POWER ELECTRO

By the same author

- Electrical Machinery
- Generalized Theory of Electrical Machines

KHANNA PUBLISHERS

Operational Office: 4575/15, Onkar House, Room No. 3-4, Ground Floor,
Darya Ganj, New Delhi-110002
Phone: 23243042 ¹ Fax: 23243043 ¹ Mobile: 9811541460

Despatch Office: 11, Community Centre, Ashok Vihar, Phase 2
Delhi-110052. Phone : 27224179

Regd. Office: 2-B, Nath Market, Nai Sarak, Delhi-110006.
Ph. 23912380

POWER ELECTRONICS

Dr. P. S. Bimbhra
Ph. D., M.E. (Hons.), F.I.E. (India), M.I.S.T.E.
Ex-Dean, Ex-Prof. and Head of Electrical & Electronics Engg. Dept.
Thapar Institute of Engineering and Technology
PATIALA-147004

Visit us at :
www.khannapublishers.in

KHANNA PUBLISHERS
2-B, Nath Market, Nai Sarak
Delhi-110006

Published by :
Romesh Chander Khanna
for **KHANNA PUBLISHERS**

Visit us at :
www.khannapublishers.in

ISBN No. 978-81-7409-215-3

Fourth Edition
23rd Reprint : 2012

Price : Rs. 250.00

Computer Typeset by :
Soft Serve Computers,
Krishna Nagar, Delhi—110051

Printed at : New A.S. Offset Press,

PREFACE

Power electronics blends the three major areas of Electrical Engineering-power, electronics and control. Under controlled power conditions, load performs better. So there has always been a popular demand to have power modulators. It is power electronics that has made possible the availability of a wide variety of controlled power converters. Power electronics has really revlutionized the art of power conversion and its control. The advent of power semiconductor device, 'thyristor', in 1957 has been the most exciting breakthrough, because its launch gave a boost to the art of power conversion and its control, and took this art to its fore-front. As a result of technological evolution, many more semiconductor devices such as triacs, asymmetrical thyristors, gate turn off thyristors, power MOSFETs, insulated gate bipolar transistors, SITs, SITHs and MOS-controlled thyristors are now available. The use of these semiconductor devices has pervaded the industrial applications relating to the field of Electrical, Electronics, Instrumentation and Control Engineering. In other words, power-electronic components find their use in low as well as high-power applications.

The purpose of this book is to provide a good understanding of the power-electronic components and the behaviour of power-electronic converters by presenting systematically all important aspects of semiconductor devices and the common type of electric-power controllers. The book begins with the study of salient features of power diodes, power transistors, MOS-controlled thyristor, silicon controlled rectifier and other members of thyristor family. Then their applications in the different types of power-converter configurations are presented in a lucid detail. In other words, this book follows the bottom-down approach (device characteristics first and then their applications). Major part of the book is intended to serve as an introductory course in power-electronics to the undergraduate students of Electrical, Electronics, Instrumentation and Control disciplines. It is presumed that the reader is familiar with the basics of elementary electronics and circuit theory. The material presented here can be covered in one semester with the omission of some topics. The instructor, after browsing through the book for some time, can plan the course contents and its sequence without loss of continuity.

The book contains thirteen chapters. Chapter 1 gives an overview of merits and demerits of power-electronic controllers and briefly discusses the topics covered in this book. This chapter also touches upon the significance of power electronics. Chapter 2 describes the characteristics of power diodes, power transistors and MOS-controlled thyristors. In Chapter 3 are presented diode characteristics, rectifiers, performance parameters and filters. Chapter 4 explains the characteristics of thyristors in detail and of Triacs, GTOs etc. Thyristor commutation techniques are given in Chapter 5. In Chapter 6, the principles of conversion from ac to dc involving single-phase as well as three-phase converters are presented. Chapters 7 to 10 pertain to the treatment of dc choppers, inverters, ac voltage controllers and cycloconverters respectively. While Chapter 11 gives study of several applications of power electronics, Chapter 12 discusses electric drives. Power factor improvement and the methods of reactive power compensation are detailed in Chapter 13. A large number of illustrative diagrams and a wide variety of worked examples add to the clarity of the subject matter. The material given in this book is class-room tested. In the appendices ; Fourier analysis, Laplace transform and a large number of objective-type questions relating to Chapters 2 to 13 are given.

The material added in the present edition includes :

    (i) performance parameters of uncontrolled and controlled rectifiers and filters,

    (ii) structural modifications in thyristors and GTOs to make than more efficient,

    (iii) SIT and SITH and (iv) Chapter 13 on power factor improvement.

Some topics have been re-written to make the presentation more lucid. Many more illustrative examples to reinforce the understanding of the subject matter are also included. Objective-type questions are thoroughly updated. It is hoped that the book in its present form will serve the purpose for the courses on power electronics of all Indian as well as foreign universities.

The author is grateful to all those students who had interacted with the author, in the class-room or outside, during the teaching of this subject. This interaction has greatly influenced the author's style of teaching and writing to a large extent and every effort has gone into making the subject matter presentation as easily comprehensible as possible. Discussion with several instructors has also been of immense help and inspiration. The author is beholden to all of them. The author, however, regrets he cannot name them all, for it is a voluminous task.

Finally, the author expresses his gratitude to his wife for her perennial encouragement, understanding and patience during the preparation of this book. The author would also thank his two sons and daughter-in-law for consistently boosting the author's morale, much needed during the revision of the book.

Suggestions leading to the improvement of the book will be gratefully acknowledged.

Dr. P.S. Bimbhra

CONTENTS

Chapter 1

Introduction

The object of this chapter is to discuss briefly the concept of power electronics, applications of power electronics and the types of power converters described in this book.

1.1. CONCEPT OF POWER ELECTRONICS

Power electronics belongs partly to power engineers and partly to electronics engineers [2]. Power engineering is mainly concerned with generation, transmission, distribution and utilization of electric energy at high efficiency. Electronics engineering, on the other hand, is guided by distortionless production, transmission and reception of data and signals of very low power level, of the order of a few watts, or milliwatts, without much consideration to the efficiency. In addition, apparatus associated with power engineering is based mainly on electromagnetic principles whereas that in electronics engineering is based upon physical phenomena in vacuum, gases/vapours and semiconductors.

Power electronics is a subject that concerns the application of electronic-principles into situations that are rated at power level rather than signal level. It may also be defined as a subject that deals with the apparatus and equipment working on the principle of *electronics* but rated at *power* level rather than signal level. For example, semiconductor power switches such as thyristors, GTOs etc. work on the principle of electronics (movement of holes and electrons), but have the name power attached to them only as a description of their power ratings. Similarly, diodes, mercury-arc rectifiers and thyratrons (gas-filled triode), high-power level devices, form a part of the subject power electronics ; because their working is based on the physical phenomena in gases and vapours, an electronic process. As the inclusion of all such power-rated electronic equipments would be a voluminous task, the present book is devoted to the study of semi-conductor-based power-electronic components and systems only. It should be understood that the techniques used in the design of high-efficiency and high-energy level power electronic circuits are quite different from those employed in the design of low-efficiency electronic circuits at signal levels.

1.2. APPLICATIONS OF POWER ELECTRONICS

The era of modern power electronics began with the invention of silicon- controlled-rectifier (SCR) by Bell Laboratories in 1956. Its prototype was introduced by GEC in 1957 and subsequently, GEC introduced SCR-based systems commercially in 1958. Since then, there have been emergence of many new power semiconductor devices. Power electronic systems today incorporate power semiconductor devices as well as microelectronic integrated circuits.

The term, 'converter system', in general, is used to denote a static device that converts ac to dc, dc to ac, dc to dc or ac to ac. Conventional power controllers based on thyratrons, mercury-arc rectifiers, magnetic amplifiers, rheostatic controllers etc. have been replaced by power electronic controllers using semiconductor devices in almost all applications. The development of new power-semiconductor devices, new circuit topologies with their improved performance and their fall in prices have opened up wide field for the new applications of power electronic converters. A judicious use of power-semiconductor devices in conjunction with microprocessors or microcomputers has further enhanced the control strategies and synthesizing capabilities of the power electronic converters. It is said that power semiconductor devices can be regarded as the muscle and the microelectronics as the intelligent brain in the modern power electronic systems.

For controlling the power flow to load, all power semiconductor devices, used in a power electronic converter, are either fully-on or fully-off. In other words, all semiconductor devices in power-electronic converter operate as switches. When the switch is fully-on, power semiconductor device handles large current (divided by the load impedance) and negligible voltage drop across it. When the switch is off, the device handles negligible current with full- voltage across it. Therefore, a power semiconductor device, during on and off periods, has very low power loss in it as compared to the power delivered by the source to load. This results in higher energy efficiency of the power electronic converter system. At the same time, low energy loss in the semiconductor device can easily be removed by its efficient cooling. This all has contributed to the widespread use of power electronic converters in the power conversion and control systems.

Table 1.1 lists various applications of power electronics. This list is however not exhaustive. No boundaries can be earmarked for the applications of power electronics, especially with the present trend of integrated design of power-semiconductor devices, microprocessors and the controlled equipment. The power ratings of power-electronic systems range from a few watts in lamps to several hundred megawatts in HVDC transmission systems. It is believed that in the early twenty-first century, 60 to 80% of the electric power consumed in utility systems will pass through power-electronics and this figure will eventually reach 100% in the future.

Table 1.1. Some Applications of Power Electronics

1. Aerospace :

Space shuttle power supplies, satellite power supplies, aircraft power systems.

2. Commercial :

Advertising, heating, airconditioning, central refrigeration, computer and office equipment, uninterruptible power supplies, elevators, light dimmers and flashers.

3. Industrial :

Arc and industrial furnaces, blowers and fans, pumps and compressors, industrial lasers, transformer-tap changers, rolling mills, textile mills, excavators, cement mills, welding.

4. Residential :

Airconditioning, cooking, lighting, space heating, refrigerators, electric-door openers, dryers, fans, personal computers, other entertainment equipment, vacuum cleaners, washing and sewing machines, light dimmers, food mixers, electric blankets, food-warmer trays.

5. Telecommunication :

Battery chargers, power supplies (dc and UPS).

6. Transportation :

Battery chargers, traction control of electric vehicles, electric locomotives, street cars, trolley buses, subways, automotive electronics.

7. Utility systems :

High voltage dc transmission (HVDC), excitation systems, VAR compensation, static circuit breakers, fans and boiler-feed pumps, supplementary energy systems (solar, wind).

1.3. ADVANTAGES AND DISADVANTAGES OF POWER-ELECTRONIC CONVERTERS

The advantages possessed by power-electronic systems are as under :

(*i*) High efficiency due to low loss in power-semiconductor devices.

(*ii*) High reliability of power-electronic converter systems.

(*iii*) Long life and less maintenance due to the absence of any moving parts.

(*iv*) Fast dynamic response of the power-electronic systems as compared to electromechanical converter systems.

(*v*) Small size and less weight result in less floor space and therefore lower installation cost.

(*vi*) Mass production of power-semiconductor devices has resulted in lower cost of the converter equipment.

Systems based on power electronics, however, suffer from the following disadvantages :

(*a*) Power-electronic converter circuits have a tendency to generate harmonics in the supply system as well as in the load circuit.

In the load circuit, the performance of the load is influenced, for example, a high harmonic content in the load circuit causes commutation problems in dc machines, increased motor heating and more accoustical noise in both dc and ac machines. So steps must be taken to filter these out from the output side of a converter.

In the supply system, the harmonics distort the voltage waveform and seriously influence the performance of other equipment connected to the same supply line. In addition, the harmonics in the supply line can also cause interference in audio- and video-equipment (called radio interference). It is, therefore, necessary to insert filters on the input side of a converter.

(*b*) Ac to dc and ac to ac converters operate at a low input power factor under certain operating conditions. In order to avoid a low pf, some special measures have to be adopted.

(*c*) Power-electronic controllers have low overload capacity. These converters must, therefore, be rated for taking momentary overloads. As such, cost of power electronic controller may increase.

(*d*) Regeneration of power is difficult in power electronic converter systems.

The advantages possessed by power electronic converters far outweigh their disadvantages mentioned above. As a consequence, semiconductor-based converters are being extensively employed in systems where power flow is to be regulated. As already stated, conventional power controllers used in many installations have already been replaced by semiconductor-based power electronic controllers.

1.4. POWER ELECTRONIC SYSTEMS

The major components of a power electronic system are shown in the form of a block diagram in Fig. 1.1. Main power source may be an ac supply system or a dc supply system.

The output from the power electronic circuit may be variable dc, or ac voltage, or it may be a variable voltage and frequency. In general, the output of a power electronic convertor circuit depends upon the requirements of the load. For example, if the load is a dc motor, the convertor output must be adjustable direct voltage. In case the load is a 3-phase induction motor, the converter may have adjustable voltage and frequency at its output terminals.

The feedback component in Fig. 1.1 measures a parameter of the load, say speed in case of a rotating machine, and compares it with the command. The difference of the two, through the digital circuit components, controls the instant of turn-on of semiconductor devices forming the

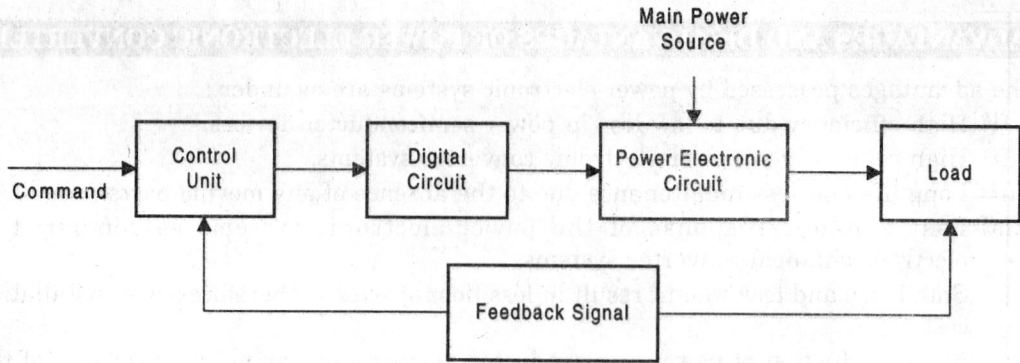

Fig. 1.1. Block diagram of a typical power electronic system.

solid-state power converter system. In this manner, behaviour of the load circuit can be controlled, as desired, over a wide range with the adjustment of the command.

1.5. POWER SEMICONDUCTOR DEVICES

Silicon controlled rectifier (SCR) was first introduced in 1957 as a power semiconductor device. Since then, several other power semiconductor devices have been developed. Most of these semiconductor devices are listed below in Table 1.2 along with their circuit, or device, symbol and present maximum ratings.

Table 1.2. Maximum ratings of power semiconductor devices

S.No.	Device	Circuit symbol	Voltage / current ratings	Upper operating freq. (kHz)
1.	Diode		5000 V/5000 A	1.0
2.	Thyristors			
	(a) SCR		7000 V/5000 A	1.0
	(b) LASCR		6000 V/3000 A	1.0
	(c) ASCR/RCT		2500 V/400 A	2.0
	(d) GTO		5000 V/3000 A	2.0

S.No.	Device	Circuit symbol	Voltage/current ratings	Upper operating freq. (kHz)
	(e) SITH	A o——⊳⊢——o K, o G	2500 V/500 A	100.0
	(f) MCT	A, G, K	1200 V/40 A	20.0
	(g) Triac	MT2, MT1, G	1200 V/1000 A	0.50
3.	Transistors			
	(a) BJT	C, B, E, npn; C, B, E, pnp	1400 V/400 A	10.0
	(b) MOSFET (n-channel)	D, G, S	1000 V/50A	100.0
	(c) SIT	D, G, S	1200 V/300 A	100.0
	(d) IGBT	C, G, E	1200 V/500 A	50.0

In the above table, the various abbreviations are; SCR (silicon controlled rectifier), LASCR (light-activated SCR), ASCR (asymmetrical SCR), RCT (reverse conducting thyristor), GTO (gate-turn off thyristor), SITH (static induction thyristor), MCT (MOS controlled thyristor), BJT (bipolar junction transistor), MOSFET (metal-oxide semiconductor field effect transistor), SIT (static induction transistor) and IGBT (insulated gate bipolar transistor).

Based on (*i*) turn-on and turn-off characteristics, (*ii*) gate signal requirements and (*iii*) degree of controllability, the power semiconductor devices can be classified as under :

(*a*) **Diodes.** These are uncontrolled rectifying devices. Their on and off states are controlled by power supply.

(*b*) **Thyristors.** These have controlled turn-on by a gate signal. After thyristors are on, they remain latched-in on-state due to internal regenerative action and gate loses control. These can be turned-off by the power circuit.

(*c*) **Controllable switches.** These devices are turned-on and turned-off by the application of control signals. The devices which behave as controllable switches are BJT, MOSFET, GTO, SITH, IGBT, SIT and MCT.

Triac and RCT possess bi-directional current capability whereas all other remaining devices (diode, SCR, GTO, BJT, MOSFET, IGBT, SIT, SITH and MCT) are unidirectional current devices.

1.6. TYPES OF POWER ELECTRONIC CONVERTERS

A power electronic system consists of one or more power electronic converters. A power electronic converter is made up of some power semiconductor devices controlled by integrated circuits. The switching characteristics of power semiconductor devices permit a power electronic converter to shape the input power of one form to output power of some other form. Static power converters perform these functions of power conversion very efficiently. Broadly speaking, power electronic converters (or circuits) can be classified into six types as under:

1. Diode Rectifiers. A diode rectifier circuit converts ac input voltage into a fixed dc voltage. The input voltage may be single-phase or three phase. Diode rectifiers find wide use in electric traction, battery charging, electroplating, electrochemical processing, power supplies, welding and uninterruptible power supply (UPS) systems.

2. Ac to dc converters (Phase-controlled rectifiers). These convert constant ac voltage to variable dc output voltage. These rectifiers use line voltage for their commutation, as such these are also called line-commutated or naturally-commutated ac to dc converters. Phase-controlled converters may be fed from 1-phase or 3-phase source. These are used in dc drives, metallurgical and chemical industries, excitation systems for synchronous machines etc.

3. DC to dc converters (DC Choppers). A dc chopper converts fixed dc input voltage to a controllable dc output voltage. The chopper circuits require forced, or load, commutation to turn-off the thyristors. For lower power circuits, thyristors are replaced by power transistors. Classification of chopper circuits is dependent upon the type of commutation and also on the direction of power flow. Choppers find wide applications in dc drives, subway cars, trolley trucks, battery-driven vehicles etc.

4. DC to ac converters (inverters). An inverter converts fixed dc voltage to a variable ac voltage. The output may be a variable voltage and variable frequency. These converters use line, load or forced commutation for turning-off the thyristors. Inverters find wide use in induction-motor and synchronous-motor drives, induction heating, UPS, HVDC transmission etc. At present, conventional thyristors are also being replaced by GTOs in high-power applications and by power transistors in low-power applications.

5. AC to ac converters. These convert fixed ac input voltage into variable ac output voltage. These are of two types as under :

(a) *AC voltage controllers* (AC voltage regulators). These converter circuits convert fixed ac voltage directly to a variable ac voltage at the same frequency. AC voltage controller employ two thyristors in antiparallel or a triac. Turn-off of both the devices is obtained by line commutation. Output voltage is controlled by varying the firing angle delay. AC voltage controllers are widely used for lighting control, speed control of fans, pumps etc.

(b) *Cycloconverters.* These circuits convert input power at one frequency to output power at a different frequency through one-stage conversion. Line commutation is more common in these converters, though forced and load commutated cycloconverters are also employed. These are primarily used for slow-speed large ac drives like rotary kiln etc.

6. Static switches. The power semiconductor devices can operate as static switches or contactors. Static switches possess many advantages over mechanical and electromechanical circuit breakers. Depending upon the input supply, the static switches are called ac static switches or dc static switches.

1.7. POWER ELECTRONIC MODULES

A power electronic converter may require two, four or more semiconductor devices depending upon the circuit configuration. For example, a single-phase half-bridge inverter requires a power module consisting of two power semiconductor devices; a full-converter (or *H*-bridge converter) requires a power module having four semiconductor devices; a three phase full converter needs a power module having six semiconductor devices. Thus, a power electronic converter can be assembled from power modules instead of from individual semiconductor devices. A power module has better performance characteristics as compared to conventional devices so far as their switching characteristics, operating speed and losses are concerned. Gate drive circuits for individual devices or power modules are also commercially available. As a result of these developments, now intelligent modules have come in the market.

Intelligent module, also called *smart-power,* is state-of-the-art power electronics and it consists of power module and a peripheral circuit. The peripheral circuit comprises of interfacing of power module with the input/output through proper isolation from low-voltage signal and from high-voltage power circuit, a drive circuit, protection and diagnostic circuitry against maloperation like excess current, over voltage etc, microcomputer control and controlled power supply. The user has merely to connect the existing supply and the load terminals to the smart-power. At present, intelligent modules are being used extensively in power electronics. It is reported that there are more than twenty manufacturers of intelligent modules.

SUMMARY

Power semiconductor devices form the heart of modern power electronics. A power electronics engineer must understand the device thoroughly for efficient, reliable and cost-effective design of power converters. For this purpose, chapter 2 is devoted to the study of power semiconductor diodes, transistors and MCT. Chapter 3 deals with diode circuits and rectifiers. In chapter 4, are discussed in detail the thyristor characteristics and its control strategies. Thyristor commutation techniques are described in chapter 5. Other power electronic converters mentioned in this chapter are described in detail in chapter 6 onwards.

PROBLEMS

1.1. (a) What is power electronics ? Discuss briefly the concept of power electronics.

(b) What is a converter ? Illustrate your answer with examples.

(c) Give the reasons leading to the widespread use of power electronic converters.

(d) Enumerate at least ten applications of power electronics.

1.2. (a) Give the advantages and disadvantages of power electronic converters.

(b) Describe a power electronic system with its general block diagram.

(c) List the following power semiconductor devices along with their circuit symbols and maximum ratings :

Diode, SCR, GTO, SITH, MCT, BJT, SIT, IGBT.

1.3. (a) Discuss the various types of power electronic converters.

(b) Compare a diode with a thyristor.

(c) List the semiconductor devices which possess the capability of withstanding
(i) bidirectional current and (ii) unidirectional current.

(d) Give the differences between a triac and a thyristor.

1.4. (a) Give the differences between an ac voltage controller and a cycloconverter.

(b) What is the difference between thyristors and controllable switches ? Make a list of uncontrolled and controllable switches.

(c) How many semiconductor devices are required for (i) H-bridge converter and (ii) three-phase full converter.

(d) What is power electronic module ? Describe smart power.

Chapter 2
Power Semiconductor Diodes and Transistors

In this Chapter

- The *p-n* Junction
- Basic Structure of Power Diodes
- Characteristics of Power Diodes
- Types of Power Diodes
- Power Transistors
- Power Mosfets
- Insulated Gate Bipolar Transistor
- Static Induction Transistor (SIT)
- Mos-controlled Thyristor (MCT)
- New Semiconducting Materials

A low-power diode, called signal diode, is a *pn*-junction device. A high-power diode, called power diode, is also a *pn*-junction device but with constructional features somewhat different from a signal diode. Lifewise, power transistors also differ in construction from signal transistors.

The voltage, current and power ratings of power diodes and transistors are much higher than the corresponding ratings for signal devices. In addition, power devices operate at lower switching speeds whereas signal diodes and transistors operate at higher switching speeds.

Power semiconductor devices are used extensively in power-electronic circuits. Some applications of power diodes include their use as freewheeling diodes, for ac to dc conversion, for recovery of trapped energy etc. Power transistors, used as a switching device in power-electronic circuits, must operate in the saturation region in order that their on-state voltage drop is low. Their applications as switching elements include dc choppers and inverters.

The object of this chapter is to describe power diodes, power transistors and MOS-controlled thyristor (MCT). A thyristor is more important component of power semiconductor devices, it is, therefore, discussed in detail in Chapter 4.

2.1. THE *p-n* JUNCTION

A *p-n* junction forms the basic building block of all power semiconductor devices. It is, therefore, worthwhile here to review this junction at an introductory level.

A *p-n* junction is formed when *p*-type semiconductor is brought in metallurgical, or physical, contact with *n*-type semiconductor. A *p*-region has greater concentration of holes whereas *n*-region has more electron-concentration. In *p*-region, free holes are called majority carriers and free electrons minority carriers. In n-region, free electrons are called majority carriers whereas free holes are called minority carriers.

Doping densities in p and n type semiconductors may be different. As such, p-type material may be designated p^+, p or p^-; similarly n-type material as n^+, n^- etc. Rough guidelines for labelling of p as p^+, p^- etc and n as n^-, n^+ etc are as under :

(a) If doping (or acceptor) density in p-type semiconductor = doping (or donor) density in n-type semiconductor, then it is called p-n junction. For example, if doping density in both p and n layers is about 10^{16} cm^{-3} to 10^{17} cm^{-3}, junction is termed p-n junction.

(b) If doping density in p-region is much greater than that in n-region, it is called p^+n junction. For example, if doping densities are 10^{19} cm^{-3} in p layer and 10^{17} cm^{-3} in n layer, then it is termed $p^+ n$ junction.

(c) If doping density in n-type is less than that given in part (b), the junction is called p^+n^- junction. For example, if doping densities are 10^{19} cm^{-3} and 10^{13} cm^{-3} for p and n types respectively, then $p^+ n^-$ junction is formed.

(d) If both p and n-layers are heavily doped, it is called $p^+ n^+$ junction and if very lightly doped, a $p^- n^-$ junction is formed. For example, if density is 10^{19} cm^{-3} in both p and n layers, $p^+ n^+$ junction is formed.

In general, p^+ indicates highly doped p region, n^- lightly doped n region and so on.

2.1.1. Depletion Layer

When physical contact between p and n regions is made, free electrons in n material diffuse across the junction into p material, Fig. 2.1 (a). Diffusion of each electron from n to p, leaves a positive charge behind in the n-region near the junction. Similarly, diffusion of each hole from p to n, leaves a negative charge behind in the p region near the junction. As a result of this diffusion, n region near the junction becomes positively charged and p region in the vicinity of junction becomes negatively charged, Fig. 2.1 (b). These charges establish an electric field across the junction. When this field grows strong enough, it stops further diffusion. Some electrons, as these diffuse from n to p, recombine with holes in p-region and disappear. Similar recombination occurs in n-region.

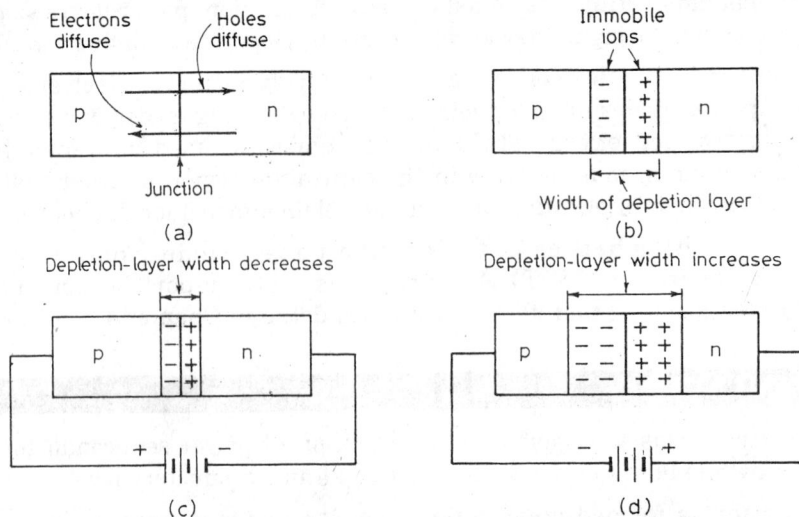

Fig. 2.1. A p-n junction showing (a) direction of holes and electrons diffusion (b) depletion region (c) effect of forward biasing and (d) effect of reverse biasing.

When electric field stops further diffusion, charge carriers (holes and electrons) don't move. As a consequence, opposite charges on each side of the junction produce immobile ions, Fig. 2.1 (*b*). The region extending into both *p* and *n* semiconductor layers is called *depletion region* or *space-charge region*. The width of depletion region, or *depletion layer*, is of the order of 5×10^{-4} mm. In equilibrium, there is a potential difference of 0.7 V across the depletion region in silicon and 0.3 V across the depletion region in germanium. This potential difference across the depletion layer is called *barrier potential*.

When positive terminal of a battery is connected to *p*-type material and negative terminal to *n*-type material, Fig. 2.1 (*c*), the *p-n* junction is forward biased. Positive terminal of the battery sucks electrons from *p* material leaving holes there. These holes travel through *p* material towards the negative charge at *p-n* junction and thus neutralize partly this negative charge. Similarly, negative terminal of the battery injects electrons into *n* layer. These electrons move through *n* material, reach the *p-n* junction thereby neutralizing partly the positive charge. As a result, width of depletion region gets reduced.

In case *p* material is connected to negative terminal of the battery and *n* material to positive terminal of battery, then it can be deduced that width of depletion layer increases, Fig. 2.1 (*d*).

A rise in junction temperature also decreases width of depletion layer. As the barrier potential depends on width of the depletion layer, the barrier potential decreases with rise in junction temperature.

For power semiconductor devices, it should be kept in mind that (*i*) a junction with lightly doped layer on its one side requires large breakdown voltage and (*ii*) a junction with highly doped layers on its both sides requires low breakdown voltage.

2.2. BASIC STRUCTURE OF POWER DIODES

Power diodes differ in structure from signal diodes. A signal diode constitutes a simple *p-n* junction as shown in Fig. 2.1. The intricacies in constructing power diodes arise from the need to make them suitable for high-voltage and high-current applications.

The practical realization and the resulting structure of a power diode is shown in Fig. 2.2 (*a*). It consists of heavily doped n^+ substrate*. On this substrate, a lightly doped n^- layer is epitexially grown. Now a heavily doped p^+ layer is diffused into n^- layer to form the anode of power diode, Fig. 2.2 (*a*). This shows that n^- layer is the basic structural feature not found in signal diodes. The function of n^- layer is to absorb the depletion layer of the reverse biased p^+n^- junction J_1. The break-down voltage needed in a power diode governs the thickness of n^- layer ; greater the breakdown voltage, more the n^- layer thickness. The drawback of n^- layer is to add significant ohmic resistance to the diode when it is conducting a forward current. This leads to large power dissipation in the diode ; so proper cooling arrangements in large diode ratings are essential.

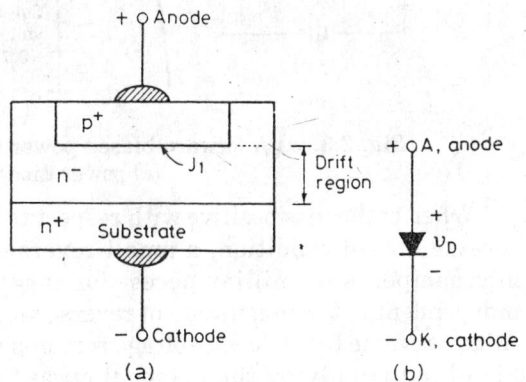

Fig. 2.2. (*a*) Structural features of power diode and (*b*) its circuit symbol.

_____

* material on which something grows.

The circuit symbol of a power diode, shown in Fig. 2.2 (b), is the same as that for a signal diode.

The modifications in the context of diode, presented above, makes them appropriate for high-power applications. As diode, or p-n junction, is the basic building block of all other power semiconductor devices; same basic modifications should be implemented in all low-power semiconductor devices in order to raise their power-handling capabilities.

2.3. CHARACTERISTICS OF POWER DIODES

As stated before, power diode is a two-terminal, p-n semiconductor device. The two terminals of diode are called anode and cathode, Fig. 2.2 (b) and Fig. 2.3 (a). Two important characteristics of power diodes are now described.

2.3.1. Diode i-v Characteristics[*]

When anode is positive with respect to cathode, diode is said to be *forward biased*. With increase of the source voltage V_s from zero value, initially diode current is zero. From $V_s = 0$ to cut-in voltage, the forward-diode current is very small. *Cut-in voltage* is also known as *threshold voltage* or *turn-on voltage*. Beyond cut-in voltage, the diode current rises rapidly and the diode is said to conduct. For silicon diode, the cut-in voltage is around 0.7 V. When diode conducts, there is a forward voltage drop of the order of 0.8 to 1 V.

For low-power diodes, current in the forward direction increases first exponentially with voltage and then becomes almost linear as shown in Fig. 2.3 (b). For power diodes, the forward current grows almost linearly with voltage, Fig. 2.3 (c). The high magnitude of current in a power diode leads to ohmic drops that hide the exponential part of i-v curve. The n^- region, or drift region, forms a considerable drop in the ohmic resistance of power diodes.

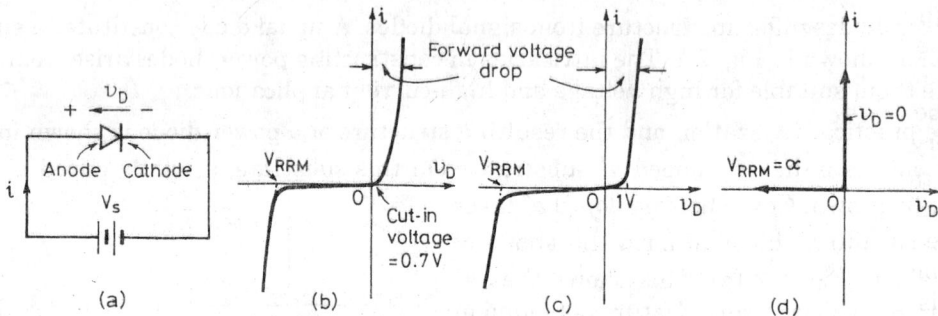

Fig. 2.3. (a) A forward-biased power diode. i-v characteristics of (b) signal diode (c) power diode and (d) ideal diode.

When cathode is positive with respect to anode, the diode is said to be *reverse biased*. In the reverse biased condition, a small reverse current called leakage current, of the order of microamperes or milliamperes (for large diodes) flows. The leakage current is almost independent of the magnitude of reverse voltage until this voltage reaches breakdown voltage. At this reverse breakdown, voltage remains almost constant but reverse current becomes quite high-limited only by the external circuit resistance. A large reverse breakdown voltage, associated with high reverse current, leads to excessive power loss that may destroy the diode. This shows that reverse breakdown of a power diode must be avoided by operating it below the specific peak reverse repetitive voltage V_{RRM}. Fig. 2.3 (c) illustrates the i-v characteristics of

---

[*]Some authors write v-i characteristics

power diode and V_{RRm}. For an ideal diode, the i–v characteristics are shown in Fig. 2.3 (d). Here, voltage drop across conducting diode, $v_D = 0$, reverse leakage current = 0, cut-in voltage = 0 and reverse breakdown voltage V_{RRM} is infinite.

Diode manufacturers also indicate the value of peak inverse voltage (PIV) of a diode. This is the largest reverse voltage to which a diode may be subjected during its working. PIV is the same as V_{RRM}.

The power diodes are now available with forward current ratings of 1A to several thousand amperes and with reverse voltage ratings of 50 V to 5000 V or more.

2.3.2. Diode Reverse Recovery Characteristics

After the forward diode current decays to zero, the diode continues to conduct in the reverse direction because of the pressure of stored charges in the depletion region and the semiconductor layers. The reverse current flows for a time called reverse recovery time t_{rr}. The diode regains its blocking capability until reverse recovery current decays to zero. The *reverse recovery time* t_{rr} is defined as the time between the instant forward diode current becomes zero and the instant reverse recovery current decays to 25% of its reverse peak value I_{RM} as shown in Fig. 2.4 (a).

The reverse recovery time is composed of two segments of time t_a and t_b, i.e. $t_{rr} = t_a + t_b$. Time t_a is the time between zero crossing of forward current and peak reverse current I_{RM}. During the time t_a, charge stored in depletion layer is removed. Time t_b is measured from the instant of reverse peak value I_{RM} to the instant when $0.25\,I_{RM}$ is reached, Fig. 2.4 (a). During t_b, charge from the semiconductor layers is removed. The shaded area in Fig. 2.4 (a) represents the stored charge, or reverse recovery charge, Q_R which must be removed during the reverse recovery time t_{rr}. The ratio t_b/t_a is called the *softness factor* or *S*-factor. This factor is a measure of the voltage transients that occur during the time diode recovers. Its usual value is unity and this indicates low oscillatory reverse recovery process. In case *S*-factor is small, diode has large oscillatory over voltages. A diode with *S*-factor equal to one is called *soft-recovery* diode and a diode with *S*-factor less than one is called

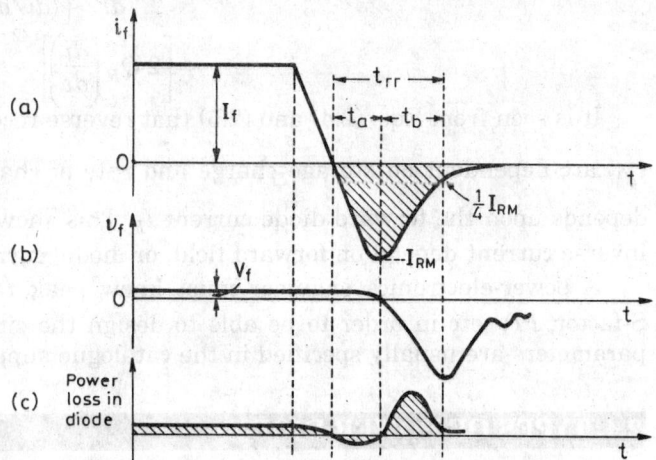

Fig. 2.4. Reverse recovery characteristics
(a) variation of forward current i_f
(b) forward voltage drop v_f and (c) power loss in a diode.

snappy-recovery diode or *fast-recovery* diode. In Fig. 2.4 (b) is shown the waveform of forward-voltage drop v_f across the diode. The product of v_f and i_f gives the power loss in a diode. Its variation is shown in Fig. 2.4 (c). The average value of $v_f i_f$ gives the total power loss in a diode. Fig. 2.4 (c) reveals that major power loss in a diode occurs during the period t_b.

It is noticed from Fig. 2.4 (a) that peak inverse current I_{RM} can be expressed as

$$I_{RM} = t_a \frac{di}{dt} \qquad \qquad ...(2.1)$$

where $\dfrac{di}{dt}$ is the rate of change of reverse current. The reverse recovery characteristics of Fig. 2.4 (a) can be taken to be triangular. Under this assumption, storage charge Q_R, from Fig. 2.4 (a), is given by

$$Q_R = \frac{1}{2} I_{RM} \cdot t_{rr}$$

or
$$I_{RM} = \frac{2\,Q_R}{t_{rr}} \qquad\qquad\qquad ...(2.2)$$

If $t_{rr} \cong t_a$, then from Eq. (2.1),

$$I_{RM} = t_{rr} \cdot \frac{di}{dt} \qquad\qquad\qquad ...(2.3)$$

From Eqs. (2.2) and (2.3), we get

$$t_{rr} \cdot \frac{di}{dt} = \frac{2\,Q_R}{t_{rr}}$$

or
$$t_{rr} = \left[\frac{2\,Q_R}{(di/dt)} \right]^{1/2} \qquad\qquad ...(2.4)$$

From Eq. (2.1), with $t_a \cong t_{rr}$, we get

$$I_{RM} = t_{rr} \cdot \frac{di}{dt} = \left[\frac{2\,Q_R}{(di/dt)} \right]^{1/2} \cdot \frac{di}{dt}$$

$$= \left[2\,Q_R \left(\frac{di}{dt} \right) \right]^{1/2} \qquad\qquad ...(2.5)$$

It is seen from Eqs. (2.4) and (2.5) that reverse recovery time t_{rr} and peak inverse current I_{RM} are dependent on storage charge and rate of change of current $\dfrac{di}{dt}$. The storage charge depends upon the forward diode current I_F. This shows that reverse recovery time and peak inverse current depend on forward field, or diode, current.

A power-electronics engineer must know peak reverse current I_{RM}, stored charge Q_R, S-factor, *PIV* etc in order to be able to design the circuitry employing power diodes. These parameters are usually specified in the catalogue supplied by the diode manufacturers.

2.4. TYPES OF POWER DIODES

Diodes are classified according to their reverse recovery characteristics. The three types of power diodes are as under :

    (*i*) General purpose diodes
    (*ii*) Fast recovery diodes
    (*iii*) Schottky diodes.

These are now described briefly.

2.4.1. General-purpose Diodes

These diodes have relatively high reverse recovery time, of the order of about 25 μs. Their current ratings vary from 1 A to several thousand amperes and the range of voltage rating is from 50 V to about 5 kV. Applications of power diodes of this type include battery charging, electric traction, electroplating, welding and uninterruptible power supplies (UPS).

2.4.2. Fast-recovery Diodes

The diodes with low reverse recovery time, of about 5 µs or less, are classified as fast-recovery diodes. These are used in choppers, commutation circuits, switched mode power supplies, induction heating etc. Their current ratings vary from about 1 A to several thousand amperes and voltage ratings from 50 V to about 3 kV.

For voltage ratings below about 400 V, the epitaxial process is used for diode fabrication. These diodes have fast recovery time, as low as 50 ns.

For voltage ratings above 400 V, diffusion technique is used for the fabrication of diodes. In order to shorten the reverse-recovery time, platinum or gold doping is carried out. But this doping may increase the forward voltage drop in a diode.

2.4.3. Schottky Diodes

This class of diodes use metal-to-semiconductor junction for rectification purposes instead of p-n junction. The metal is usually aluminium and semiconductor is silicon. Therefore, a Schottky diode has aluminium-silicon junction. The silicon is n-type.

When Schottky diode is forward biased, free electrons in n material move towards the Al-n junction and then travel through the metal (aluminium) to constitute the flow of forward current. Since metal does not have any holes, this forward current is due to the movement of *electrons* only. As the metal has no holes, there is no storage charge and no-reverse recovery time. It can, therefore, be said that rectified current flow in a Schottky diode is by the movement of majority carriers (electrons) only and the turn-off delay caused by recombination is avoided. As such, Schottky diode can switch off much faster than p-n junction diode.

As compared to p-n junction diode, a Schottky diode has (*i*) lower cut-in voltage, (*ii*) higher reverse leakage current and (*iii*) higher operating frequency. Their reverse voltage ratings are limited to about 100 V and forward current ratings vary from 1 A to 300 A. Applications of Schottky diode include high-frequency instrumentation and switching power supplies.

The electrical and thermal characteristics of power diodes are similar to those of thyristors which are described in chapter 4.

2.5. POWER TRANSISTORS

Power diodes are uncontrolled devices. In other words, their turn-on and turn-off characteristics are not under control. Power transistors, however, possess controlled characteristics. These are turned on when a current signal is given to base, or control, terminal. The transistor remains in the on-state so long as control signal is present. When this control signal is removed, a power transistor is turned off.

Power transistors are of four types as under :

    (*i*) Bipolar junction transistors (BJTs)
    (*ii*) Metal-oxide-semiconductor field-effect transistors (MOSFETs)
    (*iii*) Insulated gate bipolar transistors (IGBTs) and
    (*iv*) Static induction transistors (SITs).

These four types are now described one after the other.

2.5.1. Bipolar Junction Transistors

A bipolar transistor is a three-layer, two junction npn or pnp semiconductor device. With one p-region sandwiched by two n-regions, Fig. 2.5 (*a*), npn transistor is obtained. With two p-regions sandwiching one n-region, Fig. 2.5 (*b*), pnp transistor is obtained. The term ' bipolar denotes that the current flow in the device is due to the movement of both holes and electrons. A BJT has three terminals named collector (C), emitter (E) and base (B). An emitter is indicated by an arrowhead indicating the direction of emitter current. No arrow is associated with base or collector. Power transistors of *npn* type are easy to manufacture and are cheaper also. Therefore, use of power *npn* transistors is very wide in high-voltage and high-current applications. Hereafter, *npn* transistors would only be considered.

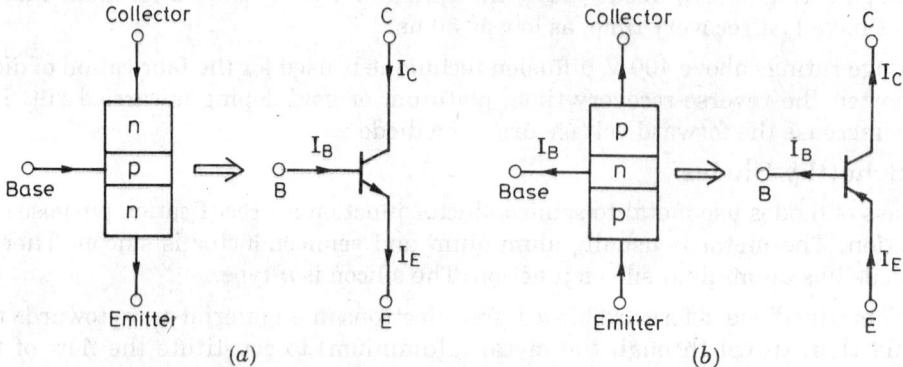

Fig. 2.5. Bipolar junction transistors (a) *npn* type and (b) *pnp* type.

2.5.1.1. Steady-state Characteristics. Out of the three possible circuit configurations for a transistor, common-emitter arrangement is more common in switching applications. So, a common emitter *npn* circuit for obtaining its characteristics is considered as shown in Fig. 2.6 (a).

Input characteristics. A graph between base current I_B and base-emitter voltage V_{BE} gives input characteristics. As the base-emitter junction of a transistor is like a diode, I_B versus V_{BE} graph resembles a diode curve. When collector-emitter voltage V_{CE2} is more than V_{CE1}, base current, for the same V_{BE}, decreases as shown in Fig. 2.6 (b).

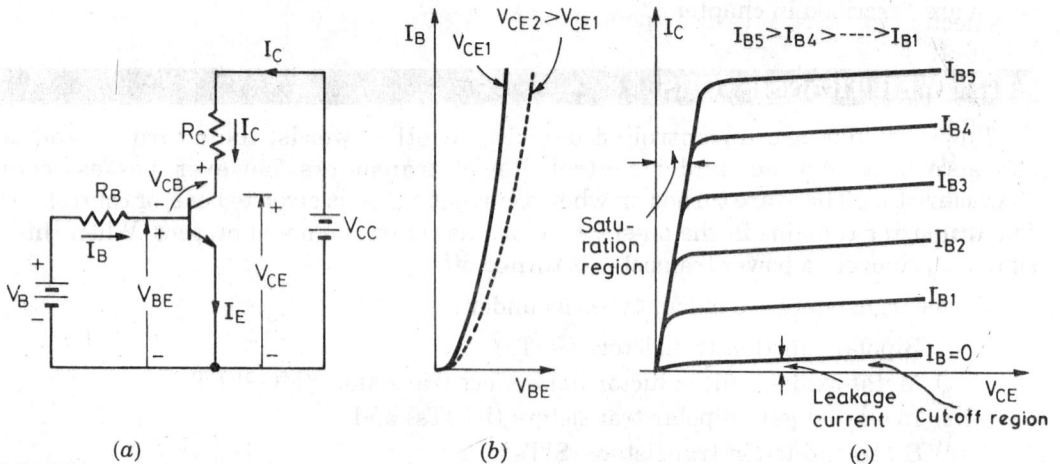

Fig. 2.6. (a) *npn* transistor circuit characteristics, (b) input characteristics and (c) output characteristics.

Output characteristics. A graph between collector current I_c and collector-emitter voltage V_{CE} gives output characteristics of a transistor. For zero base current, *i.e.* for $I_B = 0$, as V_{CE} is increased, a small leakage (collector) current exists as shown in Fig. 2.6 (c). As the base current is increased from $I_B = 0$ to I_{B1}, I_{B2} etc, collector current also rises as shown in Fig. 2.6 (c).

Fig. 2.7 (a) shows two of the output characteristic curves, 1 for $I_B = 0$ and 2 for $I_B \neq 0$. The initial part of curve 2, characterised by low V_{CE}, is called the saturation region. In this region, the transistor acts like a switch. The flat part of curve 2, indicated by increasing V_{CE} and almost constant I_C, is the active region. In this region, transistor acts like an amplifier. Almost vertically rising curve is the breakdown region which must be avoided at all costs.

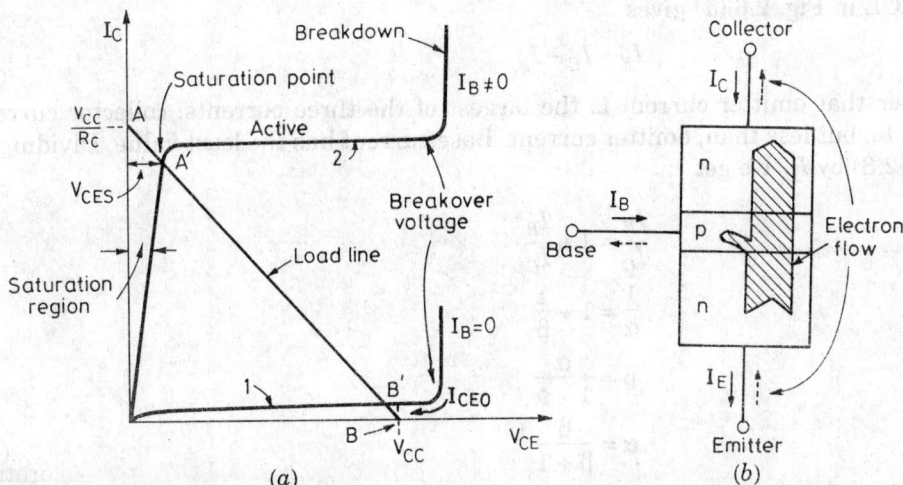

Fig. 2.7. (a) Output characteristics and load line for *npn* transistor and (b) electron flow in an *npn* transistor.

For load resistor R_C, Fig. 2.6 (a), the collector current I_C is given by

$$I_C = \frac{V_{CC} - V_{CE}}{R_C}$$

This is the equation of load line. It is shown as line AB in Fig. 2.7 (a). A load line is the locus of all possible operating points. Ideally, when transistor is on, V_{CE} is zero and $I_C = V_{CC}/R_C$. This collector current is shown by point A on the vertical axis. When the transistor is off, or in the cut-off region, V_{CC} appears across collector-emitter terminals and there is no collector current. This value is indicated by point B on the horizontal axis. For the resistive load, the line joining points A and B is the load line.

Relation between α and β. Most of the electrons, proportional to I_E, given out by emitter, reach the collector as shown in Fig. 2.7 (b). In other words, collector current I_C, though less than emitter current I_E, is almost equal to I_E. A symbol α is used to indicate how close in value these two currents are. Here α, called *forward current gain*, is defined as

$$\alpha = \frac{I_C}{I_E} \qquad \qquad ...(2.6)$$

As $I_C < I_E$, value of α varies from 0.95 to 0.99.

In a transistor, base current is effectively the input current and collector current is the output current. The ratio of collector (output) current I_C to base (input) current I_B is known as the *current gain* β.

$$\therefore \qquad \beta = \frac{I_C}{I_B} \qquad \qquad ...(2.7)$$

As I_B is much smaller, β is much more than unity ; its value varies from 50 to 300. In another system of analysis, called h parameters, h_{FE} is used in place of β.

$$\therefore \qquad \beta = h_{FE} = \frac{I_C}{I_B} \qquad\qquad ...(2.7)$$

Use of KCL in Fig. 2.6 (a) gives

$$I_E = I_C + I_B \qquad\qquad ...(2.8)$$

Remember that emitter current is the largest of the three currents, collector current is almost equal to, but less than, emitter current. Base current has the least value. Dividing both sides of Eq. (2.8) by I_C, we get

$$\frac{I_E}{I_C} = 1 + \frac{I_B}{I_C}$$

$$\frac{1}{\alpha} = 1 + \frac{1}{\beta}$$

or

$$\beta = \frac{\alpha}{1 - \alpha} \qquad\qquad ...(2.9)$$

and

$$\alpha = \frac{\beta}{\beta + 1} \qquad\qquad ...(2.10)$$

Transistor Switch. Transistor operation as a switch means that transistor operates either in the saturation region or in the cut-off region and nowhere else on the load line. As an ideal switch, the transistor operates at point A in the saturated state as closed switch with $V_{CE} = 0$ and at point B in the cut-off state as an open switch with $I_C = 0$, Fig. 2.7 (a). In practice, the large base current will cause the transistor to work in the saturation region at point A' with small saturation voltage V_{CES}. Here subscript S is used to denote saturated value. Voltage V_{CES} represents on-state voltage drop of the transistor which is of the order of about 1 V. When the control, or base, signal is reduced to zero, the transistor is turned off and its operation shifts to B' in the cut-off region, Fig. 2.7 (a). A small leakage current I_{CEO} flows in the collector circuit when the transistor is off.

For Fig. 2.6 (a), KVL for the circuit consisting of V_B, R_B and emitter gives

$$V_B - R_B I_B - V_{BE} = 0$$

or

$$I_B = \frac{V_B - V_{BE}}{R_B} \qquad\qquad ...(2.11)$$

Also, from Fig. 2.6 (a), $V_{CC} = V_{CE} + I_C R_C$

or

$$V_{CE} = V_{CC} - I_C R_C = V_{CC} - \beta I_B R_C$$

$$= V_{CC} - \frac{\beta R_C}{R_B} (V_B - V_{BE}) \qquad\qquad ...(2.12)$$

Also $V_{CE} = V_{CB} + V_{BE}$

or $V_{CB} = V_{CE} - V_{BE} \qquad\qquad ...(2.13)$

If V_{CES} is the collector-emitter saturation voltage, then collector current I_{CS} is given by

$$I_{CS} = \frac{V_{CC} - V_{CES}}{R_C} \qquad\qquad ...(2.14)$$

and the corresponding value of minimum base current, that produces saturation, is

$$I_{BS} = \frac{I_{CS}}{\beta} \qquad \ldots(2.15)$$

If base current is less than I_{BS}, the transistor operates in the active region, *i.e*. somewhere between the saturation and cut-off points. If base current is more than I_{BS}, V_{CES} is almost zero and collector current from Eq. (2.14) is given by $I_{CS} = V_{CC}/R_C$. This shows that collector current at saturation remains substantially constant even if base current is increased.

With base current more than I_{BS}, *hard drive* of transistor is obtained. With hard saturation, on-state losses of transistor increase. Normally, the practical circuit is designed for hard-drive of transistor and therefore, base current I_B is greater than I_{BS} given by Eq. (2.15). The ratio of I_B and I_{BS} is defined as the overdrive factor (ODF).

$$\therefore \qquad ODF = \frac{I_B}{I_{BS}} \qquad \ldots(2.16)$$

ODF may be as high as 4 or 5.

The ratio of I_{CS} to I_B is called *forced current gain* β_f where

$$\beta_f = \frac{I_{CS}}{I_B} < \text{natural current gain } \beta \text{ or } h_{FE} \qquad \ldots(2.17)$$

The total power loss in the two junctions of a transistor is

$$P_T = V_{BE} I_B + V_{CE} I_C \qquad \ldots(2.18)$$

Under saturated state, V_{BES} is greater than V_{CES}, this means BEJ is forward biased. Further Eq. (2.13) shows that V_{CB} is negative under saturated conditions, therefore, CBJ is also forward biased. In other words, under saturated conditions, both junctions in a power transistor are forward biased.

Example 2.1. *A bipolar transistor shown in Fig. 2.6 (a) has current gain $\beta = 40$. The load resistance $R_C = 10\ \Omega$, dc supply voltage $V_{CC} = 130\ V$ and input voltage to base circuit, $V_B = 10\ V$. For $V_{CES} = 1.0\ V$ and $V_{BES} = 1.5\ V$, calculate.*

(a) *the value of R_B for operation in the saturated state,*

(b) *the value of R_B for an over drive factor 5,*

(c) *forced-current gain and*

(d) *power loss in the transistor for both parts (a) and (b).*

Solution. Here $\beta = 40$, $R_C = 10\ \Omega$, $V_{CC} = 130\ V$, $V_B = 10\ V$, $V_{CES} = 1.0\ V$ and $V_{BES} = 1.5\ V$

(a) From Eq. (2.14), for operation in the saturated state,

$$I_{CS} = \frac{V_{CC} - V_{CES}}{R_C} = \frac{130 - 1.0}{10} = 12.90\ A$$

From Eq. (2.15), base current that produces saturation,

$$I_{BS} = \frac{I_{CS}}{\beta} = \frac{12.90}{40} = 0.3225\ A$$

Value of R_B for $I_{BS} = 0.3225\ A$ is given by Eq. (2.11) as,

$$R_B = \frac{V_B - V_{BES}}{I_{BS}} = \frac{10 - 1.5}{0.3225} = 26.357\ \Omega$$

(b) Base current with overdrive, from Eq. (2.16), is

$$I_B = ODF \times I_{BS} = 5 \times 0.3225 = 1.6125 \text{ A}$$

$$\therefore \quad R_B = \frac{10 - 1.5}{1.6125} = 5.27 \ \Omega$$

(c) Forced current gain, from Eq. (2.17), is

$$\beta_f = \frac{I_{CS}}{I_B} = \frac{12.90}{1.6125} = 8, \text{ which is less than the natural current gain } \beta = 40.$$

(d) Power loss in transistor, from Eq. (2.18), is

$$P_T = V_{BES} I_{BE} + V_{CES} I_{CS}$$

For normal base drive, $P_T = 1.5 \times 0.3225 + 1.0 \times 12.9 = 13.384$ W

With overdrive, $P_T = 1.5 \times 1.6125 + 1.0 \times 12.9 = 15.32$ W

It is seen from above that power loss with hard drive of transistor is more.

2.5.1.2. BJT Switching Performance. When base current is applied, a transistor does not turn on instantly because of the presence of internal capacitances. Fig. 2.9 shows the various switching waveforms of an *npn* power transistor with resistive load between collector and emitter, Fig. 2.8.

When input voltage v_B to base circuit is made $-V_2$ at t_0, junction *EB* or *EBJ* is reverse biased, $v_{BE} = -V_2$, the transistor is off, $i_B = I_C = 0$ and $v_{CE} = V_{CC}$, Fig. 2.9. At time t_1, input voltage v_B is made $+V_1$ and i_B rises to I_{B1} as shown in Fig. 2.9. After t_1, base-emitter voltage v_{BE} begins to rise gradually from $-V_2$ and collector current i_c begins to rise from zero (actually a small leakage current I_{CEO}

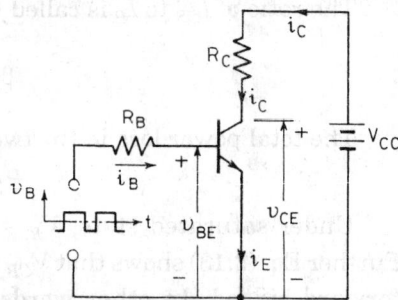

Fig. 2.8. *npn* transistor with resistive load.

exists as shown in Fig. 2.7 (a)) and collector- emitter voltage v_{CE} starts falling from its initial value V_{CC}. After some time delay t_d, called delay time, the collector current rises to 0.1 I_{CS}, v_{CE} falls from V_{CC} to 0.9 V_{CC} and v_{BE} reaches $V_{BES} = 0.7$ V. This delay time is required to charge the base-emitter capacitance to $V_{BES} = 0.7$ V. Thus, *delay time t_d* is defined as the time during which the collector current rises from zero to 0.1 I_{CS} and collector-emitter voltage falls from V_{CC} to 0.9 V_{CC}.

After delay time t_d, collector current rises from 0.1 I_{CS} to 0.9 I_{CS} and v_{CE} falls from 0.9 V_{CC} to 0.1 V_{CC} in time t_r. This time t_r is known as rise time which depends upon transistor junction capacitances. *Rise time t_r* is defined as the time during which collector current rises from 0.1 I_{CS} to 0.9 V_{CC} and collector-emitter voltage falls from 0.9 V_{CC} to 0.1 V_{CC}. This shows that total turn-on time $t_{on} = t_d + t_r$. Value of t_{on} is of the order of 30 to 300 nano seconds. The transistor remains in the on, or saturated, state so long as input voltage stays at V_1, Fig. 2.9 (a).

In case transistor is to be turned off, then input voltage v_B and input base current i_B are reversed. At time t_2, input voltage v_B to base circuit is reversed from V_1 to $-V_2$. At the same time, base current changes from I_{B1} to $-I_{B2}$ as shown in Fig. 2.9 (b). Negative base current I_{B2} removes excess carriers from the base. The time t_s required to remove these excess carriers is

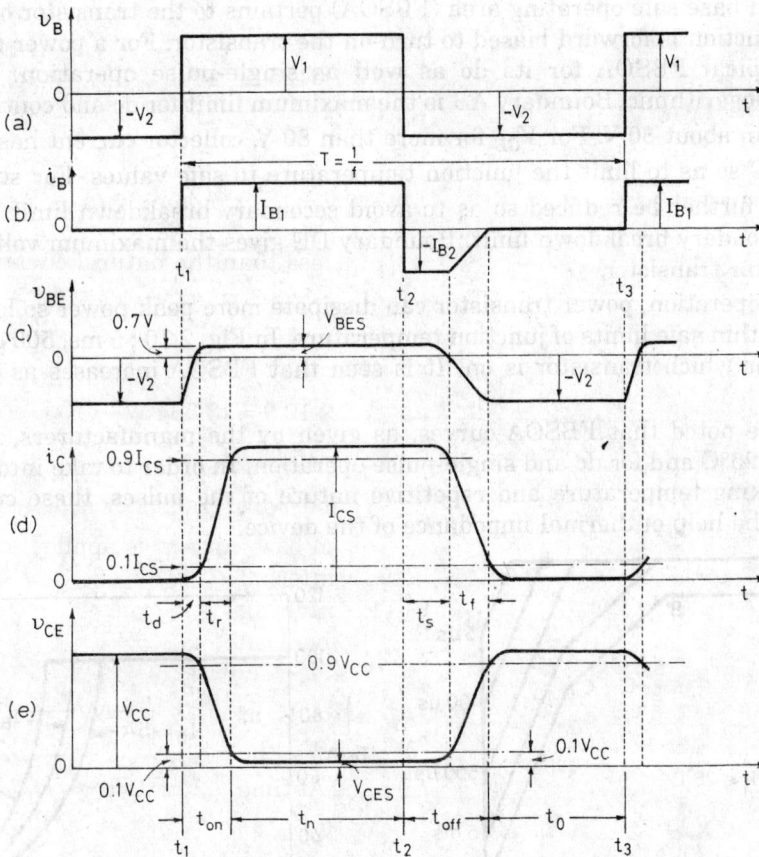

Fig. 2.9. Switching waveforms for *npn* power transistor of Fig. 2.8.

called storage time and only after t_s, base current I_{B2} begins to decrease towards zero. Transistor comes out of saturation only after t_s. *Storage time* t_s is usually defined as the time during which collector current falls from I_{CS} to 0.9 I_{CS} and collector-emitter voltage v_{CE} rises from V_{CES} to 0.1 V_{CC}, Fig. 2.9 (d) and (e). Negative input voltage enhances the process of removal of excess carriers from base and hence reduces the storage time and therefore, the turn-off time.

After t_s, collector current begins to fall and collector-emitter voltage starts building up. Time t_f, called *fall time*, is defined as the time during which collector current drops from 0.9 I_{CS} to 0.1 I_{CS} and collector-emitter voltage rises from 0.1 V_{CC} to 0.9 V_{CC}, Fig. 2.9 (d) and (e). Sum of storage time and fall time gives the transistor turn-off time t_{off} i.e. $t_{off} = t_s + t_f$. The various waveforms during transistor switching are shown in Fig. 2.9. In this figure, t_n = conduction period of transistor, t_o = off period, $T = 1/f$ is the periodic time and f is the switching frequency.

2.5.1.3. Safe Operating Area. The safe operating area (SOA or SOAR) of a power transistor specifies the safe operating limits of collector current I_C versus collector-emitter voltage V_{CE}. For reliable operation of the transistor, the collector current and voltage must always lie within this area. Actually, two types of safe operating areas are specified by the manufacturers, FBSOA and RBSOA.

The forward-base safe operating area (FBSOA) pertains to the transistor operation when base-emitter junction is forward biased to turn-on the transistor. For a power transistor, Fig. 2.10 shows typical FBSOA for its dc as well as single-pulse operation. The scale for I_C and V_{CE} are logarithmic. Boundary AB is the maximum limit for dc and continuous current for V_{CE} less than about 80 V. For V_{CE} for more than 80 V, collector current has to be reduced to boundary BC so as to limit the junction temperature to safe values. For still higher V_{CE}, current should further be reduced so as to avoid secondary breakdown limit. Boundary CD defines this secondary breakdown limit. Boundary DE gives the maximum voltage capability for this particular transistor.

For pulsed operation, power transistor can dissipate more peak power so long as average power loss is within safe limits of junction temperature. In Fig. 2.10 ; 5 ms, 500 μs etc. indicate pulse widths for which transistor is on. It is seen that FBSOA increases as pulse-width is decreased.

It should be noted that FBSOA curves, as given by the manufacturers, are for a case temperature of 25°C and for dc and single-pulse operation. In order to take into consideration the actual working temperature and repetitive nature of the pulses, these curves must be modified with the help of thermal impedance of the device.

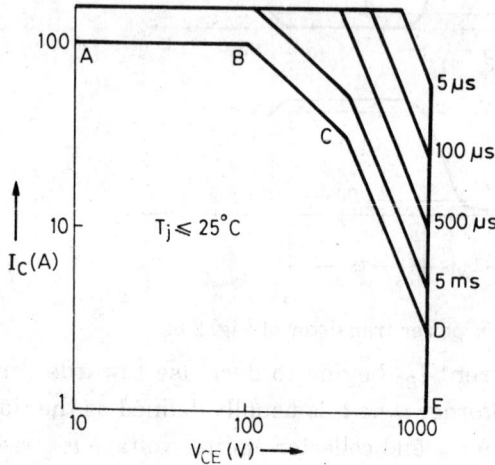

Fig. 2.10. Typical forward biased safe operating area (FBSOA) for a power transistor (logarithmic scale)

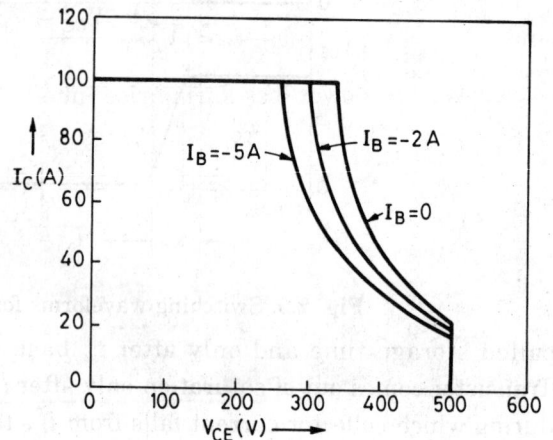

Fig. 2.11. Typical reverse-block safe operating area (RBSOA) for a power transistor.

During turn-off, a transistor is subjected to high current and high voltage with base-emitter junction reverse biased. Safe operating area for transistor during turn-off is specified as reverse blocking safe operating area (RBSOA). This RBSOA is a plot of collector current versus collector-emitter voltage as shown in Fig. 2.11. RBSOA specifies the limits of transistor operation at turn-off when the base current is zero or when the base-emitter junction is reverse biased (*i.e.* with base current negative). With increased reverse bias, area RBSOA decreases in size as shown in Fig. 2.11.

Example 2.2. *For a power transistor, typical switching waveforms are shown in Fig. 2.12(a). The various parameters of the transistor circuit are as under :*

$V_{CC} = 220$ V, $V_{CES} = 2$ V, $I_{CS} = 80$ A, $t_d = 0.4$ μs, $t_r = 1$ μs, $t_n = 50$ μs,

$t_s = 3$ μs, $t_f = 2$ μs, $t_o = 40$ μs, $f = 5$ kHz. *Collector to emitter leakage current itlaic= 2 mA.*

Determine average power loss due to collector current during t_{on} and t_n. Find also the peak instantaneous power loss due to collector current during turn-on time.

Solution. During delay time, the time limits are $0 \le t \le t_d$. Fig. 2.12 (a) shows that in this time, $i_C(t) = I_{CEO}$ and $v_{CE}(t) = V_{CC}$.

∴ Instantaneous power loss during delay time is

$$P_d(t) = i_C v_{CE} = I_{CEO} V_{CC} = 2 \times 10^{-3} \times 220 = 0.44 \text{ W}$$

Average power loss during delay time with $0 \le t \le t_d$ is given by

$$P_d = \frac{1}{T} \int_0^{t_d} i_C(t) \cdot v_{CE}(t) \, dt$$

$$= \frac{1}{T} \int_0^{t_d} I_{CEO} \cdot V_{CC} \, dt = f \cdot I_{CEO} \cdot V_{CC} \cdot t_d$$

$$= 5 \times 10^3 \times 2 \times 10^{-3} \times 220 \times 0.4 \times 10^{-6} = 0.88 \text{ mW}$$

where $f = \frac{1}{T}$ = frequency of transistor switching

During rise time, $0 \le t \le t_r$,

$$i_C(t) = \frac{I_{CS}}{t_r} \cdot t$$

and

$$v_{CE}(t) = \left[V_{CC} - \frac{V_{CC} - V_{CES}}{t_r} \cdot t \right]$$

∴ Average power loss during rise time is

$$P_r = \frac{1}{T} \int_0^{t_r} \frac{I_{CS}}{t_r} \cdot t \left[V_{CC} - \frac{V_{CC} - V_{CES}}{t_r} \cdot t \right] dt$$

$$= f \cdot I_{CS} \cdot t_r \left[\frac{V_{CC}}{2} - \frac{V_{CC} - V_{CES}}{3} \right]$$

$$= 5 \times 10^3 \times 80 \times 1 \times 10^{-6} \left[\frac{220}{2} - \frac{220 - 2}{3} \right] = 14.933 \text{ W}$$

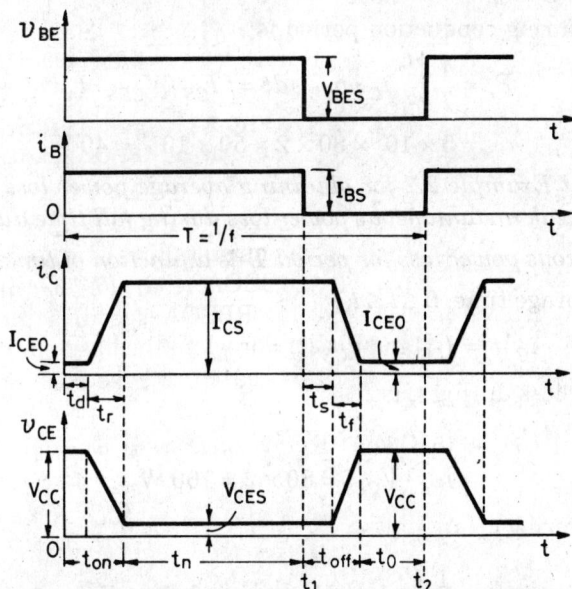

Fig. 2.12. (a) Switching waveforms for Examples 2.2 and 2.3.

Instantaneous power loss during rise time is

$$P_r(t) = \frac{I_{CS}}{t_r} \cdot t \left\{ V_{CC} - \frac{V_{CC} - V_{CES}}{t_r} \cdot t \right\}$$

$$= \frac{I_{CS} \cdot t}{t_r} V_{CC} - \frac{I_{CS} \cdot t^2}{t_r^2} [V_{CC} - V_{CES}] \qquad \ldots(i)$$

$\dfrac{d P_r(t)}{dt} = 0$ gives time t_m at which instantaneous power loss during t_r would be maximum.

It is seen from Eq. (i) that

$$t_m = \frac{V_{CC} \cdot t_r}{2 [V_{CC} - V_{CES}]} = \frac{220 \times 1 \times 10^{-6}}{2 [220 - 2]} = 0.5046 \; \mu s$$

Peak instantaneous power loss P_{rm} during rise time is obtained by substituting the value of $t = t_m$ in Eq. (i).

$\therefore$

$$P_{rm} = \frac{I_{CS}}{t_r} \cdot \frac{V_{CC}^2 \cdot t_r}{2 [V_{CC} - V_{CES}]} - \frac{I_{CS}}{t_r^2} \frac{(V_{CC} \cdot t_r)^2 [V_{CC} - V_{CES}]}{4 [V_{CC} - V_{CES}]^2}$$

$$= \frac{I_{CS} \cdot V_{CC}^2}{4 [V_{CC} - V_{CES}]} = \frac{80 \times 220^2}{4 [220 - 2]} = 4440.4 \; W$$

Total average power loss during turn-on

$$P_{on} = P_d + P_r = 0.00088 + 14.933 = 14.9339 \; W$$

During conduction time, $0 \leq t \leq t_n$

$$i_C(t) = I_{CS} \text{ and } V_{CE}(t) = V_{CES}$$

Instantaneous power loss during t_n is

$$P_n(t) = i_C \cdot v_{CE} = I_{CS} \cdot V_{CES} = 80 \times 2 = 160 \; W$$

Average power loss during conduction period is

$$P_n = \frac{1}{T} \int_0^{t_n} i_C \cdot v_{CE} \cdot dt = f I_{CS} \cdot V_{CES} \cdot t_n$$

$$= 5 \times 10^3 \times 80 \times 2 \times 50 \times 10^{-6} = 40 \; W.$$

Example 2.3. *Repeat Example 2.2 for obtaining average power loss during turn-off time and off-period, and also peak instantaneous power loss during fall time due to collector current.*

Sketch the instantaneous power loss for period T as a function of time.

Solution. During storage time, $0 \leq t \leq t_s$,

$$i_C(t) = I_{CS} \text{ and } v_{CE}(t) = V_{CES}$$

Instantaneous power loss during t_s is

$$P_s(t) = i_C(t) \, v_{CE}(t)$$

$$= I_{CS} \cdot V_{CES} = 80 \times 2 = 160 \; W$$

Average power loss during t_s is

$$P_s = \frac{1}{T} \int_0^{t_s} I_{CS} \cdot V_{CES} \cdot dt = f \cdot I_{CE} \cdot V_{CES} \cdot t_s$$

$$= 5 \times 10^3 \times 80 \times 2 \times 3 \times 10^{-6} = 2.4 \; W$$

During fall time, $0 \leq t \leq t_f$, $i_C(t) = \left[I_{CS} - \dfrac{I_{CS} - I_{CEO}}{t_f} \cdot t \right]$

During t_f, I_{CEO} is negligibly small in comparison with I_{CS},

$$\therefore \qquad i_C(t) = I_{CS} \left[1 - \frac{t}{t_f} \right]$$

and

$$v_{CE}(t) = \frac{V_{CC} - V_{CES}}{t_f} \cdot t$$

Average power loss during fall time is

$$P_f = \frac{1}{T} \int_0^{t_f} I_{CS} \left(1 - \frac{t}{t_f} \right) \left[\frac{V_{CC} - V_{CES}}{t_f} \cdot t \right] dt$$

$$= f(V_{CC} - V_{CES}) \cdot t_f \left[\frac{I_{CS}}{2} - \frac{I_{CS}}{3} \right]$$

$$= f \cdot t_f \cdot \frac{I_{CS}}{6} [V_{CC} - V_{CES}]$$

$$= 5 \times 10^3 \times 3 \times 10^{-6} \times 80 \times \frac{1}{6} \times (220 - 2) = 43.6 \text{ W}$$

Instantaneous power loss during fall time is

$$P_f(t) = I_{CS} \left[1 - \frac{t}{t_f} \right] \left[\frac{V_{CC} - V_{CES}}{t_f} t \right]$$

$$= \frac{I_{CS}(V_{CC} - V_{CES}) \cdot t}{t_f} - I_{CS}(V_{CC} - V_{CES}) \frac{t^2}{t_f^2}$$

$\dfrac{dP_f(t)}{dt} = 0$ gives time t_m at which instantaneous power loss during t_r would be maximum. Here $t_m = t_f/2$.

$\therefore$ Peak instantaneous power dissipation during t_f is

$$P_{fm} = I_{CS} \left(1 - \frac{1}{2} \right) \left(\frac{V_{CC} - V_{CES}}{2} \right) = \frac{I_{CS}(V_{CC} - V_{CES})}{4}$$

$$= \frac{80(220 - 2)}{4} = 4360 \text{ W}$$

Total average power loss during turn-off process is
$$P_{off} = P_s + P_f = 2.4 + 43.6 = 46 \text{ W}$$

During off-period, $0 \leq t \leq t_0$,
$$i_C(t) = I_{CEO} \text{ and } v_{CE}(t) = V_{CC}$$

Instantaneous power loss during t_o is
$$P_0(t) = i_C \cdot v_{CE} = I_{CEO} \cdot V_{CC} = 2 \times 10^{-3} \times 220 = 0.44 \text{ W}$$

Average power loss during t_o is

$$P_0 = \frac{1}{T} \int_0^{t_0} P_0(t) dt = f I_{CEO} \cdot V_{CC} \cdot t_0$$

$$= 5 \times 10^3 \times 2 \times 10^{-3} \times 220 \times 40 \times 10^{-6} = 0.088 \text{ W}$$

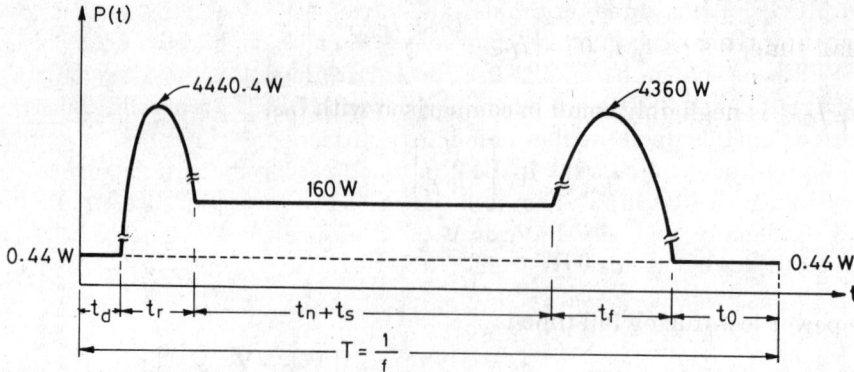

Fig. 2.12. (*b*) Sketch of instantaneous power loss in a transistor for Examples 2.2 and 2.3.

Total average power loss in power-transistor due to collector current over a period T is

$$P_T = P_{on} + P_n + P_{off} + P_0 = 14.9339 + 40 + 46 + 0.088 = 101.022 \text{ W}.$$

From the data obtained in Examples 2.2 and 2.3, the power loss variation as a function of time, over a period T, is sketched in Fig. 2.12(*b*).

Example 2.4. *A power transistor has its switching waveforms as shown in Fig. 2.13. If the average power loss in the transistor is limited to 300 W, find the switching frequency at which this transistor can be operated.*

Solution.

Energy loss during turn-on

$$= \int_0^{t_{on}} i_C \cdot v_{CE} \, dt$$

$$= \int_0^{t_{on}} \left(\frac{I_{CS}}{50} \times 10^6 \, t \right) \left(V_{CC} - \frac{V_{CC}}{40} \times 10^6 \, t \right) dt$$

$$= \int_0^{t_{on}} (2 \times 10^6 t)(200 - 5 \times 10^6 t) \, dt$$

$$= 0.1067 \text{ watt–sec}$$

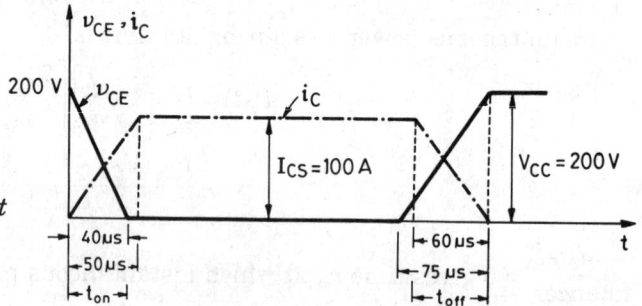

Fig. 2.13. Switching waveform for Example 2.4.

Energy loss during turn-off $= \int_0^{t_{off}} \left(100 - \frac{100}{60} \times 10^6 \, t \right) \left(\frac{200}{75} \times 10^6 \, t \right) dt$

$$= 0.1603 \text{ watt–sec}$$

Total energy loss in one cycle

$$= 0.1067 + 0.1603 = 0.267 \text{ W-sec}$$

Average power loss in transistor

$$= \text{switching frequency} \times \text{energy loss in one cycle}$$

∴ Allowable switching frequency,

$$f = \frac{300}{0.267} = 1123.6 \text{ Hz}$$

2.6. POWER MOSFETs

A metal-oxide-semiconductor field-effect transistor (MOSFET) is a recent device developed by combining the areas of field-effect concept and MOS technology.

A power MOSFET has three terminals called drain (D), source (S) and gate (G) in place of the corresponding three terminals collector, emitter and base for BJT. The circuit symbol of power MOSFET is as shown in Fig. 2.14 (a). Here arrow indicates the direction of electron flow. A BJT is a current controlled device whereas a power MOSFET is a voltage-controlled device. As its operation depends upon the flow of majority carriers only, MOSFET is a unipolar device. The control signal, or base current in BJT is much larger than the control signal (or gate current) required in a MOSFET. This is because of the fact that gate circuit impedance in MOSFET is extremely high, of the order of 10^9 ohm. This large impedance permits the MOSFET gate to be driven directly from microelectronic circuits. BJT suffers from second breakdown voltage whereas MOSFET is free from this problem. Power MOSFETs are now finding increasing applications in low-power high frequency converters.

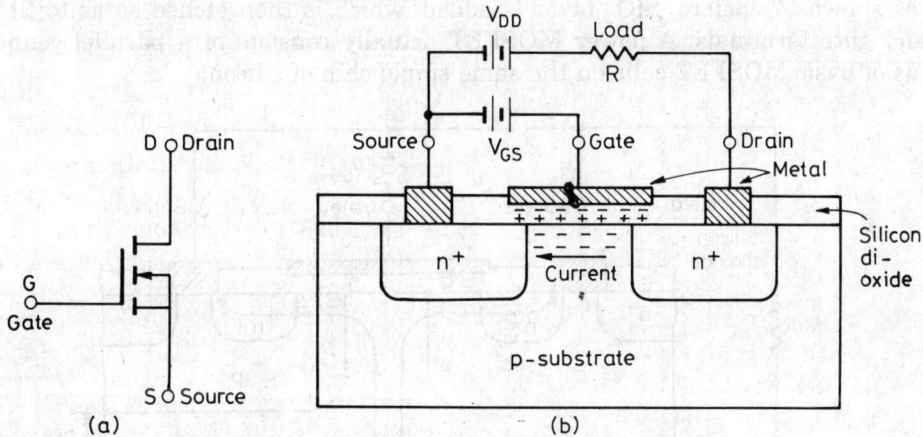

Fig. 2.14. *N*-channel enhancement power MOSFET
(*a*) circuit symbol and (*b*) its basic structure.

Power MOSFETs are of two types ; *n*-channel enhancement MOSFET and *p*-channel enhancement MOSFET. Out of these two types, *n*-channel enhancement MOSFET is more common because of higher mobility of electrons. As such, only this type of MOSFET is studied in what follows.

A simplified structure of *n*-channel planar MOSFET of low power rating is shown in Fig. 2.14 (*b*). On *p*-substrate (or body), two heavily doped n^+ regions are diffused as shown. An insulating layer of silicon dioxide (SiO_2) is grown on the surface. Now this insulating layer is etched in order to embed metallic source and drain terminals. Note that n^+ regions make contact with source and drain terminals as shown. A layer of metal is also deposited on SiO_2 layer so as to form the gate of MOSFET in between source and drain terminals, Fig. 2.14 (*b*).

When gate circuit is open, junction between n^+ region below drain and *p*-substrate is reverse biased by input voltage V_{DD}. Therefore, no current flows from drain to source and load. When gate is made positive with respect to source, an electric field is established as shown in Fig. 2.14 (*b*). Eventually, induced negative charges in the *p*-substrate below SiO_2 layer are formed thus causing the *p* layer below gate to become an induced *n* layer. These negative charges, called electrons, form *n*-channel between two n^+ regions and current can flow from drain to source as shown by the arrow. If V_{GS} is made more positive, induced *n*-channel becomes more deep and therefore more current flows from *D* to *S*. This shows that drain current I_D is enhanced by the gradual increase of gate voltage, hence the name enhancement MOSFET.

The main disadvantage of n-channel planar MOSFET of Fig. 2.14 (b) is that conducting n-channel in between drain and source gives large on-state resistance. This leads to high power dissipation in n-channel. This shows that planar MOSFET construction of Fig. 2.14 (b) is feasible only for low-power MOSFETs.

The constructional details of high power MOSFET are illustrated in Fig. 2.15. In this figure is shown a planar diffused metal-oxide-semiconductor (DMOS) structure for n-channel which is quite common for power MOSFETs. On n^+ substrate, high resistivity n^- layer is epitaxially[*] grown. The thickness of n^- layer determines the voltage blocking capability of the device. On the other side of n^+ substrate, a metal layer is deposited to form the drain terminal. Now p regions are diffused in the epitaxially grown n^- layer. Further, n^+ regions are diffused in p regions as shown. As before, SiO_2 layer is added, which is then etched so as to fit metallic source and gate terminals. A power MOSFET actually consists of a parallel connection of thousands of basic MOSFET cells on the same single chip of silicon.

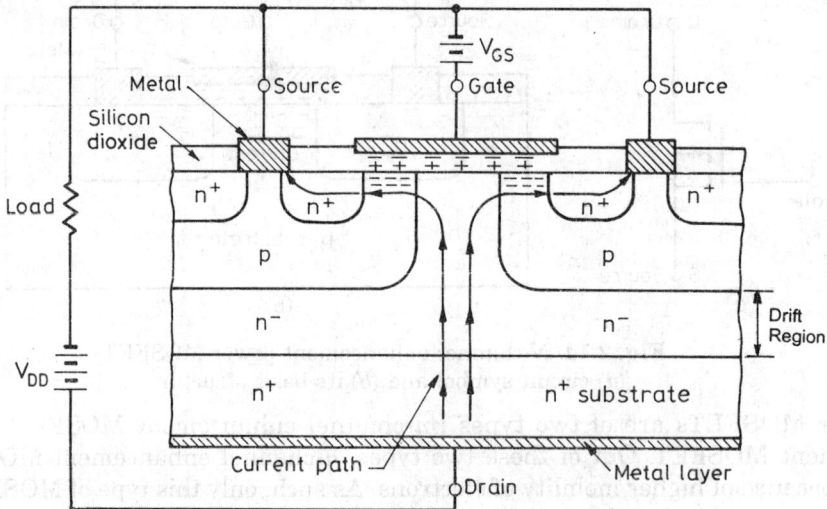

Fig. 2.15. Basic structure of a n-channel DMOS power MOSFET.

When gate circuit voltage is zero, and V_{DD} is present, $n^- - p^-$ junctions are reverse biased and no current flows from drain to source. When gate terminal is made positive with respect to source, an electric field is established and electrons form n-channel in the p^- regions as shown. So a current from drain to source is established as indicated by arrows. With gate voltage increased, current I_D also increases as expected. Length of n-channel can be controlled and therefore on-resistance can be made low if short length is used for the channel.

An examination of the basic structure of n-channel DMOS power MOSFET (PMOSFET) reveals that a parasitic npn bipolar junction transistor exists between the source and drain as shown in Fig. 2.16. The p body acts as the base, n^+ layer as the emitter (or source) and n^- layer as the collector (or drain) of this BJT. Since source is connected to both base and emitter of parasitic BJT, the source short circuits both base and emitter. As a result, potential difference between base and emitter of the parasitic BJT is zero and therefore, BJT is always in the cut-off state.

---

[*] A mixture of silicon atoms and pentavalent atoms, deposited on wafer, forms a layer of n-type semiconductor on heated surface. This layer is called epitaxial layer.

Also, vertical travel from source to drain indicates the existence of a parasitic diode as shown on the right in Fig. 2.16. The parasitic diode, with source acting as anode and drain as cathode may be used in half-bridge or full-bridge rectifiers. The parasitic diode also shows that reverse voltage blocking capability of PMOSFET is almost zero. This in-built diode is an advantage in inverter circuits.

In Fig. 2.15, source is negative and drain is positive. Therefore, electrons flow from source to n^+ layer, then through n-channel of p layer and further through n^- and n^+ layers to drain. The current must flow opposite to the

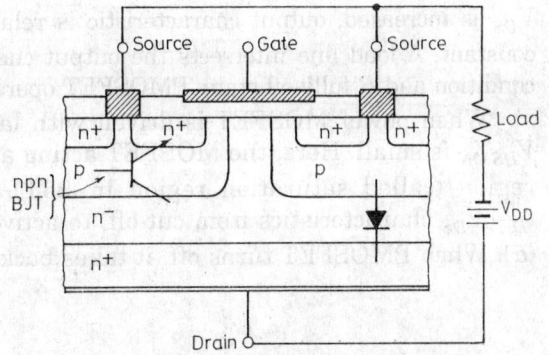

Fig. 2.16. PMOSFET showing parasitic BJT and parasitic diode.

flow of electrons as indicated in Fig. 2.15. Since the conduction of current is due to the movement of electrons only, PMOSFET is a majority carrier device. Hence, time delays caused by removal or recombination of minority carriers are eliminated during the turn-off process of this device. PMOSFET with a turn-off time of 100 ns are available. Owing to its low turn-off time, PMOSFET can be operated in a frequency range of 1 to 10 MHz.

2.6.1. PMOSFET Characteristics

The static characteristics of power MOSFET are now described briefly. The basic circuit diagram for n-channel PMOSFET is shown in Fig. 2.17 where voltage and currents are as indicated. The source terminal S is taken as common terminal, as usual, between the input and output of a MOSFET.

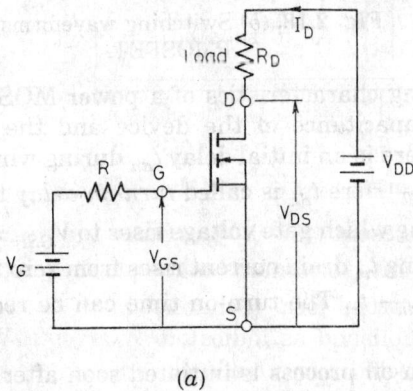

Fig. 2.17. N-channel power MOSFET (a) circuit diagram and (b) its typical transfer characteristic.

(a) **Transfer Characteristics.** This characteristic shows the variation of drain current I_D as a function of gate-source voltage V_{GS}. Fig. 2.17 (b) shows typical transfer characteristics for n-channel PMOSFET. Threshold voltage V_{GST} is an important parameter of MOSFET. V_{GST} is the minimum positive voltage between gate and source to induce n-channel. Thus, for threshold voltage below V_{GST}, device is in the off-state. Magnitude of V_{GST} is of the order of 2 to 3 V.

(b) **Output Characteristics.** PMOSFET output characteristics, shown in Fig. 2.18 (a), indicate the variation of drain current I_D as a function of drain-source voltage V_{DS}, with gate-source voltage V_{GS} as a parameter. For low values of V_{DS}, the graph between $I_D - V_{DS}$ is almost linear; this indicates a constant value of on-resistance $R_{DS} = V_{DS}/I_D$. For given V_{GS}, if

V_{DS} is increased, output characteristic is relatively flat, indicating that drain current is nearly constant. A load line intersects the output characteristics at A and B. Here A indicates fully-on condition and B fully-off state. PMOSFET operates as a switch either at A or at B just like a BJT.

When power MOSFET is driven with large gate-source voltage, MOSFET is turned on, $V_{DS.ON}$ is small. Here, the MOSFET acting as a closed switch, is said to be driven into ohmic region (called saturation region in BJT). When device turns on, PMOSFET traverses $i_D - V_{DS}$ characteristics from cut-off, to active region and then to the ohmic region, Fig. 2.18 (a). When PMOSFET turns off, it takes backward journey from ohmic region to cut-off state.

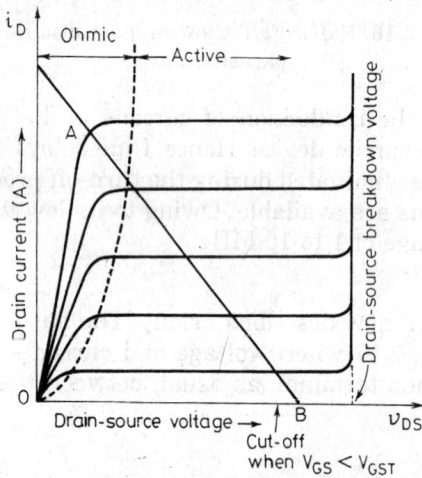

Fig. 2.18. (a) Output characteristics of PMOSFET.

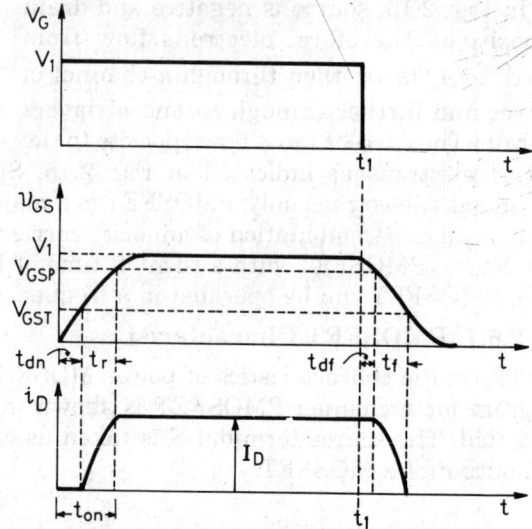

Fig. 2.18. (b) Switching waveforms for PMOSFET.

(c) **Switching characteristics.** The switching characteristics of a power MOSFET are influenced to a large extent by the internal capacitance of the device and the internal impedance of the gate drive circuit. At turn-on, there is an initial delay t_{dn} during which input capacitance charges to gate threshold voltage V_{GST}. Here t_{dn} is called *turn-on delay* time.

There is further delay t_r, called *rise time*, during which gate voltage rises to V_{GSP}, a voltage sufficient to drive the MOSFET into on state. During t_r, drain current rises from zero to full-on current I_D. Thus, the total turn-on-time is $t_{on} = t_{dn} + t_r$. The turn-on time can be reduced by using low-impedance gate-drive source.

As MOSFET is a majority carrier device, turn-off process is initiated soon after removal of gate voltage at time t_1. The turn-off delay time, t_{df}, is the time during which input capacitance discharges from overdrive gate voltage V_1 to V_{GSP}. The *fall time, t_f,* is the time during which input capacitance discharges from V_{GSP} to threshold voltage. During t_f, drain current falls from I_D to zero. So when $V_{GS} \leq V_{GST}$, PMOSFET turn-off is complete. Switching waveforms for a power MOSFET are shown in Fig. 2.18 (b).

2.6.2. PMOSFET Applications

The on-resistance of MOSFET increase with voltage rating ; this makes the device very lossy at high-current applications. Since the on-resistance has positive temperature coefficient, parallel operation of PMOSFETs is relatively easy. The positive temperature coefficient also reduces the second breakdown effect in PMOSFETs.

PMOSFETs find applications in high-frequency switching applications, varying from a few watts to few kWs. The device is very popular in switched-mode power supplies and inverters. These are, at present available with 500 V, 140 A ratings.

2.6.3. Comparison of PMOSFET with BJT

The three terminals in a PMOSFET are designated as gate, source and drain. In a BJT, the corresponding three terminals are base, emitter and collector. A PMOSFET has several features different from those of BJT. These are outlined below :

(i) BJT is a bipolar device whereas PMOSFET is a unipolar device.

(ii) A PMOSFET has high input impedance (mega ohm) whereas input impedance of BJT is low (a few kilo-ohm).

(iii) PMOSFET has lower switching losses but its on-resistance and conduction losses are more. A BJT has higher switching losses but lower conduction loss. So, at high frequency applications, PMOSFET is the obvious choice. But at lower operating frequencies (less than about 10 to 20 kHz), BJT is superior.

(iv) PMOSFET is voltage controlled device whereas BJT is current controlled device.

(v) PMOSFET has positive temperature coefficient for resistance. This makes parallel operation of PMOSFETs easy. If a PMOSFET shares increased current initially, it heats up faster, it resistance rises and this increased resistance causes this current to shift to other devices in parallel. A BJT has negative temperature coefficient, so current sharing resistors are necessary during parallel operation of BJTs.

(vi) In PMOSFETs, secondary breakdown does not occur, because it has positive temperature coefficient. As BJT has negative temperature coefficient, secondary breakdown does occur. In BJT, with decrease in resistance with rise in temperature, the current increases. This increased current over the same area results in hot spots and breakdown of the BJT.

(vii) PMOSFETs in higher voltage ratings have more conduction loss.

(viii) The state of the art PMOSFETs are available with ratings upon 500 V, 140 A whereas BJTs are available with ratings upto 1200 V, 800 A.

2.7. INSULATED GATE BIPOLAR TRANSISTOR (IGBT)

IGBT has been developed by combining into it the best qualities of both BJT and PMOSFET. Thus an IGBT possesses high input impedance like a PMOSFET and has low on-state power loss as in a BJT. Further, IGBT is free from second breakdown problem present in BJT. All these merits have made IGBT very popular amongst power-electronics engineers. IGBT is also known as metal oxide insulated gate transistor (MOSIGT), conductively-modulated field effect transistor (COMFET) or gain-modulated FET (GEMFET). It was also initially called insulated gate transistor (IGT).

2.7.1. Basic Structure

Fig. 2.19 illustrates the basic structure of an IGBT. It is constructed virtually in the same manner as a power MOSFET. There is, however, a major difference in the substrate. The n^+ layer substrate at the drain in a PMOSFET is now substituted in the IGBT by a p^+ layer substrate called collector C. Like a power MOSFET, an IGBT has also thousands of basic structure cells connected appropriately on a single chip of silicon.

In IGBT, p^+ substrate is called *injection layer* because it injects holes into n^- layer. The n^- layer is called drift region. As in other semiconductor devices, thickness of n^- layer determines the voltage blocking capability of IGBT. The p layer is called body of IGBT. The n^- layer in between p^+ and p regions serves to accommodate the depletion layer of pn^- junction, *i.e.* junction J_2.

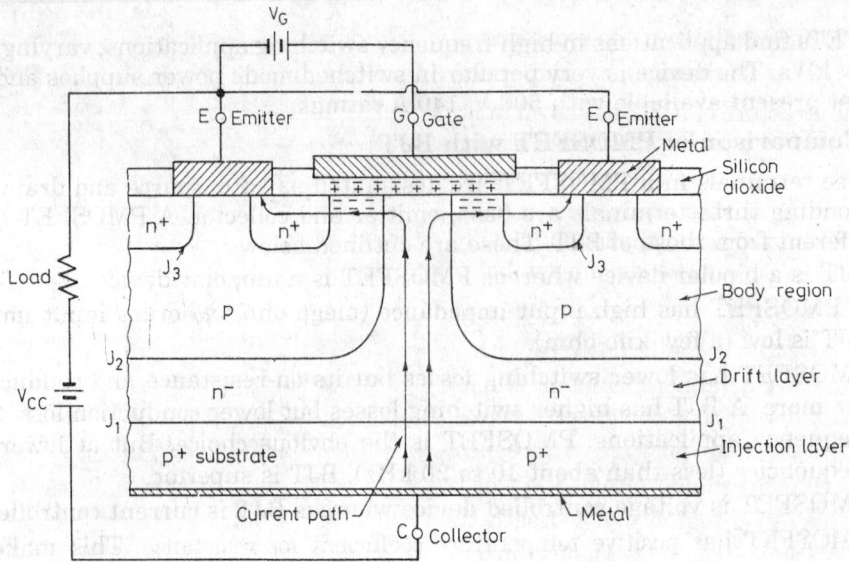

Fig. 2.19. Basic structure of an insulated gate bipolar transistor (IGBT)

2.7.2. Equivalent Circuit

An examination of Fig. 2.19 reveals that if we move vertically up from collector to emitter, we come across p^+, n^-, p layers. Thus, IGBT can be thought of as the combination of MOSFET and p^+n^-p transistor Q_1 as shown in Fig. 2.20 (b). Here R_d is resistance offered by n^- drift region. Fig. 2.20 (b) gives an approximate equivalent circuit of an IGBT.

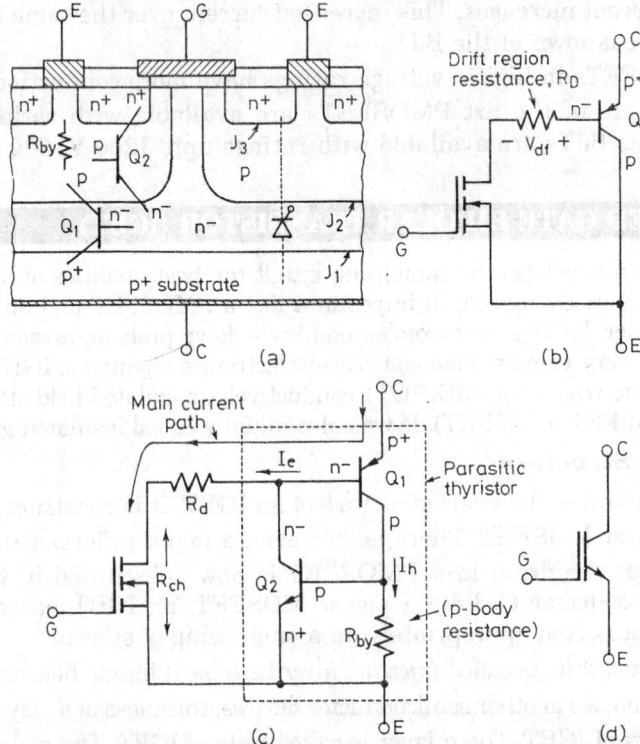

Fig. 2.20. IGBT (a) basic structure showing parasitic transistors and thyristor (b) approximate equivalent circuit (c) exact equivalent circuit and (d) circuit symbol.

Fig. 2.20 (a) also shows the existence of another path from collector to emitter; this path is collector, p^+, n^-, p (n-channel), n^+ and emitter. There is, thus, another inherent transistor Q_2 as n^-pn^+ in the structure of IGBT as shown in Fig. 2.20(a). The interconnection between two transistors Q_1 and Q_2 is shown in Fig. 2.20 (c). This figure gives the complete equivalent circuit of an IGBT. Here R_{by} is the resistance offered by p region to the flow of hole current I_h.

The two transistor equivalent circuit shown in Fig. 2.20 (c) illustrates that an IGBT structure has a parasitic thyristor in it. Parasitic thyristor is also shown dotted in Fig. 2.20 (a). Fig. 2.20 (d) gives the circuit symbol of an IGBT.

2.7.3. Working

When collector is made positive with respect to emitter, IGBT gets forward biased. With no voltage between gate and emitter, two junctions between n^- region and p region (i.e. junction J_2) are reverse biased; so no current flows from collector to emitter, Fig. 2.19.

When gate is made positive with respect to emitter by voltage V_G, with gate-emitter voltage more than the threshold voltage V_{GET} of IGBT, an n-channel or *inversion layer*, is formed in the upper part of p region just beneath the gate, as in PMOSFET, Fig. 2.19. This n-channel short-circuits the n^- region with n^+ emitter regions. Electrons from the n^+ emitter begin to flow to n^- drift region through n-channel. As IGBT is forward biased with collector positive and emitter negative, p^+ collector region injects holes into n^- drift region. In short, n^- drift region is flooded with electrons from p-body region and holes from p^+ collector region. With this, the injection carrier density in n^- drift region increases considerably and as a result, conductivity of n^- region enhances significantly. Therefore, IGBT gets turned on and begins to conduct forward current I_C.

Current I_C, or I_E, consists of two current components : (i) hole current I_h due to injected holes flowing from collector, p^+n^-p transistor Q_1, p-body region resistance R_{by} and emitter and (ii) electronic current I_e due to injected electrons flowing from collector, injection layer p^+, drift region n^-, n-channel resistance R_{ch}, n^+ and emitter. This means that collector, or load, current I_C = emitter current $I_E = I_h + I_e$.

Major component of collector current is electronic current I_e, i.e. main current path for collector, or load, current is through p^+, n^-, drift resistance R_d and n-channel resistance R_{ch} as shown in Fig. 2.20 (c). Therefore, the voltage drop in IGBT in its on-state is

$$V_{CE.on} = I_C \cdot R_{ch} + I_C \cdot R_d + V_{j1}$$

$$= \text{Voltage drop [in } n\text{-channel + across drift in } n^- \text{ region}$$
$$+ \text{ across forward biased } p^+n^- \text{ junction } J_1]$$

Here V_{j1} is usually 0.7 to 1 V as in a p-n diode. The voltage drop $I_c \cdot R_{ch}$ is due to n- channel resistance, almost the same as in a PMOSFET. The voltage drop $V_{df} = I_c \cdot R_d$ in IGBT is much less than that in PMOSFET. It is due to substantial increase in the conductivity caused by injection of electrons and holes in n^- drift region. The conductivity increase is the main reason for low on-state voltage drop in IGBT than it is in PMOSFET.

2.7.4. Latch-up in IGBT

It is seen from Fig. 2.20 (a) and (c) that IGBT structure has two inherent transistors Q_1 and Q_2, which constitute a parasitic thyristor. When IGBT is on, the hole-current flows through transistor p^+n^-p and p-body resistance R_{by}. If load current I_c is large, hole component of current I_h would also be large. This large current would increase the voltage drop $I_h \cdot R_{by}$ which may forward bias the base p-emitter n^+ junction of transistor Q_2. As a consequence, parasitic transistor Q_2 gets turned on which further facilitates in the turn-on of parasitic transistor p^+n^-p labelled Q_1. The parasitic thyristor, consisting of Q_1 and Q_2, eventually latches on through regenerative action, when sum of their current gains $\alpha_1 + \alpha_2$ reaches unity as in a conventional thyristor (discussed in chapter 4). With parasitic thyristor on, IGBT latches up and after this, collector emitter current is no longer under the control of gate terminal. The only way now to turn-off the latched up IGBT is by forced commutation of current as is done in a conventional thyristor. If this latch up is not aborted quickly, excessive power dissipation may destroy the IGBT. The latch up discussed here occurs when the collector current I_{CE} exceeds a certain critical value. The device manufactures always specify the maximum permissible value of load current I_{CE} that IGBT can handle without latch up.

At present, several modifications in the fabrication techniques are listed in the literature which are used to avoid latch-up in IGBTs. As such, latch-up free IGBTs are available.

2.7.5. IGBT Characteristics

The circuit of Fig. 2.21 (a) shows the various parameters pertaining to IGBT characteristics.

Static I-V or output characteristics of an IGBT (n-channel type) show the plot of collector current I_c versus collector-emitter voltage V_{CE} for various values of gate-emitter voltages V_{GE1}, V_{GE2} etc. These characteristics are shown in Fig. 2.21 (b). In the forward direction, the shape of the output characteristics is similar to that of BJT. But here the controlling parameter is gate-emitter voltage V_{GE} because IGBT is a voltage-controlled device. When the device is off, junction J_2 blocks forward voltage and in case reverse voltage appears across collector and emitter, junction J_1 blocks it. In Fig. 2.21 (b), V_{RM} is the maximum reverse breakdown voltage.

The transfer characteristic of an IGBT is a plot of collector current I_C versus gate-emitter voltage V_{GE} as shown in Fig. 2.21 (c). This characteristic is identical to that of power MOSFET. When V_{GE} is less than the threshold voltage V_{GET}, IGBT is in the off-state.

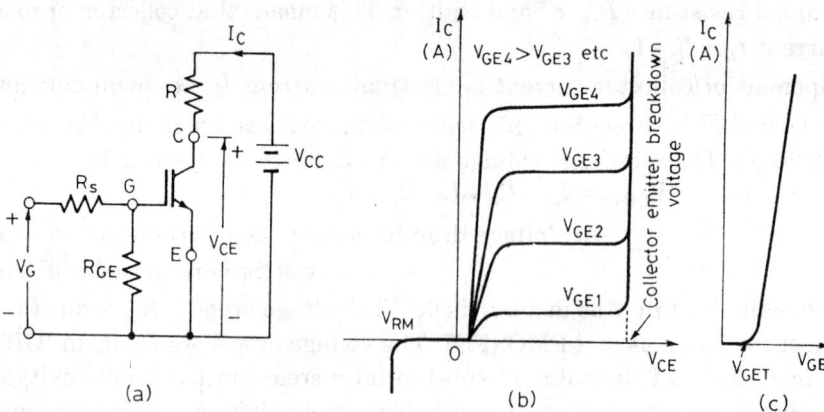

Fig. 2.21. IGBT (a) circuit diagram (b) static I-V characteristics and (c) transfer characteristics.

2.7.6. Switching Characteristics

Switching characteristics of an IGBT during turn-on and turn-off are sketched in Fig. 2.22. The turn-on time is defined as the time between the instants of forward blocking to forward on-state (7). Turn-on time is composed of delay time t_{dn} and rise time t_r, i.e. $t_{on} = t_{dn} + t_r$. The *delay time* is defined as the time for the collector-emitter voltage to fall from V_{CE} to 0.9 V_{CE}. Here V_{CE} is the initial collector-emitter voltage. Time t_{dn} may also be defined as the time for the

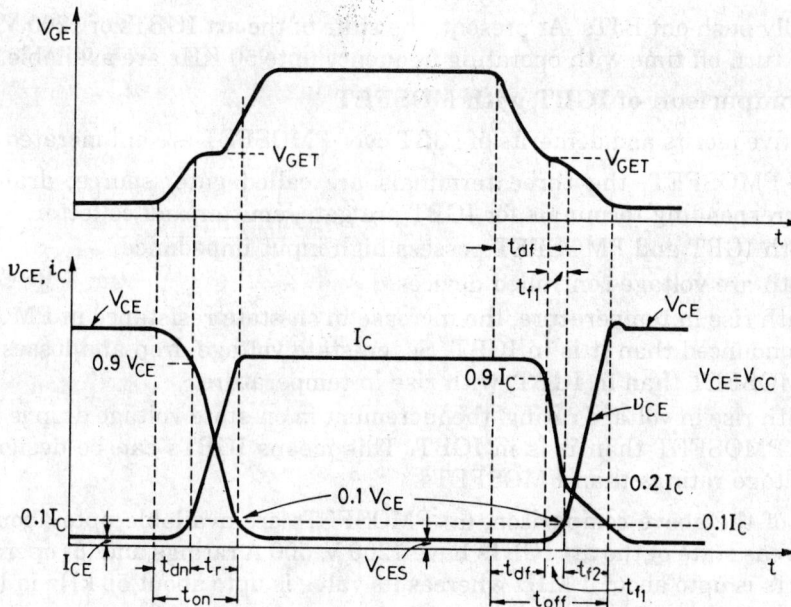

Fig. 2.22. IGBT turn-on and turn-off characteristics.

collector current to rise from its initial leakage current I_{CE} to $0.1\,I_C$. Here I_C is the final value of collector current.

The *rise time* t_r is the time during which collector-emitter voltage falls from $0.9\,V_{CE}$ to $0.1\,V_{CE}$. It is also defined as the time for the collector current to rise from $0.1\,I_C$ to its final value I_C. After time t_{on}, the collector current is I_C and the collector-emitter voltage falls to small value called conduction drop $= V_{CES}$ where subscript S denotes saturated value.

The turn-off time is somewhat complex. It consists of three intervals : (*i*) delay time, t_{df} (*ii*) initial fall time, t_{f1} and (*iii*) final fall time, t_{f2} ; *i.e.* $t_{off} = t_{df} + t_{f1} + t_{f2}$. The delay time is the time during which gate voltage falls from V_{GE} to threshold voltage V_{GET}. As V_{GE} falls to V_{GET} during t_{df}, the collector current falls from I_C to $0.9\,I_C$. At the end of t_{df}, collector-emitter voltage begins to rise. The first fall time t_{f1} is defined as the time during which collector current falls from 90 to 20% of its initial value I_C, or the time during which collector-emitter voltage rises from V_{CES} to $0.1\,V_{CE}$.

The final fall time t_{f2} is the time during which collector current falls from 20 to 10% of I_C, or the time during which collector-emitter voltage rises from $0.1\,V_{CE}$ to final value V_{CE}, see Fig. 2.22.

2.7.7. Application of IGBTs

IGBTs are widely used in medium power applications such as dc and ac motor drives, UPS systems, power supplies and drives for solenoids, relays and contactors. Though IGBTs are somewhat more expensive than BJTs, yet they are becoming popular because of lower gate-drive requirements, lower switching losses and smaller snubber circuit requirements. IGBT converters are more efficient with less size as well as cost, as compared to converters based on BJTs. Recently, IGBT inverter induction-motor drives using 15-20 kHz switching frequency are finding favour where audio-noise is objectionable. In most applications, IGBTs will eventually push out BJTs. At present, the state of the art IGBTs of 1200 V, 500 A ratings, 0.25 to 20 μs turn-off time with operating frequency upto 50 KHz are available.

2.7.8. Comparison of IGBT with MOSFET

The relative merits and demerits of IGBT over PMOSFET are enumerated below :

(i) In PMOSFET, the three terminals are called gate, source, drain whereas the corresponding terminals for IGBT are gate, emitter and collector.

(ii) Both IGBT and PMOSFET possess high input impedance.

(iii) Both are voltage-controlled devices.

(iv) With rise in temperature, the increase in on-state resistance in PMOSFET is much pronounced than it is in IGBT. So, on-state voltage drop and losses rise rapidly in PMOSFET than in IGBT, with rise in temperature.

(v) With rise in voltage rating, the increment in on-state voltage drop is more dominant in PMOSFET than it is in IGBT. This means IGBTs can be designed for higher-voltage ratings than PMOSFETs.

In view of the above comparison, (a) PMOSFETs are available upto about 500 V, 140 A ratings whereas state of the art IGBTs have 1200 V, 500 A ratings and (b) operating frequency in PMOSFETs is upto about 1 MHz whereas its value is upto about 50 kHz in IGBTs.

2.8. STATIC INDUCTION TRANSISTOR (SIT)

SIT is a high-power high-frequency semiconductor device. It is the solid-state version of triode vacuum tube. SIT was commercially introduced by Tokin Corporation of Japan in 1987.

Basic structure of SIT is shown in Fig. 2.23 (a) and its symbol in Fig. 2.23 (b). It is basically n^+n^-n device with a buried grid-like p^+ gate structure. It has short n-channel structure and p^+ gate electrodes are buried in n^-n epi-layers as shown. The buried gate structure gives lower gate-source channel resistance, lower gate source capacitance and lower thermal resistance.

SIT is normally an on device, i.e. if $V_{GS} = 0$ with V_{DS} present, electrons (majority carriers) would flow from source s to n, pass through gate p^+ electrodes and would then continue their

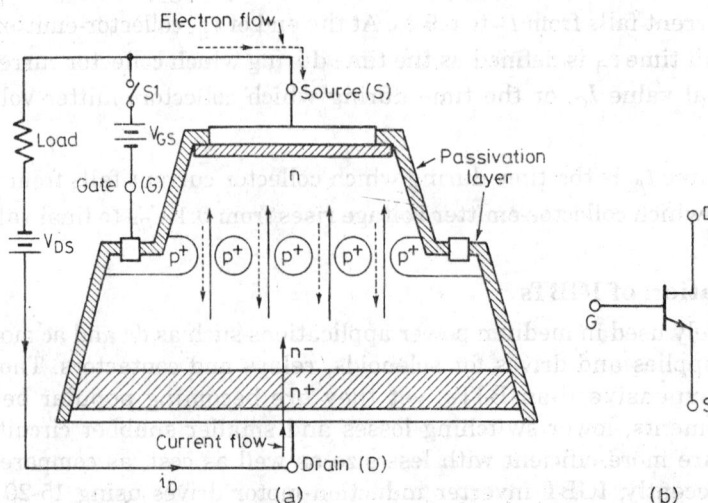

Fig. 2.23. (a) Basic structure of SIT (b) device symbol

journey through n^-, n^+ and reach drain as shown in Fig. 2.23. (a). The drain current I_D would flow from D to S as shown. If V_{GS} is negative, $p^+ n$ junctions get reverse biased. As a result, depletion layer is formed around p^+ electrodes and this reduces the current flow from its value when $V_{GS} = 0$, Fig. 2.24 (a). At some higher value of reverse bias voltage V_{GS}, the depletion layer would grow to such an extent as to cut-off the channel completely, Fig. 2.24 (b) and load current i_D would, therefore, be zero.

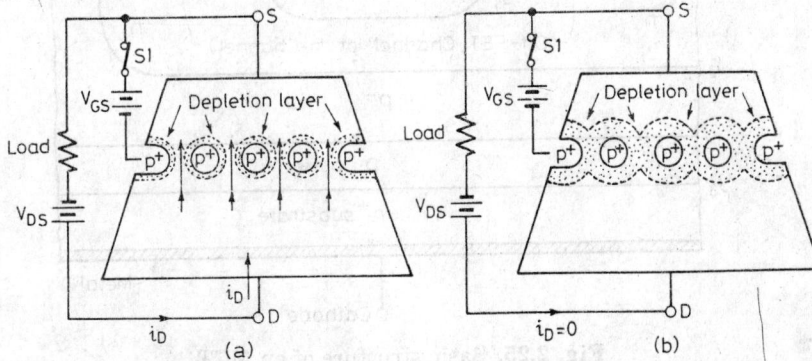

Fig. 2.24. (a) Lower reverse bias, load current i_D reduced due to depletion layer
(b) higher reverse bias, expanded depletion layer stops current flow.

Although, the device conduction drop is lower than that of equivalent series-parallel operation of PMOSFETs, the essentially large drop in SIT makes it unsuitable for general power-electronic applications. For example, a 1500 V, 180 A SIT has a channel resistance of 0.5 Ω giving 90 V conduction drop at 180 A. An equivalent thyristor or GTO drop may be around 2 V [11]. Though conduction drop in SIT is abnormally high, the turn-on and turn-off times of the device are very low. For the SIT cited above, typical t_{on} and t_{off} are around 0.35 μs. High conduction drop associated with very low turn-on and turn-off times result in low on-off energy losses. Thus, SIT is being used in high-power, high-frequency applications such as AM/FM transmitters, induction heaters, high-voltage low-current power supplies, ultrasonic generators etc. SITs with 1200 V, 300 A rating with t_{on} and t_{off} around 0.25 to 0.35 μs and 100 kHz operating frequency are available.

SIT is a majority carrier (electrons only) device, therefore SOA is limited by junction temperature. As channel resistance rises with temperature, parallel operation of SITs is easy.

SIT is normally-on device, normally-off device is under development.

2.9. MOS-CONTROLLED THYRISTOR (MCT)

An MCT is a new device in the field of semiconductor-controlled devices. It is basically a thyristor with two MOSFETs built into the gate structure. One MOSFET is used for turning on the MCT and the other for turning off the device. An MCT is a high-frequency. high-power, low-conduction drop switching device.

An MCT combines into it the features of both conventional four-layer thyristor having regenerative action and MOS-gate structure. However, in MCT, anode is the reference with respect to which all gate signals are applied. In a conventional SCR, cathode is the reference terminal for gate signals.

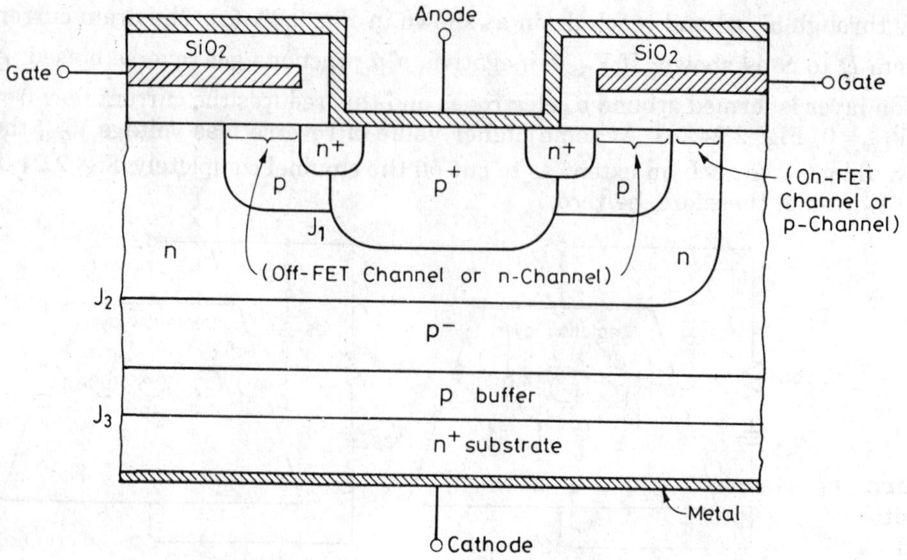

Fig. 2.25. Basic structure of an MCT.

The basic structure of an MCT cell is shown in Fig. 2.25. A practical MCT consist of thousands of these basic cells connected in parallel, just like a PMOSFET (7, 8). This is done in order to achieve a high-current carrying capacity of the device.

The equivalent circuit of MCT is shown in Fig. 2.26 (a). It consists of one on-FET, one off-FET and two transistors. One on-FET, a p-channel MOSFET and the other off-FET, an n channel MOSFET, represent MOS-gate structure of MCT. The npnp structure of MCT is represented by two transistors npn and pnp as shown in Fig. 2.26 (a). An arrow towards the gate terminal indicates n-channel MOSFET and the arrow away from the gate terminal as the p-channel MOSFET. The two transistors in the equivalent circuit indicate that there is regenerative feedback in the MCT just as it is in an ordinary thyristor. Fig. 2.26 (b) gives the circuit symbol of an MCT.

An MCT is turned-on by a negative voltage pulse at the gate with respect to the anode and is turned-off by a positive voltage pulse. Working of MCT can be understood better by referring to the equivalent circuit of Fig. 2.26 (a).

Turn-on Process. As stated above, MCT is turned on by applying a negative voltage pulse at the gate with respect to anode. In other words, for turning-on MCT, gate is made negative with respect to anode by the voltage pulse between gate and anode. Obviously, MCT must be initially forward biased and then only a negative voltage be applied. With the application of this negative voltage pulse, on-FET (p-channel) gets turned on whereas off-FET is already off. With on-FET on, current begins to flow from anode A, through on-FET and then as the base current and emitter current of npn transistor and then to cathode K. This turns on npn transistor.

As a result, collector current begins to flow in npn transistor. As off-FET is off, this collector current of npn transistor acts as the base current of pnp transistor. Subsequently, pnp transistor is also turned on. On both the transistors are on, regenerative action of the connection scheme takes place and the thyristor or MCT is turned on.

Note that on-FET and pnp transistor are in parallel when MCT is in conduction state. During the time MCT is on, base current of npn transistor flows mainly through pnp transistor because of its better conducting property.

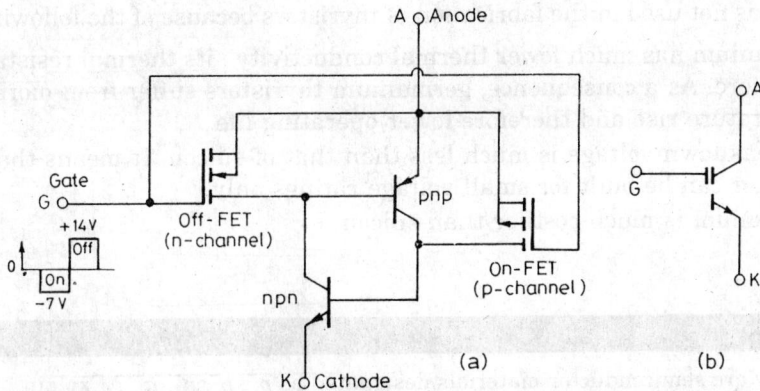

Fig. 2.26. MCT (a) equivalent circuit and (b) circuit symbol

Turn-off process. For turning-off the MCT, off-FET (or n-channel MOSFET) is energized by positive voltage pulse at the gate. With the application of positive voltage pulse, off-FET is turned on and on-FET is turned off. After off-FET is turned on, emitter-base terminals of pnp transistor are short circuited by off-FET. So now anode current begins to flow through off-FET and therefore base current of pnp transistor begins to decrease. Further, collector current of pnp transistor that forms the base current of npn transistor also begins to decrease. As a consequence, base currents of both pnp and npn transistors, now devoid of stored charge in their n and p bases respectively, begin to decay. This regenerative action eventually turns off the MCT.

An MCT has the following merits :

    (i) low forward conduction drop,

    (ii) fast turn-on and turn-off times,

    (iii) low switching losses and

    (iv) high gate input impedance, which allows simpler design of drive circuits.

Main disadvantage of MCT is its low reverse voltage blocking capability.

MCT was commercially introduced in 1992. At that time, it was predicted that its use as a power-semiconductor device would be so vast that it might challenge the existence of most of the other devices like SCR, BJT, GTO, IGBT etc. This has, however, not happened because an MCT has (i) limited reverse-biased SOA and (ii) its switching frequency is much inferior to IGBT. At present, MCTs are being promoted for their use in soft switched converter topologies, where these inferiorities do not inhibit their use.

2.10. NEW SEMICONDUCTING MATERIALS

At present, silicon enjoys monopoly as a semiconductor material for the commercial production of power-control devices. This is because silicon is cheaply available and semiconductor devices of any size can be easily fabricated on a single silicon chip. There are, however, new types of materials like gallium assenic (GaAs), silicon carbide and diamond which possess the desirable properties required for switching devices. At present, state-of-the-art technology for these materials is primitive compared with silicon, and many more years of research investment are required before these materials become commercially viable for the production of power-controlled devices. Diode, power MOSFET and thyristor made from silicon carbide have been established in the laboratory and are expected to be commercially available very soon. Superconductive materials may also be used in the manufacture of such devices, but work in this direction has not yet been reported.

Germanium is not used in the fabrication of thyristors because of the following reasons:

(*i*) Germanium has much lower thermal conductivity ; its thermal resistance is, therefore, more. As a consequence, germanium thyristors suffer from more losses, more temperature rise and therefore lower operating life.

(*ii*) Its breakdown voltage is much less than that of silicon. It means that germanium thyristor can be built for small voltage ratings only.

(*iii*) Germanium is much costlier than silicon.

PROBLEMS

2.1. (*a*) Why are semiconductor materials designated as p^+, p^-, n^-, n^+ ? Explain.

(*b*) What is *p-n* junction ? Discuss the formation of depletion layer in *p-n* junction.

(*c*) What is barrier potential ? How are depletion layer and barrier potential effected by temperature ?

2.2. (*a*) Explain the effect of forward bias and reverse bias on the depletion layer in a *p-n* junction.

(*b*) How is the magnitude of breakdown voltage effected if a junction has highly doped (*i*) layers on its both sides and (*ii*) layer on its one side only.

(*c*) Describe the structural features of power diodes. How do these differ from signal diodes ?

2.3. (*a*) What is a diode ? Discuss *i–v* characteristics of power, signal and ideal diodes.

(*b*) Describe reverse recovery characteristics of diodes. Show that reverse recovery time and peak inverse current are dependent upon storage charge and rate of change of current.

2.4. (*a*) Describe the various types of power diodes indicating clearly the differences amongst them.

(*b*) What is cut-in voltage in a diode ? What are other terms used for cut in voltage ?

(*c*) Discuss the following terms for diodes :
Softness factor, PIV, reverse recovery time, reverse recovery current.

(*d*) For a power diode, the reverse recovery time is 3.9 µs and the rate of diode-current decay is 50 A/µs. For a softness factor of 0.3, calculate the peak inverse current and the storage charge. **[Ans.** (*d*) 150 A, 292.5 µC]

2.5. (*a*) Discuss the power loss in a diode during the reverse recovery transients.

(*b*) The forward characteristic of a power diode can be represented by $v_f = 0.88 + 0.015\,i_f$. Determine the average power loss and rms current for a constant current of 50 A for 2/3 of a cycle.

$$\left[\textbf{Hint.}\ (b)\ \text{With } T \text{ as the time of a cycle, average power loss} = \frac{1}{T}\int_0^{2\,T/3} v_f \cdot I_f\,dt = \frac{2}{3} \cdot v_f I_f \text{ etc} \right]$$

[Ans. (*b*) 54.33 W, 40.825 A]

2.6. (*a*) Enumerate the types of power transistors along with their circuit symbols.

(*b*) What is a bipolar junction transistor ? Why is it so called ?
Describe the types of BJTs with their circuit symbols.

(*c*) Define α and β for BJT and develop a relation between the two. Why is α less than 1 and β more than 1 ?

(*d*) Why is it preferable to use hard drive for BJT ?

2.7. (*a*) What is the difference between β and forced β_f for BJTs ?

(*b*) What are the conditions under which a transistor operates as a switch ?
Discuss hard-drive and overdrive factor for BJT.

(*c*) Show that collector current at saturation remains substantially constant even if base **current is increased.**

2.8. A bipolar transistor, with current gain $\beta = 50$, has load resistance $R_C = 10\ \Omega$, dc supply voltage $V_{CC} = 120$ V and input voltage to base circuit, $V_B = 10$ V. For $V_{CES} = 1.2$ V and $V_{BES} = 1.6$ V, calculate

(a) the value of R_B for operation in the saturated state

(b) the value of R_B for an or drive factor 6

(c) forced current gain and

(d) power loss in the transistor for both parts (a) and (b).

[**Ans.** (a) 35.354 Ω (b) 5.892 Ω (c) 8.33 (d) 14.6362 W, 16.537 W]

2.9. (a) Explain the switching performance of BJT with relevant waveforms. Indicate clearly turn-on and turn-off times and their components.

(b) Describe FBSOA and RBSOA for BJTs.

2.10. (a) Describe the input and output characteristic for a BJT. Show the region of the transistor characteristic where it acts like a switch.

(b) Typical switching waveforms for a power transistor are shown in Fig. 2.27. Show that switch-on energy loss is given by

$$\frac{V_{CC} \cdot I_{CS}}{6} t_{on}$$

Also obtain an expression for the average value of switch-on loss.

(c) Derive expressions for the switch-off energy loss and also for its average value for the waveforms shown in Fig. 2.27.

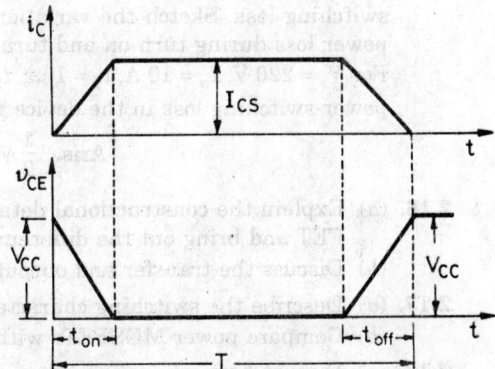

Fig. 2.27. Pertaining to Prob. 2.10 (b).

$$\left[\textbf{Ans}\ (b)\ \frac{V_{CC} \cdot I_{CS}}{6} f \cdot t_{on}\ (c)\ \frac{V_{CC} \cdot I_{CS}}{6} t_{off},\ \frac{V_{CC} \cdot I_{CS}}{6} f \cdot t_{off}\right]$$

2.11. In case $I_{CS} = 80$ A, $V_{CC} = 220$ V, $t_{on} = 1.5\ \mu s$ and $t_{off} = 4\ \mu s$ for the switching waveforms shown in Fig. 2.27, find the energy loss during switch-on and switch-off intervals. Find also the average power loss in the power transistor for a switching frequency of 2 kHz.

Derive the expressions used. [**Ans** 4.4 mWs, 11.73 mWs, 32.267 W]

2.12. (a) For the typical switching waveforms shown in Fig. 2.27 for a power transistor, find expressions that give peak instantaneous power loss during t_{on} and t_{off} intervals respectively.

(b) In case $I_{CS} = 80$ A, $V_{CC} = 220$ V, $t_{on} = 1.5\ \mu s$ and $t_{off} = 4\ \mu s$, find the peak value of instantaneous power loss during t_{on} and t_{off} intervals respectively.

$$\left[\textbf{Ans.}\ (a)\ \frac{I_{CS} \cdot V_{CC}}{4},\ \frac{I_{CS} \cdot V_{CC}}{4}\ (b)\ 4400\ W,\ 4400\ W\right]$$

2.13. A power transistor is used as a switch and typical waveforms are shown in Fig. 2.12(a). The parameters for the transistor circuit are as under :

$V_{CC} = 200$ V, $V_{CES} = 2.5$ V, $I_{CS} = 60$ A, $t_d = 0.5\ \mu s$, $t_r = 1\ \mu s$,

$t_n = 40\ \mu s$, $t_s = 4\ \mu s$, $t_f = 3\ \mu s$, $t_0 = 30\ \mu s$, $f = 10$ kHz.

Collector to emitter leakage current = 1.5 mA.

Determine average power loss due to collector current during t_{on} and t_n . Find also the peak instantaneous power loss due to collector current during turn-on time.

Sketch the instantaneous power loss during t_{on} and t_n. [**Ans.** 20.5015 W, 60 W, 3037.97 W]

2.14. Repeat Prob. 2.13 for obtaining average power loss during turn-off time and off-period, and also peak instantaneous power loss during fall time due to collector current.

Sketch the instantaneous power loss during turn-off time and off-period.

2.15. Fig. 2.28 shows the switching characteristics for a power semiconductor device. Derive the expressions for energy loss during turn-on and turn-off periods, and also for the average switching loss. Sketch the variation of power loss during turn-on and turn-off periods.

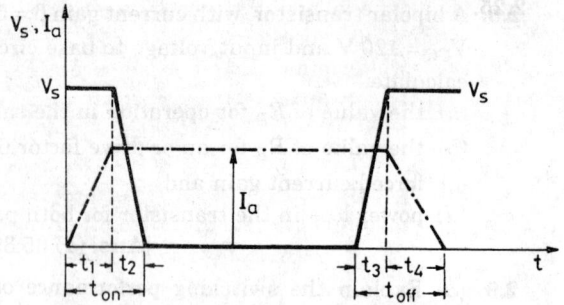

Fig. 2.28. Pertaining to Prob. 2.15.

For $V_s = 220$ V, $I_a = 10$ A, $t_1 = 1$ μs, $t_2 = 2$ μs, $t_3 = 1.5$ μs and $t_4 = 3$ μs, find the average value of power-switching loss in the device for a switching frequency of 1 kHz.

$$\left[\textbf{Ans.} \ \ \frac{1}{2} V_s \cdot I_a \ (t_1 + t_2), \ \frac{1}{2} V_s I_a \ (t_3 + t_4), \ \frac{1}{2} \ V_s I_a f \ (t_{on} + t_{off}), \ 7.5 \ \text{W} \right]$$

2.16. (a) Explain the constructional details and working of low-power MOSFET and power MOSFET and bring out the differences between the two.

(b) Discuss the transfer and output characteristics of power MOSFETs.

2.17. (a) Describe the switching characteristics of power MOSFETs.

(b) Compare power MOSFETs with BJTs.

2.18. (a) Discuss how conduction takes place in PMOSFET of n-channel type.

(b) Explain the formation of parasitic BJT and parasitic diode in a PMOSFET. Can parasitic diode be used in some power electronic applications ?

2.19. (a) What is an IGBT ? What are its other names ? Give its basic structural features. How does it differ in structure from PMOSFET ?

(b) Derive the approximate and exact equivalent circuits of an IGBT from its structural details. Also describe its output and transfer characteristics.

2.20. (a) Describe the working of an IGBT. How does latch-up occur in an IGBT ?

(b) Give a comparison between an IGBT and a PMOSFET.

2.21. (a) Explain switching characteristics of an IGBT.

(b) Discuss why PMOSFET has no reverse blocking voltage whereas an IGBT has.

(c) Why are IGBTs becoming popular in their applications to controlled converters ? Enumerate some applications of IGBTs.

2.22. (a) What is SIT ? Give its basic structural details. Explain its working with relevant diagrams.

(b) Though SIT is not suitable for general power-electronic applications, yet it is being used in some specific applications. Explain.

2.23. (a) Describe the basic structure of MOS controlled thyristor (MCT). Give its equivalent circuit and explain the turn-on and turn-off processes.

(b) Give the merits and demerits of MCTs. In what type of applications are MCTs being promoted at present ?

(c) Discuss briefly about the new semiconducting materials.

2.24. Deduce to show that the energy loss during turn-on of a power transistor is given by $(VI/6) \, T$ joules, where V = off-state voltage, I = on-state current and T = turn-on time. Assume the change of V and I to be linear over T.

Hence, calculate the turn-on loss of a power transistor for which the voltage and current, during the process of turn-on, change linearly from 300 V to zero V and zero A to 200 A respectively in 2 μs.   **[Ans.** 10 mW-s or 10×10^{-3} J]

2.25. Read the following statements carefully and indicate the power semiconductor device's each statement represents.

(*a*) two-terminal three-layer device   (*b*) majority carrier devices

(*c*) bipolar devices   (*d*) negative pulse turn-on device

(*e*) on operation in ohmic region   (*f*) normally on device

(*g*) on-state in saturation region   (*h*) two-terminal two-layer device

(*i*) uncontrolled turn-on and turn-off device

(*j*) controlled turn-on and turn-off devices.

[**Ans.** (*a*) Power diode (*b*) PMOSFET, SIT (*c*) Diode, BJT, IGBT, MCT (*d*) MCT
(*e*) PMOSFET (*f*) SIT (*g*) BJT (*h*) signal diode (*i*) Diode
(*j*) BJT, MOSFET, IGBT, SIT, MCT]

Chapter 3

Diode Circuits and Rectifiers

A rectifier is a circuit that converts ac input voltage to dc output voltage. Semiconductor diodes are used extensively in power electronic circuits for the conversion of power from ac to dc. A rectifier employing diodes is called an *uncontrolled rectifier*, because its average output voltage is a fixed dc voltage.

In this chapter, first diode circuits involving different combinations of R, L and C are studied, and then diode rectifiers are described. For simplicity, the diodes are considered as ideal switches. An ideal diode has no forward voltage drop and reverse recovery time is negligible.

3.1. DIODE CIRCUITS WITH DC SOURCE

In this section, the effect of switching a dc source to a circuit consisting of diode and different circuit parameters is examined. The conclusions arrived at can then be applied to similar situations encountered later in power-electronic circuits.

3.1.1. Resistive Load

In the circuit of Fig. 3.1 (*a*), when switch S is closed, the current rises instantaneously to V_s/R as shown in Fig. 3.1 (*b*). Here V_s is the dc source voltage and R is the load resistance. When switch S is opened at t_1, the current at once falls to zero, Fig. 3.1 (*b*). Voltage v_D across diode is zero during the time diode conducts and is equal to $+V_s$ after diode stops conducting.

3.1.2. RC Load

A circuit with dc source, diode and RC load is shown in Fig. 3.2 (*a*). When switch S is closed at $t = 0$, KVL gives

$$Ri + \frac{1}{C} \int i \, dt = V_s$$

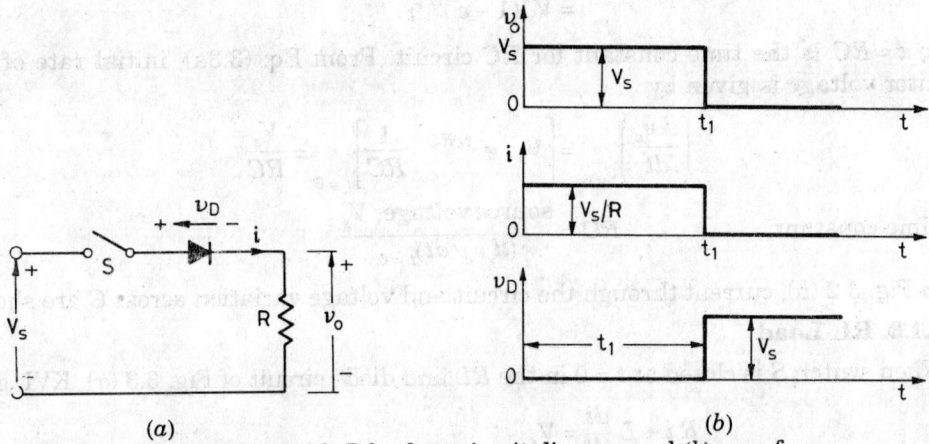

Fig. 3.1. Diode circuit with R load (a) circuit diagram and (b) waveforms.

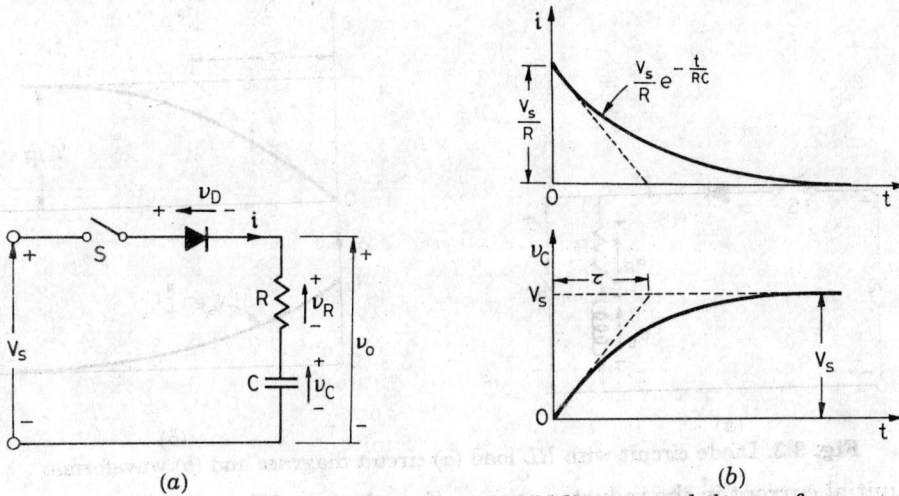

Fig. 3.2. Diode circuit with RC load (a) circuit diagram and (b) waveforms.

Its Laplace transform is $R\,I(s) + \dfrac{1}{C}\left[\dfrac{I(s)}{s} + \dfrac{q(o)}{s}\right] = \dfrac{V_s}{s}$ ...(3.1)

As the initial voltage across C is zero, $q(o) = 0$. With this, Eq. (3.1) becomes

$$I(s)\left[R + \frac{1}{Cs}\right] = \frac{V_s}{s}$$

or

$$I(s) = \frac{CV_s}{RC\left(s + \dfrac{1}{RC}\right)} = \frac{V_s}{R}\cdot\frac{1}{s + \dfrac{1}{RC}}$$

Its Laplace inverse is $\quad i(t) = \dfrac{V_s}{R}\cdot e^{-t/RC}$ ...(3.2)

The voltage across capacitor is

$$v_c(t) = \frac{1}{C}\int_0^t i\,dt = \frac{V_s}{RC}\int_0^t e^{-t/RC}\,dt$$

$$= V_s\left(1 - e^{-t/RC}\right)$$ ...(3.3a)

$$= V_s \left(1 - e^{-t/\tau}\right) \qquad \qquad ...(3.3b)$$

where $\tau = RC$ is the time constant for RC circuit. From Eq. (3.3a), initial rate of change of capacitor voltage is given by

$$\left(\frac{d\,v_c}{dt}\right)_{t=0} = \left[V_s \cdot e^{-t/RC} \cdot \frac{1}{RC}\right]_{t=0} = \frac{V_s}{RC} \qquad ...(3.4)$$

Time constant, $\qquad \qquad RC = \dfrac{\text{source voltage, } V_s}{(d\,v_c/dt)_{t=0}}$

In Fig. 3.2 (b), current through the circuit and voltage variation across C are shown.

3.1.3. RL Load

When switch S is closed at $t = 0$ in the RL and diode circuit of Fig. 3.3 (a), KVL gives

$$R\,i + L\,\frac{di}{dt} = V_s \qquad \qquad ...(3.5)$$

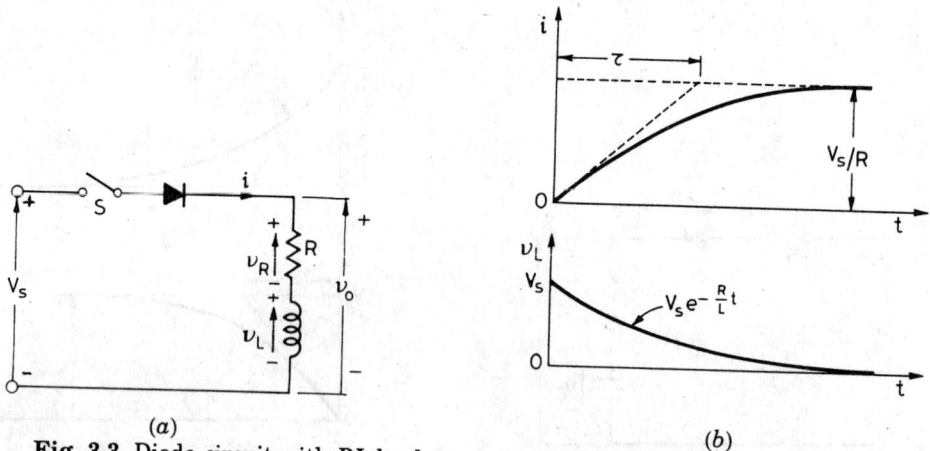

Fig. 3.3. Diode circuit with RL load (a) circuit diagram and (b) waveforms.

With initial current in the inductor as zero, the solution of Eq. (3.5) gives

$$i(t) = \frac{V_s}{R}\left(1 - e^{-\frac{R}{L}t}\right) \qquad \qquad ...(3.6)$$

Initial rate of rise of current is

$$\left.\frac{di}{dt}\right|_{t=0} = \left(\frac{V_s}{L} \cdot e^{-\frac{R}{L}t}\right)_{t=0} = \frac{V_s}{L} \qquad \qquad ...(3.7)$$

The voltage across L is $\quad v_L(t) = L\,\dfrac{di}{dt} = V_s \cdot e^{-\frac{R}{L}t} \qquad \qquad ...(3.8)$

For RL circuit, $\dfrac{L}{R} = \tau$ is the time constant. The waveforms of current through the circuit and voltage across inductance L are sketched in Fig. 3.3 (b).

It must be *noted* that the behaviour of circuits in Figs. 3.1 to 3.3 is not affected whether a diode is used or not. It is because, in these circuits, the current i does not have a tendency to reverse; *i.e.* current i remains unidirectional.

3.1.4. LC Load

A diode circuit with dc source voltage V_s, switch S and load LC is shown in Fig. 3.4 (a). When switch S is closed at $t = 0$, the voltage equation governing its performance is given by

$$L \frac{di}{dt} + \frac{1}{C} \int i \, dt = V_s$$

Its Laplace transform is $\quad L \left[s \, I(s) - i(0) \right] + \frac{1}{C} \left[\frac{I(s)}{s} + \frac{q(0)}{s} \right] = \frac{V_s}{s}$

As the circuit is initially relaxed, $i(0) = 0$ and $v_C(0) = 0$ or $q(0) = C \cdot v_C(0) = 0$

$\therefore \qquad\qquad I(s) \left[sL + \frac{1}{sC} \right] = \frac{V_s}{s}$

or $\qquad\qquad\qquad I(s) = \frac{V_s}{L} \cdot \frac{1}{s^2 + \dfrac{1}{LC}}$

Let $\omega_0 = \dfrac{1}{\sqrt{LC}}$. This gives $I(s) = \dfrac{V_s}{L \cdot \omega_0} \cdot \dfrac{\omega_0}{s^2 + \omega_0^2} = V_s \cdot \sqrt{\dfrac{C}{L}} \cdot \dfrac{\omega_0}{s^2 + \omega_0^2}$

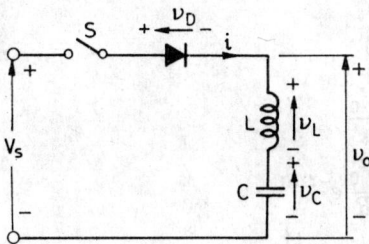

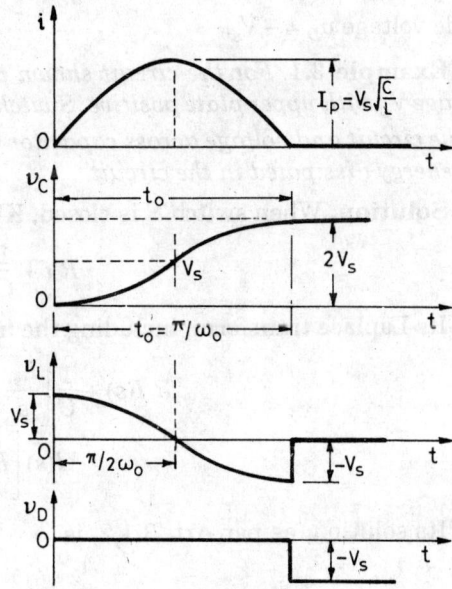

Fig. 3.4. Diode circuit with LC load (a) circuit diagram and (b) waveforms.

Its Laplace inverse is $\qquad i(t) = V_s \cdot \sqrt{\dfrac{C}{L}} \, \sin \omega_0 t$ \qquad\qquad ...(3.9)

Here $\omega_0 = \dfrac{1}{\sqrt{LC}}$ is called *resonant frequency* of the circuit. Capacitor voltage is given by

$$v_C(t) = \frac{1}{C} \int_0^t i(t) \cdot dt = \frac{1}{C} \int_0^t V_s \cdot \sqrt{\frac{C}{L}} \, \sin \omega_0 t \cdot dt$$

$$= V_s \, (1 - \cos \omega_0 t) \qquad\qquad ...(3.10a)$$

Voltage across inductance is given by

$$v_L(t) = L \frac{di(t)}{dt} = V_s \cos \omega_0 t \qquad \qquad ...(3.10b)$$

When $\omega_0 t_0 = \pi$ or when $t_0 = \pi/\omega_0$, from Eq. (3.9), $i(t_0) = 0$ and from Eq. (3.10a), $v_C(t_0) = 2 V_s$ and $v_L(t_0) = -V_s$

Here $t_0 = \pi/\omega_0 = $ conduction time of diode $= \pi \sqrt{LC}$

From Eq. (3.9), circuit or diode current at $t_0/2 = \dfrac{\pi}{2\,\omega_0}$ attains a peak value of $I_p = V_s \cdot \sqrt{C/L}$ as shown in Fig. 3.4 (b). Voltage across diode, soon after diode stops conduction at t_0 is given by

$$v_D = -v_L - v_C + V_s = 0 - 2 V_s + V_s = -V_s.$$

Waveforms of $i(t)$, v_C, v_L and v_D are sketched in Fig. 3.4 (b). It is seen that at $t_0/2 = \dfrac{\pi}{2\,\omega_0}$, diode current reaches peak value, $v_C = V_s$ and $v_L = 0$. Also at $t_0 = \pi/\omega_0 = \pi (\sqrt{LC})$, diode current decays to zero and capacitor is charged to voltage $2V_s$. Soon after t_0, voltage across L is zero and diode voltage $v_D = -V_s$.

Example 3.1. *For the circuit shown in Fig. 3.5 (a), the capacitor is initially charged to a voltage V_0 with upper plate positive. Switch S is closed at $t = 0$. Derive expressions for the current in the circuit and voltage across capacitor C. What is the peak value of diode current? Find also the energy dissipated in the circuit.*

Solution. When switch S is closed, KVL gives

$$R\,i + \frac{1}{C} \int i\,dt = 0$$

Its Laplace transform, including the initial voltage across capacitor, is

$$R\,I(s) + \frac{1}{C}\left[\frac{I_{(s)}}{s} - \frac{CV_0}{s} \right] = 0$$

or

$$I(s)\left[R + \frac{1}{Cs} \right] = \frac{V_0}{s}$$

Its solution, as per Art. 3.1.2, is

$$i(t) = \frac{V_0}{R} e^{-t/RC}$$

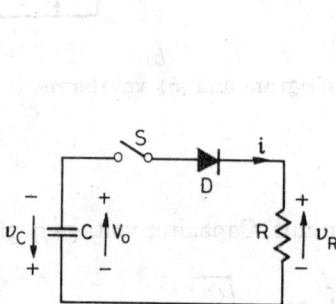

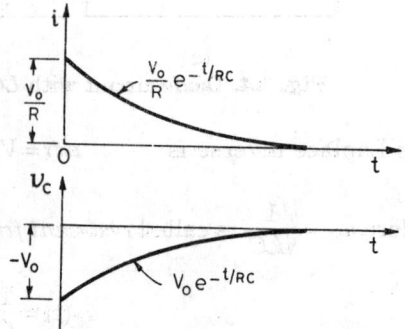

(a)   (b)

Fig. 3.5. Pertaining to Example 3.1 (a) circuit diagram and (b) waveforms.

$\therefore$ Peak diode current $= \dfrac{V_0}{R}$

Capacitor voltage,
$$v_C(t) = \frac{1}{C} \int_0^t idt - V_0$$

$$= \frac{1}{C} \int_0^t \frac{V_0}{R} \cdot e^{-t/RC} \cdot dt - V_0$$

$$= -V_0\, e^{-t/RC}$$

Current $i(t)$ and voltage $v_C(t)$ are sketched in Fig. 3.5 (b).

Energy dissipated in the circuit $= \dfrac{1}{2} CV_0^2$ Joules

Example 3.2. *In the diode and LC network shown in Fig. 3.6 (a), the capacitor is initially charged to voltage V_0 with upper plate positive. Switch S is closed at $t = 0$. Derive expressions for current through and voltage across C.*

Find the conduction time of diode, peak current through the diode and final steady-state voltage across C in case $V_s = 400$ V, $V_0 = 100$ V, $L = 100$ μH and $C = 30$ μF. Determine also the voltage across diode after it stops conduction.

Soution. When switch S is closed, KVL for Fig. 3.6 (a) gives

$$L \frac{di}{dt} + \frac{1}{C} \int idt = V_s$$

Its Laplace transform gives

$$L\,[s\,I(s) - i(0)] + \frac{1}{C}\left[\frac{I(s)}{s} + \frac{C \cdot V_0}{s}\right] = \frac{V_s}{s}$$

As initially $i(0) = 0$, the above equation becomes

$$I(s)\left[sL + \frac{1}{sC}\right] = \frac{V_s - V_0}{s}$$

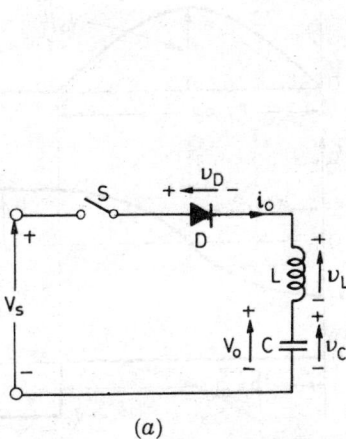

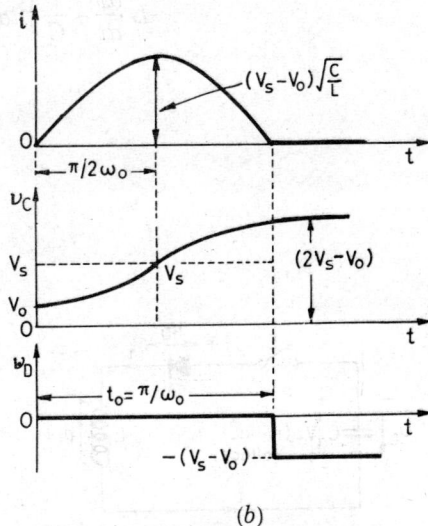

Fig. 3.6. Pertaining to Example 3.2 (a) circuit diagram and (b) waveforms.

This equation in *s*-domain can be solved as in Art 3.1.4. Its solution is

$$i(t) = (V_s - V_0) \cdot \sqrt{\frac{C}{L}} \sin \omega_0 t$$

$$v_C(t) = \frac{1}{C} (V_s - V_0) \sqrt{\frac{C}{L}} \int_0^t \sin \omega_0 t \, dt + V_0$$

$$= (V_s - V_0)(1 - \cos \omega_0 t) + V_0$$

At $\omega_0 t = 0$, $v_C(t) = V_0$

At $\omega_0 t = \pi/2$, $v_C(t) = V_s$ and at $\omega_0 t = \pi$, $v_C(t) = 2(V_s - V_0) + V_0 = 2V_s - V_0$.

Diode conduction time, $t_0 = \dfrac{\pi}{\omega_0} = \pi \sqrt{LC} = \pi\sqrt{30 \times 100} \times 10^{-6} = 54.77 \, \mu s$

Peak current through diode, $I_p = (V_s - V_0) \sqrt{\dfrac{C}{L}}$

$$= 300 \sqrt{\frac{30}{100}} = 164.32 \, A$$

Steady state voltage across *C* occurs when $\omega_0 t_0 = \pi$.

∴ $V_C = 2(V_s - V_0) + V_0 = 2V_s - V_0 = 2 \times 400 - 100 = 700 \, V$

Voltage across diode, after it stops conducting, is given by

$$v_D = -v_L - v_C + V_s = 0 - (2V_s - V_0) + V_s = -V_s + V_0 = -400 + 100$$

$$= -300 \, V.$$

Example 3.3. *In the circuit shown in Fig. 3.7 (a), the capacitor has initial voltage V_0 with upper plate positive. The circuit is switched at t = 0. Derive expressions for current and voltage across capacitor. Find the conduction time for diode and steady-state capacitor voltage.*

Solution. The voltage equation for the circuit of Fig. 3.7 (a), after switch *S* is closed at *t* = 0, is

$$L \frac{di}{dt} + \frac{1}{C} \int i \, dt = 0$$

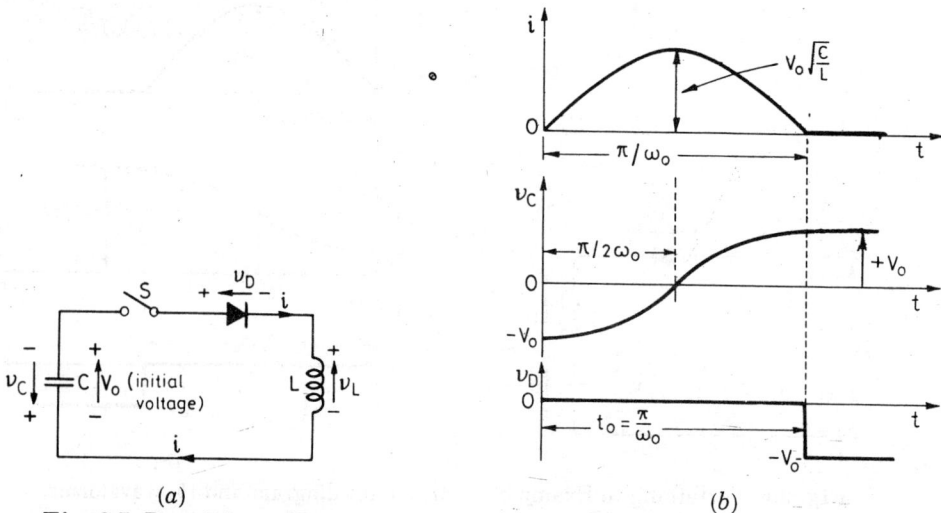

(a)   (b)

Fig. 3.7. Pertaining to Example 3.3 (a) circuit diagram and (b) waveforms.

Its Laplace transform, including initial voltage across capacitor, is

$$I(s) \cdot sL + \frac{1}{C}\left[\frac{I(s)}{s} - \frac{CV_0}{s}\right] = 0$$

$$I(s)\left[sL + \frac{1}{sC}\right] = \frac{V_0}{s}$$

Here minus sign is put before V_0, because for the direction of positive current flow, polarity of V_0 is opposite.

Solution of above s-domain equation, from Art. 3.1.4, is

$$i(t) = V_0 \sqrt{\frac{C}{L}} \sin \omega_0 t$$

Voltage across C is
$$v_C(t) = \frac{1}{C} \cdot V_0 \sqrt{\frac{C}{L}} \int_0^t \sin \omega_0 t \, dt - V_0 = -V_0 \cos \omega_0 t$$

Diode conduction time,
$$t_0 = \frac{\pi}{\omega_0} = \pi \sqrt{LC}$$

Steady-state capacitor voltage $= -V_0 \cos \pi = +V_0$

Voltage across diode, $v_D = -V_0$.

Waveforms for i, V_C and v_D are sketched in Fig. 3.7 (b).

Example 3.4. *In the circuit shown in Fig. 3.8 (a), capacitor is initially charged to voltage V_o with upper plate positive. Sketch waveforms of i, v_C, v_L and i_D after the switch S is closed.*

Solution. When switch S is closed, capacitor C begins to discharge through L and C. For obtaining i, v_C expressions, refer to Example 3.3.

Therefore,
$$i = V_0 \sqrt{\frac{C}{L}} \sin \omega_0 t, \quad v_c(t) = -V_0 \cos \omega_0 t$$

and
$$v_L = L\frac{di}{dt} - L \cdot V_0 \cdot \sqrt{\frac{C}{L}} \cdot \frac{d}{dt} \sin \omega_0 t - V_0 \cos \omega_0 t = -v_c.$$

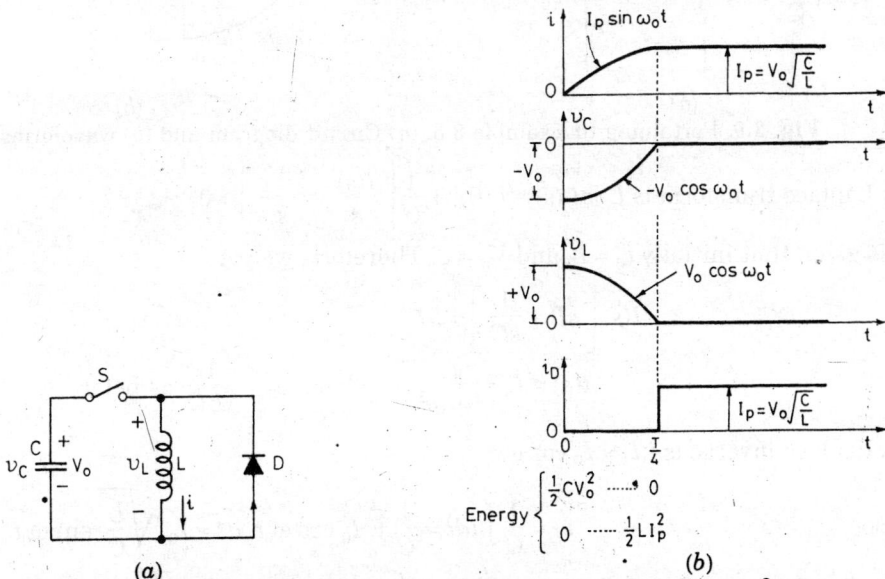

Fig. 3.8. Pertaining to Example 3.4 (a) circuit diagram and (b) waveforms.

The waveforms for i, v_C, v_L and i_D are shown in Fig. 3.8 (b). At $t = 0$, energy stored in C is $\frac{1}{2} CV_0^2$ and $i = 0$. At $t = T/4$, current i reaches peak value $V_0 \sqrt{\frac{C}{L}} = I_p$, $v_C = v_L = 0$ and energy stored in C gets transferred to L as $\frac{1}{2} LI_p^2$. Soon after $T/4$ ($\omega_0 T/4 = \pi/2$), as v_L tends to reverse, diode D gets forward biased. Current I_p now begins to flow as i_D through D and as i through L. If there were no resistance in this closed path, current I_p would continue flowing unabated. In practice, inherent resistance in the closed path would cause this current to decay exponentially.

Example 3.5. *In the circuit of Fig. 3.9 (a), current in inductor L is I_0 before $t = 0$. Sketch the waveforms of i, v_c and v_L after switch S is opened at $t = 0$.*

Solution. When switch S is opened at $t = 0$, current I_0 begins to flow in the path D, C and L. KVL for this path is

$$L \frac{di}{dt} + \frac{1}{C} \int i \, dt = 0$$

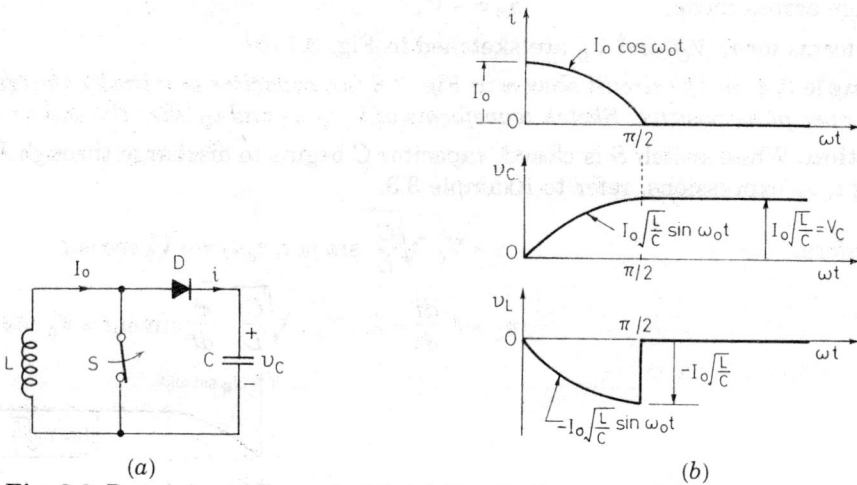

Fig. 3.9. Pertaining to Example 3.5. (a) Circuit diagram and (b) waveforms.

Its Laplace transform is $L [sI (s) - i (0)] + \frac{1}{C} \left[\frac{I(s)}{s} - \frac{CV_0}{s} \right] = 0$

It is given that initially $i_0 = I_0$ and $V_0 = 0$. Therefore, we get

$$I(s) \left[sL + \frac{1}{sC} \right] = LI_0$$

or
$$I(s) = I_0 \frac{s}{s^2 + \omega_0^2}, \quad \text{where } \omega_0 = \frac{1}{\sqrt{LC}} \text{ as before.}$$

Its Laplace inverse is $i(t) = I_0 \cos \omega_0 t$

Also
$$v_c = \frac{1}{C} \int i \, dt = \frac{1}{C} \int_0^t I_0 \cos \omega_0 t . \, dt = I_0 \sqrt{\frac{L}{C}} \sin \omega_0 t$$

and
$$v_L = L \frac{di}{dt} = L \frac{d}{dt} (I_0 \cos \omega_0 t) = - I_0 \sqrt{\frac{L}{C}} \sin \omega_0 t$$

At $\omega t = \dfrac{\pi}{2}$, current i tends to reverse, but diode D blocks this current reversal. Also, at $\omega t = \dfrac{\pi}{2}$, capacitor is changed to $I_0 \sqrt{\dfrac{L}{C}} \sin \dfrac{\pi}{2} = I_0 \sqrt{\dfrac{L}{C}} = V_C$ and voltage across inductance is $V_L = -I_0 \sqrt{\dfrac{L}{C}}$.

Thus, after $\omega t = \dfrac{\pi}{2}$, capacitor voltage remains constant at $I_0 \sqrt{\dfrac{L}{C}}$ whereas voltage across L becomes zero because current is now zero. Energy stored in inductance as $\dfrac{1}{2} L I_0^2$ at $\omega t = 0$ gets transferred to C at $\omega t = \pi/2$ as $\dfrac{1}{2} CV_c^2 = \dfrac{1}{2} C \left(I_0 \sqrt{\dfrac{L}{C}} \right)^2$.

3.1.5. RLC Load

A diode in series with RLC circuit is shown in Fig. 3.10 (a). KVL for this circuit, when switch S is closed at $t = 0$, is given by

$$Ri + L \frac{di}{dt} + \frac{1}{C} \int i\, dt = V_s$$

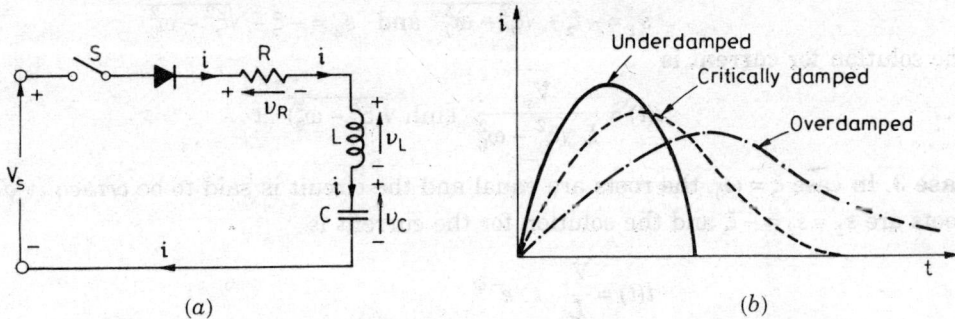

Fig. 3.10. Diode circuit with RLC load (a) circuit diagram and (b) waveforms.

With zero initial conditions, the Laplace transform of above equation is

$$I(s) \left[R + sL + \frac{1}{sC} \right] = \frac{V_s}{s}$$

or

$$I(s) = \frac{V_s}{L} \cdot \frac{1}{s^2 + \dfrac{R}{L} s + \dfrac{1}{LC}}$$

Here $s^2 + \dfrac{R}{L} s + \dfrac{1}{LC} = 0$ is the characteristic equation in s-domain. The roots of this equation are

$$s = -\frac{R}{2L} \pm \sqrt{\left(\frac{R}{2L} \right)^2 - \frac{1}{LC}}$$

or

$$s = -\xi \pm \sqrt{\xi^2 - \omega_0^2} \qquad \ldots(3.11)$$

where

$$\xi = \frac{R}{2L} \qquad \ldots(3.12)$$

is called the *damping factor.* $\qquad \omega_0 = \dfrac{1}{\sqrt{LC}}$ $\qquad$ ...(3.13)

is called resonant frequency in rad/sec

and $\qquad \omega_r = \sqrt{\dfrac{1}{LC} - \left(\dfrac{R}{2L}\right)^2}$ $\qquad$ ...(3.14)

$\qquad = \sqrt{\omega_0^2 - \xi^2} = $ ringing frequency in rad /sec.

Also $\qquad \omega_0 = \sqrt{\omega_r^2 + \xi^2}$

Depending upon the values of ξ and ω_0, the solution for the current can have three possible solutions.

Case 1. In case $\xi < \omega_0$, it is seen from Eq. (3.11) that the roots are complex and the circuit is said to be *underdamped.* The two roots are

$$s_1 = -\xi + j\,\omega_r \quad \text{and} \quad s_2 = -\xi - j\,\omega_r$$

and the current is given by

$$i(t) = \frac{V_s}{\omega_r L} \cdot e^{-\xi t} \sin \omega_r t \qquad \text{...(3.15)}$$

Case 2. If $\xi > \omega_0$, the two roots are real and the circuit is said to be *overdamped.* The two roots are

$$s_1 = -\xi + \sqrt{\xi^2 - \omega_0^2} \quad \text{and} \quad s_2 = -\xi - \sqrt{\xi^2 - \omega_0^2}$$

and the solution for current is

$$i(t) = \frac{V_s}{L\sqrt{\xi^2 - \omega_0^2}} \cdot \sinh \sqrt{(\xi^2 - \omega_0^2)} \cdot t \qquad \text{...(3.16)}$$

Case 3. In case $\xi = \omega_0$, the roots are equal and the circuit is said to be *critically damped.* The roots are $s_1 = s_2 = -\xi$ and the solution for the current is

$$i(t) = \frac{V_s}{L} \cdot t \cdot e^{-\xi t} \qquad \text{...(3.17)}$$

Waveforms of current for the three different levels of damping are sketched in Fig. 3.10 (b).

Example 3.6. *For the circuit of Fig. 3.10 (a), the data is as under :*

$$R = 10\Omega, \ L = 1\,mH, \ C = 5\,\mu F, \ V_s = 230\,V$$

The circuit is initially relaxed. With switch closed at t = 0, determine (a) current i(t) (b) conduction time of diode (c) rate of change of current at t = 0.

Solution. (a) From Eq. (3.12),

$$\xi = \frac{10 \times 1000}{2 \times 1} = 5000$$

From Eq. (3.13), $\qquad \omega_0 = \dfrac{1}{\sqrt{LC}} = \dfrac{1}{\left[1 \times 10^{-3} \times 5 \times 10^{-6}\right]^{1/2}} = \dfrac{10^5}{\sqrt{50}} = 14142.136 \text{ rad/s}$

From Eq. (3.14), $\qquad \omega_r = \left[\dfrac{10^{10}}{50} - (5000)^2\right]^{1/2} = 13228.76 \text{ rad/s}$

Here as $\xi < \omega_0$, the circuit is underdamped. The current is, therefore, given by Eq. (3.15).

$$i(t) = \frac{230 \times 1000}{13228.76 \times 1} \cdot e^{-5000t} \cdot \sin (13228.76)t$$

$$= 17.3864 \cdot e^{-5000t} \cdot \sin (13228.76\ t)$$

(b) Diode stops conducting when $\omega_r\ t_1 = \pi$

$\therefore$ Conduction time of diode,

$$t_1 = \frac{\pi}{\omega_r} = \frac{\pi}{13228.76} = 237.482\ \mu s$$

(c) From Eq. (3.15), $\quad \dfrac{di}{dt} = \dfrac{V_s}{\omega_r L} \left[e^{-\xi t} \cdot \omega_r \cos \omega_r t - \sin \omega_r t \cdot (-\xi) e^{-\xi t} \right]$

$$\frac{di}{dt} \bigg|_{t=0} = \frac{V_s}{L} = \frac{230 \times 1000}{1} = 230{,}000\ \text{A/s}.$$

3.2. FREEWHEELING DIODES

In Fig. 3.3 (a), steady state current, after switch S is closed, is equal to V_s/R. Energy stored in inductance L is $\frac{1}{2} \cdot L\ (V_r/R)^2$. If the switch S is now opened, current V_s/R would eventually decay to zero. As the current V_s/R tends to decay with the opening of switch S, a high reverse voltage appears across switch as well as the diode. High voltage across switch leads to spark across the switch contacts, thus dissipating the stored energy. In the process, the diode, subjected to high reverse voltage, may get damaged. In order to avoid such an occurrence, a diode FD, called *freewheeling, or flywheel,* diode, is connected across load RL as shown in Fig. 3.11. (a). For understanding how FD comes into play, the working of circuit of Fig. 3.11 (a) is divided into two modes.

Mode I : When switch S is closed in Fig. 3.11 (a) at $t = 0$, current flows through V_s, S, D, R and L as shown in Fig. 3.11 (b). In this circuit, current i is given by

$$i = \frac{V_s}{R} (1 - e^{-\frac{R}{L}t}) \qquad \qquad ...(3.18)$$

Final value of current, $\quad I = \dfrac{V_s}{R}$

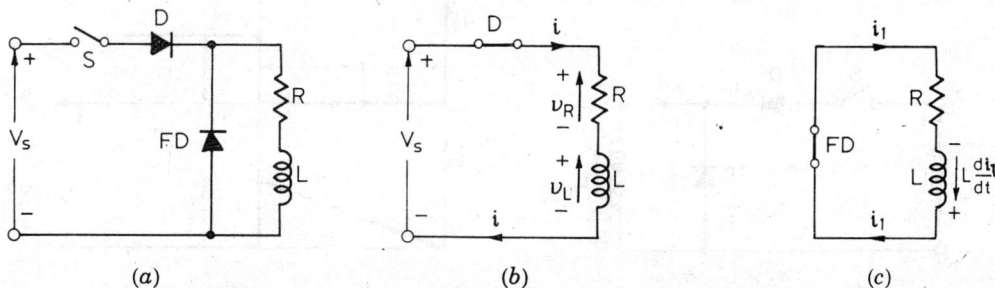

(a)   (b)   (c)

Fig. 3.11. Circuit of Fig. 3.3 with freewheeling diode.

Mode II : When switch S is opened at $t = 0$, current in the circuit tends to decay and so a voltage $L\dfrac{di}{dt}$ is induced in L which forward biases freewheeling diode. The current is, therefore, transferred to the circuit consisting of FD, R and L as shown in Fig. 3.11 (c). In this circuit, current is given by

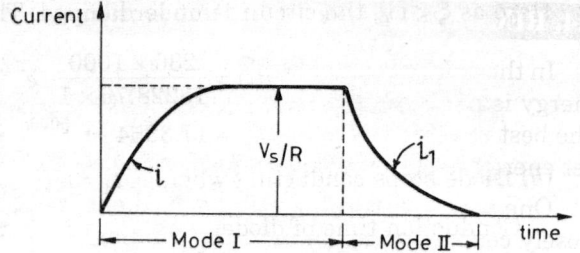

Fig. 3.11. (d) Current variation in the circuit of Fig. 3.11.

$$i_1 = \frac{V_s}{R} \cdot e^{-\frac{R}{L}t} \qquad\qquad ...(3.19)$$

The current i_1 will eventually decay to zero exponentially in mode II of Fig. 3.11 (c). The current build up during mode I and current decay during mode II are shown in Fig. 3.11 (d).

3.3. DIODE AND L CIRCUIT

Consider the circuit of Fig. 3.12 (a) where dc source feeds L through diode D. A freewheeling diode FD is connected across L. When switch S is closed at $t = 0$, KVL gives

$$V_s = L\frac{di}{dt}$$

or
$$i = \frac{V_s}{L} t \qquad\qquad ...(3.20)$$

This shows that current i rises linearly with time t. In case switch S is opened at t_1, load current $\dfrac{V_s}{L} t_1$ begins to flow through FD. As there is no resistance in the circuit formed by L and FD, current continues to flow at its constant value of $\dfrac{V_s}{L} t_1$. Energy stored in the inductance is $\dfrac{1}{2}\left(\dfrac{V_s}{L} t_1\right)^2 \cdot L = \dfrac{1}{2} \cdot \dfrac{V_s^2}{L} t_1$ joules. Current waveforms are shown in Fig. 3.12 (b).

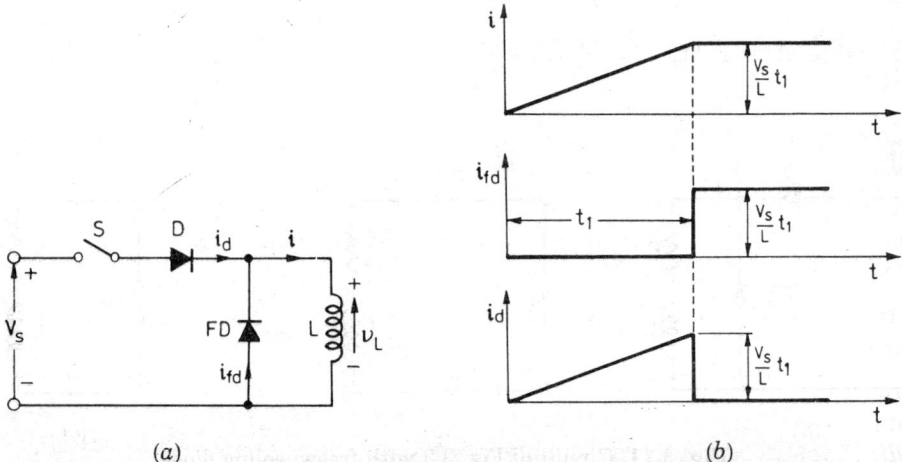

(a)   (b)
Fig. 3.12. Diode circuit with FD and L load (a) circuit diagram and (b) waveforms.

3.4. RECOVERY OF TRAPPED ENERGY

In the ideal circuit of Fig. 3.12 (a), the energy stored in the inductor is trapped. This trapped energy is not dissipated even when *FD* conducts because circuit does not contain resistance. The best way of utilization of this trapped energy is to return it to the source. In this manner, net energy taken from the source is reduced and the system efficiency improves.

One way of returning this trapped energy back to the source is to add a second winding closely coupled with the inductor winding as shown in Fig. 3.13. A diode *D* is also placed in series with the second winding. The inductor now behaves like a transformer. The two windings are so arranged that their polarity markings are opposite to each other.

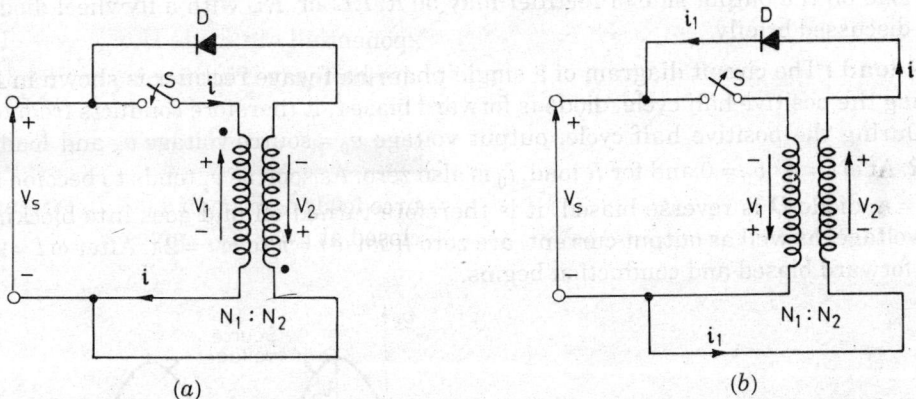

Fig. 3.13. Energy-recovery circuit (a) switch S closed and (b) switch S opened.

When switch *S* is closed, current *i* begins to flow and energy is stored in the inductance of primary winding with N_1 turns. The polarity of the secondary winding voltage V_2 is as shown. The diode *D* is reverse biased by voltage $(V_2 + V_s)$.

When switch *S* is opened, polarities of voltages V_1 and V_2 get reversed, the diode is now forward biased by voltage $(V_2 - V_s)$. As a result, diode begins to conduct a current i_1 into the positive terminal of source voltage V_s and so the trapped energy is fed back to the source.

Energy fedback to dc source = $V_s \times$ current i_1 dependent upon $(V_2 - V_s)$.

The energy stored in *L* of N_1 turns is transferred to secondary winding of N_2 turns from where it is fed back into the dc source.

3.5. SINGLE-PHASE DIODE RECTIFIERS

Rectification is the process of conversion of alternating input voltage to direct output voltage. As stated before, a rectifier converts ac power to dc power. In diode-based rectifiers, the output voltage cannot be controlled.

In this section, uncontrolled single-phase rectifiers are studied. The diode is assumed ideal as before.

A rectifier may be half-wave type or full-wave type. A half-wave rectifier is one in which current in any one line, connected to ac source, is *unidirectional*. However, a full-wave rectifier has *bidirectional* current in any one line connected to *ac* surface.

A rectifier may be one-pulse, two-pulse, three-pulse or n-pulse type. The number of pulses in any rectifier-configuration is obtained as under :

> pulse number = number of load current (or voltage) pulses during one
> cycle of ac source voltage.

3.5.1. Single-Phase Half-wave Rectifier

This is the simplest type of uncontrolled rectifier. It is never used in industrial applications because of its poor performance. Its study is, however, useful in understanding the principle of rectifier operation.

In a single-phase half-wave rectifier, for one cycle of supply voltage, there is one half-cycle of output, or load, voltage. As such, it is also called *single-phase one-pulse rectifier.*

The load on the output side of rectifier may be R, RL or RL with a flywheel diode. These are now discussed briefly.

(a) **R load :** The circuit diagram of a single-phase half-wave rectifier is shown in Fig. 3.14 (a). During the positive half cycle, diode is forward biased, it therefore conducts from $\omega t = 0°$ to $\omega t = \pi$. During the positive half cycle, output voltage v_0 = source voltage v_s and load current $i_0 = v_0/R$. At $\omega t = \pi$, $v_0 = 0$ and for R load, i_0 is also zero. As soon as v_s tends to become negative after $\omega t = \pi$, diode D is reverse biased, it is therefore turned off and goes into blocking state. Output voltage, as well as output current, are zero from $\omega t = \pi$ to $\omega t = 2\pi$. After $\omega t = 2\pi$, diode is again forward biased and conduction begins.

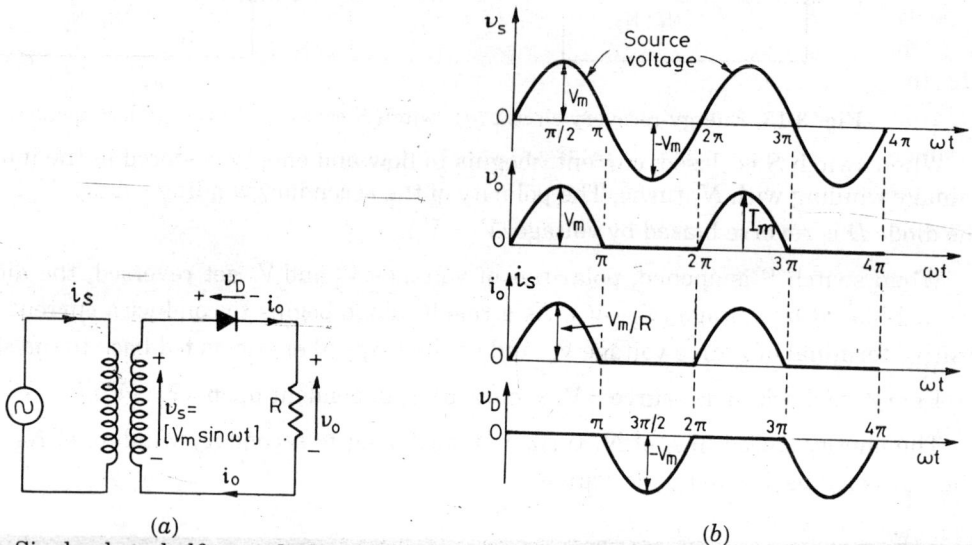

(a)   (b)

Fig. 3.14. Single-phase half-wave diode rectifier with R load (a) circuit diagram and (b) waveforms.

For a resistive load, output current i_0 has the same waveform as that of the output voltage v_0. Diode voltage v_D is zero when diode conducts. Diode is reverse biased from $\omega t = \pi$ to $\omega t = 2\pi$ as shown. The waveforms of v_s, v_0, i_0 and v_D are sketched in Fig. 3.14 (b). Here source voltage is sinusoidal *i.e.* $v_s = V_m \sin \omega t$. KVL for the circuit of Fig. 3.14 (a) gives $v_s = v_0 + v_D$.

Average value of output (or load) voltage,

$$V_0 = \frac{1}{2\pi}\left[\int_0^\pi V_m \sin \omega t \, d(\omega t)\right]$$

$$= \frac{V_m}{2\pi} \mid -\cos \omega t \mid_0^\pi = \frac{V_m}{\pi} \qquad \qquad ...(3.21)$$

Rms value of output voltage, $V_{or} = \left[\frac{1}{2\pi} \int_0^\pi V_m^2 \sin^2 \omega t \cdot d(\omega t) \right]^{1/2}$

$$= \frac{V_m}{\sqrt{2\pi}} \left[\int_0^\pi \frac{1 - \cos 2\omega t}{2} \cdot d(\omega t) \right]^{1/2}$$

$$= \frac{V_m}{2} \qquad \qquad ...(3.22)$$

Here the subscript 'r' is used to denote rms value.

Average value of load current,

$$I_0 = \frac{V_0}{R} = \frac{V_m}{\pi R} \qquad \qquad ...(3.23)$$

Rms value of load current, $\qquad I_{or} = \frac{V_{or}}{R} = \frac{V_m}{2R} \qquad \qquad ...(3.24)$

Peak value of load, or diode, current

$$= \frac{V_m}{R} \qquad \qquad ...(3.25)$$

Peak inverse voltage, PIV, is an important parameter in the design of rectifier circuits. *PIV* is the maximum voltage that appears across the device (here diode) during its blocking state. In Fig. 3.14, PIV $= V_m = \sqrt{2} \cdot V_s = \sqrt{2}$ (rms value of transformer secondary voltage). It is seen from the waveform of source current i_s (or i_o) that the transformer has to handle *dc* component of i_s. It leads to magnetic saturation of the transformer case, therefore more iron losses, more transformer heating and reduced efficiency.

Power delivered to resistive load $= (rms$ load voltage$)$ $(rms$ load current $)$

$$= V_{or} \; I_{or} = \frac{V_m}{2} \cdot \frac{V_m}{2R} = \frac{V_m^2}{4R} = \frac{V_s^2}{2R} = I_{or}^2 R \qquad ...(3.26)$$

Input power factor $\qquad = \dfrac{\text{Power delivered to load}}{\text{Input } VA}$

$$= \frac{V_{or} \cdot I_{or}}{V_s \cdot I_{or}} = \frac{V_{or}}{V_s} = \frac{\sqrt{2} V_s}{2 V_s} = 0.707 \text{ lag.}$$

(b) **L load :** Single-phase half-wave diode rectifier with L load is shown in Fig. 3.15 (a). When switch S is closed at $\omega t = 0$, diode starts conducting. *KVL* for this circuit gives

$$v_s = v_0 = L \frac{di_o}{dt} = V_m \sin \omega t$$

or $\qquad\qquad i_0 = \frac{V_m}{L} \int \sin \omega t \cdot dt$

$$= - \frac{V_m}{\omega L} \cos \omega t + A \qquad ...(3.27\ a)$$

At $\omega t = 0$, $i_0 = 0$, $\therefore \qquad 0 = - \frac{V_m}{\omega L} + A$

or $\qquad\qquad A = V_m / \omega L$

Substitution of the value of A in Eq. (3.27 a) gives

$$i_0 = \frac{V_m}{\omega L}(1 - \cos \omega t) \qquad \qquad ...(3.27\ b)$$

Output voltage, $\qquad v_0 = L\frac{di_0}{dt} = L\frac{V_m}{\omega L}[\sin \omega t]\ \omega = V_m \sin \omega t = v_s$

Source voltage v_s and both output voltage v_0 and output current i_0 are plotted in Fig. 3.15 (b).

Average value of output voltage, $V_0 = 0$

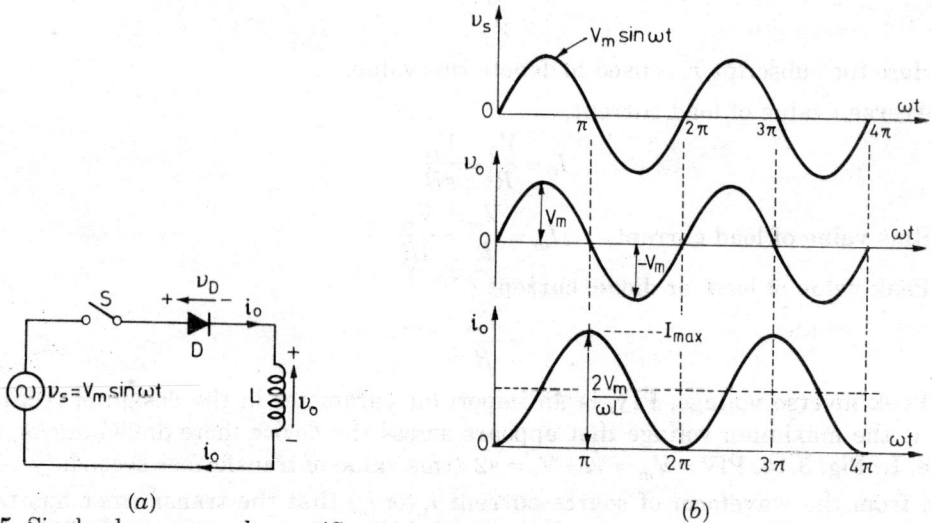

(a) (b)

Fig. 3.15. Single-phase one-pulse rectifier with L load (a) circuit diagram and (b) waveforms.

The output current i_0 consists of dc component and fundamental frequency component of frequency ω.

Peak value of current I_{max} occurs at $\omega t = \pi$

$$\therefore \qquad I_{max} = \frac{V_m}{\omega L}(1 + 1) = \frac{2V_m}{\omega L} \qquad \qquad ...(3.28)$$

Average value of current, $\qquad I_0 = \frac{1}{2\pi}\int_0^{2\pi}\frac{V_m}{\omega L}(1 - \cos \omega t)\, d(\omega t)$

$$= \frac{V_m}{\omega L} = \frac{1}{2}I_{max} \qquad \qquad ...(3.29)$$

Rms value of fundamental current, I_{1r} is given by

$$I_{1r} = \left[\frac{1}{2\pi}\left(\frac{V_m}{\omega L}\right)^2\int_0^{2\pi}(\cos \omega t)^2\, d(\omega t)\right]^{1/2}$$

$$= \frac{V_m}{\sqrt{2}\cdot \omega L} = \frac{V_s}{\omega L} = \frac{I_0}{\sqrt{2}} \qquad \qquad ...(3.30)$$

Rms value of rectified current $= \left[I_0^2 + I_{1r}^2\right]^{1/2}$

$$= \left[I_0^2 + \frac{I_0^2}{2}\right]^{1/2} = 1.225\,I_0 \qquad \qquad ...(3.31)$$

Voltage across diode, $v_D = 0$.

(c) **C Load :** In Fig. 3.16 (a), when switch S is closed at $\omega t = 0$, the equation governing the behaviour of the circuit is

$$i_0 = C \frac{dv_s}{dt} = C \frac{d}{dt} (V_s \sin \omega t)$$

$$= \omega C V_m \cos \omega t \qquad \qquad ...(3.32)$$

Output voltage, $\qquad v_0 = \frac{1}{C} \int i \, dt = V_m \sin \omega t = v_s = v_C$

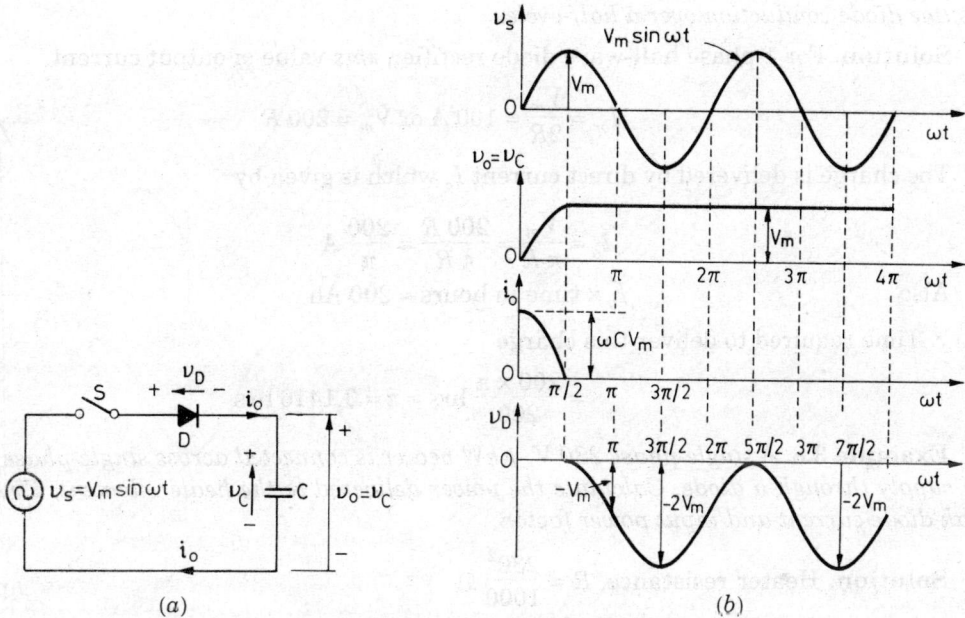

Fig. 3.16. Single-phase half-wave diode rectifier with C load (a) circuit diagram and (b) waveforms.

Capacitor is charged to voltage V_m at $\omega t = \frac{\pi}{2}$ and subsequently this voltage remains constant at V_m. This is shown as $v_0 = v_C$ in Fig. 3.16 (b).

Capacitor current or load current is maximum at $\omega t = 0$. Its value at $\omega t = 0$ is $\omega C V_m$ as shown.

The diode conducts for $\frac{\pi}{2\omega}$ seconds only from $\omega t = 0$ to $\omega t = \frac{\pi}{2}$. During this interval, diode voltage is, therefore, zero. After $\omega t = \pi/2$, diode voltage v_D is given by

$$v_D = -v_0 + v_s = -V_m + V_m \sin \omega t$$

$$= V (\sin \omega t - 1) \qquad \qquad ...(3.33)$$

For Eq. (3.33), the time origin is redefined at $\omega t = \pi/2$.

After $\omega t = \pi/2$, diode voltage is plotted as shown in Fig. 3.16 (b). At $\omega t = \frac{3\pi}{2}$, $v_D = -2 V_m$.

Average value of voltage across diode,

$$V_D = \frac{1}{2\pi} \int_0^{2\pi} V_m (\sin \omega t - 1) \, d(\omega t)$$

$$= V_m = \sqrt{2} \, V_s \qquad \qquad ...(3.34 \, a)$$

Rms value of fundamental component of voltage across diode,

$$V_{1r} = \left[\frac{1}{2\pi}\int_0^{2\pi} V_m^2 \sin^2 \omega t \, d(\omega t)\right]^{1/2} = \frac{V_m}{\sqrt{2}} \qquad ...(3.34\ b)$$

Rms value of voltage across diode

$$= \sqrt{V_D^2 + V_{1r}^2} = 1.225\, V_m \qquad ...(3.35)$$

Example 3.7. *Find the time required to deliver a charge of 200 Ah through a single-phase half-wave diode rectifier with an output current of 100 A rms and with sinusoidal input voltage. Assume diode conduction over a half-cycle.*

Solution. For 1-phase half-wave diode rectifier, *rms* value of output current,

$$I_{or} = \frac{V_m}{2R} = 100\ A \text{ or } V_m = 200\ R$$

The charge is delivered by direct current I_o which is given by

$$I_o = \frac{V_m}{\pi R} = \frac{200\ R}{\pi R} = \frac{200}{\pi}\ A$$

Also $I_o \times$ time in hours $= 200$ Ah

∴ Time required to deliver this charge

$$= \frac{200 \times \pi}{200}\ \text{hrs} = \pi = 3.1416\ \text{hrs}$$

Example 3.8. *A single-phase 230 V, 1 kW heater is connected across single-phase 230 V, 50 Hz supply through a diode. Calculate the power delivered to the heater element. Find also the peak diode current and input power factor.*

Solution. Heater resistance, $R = \dfrac{230^2}{1000}\ \Omega$

Rms value of output voltage, from Eq. (3.22), is

$$V_{or} = \frac{\sqrt{2} \times 230}{2}$$

Power absorbed by heater element

$$= \frac{V_{or}^2}{R} = \frac{2 \times 230^2}{4} \times \frac{1000}{230^2} = 500\ W$$

Peak value of diode current, from Eq. (3.25), is given by

$$\frac{\sqrt{2} \times 230}{230^2} \times 1000 = 6.1478\ A$$

Input power factor $= \dfrac{V_{or}}{V_s} = \dfrac{\sqrt{2} \times 230}{2} \times \dfrac{1}{230} = 0.707$ lag.

(d) **RE Load :** Single-phase half-wave diode rectifier with load resistance R and load counter emf E is shown in Fig. 3.17 (a). If the switch S is closed at $\omega t = 0°$ or when $v_s = 0$, then diode would not conduct at $\omega t = 0$ because diode is reverse biased until source voltage v_s equals E. When $V_m \sin \theta_1 = E$, diode D starts conducting and the turn-on angle θ_1 is given by

$$\theta_1 = \sin^{-1}\left(\frac{E}{V_m}\right) \qquad ...(3.36)$$

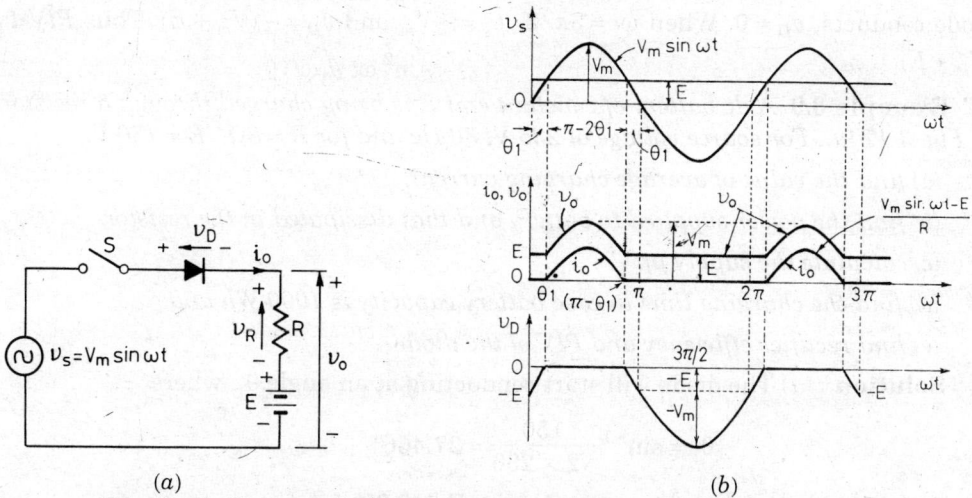

(a)   (b)

Fig. 3.17. Single-phase half-wave diode rectifier with RE load
(a) circuit diagram and (b) wave forms.

The diode now conducts from $\omega t = \theta_1$ to $\omega t = (\pi - \theta_1)$, *i.e.*, conduction angle for diode is $(\pi - 2\theta_1)$ as shown in Fig. 3.17 (b). During the conduction period of diode, the voltage equation for the circuit is

$$V_m \sin \omega t = E + i_0 R$$

or
$$i_0 = \frac{V_m \sin \omega t - E}{R} \qquad \qquad ...(3.37)$$

Average value of this current is given by

$$I_0 = \frac{1}{2\pi R}\left[\int_{\theta_1}^{\pi - \theta_1}(V_m \sin \omega t - E)\, d(\omega t)\right]$$

$$= \frac{1}{2\pi R}\left[2 V_m \cos \theta_1 - E\,(\pi - 2\theta_1)\right] \qquad \qquad ...(3.38)$$

Rms value of the load current of Eq. (3.37) is

$$I_{or} = \left[\frac{1}{2\pi}\int_{\theta_1}^{\pi - \theta_1}\left(\frac{V_m \sin \omega t - E}{R}\right)^2 \cdot d(\omega t)\right]^{1/2}$$

$$= \left[\frac{1}{2\pi R^2}\int_{\theta_1}^{\pi - \theta_1}(V_m^2 \sin^2 \omega t + E^2 - 2 V_m E \sin \omega t)\, d(\omega t)\right]^{1/2}$$

$$= \left[\frac{1}{2\pi R^2}\left\{(V_s^2 + E^2)\,(\pi - 2\theta_1) + V_s^2 \sin 2\theta_1 - 4 V_m E \cos \theta_1\right\}\right]^{1/2} \qquad ...(3.39)$$

Power delivered to load,

$$P = E I_0 + I_{or}^2 R \text{ watts} \qquad \qquad ...(3.40)$$

Supply pf
$$= \frac{\text{Power delivered to load}}{(\text{Source voltage}) (\text{rms value of source current})}$$

$$= \frac{E I_0 + I_{or}^2 R}{V_s \cdot I_{or}} \qquad \qquad ...(3.41)$$

It is seen from Fig. 3.17 (a) that at $\omega t = 0°$, $v_D = -E$ and at $\omega t = \theta_1$, $v_D = 0$. During the period diode conducts, $v_D = 0$. When $\omega t = 3\pi/2$, $v_s = -V_m$ and $v_D = -(V_m + E)$. Thus PIV for diode is $(V_m + E)$.

Example 3.9. *A dc battery of constant emf E is being charged through a resistor as shown in Fig. 3.17 (a). For source voltage of 235 V, 50 Hz and for R = 8Ω, E = 150 V,*

(a) *find the value of average charging current,*

(b) *find the power supplied to battery and that dissipated in the resistor,*

(c) *calculate the supply pf,*

(d) *find the charging time in case battery capacity is 1000 Wh and*

(e) *find rectifier efficiency and PIV of the diode.*

Solution : (a) The diode will start conducting at an angle θ_1, where

$$\theta_1 = \sin^{-1} \frac{150}{\sqrt{2} \times 230} = 27.466°$$

Average value of charging current, from Eq. (3.38), is

$$I_0 = \frac{1}{2\pi \times 8}\left[2 \cdot \sqrt{2} \times 230 \cos 27.466° - 150\left(\pi - \frac{2 \times 27.466 \times \pi}{180}\right)\right]$$

$$= 4.9676 \text{ A}$$

(b) Power delivered to battery

$$= E\, I_0 = 150 \times 4.9676 = 745.14 \text{ W}$$

Rms value of charging current, from Eq. (3.39), is

$$I_{or} = \left[\frac{1}{2\pi \times 64}\left\{(230^2 + 150^2)\left(\pi - 2 \times 27.466 \times \frac{\pi}{180}\right) + 230^2 \sin 2 \times 27.466° \right.\right.$$
$$\left.\left. - 4\sqrt{2} \times 230 \times 150 \cos 27.466°\right\}\right]^{1/2} = 9.2955 \text{ A}$$

Power dissipated in resistor

$$= I_{or}^2\, R = (9.2955)^2 \times 8 = 691.25 \text{ W}$$

(c) From Eq. (3.41), the supply

$$pf = \frac{745.14 + 691.25}{230 \times 9.2955} = 0.672 \text{ lag}$$

(d) (Power delivered to battery) (charging time in hours)

= Battery capacity in Wh.

∴ Charging time $= \dfrac{1000}{745.14} = 1.342 \text{ h}$

(e) Rectifier efficiency $= \dfrac{\text{Power delivered to battery}}{\text{Total input power}}$

$$= \frac{745.14}{745.14 + 691.25} \times 100 = 51.876\%$$

(f) PIV of diode $= V_m + E = \sqrt{2} \times 230 + 150 = 475.22 \text{ V}.$

(e) **RL Load :** A single-phase one-pulse diode rectifier feeding RL load is shown in Fig. 3.18 (a). Current i_0 continues to flow-even after source voltage v_s has become negative ; this is because of the **presence of inductance L in the load circuit**. After positive half cycle of source

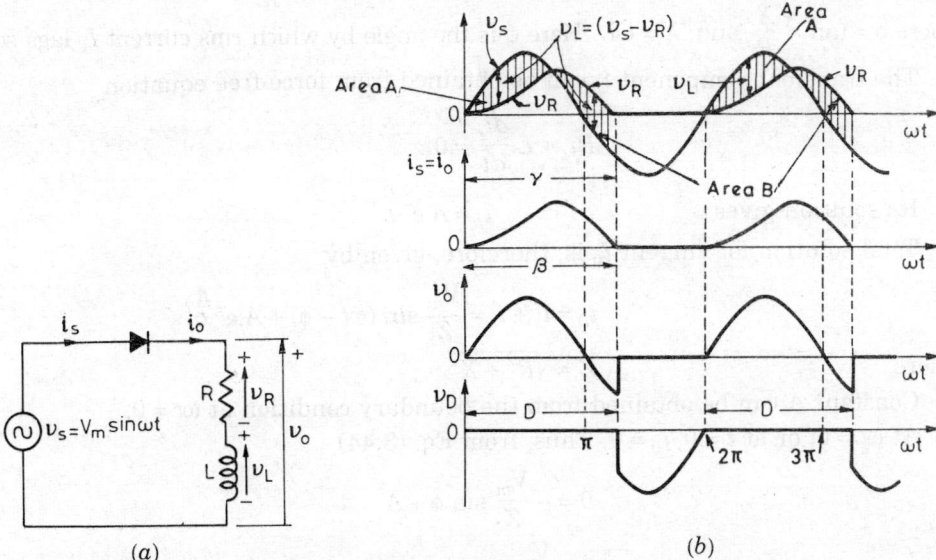

Fig. 3.18. Single-phase half-wave diode rectifier with RL load
(a) circuit diagram and (b) waveforms.

voltage, diode remains on, so the negative half cycle of source voltage appears across load until load current i_o decays to zero at $\omega t = \beta$. Voltage $v_R = i_o R$ has the same waveshape as that of i_o. Inductor voltage $v_L = v_s - v_R$ is also shown. The current i_0 flows till the two areas A and B are equal. Area A (where $v_s > v_R$) represents the energy stored by L and area B (where $v_s < v_R$) the energy released by L. It must be noted that average value of voltage v_L across inductor L is zero.

When $i_0 = 0$ at $\omega t = \beta$; $v_L = 0$, $v_R = 0$ and voltage v_s appears as reverse bias across diode D as shown. At β, voltage v_D across diode jumps from zero to $V_m \sin \beta$ where $\beta > \pi$. Here $\beta = \gamma$ is also the conduction angle of the diode.

Average value of output voltage,

$$V_0 = \frac{1}{2\pi} \int_0^\beta V_m \sin \omega t \cdot d(\omega t)$$

$$= \frac{V_m}{2\pi} (1 - \cos \beta) \qquad \qquad ...(3.42)$$

Average value of load or output current

$$I_0 = \frac{V_0}{R} = \frac{V_m}{2\pi R} (1 - \cos \beta) \qquad \qquad ...(3.43)$$

A general expression for output current i_0 for $0 < \omega t < \beta$ can be obtained as under :

When diode is conducting, KVL for the circuit of Fig. 3.18 (a) gives

$$R i_0 + L \frac{di_0}{dt} = V_m \sin \omega t$$

The load, or output, current i_0 consists of two components, one steady state component i_s and the other transient component i_t. Here i_s is given by

$$i_s = \frac{V_m}{\sqrt{R^2 + X^2}} \sin (\omega t - \phi)$$

where $\phi = \tan^{-1}\dfrac{X}{R}$ and $X = \omega L$. Here ϕ is the angle by which rms current I_s lags source voltage V_s. The transient component i_t can be obtained from force-free equation

$$Ri_t + L\frac{di_t}{dt} = 0$$

Its solution gives $\qquad\qquad i_t = A\,e^{-\frac{R}{L}t}$

Total solution for current i_0 is, therefore, given by

$$i_0 = i_s + i_t = \frac{V_m}{Z}\sin(\omega t - \phi) + A\,e^{-\frac{R}{L}t} \qquad\qquad ...(3.44)$$

where $\qquad\qquad\qquad\qquad Z = \sqrt{R^2 + X^2}$

Constant A can be obtained from the boundary condition at $\omega t = 0$.

At $\omega t = 0$, or at $t = 0$, $i_0 = 0$. Thus, from Eq. (3.44)

$$0 = -\frac{V_m}{Z}\sin\phi + A$$

$\therefore \qquad\qquad\qquad\qquad A = \dfrac{V_m}{Z}\sin\phi$

Substitution of A in Eq. (3.44) gives

$$i_0 = \frac{V_m}{Z}\left[\sin(\omega t - \phi) + \sin\phi \cdot e^{-\frac{R}{L}t}\right] \qquad\qquad ...(3.45)$$

for $0 \le \omega t \le \beta$

It is also seen from the waveform of i_0 in Fig. 3.18 (b) that when $\omega t = \beta$, $i_0 = 0$. With this condition, Eq. (3.45) gives

$$\sin(\beta - \phi) + \sin\phi \cdot \exp\left[-\frac{R}{\omega L}\beta\right] = 0$$

The solution of this transcendental equation can give the value of extinction angle β.

(f) **RL load with freewheeling diode***: Performance of single-phase one-pulse diode rectifier with RL load can be improved by connecting a freewheeling diode across the load as shown in Fig. 3.19 (a). Output voltage is $v_0 = v_s$ for $0 \le \omega t \le \pi$. At $\omega t = \pi$, source voltage v_s is zero, but output current i_0 is not zero because of L in the load circuit. Just after $\omega t = \pi$, as v_s tends to reverse, negative polarity of v_s reaches cathode of FD through conducting diode D, whereas positive polarity of v_s reaches anode of FD direct. Freewheeling (or flywheel) diode FD, therefore, gets forward-biased. As a result, load current i_0 is immediately transferred from D to FD as v_s tends to reverse. After $\omega t = \pi$, diode, or source, current $i_s = 0$ and diode D is subjected to reverse voltage with PIV equal to V_m at $\omega t = \dfrac{3\pi}{2}, \dfrac{7\pi}{2}$ etc.

After $\omega t = \pi$, current freewheels through circuit R, L and FD. The energy stored in L is now dissipated in R. When energy stored in L = energy dissipated in R, current falls to zero at $\omega t = \beta < 2\pi$. Depending upon the value of R and L, the current may not fall to zero even when $\omega t = 2\pi$, this is called continuous conduction. But in Fig. 3.19 (b), load current decays to zero before $\omega t = 2\pi$; load current is therefore discontinuous.

---

*Freewheeling diode is also called bypass diode or commutating diode.

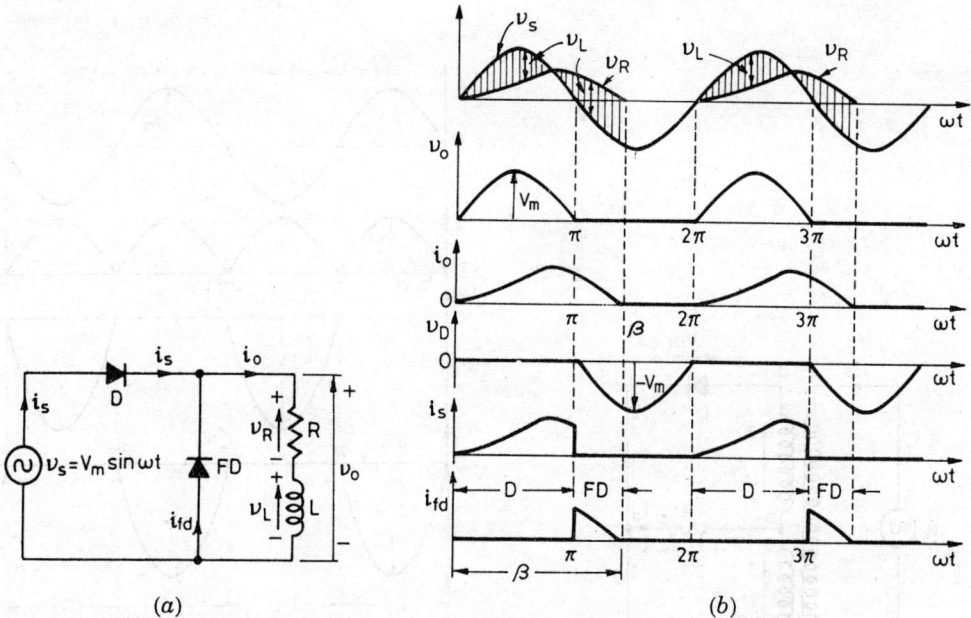

Fig. 3.19. Single-phase one-pulse diode rectifier with RL load and freewheeling diode
(a) circuit diagram and (b) waveforms.

The effects of using freewheeling diode are as under :

(i) It prevents the output (or load) voltage from becoming negative.

(ii) As the energy stored in L is transferred to load R through FD, the system efficiency is improved.

(iii) The load current waveform is more smooth, the load performance, therefore, gets better.

The waveforms for v_s, v_0, i_0, v_D, i_s and i_{fd} are drawn in Fig. 3.19 (b).

The expression for the load current i_0 can be obtained from Art. 6.1.2 if required. It is seen from Fig. 3.19 (b) that

average output voltage, $$V_0 = \frac{1}{2\pi} \int_0^\pi V_m \sin \omega t \, d(\omega t) = \frac{V_m}{\pi} \qquad \ldots (3.46)$$

and average load current, $$I_0 = \frac{V_m}{\pi R} \qquad \ldots (3.47)$$

(g) **Single-phase full-wave diode rectifier :** There are two types of full-wave diode rectifiers, one is centre-tapped (or mid-point) full-wave diode rectifier and the other is full-wave diode bridge rectifier. These are now described briefly.

(i) **Single-phase full-wave mid-point diode rectifier :** Fig. 3.20 (a) illustrates a single-phase full-wave mid-point rectifier using diodes. The turns ratio from each secondary to primary is taken as unity for simplicity. When 'a' is positive with respect to 'b', or mid-point O, diode D1 conducts for π radians. In the next half cycle, 'b' is positive with respect to 'a', or mid-point O, and therefore diode D2 conducts. The output voltage is shown as v_0 in Fig. 3.20 (b). The waveform for output current i_0 (not shown in the figure) is similar to v_0 waveform.

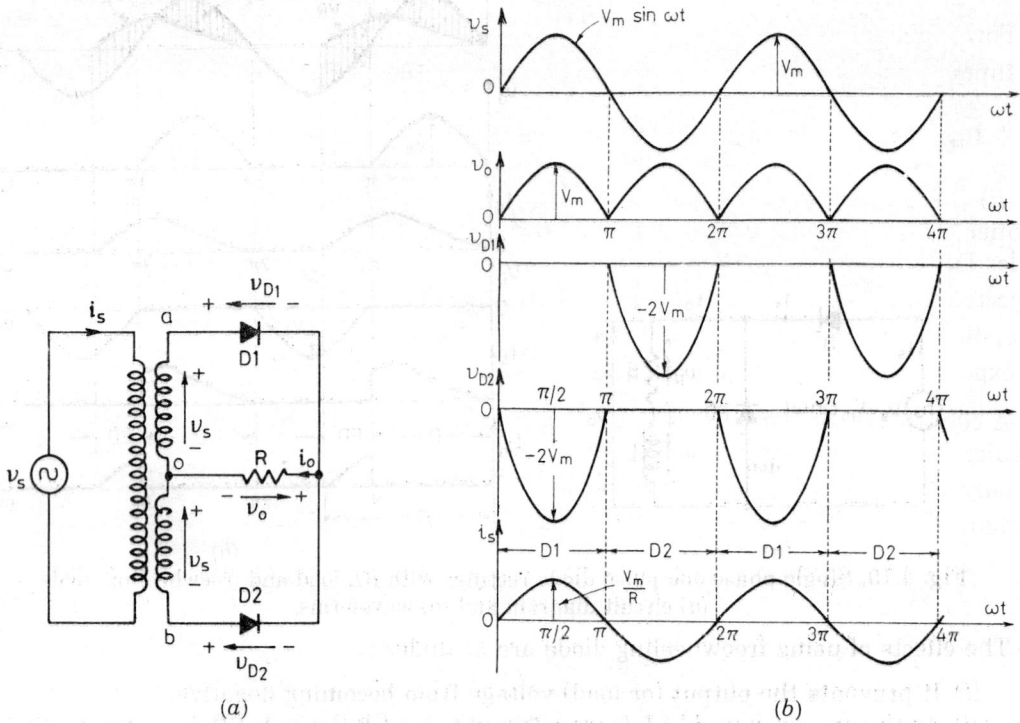

Fig. 3.20. Single-phase full-wave mid-point diode rectifier
(a) circuit diagram and (b) waveforms.

When 'a' is positive with respect to 'b', diode D2 is subjected to a reverse voltage of $2v_s$. In the next half cycle, diode D1 experiences a reverse voltage of $2v_s$. This is shown in Fig. 3.20 (b) as v_{D1} and v_{D2}. Thus, for diodes D1 and D2, peak inverse voltage is $2V_m$. Waveforms of Fig. 3.20 (b) show that for one cycle of source voltage, there are two pulses of output voltage. So single-phase full-wave diode rectifier can also be called *single-phase two-pulse* diode rectifier.

Source current waveform i_s is also shown in Fig. 3.20 (b). When D1 conducts, current in secondary flows upward from o to a and therefore, primary current i_s must flow *downward* to balance the secondary mmf from 0 to π rad. When D2 conducts, secondary current flows downward from o to b, therefore, primary current i_s must flow *upward* to balance the secondary mmf from π to 2π and so on.

Average output voltage,
$$V_0 = \frac{1}{\pi} \int_0^\pi V_m \sin \omega t \, d(\omega t) = \frac{2V_m}{\pi} \qquad \ldots (3.48\ a)$$

Average output current,
$$I_0 = \frac{V_0}{R}$$

Rms value of output voltage,
$$V_{or} = \left[\frac{1}{\pi} \int_0^\pi V_m^2 \sin^2 \omega t \, d(\omega t) \right]^{1/2}$$

$$= \frac{V_m}{\sqrt{2}} = V_s \qquad \ldots (3.48\ b)$$

Rms value of load current, $I_{or} = \dfrac{V_s}{R}$

Power delivered to load $= V_{or} \cdot I_{or} = I_{or}^2 \cdot R$

Input voltamperes $= V_s \cdot I_{or}$

∴ Input $pf = \dfrac{V_{or} \cdot I_{or}}{V_s \cdot I_{or}} = 1$

(*ii*) **Single-phase full-wave diode bridge rectifier :** A single-phase full-wave bridge rectifier employing diodes is shown in Fig. 3.21 (*a*). When '*a*' is positive with respect to '*b*', diodes D1, D2 conduct together so that output voltage is v_{ab}. Each of the diodes D3 and D4 is subjected to a reverse voltage of v_s as shown in Fig. 3.21 (*b*). When '*b*' is positive with respect to '*a*', diodes D3, D4 conduct together and output voltage is v_{ba}. Each of the two diodes D1 and D2 experience a reverse voltage of v_s as shown.

A comparison of Figs. 3.20 (*b*) and 3.21 (*b*) reveals that a diode in mid-point full-wave rectifier is subjected to PIV of $2V_m$ whereas a diode in full-wave bridge rectifier has PIV of V_m only. However, average and rms values of output voltage are the same for both rectifier configurations.

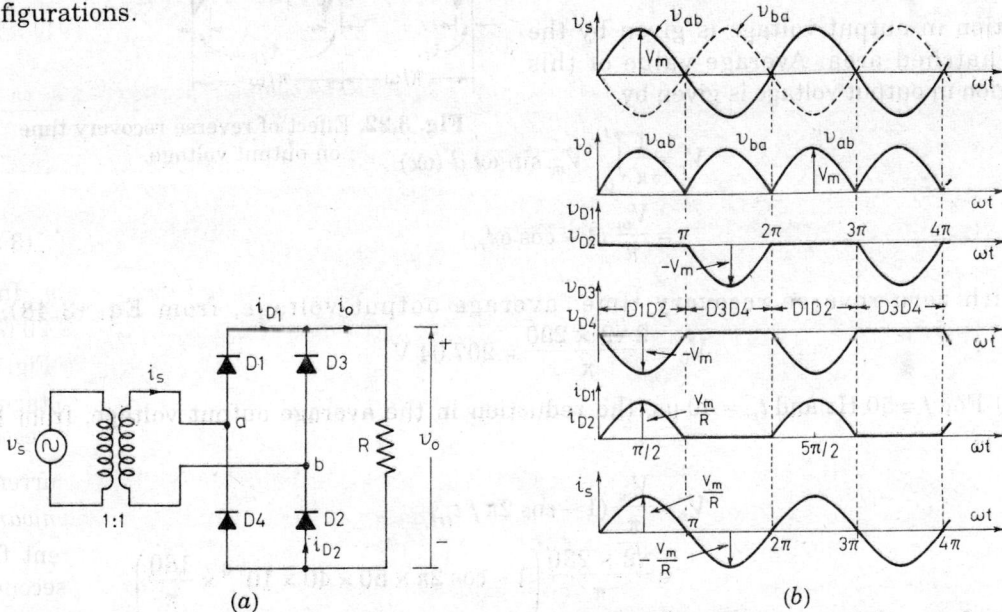

Fig. 3.21. Single-phase full-wave diode bridge rectifier (*a*) circuit diagram and (*b*) waveforms.

For the waveforms of diode current i_{D1} or i_{D2} in Fit 3.21 (*b*) and also for i_{D3}, i_{D4} for the circuit of Fig. 3.20 (*a*) (not shown in Fig. 3.20 (*b*)), the average and *rms* values for diode current are obtained as under :

Average value of diode current, $I_{DA} = \dfrac{1}{2\pi} \displaystyle\int_o^\pi I_m \sin \omega t \cdot d\,(\omega t) = \dfrac{I_m}{\pi}$ ..(3.49 *a*)

Rms value of diode current, $I_{Dr} = \left[\dfrac{1}{2\pi} \displaystyle\int_o^\pi I_m^2 \sin^2 \omega t\, d\,(\omega t) \right]^{1/2} = \dfrac{I_m}{2}$ ...(3.49 *b*)

Peak repetitive diode current, $I_m = \dfrac{V_m}{R}$ ...(3.49 *c*)

It can similarly be shown that average value of voltage across each diode in Fig. 3.20 (b) is $\frac{2 V_m}{\pi}$ and that in Fig. 3.21 (b) is $\frac{V_m}{\pi}$. The corresponding *rms* values of voltage across each diode is $V_m = \sqrt{2}\, V_s$ in Fig. 3.20 (b) and $\frac{V_m}{\sqrt{2}} = V_s$ in Fig. 3.21 (b).

Three-phase rectifiers using diodes are discussed in Art. 3.9. Example 3.10 is formulated to illustrate the effect of reverse recovery time on the average output voltage.

Example 3.10. *In a single-phase full-wave diode bridge rectifier, the diodes have a reverse recovery time of 40 μs. For an ac input voltage of 230 V, determine the effect of reverse recovery time on the average output voltage for a supply frequency of (a) 50 Hz and (b) 2.5 kHz.*

Solution. Single-phase full-wave diode bridge rectifier is shown in Fig. 3.21 (a) and output voltage v_0 is shown in Fig. 3.21 (b). If reverse recovery time is taken into consideration, the diodes D1 and D2 will not be off at $\omega t = \pi$ in Fig. 3.21 (b), but will continue to conduct until $t = \frac{\pi}{\omega} + t_{rr}$ as depicted in Fig. 3.22. The reduction in output voltage is given by the cross-hatched area. Average value of this reduction in output voltage is given by

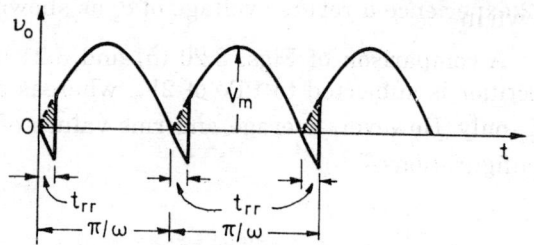

Fig. 3.22. Effect of reverse recovery time on output voltage.

$$V_r = \frac{1}{\pi} \int_0^{t_{rr}} V_m \sin \omega t\, d\,(\omega t)$$

$$= \frac{V_m}{\pi} (1 - \cos \omega t_{rr}) \qquad \ldots(3.50)$$

With zero reverse recovery time, average output voltage, from Eq. (3.48), is

$$V_0 = \frac{2\sqrt{2} \times 230}{\pi} = 207.04 \text{ V}$$

(a) For $f = 50$ Hz and $t_{rr} = 40$ μs, the reduction in the average output voltage, from Eq. (3.50), is

$$V_r = \frac{V_m}{\pi} (1 - \cos 2\pi f t_{rr})$$

$$= \frac{\sqrt{2} \times 230}{\pi} \left(1 - \cos 2\pi \times 50 \times 40 \times 10^{-6} \times \frac{180}{\pi} \right)$$

$$= 8.174 \text{ mV}$$

Percentage reduction in average output voltage

$$= \frac{8.174 \times 10^{-3}}{207.04} \times 100 = 3.948 \times 10^{-3} \%$$

(b) For $f = 2500$ Hz, the reduction in the average output voltage, from Eq. (3.50), is

$$V_r = \frac{\sqrt{2} \times 230}{\pi} \left(1 - \cos 2\pi \times 2500 \times 40 \times 10^{-6} \times \frac{180}{\pi} \right)$$

$$= 19.77 \text{ V}$$

Percentage reduction in average output voltage $= \dfrac{19.77}{207.04} \times 100 = 9.594\%$.

It is seen from above that the effect of reverse recovery time is negligible for diode operation at 50 Hz, but for high-frequency operation of diodes, the effect is noticeable.

Example 3.11. *A single-phase full bridge diode rectifier is supplied from 230 V, 50 Hz source. The load consists of R = 10 Ω and a large inductance so as to render the load current constant. Determine*

(a) average values of output voltage and output current,

(b) average and rms values of diode currents,

(c) rms values of output and input currents, and supply pf.

Solution. The circuit diagram and relevant waveforms for this uncontrolled rectifier are shown in Fig. 3.23.

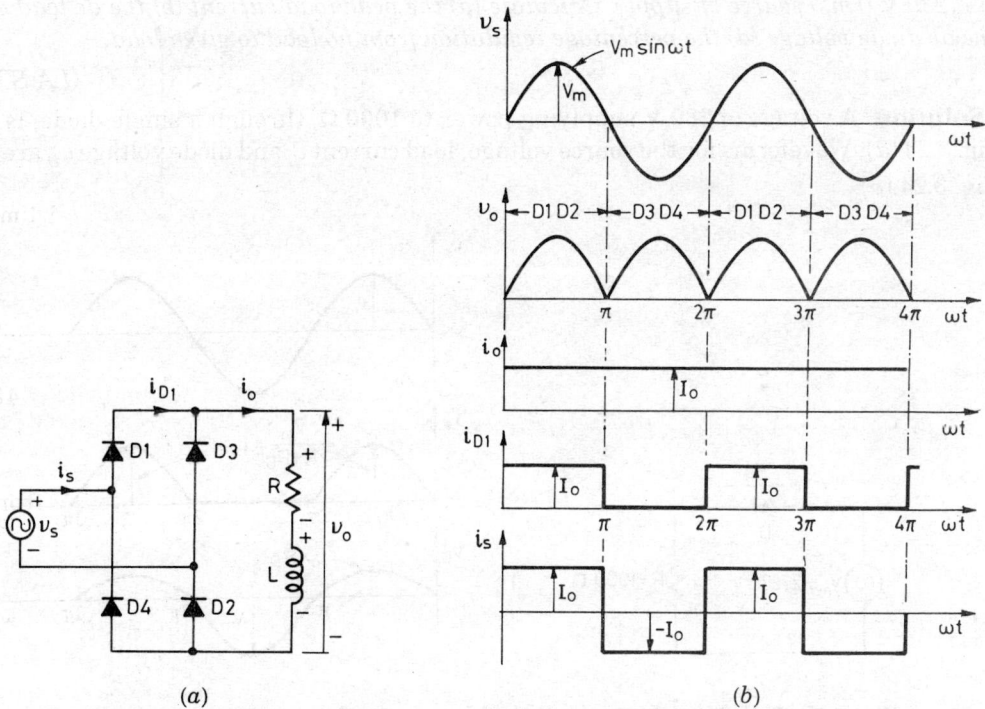

Fig. 3.23. Pertaining to Example 3.11 (a) circuit diagram and (b) waveforms.

(a) Average value of output voltage,

$$V_0 = \frac{2V_m}{\pi} = \frac{2\sqrt{2} \times 230}{\pi} = 207.04 \text{ V}$$

Average value of output current,

$$I_0 = \frac{V_0}{R} = \frac{207.04}{10} = 20.704 \text{ A}$$

(b) Average value of diode current,

$$I_{DA} = \frac{I_0 \cdot \pi}{2\pi} = \frac{I_0}{2} = \frac{20.704}{2} = 10.352 \text{ A}$$

Rms value of diode current, $I_{Dr} = \sqrt{\dfrac{I_0^2\,\pi}{2\pi}} = \dfrac{I_0}{\sqrt{2}} = \dfrac{20.704}{\sqrt{2}} = 14.642$ A

As load, or output, current is ripple free, rms value of output current

$$= \text{average value of output current} = I_0 = 20.704 \text{ A}$$

Rms value of source current, $I_s = \sqrt{\dfrac{I_0^2\,\pi}{\pi}} = I_0 = 20.704$ A

Load power $= V_0 I_0 = 207.04 \times 20.704$ W

Input power $= V_s I_s \cos\phi$

∴ $230 \times 20.704 \times \cos\phi = 207.04 \times 20.704$

∴ Supply pf $= \cos\phi = \dfrac{207.04}{230} = 0.90$ lagging.

Example 3.12. *A diode whose internal resistance is 20 Ω is to supply power to a 1000 Ω load from a 230 V (rms) source of supply. Calculate (a) the peak load current (b) the dc load current (c) the dc diode voltage (d) the percentage regulation from no load to given load.*

(I.A.S., 1983)

Solution. A voltage of 230 V supplying power to 1000 Ω, through a single diode, is shown in Fig. 3.24 (a). Waveforms for the source voltage, load current i_0 and diode voltage v_D are shown in Fig. 3.24 (b).

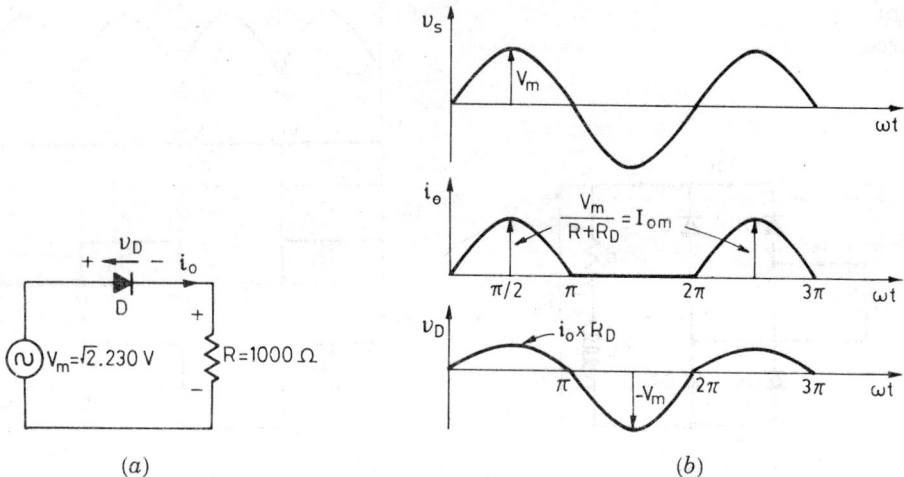

(a)   (b)

Fig. 3.24. Pertaining to Example 3.12 (a) circuit diagram and (b) waveforms.

(a) It is seen from the waveform of i_0 that peak load current I_{om} is given by

$$I_{om} = \frac{V_m}{R + R_D} = \frac{\sqrt{2} \times 230}{1020} = 0.3189 \text{ A}$$

Here R = load resistance and R_D = internal resistance of diode.

(b) DC load current, $I_0 = \dfrac{1}{2\pi}\displaystyle\int_0^\pi I_{om} \sin\omega t\, d(\omega t)$

$$= \frac{I_{om}}{\pi} = 0.10151 \text{ A}$$

(c) DC diode voltage, $\qquad V_D = I_0 R_D - \dfrac{1}{2\pi} \displaystyle\int_0^\pi 230\sqrt{2}\ \sin \omega t\ d(\omega t)$

$$= I_0 R_D - \frac{V_m}{\pi} = 0.10151 \times 20 - \frac{230\sqrt{2}}{\pi} = -101.5\ \text{V}$$

(d) At no load, load voltage, $\quad V_{on} = \dfrac{V_m}{\pi} = \dfrac{\sqrt{2} \times 230}{\pi} = 103.521\ \text{V}$

At given load, load voltage, $\quad V_{01} = \dfrac{230\sqrt{2}}{\pi} \times \dfrac{1000}{1020} = 101.491\ \text{V}$

$\therefore$ Voltage regulation $\qquad = \dfrac{V_{on} - V_{01}}{V_{on}} \times 100 = \dfrac{103.521 - 101.491}{103.521} = 1.961\%.$

3.6. ZENER DIODES

Zener diodes are specially constructed to have accurate and stable reverse breakdown voltage.

Circuit symbol for Zener diode is shown in Fig. 3.25 (a). When it is forward biased, it behaves as a normal diode. When reverse biased, a small leakage current flows. If the reverse voltage across Zener diode is increased, a value of voltage is reached at which reverse breakdown occurs. This is indicated by a sudden increase of Zener current, Fig. 3.25 (b). The voltage after reverse breakdown remains practically constant over a wide range of Zener current. This makes it suitable for use as a voltage regulator to furnish constant voltage from a source whose voltage may vary noticeably.

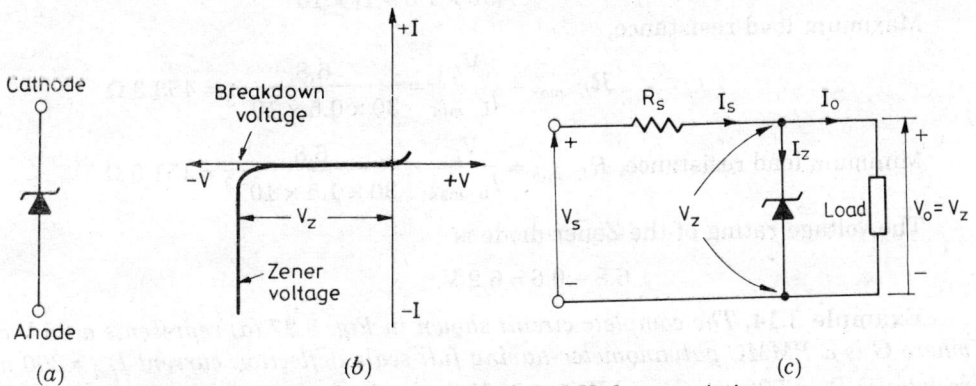

Fig. 3.25. Zener diode (a) circuit symbol (b) I-V characteristics (c) use as a voltage regulator.

For the operation of Zener diode as a voltage regulator, (i) it must be reverse biased with a voltage greater than its breakdown, or Zener, voltage and (ii) a series resistor R_s, Fig. 3.25 (c) is necessary to limit the reverse current through the diode below its rated value.

If V_z = voltage across Zener diode, then it is seen from Fig. 3.25 (c) that source current I_s is

$$I_s = \frac{V_s - V_z}{R_s}$$

Load, or output, current, $I_0 = \dfrac{V_z}{R}$ where R = load resistance. Current through Zener diode,

$$I_z = I_s - I_0$$

Power rating of a Zener diode is $V_Z \cdot I_Z$. These are available in a voltage range from few volts to about 280 V.

Example 3.13. *Design a Zener voltage regulator, shown in Fig. 3.26, to meet the following specifications :*

Load voltage = 6.8 V, Source voltage V_s is 20 V ± 20% and load current is 30 mA ± 50%.

The Zener requires a minimum current of 1 mA to breakdown. The diode D has a forward voltage drop of 0.6 V.

Solution. When source voltage is maximum and load current is minimum, then source resistance should be maximum.

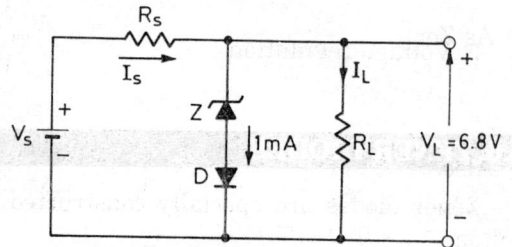

Fig. 3.26. Pertaining to Example 3.13.

$$\therefore \qquad V_{s \cdot max} = V_L + (I_{L \cdot min} + I_z)\, R_{s \cdot max}$$

$$\therefore \qquad R_{s \cdot min} = \frac{(20 \times 1.2) - 6.8}{[30 \times 0.5 + 1] \times 10^{-3}} = 1075\ \Omega$$

Similarly,

$$V_{s \cdot min} = V_L + (I_{L \cdot max} + I_z)\, R_{s \cdot min}$$

$$\therefore \qquad R_{s \cdot min} = \frac{(20 \times 1.2) - 6.8}{[30 \times 1.5 + 1] \times 10^{-3}} = 200\ \Omega$$

Maximum load resistance,

$$R_{L \cdot max} = \frac{V_L}{I_{L \cdot min}} = \frac{6.8}{30 \times 0.5 \times 10^{-3}} = 453.3\ \Omega$$

Minimum load resistance, $R_{L \cdot min} = \dfrac{V_L}{I_{L \cdot max}} = \dfrac{6.8}{30 \times 1.5 \times 10^{-3}} = 151.5\ \Omega$

The voltage rating of the Zener diode is

$$6.8 - 0.6 = 6.2\ \text{V}.$$

Example 3.14. *The complete circuit shown in Fig. 3.27 (a) represents a 25 V dc voltmeter where G is a PMMC galvanometer having full-scale deflection current $I_{fsd} = 200$ micro-A and resistance $R_G = 500$ ohms, and D is a 20-V Zener diode. Find R_1 and R_2. What is the function of the diode D in this circuit ?*

(GATE, 1990)

Solution. Current through galvanometer,

$$I_{fsd} = I_2 = \frac{\text{Zener voltage}}{R_2 + R_G}$$

or

$$\frac{20}{R_2 + 500} = 200 \times 10^{-6}$$

or

$$R_2 = \frac{20 \times 10^6}{200} - 500 = 99.5\ \text{k}\ \Omega$$

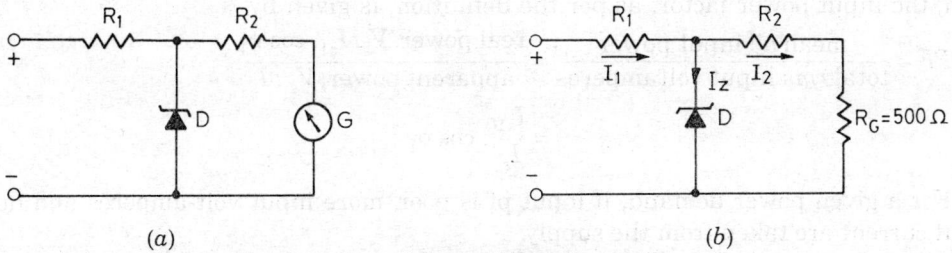

(a)   (b)

Fig. 3.27. Pertaining to Example 3.14.

As Zener diode current is not specified, let it be assumed zero. Therefore, from Fig. 3.27 (b),

∴ $$I_1 - I_2 = I_z = 0 \quad \text{or} \quad I_1 = I_2 = 200 \ \mu A$$

Also $$I_1 = \frac{25 - 20}{R_1} = 200 \times 10^{-6}$$

or $$R_1 = \frac{5 \times 10^6}{200} = 25 \text{ k}\Omega$$

Function of Zener diode is to provide a constant voltage to the galvanometer circuit. Whenever voltage across this diode exceeds 20 V, it conducts and the excess current is shunted away from galvanometer G. So here diode D prevents overloading of the PMMC galvanometer.

3.7. PERFORMANCE PARAMETERS

The input voltage to rectifiers is usually sinusoidal. It is desired that the output voltage from a rectifier should be constant with no ripples in it. This, however, is not the case. This shows that the rectified output voltage is made up of constant *dc* voltage plus harmonic components. The waveform of input and output currents depend on the nature of load and the rectifier configuration. In order to evaluate the overall performance of rectifier-load combinations, certain performance parameters relating to their input and output must be known. The object of this article is to define the various performance parameters (or indices) relating to input as well as output voltages and currents.

3.7.1. Input Performance Parameters

The various parameters relating to the source (or input) side of the converter-load combination are defined below :

(*i*) **Input power-factor.** Input voltage taken from power-supply undertaking is generally sinusoidal. However, *ac* input current is usually non-sinusoidal. Under such a condition, only the fundamental component of input current takes part in extracting mean *ac* input power from the source.

The input power factor PF is defined as the ratio of mean input power (real power) to the total *rms* input voltamperes (apparent power) given to the converter (or rectifier) system.

If V_s = *rms* value of supply phase voltage.

I_s = *rms* value of supply phase current including fundamental and harmonics

I_{s1} = *rms* value of fundamental component of supply current I_s and

ϕ_1 = phase angle between supply voltage V_s and fundamental component I_{s1} of supply current I_s ; see Fig. 3.28 ;

then, the input power factor, as per the definition, is given by

$$PF = \frac{\text{mean } ac \text{ input power}}{\text{total } rms \text{ input voltamperes}} = \frac{\text{real power, } V_s. I_{s1}. \cos \phi_1}{\text{apparent power, } V_s . I_s}$$

$$= \frac{I_{s1}}{I_s} . \cos \phi_1 \qquad \qquad ...(3.51)$$

For a given power demand, if input pf is poor, more input volt-amperes and hence more input current are taken from the supply.

(*ii*) **Input displacement factor (DF).** As stated above, the phase angle between sinusoidal supply voltage V_s and fundamental component I_{s1} of supply current I_s is ϕ_1. This angle ϕ_1, shown in Fig. 3.28, is usually known as *input displacement angle*. Its cosine is called the input displacement factor DF.

$\therefore$ $$DF = \cos \phi_1 \qquad \qquad ...(3.52)$$

DF is also called *fundamental power factor*.

(*iii*) **Input current distortion factor (CDF).** It is defined as the ratio of the rms value of fundamental component I_{s1} of the input current to the *rms* value of input, or supply, current I_s.

$\therefore$ $$CDF = \frac{I_{s1}}{I_s} \qquad \qquad ...(3.53)$$

It is seen from Eqs. (3.51) to (3.53) that $PF = (CDF) \times (DF)$

or input power factor = (input current distortion factor) × (input displacement factor)

(*iv*) **Input current harmonic factor (HF).** Non-sinusoidal input, or supply, current is made up of fundamental current plus current components of higher frequencies. The harmonic factor (HF) is equal to the *rms* value of all the harmonics divided by the rms value of fundamental component of the input current.

If I_h = *rms* value of all the harmonic-components combined

$$= \sqrt{I_s^2 - I_{s1}^2}$$

then, as per the definition, $$HF = \frac{I_h}{I_{s1}} = \frac{\sqrt{I_s^2 - I_{s1}^2}}{I_{s1}} = \frac{[\sum_{n=2}^{\infty} I_{sn}]}{I_{s1}} \qquad \qquad ...(3.54)$$

where I_{sn} = *rms* value of nth harmonic content.

Harmonic factor is a measure of the harmonic content in the input supply current. HF is also known as total harmonic distortion (THD). Greater the value of HF (or THD), greater is the harmonic content and hence greater is the distortion of input supply current.

Also, $$HF = \sqrt{\left(\frac{I_s}{I_{s1}}\right)^2 - 1} = \sqrt{\frac{1}{CDF^2} - 1} \qquad \qquad ...(3.55)$$

Higher value of input distortion factor CDF indicates lower magnitude of harmonic content in the source current.

Non-sinusoidal input current can be resolved into Fourier series as under :

$$i = \frac{a_o}{2} + \sum_{n=1, 2, 3,...}^{\infty} (a_n \cos n\omega t + b_n \sin n\omega t) \qquad \qquad ...(3.56)$$

$$= \frac{a_o}{2} + \sum_{n = 1,2,3,\ldots}^{\infty} C_n \sin (n\omega t + \phi_n) \qquad \ldots(3.57)$$

where $a_o = \frac{2}{T} \int_o^T i(t).dt$, $a_n = \frac{2}{T} \int_0^T i(t). \cos n\omega t. dt$ and $b_n = \frac{2}{T} \int_o^T i(t). \sin n\omega t. dt$

$$C_n = \left[\frac{a_n^2 + b_n^2}{2} \right]^{1/2} \text{ and } \phi_n = \tan^{-1} \left(\frac{a_n}{b_n} \right) \qquad \ldots(3.58)$$

(v) **Crest Factor (CF).** Crest factor for input current is defined as the ratio of peak input current I_{sp} to its *rms* value I_s.

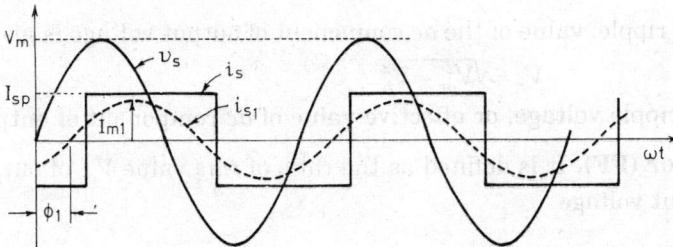

Fig. 3.28. Waveforms for source voltage v_s, source current i_s, fundamental component i_{s1} of source current and ϕ_1 = phase angle between v_s and i_{s1}.

$$\therefore \qquad CF = \frac{I_{sp}}{I_s} \qquad \ldots(3.59)$$

CF is used for specifying the current ratings of power semiconductor devices and other components.

3.7.2. Output performance Parameters

The load, or output, voltage and the load (or output) current at the output terminals of *ac* to *dc* converters are unidirectional but pulsating in nature. Fouriers series is used to express these output quantities in terms of its two components, namely (i) average (or *dc*) value and (ii) *ac* component superimposed on *dc* value as under :

In general, average value of output quantity y is, $Y_o = Y_{dc} = \frac{1}{T} \int_{t_1}^{t_1 + T} y. dt$

and its *rms* value is, $Y_{or} = \left[\frac{1}{T} \int_{t_1}^{t_1 + T} y^2 dt \right]^{1/2}$

where y = instantaneous value of the function in terms of t
and T = time period for one cycle of y variation.

$\therefore$ Output *dc* power, P_{dc} = (average output voltage, V_o) × (average output current, I_o)

$$= V_o I_o \qquad \ldots(3.60)$$

where subscript 'o' denotes output *dc* values.

Output *ac* power $P_{ac} = V_{or}. I_{or}$

where subscript "or" denotes *rms* value of output quantities.

The various output parameters are now defined below.

(i) **Rectification ratio** η. Rectification ratio, also called efficiency of a converter, is defined as the ratio of *dc* output power P_{dc} to *ac* output power P_{ac}.

$$\therefore \qquad \eta = \frac{P_{dc}}{P_{ac}} \qquad \qquad ...(3.61)$$

Rectifier ratio is also known as *rectifier efficiency* or *figure of merit*. In case R_d = forward rectifier resistance, then

$$\eta = \frac{P_{dc}}{P_{ac} + I_{or}^2 R_d} \qquad \qquad ...(3.62)$$

(ii) **Effective, or ripple, value of the** *ac* component of output voltage is given by

$$V_r = \sqrt{V_{or}^2 - V_o^2} \qquad \qquad ...(3.63)$$

where V_r is called ripple voltage, or effective value of *ac* component of output voltage.

(iii) **Form factor (FF).** It is defined as the ratio of *rms* value V_{or} of output voltage to the *dc* value V_o of output voltage.

$$\therefore \qquad FF = \frac{V_{or}}{V_o} \qquad \qquad ...(3.64)$$

FF is a measure of the shape of the output voltage. The closer FF is to unity, the better is the *dc* output voltage waveform. For constant *dc* output voltage, *rms* value of output voltage, V_{or} = average value of output voltage, V_o.

(iv) **Voltage ripple factor (VRF).** It is defined as the ratio of ripple voltage V_r to the average output voltage V_o.

$$\therefore \qquad VRF = \frac{V_r}{V_o} \qquad \qquad ...(3.65)$$

Substituting the value of V_r from Eq. (3.63) in Eq. (3.65) gives

$$VRF = \left[\left(\frac{V_{or}}{V_o} \right)^2 - 1 \right]^{1/2} = \sqrt{FF^2 - 1} \qquad \qquad ...(3.66\ a)$$

or
$$FF = \sqrt{VRF^2 + 1} \qquad \qquad ...(3.66\ b)$$

(v) **Per-unit average output voltage.** It is defined as the ratio of the average output voltage V_o for any value of triggering angle to the average output voltage V_{om} for zero-degree firing angle.

$$\therefore \qquad V_{o.pu} = \frac{V_o}{V_{om}} \qquad \qquad ...(3.67)$$

(vi) **Current ripple factor (CRF).** It is defined as the ratio of *rms* value of all harmonic components of output current to the *dc* component I_o of the output current.

$$\therefore \qquad CRF = \frac{I_r}{I_o} = \frac{\sqrt{I_{or}^2 - I_o^2}}{I_o} = \left[\left(\frac{I_{or}}{I_o} \right)^2 - 1 \right]^{1/2} \qquad \qquad ...(3.68)$$

Here I_{or} = rms value of output current including dc and harmonics,

I_r = rms value of all harmonic components of output current

I_o = dc component of output current.

Note that $I_{or}^2 = I_o^2 + I_r^2$.

(vii) **Transformer utilization factor (TUF).** If $V_2 (= V_s)$ and $I_2 (= I_s)$ are respectively the rms voltage and rms current ratings of the secondary winding of a transformer, then TUF is defined as

$$TUF = \frac{P_{dc}}{V_2 I_2} = \frac{P_{dc}}{V_s I_s} \qquad \qquad ...(3.69)$$

∴ Transformer VA rating $= \dfrac{P_{dc}}{TUF}$ ...(3.70)

Lower the TUF, higher is the transformer VA rating required.

It is desirable that a rectifier produces a perfect dc output voltage so that (i) rms value = dc value (ii) FF = 1.0 (iii) ac component of output voltage = 0 (iv) HF = 0 (vi) PF = 1.0 and TUF = 1.

3.8. COMPARISON OF SINGLE-PHASE DIODE RECTIFIERS

In this article, the performance parameters of single-phase diode rectifiers feeding resistive loads are evaluated. The rectifier types discussed are 1-phase half-wave rectifier and 1-phase full-wave mid-point and bridge types. The performance parameters are then collated in tabular form.

3.8.1. Single-phase Half-wave Rectifier

This rectifier, when feeding a resistive load, Fig. 3.14 (a), has waveforms for source voltage v_s, output voltage v_o and output current i_o in Fig. 3.14 (b). Its various performance parameters are obtained as under :

From Eq. (3.21), dc output voltage, $V_o = \dfrac{V_m}{\pi}$

and dc output current, $\qquad\qquad\qquad I_o = \dfrac{V_m}{\pi . R} = \dfrac{I_m}{\pi}$

where $I_m = \dfrac{V_m}{R}$ = maximum values of dc current as shown in Fig. 3.14 (b).

Output dc power, $\qquad\qquad\qquad P_{dc} = V_o I_o = \dfrac{V_m \cdot I_m}{\pi^2}$ ...(i)

From Eq. (3.22), rms output voltage, $\quad V_{or} = \dfrac{V_m}{2}$

and rms output current, $\qquad\qquad\quad I_{or} = \dfrac{V_m}{2.R} = \dfrac{I_m}{2}$

Output power, $\qquad\qquad\qquad\quad P_{ac} = V_{or} I_{or} = \dfrac{V_m I_m}{4}$ ...(ii)

Rectifier efficiency, $\qquad\quad \eta = \dfrac{P_{dc}}{P_{ac}} = \dfrac{V_m I_m}{\pi^2} \cdot \dfrac{4}{V_m I_m} = \dfrac{4}{\pi^2} = 0.4053$ or 40.53%

Form factor, $\qquad\qquad\quad FF = \dfrac{V_{or}}{V_o} = \dfrac{V_m}{2} \cdot \dfrac{\pi}{V_m} = \dfrac{\pi}{2} = 1.5708$

Ripple voltage, $\qquad V_r = \sqrt{V_{or}^2 - V_o^2} = \sqrt{\left(\dfrac{V_m}{2}\right)^2 - \left(\dfrac{V_m}{\pi}\right)^2} = 0.3856\, V_m$

Voltage ripple factor, $\qquad VRF = \dfrac{V_r}{V_o} = \dfrac{0.3856\, V_m \cdot \pi}{V_m} = 1.211.$

Also, $\qquad\qquad\qquad\quad VRF = \sqrt{FF^2 - 1} = \sqrt{1.5708^2 - 1} = 1.211$

Since source voltage v_i is a sine wave, its rms value, $V_s = \dfrac{V_m}{\sqrt{2}}$

Load current i_o waveform is the same as that of source current i_s.

$\therefore$ Rms value of source current, I_s = rms value of output current, $I_{or} = \dfrac{I_m}{2}$.

VA rating of transformer $\qquad = V_s\, I_s = \dfrac{V_m}{\sqrt{2}} \times \dfrac{I_m}{2} = \dfrac{V_m I_m}{2\sqrt{2}}$

$\therefore$ Transformer utilization factor, $TUF = \dfrac{P_{dc}}{V_s\, I_s} = \dfrac{V_m I_m}{\pi^2} \times \dfrac{2\sqrt{2}}{V_m I_m} = 0.2865$

A $TUF = 0.2865$ means that VA rating of transformer is $\dfrac{1}{TUF}$ times the dc power output.

For a load of 100 watt, a transformer having a rating of $\dfrac{100}{0.2865} = 349.6\ VA$ would be required.

$$PIV = \sqrt{2}\ V_s = V_m$$

Peak value of source current, $\qquad I_{sp} = I_m$

Rms value of source current, $\qquad I_s = I_{or} = \dfrac{I_m}{2}$

Crest factor, $\qquad\qquad\qquad CF = \dfrac{I_{sp}}{I_s} = \dfrac{I_m}{I_m} \times 2 = 2$

3.8.2. Single-phase Full-wave Mid-point Rectifier

Its circuit diagram and various waveforms are shown in Fig. 3.20. Its different performance parameters are obtained as under.

From Eq. (3.48), dc output voltage, $V_o = \dfrac{2V_m}{\pi}$

and dc output current, $\qquad\qquad I_0 = \dfrac{2V_m}{\pi R} = \dfrac{2}{\pi} I_m$

where I_m = maximum value of load current $= \dfrac{V_m}{R}$

Output dc power, $\qquad\qquad P_{dc} = V_o I_o = \dfrac{2V_m}{\pi} \cdot \dfrac{2}{\pi} I_m = \left(\dfrac{2}{\pi}\right)^2 \cdot V_m I_m$

Rms output voltage, $\qquad\qquad V_{or} = \dfrac{V_m}{\sqrt{2}} = V_s$

Rms output current, $\qquad\qquad I_{or} = \dfrac{V_m}{\sqrt{2}} \times \dfrac{1}{R} = \dfrac{1}{\sqrt{2}}\left(\dfrac{V_m}{R}\right) = \dfrac{1}{\sqrt{2}} I_m = I_s$

Output ac power, $\qquad\qquad P_{ac} = V_{or} \cdot I_{or} = \dfrac{V_m}{\sqrt{2}} \cdot \dfrac{I_m}{\sqrt{2}} = \dfrac{V_m I_m}{2}$

Rectifier efficiency, $\qquad \eta = \dfrac{P_{dc}}{P_{ac}} = \dfrac{4}{\pi^2} V_m I_m \cdot \dfrac{2}{V_m I_m} = \dfrac{8}{\pi^2} = 0.8106$

Form factor, $\qquad FF = \dfrac{V_{or}}{V_o} = \dfrac{V_m}{\sqrt{2}} \cdot \dfrac{\pi}{2V_m} = \dfrac{\pi}{2\sqrt{2}} = 1.11$

Ripple voltage, $\qquad V_r = \sqrt{V_{or}^2 - V_o^2} = \left[\left(\dfrac{V_m}{\sqrt{2}} \right)^2 - \left(\dfrac{2V_m}{\pi} \right)^2 \right]^{1/2} = 0.3077\, V_m$

Voltage ripple factor, $\qquad VRF = \dfrac{V_r}{V_o} = 0.3077\, V_m \times \dfrac{\pi}{2V_m} = 0.483$

Also, $\qquad VRF = \sqrt{FF^2 - 1} = \sqrt{1.11^2 - 1} = 0.482$

TUF can be obtained as under :

Rms value of voltage for each secondary winding $= \dfrac{V_m}{\sqrt{2}}$

Note that current in each secondary winding flows for half cycle only.

$\therefore$ Rms value of current in each secondary winding $= \dfrac{I_m}{2}$

VA rating of secondary winding = 2 [voltage rating of each secondary winding]
$$\times \text{[current rating of each secondary winding]}$$
$$= 2 \times \dfrac{V_m}{\sqrt{2}} \times \dfrac{I_m}{2} = \dfrac{V_m I_m}{\sqrt{2}} = 0.707\, V_m I_m$$

Primary winding current is, however, made up of both positive and negative half cycles.

$\therefore$ Primary *rms* current $\qquad = \dfrac{I_m}{\sqrt{2}}$

Primary *rms* voltage $\qquad = \dfrac{V_m}{\sqrt{2}}$

Primary VA rating $\qquad = \dfrac{V_m I_m}{2} = 0.5\, V_m I_m$

$\therefore$ Average VA rating of transformer $= \dfrac{0.5 + 0.707}{2} \cdot V_m I_m = 0.6035\, V_m I_m$

$$TUF = \dfrac{P_{dc}}{\text{average } VA \text{ rating of transformer}} = \dfrac{4}{\pi^2} \cdot V_m I_m \times \dfrac{1}{0.6035\, V_m I_m} = 0.672$$

PIV for each diode $= 2V_m$

Peak value of source current, $I_{sp} = I_m$

Rms value of source current, $\quad I_s = \dfrac{I_m}{\sqrt{2}}$

$\therefore$ CF of input current $\qquad = \dfrac{I_{sp}}{I_s} = \dfrac{I_m}{I_m} \sqrt{2} = \sqrt{2} = 1.414.$

3.8.3. Single-phase full-wave Bridge Rectifier

Its circuit diagram is given in Fig. 3.21 (*a*). It is seen from Fig. 3.20 (*b*) and 3.21 (*b*) that waveform of output (or load) type v_o and out current i_o are identical in both $M - 2$ and $B - 2$ types of diode rectifiers. Therefore, in single-phase $B - 2$ diode rectifier also,

$$V_o = \frac{2\,V_m}{\pi}, I_o = \frac{2\,I_m}{\pi}, V_{or} = \frac{V_m}{\sqrt{2}} \text{ and } I_{or} = \frac{I_m}{\sqrt{2}}$$

This shows that the rectifier efficiency, FF, ripple voltage V_r, VRF are the same for both types of diode rectifiers. However, PIV of diode in single-phase $B-2$ rectifier is V_m whereas it is $2V_m$ in 1-phase $M-2$ rectifier.

TUF : Rms value of source voltage $V_s = \dfrac{V_m}{\sqrt{2}}$

Rms value of source current, $\qquad I_s = \dfrac{I_m}{\sqrt{2}}$

VA rating of transformer $\qquad\qquad = V_s I_s = \dfrac{V_m I_m}{2}$

$$P_{dc} = V_o I_o = \frac{2\,V_m}{\pi} \cdot \frac{2\,I_m}{\pi} = \left(\frac{2}{\pi}\right)^2 \cdot V_m I_m$$

$\therefore \qquad TUF = \dfrac{P_{dc}}{VA \text{ rating of transformer}} = \dfrac{4}{\pi^2} V_m I_m \times \dfrac{2}{V_m I_m} = \dfrac{8}{\pi^2} = 0.8106$

Source current waveforms for both types are identical, therefore $CF = \sqrt{2}$.

A comparison of three types of 1-phase diode rectifiers discussed above is given in the table below where $V_m = \sqrt{2}\,V_s$. Here $V_s =$ rms value of sinusoidal source voltage and $f =$ source frequency in Hz.

S.No.	Parameters	Half-wave (or one-pulse)	Full-wave (or Two pulse)	
			Centre-tap (M2)	Bridge (B- 2)
1.	DC output voltage, V_o	$\dfrac{V_m}{\pi}$	$\dfrac{2\,V_m}{\pi}$	$\dfrac{2\,V_m}{\pi}$
2.	Rms value of output voltage, V_{or}	$\dfrac{V_m}{2}$	$\dfrac{V_m}{\sqrt{2}}$	$\dfrac{V_m}{\sqrt{2}}$
3.	Ripple voltage, V_r	$0.3856\,V_m$	$0.3077\,V_m$	$0.3077\,V_m$
4.	Voltage ripple factor, VRF	1.211	0.482	0.482
5.	Rectification efficiency, η	40.53%	81.06%	81.06%
6.	Transformer utilization factor, TUF	0.2865	0.672	0.8106
7.	Peak inverse voltage, PIV	V_m	$2V_m$	V_m
8.	Crest factor, CF	2	$\sqrt{2}$	$\sqrt{2}$
9.	Number of diodes	1	2	4
10.	Ripple frequency	f	$2f$	$2f$

It is seen from the above table that both full-wave diode rectifiers

(i) are better than the half-wave rectifier in so far as voltage ripple factor, rectification efficiency, TUF and crest factor are concerned, (*ii*) have average output voltage double of that of the half-wave rectifier (for the **same** input voltage),

    (*iii*) have ripple frequency double of that of half-wave rectifier.

For both the full-wave rectifiers, the following is observed from the table.

    (*i*) *TUF* of $B-2$ rectifier is superior than the $M-2$ type. Therefore, transformer required in $M-2$ configuration is bulky and weighty.

    (*ii*) PIV of diodes in $B-2$ rectifier is half of that of the diodes used in $M-2$ rectifier.

    (*iii*) $B-2$ rectifier requires four diodes whereas $M-2$ requires only two diodes

    (*iv*) Overall, a bridge rectifier using four diodes is more economical.

Example 3.15. *A load of $R = 60\ \Omega$ is fed from 1-phase, 230 V, 50 Hz supply through a step-up transformer and then one diode. The transformer turns ratio is two. Find the VA rating of transformer.*

Solution. The half-wave diode rectifier uses a step-up transformer therefore, *ac* voltage applied to rectifier $= 230 \times 2 = 460\ V = V_s$

Average value of load voltage, $V_o = \dfrac{V_m}{\pi} = \dfrac{\sqrt{2} \times 460}{\pi} = 207.04\ V$

Output *dc* power, $\qquad P_{dc} = \dfrac{V_o^2}{R} = \dfrac{207.04^2}{60} = 714.43\ W$

It is seen from the table that *TUF* for 1-phase half-wave diode rectifier is 0.2865.

$\therefore$ VA rating of transformer $\qquad = \dfrac{P_{dc}}{TUF} = \dfrac{714.43}{0.2865} = 2493.65\ VA$

So choose a transformer with 2.5 kVA (next round figure) rating.

Example 3.16. *A 230 V, 50 Hz supply is connected to a 1-phase transformer which feeds a diode bridge as shown in Fig. 3.23 (a). Primary to secondary turns ratio for transformer is 0.5 and load RL has a ripple free current $I_o = 10\ A$. Determine (i) average value of output voltage (ii) input current distortion factor (iii) input displacement factor DF (iv) input power factor (v) input current harmonic factor HF (or THD) and (vi) crest factor.*

Solution. Waveforms for supply voltage v_s, constant load current $i_o = I_o = 10\ A$ and source current i_s are shown in Fig. 3.23 (*b*).

Rms value of input voltage to bridge rectifier, $\dfrac{N_1}{N_2} = \dfrac{230}{V_s} = 0.5$

$\therefore \qquad\qquad\qquad\qquad V_s = \dfrac{230}{0.5} = 460\ V$

The source current, or input current, i_s can be expressed in Fourier series as under :

$$i_s = I_{dc} + \sum_{n = 1,3,5}^{\infty} (a_n \cos n\omega t + b_n \sin n\omega t) \qquad\qquad ...(3.56)$$

Here $I_{dc} = dc$ value of source current

$$= \frac{1}{2\pi} \int_o^{2\pi} i_s \cdot d\,(\omega t) = \frac{1}{2\pi} \left[\int_o^\pi I_o \cdot d(\omega t) - \int_\pi^{2\pi} I_o\, d\,(\omega t) \right] = 0$$

It can also be stated from the waveform of i_s that as the area of positive and negative half cycle are equal, average value of i_s i.e. $I_{dc} = 0$.

$$a_n = \frac{1}{\pi} \int_o^{2\pi} i_o(t) \cdot \cos n\omega t \cdot \cos n\omega t \cdot d(\omega t)$$

$$= \frac{2}{\pi} \int_o^\pi I_o \cos n\omega t \cdot d(\omega t) = \frac{2I_o}{n\pi} \mid \sin n\omega t \mid_o^\pi = 0 \text{ for all } n.$$

$$b_n = \frac{1}{\pi} \int_o^{2\pi} i_o(t) \sin n\omega t \, d(\omega t) = \frac{2}{\pi} \int_o^\pi I_o \cdot \sin n\omega t \cdot d(\omega t)$$

$$= \frac{2 I_o}{n\pi} [-\cos n\omega t]_o^\pi = \frac{2I_o}{n \cdot \pi} [1 - \cos n\pi]$$

$$= \frac{4I_o}{n\pi} \text{ for } n = 1, 3, 5\ldots\ldots(\text{for odd values of } n)$$

and $\qquad\qquad b_n = 0 \qquad$ for $= 2, 4, 6\ldots\ldots(\text{for even values of } n)$

Substituting the values of I_{dc}, a_n and b_n in Eq. (3.56), we get

$$i_s = \frac{4I_o}{n\pi} \sin n\omega t \text{ and } \phi_n = \tan^{-1}\left[\frac{0}{b_n}\right] = 0$$

$$\therefore \qquad i_s = \frac{4I_o}{\pi}\left[\sin \omega t + \frac{1}{3} \sin 3\omega t + \frac{1}{5} \sin 5\omega t + \frac{1}{7} \sin 7\omega t + \ldots\right]$$

(i) Average value of output voltage, $V_o = \dfrac{2V_m}{\pi} = \dfrac{2\sqrt{2} \times 460}{\pi} = 414.08$ V

(ii) Since fundamental component of input source current $\dfrac{4I_o}{\pi} \sin \omega t$ is in phase with source voltage $V_m \sin \omega t$, the displacement angle = 0. Also, from above, $\phi_1 = 0$.

∴ Input displacement factor, $DF = \cos \phi_1 = \cos 0° = 1$.

(iii) Rms value of fundamental component of source current, $I_{s1} = \dfrac{4I_o}{\pi} \times \dfrac{1}{\sqrt{2}}$. A. Rms value of source current, $I_s = \left[\dfrac{I_o^2 \times \pi}{\pi}\right]^{1/2} = I_o = 10$ A.

Input current distortional factor, $CDF = \dfrac{I_{s1}}{I_s} = \dfrac{4I_o}{\pi \cdot \sqrt{2}} \times \dfrac{1}{I_o} = \dfrac{\sqrt{2} \times 2}{\pi} = 0.9$

(iv) Input $pf = CDF \times DF = 0.9 \times 1 = 0.90$ (lagging)

(v) $HF = THD = \left[\left(\dfrac{I_s}{I_{s1}}\right)^2 - 1\right]^{1/2} = \left[\left(\dfrac{1}{0.9}\right)^2 - 1\right]^{1/2} = 0.4843$ or 48.43%

(vi) Crest factor. Here $I_{sp} = I_o = 10$ A and $I_s = 10$ A.

∴ $\qquad\qquad\qquad CF = \dfrac{10}{10} = 1.00$

Example 3.17. *A single-phase B-2 diode rectifier is required to supply a dc output voltage of 230 V to a load of R = 10 Ω. Determine the diode ratings and transformer rating required for this configuration.*

Solution.

Average, or *dc* output voltage, $\qquad V_o = \dfrac{2V_m}{\pi} = \dfrac{2\sqrt{2} \cdot V_s}{\pi} = 230$ V

$\therefore$ Rms value of input voltage to rectifier = transformer secondary voltage, V_s

$$= \frac{230 \times \pi}{2\sqrt{2}} = 255.5 \text{ V}$$

Average load current, $\qquad I_o = \frac{V_o}{R} = \frac{230}{10} = 23 \text{ A}$

Maximum value of diode current, $I_m = \frac{V_m}{R} = \frac{\sqrt{2} \times 255.5}{10} = 36.13 \text{ A}$

It is seen from the waveform of diode current i_{D1} from Fig. 3.23 (b) that average value of diode current is

$$I_{DAV} = \frac{1}{2\pi} \int_o^\pi I_m \sin \omega t. \, d(\omega t) = \frac{I_m}{\pi} = \frac{36.13}{\pi} = 11.50 \text{ A}$$

and *rms* value of diode current, $\quad I_{Dr} = \left[\frac{1}{2\pi} \int_o^\pi I_m^2 \sin^2 \omega t \, . \, d(\omega t) \right]^{1/2} = \frac{I_m}{2} = \frac{36.13}{2} = 18.07 \text{ A}$

$$PIV = \sqrt{2} V_s = \sqrt{2} \times 255.5 = 361.3 \text{ V}$$

Transformer secondary current $\qquad = \frac{I_m}{\sqrt{2}} = \frac{36.13}{\sqrt{2}} = 25.55 \text{ A} = I_s$

Transformer rating $\qquad = V_s I_s = 255.5 \times 25.55 = 6528 \text{ VA} = 6.528 \text{ (kVA)}$

[Check : $\qquad P_{dc} = V_o I_o = 230 \times 23 = 5290 \text{ W}$

$\therefore$ Transformer rating $\qquad = \frac{P_{dc}}{TUF} = \frac{5290}{0.81} = 6.530 \text{ VA} = 6.53 \text{ kVA}]$

Thus, diode ratings are : $\qquad I_{DAV} = 11.50 \text{ A}, I_{Dr} = 18.07 \text{ A}$

Peak diode current, $I_m = 36.13$ A and PIV = 361.3 V and transformer rating = 6.528 kVA.

3.9. THREE-PHASE RECTIFIERS

The highest possible value of average output voltage from a single-phase full wave rectifier is $2V_m/\pi = 0.63662 \, V_m$. At the same time, single-phase rectifiers are suitable up to power loads of about 15 kW. For higher power demands, three-phase rectifiers are preferred due to the following reasons :

(i) Higher dc voltage

(ii) Better TUF

(iii) Better input pf

(iv) Less ripple content in output current; therefore better load performance and

(v) lower size of filter circuit parameters because of higher ripple frequency.

Three-phase rectifier are classified as under :

(a) Three-phase half-wave rectifier

(b) Three-phase mid-point 6-pulse rectifier

(c) Three-phase bridge rectifier and

(d) Three-phase 12-pulse rectifier.

These are now described one after the other.

3.9.1. Three-Phase Half-wave Diode Rectifier

Circuit diagram of a three-phase half-wave rectifier using three diodes is shown in Fig. 3.29. It uses a 3-phase transformer with primary in delta and secondary in star. The primary in delta provides a path for the triplen harmonic currents. This stabilizes the voltages on the

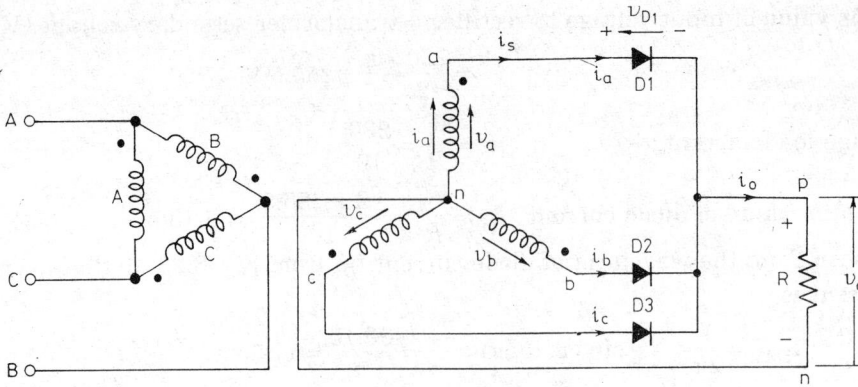

Fig. 3.29. Three-phase half-wave diode rectifier with common cathode arrangement.

secondary star. The three diodes D1, D2 and D3, one in each phase, have their cathodes connected together to common load R. Neutral is used to complete the path for the return of load current. As the cathodes of three diodes are connected together, circuit of Fig. 3.29 is also known as *common-cathode circuit* for a 3-phase half-wave rectifier. The three-phase supply voltage is shown as v_a (= v_{an}, voltage between a and n), v_b, v_c in Fig. 3.30 (a).

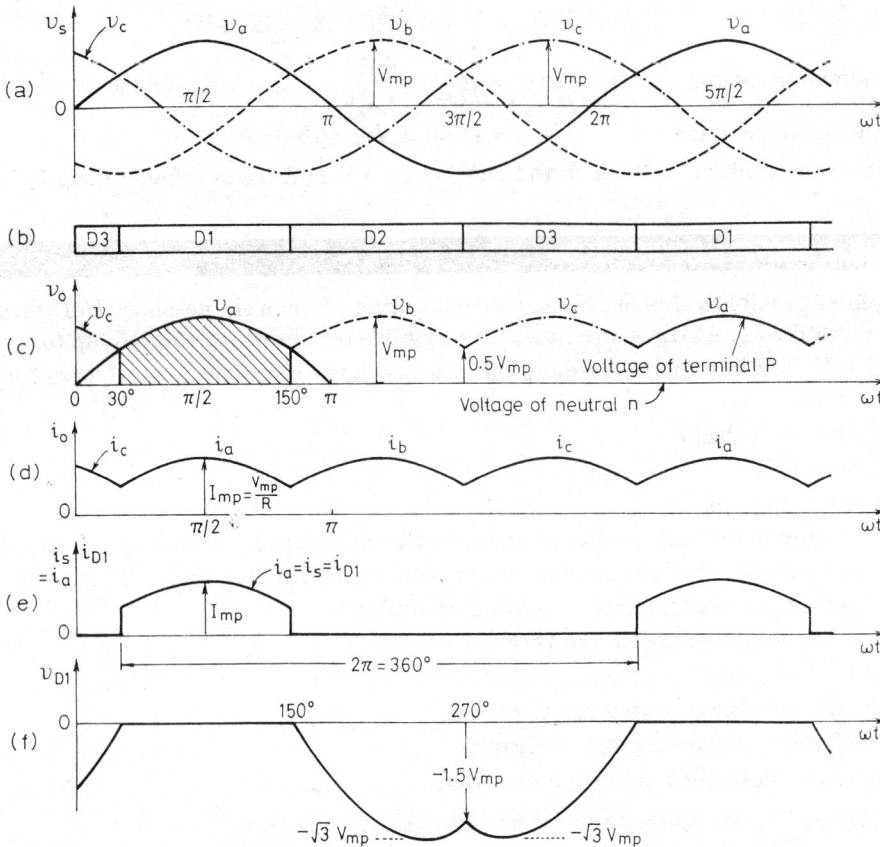

Fig. 3.30 (a) Line to neutral source voltages (b) diode conduction (c) load voltage (d) load current (e) **source current and** (f) voltage across diode D1.

The rectifier element connected to the line at the highest positive instantaneous voltage can only conduct. In Fig. 3.29, a diode with the highest positive voltage will begin to conduct at the cross-over points of the three-phase supply. It is seen from Fig. 3.30 (a) that diode D1 will conduct for $\omega t = 30°$ to $\omega t = 150°$ as this diode senses the most positive voltage v_a, as compared to the other two diodes, during the interval. Diode D2 will conduct from $\omega t = 150°$ to 270° and diode D3 from $\omega t = 270°$ to 390°. The conduction of diodes in proper sequence is shown in Fig. 3.30 (b). When a diode is conducting, the common cathode terminal P rises to the highest positive voltage of that phase and the other two blocking diodes are reverse biased. The voltage v_o across the load follows the positive supply voltage envelope and has the waveform as shown in Fig. 3.30 (c). It should be noted that voltage of the neutral point 'n' is taken as zero and is given by the reference line ωt. The voltage of point P in Fig. 3.29 is shown by v_a, v_b, v_c etc above the reference line in Fig. 3.30 (c). The dc load voltage v_o varies between V_{mp} (= maximum phase voltage) and $0.5\,V_{mp}$. It is observed that for one cycle of supply voltage, output voltage has three pulses, the circuit of Fig. 3.29 can therefore be called a *3-phase 3-pulse diode rectifier* or 3-phase half-wave diode rectifier.

Voltage variation across diode D1 can be obtained by applying KVL to the loop consisting of D1, phase 'a' winding and load R. So $v_{D1} - v_a + v_o = 0$ or $v_{D1} = v_a - v_o$

When diode D1 conducts, $v_o = v_o$, $\therefore$ $v_{D1} = v_a - v_a = 0$. This is shown from $\omega t = 30°$ to $\omega t = 150°$ in Fig. 3.30 (f).

When diode D2 conducts, $v_o = v_b$, $\therefore$ $v_{D1} = v_a - v_b$

At $\omega t = 180°$, $v_b = 0.866\,V_{mp}$ and $v_a = 0$, $\therefore v_{D1} = 0 - 0.866\,V_{mp} = -0.866\,V_{mp}$

At $\omega t = 210°$, $v_b = V_{mp}$ and $v_a = -0.5\,V_{mp}$, $\therefore v_{D1} = (-0.5 - 1.0)\,V_{mp} = -1.5\,V_{mp}$

At $\omega t = 240°$, $v_b = 0.866\,V_{mp}$ and $v_a = -0.866\,V_{mp}$, $\therefore v_{D1} = (0 - 0.866 - 0.866)\,V_{mp} = -\sqrt{3}\,V_{mp}$

At $\omega t = 270°$, $v_b = 0.5\,V_{mp}$ and $v_a = -V_{mp}$, $\therefore v_{D1} = (-1 - 0.5) = -1.5\,V_{mp}$.

When D3 conducts, $v_{D1} = v_a - v_c$ and variation of voltage v_{D1} from $\omega t = 270°$ to $\omega t = 390°$ is obtained as outlined before. It is seen from Fig. 3.30 (f) that peak inverse voltage across diode D1 is $\sqrt{3}\,V_{mp}$; in general PIV = $\sqrt{3}\,V_{mp}$ for each of the three diodes D1, D2 and D3.

As in single-phase rectifiers, the average output voltage in a 3-phase diode rectifier can be obtained by considering the output voltage over one periodic cycle.

For a 3-phase diode rectifier of Fig. 3.29, the periodicity is 120° or $2\pi/3$ radians as per Fig. 3.30 (c). Here the output voltage comprises of phase voltages v_a, v_b, v_c and its average (or mean) value V_o is given by

$$V_o = \frac{\text{Area over one periodicity (shown cross-hatched in Fig. 3.30 (c))}}{\text{Periodicity}}$$

$$= \frac{1}{\text{Periodicity}} \int_{\alpha_1}^{\alpha_2} v_a \cdot d\,(\omega t)$$

In the above expression, v_a is zero at $\omega t = 0$; therefore, $V_{mp} \sin \omega t$ is written for v_a. Further v_a appears from $\omega t = 30°$ to 150° in the output voltage waveform; these are, therefore, the limits of integration and periodicity is 120° = $2\pi/3$ radians.

$$\therefore \qquad V_o = \frac{3}{2\pi} \int_{\pi/6}^{5\pi/6} V_{mp} \sin \omega t \cdot d\,(\omega t) = \frac{3\sqrt{3}}{2\pi}\,V_{mp} = \frac{3\sqrt{6}}{2\pi}\,V_{ph} = \frac{3}{2\pi} \cdot V_{ml} \qquad ...(3.71)$$

where V_{mp} = maximum value of phase voltage, $V_{ph} = \sqrt{2}\,V_{ph}$

and V_{ml} = maximum value of line voltage, $V_l = \sqrt{3} \cdot V_{mp} = \sqrt{6} \cdot V_{ph}$

Rms value of output voltage, $V_{or} = \left[\dfrac{1}{2\pi/3} \displaystyle\int_{\pi/6}^{5\pi/6} (V_{mp} \sin \omega t)^2 . \, d(\omega t) \right]^{1/2}$

$$= \left[\dfrac{3\,V_{mp}^2}{2\pi \times 2} . \left| \omega t - \dfrac{\sin 2\omega t}{2} \right|_{\pi/6}^{5\pi/6} \right]^{1/2} = 0.84068\,V_{mp} \quad ...(3.72)$$

Ripple voltage, $\quad V_r = \sqrt{V_{or}^2 - V_o^2} = V_{mp} \sqrt{0.84068^2 - 0.827^2} = 0.151\,V_{mp}$

$$VRF = \dfrac{V_r}{V_o} = \dfrac{0.151}{0.827} = 0.1826 \text{ or } 18.26\%$$

$$FF = \dfrac{V_{or}}{V_o} = \dfrac{0.84068}{0.827} = 1.0165$$

Rms value of output current, $I_{or} = \dfrac{V_{or}}{R} = \dfrac{0.84068}{R} V_{mp} = 0.84068\,I_{mp}$

where $I_{mp} = \dfrac{V_{mp}}{R} =$ peak value of load, or output, current.

$$P_{dc} = V_o\,I_o = \dfrac{3\sqrt{3}}{2\pi} V_{mp} . \dfrac{3\sqrt{3}}{2\pi} I_{mp}$$

$$P_{ac} = V_{or} . I_{or} = (0.84068)^2\,V_{mp}\,I_{mp}$$

Rectifier efficiency $\quad = \dfrac{P_{dc}}{P_{ac}} = \left(\dfrac{3\sqrt{3}}{2\pi} \right)^2 \times \dfrac{1}{0.84068^2} = 0.96765 \quad$ or $\quad 96.765\%$

During the interval $\omega t = 30°$ to $\omega t = 150°$, source current $i_s = i_a$ and the periodicity of source current is 2π radians as shown in Fig. 3.30 (e).

$\therefore$ RMS value of source current, $I_s = \left[\dfrac{1}{2\pi} \displaystyle\int_{\pi/6}^{5\pi/6} (I_{mp} \sin \omega t)^2 \, d\,(\omega t) \right]^{1/2} = 0.4854\,I_{mp}$

Rms value of source voltage, $V_s = \dfrac{V_{mp}}{\sqrt{2}} = 0.707\,V_{mp}$

Transformer has three phases, therefore,

VA rating of transformer $\quad = 3\,V_s\,I_s = 3 \times 0.707\,V_{mp} \times 0.4854\,I_{mp} = 1.0295\,V_{mp}\,I_{mp}$

DC output power, $\quad\quad P_{dc} = \left(\dfrac{3\sqrt{3}}{2\pi} \right)^2 \times V_{mp}\,I_{mp} = 0.684\,V_{mp}\,I_{mp}$

$\therefore \quad\quad\quad\quad TUF = \dfrac{P_{dc}}{\text{Transformer VA rating}} = \dfrac{0.684\,V_{mp} . I_{mp}}{1.0295\,V_{mp} . I_{mp}} = 0.6644$

As stated before, PIV for each diode $= \sqrt{3}\,V_{mp}$

Following observations can be made from the above analysis.

(i) Each diode conducts for 120° only.

(ii) These are three pulses of output voltage, or output current, during one cycle of input voltage. It is, therefore, called *3-phase three-pulse diode rectifier*.

(iii) Current in the transformer secondary is unidirectional, therefore, *dc* exists in the transformer secondary current. As a result, transformer core gets saturated leading to more iron losses and reduced efficiency.

3.9.2. Three-phase Mid-point 6-pulse Diode Rectifier

This rectifier is also called *six-phase half-wave diode rectifier* or *three-phase M-6 diode rectifier*. It is seen from the previous section that the performance of three-phase three-pulse diode rectifier is better than the single-phase two-pulse diode rectifier so far as magnitudes of average output voltage and ripple content are concerned. From this, it appears to be logical to think that a rectifier with more number of pulses per cycle would give an overall improved performance.

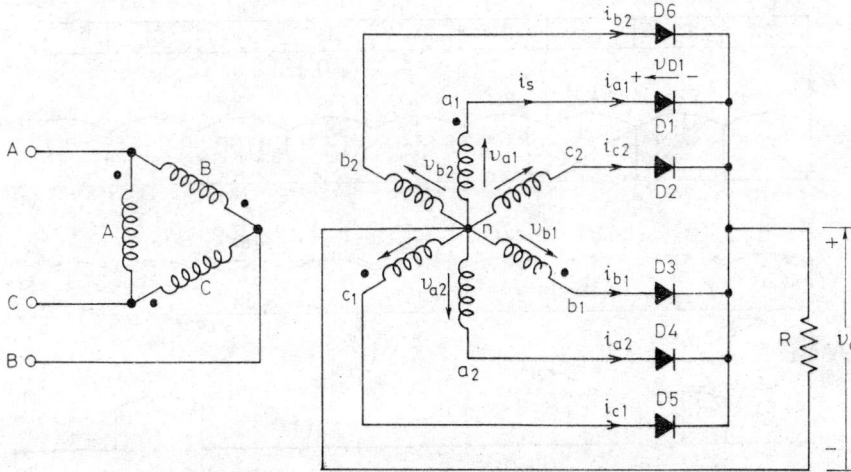

Fig. 3.31. Three-phase mid-point 6-pulse diode rectifier.

Fig. 3.31 shows a three-phase mid-point 6-pulse rectifier using six diodes. A three-phase transformer with primary in delta and secondary in double-star is used. One diode in each phase is connected as shown. Note that secondary of each phase winding is in two halves. The mid-points of the three secondary windings are connected to form the neutral n. Six-phase supply is available from six terminals a_1, c_2, b_1, a_2, c_1 and b_2. Phase voltages V_{a1}, V_{b1}, V_{c1} are phase-displaced by 120°, similarly V_{a2}, V_{b2}, V_{c2} are displaced by 120°. But voltages V_{a1} and V_{a2} are out of phase by 180°. However; V_{a1}, V_{c2}, are out of phase by 60° as shown in Fig. 3.32.

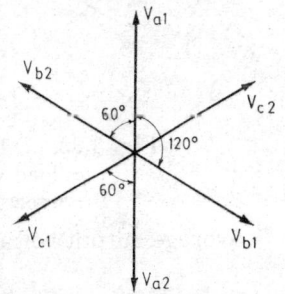

Fig. 3.32. Six-phase voltages for Fig. 3.31.

Therefore, if $v_{a1} = V_{mp} \sin \omega t$, then $v_{c2} = V_{mp} \sin (\omega t - 60°)$

$$v_{b1} = V_{mp} \sin (\omega t - 120°), \, v_{a2} = V_{mp} \sin (\omega t - 180°) = -v_{a1}$$

$$v_{c1} = V_{mp} \sin (\omega t - 240°) = -V_{mp} \sin (\omega t - 60°) = -v_{c2}$$

and $\qquad v_{b2} = V_{mp} \sin (\omega t - 300°) = -v_{b1}$

Here V_{mp} = maximum value of per phase voltage

The waveform of six-phase voltages, v_{a1}, v_{c2}, v_{b1} .... are sketched in Fig. 3.33 (a). As before, a diode sensing the highest positive anode potential gets forward biased and conducts. Therefore, from $\omega t = 0°$ to $\omega t = 60°$, voltage v_{b2} is the highest positive, therefore, diode D6 conducts; from $\omega t = 60°$ to $\omega t = 120$, diode. D1 conducts and so on, Fig. 3.33 (a) and (b). Each diode conducts for 60°. It is seen from Fig. 3.33 (c) that load voltage v_0 is made up of v_{b2} from $\omega t = 0°$ to 60° ; v_{a1} from $\omega t = 60°$ to 120° and so on. Also, v_o varies between V_{mp} and 0.866 V_{mp}. Periodicity of output voltage v_o is 60°.

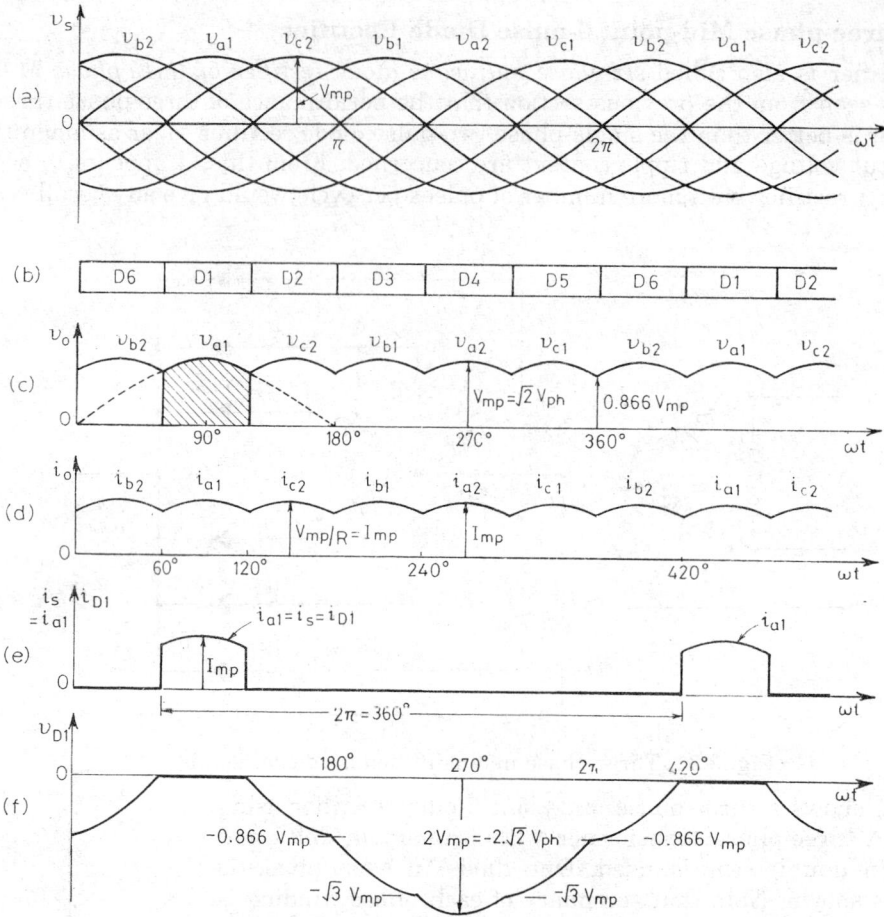

Fig. 3.33. (a) Line to neutral (or phase) source voltages (b) diode conduction
(c) load voltage (d) load current (e) source current and (f) voltage
across diode D1 for 3-phase $M-6$ diode rectifier of Fig. 3.31.

Average output voltage,

$$V_o = \frac{1}{\pi/3} \int_{\pi/3}^{2\pi/3} V_{mp} \sin \omega t \cdot d(\omega t) = \frac{3 V_{mp}}{\pi} \qquad \qquad ...(3.73)$$

Rms value of output voltage,

$$V_{or} = \left[\frac{1}{\pi/3} \int_{\pi/3}^{2\pi/3} (V_{mp} \sin \omega t)^2 \, d(\omega t) \right]^{1/2}$$

$$= \left[\frac{3 V_{mp}}{2\pi} \left\{ \frac{\pi}{3} - \frac{\sin 240° - \sin 120°}{2} \right\} \right]^{1/2} = 0.9558 \, V_{mp} \qquad ...(3.74)$$

Ripple voltage, $V_r = \sqrt{V_{or}^2 - V_o^2} = V_{mp} \left[0.9558^2 - \left(\frac{3}{\pi} \right)^2 \right]^{1/2} = 0.0408 \, V_{mp}$

$$VRF = \frac{V_r}{V_o} = \frac{0.0408 \times \pi}{3} = 0.043 \text{ or } 4.3\%$$

$$FF = \frac{V_{or}}{V_o} = \frac{0.9558 \times \pi}{3} = 1.009.$$

Rms value of output current, $I_{or} = \dfrac{V_{or}}{R} = \dfrac{0.9558}{R} V_{mp} = 0.9558\, I_{mp}$

$$P_{dc} = V_o\, I_o = \left(\dfrac{3}{\pi}\right)^2 V_{mp}\, I_{mp}.$$

$$P_{ac} = V_{or}\, I_{or} = (0.9558)\, V_{mp}\, I_{mp}$$

Rectifier efficiency $= \left(\dfrac{3}{\pi}\right)^2 \times \dfrac{1}{0.9558^2} = 0.9982$ or 99.82%

The source current i_s in phase a_1 has the same waveform as for the current i_{a1}, Fig. 3.33 (d) and (e). It is seen that periodicity of i_s is 2π radians.

$\therefore$ Rms value of source current, $I_s = \left[\dfrac{1}{2\pi} \displaystyle\int_{\pi/3}^{2\pi/3} (I_{mp} \sin \omega t)^2.\, d\,(\omega t) \right]^{1/2} = 0.39\, I_{mp}$

VA rating of transformer $= 6\, V_s\, I_s = 6 \dfrac{V_{mp}}{\sqrt{2}} \times 0.39\, I_{mp} = 1.655\, V_{mp}\, I_{mp}$

DC output power, $P_{dc} = V_o\, I_o = \dfrac{3\, V_{mp}}{\pi} \times \dfrac{3}{\pi} I_{mp} = 0.912\, V_{mp}\, I_{mp}$

$$TUF = \dfrac{P_{dc}}{VA \text{ rating of transf.}} = \dfrac{0.912}{1.655} = 0.551 \text{ or } 55.1\%.$$

Voltage variation across any diode, say D1, can be obtained as done in 3-phase 3-pulse diode rectifier. Therefore, $v_{D1} = v_{a1} - v_o$

When D1 conducts, $v_o = v_{a1}$, $\therefore v_{D1} = v_{a1} - v_{a1} = 0$ from $\omega t = 60°$ to $120°$.

When D2 conducts from $\omega t = 120°$ to $180°$, $v_o = v_{c2}$, $\therefore v_{D1} = v_{a1} - v_{c2}$

At $\omega t = 150°$, $v_{c2} = V_{mp}$ and $v_{a1} = 0.5\, V_{mp}$; $\therefore v_{D1} = (0.5 - 1.0)\, V_{mp} = -0.5\, V_{mp}$

At $\omega t = \pi$, $v_{c2} = 0.866\, V_{mp}$ and $v_{a1} = 0$, $\therefore v_{D1} = -0.866\, V_{mp}$

When D3 conducts from $\omega t = 180°$ to $240°$, $v_o = v_{b1}$, $\therefore v_{D1} = v_{a1} - v_{b1}$

At $\omega t = 240°$, $v_{b1} = 0.866\, V_{mp}$ and $v_{a1} = -0.866\, V_{mp}$, $\therefore v_{D1} = (-0.866 - 0.866)\, V_{mp} = -\sqrt{3}\, V_{mp}$

When D4 conducts from $\omega t = 240°$ to $300°$ $v_o = v_{a2}$ $\therefore v_{D1} = v_{a1} - v_{a2}$

At $\omega t = 270°$, $v_{a1} = -V_{mp}$ and $v_{a2} = V_{mp}$, $\therefore v_{D1} = (-1 - 1)\, V_{mp} = -2\, V_{mp}$

Similarly, v_{D1} waveform can be obtained when diodes D5, D6 conduct, this is shown in Fig. 3.33 (f). It is seen from v_{D1} waveform that PIV for each diode $= 2\, V_{mp}$.

The above analysis reveals the following.

> (i) Quality of output is superior as compared to 3 pulse rectifier, because RF is 4.3% and FF is close to unity.

> (ii) TUF is poor as compared to 3 pulse rectifier ; it is because of lower value of conduction angle (= 60°) for each phase and diode of this rectifier.

> (iii) Output frequency is $6f$; size of filter, if required, is therefore reduced.

It may be observed from Fig. 3.33 that each phase winding carries undirectional current and there are six pulses during one cycle of source voltage. That is why it is called six-phase half-wave diode rectifier.

At any time, only one secondary phase winding, say phase a_1, carries current, this gets reflected downward in the primary delta for *mmf* balance. After further $\omega t = 180°$, secondary-phase winding a_2 carries current, this gets reflected upward in the primary delta. This shows that both secondary and primary windings handle alternating current during one cycle of source voltage ; there is therefore no magnetic saturation of the transformer core as it is in the transformer used in 3-phase half-wave rectifier.

3.9.3. Multiphase Diode Rectifier

For three-phase three-pulse rectifier, or three-phase half-wave rectifier, each phase conducts for $2\pi/3$ radians of a cycle of 2π radians. Three-phase M-6 rectifier may be considered as 6-phase half-wave rectifier and it is seen that each phase of this type conducts for $2\pi/6$ rad. In general, in m-phase half-wave diode rectifier, each phase and diode would conduct for $2\pi/m$ rad and number of output voltage pulses p would be equal to number of phases m.

For m more than three, an m-phase diode rectifier would have delta-connected primary and the secondary would have mid-tapped $m/2$ windings. The number of diodes is equal to m. Fig. 3.34 shows a few pulses of output voltage waveform for m-pulse half-wave diode rectifier. Each phase conducts for $2\pi/m$ or $2\pi/p$ radians, because number of pulses p = number of phases m for half-wave rectifiers. With time origin AA' taken at the peak value of output voltage in Fig. 3.34, the instantaneous phase voltage is

$$v = V_{mp} \cos \omega t = \sqrt{2}\, V_{ph}. \cos \omega t$$

where V_{ph} = rms value of per-phase supply voltage.

Waveform of output voltage v_o in Fig. 3.34 shows that in m-phase half-wave diode rectifier, conduction occurs from $-\dfrac{\pi}{m}$ to $\dfrac{\pi}{m}$, or from $-\dfrac{\pi}{p}$ to π/p with time origin at AA' and periodicity is $2\pi/m$, or $2\pi/p$, radians.

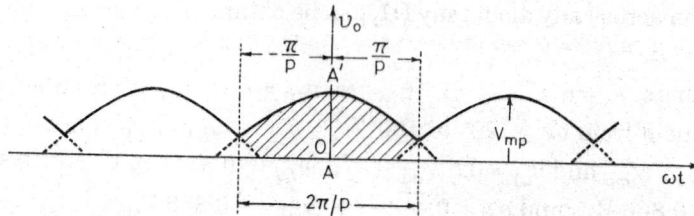

Fig. 3.34. Output voltage waveform for m-phase half-wave diode rectifier.

$\therefore$ Average value of output voltage, $V_o = \dfrac{1}{2\pi/p} \displaystyle\int_{-\pi/p}^{\pi/p} V_{mp} \cos \omega t\, d\,(\omega t)$

$$= V_{mp}. \frac{p}{\pi} \sin \frac{\pi}{p} \qquad\qquad\qquad \dots(3.75)$$

Rms value of output voltage, $\quad V_{or} = \left[\dfrac{p}{2\pi} \displaystyle\int_{-\pi/p}^{\pi/p} (V_{mp} \cos \omega t)^2. \, d\,(\omega t) \right]^{1/2}$

$$= \left[\frac{p.V_{mp}^2}{4\pi} \int_{-\pi/p}^{\pi/p} (1 + \cos 2\omega t)\, d\,(\omega t) \right]$$

$$= V_{mp} \left[\frac{p}{2\pi} \left(\frac{\pi}{p} + \frac{1}{2} \sin \frac{2\pi}{p} \right) \right]^{1/2} \qquad \dots(3.76)$$

Maximum value of load current $\quad = \dfrac{V_{mp}}{R} = I_{mp}$

Average value of diode current, $\quad I_D = \dfrac{1}{2\pi} \displaystyle\int_{-\pi/p}^{\pi/p} I_{mp} \cos \omega t. \, d\,(\omega t) = \dfrac{I_{mp}}{\pi} \sin \dfrac{\pi}{p} \qquad \dots(3.77)$

Rms value of diode current, $I_{Dr} = \left[\dfrac{1}{2\pi} \displaystyle\int_{-\pi/p}^{\pi/p} (I_{mp} \cos \omega t)^2 \, d\,(\omega t) \right]^{1/2}$

$$= I_{mp} \left[\dfrac{1}{2\pi} \left(\dfrac{\pi}{p} + \dfrac{1}{2} \sin \dfrac{2\pi}{p} \right) \right]^{1/2} \qquad \qquad ...(3.78)$$

Example 3.18. *A step-down delta-star transformer, with per-phase turns ratio 5, is fed from 3-phase, 1100 V, 50 Hz source. The secondary of this transformer, through a rectifier, feeds a load R = 10 Ω. Calculate the average value of output voltage, average and rms values of diode current and power delivered to load in case the rectifier is (a) 3-phase, 3-pulse type and (b) 3-phase M-6 type.*

Solution.

(*a*) *3-phase three-pulse type* : Per-phase secondary voltage

$$V_{ph} = \dfrac{1100}{5} = 220 \text{ V and } V_{mp} = \sqrt{2} \times 220 \text{ V}.$$

From Eq. (3.71), or from Eq. (3.75) with $p = 3$, average value of output voltage,

$$V_o = \dfrac{3\sqrt{3}}{2\pi} V_{mp} = \dfrac{3\sqrt{3}}{2\pi} \times \sqrt{2} \times 220 = 257.3 \text{ V}$$

Maximum value of **load current**,

$$I_{mp} = \dfrac{V_{mp}}{R} = \dfrac{\sqrt{2} \times 220}{10} = \sqrt{2} \times 22 \text{ A}$$

From Eq. (3 77), average value of diode current, $I_D = \dfrac{22 \times \sqrt{2}}{\pi} \sin \dfrac{\pi}{3} = 8.575$ A

From Eq. (3.78), rms value of diode current,

$$I_{Dr} = 22 \times \sqrt{2} \left[\dfrac{1}{2\pi} \left(\dfrac{\pi}{3} + \dfrac{\sin 120°}{2} \right) \right]^{1/2} = 15.10 \text{ A}$$

From Eq. (3.76), rms value of output voltage,

$$V_{or} = 220 \times \sqrt{2} \left[\dfrac{3}{2\,\pi} \left(\dfrac{\pi}{3} + \dfrac{\sin 120°}{2} \right) \right]^{1/2} = 2.61.52 \text{ V}$$

Power delivered to load $= \dfrac{V_{or}^2}{R} = \dfrac{261.52^2}{10} = 6839.3$ watts

(*b*) *3-phase M-6 type* : Per-phase secondary voltage,

$$V_{ph} = \dfrac{220}{2} = 110 \text{ V and } V_{mp} = \sqrt{2} \times 110 \text{ V}$$

From Eq. (3.73), or from Eq. (3.75) with $p = 6$, average output voltage is

$$V_o = \dfrac{3}{\pi} \times \sqrt{2} \times 110 = 148.53 \text{ V}$$

Maximum value of load current,

$$I_{mp} = \dfrac{V_{mp}}{R} = \dfrac{\sqrt{2} \times 110}{10} = \sqrt{2} \times 11 \text{ A}$$

From Eq. (3.77), average value of diode current, $I_D = \dfrac{\sqrt{2} \times 11}{\pi} \sin \pi/6 = 2.4755$ A

From Eq. (3.78), rms value of diode current, $I_{Dr} = 11 \times \sqrt{2} \left[\dfrac{1}{2\pi} \left(\dfrac{\pi}{6} + \dfrac{\sin 60}{2} \right) \right]^{1/2} = 6.069$ A

From Eq. (3.76), rms value of output voltage,

$$V_{or} = 110 \times \sqrt{2} \left[\frac{6}{2\pi} \left(\frac{\pi}{6} + \frac{\sin 60°}{2} \right) \right]^{1/2} = 148.66 \text{ V}$$

Power delivered to load $\quad = \dfrac{V_{or}^2}{R} = \dfrac{148.66^2}{10} = 2209.98 = 2210$ W

3.9.4. Evolution of Three-phase Bridge Rectifier

Before studying the 3-phase bridge rectifier, let us first examine the evolution of this rectifier type.

Three-phase half-wave diode rectifier, with common cathode configuration, has already been discussed in Art. 3.9.1.

For the circuit of Fig. 3.29, the conduction of diodes D1, D2, D3 indicated in Fig. 3.30 (b), is again shown in Fig. 3.36 (b), just for the sake of convenience in understanding the evolution of 3-phase bridge rectifier.

Consider now a three-phase half-wave diode rectifier with *common-anode* arrangement as shown in Fig. 3.35. In this circuit, a diode will conduct only during the most negative part of the supply voltage cycle. This means that a diode will conduct when the neutral is positive with respect to terminal a, b or c. Therefore, for the supply waveform of Fig. 3.36 (a), diode D5 would conduct from $\omega t = 0°$ to $\omega t = 90°$ as the voltage v_b is the most negative for this interval. From $\omega t = 90°$ to $\omega t = 210°$, voltage v_c is the most negative, therefore diode D6 must conduct during this interval as shown in Fig. 3.36 (c). Similarly, diode D4 would conduct from $\omega t = 210°$ to $\omega t = 330°$ and so on. Each diode conducts for 120° (as in common-cathode configuration). The load voltage waveform v_o, derived from Fig. 3.36 (a), follows the negative supply voltage envelope as shown in Fig. 3.36 (d) for the diode configuration of Fig. 3.35. Voltage of neutral n is fixed at zero by the reference line ωt in Fig. 3.36 (d). The voltage of terminal Q of Fig. 3.35 is shown by v_b, v_c, v_a etc below the reference line in Fig. 3.36 (d).

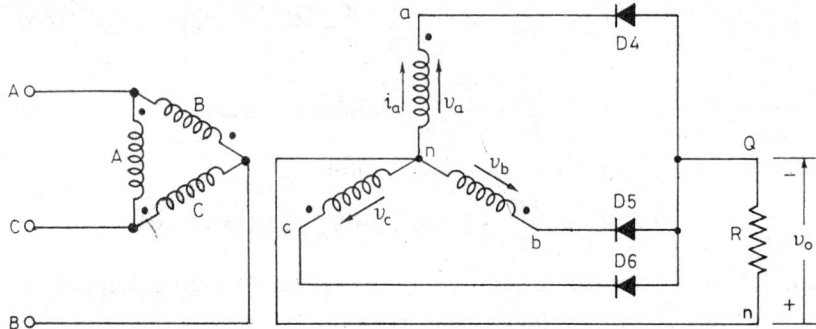

Fig. 3.35. Three-phase half-wave diode rectifier with common anode arrangement.

The three-phase half-wave rectifier circuits of Figs. 3.29 and 3.35 can be connected in series as shown in Fig. 3.37 (a). An examination of this series connected circuit reveals that load current can exist even without neutral wire n. For example, when diode D1 is conducting from $\omega t = 30°$ to $\omega t = 150°$ in Fig. 3.36 (b), the return path for the current is through diode D5 from $\omega t = 30°$ to 90° and through diode D6 from $\omega t = 90°$ to 150°, see Fig. 3.36 (b) and (c). Supply point

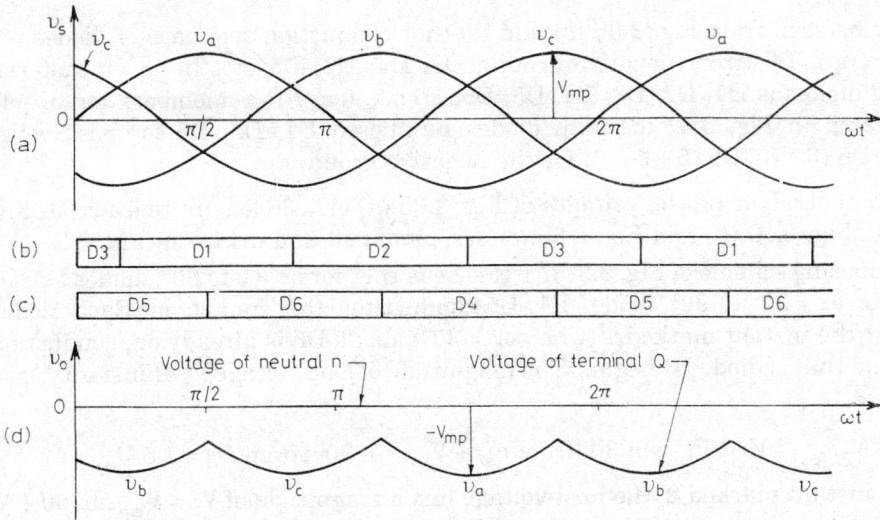

Fig. 3.36. (a) Line to neutral source voltages (b) diode conduction for
Fig. 3.29, (c) diode conduction for Fig. 3.35, (d) load voltage.

'a' connected to the anode of D1 is the same as that connected to the cathode of diode D4. The neutral wire can thus be eliminated and cathode terminal of D4 can be connected to anode of D1. Thus, the circuit of Fig. 3.37 (a) can be redrawn as shown in Fig. 3.37 (b). This circuit can further be rearranged to that shown in Fig. 3.37 (c). The only difference between Fig. 3.37 (a) and Fig. 3.37 (b) and (c) is that load voltage is equal to line to neutral voltage in Fig. 3.37 (a) and it is line to line voltage in Fig. 3.37 (b) and (c). The circuit configuration shown in Fig. 3.37 (c) is called *3-phase full wave bridge rectifier*, or *3-phase six-pulse bridge rectifier*. Note that diodes D1, D2, D3 of the bridge would conduct when supply voltage is the most positive, whereas diodes D4, D5, D6 would conduct when supply voltage is the most negative. Diodes D1, D2, D3 may therefore be called a *positive diode group* and D4, D5, D6 a *negative diode group*. The voltage across load would always be the direct emf with the polarity of P positive and that of Q negative as indicated.

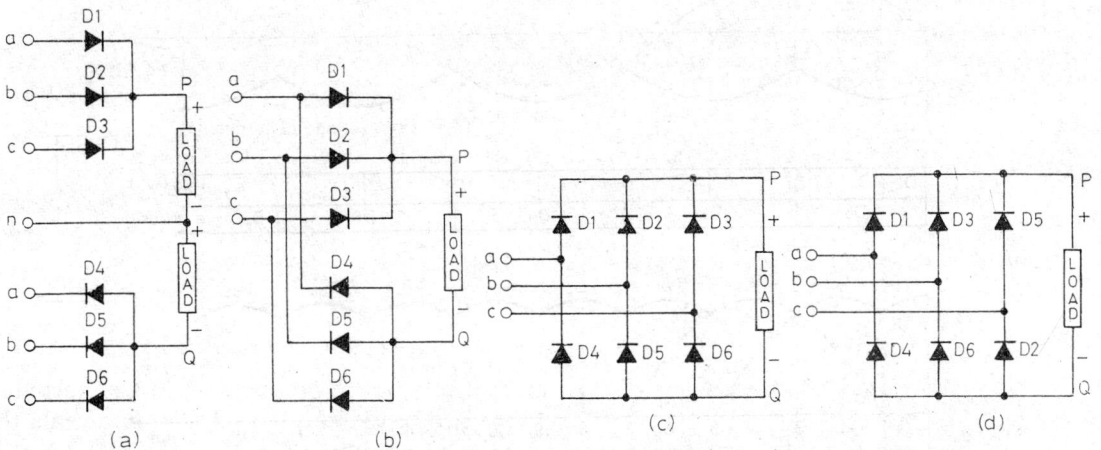

Fig. 3.37. Evolution of 3-phase six-pulse rectifier (a) circuits of Figs. 3.29 and 3.35
connected in series, (b) circuit obtained from (a), (c) circuit of
(b) rearranged and (d) diode numbering scheme altered.

It may be seen from Fig. 3.36 (b) and (c) that conduction sequence of diodes is D1 (from positive group), D6 (from negative group), D2, D4, D3, D5 etc. In order that sequence of conducting diodes is D1, D2, D3, D4, D5, D6, D1. .... (easy to remember), circuit of Fig. 3.37 (c) is redrawn in Fig. 3.37 (d) with diodes numbered D1, D3, D5 for positive group and D4 (1 + 3), D6 (3 + 3), D2 (5 + 3 − 6) for the negative group.

Line to neutral, or phase, voltages of Fig. 3.30 (a) or 3.36 (a) are redrawn in Fig. 3.38 (a) as v_a, v_b, v_c. Fig. 3.36 (b) and Fig. 3.36 (c) are combined and drawn in Fig. 3.38 (b) but with diode-numbering scheme of Fig. 3.37 (d). It is seen that for $\omega t = 0$ to 30°, diodes D5, D6 conduct together, for $\omega t = 30°$ to 90°, diodes D1, D6 conduct together and so on. Each diode conducts for 120°. At the instant marked 1 (when $\omega t = 30°$), diode D6 is already on, conduction of diode D5 stops and that of diode D1 begins. The magnitude of load voltage V_1 at instant 1 is, therefore, given by

$$V_1 = V_{mp} \sin 30 \text{ (from } v_a) + V_{mp} \sin 90° \text{ (from } v_b) = 1.5 \, V_{mp}$$

At the instant marked 2, the load voltage has a magnitude of $V_2 = V_{mp} \sin 60 + V_{mp} \sin 60$ $= \sqrt{3} \, V_{mp}$.

At the instant marked 3, $V_3 = 1.5 \, V_{mp}$

Here V_{mp} = maximum value of phase (or line to neutral) voltage.

The voltage of the load terminals P and Q of Fig. 3.37 (c) or (d) is shown in Fig. 3.38 (a). This figure also reveals that at the instants marked 2, 4, 6, 8, 10 etc, the load voltage has a magnitude of $\sqrt{3} \, V_{mp}$. At the instants marked 1, 3, 5, 7, 9, 11 etc, the magnitude of load voltage is $1.5 \, V_{mp}$. The load voltage, or the rectified output voltage, v_o can therefore, be plotted as shown in Fig. 3.38 (c). In this figure, voltage of terminal Q is shown at zero potential by straight reference line ωt, whereas the potential of terminal P is shown by line voltages v_{cb}, v_{ab}, v_{ac} etc. In fact, if voltage waveform of terminal Q in Fig. 3.38 (a) is made a straight line, Fig. 3.38 (c)

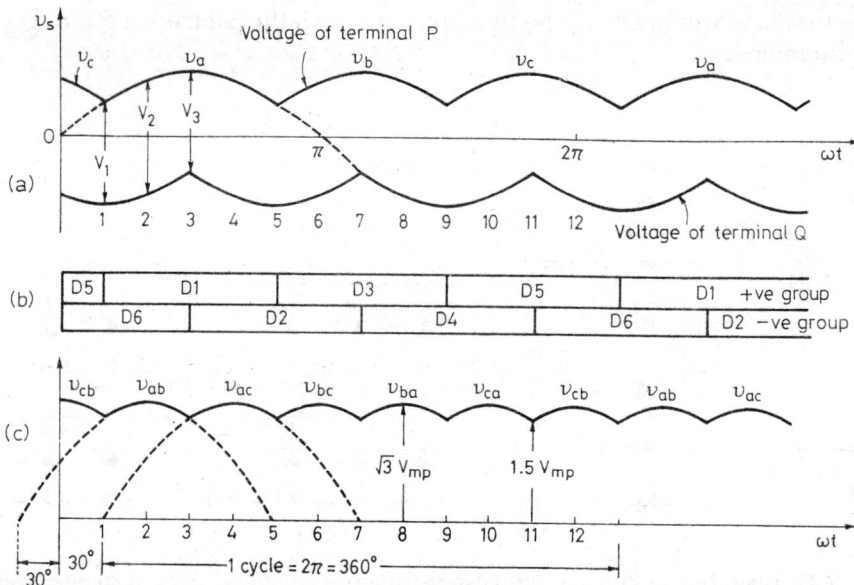

Fig. 3.38. (a) 3-phase input voltage waveforms (b) conduction sequence of positive and negative group of diodes (c) output voltage waveform of 3-phase six-pulse diode bridge of Fig. 3.37 (d).

is obtained. It should be remembered that in Fig. 3.38 (c); v_{ab}, v_{ac}, v_{bc} etc are line voltages, whereas in Fig. 3.38 (a) ; v_a, v_b, v_c are phase voltages. The dual subscript ab in v_{ab} should be taken to denote that as per the *first* subscript 'a', diode connected to phase terminal 'a' from *positive* group, *i.e.* D1 conducts. As per the *second* subscript 'b', diode connected to phase terminal 'b' from the *negative* group, *i.e.* D6 conducts. For example, for output voltage v_{cb}, diode D5 from positive group and diode D6 from negative group conduct. Note that each diode conducts for 120°.

Fig. 3.38 (c) reveals that there are six pulses for one cycle of supply voltage. Thus, three-phase bridge rectifier of Fig. 3.37 (d) can be called 3-phase six-phase diode rectifier or *3-phase B-6 diode rectifier.* Here B denotes bridge and 6 denotes the number of output-voltage pulses per cycle of source voltage.

It is thus seen from above that when two 3-phase 3-pulse rectifiers are connected in antiparallel, a 3-phase 6 pulse rectifier (or a 3-phase bridge rectifier) is evolved.

3.9.5. Three-phase Bridge Rectifier

Power circuit diagram for a 3-phase bridge rectifier using six diodes in shown in Fig. 3.39. The diodes are arranged in three legs. Each leg has two series-connected diodes. Upper diodes D1, D3, D5 constitute the positive group of diodes. The lower diodes D2, D4, D6 form the negative group of diodes. The three-phase transformer feeding the bridge is connected in delta-star. This rectifier is also called 3-phase 6-pulse diode rectifier, 3-phase full-wave diode rectifier, or three-phase B-6 diode rectifier.

Positive group of diodes conduct when these have the *most positive anode*. Similarly, negative group of diodes would conduct if these have the most negative anode. In other words, diodes D1, D3, D5, forming positive group, would conduct when these experience the highest positive voltage. Likewise, diodes D2, D4, D6 would conduct when these are subjected to the most negative voltage.

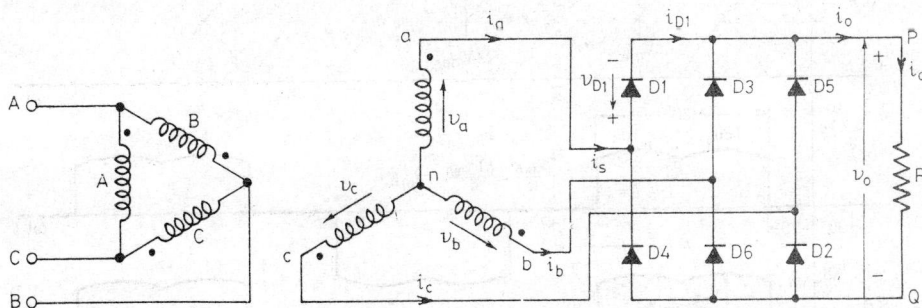

Fig. 3.39. Three-phase bridge rectifier using diodes.

It is seen from the source voltage waveform v_s of Fig. 3.40 (a) that from $\omega t = 30°$ to 150°, voltage v_a is more positive than the voltages v_b, v_c. Therefore, diode D1 connected to line 'a' (as per subscript 'a' in v_a) counducts during the interval $\omega t = 30°$ to 150°. Likewise, from $\omega t = 150°$ to 270°, voltage v_b is more positive as compared to v_a, v_c; therefore, diode D3 connected to line 'b' (as per the subscript 'b' in v_b) conducts during this interval. Similarly, diode D5 from the positive group conducts from $\omega t = 270°$ to 390° and so on. Note also that from $\omega t = 0$ to 30°, v_c is the most positive, therefore, diode D5 from the positive group conducts for this interval. Conduction of positive group diodes is shown in Fig. 3.40 (*b*) as D5, D1, D3, D5, D1 etc.

Voltage v_c is the most negative from $\omega t = 90°$ to $210°$. Therefore, negative group diode D2 connected to line '*c*' (as per subscript '*c*' in v_c) conducts during this interval. Similarly, diode D4 conducts from $210°$ to $330°$ and diode D6 from $330°$ to $450°$ and so on. Note also that from $\omega t = 0°$ to $90°$, v_b is the most negative, therefore diode D6 conducts during this interval. Conduction of negative group diodes is shown as D6, D2, D4, D6, etc in Fig. 3.40 (*b*).

During the interval $\omega t = 0°$ to $30°$, it is seen from Fig. 3.40 (*b*) that diode D5 and D6 conduct. Fig. 3.39 shows that conduction of D5 connects load terminal *P* to line terminal *c*; similarly, conduction of D6 connects load terminal *Q* to line terminal *b*. As a result, load voltage is $v_{pq} = v_o =$ line voltage v_{cb} (first positive subscript corresponding to *c* and second negative subscript corresponding to *b*) from $\omega t = 0°$ to $\omega t = 30°$. Likewise, during $\omega t = 30°$ to $90°$, diodes D1 and D6 conduct. Conduction of diode D1 connects *P* to *a* and D6 to *b*. Therefore, load voltage

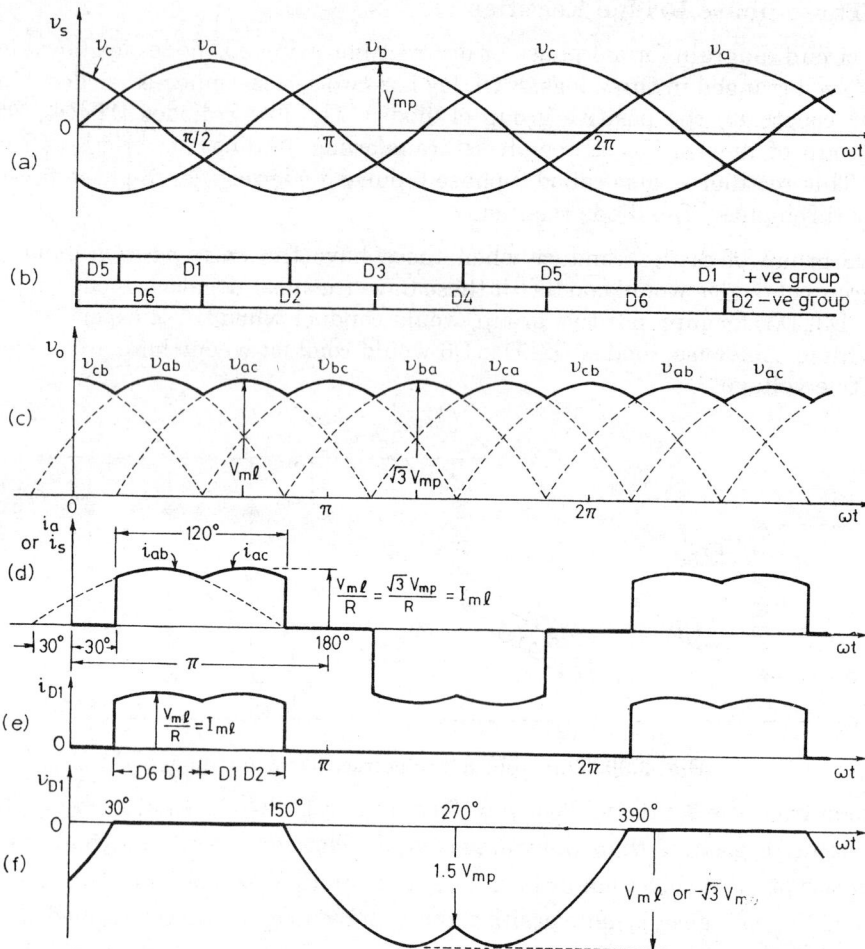

Fig. 3.40. Three-phase diode bridge rectifier (*a*) 3-phase input voltage waveform (*b*) conduction sequence of diodes (*c*) output voltage waveform (*d*) input current waveform (*e*) diode current waveform through D1 and (*f*) voltage variation across diode D1.

during this interval is v_o = line voltage v_{ab}. Similarly, for interval 90° to 150°, diodes D1 and D2 conduct and v_o = line voltage v_{ac}; for interval 150° to 210°, diodes $D3$ and D2 conduct and v_o = line voltage v_{bc} and so on. Output, or load, voltage waveform is drawn by a thick curve in Fig. 3.40 (c).

Average value of load voltage, $V_o = \dfrac{1}{\text{periodicity}} \displaystyle\int_{\alpha_1}^{\alpha_2} v_{ab} \cdot d\,(\omega t)$

It is seen from Fig. 3.40 (c) that the value of v_{ab} at $\omega t = 0$ is $V_{ml} \cdot \sin 30°$ and its periodicity is 60° or $\pi/3$ rad.

$$\therefore \qquad V_o = \frac{3}{\pi} \int_{\pi/6}^{\pi/2} V_{ml} \sin(\omega t + 30°)\, d\,(\omega t)$$

$$= \frac{3 \cdot V_{ml}}{\pi} = \frac{3\sqrt{2}\,V_l}{\pi} = \frac{3 \cdot \sqrt{6}\,V_p}{\pi} \qquad \text{...(3.79)}$$

where V_{ml} = maximum value of line voltage

V_l = rms value of line voltage and

V_p = rms value of phase voltage.

Average value of output voltage, V_o, can also be obtained as under :

(i) Take any sinusoidal wave and integrate it from 60° to 120°. It is because the voltage-pulse area required extends from $\omega t = 60°$ to $\omega t = 120°$ for the sin ωt function.

$$\therefore \qquad V_o = \frac{3}{\pi} \int_{\pi/3}^{2\pi/3} V_{ml} \cdot \sin \omega t \cdot d\,(\omega t) = \frac{3\,V_{ml}}{\pi} \qquad \text{...(3.79)}$$

(ii) For a cosine function cos ωt, voltage pulse of 60° duration extends 30° to the left of its peak and 30° to the right of its peak.

$$\therefore \qquad V_o = \frac{3}{\pi} \int_{-\pi/6}^{\pi/6} V_{ml} \cos \omega t \cdot d\,(\omega t) = \frac{3V_{ml}}{\pi} \qquad \text{...(3.79)}$$

Rms value of output voltage, $V_{or} = \left[\dfrac{3}{\pi} \displaystyle\int_{\pi/3}^{2\pi/3} V_{ml}^2 \sin^2 \omega t \cdot d\,(\omega t) \right]^{1/2}$

$$= \left[\frac{3V_{ml}^2}{2\pi} \left| \omega t - \frac{\sin 2\omega t}{2} \right|_{\pi/3}^{2\pi/3} \right]^{1/2} = 0.9558\, V_{ml} \qquad \text{...(3.80)}$$

Ripple voltage, $\qquad V_r = \sqrt{V_{or}^2 - V_o^2} = \left[0.9558^2 - \left(\dfrac{3}{\pi} \right)^2 \right]^{1/2} V_{ml} = 0.0408\, V_{ml}$

Voltage ripple factor, $\qquad \text{VRF} = \dfrac{V_r}{V_o} = \dfrac{0.0408}{3/\pi} = 0.0427 \text{ or } 4.27\%$

$$FF = \frac{V_{or}}{V_o} = \frac{0.9558\, V_{ml}}{\dfrac{3}{\pi} V_{ml}} = 1.0009$$

Rms value of output current, $I_{or} = \dfrac{V_{or}}{R} = \dfrac{0.9558}{R} V_{ml} = 0.9558\, I_{ml}$

$$P_{dc} = V_o I_o = \left(\frac{3}{\pi}\right)^2 V_{ml} \cdot I_{ml}$$

$$P_{ac} = V_{or} \cdot V_{or} = (0.9558) \, V_{ml} I_{ml}$$

Rectifier efficiency $= \left(\frac{3}{\pi}\right)^2 \times \frac{1}{0.9558} = 0.9982$ or 99.82%

For a resistive load, peak current through each diode is $i_{ml} = \dfrac{V_{ml}}{R} = \dfrac{\sqrt{3} \, V_{mp}}{R}$. It is seen from the waveform of line current i_a (or transformer secondary current i_s) that (i) periodicity of this current is π rad. and (ii) this current has two pulses, each of $60°$ duration, for each periodicity of π rad.

$\therefore$ Rms value of line current = rms value of transformer secondary current

$$I_s = \left[\frac{2}{\pi} \int_{\pi/3}^{2\pi/3} I_{ml}^2 \sin^2 \omega t. \, d\,(\omega t)\right]^{1/2} = 0.7804 \, I_{ml} \qquad \ldots(3.81)$$

Transformer VA rating $= 3 V_s I_s = 3 \times \dfrac{V_{ml}}{\sqrt{6}} \times 0.7804 \, I_{ml}$

$\therefore$

$$TUF = \frac{P_{dc}}{VA \text{ rating of transformer}} = \left(\frac{3}{\pi}\right)^2 \times \frac{\sqrt{6}}{3 \times 0.7804} = 0.9541$$

It is seen from diode-current waveform of Fig. 3.40 (d) that average value of diode current is

$$I_D = \frac{2}{2\pi} \int_{\pi/3}^{2\pi/3} I_{ml} \cdot \sin \omega t \, d(\omega t) = \frac{I_{ml}}{\pi} \qquad \ldots(3.82)$$

Rms value of diode current, $I_{Dr} = \left[\dfrac{2}{2\pi} \displaystyle\int_{\pi/3}^{2\pi/3} I_{ml}^2 \sin^2 \omega t \, . \, d\,(\omega t)\right]^{1/2} = 0.582 \, I_{ml} \qquad \ldots(3.83)$

A conducting diode has zero voltage drop across it. Let it be required to sketch in Fig. 3.40 (f), the variation of voltage across diode D1 belonging to positive group in Fig. 3.39.

For dc output voltage v_{ab}, v_{ac} ; diode D1 conducts, therefore, voltage across diode D1 is zero, $i.e.$ $v_{D1} = 0$ from $\omega t = 30°$ to $150°$ and this is shown in Fig. 3.40 (f). After $\omega t \approx 150°$, diode D3 conducts for $120°$ with output voltage v_{bc}, v_{ba}. Now cathode of D1 gets connected to supply terminal b through conducting diode D3 for a period of $120°$, whereas its anode is already connected to supply terminal 'a'. Therefore, voltage across D1 from $\omega t = 150°$ to $270°$ is $v_{D1} = v_a - v_b$ as per KVL $(v_{D1} - v_a + v_b = 0)$. This voltage v_{ab} during interval of $\omega t = 150°$ to $270°$ reverse biases diode D1 and therefore, voltage v_{ab} is shown below the reference line in Fig. 3.40 (f). Examination of Fig. 3.40 (c) reveals that waveform v_{ba} from $\omega t = 150°$ to $270°$, when reversed, becomes v_{ab} in Fig. 3.40 (f).

After $\omega t = 270°$, diode D5 conducts for $120°$ and now voltage $v_{D1} = v_{ac}$ from $\omega t = 270$ to $390°$. This voltage is shown as v_{ac} and as a reverse bias across diode D1, so v_{ac} is sketched below the reference line in Fig. 3.40 (f). As before, voltage waveform v_{ca} from $\omega t = 270°$ to $390°$ in Fig. 3.40 (c), when reversed, becomes v_{ac} in Fig. 3.40 (f).

Fig. 3.40 (f) reveals that PIV for diode D1, or any other diode, is $\sqrt{3} \, V_{mp} = V_{ml} =$ maximum value of line voltage.

An examination of waveforms in Fig. 3.40 (*a*) and (*c*) reveals that line voltage v_{ab} leads phase voltage v_a by 30°. Similarly, v_{bc} leads v_b by 30° and v_{ca} leads v_c by 30°. A phasor diagram showing 3-phase phase voltages V_a, V_b, V_c and the corresponding 3-phase line voltages $V_{ab} (= \overline{V}_a - \overline{V}_b), V_{bc} (= \overline{V}_b - \overline{V}_c), V_{ca} (= \overline{V}_c - \overline{V}_a)$ is drawn in Fig. 3.41. Line voltage V_{ba} is 180° away from V_{ab} as shown. Similarly, line voltages V_{cb} and V_{ac} are shown in Fig. 3.41, thus resulting in a six-phase system of line voltages $V_{ab}, V_{ac}, V_{bc}, V_{ba}, V_{ca}, V_{cb}$. Phasor diagram of Fig. 3.41 reveals that line voltage V_{ab} leads phase voltage V_a by 30°, similarly V_{bc} leads V_b by 30° and V_{ca} leads V_c by 30° ; this matches with the waveforms in Fig. 3.40 as expected.

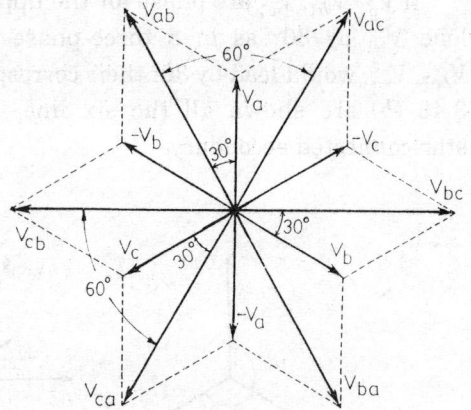

Fig. 3.41. Six-phase line voltage V_{ab}, V_{ac}, V_{bc} etc for secondary winding of delta-star transformer of Fig. 3.39.

Note that $V_a = V_b = V_c$ = phase voltage, V_{ph} and $V_{ab} = V_{bc} = V_{ca} = V_{ba} = V_{cb} = V_{ac}$ = line voltage, V_l.

3.9.6. Three-phase Twelve-pulse Rectifier

It has been stated before that as the number of pulses per cycle are increased, the output *dc* waveform gets improved. So, with twelve pulses per cycle, the quality of output voltage waveform would definitely be improved with low ripple content.

Fig. 3.42 shows the circuit diagram for a 3-phase twelve-pulse rectifier using a total of twelve diodes. A 3-phase transformer with two secondaries and one delta-connected primary feeds the diode rectifier circuit. One secondary winding is connected in star and the other is in delta. Star-connected secondary feeds the upper 3-phase diode bridge rectifier 1, whereas the delta-connected secondary is connected to lower 3-phase diode bridge rectifier 2. Each bridge rectifier uses six diodes as shown. The two bridges are series connected so that net output, or load, voltage v_o = output voltage of upper rectifier, v_{01} + output voltage of lower rectifier v_{02}.

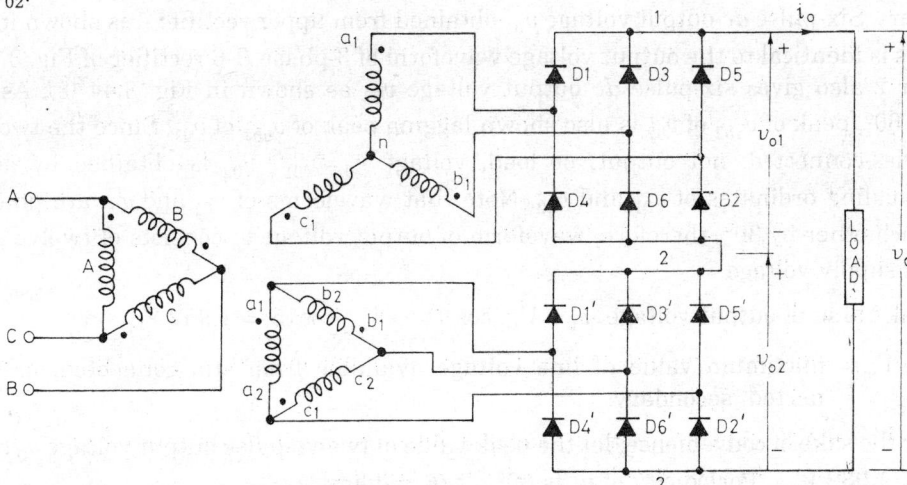

Fig. 3.42. Three-phase twelve-pulse rectifier.

If V_{a1}, V_{b1}, V_{c1} are phase for the upper star, then upper line voltage V_{ab1} ($= \overline{V}_{a1} - \overline{V}_{b1}$) would lead V_{a1} by 30° as in a three-phase bridge rectifier of Fig 3.39. Similarly, line voltages V_{bc1}, V_{ca1} would lead by 30° their corresponding phase voltages V_{b1} and V_{c1} respectively. In Fig. 3.43 (b) are shown all the six line voltages V_{ab1}, V_{ac1}, V_{bc1}, V_{ba1}, V_{ca1}, V_{cb1} for the upper star-connected secondary.

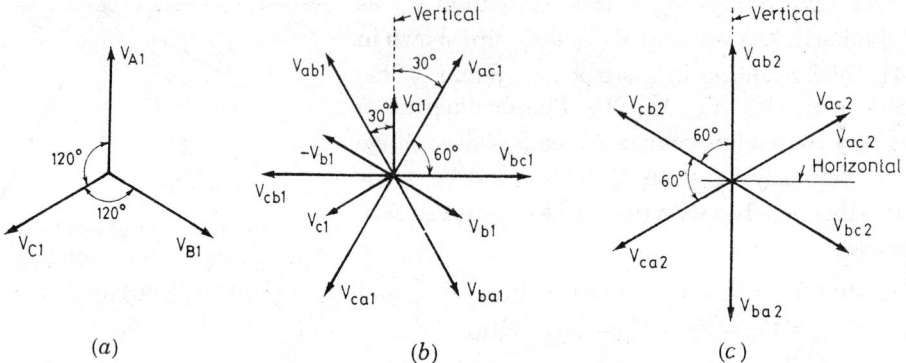

Fig. 3.43. Voltage phasor diagram for (a) primary line voltages (b) line voltages for secondary star and (c) line voltages secondary delta.

Phase voltage V_{a1} of upper star must be in phase with primary line, or phase, voltage V_{A1} as per the transformer principle. Likewise, line, or phase, voltage V_{ab2} of the delta-connected secondary must be in phase with V_{A1}. All the secondary line voltages V_{ab2}, V_{ac2}, V_{bc2}, V_{ba2}, V_{ca2}, V_{cb2} for the delta-connected secondary are shown in Fig. 3.43 (c). The line voltages V_{ab1} for upper rectifier and V_{ab2} for lower rectifier are phase-displaced by 30° with V_{ab1} leading V_{ab2} by 30°. In case input line voltages to upper rectifier 1 and lower rectifier 2 are superimposed; line voltages V_{ab1}, V_{ab2}, V_{ac1}, V_{ac2} etc would be obtained; these line voltages would be phase-displaced from each other by 30°.

In Fig. 3.44 (a) are sketched waveforms for voltages available across upper-star-connected secondary. Six-pulse dc output voltage v_{01} obtained from upper rectifier 1 is shown in Fig. 3.44 (b) ; this is identical to the output voltage waveform of 3-phase B-6 rectifier of Fig. 3.39. Lower rectifier 2 also gives six-pulse dc output voltage v_{02} as shown in Fig. 3.44 (c). As V_{ab2} lags V_{ab1} by 30°, peak of v_{ab2} of v_{02} is also shown lagging peak of v_{ab1} of v_{01}. Since the two rectifiers are series-connected, net output, or load, voltage $v_0 = v_{01} + v_{02}$ is obtained by adding the corresponding ordinates of v_{01} and v_{02}. Note that waveforms of v_{01} and v_{02} are phase-shifted from each other by 30°, therefore, waveform of output voltage v_o consists of twelve pulses per cycle of supply voltage.

Peak value of output voltage, $v_0 = V_{ml} \cos 15° + V_{ml} \cos 15° = 1.932 \, V_{ml}$

where V_{ml} = maximum value of line voltage available from star-connected, or delta-connected, secondary.

For the sake of convenience, let the peak value of twelve-pulse output voltage v_0 be denoted by V_p ($= 1.932 \, V_{ml}$). Periodicity of v_0 is 30° $= \pi/6$ radians.

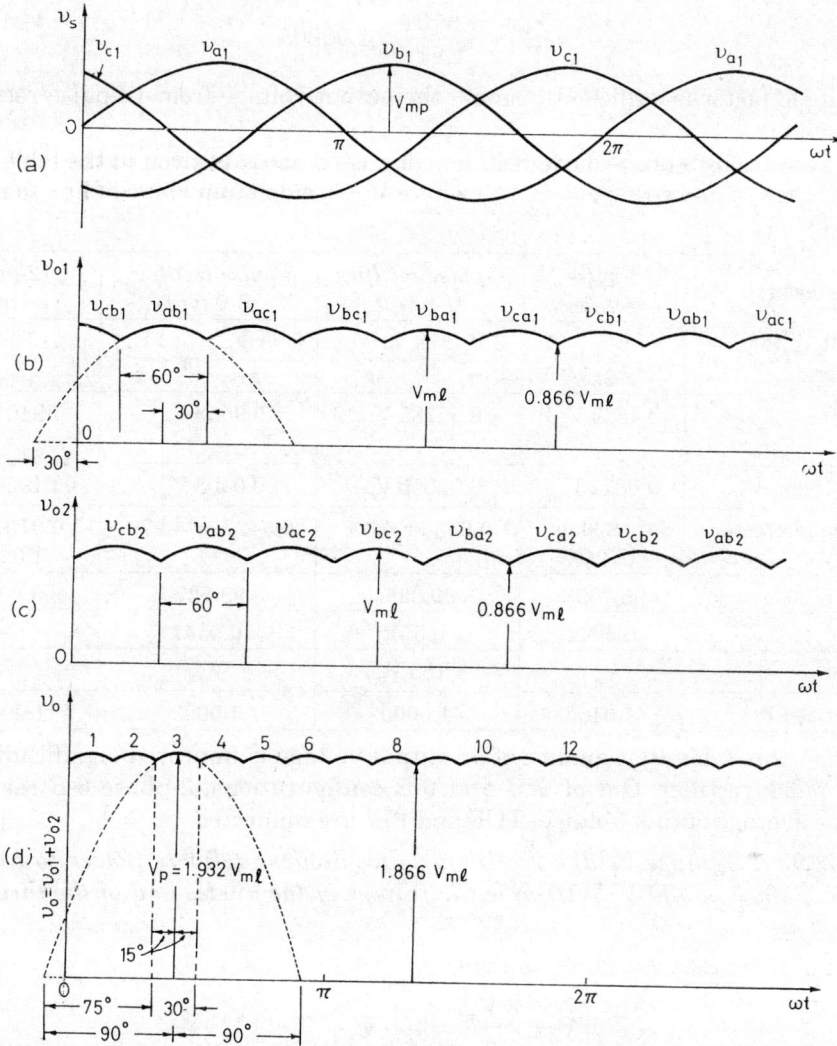

Fig. 3.44. Waveforms for (a) voltages across star-connected secondary (b) six-pulse output voltage v_{01} from upper bridge 1 (c) six-pulse output voltage v_{02} from lower bridge 2 and (d) resultant twelve-pulse output voltage v_o obtained from $v_{01} + v_{02}$.

$\therefore$ Average value of output voltage, $V_0 = \dfrac{6}{\pi} \displaystyle\int_{75°}^{105°} V_p \sin \omega t. \, d\,(\omega t) = 0.988616 \, V_p$

$$= 0.988616 \times 1.932 \, V_{ml} = 1.91 \, V_{ml} \qquad \qquad ...(3.83)$$

Rms value of output voltage, $\quad V_{or} = \left[\dfrac{6}{\pi} \displaystyle\int_{75°}^{105°} V_p^2 \sin^2 \omega t \, d\,(\omega t)\right]^{1/2} = 0.988668 \, V_p$

$$= 0.988668 \times 1.932 \, V_{ml} = 1.9101 \, V_{ml} \qquad \qquad ...(3.84)$$

Ripple voltage, $\qquad V_r = \sqrt{V_{or}^2 - V_o^2} = [1.9101^2 - 1.91^2]^{1/2} \, V_{ml} = 0.019545 \, V_{ml}$

Voltage ripple factor, $\qquad VRF = \dfrac{V_r}{V_o} = \dfrac{0.019545 \,.\, V_{ml}}{1.91 \times V_{ml}} = 0.01023 \text{ or } 1.023\%$

Form factor,
$$FF = \frac{V_{or}}{V_o} = \frac{1.9101}{1.91} = 1.00005$$

As voltage ripple factor is sufficiently small, the output voltage from 12-pulse rectifier is almost pure *dc* voltage.

A comparison of various 3-phase diode rectifiers discussed above is given in the table below : [V_{ml} = maximum value of line voltage = $\sqrt{3}\ V_{mp}$ where V_{mp} = maximum value of line to neutral, or phase, voltage].

S.No.	Parameters	3-pulse rectifier	6-pulse rectifier M-6 type	6-pulse rectifier B-6 type	12-pulse rectifier
1.	DC output voltage, V_o	$\dfrac{3V_{ml}}{2\pi}$	$\dfrac{3\ V_{mp}}{\pi}$ or $\dfrac{\sqrt{3}\ V_{ml}}{\pi}$	$\dfrac{3\sqrt{3}\ V_{mp}}{\pi}$ or $\dfrac{3\ V_{ml}}{\pi}$	$1.91\ V_{ml}$
2.	Rms output voltage, V_{or}	$0.4854\ V_{ml}$	$0.55185\ V_{ml}$	$0.9558\ V_{ml}$	$1.9101\ V_{ml}$
3.	Ripple voltage, V_r	$0.0872\ V_{ml}$	$0.02356\ V_{ml}$	$0.0408\ V_{ml}$	$0.019545\ V_{ml}$
4.	Voltage ripple factor, VRF	0.1826 or 18.26%	0.043 or 4.3%	0.0427 or 4.27%	0.01023 or 1.023%
5.	Rectifier efficiency, η	96.765%	99.82%	99.82%	—
6.	TUF	0.6644	0.551	0.9541	—
7.	PIV	V_{ml}	$1.155\ V_{ml}$	V_{ml}	—
8.	Form factor, FF	1.0165	1.0009	1.0009	1.00005

It is seen from this table that quality of *dc* output voltage is improved significantly with 3-phase twelve-pulse rectifier. Out of 6-*M* and 6-*B* configurations, 3-phase 6-*B* rectifier is superior so far as average output voltage, TUF and PIV are connected.

Example 3.19. *A 3-phase bridge rectifier, using diodes, delivers power to a load of R = 10 Ω at a dc voltage of 400 V. Determine the ratings of the diodes and of the three-phase delta-star transformer.*

Solution. It is given that *dc* output voltage,
$$V_o = \frac{3\ V_{ml}}{\pi} = 400 \text{ V}$$

∴ Maximum value of line voltage, $V_{ml} = \dfrac{400 \times \pi}{3} = 418.88$ V

∴ Rms value of phase voltage for share-secondary,
$$V_s = \frac{V_{ml}}{\sqrt{2} \times \sqrt{3}} = \frac{418.88}{\sqrt{6}} = 171.0 \text{ V}$$

Maximum value of load current, $I_m = \dfrac{V_{ml}}{R} = \dfrac{418.88}{10} = 41.89$ A

From Eq. (3.81), rms value of phase current,
$$I_s = 0.7804\ I_m = 0.7804 \times 41.89 = 32.691 \text{ A}$$

Transformer rating
$$= 3\ V_{ph} \cdot I_{ph} = 3\ V_s \cdot I_s = 3 \times 171 \times 32.691 = 16770.5 \text{ VA}$$

Rms value of diode current,
$$I_{Dr} = \left[\frac{2}{2\pi} \int_{\pi/3}^{2\pi/3} I_m^2 \sin^2 \omega t \cdot d(\omega t) \right]^{1/2}$$

$$= \left[\frac{I_m^2}{2\pi} \left(\frac{\pi}{3} + \frac{\sqrt{3}}{2} \right) \right]^{1/2} = 0.5518 \, I_m$$

$$= 0.5518 \times 41.89 = 23.115 \, \text{A}$$

Average value of diode current, $I_D = \frac{1}{\pi} \int_{\pi/3}^{2\pi/3} I_m \sin \omega t \, d \, (\omega t) = \frac{I_m}{\pi} = \frac{41.89}{\pi} = 13.334 \, \text{A}$

Peak diode current $= I_m = 41.89 \, \text{A}, \, PIV = V_{ml} = 418.88 \, \text{V}$

Check : $P_d = V_o \, I_o = 400 \times \dfrac{400}{10} = 16000 \, \text{W}$

∴ Transformer rating $= \dfrac{P_d}{TUF} = \dfrac{16000}{0.9541} = 16769.73 \, \text{VA}$

Example 3.20. *A 3-phase bridge rectifier charges a 240-V battery. Input voltage to rectifier is 3-phase, 230 V, 50 Hz. Current limiting resistance in series with battery is 8Ω and an inductor makes the load current almost ripple free. Determine (a) power delivered to battery and the load (b) input displacement factor (c) current distortion factor (d) input power factor (e) input HF or THD (f) transformer rating.*

Solution.

(a) Here $V_{ml} = \sqrt{2} \times V_l = \sqrt{2} \times 230 \, \text{V}$

Arrange output voltage, $V_o = \dfrac{3 \, V_{ml}}{\pi} = \dfrac{3 \sqrt{2} \times 230}{\pi} = 310.56 \, \text{V}$

But $V_o = E + I_o R$

∴ Average value of battery charging current, $I_o = \dfrac{V_o - E}{R} = \dfrac{310.56 - 240}{8} = 8.82 \, \text{A}$

Power delivered to battery $= EI_o = 240 \times 8.82 = 2116.8 \, \text{W}$

Power delivered to load, $P_d = EI_o + I_{or}^2 \, R$

Since load current is ripple free, $I_{or} = I_o = 8.82 \, \text{A}$

∴ $P_d = 240 \times 8.82 + 8.82^2 \times 8 = 2739.16 \, \text{W}$

(b) For ripple free load current, Fig. 3.40 (d) shows that phase-a current i_a, or transformer secondary current i_s, would be constant at $I_o = 8.82 \, \text{A}$ from $\omega t = 30°$ to $150°$ and $- I_o$ from $210°$ to $330°$ and so on. As positive and negative half cycles are identical, average value of $i_s = 0$, i.e. $I_{dc} = 0$.

$$a_n = \frac{2}{\pi} \int_{\pi/6}^{5\pi/6} I_0 \cos n\omega t \, . \, d \, (\omega t)$$

or $\quad a_1 = \dfrac{2}{\pi} \displaystyle\int_{\pi/6}^{5\pi/6} I_0 \cos \omega t . \, d \, (\omega t) = \dfrac{2I_o}{\pi} [\sin 150° - \sin 30°] = 0$

$\quad b_1 = \dfrac{2}{\pi} \displaystyle\int_{\pi/6}^{5\pi/6} I_0 \sin \omega t . \, d \, (\omega t) = \dfrac{2I_o}{\pi} [- \cos 150° + \cos 30°] = \dfrac{2\sqrt{3}}{\pi} I_o$

From Eq. (3.56), fundamental component of source current is given by

$$i_{s1} = \frac{2\sqrt{3}}{\pi} I_o \sin \omega t \quad \text{and} \quad \phi_1 = \tan^{-1} \left[\frac{0}{b_1} \right] = 0°$$

Input displacement factor, $DF = \cos \phi_1 = 1$

(c) Rms value of fundamental component of source current, $I_{s1} = \dfrac{2\sqrt{3}}{\pi} \times \dfrac{I_o}{\sqrt{2}}$

Rms value of source current, $I_s = \left[\dfrac{I_o^2 \times 2\pi}{\pi \times 3} \right]^{1/2} = \sqrt{\dfrac{2}{3}} \cdot I_o$

Current distortion factor, CDF $= \dfrac{I_{s1}}{I_s} = \dfrac{2\sqrt{3} \times I_o}{\pi \cdot \sqrt{2}} \times \dfrac{\sqrt{3}}{\sqrt{2} \cdot I_o} = \dfrac{3}{\pi} = 0.955$

(d) Input $pf = CDF \times DF = 0.955 \times 1 = 0.955$ (lagging)

(e) HF = THD $= \left[\left(\dfrac{I_s}{I_{s1}} \right)^2 - 1 \right]^{1/2} = \left[\left(\dfrac{1}{0.0955} \right)^2 - 1 \right]^{1/2} = 0.3106$

(f) Transformer rating $= \sqrt{3}\, V_s \cdot I_s = \sqrt{3} \times 230 \times \sqrt{\dfrac{2}{3}} \times 8.82 = 2868.4$ VA

Also, transformer rating $= \dfrac{P_d}{TUF} = \dfrac{2739.14}{0.9541} = 2876.92$ VA

3.10. FILTERS

A rectifier should provide an output voltage that should be as smooth as possible. In practice, however, output voltage from rectifiers consists of *dc* component plus *ac* component, or *ac* ripples. The *ac* component is made up of several dominant harmonics. It is more so in single-phase rectifiers with *R* load. The *ac* component does no useful work. For example, in a *dc* motor, it is the *dc* current that produces the required torque; in a battery, energy is stored due to *dc* current only. AC ripples in rectifier output current do not contribute to motor torque, or to the energy stored in the battery. AC component merely causes more ohmic losses in the circuit leading to reduced efficiency of the system. This shows that it is of paramount importance to filter out the unwanted *ac* component present in the rectifier output. For this purpose, filters are used. When used on the rectifier output side, these are called *dc filters*; these tend to make the *dc* output voltage and current as level as possible. The more common *dc* filters are of *L, C* and *LC* type as shown in Fig. 3.45 (*a*), (*b*) and (*c*).

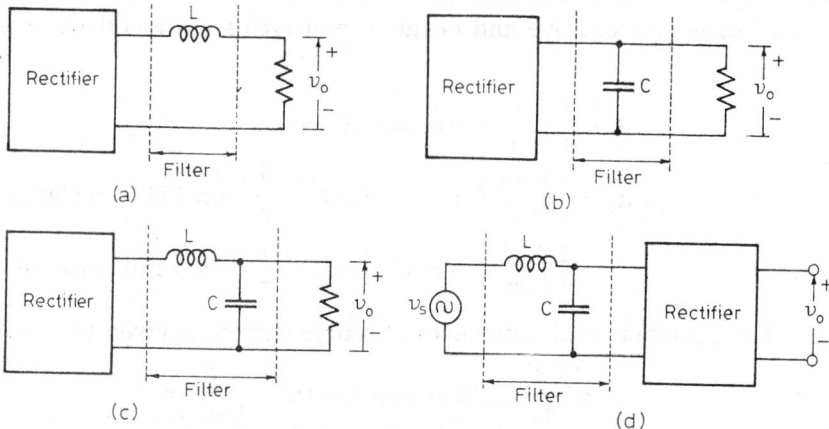

Fig. 3.45. (*a*), (*b*) and (*c*) *dc* filters, (*d*) ac filter.

The non-sinusoidal output current in rectifier circuit causes the supply line current to contain harmonics. For reducing these harmonics in the supply current, *ac filters* are used at the output terminals of rectifier circuits. Fig. 3.45 (*d*) shows an *ac* filter of LC type.

An inductor *L* in series with load *R*, Fig. 3.45 (*a*), reduces the *ac* component, or *ac* ripples, considerably. It is because *L* in series with *R* offers high impedance to *ac* component but very low resistance to *dc*. Thus *ac* component gets attenuated considerably. A capacitor *C* across load *R*, Fig. 3.45 (*b*), offers direct short circuit to *ac* component, these are therefore not allowed to reach the load. However, *dc* gets stored in the form of energy in *C* and this allows the maintenance of almost constant *dc* output voltage across the load.

In this article, a simple design of *L*, *C* and *LC* type *dc* filters is presented.

3.10.1. Capacitor Filter (*C*-Filter)

Fill-wave, or two-pulse, rectifier is more often used than a half-wave, or one-pulse, rectifier. In the present discussion, therefore, single-phase full-wave diode rectifier is only examined. Its ripple frequency is $2f$, where f is the supply frequency. A capacitor C directly connected across the load, as shown in Fig. 3.46, serves to smoothen out the *dc* output wave.

Fig. 3.47 shows the steady state waveforms pertaining to Fig. 3.46. Fig. 3.48 (*a*) gives the circuit model of Fig. 3.46. Source voltage $v_s = V_m \sin \omega t$ is sketched in Fig. 3.47 (*a*). Load voltage v_o is shown in Fig. 3.47 (*b*). In this figure, from $\omega t = 0$ to $\omega t = \theta$, source voltage v_s is less

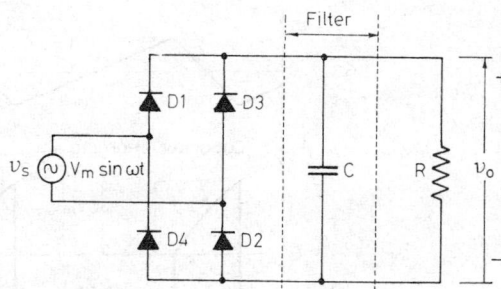

Fig. 3.46. Single-phase full-wave diode rectifier with capacitor filter.

than capacitor voltage $v_c = v_o$; therefore diodes D1, D2 are reverse biased and cannot conduct. During this interval, *i.e.* from $\omega t = 0°$ to $\omega t = \theta$, capacitor discharges through load resistance R. At $\omega t = \theta$, $v_o = v_c = V_2$ as shown in Fig. 3.47 (*b*). Soon after $\omega t = \theta$, source voltage v_s exceeds $v_o (= v_c)$, diodes D1, D2 get forward biased and begin to conduct. As a result, source voltage charges capacitor from V_2 to V_m at $\omega t = \pi/2$, Fig. 3.47 (*b*) and Fig. 3.48 (*b*). Soon after $\omega t = \pi/2$, source voltage begins to decrease faster than the capacitor voltage; it is because capacitor discharges gradually through R. Therefore, after $\omega t = \pi/2$, diodes D1, D2 are reverse biased and capacitor discharges through R as shown in Fig. 3.48 (*c*). The capacitor voltage falls exponentially, Fig. 3.47 (*b*). In the next half cycle, $v_c = v_o = V_2$ at $\omega t = (\pi + \theta)$. Just after $\omega t = (\pi + \theta)$, $v_s > v_c$, diodes D3, D4 get forward biased and begin to conduct. The capacitor voltage rises from V_2 to V_m at $\omega t = 3\pi/2$, Fig. 3.47 (*b*). It is seen from this figure that voltage drop from maximum to minimum is $V_m - V_2$, or peak to peak ripple voltage, $V_{rp.p} = V_m - V_2$.

In Fig. 3.47 (*c*) is drawn the profile of ripple voltage with the help of Fig. 3.47 (*b*). A horizontal line at a height $\frac{1}{2}(V_m + V_2)$, from reference line ωt in Fig. 3.47 (*b*) is now taken as the reference line in Fig. 3.47 (*c*) for plotting voltage profile v_r. As stated before, peak to peak ripple voltage is $V_{rpp} = V_m - V_2$ and peak ripple voltage $V_{rp} = \frac{1}{2}(V_m - V_2)$ as shown in Fig. 3.47 (*c*). Ripple voltage is seen to be almost triangular in shape.

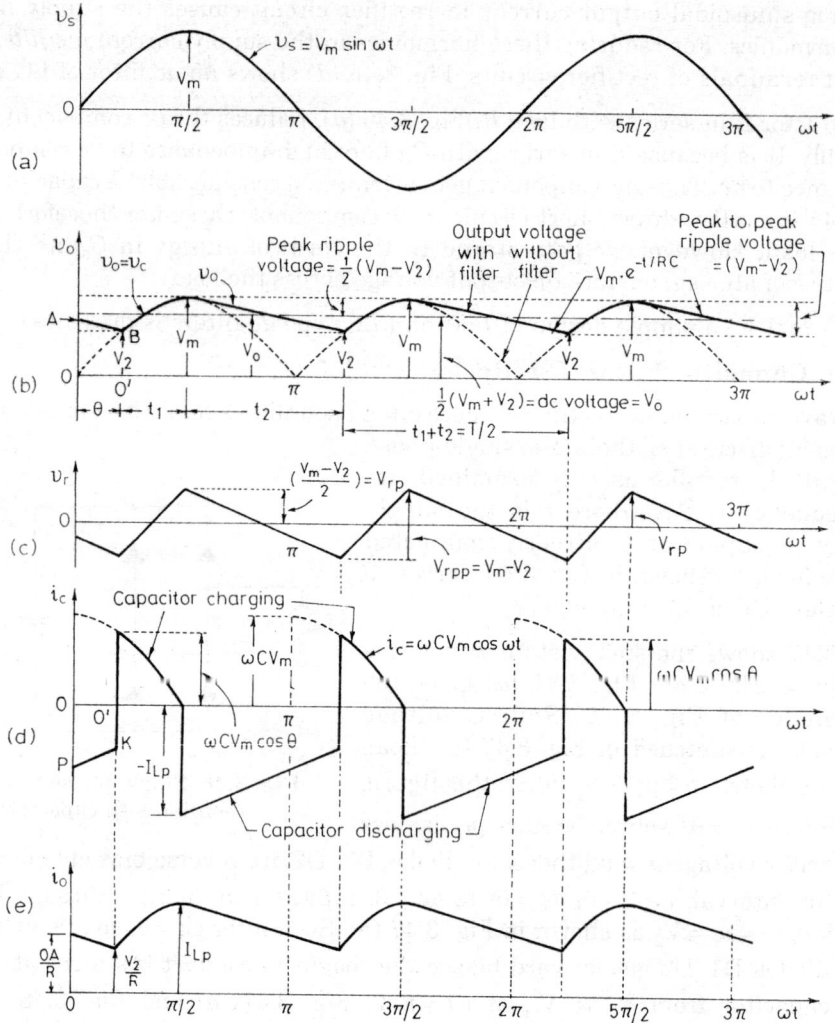

Fig. 3.47. Waveforms for (a) source voltage (b) load voltage with and without filter (c) ripple voltage (d) capacitor current and (e) load current for the circuit of Fig. 3.46.

Charging of Capacitor. From $\omega t = \theta$ to $\pi/2$, capacitor charges from V_2 to V_m. The equivalent circuit for capacitor charging of Fig. 3.48 (b) gives the charging current i_c as under :

$$i_c = C \frac{dv_s}{dt} = C \frac{d}{dt} (V_m \sin \omega t) = \omega C V_m \cos \omega t \qquad \qquad ...(3.85)$$

The charging current i_c at $\omega t = \pi/2$ is $\omega C V_m \cos 90° = 0$, but $v_c = V_m$. Therefore, energy stored in C at $\omega t = \pi/2$ is $\frac{1}{2} C V_m^2$.

Discharging of capacitor. KVL for the circuit model of Fig. 3.48 (c) for capacitor discharging gives

$$\frac{1}{C} \int i \, dt + Ri = 0$$

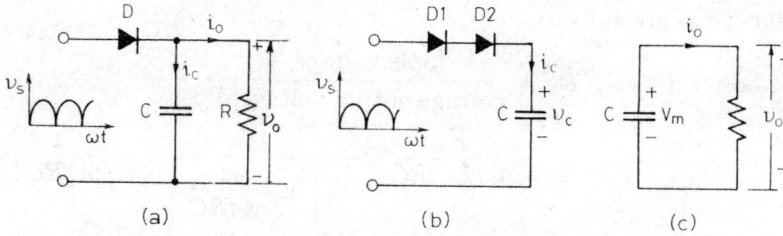

Fig. 3.48. Single-phase full-bridge diode rectifier (a) circuit model (b) charging and (c) discharging.

In this equation, time origin is taken at $\omega t = \pi/2$. Laplace transform of this equation is

$$\frac{1}{C}\left[\frac{I(s)}{s} - \frac{C.V_m}{s}\right] + R \cdot I(s) = 0$$

or

$$I(s) = \frac{V_m}{R}\frac{1}{s + \dfrac{1}{RC}}$$

$\therefore$

$$i(t) = \frac{V_m}{R} e^{-t/RC} = \frac{V_m}{R} e^{-t/\tau} \qquad \text{where } \tau = RC$$

Load voltage, $\qquad v_o = R \cdot i(t) = V_m e^{-t/\tau} = V_m e^{-t/RC}$

Peak to peak value of ripple voltage, $V_{rpp} = V_m - V_2 = V_m - [v_o \text{ at } t = t_2] = V_m - V_m e^{-t_2/\tau}$

It is known that $e^{-x} = 1 - x$.

$\therefore$

$$V_{rpp} = V_m - V_m\left(1 - \frac{t_2}{RC}\right) = \frac{V_m t_2}{RC}$$

Charging time t_1 is usually small, it can therefore be neglected. As a result, $t_2 = \dfrac{T}{2}$. But $T = \dfrac{1}{f}$, therefore $t_2 = \dfrac{1}{2f}$ and peak to peak ripple voltage, $V_{rpp} = \dfrac{V_m}{2\,fRC}$

Peak value of ripple voltage, $V_{rp} = \dfrac{V_{rpp}}{2} = \dfrac{V_m}{4\,fRC}$

Variation of ripple voltage, v_r, shown in Fig. 3.47 (c), is not a sine wave, therefore rms value of V_r of ripple voltage can be approximately found from the relation

$$V_r = \frac{V_{rp}}{\sqrt{2}} = \frac{V_m}{4\sqrt{2}.fRC} \qquad \qquad ...(3.86)$$

It is seen from the waveform of output voltage v_o, Fig. 3.47 (b), that variation of ripple voltage v_r shown in Fig. 3.47 (c) is almost triangular and average value V_o of output voltage is usually taken as

V_o = maximum value of V_m-peak value of triangular ripple voltage

$$= V_m - V_{rp} = V_m - \frac{V_m}{4\,fRC} = V_m\left[1 - \frac{1}{4\,fRC}\right] \qquad ...(3.87)$$

Minimum value of load voltage, $V_2 = V_m - V_{rpp} = V_m\left[1 - \dfrac{1}{2\,fRC}\right]$.

Ripple factor $\overline{RF}$ is given by

$$\overline{RF} = \frac{\text{ripple voltage, } V_r}{\text{average output voltage, } V_o}$$

$$= \frac{V_m}{4\sqrt{2} \cdot fRC} \times \frac{1}{V_m\left[1 - \frac{1}{4\,fRC}\right]} = \frac{1}{\sqrt{2}\,[4\,fRC - 1]} \qquad \text{...(3.88)}$$

Its simplification gives

$$C = \frac{1}{4\,fR}\left[1 + \frac{1}{\sqrt{2} \cdot RF}\right] \qquad \text{...(3.89)}$$

Waveform of capacitor current is as under :

At $\omega t = 0$, $i_c = OP = \dfrac{v_c}{R} = \dfrac{OA}{R}$

At $\omega t = \theta$, $i_c = O'K = \dfrac{O'B}{R} = \dfrac{V_2}{R}$; both OP and $O'K$ are shown negative because capacitor discharges from $\omega t = 0$ to $\omega t = \theta$.

During charging time t_1, i_c ($= \omega\, CV_m \cos \omega t$) follows the cosine wave with peak value $\omega CV_m \cos \theta$ at $\omega t = \theta$ and $i_c = 0$ at $\omega t = \pi/2$.

Current i_c is positive as the capacitor is getting charged. Soon after the maximum value of V_m at $\omega t = \pi/2$, capacitor begins to discharge and $i_c = -I_{Lp}$ where $I_{Lp} = \dfrac{V_m}{R}$. At $\omega t = (\pi + \theta)$, v_c decays to V_2 and $i_c = O'K$ as before.

Waveform of load current i_o in Fig. 3.47 (e) is identical with the waveform of v_o in Fig. 3.47 (b). At $\omega t = 0$, $i_o = \dfrac{OA}{R}$, at $\omega t = \theta$, $i_o = \dfrac{V_2}{R}$, at $\omega t = \pi/2$, $i_o = \dfrac{V_m}{R}$ and so on.

Eq. (3.86) shows that if C is increased, ripple voltage gets reduced. But high value of C increases the amplitude of charging current as per Eq. (3.85). A high charging current would entail higher current rating of diodes. This leads to increased cost of the rectifier-filter circuit. Thus, a compromise between the value of C and the magnitude of ripple voltage must be made.

Example 3.21. *A single-phase diode B-2 rectifier is fed from 230 V, 50 Hz source and is connected to a load of $R = 400\ \Omega$.*

   (a) *Design a capacitor-filter so that the ripple factor of the output voltage is less than 5%.*

   (b) *With the value of C obtained in part (a), determine the average value of output voltage.*

   (c) *Determine the average value of output voltage without C- filter.*

Solution. (a) From Eq. (3.39), the value of capacitor C to limit the ripple factor $\overline{RF}$ to 5% is

$$C = \frac{1}{4\,fR}\left[1 + \frac{1}{\sqrt{2} \cdot RF}\right] = \frac{1}{4 \times 50 \times 400}\left[1 + \frac{1}{\sqrt{2} \times 0.05}\right] = 189.3\ \mu F.$$

(b) From Eq. (3.87), average value V_o of the output voltage, with filter C, is

$$V_o = V_m \left[1 - \frac{1}{4fRC} \right] = 230\sqrt{2} \left[1 - \frac{10^6}{4 \times 50 \times 400 \times 189.3} \right] = 303.745 \text{ V}$$

(c) Average value of output voltage without C-filter is

$$V_o = \frac{2V_m}{\pi} = \frac{2\sqrt{2} \times 230}{\pi} = 207.04 \text{ V}$$

It is seen from this example that use of C filter has reduced the $\overline{RF}$ from 48.2% to 5% and at the same time, average output voltage V_o has increased from 207.04 V to 303.754V.

3.10.2. Inductor Filter (L-filter)

An inductor filter connected in series with the resistive load serves to provide the requisite filtering. In a resistive load, current waveform is identical with voltage waveform. An inductance in series with R load does not allow sudden changes in the load current. As a consequence, load current profiles becomes noticeably smooth. This has the effect of reducing the current ripple factor.

Fig. 3.49 (a) shows an inductor filter L connected in series with R-load in a 1-phase two-pulse diode rectifier. Fig. 3.49 (b) shows the rectified voltage v_o and the load current i_o. It is seen that inductor has a smoothing effect on the load current profile.

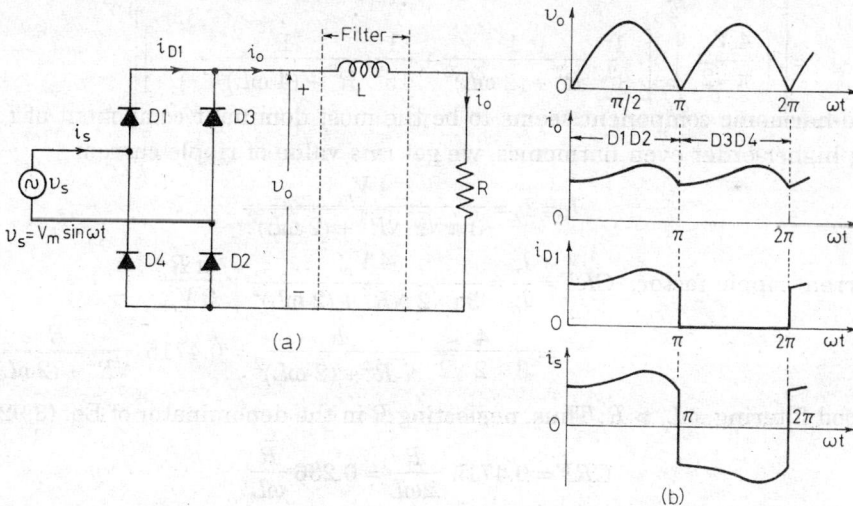

Fig. 3.49. Single-phase diode bridge rectifier (a) circuit diagram and (b) waveforms for load voltage, load current, diode D1 current and source current.

From $\omega t = 0$ to $\omega t = \pi$; diodes D1, D2 conduct and from $\omega t = \pi$ to 2π; diodes D3, D4 conduct. The load current is continuous.

The output voltage v_o of the 2-pulse rectifier can be analysed into Fourier series as (see appendix A),

$$v_o = \frac{2V_m}{\pi} - \frac{4V_m}{3\pi}\cos 2\omega t - \frac{4V_m}{15 \cdot \pi}\cos 4\omega t - \frac{4V_m}{35\pi}\cos 6\omega t \qquad ...(3.90)$$

where V_m = maximum value of source voltage.

Average load voltage, $V_o = \dfrac{2V_m}{\pi}$

Average load current, $I_o = \dfrac{V_o}{R} = \dfrac{2\,V_m}{\pi R}$

Load impedance for nth harmonic, $Z_n = \sqrt{R^2 + (n\,\omega L)^2}$

Magnitude of second harmonic load current, $I_2 = \dfrac{4\,V_m}{3\pi.\,\sqrt{2}\,\sqrt{R^2 + (2\,\omega L)^2}}$

Similarly, fourth harmonic current, $I_4 = \dfrac{4\,V_m}{15\,\pi.\,\sqrt{2}\,\sqrt{R^2 + (4\,\omega L)^2}}$

Load current i_o can, therefore, be written as

$$i_o = \frac{2V_m}{\pi R} - \frac{4V_m}{3\pi\sqrt{R^2 + (2\omega L)^2}}\cos(2\omega t - \theta_2) - \frac{4\,V_m}{15\pi\sqrt{R^2 + (4\,\omega L)^2}}\cos(4\,\omega t - \theta_4)$$

where first, second and third terms of above current expression are dc, second-harmonic and fourth-harmonic components of i_o. Rms value of harmonic-current components, or ripple current, I_r is

$$I_r = \left[\left(\frac{4\,V_m}{3\pi\sqrt{2}}\right)^2 \frac{1}{R^2 + (2\,\omega L)^2} + \left(\frac{4\,V_m}{15\,\pi\sqrt{2}}\right)^2 \frac{1}{R^2 + (4\,\omega L)^2} + \dots\right]^{1/2}$$

$$= \left[\left(\frac{4\,V_m}{\pi\sqrt{2}}\right)^2 \left[\frac{1}{3^2}.\frac{1}{R^2 + (2\,\omega L)^2} + \frac{1}{15^2}.\frac{1}{R^2 + (4\,\omega L)^2} + \dots\right]\right]^{1/2} \qquad \dots(3.91)$$

Second-harmonic component seems to be the most dominant component of I_r. Therefore, neglecting higher-order even harmonics, we get *rms* value of ripple current,

$$I_r = I_2 = \frac{4\,V_m}{3\,\pi\,\sqrt{2}\,\sqrt{R^2 + (2\,\omega L)^2}} \qquad \dots(3.91\,a)$$

$\therefore$ Current-ripple factor, $CRF = \dfrac{I_r}{I_o} = \dfrac{4\,V_m}{3\pi\sqrt{2}\,\sqrt{R^2 + (2\,\omega L)^2}} \times \dfrac{\pi\,R}{2\,V_m}$

$$= \frac{4}{3 \times 2\sqrt{2}}.\frac{R}{\sqrt{R^2 + (2\,\omega L)^2}} = 0.4715.\frac{R}{\sqrt{R^2 + (2\omega L)^2}} \qquad \dots(3.92)$$

For good filtering, $\omega L \gg R$. Thus, neglecting R in the denominator of Eq. (3.92), we get

$$CRF = 0.4715.\frac{R}{2\omega L} = 0.236\,\frac{R}{\omega L}$$

For 50 Hz supply, $CRF = 0.236\,\dfrac{R}{(2\pi \times 50)\,L} = 7.51 \times 10^{-4}\,\dfrac{R}{L}$

It is seen from **above that**
   (*i*) reduced current ripple requires large value of L,
   (*ii*) as the load is reduced (or R increased), ripple current increases,
   (*iii*) inductive filter is preferred where load resistance R is consistently low or load current is invariably high,
   (*iv*) inductor in the load circuit introduces time delay of the load current with respect to the load voltage,
   (*v*) as the current profile becomes more smooth, transformer utilization factor is improved.

Comparison between C and L filters. It is worthwhile at this stage to compare a C-filter with an L-filter.

(a) If load resistance is low, ripple factor for L-filter is low and high for C-filter.

(b) In both C-filter and L-filter, time constant should be large for better waveform, i.e. for low ripple factor, τ should be high.

(c) For C-filter, if R is increased, $\tau\,(=RC)$ increases, and therefore, ripple factor gets reduced.

(d) For L-filter, if R is lowered, $\tau\,(=L/R)$ increases, therefore ripple factor becomes low. This shows that C-filter is suitable for loads having low current (high load resistance) consistently. L-filter is suited for loads requiring high load current (low R) consistently.

(e) A high value of C reduces ripple factor but increases the charging current and the diode-current rating.

(f) A high value of L reduces the ripple factor, but a delay is introduced in the response.

Inductor filter is bulky, weighty, expensive and causes extra ohmic loss as compared to C-filter. Besides, L-filter is noisy in nature.

Example 3.22. *A single-phase two-pulse diode rectifier has input supply of 230 V, 50 Hz and the load resistance $R = 300\ \Omega$. Calculate the value of inductance to be connected in series with R so as to limit the current ripple factor to 5%. Find the value of L in case $R = 30\ \Omega$. Determine also the value of CRF without L.*

Solution. From Eq. (3.92), current ripple factor (CRF) with L is

$$CRF = 0.4715 . \frac{R}{\sqrt{R^2 + (2\omega L)^2}}$$

For $R = 300\ \Omega$,
$$0.05 = 0.4715 \frac{300}{[300^2 + (2 \times 2\pi \times 50 \times L)^2]^{1/2}}$$

$\therefore$
$$L = 4.4755\ \text{H}$$

For $R = 30\ \Omega$,
$$0.05 = 0.4715 \frac{30}{\sqrt{30^2 + (2 \times 2\pi \times 50 \times L)^2}}$$

$\therefore$
$$L = 0.4477\ \text{H}$$

CRF without filter L can be obtained by putting $L = 0$ in Eq. (3.92). This gives CRF = 0.4715 without L.

3.10.3. L-C Filter

An L-C filter consists of inductor L in series with the load and capacitor C across the load. This filter possesses the advantages of both L-filter and C-filter. In addition, ripple factor in L-C filter has lower value than that obtained by either L-filter or C-filter for the same values of L and C.

Fig. 3.50 (a) shows the use of L-C filter for reducing the ripple from the output voltage of a single-phase full-wave diode rectifier. Its equivalent circuit is given in Fig. 3.50 (b). The inductor L blocks the dominant harmonics. Capacitor C provides an easy path to the nth harmonic ripple currents.

In order that capacitor yields an easy path for harmonics, load impedance R must be much greater than nth harmonic capacitive reactance; i.e. $R \gg \dfrac{1}{n\omega C}$. It has been found in practice that capacitor provides effective filtering if

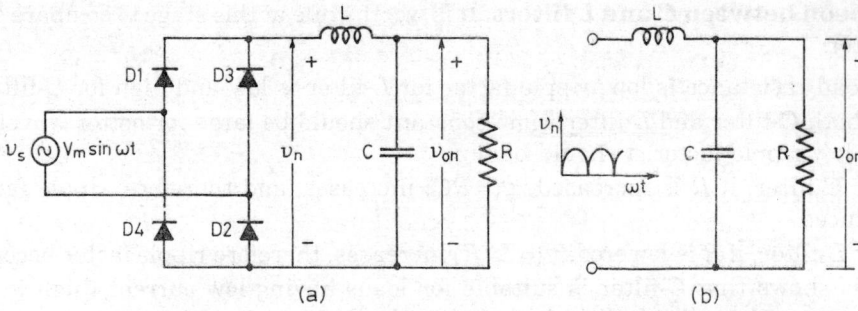

Fig. 3.50. Single-phase diode bridge rectifier with $L - C$ filter
(a) circuit diagram and (b) its equivalent circuit.

$$R = \frac{10}{n\,\omega C} \qquad \qquad ...(3.93\,a)$$

In case load consists of R and L_L in series, then Eq. (3.93 a) becomes

$$\sqrt{R^2 + (n\omega L_L)^2} = \frac{10}{n\,\omega C} \qquad \qquad ...(3.93\,b)$$

Under the condition of Eq. (3.93), effect of load R, or R and L_L, can be ignored during further analysis. Therefore, nth harmonic current in Fig. 3.50 is

$$I_n = \frac{V_n}{n\omega L - \dfrac{1}{n\omega C}}$$

where $V_n = rms$ value of nth harmonic of rectifier output voltage.

Thus, nth harmonic component of load voltage V_{on} across filter C in Fig. 3.50 (b) is

$$V_{on} = \left[\frac{-(1/n\omega C)}{n\omega L - \dfrac{1}{n\omega C}}\right] V_n = \left[\frac{-1}{(n\omega)^2 LC - 1}\right].V_n \qquad ...(3.94)$$

Total ripple voltage due to all harmonics is

$$V_r = \left[\sum_{n\,=\,2,4,6,}^{\infty} .V_{on}^2\right]^{1/2} \qquad ...(3.95)$$

The Fourier series analysis of output voltage of 1-phase full-wave rectifier is given in Eq. (3.90). The average value of output voltage $V_o = 2V_m/\pi$. It is seen from Eq. (3.90) that second harmonic is the most dominant component. Therefore, other harmonic components can be neglected.

From Eq. (3.90), the rms value V_2 of second harmonic voltage is

$$V_2 = \frac{4\,V_m}{3\pi.\sqrt{2}}$$

From Eq. (3.94) for $n = 2$, ripple voltage is

$$V_{02} = V_r = \left[\frac{-1}{(2\omega)^2 LC - 1}\right]^{1/2} \qquad ...(3.96)$$

The value of filter capacitor C can be obtained from Eq. (3.93) as

$$C = \frac{10}{2\omega R} \qquad \qquad ...(3.97a)$$

or

$$C = \frac{10}{2\omega \sqrt{R^2 + (2\omega L_L)^2}} \qquad \qquad ...(3.97b)$$

The VRF is defined as

$$VRF = \frac{V_r}{V_o} = \frac{V_2}{V_o} \cdot \frac{1}{(2\omega)^2 LC - 1} = \left(\frac{4 V_m}{3 \pi \cdot \sqrt{2}} \right) \cdot \left(\frac{\pi}{2V_m} \right) \times \frac{1}{(2\omega)^2 LC - 1}$$

$$= \frac{\sqrt{2}}{3} \left[\frac{1}{(2\omega)^2 LC - 1} \right] \qquad \qquad ...(3.98)$$

Once filter C is obtained from Eq. (3.97), the value of filter inductor L can be calculated from Eq. (3.98) for a specified value of VRF.

Example 3.23. *A single-phase two-pulse diode rectifier has input supply of 230 V, 50 Hz and a load resistance R = 50 Ω and load inductance L_L = 10 mH. An L-C filter is to be used on the output side so as to reduce the output voltage ripple to 10%. Design the LC filter.*

Solution. From Eq. (3.97 b), the value of filter capacitor C is

$$C = \frac{10}{2 \times 2\pi 50 \sqrt{50^2 + (200 \pi \times 10 \times 10^{-3})^2}} = 315.83 \, \mu F$$

From Eq. (3.98),

$$VRF = 0.1 = \frac{\sqrt{2}}{3} \cdot \left[\frac{1}{(200 \pi)^2 \cdot L \times 315.83 \times 10^{-6} - 1} \right]$$

or

$$(200\pi)^2 \times 315.83 \times L = \frac{\sqrt{2}}{3} \times \frac{1}{0.1} + 1$$

or

$$L = 0.045822 \, H \text{ or } 45.822 \, mH.$$

PROBLEMS

3.1. Capacitor in the circuit of Fig. 3.2 (a) is initially charged with (a) V_0 volts and (b) – V_0 volts. For both these parts, determine the expressions for current in the circuit and voltage across capacitor. Sketch the waveforms for current as well as capacitor voltage.

What is the final value of voltage across capacitor in each case ?

$$\left[\textbf{Ans.} \quad (a) \; \frac{V_s - V_0}{R} e^{-t/RC}, \; V_0 + (V_s - V_0)(1 - e^{-t/RC}), \; V_s \right.$$

$$\left. (b) \; \frac{V_s + V_0}{R} e^{-t/RC}, \; - V_0 + (V_s + V_0)(1 - e^{-t/RC}), \; V_s \right]$$

3.2. A diode is connected in series with LC circuit. If this circuit is switched on to dc source of voltage V_s at $t = 0$, derive expressions for current through and voltage across capacitor. The capacitor is initially charged to a voltage of – V_0. Sketch waveforms for i, v_C, v_L and v_D.

In case this circuit has $V_s = 230$ V, $V_0 = 50$ V, $L = 0.2$ mH and $C = 10$ μF, determine the diode conduction time, diode peak current and final steady state voltage across the capacitor and diode.

$$\left[\textbf{Ans} \; i_{(t)} = (V_s + V_0)\sqrt{\frac{C}{L}} \sin \omega_0 t, \; v_C(t) = (V_s + V_0)(1 - \cos \omega_0 t) \right.$$

$$\left. - V_0, \; 140.496 \, \mu s, \; 62.61 \, A \; 510 \, V, \; - 280 \, V \right]$$

3.3. (a) For the circuit shown in Fig. 3.4 (a), the circuit is initially relaxed. If switch S is closed at $t = 0$, sketch the variations of i, v_L, v_C and v_D as a function of time. Derive the expressions describing these functions.

   (b) For part (a), $V_s = 220$ V, $L = 4$ mH, $C = 5$ μF. Find the diode conduction time and peak diode current. Determine also v_C, v_L and v_D after diode stops conducting.

                                        [**Ans.** (b) 0.444 ms, 7.778 A, 440 V, 0, – 220 V]

3.4. For the circuit shown in Fig. 3.5 (a) ; $V_0 = 230$ V, $R = 25$ Ω and $C = 10$ μF. If switch S is closed at $t = 0$, determine expressions for the current in the circuit and voltage across capacitor C. Find the peak value of diode current and energy lost in the circuit.

   Derive the expressions used.   [**Ans.** $9.2\,e^{-4000t}$, $-230\,e^{-4000t}$, 9.2 A, 0.2645 watt–sec]

3.5. In the circuit shown in Fig. 3.51, switch S is open and a current of 20 A is flowing through the freewheeling diode, R and L. If switch S is closed at $t = 0$, determine the expression for the current through the switch.   [**Ans** $i_{(t)} = 22 - 2\,e^{-1000t}$]

3.6. (a) Describe how the energy trapped in an inductor can be recovered and returned to the source.

   (b) A 230 V, 1 kW heater, fed through single-phase half-wave diode rectifier, has rated voltage at its terminals. Find the ac input voltage. Find also PIV of diode and peak-diode current.

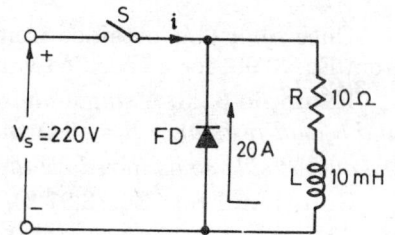

Fig. 3.51. Pertaining to Prob. 3.5.

                                                   [**Ans.** (b) 325.32 V, 460 V, 8.696 A]

3.7. (a) In the circuit shown in Fig. 3.52, a PMMC ammeter is placed in series with diode and a PMMC voltmeter across the diode. Take PMMC instruments ideal. Find the readings on these instruments. Derive the expressions used for obtaining these readings.

   (b) If PMMC ammeter is replaced by MI ammeter, find its reading.

                                        [**Ans.** (a) 10.352 A, 0 V (b) 12.6812 A]

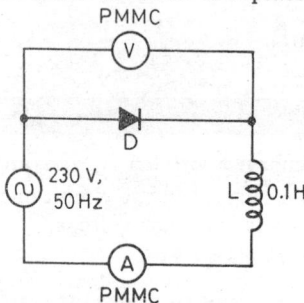

Fig. 3.52. Pertaining to Prob. 3.7.

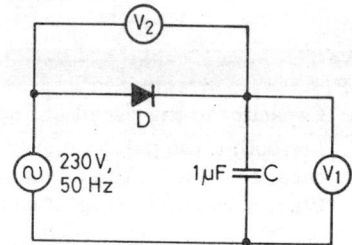

Fig. 3.53. Pertaining to Prob. 3.8.

3.8. (a) In the circuit of Fig. 3.53, ideal PMMC voltmeters are placed, one across capacitor and another across diode as shown. Find the voltmeter readings. Obtain the expressions used for determining these readings.

   (b) In case PMMC voltmeter 2 is replaced by MI voltmeter, find its reading.

                                        [**Ans.** (a) 325.22 V, 325.22 V (b) 398.394 V]

3.9. A battery is charged by a single-phase half-wave diode rectifier. The supply is 30 V, 50 Hz and the battery emf is constant at 6 V. Find the resistance to be inserted in series with the battery to limit the charging current to 4 A. Take a voltage drop of 1 V across diode. Derive the expression used.

   Draw waveform of voltage across diode and find its PIV. $\left[\textbf{Hint.}\ \ \theta_1 = \sin^{-1}\dfrac{7}{\sqrt{2} \times 30}\ \text{etc.}\right]$

$$\left[\textbf{Ans.}\ \ \frac{1}{2\pi R}\,[2V_m \cos\theta_1 - (E + 1)\,(\pi - 2\,\theta_1)],\ \ 2.5467\ \Omega,\ \ 49.42\ \text{V}\right]$$

3.10. (a) A single-phase half-wave uncontrolled rectifier is connected to RL load. Derive an expression for the load current in terms of V_m, Z, ω etc.

(b) For part (a), $V_s = 230$ V at 50 Hz, $R = 10\ \Omega$, $L = 5$ mH, extinction angle = 210°. Find average values of output voltage and output current. **[Ans.** (b) 193.172 V, 19.3172 A**]**

3.11. (a) A single-phase half-wave diode rectifier feeds power to (i) RL load and (ii) RL load with freewheeling diode across it. Describe the working of this rectifier for both these parts with relevant waveforms and bring out the differences if any. Hence point out the effect of using a freewheeling diode.

(b) For part (a), $V_s = 230$ V at 50 Hz, $R = 20\ \Omega$, $L = 1$ H. Find the average values of the output voltage and output current with and without the use of a flywheeling diode.

[Ans. (b) With freewheeling diode : $V_0 = 103.52$ V and $I_0 = 5.176$ A

Without freewheeling diode : Extinction angle β not known, so V_0, I_0 cannot be calculated]

3.12. (a) For the circuit shown in Fig. 3.54, the output current i_0 is considered constant at I_0 because of large L. Sketch the waveforms of v_s, i_0, v_0, i_D, i_{fd}, and i_s.

(b) For the above circuit, find

(i) average values of output voltage and output current,

(ii) average and rms values of freewheeling diode current,

(iii) supply pf. **[Ans.** (b) (i) 103.52 V, 26.76 A (ii) 13.38 A, 18.925 A (iii) 0.6364 lag**]**

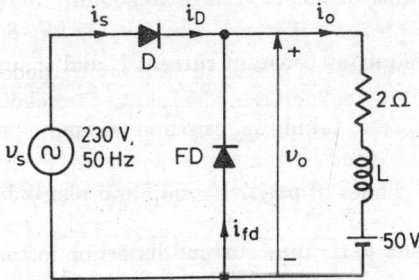

Fig. 3.54. Pertaining to Prob. 3.12.

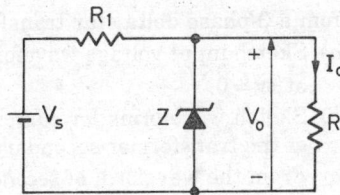

Fig. 3.55. Pertaining to Prob. 3.13.

3.13. For the circuit shown in Fig. 3.55, $V_s = 160$ V, $V_z = 40$ V and zener diode current varies from 4 to 40 mA. Find the minimum and maximum values of R_1 so as to allow voltage regulation for output current $I_0 = $ zero to its maximum value I_{0m}. Also calculate I_{0m}.

[Ans. 3k Ω, 30k Ω, 36 mA**]**

3.14. (a) Enumerate the input performance parameters of a rectifier. Discuss how the performance of a rectifier circuit is influenced in case these parameters have low, or high, value.

(b) Define input power factor, displacement factor DF and current distortion factor CDF for a rectifier system and show that input power factor = $CDF \times DF$.

3.15. (a) Define input current harmonic factor (HF) and crest factor. Express (HF) in terms of current distortion factor. If HF is more, what does it indicate in a rectifier system.

(b) Define the following terms :
Rectification ratio, ripple voltage, form factor, voltage ripple factor, current ripple factor and transformer utilization factor.

3.16. For a single-phase half-wave diode rectifier feeding a resistive load R, find the values of rectifier efficiency, form factor, voltage ripple factor, transformer utilization factor and crest factor.

3.17. A single-phase half-wave diode rectifier is designed to supply *dc* output voltage of 230 V to a load of $R = 10\ \Omega$. Calculate the ratings of diode and transformer for this circuit arrangement.

[Ans. $I_{DAV} = 23$ A, $I_{Dr} = 36.13$ A, $PIV = 722.6$ V, Trans. rating = 18.462 kVA**]**

3.18. A single-phase full-wave mid-point diode rectifier feeds resistive load R. For this circuit, determine rectifier efficiency, form factor, voltage ripple factor, transformer utilization factor and crest factor.

How does this rectifier circuit differ from single-phase full-wave bridge rectifier ?

3.19. (a) Why are three-phase rectifiers preferred over single-phase rectifiers ?
(b) For a 3-phase half-wave diode rectifier feeding load R, obtain the following :
Average output voltage, rms output voltage, VRF, FF, TUF and PIV

3.20. Describe the evolution of three-phase six-pulse diode rectifier from 3-phase three-pulse diode rectifiers with appropriate circuits and waveforms. Hence, derive an expression for the average output voltage of 3-phase six-pulse diode rectifier.

3.21. Describe a 3-phase M-6 diode rectifier with a circuit diagram and relevant waveforms for resistive load R.

Hence, derive expressions for average and rms values of output voltage and obtain therefrom VRF, FF, rectifier efficiency and TUF.

3.22. A 3-phase mid-point 6-pulse diode rectifier feeds a load of 10 Ω at a *dc* voltage of 400 V. Find the ratings of diodes and the three-phase transformer.

[**Ans.** $I_{DAV} = 6.667$ A, $I_{Dr} = 16.337$ A, PIV $= 837.76$ V, Trans. rating $= 29.038$ kVA]

3.23. Describe a 3-phase full-wave diode-bridge rectifier with a circuit diagram and relevant waveforms for load R.

Hence, derive expressions for average and rms values of output voltage and obtain there from VRF, FF, rectifier efficiency and TUF.

3.24. A 3-phase full-wave diode rectifier feeds a load requiring constant current I_o and is supplied from a 3-phase delta-star transformer.
(a) Sketch input voltage waveforms for v_{ab}, v_{ac}, v_{bc} etc., taking v_{ab} zero and becoming positive at $\omega t = 0$.
(b) Sketch waveforms for currents for the three diodes of positive group and phase current of the transformer secondary.
(c) From the waveform of secondary phase current, determine current distortion factor CDF and THD.

[**Hint.** Here $a_1 = -\dfrac{\sqrt{3}}{\pi} I_o$, $b_1 = \dfrac{3}{\pi} I_o$, etc.]

[**Ans.** (c) 0.955, 0.3106]

3.25. A 3-phase full-wave diode rectifier delivers power to an inductive load which takes ripple-free current of 120 A. The source voltage is 3-phase, 400 V, 50 V, 50 Hz. Determine the ratings of diodes, power delivered to load and the rms value of source current.

[**Ans.** $I_{DAV} = 40$ A, $I_{Dr} = 69.284$ A, PIV $= 565.6$ V, 64813.2 W, 97.98 A]

3.26. Describe a three-phase 12-pulse diode rectifier with circuit diagram and appropriate waveforms. Hence derive expressions for average and rms values of output voltage. From these, obtain VRF and FF.

3.27. (a) What are the advantages of 3-phase bridge rectifier over 3-phase M-6 rectifier.
(b) For a 3-phase p-pulse diode rectifier, prove the following:

Average output voltage, $V_o = V_{mp}.\dfrac{p}{\pi} \sin \dfrac{\pi}{p}$

and *rms* output voltage, $V_{or} = V_{mp}\left[\dfrac{p}{2\pi}\left(\dfrac{\pi}{p} + \dfrac{1}{2} \sin \dfrac{2\pi}{p} \right) \right]^{1/2}$

where V_{mp} = maximum value of per-phase supply voltage.

3.28. (a) What are the functions of filters in rectifier circuits ? Distinguish between *dc* and *ac* filters.
(b) Explain how the inductance L and capacitance C play their role in reducing the harmonic contents in rectifier circuits.

3.29. A single-phase full-wave diode rectifier feeds R with a capacitor C directly connected across load. Describe the operation of C as a filter with relevant voltage and current waveforms. Show that peak ripple voltage is $\frac{1}{2}(V_m - V_2)$.

3.30. A single-phase two-pulse diode rectifier feeds R and C in parallel. Explain charging and discharging of capacitor C and derive expressions for ripple factor and the value of filter capacitor C.

3.31. A single-phase diode bridge rectifier is fed at 230 V, 50 Hz. The load is $R = 200\ \Omega$ shunted by a capacitance of 300 µF. Neglecting all losses, determine the average value of load voltage, VRF, maximum and minimum value of load current, peak capacitor current and average load current. **[Ans.** 298.12 V, 0.0589, $I_{max} = 1.6261$ A, $I_{min} = 1.3551$ A, 30.651 A 1.4911 A**]**

3.32. A single-phase two-pulse diode rectifier feeds load R with an inductor L in series with it. Describe the working of L as filter with relevant voltage and current waveforms. Derive expression for current ripple factor and show that for 50 Hz supply, $CRF = 7.51 \times 10^{-4} \cdot \frac{R}{L}$.

3.33. A single-phase full-wave diode rectifier has mean output voltage of 200 V and the load resistance is 400 Ω. Determine the inductance required to limit the amplitude of second-harmonic current in the load to 0.06 A. **[Ans.** $L = 3.48$ H**]**

3.34. A single-phase full-wave diode rectifier with L-C filter feeds load R. Describe its working and derive expressions from which the parameters of L-C filter can be obtained.

3.35. (a) Compare C-filter with L-filter. In what type of applications are the two types usually preferred ?

(b) A single-phase two-pulse diode rectifier is fed from 230 V. The load is $R = 200\ \Omega$. Design an LC filter so as to get voltage ripple factor of 5.89%. Find the *rms* value of ripple voltage.
[Ans. $C = 79.58$ µF, $L = 0.28654$ H, 12.195 V**]**

FOUR

Thyristors

....

In this Chapter

- Terminal Characteristics of Thyristors
- Thyristor Turn-on Methods
- Switching Characteristics of Thyristors
- Thyristor Gate Characteristics
- Two-Transistor Model of a Thyristor
- Thyristor Ratings
- Thyristor Protection
- Improvement of Thyristor Characteristics
- Heating, Cooling and Mounting of Thyristors
- Series and Parallel Operation of Thyristors
- Other Members of the Thyristor Family
- Gate Turn off (G.T.O.) Thyristor
- Static Induction Thyristor
- Firing Circuits for Thyristors
- Pulse Transformer in Firing Circuits
- Triac Firing Circuit

....

As stated before, Bell Laboratories were the first to fabricate a silicon-based semiconductor device called thyristor. Its first prototype was introduced by GEC (USA) in 1957. This company did a great deal of pioneering work about the utility of thyristors in industrial applications. Later on, many other devices having characteristics similar to that of a thyristor were developed. These semiconductor devices, with their characteristics identical with that of a thyristor, are triac, diac, silicon-controlled switch, programmable unijunction transistor (PUT), GTO, RCT etc. This whole family of semiconductor devices is given the name thyristor. Thus the term thyristor denotes a family of semiconductor devices used for power control in dc and ac systems. One oldest member of this thyristor family, called silicon-controlled rectifier (SCR), is the most widely used device. At present, the use of SCR is so vast that over the years, the word thyristor has become synonymous with SCR. It appears that the term thyristor is now becoming more common than the actual term SCR. In this book, the term SCR and thyristor have been used at random for the same device SCR. Other members of thyristor family are also discussed in this chapter.

A thyristor has characteristics similar to a thyratron tube. But from the construction view point, a thyristor (a *pnpn* device) belongs to transistor (*pnp or npn* device) family. The name 'thyristor', is derived by a combination of the capital letters from THYRatron and transISTOR. This means that thyristor is a solid state device like a transistor and has characteristics similar to that of a thyratron tube. The present-day reader may not be familiar with thyratron tube as this is not being taught these days. Actually, the name 'thyristor' came into existence after a formal decision taken at a conference held by IEC (International Electrotechnical Commission) in 1963. Prior to that, it was called silicon controlled rectifier, or SCR. It appears that commission must have evolved the name 'thyristor' as discussed above.

At this conference, the definition of thyristor was decided as under :

(i) It constitutes three or more p-n junctions.

(ii) It has two stable states, an ON-state and an OFF-state and can change its state from one to another.

As per this definition, thyristor now includes a large variety of semiconductor devices having similar basic characteristics.

The object of this chapter is to discuss the thyristor characteristics and other related topics useful for their industrial applications.

4.1. TERMINAL CHARACTERISTICS OF THYRISTORS

Thyristor is a four layer, three-junction, p-n-p-n semiconductor switching device. It has three terminals ; anode, cathode and gate. Fig. 4.1 (a) gives constructional details of a typical thyristor. Basically, a thyristor consists of four layers of alternate p-type and n-type silicon semiconductors forming three junctions J_1, J_2 and J_3 as shown in Fig. 4.1 (a). The threaded portion is for the purpose of tightening the thyristor to the frame or heat sink with the help of a nut. Gate terminal is usually kept near the cathode terminal, Fig. 4.1 (a). Schematic diagram and circuit symbol for a thyristor are shown respectively in Figs. 4.1 (b) and (c). The terminal connected to outer p region is called anode (A), the terminal connected to outer n region is called cathode and that connected to inner p region is called the gate (G). For large current applications, thyristors need better cooling ; this is achieved to a great extent by mounting them onto heat sinks. SCR rating has improved considerably since its introduction in 1957. Now SCRs of voltage rating 10 kV and an rms current rating of 3000 A with corresponding power-handling capacity of 30 MW are available. Such a high power thyristor can be switched on by a low voltage supply of about 1 A and 10 W and this gives us an idea of the immense power amplification capability (= 3×10^6) of this device. As SCRs are solid state devices, they are compact, possess high reliability and have low loss. Because of these useful features, SCR is almost universally employed these days for all high power-controlled devices.

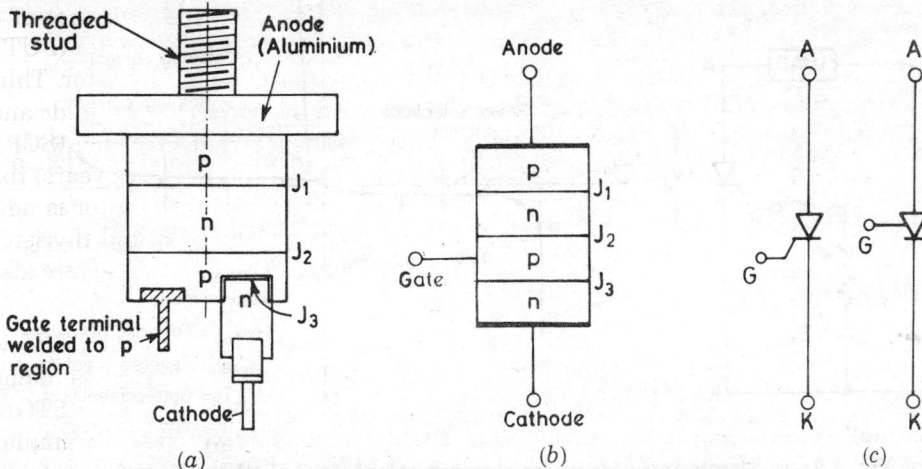

Fig. 4.1. (a) Constructional details (b) Schematic diagram and (c) circuit symbol of a thyristor.

An SCR is so called because *silicon* is used for its construction and its operation as a *rectifier* (very low resistance in the forward conduction and very high resistance in the reverse direction) can be *controlled*. Like the diode, an SCR is an unidirectional device that blocks the current flow from cathode to anode. Unlike the diode, a thyristor also blocks the current flow from anode to cathode until it is triggered into conduction by a proper gate signal between gate and cathode terminals.

For engineering applications of thyristors, their terminal characteristics must be known. In this article, their static *I-V* characteristics, dynamic characteristics during turn-on and turn-off processes and their gate characteristics are discussed.

4.1.1. Static I-V Characteristics of a Thyristor

An elementary circuit diagram for obtaining static *I-V* characteristics of a thyristor is shown in Fig. 4.2 (*a*). The anode and cathode are connected to main source through the load. The gate and cathode are fed from a source E_s which provides positive gate current from gate to cathode.

Fig. 4.2 (*b*) shows static *I-V* characteristics of a thyristor. Here V_a is the anode voltage across thyristor terminals A, K and I_a is the anode current. Typical SCR *I-V* characteristic shown in Fig. 4.2 (*b*) reveals that a thyristor has three basic modes of operation ; namely, reverse blocking mode, forward blocking (off-state) mode and forward conduction (on-state) mode. These three modes of operation are now discussed below :

Reverse Blocking Mode. When cathode is made positive with respect to anode with switch S open, Fig. 4.2 (*a*), thyristor is reverse biased as shown in Fig. 4.3 (*a*). Junctions J_1, J_3 are seen to be reverse biased whereas junction J_2 is forward biased. The device behaves as if two diodes are connected in series with reverse voltage applied across them. A small leakage current of the order of a few milliamperes (or a few microamperes depending upon the SCR rating) flows. This is reverse blocking mode, called the off-state, of the thyristor. In Fig. 4.2 (*b*), reverse blocking mode is shown by OP. If the reverse voltage is increased, then at a critical

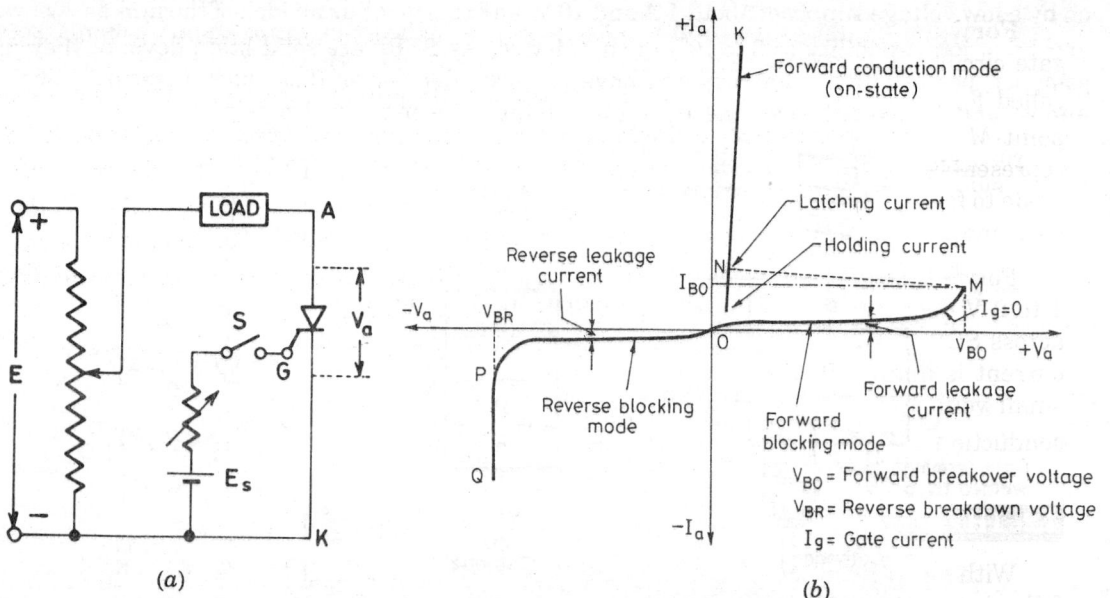

Fig. 4.2. (*a*) Elementary circuit for obtaining thyristor I-V characteristics
(*b*) Static I-V characteristics of a thyristor.

breakdown level, called reverse breakdown voltage V_{BR}, an avalanche occurs at J_1 and J_3 and the reverse current increases rapidly. A large current associated with V_{BR} gives rise to more losses in the SCR. This may lead to thyristor damage as the junction temperature may exceed its permissible temperature rise. It should, therefore, be ensured that maximum working reverse voltage across a thyristor does not exceed V_{BR}. In Fig. 4.2 (b), reverse avalanche region is shown by PQ. When reverse voltage applied across a thyristor is less than V_{BR}, the device offers a high impedance in the reverse direction. The SCR in the reverse blocking mode may therefore be treated as an open switch.

Note that I-V characteristic after avalanche breakdown during reverse blocking mode is applicable only when load resistance is zero, Fig. 4.2

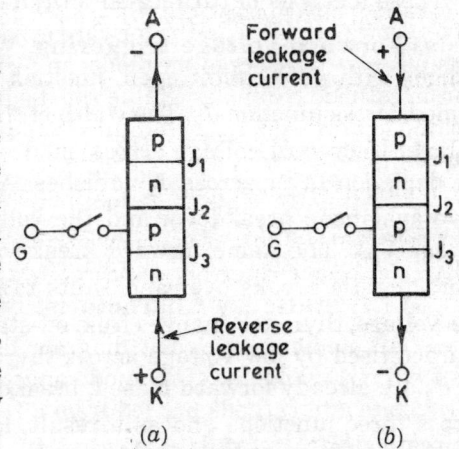

Fig. 4.3. (a) J_2 forward biased and J_1, J_3 reverse biased (b) J_2 reverse biased and J_1, J_3 forward biased.

(b). In case load resistance is present, a large anode current associated with avalanche breakdown at V_{BR} would cause substantial voltage drop across load and as a result, I-V characteristic in third quadrant would bend to the right of vertical line drawn at V_{BR}.

Forward Blocking Mode : When anode is positive with respect to the cathode, with gate circuit open, thyristor is said to be forward biased as shown in Fig. 4.3 (b). It is seen from this figure that junctions J_1, J_3 are forward biased but junction J_2 is reverse biased. In this mode, a small current, called forward leakage current, flows as shown in Figs. 4.2 (b) and 4.3 (b). In Fig. 4.2 (b), OM represents the forward blocking mode of SCR. As the forward leakage current is small, SCR offers a high impedance. Therefore, a thyristor can be treated as an open switch even in the forward blocking mode.

Forward Conduction mode. When anode to cathode forward voltage is increased with gate circuit open, reverse biased junction J_2 will have an avalanche breakdown at a voltage called forward breakover voltage V_{BO}. After this breakdown, thyristor gets turned on with point M at once shifting to N and then to a point anywhere between N and K. Here NK represents the *forward conduction mode*. A thyristor can be brought from forward blocking mode to forward conduction mode by turning it on by applying (i) a positive gate pulse between gate and cathode or (ii) a forward breakover voltage across anode and cathode.

Forward conduction mode NK shows that voltage drop across thyristor is of the order of 1 to 2 V depending upon the rating of SCR. It may also be seen from NK that voltage drop across SCR increases slightly with an increase in anode current. In conduction mode, anode current is limited by load impedance alone as voltage drop across SCR is quite small. This small voltage drop V_T across the device is due to ohmic drop in the four layers. In forward conduction mode, thyristor is treated as a closed switch.

4.2. THYRISTOR TURN-ON METHODS

With anode positive with respect to cathode, a thyristor can be turned on by any one of the following techniques : (a) Forward voltage trigging (b) gate triggering (c) dv/dt triggering (d) temperature triggering and (e) light triggering.

These methods of turning-on a thyristor are now discussed one after the other.

(a) **Forward voltage triggering.** When forward voltage is applied between anode and cathode with gate circuit open, junction J_2 is reverse biased. As a result, depletion layer is formed across junction J_2. The width of this layer decreases with an increase in anode-cathode voltage. If forward voltage across anode-cathode is gradually increased, a stage comes when the depletion layer across J_2 vanishes. At this moment, reverse biased junction J_2 is said to have avalanche breakdown and the voltage at which it occurs is called *forward breakover voltage* V_{BO}. The name forward breakover voltage is given because at this voltage V_{BO}, i-v characteristic breaks over and shifts to its on-state position with breakover current I_{BO}. At this voltage, thyristor changes from off-state (high voltage with low leakage current) to on-state characterised by low voltage across thyristor with large forward current. As other junctions J_1, J_3 are already forward biased, breakdown of junction J_2 allows free movement of carriers across three junctions and as a result, large forward anode-current flows. As stated before, this forward current is limited by the load impedance. In practice, the transition from off-state to on-state obtained by exceeding V_{BO} is never employed as it may destroy the device.

The magnitudes of forward breakover and reverse breakdown voltages are nearly the same and both are temperature dependent. In practice, it is found that V_{BR} is slightly more than V_{BO}. Therefore, forward breakover voltage is taken as the final voltage rating of the device during the design of SCR applications.

After the avalanche breakdown, junction J_2 loses its reverse blocking capability. Therefore, if the anode voltage is reduced below V_{BO}, SCR will continue conduction of the current. The SCR can now be turned off only by reducing the anode current below a certain value called holding current (defined later).

(b) **Gate Triggering.** Turning on of thyristors by gate triggering is simple, reliable and efficient, it is therefore the most usual method of firing the forward biased SCRs. A thyristor with forward breakover voltage (say 800 V) higher than the normal working voltage (say 400 V) is chosen. This means that thyristor will remain in forward blocking state with normal working voltage across anode and cathode and with gate open. However, when turn-on of a thyristor is required, a positive gate voltage between gate and cathode is applied. With gate current thus established, charges are injected into the inner p layer and voltage at which forward breakover occurs is reduced. The forward voltage at which the device switches to on-state depends upon the magnitude of gate current. Higher the gate current, lower is the forward breakover voltage.

When positive gate current is applied, gate p layer is flooded with electrons from the cathode. This is because cathode n layer is heavily doped as compared to gate p layer. As the thyristor is forward biased, some of these electrons reach junction J_2. As a result, width of depletion layer near junction J_2 is reduced. This causes the junction J_2 to breakdown at an applied voltage lower than the forward breakover voltage V_{BO}. If magnitude of gate current is increased, more electrons would reach junction J_2, as a consequence thyristor would get turned on at a much lower forward applied voltage.

Fig. 4.4 (a) shows that for gate current $I_g = 0$, forward breakover voltage is V_{BO}. For gate current I_{g1}, forward breakover, or turn-on voltage is V_1 which is less than V_{BO}. For $I_{g2} > I_{g1}$, forward breakdover voltage is further reduced to $V_2 < V_1$. For $I_{g3} > I_{g2}$, the forward breakover voltage is $V_3 < V_2$, Fig. 4.4 (a). The effect of gate current on the forward breakover voltage of a

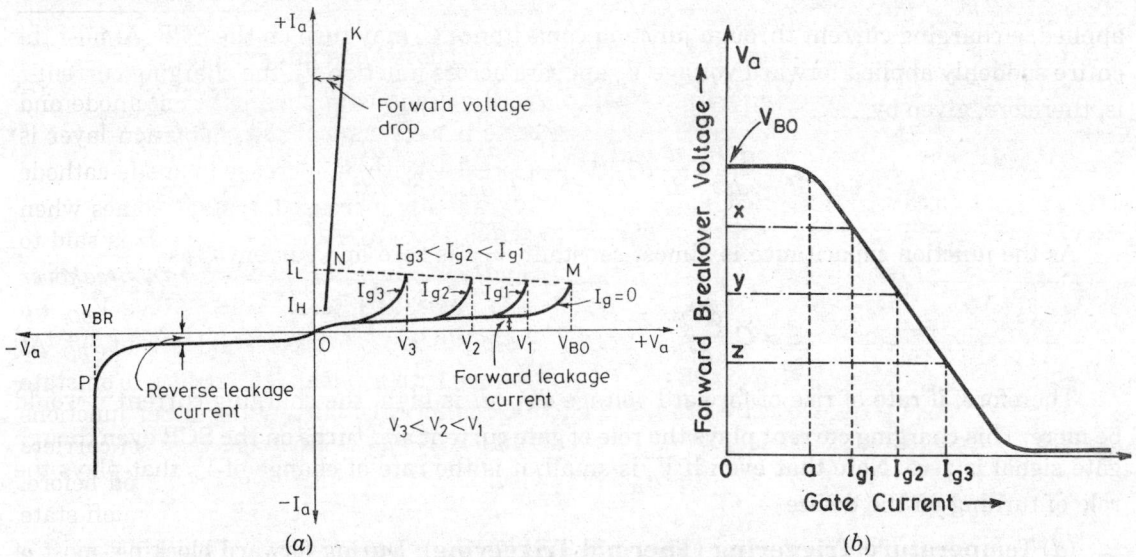

Fig. 4.4. Effect of gate current on forward breakover voltage.

thyristor can also be illustrated by means of a curve as shown in Fig. 4.4 (b). For $I_g < oa$, forward breakover voltage remains almost constant at V_{BO}. For gate currents I_{g1}, I_{g2} and I_{g3}, the magnitudes of forward breakover voltages are $ox = V_1$, $oy = V_2$ and $oz = V_3$ respectively as shown in Fig. 4.4 (a) and (b). In Fig. 4.4 (a), the curve marked $I_g = 0$ is actually for gate current less than oa. In practice, the magnitude of gate current is more than the minimum gate current required to turn on the SCR. Typical gate current magnitudes are of the order of 20 to 200 mA.

Once the SCR is conducting a forward current, reverse biased junction J_2 no longer exists. As such, no gate current is required for the device to remain in on-state. Therefore, if the gate current is removed, the conduction of current from anode to cathode remains unaffected. However, if gate current is reduced to zero before the rising anode current attains a value, called the latching current, the thyristor will turn-off again. The gate pulse width should therefore be judiciously chosen to ensure that anode current rises above the latching current. Thus *latching current* may be defined as the minimum value of anode current which it must attain during turn-on process to maintain conduction when gate signal is removed.

Once the thyristor is conducting, gate loses control. The thyristor can be turned-off (or the thyristor can be returned to forward blocking state) only if the forward current falls below a low-level current called the holding current. Thus *holding current* may be defined as the minimum value of anode current below which it must fall for turning-off the thyristor. The latching current is higher than the holding current. Note that latching current is associated with turn-on process and holding current with turn-off process. It is usual to take latching current as two to three times the holding current [1]. In industrial applications, holding current (typically 10 mA) is almost taken as zero.

(c) $\dfrac{dv}{dt}$ **Triggering.** With forward voltage across the anode and cathode of a thyristor, the two outer junction J_1, J_3 are forward biased, but inner junction J_2 is reverse biased. This reverse biased junction J_2, Fig. 4.3 (b), has the characteristics of a capacitor due to charges existing across the junction. In other words, space-charges exist in the depletion region near junction J_2 and therefore junction J_2 behaves like a capacitance. If forward voltage is suddenly applied, a charging current through junction capacitance C_j may turn on the SCR. Almost the entire suddenly applied forward voltage V_a appears across junction J_2, the charging current i_c is, therefore, given by

$$i_c = \frac{dQ}{dt} = \frac{d}{dt}(C_j \cdot V_a) = C_j \frac{dV_a}{dt} + V_a \cdot \frac{dC_j}{dt} \qquad ...(4.1\ a)$$

As the junction capacitance is almost constant, $\dfrac{dC_j}{dt}$ is zero and current i_c as

$$i_c = C_j \frac{dV_a}{dt} \qquad ...(4.1\ b)$$

Therefore, if rate of rise of forward voltage dV_a/dt is high, the charging current i_c would be more. This charging current plays the role of gate current and turns on the SCR even though gate signal is zero. Note that even if V_a is small, it is the rate of change of V_a that plays the role of turning-on the device.

(d) **Temperature Triggering (Thermal Triggering).** During forward blocking, most of the applied voltage appears across reverse biased junction J_2. This voltage across, J_2, associated with leakage current, would raise the temperature of this junction. With increase in temperature, width of depletion layer decreases. This further leads to more leakage current and therefore, more junction temperature. With the cumulative process, at some high temperature (within the safe limits), depletion layer of reverse biased junction vanishes and the device gets turned on.

(e) **Light Triggering.** For light-triggered SCRs, a recess (or niche) is made in the inner p-layer as shown in Fig. 4.5 (a). When this recess is irradiated, free charge carriers (pairs of holes and electrons) are generated just like when gate signal is applied between gate and cathode. The pulse of light of appropriate wavelength is guided by optical fibres for irradiation. If the intensity of this light thrown on the recess exceeds a certain value, forward-biased SCR is turned on. Such a thyristor is known as light-activated SCR (LASCR).

LASCR may be triggered with a light source or with a gate signal. Sometimes a combination of both light source and gate signal is used to trigger an SCR. For this, the gate is biased with voltage or current slightly less than that required to turn it on, now a beam of light directed at the inner p-layer junction turns on the SCR. The light intensity required to turn-on the SCR depends upon the voltage bias given to the gate. Higher the voltage (or current) bias, lower the light intensity required.

Light-triggered thyristors have now been used in high-voltage direct current (HVDC) transmission systems. In these several SCRs are connected in series-parallel combination and their light-triggering has the advantage of electrical isolation between power and control circuits.

Example 4.1. *Discuss what would happen if gate is made positive with respect to cathode during the reverse blocking of a thyristor.*

Solution. Before answering this question, it is worthwhile to know a little more about the internal details of a thyristor.

Fig. 4.5 (b) shows cross-section of a conventional centre-gate thyristor. In this figure, approximate doping densities in number per cubic centimeter are also indicated for all the four layers. For example, for p_1 layer, the doping density is 10^{19} per cm^3.

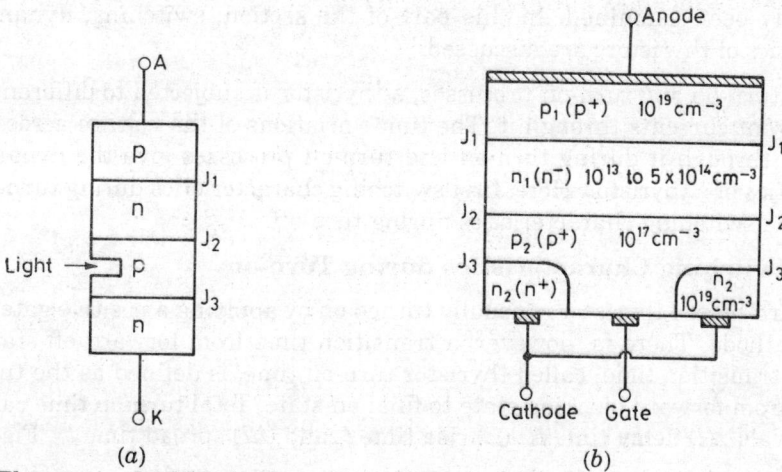

Fig. 4.5. (*a*) Elementary LASCR (*b*) Structural details of conventional centre-gate thyristor.

For semiconductor devices, it should be kept in mind that

(*i*) junction with lightly doped layers (at least on one side of the junction) requires large breakdown voltage,

(*ii*) junction with highly doped layers on both sides requires low breakdown voltage.

When thyristor is in forward blocking state, junctions J_1, J_3 are forward biased whereas junction J_2 is reverse biased. As layer n_1 is lightly doped around junction J_2, depletion region of junction J_2 extends mainly into n_1 layer. Therefore, n_1 layer is made to have larger width to withstand the high voltage during forward blocking state.

For reverse voltage on the device ; junctions J_1, J_3 are reverse biased and J_2 is forward biased. As layers p_2, n_2 across junction J_3 are heavily doped, J_3 has low breakdown voltage. Layer n_1 being lightly doped as compared to layer p_1, junction J_1 has large breakdown voltage. As a consequence, during the reverse blocking of a thyristor, junction J_1 supports most of the reverse voltage. Even during blocking, the depletion region extends into the n_1 layer. This shows that width of layer n_1 absorbs most of the voltage during forward blocking mode and also during the reverse blocking mode of a thyristor.

If positive gate voltage is applied between gate and cathode during the reverse blocking of a thyristor, blocking property of junction J_3 disappears as J_3 has low breakdown voltage. As a result, reverse voltage appears across junction J_1. Positive charge carriers are now injected into the n_1 layer of reverse biased junction J_1. This causes an increase in the reverse leakage current. The flow of large leakage current associated with high reverse voltage results in increased power loss across junction J_1 and heat thus generated may raise the junction temperature above the allowable maximum and this may destroy the SCR. Such an happening can be avoided if no positive gate voltage is applied between gate and cathode during the reverse blocking of SCR. Some manufacturers do specify the maximum positive voltage (usually less than 0.25 V) that can exist between gate and cathode during the reverse blocking of a thyristor.

4.3. SWITCHING CHARACTERISTICS OF THYRISTORS

Static and switching characteristics of thyristors are always taken into consideration for economical and reliable design of converter equipment. Static characteristics of a thyristor have already been examined. In this part of the section; switching, dynamic or transient, characteristics of thyristors are discussed.

During turn-on and turn-off processes, a thyristor is subjected to different voltages across it and different currents through it. The time variations of the voltage across a thyristor and the current through it during turn-on and turn-off processes give the dynamic or switching characteristics of a thyristor. Here, first switching characteristics during turn-on are described and then the switching characteristics during turn-off.

4.3.1. Switching Characteristics during Turn-on

A forward-biased thyristor is usually turned on by applying a positive gate voltage between gate and cathode. There is, however, a transition time from forward off-state to forward on state. This transition time, called thyristor turn-on time, is defined as the time during which it changes from forward blocking state to final on-state. Total turn-on time can be divided into three intervals ; (i) delay time t_d, (ii) rise time t_r and (iii) spread time t_p, Fig. 4.8.

(i) *Delay time* t_d : The delay time t_d is measured from the instant at which gate current reaches 0.9 I_g to the instant at which anode current reaches $0.1 I_a$. Here I_g and I_a are respectively the final values of gate and anode currents. The delay time may also be defined as the time during which anode voltage falls from V_a to $0.9 V_a$ where V_a = initial value of anode voltage. Another way of defining delay time is the time during which anode current rises from forward leakage current to 0.1 I_a where I_a = final value of anode current. With the thyristor initially in the forward blocking state, the anode voltage is OA and anode current is small leakage current as shown in Fig. 4.8. Initiation of turn-on process is indicated by a rise in anode current from small forward leakage current and a fall in anode-cathode voltage from forward blocking voltage OA. As gate current begins to flow from gate to cathode with the application of gate signal, the gate current has non-uniform distribution of current density over the cathode surface due to the p layer. Its value is much higher near the gate but decreases rapidly as the distance from the gate increases, see Fig. 4.6 (a). This shows that during delay time t_d, anode current flows in a narrow region near the gate where gate current density is the highest.

The delay time can be decreased by applying high gate current and more forward voltage between anode and cathode. The delay time is fraction of a microsecond.

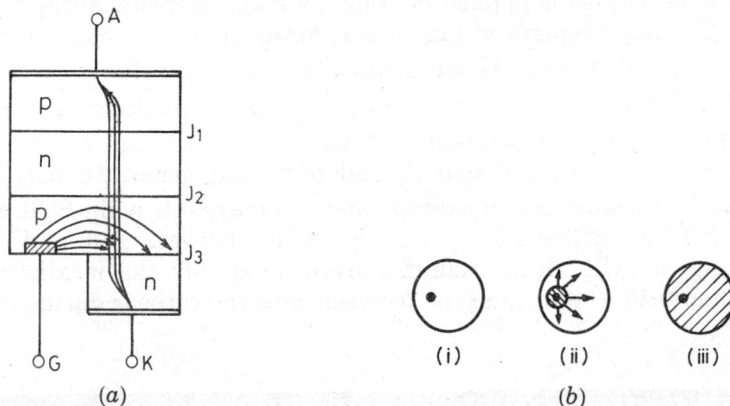

Fig. 4.6. (a) Distribution of gate and anode currents during delay time
(b) Conducting area of cathode (i) during t_d (ii) after t_r (iii) after t_p.

(ii) *Rise time* t_r : The rise time t_r is the time taken by the anode current to rise from 0.1 I_a to 0.9 I_a. The rise time is also defined as the time required for the forward blocking off-state voltage to fall from 0.9 to 0.1 of its initial value OA. The rise time is inversely proportional to the magnitude of gate current and its build up rate. Thus t_r can be reduced if high and steep current pulses are applied to the gate. However, the main factor determining t_r is the nature of anode circuit. For example, for series RL circuit, the rate of rise of anode current is slow, therefore, t_r is more. For RC series circuit, di/dt is high, t_r is therefore, less.

From the beginning of rise time t_r, anode current starts spreading from the narrow conducting region near the gate. The anode current spreads at a rate of about 0.1 mm per microsecond [2]. As the rise time is small, the anode current is not able to spread over the entire cross-section of cathode. Fig. 4.6 (b) illustrates how anode current expands over cathode surface area during turn-on process of a thyristor. Here the thyristor is taken to have single gate electrode away from the centre of p-layer. It is seen that anode current conducts over a small conducting channel even after t_r –this conducting channel area is however, greater than that during t_d.

Fig. 4.7. Typical waveform for gate current.

During rise time, turn-on losses in the thyristor are the highest due to high anode voltage (V_a) and large anode current (I_a) occurring together in the thyristor as shown in Fig. 4.8. As these losses occur only over a small conducting region, local hot spots may be formed and the device may be damaged.

(iii) *Spread time* t_p : The spread time is the time taken by the anode current to rise from 0.9 I_a to I_a. It is also defined as the time for the forward blocking voltage to fall from 0.1 of its initial value to the on-state voltage drop (1 to 1.5 V). During this time, conduction spreads over the entire cross-section of the cathode of SCR. The spreading interval depends on the area of cathode and on gate structure of the SCR. After the spread time, anode current attains steady state value and the voltage drop across SCR is equal to the on-state voltage drop of the order of 1 to 1.5 V, Fig. 4.8.

Total turn-on time of an SCR is equal to the sum of delay time, rise time and spread time. Thyristor manufacturers usually specify the rise time which is typically of the order of 1 to 4 μ sec. Total turn-on time depends upon the anode circuit parameters and the gate signal waveshapes.

During turn-on, SCR may be considered to be a charge controlled device. A certain amount of charge must be injected into the gate region for the thyristor conduction to begin. This charge is directly proportional to the value of gate current. Therefore, higher the magnitude of gate current, the lesser time it takes to inject this charge. The turn-on time can therefore be reduced by using higher values of gate currents. The magnitude of gate current is usually 3 to 5 times the minimum gate current required to trigger an SCR.

When gate current is several times higher than the minimum gate current required, a thyristor is said to be hard-fired or overdriven. *Hard-firing* or *overdriving* of a thyristor reduces its turn-on time and enhances its di/dt capability. A typical waveform for gate current, that is widely used, is shown in Fig. 4.7. This waveform has higher initial value of gate current with a very fast rise time. The initial high value of gate current is then reduced to a lower value where it stays for several microseconds in order to avoid unwanted turn-off of the device.

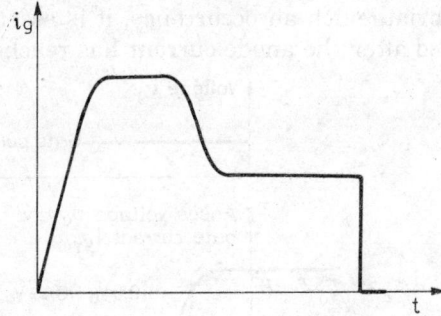

4.3.2. Switching Characteristics during Turn-off

Thyristor turn-off means that it has changed from on to off state and is capable of blocking the forward voltage. This dynamic process of the SCR from conduction state to forward blocking state is called *commutation process or turn-off process*.

Once the thyristor is on, gate loses control. The SCR can be turned off by reducing the anode current below holding current[*]. If forward voltage is applied to the SCR at the moment its anode current falls to zero, the device will not be able to block this forward voltage as the carriers (holes and electrons) in the four layers are still favourable for conduction. The device will therefore go into conduction immediately even though gate signal is not applied. In order to obviate such an occurrence, it is essential that the thyristor is reverse biased for a finite period after the anode current has reached zero.

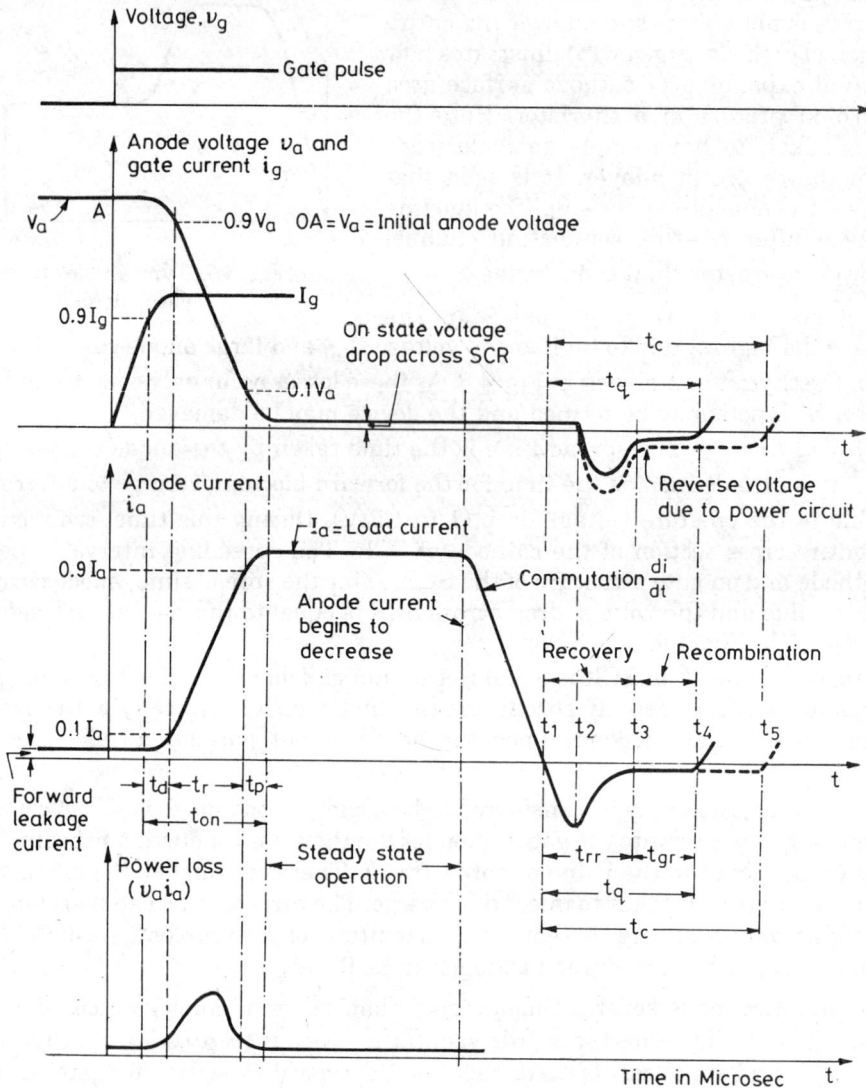

Fig. 4.8. Thyristor voltage and current waveforms during turn-on and turn-off processes.

---

[*] This can be achieved through natural commutation or forced commutation.

The *turn-off time t_q* of a thyristor is defined as the time between the instant anode current becomes zero and the instant SCR regains forward blocking capability. During time t_q, all the excess carriers from the four layers of SCR must be removed. This removal of excess carriers consists of sweeping out of holes from outer p-layer and electrons from outer n-layer. The carriers around junction J_2 can be removed only by recombination. The turn-off time is divided into two intervals ; reverse recovery time t_{rr} and the gate recovery time t_{gr}; i.e. $t_q = t_{rr} + t_{gr}$. The thyristor characteristics during turn-on and turn-off processes are shown in one Fig. 4.8 so as to gain insight into these processes.

At instant t_1, anode current becomes zero. After t_1, anode current builds up in the reverse direction with the same di/dt slope as before t_1. The reason for the reversal of anode current after t_1 is due to the presence of carriers stored in the four layers. The reverse recovery current removes excess carriers from the end junctions J_1 and J_3 between the instants t_1 and t_3. In other words, reverse recovery current flows due to the sweeping out of holes from top p-layer and electrons from bottom n-layer. At instant t_2, when about 60% of the stored charges are removed from the outer two layers, carrier density across J_1 and J_3 begins to decrease and with this reverse recovery current also starts decaying. The reverse current decay is fast in the beginning but gradual thereafter. The fast decay of recovery current causes a reverse voltage across the device due to the circuit inductance. This reverse voltage surge appears across the thyristor terminals and may therefore damage it. In practice, this is avoided by using protective RC elements across SCR. At instant t_3, when reverse recovery current has fallen to nearly zero value, end junctions J_1 and J_3 recover and SCR is able to block the reverse voltage. For a thyristor, reverse recovery phenomenon between t_1 and t_3 is similar to that of a rectifier diode.

At the end of reverse recovery period $(t_3 - t_1)$, the middle junction J_2 still has trapped charges, therefore, the thyristor is not able to block the forward voltage at t_3. The trapped charges around J_2, i.e. in the inner two layers, cannot flow to the external circuit, therefore, these trapped charges must decay only by recombination. This recombination is possible if a reverse voltage is maintained across SCR, though the magnitude of this voltage is not important. The rate of recombination of charges is independent of the external circuit parameters. The time for the recombination of charges between t_3 and t_4 is called *gate recovery time t_{gr}*. At instant t_4, junction J_2 recovers and the forward voltage can be reapplied between anode and cathode. The thyristor turn-off time t_q is in the range of 3 to 100 µsec. The turn-off time is influenced by the magnitude of forward current, di/dt at the time of commutation and junction temperature. An increase in the magnitude of these factors increases the thyristor turn-off time. If the value of forward current before commutation is high, trapped charges around junction J_2 are more. The time required for their recombination is more and therefore turn-off time is increased. But turn-off time decreases with an increase in the magnitude of reverse voltage, particularly in the range of 0 to – 50 V. This is because high reverse voltage sucks out the carriers out of the junctions J_1, J_3 and the adjacent transition regions at a faster rate. It is evident from above that turn-off time t_q is not a constant parameter of a thyristor.

The thyristor turn-off time t_q is applicable to an individual SCR. In actual practice, thyristor (or thyristors) form a part of the power circuit. The turn-off time provided to the thyristor by the practical circuit is called *circuit turn-off time t_c*. It is defined as the time between the instant anode current becomes zero and the instant reverse voltage due to practical circuit reaches zero, see Fig. 4.8. Time t_c must be greater than t_q for reliable turn-off, otherwise the device may turn-on at an undesired instant, a process called *commutation failure*.

Thyristors with slow turn-off time (50 – 100 μsec) are called *converter grade* SCRs and those with fast turn-off time (3 – 50 μsec) are called inverter-grade SCRs. Converter-grade SCRs are cheaper and are used where slow turn-off is possible as in phase-controlled rectifiers, ac voltage controllers, cycloconverters etc. Inverter-grade SCRs are costlier and are used in inverters, choppers and force-commutated converters.

4.4. THYRISTOR GATE CHARACTERISTICS

The forward gate characteristics of a thyristor are shown in Fig. 4.9 in the form of a graph between gate voltage and gate current. Here positive gate to cathode voltage V_g and positive gate to cathode current I_g represent dc values. As gate-cathode circuit of a thyristor is a *p-n* junction, gate characteristics of the device are similar to that of a diode. For a particular type of SCRs, V_g–I_g characteristic has a spread between two curves 1 and 2 as shown in Fig. 4.9. This spread, or scatter, of gate characteristics is due to inadvertent difference in the doping levels of *p* and *n* layers. The gate trigger circuitry must be suitably designed to take care of this unavoidable scatter of characteristics. In Fig. 4.9, curve 1 represents the lowest voltage values that must be applied to turn-on the SCR. Curve 2 gives the highest possible voltage values that can be safely applied to gate circuit.

Each thyristor has maximum limits as V_{gm} for gate voltage and I_{gm} for gate current. There is also rated (average) gate power dissipation P_{gav} specified for each SCR. These limits should not be exceeded in order to avoid permanent damage of junction J_3, Fig. 4.3. There are also minimum limits for V_g and I_g for reliable turn-on, these are represented by *oy* and *ox* respectively in Fig. 4.9. As stated before, if V_{gm}, I_{gm} and P_{gav} are exceeded, the thyristor can be destroyed. This shows that preferred gate drive area for an SCR is *bcdefghb* as shown in Fig. 4.9.

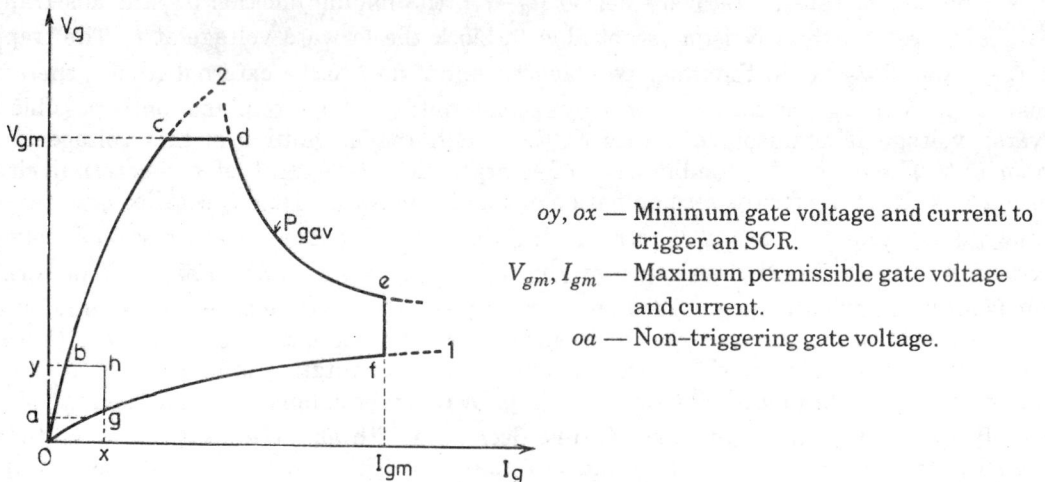

oy, *ox* — Minimum gate voltage and current to trigger an SCR.

V_{gm}, I_{gm} — Maximum permissible gate voltage and current.

oa — Non–triggering gate voltage.

Fig. 4.9. Forward gate characteristics of thyristor.

A non-triggering gate voltage is also prescribed by the manufacturers of SCRs. This is indicated by *oa* in Fig. 4.9. If firing circuit generates positive gate signal prior to the desired instant of triggering the SCR, it should be ensured that this unwanted signal is less than the non-triggering gate voltage *oa*. At the same time, all spurious or noise signals should be less than the voltage *oa*.

The design of the firing circuit can be carried out with the help of Figs. 4.10 and 4.11. In Fig. 4.10 (*a*) is shown a trigger circuit feeding power to gate-cathode circuit. For this circuit,

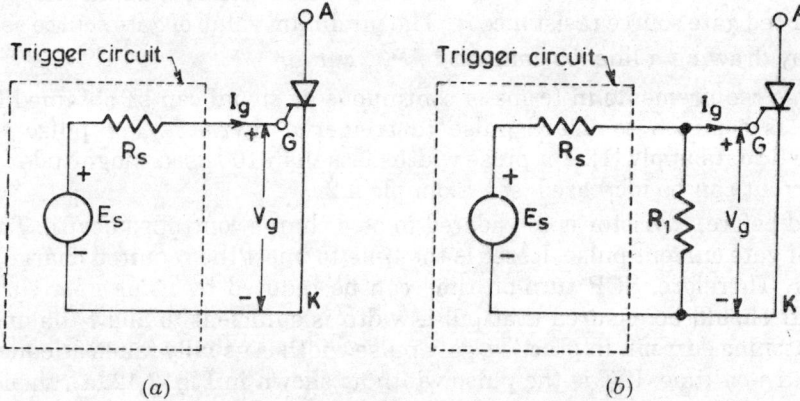

Fig. 4.10. Trigger circuit connected to gate-cathode circuit of an SCR.

$$E_s = V_g + I_g R_s \qquad \qquad ...(4.2a)$$

where E_s = gate source voltage

V_g = gate–cathode voltage

I_g = gate current

and R_s = gate–source resistance

The internal resistance R_s of trigger source should be such that current (E_s/R_s) is not harmful to the source as well as to the gate circuit when SCR is turned on. In case R_s is low, an external resistance in series with R_s must be connected.

A resistance R_1 is also connected across gate-cathode terminals, Fig. 4.10 (b), so as to provide an easy path to the flow of leakage current between SCR terminals. If I_{gmn} and V_{gmn} are the minimum gate current and gate voltage to turn-on SCR, then it is seen from Fig. 4.10 (b) that current through R_1 is V_{gmn}/R_1 and the trigger source voltage E_s is given by

$$E_s = \left(I_{gmn} + \frac{V_{gmn}}{R_1} \right) R_s + V_{gmn} \qquad ...(4.2b)$$

For low-power circuits, it is customary to obtain the operating point by utilizing the V-I characteristics of both source and the device. In view of this, for selecting the operating point for the circuit of Fig. 4.10, a load line of the gate source voltage $E_s = OA$ is drawn as AD in Fig. 4.11. Here OD = trigger circuit short circuit current = E_s/R_s. Let us consider a thyristor whose V_g-I_g characteristic is given by curve 3. Intersection of load line AD and V_g-I_g curve 3 gives the operating point S. Thus, for this SCR, gate voltage = PS and gate current = OP. In order to minimise turn-on time and jitter (unreliable turn-on), the load line and hence the operating point S, which may change from S_1 to S_2, must be as close to the P_{gav} curve as possible. At the same time, the operating point

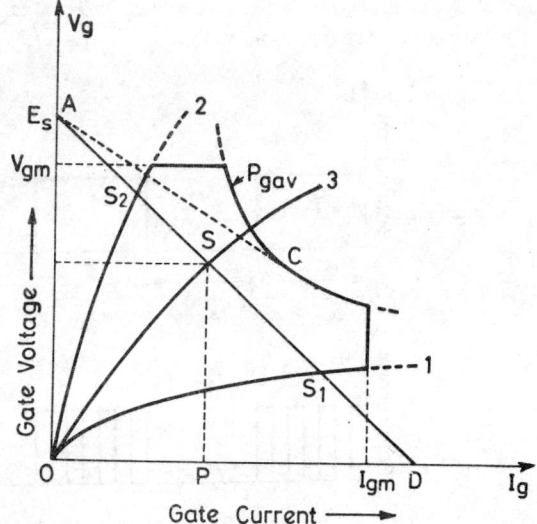

Fig. 4.11. Choice of gate circuit parameters.

S must lie within the limit curves 1 and 2. The gradient of the load line AD $(= OA/OD)$ will give the required gate source resistance R_s. The minimum value of gate source series resistance is obtained by drawing a line AC tangent to P_{gav} curve.

Gate drive requirements in terms of continuous dc signal can be obtained from Fig. 4.11. However, it is common to use a pulse to trigger a thyristor. For pulse widths beyond 100 μsec, the dc data apply [1]. For pulse widths less than 100 μsec, magnitudes of gate voltage and gate current can be increased, see Example 4.2.

As stated before, thyristor is considered to be a charge controlled device. Thus, higher the magnitude of gate current pulse, lesser is the time to inject the required charge for turning-on the thyristor. Therefore, SCR turn-on time can be reduced by using gate current of higher magnitude. It should be ensured that pulse width is sufficient to allow the anode current to exceed the latching current. In practice, gate pulse width is usually taken as equal to, or greater than, SCR turn-on time. If T is the pulse width as shown in Fig. 2.12 (a), then $T \geq t_{on}$

With pulse triggering, greater amount of gate power dissipation can be allowed ; this should, however, be less than the peak instantaneous gate power dissipation P_{gm} as specified by the manufacturers. Frequency of firing (or pulse width) for trigger pulses can be obtained by taking pulse of (i) amplitude P_{gm} (ii) pulse width T and (iii) periodicity T_1. Therefore,

$$\frac{P_{gm} T}{T_1} \geq P_{gav} \qquad \text{or} \qquad P_{gm} \cdot T \cdot f \geq P_{gav}$$

or
$$\frac{P_{gav}}{f T} \leq P_{gm} \qquad\qquad\qquad ...(4.3a)$$

where $f = \dfrac{1}{T_1}$ = frequency of firing, or pulse repetition rate, in Hz,

and T = pulse width in sec.

In the limiting case, $\dfrac{P_{gav}}{fT} = P_{gm}$ or $f = \dfrac{P_{gav}}{T \cdot P_{gm}}$

A duty cycle is defined as the ratio of pulse-on period to periodic time of pulse. In Fig. 4.12 (a), pulse-on period is T and periodic time is T_1. Therefore, duty cycle δ is given by

$$\delta = \frac{T}{T_1} = fT$$

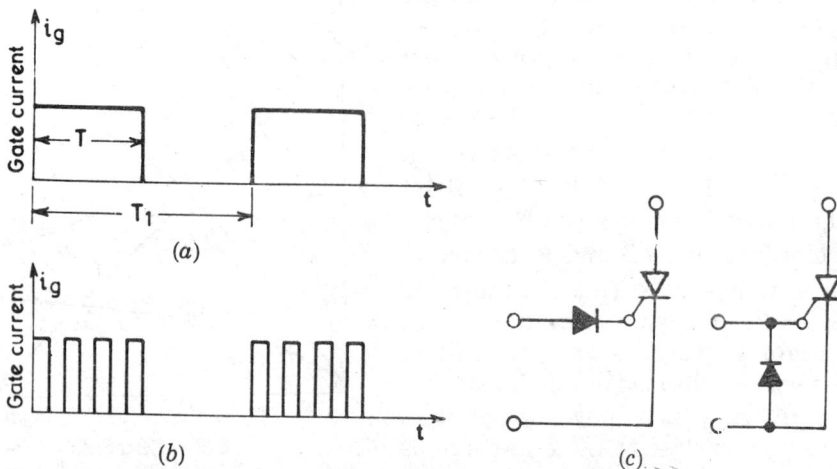

Fig. 4.12. (a) Pulse gating and (b) high-frequency carrier gating of SCRs, (c) Thyristor protection against reverse overvoltages.

From Eq. (4.3a), $\quad \dfrac{P_{gav}}{\delta} \le P_{gm}$ or $\dfrac{P_{gav}}{\delta} = P_{gm}$ $\qquad\qquad$ ...(4.3 b)

Sometimes the pulses of Fig. 4.12 (a) are modulated to generate a train of pulses as shown in Fig. 4.12 (b). This technique of firing the thyristor is called *high-frequency carrier gating*. The advantages offered by this method of firing the SCRs are lower rating, reduced dimensions and therefore an overall economical design of the pulse transformer needed for isolating the low power circuit from the main power circuit.

For an SCR, V_{gm} and I_{gm} are specified separately. If both of these are used for pulse firing, then P_{gm} may be exceeded and the thyristor would be damaged. For example, GE-C35 thyristor has $V_{gm} = 10$ V and $I_{gm} = 2$ A. If both these limits are placed on C35, the power dissipation is 20 W. But this is far excess of the specified $P_{gm} = 5$ W. It should be ensured that (pulse voltage amplitude) (pulse current amplitude) $< P_{gm}$.

There is also prescribed a peak reverse voltage (gate negative with respect to cathode) that can be applied across gate-cathode terminals. Any voltage signal, given by the trigger circuit (or by any interference), exceeding this prescribed limit of about 5 to 20 V may damage the gate circuit. For preventing the occurrence of such hazards, a diode is connected either in series with the gate circuit or across the gate-cathode terminals as shown in Fig. 4.12 (c). Diode across the gate-cathode terminals, called clamping diode, prevents the gate-cathode voltage from becoming more than about 1 V. Diode in series with gate circuit prevents the flow of negative gate source current from becoming more than small reverse leakage current.

The magnitude of gate voltage and gate current for triggering an SCR is inversely proportional to junction temperature. Thus, at very low temperatures, gate voltage and gate current must have high values in order to ensure turn-on. But P_{gm} should not be exceeded in any case.

The resistor R_1, connected across gate-cathode terminals, Fig. 4.10 (b), also serves to bypass part of the thermally-generated leakage current across junction J_2 when SCR is in the forward blocking mode ; this improves the thermal stability of SCR.

Example 4.2. (a) *The average gate power dissipation for an SCR is 0.5 W. The allowable gate voltage variation is from a minimum of 2 V to a maximum of 10 V. Taking average gate power dissipation constant, plot allowable gate voltage as a function of gate current.*

(b) *If SCR of part (a) is triggered with gate pulses of duty cycle 0.5, find the new value of average gate power dissipation.*

Solution. (a) Here $\qquad V_g I_g = 0.5$ W

For $\qquad\qquad\qquad V_g = 2$ V, $I_g = 0.5/2 = 0.25$ A

For $\qquad\qquad\qquad V_g = 10$ V, $I_g = 0.5/10$

$\qquad\qquad\qquad\qquad\qquad = 0.05$ A

For other values of gate voltage V_g in between 2 and 10 V, gate current I_g is obtained and plotted in Fig. 4.13 showing the variation of V_g as a function of I_g for constant P_{gav}.

(b) For this example, $T_1 = 2T$ in Fig. 4.12 (a) so that $\delta = 0.5$. For dc values, $V_g I_g = 0.5$ W.

For pulse firing, Fig. 4.12 (a), the average gate power dissipation can be obtained from the relation

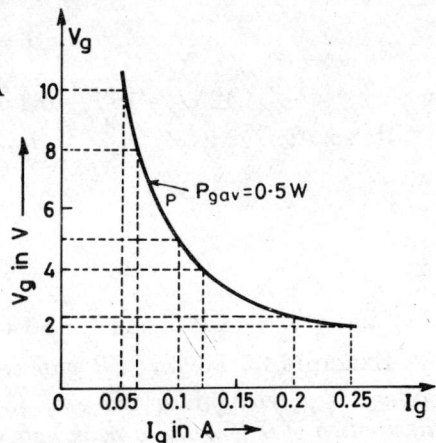

Fig. 4.13. Pertaining to Example 4.2.

$$\frac{1}{T_1} \int_0^T v_g i_g$$

where v_g, i_g are the instantaneous values of gate voltage and gate current. Therefore, for this example, average gate power dissipation is given by

$$V_g \cdot I_g \cdot \frac{T}{2T} = (0.5)\frac{1}{2} = 0.25 \text{ W}.$$

As this is less than the allowable P_{gav}, higher values of v_g, i_g can be used for the pulse firing of SCRs.

Example 4.3. *For an SCR, the gate-cathode characteristic has a straight-line slope of 130. For trigger source voltage of 15 V and allowable gate power dissipation of 0.5 watts, compute the gate-source resistance.*

Solution. Here $V_g I_g = 0.5 \text{ W}$

and $\dfrac{V_g}{I_g} = 130$

∴ $130 I_g^2 = 0.5$

This gives $I_g = [0.5/130]^{1/2} = 0.062 = 62 \text{ mA}$

∴ Gate voltage, $V_g = 130 \times 62 \times 10^{-3} = 8.06 \text{ V}$

For the gate circuit, $E_s = I_g R_s + V_g = 0.062 R_s + 8.06 = 15$

or $R_s = \dfrac{15 - 8.06}{0.062} = 111.94 \Omega.$

Example 4.4. *The trigger circuit of a thyristor has a source voltage of 15 V and the load line has a slope of – 120 V per ampere. The minimum gate current to turn-on the SCR is 25 mA. Compute*

(a) source resistance required in the gate circuit,

(b) the trigger voltage and trigger current for an average gate power dissipation of 0.4 watts.

Solution. (a) The slope of load line gives the required gate source resistance. From the load line, series resistance required in the gate circuit is 120 Ω.

(b) Here $V_g I_g = 0.4 \text{ W}$

For the gate circuit, $E_s = R_s I_g + V_g$

∴ $15 = 120 I_g + \dfrac{0.4}{I_g}$

or $120 I_g^2 - 15 I_g + 0.4 = 0$

Its solution gives $I_g = 38.56 \text{ mA or } 86.44 \text{ mA}$

∴ $V_g = \dfrac{0.4 \times 10^3}{38.56} = 10.37 \text{ V}$

or $V_g = \dfrac{0.4 \times 10^3}{86.44} = 4.627 \text{ V}.$

Choose $V_g = 4.627$ V and $I_g = 86.44$ mA for minimum gate current of 25 mA.

Example 4.5. *For an SCR, gate-cathode characteristic is given by $V_g = 1 + 10 I_g$. Gate source voltage is a rectangular pulse of 15 V with 20 μ sec duration. For an average gate power dissipation of 0.3 W and a peak gate-drive power of 5 W, compute*

(a) the resistance to be connected in series with the SCR gate,

(b) the triggering frequency and

(c) the duty cycle of the triggering pulse.

Solution. (*a*) Here $\qquad V_g = 1 + 10\, I_g.$

For pulse-triggering of SCRs,

(Peak gate voltage) (peak gate current) during pulse-on period
$$= \text{peak gate drive power, } P_{gm}.$$

As the gate pulse width is 20 μ sec (less than 100 μ sec), the dc data does not apply. Had the gate pulse width been more than 100 μsec, the relation $(1 + 10\, I_g)\, I_g = 0.3$ W will hold good. But as the dc data does not apply, we have here

$$(1 + 10\, I_g)\, I_g = 5 \text{ W}$$

or $\qquad\qquad\qquad\qquad 10\, I_g^2 + I_g - 5 = 0$

Its solution gives, $\qquad\qquad\qquad I_g = 0.659$ A.

∴ Amplitude of current pulse $\qquad = 0.659$ A

During the pulse-on period, $\qquad E_s = R_s\, I_g + V_g$

or $\qquad\qquad\qquad\qquad 15 = R_s\, I_g + 1 + 10\, I_g$

$\Big\{$ ∴ $\qquad\qquad\qquad\qquad R_s = \dfrac{15 - 1}{0.659} - 10 = 11.244\,\Omega$

(*b*) $\qquad\qquad\qquad\qquad P_{gm} = \dfrac{P_{gav}}{fT}.$ Here $T = 20$ μsec

∴ Triggering frequency, $\qquad f = \dfrac{0.3 \times 10^6}{5 \times 20} = 3$ kHz

(*c*) Duty cycle, $\qquad\qquad \delta = fT = 3 \times 10^3 \times 20 \times 10^{-6} = 0.06.$

Example 4.6. *Latching current for an SCR, inserted in between a dc voltage source of 200 V and the load, is 100 mA. Compute the minimum width of gate-pulse current required to turn-on this SCR in case the load consists of (a) L = 0.2 H, (b) R = 20 Ω in series with L = 0.2 H and (c) R = 20 Ω in series with L = 2.0 H.*

Solution. (*a*) When load consists of pure inductance L, the voltage equation is

$$E = L \cdot \frac{di}{dt} \qquad \text{or} \quad di = \frac{E}{L}\, t \quad \text{or} \quad i = \frac{E}{L}\, t$$

∴ $\qquad\qquad 0.100 = \dfrac{200}{0.2}\, t \qquad \text{or} \quad t = \dfrac{0.1 \times 0.2}{200} = 100$ μsec

Thus, minimum gate-pulse is 100 μsec

(*b*) The voltage equation for R-L load is

$$E = Ri + L\frac{di}{dt}$$

or $\qquad\qquad i = \dfrac{E}{R}\left(1 - e^{-\frac{R}{L}t}\right) \qquad \text{or} \qquad 0.100 = \dfrac{200}{20}\,(1 - e^{-100\, t})$

or $\qquad\qquad t = 100.503$ μsec

∴ Minimum gate-pulse width is 100.503 μsec

(c)
$$i = \frac{E}{R}\left(1 - e^{-\frac{R}{L}t}\right)$$

or
$$0.1 = \frac{200}{20}(1 - e^{-10t}) \quad \text{or} \quad t = 1005.03 \ \mu sec.$$

This example shows that if load resistance is increased from zero to 20 Ω, the gate-pulse width remains almost unaffected. But with an increase in inductance from 0.2 H to 2 H, the gate-pulse width becomes 10 times its previous value.

Example 4.7. *The gate current of a forward biased SCR is gradually increased from zero until the device is turned on. It is observed that gate current, just prior to the instant of turn-on, is 1 mA and soon after SCR goes into conduction, gate current decays to about 0.3 mA. Discuss how it happens.*

Solution. When anode of an SCR is made positive with respect to cathode, a small voltage E'_g generated internally, appears across the gate-cathode terminals, Fig. 4.14 (a). The magnitude of E'_g depends upon applied anode voltage and the device geometry. In the gate-cathode equivalent circuit of Fig. 4.14 (a), R is the static non-linear gate resistance.

If the SCR is turned on by applying a positive gate signal, then the equivalent circuit for the trigger circuit is as shown in Fig. 4.14 (b). Here E_g is the gate voltage generated internally due to the flow of anode current. The magnitude of E'_g is much smaller as compared to E_g. For a typical SCR, $E'_g = 0.05$ V and $E_g = 0.7$ V.

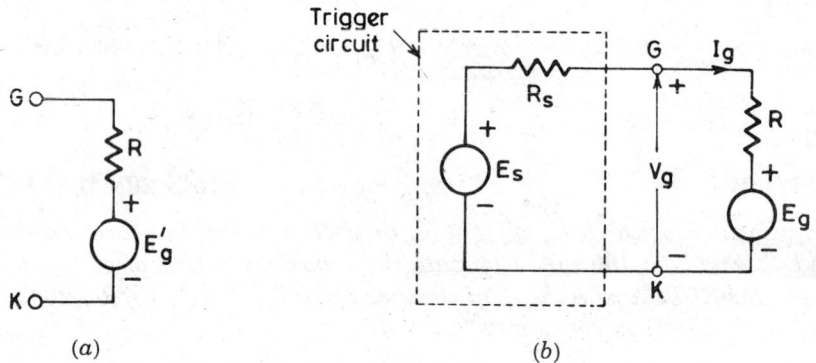

Fig. 4.14. Pertaining to Example 4.7.

Before the SCR starts conducting, gate current $I'_g = \dfrac{E_s - E'_g}{R + R_s}$. As E'_g is very small,

$I'_g \cong E_s/(R + R_s)$. After SCR goes into conduction, $I_g = \dfrac{E_s - E_g}{R + R_s}$. Voltage E_g is quite large as compared to E'_g, therefore, gate current is reduced from a higher value of I'_g to a lower value of I_g.

In case E_s is reduced to zero, gate current becomes negative with its value equal to $I''_g = \dfrac{-E_g}{R + R_s}$. Under this condition of $E_s = 0$, the voltage appearing across the gate-cathode terminals is $E_g - I''_g R$.

Example 4.8. *Gate-cathode characteristics of a thyristor have a spread given by the following two relations :*

$$I_g = 2.1 \times 10^{-3} V_g^2 \text{ and } I_g = 2.1 \times 10^{-3} V_g^{1.5}$$

The gate source voltage is 16 V and load line has a slope of – 128 V/A. Calculate the trigger voltage and trigger current for an average gate power dissipation of 0.5 W.

Are the values of V_g, I_g obtained here justified ? Discuss.

Solution. Slope of load line gives gate-source resistance, $R_s = 128\ \Omega$. Here $V_g I_g = 0.5$ W and $E_s = I_g R_s + V_g$

$$\therefore \qquad 16 = I_g \times 128 + \frac{0.5}{I_g}$$

or $\qquad 128\ I_g^2 - 16\ I_g + 0.5 = 0$

It solution gives $I_g = 62.5$ mA and $V_g = 8$ V

So point S in Fig. 4.11 has $V_g = 8$ V and $I_g = 62.5$ mA.

For the same V_g; $I_g = 2.1 \times 10^{-3}\ V_g^2$ gives more I_g, therefore it represents curve 1 of Fig. 6.11. Point S_1 on this curve can be obtained from $V_g I_g = 0.5$ W and $I_g = 2.1 \times 10^{-3}\ V_g^2$

$$\therefore \qquad I_g = 2.1 \times 10^{-3} \left(\frac{0.5}{I_g} \right)^2 \text{ or } I_g = 80.67 \text{ mA and } V_g = 6.198 \text{ V}$$

Point S_2 can be obtained from $V_g I_g = 0.5$ W and $I_g = 2.1 \times 10^{-3}\ V_g^{1.5}$

$$\therefore \qquad I_g = (2.1) \times 10^{-3} \left(\frac{0.5}{I_g} \right)^{1.5} \text{ or } I_g = 56.01 \text{ mA and } V_g = 8.93 \text{ V}.$$

Since point S (8 V, 62.5 mA) lies in between S_1 (6.198 V, 80.67 mA) and S_2 (8.93 V, 56.01 mA) as desired, the calculated values of $V_g = 8$ V and $I_g = 62.5$ mA are justified.

4.5. TWO-TRANSISTOR MODEL OF A THYRISTOR

The principle of thyristor operation can be explained with the use of its two-transistor model (or two-transistor analogy). Fig. 4.15 (*a*) shows schematic diagram of a thyristor. From this figure, two-transistor model is obtained by bisecting the two middle layers, along the dotted line, in two separate halves as shown in Fig. 4.15 (*b*). In this figure, junctions $J_1 - J_2$ and $J_2 - J_3$ can be considered to constitute *pnp* and *npn* transistors separately. The circuit representation of the two-transistor model of a thyristor is shown in Fig. 4.15 (*c*).

In the off-state of a transistor, collector current I_C is related to emitter current I_E as

$$I_C = \alpha\, I_E + I_{CBO}$$

where α is the *common-base current gain* and I_{CBO} is the common-base leakage current of collector-base junction of a transistor.

For transistor Q_1 in Fig. 4.15 (*c*), emitter current I_E = anode current I_a and I_C = collector current I_{C1}. Therefore, for Q_1,

$$I_{C1} = \alpha_1 I_a + I_{CBO1} \qquad \qquad ...(4.4)$$

where $\qquad \alpha_1$ = common–base current gain of Q_1

and $\qquad I_{CBO1}$ = common–base leakage current of Q_1.

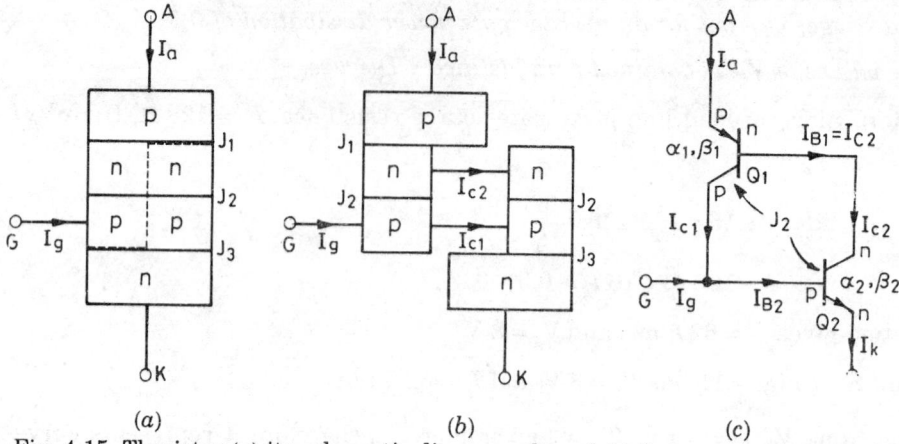

(a)   (b)   (c)

Fig. 4.15. Thyristor (a) its schematic diagram, (b) and (c) its two-transistor model.

Similarly, for transistor Q_2, the collector current I_{C2} is given by

$$I_{C2} = \alpha_2 I_k + I_{CBO2} \qquad \qquad \text{...(4.5)}$$

where α_2 = common–base current gain of Q_2

I_{CBO2} = common–base leakage current of Q_2

and I_k = emitter current of Q_2.

The sum of two collector currents given by Eqs. (4.4) and (4.5) is equal to the external circuit current I_a entering at anode terminal A.

$$\therefore \quad I_a = I_{C1} + I_{C2}$$

or $$I_a = \alpha_1 I_a + I_{CBO1} + \alpha_2 I_k + I_{CBO2} \qquad \qquad \text{.(4.6)}$$

When gate current is applied, then $I_k = I_a + I_g$. Substituting this value of I_k in Eq. (4.6) gives

$$I_a = \alpha_1 I_a + I_{CBO1} + \alpha_2 (I_a + I_g) + I_{CBO2}$$

or $$I_a = \frac{\alpha_2 I_g + I_{CBO1} + I_{CBO2}}{1 - (\alpha_1 + \alpha_2)} \qquad \qquad \text{...(4.7)}$$

For a silicon transistor, current gain α is very low at low emitter current. With an increase in emitter current, α builds up rapidly as shown in Fig. 4.16. With gate current $I_g = 0$ and with thyristor forward biased, $(\alpha_1 + \alpha_2)$ is very low as per Fig. 4.16. Under these conditions, Eq. (4.7) shows that forward leakage current somewhat more than $(I_{CBO1} + I_{CBO2})$ flows. If, by some means, the emitter current of two component transistors can be increased so that $\alpha_1 + \alpha_2$ approaches unity, then as per Eq. (4.7), I_a would tend to become infinity thereby turning-on the device. Actually, external load limits the anode current to a safe value after the thyristor begins conduction. The methods of turning-on a thyristor, in fact, are the methods of making $\alpha_1 + \alpha_2$ to approach unity. These various

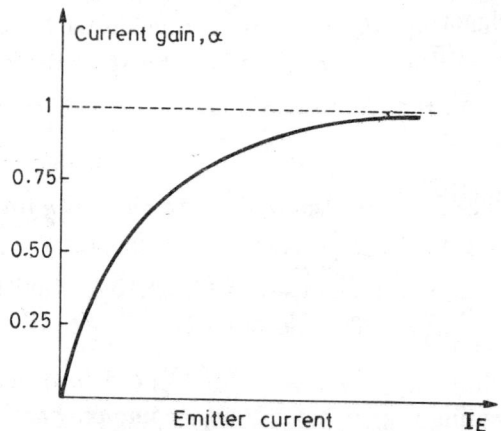

Fig. 4.16. Typical variation of current gain with emitter current of a thyristor.

mechanisms for turning-on a thyristor are now discussed below :

(i) *GATE Triggering* : With anode positive with respect to cathode and with gate current $I_g = 0$, Eq. (4.6) shows that anode current, equal to the forward leakage current, is somewhat more than $I_{CBO1} + I_{CBO2}$. Under these conditions, the device is in the forward blocking state.

Now a sufficient gate-drive current between gate and cathode of thyristor, or the transistor of Fig. 4.15 (c) is applied. This gate-drive current is equal to base current $I_{B2} = I_g$ and emitter current I_k of transistor Q_2. With the establishment of emitter current I_k of Q_2, current gain α_2 of Q_2 increases and base current I_{B2} causes the existence of collector current $I_{C2} = \beta_2 I_{B2} = \beta_2 I_g$. This amplified current I_{C2} serves as the base current I_{B1} of transistor Q_1. With the flow of I_{B1}, collector current $I_{C1} = \beta_1 I_{B1} = \beta_1 \beta_2 I_g$ of Q_1 comes into existence. Currents I_{B1} and I_{C1} lead to the establishment of emitter current I_a of Q_1 and this causes current gain α_1 to rise as desired. Now current $I_g + I_{C1} = (1 + \beta_1\beta_2) I_g$ acts as the base current of Q_2 and therefore its emitter current $I_k = I_{C1} + I_g$ rise. With the rise in emitter current I_k, α_2 of Q_2 increases and this further causes $I_{C2} = \beta_2 (1 + \beta_1\beta_2) I_g$ to rise. As amplified collector current I_{C2} is equal to the base current of Q_1, current gain α_1 eventually rises further. There is thus established a regenerative action internal to the device. This regenerative or positive feedback effect causes $\alpha_1 + \alpha_2$ to grow towards unity. As a consequence, anode current begins to grow towards a larger value limited only by load impedance external to the device. When regeneration has grown sufficiently, gate current can be withdrawn. Even after I_g is removed, regeneration continues. This characteristic of the thyristor makes it suitable for pulse triggering. Note that thyristor is a latching device.

After thyristor is turned on, all the four layers are filled with carriers and all junctions are forward biased. Under these conditions, thyristor has very low impedance and is in the forward on-state.

(ii) *Forward-voltage triggering* : If the forward anode to cathode voltage is increased, the collector to emitter voltages of both the transistors are also increased. As a result, the leakage current at the middle junction J_2 of thyristor increases, which is also the collector current of Q_2 as well as Q_1. With increase in collector currents I_{C1} and I_{C2} due to avalanche effect, the emitter currents of the two transistors also increase causing $\alpha_1 + \alpha_2$ to approach unity. This leads to switching action of the device due to regenerative action. The forward-voltage triggering for turning-on a thyristor may be destructive and should therefore be avoided.

(iii) *dv/dt triggering* : The reversed biased junction J_2 behaves like a capacitor because of the space-charge present there. Let the capacitance of this junction be C_j. For any capacitor, $i = C \dfrac{dv}{dt}$. In case it is assumed that entire forward voltage v_a appears across reverse biased junction J_2, then charging current across the junction is given by

$$i = C_j \frac{dv_a}{dt}$$

This charging or displacement current across junction J_2 is collector currents of Q_2 and Q_1. Currents I_{C2}, I_{C1} will induce emitter current in Q_2, Q_1. In case rate of rise of anode voltage is large, the emitter currents will be large and as a result, $\alpha_1 + \alpha_2$ will approach unity leading to eventual switching action of the thyristor.

(iv) *Temperature triggering* : At high temperature, the forward leakage current across junction J_2 rises. This leakage current serves as the collector junction current of the component transistors Q_1 and Q_2. Therefore, an increase in leakage current I_{C1}, I_{C2} leads to an increase in the emitter currents of Q_1, Q_2. As a result, $(\alpha_1 + \alpha_2)$ approaches unity. Consequently, switching action of thyristor takes place.

(v) *Light triggering* : When light is thrown on silicon, the electron-hole pairs increase. In the forward-biased thyristor, leakage current across J_2 increases which eventually increases $\alpha_1 + \alpha_2$ to unity as explained before and switching action of thyristor occurs.

As stated before, gate-triggering is the most common method for turning-on a thyristor. Light-triggered thyristors are used in HVDC applications.

4.6. THYRISTOR RATINGS

Thyristor ratings indicate voltage, current, power and temperature limits within which a thyristor can be used without damage or malfunction. Ratings and specifications serve as a link between the designer and the user of SCR systems.

For reliable operation of a thyristor, it should be ensured that its current and voltage ratings are not exceeded during its working. One of the major disadvantages of thyristors is that they have low thermal time constant. If a thyristor handles voltage, current and power greater than its specified ratings, the junction temperature may rise above the safe limit and as a result, thyristor may get damaged. Therefore, when SCRs are selected, some safety margin must be kept in the form of choosing device ratings somewhat higher than their normal working values. The manufacturers of thyristors make a comprehensive list of the voltage, current, power and temperature ratings after carefully testing the device. If SCRs are operated under these specified conditions, no damage will be done to SCRs. The object of this section is to discuss the various SCR ratings.

A thyristor has several ratings such as voltage, current, power, dv/dt, di/dt, turn-on time, turn-off time etc. For correct application of the device in thyristor circuits, a knowledge of these ratings is desirable.

Some subscripts are associated with voltage and current ratings for convenience in identifying them. First subscript letter indicates the direction or the state :

$D \rightarrow$forward-blocking region with gate circuit open ; $T \rightarrow$on-state ; $R \rightarrow$reverse ; $F \rightarrow$forward.

Except for the gate G, second subscript letter denotes the operating values.

$W \rightarrow$working value ; $R \rightarrow$repetitive value ; $S \rightarrow$surge or non-repetitive value ; $T \rightarrow$trigger Third subscript letter M indicates the maximum or peak value.

Ratings with less than three subscripts may not follow these rules. Gate ratings involve the subscript G. Subscript A usually stands for anode and subscript AV for average.

4.6.1. Anode Voltage Ratings

A thyristor is made up of four layers and three junctions as shown in Fig. 4.1 (*b*). The middle junction J_2 blocks the forward voltage whereas the two end junctions J_1, J_3 block the reverse voltage. The anode voltage ratings indicate the values of maximum voltages that a thyristor can withstand without a breakdown of the junction area with gate circuit open.

For ac systems, the supply voltage may not be a smooth sine wave. The voltage transients may occur regularly or at random as shown in Fig. 4.17 (*a*). The different anode voltage ratings are as under :

(*i*) V_{DWM} —*Peak working forward-blocking voltage.* It specifies the maximum forward-blocking voltage that a thyristor can withstand during its working. Fig. 4.17 (*a*) shows that V_{DWM} is equal to the maximum value of the sine voltage wave.

(*ii*) V_{DRM}—*Peak repetitive forward-blocking voltage.* It refers to the peak transient voltage that a thyristor can withstand repeatedly or periodically in its forward-blocking mode. The rating is specified at a maximum allowable junction temperature with gate circuit open or with a specified biasing resistance between gate and cathode.

Voltage V_{DRM} is encountered when a thyristor is commutated or turned-off. It may be recalled that during turn-off process, an abrupt change in reverse recovery current is accompanied by a spike voltage $L \dfrac{di}{dt}$; this is responsible for the appearance of V_{DRM} across thyristor terminals.

(*iii*) V_{DSM} —*Peak surge (or non-repetitive) forward-blocking voltage.* It refers to the peak value of the forward surge voltage that does not repeat. Its value is about 130% of V_{DRM}, but V_{DSM} is less than forward breakover voltage V_{BO} as shown in Fig. 4.17 (*b*).

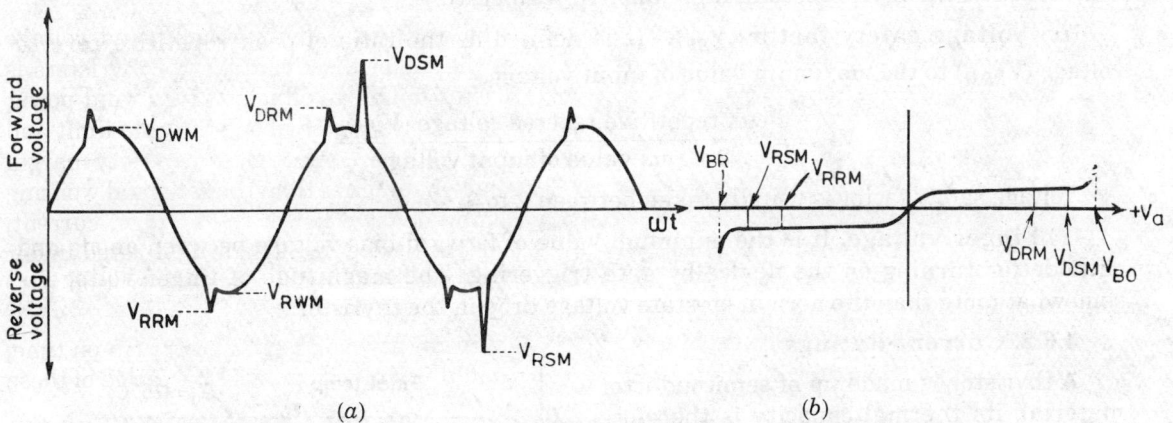

Fig. 4.17. Anode voltage ratings during the blocking state of a thyristor.

(*iv*) V_{RWM} —*Peak working reverse voltage.* It is the maximum reverse voltage that a thyristor can withstand repeatedly. Actually, it is equal to the peak negative value of a sine voltage wave, Fig. 4.17 (*a*).

(*v*) V_{RRM} —*Peak repetitive reverse voltage.* It specifies the peak reverse transient voltage that may occur repeatedly in the reverse direction at the allowable maximum junction temperature. The transient lasts for a fraction of the time of one cycle, Fig. 4.17 (*a*). The reason for the periodic appearance of V_{RRM} is the same as for V_{DRM}.

(*vi*) V_{RSM} —*Peak surge (or non-repetitive) reverse voltage.* It represents the peak value of the reverse surge voltage that does not repeat. Its value is about 130% of V_{RRM}. But V_{RSM} is less than reverse breakover voltage V_{BR} as shown in Fig. 4.17 (*b*).

Both V_{DSM} and V_{RSM} ratings can be increased by connecting a diode in series with a thyristor. The anode voltage ratings listed above from (*i*) to (*iii*) pertain to forward blocking voltages whereas from (*iv*) to (*vi*) belong to reverse blocking voltages ; a thyristor must be able to support these voltages safely with gate circuit open.

(*vii*) V_T —*On-state voltage drop.* It is the voltage drop between anode and cathode with specified forward on-state current and junction temperature. Its value is of the order of 1 to 1.5 V.

(*viii*) *Forward dv/dt rating.* If rate of rise of forward anode-to-cathode voltage is high, thyristor may turn on even when

(a) there is no gate signal and

(b) anode-to-cathode voltage is less than forward breakover voltage.

When a thyristor is in the forward blocking mode, the applied voltage appears across junction J_2 as junctions J_1 and J_3 are forward-biased. The reverse biased junction behaves like a capacitor. When forward voltage is suddenly applied to the device, a charging current $C_j \, dv/dt$ begins to flow which may turn on SCR as explained in Art. 4.2. A high value of dv/dt, at which a thyristor just gets turned on is called *critical rate of rise of anode voltage* or *forward dv/dt rating* of the device. If applied dv/dt exceeds this critical value, thyristor gets turned on. For applied dv/dt lower than forward dv/dt rating, thyristor remains in forward blocking mode.

The forward dv/dt rating depends on the junction temperature ; higher the junction temperature, lower the forward dv/dt rating of the device. In practice, dv/dt triggering is never employed as it gives random turn-on of a thyristor. This type of triggering also leads to destruction of the device through high junction temperature.

(*ix*) **Voltage safety factor (V_{SF}).** It is defined as the ratio of peak repetitive reverse voltage (V_{RRM}) to the maximum value of input voltage.

$$\therefore \qquad V_{SF} = \frac{\text{Peak repetitive reverse voltage } (V_{RRM})}{\sqrt{2} \times rms \text{ value of input voltage}}$$

Voltage safety factor is usually taken between 2 to 3.

(*x*) **Finger voltage.** It is the minimum value of forward bias voltage between anode and cathode for turning-on the device by gate triggering. The magnitude of finger voltage is somewhat more than the normal on-state voltage drop in the thyristor.

4.6.2. Current Ratings

A thyristor is made up of semiconductor material, its thermal capacity is therefore quite small. Even for short overcurrents, the junction temperature may exceed the rated value and the device may be damaged. As the junction temperature is dependent on the current handled by a thyristor, a correct choice of current ratings is essential for a long working life of the device. In this part of the article, current ratings of SCRs are discussed for both repetitive and non-repetitive type of current waveforms.

Average on-state current (I_{TAV}). The forward voltage drop across conducting SCR is low, therefore power loss in a thyristor depends primarily on forward average on-state current I_{TAV}. For the purpose of illustrating the significance of average on-state current, consider a continuous dc current OA flowing through the SCR, Fig. 4.18 (*a*). After the application of this current at $t = 0$, junction temperature begins to rise

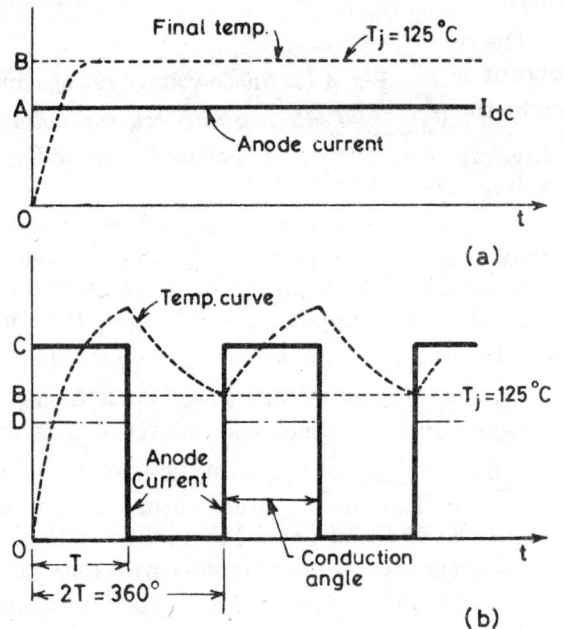

Fig. 4.18. Variation of junction temperature with constant anode current i_a and with rectangular wave of i_a.

until finally it reaches its rated value $T_j = 125°C$. As the SCR has low thermal time constant, final temperature of 125°C is reached in a relatively short time. Suppose now that anode current is of rectangular waveshape with conduction angle $180° \left(= \dfrac{T}{2T} \times 360° \right)$, as shown in Fig. 4.18 (b). If the rectangular wave has average value equal to the constant current OA in Fig. 4.18 (a), then current amplitude of rectangular wave in Fig. 4.18 (b) is $OC = 2$ times OA. As the SCR has short time constant, junction temperature in Fig. 4.18 (b) is likely to exceed the allowable temperature of 125°C and this is not desirable. In order to limit the temperature to 125°C for rectangular waveform of anode current; there are two techniques, (i) provide better cooling to the thyristor or (ii) reduce the pulse amplitude from OC.

As per the second method, pulse amplitude of anode current is reduced from OC to some lower value OD (say), so that junction temperature remains within limits, Fig. 4.18 (b). But a reduction in the amplitude of rectangular wave would result in a lower value of average anode current. This means that for the temperature rise to remain within limits, SCR must be rated at a lower value of average forward current I_{TAV} when it is conducting a pulsed anode current than when it is carrying a constant dc. This shows that thyristor is derated when it handles rectangular or square wave of anode current. The effect of conduction angle on anode current I_{TAV} is depicted in Fig. 4.19 (a) for rectangular waves. The average on-state power loss P_{av} in this figure is approximately given by

$$P_{av} = (\text{forward on–state voltage across a thyristor}) \times I_{TAV}$$

It can be obtained more accurately from the relation

$$P_{av} = \frac{1}{T} \int (\text{instantaneous voltage across SCR}) (\text{instantaneous current through SCR}) \, dt$$

where T = periodic time of the anode current waveform.

The rms current for an SCR is constant whatever the conduction angle may be. But average current is given by (I_{rms}/FF) where FF is the form factor of the current waveform. The conduction angle for sine wave is defined in Fig. 4.19 (b). For the same conduction angle, the

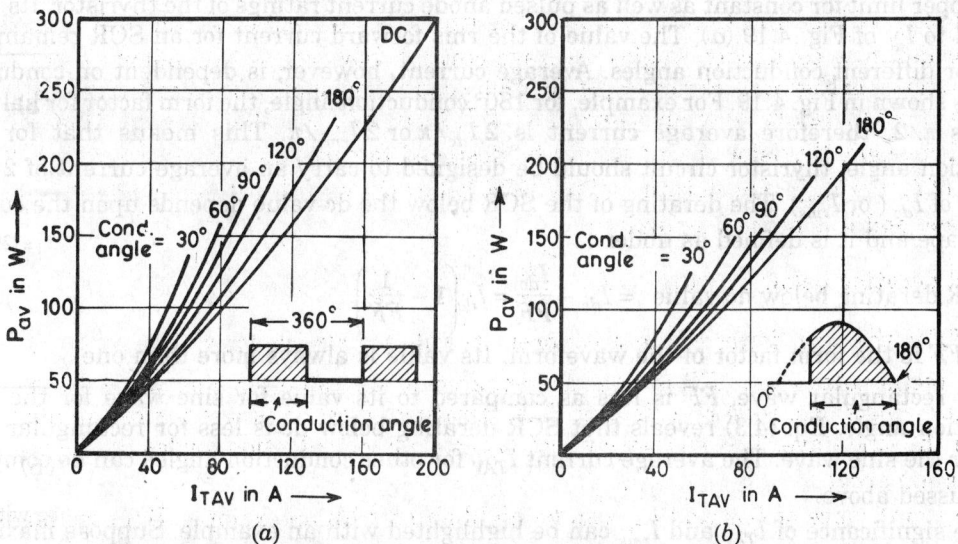

Fig. 4.19. Average on-state power dissipation P_{av} as a function of I_{TAV} for
(a) rectangular wave and (b) half-wave sinusoid.

form factor for sine wave is higher than, for the rectangular wave (see Examples 4.8 and 4.9). This means that average current for sine wave will be lower than it is for the rectangular wave for the same dc (or rms) current. The derating of the SCR is therefore more for sine waves than for the square or rectangular waves. The effect of conduction angle on average current is depicted in Fig. 4.19 (b) for sine wave. For 180° conduction angle, the anode current in Fig. 4.19 (b) is less than that in Fig. 4.19 (a). This diagram is applicable for 1-phase half-wave circuit (or 1-phase one-way, one pulse circuit).

Curves shown in Fig. 4.19 are supplied by the manufacturers of thyristors and are valid for supply frequency range of 50 to 400 Hz. The curve marked dc in Fig. 4.19 (a) is applicable when anode current is continuous dc. The current for different conduction angles are terminated at different values of average current in Fig. 4.19. For example, for 30° conduction angle, I_{TAV} terminates at $[I_{dc}/(\text{form factor})] = (I_{dc}/3.464)$ in Fig. 4.19 (a) for rectangular wave and at $(I_{dc}/3.979)$ in Fig. 4.19 (b) for sine wave. At these terminal points, maximum rms current ratings of the device is reached. Table 4.1 gives different values of form-factor for different conduction angles of the half-wave sine waveforms.

Table 4.1. Form Factor for Sine Waves

Conduction angle	15°	30°	45°	60°	90°	120°	180°
Form factor	5.650	3.9812	3.233	2.7781	2.2214	1.878	1.5708

Curves of Fig. 4.19 are applicable when load is purely resistive. In case load is inductive in nature, these curves should be modified. With an improvement in the waveform, i.e. with waveform becoming more smooth, the form factor decreases and as a consequence, higher average on-state current I_{TAV} can be handled by the device.

RMS on-state current (I_{RMS}). By definition, for direct current, rms value I_{RMS} or I_{rms} = average or dc value, I_{dc}. Heating of the resistive elements of a thyristor, such as metallic joints, leads and interfaces depends on the forward rms current I_{rms}. The rms current rating is used as an upper limit for constant as well as pulsed anode current ratings of the thyristor. Its value is equal to I_{dc} of Fig. 4.19 (a). The value of the rms forward current for an SCR remains the same for different conduction angles. Average current, however, is dependent on conduction angle as shown in Fig. 4.19. For example, for 180° conduction angle, the form factor for half-sine wave is $\pi/2$, therefore average current is $2 I_{dc}/\pi$ or $2 I_{rms}/\pi$. This means that for 180° conduction angle, thyristor circuit should be designed to carry an average current of $2 I_{dc}/\pi$ instead of I_{dc} (or I_{rms}). The derating of the SCR below the dc value depends upon the current waveshape and it is defined as under :

$$\text{SCR derating below dc value} = I_{dc} - \frac{I_{dc}}{FF} = I_{dc}\left(1 - \frac{1}{FF}\right) \qquad \qquad ...(4.3)$$

where FF is the form factor of the waveform. Its value is always more than one.

For rectangular wave, FF is less as compared to its value for sine wave for the same conduction angle. Eq. (4.3) reveals that SCR derating below dc is less for rectangular wave than for the sine wave. The average current I_{TAV} for other conduction angles can be computed as discussed above.

The significance of I_{TAV} and I_{rms} can be highlighted with an example. Suppose maximum rms current for a thyristor is 35 A. For 120° conduction angle for sine wave, $I_{TAV} = \dfrac{35}{1.875}$

= 18.637 A. This means that thyristor can handle an average current of 18.637 A for 120° conduction angle and its temperature will remain within limits. Suppose an ammeter is placed in series with the SCR for measuring the average current. Now decrease the conduction angle to 30° but with average current as measured by the ammeter remaining unchanged at 18.637 A. But an average current of 18.637 A at 30° conduction angle would require an rms current of $I_{rms} = 18.637 \times 3.9812 = 74.1976$ A. But such a large value of rms current would cause large ohmic losses and is, therefore, certainly going to destroy the SCR. This shows that as conduction angle is reduced, I_{TAV} must be lowered accordingly so that rms current is not exceeded beyond its rated value and the SCR is not damaged.

The current ratings I_{TAV} and I_{rms} are of repetitive type. They are dependent on maximum junction temperature. If better cooling is provided to a thyristor body, these ratings can be upgraded.

As stated above, power loss in a thyristor and its heating is dependent upon the rms current. Manufacturers also provide curves showing the variation of case temperature T_c with

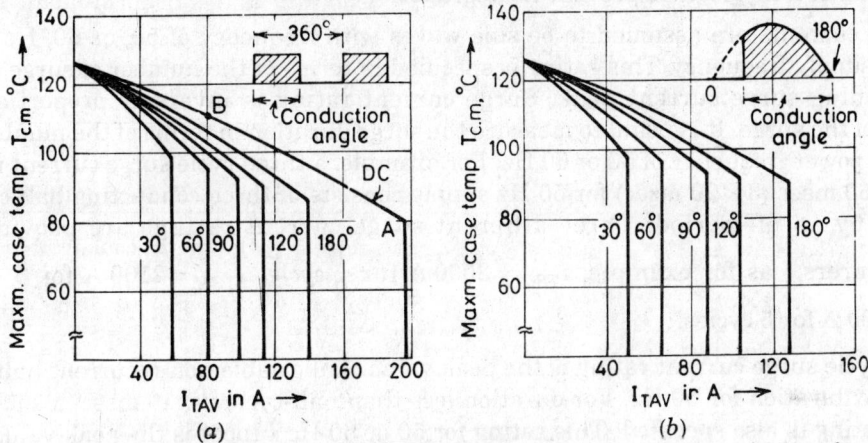

Fig. 4.20. Maximum allowable case temperature T_{cm} as a function of I_{TAV} for (a) rectangular wave and (b) for half-wave sinusoid.

average on-state current I_{TAV}, Fig. 4.20. These curves can be obtained from Fig. 4.19 provided θ_{jc} (thermal resistance between junction and thyristor case)* in °C/W is known. If T_j is the junction temperature, then

$$T_j - T_c = \theta_{jc} \cdot P_{av}$$

For SCRs, T_j is usually 125°C. Taking $\theta_{jc} = 0.15$°C/W for dc current of 200 A; $P_{av} = 300$ W, from Fig. 4.19 (a), is obtained for $I_{TAV} = 200$ A.

$$\therefore \qquad 125 - T_c = 0.15 \times 300 \quad \text{or} \quad T_c = 80°C$$

This point is plotted in Fig. 4.20 (a) as A. For 80 A dc, $P_{av} = 100$ W.

$$\therefore \qquad 125 - T_c = 0.15 \times 100 \quad \text{or} \quad T_c = 110°C$$

This point is plotted as B in Fig. 4.20 (a). For 180° conduction angle, for $I_{TAV} = 140$ A, $P_{av} = 225$ W from Fig. 4.19 (a).

$$\therefore \qquad T_c = 125 - 0.15 \times 225 = 91.25°C$$

---

*For understanding the term thermal resistance read Art. 4.8.

This point is plotted as C in Fig. 4.20 (a). Other points can be plotted accordingly for rectangular as well as half-wave sinusoids to obtain the curves of Fig. 4.20. These curves indicate that for junction temperature $T_j = 125°C$, lower the average on-state current I_{TAV}, greater is the case temperature that can be allowed for the same conduction angle. For example, for sine wave with 180° conduction angle, for $I_{TAV} = 120$ A the case temperature $T_{cm} = 91°C$; for $I_{TAV} = 80$ A the case temperature $T_{cm} = 104°C$ and so on.

Surge Current Rating. When a thyristor is working under its repetitive voltage and current ratings, its permissible junction temperature is never exceeded. However, a thyristor may be subjected to abnormal operating conditions due to faults or short circuits. In order to accommodate these unusual working conditions, surge current rating, I_{TSM} (peak non-repetitive on-state current), of thyristors is also specified. A surge current rating indicates the maximum possible non-repetitive, or surge, current which the device can withstand. Higher currents caused by non-repetitive faults or short circuits should occur once in a while during the life span of a thyristor to prevent its degradation.

Surge currents are assumed to be sine waves with frequency of 50, or 60, Hz depending upon the supply frequency. This rating is specified in terms of the number of surge cycles with corresponding surge current peak. Surge current rating is inversely proportional to the duration of the surge. It is usual to measure the surge duration in terms of the number of cycles of normal power frequency of 50 or 60 Hz. For example, a three-cycle surge current rating for a period of 60 msec (3×20 msec) for 50 Hz supply consists of three conducting half-cycles, each followed by an off-period. Three different surge current ratings are provided by the manufacturers ; as for example, $I_{TSM} = 3000$ A for $\frac{1}{2}$ cycle, $I_{TSM} = 2100$ A for 3 cycles and $I_{TSM} = 1800$ A for 5 cycles.

One cycle surge current rating is the peak value of allowable non-recurrent half-sine wave of 10 msec duration for 50 Hz. For duration less than half-cycle *i.e.* 10 msec, a subcycle surge current rating is also specified. This rating for 50 or 60 Hz supply is the peak value for a part of the half-sine wave. The subcycle surge current rating I_{sb} can be determined by equating the energies involved in one cycle surge and one subcycle surge as follows :

$$I_{sb}^2 \cdot t = I^2 \cdot T$$

or
$$I_{sb} = I \sqrt{\frac{T}{t}}$$ ...(4.8 a)

where T = time for one half–cycle of supply frequency, sec
I = one–cycle surge current rating, A
I_{sb} = subcycle surge current rating, A
t = duration of subcycle surge, sec
For 50 Hz supply, $T = 10$ msec

$\therefore$ $I_{sb} = \frac{I}{10} \cdot \frac{1}{\sqrt{t}}$ ...(4.8 b)

I^2t **rating.** This rating is employed in the choice of a fuse or other protective equipment for thyristors. The rating in terms of amp^2-sec specifies the energy that the device can absorb for a short time before the fault is cleared. It is usually specified for overloads lasting for less than, or equal to, one-half cycle of 50 or 60 Hz supply. The I^2t rating is given by the relation.

(rms value of one–cycle surge current $)^2 \times$ time for one cycle   ...(4.9)

As an example, I^2t rating for 4 A (rms) SCR is 10 amp^2-sec and for 35 A SCR is 100 amp^2-sec. In order that a fuse (or other protective equipment) protects a thyristor reliably, the I^2t rating of fuse must be less than the I^2t rating of the series-connected thyristor.

di/dt rating. This rating of a thyristor indicates the maximum rate of rise of current from anode to cathode without any harm to the device. When a thyristor is turned on, conduction starts at a place near the gate. This small area of conduction spreads to the whole area of junction. If the rate of rise of anode current (di/dt) is large as compared to the spreading velocity of carriers across the cathode junction, local hot spots will be formed near the gate connection on account of high current density. This causes the junction temperature to rise above the safe limit and as a consequence, SCR may be damaged permanently. Therefore, a limit on the value of di/dt at turn-on is specified in amperes per microsecond for all SCRs. Typical values of di/dt are 20 to 500 A/µ sec.

Other ratings. In addition to the voltage and current ratings of thyristors discussed above, there are some other ratings as under :

(a) Latching and holding currents,

(b) Turn-on and turn-off times,

(c) Gate circuit voltage, current and power ratings.

These ratings have already been discussed in Art. 4.1 to Art. 4.5.

Detailed ratings of any SCR during on-state and off-state can be obtained from the manufacturers by quoting the specification sheet number.

Example 4.9. *The specification sheet for an SCR gives maximum rms on-state current as 35 A. If this SCR is used in a resistive circuit, compute average on-state current rating for half-sine wave current for conduction angles of (a) 180° (b) 90° and (c) 30°.*

Solution. For half-sine wave of current as shown in Fig. 4.21 (a),

$$I_{av} = \frac{1}{2\pi} \int_{\theta_1}^{\pi} I_m \sin\theta \, d\theta = \frac{I_m}{2\pi} (1 + \cos\theta_1)$$

$$I_{rms} = \left[\frac{1}{2\pi} \int_{\theta_1}^{\pi} I_m^2 \sin^2\theta \, d\theta \right]^{1/2} = \left[\frac{I_m^2}{2\pi} \int_{\theta_1}^{\pi} \left\{ \frac{1}{2} - \frac{\cos 2\theta}{2} \right\} \right]^{1/2}$$

$$= \left[\frac{I_m^2}{2\pi} \left| \frac{\theta}{2} - \frac{\sin 2\theta}{4} \right|_{\theta_1}^{\pi} \right]^{1/2}$$

$$= \left[\frac{I_m^2}{2\pi} \left\{ \frac{\pi - \theta_1}{2} + \frac{1}{4} \sin 2\theta_1 \right\} \right]^{1/2}$$

(a) For 180° conduction angle, $\theta_1 = 0°$

$$\therefore \qquad I_{av} = \frac{I_m}{2\pi} (1 + \cos 0°) = \frac{I_m}{\pi}$$

and

$$I_{rms} = \left[\frac{I_m^2}{2\pi} \left\{ \frac{\pi}{2} + \frac{1}{4} (0) \right\} \right]^{1/2} = \frac{I_m}{2}$$

$$\therefore \qquad \text{Form factor, } (FF) = \frac{I_{rms}}{I_{av}} = \frac{I_m}{2} \cdot \frac{\pi}{I_m} = \frac{\pi}{2}$$

$$\therefore \qquad I_{TAV} = \frac{I_{rms}}{FF} = \frac{35 \times 2}{\pi} = 22.282 \text{ A}$$

(b) For 90° conduction angle, $\theta_1 = 90°$

$$\therefore \qquad I_{av} = \frac{I_m}{2\pi}[1 + \cos 90] = \frac{I_m}{2\pi}$$

and

$$I_{rms} = \left[\frac{I_m^2}{2\pi}\left\{\frac{\pi}{4} + \frac{1}{4}(0)\right\}\right]^{1/2} = \frac{I_m}{2\sqrt{2}}$$

$$\therefore \qquad \text{Form factor} = \frac{I_m}{2\sqrt{2}} \cdot \frac{2\pi}{I_m} - \frac{\pi}{\sqrt{2}}$$

$$\therefore \qquad I_{TAV} = \frac{35 \times \sqrt{2}}{\pi} = 15.755 \text{ A}$$

(c) For 30° conduction angle, $\theta_1 = 150°$

$$\therefore \quad I_{av} = \frac{I_m}{2\pi}[1 + (-0.866)] = 0.0213227\, I_m$$

$$I_{rms} = \left[\frac{I_m^2}{2\pi}\left\{\frac{\pi}{12} + \frac{1}{4}(-0.866)\right\}\right]^{1/2} = 0.0849035\, I_m$$

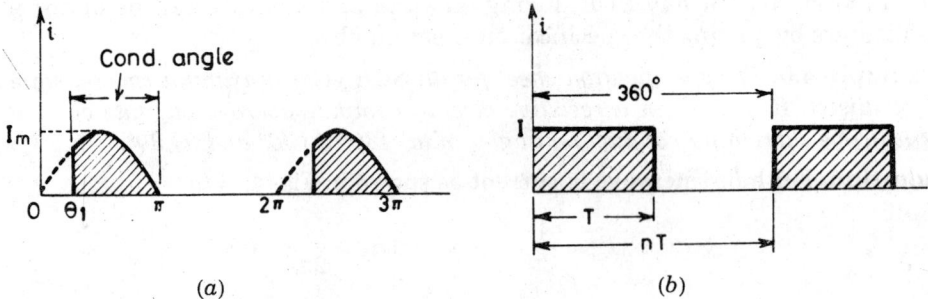

(a)   (b)

Fig. 4.21. Pertaining to (a) Example 4.9 and (b) Example 4.10.

$$\therefore \qquad \text{Form factor} = \frac{0.0849035\, I_m}{0.0213227\, I_m} = 3.9818363$$

$$\therefore \qquad I_{TAV} = \frac{35}{3.98184} = 8.7899 \text{ A.}$$

Example 4.10. *Repeat Example 4.9 in case the current has rectangular waveshape.*

Solution. For the rectangular waveform of current shown in Fig. 4.21 (b),

Conduction angle $$= \frac{T}{nT} \times 360$$

or $$n = \frac{360°}{\text{Conduction angle}}$$

Here $$I_{av} = \frac{I \times T}{nT} = \frac{I}{n}$$

$$I_{rms} = \left[\frac{I^2 \times T}{nT}\right]^{1/2} = \frac{I}{\sqrt{n}}$$

(a) For 180° conduction angle, $n = \dfrac{360}{180} = 2$

$\therefore$ $I_{av} = \dfrac{I}{2}$ and $I_{rms} = \dfrac{I}{\sqrt{2}}$

$\therefore$ Form factor $= \dfrac{I}{\sqrt{2}} \cdot \dfrac{2}{I} = \sqrt{2}$

$$I_{TAV} = \dfrac{35}{\sqrt{2}} = 24.7487 \text{ A}$$

(b) For 90° conduction angle, $n = \dfrac{360}{90} = 4$

$\therefore$ $I_{av} = \dfrac{I}{4}$ and $I_{rms} = \dfrac{I}{\sqrt{4}} = \dfrac{I}{2}$

$\therefore$ Form factor $= \dfrac{I}{2} \cdot \dfrac{4}{I} = 2$

$$I_{TAV} = \dfrac{35}{2} = 17.5 \text{ A}$$

(c) For 30° conduction angle, $n = \dfrac{360}{12} = 12$

$$I_{av} = \dfrac{I}{12} \quad \text{and} \quad I_{rms} = \dfrac{I}{\sqrt{12}}$$

Form factor $= \dfrac{I}{\sqrt{12}} \cdot \dfrac{12}{I} = \sqrt{12}$

$$I_{TAV} = \dfrac{35}{\sqrt{12}} = 10.1036 \text{ A}.$$

Example 4.11. *An SCR has half-cycle surge current rating of 3000 A for 50 Hz supply. Calculate its one-cycle surge current rating and $I^2 t$ rating.*

Solution. Let I and I_{sb} be the one-cycle and sub-cycle surge current ratings of the SCR respectively. Then equating the energies involved in them, we get

$$I^2 T = I_{sb}^2 \cdot t$$

or $\qquad I^2 \times \dfrac{1}{100} = (3000)^2 \times \dfrac{1}{200}$ or $I = \dfrac{3000}{\sqrt{2}} = 2121.32 \text{ A}$

From Eq. (4.9), $\qquad I^2 t$ rating $= I^2 \times \dfrac{1}{2f} = \left(\dfrac{3000}{\sqrt{2}}\right)^2 \times \dfrac{1}{100} = 45000 \text{ Amp}^2 \cdot \text{sec.}$

Thus the SCR has one-cycle surge current rating of 2121.32 A and $I^2 t$ rating of 45000 amp²-sec.

Example 4.12. *In the circuit of Fig. 4.22, the thyristor is gated with a pulse width of 40 microsec. The latching current of thyristor is 36 mA. For a load of 60 Ω and 2 H, will the thyristor get turned on ? Check. If the answer is negative, how this difficulty can be overcome for the given load. Find the maximum value of the remedial parameter shown dotted.*

Solution. The current through load and thyristor is

$$i_T = \dfrac{V_s}{R}\left(1 - e^{-\frac{R}{L}t}\right)$$

For the circuit shown, $\qquad i_T = \dfrac{300}{60}\left(1 - e^{-\frac{60}{2} \times 40 \times 10^{-6}}\right) = 5.996 \times 10^{-3} = 5.996 \text{ mA}$

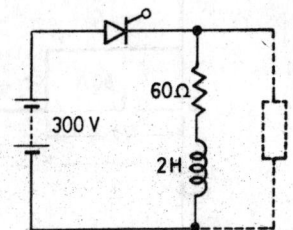

Fig. 4.22. Pertaining to Example 4.12.

This shows that for a pulse width of 40 μs, the anode current rises to 5.996 mA which is far less than the latching current of 36 mA. So thyristor will not get turned on.

The remedial parameter, shown dotted in Fig. 4.22, should be resistance, say R_1, because current can rise in resistance without any time delay. The value of R_1 can be obtained as under :

$$i_T = 36 \times 10^{-3} = \frac{300}{R_1} + \frac{300}{60}(1 - e^{-0.0012})$$

or
$$R_1 = \frac{300}{30.004} \times 10^3 = 9998 \ \Omega = 9.998 \ \text{k}\ \Omega.$$

Example 4.13. *During forward conduction, a thyristor has static I-V characteristic as shown by a straight line in Fig. 4.23. Find the average power loss in the thyristor and its rms current rating for the following load conditions :*

(a) A constant current of 80 A for one-half cycle.

(b) A constant current of 30 A for one-third cycle.

(c) A half-sine wave of peak value 80 A.

Solution. It is seen from Fig. 4.23 that for any current i_a, the voltage drop across thyristor is

$$v_T = 0.8 + \frac{2.0 - 0.8}{100} \times i_a = 0.8 + 0.012 \ i_a$$

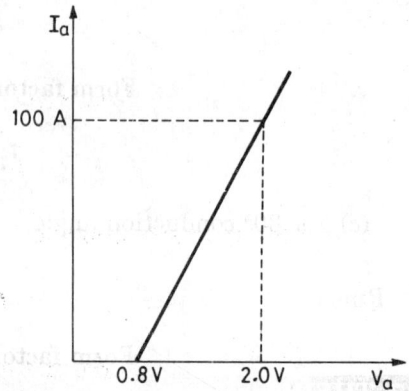

Fig. 4.23. Pertaining to Example 4.13.

(a) Constant current of 80 A for one-half cycle is shown in Fig. 4.24 (a). For $i_a = 80$ A, the voltage drop across thyristor is $v_T = 0.8 + 0.012 \times 80 = 1.76$ V. From the waveforms of i_a, v_T shown in Fig. 4.24 (a), the average on-state power loss in thyristor is

$$P_{av} = \frac{1}{T}\int_0^{T/2} v_T \cdot i_a \cdot dt$$

$$= \frac{1}{T}\int_0^{T/2} 1.76 \times 80 \ dt = \frac{1.76 \times 80 \times T}{2T} = 70.4 \ \text{W}$$

Waveform of i_a gives the rms current rating of thyristor as

$$\sqrt{\frac{80^2 \times T}{2T}} = 56.577 \ \text{A}$$

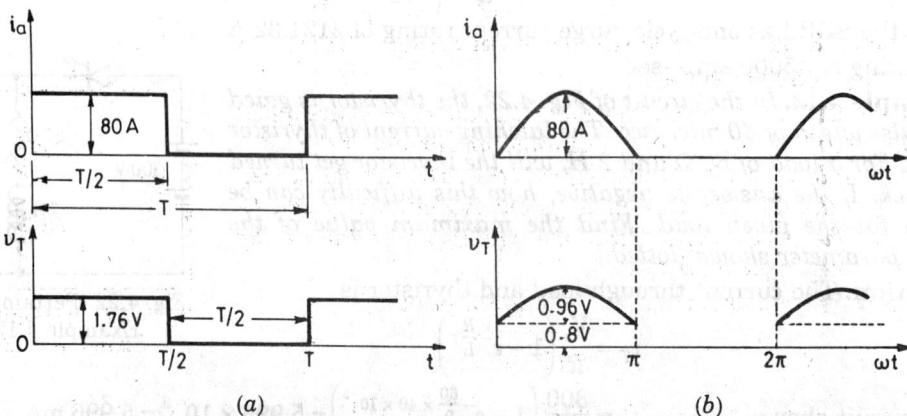

(a)   (b)

Fig. 4.24. Current and voltage waveforms pertaining to Example 4.13.

(b) Here $\qquad v_T = 0.8 + 0.012 \times 30 = 1.16$ V

$$\therefore \qquad P_{av} = \frac{1.16 \times 30 \times T}{3T} = 11.6 \text{ W}$$

Rms current rating $\quad = 30 \times \dfrac{1}{\sqrt{3}} = 17.321$ A

(c) Half-sine wave of peak value of 80 A, Fig. 4.24 (b), can be expressed as $i_a = 80 \sin \omega t$.

$$\therefore \qquad v_T = 0.8 + 0.012 \times 80 \sin \omega t = 0.8 + 0.96 \sin \omega t$$

From the waveforms for i_a and v_T shown in Fig. 4.24 (b), the average on-state power loss is given by

$$P_{av} = \frac{1}{2\pi} \int_0^\pi (0.8 + 0.96 \sin \omega t)\,(80 \sin \omega t)\,d(\omega t)$$

$$= \frac{1}{2\pi} \int_0^\pi 64 \sin \omega t \cdot d(\omega t) + \frac{1}{2\pi} \int_0^\pi 76.8 \sin^2 \omega t \cdot d(\omega t)$$

$$= \frac{1}{2\pi} \times 64 \left| - \cos \omega t \right|_0^\pi + \frac{76.8}{4\pi} \left| \omega t - \frac{\sin 2\omega t}{2} \right|_0^\pi$$

$$= 20.372 + 19.2 = 39.572 \text{ W}$$

Rms current rating $\qquad = \dfrac{I_{max}}{2} = \dfrac{80}{2} = 40$ A.

4.7. THYRISTOR PROTECTION

Reliable operation of a thyristor demands that its specified ratings are not exceeded. In practice, a thyristor may be subjected to overvoltages or overcurrents. During SCR turn-on, di/dt may be prohibitively large. There may be false triggering of SCR by high value of dv/dt. A spurious signal across gate-cathode terminals may lead to unwanted turn-on. A thyristor must be protected against all such abnormal conditions for satisfactory and reliable operation of SCR circuit and the equipment. SCRs are very delicate devices, their protection against abnormal operating conditions is, therefore, essential. The object of this section is to discuss various techniques adopted for the protection of SCRs.

(a) *di/dt protection.* When a thyristor is forward biased and is turned on by a gate pulse, conduction of anode current begins in the immediate neighbourhood of the gate-cathode junction, Fig. 4.6 (a). Thereafter, the current spreads across the whole area of junction. The thyristor design permits the spread of conduction to the whole junction area as rapidly as possible. However, if the rate of rise of anode current, *i.e. di/dt*, is large as compared to the spread velocity of carriers, local hot spots will be formed near the gate connection on account of high current density. This localised heating may destroy the thyristor. Therefore, the rate of rise of anode current at the time of turn-on must be kept below the specified limiting value. The value of *di/dt* can be maintained below acceptable limit by using a small inductor, called *di/dt inductor*, in series with the anode circuit. Typical *di/dt* limit values of SCRs are 20–500 A/μ sec. The method of determining inductance of *di/dt* inductor is illustrated in Example 4.14.

Local spot heating can also be avoided by ensuring that the conduction spreads to the whole area as rapidly as possible. This can be achieved by applying a gate current nearer to (but never greater than) the maximum specified gate current.

(b) *dv/dt protection.* It has already been discussed in Art.4.2 that if rate of rise of suddenly applied voltage across thyristor is high, the device may get turned on. Such phenomena of turning-on a thyristor, called *dv/dt turn-on* must be avoided as it leads to false operation of the thyristor circuit. For controllable operation of the thyristor, the rate of rise of forward anode to cathode voltage dV_a/dt must be kept below the specified rated limit. Typical values of *dv/dt* are 20 – 500 V/μsec. False turn-on of a thyristor by large *dv/dt* can be prevented by using a snubber circuit in parallel with the device.

4.7.1. Design of Snubber Circuits

A snubber circuit consists of a series combination of resistance R_s and capacitance C_s in parallel with the thyristor as shown in Fig. 4.25. Strictly speaking, a capacitor C_s in parallel with the device is sufficient to prevent unwanted *dv/dt* triggering of the SCR. When switch S is closed, a sudden voltage appears across the circuit. Capacitor C_s behaves like a short circuit, therefore voltage across SCR is zero. With the passage of time, voltage across C_s builds up at a slow rate such that *dv/dt* across C_s and therefore across SCR is less than the specified maximum *dv/dt* rating of the device. Here the question arises that if C_s is enough to prevent accidental turn-on of the device by *dv/dt*, what is the need of putting R_s in series with C_s? The answer to this is as under.

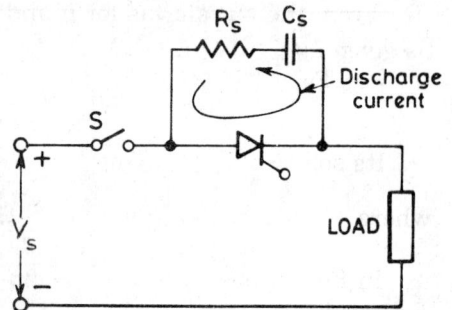

Fig. 4.25. Snubber circuit across SCR.

Before SCR is fired by gate pulse, C_s charges to full voltage V_s. When the SCR is turned on, capacitor discharges through the SCR and sends a current equal to $V_s/$ (resistance of local path formed by C_s and SCR). As this resistance is quite low, the turn-on *di/dt* will tend to be excessive and as a result, SCR may be destroyed. In order to limit the magnitude of discharge current, a resistance R_s is inserted in series with C_s as shown in Fig. 4.25. Now when SCR is turned on, initial discharge current V_s/R_s is relatively small and turn-on *di/dt* is reduced.

In actual practice ; R_s, C_s and the load circuit parameters should be such that *dv/dt* across C_s during its charging is less than the specified *dv/dt* rating of the SCR and discharge current at the turn-on of SCR is within reasonable limits. Normally, R_s, C_s and load circuit parameters form an underdamped circuit so that *dv/dt* is limited to acceptable values.

The design of snubber circuit parameters is quite complex. Here only an approximate method of their calculation is presented in Example 4.14. In practice, designed snubber parameters are adjusted up or down in the final assembled power circuit so as to obtain a satisfactory performance of the power electronics system.

Example 4.14. *Fig. 4.26 (a) shows a thyristor controlling the power in a load resistance R_L. The supply voltage is 240 V dc and the specified limits for di/dt and dv/dt for the SCR are 50 A/μsec and 300 V/μsec respectively. Determine the values of the di/dt inductance and the snubber circuit parameters R_s and C_s.*

Solution. Snubber circuit parameters R_s and C_s are connected across SCR and *di/dt* inductor L in series with anode circuit as shown in Fig. 4.26 (b). When switch S is closed, the capacitor behaves like a short circuit and SCR in the forward blocking state offers a very high resistance. Therefore, the equivalent circuit soon after the instant of closing the switch S is as shown in Fig. 4.26 (c). For this circuit, the voltage equation is

$$V_s = (R_s + R_L)\, i + L\, \frac{di}{dt} \qquad \qquad ...(4.10\ a)$$

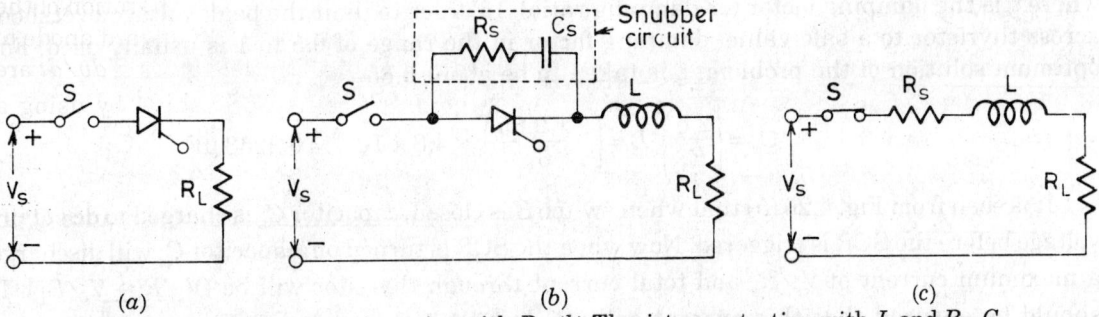

Fig. 4.26. (a) Thyristor in series with R_L (b) Thyristor protection with L and R_s, C_s
(c) Equivalent circuit of Fig. 4.26 (b) at the instant switch S is closed.

Its solution gives, $\qquad i = I\,(1 - e^{-t/\tau})$

where $\qquad\qquad I = \dfrac{V_s}{R_s + R_L}$ and $\tau = \dfrac{L}{R_s + R_L}$

In Eq. (4.10 a), t is the time in seconds measured from the instant of closing the switch. From this equation,

$$\frac{di}{dt} = I \cdot e^{-t/\tau} \cdot \frac{1}{\tau} = \frac{V_s}{R_s + R_L} \cdot \frac{R_s + R_L}{L} e^{-t/\tau}$$

$$= \frac{V_s}{L} e^{-t/\tau}$$

The value of di/dt is maximum when $t = 0$.

$\therefore \qquad\qquad \left(\dfrac{di}{dt}\right)_{max} = \dfrac{V_s}{L} \qquad\qquad\qquad\qquad$ ...(4.10 b)

or $\qquad\qquad L = \dfrac{V_s}{(di/dt)_{max}} = \dfrac{240 \times 10^{-6}}{50} = 4.8\ \mu H$

The voltage across SCR is given by, $v_a = R_s \cdot i$

or $\qquad\qquad\qquad \dfrac{dv_a}{dt} = R_s \cdot \dfrac{di}{dt}$

or $\qquad\qquad\qquad \left(\dfrac{dv_a}{dt}\right)_{max} = R_s \cdot \left(\dfrac{di}{dt}\right)_{max} \qquad\qquad$ ...(4.11)

From Eq. (4.10 b) and (4.11),

$$\left(\frac{dv_a}{dt}\right)_{max} = \frac{R_s \cdot V_s}{L} \qquad\qquad\qquad\qquad \text{...(4.12)}$$

or $\qquad\qquad R_s = \dfrac{L}{V_s}\left(\dfrac{dv_a}{dt}\right)_{max} = \dfrac{4.8}{240} \times 300 = 6\ \Omega$

The circuit of Fig. 4.26, consisting of R, L, C, should be fully analysed to determine the optimum values of snubber circuit parameters R_s, C_s. The analysis of this circuit shows that resistance R_s can be obtained from the relation [9]

$$R_s = 2\,\xi\,\sqrt{\frac{L}{C_s}}$$

where ξ is the damping factor (or damping ratio). In order to limit the peak voltage overshoot across thyristor to a safe value, damping factor in the range of 0.5 to 1 is usually used. For optimum solution of the problem, ξ is taken to be about 0.65.

$$\therefore \qquad C_s = \left(\frac{2\,\xi}{R_s}\right)^2 L = \left(\frac{2 \times 0.65}{6}\right)^2 \times 4.8 \times 10^{-6} = 0.2253 \; \mu F$$

It is seen from Fig. 4.26 (b) that when switch S is closed, capacitor C_s is charged to dc supply voltage before the SCR is triggered. Now when the SCR is turned on, capacitor C_s will discharge a maximum current of V_s/R_s and total current through thyristor will be $(V_s/R_s + V_s/R_L)$. It should be ensured that this current spike is less than the peak repetitive current rating (I_{TRM}) of the SCR. Thus if R_s is small, the current spike contributed by the discharge of C_s will be large. In order to reduce this spike, R_s is normally taken greater than what is required to limit dv/dt. At the same time, value of C_s is also reduced so that energy stored in C_s is small and the snubber discharge does not harm SCR when it is turned on. Thus, in the present case, R_s may be chosen somewhat higher than 6Ω, say 10Ω and C_s somewhat less than $0.2253 \; \mu F$, say $0.15 \; \mu F$. The adoption of the new value of R_s demands a new value of L. From Eq. (4.12),

$$L = \frac{R_s \cdot V_s}{(dv_a/dt)_{max}} = \frac{10 \times 240}{300} = 8 \; \mu H$$

This value of inductance is more than that required to limit di/dt to 50 A/μ sec.

For ac circuits, maximum value of input voltage (V_m) can be used in place of V_s in Eq. (4.12) for computing R_s.

Example 4.15. *A thyristor operating from a peak supply voltage of 400 V has the following specifications :*

Repetitive peak current, $I_p = 200$ A, $(di/dt)_{max} = 50$ A/μs, $\left(\dfrac{dv}{dt}\right)_{max} = 200$ V/μs.

Choosing a factor of safety of 2 for I_p, $\left(\dfrac{di}{dt}\right)_{max}$ and $\left(\dfrac{dv}{dt}\right)_{max}$, design a suitable snubber circuit. The minimum value of load resistance is 10 Ω.

Solution. For a factor of safety of 2, the permitted values are $I_p = \dfrac{200}{2} = 100$ A, $\left(\dfrac{di}{dt}\right)_{max} = \dfrac{50}{2} = 25$ A/μs, $(dv/dt)_{max} = \dfrac{200}{2} = 100$ V/μs.

In order to restrict the rate of rise of current beyond specified value, (di/dt) inductor must be inserted in series with thyristor. From Example 4.14,

$$L = \frac{V_s}{(di/dt)_{max}} = \frac{400 \times 10^{-6}}{25} = 16 \; \mu H$$

$$R_s = \frac{L}{V_s} \cdot \left(\frac{dv}{dt}\right)_{max} = \frac{16 \times 10^{-6}}{400} \times \frac{100}{10^{-6}} = 4 \; \Omega$$

Before thyristor is turned on, C_s is charged to 400 V. When thyristor is turned on, the peak current through the thyristor is

$$\frac{400}{10} + \frac{400}{4} = 140 \; A.$$

As this peak current through SCR is more than the permissible peak current of 100 A, the magnitude of R_s must be increased. Taking R_s as 8 Ω, the peak current through the SCR

$$= \frac{400}{10} + \frac{400}{8} = 90 \text{ A, less than the allowable peak current. So choose } R_s = 8 \text{ Ω.}$$

Also

$$C_s = \left(\frac{2\,\xi}{R_s}\right)^2 L = \left(\frac{1.3}{8}\right)^2 \times 16 \times 10^{-6} = 0.4225 \text{ μF}$$

The value of C_s may be lowered as discussed in the previous example, so C_s may be taken as 0.30 μF.

At the instant switch S is closed, Fig. 4.26, thyristor is open circuited and current through C_s is given by

$$C_s \frac{dv}{dt} \cong \frac{V_s}{R_s + R_L}$$

or

$$0.3 \times 10^{-6} \frac{dv}{dt} = \frac{400}{10 + 8}$$

or

$$\frac{dv}{dt} = \frac{400}{18} \times \frac{1}{0.3 \times 10^{-6}} = 74.07 \text{ V/μs}$$

Since designed value of (dv/dt) is less than the specified maximum value of 100 $V/\mu s$, value of C_s chosen is correct. So choose $L = 10$ μH, $R_s = 8$ Ω and $C_s = 0.3$ μF.

Example 4.16. *Thyristor shown in Fig. 4.27 has $I^2 t$ rating of 20 $A^2 s$. If terminal A gets short-circuited to ground, calculate the fault clearance time so that SCR is not damaged.*

Solution. The worst possible fault current should be considered for calculating the fault clearance time. Maximum fault current occurs when source voltage is at its peak = 230√2 V.

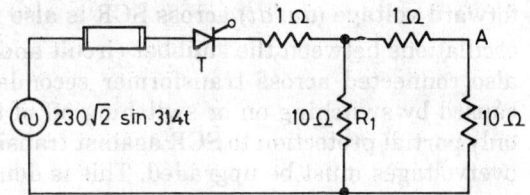

Fig. 4.27. Pertaining to Example 4.16.

When terminal A gets short-circuited to ground, the resistance offered to source

$$= 1 + \frac{10 \times 1}{10 + 1} = 21/11 \text{ Ω. Assuming maximum fault current} = \frac{230\sqrt{2} \times 11}{21} \text{ A to remain constant}$$

during the short clearance time t_c, we get

$$\int_0^{t_c} i^2 \cdot dt = \int_0^{t_c} \left(\frac{230\sqrt{2} \times 11}{21}\right)^2 \cdot dt = 20 \text{ A}^2\text{s}$$

$$\therefore \qquad t_c = 20 \left[\frac{21}{230\sqrt{2} \times 11}\right]^2 \times 1000 \text{ ms} = 0.6892 \text{ ms.}$$

4.7.2. Overvoltage Protection

Thyristors are very sensitive to overvoltages just as other semi-conductor devices are. Overvoltage transients are perhaps the main cause of thyristor failure. Transient overvoltages cause either maloperation of the circuit by unwanted turn-on of a thyristor or permanent damage to the device due to reverse breakdown. A thyristor may be subjected to internal or external overvoltages ; the former is caused by the thyristor operation whereas the latter comes from the supply lines or the load circuit.

(*i*) *Internal overvoltages.* Large voltages may be generated internally during the commutation of a thyristor. After thyristor anode current reduces to zero, anode current reverses due to stored charges. This reverse recovery current rises to a peak value at which time the SCR begins to block. After this peak, reverse recovery current decays abruptly with large di/dt. Because of the series inductance L of the SCR circuit, large transient voltage $L\dfrac{di}{dt}$ is produced. As this internal overvoltage may be several times the breakover voltage of the device, the thyristor may be destroyed permanently.

(*ii*) *External overvoltages.* External overvoltages are caused due to the interruption of current flow in an inductive circuit and also due to lightning strokes on the lines feeding the thyristor systems. When a thyristor converter is fed through a transformer, voltage transients are likely to occur when the transformer primary is energised or de-energised. Such overvoltages may cause random turn on of a thyristor. As a result, the overvoltages may appear across the load causing the flow of large fault currents. Overvoltages may also damage the thyristor by an inverse breakdown. For reliable operation, the overvoltages must be suppressed by adopting suitable techniques.

4.7.2.1. Suppression of overvoltages. In order to keep the protective components to a minimum, thyristors are chosen with their peak voltage ratings of 2.5 to 3 times their normal peak working voltage. The effect of overvoltages is usually minimised by using RC circuits and non-linear resistors called *voltage clamping devices.*

The RC circuit, called snubber circuit, is connected across the device to be protected, see Fig. 4.30. It provides a local path for internal overvoltages caused by reverse recovery current. Snubber circuit is also helpful in damping overvoltage transient spikes and for limiting dv/dt across the thyristor. The capacitor charges at a slow rate and thus the rate of rise of forward voltage (dv/dt) across SCR is also reduced. The resistance R_s damps out the ringing oscillations between the snubber circuit and the stray circuit inductance. Snubber circuits are also connected across transformer secondary terminals to suppress overvoltage transients caused by switching on or switching off of the primary winding. As snubber circuits provide only partial protection to SCR against transient overvoltages, thyristor protection against such overvoltages must be upgraded. This is done with the help of voltage-clamping devices.

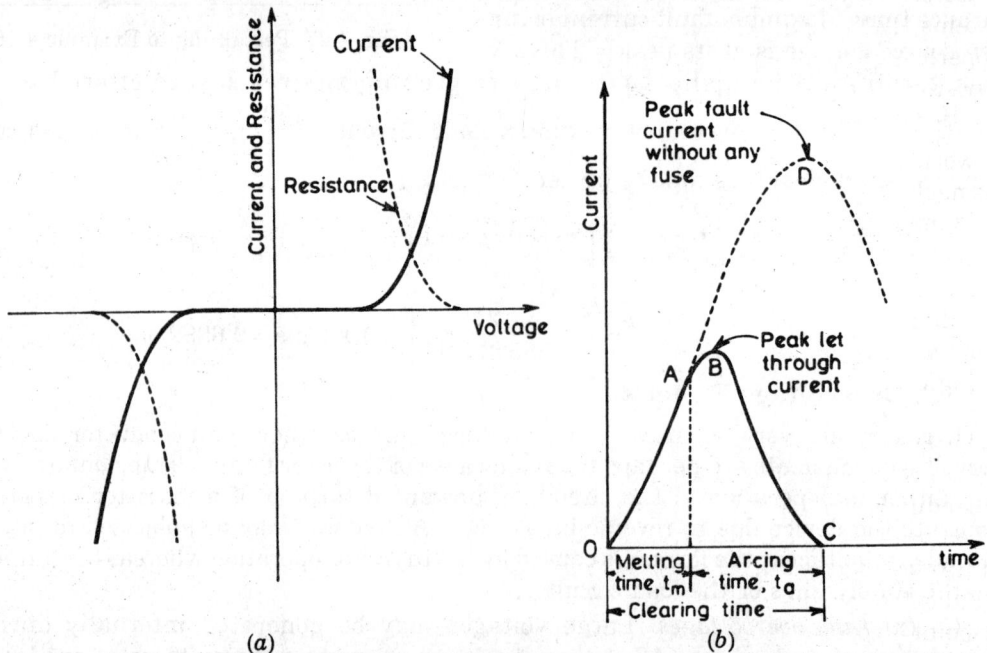

Fig. 4.28. (*a*) Volt-ampere and volt-resistance characteristics of voltage-clamping device
(*b*) Action of current-limiting fuse in an ac circuit.

A voltage-clamping (V.C.) device is a non-linear resistor connected across SCR as shown in Fig. 4.30. The V.C. device has falling resistance characteristic with increasing voltage, Fig. 4.28 (a). Under normal working conditions of voltage below the clamping level, the device has a high resistance and draws only a small leakage current. When a voltage surge appears, the V.C. device operates in the low resistance region and produces a virtual short circuit across the SCR. The increased current associated with virtual short circuit produces an increased voltage drop in the source and line impedances and as a result, voltage across SCR is clamped to a safe value. After the surge energy is dissipated in the non-linear resistor, the operation of the V.C. device returns to its high resistance region. Selenium thyrector diodes, metal oxide varistors or avalanche diode suppressors are commonly employed for protecting the thyristor circuit against overvoltages. As the voltage clamping ability of a thyrector is inferior to those of metal oxide varistor and avalanche-diode suppressor, use of thyrector is on the decline.

It has already been stated that RC snubber is not enough for overvoltage protection of SCR. In practic, therefore, a combined protection consisting of RC snubber and V.C. device is provided to thyristors as shown in Fig. 4.30.

4.7.3. Overcurrent Protection

Thyristors have small thermal time constants. Therefore, if a thyristor is subjected to overcurrent due to faults, short circuits or surge currents ; its junction temperature may exceed the rated value and the device may be damaged. There is thus a need for the overcurrent protection of SCRs. As in other electrical systems, overcurrent protection in thyristor circuits is achieved through the use of circuit breakers and fast-acting fuses as shown in Fig. 4.30.

The type of protection used against overcurrent depends upon whether the supply system is weak or stiff. In a weak supply network, fault current is limited by the source impedance below the multi-cycle surge current rating of the thyristor. In machine tool and excavator drives, if the motor stalls due to overloads, the current is limited by the source and motor impedances. The filter inductance commonly employed in dc and ac drives may limit the rate of rise of fault current below the multicycle surge current rating of the thyristor. For all such systems, overcurrent can be interrupted by conventional fuses and circuit breakers. However, proper co-ordination is essential to guarantee that (i) fault current is interrupted before the thyristor is damaged and (ii) only faulty branches of the network are isolated.

Conventional protective methods are, however, inadequate in electrical stiff supply networks. In such systems, magnitude and rate of rise of current is not limited because source has negligible impedance. As such, fault current and therefore junction temperature rise within a few milliseconds. Special fast-acting current-limiting fuses are, therefore, required for the protection of thyristors in these stiff supply networks.

The operation of fast-acting current-limiting fuse is illustrated in Fig. 4.28 (b). These fuses and thyristors are found to have similar thermal properties, their co-ordination is therefore simpler. The current-limiting fuse consists of one or more fine silver ribbons having very short fusing time. In Fig. 4.28 (b), fault is shown to occur at zero crossing of the ac sine wave, i.e. at $t = 0$. Without fuse, the fault current would rise upto A and then would follow dotted curve, reach peak value D and then decrease as shown. A properly selected current limiting fuse melts at A. An arc is then struck. For a brief interval after A, the current continues to rise depending upon the circuit parameters and the fuse design. This current reaches a peak value, called *peak let through current,* which is indicated by point B in Fig. 4.28 (b). Note that peak let through current is considerably less than the peak fault current without the fuse, the latter is indicated by point D. After the point B, arc resistance increases and fault current decreases. At point C, arcing stops and the fault current is cleared. The total clearing time t_c is the sum of melting time t_m and arcing time t_a, i.e. $t_c = t_m + t_a$.

Proper co-ordination between fast-acting current-limiting fuse and thyristor is essential. A fuse carries the thyristor current as both are placed in series. Therefore, the fuse must be rated to carry full-load current plus a marginal overload current for an indefinite period. But the peak let through current of fuse must be less than the subcycle surge current rating of the SCR. The voltage across the fuse during arcing period is known as arcing, or recovery, voltage. This voltage is equal to the sum of source voltage and the emf induced in the circuit inductance during arcing time t_a. If the fuse current is interrupted abruptly, induced e.m.f. $L\dfrac{di}{dt}$ may be high ; as a result arcing voltage would be excessive. It should therefore be ensured during fuse design and co-ordination that arcing voltage is limited to less than twice the peak supply voltage. In case voltage rating of the fuse is far in excess of circuit voltage, an abrupt current interruption would lead to dangerous overvoltages.

When both circuit breaker and fast-acting current-limiting fuse are used for overcurrent protection of SCR, Fig. 4.30, the faulty circuit must be cleared before any damage is done to the device. A circuit breaker has long tripping time, it is therefore generally used for protecting the semiconductor device against the continuous overloads or against surge currents of long duration. A fast-acting C.L. fuse is used for protecting thyristors against large surge currents of very short duration. The tripping time of the circuit breaker, the fusing-time of the fast-acting fuse must be properly co-ordinated with the rating of a thyristor. In order that fuse protects the thyristor reliably, the I^2t rating of the fuse must be less than that of the SCR.

Electronic crowbar protection. As thyristor possesses high surge current capability, it can be used in an electronic crowbar circuit for overcurrent protection of power converters using SCRs. An electronic crowbar protection provides rapid isolation of the power converter before any damage occurs.

Fig. 4.29 illustrates the basic principle of electronic crowbar protection. A crowbar thyristor is connected across the input dc terminals. A current sensing resistor detects the value of converter current. If it exceeds preset value, gate circuit provides the signal to crowbar SCR and turns it on in a few microseconds. The input terminals are then short-circuited by crowbar SCR

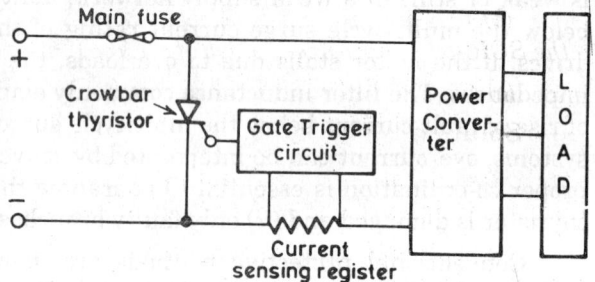

Fig. 4.29. Elementary electronic crowbar circuit.

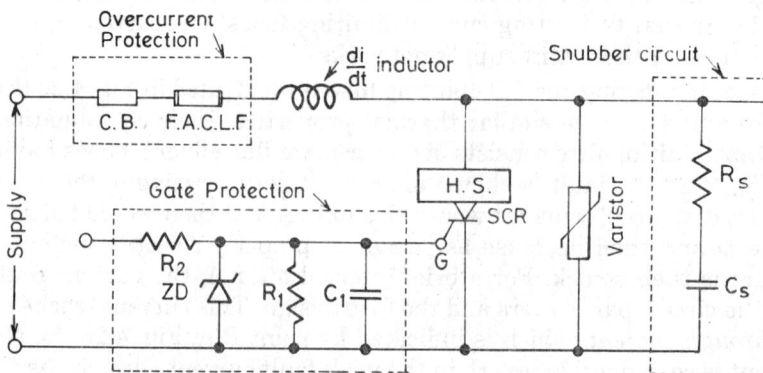

Fig. 4.30. Circuit components showing the thyristor protection.
C.B.—Circuit breaker ; F.A.C.L.F.—Fast acting current limiting fuse ; H.S.—Heat sink ; ZD—Zener diode.

and it shunts away the converter overcurrent. The crowbar thyristor current depends upon the source voltage and its impedance. After some time, main fuse interrupts the fault current. The fuse may be replaced by a circuit breaker if SCR has adequate surge current rating.

4.7.4. Gate Protection

Gate circuit should also be protected against overvoltages and overcurrents. Overvoltages across the gate circuit can cause false triggering of the SCR. Overcurrent may raise junction temperature beyond specified limit leading to its damage. Protection against over-voltages is achieved by connecting a zener diode ZD across the gate circuit. A resistor R_2 connected in series with the gate circuit provides protection against overcurrents.

A common problem in thyristor circuits is that they suffer from *spurious, or noise, firing*. Turning-on or turning-off of an SCR may induce trigger pulses in a nearby SCR. Sometimes transients in a power circuit may also cause unwanted signal to appear across the gate of a neighbouring SCR. These undesirable trigger pulses may turn on the SCR leading to false operation of the main SCR. Gate protection against such spurious firing is obtained by using shielded cables or twisted gate leads. A varying flux caused by nearby transients cannot pass through twisted gate leads or shielded cables. As such no e.m.f. is induced in these cables and spurious firing of thyristors is thus minimised. A capacitor and a resistor are also connected across gate to cathode to bypass the noise signals, Fig. 4.30. The capacitor should be less than $0.1\ \mu F$ and must not deteriorate the waveshape of the gate pulse.

Example 4.17. *For the circuit shown in Fig. 4.31,*

(a) calculate the maximum values of di/dt and dv/dt for the SCR,

(b) find the rms and average current ratings of the SCR for firing angle delays of 90°and 150° and

(c) suggest a suitable voltage rating of the SCR.

Solution. (*a*) From Eq. (4.10),

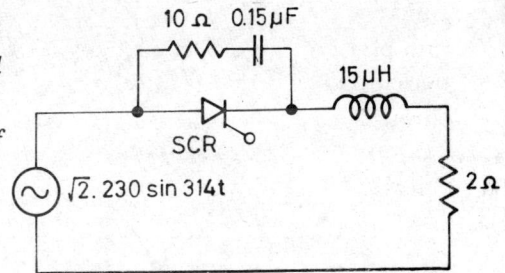

Fig. 4.31. Pertaining to Example 4.17.

$$\left(\frac{di}{dt}\right)_{max} = \left(\frac{V_s}{L}\right)$$

$$= \frac{\sqrt{2}\cdot 230}{15\times 10^{-6}} = 21.685\ \text{A/μsec.}$$

From Eq. (4.11),
$$\left(\frac{dv}{dt}\right)_{max} = R_s\left(\frac{di}{dt}\right)_{max} = 10\times 21.685 = 216.85\ \text{V/μsec.}$$

(*b*) For $15\ \mu H$, $X_L = 314\times 15\times 10^{-6} = 0.00471\ \Omega$. As this value of X_L is much lower than $R = 2\ \Omega$, the current is primarily limited by $2\ \Omega$.

$$\therefore \qquad I_{max} = \frac{\sqrt{2}\cdot 230}{2} = 115\cdot\sqrt{2}$$

For firing angle delays of 90° and 150°, the conduction angles are 90° and 30° respectively and from Example 4.9, the respective values of form factors are $\pi/\sqrt{2}$ and 3.98184.

$$\therefore\ \text{For firing angle delay of 90°,}\quad I_{TAV} = \frac{\sqrt{2}\cdot 115\cdot\sqrt{2}}{\pi} = 73.211\ \text{A}$$

and for firing angle delay of 150°, $\quad I_{TAV} = \frac{\sqrt{2}\cdot 115}{3.98184} = 40.844\ \text{A}$

RMS current rating of the thyristor is $115\sqrt{2} = 162.634$ A for any conduction angle, but average currents are 73.211 A for conduction angle of 90° and 40.844 A for conduction angle of 30°.

(c) Voltage rating of the SCR = (2.5 to 3) times the peak working voltage

$$= (2.5 \text{ to } 3) \times \sqrt{2} \cdot 230 = 813.173 \text{ V to } 975.807 \text{ V}.$$

So a voltage rating of about 900 V may be chosen for the SCR.

Example 4.18. *For the circuit shown in Fig. 4.32 (a), the initial voltage across capacitor is $v_c(0) = -100$ V. Sketch the time variations of i, v_L, v_C, i_D and i_L after the thyristor is turned on at $t = 0$.*

Solution. When the thyristor is turned on at $t = 0$, the voltage equation for the circuit is

$$L\frac{di}{dt} + \frac{1}{C}\int i\,dt = V_s$$

Its Laplace transform is

$$sL \cdot I(s) + \frac{1}{C}\left[\frac{I(s)}{s} + \frac{C \cdot v_c(0)}{s}\right] = \frac{V_s}{s}$$

or

$$I(s) = \frac{V_s - v_{co}}{L} \cdot \frac{1}{s^2 + \frac{1}{LC}} = \frac{300}{L} \cdot \frac{1/\sqrt{LC}}{\frac{1}{\sqrt{LC}}\left(s^2 + \frac{1}{LC}\right)}$$

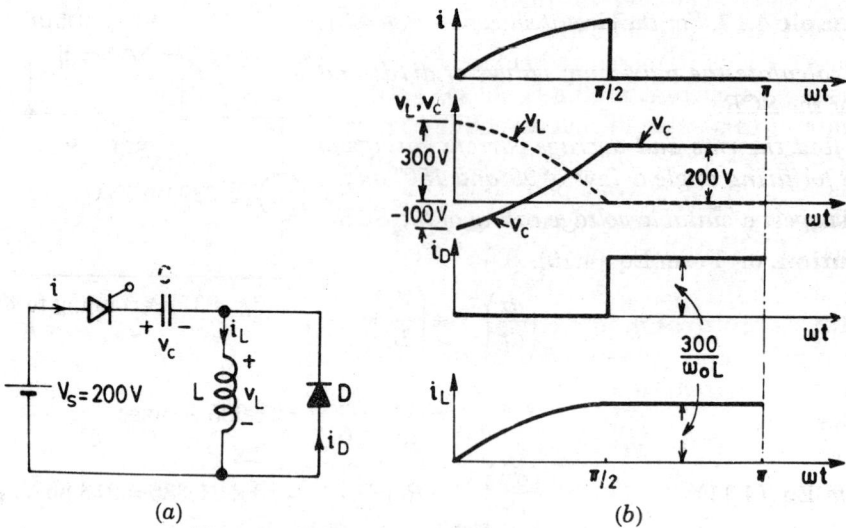

Fig. 4.32. Pertaining to Example 4.18.

Its Laplace inverse is $i(t) = \dfrac{300}{\omega_0 L}\sin\omega_0 t$ where $\omega_0 = \dfrac{1}{\sqrt{LC}}$

$\therefore$

$$v_L = L\frac{di(t)}{dt} = 300\cos\omega_0 t$$

$$v_C = V_s - v_L = 200 - 300\cos\omega_0 t$$

The current and voltage waveforms are as shown in Fig. 4.32 (b) At $\pi/2$, v_L tends to reverse and as a result, diode D gets forward biased and current i_L starts flowing through D as $i_D \cdot v_L$ is therefore zero from $\pi/2$ to π. Voltage v_c remains 200 V and current i zero from $\pi/2$ to π as shown in Fig. 4.32 (b).

4.8. IMPROVEMENT OF THYRISTOR CHARACTERISTICS

It has been explained in Art. 4.3.1 that rate of growth of anode current during rise time of t_{on} is high, but cathode-conduction area is small. High rise of anode current in a thyristor, associated with high anode voltage, causes more losses. These high losses, occurring over a small cathode-conduction area during rise time, may result in hot spots leading to the destruction of the device. A high value of dv/dt may turn on the thyristor at an unwanted instant which is undesirable. This all prompts us for an improvement in di/dt as well as dv/dt ratings of thyristors. A boost in these ratings can be made by doing some structural modifications in thyristors; this is explained below.

4.8.1. Improvement in di/dt Rating

The rate of rise of anode current (di/dt) in a thyristor depends primarily on the initial area of cathode conduction during rise time. This implies that if initial cathode conduction area is increased, the di/dt rating also gets improved. There are two methods of doing this, (i) by using a higher-gate current (ii) by intermixing the gate-cathode regions.

4.8.1.1. Higher-gate current. At the start of turn on, if higher-gate current is applied, turned-on area of cathode surface has to be more for handling this higher-gate current. As a consequence, initial cathode-conduction area for allowing anode current to pass through it, increases, and this is what is desired. The widely used gate current profile is shown in Fig. 4.7. However, higher gate current should not be obtained from a gate drive circuit. The usual way of accomplishing this goal of higher-gate current is through the use of a pilot thyristor shown in Fig. 4.33. When pilot thyristor is turned on, a high value of gate current

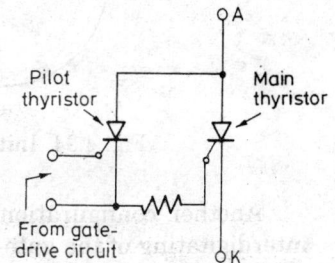

Fig. 4.33. Main thyristor turn on by pilot thyristor.

flows from anode A, pilot thyristor and gate-cathode terminals of main thyristor for switching it on.

4.8.1.2. Structural modification of the device. As stated above, the di/dt rating of a thyristor can be improved by having more cathode-conduction area during delay and rise time of t_{on}. This can be achieved by higher-gate current (already discussed) and by modifying the gate-cathode geometry. This alteration consists of intermixing, or interdigitating gate and cathode regions. The effect of this structural change can be realized by examining the initial conduction process first (i) in side-gate thyristor and then (ii) in centre-gate thyristor.

A side-gate thyristor is shown in Fig. 4.34 (a). When gate current is applied, the gate-current density is higher near the gate terminal. As a result, cathode-conduction area is small during delay and rise time. This shows that initial conduction occurs over a narrow channel near the gate terminal as shown in Fig. 4.34 (a). Reference to Fig. 4.6 and its relevant write-up is also helpful.

A centre-gate thyristor is shown in Fig. 4.34 (b). When positive gate pulse is applied, the gate current flows from gate to cathode in all possible directions covering a ring-shaped area on the cathode surface as indicated in Fig. 4.34 (b). Examination of Fig. 4.34 (a) and (b) shows that initial area of cathode conduction is very large in centre-gate thyristor as compared to that in side-gate thyristor. This illustrates that initial area of cathode conduction can be enhanced significantly by intermixing the gate-cathode regions appropriately.

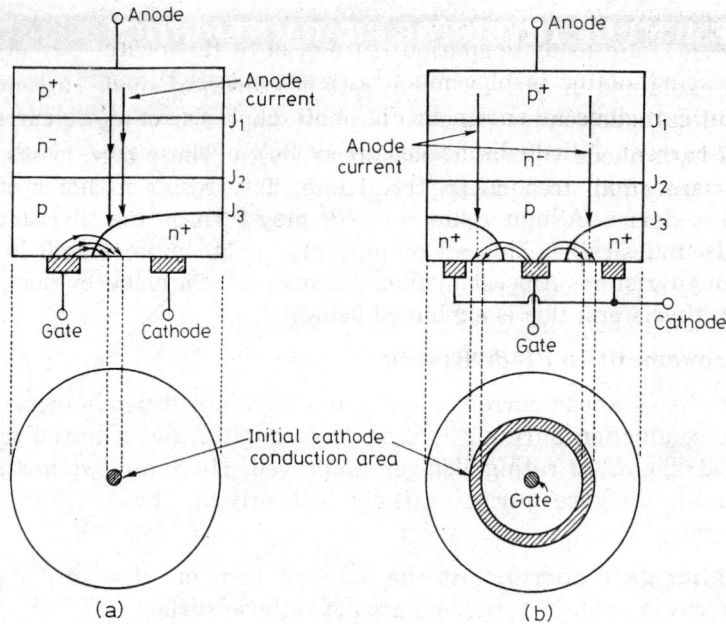

Fig. 4.34. Initial cathode-conduction area in (a) side-gate thyristor and (b) centre-gate thyristor.

Another configuration indicating the intermixing, or interdigitating of the gate-cathode regions is illustrated in Fig. 4.35.

4.8.2. Improvement in dv/dt rating

It has already been discussed in Art. 4.2 (dv/dt triggering) that when dv/dt is large, high charging currents flow through the reversed biased junction J_2 which may turn on the thyristor. The effect of capacitor-charging current, or dv/dt, can be minimised by using *cathode-short structure* shown in Fig. 4.36 (a). Cathode-shorts are realized by overlapping metal on cathode n^+ regions with a narrow p-region in between. Fig. 4.36 (a) shows metallization M and N which form the cathode.

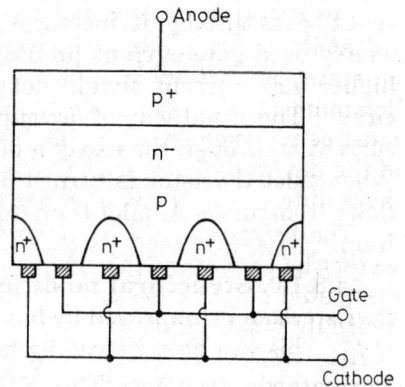

Fig. 4.35. Interdigitating of gate-cathode regions in a thyristor.

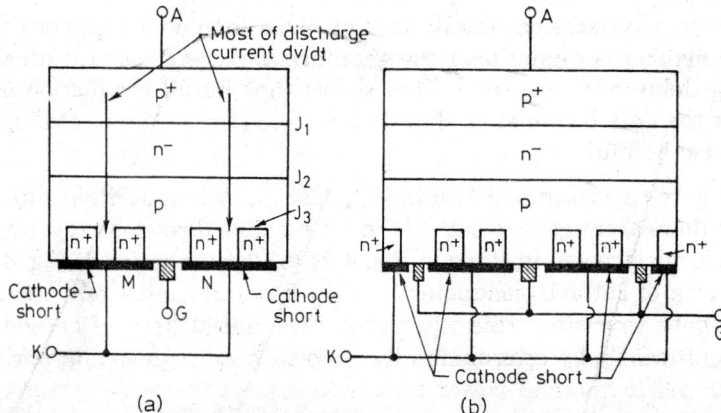

Fig. 4.36. Thyristor cathode-shorts (a) elementary intermixing (b) advanced intermixing.

In normal structure, discharge current dv/dt (acting as gate current) flows through J_3 junction, Fig. 4.3 (a), and leads to spurious turn of of SCR. In cathode-short structure, most of the discharge current (or displacement current) passes through narrow p channels in between cathode n^+ regions as shown in Fig. 4.36 (a). Junction J_3 shares only a negligible amount of dv/dt current. A little discharge current flowing through J_3 junction (and acting as gate current) is too small to turn on the device. Thus, higher valuos of dv/dt are now permissible with cathode-short structure.

The thermally generated leakage current across junction J_2 also does not pass through gate-cathode junction J_3. Therefore, current injection across gate-cathode, or J_3, junction is drastically reduced, hence the total discharge current, or dv/dt, can be larger without turning-on the device. It can, therefore, be inferred that cathode-short structure improves dv/dt rating of the thyristor.

4.9. HEATING, COOLING AND MOUNTING OF THYRISTORS

Some power loss occurs in a thyristor during its working. The various components of this power loss in the junction region of a thyristor are as under :

    (i) Forward conduction loss

    (ii) Loss due to leakage current during forward and reverse blocking

    (iii) Switching losses at turn-on and turn-off

    (iv) Gate triggering loss

At industrial power frequencies between zero and 400 Hz, the forward conduction loss, or on-state conduction loss, is usually the major component. But switching losses become dominant at high operating frequencies. These electrical losses produce thermal heat which must be removed from the junction region. The thermal losses and hence the temperature rise of the device increase with the thyristor rating. The cooling of thyristors, therefore, becomes more difficult as the SCR rating increases.

The heat produced in a thyristor by electrical loss is dissipated to ambient fluid (air or water) by mounting the device on a heat sink. When heat due to losses is equal to that dissipated by the heat sink, steady junction temperature is reached. Thyristor heating and hence its junction temperature rise is dependent primarily on current handled by the device during its working. As such, current rating of thyristors is often based on thermal considerations.

4.9.1. Thermal Resistance

Thermal energy, or heat, flows from a region of higher temperature to a region of lower temperature. This is similar to the flow of current from higher to lower potential in an electric circuit. There is thus an analogy between thermal-power flow and current flow as given in tabular form below :

	Thermal quantities		Electrical quantities
1.	Heat, J or Ws	1.	Charge, C or As
2.	Temperature difference, °C	2.	Potential difference, V
3.	Thermal power, or rate of heat transfer, W	3.	Current, or rate of charge transfer, A
4.	Thermal resistance, °C/W	4.	Electrical resistance, V/A or ohms.

It is seen from above that thermal resistance, analogous to electrical resistance, is the resistance offered to thermal power flow. Thermal resistance is denoted by θ. If power loss,

P_{av} in watts, causes the temperature of two points to be at $T_1 \,°C$ and $T_2 \,°C$ where $T_1 > T_2$, then thermal resistance is given by

$$\theta_{12} = \frac{T_1 - T_2}{P_{av}} \,°C/W \qquad \qquad ...(4.13)$$

The heat generated in a thyristor due to internal losses is taken to be developed at a junction within the semiconductor material. A simple arrangement of thyristor, its case and heat sink is shown in Fig. 4.37 (a). Various temperatures and thermal resistance are also indicated in this figure. The heat flow from thyristor junction to ambient fluid is as under:

   (i) from the junction to thyristor case, thermal resistance θ_{jc}

   (ii) from the thyristor case to heat sink, thermal resistance θ_{cs} and

   (iii) from the heat sink to the surrounding ambient fluid (air or water), thermal resistance θ_{SA}.

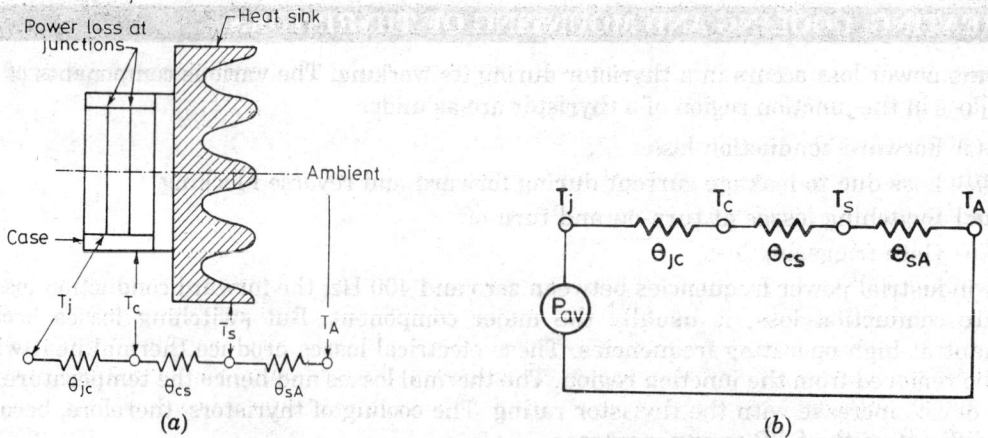

Fig. 4.37. (a) Thyristor with its case and heat sink
(b) thermal equivalent circuit for a thyristor.

There is thus thermal resistance θ_{jc} between junction temperature T_j and case temperature T_c. Similarly, there is thermal resistance θ_{cs} between T_c and sink temperature T_s and θ_{sA} between T_s and ambient temperature T_A. Using the electrical analogy, a thermal equivalent circuit depicting the flow of heat from junction to ambient fluid can be drawn as shown in Fig. 4.37 (b). Here P_{av} is the average rate of heat generated at a thyristor junction and is analogous to constant-current source. Here

$$P_{av} = \frac{T_j - T_c}{\theta_{jc}} = \frac{T_c - T_s}{\theta_{cs}} = \frac{T_s - T_A}{\theta_{sA}}$$

$$= \frac{T_j - T_A}{\theta_{jA}} \qquad \qquad ...(4.14)$$

where $\qquad \qquad \theta_{jA} = \theta_{jc} + \theta_{cs} + \theta_{sA},$

is the total thermal resistance between junction and ambient.

The junction-to-case thermal resistance θ_{jc} is specified in the thyristor data sheet. The case-to-sink thermal resistance θ_{cs} depends on the size of the device case, flatness of the case surface, the clamping pressure and the use of conducting grease between the interfaces. Usual value of θ_{cs} varies between 0.05°C/W and 0.5°C/W. In addition to θ_{jc}, thyristor data sheet also specify θ_{cs} assuming correct installation procedure and use of the interface thermal lubrication. The sink-to-ambient thermal resistance θ_{sA} is independent of the thyristor configuration. The parameters on which θ_{sA} depends are heat sink material, surface area and finish of the heat sink, volume occupied by heat sink and the type of cooling (air cooling or water cooling). For naturally cooled heat sink, θ_{sA} may be equal to 0.5°C/W and this value would be lower for better cooled heat sinks.

The difference in temperature between junction and ambient can be written from Fig. 4.37 as

$$T_j - T_A = P_{av}(\theta_{jc} + \theta_{cs} + \theta_{sA}) \qquad ...(4.15)$$

Eq. (4.15) shows that for maximum value of T_j (= 125°C), P_{av} can be increased by reducing θ_{sA}. This means that by providing efficient cooling system to the SCR, the power dissipation capability of the device can be increased.

4.9.2. Heat Sink Specifications

The thyristor data sheet specifies maximum junction temperature T_j, thermal resistances θ_{jc} and θ_{cs}. The manufacturers of heat sinks provide catalogs in which sufficient data on heat sink is available. Fig. 4.38 gives typical data in the form of curves for standard heat sinks of aluminium extrusions. These curves relate temperature difference $(T_s - T_A)$ in °C between heat sink and ambient *versus* average power dissipation P_{av} in watts.

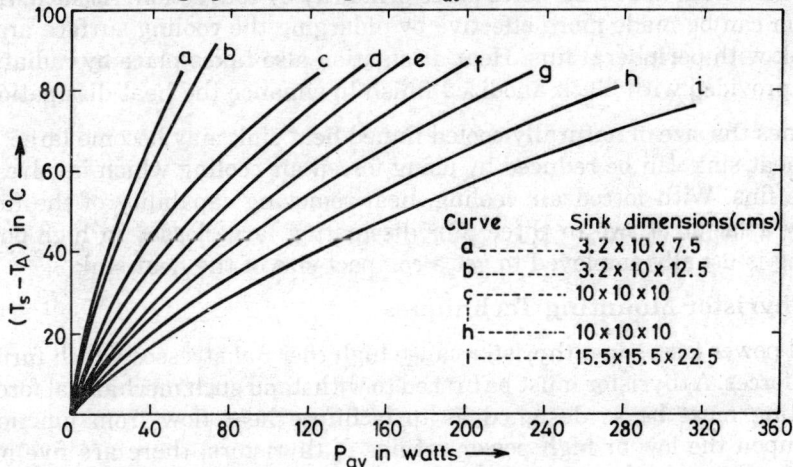

Fig. 4.38. Standard heat sink ratings of aluminium extrusions.

In order to illustrate the use of these curves, choose a particular heat sink and read P_{av} and $(T_s - T_A)$. Then the thermal resistance of the heat sink to ambient is calculated as

$$\theta_{sA} = \frac{T_s - T_A}{P_{av}} \qquad ...(4.16)$$

For maximum specified temperature T_j (usually 125°C) and a known ambient temperature T_A, the permissible value of P_{av} (with θ_{jc} and θ_{cs} already known) is calculated from Eq. (4.15) with θ_{sA} computed from Eq. (4.16). If this P_{av} is different from that chosen earlier from Fig. 4.38, another heat sink with other values of P_{av} and $(T_s - T_A)$ is tried until Eqs. (4.15) and (4.16) are satisfied. After deciding the value of P_{av}, use this value of P_{av} in Fig. 4.19 to obtain permissible value of average current rating for a given conduction angle and current waveform.

In the second method of heat sink design, first average armature current is determined from a known current waveform and conduction angle. Corresponding to this average current, P_{av} is read off from Fig. 4.19. For this P_{av}, thermal resistance θ_{sA} is determined from Eq. (4.15) as temperatures T_j (= 125°C), T_A and θ_{jc}, θ_{cs} are already known. This computed value of θ_{sA} is used to obtain temperature difference $(T_s - T_A)$ between heat sink and the ambient from Eq. (4.16). Using these values of P_{av} and $(T_s - T_A)$, an appropriate heat sink is selected from Fig. 4.38, the details of which are usually supplied by the manufacturers.

In the third method of heat sink selection, first compute average armature current as done in the second method. For this value of average current, obtain P_{av} from Fig. 4.19 and case temperature T_c from Fig. 4.20 for the known current waveform and conduction angle. An examination of Fig. 4.38 reveals that the sink temperature T_s in terms of case temperature T_c is given by

$$T_s = T_c - P_{av} \cdot \theta_{cs} \qquad \qquad ...(4.17)$$

As ambient temperature is known, $(T_s - T_A)$ can be calculated. Now, with the knowledge of P_{av} and $(T_s - T_A)$, a choice of suitable heat sink can be made from Fig. 4.38.

Heat sinks are made from metal with high thermal conductivity. Aluminium is the most commonly used metal. Copper, being a costly metal, is seldom used as a heat sink material. Heat dissipation from heat sink takes place primarily by convection. As such, thyristor cooling by convection can be made more effective by enlarging the cooling surface area by providing the heat sink with peripheral fins. Heat dissipation also takes place by radiation. Heat sinks are usually provided with black anodized finish to enhance the heat dissipation by radiation.

Sometimes the size of naturally-cooled finned heat sink may become large. In such a case, size of the heat sink can be reduced by using forced air cooling which involves a fan blowing air over the fins. With forced air cooling, heat-removing capability of the finned heat sink increases by a factor of two to three. For dissipating large losses in high-power thyristors, water-cooling is usually employed to get a compact size of the heat sink.

4.9.3. Thyristor Mounting Techniques

Internal power losses in a thyristor cause high thermal stresses which further give rise to mechanical forces. A thyristor must be braced to withstand such mechanical forces. In addition, SCR mounting must be so designed as to facilitate heat flow from junction to the case. Depending upon the low or high power ratings of thyristors, there are five major mounting techniques for SCRs as described below :

(*a*) *Lead-mounting.* For load-current rating of about one ampere, lead-mounted SCRs are used, Fig. 4.39 (*a*). Such SCRs do not require any additional cooling or heat sink. Their housings dissipate sufficient heat by radiation and convection.

(*b*) *Stud-mounting.* This type of construction shown in Fig. 4.39 (*b*) is very widely used due to its flexibility and ruggedness. The threaded stud forms the anode. The SCR is attached to a heat sink by means of threaded stud and nut. Thus anode gets electrically connected to the heat sink. If electrical connection between anode and heat sink is undesirable, then mica or PTFE washers are used in between the joining surfaces. Both mica and PTFE conduct heat easily but act as insulators to electricity.

(*c*) *Bolt-down mounting.* This is also called flat-pack mounting. This type of device **mounting** has tabs with one or more holes. Sometimes the hole is provided in the middle as

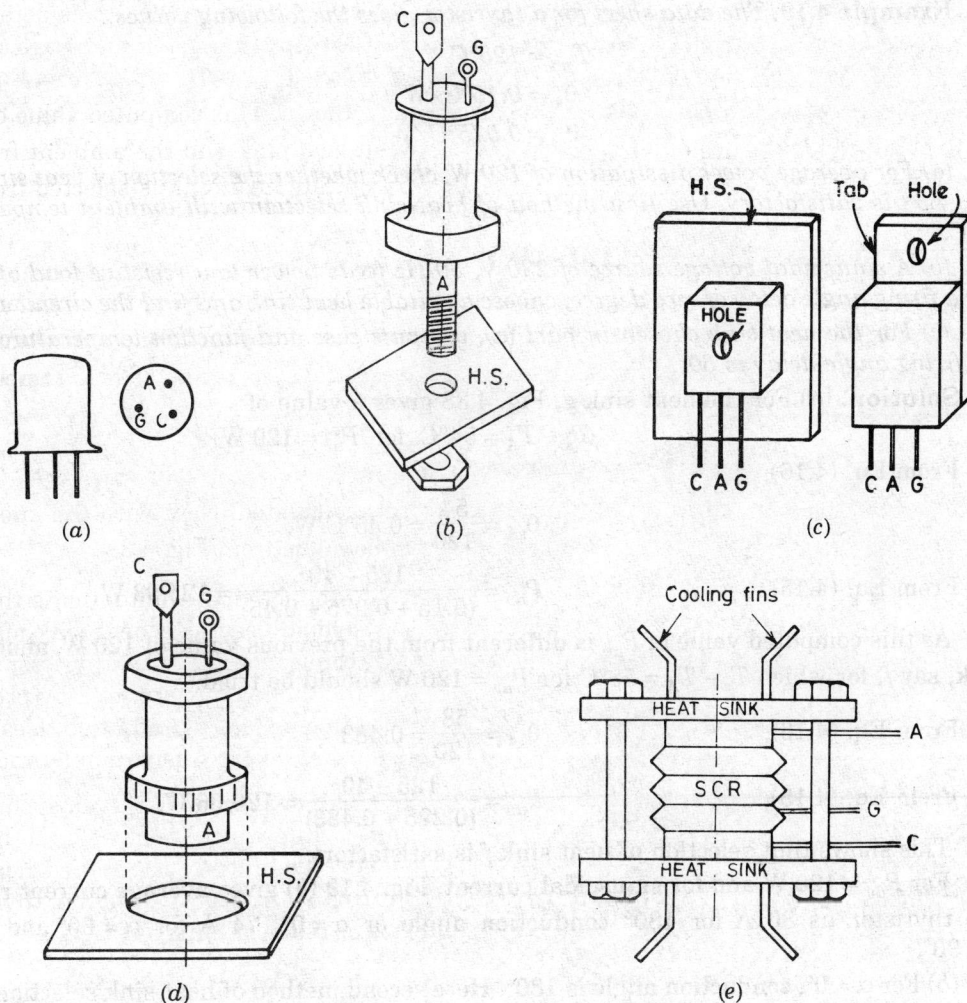

Fig. 4.39. Different SCR mountings and heat sink.

shown in Fig. 4.39 (c). Bolts are pushed through these holes so as to mount the device on to heat sink with nuts etc. In case the device is to be insulated from the heat sink, a thin insulating mica or PTFE washer is used between the device and heat sink and the bolt is made up of nylon. This type of mounting is used for small and medium ratings.

(d) *Press-fit mounting.* Press-fit (or pressure-fit) package is designed for insertion into an appropriate sized hole in the heat sink. The insertion may be done by using a vice and pressing the device into the hole using wooden block etc. For large sizes, the insertion is carried out by means of a hydraulic ram. This type of mounting is used for large rated thyristors. Fig. 4.39 (d) illustrates press-fit mounting of SCRs.

(e) *Press-pak mounting.* This type of mounting is also called "disc" or "hockey-puck" mounting because of its shape. The SCR is clamped between two heat sinks, Fig. 4.39 (e) and external pressure is applied evenly so that there is no deformation of any part. The heat sinks may be air, water or oil cooled. Such type of mounting is used for thyristors of very high current ratings.

Example 4.19. *The data sheet for a thyristor gives the following values :*

$$T_{jm} = 125°C$$
$$\theta_{jc} = 0.15°C/W$$
$$\theta_{cs} = 0.075°C/W$$

(a) *For average power dissipation of 120 W, check whether the selection of heat sink g from Fig. 4.38 is satisfactory. Use first method of heat sink selection with ambient temperature of 40°C.*

(b) *A sinusoidal voltage source of 230 V, 50 Hz feeds power to a resistive load of $R = 2\,\Omega$. For a firing angle delay of zero degree, choose a suitable heat sink and find the circuit efficiency.*

(c) *For the heat sink chosen in part (a), compute case and junction temperatures in case the firing angle delay is 60°.*

Solution. (a) For the heat sink g, Fig. 4.38 gives a value of

$$T_s - T_A = 54°C \text{ for } P_{av} = 120 \text{ W}.$$

From Eq. (4.16),

$$\theta_{sA} = \frac{54}{120} = 0.45°C/W.$$

From Eq. (4.15),

$$P_{av} = \frac{125 - 40}{(0.15 + 0.075 + 0.45)} = 125.93 \text{ W}.$$

As this computed value of P_{av} is different from the previous value of 120 W, another heat sink, say *f*, for which $T_s - T_A = 58°C$ for $P_{av} = 120$ W should be tried.

From Eq. (4.16),

$$\theta_{sA} = \frac{58}{120} = 0.483$$

From Eq. (4.15),

$$P_{av} = \frac{125 - 40}{(0.225 + 0.483)} = 120.06 \text{ W}.$$

This shows that selection of heat sink *f* is satisfactory.

For $P_{av} = 120$ W and for sinusoidal current, Fig. 4.19 (*b*) gives average current rating for the thyristor as 80 A for 180° conduction angle or $\alpha = 0°$, 74 A for $\alpha = 60°$ and 68 A for $\alpha = 90°$.

(b) For $\alpha = 0°$, conduction angle is 180°. Here second method of heat-sink selection is used.

$$I_{TAV} = \frac{1}{2\pi} \int_0^\pi \frac{V_m}{R} \sin \omega t \cdot d(\omega t) = \frac{V_m}{\pi R} = \frac{\sqrt{2} \times 230}{\pi \times 2} = 51.77 \cong 52 \text{ A}$$

For this current, P_{av} from Fig. 4.19 (*b*) is 90 W.

From Eq. (4.15),

$$\theta_{sA} = \frac{125 - 40}{90} - (0.225) = 0.7194° \text{ C/W}$$

From Eq. (4.16),

$$T_s - T_A = 90 \times 0.7194 = 64.75°C$$

For $T_s - T_A = 64.75°C$ and $P_{av} = 90$ W, Fig. 4.38 shows that heat sink *c* should be selected.

The use of third method of heat-sink selection is also demonstrated. First I_{TAV} is calculated as in the second method. For this current, $P_{av} = 90$ W from Fig. 4.19 (*b*) and $T_c = 112°C$ from Fig. 4.20 (*b*). Now sink temperature from Eq. (4.17) is

$$T_s = 112 - 90 \times 0.075 = 105.25°C$$

and

$$T_s - T_A = 105.25 - 40 = 65.25°C.$$

For $T_s - T_A = 65.25°C$ and $P_{av} = 90$ W, Fig. 4.38 shows that heat sink *c* should be chosen. As expected, this agrees with the choice made by the use of second method.

Power delivered to load, $P = I_r^2 R = \dfrac{V_r^2}{R}$

where V_r = rms value of load voltage

$$\therefore \quad V_r = \left[\frac{1}{2\pi} \int_0^\pi V_m^2 \sin^2 \omega t \, d(\omega t) \right]^{1/2} = \frac{V_m}{2}$$

$$\therefore \quad P = \left(\frac{V_m}{2} \right)^2 \frac{1}{R} = \left(\frac{230\sqrt{2}}{2} \right)^2 \frac{1}{2} = 13225 \text{ W}.$$

$$\therefore \text{ Circuit efficiency} \qquad = \frac{13225}{13225 + 90} = 0.993 \text{ pu or } 99.3\%.$$

(c) For $\alpha = 60°$, conduction angle is $180° - 60° = 120°$

$$\therefore \quad I_{TAV} = \frac{1}{2\pi} \int_\alpha^\pi \frac{V_m}{R} \sin \omega t \cdot d(\omega t) = \frac{V_m}{2\pi R} (1 + \cos \alpha)$$

$$= \frac{230\sqrt{2}}{2\pi \cdot 2} (1 + \cos 60°) = 38.82 \text{ A}.$$

For $I_{TAV} = 38.82$ A and conduction angle of $120°$, P_{av} from Fig. 4.19 (b) is 52 W and from Fig. 4.38 for heat sink c.

$$T_s - T_A = 46°C$$
$$\therefore \qquad\qquad T_s = 40 + 46 = 86°C.$$

From Eq. (4.17), case temperature,

$$T_c = T_s + P_{av} \cdot \theta_{cs} = 86 + 90 \times 0.075 = 92.75°C$$

and junction temperature, $\quad T_j = T_c + P_{av} \cdot \theta_{jc} = 92.75 + 90 \times 0.15 = 106.25°C.$

This example demonstrates that selection of heat sink by second and third methods is more simpler than by the first method.

Example 4.20. *For a thyristor, maximum junction temperature is 125°C. The thermal resistances for the thyristor-sink combination are $\theta_{jc} = 0.16$ and $\theta_{cs} = 0.08°C/W$. For a heat-sink temperature of 70°C, compute the total average power loss in the thyristor-sink combination.*

In case the heat sink temperature is brought down to 60°C by forced cooling, find the percentage increase in the device rating.

Solution. From the equivalent circuit of Fig. 4.37 (b)

$$T_j = T_s + P_{av} (\theta_{jc} + \theta_{cs})$$

$$P_{av1} = \frac{125 - 70}{0.16 + 0.08} = 229.17 \text{ W}$$

Thus total average power loss in the thyristor-sink combination is 229.17 W. With improved cooling,

$$P_{av2} = \frac{125 - 60}{0.24} = 270 \ 83 \text{ W}.$$

Thyristor rating is proportional to the square root of average power loss.

$\therefore$ Percentage increase in thyristor rating

$$= \frac{\sqrt{270.83} - \sqrt{229.17}}{\sqrt{229.17}} \times 100 = 8.71\%.$$

4.10. SERIES AND PARALLEL OPERATION OF THYRISTORS

SCR ratings have improved considerably since its introduction in 1957. Presently, SCRs with voltage and current ratings of 10 kV and 3 kA are available. However, for some industrial applications, the demand for voltage and current ratings is so high that a single SCR cannot fulfil such requirements. In such cases, SCRs are connected in series in order to meet the h.v. demand and in parallel for fulfilling the high current demand. For series or parallel connected SCRs, it should be ensured that each SCR rating is fully utilized and the system operation is satisfactory. *String efficiency* is a term that is used for measuring the degree of utilization of SCRs in a string. String efficiency of SCRs connected in series/parallel is defined as

string efficiency

$$= \frac{\text{Actual voltage/current rating of the whole string}}{[\text{Individual voltage/current rating of one SCR}]\ [\text{Number of SCRs in the string}]}$$

In practice, this ratio is less than one. For obtaining highest possible string efficiency, the SCRs connected in series/parallel string must have identical *I-V* characteristics. As SCRs of the same ratings and specifications do not have identical characteristics, unequal voltage/current sharing is bound to occur for all SCRs in a string. As a consequence, string efficiency can never be equal to one. However, unequal voltage/current sharing by the SCRs in a string can be minimised to a great extent by using external equalizing circuits.

Even in a string provided with external equalizing circuits, the string efficiency is less than unity. For a given system, if one extra unit is added to the series/parallel string, the voltage/current shared by each device would become lower than its normal rating. The use of this extra unit will certainly improve the reliability of the string though at an increased cost. A measure of the reliability of string is given by a factor called *derating factor* DRF defined as under :

$$DRF = 1 - \text{string efficiency}$$

For example, for a string voltage of 3300 V, let there be six series-connected SCRs, each of 600 V rating.

$\therefore$ String efficiency $\qquad = \dfrac{3300}{600 \times 6} = 0.917$ or 91.7%

and $\qquad\qquad\qquad\qquad DRF = 1 - 0.917 = 0.083$ or 8.3%

If one extra unit is connected in series with the same system voltage, then string efficiency $= \dfrac{3300}{600 \times 7} = 0.786$ or 78.6% and DRF $= 1 - 0.786 = 0.214$ or 21.4%.

With the addition of extra SCR, DRF has increased from 8.3% to 21.4%, indicating higher reliability of the string, though at an extra cost.

The object of this section is to study the problems concerning the series/parallel operation of SCRs and to discuss the measures adopted to overcome these problems.

4.10.1. Series Operation

When system voltage is more than the voltage rating of a single thyristor, SCRs are connected in series in a string. As stated before, these SCRs should have their *I – V* characteristics as close as possible. On account of inherent variations in their characteristics, the voltage shared by each SCR may not be equal. For instance, consider two SCRs with their static *I – V* characteristics as shown in Fig. 4.40. For SCR1, leakage resistance ($= V_1/I_0$) is high whereas for SCR2, it is low (V_2/I_0). For the same leakage current I_0 in the series connected

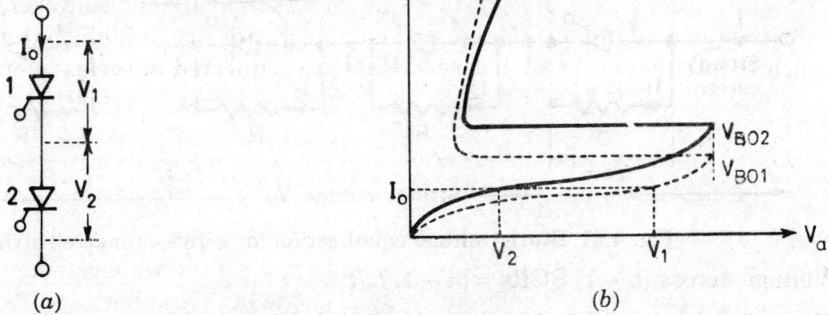

Fig. 4.40. Series connected SCRs.

SCRs, SCR1 supports rated voltage V_1 whereas SCR2 supports voltage $V_2 < V_1$. Each SCR in Fig. 4.40 is rated for a forward blocking voltage of V_1 volts which is always less than its forward breakover voltage. Here V_{BO1} and V_{BO2} are the forward breakover voltages for thyristors 1 and 2 respectively. It is seen from Fig. 4.40 that two SCRs can support a maximum voltage of $V_1 + V_2$ and not the rated blocking voltage $2V_1$. The string efficiency for two series connected SCRs of Fig. 4.40 is,

Therefore,
$$\frac{V_1 + V_2}{2V_1} = \frac{1}{2}\left(1 + \frac{V_2}{V_1}\right)$$

This shows that even though SCRs have identical ratings, voltage shared by each is not the same and string efficiency is therefore less than one.

A uniform voltage distribution in steady state can be achieved by connecting a suitable resistance across each SCR such that each parallel combination has the same resistance. This will require different value of resistance for each SCR which is a difficult proposition. A more practical way of obtaining a reasonably uniform voltage distribution during steady state working of series-connected SCRs is to connect the same value of shunt resistance R across each SCR as shown in Fig. 4.41. This shunt resistance R is called the *static equalizing circuit*. Magnitude of parallel resistance R can be obtained as follows [1].

Consider n thyristors connected in series as shown in Fig. 4.41. Let SCR1 has minimum leakage current I_{bmn} and each of the remaining $(n-1)$ SCRs have the same leakage current $I_{bmx} > I_{bmn}$. An examination of Fig. 4.40 (b) reveals that an SCR with lower leakage current blocks more voltage.

As SCR1 has lower leakage current, it will block voltage V_{bm} (say) which is more than that shared by each of the other $(n-1)$ SCRs. Here V_{bm} is the maximum permissible blocking voltage of SCR 1. It is seen from Fig. 4.41 that

$$I_1 = I - I_{bmn} \quad \text{and} \quad I_2 = I - I_{bmx}$$

where I = total string current

Voltage across SCR1 is $V_{bm} = I_1 R$

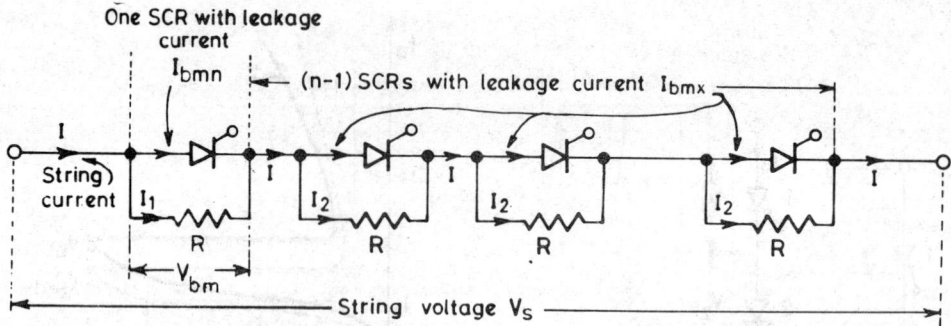

Fig. 4.41. Static-voltage equalization for series-connected string.

Voltage across $(n-1)$ SCRs $= (n-1) I_2 R$

For a string voltage V_s, the voltage equation for the series circuit of Fig. 4.41 is

$$V_s = I_1 R + (n-1) R I_2 = V_{bm} + (n-1) R (I - I_{bmx})$$
$$= V_{bm} + (n-1) R [I_1 - (I_{bmx} - I_{bmn})]$$
$$= V_{bm} + (n-1) R I_1 - (n-1) R \cdot \Delta I_b$$

where $\Delta I_b = I_{bmx} - I_{bmn}$

As $R I_1 = V_{bm}, \quad V_s = n V_{bm} - (n-1) R \cdot \Delta I_b$

or $$R = \frac{n V_{bm} - V_s}{(n-1) \cdot \Delta I_b} \qquad \qquad ...(4.18)$$

The SCR data sheet contains only maximum blocking current I_{bmx} and rarely ΔI_b. In such a case, it is usual to assume $\Delta I_b = I_{bmx}$ with $I_{bmn} = 0$. With this, the value of R calculated from Eq. (4.18) is low than what is actually required. The value of minimum leakage, or blocking, current I_{bm} may be acquired from manufacturers if required, but data sheet does not give its value.

Once the value of R is calculated, its power rating is given by

$$P_R = \frac{V_r^2}{R}$$

where V_r = rms voltage across R.

It is likely that SCRs do not have identical dynamic characteristics. In such a case, series-connected SCRs will have unequal voltage distribution during the transient conditions of turn-on, turn-off and high frequency operation. The dynamic characteristics of two SCRs during turn-on are shown in Fig. 4.42 (a) where it is assumed that turn-on time of SCR2 is more than that of SCR1 by Δt_d. Before these two SCRs are gated, string voltage V_s is shared as $V_s/2$ by each thyristor as shown. Now both SCRs are gated at the same time. As SCR1 has less turn-on time, it gets turned-on at instant t_1, whereas SCR2 is yet off. Voltage across SCR1 drops from $V_s/2$ to almost zero. At the same instant t_1, voltage across off SCR2 will boost from $V_s/2$ to almost full V_s. Thus, the voltage shared by two SCRs are unequal. After instant t_1, voltage V_s across SCR2 may turn it on in case V_s is greater than its breakover voltage. SCR2 will, however, get turned on at time $(t_1 + \Delta t_d)$ as assumed, Fig. 4.42 (a).

During turn-off, thyristor characteristics are shown in Fig. 4.42 (b). SCR1 is assumed to have less turn-off time t_{q1} than that of SCR2, i.e. $t_{q1} < t_{q2}$. At instant t_2, SCR1 has recovered and is passing through zero voltage whereas SCR2 is developing reverse recovery voltage xy. At

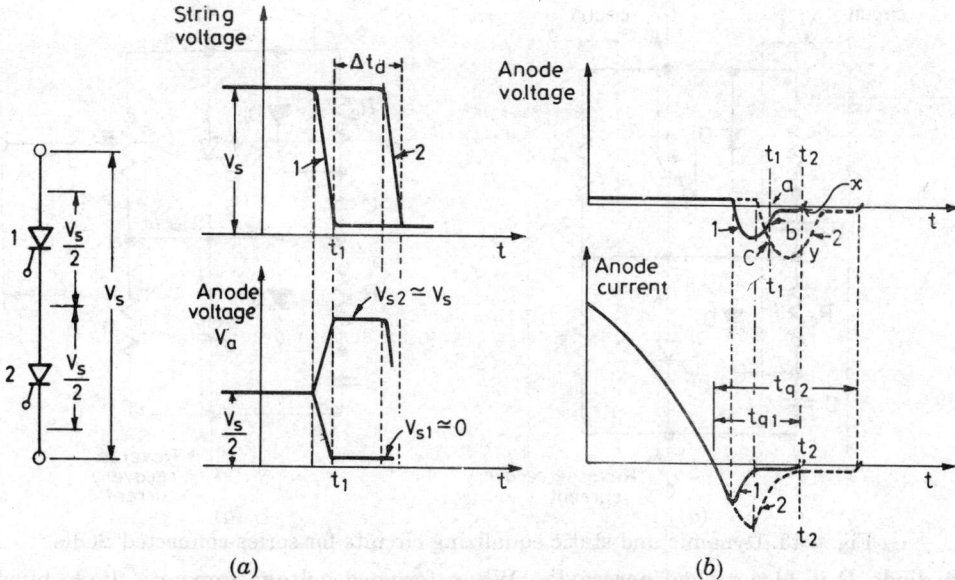

Fig. 4.42. Unequal voltage distribution for two series connected SCRs during
(a) turn-on and (b) turn-off.

instant t_1 in Fig. 4.42 (b), both SCRs are developing different reverse recovery voltages given by ab for SCR1 and ac for SCR2 as shown, so the two SCRs have unequal voltages across them at t_1. It is thus seen that SCRs with different characteristics during turn-off time suffer from unequal voltage distribution during their turn-off processes. It may thus be concluded from above that series-connected SCRs do suffer from unequal voltage distribution across them during their turn-on and turn-off processes and also during their high-frequency operation which means more frequent turning-on and turning-off of the devices.

A simple resistor as shown in Fig. 4.41 for static voltage equalization cannot maintain equal voltage distribution under transient condition. During turn-on and turn-off, the capacitance of the reverse biased junctions determines the voltage distribution across SCRs in a series connected string. As reverse biased junctions are likely to have different capacitances, called self-capacitances, the voltage distribution during turn-on and turn-off periods would be unequal. Voltage equalization under these conditions can, however, be achieved by employing shunt capacitors as shown in Fig. 4.43. This capacitance has the effect of removing the inequalities in thyristor self-capacitances. In other words, during turn-on and turn-off periods, the resultant of shunt capacitance and self-capacitance of each SCR tend to be equal for each of the series connected SCRs. Thus the shunt capacitors play a dominant role in equalizing the voltage distribution across the series-connected thyristors during their turn-on and turn-off processes.

When any SCR is in the forward blocking state, the capacitor connected across it gets charged to a voltage existing across that SCR. When this SCR is turned on, capacitor discharges heavy current through this SCR. For limiting this discharge current spike, a damping resistor R_C is used in series with capacitor C as shown in Fig. 4.43. Resistor R_C also damps out the high frequency oscillations that may arise due to the series combination of R_C, shunt capacitor and circuit inductance. Combination of R_C and C is called the *dynamic equalizing circuit* and is shown in Fig. 4.43 (a). Note that the function of R_C and C used in Fig. 4.43 is to equalize the voltage during dynamic (or transient) conditions and to protect the thyristors against high dv/dt.

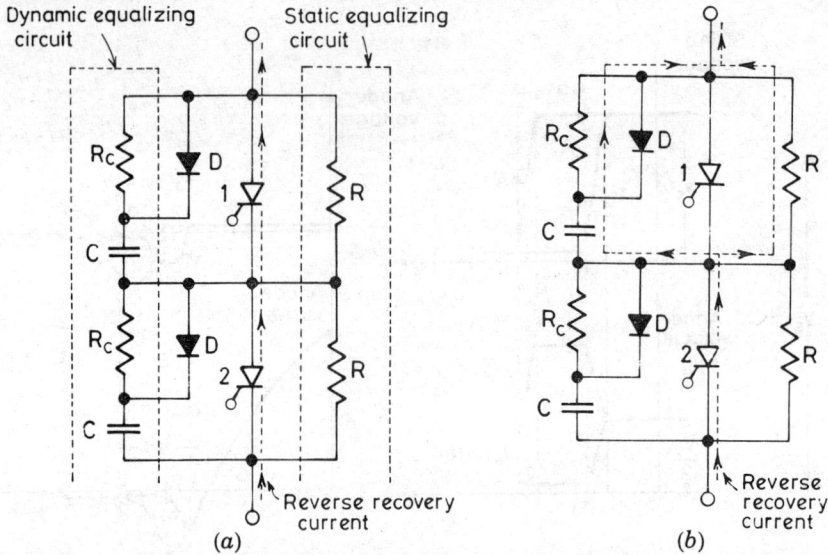

Fig. 4.43. Dynamic and static equalizing circuits for series-connected SCRs.

A diode D is also placed across R_C. When forward voltage appears, diode bypasses R_C during charging time of the capacitor C. This makes the capacitor more effective in voltage equalization and for limiting dv/dt across SCR. However, during capacitor discharge, R_C comes into play for limiting the current spike and rate of change of current di/dt. During turn-off period, when all SCRs are developing reverse voltage, the reverse recovery current i_r flows through all series connected SCRs as shown in Fig. 4.43 (a). However, if one SCR recovers early, it will not allow the passage of i_r from the other SCRs. If SCR1 is assumed to recover fully and earlier than other SCRs, then reverse recovery current due to other SCRs can pass through R connected across SCR1 as shown in Fig. 4.43 (b). In this figure, i_r may flow through C, R_C also in case the conditions are favourable. For simplicity, only two SCRs are shown in Fig. 4.43. The existence of reverse recovery current is desirable as it facilitates the turning-off process of the series-connected SCR string.

Value of capacitance C shown in Fig. 4.43 can be obtained as under :

In series connected SCRs, voltage unbalance during turn-off time is more predominant than it is during turn-on time, Fig. 4.42. Therefore, choice of capacitor C is based on the reverse recovery characteristics of SCRs. In Fig. 4.44 (b) are shown reverse recovery characteristics for two SCRs of Fig. 4.44 (a). SCR1 is assumed to have short reverse recovery time as compared to SCR 2. Shaded area ΔQ, proportional to the product of current and time, is the difference in the reverse recovery charges of two thyristors 1 and 2. Under this assumption, SCR1 recovers first ; it, therefore, goes into blocking state and does not allow the passage of excess charge ΔQ left on SCR2. This charge ΔQ can, however, pass through C as shown in Fig. 4.44 (a).

The voltage induced by ΔQ in the capacitor C, connected across SCR1, is $\Delta Q/C$; whereas no voltage is induced by $\Delta Q\ (= Q_2 - Q_1)$ in C connected across SCR2. There is thus a difference in voltages, equal to $\dfrac{Q_2 - Q_1}{C} = \dfrac{\Delta Q}{C}$, to which the two shunt capacitors are charged. The thyristor with the least reverse recovery time will share the highest transient voltage, say V_{bm}. As stated above, the voltage difference to which the two shunt capacitors are charged

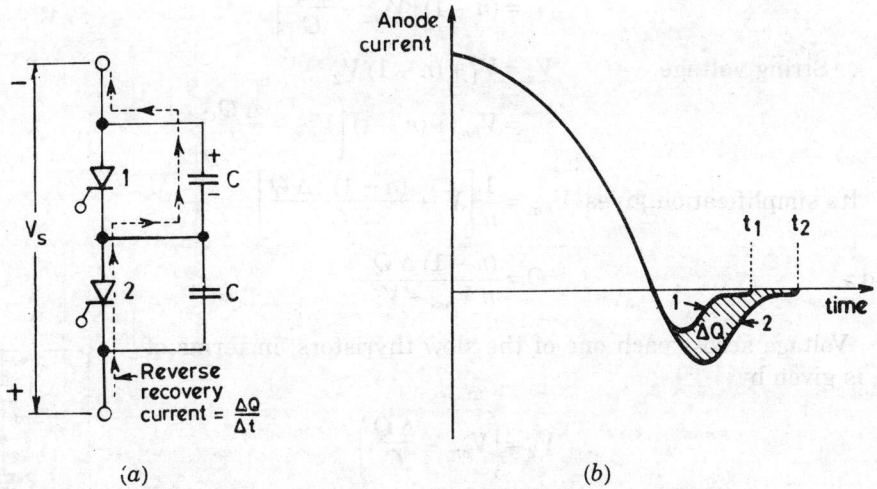

Fig. 4.44. (a) Flow of reverse recovery current if SCR1 recovers first
(b) Variation of reverse recovery characteristics for two SCRs of Fig. 4.44 (a).

during reverse recovery time is $\Delta Q/C$, the transient voltage shared by slow thyristor 2 must be $V_{bm} - \dfrac{\Delta Q}{C}$ (less than V_{bm} shared by fast thyristor 1). Thus, in Fig. 4.44 (a),

voltage across fast top thyristor 1 , $V_1 = V_{bm}$

and voltage across slow bottom thyristor 2, $V_2 = V_{bm} - \dfrac{\Delta Q}{C}$

$\therefore$ String voltage, $\qquad V_s = V_1 + V_2 = V_{bm} + V_{bm} - \dfrac{\Delta Q}{C}$

or $\qquad V_{bm} = \dfrac{1}{2}\left(V_s + \dfrac{\Delta Q}{C}\right)$

and $\qquad V_2 = V_{bm} - \dfrac{\Delta Q}{C} = \dfrac{1}{2}\left(V_s - \dfrac{\Delta Q}{C}\right)$

In order to aid the reverse recovery process of the thyristors in a string, the string voltage reverses in polarity as shown in Fig. 4.44 (a).

Now consider that there are n series-connected SCRs in a string as shown in Fig. 4.45. If top SCR has characteristics similar to SCR1 of Fig. 4.44 (b) and the remaining $(n-1)$ SCRs have characteristics similar to SCR2 of Fig. 4.44 (b), then SCR1 would recover first and support a voltage V_{bm}. The charge $(n-1)\,\Delta Q$ from the remaining $(n-1)$ thyristors would pass through C connected across top fast SCR1 and as a result, a voltage $(n-1)\,\Delta Q/C$ would be induced in C. As before, excess charge contributed by each one of the $(n-1)$ thyristors is ΔQ, therefore, the voltage across each one of the slow thyristors is $\left(V_{bm} - \dfrac{\Delta Q}{C}\right)$ as shown in Fig. 4.45. Thus, for a string of n series-connected thyristors, voltage across fast top thyristor $1, V_1 = V_{bm}$

voltage across each one of the slow thyristors, V_2 is

$$V_2 = V_{bm} - \dfrac{\Delta Q}{C}$$

and voltage across $(n-1)$ slow thyristors $= (n-1)\,V_2$

$$= (n-1)\left(V_{bm} - \frac{\Delta Q}{C}\right)$$

$\therefore$ String voltage, $V_s = V_1 + (n-1)\,V_2$

$$= V_{bm} + (n-1)\left(V_{bm} - \frac{\Delta Q}{C}\right)$$

Its simplification gives $V_{bm} = \dfrac{1}{n}\left[V_s + \dfrac{(n-1)\cdot \Delta Q}{C}\right]$

and $C = \dfrac{(n-1)\,\Delta Q}{n\,V_{bm} - V_s}$ ...(4.19)

Voltage across each one of the slow thyristors, in terms of V_s, is given by

$$V_2 = \left(V_{bm} - \frac{\Delta Q}{C}\right)$$

$$= \frac{V_s}{n} + \frac{(n-1)\,\Delta Q}{nC} - \frac{\Delta Q}{C}$$

or $V_2 = \dfrac{V_s - \dfrac{\Delta Q}{C}}{n}$ ...(4.20)

During the turn-off process, the source voltage V_s must reverse to aid the reverse recovery current. The transient voltage which each SCR must be able to withstand is V_{bm}. However, total voltage acting across the circuit consisting of V_s, thyristors n, 4, 3, 2 and top C, and per KVL, is $V_s + \dfrac{(n-1)\cdot\Delta Q}{C}$ and this must be supported by all SCRs together, which is equal to $n.V_{bm}$.

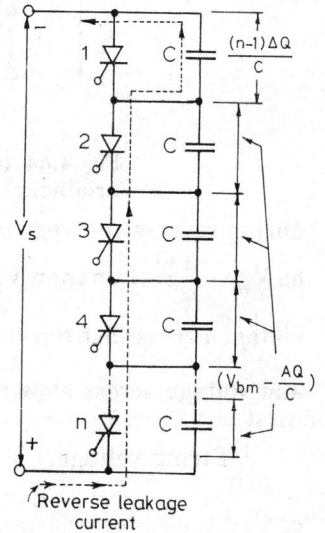

Fig. 4.45. String having n-series connected thyristors.

$\therefore$ $n.V_{bm} = V_s + \dfrac{(n-1)\cdot\Delta Q}{C}$

or $V_{bm} = \dfrac{1}{n}\left[V_s + \dfrac{(n-1)\cdot\Delta Q}{C}\right]$

or $C = \dfrac{(n-1)\cdot\Delta Q}{n.V_{bm} - V_s}$ ...(4.19)

4.10.2. Parallel Operation

When current required by the load is more than the rated current of a single thyristor, SCRs are connected in parallel in a string. For equal sharing of currents, I-V characteristics of SCRs during forward conduction must be identical as far as possible.

In Fig. 4.46 (a) are shown two SCRs in parallel and their characteristics during forward conduction are shown in Fig. 4.46 (b). For parallel-connected SCRs, voltage drop V_T across them must be equal. Fig. 4.46 (b) shows that for the same voltage drop V_T, SCR1 shares a rated current I_1 whereas SCR2 carries current I_2 much less than the rated current I_1. The total current carried by the unit is $I_1 + I_2$ and not the rated current $2I_1$ as required. Therefore, string efficiency is given by

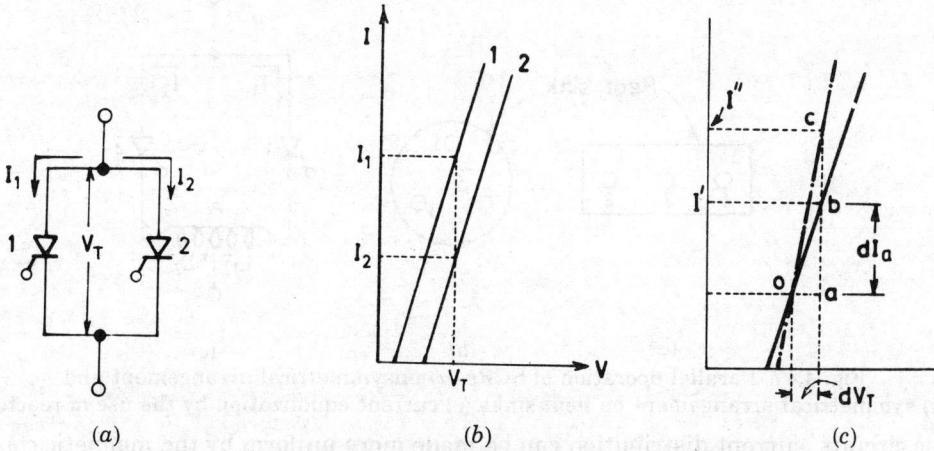

Fig. 4.46. (a) and (b) Parallel operation of two thyristors
(c) Dynamic resistance decreases as junction temperature rises.

$$\frac{I_1 + I_2}{2I_1} = \frac{1}{2}\left(1 + \frac{I_2}{I_1}\right)$$

Now consider n parallel connected SCRs. For satisfactory operation of these SCRs, they should get turned on at the same moment. The importance of their simultaneous turn on can be explained with an example. Consider that SCR1 has large turn-on time whereas the remaining $(n-1)$ SCRs have low turn-on time. Under this assumption, $(n-1)$ SCRs will turn on first but one SCR1 with longer turn-on time is likely to remain off. The voltage drop across $(n-1)$ SCRs falls to a low value and SCR1 is therefore subjected to this low voltage. For a given gate drive power, anode to cathode must have some minimum forward voltage, called *finger voltage,* for a thyristor to turn-on. If voltage across SCR1 drops to a value less than its finger voltage, then this thyristor will not turn on. As a consequence, the remaining $(n-1)$ SCRs, which are already on, will have to share the entire load current. As such, these SCRs may be overloaded and damaged because of heating caused by overcurrents.

If one SCR1 in a parallel unit carries more current than other SCRs, then this SCR1 will have greater junction temperature rise. As a result, its dynamic resistance $(= dV_T/dI_a)$ during forward conduction, Fig. 4.46 (c) decreases and this further increases the current shared by this SCR. In Fig. 4.46 (c), dynamic resistance is oa/ab and current shared is I'. Because of junction temperature rise, its dynamic resistance decreases to oa/ac and current shared by SCR1 increases to I''. This process of anode current rise becomes cumulative and subsequently the junction temperature of SCR1 exceeds its rated value ; as a result SCR1 is damaged. This sequence of events may engulf another SCR and in this manner all SCRs in the string may be destroyed permanently. Therefore, when SCRs are to be operated in parallel, it should be ensured that they operate at the same temperature. This can be achieved by mounting the parallel unit on one common heat sink.

Unequal current distribution in a parallel unit is also caused by the inductive effect of current carrying conductors. When SCRs are arranged unsymmetrically as shown in Fig. 4.47 (a), the middle conductor will have more inductance because of more flux linkages from two nearby conductors. As a consequence, less current flows through the middle SCR as compared to outer two SCRs. This unequal current distribution can be avoided by mounting the SCRs symmetrically on the heat sink as shown in Fig. 4.47 (b).

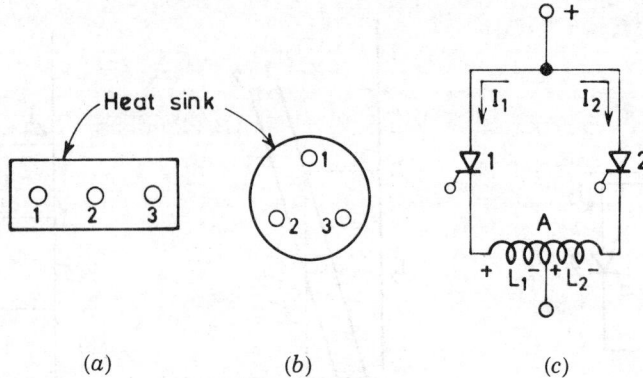

(a)   (b)   (c)

Fig. 4.47. Parallel operation of SCRs (a) unsymmetrical arrangement and
(b) symmetrical arrangement on heat sinks (c) current equalization by the use of reactor.

In ac circuits, current distribution can be made more uniform by the magnetic coupling of the parallel paths as shown in Fig. 4.47 (c). The tapped point A is the mid point of the reactor. If anode currents are such that $I_1 = I_2$, then flux produced by two halves of the reactor oppose each other. As A is the mid point, opposing flux linkages cancel and there is therefore no voltage drop in the reactor. If currents I_1 and I_2 are unequal, say $I_1 > I_2$, then resultant flux linkages are not zero. These flux linkages induce emfs in $L1$ and $L2$ as shown. Emf across reactor $L1$ opposes the flow of I_1 whereas that across $L2$ aids the flow of I_2. There is thus a *tendency* to buck I_1 and boost I_2 so as to minimise the unbalance of currents in the parallel unit.

When three or more SCRs are connected in parallel, reactors can be arranged accordingly so as to minimise the current unbalance.

Example 4.21. *A string of four series-connected thyristors is provided with static and dynamic equalizing circuits. This string has to withstand an off-state voltage of 10 kV. The static equalizing resistance is 25000 Ω and the dynamic equalizing circuit has $R_C = 40$ Ω and $C = 0.08$ μF. The leakage currents for four thyristors are 21 mA, 25 mA, 18 mA and 16 mA respectively. Determine voltage across each SCR in the off-state and the discharge current of each capacitor at the time of turn-on.*

Solution. Let I be the string current in the off-state. Then current through static-equalizing resistance R of 25000 Ω is (I-leakage current), current through each SCR is its own leakage current and no current flows through series combination of R_C and C.

∴ Voltage across R = voltage across each SCR

Voltage across SCR1 $= (I - 0.021) \times 25000 = V_1$

Voltage across SCR2 $= (I - 0.025) \times 25000 = V_2$

Voltage across SCR3 $= (I - 0.018) \times 25000 = V_3$

Voltage across SCR4 $= (I - 0.016) \times 25000 = V_4$

The sum of V_1, V_2, V_3 and V_4 gives

$$25000 \, (4I - 0.08) = V_1 + V_2 + V_3 + V_4 = \text{string voltage,} \ \ 10000 \text{ V}$$

or $I = 0.12$ A

From above, voltage across SCR1

$$= (0.12 - 0.021) \times 25000 = 2475 \text{ V}$$

Similarly $V_2 = 2375$ V, $V_3 = 2550$ V and $V_4 = 2600$ V.

Discharge current through SCR1 at the time of turn on

$$= \frac{V_1}{R_C} = \frac{2475}{40} = 61.875 \text{ A}$$

Similarly, discharge currents through thyristors 2, 3 and 4 are respectively 59.375 A, 63.75 A and 65 A.

Example 4.22. *SCRs with a rating of 1000 V and 200 A are available to be used in a string to handle 6 kV and 1 kA. Calculate the number of series and parallel units required in case derating factor is (a) 0.1 and (b) 0.2.*

Solution. (*a*) Derating factor, $DRF = 1 -$ string efficiency

$$\therefore \qquad\qquad 0.1 = 1 - \frac{6000}{n_s \times 1000} = 1 - \frac{1000}{n_p \times 200}$$

∴ Number of series-connected SCRs,

$$n_s = \frac{6000}{1000 \times 0.9} = 6.6 \cong 7$$

Number of parallel-connected SCRs,

$$n_p = \frac{1000}{200 \times 0.9} = 5.5 \cong 6$$

(*b*) As above, number of series-connected SCRs,

$$n_s = \frac{6000}{1000 \times 0.8} = 7.5 \cong 8$$

and number of parallel-connected SCRs,

$$n_p = \frac{1000}{200 \times 0.8} = 6.25 \cong 7$$

With higher value of DRF, more SCRs are required and therefore voltage and current shared by each device are lower than their normal rating. This increases the string reliability though at an increased investment.

Example 4.23. *It is required to operate 250-A SCR in parallel with 350-A SCR with their respective on-state voltage drops of 1.6 V and 1.2 V. Calculate the value of resistance to be inserted in series with each SCR so that they share the total load of 600 A in proportion to their current ratings.*

Solution.

Dynamic resistance of 250-A SCR1 $= \dfrac{1.6}{250} \ \Omega$

Dynamic resistance of 350-A SCR2 $= \dfrac{1.2}{350} \ \Omega$

Let R be the resistance inserted in series with each SCR. With this, current shared by

$$\text{SCR1} = 600 \ \frac{\dfrac{1.2}{350} + R}{\text{Total resistance}} \propto 250$$

and current shared by $\qquad\qquad \text{SCR2} = 600 \ \dfrac{\dfrac{1.6}{250} + R}{\text{Total resistance}} \propto 350$

From above,
$$\frac{\dfrac{1.2}{350} + R}{\dfrac{1.6}{250} + R} = \frac{250}{350} = \frac{5}{7}.$$

Its simplification gives $R = 0.004 \ \Omega$.

Thus the resistance to be inserted in series with each SCR is $0.004 \ \Omega$.

Example 4.24. *Discuss the conditions which must be satisfied for turning-on an SCR with a gate signal.*

Solution. Conditions which must be satisfied for turning-on SCR with a gate signal are as under :

(*a*) An SCR must be forward-biased. It means that anode must be positive with respect to cathode.

(*b*) Gate pulse width must be more than the turn-on time of an SCR. This will ensure that anode current exceeds the latching current before gate signal is removed.

(*c*) Anode to cathode voltage must be more than finger voltage. A finger voltage is that voltage below which an SCR cannot be turned on with a gate signal.

(*d*) Magnitude of gate current must be more than the minimum gate current required to turn-on a thyristor, otherwise the thyristor turn-on will not be reliable.

(*e*) Magnitude of gate current must be less than the maximum gate current allowed, otherwise gate circuit may be damaged.

(*f*) The gate triggering must synchronize with the ac supply.

4.11. OTHER MEMBERS OF THE THYRISTOR FAMILY

The term thyristor includes all four-layer *p-n-p-n* devices used for the control of power in ac and dc systems. The silicon controlled rectifier is the most popular member of thyristor family. There are several other members of thyristor family like PUT, SUS, SCS, triac, diac etc. All these devices, except triac, are low power devices. Several new devices have been developed and added to the thyristor family. These recently developed thyristor devices are asymmetric thyristor (ASCR), reverse conducting thyristor (RCT), static induction thyristor (SITH), gate-assisted turn-off thyristor and gate turn-off (GTO) thyristor. MOS-controlled thyristor (MCT) has already been described in Chapter 2. The object of this section is to discuss other members of the thyristor family.

4.11.1. PUT (Programmable Unijunction Transistor)

It is a *pnpn* device like an SCR. But the major difference is that gate is connected to *n*-type material near the anode as shown in Fig. 4.48 (*a*). PUT is used mainly in time-delay, logic and SCR trigger circuits. Its largest rating is about 200 V and 1 A. Circuit symbol and $I - V$ characteristics of a PUT are shown in Fig. 4.48 (*b*) and (*c*) respectively.

In a PUT, G is always biased positive with respect to cathode. When anode voltage exceeds the gate voltage by about 0.7 V, junction J_1 gets forward biased and PUT turns on. When anode voltage becomes less than gate voltage, PUT is turned off.

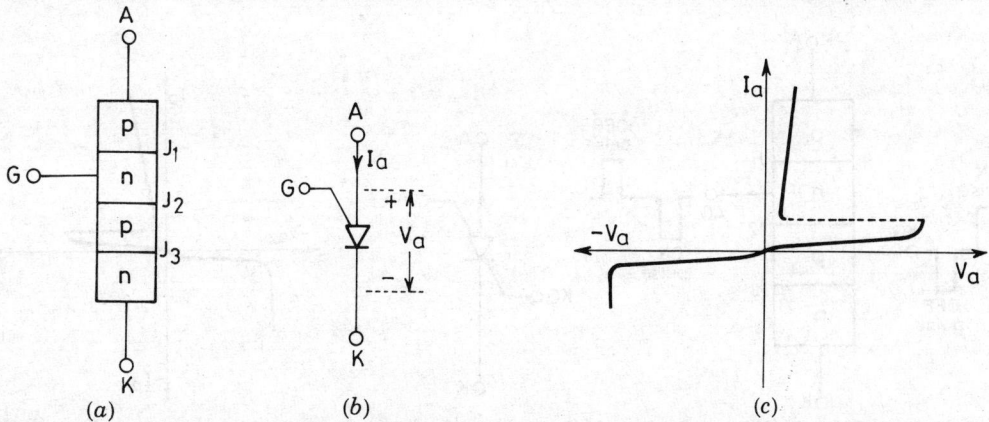

Fig. 4.48. (a) Schematic diagram (b) circuit symbol and (c) I-V characteristics of a PUT.

4.11.2. SUS (Silicon Unilateral Switch)

A SUS is similar to a PUT but with an inbuilt low-voltage avalanche diode between gate and cathode as shown in Fig. 4.49 (a). Because of the presence of diode, SUS turns on for a fixed anode-to-cathode voltage unlike an SCR whose trigger voltage and/or current vary widely with changes in ambient temperature. SUS is used mainly in timing, logic and trigger circuits. Its ratings are about 20 V and 0.5 A. Circuit symbol, equivalent circuit and $I - V$ characteristic of an SUS are shown in Fig. 4.49 (b), (c) and (d) respectively.

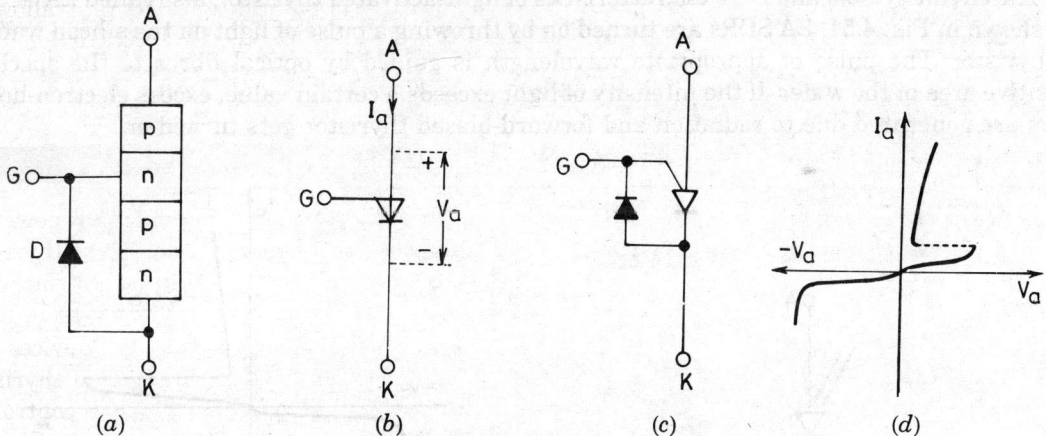

Fig. 4.49. (a) Schematic diagram (b) circuit symbol
(c) equivalent circuit and (d) I-V characteristics of an SUS.

4.11.3. SCS (Silicon Controlled Switch)

SCS is a tetrode, *i.e.* four electrode thyristor. It has two gates, one anode gate (AG) like a PUT and another cathode gate (KG) like an SCR. In other words, SCS is a four layer, four terminal *pnpn* device ; with anode A, cathode K, anode gate AG and cathode gate KG, Fig. 4.50 (a). SCS can be turned on by either gate. Circuit symbol and $I - V$ characteristic of an SCS are shown in Fig. 4.50 (b) and (c) respectively.

When a negative pulse is applied to gate AG, junction J_1 is forward biased and SCS is turned on. A positive pulse at AG will reverse bias junction J_1 and turns off the SCS.

A positive pulse at gate KG turns on the device (just like an SCR) and a negative pulse at KG turns it off (just like a G.T.O.).

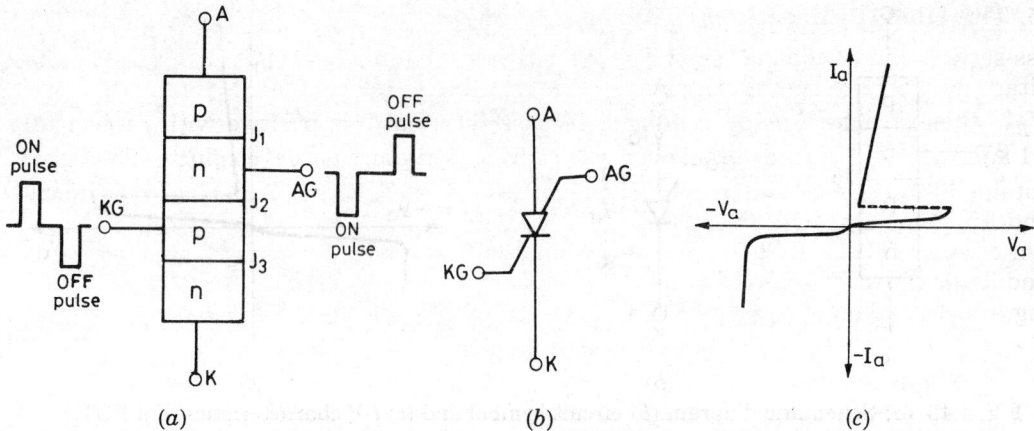

(a)   (b)   (c)

Fig. 4.50. (a) Schematic diagram (b) circuit symbol and (c) I-V characteristic of an SCS.

Its ratings are about 100 V and 200 mA. This can be operated like an *OR* gate. Its applications include :

(*i*) timing, logic and triggering circuits   (*ii*) pulse generators
(*iii*) voltage sensors   (*iv*) oscillators etc.

4.11.4. Light Activated Thyristors

The circuit symbol and $I - V$ characteristics of light-activated thyristor, also called LA SCR, are shown in Fig. 4.51. LA SCRs are turned on by throwing a pulse of light on the silicon wafer of thyristor. The pulse of appropriate wavelength is guided by optical fibres to the special sensitive area of the wafer. If the intensity of light exceeds a certain value, excess electron-hole pairs are generated due to radiation and forward-biased thyristor gets turned on.

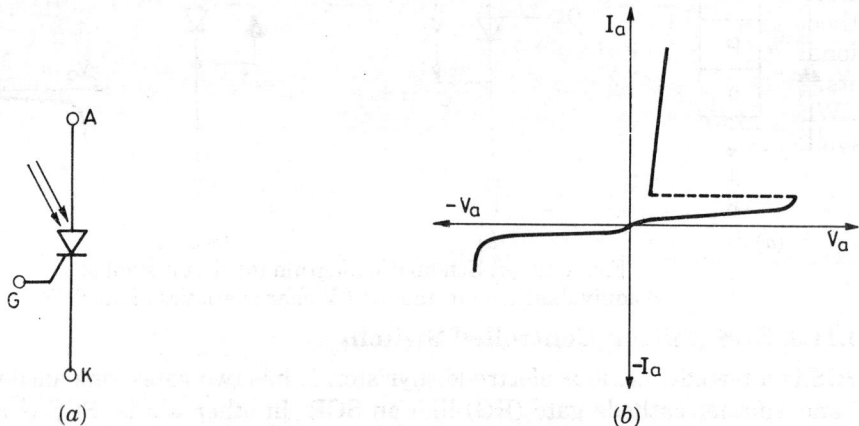

(a)   (b)

Fig. 4.51. (a) Circuit symbol and (b) I-V characteristic of LASCR

The primary use of light-fired thyristors is in high-voltage high-current applications, static reactive-power compensation etc. A light-fired thyristor has complete electrical isolation between the light-triggering source and the high-voltage anode-cathode circuit. Light-activated thyristors are available up to 6 kV and 3.5 kA, with on-state voltage drop of about 2 V and with light-triggering requirements of 5 mW.

4.11.5. The Diac (Bidirectional Thyristor Diode)

A cross-sectional view of a diac showing all its layers and junctions is depicted in Fig. 4.52 (a). If voltage V_{12}, with terminal 1 positive with respect to terminal 2, exceeds break-over voltage V_{B01}, then structure $pn\,pn$ conducts. In case terminal 2 is positive with respect to terminal 1 and when V_{21} exceeds breakover voltage V_{B02}, structure $pn\,pn'$ conducts. The term 'diac' is obtained from capital letters, DIode that can work on AC. Fig. 4.52 (b) gives the circuit symbol and Fig. 4.52 (c) the I-V characteristics of a diac. It is seen that diac has symmetrical breakdown characteristics. Its leads are interchangeable. Its turn-on voltage is about 30 V. When conducting, it acts like a low resistance with about 3 V drop across it. When not conducting, it acts like an open switch. A diac is sometimes called a gateless triac.

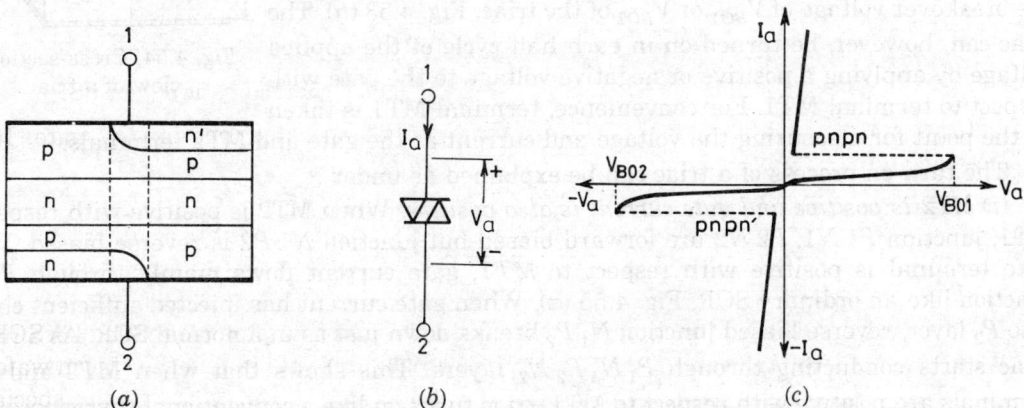

Fig. 4.52. (a) Cross-sectional view (b) circuit symbol and (c) I-V characteristics of a diac.

4.11.6. The Triac

An SCR is a unidirectional device as it can conduct from anode to cathode only and not from cathode to anode. A triac can, however, conduct in both the directions. A triac is thus a bidirectional thyristor with three terminals. It is used extensively for the control of power in ac circuits. Triac is the word derived by combining the capital letters from the words TRIode and AC. When in operation, a triac is equivalent to two SCRs connected in antiparallel. The circuit symbol and its characteristics are shown in Fig. 4.53 (a) and (b) respectively. As the

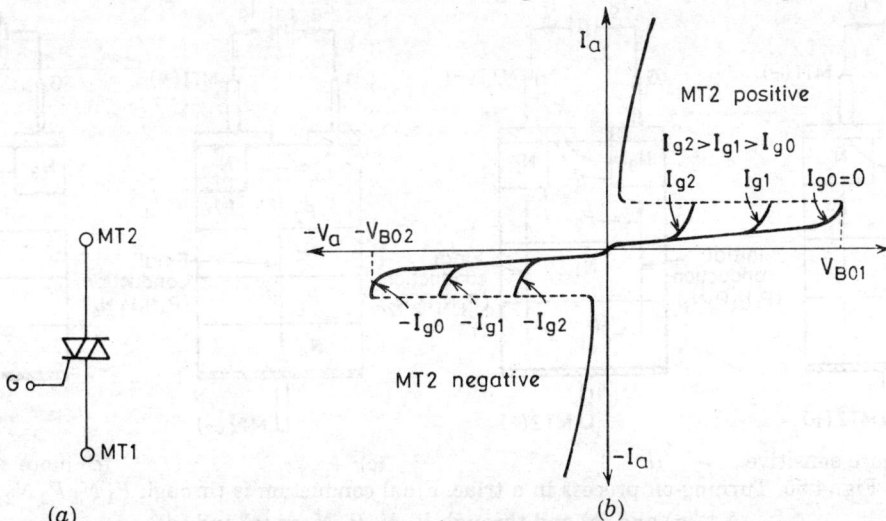

Fig. 4.53. (a) Circuit symbol and (b) static I-V characteristics of a triac.

triac can conduct in both the directions, the terms anode and cathode are not applicable to triac. Its three terminals are usually designated as MT1 (main terminal 1), MT2 and the gate by G as in a thyristor. For understanding the operation of the triac, its cross-sectional view showing all the layers and junctions is sketched in Fig. 4.54. The gate G is near terminal MT1. The cross-hatched strip shows that G is connected to N_3 as well as P_2. Similarly, terminal MT1 is connected to P_2 and N_2 ; terminal MT2 to P_1 and N_4.

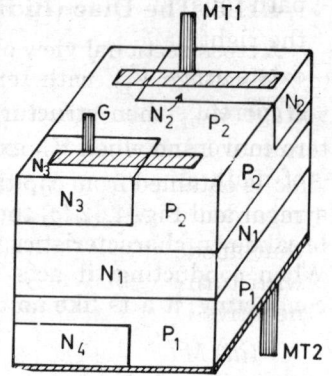

Fig. 4.54. Cross-sectional view of a triac.

With no signal to gate, the triac will block both half cycles of the ac applied voltage in case peak value of this voltage is less than the breakover voltage of V_{BO1} or V_{BO2} of the triac, Fig. 4.53 (b). The triac can, however, be turned on in each half cycle of the applied voltage by applying a positive or negative voltage to the gate with respect to terminal MT1. For convenience, terminal MT1 is taken as the point for measuring the voltage and current at the gate and MT2 terminals.

The turn-on process of a triac can be explained as under :

(i) *MT2 is positive and gate current is also positive.* When MT2 is positive with respect to MT1, junction $P1\,N1$, $P2\,N2$ are forward biased but junction $N1\,P2$ is reverse biased. When gate terminal is positive with respect to $MT1$, gate current flows mainly through $P2\,N2$ junction like an ordinary SCR, Fig. 4.55 (a). When gate current has injected sufficient charge into P_2 layer, reverse biased junction $N_1\,P_2$ breaks down just as in a normal SCR. As a result, triac starts conducting through $P_1\,N_1\,P_2\,N_2$ layers. This shows that when MT2 and gate terminals are positive with respect to MT1, triac turns on like a conventional thyristor. Under this condition, triac operates in the first quadrant of Fig. 4.53 (b). The device is *more sensitive* in this mode. It is recommended method of triggering if the conduction is desired in the first quadrant.

(ii) *MT2 is positive but gate current is negative.* When gate terminal is negative with respect to MT1, gate current flows through $P_2\,N_3$ junction, Fig. 4.55 (b) and reverse biased junction $N_1\,P_2$ is forward biased as in a normal thyristor. As a result, triac starts conducting through $P_1\,N_1\,P_2\,N_3$ layers initially. With the conduction of $P_1\,N_1\,P_2\,N_3$, the voltage drop across this

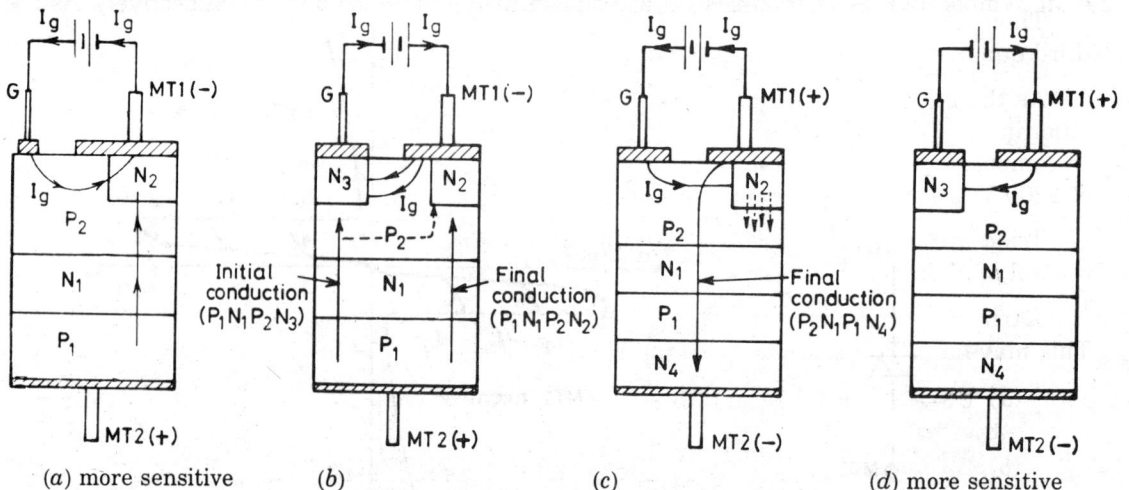

(a) more sensitive   (b)   (c)   (d) more sensitive

Fig. 4.55. Turning-on process in a triac. Final conduction is through $P_1\,N_1\,P_2\,N_2$ in (a) and (b) and through $P_2\,N_1\,P_1\,N_4$ in (c) and (d).

path falls but potential of layer between $P_2 N_3$ rises towards the anode potential of MT2. As the right hand portion of P_2 is clamped at the cathode potential of MT1, a potential gradient exists across layer P_2, its left hand region being at higher potential than its right hand region. A current shown dotted is thus established in layer P_2 from left to right. This current is similar to conventional gate current of an SCR. As a consequence, right-hand part of triac consisting of main structure $P_1 N_1 P_2 N_2$ begins to conduct. The device structure $P_1 N_1 P_2 N_3$ may be regarded as pilot SCR and the structure $P_1 N_1 P_2 N_2$ as the main SCR. It can then be stated that anode current of pilot SCR serves as the gate current for the main SCR. As compared with turn-on process discussed in (i) above, the device with MT2 positive but gate current negative is *less sensitive* and therefore, more gate current is required.

(iii) *MT2 is negative but gate current is positive.* The gate current I_g forward biases $P_2 N_2$ junction Fig. 4.55 (c). Layer N_2 injects electrons into P_2 layer as shown by dotted arrows. As a result, reverse biased junction $N_1 P_1$ breaks down as in a conventional thyristor. Eventually the structure $P_2 N_1 P_1' N_4$ is completely turned on. As usual, the current after turn-on is limited by the external load. As the triac is turned on by remote gate N_2, the device is *less sensitive* in the third quadrant with positive gate current.

(iv) *Both MT2 and gate current are negative.* In this mode, N_3 acts as a remote gate, Fig. 4.55 (d). The gate current I_g flows from P_2 to N_3 as in a normal thyristor. Reverse-biased junction $N_1 P_1$ is broken and finally, the structure $P_2 N_1 P_1 N_4$ is turned on completely. Though the triac is turned on by remote gate N_3 in third quadrant, yet the device is *more sensitive* under this condition compared with turn-on action with positive gate current discussed in (iii) above.

It can, therefore, be concluded from above that :

(i) sensitivity of the triac is greatest in the first quadrant when turned on with positive gate current and also in the third quadrant when turned on with negative gate current,

(ii) sensitivity of the triac is low in the first quadrant when turned on with negative gate current and also in the third quadrant when turned-on with positive gate current.

Thus the triac is rarely operated in first quadrant with negative gate current and in the third quadrant with positive gate current.

As the two conducting paths from MT1 to MT2 or from MT2 to MT1 interact with each other in the structure of the triac ; their voltage, current and frequency ratings are much lower as compared with conventional thyristors. At present, triacs with voltage and current ratings of 1200 V and 300 A (rms) are available.

Triacs are used extensively in residential lamp dimmers, heat control and for the speed control of small single-phase series and induction motors.

A triac may sometimes operate in the rectifier mode rather than in the bidirectional mode. This may happen due to the following reasons :

(a) For a given value of positive gate current, a triac may turn on with MT2 positive in first quadrant but may fail to turn on with MT2 negative.

(b) With constant negative gate current, the triac may turn on with MT2 negative in third quadrant but may not turn on with MT2 positive.

The rectifier-mode can be overcome by increasing the value of gate current.

4.11.7. Asymmetrical Thyristor (ASCR)

A conventional thyristor is able to block a large reverse voltage, but this blocking capability is not required in several industrial applications. For example, in voltage source inverters converting dc to ac and in some chopper circuits, a freewheeling diode is usually connected in antiparallel across each thyristor. This freewheeling diode clamps the thyristor voltage to 1 to 2 V under steady state conditions. An asymmetrical thyristor, or ASCR, is specially fabricated to have limited reverse voltage capability ; this permits a reduction in turn-on time, turn-off time and on-state voltage drop in ASCR. A typical ASCR may have reverse blocking capability of 20 to 30 V and forward blocking voltage of 400 to 2000 V. ASCRs with turn-off time half of that of a similar rated conventional SCRs have been developed. Fast turn-off ASCRs minimize the size, weight and cost of commutating components and permit high frequency operation (20 KHz or more) with improved efficiency.

4.11.8. Reverse Conducting Thyristor (RCT)

A reverse conducting thyristor is a special case asymmetrical thyristor with a monolithically integrated antiparallel diode on the same silicon chip. This construction reduces to zero the reverse blocking capability of RCT. A current pulse through the diode part of the chip turns off RCT. The arrangement of ASCR and diode in a single device reduces the heat sink size and leads to compactness of the converter. The undesirable stray loop inductance between ASCR and diode is also eliminated and unwanted reverse voltage transients across ASCR are avoided ; this leads to better turn off behaviour of RCT. RCTs with 2000 V and 500 A ratings are available. For high-performance inverter and chopper circuits, RCTs can now be tailor-made.

Fig. 4.56.
Reverse
conducting
thyristor

4.11.9. Other Thyristor Devices

Gate-assisted turn-off thyristor (GAT) is a normal four-layer thyristor, but its turn-off is achieved by applying a negative gate drive across gate-cathode terminals. In order to reduce the turn-off time appreciably, the gate-cathode junction is highly interdigitated so that stored charges can be removed more effectively from the base region. GAT thyristors are extensively employed in TV deflection circuits at frequencies around 20 kHz with turn-off times as low as 2.5 μ sec for 200-V devices.

Gate-turn-off thyristor (GTO) and static induction thyristor are described in the next two articles. The latest semiconductor device to enter the family of thyristors is integrated-gate commutated thyristor (IGCT).

IGCT is basically a hard-switched GTO. IGCT with 4500 V, 3000 A ratings are available. Its advantages over GTO are (i) lower conduction drop, (ii) faster switching speed, (iii) monolithic by-pass diode, (iv) snubberless operation and (v) ease of series operation [12].

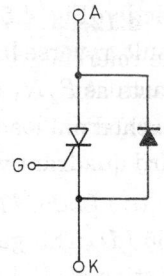

4.12. GATE TURN OFF THYRISTOR (GTO)

Conventional thyristors (CTs) are nearly ideal switches for their use in power-electronic applications. These can easily be turned on by positive gate current. Once in the on-state, gate loses control. CTs can now be turned off by expensive and bulky commutation circuitry. This shortcoming of thyristors limit their use up to about 1 kHz applications. These drawbacks in thyristors has led to the development of GTOs.

A GTO is a more versatile power-semiconductor device. It is like a CT but with added features in it. A GTO can easily be turned off by a negative gate pulse of appropriate amplitude. Thus, a GTO is a *pn pn* device that can be turned-on by a positive gate current and turned-off by a negative gate current at its gate cathode terminals.

Self-turn off capability of GTO makes it the most suitable device for inverter and chopper applications.

4.12.1. Basic Structure

A GTO is *pn pn*, there terminal device with anode (A), cathode (K) and gage (G), Fig. 4.57 (*a*). The four layers are $p^+ n p^+ n^+$ as shown. In CT, anode consists of p^+ layer, but in a GTO, anode is made up of n^+ type fingers diffused into p^+ layer.

Fig. 4.57 (*c*) gives two alternate circuit symbols for GTO. Since GTO is a four layer *pn pn* device just like CT, it can also be modelled by two-transistor analogy as shown in Fig. 4.57 (*b*). The four layers have different doping levels indicated by $p^+ n p^+ n^+$. Transistor Q_1 is $p^+ n p^+$ type and transistor Q_2 is $n p^+ n^+$ type, with p^+ emitter of Q_1 as anode A and n^+ emitter of Q_2 as cathode K.

4.12.1.1. Turn-on Process. A GTO is turned on by applying a positive gate current I_g in the reference direction shown in Fig. 4.57 (*b*). As GTO is forward biased, regeneration process starts as in a CT. Current gains α_1, α_2 begin to rise and when $\alpha_1 + \alpha_2 = 1$, saturation level is reached and GTO is turned on. The anode current I_a is then limited by load impedance.

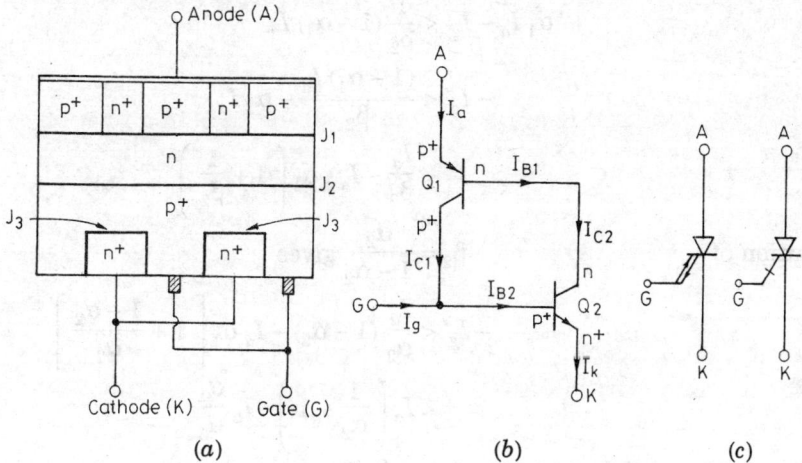

Fig. 4.57. Gate turn-off thyristor (*a*) basic structure
(*b*) two-transistor analogy and (*c*) circuit symbol.

4.12.1.2. Turn-off process

The turn-off process in GTO is quite different from that in a CT. The two-transistor model is analysed for understanding the turn-off process in a GTO. For convenience, the two-transistor model of GTO is redrawn in Fig. 4.58.

From Eq. (2.7), $I_{C2} = \beta_2 \cdot I_{B2}$

and $I_{C1} = \beta_1 \cdot I_{B1}$

From Eq. (2.6), $I_{C1} = \alpha_1 I_{E1}$

and $I_{C2} = \alpha_2 \cdot I_{E2}$

As stated before, for initiating the turn-off process in a GTO, a negative gate current I_g' is applied across gate-cathode terminals as shown in Fig. 4.58.

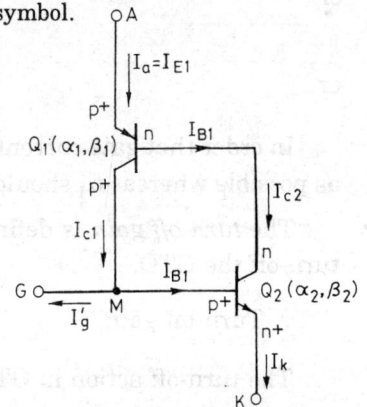

Fig. 4.58. Two-transistor model for GTO with negative gate current I_g'.

Now KCL at anode M in Fig. 4.58 gives

$$I_{C1} - I_g' - I_{B2} = 0$$

or

$$I_{B2} = I_{C1} - I_g' = \alpha_1 I_a - I_g' \qquad ...(4.20)$$

Fig. 4.58 also reveals that $I_a = I_{C1} + I_{C2}$

or

$$I_{c2} = I_a - I_{c1} = (1 - \alpha_1) I_a \qquad ...(4.21)$$

When saturation in Q_2 has occurred, $I_{B2} = \dfrac{I_{C2}}{\beta_2}$. For initiating the turn-off process, Q_z must be brought out of saturation. This can be accomplished only if I_{B2} is made less than I_{C2}/β_2. So, when $I_{B2} < (I_{C2}/\beta_2)$, Q_2 would shift to active region and regenerative action would eventually turn-off the GTO.

$\therefore$ For turning off of Q_2 (or GTO), $I_{B2} < \dfrac{I_{C2}}{\beta_2}$

Substituting the value of I_{B2} from Eq. (4.20) and I_{C2} from Eq. (4.21) we get

$$\alpha_1 I_a - I_g' < \frac{1}{\beta_2} (1 - \alpha_1) I_a$$

$$-I_g' < \frac{(1 - \alpha_1) I_a}{\beta_2} - \alpha_1 I_a$$

$$< \frac{I_a}{\beta_2} - I_a \cdot \alpha_1 \left(1 + \frac{1}{\beta_2} \right)$$

Substitution of

$$\beta_2 = \frac{\alpha_2}{1 - \alpha_2} \text{ gives}$$

$$-I_g' < \frac{I_a}{\alpha_2} (1 - \alpha_2) - I_a \alpha_1 \left[1 + \frac{1 - \alpha_2}{\alpha_2} \right]$$

$$< I_a \left(\frac{1}{\alpha_2} - 1 \right) - I_a \frac{\alpha_1}{\alpha_2}$$

or

$$-I_g' < I_a \left[\frac{1 - \alpha_1 - \alpha_2}{\alpha_2} \right]$$

or

$$I_g' > I_a \left[\frac{\alpha_1 + \alpha_2 - 1}{\alpha_2} \right] \qquad ...(4.22)$$

In order that gate current I_g' for turning-off GTO is low, α_2 should be made as near to unity as possible whereas α_1 should be made small.

The *turn-off gain* is defined as the ratio of anode current I_a to gate current I_g' needed to turn-off the GTO.

$\therefore$ Turn-off gain,

$$\beta_{off} = \frac{I_a}{I_g'} = \frac{\alpha_2}{\alpha_1 + \alpha_2 - 1} \qquad ...(4.23)$$

The turn-off action in GTO can now be explained as under :

(i) Fig. 4.58 shows that $I_a = I_k$ and Eq. (4.22) gives I_g' more than I_k. So when negative gate current I_g' flows between gate-cathode terminals, net base current $(I_{B2} - I_g')$ is reversed, excess carriers are drawn from base p^+ region of Q_2 and collector current I_{C1} of Q_1 is diverted into the external gate circuit. This removes base drive of transistor Q_2. This further removes base current I_{B1} of transistor Q_1 and the GTO is eventually turned off.

(ii) As stated above, a low value of negative gate current requires low value of α_1 and high value of α_2.

Low value of current gain α_1 of Q_1 can be achieved (a) by diffusing gold or other heavy metal n base of Q_1 transistor (b) or by introducing short-circuiting n^+ fingers in the anode p^+ layer as shown in Fig. 4.57 (a), (c) or by a combination of both the techniques listed in (a) and (b) here. Techniques (a) and (b) are described below.

(I) **Gold-doped GTO.** A gold-doped GTO retains its reverse blocking capability. Gold-doping also reduces turn-off time, therefore, these GTOs are suitable for high-frequency operation. However, gold-doped GTOs suffer from more on-state voltage drop for a given current than a similar CT.

(II) **Anode-shorted GTO.** The short-circuiting fingers, also called *anode-shorts*, leads to short-circuit of the emitter P^+ (anode A) with base n of Q_1 transistor as shown in Fig. 4.59. For anode or emitter current I_a, effective emitter current I_{E1} is reduced because of anode short. This further decreases collector current I_{C1}. Therefore, effective current gain α_1 of Q_1, now given by I_{C1}/I_a gets reduced. So by anode-short structure, α_1 is reduced but α_2 remains unchanged as desired.

Anode-short, however, reduces reverse voltage blocking capability. With reverse biased GTO, junction that blocks reverse voltage is J_3 only. Junction J_1 blocks no voltage because of n^+ fingers in between p^+ anode. As J_3 junction has large doped layers p^+, n^+ on its two sides, J_3 has lower breakdown voltage, of the order of about 20 to 30 V.

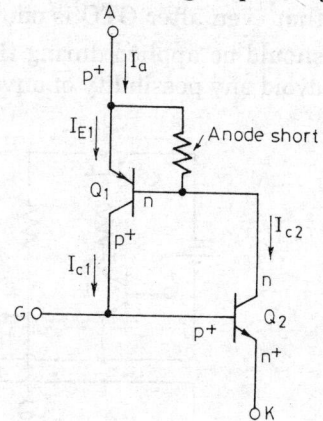

Fig. 4.59. Two-transistor model of GTO with anode-short.

The above is summarised below :

Gold-doped GTO		Anode-shorted GTO	
1.	More on-state voltage drop	1.	Low on-state voltage drop
2.	High reverse-voltage blocking capability	2.	Low reverse-voltage blocking capability.
3.	Suitable for high-frequency operation	3.	Suitable for low-frequency operation

4.12.2. Static I-V Characteristics

The static I-V characteristics of a GTO is identical with that of a conventional thyristor. Latching current for GTO is, however, several amperes, say 2A, as compared to 100-500 mA for a conventional thyristor of the same rating. If gate current is not able to turn on the GTO, it behaves like a high-voltage, low gain transistor with considerable anode current. This leads to a noticeable power loss under such conditions.

In the reverse mode, reverse-voltage blocking capability of GTO is low, typically 20 to 30 V, because of (i) anode shorts and (ii) large doping densities on both sides of reverse blocking junction J_3, Fig. 4.57 (a).

4.12.3. Switching Performance

A basic gate drive circuit for a GTO is shown in Fig. 4.60 (a). For turning-on a GTO, first transistor TR1 is turned on, this in turn switches on TR2 to apply a positive gate-current pulse to turn on GTO. For turning off the GTO, the turn-off circuit should be capable of outputting a high peak current. Usually, a thyristor is used for this purpose. In Fig. 4.60 (*a*), turn-off process is initiated by gating thyristor T1. When T1 is turned on, a large negative gate current pulse turns off the GTO.

4.12.3.1. Gate turn-on. The turn-on process in a GTO is similar to that of a conventional thyristor. Gate turn-on time for GTO is made up of delay time, rise time and spread time like a CT. Further, turn-on time in a GTO can be decreased by increasing its forward gate current as in a thyristor.

In Fig. 4.60 (*b*), a steep-fronted gate pulse is applied to turn-on GTO. Gate drive can be removed once anode current exceeds latching current. However, some manufacturers advise that even after GTO is on, a continuous gate current, called *back porch current* I_{gb} as shown, should be applied during the entire on-period of GTO. The aim of this recommendation is to avoid any possibility of unwanted turn-off of the GTO.

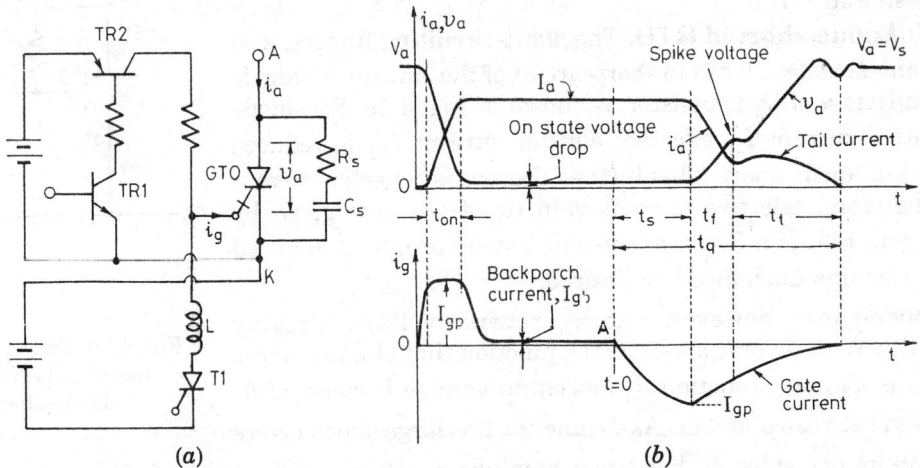

Fig. 4.60. Gate turn-off thyristor (*a*) basic gate drive circuit and (*b*) switching characteristics.

4.12.3.2. Gate turn-off. The turn-off characteristics of a GTO are different from those of an SCR. Before the initiation of turn-off process, a GTO carries a steady current I_a, Fig. 4.60 (*b*). This figure shows a typical dynamic turn-off characteristic for a GTO. The total turn-off time t_q is subdivided into three different periods; namely the storage period (t_s), the fall period (t_f) and the tail period (t_t). In other words,

$$t_q = t_s + t_f + t_t$$

Initiation of turn-off process starts as soon as negative gate current begins to flow after $t = 0$ at instant A. The rate of rise of this gate current depends upon the gate circuit inductance L and the gate voltage applied. During the storage period, anode current I_a and anode voltage (equal to on-state voltage drop) remain constant. Termination of the storage period is indicated by a fall in I_a and rise in V_a. During t_s, excess charges, *i.e.* holes, in p^+ base are removed by negative gate current and the centre junction comes out of saturation. In other words, during storage time t_s, the negative gate current rises to a particular value and prepares the GTO for turning-off (or commutation) by flushing out the stored carriers. After t_s, anode current begins to fall rapidly and anode voltage starts rising. As shown in Fig. 4.60 (*b*), the anode current falls to a certain value and then abruptly changes its rate of fall. This interval during which anode current falls rapidly is the *fall time t_f*, Fig. 4.60 (*b*) and is of the order of 1 μsec [4]. The fall period t_f is measured from the instant gate current is maximum negative to the instant anode current falls to its tail current.

At the time $t = t_s + t_f$, there is a spike in voltage due to abrupt change in anode current. After t_f, anode current i_a and anode voltage v_a keep moving towards their turn-off values for a time t_t called *tail time*. After t_t, anode current reaches zero value and v_a undergoes a transient overshoot due to the presence of R_s, C_s and then stabilizes to its off-state value equal to the source voltage applied to the anode circuit. Here R_s and C_s are the snubber circuit parameters. The turn-off process is complete when tail current reaches zero. The over shoot voltage and tail current can be decreased by increasing the size of C_s, but a compromise with snubber loss must be made. The duration of t_t depends upon the device characteristics [4].

4.12.4. Comparison between GTO and thyristor

A GTO has the following disadvantages as compared to a conventional thyristor :

(*i*) Magnitude of latching and holding currents is more in a GTO.

(*ii*) On state voltage drop and the associated loss is more in a GTO.

(*iii*) Due to the multicathode structure of GTO, triggering gate current is higher than that required for a conventional SCR.

(*iv*) Gate drive circuit losses are more.

(*v*) Its reverse-voltage blocking capability is less than its forward-voltage blocking capability. But this is no disadvantage so far as inverter and chopper circuits are concerned.

In spite of all these demerits, GTO has the following advantages over an SCR :

(*i*) GTO has faster switching speed.

(*ii*) Its surge current capability is comparable with an SCR.

(*iii*) It has more di/dt rating at turn-on.

(*iv*) GTO circuit configuration has lower size and weight as compared to thyristor circuit unit.

(*v*) GTO unit has higher efficiency because an increase in gate-drive power loss and on-state loss is more than compensated by the elimination of forced-commutation losses.

(*vi*) GTO unit has reduced acoustical and electromagnetic noise due to elimination of commutation chokes.

4.12.5. Application of GTOs

In view of the above facts, GTO device are now being used in (*a*) high-performance drive systems, such as the field-oriented control scheme used in rolling mills, robotics and machine tools [4], (*b*) traction purposes because of their lighter weight and (*c*) adjustable-frequency inverter drives. At present, GTOs with ratings up to 5000 V and 3000 A are available.

4.13. STATIC INDUCTION THYRISTOR (SITH)

A static induction thyristor, or SITH, is a three terminal self-controlled device just like a GTO. It was commercially introduced in Japan in 1988. A similar device, known as field-controlled thyristor (FCT), or field-controlled diode (FCD), was developed earlier by General Electric but could not be commercially launched. However, commercial use of SITH is being promoted by Japanese universities and industries.

4.13.1. Basic Structure

The basic structure of SITH is shown in Fig. 4.61 (a) and the device in Fig. 4.61 (b), It is primarily a p^+nn^+ diode, with p^+ grid-like gate electrodes buried in n layer. Comparison of Fig. 2.23 (a) and Fig. 4.61 (a) reveals that device structure of SITH is analogous to SIT except that a p^+ layer is added on the anode side. In addition, n^+ type fingers are diffused in p^+ anode layer just as in a GTO.

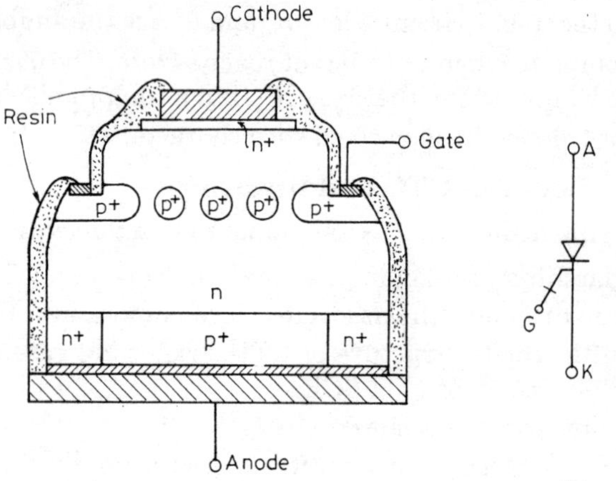

(a) (b)

Fig. 4.61. Static induction thyristor (a) basic structure and (b) device symbol.

4.13.2. Turn-on and Turn-off processes

A simplified structure shown in Fig. 4.62 is employed for explaining the turn on and the turn-off processes in a SITH. When anode is forward biased with gate-cathode voltage V_g equal to zero, the device behaves like a diode. Load current i_a flows from anode to cathode as p^+n junction is forward biased, Fig. 4.62 (a). This shows that SITH is a *normally-on* device just like SIT. When gate is reverse biased with respect to cathode, *i.e.* when V_g is negative, a depletion layer would be formed as shown in Fig. 4.62 (b). This depletion layer blocks the flow of anode current. With varying negative gate bias, the magnitude of anode current can be controlled.

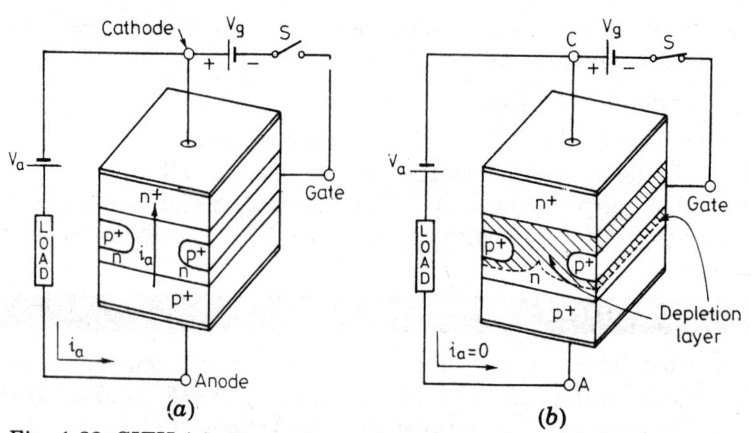

(a) (b)

Fig. 4.62. SITH (a) on-condition when gate voltage V_g is zero and (b) off-condition when V_g is negative.

If SITH is reverse biased, with cathode positive and anode negative, electrons can flow from anode, intermixed n^+ layer, n, through p^+ grid, n^+ and finally to cathod, Fig. 4.61 (a). Thus, reverse current from cathode to anode can exist when SITH is reverse biased. This shows that SITH does not have any reverse blocking capability due to emitter-shorting (p^+ layer interdigitated with n^+ layers at the anode).

4.13.3. Application of SITH and comparison with GTO

At present, SITHs with 2500 V/500 A ratings and 100 kHz operating frequency are available. SITHs with higher power ratings and with normally-off characteristics are likely to be developed in the near future. Their use in induction heating, high frequency link dc–dc converters and HVDC converters is being promoted by Japanese organisations.

When compared with a GTO, a SITH (i) is normally-on device unlike GTO (ii) has higher conduction drop (iii) has lower turn-off current gain, typically 1 to 3 instead of 4 to 5 for GTO (iv) has higher switching frequency (v) the dv/dt and dt/dt ratings are higher and (vi) has more SOA, (safe operating area).

4.14. FIRING CIRCUITS FOR THYRISTORS

An SCR can be switched from off-state to on-state in several ways ; these are forward-voltage triggering, dv/dt triggering, temperature triggering, light triggering and gate triggering, see Art. 4.2. The instant of turning on the SCR cannot be controlled by the first three methods listed above. Light triggering is used in some applications, particularly in a series-connected string. Gate triggering is, however, the most common method of turning on the SCRs, because this method lends itself accurately for turning on the SCR at the desired instant of time. In addition, gate triggering is an efficient and reliable method. In this section, firing circuits for thyristors are studied in detail.

4.14.1. Main Features of Firing Circuits

As stated above, the most common method for controlling the onset of conduction in an SCR is by means of gate voltage control. The gate control circuit is also called firing, or triggering, circuit. These gating circuits are usually low-power electronic circuits. A firing circuit should fulfil the following two functions.

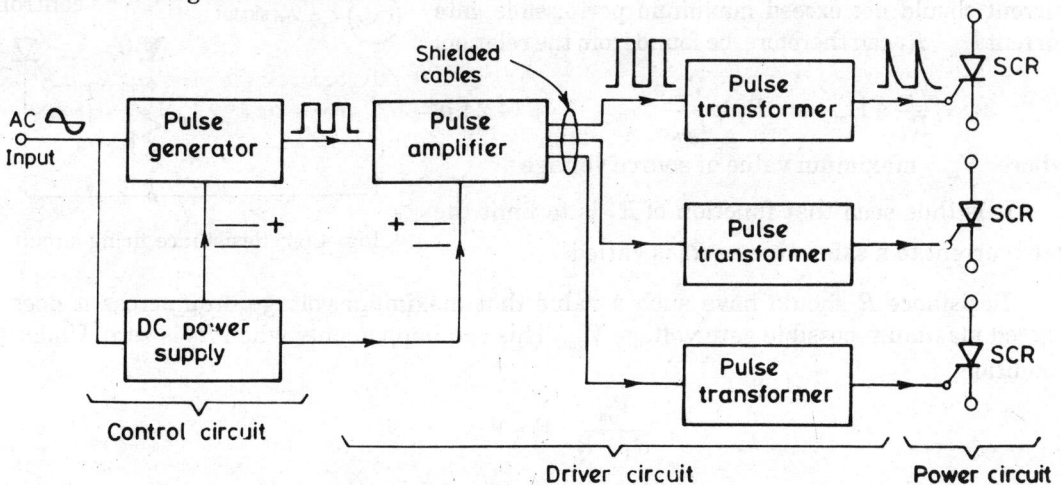

Fig. 4.63. A general layout of the firing circuit scheme for SCRs.

(*i*) If power circuit has more than one SCR, the firing circuit should produce gating pulses for each SCR at the desired instant for proper operation of the power circuit. These pulses must be periodic in nature and the sequence of firing must correspond with the type of thyristorised power controller. For example, in a single-phase semiconverter using two SCRs, the triggering circuit must produce one firing pulse in each half cycle; in a 3-phase full converter using six SCRs, gating circuit must produce one trigger pulse after every 60° interval.

(*ii*) The control signal generated by a firing circuit may not be able to turn-on an SCR. It is therefore common to feed the voltage pulses to a *driver circuit* and then to gate-cathode circuit. A driver circuit consists of a pulse amplifier and a pulse transformer.

A firing circuit scheme, in general, consists of the components shown in Fig. 4.63. A regulated dc power supply is obtained from an alternating voltage source. Pulse generator, supplied from both ac and dc sources, gives out voltage pulses which are then fed to pulse amplifier for their amplification. Shielded cables transmit the amplified pulses to pulse transformers. The function of pulse transformer is to isolate the low-voltage gate-cathode circuit from the high-voltage anode-cathode circuit. Some firing circuit schemes are described in this section.

4.14.2. Resistance and Resistance-Capacitance Firing Circuits

R and RC firing circuits are not in commercial use these days. These are presented here for the sake of highlighting the basic principles of triggering the SCRs. They offer simple and economical firing circuits [3].

(*a*) *Resistance firing circuits.* As stated above, resistance trigger circuits are the simplest and most economical. They however, suffer from a limited range of firing angle control (0° to 90°), great dependence on temperature and difference in performance between individual SCRs.

Fig. 4.64 shows the most basic resistance triggering circuit. R_2 is the variable resistance, R is the stabilizing resistance. In case R_2 is zero, gate current may flow from source, through load, R_1, D and gate to cathode. This current should not exceed maximum permissible gate current I_{gm}. R_1 can therefore, be found from the relation,

$$\frac{V_m}{R_1} \le I_{gm} \quad \text{or} \quad R_1 \ge \frac{V_m}{I_{gm}} \qquad ...(4.24\ a)$$

where V_m = maximum value of source voltage

It is thus seen that function of R_1 is to limit the gate current to a safe value as R_2 is varied.

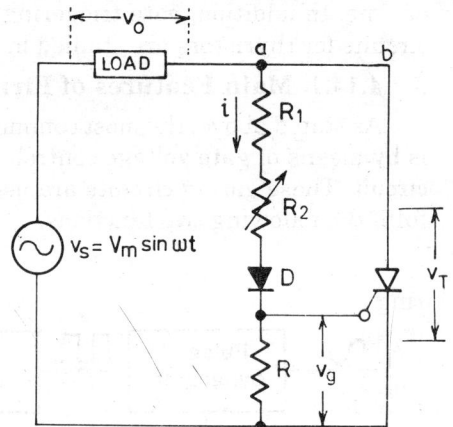

Fig. 4.64. Resistance firing circuit.

Resistance R should have such a value that maximum voltage drop across it does not exceed maximum possible gate voltage V_{gm}. This can happen only when R_2 is zero. Under this condition,

$$\frac{V_m}{R_1 + R} \cdot R \le V_{gm}$$

or

$$R \le \frac{V_{gm} \cdot R_1}{V_m - V_{gm}} \qquad ...(4.24\ b)$$

As resistances R_1, R_2 are large, gate trigger circuit draws a small current. Diode D allows the flow of current during positive half cycle only, *i.e.* gate voltage v_g is half-wave dc pulse. The amplitude of this dc pulse can be controlled by varying R_2.

The potentiometer setting R_2 determines the gate voltage amplitude. When R_2 is large, current i is small and the voltage across R, *i.e.* $v_g = iR$ is also small as shown in Fig. 4.65 (a). As V_{gp} (peak of gate voltage v_g) is less than V_{gt} (gate trigger voltage), SCR will not turn on. Therefore, load voltage $v_0 = 0$, $i_0 = 0$ and supply voltage v_s appears as v_T across SCR as shown in Fig. 4.65 (a)[*]. Note that trigger circuit consists of resistances only, v_g is therefore in phase with source voltage v_s. In Fig. 4.65 (b), R_2 is adjusted such that $V_{gp} = V_{gt}$. This gives the value of

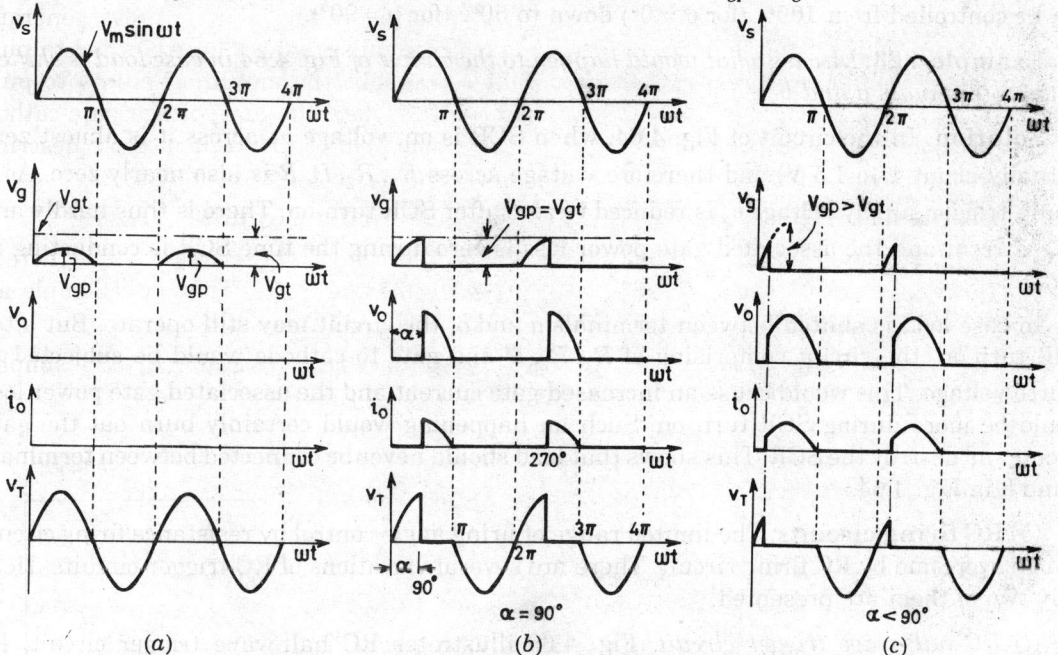

Fig. 4.65. Resistance firing of an SCR in a half-wave circuit with dc load
(a) No triggering of SCR (b) $\alpha = 90°$ (c) $\alpha < 90°$.

firing angle as 90°. The various current and voltage waveforms are shown in Fig. 4.65 (b). In Fig. 4.65 (c), $V_{gp} > V_{gt}$. As soon as v_g becomes equal to V_{gt} for the first time SCR is turned on. The resistance triggering cannot give firing angle beyond 90°. Increasing v_g above V_{gt} turns on the SCR at firing angles less than 90°. When v_g reaches V_{gt} for the first time, SCR fires, gate loses control and v_g is reduced to almost zero (about 1 V) value as shown. It may also be seen that firing angle can never be equal to zero degree however large V_{gp} may be ; it can, of course, be brought nearer (2°–4°) to zero degree firing angle. A relationship between peak gate voltage V_{gp} and gate trigger voltage V_{gt} may be expressed as follows :

$$V_{gp} \sin \alpha = V_{gt} \qquad \text{or} \qquad \alpha = \sin^{-1}(V_{gt}/V_{gp})$$

---

[*] Some students argue that in every positive cycle of source, diode circuit will be active and will therefore draw current from source. The current will cause voltage drop v_o across load and therefore, v_o and i_o should be shown in Fig. 4.65 (a). Actually, load resistance (a few ohms) in comparison with $R_1 + R_2 + R$ (kΩ) is quite small. Therefore, current during the positive cycle of source is negligibly small and likewise v_o across the load.

Since
$$V_{gp} = \frac{V_m R}{R_1 + R_2 + R}$$

$$\alpha = \sin^{-1}\left[\frac{V_{gt} \cdot (R_1 + R_2 + R)}{V_m R}\right]$$

As V_{gt}, R_1, R and V_m are fixed, $\alpha \propto \sin^{-1}(R_2)$ or $\alpha \propto R_2$.

This shows that firing angle is proportional to R_2. As R_2 is increased from small value (*i.e.* small α), firing angle increases. In any case, α can never be more than 90°.

As the firing angle control is from 0° (approximately) to 90°, the half-wave power output can be controlled from 100% (for $\alpha = 0°$) down to 50% (for $\alpha = 90°$).

Example 4.23. *Discuss what would happen to the circuit of Fig. 4.64 in case load is shifted between terminals a and b.*

Solution. In the circuit of Fig. 4.64, when SCR is on, voltage v_T across it is almost zero (actually about 1 to 1.5 V) and therefore voltage across R_1, R_2, D, R is also nearly zero. As a result, trigger supply voltage v_g is reduced to zero after SCR turn-on. There is thus hardly any gate current and the associated gate power loss is zero during the time SCR is conducting in Fig. 4.64.

In case load is shifted between terminals a and b, the circuit may still operate. But after SCR turn-on, the circuit comprising of R_1, R_2, D and gate to cathode would be subjected to source voltage. This would cause an increased gate current and the associated gate power loss would be more during SCR turn on. Such an happening would certainly burn out the gate circuit and destroy the SCR. This shows that load should never be connected between terminals a and b in Fig. 4.64.

(*b*) **RC firing circuits.** The limited range of firing angle control by resistance firing circuit can be overcome by RC firing circuit. There are several variations of RC trigger circuits. Here only two of them are presented.

(*i*) *RC half-wave trigger circuit.* Fig. 4.66 illustrates RC half-wave trigger circuit. By varying the value of R, firing angle can be controlled from 0° to 180°. In the negative half cycle, capacitor C charges through $D2$ with lower plate positive to the peak supply voltage V_m at $\omega t = -90°$. After $\omega t = -90°$, source voltage v_s decreases from $-V_m$ at $\omega t = -90°$ to zero at $\omega t = 0°$. During this period, capacitor voltage v_C may fall from $-V_m$ at $\omega t = -90°$ to some lower value $-oa$ at $\omega t = 0°$ as shown in Fig. 4.67. Now, as the SCR anode voltage passes through zero and becomes positive, C begins to charge through variable resistance R from the initial voltage $-oa$. When capacitor charges to positive voltage equal to gate trigger voltage V_{gt}, SCR is fired and after this, capacitor holds to a small positive voltage, Fig. 4.67. Diode D1 is used to prevent the breakdown of cathode to gate junction through D2 during the negative half cycle. An examination of Fig. 4.67 reveals that firing angle can never be zero and 180°.

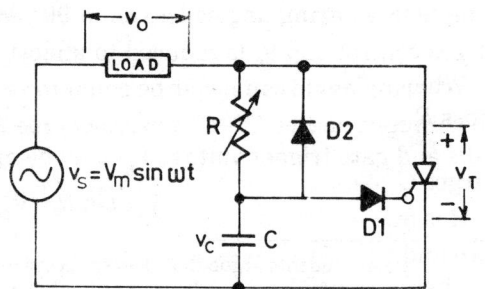

Fig. 4.66. RC half-wave trigger circuit.

In the range of power frequencies, it may be empirically shown [3] that RC for zero output voltage is given by

$$RC \geq \frac{1.3\,T}{2} \cong \frac{4}{\omega} \qquad \qquad ...(4.25)$$

where $\qquad\qquad T = \frac{1}{f}$ = period of ac line frequency in seconds.

The SCR will trigger when $v_c = V_{gt} + v_d$, where v_d is the voltage drop across diode D1. At the instant of triggering, if v_c is assumed constant, the current I_{gt} must be supplied by voltage source through R, D1 and gate to cathode circuit. Hence the maximum value of R is given by

$$v_s \geq RI_{gt} + v_c$$

or $\qquad\qquad v_s \geq RI_{gt} + V_{gt} + v_d$

or $\qquad\qquad R \leq \dfrac{v_s - V_{gt} - v_d}{I_{gt}} \qquad\qquad ...(4.26)$

where v_s is the source voltage at which thyristor turns on. Approximate values of R and C can be obtained from Eqs. (4.25) and (4.26).

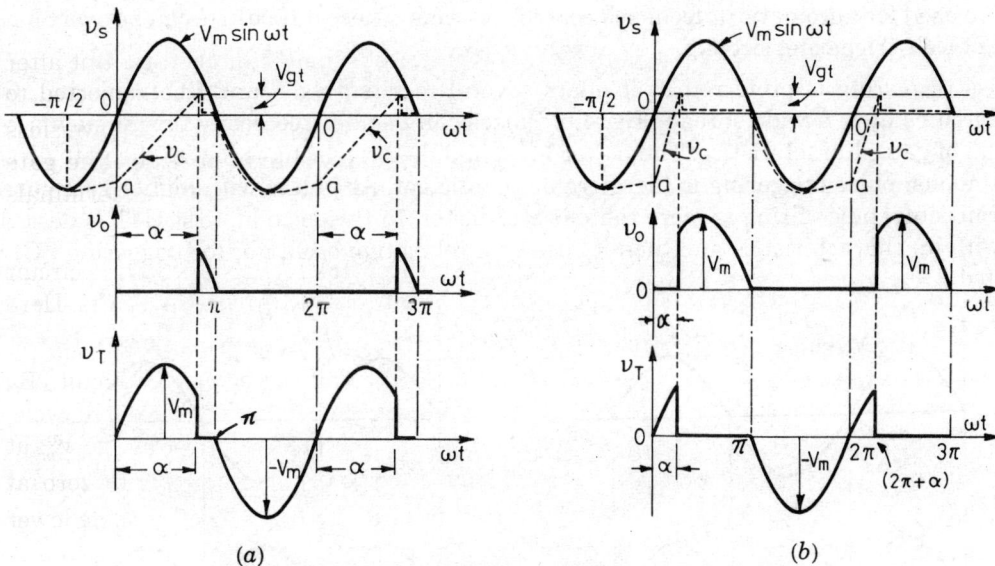

(a)   (b)

Fig. 4.67. Waveforms for RC half-wave trigger circuit of
Fig. 4.66 (a) high value of R (b) low value of R.

When SCR triggers, voltage drop across it falls to 1 to 1.5 V. This, in turn, lowers the voltage across R and C to this low value of 1 to 1.5 V. Low voltage across SCR during conduction period keeps C discharged in positive half cycle until negative voltage cycle across C appears. This charges C to maximum negative voltage $- V_m$ as shown in Fig. 4.67 by dotted line. In Fig. 4.67 (a), R is more, the time taken for C to charge from $- oa$ to $(V_{gt} + v_d) \cong V_{gt}$ is more, firing angle is more and therefore average output voltage is low. In Fig. 4.67 (b), R is less, firing angle is low and therefore average output voltage is more.

(ii) RC full-wave trigger circuit. A simple *RC* trigger circuit giving full-wave output voltage is shown in Fig. 4.68. Diodes D1–D4 form a full-wave diode bridge. In this circuit, the initial

voltage from which the capacitor C charges is almost zero. The capacitor C is set to this low positive voltage (upper plate positive) by the clamping action of SCR gate. When capacitor charges to a voltage equal to V_{gt}, SCR triggers and rectified voltage v_d appears across load as v_0. The value of RC is calculated by the empirical relation [3],

$$RC \geq 50 \frac{T}{2} \cong \frac{157}{\omega} \qquad ...(4.27)$$

As per Eq. (4.26), the value of R is given by

$$R \ll \frac{v_s - V_{gt}}{I_{gt}}$$

where v_s is the source voltage at which thyristor turns on. In Fig. 4.69 (a), firing angle α is more than 90° and in Fig. 4.69 (b), $\alpha < 90°$.

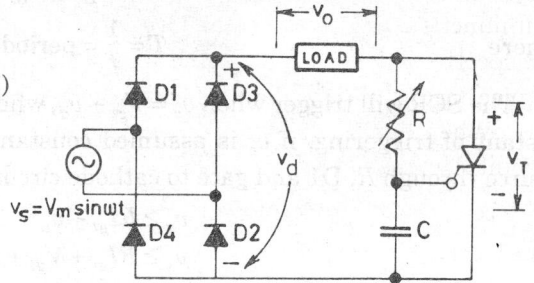

Fig. 4.68. RC full-wave trigger circuit.

4.14.3. Unijunction Transistor (UJT)

Resistance and RC triggering circuits described above give prolonged pulses. As a result, power dissipation in the gate circuit is large. At the same time, R and RC triggering circuits cannot be used for automatic or feedback control systems. These difficulties can be overcome by the use of UJT triggering circuits.

Pulse triggering is preferred as it offers several merits over R and RC triggering. Gate characteristics have a wide spread, Fig. 4.9. Pulses can be adjusted easily to suit such a wide spectrum of gate characteristics. The power level in pulse triggering is low as the gate drive is discontinuous, pulse triggering is therefore more efficient. As pulses with higher gate current are permissible, pulse firing is more reliable and faster. In this section, first UJT is described along with its characteristics and then its use as a relaxation oscillator for triggering SCRs is presented.

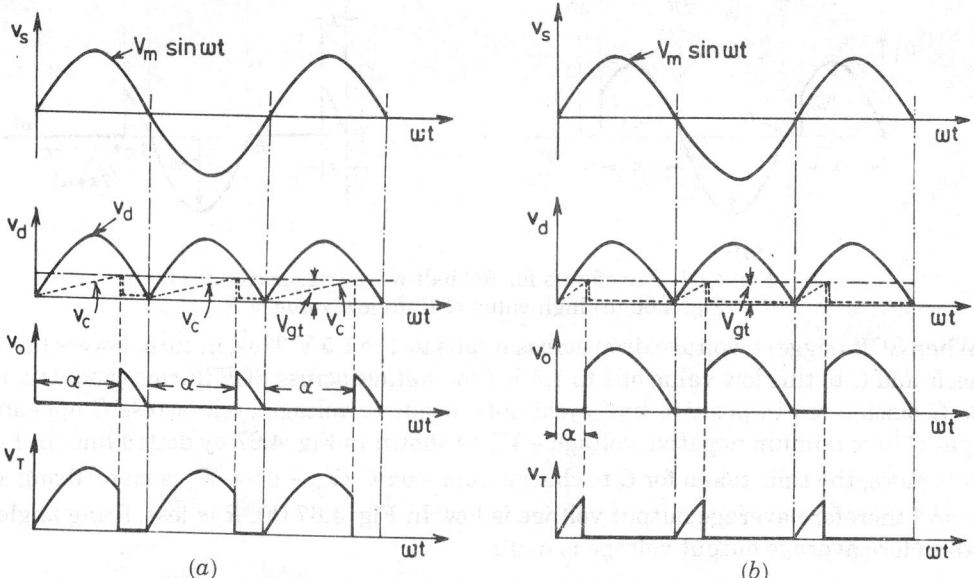

Fig. 4.69. Waveforms for RC half-wave trigger circuit of
Fig. 4.68 (a) high value of R (b) low value of R.

An UJT is made up of an n-type silicon base to which p-type emitter is embedded, Fig. 4.70 (a). The n-type base is lightly doped whereas p-type is heavily doped. The two ohmic contacts provided at each end are called base-one B_1 and base-two B_2. So, an UJT has three terminals, namely the emitter E, base-one B_1 and base-two B_2. Between bases B_1 and B_2, the unijunction behaves like an ordinary resistance. R_{B1} and R_{B2} are the internal resistances respectively from bases B_1 and B_2 to eta point A, Fig. 4.70 (a). Its symbolic representation is given in Fig. 4.70 (b) and its equivalent circuit in Fig. 4.70 (c).

When a voltage V_{BB} is applied across the two base terminals B_1 and B_2, the potential of point A with respect to B_1 is given by

$$V_{AB1} = \frac{V_{BB}}{R_{B1} + R_{B2}} \cdot R_{B1} = \frac{R_{B1}}{R_{B1} + R_{B2}} \cdot V_{BB} = \eta \, V_{BB}$$

where $\eta = \dfrac{R_{B1}}{R_{B1} + R_{B2}}$ is called the *intrinsic stand-off ratio*. Typical values of η are 0.51 to 0.82.

Interbase resistance $R_{BB} = R_{B1} + R_{B2}$ is of the order of 5–10 kΩ.

This resistance R_{BB} can easily be measured by a multimeter with emitter open. As stated before, R_{BB} is broken up into two resistances, R_{B1} between emitter and base B_1, and R_{B2} in between emitter and base B_2. Since emitter is nearer to B_2, resistance R_{B2} is less than the resistance R_{B1}.

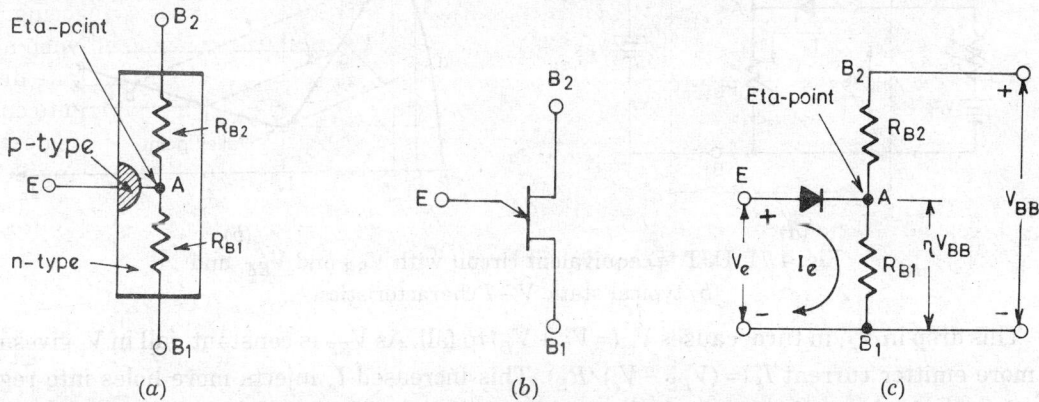

Fig. 4.70. (a) Basic structure of UJT (b) symbolic representation and (c) its equivalent circuit.

The operation of UJT can be understood with its equivalent circuit of Fig. 4.70 (c). As UJT is usually operated with both B_2 and E biased positive with respect to reference base terminal B_1. DC voltage source V_{BB} between B_2 and B_1 is constant. DC source V_{EE} in series with resistance R_E is considered as input to the UJT. Both V_{BB} and V_{EE} are shown in Fig. 4.71 (a) where UJT equivalent circuit is shown inside the dotted rectangle. As before

$$V_{AB1} = V_A = \frac{V_{BB}}{R_{B1} + R_{B2}} \cdot R_{B1} = \eta \cdot V_{BB}$$

The magnitude of voltage V_e can be varied by regulating external resistance R_E. As long as emitter voltage $V_e < \eta \cdot V_{BB}$, the $E - B_1$ unijunction (or $p - n$ junction) is reverse biased and emitter current I_e is negative as shown by curve PS in Fig. 4.71 (b). The region PS of very low current is treated as 'off' state of UJT. The resistance between $E - B_1$ junction is therefore very high. At point S, $I_e = 0$, drop across R_E is zero, therefore V_e = source voltage, *i.e.* $OS = V_e = V_{EE}$.

Actually, off-state of UJT extends to a point where emitter voltage V_e exceeds V_A, or $\eta.V_{BB}$, by diode voltage V_D in $E - B_1$ junction. So when $V_e = \eta . V_{BB} + V_D$, point B is reached and $E - B_1$ junction gets forward biased to allow forward current through the diode. Here V_D is the forward voltage drop across $E - B_1$ junction (usually 0.5 V).

Point B is called the *peak point*. Voltage V_p and current I_p pertaining to point B are called *peak-point voltage* and *peak-point current* respectively. By varying R_E, V_e is increased till V_e approaches V_p. At this peak point, $V_e = V_p = \eta.V_{BB} + V_D$, the p-emitter begins to inject holes from the heavily doped emitter E into the lower base region B_1. As n type base is lightly doped, the holes rarely get any chance to recombine. The lower base region B_1 is, therefore, filled up with additional current carriers (holes). As a result, resistance R_{B1} of $E - B1$ junction decreases. The fall in R_{B1} causes potential of eta point A to drop.

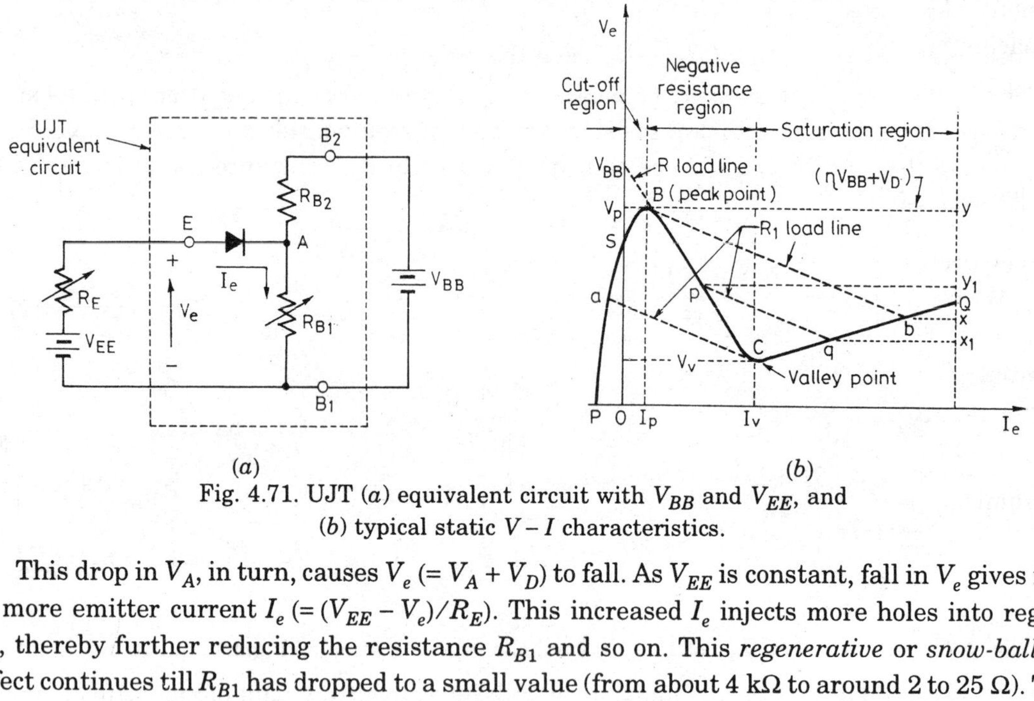

Fig. 4.71. UJT (a) equivalent circuit with V_{BB} and V_{EE}, and
(b) typical static $V - I$ characteristics.

This drop in V_A, in turn, causes V_e ($= V_A + V_D$) to fall. As V_{EE} is constant, fall in V_e gives rise to more emitter current I_e ($= (V_{EE} - V_e)/R_E$). This increased I_e injects more holes into region B_1, thereby further reducing the resistance R_{B1} and so on. This *regenerative* or *snow-balling* effect continues till R_{B1} has dropped to a small value (from about 4 kΩ to around 2 to 25 Ω). The emitter current, limited by external resistance R_E, is then given by

$$I_e = \frac{V_{EE} - V_D}{R_{B1} + R_E}$$

When R_{B1} has dropped to a very small value, indicated by point C in Fig. 4.71 (b), the UJT has reached 'on' state. At point C, entire base region B_1 is saturated and resistance R_{B1} cannot decrease any more. This point C is called the *valley point*; V_v and I_v are the corresponding emitter potential and current. After UJT is on, or after valley point is reached, an increase in V_e is accompanied by an increase in I_e; this is indicated by curve CQ. At point Q, V_e is a little more than its valley point voltage V_v. Between points B and C, emitter voltage V_e falls as I_e increases; UJT, therefore, exhibits negative resistance between these two points. The negative resistance region between peak and valley points in Fig. 4.71 (*b*) gives UJT the switching characteristics for use in SCR triggering circuits.

At the valley point, the current is given by V_v/R_{B1}. Valley-point current, also called holding current, keeps UJT on. When emitter current I_e falls below I_v, UJT turns off.

UJT oscillator triggering. The unijunction transistor is a highly efficient switch ; its switching time is in the range of nanoseconds. Since UJT exhibits negative resistance characteristics, it can be used as a relaxation oscillator. Fig. 4.72 (*a*) shows a circuit diagram with UJT working in the oscillator mode. The external resistances R_1, R_2 are small in comparison with the internal resistances R_{B1}, R_{B2} of UJT bases. The charging resistance R should be such that its load line intersects the device characteristics only in the negative resistance region.

In Fig. 4.72 (*a*), when source voltage V_{BB} is applied, capacitor C begins to charge through R exponentially towards V_{BB}. During this charging, emitter circuit of UJT is an open circuit. The capacitor voltage v_c, equal to emitter voltage v_e, is given by

$$v_c = v_e = V_{BB}\left(1 - e^{-t/RC}\right)$$

The time constant of the charge circuit is $\tau_1 = RC$.

When this emitter voltage v_e (or v_c) reaches the peak-point voltage V_p ($= \eta\, V_{BB} + V_D$), the unijunction between $E - B_1$ breaks down. As a result, UJT turns on and capacitor C rapidly discharges through low resistance R_1 with a time constant $\tau_2 = R_1 C$. Here τ_2 is much smaller than τ_1. When the emitter voltage decays to the valley-point voltage V_v, emitter current ($V_v/(R_{B1} + R_1)$) falls below I_v and UJT turns off. The time T required for capacitor C to charge from initial voltage V_v to peak-point voltage V_p, through large resistance R, can be obtained as under :

$$V_p = \eta\, V_{BB} + V_D = V_v + V_{BB}\,(1 - e^{-T/RC})$$

Assuming

$$V_D = V_v, \quad \eta = (1 - e^{-T/RC})$$

or

$$T = \frac{1}{f} = RC\ \ln\left(\frac{1}{1-\eta}\right) \qquad \qquad ...(4.28)$$

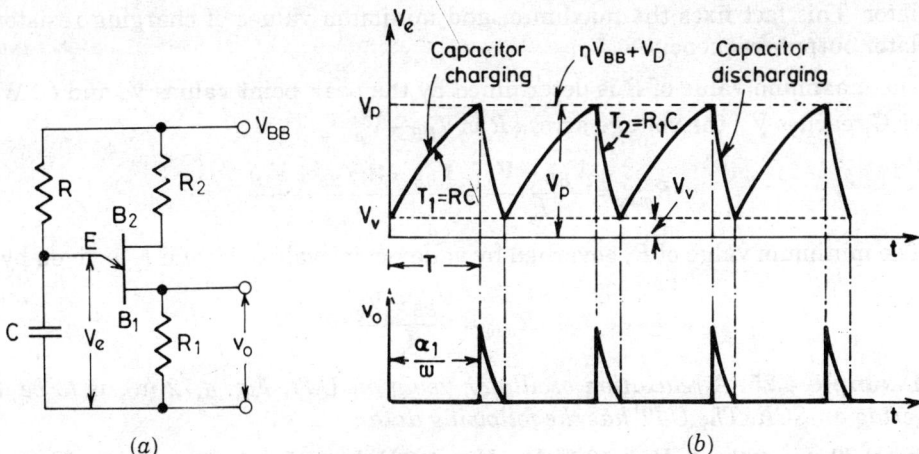

(*a*)   (*b*)

Fig. 4.72. UJT oscillator (*a*) Connection diagram and (*b*) Voltage waveforms.

In case T is taken as the time period of output pulse duration (neglecting small discharge time), then the value of firing angle α_1 is given by

$$\alpha_1 = \omega T = \omega RC \ \ln \ \frac{1}{1-\eta} \qquad \qquad ...(4.29)$$

where ω is the angular frequency of UJT oscillator.

The amplitude of pulse voltage is obtained by drawing a load line $B\,b$ for R_1 as shown in Fig. 4.71 (b). The vertical projection of Bb, equal to xy, gives the voltage pulse amplitude. With the discharge of capacitor, the operating points B and b move towards C. For points p and q, the pulse amplitude is x_1y_1. Eventually, point C is reached at which pulse voltage is zero, then the operating point shifts to a, Fig. 4.71 (b). The potential of eta point A is ηV_{BB}, but that of the emitter is V_v which is less than ηV_{BB}. As a result, $E - B_1$ unijunction is reverse biased and ceases to conduct, the UJT turns off and goes into blocking mode. Capacitor C now again charges from $V_e = V_v$ to voltage $\eta V_{BB} + V_D$, $E - B_1$ unijunction breaks down and the above cycle repeats.

If the output voltage pulses are used for triggering an SCR, resistance R_1 should be sufficiently small so that normal leakage current drop across R_1, when UJT is off, is not able to trigger the SCR. In other words,

$$\frac{V_{BB} \cdot R_1}{R_{BB} + R_1 + R_2} < \text{SCR trigger voltage } V_{gt}$$

where $R_{BB} = R_{B1} + R_{B2}$

The emitter-diode forward characteristics vary with temperature in such a manner that V_D decreases and R_{BB} increases with temperature. In order to provide compensation against this thermal effect, the value of R_2 used in Fig. 4.72 should be calculated from the relation

$$R_2 = \frac{10^4}{\eta V_{BB}} \qquad \qquad ...(4.30)$$

The width of triggering pulse is sometimes taken equal to R_1C.

In case load line for R intersects the UJT characteristics in the region CQ, Fig. 4.71 (b), the intersecting point will result in stable operating point and the circuit then cannot work as an oscillator. This fact fixes the maximum and minimum values of charging resistor R and the oscillator output frequency.

The maximum value of R is determined by the peak-point values V_p and I_p. When voltage across C reaches V_p, the voltage across R is $V_{BB} - V_p$.

$\therefore$ $$R_{max} = \frac{V_{BB} - V_p}{I_p} = \frac{V_{BB} - (\eta V_{BB} + V_D)}{I_p} \qquad \qquad ...(4.31a)$$

The minimum value of R, governed by valley-point values V_v and I_v is given by

$$R_{min} = \frac{V_{BB} - V_v}{I_v} \qquad \qquad ...(4.31b)$$

Example 4.25. *A relaxation oscillator using an UJT, Fig. 4.72 (a), is to be designed for triggering an SCR. The UJT has the following data :*

$\eta = 0.72$, $I_p = 0.6$ mA, $V_p = 18.0$ V, $V_v = 1.0$ V, $I_v = 2.5$ mA, $R_{BB} = 5$ kΩ, Normal leakage current with emitter open = 4.2 mA.

The firing frequency is 2 kHz. For C = 0.04 µF, compute the values of R, R_1 and R_2.

Solution. The value of charging resistor R, from Eq. (4.28), is

$$R = \frac{T}{C \ln \dfrac{1}{1-\eta}} = \frac{1}{fC \ln \dfrac{1}{1-\eta}} = \frac{10^6}{2000 \times 0.04 \ln \dfrac{1}{0.28}} = 9.82 \text{ k}\Omega$$

As V_D is not given, $V_p = \eta V_{BB}$

$$\therefore \qquad V_{BB} = \frac{V_p}{\eta} = \frac{18.00}{0.72} = 25 \text{ V}$$

From Eq. (4.30), $\qquad R_2 = \dfrac{10^4}{0.72 \times 25} = 555.55 \ \Omega$

With emitter open, $\qquad V_{BB} = \text{Leakage current } (R_1 + R_2 + R_{BB})$

$$\therefore \qquad R_1 = \frac{25}{4.2 \times 10^{-3}} - 5000 - 555.55 \ \Omega = 396.83 \cong 397 \ \Omega$$

Example 4.26. *If the firing frequency of the SCR in Example 4.25 is changed by varying charging resistor R, obtain the maximum and minimum values of R and the corresponding frequencies.*

Solution.

From Eq. (4.31), $\qquad R_{max} = \dfrac{25 \ (1 - 0.72)}{0.6 \times 10^{-3}} = 11.67 \text{ k}\Omega$

$$R_{min} = \frac{25.0 - 1.0}{2.5 \times 10^{-3}} = 9.6 \text{ k}\Omega$$

From Eq. (4.28), $\qquad f_{min} = \dfrac{1}{T_{\max}} = \dfrac{1}{R_{max} \, C \ln \dfrac{1}{1-\eta}}$

$$= \frac{10^3}{11.67 \times 0.04 \ln \dfrac{1}{0.28}} = 1682.8 \text{ Hz} \cong 1.683 \text{ kHz}$$

and $\qquad f_{max} = \dfrac{10^3}{9.6 \times 0.04 \ln \dfrac{1}{0.28}} = 2045.7 \text{ Hz} \cong 2.05 \text{ kHz}$

Synchronized UJT triggering (or Ramp triggering). A synchronized UJT trigger circuit using an UJT is shown in Fig. 4.73. Diodes $D_1 - D_4$ rectify ac to dc. Resistor R_1 lowers V_{dc} to a suitable value for the zener diode and UJT. Zener diode Z functions to clip the rectified voltage to a standard level V_z, which remains constant except near the V_{dc} zero, Fig. 4.74. This voltage V_z is applied to the charging circuit RC. Current i_1 charges capacitor C at a rate determined by R. Voltage across capacitor is marked by v_c in Figs. 4.73 and 4.74. When voltage v_c reaches the unijunction threshold voltage ηV_z, the $E - B_1$ junction of UJT breaks down and the capacitor C discharges through primary of pulse transformer sending a current i_2 as shown in Fig. 4.73.

As the current i_2 is in the form of pulse, windings of the pulse transformer have pulse voltages at their secondary terminals. Pulses at the two secondary windings feed the same in-phase pulse to two SCRs of a full-wave circuit. SCR with positive anode voltage would turn

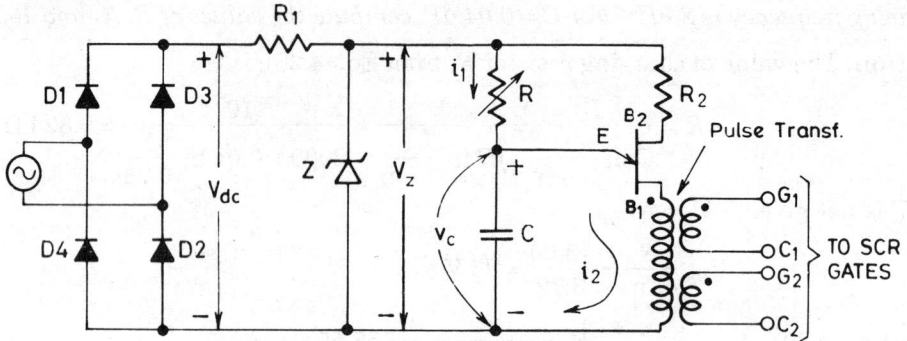

Fig. 4.73. Synchronised UJT trigger circuit.

on. As soon as the capacitor discharges, it starts to recharge as shown. Rate of rise of capacitor voltage can be controlled by varying R. The firing angle can be controlled up to about 150°. This method of controlling the output power by varying charging resistor R is called *ramp control, open-loop control* or *manual control.*

As the zener diode voltage V_z goes to zero at the end of each half cycle, the synchronization of the trigger circuit with the supply voltage across SCRs is achieved. Thus the time t, equal to α/ω, when the pulse is applied to SCR for the first time, will remain constant for the same value of R. Small variations in the supply voltage and frequency are not going to effect the circuit operation.

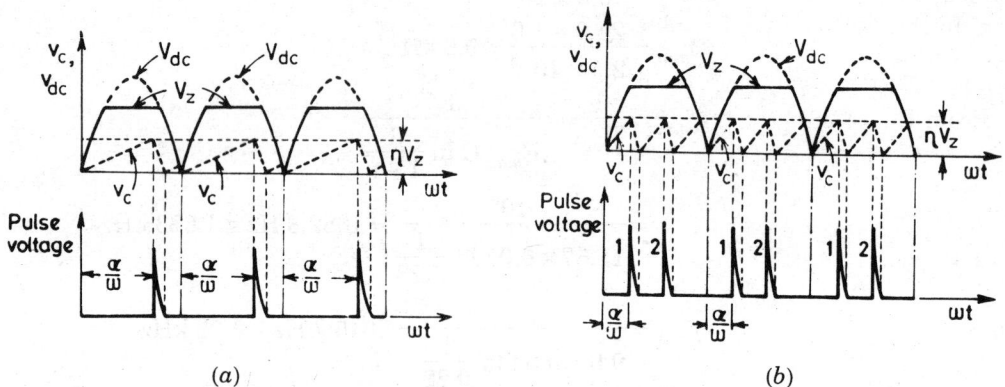

Fig. 4.74. Generation of output pulses for the circuit of Fig. 4.73. Here, $t = \alpha/\omega$.

In case R is reduced so that v_c reaches UJT threshold voltage twice in each half cycle as shown in Fig. 4.74 (b), then there will be two pulses in each half cycle. As the first pulse will be able to turn-on the SCR, second pulse in each cycle is redundant.

Ramp-and-pedestal triggering. Ramp and pedestal triggering is an improved version of synchronized-UJT-oscillator triggering. Fig. 4.75 shows the circuit for ramp-and-pedestal triggering of two SCRs connected in antiparallel for controlling power in an ac load. This trigger circuit can also be used for triggering the thyristors in a single-phase semiconverter or a single-phase full converter. The various voltage waveforms are shown in Fig. 4.76.

Zener diode voltage V_z is constant at its thresh-hold voltage. R_2 acts as a potential divider. Wiper of R_2 controls the value of pedestal voltage V_{pd}. Diode D allows C to be quickly charged to V_{pd} through the low resistance of the upper portion of R_2. The setting of wiper on R_2 is such

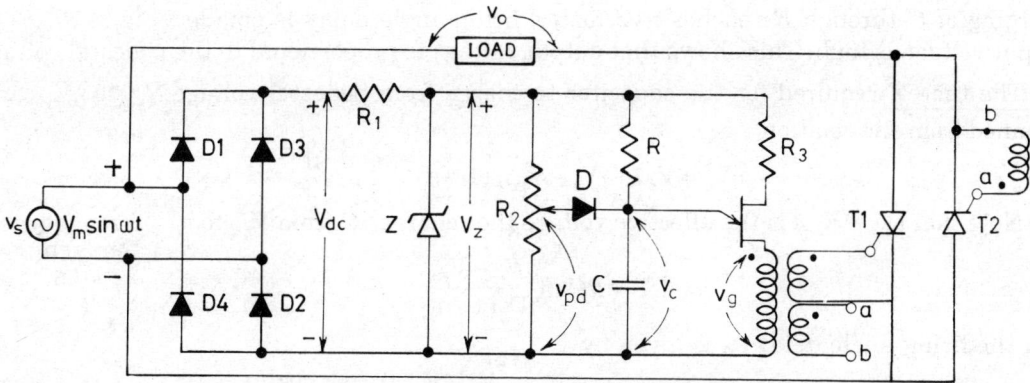

Fig. 4.75. Ramp and pedestal trigger circuit for ac load.

that this value of V_{pd} is always less than the UJT firing point voltage ηV_z. When wiper setting is such that V_{pd} is small, Fig. 4.76 (a), voltage V_z charges C through R. When this ramp voltage v_c reaches ηV_z, UJT fires and voltage v_g, through the pulse transformer, is transmitted to the gate circuits of both SCRs T1 and T2. The forward biased SCR T1 is turned on. After this, v_c reduces to V_{pd} and then to zero at $\omega t = \pi$. As v_c is more than V_{pd}, during the charging of capacitor C through charging resistor R, diode D is reverse biased and turned off. Thus V_{pd} does not effect in any way the discharge of C through UJT emitter and primary of pulse transformer. From 0 to π, T1 is forward biased and is turned on. From π to 2π, T2 is forward biased and is turned on. In this manner, load is subjected to alternating voltage v_0 as shown in Fig. 4.76.

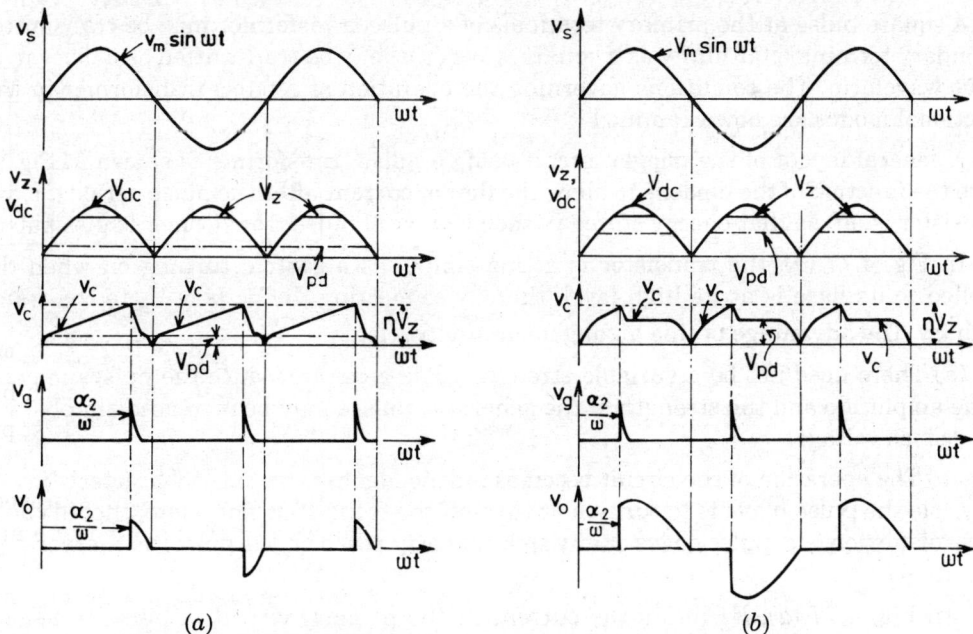

(a) (b)

Fig. 4.76. Waveforms for ramp-and-pedestal circuit of Fig. 4.75.

With the setting of wiper on R_2, pedestal voltage V_{pd} on C can be adjusted. With *low* pedestal voltage across C, ramp charging of C to ηVz takes longer time, Fig. 4.76 (a) and firing angle delay is therefore more and output voltage is *low*. With high pedestal on C, voltage-ramp charging of C through R reaches ηV_z faster, firing angle delay is smaller, Fig. 4.76 (b) and output voltage is high. This shows that output voltage is proportional to the pedestal voltage.

The time T required for the capacitor to charge from pedestal voltage V_{pd} to ηV_z can be obtained from the relation

$$\eta V_z = V_{pd} + (V_z - V_{pd})(1 - e^{-T/RC})$$

Note that $(V_z - V_{pd})$ is the effective voltage that charges C from V_{pd} to ηV_z. From above

$$T = RC \ln \frac{V_z - V_{pd}}{V_z (1 - \eta)} \qquad \qquad ...(4.32)$$

and the firing angle delay α_2 is given by

$$\alpha_2 = \omega RC \ln \frac{V_z - V_{pd}}{V_z (1 - \eta)} \qquad \qquad ...(4.33)$$

4.15. PULSE TRANSFORMER IN FIRING CIRCUITS

Pulse transformers are used quite often in firing circuits for SCRs and GTOs. This transformer has usually two secondaries. The turn ratio from primary to two secondaries is 2 : 1 : 1 or 1 : 1 : 1. These transformers are designed to have low winding resistance, low leakage reactance and low inter-winding capacitance. The advantages of using pulse transformers in triggering semiconductor devices are :

(i) the isolation of low-voltage gate circuit from high-voltage anode circuit and

(ii) the triggering of two or more devices from the same trigger source.

A square pulse at the primary terminals of a pulse transformer may be transmitted at its secondary terminals faithfully a square wave or it may be transmitted as a derivative of the input waveform. The conditions governing the operation of a pulse transformer in these two functional modes are now examined.

A general layout of the trigger circuit using a pulse transformer is shown in Fig. 4.77 (a). Here the function of the diode is to allow the flow of current after the pulse period (i.e. when the transistor is off) so that energy stored in the primary of pulse transformer is dissipated.

In Fig. 4.77 (a), the transistor is acting simply as a switch, turning on when the pulse applied to its base is at its high level, thereby connecting the dc bias V_B to the transformer primary. The advantage of this arrangement are two fold :

(a) There need not be a variable strength pulse generator since the pulses may be of the same amplitude and the strength of the generated pulses may be increased simply by varying the dc bias voltage.

(b) The operation of the circuit becomes independent of the pulse characteristics since the only role the pulse plays is to turn-on or turn-off the transistor. Therefore, there is no effect of pulse distortion (e.g. pulse edges or any spike superimposed on the pulse) on the working of this circuit.

In Fig. 4.77 (a), R_L limits the current in the primary circuit of pulse transformer. Its equivalent circuit is drawn in Fig. 4.77 (b), where L is the magnetizing inductance of the pulse transformer and R_g is the resistance of gate-cathode circuit of an SCR. Fig. 4.77 (c) shows the

transfer of R_g to pulse transformer primary as $R_1 = \left(\dfrac{N_1}{N_2}\right)^2 R_g$. This circuit can be analysed by

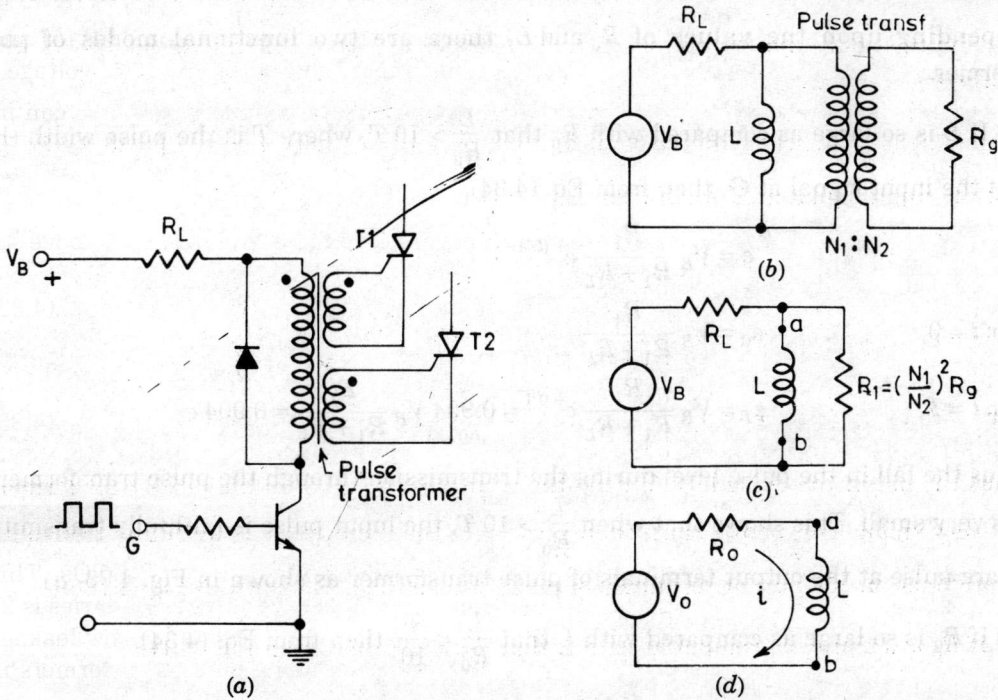

Fig. 4.77. (a) Pulse transformer trigger circuit (b), (c) and (d) its equivalent circuits.

applying Thevenin's theorem at the terminals a b. Fig. 4.77 (d) is the Thevenin's equivalent circuit, where

$$V_0 = V_B \frac{R_1}{R_1 + R_L} \quad \text{and} \quad R_0 = \frac{R_1 R_L}{R_1 + R_L}$$

The voltage equation for Fig. 4.77 (d) is

$$V_0 = R_0 i + L \frac{di}{dt}$$

or

$$V_B \frac{R_1}{R_1 + R_L} = \frac{R_1 R_L}{R_1 + R_L} i + L \frac{di}{dt}$$

or

$$V_B = R_L i + L \left(\frac{R_1 + R_L}{R_1} \right) \frac{di}{dt}$$

Its solution is given by

$$i = \frac{V_B}{R_L} \left[1 - e^{-\frac{R_1 R_L}{L(R_1 + R_L)} t} \right]$$

The voltage across L appears as the output voltage. The magnitude of this voltage from pulse transformer is

$$e = L \frac{di}{dt} = \frac{V_B \cdot R_1}{R_1 + R_L} e^{-\frac{R_1 R_L}{L(R_1 + R_L)} t}$$

or

$$e = \frac{V_B \cdot R_1}{R_1 + R_L} e^{-(R_0/L) t} \qquad \qquad ...(4.34)$$

where

$$R_0 = \frac{R_1 R_L}{R_1 + R_L}$$

Depending upon the values of R_0 and L, there are two functional modes of pulse transformer.

(a) If L is so large as compared with R_0 that $\dfrac{L}{R_0} > 10\,T$, where T is the pulse width (Fig. 4.78) of the input signal at G, then from Eq. (4.34),

$$e = V_B \frac{R_1}{R_1 + R_L} e^{-(t/10\,T)}$$

For $t = 0$,

$$e_0 = V_B \frac{R_1}{R_1 + R_L}$$

and for $t = T$,

$$e_T = V_B \frac{R_1}{R_1 + R_L} e^{-0.1} = 0.904\,V_B \frac{R_1}{R_1 + R_L} = 0.904\,e_0$$

Thus the fall in the pulse level during the transmission through the pulse transformer at $t = T$ is very small. This shows that when $\dfrac{L}{R_0} > 10\,T$, the input pulse is faithfully transmitted as square pulse at the output terminals of pulse transformer as shown in Fig. 4.78 (a).

(b) If R_0 is so large as compared with L that $\dfrac{L}{R_0} < \dfrac{T}{10}$, then from Eq. (4.34).

$$e = V_B \frac{R_1}{R_1 + R_L} \cdot e^{-(10/T)\,t}$$

For $t = 0$,

$$e_0 = V_B \frac{R_1}{R_1 + R_L}$$

and for $t = T$,

$$e_T = V_B \frac{R_1}{R_1 + R_L} e^{-10} = 0.0000453 \cdot V_B \frac{R_1}{R_1 + R_L} = 0.0000453\,e_0.$$

This shows that for $\dfrac{L}{R_0} < \dfrac{T}{10}$, the input pulse is transmitted in the form of exponentially decaying pulses as shown in Fig. 4.78 (b). It is seen that for a step rise in input voltage, the pulse transformer output is a positive pulse. In other words, the input signal is transmitted as a derivative of the input waveform for a step rise. Likewise, for a step fall in input voltage, a negative pulse appears at the pulse transformer output. Fig. 4.78 (b). The operation of the pulse transformer in this mode can be achieved by using a small value of L, i.e. by using an air core for the pulse transformer.

It can thus be inferred from above that the deciding factor in the waveshape of the output pulses from a pulse transformer is its inductance. If the pulse transformer has large inductance, the pulses are faithfully reproduced and if the inductance is small, the pulses are exponentially decaying pulses.

The negative going pulses can be easily removed by using a clipper.

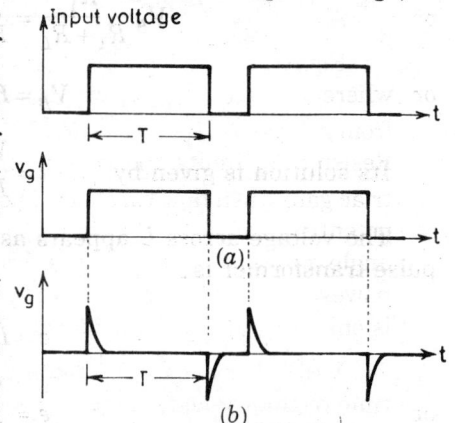

Fig. 4.78. Output voltage waveform of a pulse transformer for

(a) $\dfrac{L}{R_0} > 10T$ and (b) $\dfrac{L}{R_0} < \dfrac{T}{10}$

The amplitude of the trigger voltage at the secondary terminals of pulse transformer is

$$V_g = \frac{N_2}{N_1} \cdot V_B \frac{R_1}{R_1 + R_L}$$

The magnitude of V_B should be large enough to produce trigger voltage V_{gt} at the gate circuit of SCR for its reliable turn on, *i.e.*

$$\frac{N_2}{N_1} \cdot \frac{V_B R_1}{R_1 + R_L} \geq V_{gt}$$

or

$$V_B \geq V_{gt} \frac{N_1}{N_2} \left(1 + \frac{R_L}{R_1}\right)$$

But

$$R_1 = \left(\frac{N_1}{N_2}\right)^2 R_g$$

$$V_B \geq V_{gt} \frac{N_1}{N_2} \left[1 + \left(\frac{N_2}{N_1}\right)^2 \frac{R_L}{R_g}\right] \qquad \qquad ...(4.35)$$

In practice, exponentially decaying trigger pulses of Fig. 4.78 (b) are preferred due to the following reasons :

(*i*) This pulse waveform is suitable for injecting a large charge in the gate circuit for reliable turn on.

(*ii*) The duration of this pulse is small, therefore no significant heating of the gate circuit is observed.

(*iii*) For the same gate-cathode power, it is permissible to raise V_B to a suitable high value so that a hard-drive of SCR is obtained. A device with a hard-drive can withstand high di/dt at the anode circuit which is desirable.

(*iv*) The size of the pulse transformer is reduced. For an extended pulse, large L (with iron-core) is required which increases size and cost of the pulse transformer.

4.16. TRIAC FIRING CIRCUIT

A triggering circuit for a triac using a diac is discussed in this section.

Fig. 4.79 shows a triac firing circuit employing a diac. In this circuit, resistor R is variable whereas resistor R_1 has constant resistance. When R is zero, R_1 protects the diac and triac gate from getting exposed to almost full supply voltage. Resistor R_2 limits the current in the diac and triac gate when diac turns on. The value of C and potentiometer R are so selected as to give a firing angle range of nearly 0° and 180°. In practice, however, a triggering angle range of 10° to 170° is only possible by the firing circuit of Fig. 4.79.

Variable resistor R controls the charging time of the capacitor C and therefore the firing angle of the triac. When R is small, the charging time constant, equal to $(R_1 + R) C$, is small. Therefore, source voltage charges capacitor C to

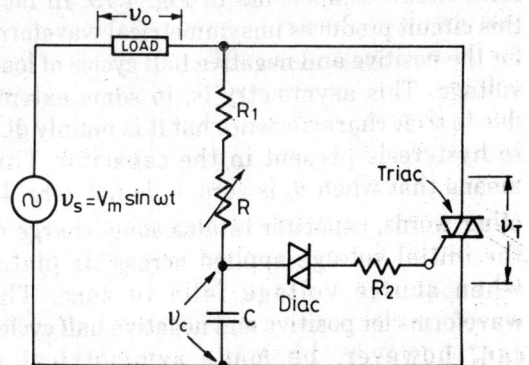

Fig. 4.79. Firing circuit for a triac using a diac.

diac trigger voltage earlier and firing angle for triac is small. Likewise, when R is high, firing angle of triac is large.

When capacitor C (with upper plate positive) charges to breakdown voltage V_{dt} of diac, diac turns on. As a consequence, capacitor discharges rapidly thereby applying capacitor voltage v_c in the form of pulse across the triac gate to turn it on. After triac turn-on at firing angle α, source voltage v_s appears across the load during the positive half cycle for $(\pi - \alpha)$ radians. When v_s becomes zero at $\omega t = \pi$, triac turns off. After $\omega t = \pi$, v_s becomes negative, the capacitor C now charges with lower plate positive. When v_c reaches V_{dt} of diac, diac and triac turn on and v_s appears across the load during the negative half cycle for $(\pi - \alpha)$ radians. At $\omega t = 2\pi$, triac turns off again and the above process repeats.

The waveforms for v_s, v_c, v_T and v_0 are shown in Fig. 4.80 (a) for minimum R and in Fig. 4.80 (b) for maximum R. Here v_s is the source voltage, v_c is the voltage across capacitor, v_T is the voltage across triac and v_0 is the output or load voltage. After triac turn-on, capacitor C holds to a small positive voltage.

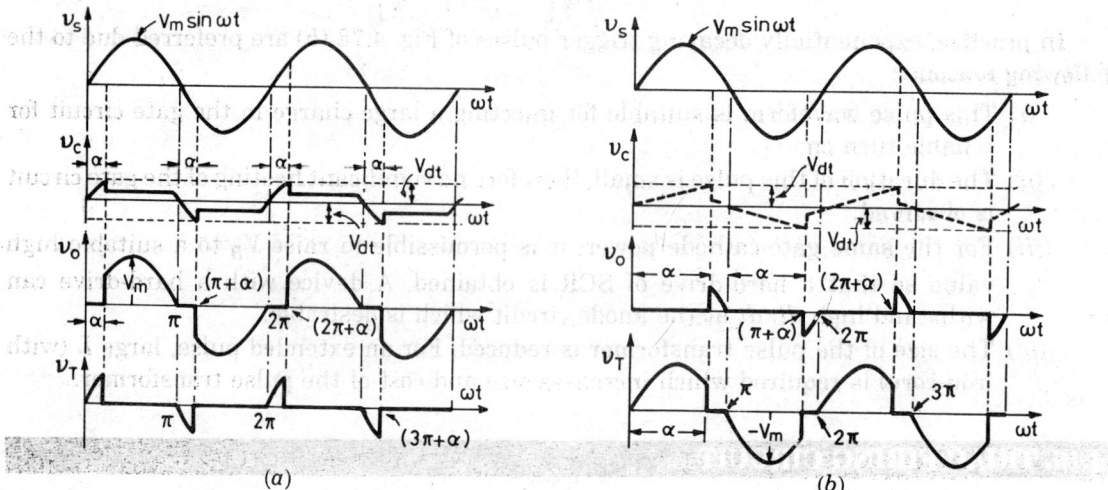

(a)   (b)

Fig. 4.80. Waveforms for triact firing circuit using a diac with
(a) pot. R adjusted to minimum and (b) pot. R adjusted to maximum.

The waveforms shown in Fig. 4.80 are for ideal circuit components in Fig. 4.79. In fact, this circuit produces unsymmetrical waveform for the positive and negative half cycles of load voltage. This asymmetry is, to some extent, due to triac characteristics but it is mainly due to hysteresis present in the capacitor. This means that when v_s is zero, v_c is not zero. In other words, capacitor retains some charge of the initial voltage applied across its plates when source voltage falls to zero. The waveforms for positive and negative half cycles can, however, be made symmetrical if

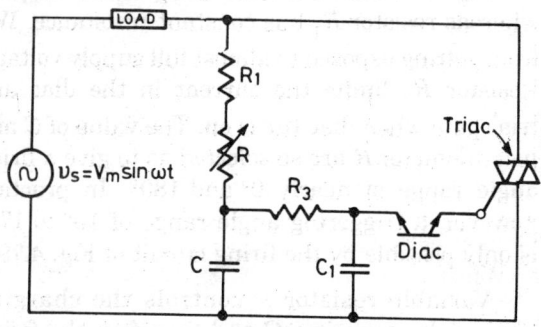

Fig. 4.81. Commercial triac firing circuit
using a diac.

additional resistance R_3 and capacitor C_1 are employed as shown in Fig. 4.81. This circuit is commercially used for controlling the power in lamp dimmers, heat convertors, speed control of fans etc. For inductive loads, snubber circuit must be used across the triac.

Example 4.27. *The firing circuit for a triac using a diac, Fig. 4.79, has the following data :*

$$R_1 = 1000 \ \Omega, \quad R = zero \ to \quad 25000 \ \Omega, \quad C = 0.5 \ \mu F,$$
$$V_s = 230 \ V \ at \ 50 \ Hz, \quad Diac \ breakdown \ voltage = 30 \ V.$$

Find the magnitude of maximum and minimum firing-angle delays for the triac. The effect of load impedance may be neglected.

Solution. When the diac is not conducting, the current through R_1, R and C is given by

$$I_1 = \frac{V_s}{Z} \qquad \text{where} \qquad Z = \left[(R_1 + R)^2 + \left(\frac{1}{\omega C} \right)^2 \right]^{1/2}$$

When $R = 0$:
$$Z = \left[1000^2 + \left(\frac{10^6}{2\pi \times 50 \times 0.5} \right)^2 \right]^{1/2}$$

$$= [1000^2 + 6366.2^2]^{1/2} = 6444.3 \ \Omega$$

I_1 leads V_s by an angle $\quad \phi = \tan^{-1} \dfrac{1/\omega C}{R_1 + R} = \tan^{-1} \dfrac{6366.2}{1000} = 81.07°$

$\therefore \qquad \bar{I}_1 = \dfrac{230 \angle 0°}{6444.3 \angle - 81.07°}$

$\therefore$ Voltage across capacitor $\quad \bar{V}_c = \bar{I}_1 \cdot \bar{X}_c$

$$= \frac{230 \angle 81.07°}{6444.3} \times 6366.2 \angle - 90° = 227.2 \angle - 8.93°$$

or $\qquad v_c = \sqrt{2} \cdot (227.2) \sin (\omega t - 8.93°)$

When capacitor voltage v_c reaches the breakdown voltage of the diac, the triac firing angle α_1 is given by

$$v_c = \sqrt{2} \cdot (227.2) \sin (\alpha_1 - 8.93°) = 30 \ V$$

or $\qquad \alpha_1 = \sin^{-1} \dfrac{30}{\sqrt{2} \times 227.2} + 8.93° = 14.3°$

When $R = 25000 \ \Omega$: $\qquad Z = [26000^2 + 6366.2^2]^{1/2} = 26768 \ \Omega$

$$\phi = \tan^{-1} \left[\frac{6366.2}{26000} \right] = 13.76°$$

$\therefore \qquad \bar{I}_1 = \dfrac{230 \angle 0°}{26768 \angle - 13.76°}$

$$V_c = \frac{230 \angle 13.76°}{26768} \times 6366.2 \angle - 90° = 54.7 \angle - 76.24°$$

or $\qquad v_c = \sqrt{2} \times 54.7 \sin (\omega t - 76.24°)$

When v_c equal 30 V, let the firing angle be α_2

$\therefore \qquad \sqrt{2} \times 54.7 \sin (\alpha_2 - 76.24°) = 30 \ V$

or $\qquad \alpha_2 = \sin^{-1} \left(\dfrac{30}{\sqrt{2} \times 54.7} \right) + 76.24° = 99.06°$

Thus the maximum and minimum values of firing-angle delays are 99.06 ° and 14.3 ° respectively.

4.17. GATING CIRCUITS FOR SINGLE-PHASE CONVERTERS

A gate trigger circuit for thyristors in phase-controlled rectifiers should possess the following :

   (*i*) A circuit for the detection of zero crossing of the input voltage.
   (*ii*) Generation of trigger pulses of required waveshape.
   (*iii*) DC power supply for pulse amplifier.
   (*iv*) Gate trigger circuit isolation from the line potential by means of pulse transformers or optocouplers.

A general block diagram for gate trigger circuit for single-phase converter is shown in Fig. 4.82. The gating circuit consists of synchronizing transformer, diode rectifier, zero crossing detector, firing-angle delay block, pulse amplifier, gate-pulse isolation transformer and power circuit for the converter.

Synchronizing mid-tapped transformer steps down the supply voltage suitable for zero crossing detector and for delivering dc supply V_{CC} to gate trigger circuit. The zero crossing

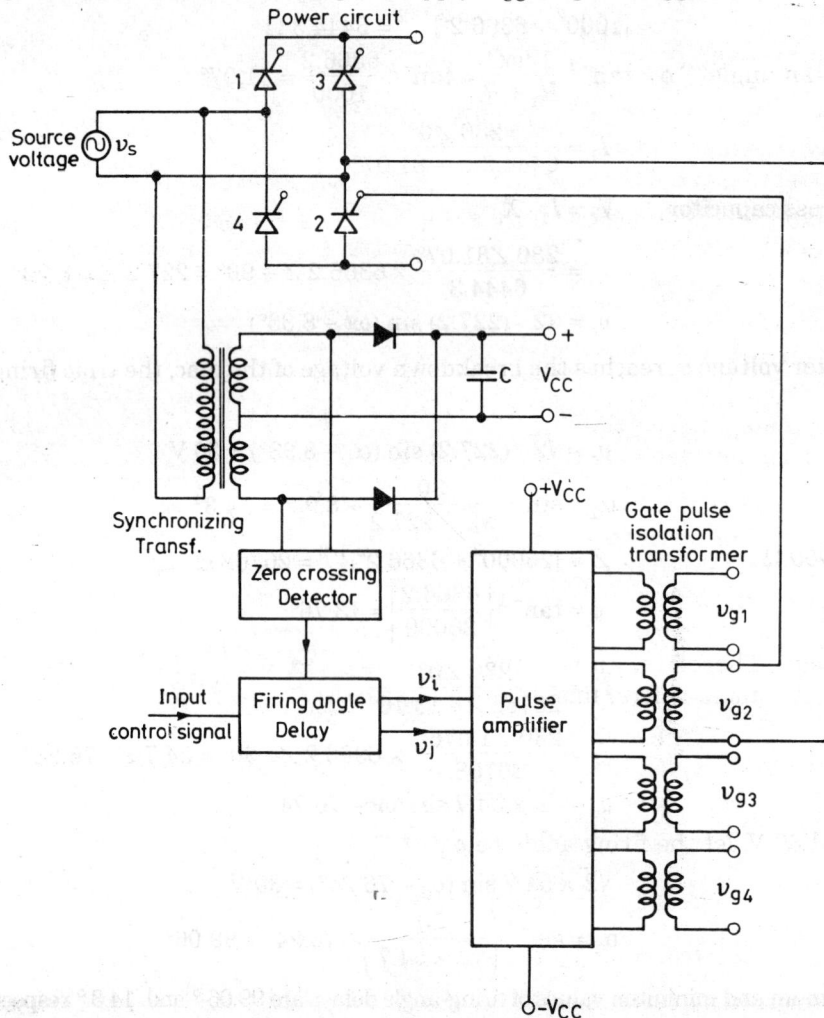

Fig. 4.82. Block diagram of a thyristor gating circuit.

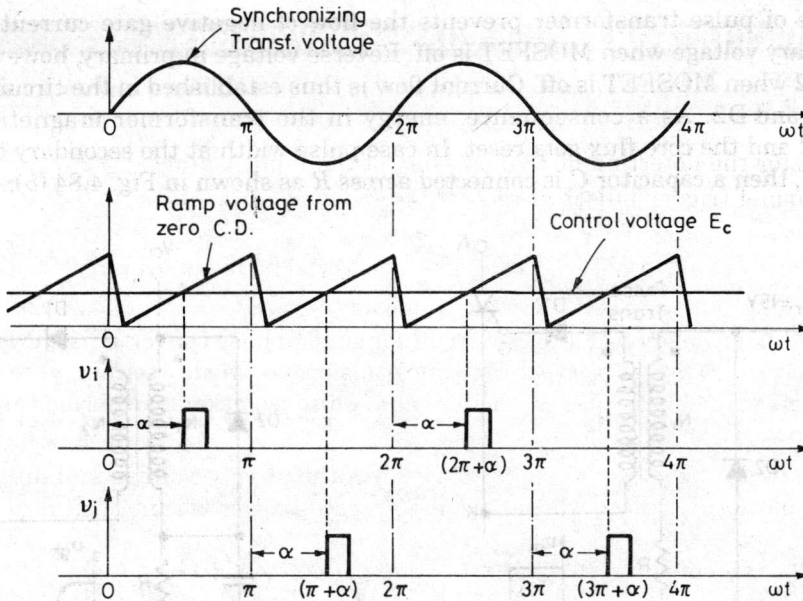

Fig. 4.83. Waveforms for the circuit of Fig. 4.82.

detector converts ac synchronizing input voltage into ramp voltage and synchronizes this ramp voltage with the zero crossing of the ac supply voltage as shown in Fig. 4.83. In the firing-angle delay block, the constant amplitude ramp voltage is compared with control voltage E_C. When rising ramp voltage equals control voltage E_C, a pulse signal of controlled duration is generated as shown in Fig. 4.83. These signals are indicated as v_i for thyristors 1 and 2 and v_j for thyristors 3 and 4 for the power circuit of Fig. 4.82. If E_C is lowered, firing angle decreases and in case E_C is raised, firing angle increases. This shows that firing-delay angle is directly proportional to the control signal voltage. The pulse output from the firing-delay angle block are next fed to a pulse amplifier circuit. The amplified pulses are then used for triggering thyristors 1, 2, 3 and 4 through gate-pulse isolation transformers as shown.

4.17.1. Gate Pulse Amplifiers

Pulse output from integrated circuits (ICs) may be directly fed to gate-cathode circuit of a low-power thyristor to turn it on. But in high-power thyristors, trigger-current requirement is high. Therefore, pulses derived from ICs must be amplified and then fed to thyristor for its reliable turn on. In a thyristor, anode circuit is subjected to high voltage whereas gate circuit works at a low voltage. Therefore, an isolation is essential between a thyristor and the gate-pulse generator. As stated before, this isolation is provided by an optocoupler or a pulse transformer.

A pulse-amplifier circuit for amplifying the input pulses is shown in Fig. 4.84. It consists of a MOSFET (or a transistor), a pulse transformer for isolation and diodes D1, D2. When a voltage of appropriate level is applied to the gate of MOSFET, it gets turned on. As a result, most of the dc voltage V_{CC} appears across transformer primary and corresponding pulse voltage is induced in the transformer secondary. This amplified pulse on the secondary side is applied to gate and cathode of a thyristor to turn it on. When pulse signal applied to the gate of MOSFET goes to zero, MOSFET turns off. The primary current due to V_{CC} tends to fall and likewise flux in core also tends to decrease. Due to this tendency, a voltage of opposite polarity is induced in both primary and secondary windings of pulse transformer. Diode D1 on the secondary side of pulse transformer prevents the flow of negative gate current due to the reverse secondary voltage when MOSFET is off. Reverse voltage in primary, however, forward biases diode D2 when MOSFET is off. Current flow is thus established in the circuit consisting of primary, R and D2. As a consequence, energy in the transformer magnetic core gets dissipated in R and the core flux gets reset. In case pulse width at the secondary terminals is to be increased, then a capacitor C is connected across R as shown in Fig. 4.84 (b).

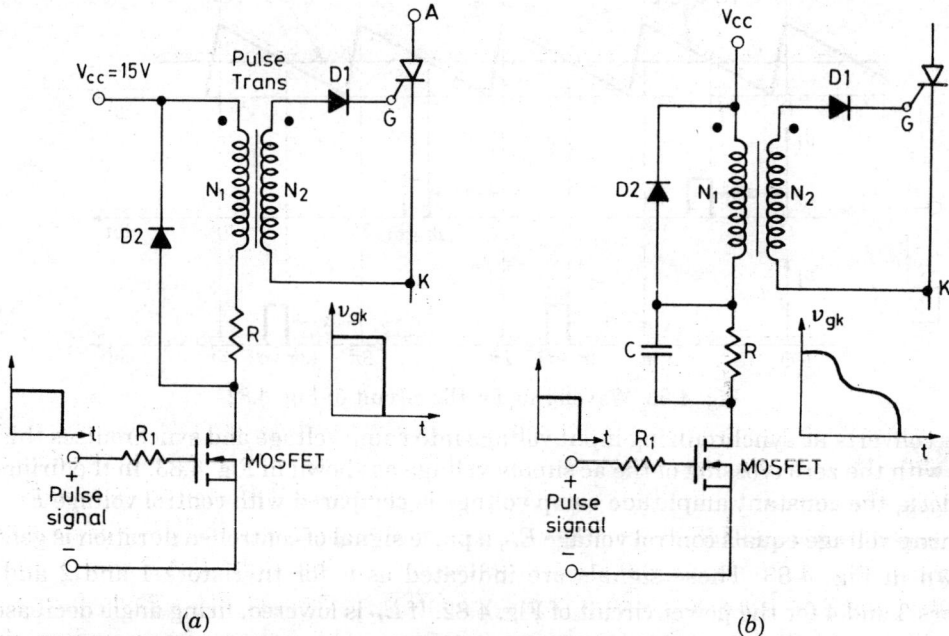

Fig. 4.84. Pulse amplifier circuit using a MOSFET for a thyristor trigger circuit
(a) short-pulse output (b) long-pulse output.

4.17.2. Pulse Train Gating

Pulse gating is not suitable for inductive, i.e. RL loads, because initiation of thyristor conduction is not well defined for these types of loads. This difficulty for such situations can be overcome by triggering the thyristor continuously. Continuous gating, however, suffers from some disadvantages like increased thyristor losses and distortion of output pulse due to saturation of pulse transformer by continuous pulse. In order to overcome these shortcomings of continuous gate signal, a train of firing pulses is used to turn on a thyristor. A pulse train of gating signal is also called high-frequency carrier gating. A pulse train can be generated by modulating the pulse width at a high frequency (10 to 30 kHz) as shown in Fig. 4.85.

A circuit for generating a pulse train is shown in Fig. 4.85 (a). This circuit consists of an AND-logic gate, 555 timer, MOSFET, isolation pulse transformer and diodes D1, D2. The pulse signal v_i, obtained from the thyristor trigger circuit and shown in the top of Fig. 4.85 (b), is fed to AND gate. The output v_t of the timer 555 as shown is also fed to the AND gate. The duty cycle of the timer should be less than 50% in order to allow the transformer flux to reset. The pulse signal v_i and timer output v_t are processed in the AND gate to get the waveform output v_x as shown. The output from AND gate is then applied to pulse amplifier circuit to augment the amplitude of v_x to v_{gk}. The amplified output waveform v_{gk} is then applied across gate-cathode terminals of a thyristor to turn it on.

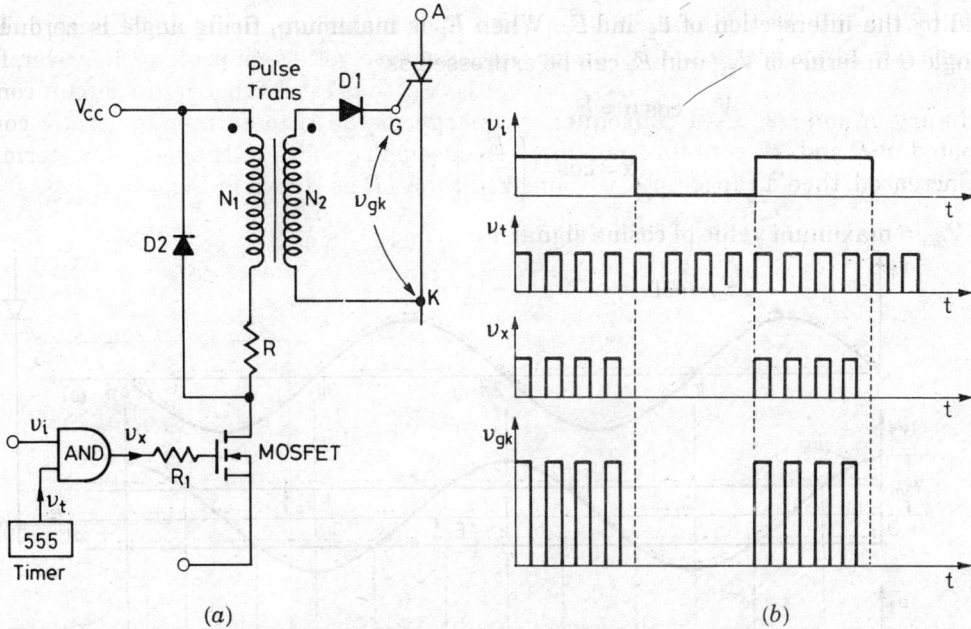

Fig. 4.85. Pulse train gating (a) circuit (b) waveforms.

4.18. COSINE FIRING SCHEME

Cosine firing scheme for thyristors in single-phase converters is shown in Fig. 4.86. The synchronizing transformer steps down the supply voltage to an appropriate level. The input to this transformer is taken from the same source from which converter circuit is energized. The output voltage v_1 of synchronizing transformer is integrated to get cosine-wave v_2. The dc control voltage E_C varies from maximum positive E_{cm} to maximum negative E_{cm} so that firing angle can be varied from zero to 180°. The cosine wave v_2 is compared in comparators 1 and 2 with E_c and $-E_c$. When E_c is high as compared to v_2, output voltage v_3 is available from comparator 1. Same is true for comparator 2. So the comparators 1 and 2 give output pulses v_3 and v_4 respectively as shown in Fig. 4.87. It is seen from this figure that firing angle is

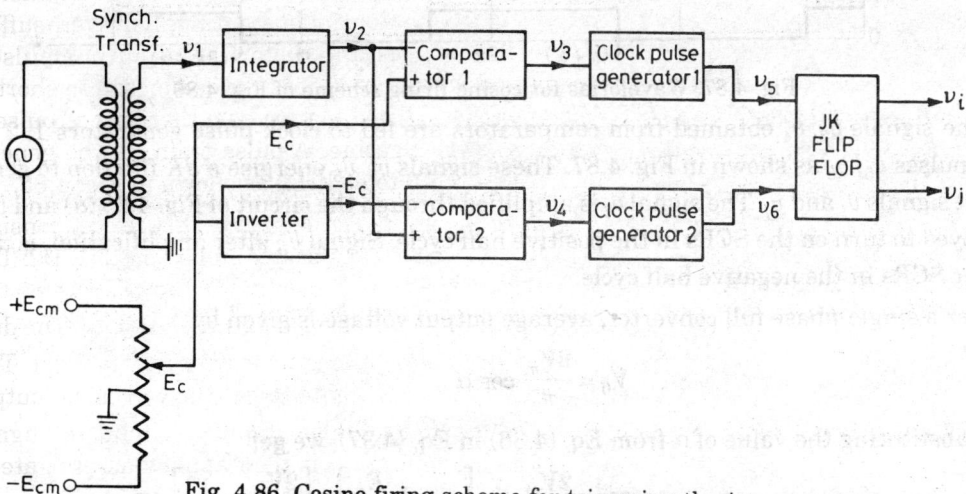

Fig. 4.86. Cosine firing scheme for triggering thyristors.

governed by the intersection of v_2 and E_c. When E_c is maximum, firing angle is zero. Thus, firing angle α in terms of V_{2m} and E_c can be expressed as

$$V_{2m} \cos \alpha = E_c$$

or

$$\alpha = \cos^{-1} \left(\frac{E_c}{V_{2m}} \right) \qquad \qquad \text{...(4.36)}$$

where V_{2m} = maximum value of cosine signal v_2.

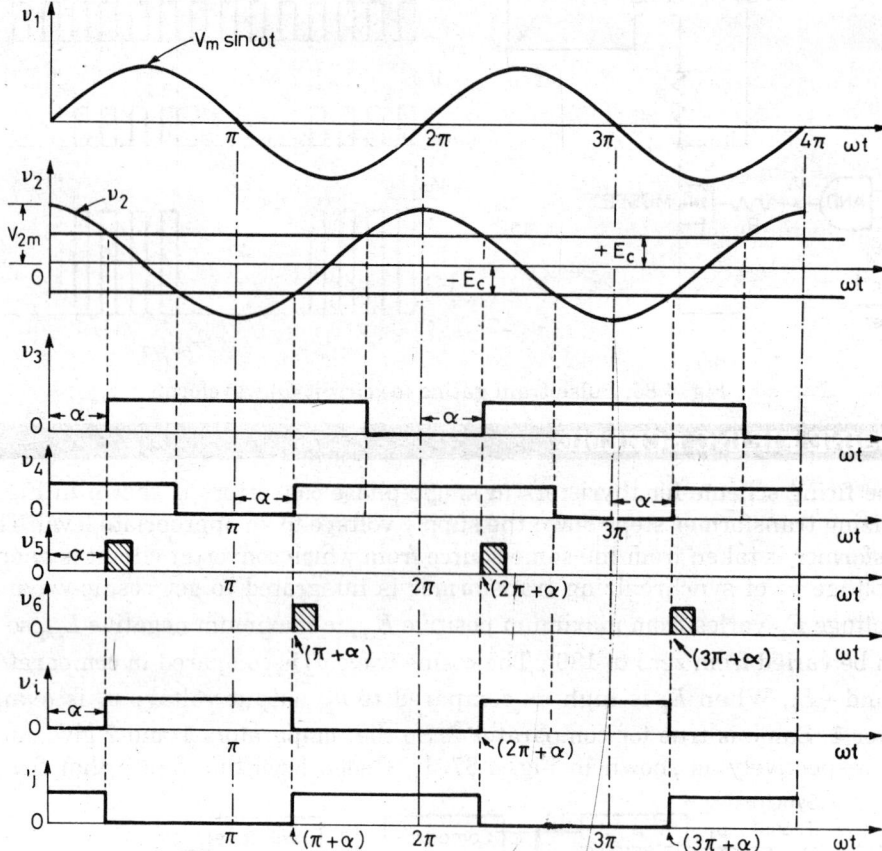

Fig. 4.87. Waveforms for cosine firing scheme of Fig. 4.86.

The signals v_3, v_4 obtained from comparators are fed to clock-pulse generators 1, 2 to get clock pulses v_5, v_6 as shown in Fig. 4.87. These signals v_5, v_6 energise a JK flip flop to generate output signals v_i and v_j. The signal v_i is amplified through the circuit of Fig. 4.85 (a) and is then employed to turn on the SCRs in the positive half cycle. Signal v_j, after amplification, is used to trigger SCRs in the negative half cycle.

For a single-phase full converter, average output voltage is given by

$$V_0 = \frac{2V_m}{\pi} \cos \alpha \qquad \qquad \text{...(4.37)}$$

Substituting the value of α from Eq. (4.36) in Eq. (4.37), we get

$$V_0 = \frac{2V_m}{\pi} \cos \left[\cos^{-1} \frac{E_c}{V_{2m}} \right] = \left[\frac{2V_m}{\pi} \cdot \frac{1}{V_{2m}} \right] \cdot E_c$$

$$V_0 = k \, E_c \qquad \qquad ...(4.38)$$

This shows that cosine firing scheme provides a linear transfer characteristic between the average output voltage V_0 and the control voltage E_c. This scheme, on account of its linear transfer characteristic, improves the closed-loop response of the converter system. This feature has made the cosine firing scheme quite popular in industrial applications.

PROBLEMS

4.1. (a) What is a thyristor ? How has this term been coined ? Name the most popular thyristor.

(b) What is the definition of thyristor as per *IEC* ?

(c) Give constructional details of side-gate thyristor. Sketch its schematic diagram and the circuit symbol.

(d) Sketch static *I-V* characteristics of a thyristor. Label the various voltages, currents and the operating modes on this sketch.

4.2. (a) Describe the different modes of operation of a thyristor with the help of its static *I–V* characteristics.

(b) Enumerate the various mechanisms by which thyristors can be triggered into conduction. Discuss the techniques which result in random turn-on of a thyristor.

4.3. (a) Describe gate-triggering of a thyristor. Does the gate-current has any effect on the forward-breakover voltage ? Discuss.

(b) How does light triggering of a thyristor differ from gate triggering ? Where are LASCRs used ?

4.4. (a) Define latching and holding currents as applicable to an SCR. Show these currents on its static *I-V* characteristics.

(b) What are the necessary conditions for turning-on of an SCR ? Discuss.

(c) Define turn-on and turn-off times for an SCR.

4.5. Sketch switching (or dynamic) characteristics of a thyristor during its turn-on and turn-off processes. Show the variation of voltage across the thyristor and current through it during these two dynamic processes. Indicate clearly the various intervals into which turn-on and turn-off times can be subdivided. Discuss briefly the nature of these curves.

4.6. (a) Can a forward voltage be applied to an SCR soon after its anode current has fallen to zero ? Explain.

(b) A forward voltage is applied to an SCR soon after reverse recovery current drops nearly to zero value. Discuss what would happen to the SCR.

(c) Discuss the importance of di/dt rating during the turn-on process of a thyristor.

4.7. (a) Discuss the conditions which must be satisfied for turning on an SCR with a gate signal.

(b) A thyristor is conducting a forward current. Discuss the basic requirements for commutating (turning-off) this thyristor.

(c) Bring out clearly how the anode current expands over the cathode surface area during the turn-on process of a thyristor.

4.8. (a) Are the turn-on and turn-off times of a thyristor constant ? On what factors do these depend ?

(b) In an SCR, the anode current rises linearly from zero to $I_1 = 100$ A whereas anode voltage across SCR falls linearly from $V_1 = 600$ V to zero during its turn-on time of $t_1 = 5$ μS. Derive an expression for the average power loss in SCR for $t_1 = 5$ μs. Derive an expression for the average power loss in SCR for a triggering frequency f. In case $f = 100$ Hz, find the average power loss in SCR. [**Ans.** $1/6 \, V_1 . I_1 . t_1 . f$; 5 watts]

4.9. (a) Justify the statement, "Higher the gate current, lower is the forward breakover voltage."

(b) What is hard-driving for a thyristor ? What are its advantages ? Sketch a typical waveform for gate current for hard-driving the thyristor.

(c) For an SCR, the gate-cathode characteristic is given by a straight line with a gradient of 20 volts per ampere passing through origin. The maximum turn-on time is 4 μs and the minimum gate current required to quick turn-on is 400 mA. If the gate source voltage is 15 V, calculate the resistance to be connected in series and the gate-power dissipation.

Given that pulse width is equal to the turn-on time and the average power dissipation is 0.2 W, compute the maximum triggering frequency that will be possible when pulse firing is used. **[Ans :** (c) 17.5 Ω, 15.625 kHz]

4.10. (a) Draw thyristor gate characteristics showing the six gate ratings as specified by the manufacturers. Discuss these ratings. Indicate clearly the preferred gate drive area. Are there any other gate ratings in addition to the six mentioned above ? If yes, describe this/these briefly.

(b) The gate-cathode characteristic of an SCR is given by $V_g = 0.5 + 8\,I_g$. For a triggering frequency of 400 Hz and duty cycle of 0.1, compute the value of resistance to be connected in series with the gate circuit. The rectangular trigger pulse applied to the gate circuit has an amplitude of 12 V. The thyristor has average gate-power loss of 0.5 watts.

$$\left[\textbf{Hint :} \ (b) \ \delta = \frac{T}{T_1} \ \text{or pulse width,} \ T = \frac{0.1}{400} = 250 \ \mu s. \ \text{As T is more than 100 μs, dc data apply} \right]$$

[Ans : (a) Peak gate–power dissipation and peak reverse gate voltage
(b) 44.23 Ω]

4.11. (a) Draw the gate input characteristics of a batch of thyristors indicating the upper and lower limit loci and explain why this variation exists.

(b) Draw the circuit model of a triggering circuit connected to the gate-cathode terminals of a thyristor. Explain the purpose of connecting a resistor across the gate circuit of an SCR.

(c) A thyristor data sheet gives 1.5 V and 100 mA as the minimum value of gate-trigger voltage and gate-trigger current respectively. A resistor of 20 Ω is connected across gate-cathode terminals. For a trigger supply voltage of 8 V, compute the value of resistance that should be connected in series with gate circuit in order to ensure turn-on of the device.

[Ans : (b) 37.143 Ω]

4.12. (a) Draw thyristor gate V–I characteristics indicating clearly the gate drive limits. Explain, with the help of these characteristics, the selection of an operating point and the choice of gate circuit parameters.

Discuss also how turn-on time and jitter can be minimised.

(b) In case gating signal for an SCR consists of a train of pulses instead of continuous dc signal, explain how the frequency of triggering and other factors are decided.

(c) A thyristor is triggered by a train of pulses of frequency 4 kHz and of duty cycle 0.2. Calculate the pulse width. In case average gate power dissipation is 1 W, find the maximum allowable gate power drive. **[Ans :** (c) 50 μs, 5 W]

4.13. (a) Discuss the function of connecting a

(i) diode across gate-cathode terminals,

(ii) diode in series with gate circuit,

(iii) a resistor across gate-cathode terminals.

(b) How are the magnitudes of gate-voltage and gate-current influenced by temperature rise in a thyristor ?

(c) During turn-off of a thyristor, idealized voltage and current waveforms are shown in Fig. 4.88. For a triggering frequency of 50 Hz, find the mean power loss due to turn-off loss. **Also obtain the reversed recovery charge.**

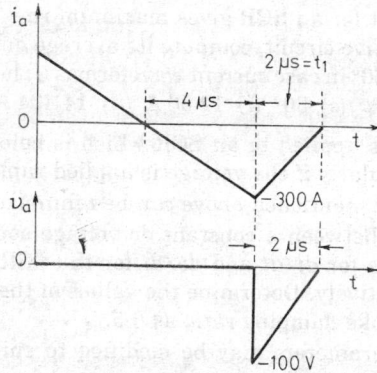

Fig. 4.88. Pertaining to Prob. 4.13.

$$\left[\text{Hint}: \quad (c)\ P_a = \frac{1}{3} f V_1 I_1 \cdot t_1\right]$$

[**Ans**: (c) 1W, 900 µC]

4.14. The spread in the gate-cathode characteristics of a thyristor is given by the following two relations :

$$I_g = 2.0 \times 10^{-3}\ V_g^2 \text{ and } I_g = 2.0 \times 10^{-3}\ V_g^{1.5}$$

For an average gate-power dissipation of 0.5 W, design the trigger circuit voltage and current for hard-drive.

[**Hint** : Here $I_g = 79.37$ mA and $V_g = 6.3$ V. Also $I_g = 47.82$ mA and $V_g = 10.456$ V. For hard-drive, choose I_g say 75 mA etc.]

[**Ans**. $V_g = 5.56$ V, $I_g = 75$ mA]

4.15. (a) Discuss the two-transistor model of a thyristor. Derive an expression for the anode current and discuss therefrom the turn-on mechanisms of a thyristor.

4.16. (a) Which current rating of an SCR is the most important ?

(b) What is the difference between repetitive-current and surge-current ratings of a thyristor ?

(c) What are V_{DRM} and V_{RRM} ? Are these ratings different from each other for a thyristor ?

(d) Is it possible to exceed rms current rating of an SCR?

(e) What are V_{DWM} and V_{DRM} ? Which rating is low ?

(f) An SCR has maximum rms current rating of 78.5 A. Find its maximum average current rating. [**Ans** : (a) Rms current (c) $V_{DRM} = V_{RRM}$ (d) No (e) V_{DWM} is low (f) 50 A]

4.17. (a) Describe the various anode voltage ratings as applicable to an SCR. Indicate these voltage ratings on a relevant voltage waveform.

(b) Discuss the significance of dv/dt in case of thyristors.

(c) Explain why an SCR is derated when it handles pulsed anode current as compared to its rating for constant dc current.

(d) The average current rating of an SCR decreases as its conduction angle is reduced. Explain.

4.18. Define the following terms relating to SCR and discuss their significance :

(i) Forward breakover voltage (ii) peak inverse voltage (iii) critical rate of rise of voltage (iv) voltage safety factor (v) on-state voltage drop (vi) finger voltage.

4.19. (a) The derating of an SCR is more for sine waves than for the square (or rectangular) waves. Explain.

Sketch the curves showing average power dissipation as a function of average forward current for different conduction angles for both sine and square waves.

(b) What is the effect on average current rating of an SCR in case inductance is inserted in the anode circuit ? Discuss.

(c) The specification sheet for an SCR gives maximum rms on-state current as 50 A. If this SCR is used in a resistive circuit, compute its average on-state current rating for conduction angles of 30° and 60° in case current waveform is (i) half-sine wave and (ii) rectangular wave.　　　　[**Ans:** (c) 30° (i) 12.56 A (ii) 14.434 A ; 60° (i) 18.00 A (ii) 20.412 A]

4.20. (a) If a forward voltage is applied to an SCR which is below its breakover voltage, it may well switch on, particularly if the voltage is applied rapidly. Explain why this is so. Discuss how the effect mentioned above can be minimized.

(b) A thyristor is placed between a constant dc voltage source of 240 V and resistive load R. The specified limits for di/dt and dv/dt for the SCR are 60 A/micro second and 300 V/micro second respectively. Determine the values of the di/dt inductor and the snubber circuit parameters. Take damping ratio as 0.5.

Discuss how these parameters may be modified to suit the working conditions in the circuit.

Derive the various expressions used.

[**Ans:** (b) Computed values : 4 µH, 5 Ω, 0.16 µF
modified values : 6.4 µH, 8 Ω, 0.12 µF]

4.21. (a) Snubber circuit for an SCR should primarily consist of capacitor only. But, in actual practice, a resistor is used in series with the capacitor. Discuss.

(b) R, L and C in an SCR circuit meant for protecting against dv/dt and di/dt are 4 Ω, 6 µH and 6 µF respectively. If the supply voltage to the circuit is 300 V, calculate permissible maximum values of dv/dt and di/dt.

[**Hint:** (b) Rate of change of voltage across the thyristor at $t = 0$ when the supply is switched on is given by

$$\frac{dv_a}{dt} = R_s \frac{di}{dt} + \frac{I_{sc}}{C} \text{ where } I_{sc} = \frac{V_s}{R_s} = \frac{300}{4} = 75 \text{ A etc]}$$ [**Ans:** (b) 50 A/µs, 212.5 V/µs]

4.22. Following are the specifications of a thyristor operating from a peak supply of 500V:
Repetitive peak current, $I_p = 250$ A

$$\left(\frac{di}{dt}\right)_{max} = 60 \text{ A/µs}, \left(\frac{dv_a}{dt}\right)_{max} = 200 \text{ V/µs}$$

Take a factor of safety of 2 for the three specifications mentioned above. Design a suitable snubber circuit if the minimum load resistance is 20 Ω. Take ξ = 0.65.

[**Ans:** 17 µH, 6 Ω, 0.5 µF]

4.23. (a) Discuss how a thyristor may be subjected to internal and external overvoltages. Describe the methods adopted for suppressing such overvoltages in thyristor systems.

(b) During the **turn-off** process in a thyristor, the reverse recovery current of 10 A is interrupted in a **time** interval of 4 µs. The thyristor is connected in series with an inductance of 6 mH with **no resistance** in the circuit. If the source voltage during turn-off process is − 300 V, calculate

(i) peak voltage across the thyristor when reverse current is interrupted and

(ii) the value of snubber circuit resistance in case snubber capacitance $C_s = 0.3$ µF and damping ratio is 0.65.　　　　[**Ans:** (b) − 15.3 kV, 183.85 Ω]

4.24. (a) Explain the methods adopted for the protection of SCRs against overcurrents.

(b) A thyristor, having maximum rms on-state current of 45 A, is used in a resistive circuit. Compute its average on-state current rating for half-sine wave for conduction angles of π/3 and π/2.　　　　[**Ans:** (b) 16.198 A, 20.26 A]

4.25. (a) Describe electronic crowbar protection scheme employed for the overcurrent protection of power converters.

(b) **Draw a circuit diagram illustrating the protection of both anode and gate circuits of an SCR. Describe briefly the function of various components used.**

4.26. (a) Enumerate the various abnormal conditions against which thyristors must be protected.

(b) Describe the significance of di/dt and dv/dt in SCRs.

(c) Describe, with the help of a circuit diagram, the function of various components used for the protection of gate circuit of a thyristor.

4.27. (a) Describe the methods employed for improving di/dt rating in a thyristor.

(b) Large dv/dt may turn on a thyristor at random. Describe how cathode-shorts in thyristors improve their dv/dt ratings.

4.28. (a) Discuss briefly the different components of power loss that occur in a thyristor during its working. Which of the power loss component/components is/are dominant at power frequencies and which at high frequencies ?

(b) Give the concept of thermal resistance. Describe the analogy between thermal and electrical quantities.

(c) Draw the thermal equivalent circuit for an SCR and discuss the various parameters involved in it.

(d) Describe any one method of designing the heat sinks for thyristors.

4.29. (a) For thyristors, various mounting techniques are based on their thermal considerations. Discuss these mounting techniques with relevant diagrams.

(b) A thyristor is rated to carry full-load current with an allowable case temperature of 100°C, for maximum allowable junction temperature of 125°C and thermal resistance between case and ambient as 0.5°C/W. Find the sink temperature for an ambient temperature of 40°C. Take thermal resistance between sink and ambient as 0.4°C/W.

[Ans : (b) 88°C]

4.30. A thyristor is rated to carry an rms current of 100 A. Its maximum allowable junction temperature is 125°C.

(a) If this thyristor is made to carry direct current continuously, find the maximum allowable current rating of the SCR.

(b) If this SCR is used in a single-phase half-wave circuit with resistive load, find the maximum allowable average current for firing angles of $\alpha_1 = 30°$ and $\alpha_2 = 120°$.

(c) For part (b), determine the sink temperatures if average powers dissipated are 200 W for α_1 and 150 W for α_2. The value of thermal impedances are :

$$\theta_{jc} = 0.15° \text{ C/W}, \quad \theta_{cs} = 0.07°\text{C/W for } \alpha_1$$

and $$\theta_{jc} = 0.16°\text{C/W}, \quad \theta_{cs} = 0.08°\text{C/W for } \alpha_2.$$

[Ans : (a) 100 A (b) 60.273 A, 25.118 A (c) 81°C and 89°C]

4.31. A thyristor string is made up of a number of SCRs connected in series and parallel. The string has voltage and current ratings of 11 kV and 4 kA respectively. The voltage and current ratings of available SCRs are 1800 V and 1000 A respectively. For a string efficiency of 90%, calculate the number of series and parallel connected SCRs.

For these SCRs, maximum off-state blocking current is 12 mA. Determine the value of static equalizing resistance for the string. Derive the formula used for this resistance.

[Ans : Series–7, Parallel–5, $R = 22.22$ kΩ]

4.32. For the thyristors of Prob. 4.31, maximum difference in their reverse recovery charge is 25 microcoulombs. Compute the value of dynamic equalizing capacitance of this string. Derive the formula used for the computation of this capacitance. **[Ans :** $C = 0.094$ µF]

4.33. Three series-connected thyristors, provided with static and dynamic equalizing circuits, have to withstand an off-state voltage of 8 kV. The static equalizing resistance is 20 kΩ and the dynamic equalizing circuit has $R_c = 40$ Ω and $C = 0.06$ µF. These three thyristors have leakage currents of 25 mA, 23 mA and 22 mA respectively. Determine voltage across each SCR in the off-state and the discharge current of each capacitor at the time of turn on.

[Ans : 2500 V, 2540 V, 2560 V; 62.5 A, 63.5 A, 64A]

4.34. In a power circuit, four SCRs are to be connected in series. Permissible difference in blocking voltage is 20 V for a maximum difference in their blocking currents of 1 mA. Difference in recovery charge is 10 μC. Design suitable equalizing circuit.

[**Ans** : Static equalizing resistance = 20 k Ω; shunt capacitance = 0.5 μF]

4.35. (a) Discuss how SCRs suffer from unequal voltage distribution across them during their turn-on and turn-off processes.

(b) A number of SCRs, each with a rating of 2000 V and 50 A, are to be used in series-parallel combination in a circuit to handle 11 kV and 400 A. For a derating factor of 0.15, calculate the number of SCRs in series and parallel units.

The maximum difference in their reverse recovery charge is 20 microcoulombs. Calculate (i) the value of dynamic equalizing capacitance and (ii) the voltage across each of the slow thyristors in case one series-connected SCR is fast.

[**Ans** : (b) $n_s = 7$, $n_p = 10$, $C = 0.04$ μF, 1500 V]

4.36. Define string efficiency for series / parallel connected SCRs. Show that string efficiency of two series connected SCRs is usually less than one.

Derive an expression for the resistance used for static voltage equalization for a series connected string.

4.37. Describe how two series connected SCRs are subjected to unequal voltage distribution during their dynamic conditions. Derive an expression for capacitance C used in the dynamic equalizing circuit for n series connected SCRs.

4.38. Show that string efficiency for two parallel connected SCRs is usually less than one.

Discuss the problems associated with the parallel operation of SCRs and how these are overcome.

4.39. (a) Describe briefly the following members of thyristor family.

PUT, SUS, SCS

Illustrate your answer with suitable diagrams.

(b) Draw the cross-sectional view of the diac and explain how it can conduct in both the directions.

(c) Give the cross-sectional view of a triac and explain its turn-on process with relevant diagrams. Hence show that a triac is rarely operated in first quadrant with negative gate current and in third quadrant with positive gate current.

4.40. (a) Describe LASCR. Give its industrial applications.

(b) Discuss how a triac may sometimes operate in the rectifier mode.

(c) Enumerate the advantages of ASCR and RCT over conventional thyristors.

4.41. (a) What is a GTO ? Describe its basic structure.

(b) The turn-off process in a GTO can be described with its two-transistor model. Explain this in detail.

Bring out clearly the difference between gold-doped GTOs and anode-shorted GTOs.

4.42. (a) Describe switching performance in a GTO with relevant voltage and current waveforms.

(b) Give the merits and demerits of GTOs as compared to conventional thyristors.

4.43. (a) Define the following terms as applicable to GTOs and discuss their significance.

Turn-off gain, backporch current.

(b) Give the application of GTOs.

4.44. (a) Describe the basic structure of a static induction thyristor (SITH).

(b) Explain the turn-on and turn-off processes in a SITH. Show that SITH is a normally-on device.

(c) Compare SITH with a GTO.

4.45. (a) Discuss the features that the firing circuits for thyristors should possess.

Give the general layout of a firing circuit scheme and explain the function of various components used in it.

(b) Describe the resistance firing circuit used for triggering SCRs. Is it possible to get a firing angle greater than 90° with resistance firing ? Illustrate your answer with appropriate waveforms.

4.46. (a) For resistance firing circuits show that firing-delay angle is proportional to the variable resistance.

(b) Resistance firing circuit is used for triggering an SCR in a laboratory. This SCR is destroyed by a batch of students inadvertently.

A new SCR with the same specification number is installed. But it is found that maximum firing angle attained is 75° only. Explain how the desired maximum firing angle of 90° can be obtained. **[Ans :** (b) Increase R_1 or R_2 or else decrease R in Fig. 4.64]

4.47. (a) Draw RC half-wave trigger circuit for one SCR and discuss the function of the various components used.

Describe, with the help of waveforms, how the output voltage is controlled by varying the resistance. Draw the voltage waveform across SCR also.

(b) Describe RC full-wave trigger circuit for one SCR when the load is (i) ac type (ii) dc type. Relevant diagrams and waveforms should be drawn to illustrate your answer.

4.48. (a) Compare an UJT firing circuit with R and RC firing circuits.

(b) A unijunction transistor, used in relaxation oscillator, has the following data :

$$\eta = 0.67, \ I_v = 10 \text{ mA}, \ V_v = 2.5 \text{ V}, \ I_p = 15 \text{ μA}$$

An oscillator, with an oscillation frequency of 1 kHz, is to be designed by using this UJT. Compute the values of charging resistor and external resistors needed in the base circuits. Take $C = 0.4$ μF and forward-voltage drop of $E - B_1$ junction as 0.5 V. Source voltage is 24 V dc and triggering pulse width is 50 μs.

[Ans : (b) $R = 2.772$ k Ω, $R_{max} = 495$ k Ω, $R_{min} = 2.15$ k Ω,
$R_2 = 621.9$ Ω, $R_1 = 125$ Ω]

4.49. (a) Explain the working of an oscillator employing an UJT. Derive expressions for the frequency of triggering and firing angle delay in terms of eta, charging resistance etc.

(b) A relaxation oscillator, using an UJT, is to be designed for triggering an SCR. The UJT has the following data :

$$\eta = 0.7, \ I_p = 0.5 \text{ mA}, \ V_p = 15.0 \text{ V}, \ V_v = 0.8 \text{ V}, \ I_v = 2 \text{ mA}, \ R_{BB} = 6 \text{ kΩ}.$$

Normal leakage current with emitter open = 3 mA.

The firing frequency is 1.5 kHz. For $C = 0.05$ μF, compute the values of charging resistor and the external resistors connected in the base circuits. Take forward-voltage drop of $E - B_1$ junction as zero.

(c) If the frequency of firing the SCR in part (b) is changed by varying charging resistor R, obtain the maximum and minimum values of R and the corresponding frequencies.

[Ans : (b) 11.074 kΩ, 476.66 Ω, 666.67 Ω
(c) 12.858 kΩ, 10.315 kΩ, 1.292 kHz, 1.611 kHz]

4.50. (a) The intrinsic stand-off ratio for an UJT is 0.65. Its interbase resistance is 10 k Ω. Calculate the values of the interbase resistances.

(b) Estimate the minimum and maximum values of charging resistor in the UJT oscillator circuit for manual trigger-angle control of α between 20° and 160° for 50 Hz supply. Assume $C = 0.4$ μF and $\eta = 0.7$. **[Ans :** (a) 6.5 k Ω, 3.5 k Ω, (b) 2.307 k Ω, 18.457 k Ω]

4.51. (a) Draw and explain the working of an UJT oscillator. Discuss how the amplitude of output voltage pulse can be estimated in this oscillator.

(b) Using a 15-V supply to an UJT, design the oscillator circuit for a frequency of 5 kHz. Data for UJT is as under :

$$\eta = 0.65 \text{ to } 0.75, \quad R_{BB} = 4.7 \text{ to } 9.1 \text{ k}\Omega$$

Take $C = 0.04$ μF. Missing data may be assumed.

[**Hint :** (b) Assume leakage current = 1.88 mA]

[**Ans :** (b) $R = 4.153$ k Ω, $R_2 = 952.4$ Ω, $R_1 = 126.3$ Ω, $R_{max} = 6$ kΩ,

$R_{min} = 3$ k Ω, $f_{min} = 3460.8$ Hz, $f_{max} = 6921.6$ Hz]

4.52. A relaxation oscillator using an UJT is fabricated to generate pulses for triggering SCRs. When the circuit is energised, the circuit fails to oscillate. What could be the plausibe causes of this failure ? How can the circuit be made functional ?

[**Ans.** More V_{BB} or less V_{BB} than required; $R < R_{min}$ or $R > R_{max}$]

4.53. An UJT of Fig. 4.71 (a) has the following parameters :

$\eta = 0.67, V_D = 0.7$ V, $I_v = 3$ mA, $V_v = 1$ V, $I_p = 12$ μA, $V_{BB} = 20$ V

(a) Find the value of V_{EE} so as to turn-on UJT if $R_E = 1$ kΩ.

(b) Find the value of V_{EE} to which it must be reduced to turn-off the UJT.

[**Hint.** (a) $V_{EE} = \eta . V_{BB} + V_D + I_p . R_E$ etc.]   [**Ans.** (a) 14.112 V (b) 4 V]

4.54. Draw synchronized UJT trigger circuit using a zener diode.

Describe it briefly with relevant voltage and current waveforms.

Explain how synchronization of the trigger circuit with the supply voltage across SCR is achieved. In case charging resistor is small so that the capacitor voltage reaches UJT threshold voltage twice in each half cycle, explain how the circuit operation is influenced.

4.55. (a) Draw a circuit diagram for the ramp-and-pedestal trigger circuit used for a single-phase semiconverter. Describe its operation with appropriate waveforms.

For this trigger circuit, derive expressions for the frequency of triggering and firing-angle delay in terms of eta, charging resistor etc.

(b) A firing circuit, using ramp-and-pedestal triggering scheme, has the following data :

Charging resistor = 4 k Ω, charging capacitor = 0.2 μF, supply frequency = 50 Hz, $\eta = 0.75$, zener–diode voltage = 15 V.

Compute the magnitude of firing angle in case pedestal voltage is (i) zero and (ii) 4 V.

[**Ans :** (b) (i) 19.963° (ii) 15.496°]

4.56. (a) Describe the use of pulse transformer in the triggering of SCRs and GTOs. With a suitable circuit, discuss the conditions under which the input pulse is faithfully transmitted or is transmitted in the form of exponentially decaying pulse. Which of these two functional modes is preferred and why?

(b) The primary of a pulse transformer is connected in series with a transistor and a current limiting resistor R_L. The data for the triggering circuit is as under :

$R_L = 500$ Ω, gate to cathode resistance = 200 Ω

Primary to secondary turns ratio = $\dfrac{1}{2}$, voltage required to trigger the SCR = 3 V.

Compute the voltage applied to the circuit consisting of transformer primary, R_L etc. Derive the expression used.   [**Ans :** (b) 16.5 V]

4.57. (a) Describe the trigger circuit for a triac using a diac.

(b) A diac with a breakdown voltage of 35 V, Fig. 4.79, is used for triggering a triac. This circuit has $R_1 = 1000$ Ω, $R =$ zero to 280 k Ω and $C = 0.1$ μF. For a supply voltage of 230 V, 50 Hz ; calculate the maximum and minimum values of firing-delay angles for the triac. The effect of load impedance may be neglected.   [**Ans :** (b) 156.5°, 7.98°]

4.58. (a) Describe a gate trigger circuit for a single-phase full converter. Discuss how the adjustment of control voltage varies the firing-delay angle.

(b) Describe a gate-pulse amplifier using a MOSFET.

4.59. (a) Why is pulse-train gating preferred over pulse gating ? Explain, with relevant circuit and waveforms, the pulse-train gating of SCRs.

(b) Why is the cosine-firing scheme so popular ? Describe a cosine-firing scheme for the triggering of thyristors.

Chapter 5

Thyristor Commutation Techniques

In this Chapter

- Class A Commutation : Load Commutation
- Class B Commutation : Resonant-pulse Commutation
- Class C Commutation : Complementary Commutation
- Class D Commutation : Impulse Commutation
- Class E Commutation : External Pulse Commutation
- Class F Commutation : Line Commutation

A thyristor is turned on by applying a signal to its gate-cathode circuit. For the purpose of power control or power conditioning, a conducting thyristor must be turned-off as desired. As stated before, the turn-off of a thyristor means bringing the device from forward-conduction state to forward-blocking state. The thyristor turn-off requires that (*i*) its anode current falls below the holding current and (*ii*) a reverse voltage is applied to thyristor for a sufficient time to enable it to recover to blocking state. *Commutation* is defined as the process of turning-off a thyristor. Once thyristor starts conducting, gate loses control over the device, therefore, external means may have to be adopted to commutate the thyristor. Several commutation techniques have been developed with the sole objective of reducing their turn-off (or commutation) time.

The use of thyristor circuits in low-power converters has declined relatively. This is because of recent advances in semiconductor power devices leading to the availability of power transistors, GTOs and IGBTs. However, for high-voltage and high-current applications above about 1 kV and 0.5 kA, thyristor circuits offer popular circuit configurations.

The classification of thyristor commutation techniques, as reported by various authors, is not the same. Here, an attempt is made to refer to all these classification techniques. Primarily, the classification of commutation techniques is based on the manner in which anode current is reduced to zero and on the configuration of the commutating circuits.

Thyristor commutation techniques use resonant LC, or underdamped RLC circuits, to force the current and / or voltage of a thyristor to zero to turn off the device. Several power-electronic converters employ the circuit configurations used for describing the thyristor commutation techniques. Therefore, a study of the various commutation techniques serves as an introduction and leads to a better understanding of the transient phenomena occuring in power-electronic converters under switching conditions.

The various commutation techniques are now described in this chapter.

5.1. CLASS A COMMUTATION : LOAD COMMUTATION

For achieving load commutation of a thyristor, the commutating components L and C are connected as shown in Fig. 5.1. Here R is the load resistance. For low value of R, L and C are

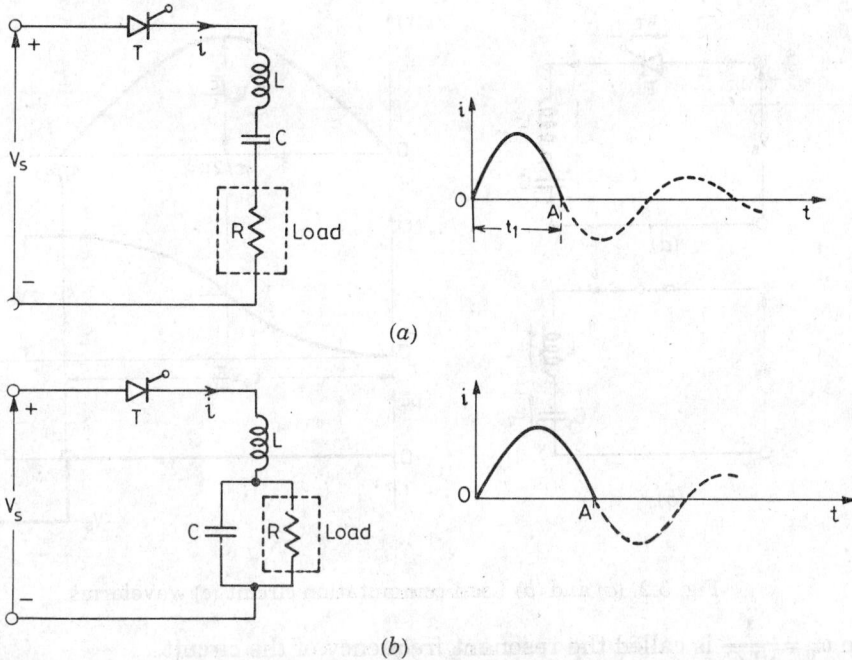

Fig. 5.1. Class A or load commutation (a) series capacitor (b) shunt capacitor.

connected in series with R, Fig. 5.1 (a). For high value of R, load R is connected across C, Fig. 5.1 (b). The essential requirement for both the circuits of Fig. 5.1 is that the overall circuit must be underdamped. When these circuits are energized from dc, current waveforms as shown on the right hand side of Fig. 5.1 are obtained. It is seen that current i first rises to maximum value and then begins to fall. When current decays to zero and tends to reverse, thyristor T in Fig. 5.1 is turned-off on its own at instant A.

Load, or class-A, commutation is prevalent in thyristor circuits supplied from a dc source. The nature of the circuit should be such that when energized from a dc source, current must have a natural tendency to decay to zero for the load commutation to occur in a thyristor circuit. Load commutation is possible in dc circuits and not in ac circuits. Class A, or load, commutation is also called *resonant commutation* or *self-commutation*. A practical circuit employing load commutation is a series inverter which is described in Chapter 8. A simple example illustrating the basic principle of load commutation is given below :

Example 5.1. *The circuit shown in Fig. 5.2 (a) is initially relaxed. The thyristor T is turned on at t = 0. Determine (a) conduction time of thyristor and (b) voltage across thyristor and capacitor after SCR is turned off. Calculate these values for L = 5 mH, C = 20 μF and V_s = 200 V.*

Solution. When thyristor is turned on, it behaves like a diode. Therefore, with SCR on, the device acts like a closed switch, Fig. 5.2 (b). KVL for this circuit gives

$$L \frac{di}{dt} + \frac{1}{C} \int i \, dt = V_s$$

Its solution, from Art. 3.1.4, is given by Eq. (3.9) which is repeated here.

$$i(t) = V_s \sqrt{\frac{C}{L}} \sin \omega_0 t \qquad \ldots(5.1)$$

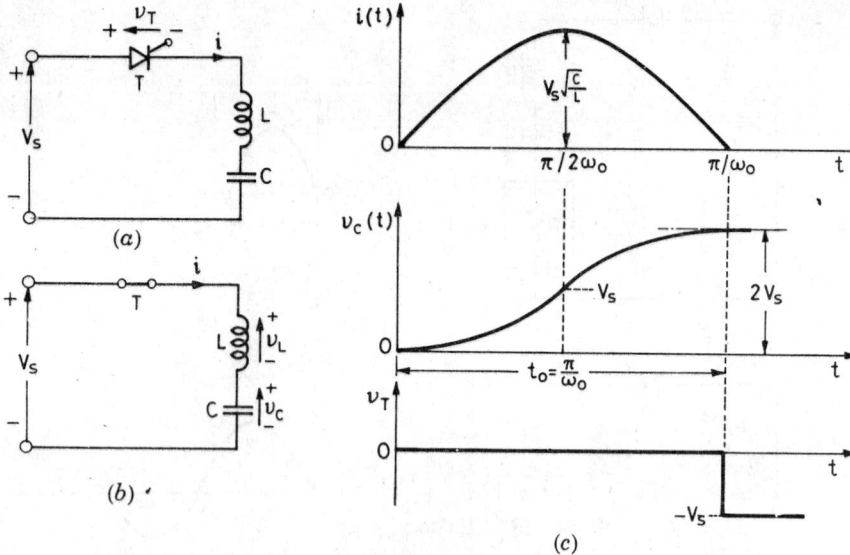

Fig. 5.2. (a) and (b) Load commutation circuit (c) waveforms.

Here $\omega_0 = \dfrac{1}{\sqrt{LC}}$ is called the resonant frequency of the circuit.

Capacitor voltage, from Eq. (3.10a) is

$$v_c(t) = V_s(1 - \cos \omega_0 t) \qquad \qquad ...(5.2)$$

It is seen from above equations that at time $t = t_0 = \pi/\omega_0$, $i(t) = 0$ and $v_c(t) = +2V_s$. This shows that π/ω_0 sec or $\pi\sqrt{LC}$ sec after thyristor is closed at $t = 0$, the charging current becomes zero, Fig. 5.2 (c) and thyristor is, therefore, turned off on its own. Here

$$t_0 = \text{conduction time of the thyristor} = \pi\sqrt{LC} \qquad \qquad ...(5.3)$$

Voltage v_T across thyristor during its conduction time t_0 is zero. When it stops conducting, $v_T = -2V_s + V_s = -V_s$. It implies that SCR is subjected to a reverse voltage of V_s which helps in its recovery.

For the circuit parameters given, the calculations are as under :

Resonant frequency of the circuit, $\omega_0 = \dfrac{1}{[5 \times 10^{-3} \times 20 \times 10^{-6}]^{1/2}} = \dfrac{10^4}{\sqrt{10}}$

$$= 3162.27 \text{ rad/s.}$$

Conduction time of thyristor, $\quad t_0 = \dfrac{\pi}{\omega_0} = \dfrac{\pi\sqrt{10}}{10^4} = 9.9346 \times 10^{-4} \text{ s}$

$$= 99.346 \text{ ms.}$$

Voltage across thyristor after it is turned off

$$= -V_s = -200 \text{ V.}$$

5.2. CLASS B COMMUTATION : RESONANT-PULSE COMMUTATION

For explaining class-B, or resonant-pulse, commutation, refer to Fig. 5.3 (a). In this figure, source voltage V_s charges capacitor C to voltage V_s with left hand plate positive as shown. Main

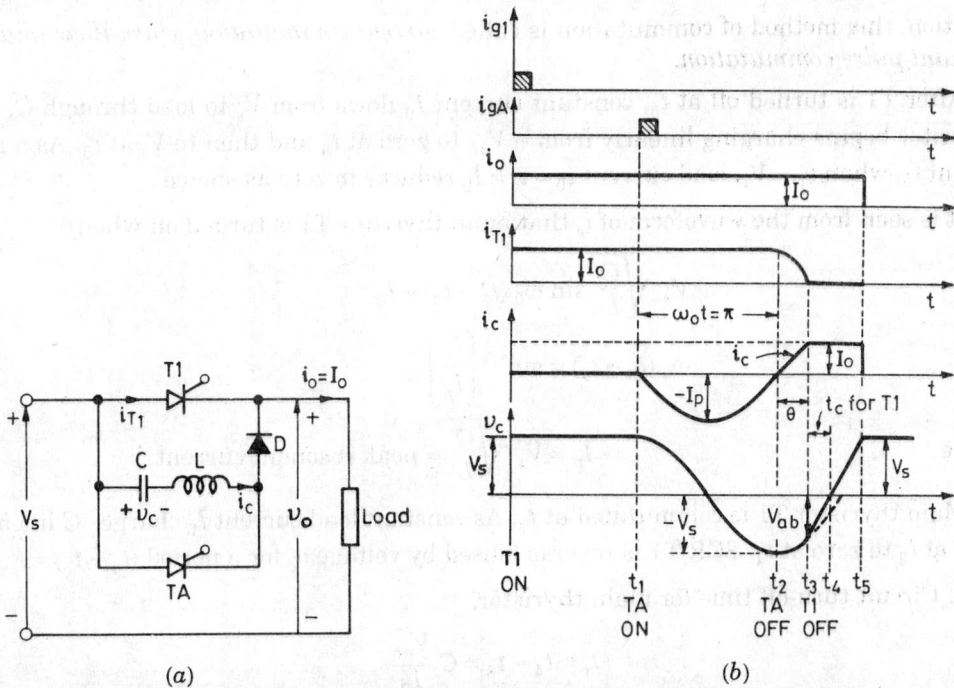

(a)   (b)

Fig. 5.3. Resonant-pulse commutation (a) circuit diagram (b) waveforms.

thyristor T1 as well as auxiliary thyristor TA are off. Positive direction of capacitor voltage v_c and capacitor current i_c are marked. When T1 is turned on at $t = 0$, a constant current I_0 is established in the load circuit. Here, for simplicity, load current is assumed constant.

Uptill time t_1, $v_c = V_s$, $i_c = 0$, $i_0 = I_0$ and $i_{T1} = I_0$, Fig. 5.3 (b). For initiating the commutation of main thyristor T1, auxiliary thyristor TA is gated at $t = t_1$. With TA on, a resonant current i_c begins to flow from C through TA, L and back to C. This resonant current, with time measured from instant t_1, is given by

$$i_c = - V_s \sqrt{\frac{C}{L}} \sin \omega_0 t = - I_p \sin \omega_0 t$$

Minus sign before $I_p \sin \omega_0 t$ is due to the fact that this current flows opposite to the reference positive direction chosen for i_C in Fig. 5.3 (a).

Capacitor voltage $\qquad v_c (t) = \dfrac{1}{C} \int i_c \, dt$

$$= V_s \cos \omega_0 t \qquad\qquad\qquad ...(5.4)$$

After half a cycle of i_c from instant t_1; $i_c = 0$, $v_c = - V_s$ and $i_{T1} = I_0$. After π radians from instant t_1, *i.e.* just after instant t_2, as i_c tends to reverse, TA is turned off at t_2. With $v_c = - V_s$, right-hand plate has positive polarity. Resonant current i_c now builds up through C, L, D and T1. As this current i_c grows opposite to forward thyristor current of T1, net forward current $i_{T1} = I_0 - i_c$ begins to decrease. Finally, when i_c in the reversed direction attains the value I_0, forward current in T1 ($i_{T1} = I_0 - I_0 = 0$) is reduced to zero and the device T1 is turned off at t_3. For reliable commutation, peak resonant current I_p must be greater than load current I_0. As thyristor is commutated by the gradual build up of resonant current in the reversed direction, this method of commutation is called *current commutation, class-B commutation* or *resonant-pulse commutation*.

After T1 is turned off at t_3, constant current I_0 flows from V_s to load through C, L and D. Capacitor begins charging linearly from $-V_{ab}$ to zero at t_4 and then to V_s at t_5. As a result, at instant t_5, when $v_c = V_s$, load current $i_0 = i_c = I_0$ reduces to zero as shown.

It is seen from the waveform of i_c that main thyristor T1 is turned off when

$$V_s \sqrt{\frac{C}{L}} \sin \omega_0 (t_3 - t_2) = I_0$$

or
$$\omega_0 (t_3 - t_2) = \sin^{-1}\left(\frac{I_0}{I_p}\right) \qquad \qquad ...(5.5)$$

where
$$I_p = V_s \sqrt{\frac{C}{L}} = \text{peak resonant current.}$$

Main thyristor T1 is commutated at t_3. As constant load current I_0 charges C linearly from $-V_{ab}$ at t_3 to zero at t_4, SCR T1 is reverse biased by voltage v_c for a period $(t_4 - t_3) = t_c$.

∴ Circuit turn-off time for main thyristor,

$$t_c = t_4 - t_3 = C \frac{V_{ab}}{I_0} \qquad \qquad ...(5.6)$$

Eq. (5.6) shows that t_c is dependent on the load current. Waveform of capacitor voltage v_c reveals that the magnitude of reverse voltage V_{ab} across main thyristor T1, when it gets commutated, is given by

$$V_{ab} = V_s \cos \omega_0 (t_3 - t_2) \qquad \qquad ...(5.7)$$

Example 5.2. *Circuit of Fig. 5.3 (a) employing resonant-pulse commutation (or class-B commutation) has $C = 20 \ \mu F$ and $L = 5 \ \mu H$. Initial voltage across capacitor is $V_s = 230$ V. For a constant load current of 300 A, calculate*

(a) conduction time for the auxiliary thyristor,

(b) voltage across the main thyristor when it gets commutated and

(c) the circuit turn-off time for the main thyristor.

Solution. Peak value of resonant current,

$$I_p = V_s \sqrt{\frac{C}{L}} = 230 \sqrt{\frac{20}{5}} = 460 \text{ A}$$

Resonant frequency, $\omega_0 = \dfrac{1}{\sqrt{LC}} = \dfrac{10^6}{\sqrt{100}} = 0.1 \times 10^6 \text{ rad/s}$

(a) Conduction time for auxiliary thyristor

$$= \frac{\pi}{\omega_0} = \frac{\pi}{0.1 \times 10^6} = 31.416 \ \mu s.$$

(b) From Eq. (5.5), $\omega_0 (t_3 - t_2) = \sin^{-1}\left(\dfrac{300}{460}\right) = 40.706$ or 0.71045 rad.

Voltage across main thyristor, when it gets turned-off, is given by Eq. (5.7).

∴
$$V_{ab} = V_s \cos \omega_0 (t_3 - t_2) = 230 \cos (40.706°) = 174.355 \text{ V}$$

(c) Circuit turn-off time for main thyristor, from Eq. (5.6), is

$$t_c = t_4 - t_3 = C \frac{V_{ab}}{I_0} = 20 \times 10^{-6} \frac{174.355}{300} = 11.624 \, \mu s.$$

5.3. CLASS C COMMUTATION : COMPLEMENTARY COMMUTATION

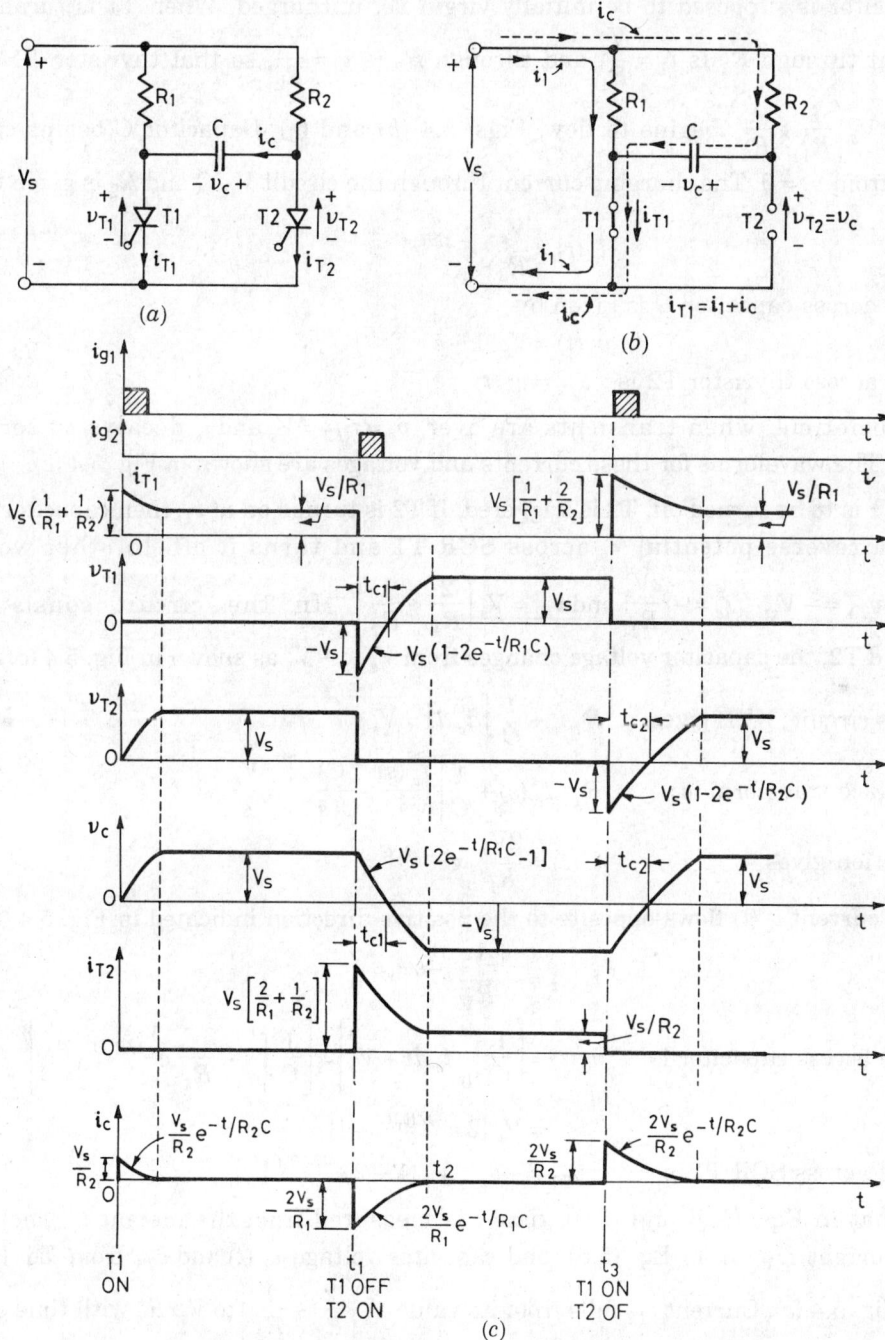

Fig. 5.4. Class-C commutation (a) and (b) circuit diagrams (c) waveforms.

In this type of commutation, a thyristor carrying load current is commutated by transferring its load current to another incoming thyristor. Fig. 5.4 (a) illustrates an arrangement employing complementary commutation. In this figure, firing of SCR T1 commutates T2 and subsequently, firing of SCR T2 would turn off T1.

Positive and negative directions of voltages and currents are marked in Fig. 5.4 (a). In this figure, capacitor is supposed to be initially virgin *i.e.* uncharged. When T1 is turned on at $t = 0$, current through R_1 is $i_1 = \dfrac{V_s}{R_1}$ and through R_2 is $i_c = \dfrac{V_s}{R_2}$, so that thyristor T1 current $i_{T1} = i_1 + i_c = V_s \left(\dfrac{1}{R_1} + \dfrac{1}{R_2} \right)$ begins to flow, Figs. 5.4 (b) and (c). Capacitor C begins charging through R_2 from $v_c = 0$. The charging current through the circuit V_s, C and R_2 is given by

$$i_c(t) = \frac{V_s}{R_2} \cdot e^{-t/R_2 C}$$

and voltage across capacitor C is given by

$$v_c(t) = V_s \left(1 - e^{-t/R_2 C} \right)$$

Voltage across thyristor T2 is $v_{T2} = v_c(t)$

After sometime, when transients are over, $v_c = v_{T2} = V_s$ and i_c decalys to zero. Also $i_{T1} = V_s/R_1$. The waveforms for these currents and voltages are shown in Fig. 5.4 (c).

When T1 is to be turned off, T2 is triggered. If T2 is turned on at t_1, then capacitor voltage v_c applies a reverse potential V_s across SCR T1 and turns it off. In other words, at t_1, $v_{T2} = 0$, $v_{T1} = -V_s$, $i_c = -\dfrac{2V_s}{R_1}$ and $i_{T2} = V_s \left(\dfrac{2}{R_1} + \dfrac{1}{R_2} \right)$. In the circuit consisting of V_s, R_1, C and T2, the capacitor voltage changes from V_s to $-V_s$ as shown in Fig. 5.4 (c).

For this circuit, KVL gives $R_1 i_c + \dfrac{1}{C} \int i_c \, dt = V_s$

Its Laplace transform is $R_1 \cdot I_c(s) + \dfrac{1}{C} \left[\dfrac{I_c(s)}{s} - \dfrac{CV_s}{s} \right] = \dfrac{V_s}{s}$

Its solution gives, $i_c(t) = \dfrac{2V_s}{R_1} \cdot e^{-t/R_1 C}$

As this current $i_c(t)$ flows opposite to the positive direction indicated in Fig. 5.4 (a),

$$i_c(t) = -\frac{2V_s}{R_1} e^{-t/R_1 C} \qquad \qquad \text{...(5.8)}$$

Voltage across capacitor is $v_c(t) = \left[\dfrac{1}{C} \displaystyle\int_0^t i_c \, dt + V_s \right] = \left[\dfrac{1}{C} \displaystyle\int_0^t - \dfrac{2V_s}{R_1} e^{-t/R_1 C} + V_s \right]$

$$= V_s \left[2 e^{-t/R_1 C} - 1 \right] \qquad \qquad \text{...(5.9a)}$$

Voltage across SCR T1 is $v_{T1} = -v_c = V_s \left[1 - 2 e^{-t/R_1 C} \right] \qquad \text{...(5.9b)}$

Note that in Eqs. (5.8) and (5.9), time t is measured from the instant t_1. The plots of capacitor current $i_c(t)$ from Eq. (5.8), and capacitor voltage $v_c(t)$ and v_{T1} from Eq. (5.9) are shown in Fig. 5.4 (c). Current i_{T2} falls from its value $V_s \left(\dfrac{2}{R_1} + \dfrac{1}{R_2} \right)$ to V_s/R_2 with time constant $R_1 C$.

When transients are over after t_1, $v_{T1} = V_s$, $v_c = -V_s$, $i_c = 0$, $v_{T2} = 0$, $i_{T2} = V_s/R_2$ and $i_{T1} = 0$. When T1 is turned on to commutate T2 at instant t_3, $i_{T2} = 0$, $i_{T1} = V_s \left(\dfrac{2}{R_2} + \dfrac{1}{R_1} \right)$, $v_{T2} = -V_s$, $v_{T1} = 0$ and $i_c = \dfrac{2V_s}{R_2}$.

With the turn on of T2 at t_1, capacitor voltage V_s suddenly appears as reverse bias across T1 to turn it off. Similarly, at t_3, capacitor voltage V_s applies a sudden reverse bias across T2 to turn it off. On account of this, class-C commutation is also called *complementary impulse commutation*.

Waveforms for voltages and currents are drawn in Fig. 5.4 (c). Waveform for v_{T1} indicates that a reverse voltage $-V_s$ to zero appears across thyristor T1 for a certain period. This period, called circuit turn-off time t_{c1} for T1 is given by

$$v_{T1} = 0 = V_s \left[1 - 2\, e^{-t_{c1}/R_1 C} \right]$$

or
$$t_{c1} = R_1 C \ln (2) \qquad \qquad \text{...(5.10 } a\text{)}$$

Similarly, circuit turn-off time for T2 is

$$t_{c2} = R_2 C \ln (2) \qquad \qquad \text{...(5.10 } b\text{)}$$

Example 5.3. *Circuit of Fig. 5.4 (a), employing class-C commutation, has $V_s = 200$ V, $R_1 = 10\ \Omega$ and $R_2 = 100\ \Omega$. Determine*

(a) peak value of current through thyristors T1 and T2

(b) value of capacitor C if each thyristor has turn-off time of 40 µs. Take a factor of safety 2.

Solution. (a) An examination of Fig. 5.4 (c) reveals that

peak value of current through T1 $= V_s \left[\dfrac{1}{R_1} + \dfrac{2}{R_2} \right] = 200 \left[\dfrac{1}{10} + \dfrac{2}{100} \right] = 24$ A

and peak value of current through T2 $= V_s \left[\dfrac{2}{R_1} + \dfrac{1}{R_2} \right]$

$$= 200 \left[\dfrac{2}{10} + \dfrac{1}{100} \right] = 42 \text{ A}$$

(b) From Eq. (5.10 a), $\qquad C = \dfrac{t_{c1}}{R_1 \ln (2)}$

$$= \dfrac{2 \times 40 \times 10^{-6}}{10 \ln (2)} = 11.542\ \mu\text{F}$$

From Eq. (5.10 b), $\qquad C = \dfrac{2 \times 40 \times 10^{-6}}{100 \ln (2)}\ 1.1542\ \mu\text{F}$

So choose a capacitor of large size of 11.542 µF.

5.4. CLASS D COMMUTATION : IMPULSE COMMUTATION

For explaining class D, or impulse, commutation, refer to the circuit of Fig. 5.5 (a). In this figure, T1 and *TA* are called main and auxiliary thyristors respectively.

Initially, main thyristor T1 and auxiliary thyristor *TA* are off and capacitor is assumed charged to voltage V_s with upper plate positive. When T1 is turned on at $t = 0$, source voltage

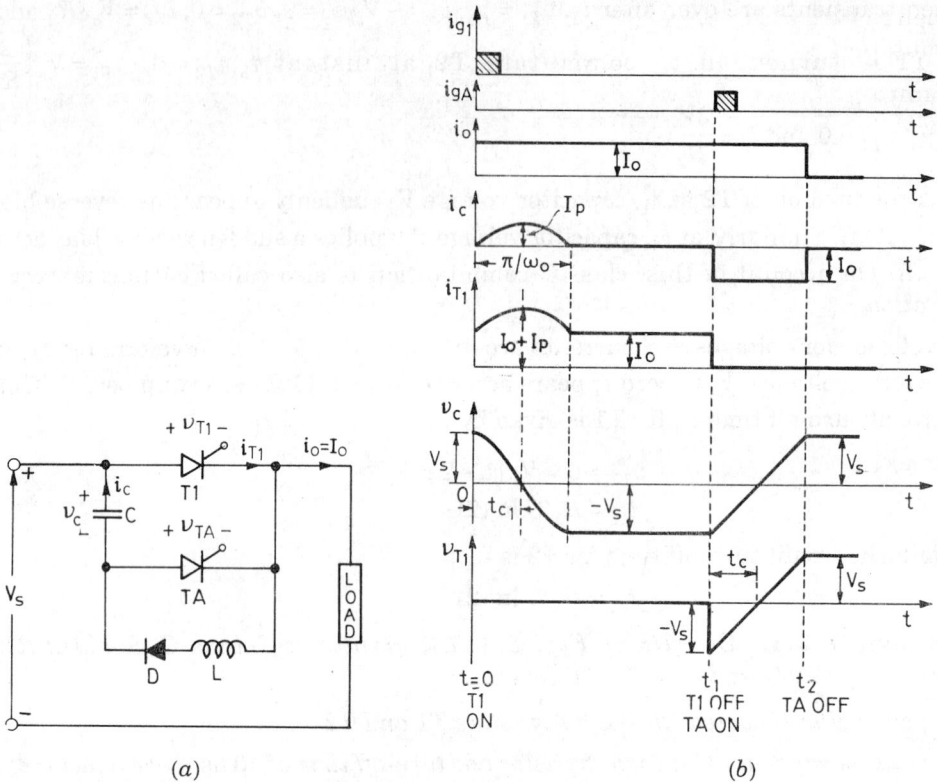

(a)   (b)

Fig. 5.5. Class-D commutation (a) circuit diagram (b) waveforms.

V_s is applied across load and load current I_0 begins to flow which is assumed to remain constant. With T1 on at $t = 0$, another oscillatory circuit consisting of C, T1, L and D is formed where the capacitor current is given by

$$i_c = V_s \sqrt{\frac{C}{L}} \sin \omega_0 t = I_p \sin \omega_0 t$$

When $\omega_0 t = \pi$, $i_c = 0$. Between $0 < t < (\pi/\omega_0)$, $i_{T1} = I_0 + I_p \sin \omega_0 t$. Capacitor voltage changes from $+V_s$ to $-V_s$ co-sinusoidally and the lower plate becomes positive. At $\omega_0 t = \pi$, $i_C = 0$, $i_{T1} = I_0$ and $v_c = -V_s$, Fig. 5.5 (b).

At t_1, auxiliary thyristor TA is turned on. Immediately after TA is on, capacitor voltage V_s applies a reverse voltage across main thyristor T1 so that $v_{T1} = -V_s$ at t_1 and SCR T1 is turned off and $i_{T1} = 0$. The load current is now carried by C and TA. Capacitor gets charged from $-V_s$ to V_s with constant load current I_0. The change is, therefore, linear from $+V_s$ to $-V_s$ as shown. When $v_c = V_s$, $i_c = 0$ at t_2, thyristor TA is turned off. During the time TA is on from t_1 to t_2, $v_c = v_{T1}, i_c = -I_0$ and $i_0 = I_0$. For main thyristor T1, circuit turn-off time is t_c as shown in Fig. 5.5 (b).

With the firing of thyristor TA, a reverse voltage V_s is suddenly applied across T1 ; this method of commutation is therefore, also called *voltage commutation*. With sudden appearance of reverse voltage across T1, its current is quenched ; in fact the current momentarily reverses to recover the stored charge of T1. As an auxiliary thyristor TA is used for turning-off the main thyristor T1, this type of commutation is also known as *auxiliary commutation*.

When thyristor *TA* is turned on, capacitor gets connected across T1 to turn it off, this type of commutation is, therefore, also called *parallel-capacitor commutation.*

Example 5.4. *Circuit of Fig. 5.5 (a) illustrates class-D commutation. For this circuit, $V_s = 230$ V, $L = 20$ µH and $C = 40$ µF. For a constant load current of 120 A, calculate.*

(a) peak value of current through capacitance and also through main and auxiliary thyristors,

(b) circuit turn-off times for main and auxiliary thyristors.

Solution. *(a)* When main thyristor T1 is turned on, an oscillatory current in the circuit C, T1, L and D is set up and it is given by

$$i_c(t) = V_s \cdot \sqrt{\frac{C}{L}} \sin \omega_0 t$$

∴ Peak value of current through capacitor

$$I_p = V_s \sqrt{\frac{C}{L}} = 230 \sqrt{\frac{40}{20}} = 325.22 \text{ A}$$

Peak value of current through main thyristor

$$T1 = I_p + I_0 = 325.22 + 120 = 445.22 \text{ A}$$

Peak value of current through auxiliary thyristor $TA = I_0 = 120$ A

(b) Waveforms for v_{T1} or v_c in Fig. 5.5 *(b)* indicate that circuit turn-off time for main thyristor T1 is the time required for v_{T1} or v_c to change linearly from $-V_s$ to zero.

$$\therefore \quad I_0 = C \frac{V_s}{t_c}$$

∴ Circuit turn-off time for main thyristor

$$t_c = C \frac{V_s}{I_0} = 40 \times 10^{-6} \frac{230}{120} = 76.67 \text{ µs}$$

An examination of Fig. 5.5 reveals that when T1 conducts and during the time upper plate of C is positive, $v_{TA} = -v_c$ *i.e.* auxiliary thyristor *TA* is reverse biased by v_c. This gives circuit turn-off time t_{c1} for $TA = \dfrac{\pi}{2\,\omega_0}$

Here

$$\omega_0 = \frac{1}{\sqrt{LC}} = \frac{10^6}{\sqrt{20 \times 40}} = \frac{10^6}{\sqrt{800}}$$

Circuit turn-off time for auxiliary thyristor,

$$t_{c1} = \frac{\pi}{2\omega_0} = \frac{\pi \sqrt{800}}{2 \times 10^6} = 44.43 \text{ µs.}$$

5.5. CLASS E COMMUTATION : EXTERNAL PULSE COMMUTATION

In this type of commutation, a pulse of current is obtained from a separate voltage source to turn off the conducting SCR. The peak value of this current pulse must be more than the load current. Fig. 5.6 shows a circuit using external-pulse commutation. Here V_s is the voltage of the main source and V_1 is the voltage of the auxiliary supply. Thyristor T1 is conducting and load is connected to source V_s. When thyristor T3 is turned on at $t = 0$; V_1, T_3, L and C form an oscillatory circuit. Therefore, C is charged to a voltage $+2V_1$ with upper plate positive at $t = \pi\sqrt{LC}$ as shown and as oscillatory current falls to zero, see Art. 3.1.4, thyristor T3 gets commutated. For turning off the main thyristor T1, thyristor T2 is turned on. With T2 on, T1 is subjected to a reverse voltage equal to $V_s - 2V_1$ and $T1$ is therefore turned off. After T1 is off, capacitor discharges through the load.

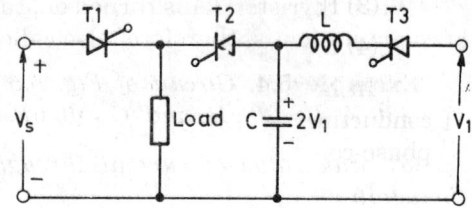

Fig. 5.6. External-pulse commutation circui.

5.6. CLASS F COMMUTATION : LINE COMMUTATION

This type of commutation is also known as *natural commutation*. This can occur only when the source is *ac*. When an SCR circuit is energised from *ac* source, current has to pass through its natural zero at the end of every positive half cycle. Then *ac* source applies a reverse bias across SCR automatically. As a result, SCR is turned off. This is called natural commutation because no external circuit is employed to turn-off the thyristor. This method of commutation is applied to phase-controlled converters, line-commutated inverters, ac voltage controllers and step-down cycloconverters.

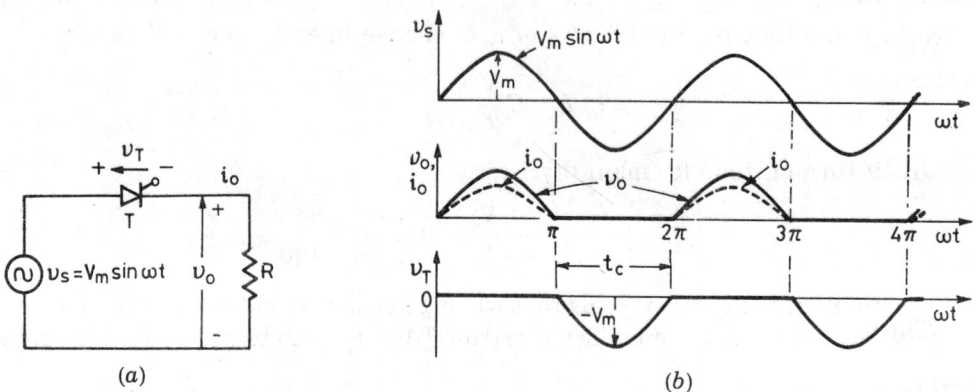

(a) (b)

Fig. 5.7. Class F commutation (a) circuit diagram (b) waveforms.

A single-phase half-wave (or one-pulse) controlled converter employing line commutation is shown in Fig. 5.7 (a). In this figure, thyristor T is fired at firing angle equal to zero, *i.e.* when $\omega t = 0$, $v_s = 0$. Load is resistive in nature. With zero degree firing-delay angle, the thyristor behaves like a diode. During the positive half-cycle, $v_0 = v_s$ and waveshape of load current i_0 is identical with the waveshape of v_0 for a resistive load. At $\omega t = \pi$, $v_s = 0$, $v_0 = 0$ and $i_0 = 0$; therefore T gets turned off at this instant. From $\omega t = \pi$ to $\omega t = 2\pi$, T is reverse biased for a period $t_c = \pi/\omega$ sec, longer than the thyristor turn-off time t_q. Here t_c is called the circuit turn-off time.

Another method of classification of thyristor commutation technique is as under :
(1) Line commutation : class F
(2) Load commutation : class A

(3) Forced commutation : class B, C and D

(4) External-pulse commutation : class E.

In line, or natural, commutation, natural reversal of ac supply voltage commutates the conducting thyristor. As stated before, line commutation is widely used in ac voltage controllers, phase-controlled rectifiers and step-down cycloconverters.

In load commutation, L and C are connected in series with the load or C in parallel with the load such that overall load circuit is under damped. Load commutation is commonly employed in series inverters.

In forced commutation, the commutating components L and C do not carry load current continuously. So class B, C and D commutation constitute forced commutation techniques. As stated before, in forced commutation, forward current of the thyristor is forced to zero by external circuitry called *commutation circuit*. Forced commutation is usually employed in dc choppers and inverters.

Example 5.5. *In the circuit shown in Fig. 5.8, SCR is forced commutated by circuitry not shown in the figure. Compute the minimum value of C so that SCR does not get turned on due to re-applied dv/dt. The SCR has minimum charging current of 5 mA to turn it on and its junction capacitance is 25 pF.*

Solution. Under steady state, SCR conducts a current $= \dfrac{V_s}{R} = \dfrac{200}{50} = 4$ A and voltage across ideal SCR = voltage v_c across $C = 0$.

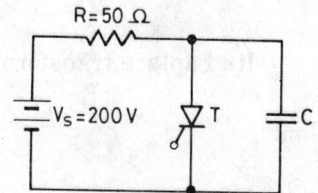

Fig. 5.8. Pertaining to Example 5.5.

When SCR is force commutated, capacitor C begins charging from source V_s through R so that capacitor voltage v_c ($= v_T$) is given by

$$v_c = V_s \left[1 - e^{-t/RC}\right]$$

$$\left[\frac{dv_c}{dt}\right] = V_s \cdot e^{-t/RC} \cdot \frac{1}{RC}$$

or $$\left[\frac{dv_c}{dt}\right]_{t=0} = \frac{V_s}{RC} \qquad \ldots(i)$$

The rate of rise of capacitor voltage v_c across SCR may be large. In case SCR charging current $C_j \cdot \left(\dfrac{dv_c}{dt}\right)_{t=0}$ happens to be equal to 5 mA, SCR will get turned on. Here C_j is the junction capacitance of SCR.

$\therefore$ $$C_j \cdot \left(\frac{dv_c}{dt}\right)_{t=0} = 5 \text{ mA}$$

Substituting the value of $\left(\dfrac{dv_c}{dt}\right)_{t=0}$ from Eq. (i) above, we get

$$C_j \frac{V_s}{RC} = 5 \times 10^{-3}$$

or $$25 \times 10^{-12} \frac{200}{50 \times C} = 5 \times 10^{-3}$$

or
$$C = \frac{25 \times 10^{-12} \times 200}{250 \times 10^{-3}} = 0.02 \ \mu F$$

In order to obviate turning on of SCR, the value of capacitance C should be less than $0.02 \ \mu F$.

Example 5.6. *For a voltage or impulse commutated thyristor circuit shown in Fig. 5.9, capacitor is initially charged to V_s with polarity as shown. Find the circuit turn-off time for the main thyristor in case $C = 10 \ \mu F$, $R = 5\Omega$ and $V_s = 200$ V dc.*

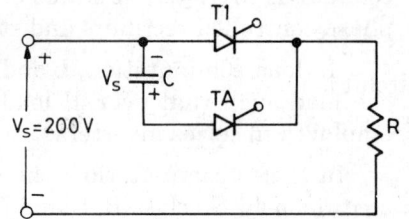

Fig. 5.9. Pertaining to Example 5.6.

Solution. When auxiliary thyristor *TA* is turned on, main thyristor T1 is turned off by means of capacitor voltage V_s appearing as reverse bias. After T1 is off, KVL for the circuit consisting of V_s, C, *TA* and R in series is given by

$$R \cdot i\,(t) + \frac{1}{C} \int i\,(t)\, dt = V_s$$

Its Laplace transforme is $\quad R \cdot I\,(s) + \dfrac{1}{C}\left[\dfrac{I\,(s)}{s} - \dfrac{C \cdot V_s}{s}\right] = \dfrac{V_s}{s}$

or
$$I\,(s)\left[R + \frac{1}{sC}\right] = \frac{2V_s}{s}$$

$$I\,(s) = \frac{2V_s}{s} \cdot \frac{sC}{(RsC + 1)} = \frac{2V_s}{R} \cdot \frac{1}{s + \dfrac{1}{RC}}$$

Its Laplace inverse is $\quad i\,(t) = \dfrac{2V_s}{R} \cdot e^{-t/RC}$

The voltage across capacitor C is

$$v_c\,(t) = \frac{1}{C} \int i\,(t)\, dt + \text{initial voltage across capacitor}$$

$$= \frac{1}{C} \int_0^t \frac{2V_s}{R}\, e^{-t/RC} - V_s = V_s\,[1 - 2\, e^{-t/RC}]$$

During the time auxiliary SCR *TA* is on, $v_c = v_{T1} = V_s\,[1 - e^{-t/RC}]$. The circuit turn-off time for T1 is the time taken by $v_c = v_{T1}$ to change from its value $-V_s$ to zero.

$\therefore$
$$0 = V_s\,[1 - 2\, e^{-t_c/RC}]$$

or
$$t_c = RC \ln(2) = 5 \times 10 \times 10^{-6} \ln(2) = 34.6574 \ \mu s.$$

Example 5.7. *For the circuit shown in Fig. 5.1 (a), commutating elements $L = 20 \ \mu H$ and $C = 40 \ \mu F$ are connected in series with load resistance $R = 1 \ \Omega$. Check whether self-commutation, or load commutation, would occur or not. Find the conduction time of the thyristor.*

Solution. It is seen from Art 3.1.5 that ringing frequency ω_r in rad/sec, from Eq (3.14), is given by

$$\omega_r = \sqrt{\frac{1}{LC} - \left(\frac{R}{2L}\right)^2} = \text{damped frequency of oscillation, } \omega_d$$

The condition for underdamping is that $\omega_d > 0$

or $\qquad \dfrac{1}{LC} - \left(\dfrac{R}{2L}\right)^2 > 0 \quad$ or $\quad R < \sqrt{\dfrac{4L}{C}}$

Here $\dfrac{4L}{C} = \dfrac{4 \times 20 \times 10^{-6}}{40 \times 10^{-6}} = 2$. Therefore $\sqrt{\dfrac{4L}{C}} = \sqrt{2} = 1.414$ and $R = 1\,\Omega$. As $R < \sqrt{\dfrac{4L}{C}}$, the circuit is underdamped.

Fig. 5.1 (a) shows that thyristor stops conducting when $\omega_r\, t_1 = \pi$.

Here $\qquad \omega_r = \left[\dfrac{10^{12}}{2 \times 40} - \left(\dfrac{1 \times 10^6}{2 \times 20}\right)^2\right]^{1/2} = 25000$ rad/sec

$\therefore$ Conduction time of thyristor, $t_1 = \dfrac{\pi}{\omega_r} = \dfrac{\pi \times 10^6}{25000}$ µs $= 125.664$ µs.

Example 5.8. (a) *For the circuit shown in Fig. 5.10 (a), dv/dt rating of thyristor T is 400 V/µs and its junction capacitance is 25 pF. Switch S is closed at t = 0. Calculate the value of C_s so that thyristor T is not turned on due to dv/dt.*

(b) *In case maximum current through thyristor of Fig. 5.10 (a) is limited to 40 A, determine the value of R_s.*

Solution. (a) When switch S is closed, the equivalent circuit for Fig. 5.10 (a) is as shown in Fig. 5.10 (b) where $C = C_j + C_s$. The voltage rise across C is given by

$$v_c = V_s\,[1 - e^{-t/\tau}] \text{ where } \tau = RC$$

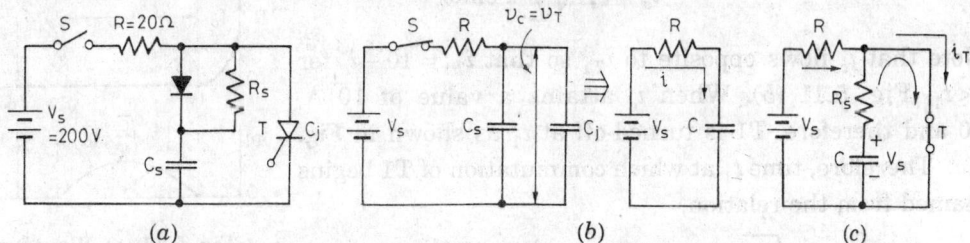

Fig. 5.10 (a) Pertaining to Example 5.8. Equivalent circuit (b) for (a) at $t = 0$ and (c) for (a) when SCR turns on.

The voltage variation across C is the same as that across C_s or C_j of thyristor T.

$\therefore \qquad v_C = v_T = V_s\left[1 - e^{-t/\tau}\right]$

$\left(\dfrac{dv_T}{dt}\right)_{t=0} = \dfrac{V_s}{RC}$ or $\dfrac{400}{10^{-6}} = \dfrac{200}{20 \times C}$, $\therefore C = \dfrac{10 \times 10^{-6}}{400} = 0.025$ µF

$\therefore \qquad C_s = C - C_j = 0.025 \times 10^{-6} - 0.025 \times 10^{-12} \simeq 0.025$ µF

(b) When switch S is closed, C_s would be charged to voltage V_s with upper plate positive as shown in Fig. 5.10 (c). Now when thyristor is turned on, current i_T at that moment through T would be given by

$$I_T = \dfrac{V_s}{R} + \dfrac{V_s}{R_s} \quad \text{or} \quad 40 = \dfrac{200}{20} + \dfrac{200}{R_s}$$

$\therefore \qquad R_s = 10\,\Omega.$

Example 5.9. *For the circuit shown in Fig. 5.11 (a), $V_s = 200V$, $L = 0.2\ mH$, $C = 20\ \mu F$, constant load current $I_o = 10\ A$ and capacitor C is initially charged to source voltage V_s, with lower plate positive. The auxiliary thyristor TA is turned on at $t = 0$ to commutate the main thyristor T1. Calculate (a) the time at which the commutation of main thyristor T1 gets initiated (b) the circuit turn-off time for T1. Comment on the conduction time of auxiliary thyristor.*

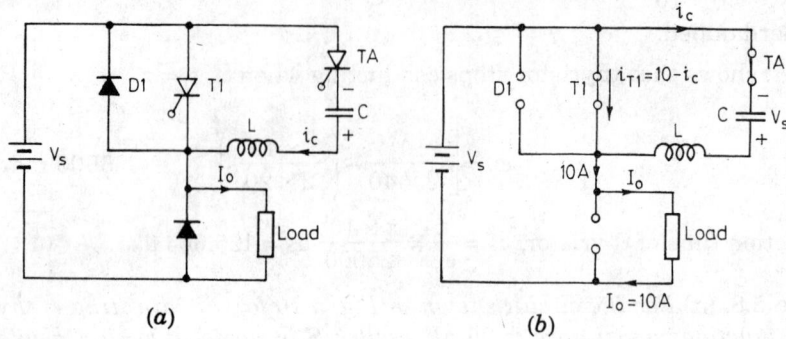

Fig. 5.11. (a) Pertaining to Example 5.9 (b) Circuit model when T1 is turned on.

Solution. (a) When auxiliary thyristor TA is turned on at $t = 0$, C begins to discharge through L, T1, TA and C. Eventually, upper plate of capacitor would become positive with $v_c = V_s$. With TA on, current i_c begins to rise as

$$i_c = V_s \sqrt{\frac{C}{L}} \cdot \sin \omega_o t$$

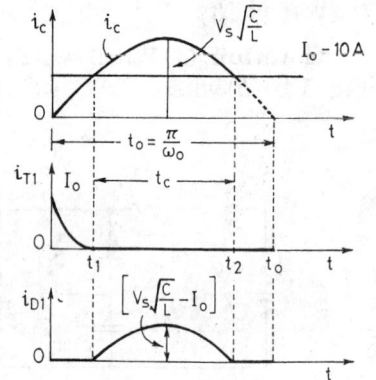

Fig. 5.11 (c). Waveforms pertaining to Example 5.9.

Note that i_c flows opposite to i_{T1} so that $i_{T1} = 10 - i_c$ for $0 < t < t_1$, Fig. 5.11 (b). When i_c attains a value of 10 A, $i_{T1} = 0$ and therefore, T1 is turned-off at t_1 as shown in Fig. 5.11 (c). Therefore, time t_1 at which commutation of T1 begins is obtained from the relation,

$$V_s \sqrt{\frac{C}{L}} \sin \omega_o t_1 = 10\ \text{A}$$

Here $\omega_o = \dfrac{1}{\sqrt{LC}} = \sqrt{\dfrac{10^9}{0.2 \times 20}} = \dfrac{10^5}{\sqrt{40}}$ and $V_s \sqrt{\dfrac{C}{L}} = 200 \sqrt{\dfrac{200 \times 10^{-6}}{0.2 \times 10^{-6}}} = 40\ \text{A}$

$\therefore \qquad 40 \sin \omega_o t_1 = 10\ \text{A}$

or $\qquad t_1 = \dfrac{1}{\omega_o} \sin^{-1}\left(\dfrac{10}{40}\right) = \dfrac{\sqrt{40}}{10^5} \sin^{-1}(0.25) = 15.981\ \mu s$

(b) After t_1, as i_c exceeds I_o, diode D1 begins to conduct till time t_2 where i_{D1} falls to I_o. Here time t_o of half cycle of i_c is given by

$$t_o = \frac{\pi}{\omega_o} = \pi \sqrt{LC} = \pi \frac{\sqrt{40}}{10^5} = 198.693\ \mu s$$

Circuit turn-off time for T1 = conduction time of diode D1 = $t_2 - t_1 = t_0 - 2 t_1 = t_c$

$$= 198.692 - 2 \times 15.981 = 166.73\ \mu s$$

The auxiliary thyristor *TA* will continue conducting till capacitor charges to V_s with upper plate positive and i_c falls to zero.

Example 5.10. *The circuit of Fig. 5.12 can be used to explain class-B commutation. With positive directions indicated for v_c, i_c, i_o, i_T and v_T, describe, with appropriate waveforms of i_o, i_c, v_c, i_T and v_T, how current-commutation is achieved in this circuit. Derive expressions for i_c, reverse voltage across SCR when it is turned off and the circuit turn-off time. State the assumptions made.*

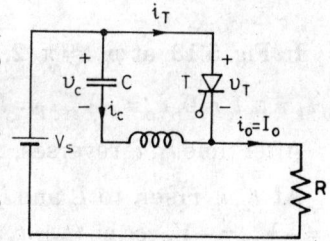

Fig. 5.12. Pertaining to Example 5.10.

Solution. The assumptions are as under :

(*i*) Thyristor is initially off.

(*ii*) Capacitor C is charged to source voltage V_s with upper plate positive.

(*iii*) Load current is constant and (*iv*) SCR is an ideal device.

Now, when T is turned on at $t = 0$, capacitor begins to discharge through C, T, L, Fig. 5.14 (*a*). Therefore, an oscillatory current $i_c = V_s \sqrt{\dfrac{C}{L}} \sin \omega_o t$ is established and $v_c = V_s \cos \omega_o t$.

Also, at $t = 0$, load current I_o is set up so that thyristor current $i_T = I_o + V_s \sqrt{\dfrac{C}{L}} \cdot \sin \omega_o t$ as shown in Fig. 5.13 and Fig. 5.14 (*a*).

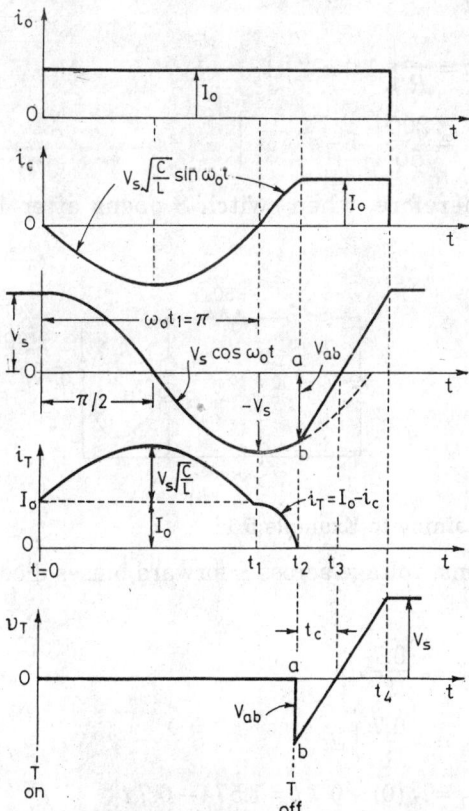

Fig. 5.13. Waveforms for the circuit of Fig. 5.12

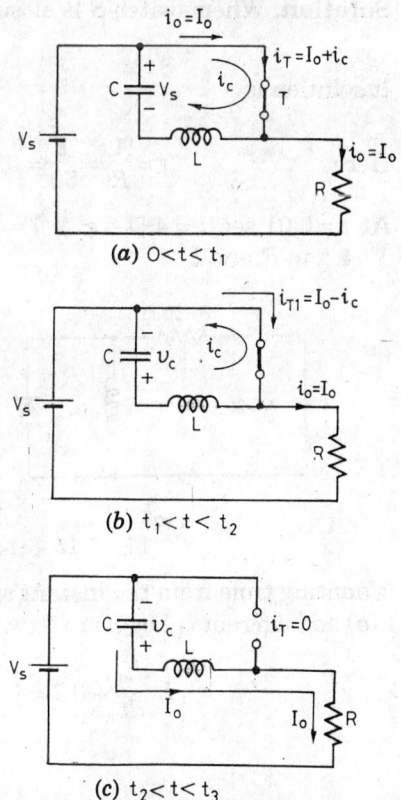

Fig. 5.14. (*a*) When T is turned on at $t = 0$, $i_T = I_o + i_c$ (*b*) $t_1 < t < t_2$ and (*c*) $t_2 < t < t_3$.

In Fig. 5.13, at $\omega_o t = \pi/2$, $i_c = -V_s\sqrt{\dfrac{C}{L}}$, $v_c = 0$, $i_T = I_o + V_s\sqrt{\dfrac{C}{L}}$ (peak value), $v_T = 0$ and at $\omega_o t_1 = \pi$, $i_c = 0$, $v_c = -V_s$, $i_T = I_o$ and $v_T = 0$.

After time t_1, i_c reverses, therefore $i_T = I_o - i_c$ begins to decrease, Fig. 5.13 and Fig. 5.14 (b).

At t_2, i_c rises to I_o and $i_{T1} = 0$, therefore T is turned off. Also at tim t_2, $i_c = I_o$, $i_T = 0$, $v_c = -V_{ab} = -V_s\cos\omega_o(t_2 - t_1)$, $v_T = -V_{ab} = -V_s\cos\omega_o(t_2 - t_1)$. Note that SCR T is subjected to reverse voltage V_{ab}. Between $(t_3 - t_2)$, current $i_c = I_o$ charges C linearly from V_{ab} at t_2 to zero at t_3 so that $v_T = 0$ at t_3, Fig. 5.13 and Fig. 5.14 (c). This gives

$$I_o = C\frac{V_{ab}}{t_3 - t_2}$$

$\therefore$ Circuit turn-off time for $T = (t_3 - t_2) = t_c = C\dfrac{V_{ab}}{I_o}$

Eventually C charges linearly from zero at t_3 to V_s at t_4 as shown in Fig. 5.13. Then i_c reduces to zero, $i_o = 0$, $v_c = V_s$ (as before),($i_T = 0$, $v_T = V_s$ after t_4.

Example 5.11. *In the circuit shown in Fig. 5.15 (a), switch S closes at t = 0 and opens after 10 ms. What will the currents in R and L, 8 ms after the switch S opens. Assume 0.7 V drop across diode whenever it conducts.*

Solution. When switch S is closed, $Ri + L\dfrac{di}{dt} = V_s$

It solution is $i = \dfrac{V_s}{R}\left[1 - e^{-t/\tau}\right]$

Here $\tau = \dfrac{1}{R} = \dfrac{1}{50} = 0.02$ sec and $i = \dfrac{200}{50}\left[1 - e^{-\frac{t}{0.02}}\right]$

At $t = 0.01$ sec, $i = 4[1 - e^{-1/2}] = 1.574$ A. Therefore, when switch S opens after 10 ms, $i_o = 1.54$ A in R and L.

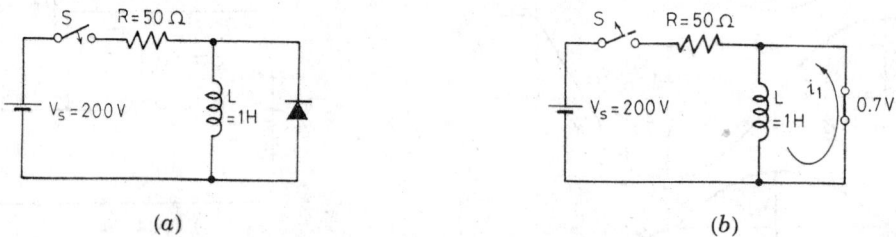

(a)   (b)

Fig. 5.15. (a) and (b). Pertaining to Example 5.11

Counting time from the instant switch S opens, voltage across L forward biases diode, Fig. 5.11 (b) and current i_1 begins to flow.

$\therefore$ $L\dfrac{di_1}{dt} + 0.7 = 0$ or $di_1 = -\dfrac{0.7}{L}dt$

or $\displaystyle\int_0^t di_1 = -\dfrac{0.7}{1}\int dt$

$i_1 - i_1(0) = -0.7t$ or $i_1 = i_1(0) - 0.7t = 1.574 - 0.7t$

When $t = 0.008$ s, current in L would be $i_1 = 1.574 - 0.7 \times 0.008 = 1.5684$ A

Current in R at $t = 0.008$ s would $i_R = 0$.

Example 5.12. *In the circuit shown in Fig. 5.16, switch S closes at t = 0 and opens after 10 ms. What will be current in R, L and voltage across C, 9 ms after switch S opens. Assume diode to be ideal.*

Solution. When switch S is closed, $Ri + L \dfrac{di}{dt} = V_s$

and $\qquad\qquad\qquad\qquad\qquad i = 4\left[1 - e^{-t/0.02} \right]$

At $t = 0.01$ s, $i = 1.574$ A as in the previous example.

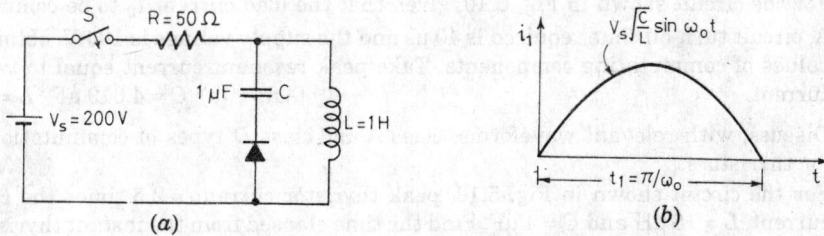

Fig. 5.16. Pertaining to Example 5.12

When switch S opens, $i_R = 0$ and voltage across L, equal to $L \dfrac{di}{dt}$, forward biases the ideal diode which begins to conduct at once. An oscillatory circuit consisting of L, C and diode is formed in which current i_1 will decay to zero, Fig. 5.16 (b), at time t_1 given by

$$t_1 = \frac{\pi}{\omega_o} = \pi \sqrt{LC} = \pi \sqrt{1 \times 1 \times 10^{-6}} = 3.141 \text{ ms}$$

Since t_1 is less than 9 ms, energy in L gets transferred to C as the current i_1 in L decays to zero.

$$\therefore \qquad\qquad \frac{1}{2} CV_c^2 = \frac{1}{2} L\, i_o^2$$

$$V_c = \sqrt{\frac{L}{C}} \times \text{Current when } S \text{ opens, } i_o$$

$$= \sqrt{\frac{1}{1 \times 10^{-6}}} \times 1.574 = 1.574 \text{ kV.}$$

PROBLEMS

5.1. (a) Explain the need of commutation in thyristor circuits. What are the different methods of commutation schemes ? Discuss one of them, involving two thyristors, with a neat schematic and waveforms.

(b) A circuit employing parallel-resonance turn-off (or class-B commutation) circuit has $C = 50$ μF, $L = 20$ μH, $V_s = 200$ V and initial voltage across capacitor is 200 V. Determine the circuit turn-off time for main thyristor for load $R = 1.5$ Ω. **[Ans :** (b) 68 μs]

5.2. (a) Distinguish clearly between voltage commutation and current commutation in thyristor circuits.

    (b) Discuss how the voltage across the commutating capacitor is reversed in a commutating circuit.

    (c) For the circuit in Fig. 5.3 (a), supply voltage $V_s = 230$ V dc, load current $I_0 = 200$ A, circuit turn-off time for main thyristor = 25 μs and reversal current is limited to 150% of I_0. Determine the values of commutating components C and L.

                                      [Ans : (c) $C = 29.166$ μF, $L = 17.143$ μH**]**

5.3. In the circuit shown in Fig. 5.9, capacitor C is initially charged to $V_s = 200$ V with polarity as indicated. Find the circuit turn-off time for main thyristor T1 after it is voltage commutated by thyristor TA. Load current is constant at 40 A and $C = 10$ μF.   **[Ans. 50μs]**

5.4. (a) Explain the merits and demerits of self-commutation of SCR and its other methods of commutation.

    (b) For the circuit shown in Fig. 5.10, given that the load current I_0 to be commutated is 10 A, circuit turn-off time required is 40 μs and the supply voltage is 100 V, obtain the proper values of commutating components. Take peak resonant current equal to twice the load current.   **[Ans :** (b) $C = 4.619$ μF, $L = 115.475$ μH**]**

5.5. (a) Discuss, with relevant waveforms, class A and class D types of commutations employed for thyristors.

    (b) For the circuit shown in Fig. 5.10, peak thyristor current = 2.5 times the constant load current, $L = 18$ μH and $C = 4$ μF. Find the time elapsed from the instant thyristor is turned on to the instant it gets turned off.   **[Ans :** (b) 32.852 μs**]**

5.6. (a) Enumerate the various commutation techniques used for thyristors.

    (b) Describe line-commutation and class-E commutation for thyristors. Name the circuit configuration where line-commutation is employed.

5.7. (a) Discuss, with relevant waveforms, class B and class E types of commutations employed for thyristor circuits.

    (b) A circuit employing resonant-pulse commutation has $C = 20$ μF and $L = 3$ μH. The initial capacitor voltage = source voltage, $V_s = 230$ V dc. Determine conduction time for auxiliary thyristor and circuit turn-off time for main thyristor in case constant load current is (i) 300 A and (ii) 60 A.   **[Ans :** (b) (i) 24.335 μs, $t_c = 13.23$ μs (ii) 24.335 μs, $t_c = 76.273$ μs**]**

5.8. (a) Describe class-C type of commutation used for thyristors with appropriate current and voltage waveforms.

    (b) An impulse-commutated circuit is shown in Fig. 5.5 (a). In this circuit, capacitor is initially charged to source voltage $V_s = 200$ V with upper plate negative. When auxiliary thyristor is turned on main thyristor gets commutated in 50 μs. Find the value of C in case load resistance is 20 Ω.

    If peak value of current through main thyristor is limited to twice the full-load current, calculate the value of commutating inductance.   **[Ans :** (b) 3.607 μF, 1.4428 mH**]**

5.9. What is complementary impulse commutation ? Describe this type of commutation with a circuit diagram and appropriate waveforms.

    Derive expressions for current through and voltage across commutating capacitor. Find also the circuit turn off times for the complementary thyristors.

5.10. (a) A capacitor C, initially charged to dc voltage V_s, is connected to inductance L through a thyristor. Determine

        (i) the peak value of current through thyristor and

        (ii) the maximum value of di/dt through SCR.

    (b) For illustrating class C commutation, circuit of Fig. 5.4 (a) is employed where $V_s = 200$ V and $R_1 = 10$ Ω. Find the value of C so that thyristor T1 is commutated in 50 μs.

It is required that SCR T2 is turned off naturally when current through it falls below the holding current of 4 mA. Find the value of R_2.

[**Hint :** (b) When C is fully charged, current through T2 = holding current = $\dfrac{V_s}{R_2}$ etc.]

$$\left[\textbf{Ans:} \ (a) \ V_s \sqrt{\frac{C}{L}}, \ \frac{V_s}{L} \ \text{Amp/sec} \ (b) \ 7.2135 \ \mu\text{F}, \ 50 \ \text{k}\Omega \right]$$

5.11. In the circuit of Fig. 5.4 (a) employing complementary commutation ; $V_s = 200$ V, $R_1 = 20 \ \Omega$ and $R_2 = 100 \ \Omega$. Determine the minimum value of C so that thyristors do not get turned on due to re-applied dv/dt. Each SCR has a minimum charging current of 4 mA to turn it on and its junction capacitance is 20 pF. **[Ans: 0.1 μF]**

5.12. For current-commutated circuit of Fig. 5.3 (a) ; $V_s = 230$ V, $L = 16 \ \mu$H and $C = 5 \ \mu$F. Capacitor is initially charged to voltage V_s with left hand plate positive. Auxiliary thyristor TA is turned on at $t = 0$. Find the total time for which capacitor current i_c exists. The peak resonant current is 1.5 times the full-load current.

$$\left[\textbf{Hint :} \ \text{In Fig. 5.3 (b)}, \ t_5 - t_3 = C \ \frac{V_s + V_{ab}}{I_0} \ \text{etc.} \right]$$ **[Ans : 58.047 μs]**

Chapter 6

Phase Controlled Rectifiers

Many industrial applications make use of controllable dc power. Examples of such applications are as follows :

(a) Steel-rolling mills, paper mills, printing presses and textile mills employing dc motor drives.

(b) Traction systems working on dc.

(c) Electrochemical and electrometallurgical processes.

(d) Magnet power supplies.

(e) Portable hand tool drives.

(f) High-voltage dc transmission.

Earlier, dc power was obtained from motor-generator (MG) sets or ac power was converted to dc power by means of mercury-arc rectifiers or thyratrons. The advent of thyristors has changed the art of ac to dc conversion. Presently, phase-controlled ac to dc converters employing thyristors are extensively used for changing constant ac input voltage to controlled dc output voltage. In an industry where there is a provision for modernization, mercury-arc rectifiers and thyratrons are being replaced by thyristors.

In phase-controlled rectifiers, a thyristor is turned off as ac supply voltage reverse biases it, provided anode current has fallen to a level below the holding current. The turning-off, or commutation, of a thyristor by supply voltage itself is called *natural,* or *line commutation.* In industrial applications, rectifier circuits make use of more than one SCR. In such circuits, when an incoming SCR is turned on by triggering, it immediately reverse biases the outgoing SCR and turns it off. As phase-controlled rectifiers need no commutation circuitry, these are simple, less expensive and are therefore widely used in industries where controlled dc power is required.

In the study of thyristor systems, SCRs and diodes are assumed ideal switches which means that (i) there is no voltage drop across them, (ii) no reverse current exists under reverse voltage conditions and (iii) holding current is zero.

Trigger circuits are not shown in SCR circuit for convenience.

In this chapter, single-phase and three-phase controlled converters are described and the effect of source inductance on their performance is examined. Basic operating features of dual converters are also presented.

6.1. PRINCIPLE OF PHASE CONTROL

The simplest form of controlled rectifier circuits consist of a single thyristor feeding dc power to a resistive load R as shown in Fig. 6.1 (a). The source voltage is $v_s = V_m \sin \omega t$, Fig. 6.1 (b). An SCR can conduct only when anode voltage is positive and a gating signal is applied. As such, a thyristor blocks the flow of load current i_0 until it is triggered. At some delay angle α, a positive gate signal applied between gate and cathode turns on the SCR. Immediately, full supply voltage is applied to the load as v_0, Fig. 6.1 (b). At the instant of delay angle α, v_0 rises from zero to $V_m \sin \alpha$ as shown. For resistive load, current i_0 is in phase with v_0. Firing angle of a thyristor is measured from the instant it would start conducting if it were replaced by a diode. In Fig. 6.1, if thyristor is replaced by diode, it would begin conduction at $\omega t = 0$, 2π, 4π etc. ; firing angle is therefore measured from these instants. A *firing angle* may thus be defined as the angle between the instant thyristor would conduct if it were a diode and the instant it is triggered.

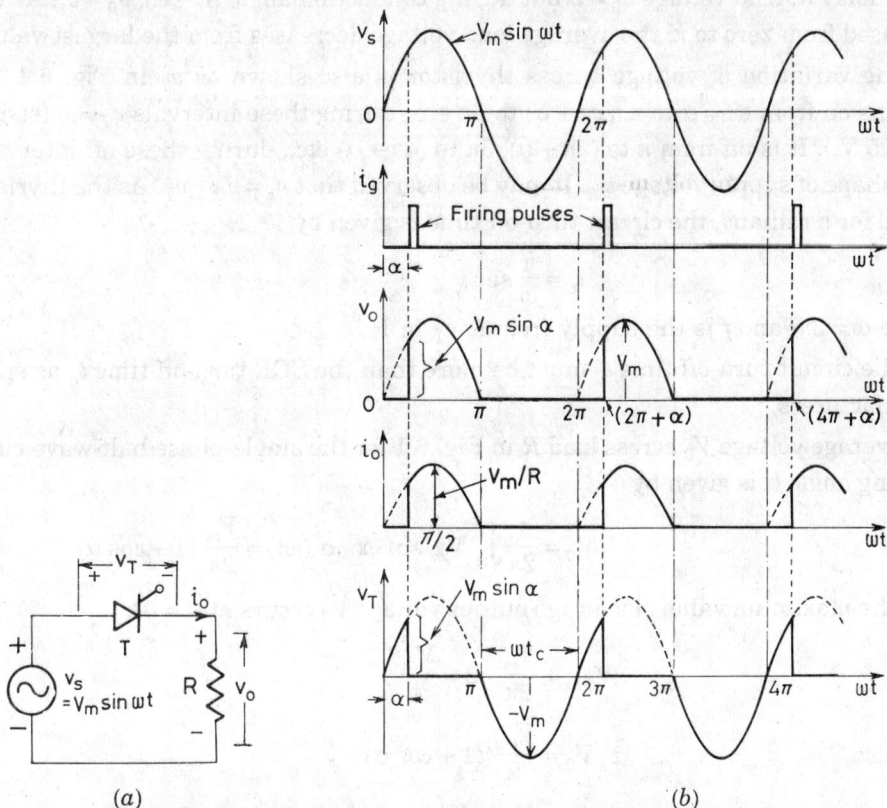

Fig. 6.1. Single-phase half-wave thyristor circuit with R load
(a) circuit diagram and (b) voltage and current waveforms.

A firing angle may also be defined as follows : *A firing angle* is measured from the angle that gives the largest average output voltage, or the highest load voltage. If thyristor in Fig. 6.1 is fired at $\omega t = 0$, 2π, 4π etc., the average load voltage is the highest ; the firing angle should thus be measured from these instants. A *firing angle* may thus be defined as the angle measured from the instant that gives the largest average output voltage to the instant it is triggered.

A critical observation of Fig. 6.1 leads to the emergence of another definition of firing angle. Thus, a *firing angle* may be defined as the angle measured from the instant SCR gets forward biased to the instant it is triggered.

Once the SCR is on, load current flows, until it is turned-off by reversal of voltage at $\omega t = \pi$, 3π etc. At these angles of π, 3π, 5π etc. load current falls to zero and soon after the supply voltage reverse biases the SCR, the device is therefore turned off. It is seen from Fig. 6.1 (*b*) that by varying the firing angle α, the *phase* relationship between the start of the load current and the supply voltage can be *controlled* ; hence the term *phase control* is used for such a method of controlling the load currents [3] .

A single-phase half-wave circuit is one which produces only *one* pulse of load current during one cycle of source voltage. As the circuit shown in Fig. 6.1 (*a*) produces only one load current pulse for one cycle of sinusoidal source voltage, this circuit represents a single-phase half-wave thyristor circuit.

In Fig. 6.1 (*b*), thyristor conducts from $\omega t = \alpha$ to π, $(2\pi + \alpha)$ to 3π and so on. Over the firing angle delay α, load voltage $v_0 = 0$ but during conduction angle $(\pi - \alpha)$, $v_0 = v_s$. As firing angle is increased from zero to π, the average load voltage decreases from the largest value to zero.

The variation of voltage across thyristor is also shown as v_T in Fig. 6.1 (*b*). Thyristor remains on from $\omega t = \alpha$ to π, $(2\pi + \alpha)$ to 3π etc., during these intervals $v_T = 0$ (strictly speaking 1 to 1.5 V). It is off from π to $(2\pi + \alpha)$, 3π to $(4\pi + \alpha)$ etc., during these off intervals v_T has the waveshape of supply voltage v_s . It may be observed that $v_s = v_0 + v_T$. As the thyristor is reverse biased for π radians, the circuit turn-off time is given by

$$t_c = \frac{\pi}{\omega} \text{ sec}$$

where $\omega = 2\pi f$ and f is the supply frequency in Hz.

The circuit turn-off time t_c must be more than the SCR turn-off time t_q as specified by the manufacturers.

Average voltage V_0 across load R in Fig. 6.1 for the single-phase half-wave circuit in terms of firing angle α is given by

$$V_0 = \frac{1}{2\pi} \int_\alpha^\pi V_m \sin \omega t \cdot d (\omega t) = \frac{V_m}{2\pi} (1 + \cos \alpha) \qquad \text{...(6.1)}$$

The maximum value of average output voltage V_0 occurs at $\alpha = 0°$.

$$\therefore \qquad\qquad V_{o \cdot m} = \frac{V_m}{2\pi} \cdot 2 = \frac{V_m}{\pi}$$

Also, $\qquad\qquad V_0 = \frac{V_{om}}{2} (1 + \cos \alpha)$

Average load current, $\qquad I_0 = \frac{V_0}{R} = \frac{V_m}{2\pi R} (1 + \cos \alpha) \qquad \text{...(6.2)}$

In some types of loads, one may be interested in rms value of load voltage V_{or}. Examples of such loads are electric heating and incandescent lamps. Rms voltage V_{or} in such cases is given by

$$V_{or} = \left[\frac{1}{2\pi} \int_\alpha^\pi V_m^2 \sin^2 \omega t \cdot d\,(\omega t) \right]^{1/2}$$

$$= \frac{V_m}{2\sqrt{\pi}} \left[(\pi - \alpha) + \frac{1}{2} \sin 2\alpha \right]^{1/2} \qquad \qquad ...(6.3)$$

The value of rms current I_{or} is

$$I_{or} = \frac{V_{or}}{R}$$

Power delivered to resistive load = (rms load voltage) (rms load current)

$$= V_{or} I_{or} = \frac{V_{or}^2}{R} = I_{or}^2 R \qquad \qquad ...(6.4)$$

Input voltamperes = (rms source voltage) (total rms line current)

$$= V_s \cdot I_{or} = \frac{\sqrt{2}\, V_s^2}{2R \sqrt{\pi}} \left[(\pi - \alpha) + \frac{1}{2} \sin 2\alpha \right]^{1/2}$$

Input power factor $= \dfrac{\text{Power delivered to load}}{\text{Input VA}} = \dfrac{V_{or} \cdot I_{or}}{V_s \cdot I_{or}} = \dfrac{V_{or}}{V_s}$

From Eq. (6.3), input $pf = \dfrac{1}{\sqrt{2\pi}} \left[(\pi - \alpha) + \dfrac{1}{2} \sin 2\alpha \right]^{1/2}$ $\qquad ...(6.5)$

6.1.1. Single-phase Half-wave Circuit with RL Load

A single-phase half-wave thyristor circuit with RL load is shown in Fig. 6.2 (a). Line voltage v_s is sketched in the top of Fig. 6.2 (b). At $\omega t = \alpha$, thyristor is turned on by gating signal (not shown). The load voltage v_0 at once becomes equal to source voltage v_s as shown. But the inductance L forces the load, or output, current i_0 to rise gradually. After some time, i_0 reaches maximum value and then begins to decrease. At $\omega t = \pi$, v_0 is zero but i_0 is not zero because of the load inductance L. After $\omega t = \pi$, SCR is subjected to reverse anode voltage but it will not be turned off as load current i_0 is not less than the holding current. At some angle $\beta > \pi$, i_0 reduces to zero and SCR is turned off as it is already reverse biased. After $\omega t = \beta$, $v_o = 0$ and $i_0 = 0$. At $\omega t = 2\pi + \alpha$, SCR is triggered again, v_0 is applied to the load and load current develops as before. Angle β is called the *extinction angle* and $(\beta - \alpha) = \gamma$ is called the *conduction angle*.

The waveform of voltage v_T across thyristor T in Fig. 6.2 (b) reveals that when $\omega t = \alpha$, $v_T = V_m \sin \alpha$; from $\omega t = \alpha$ to β, $v_T = 0$ and at $\omega t = \beta$, $v_T = V_m \sin \beta$. As $\beta > \pi$, v_T is negative at $\omega t = \beta$. Thyristor is therefore reverse biased from $\omega t = \beta$ to 2π. Thus, circuit turn-off time $t_C = \dfrac{2\pi - \beta}{\omega}$ sec. For satisfactory commutation, t_C should be more than t_q the thyristor turn-off time.

The voltage equation for the circuit of Fig. 6.2 (a), when T is on, is

$$V_m \sin \omega t = R\, i_0 + L \frac{di_0}{dt}$$

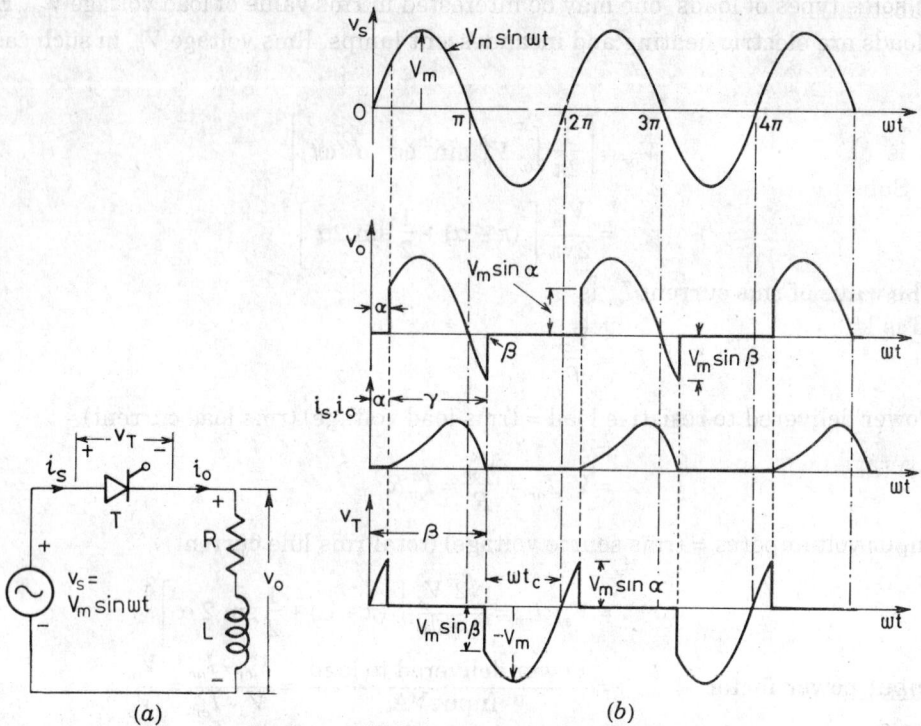

Fig. 6.2. Single-phase half-wave circuit with RL load
(a) circuit diagram and (b) voltage and current waveforms.

The load current i_0 consists of two components, one steady-state component i_s and the other transient component i_t. Here i_s is given by

$$i_s = \frac{V_m}{\sqrt{R^2 + X^2}} \sin(\omega t - \phi)$$

where $\phi = \tan^{-1} \dfrac{X}{R}$ and $X = \omega L$.   Here ϕ is the angle by which rms current I_s lags V_s.

The transient component i_t can be obtained from force-free equation

$$R i_t + L \frac{di_t}{dt} = 0$$

Its solution gives, $i_t = A e^{-(R/L)t}$

$\therefore$ $i_0 = i_s + t_t = \dfrac{V_m}{Z} \sin(\omega t - \phi) + A^{-(R/L)t}$ ...(6.6)

where $Z = \sqrt{R^2 + X^2}$

Constant A can be obtained from the boundary condition at $\omega t = \alpha$.

At this time $t = \dfrac{\alpha}{\omega}$, $i_0 = 0$. Thus, from Eq. (6.6),

$$0 = \frac{V_m}{Z} \sin(\alpha - \phi) + A e^{-R\alpha/L\omega}$$

or $A = -\dfrac{V_m}{Z} \sin(\alpha - \phi) e^{R\alpha/\omega L}$

Substitution of A in Eq. (6.6) gives

$$i_0 = \frac{V_m}{Z} \sin(\omega t - \phi) - \frac{V_m}{Z} \sin(\alpha - \phi) \exp\left\{ -\frac{R}{\omega L}(\omega t - \alpha) \right\} \qquad ...(6.7)$$

for $\qquad \alpha < \omega t < \beta$

It is also seen from the waveform of i_0 in Fig. 6.2 (b) that when $\omega t = \beta$, load current $i_0 = 0$. Substituting this in Eq. (6.7) gives

$$\sin(\beta - \phi) = \sin(\alpha - \phi). \exp\left\{ -\frac{R}{\omega L}(\beta - \alpha) \right\}$$

This transcendental equation can be solved to obtain the value of extinction angle β. In case β is known, average load voltage V_0 is given by

$$V_0 = \frac{1}{2\pi} \int_\alpha^\beta V_m \sin \omega t \, d(\omega t) = \frac{V_m}{2\pi}(\cos \alpha - \cos \beta) \qquad ...(6.8)$$

Average load current, $\quad I_0 = \frac{V_m}{2\pi R}(\cos \alpha - \cos \beta) \qquad\qquad ...(6.9)$

Rms load voltage, $\quad V_{or} = \left[\frac{1}{2\pi} \int_\alpha^\beta V_m^2 \sin^2 \omega t \cdot d(\omega t) \right]^{1/2}$

$$= \frac{V_m}{2\sqrt{\pi}} \left[(\beta - \alpha) - \frac{1}{2} \{ \sin 2\beta - \sin 2\alpha \} \right]^{1/2} \qquad ...(6.10)$$

Rms load current can be obtained from Eq. (6.7) if required.

6.1.2. Single-phase Half-wave Circuit with RL Load and Freewheeling Diode

The waveform of load current i_0 in Fig. 6.2 (b) can be improved by connecting a freewheeling (or flywheeling) diode across load as shown in Fig. 6.3 (a). A freewheeling diode is also called *by-pass* or *commutating diode*. At $\omega t = 0$, source voltage is becoming positive. At some delay angle α, forward biased SCR is triggered and source voltage v_s appears across load as v_0. At $\omega t = \pi$, source voltage v_s is zero and just after this instant, as v_s tends to reverse, freewheeling diode FD is forward biased through the conducting SCR. As a result, load current i_0 is immediately transferred from SCR to FD as v_s tends to reverse. At the same time, SCR is subjected to reverse voltage and zero current, it is therefore turned off at $\omega t = \pi$. It is assumed that during freewheeling period, load current does not decay to zero until the SCR is triggered again at $(2\pi + \alpha)$. Voltage drop across FD is taken as almost zero, the load voltage v_0 is, therefore, zero during the freewheeling period. The voltage variation across SCR is shown as v_T in Fig. 6.3 (b). It is seen from this wave-form that SCR is reverse biased from $\omega t = \pi$ to $\omega t = 2\pi$. Therefore, circuit turn-off time is

$$t_C = \frac{\pi}{\omega} \sec$$

The source current i_s and thyristor current i_T have the same waveform as shown.

Operation of the circuit of Fig. 6.3 (a) can be explained in two modes. In the first mode, called *conduction mode*, SCR conducts from α to π, $2\pi + \alpha$ to 3π and so on and FD is reverse biased. The duration of this mode is for $[(\pi - \alpha)/\omega]$ sec. Let the load current at the beginning of mode I be I_0. The expression for current i_0 in mode I can be obtained as follows :

Mode I : For conduction mode, the voltage equation is

$$V_m \sin \omega t = Ri_0 + L\frac{di_0}{dt}$$

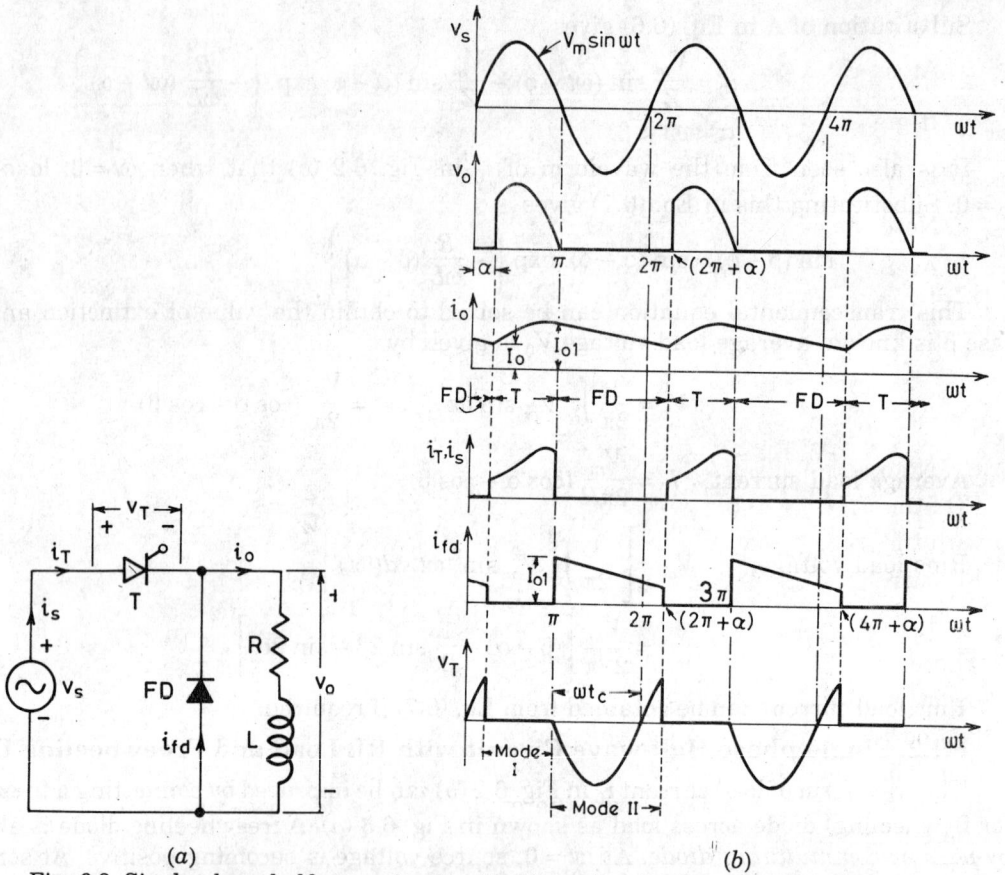

Fig. 6.3. Single-phase half-wave circuit with RL load and a freewheeling diode,
(a) circuit diagram and (b) voltage and current waveforms.

Its solution, already obtained in the previous section, is repeated here from Eq. (6.6) as

$$i_0 = \frac{V_m}{Z} \sin(\omega t - \phi) + A\, e^{-(R/L)t}$$

At $\omega t = \alpha$, $i_0 = I_0$, i.e. at $t = \dfrac{\alpha}{\omega}$, $i_0 = I_0$

$$\therefore \qquad A = \left[I_0 - \frac{V_m}{Z} \sin(\alpha - \phi) \right] e^{R\alpha/\omega L}$$

$$\therefore \qquad i_0 = \frac{V_m}{Z} \sin(\omega t - \phi) + \left[I_0 - \frac{V_m}{Z} \sin(\alpha - \phi) \right] \exp\left\{ -\frac{R}{L}\left(t - \frac{\alpha}{\omega} \right) \right\} \qquad ...(6.11)$$

Note that for mode I, $\alpha \le \omega t \le \pi$

Mode II : This mode, called *freewheeling mode*, extends from π to $2\pi + \alpha$, 3π to $4\pi + \alpha$ and so on. In this mode, SCR is reverse biased from π to 2π, 3π to 4π... as shown by voltage waveform v_T in Fig. 6.3 (b). As the load current is assumed continuous, FD conducts from π to $(2\pi + \alpha)$, 3π to $(4\pi + \alpha)$ and so on. Let the current at the beginning of mode II be I_{01} as shown. As load current is passing through FD, the voltage equation for mode II is

$$0 = Ri_0 + L \frac{di_0}{dt}$$

Its solution is $\qquad i_0 = A\, e^{-(R/L)\,t}$

At $\omega t = \pi,$ $\qquad\qquad i_0 = I_{01}.$

It gives $\qquad\qquad A = I_{01}\, e^{R\pi/\omega L}$

$\therefore \qquad\qquad i_0 = I_{01} \cdot \exp\left[-\frac{R}{L}\left(t - \frac{\pi}{\omega} \right) \right]$ ...(6.12)

Note that for mode II, $\pi < \omega t \leq (2\pi + \alpha)$

Average load voltage V_0 from Fig. 6.3 (b) is given by

$$V_0 = \frac{1}{2\pi} \int_\alpha^\pi V_m \sin \omega t \; d\,(\omega t) = \frac{V_m}{2\pi}\,(1 + \cos \alpha) \qquad ...(6.13)$$

Average load current, $\quad I_0 = \frac{V_0}{R} = \frac{V_m}{2\pi R}\,(1 + \cos \alpha)$ ...(6.14)

Note that load current i_0 is contributed by SCR from α to π, $(2\pi + \alpha)$ to 3π and so on and by FD from 0 to α, π to $(2\pi + \alpha)$ and so on. Thus the waveshape of thyristor current i_T is identical with the waveshape of i_0 for $\omega t = \alpha$ to π, $(2\pi + \alpha)$ to 3π and so on. Similarly, the wave shape of FD current i_{fd} is identical with the waveform of i_0 for $\omega t = 0°$ to α, π to $(2\pi + \alpha)$ and so on.

In Fig. 6.2, load consumes power p_1 from source for α to π (both v_0 and i_0 are positive) whereas energy stored in inductance L is returned to the source as power p_2 for π to β (v_0 is negative and i_0 is positive). As a result, net power consumed by the load is the difference of these two powers p_1 and p_2. In Fig. 6.3, load absorbs power for α to π, but for π to $(2\pi + \alpha)$, energy stored in L is delivered to load resistance R through the FD. As a consequence, power consumed by load is more in Fig. 6.3. It can, therefore, be concluded that power delivered to load, for the same firing angle, is more when FD is used. As volt-ampere input is almost the same in both Figs. 6.2 and 6.3, the input pf (= power delivered to load/input volt-ampere) with the use of FD is improved.

It is also seen from Figs. 6.2 (b) and 6.3 (b) that load current waveform is improved with FD in Fig. 6.3 (b). Thus the advantages of using freewheeling diode are

    (i) input pf is improved

    (ii) load current waveform is improved

    (iii) as a result of (ii), load performance is better and

    (iv) as energy stored in L is transferred to R during the freewheeling period, overall converter efficiency improves.

It may be seen from Fig. 6.3 (b) that freewheeling diode prevents the load voltage v_0 from becoming negative. Whenever load voltage tends to go negative, FD comes into play. As a result, load current is transferred from main thyristor to FD, allowing the thyristor to regain its forward blocking capability.

It is seen from Figs. 6.2 (b) and 6.3 (b) that supply current i_s taken from the source is unidirectional and is in the form of dc pulses. Single phase half-wave converter thus introduces a dc component into the supply line. This is undesirable as it leads to saturation of the supply transformer and other difficulties (harmonics etc.).

These shortcomings can be overcome to some extent by the use of single-phase fullwave circuits discussed in Art. 6.2.

6.1.3. Single-phase Half-wave Circuit with RLE Load

A single-phase half-wave controlled converter with RLE load is shown in Fig. 6.4 (a). The counter emf E in the load may be due to a battery or a dc motor. The minimum value of firing angle is obtained from the relation $V_m \sin \omega t = E$. This is shown to occur at an angle θ_1 in Fig. 6.4 (b), where

$$\theta_1 = \sin^{-1}(E/V_m) \qquad \qquad ...(6.15)$$

In case thyristor T is fired at an angle $\alpha < \theta_1$, then $E > V_s$, SCR is reverse biased and therefore it will not turn on. Similarly, maximum value of firing angle is $\theta_2 = \pi - \theta_1$, Fig. 6.4 (b). During the interval load current i_0 is zero, load voltage $v_0 = E$ and during the time i_0 is not zero, v_0 follows v_s curve. For the circuit of Fig. 6.4 (a) and with SCR T on, KVL gives the voltage differential equation as

$$V_m \sin \omega t = R\, i_0 + L \frac{d\, i_0}{dt} + E \qquad \qquad ...(6.16)$$

The solution of this equation is made up of two components ; namely steady-state current component i_s and the transient current component i_t. For convenience, i_s may be thought of as

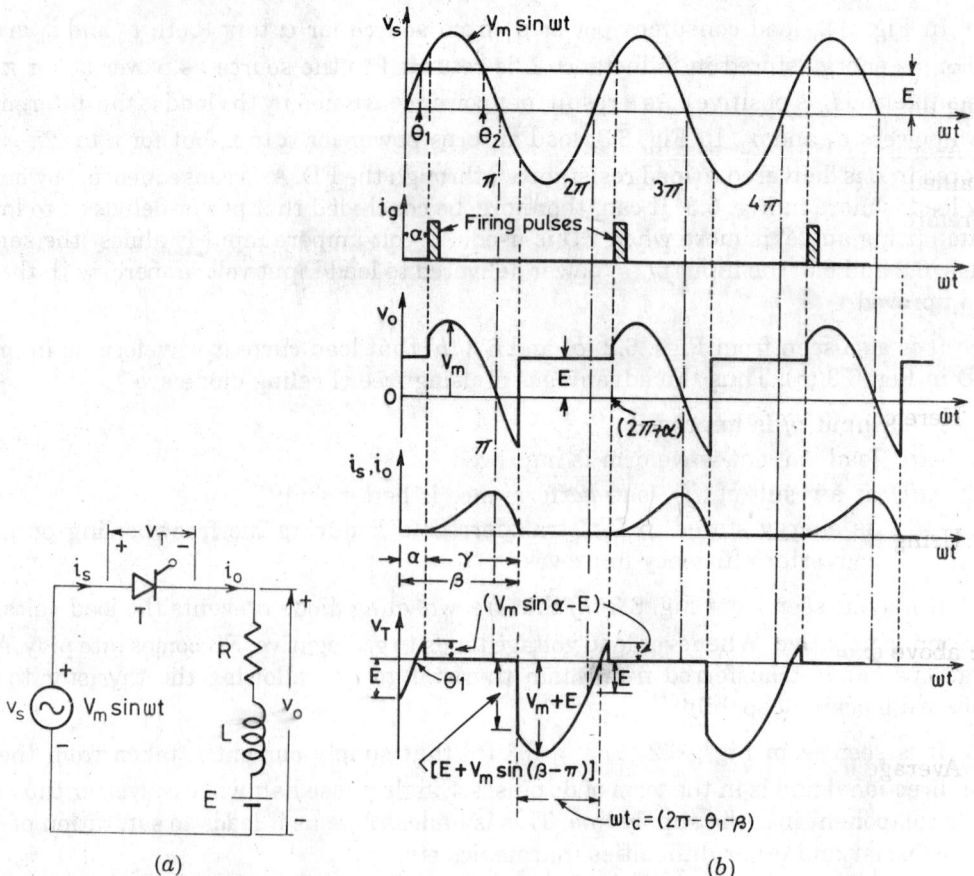

Fig. 6.4. Single-phase half-wave circuit with RLE load
(a) circuit diagram and (b) voltage and current waveforms.

the sum of i_{s1} and i_{s2}, where i_{s1} is the steady state current due to ac source voltage acting alone and i_{s2} is that due to dc counter emf E acting alone. As in the presentation leading to Eq. (6.6), i_{s1} due to source voltage $V_m \sin \omega t$ is given by

$$i_{s1} = \frac{V_m}{Z} \sin (\omega t - \phi)$$

If only E were present, then steady state current i_{s2} would be given by

$$i_{s2} = - (E/R)$$

The transient current i_t is given by $\quad i_t = A\, e^{-(R/L)t}$

Thus the total current i_0 is given by $\quad i_0 = i_{s1} + i_{s2} + i_t = \dfrac{V_m}{Z} \sin (\omega t - \phi) - \dfrac{E}{R} + A\, e^{-(R/L)t}$

At $\omega t = \alpha$, $i_0 = 0$, *i.e.* at $t = \dfrac{\alpha}{\omega}$, $i_0 = 0$. This gives $A = \left[\dfrac{E}{R} - \dfrac{V_m}{Z} \sin (\alpha - \phi) \right] e^{R\alpha/L\omega}$

$$\therefore \quad i_0 = \frac{V_m}{Z} \left[\sin (\omega t - \phi) - \sin (\alpha - \phi) \exp \left\{ -\frac{R}{\omega L} (\omega t - \alpha) \right\} \right] - \frac{E}{R} \left[1 - \exp \left\{ -\frac{R}{\omega L} (\omega t - \alpha) \right\} \right]$$

$$...(6.17)$$

Eq. (6.17) is applicable for $\alpha \le \omega t \le \beta$. The extinction angle β depends upon load emf E, firing angle α and the load impedance angle ϕ.

Average voltage across inductance is zero. Thus, average value of load current can be obtained by integrating $(V_m \sin \omega t - E)/R$ between α and β. The average load current I_0 is therefore given by

$$I_0 = \frac{1}{2\pi R} \left[\int_\alpha^\beta (V_m \sin \omega t - E)\, d(\omega t) \right]$$

$$= \frac{1}{2\pi R} [V_m (\cos \alpha - \cos \beta) - E (\beta - \alpha)] \qquad ...(6.18)$$

Here conduction angle $\gamma = \beta - \alpha$. Putting $\beta = \gamma + \alpha$ in Eq. (6.18) gives

$$I_0 = \frac{1}{2\pi R} [V_m \{\cos \alpha - \cos (\gamma + \alpha)\} - E \cdot \gamma]$$

Using the trigonometric relation,

$$\cos x - \cos y = 2 \sin \frac{x+y}{2} \sin \frac{y-x}{2}$$

the above expression for I_0 can be written as

$$I_0 = \frac{1}{2\pi R} \left[2\, V_m \sin \left(\alpha + \frac{\gamma}{2} \right) \sin \frac{\gamma}{2} - E \cdot \gamma \right] \qquad ...(6.19)$$

Average load voltage V_0 is given by

$$V_0 = E + I_0 R = E + \frac{1}{2\pi} \left[2V_m \sin \left(\alpha + \frac{\gamma}{2} \right) \sin \frac{\gamma}{2} - \gamma \cdot E \right]$$

$$= E \left(1 - \frac{\gamma}{2\pi} \right) + \frac{V_m}{\pi} \sin \left(\alpha + \frac{\gamma}{2} \right) \sin \frac{\gamma}{2} \qquad ...(6.20)$$

The above expression for the average load voltage V_0 can also be obtained as under :

For periodicity 2π, extending from α to $(2\pi + \alpha)$, we have

$$V_0 = \frac{1}{2\pi}\left[\int_\alpha^\beta V_m \sin \omega t \cdot d(\omega t) + E(2\pi + \alpha - \beta)\right]$$

$$= \frac{1}{2\pi}\left[V_m(\cos \alpha - \cos \beta) + E(2\pi + \alpha - \beta)\right]$$

In case β is made equal to $(\gamma + \alpha)$ in the above expression, Eq. (6.20) can be obtained.

If load inductance L is zero in Fig. 6.4 (a), then extinction angle β would be equal to $\theta_2 = \pi - \theta_1$, i.e. now β would be less than π. Average value of load current can still be obtained from Eq. (6.18) by substituting $\beta = \pi - \theta_1$. Therefore, average load current I_o, with $L = 0$, is

$$I_o = \frac{1}{2\pi R}\left[V_m(\cos \alpha - \cos(\pi - \theta_1)) - E(\pi - \theta_1 - \alpha)\right]$$

$$= \frac{1}{2\pi R}\left[V_m(\cos \alpha + \cos \theta_1) - E(\pi - (\theta_1 + \alpha))\right] \qquad ...(6.18)$$

Rms value of load current, with $L = 0$ in Fig. 6.4 (a), is given by

$$I_{or}^2 = \frac{1}{2\pi}\int_\alpha^\beta \left(\frac{V_m \sin \omega t - E}{R}\right)^2 d(\omega t)$$

$$= \frac{1}{2\pi R^2}\int_\alpha^\beta (V_m^2 \sin^2 \omega t + E^2 - 2 V_m E \sin \omega t)\, d(\omega t)$$

Its amplification gives

$$I_{or} = \left[\frac{1}{2\pi R^2}\left\{(V_s^2 + E^2)(\beta - \alpha) - \frac{V_s^2}{2}(\sin 2\beta - 2\sin 2\alpha) - 2 V_m E(\cos \alpha - \cos \beta)\right\}\right]^{1/2} \qquad ...(6.21)$$

Power delivered to load, $\quad P = I_{or}^2 R + I_o E.$ $\qquad ...(6.22)$

Supply power factor $\qquad = \dfrac{I_{or}^2 R + I_o \cdot E}{V_s \cdot I_{or}}$ $\qquad ...(6.23)$

The time variation of voltage across thyristor is shown as v_T in Fig. 6.4 (b). At $\omega t = 0$, $v_s = 0$, and therefore, $v_T = -E$. At $\omega t = \theta$, $v_s = E$, therefore $v_t = 0$. At $\omega t = \alpha$, $v_s = V_m \sin \alpha$, therefore $v_T = V_m \sin \alpha - E$. During the conduction angle $\gamma = (\beta - \alpha)$, $v_T = 0$. At $\omega t = \beta$, v_s has reverse polarity. Therefore, just after thyristor is turned off at $\omega t = \beta$, voltage $v_T = -[V_m \sin(\beta - \pi) + E]$. It is also possible to write $v_T = V_m \sin \beta - E$ at $\omega t = \beta$, because $V_m \sin \beta$ is negative for $\beta > \pi$. The magnitude of maximum reverse voltage is $(V_m + E)$ as shown in Fig. 6.4 (b). Fig. 6.4 (b) also reveals that circuit turn-off time is $\dfrac{2\pi + \theta_1 - \beta}{\omega}$ sec.

Example 6.1. *A single-phase 230 V, 1 kW heater is connected across 1-phase, 230 V , 50 Hz supply through an SCR. For firing angle delays of 45° and 90°, calculate the power absorbed in the heater element.*

Solution. Heater resistance $\quad R = \dfrac{(230)^2}{1000}\ \Omega$

From Eq. (6.3), the value of rms voltage for $\alpha = 45°$ is

$$V_{or} = \frac{\sqrt{2} \cdot 230}{2\sqrt{\pi}} \left[\left(\pi - \frac{\pi}{4} \right) + \frac{1}{2} \sin 90° \right]^{1/2} = 155.071 \text{ V}$$

$\therefore$ Power absorbed by heater element for $\alpha = 45°$ is

$$\frac{V_{or}^2}{R} = \left(\frac{155.071}{230} \right)^2 \times 1000 = 454.57 \text{ watts}$$

For $\alpha = 90°$, rms voltage is

$$V_{or} = \frac{\sqrt{2} \cdot 230}{2\sqrt{\pi}} \left[\left(\pi - \frac{\pi}{2} \right) + 0 \right]^{1/2} = 115 \text{ V}$$

$\therefore$ Power absorbed for $\alpha = 90°$ is

$$\frac{V_{or}^2}{R} = \left(\frac{115}{230} \right)^2 \times 1000 = 250 \text{ watts}.$$

Example 6.2. *A dc battery is charged through a resistor R as shown in Fig. 6.5 (a). Derive an expression for the average value of charging current in terms of V_m, E, R etc. on the assumption that SCR is fired continuously.*

(a) For an ac source voltage of 230 V, 50 Hz, find the value of average charging current for $R = 8 \Omega$ and $E = 150$ V.

(b) Find the power supplied to battery and that dissipated in the resistor.

(c) Calculate the supply pf.

Solution. For the circuit of Fig. 6.5 (a), the voltage equation is

$$V_m \sin \omega t = E + i_0 R$$

or

$$i_0 = \frac{V_m \sin \omega t - E}{R}$$

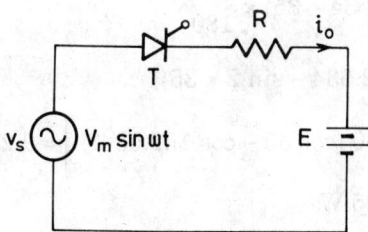

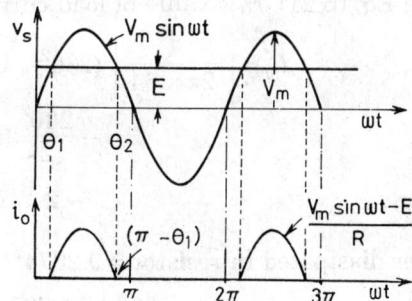

(a)   (b)

Fig. 6.5. (a) Power circuit diagram (b) various waveforms for Example 6.2.

It is seen from Fig. 6.5 that SCR is turned on when $V_m \sin \theta_1 = E$ and is turned off when $V_m \sin \theta_2 = E$, where $\theta_2 = \pi - \theta_1$. The battery charging requires only the average current I_0 given by

$$I_0 = \frac{1}{2\pi R} \left[\int_{\theta_1}^{\pi - \theta_1} (V_m \sin \omega t - E) \, d(\omega t) \right]$$

$$= \frac{1}{2\pi R} [2V_m \cos \theta_1 - E(\pi - 2\theta_1)]$$

(a) Here $\theta_1 = \sin^{-1} \dfrac{150}{\sqrt{2} \cdot 230} = 27.466°$

$\therefore$ $I_0 = \dfrac{1}{2\pi \cdot 8} \left[2 \cdot \sqrt{2} \cdot 230 \cos 27.466° - 150 \left(\pi - \dfrac{2 \times 27.496 \times \pi}{180} \right) \right] = 4.9676$ A.

(b) Power supplied to battery $= EI_0 = 150 \times 4.9676 = 745.14$ W.

For finding the power dissipated in R, rms value of charging current must by obtained. From Eq. (3.39),

$$I_{or} = \left[\frac{1}{2\pi \cdot 64} \left\{ (150^2 + 230^2) \left(\pi - 2 \times 27.466 \times \frac{\tau}{180} \right) + (230)^2 \sin 2 \times 27.466 \right. \right.$$
$$\left. \left. - 4 \cdot \sqrt{2} \cdot 230 \cdot 150 \cos 27.466° \right] \right]^{1/2} = 9.2955 \text{ A.}$$

$\therefore$ Power dissipated in resistor $= (9.2955)^2 \times 8 = 691.25$ Watts.

(c) From Eq. (6.23), supply pf $= \dfrac{691.25 + 745.14}{230 \times 9.2955} = 0.672$ lagging.

Example 6.3. *Repeat Example 6.2 in case thyristor is triggered at a firing angle of 35° in every positive half cycle.*

Solution. (a) Here $\alpha = 35°$, $\beta = \theta_2 = \pi - \theta_1 = 180 - 27.466° = 152.534°$

From Eq. (6.18), average charging current is given by

$$I_o = \frac{1}{2\pi \times 8} \left[\sqrt{2} \times 230 (\cos 35° - \cos 152.534°) - 150 (152.534 - 35°) \times \frac{\pi}{180} \right]$$

$$= 4.9192 \text{ A}$$

(b) Power delivered to battery $= EI_o = 150 \times 4.9192 = 737.88$ W

From Eq. (6.21), *rms* value of load current is

$$I_{or} = \left[\frac{1}{2\pi \times 64} \left\{ (230^2 + 150^2) (152.534 - 35) \times \frac{\pi}{180} \right. \right.$$
$$- \frac{230^2}{2} (\sin 2 \times 152.534 - \sin 2 \times 35)$$
$$\left. \left. - 2 \cdot \sqrt{2} \cdot 230 \times 150 (\cos 35 - \cos 152.534°) \right] \right] = 9.2874 \text{ A}$$

Power dissipated in resistor $= 9.2874^2 \times 8 = 690.05$ W

(c) Supply power factor $= \dfrac{690.05 + 737.88}{230 \times 9.2874} = 0.6685$ lagging

Examples 6.2 and 6.3 demonstrate that an increase in the firing angle reduces the value of average charging current, *rms* current and the supply power factor.

Example 6.4. *A 230 V, 50 Hz, one-pulse SCR controlled converter is triggered at a firing angle of 40° and the load current extinguishes at an angle of 210°. Find the circuit turn off time, average output voltage and the average load current for*

(a) $R = 5 \, \Omega$ and $L = 2mH$.

(b) $R = 5 \, \Omega$, $L = 2 \, mH$ and $E = 110 \, V$.

Solution. (a) For this part, refer to Fig. 6.2. It is seen from this figure that circuit turn off time t_c

$$= \frac{2\pi - \beta}{\omega} = \frac{(360 - 210)\,\pi}{180 \times 2\pi \times 50} = 8.333 \text{ m–sec}$$

From Eq. (6.8), average output voltage

$$V_0 = \frac{\sqrt{2} \cdot 230}{2\pi} \left[\cos 40° - \cos 210°\right] = 84.477 \text{ V}$$

Average load current $\quad I_0 = \frac{V_0}{R} = \frac{84.477}{5} = 16.8954 \text{ A.}$

(b) Fig. 6.4(b) shows that circuit turn-off time is

$$t_c = \frac{2\pi + \theta_1 - \beta}{\omega}$$

Here $\qquad \theta_1 = \sin^{-1}\frac{E}{V_m} = \sin^{-1}\frac{110}{\sqrt{2} \times 230} = 19.77°$

$\therefore \qquad t_c = \frac{(360 + 19.77 - 210)\,\pi}{180 \times 2\pi \times 50} = 9.432 \text{ ms.}$

From Eq. (6.18), the average charging current is

$$I_0 = \frac{1}{2\pi \cdot 5}\left[\sqrt{2} \cdot 230\,(\cos 40° - \cos 210°) - 110\,(210 - 40)\frac{\pi}{180}\right]$$

$$= 6.5064 \text{ A.}$$

$\therefore$ Average load voltage, $V_0 = E + I_0 R = 110 + 6.5064 \times 5 = 142.532 \text{ V.}$

Example 6.5. *A single-phase transformer, with secondary voltage of 230 V, 50 Hz, delivers power to load R = 10 Ω through a half-wave controlled rectifier circuit. For a firing-angle delay of 60°, determine (a) the rectification efficiency (b) form factor (c) voltage ripple factor (d) transformer utilization factor and (e) PIV of thyristor.*

Solution. Here $V_s = 230$ V, $f = 50$ Hz, $R = 10$ Ω $\alpha = 60°$

From Eq. (6.1), $\qquad V_o = \frac{V_m}{2\pi}\,(1 + \cos \alpha) = \frac{\sqrt{2} \times 230}{2\pi}\,(1 + \cos 60°) = 77.64 \text{ V}$

$$I_o = \frac{77.64}{10} = 7.76 \text{ A}$$

From Eq. (6.3), $\qquad V_{or} = \frac{\sqrt{2} \times 230}{2\sqrt{\pi}}\left[\pi - \frac{\pi}{3}\right) + \frac{1}{2}\sin 120°\right]^{1/2} = 145.873 \text{ V}$

$$I_{or} = \frac{145.873}{10} = 14.587 \text{ A}$$

Output *dc* power, $\qquad P_{dc} = V_o I_o = 7.64 \times 7.764 = 602.8 \text{ W}$

Output *ac* power, $\qquad P_{ac} = V_{or} I_{or} = 145.873 \times 14.587 = 2127.85 \text{ W}$

(a) Rectification efficiency $= \dfrac{P_{dc}}{P_{ac}} = \dfrac{602.8}{2127.85} = 0.2833$ or 28.33%

(b) Form factor, $\qquad FF = \dfrac{V_{or}}{V_o} = \dfrac{145.873}{77.64} = 1.879$

(c) Voltage ripple factor, VRF $= \sqrt{FF^2 - 1} = \sqrt{1.879^2 - 1} = 1.5908$

(d) $$\text{TUF} = \frac{V_o I_o}{V_s I_s} = \frac{V_o I_o}{V_s . I_{or}} = \frac{602.8}{230 \times 14.587} = 0.1797$$

(e) $$\text{PIV} = V_m = \sqrt{2} . V_s = \sqrt{2} \times 230 = 325.22 \text{ V}$$

6.2. FULL-WAVE CONTROLLED CONVERTERS

There is a large variety of SCR controlled converters (or rectifiers). One way of classifying these ac to dc converters is according to the number of supply phases on the input side. As per this classification, the ac to dc converters discussed in Figs. 6.1 to 6.4 are single-phase half-wave converters. Three-phase controlled rectifiers, as the name suggests, have three-phase supply on their input side, these are discussed later in this chapter. The other way of classification is according to the number of load current pulses per cycle of source voltage. It is seen from Art. 6.1 that single-phase half-wave controlled rectifiers produce only one pulse of load current during one cycle of source voltage, these can therefore be termed as single-phase one-pulse converters. Thus, the controlled rectifiers discussed in Figs. 6.1 to 6.4 are all single-phase one pulse converters.

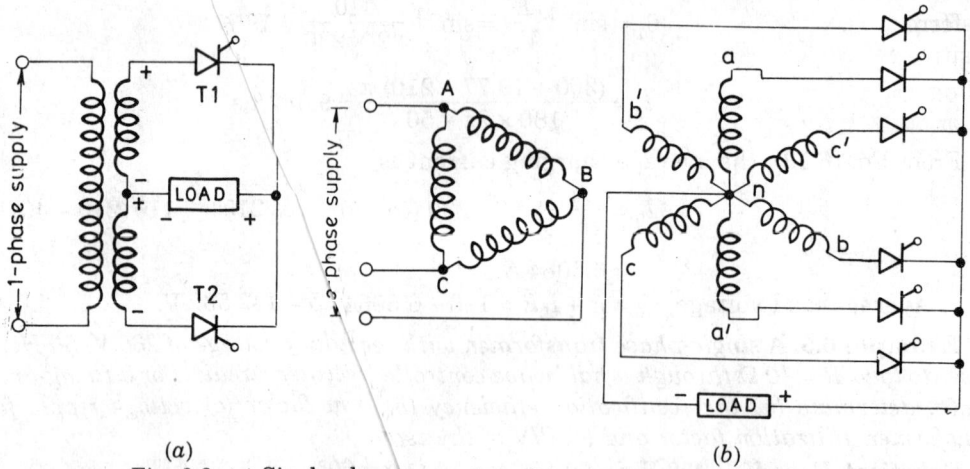

(a)   (b)

Fig. 6.6. (a) Single-phase two-pulse mid-point converter and
(b) three-phase six pulse mid-point converter.

The disadvantages of single-phase half-wave, or single-phase one-pulse converter, are minimised by the use of single-phase full wave, or single-phase two pulse, converters. In practice, there are two basic configurations for full-wave controlled converters. One configuration uses an input transformer with two windings for each input phase winding. This is called *mid-point converter*. A single-phase two-pulse mid-point SCR converter is shown in Fig. 6.6 (a) and a three-phase 6 pulse mid-point converter in Fig. 6.6 (b).

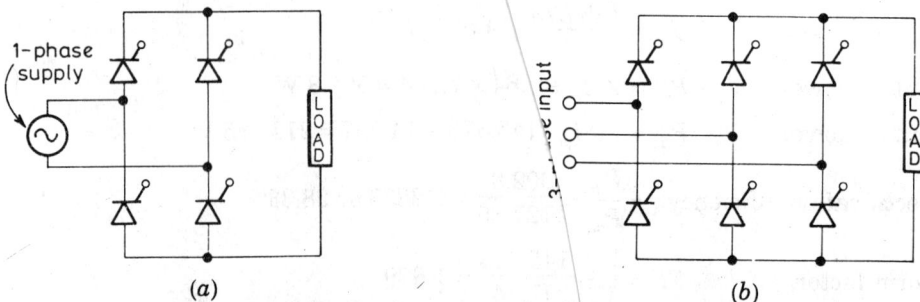

(a)   (b)

Fig. 6.7. (a) Single-phase two-pulse bridge converter and
(b) Three-phase six-pulse bridge converter.

The second configuration uses SCRs in the form of a bridge circuit. Single-phase full-wave, or two-pulse, bridge converter using four SCRs is shown in Fig. 6.7 (*a*) and a three-phase six-pulse bridge converter using six SCRs in Fig. 6.7 (*b*). A bridge converter has some advantages over mid-point converter, these will be discussed after both these configurations are studied in the next article.

6.3. SINGLE-PHASE FULL-WAVE CONVERTERS

In single-phase two-pulse (or full-wave) converters, voltage at the output terminals can be controlled by adjusting the firing angle delay of the thyristors. Mid-point or bridge-type circuits may be used for ac to dc conversion. In this section, first mid-point and then bridge-type configurations are discussed with input from single-phase source.

6.3.1. Single-phase Full-wave Mid-point Converter (M-2 Connection)

The circuit diagram of a single-phase full-wave converter using a centre-tapped transformer is shown in Fig. 6.8 (*a*). When terminal *a* is positive with respect to *n*, terminal *n* is positive with respect to *b*. Therefore, $v_{an} = v_{nb}$ or $v_{an} = -v_{bn}$ as *n* is the mid-point of secondary winding. Equivalent circuit of this arrangement is shown in Fig. 6.8 (*b*). It is assumed here that load, or output, current is continuous and turns ratio from primary to each secondary is unity.

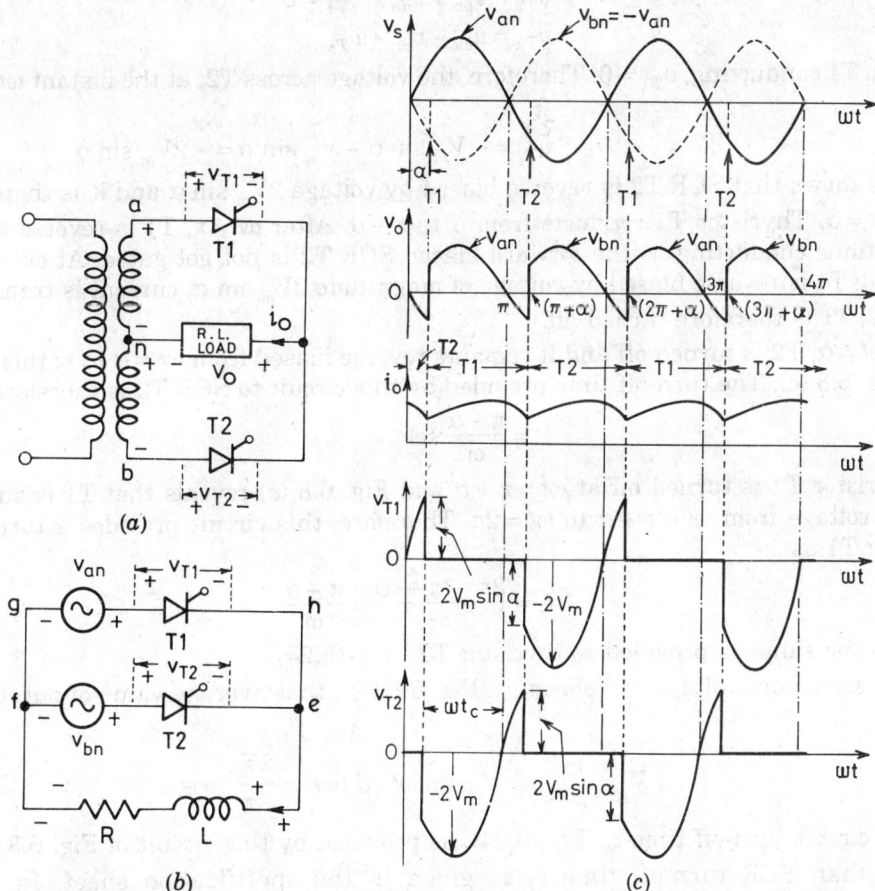

Fig. 6.8. Single-phase full-wave mid-point converter (*a*) circuit diagram (*b*) equivalent circuit and (*c*) various voltage and current waveforms.

Thyristors T1 and T2 are forward biased during positive and negative half cycles respectively ; these are therefore triggered accordingly. Suppose T2 is already conducting. After $\omega t = 0$, v_{an} is positive, T1 is therefore forward biased and when triggered at delay angle α, T1 gets turned on. At this firing angle α, supply voltage $2V_m \sin \alpha$ reverse biases T2, this SCR is therefore turned off. Here T1 is called the incoming thyristor and T2 the outgoing thyristor. As the incoming SCR is triggered, ac supply voltage applies reverse bias across the outgoing thyristor and turns it off. Load current is also transferred from outgoing SCR to incoming SCR. This process of SCR turn off by natural reversal of ac supply voltage is called *natural or line commutation.*

From the equivalent circuit of Fig. 6.8 (b), it is seen that if
$$v_{an} = V_m \sin \omega t,$$
then
$$v_{bn} = -v_{nb} = -V_m \sin \omega t$$
and
$$v_{ab} = v_{an} + v_{nb} = 2V_m \sin \omega t$$

When $\omega t = \alpha$, T1 is triggered. SCR $T2$ is subjected to a reverse voltage $v_{ab} = 2V_m \sin \alpha$ as stated before ; current is transferred from T2 to T1 and as a result T2 is turned off. The magnitude of reverse voltage across T2 can also be obtained by applying KVL to the loop *efghe* of the equivalent circuit of Fig. 6.8 (b) at the instant T1 is triggered. Thus,
$$v_{T2} - v_{bn} + v_{an} - v_{T1} = 0$$
or
$$v_{T2} = v_{bn} - v_{an} + v_{T1}$$

With T1 conducting, $v_{T1} = 0$. Therefore, the voltage across T2, at the instant $\omega t = \alpha$ is given by
$$v_{T2} = -V_m \sin \alpha - V_m \sin \alpha = -2V_m \sin \alpha$$

This shows that SCR T2 is reverse biased by voltage $2V_m \sin \alpha$ and it is therefore turned off at $\omega t = \alpha$. Thyristor T1 conducts from α to $\pi + \alpha$. After $\omega t = \pi$, T1 is reverse biased but it will continue conducting as the forward biased SCR T2 is not get gated. At $\omega t = \pi + \alpha$, T2 is triggered, T1 is reverse biased by voltage of magnitude $2V_m \sin \alpha$, current is transferred from T1 to T2, T1 is therefore turned off.

At $\omega t = \alpha$, T2 is turned off and it remains reverse biased from $\omega t = \alpha$ to π, this can be seen from Fig. 6.8 (c). The turn-off time provided by this circuit to SCR T2 is therefore given by
$$t_c = \frac{\pi - \alpha}{\omega} \sec \qquad \qquad ...(6.24)$$

Thyristor T1 is turned off at $\omega t = \pi + \alpha$ and Fig. 6.8 (c) reveals that T1 is subjected to a reverse voltage from $\omega t = \pi + \alpha$ to $\omega t = 2\pi$. Therefore, this circuit provides a turn-off time to thyristor T1 as
$$t_c = \frac{2\pi - (\pi + \alpha)}{\omega} = \frac{\pi - \alpha}{\omega}$$
which is the same as provided to thyristor T2 ; Eq. (6.24).

It is seen from voltage waveform v_0, Fig. 6.8 (c) , that average value of output voltage is given by
$$V_0 = \frac{1}{\pi} \int_{\alpha}^{\alpha + \pi} V_m \sin \omega t \cdot d\,(\omega t) = \frac{2V_m}{\pi} \cos \alpha \qquad \qquad ...(6.25)$$

The circuit turn-off time t_c, Eq. (6.24), as provided by this circuit of Fig. 6.8 (a) must be greater than SCR turn-off time t_q as given in the specification sheet. In case $t_c < t_q$, commutation failure will occur and the whole secondary winding will be short circuited. During commutation failure, if the rate of rise of fault current is high, the incoming SCR may be damaged in case protective elements do not clear the fault. Fig. 6.8 (c) reveals that each SCR is subjected to a peak voltage of $2V_m$.

The following observations can be made from the above study.

(i) When commutation of an SCR is desired, it must be reverse biased and the incoming SCR must be forward biased.

(ii) When incoming SCR is gated on, current is transferred from outgoing SCR to incoming SCR.

(iii) The circuit turn-off time must be greater than SCR turn-off time.

It is seen from above that thyristor commutation achieved by means of natural reversal of line voltage, called line or natural commutation, is simple ; it is therefore employed in all phase-controlled rectifiers, ac voltage controllers and cycloconverters.

6.3.2. Single-phase Full-wave Bridge Converters

Phase-controlled single-phase, or three-phase, full-wave converters are primarily of three types ; namely uncontrolled converters, half-controlled converters and fully-controlled converters. An *uncontrolled converter* or *rectifier* uses only diodes and the level of dc output voltage cannot be controlled. A *half-controlled converter* or semiconverter uses a mixture of diodes and thyristors and there is a limited control over the level of dc output voltage. A *fully-controlled converter* or *full converter* uses thyristors only and there is a wider control over the level of dc output voltage.

A semiconverter is *one-quadrant* converter. A one-quadrant converter has one polarity of dc output voltage and current at its output terminals, Fig. 6.9 (a). A *two-quadrant* converter is one in which voltage polarity can reverse but current direction cannot reverse because of the unidirectional nature of thyristors, Fig. 6.9 (b). In this part of the section, single-phase bridge type full converters and semiconverters are studied in detail.

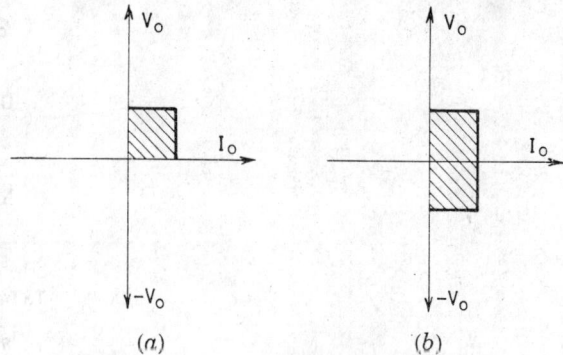

Fig. 6.9. (a) One-quadrant converter and (b) two-quadrant converter.

6.3.2.1. Single-phase full converter (B-2 Connection)

A single-phase full converter bridge using four SCRs is shown in Fig. 6.10 (a). The load is assumed to be of *RLE* type, where E is the load circuit emf. Voltage E may be due to a battery in the load circuit or may be generated emf of a dc motor. Thyristor pair T1, T2 is simultaneously triggered and π radians later, pair T3, T4 is gated together. When a is positive with respect to b, supply voltage waveform is shown as v_{ab} in Fig. 6.10 (b). When b is positive with respect to a, supply voltage waveform is shown dotted as v_{ba}. Obviously, $v_{ab} = -v_{ba}$. The current directions and voltage polarities shown in Fig. 6.10 (a) are treated as positive.

Load current or output current i_0 is assumed continuous over the working range ; this means that load is always connected to the ac voltage source through the thyristors. Between $\omega t = 0$ and $\omega t = \alpha$; T1, T2 are forward biased through already conducting SCRs T3 and T4 and block the forward voltage. For continuous current, thyristors T3, T4 conduct after $\omega t = 0$ even though these are reverse biased. When forward biased SCRs T1, T2 are triggered at $\omega t = \alpha$, they get turned on. As a result, supply voltage $V_m \sin \alpha$ immediately appears across thyristors T3, T4 as a reverse bias, these are therefore turned off by natural, or line, commutation. At the same time, load current i_0 flowing through T3, T4 is transferred to T1, T2 at $\omega t = \alpha$. Note that when T1, T2 are gated at $\omega t = \alpha$, these SCRs will get turned on only if $V_m \sin \alpha > E$. Thyristors T1, T2 conduct from $\omega t = \alpha$ to $\pi + \alpha$. In other words, $T1, T2$ conduct for π radians. Likewise, waveform of current i_{T1} through T_1 (or i_{T2} through T_2) is shown to flow π radians in Fig. 6.10 (b). At $\omega t = \pi + \alpha$, forward biased SCRs T3, T4 are triggered. The supply voltage turns off T1, T2 by natural commutation and the load current is transferred from T1, T2 to T3, T4.

Voltage across thyristors T1, T2 is shown as $v_{T1} = v_{T2}$ and that across T3, T4 as $v_{T3} = v_{T4}$. Maximum reverse voltage across T1, T2, T3 or T4 is V_m and at the instant of triggering with firing angle α, each SCR is subjected to a reverse voltage of $V_m \sin \alpha$. Source current i_s is treated as positive in the arrow direction. Under this assumption, source current is shown positive when T1, T2 are conducting and negative when T3, T4 are conducting, Fig. 6.10 (b).

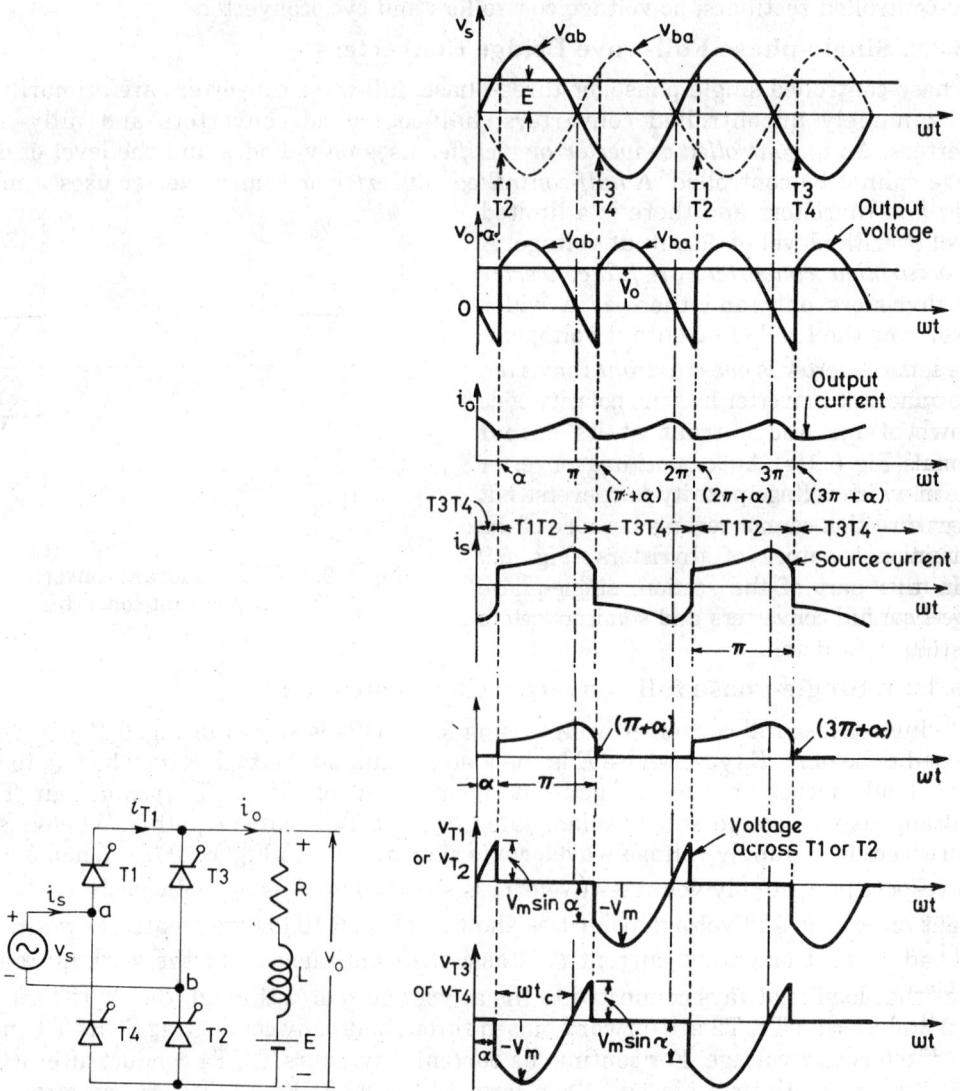

Fig. 6.10. (a) Single-phase full converter bridge with RLE load
(b) voltage and current waveforms for continuous load current.

During α to π, both v_s and i_s are positive, power therefore flows from ac source to load. During the interval π to $(\pi + \alpha)$, v_s is negative but i_s is positive, the load therefore returns some of its energy to the supply system. But the net power flow is from ac source to dc load because $(\pi - \alpha) > \alpha$ in Fig. 6.10 (b).

The load terminal voltage, or full-converter output voltage, v_0 is shown in Fig. 6.10 (b). The average value of output voltage V_0 is given by

$$V_0 = \frac{1}{\pi} \int_\alpha^{\pi+\alpha} V_m \sin \omega t \cdot d(\omega t) = \frac{2 V_m}{\pi} \cos \alpha \qquad \ldots(6.26)$$

Rms value of output voltage for single-phase M–2, or B–2, controlled converter can also be obtained as under.

$$V_{or}^2 = \frac{1}{\pi} \int_\alpha^{\pi+\alpha} V_m^2 \sin^2 \omega t \, d(\omega t)$$

$$= \frac{V_m^2}{2\pi} \left[\omega t - \frac{1}{2} \mid \sin 2 \omega t \mid_\alpha^{\pi+\alpha} \right] = \frac{V_m^2}{2} = V_s^2$$

$$\therefore \qquad V_{or} = V_s \qquad \ldots(6.27)$$

Eq. (6.26) shows that if $\alpha > 90°$, V_0 is negative. This is illustrated in Fig. 6.10 (c), where α is shown greater than 90°. In this figure, average terminal voltage V_0 is negative. If the load circuit emf E is reversed, this source E will feed power back to ac supply. This operation of full converter is known as inverter operation of the converter. The full converter with firing angle delay greater than 90° is called *line-commutated inverter*. Such an operation is used in the regenerative braking mode of a dc motor in which case then E is counter emf of the dc motor.

During 0 to α, ac source voltage v_s is positive but ac source current i_s is negative, power therefore flows from dc source to ac source. From α to π, both v_s and i_s are positive, power therefore, flows from ac source to dc source. But the net power flow is from dc source to ac source, because $(\pi - \alpha) < \alpha$ in Fig. 6.10 (c).

In converter operation, the average value of output voltage V_0 must be greater than load circuit emf E. During inverter operation, load circuit emf when inverted to ac must be more than ac supply voltage. In other words, dc source voltage E must be more than inverter voltage V_0, only then power would flow from dc source to ac supply system. But in both converter and inverter modes, thyristors must be forward

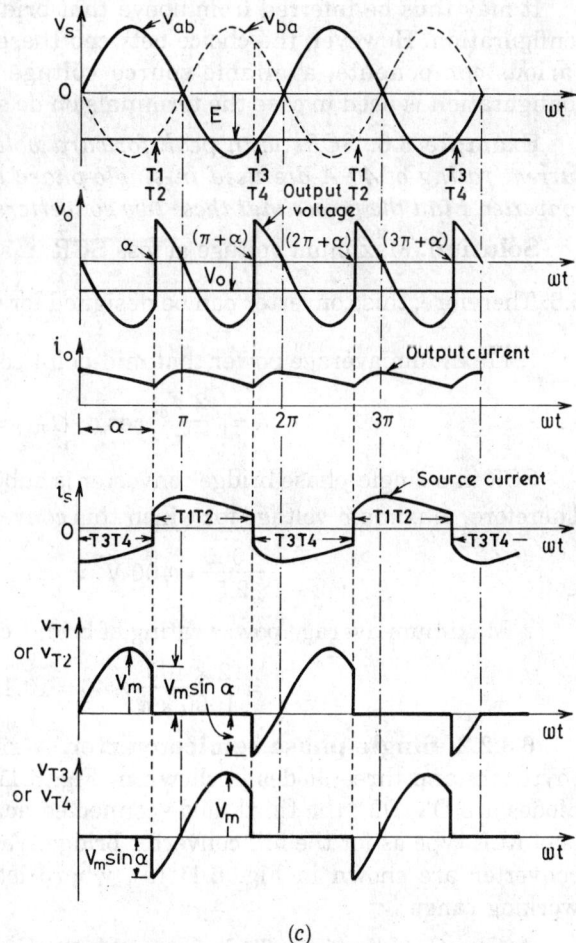

(c)

Fig. 6.10 (c) Voltage and current waveform for single-phase full converter of Fig. 6.10 (a) for $\alpha > 90°$.

biased and current through SCRs must flow in the same direction as these are unidirectional devices. This is the reason output current i_0 is shown positive in Fig. 6.10 (c). As before, source current i_s is positive when T1, T2 are conducting.

The variation of voltage across thyristors T1, T2, T3 or T4 reveals that circuit turn-off time for both converter and inverter operations is given by

$$t_c = \frac{\pi - \alpha}{\omega} \text{ sec}$$

As both the types of phase-controlled converters have been studied, the advantages of single-phase bridge converter over single-phase mid-point converter can now be stated :

(i) SCRs are subjected to a peak inverse voltage of $2\,V_m$ in mid-point converter and V_m in full converter. Thus for the same voltage and current ratings of SCRs, power handled by mid-point configuration is about half of that handled by bridge configuration, see Example 6.6.

(ii) In mid-point converter, each secondary should be able to supply the load power. As such, the transformer rating in mid-point converter is double the load rating. This, however, is not the case in single-phase bridge converter.

It may thus be inferred from above that bridge configuration is preferred over mid-point configuration. However, the choice between these two types depends primarily on cost of the various components, available source voltage and the load voltage required. Mid-point configuration is used in case the terminals on dc side have to be grounded.

Example 6.6. *SCRs with peak forward voltage rating of 1000 V and average on-state current rating of 40 A are used in single-phase mid-point converter and single-phase bridge converter. Find the power that these two converters can handle. Use a factor of safety of 2.5.*

Solution. Maximum voltage across SCR in single-phase mid-point converter is $2V_m$, Fig. 6.8. Therefore, this converter can be designed for a maximum voltage of $\dfrac{1000}{2 \times 2.5} = 200$ V.

∴ Maximum average power that mid-point converter can handle

$$= \left(\frac{2\,V_m}{\pi} \cos \alpha \right) I_{TAV} = \frac{2 \times 200}{\pi} \times 40 \times \frac{1}{1000} = 5.093 \text{ kW}$$

SCR in a single-phase bridge converter is subjected to a maximum voltage of V_m, Fig. 6.10. Therefore, maximum voltage for which this converter can be designed is

$$\frac{1000}{2.5} = 400 \text{ V}$$

∴ Maximum average power rating of bridge converter

$$= \frac{2 \times 400}{1000 \times \pi} \times 40 = 10.186 \text{ kW}.$$

6.3.2.2. Single-phase semiconverter. A single-phase semiconverter bridge with two thyristors and three diodes is shown in Fig. 6.11 (a). The two thyristors are T1, T2; the two diodes are D1, D2 ; the third diode connected across load is freewheeling diode FD. The load is of RLE type as for the full converter bridge. Various voltage and current waveforms for this converter are shown in Fig. 6.11 (b), where load current is assumed continuous over the working range.

After $\omega t = 0$, thyristor T1 is forward biased only when source voltage $V_m \sin \omega t$ exceeds E. Thus, T1 is triggered at a firing angle delay α such that $V_m \sin \alpha > E$. With T1 on, load gets

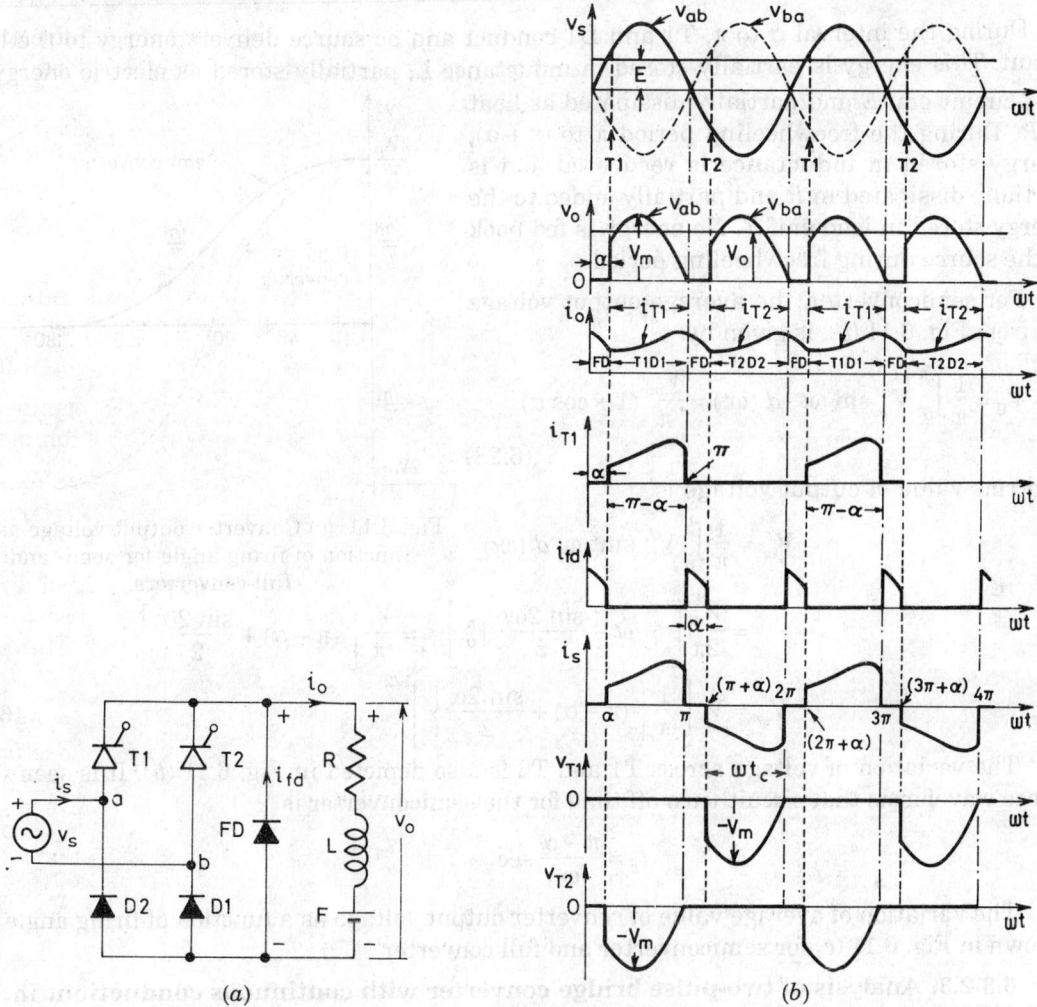

Fig. 6.11. Single-phase semiconverter bridge (a) Power-circuit diagram with RLE load and
(b) voltage and current waveforms for continuous load current.

connected to source through T1 and D1. For the period $\omega t = \alpha$ to π, load current i_0 flows through
RLE, D1, source and T1 and the load terminal voltage v_0 is of the same waveshape as the ac
source voltage v_s. Soon after $\omega t = \pi$, load voltage v_0 tends to reverse as the ac source voltage
changes polarity. Just as v_0 tends to reverse (at $\omega t = \pi +$), FD gets forward biased and starts
conducting. The load, or output, current i_0 is transferred from T1, D1 to FD. As SCR T1 is
reverse biased at $\omega t = \pi +$ through FD, T1 is turned off at $\omega t = \pi +$ The waveform of current
i_{T1} through thyristor T_1 is shown in Fig. 6.11 (b). It flows from α to π, $2\pi + \alpha$ to 3π and so on for
an interval of $(\pi - \alpha)$ radians. The load terminals are short circuited through FD, therefore load,
or output, voltage v_0 is zero during $\pi < \omega t < (\pi + \alpha)$. After $\omega t = \pi$, during the negative half cycle,
T2 will be forward biased only when source voltage is more than E. At $\omega t = \pi + \alpha$, source voltage
exceeds E, T2 is therefore triggered. Soon after $(\pi + \alpha)$, FD is reverse biased and is therefore
turned off ; load current now shifts from FD to T2, D2. At $\omega t = 2\pi$, FD is again forward biased
and output current i_0 is transferred from T2, D2 to FD as explained before. The source current
i_s is positive from α to π when T1, D1 conduct and is negative from $(\pi + \alpha)$ to 2π when T2, D2
conduct, see Fig. 6.11 (b).

During the interval α to π, T1 and D1 conduct and ac source delivers energy to the load circuit. This energy is partially stored in inductance L, partially stored as electric energy in load-circuit emf E and partially dissipated as heat in R. During the freewheeling period π to $(\pi + \alpha)$, energy stored in inductance is recovered and is partially dissipated in R and partially added to the energy stored in load emf E. No energy is fed back to the source during freewheeling period.

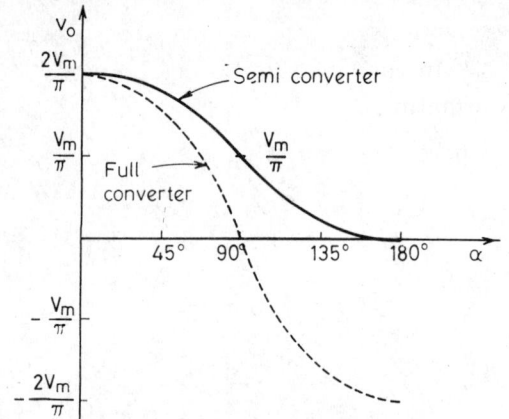

Fig. 6.11. (c) Converter output voltage as a function of firing angle for semi- and full-converters.

For semiconverter, the average output voltage V_0, from Fig. 6.11 (b), is given by

$$V_0 = \frac{1}{\pi} \int_\alpha^\pi V_m \sin \omega t \cdot d\,(\omega t) = \frac{V_m}{\pi}(1 + \cos \alpha)$$

...(6.28)

and rms value of output voltage is

$$V_{or}^2 = \frac{1}{\pi} \int_\alpha^\pi V_m^2 \sin^2 \omega t \, d\,(\omega t)$$

$$= \frac{V_m^2}{2\pi}\left[\, |\, \omega t - \frac{\sin 2\omega t}{2}\, |_\alpha^\pi \right] = \frac{V_s^2}{\pi}\left[(\pi - \alpha) + \frac{\sin 2\alpha}{2}\right]$$

$$\therefore \qquad V_{or} = V_s\left[\frac{1}{\pi}\left\{(\pi - \alpha) + \frac{\sin 2\alpha}{2}\right\}\right]^{1/2} \qquad\qquad ...(6.29)$$

The variation of voltage across T1 and T2 is also depicted in Fig. 6.11 (b). It is seen from these waveforms that circuit-turn off time for the semiconverter is

$$t_c = \frac{\pi - \alpha}{\omega} \text{ sec}$$

The variation of average value of converter output voltage as a function of firing angle α is shown in Fig. 6.11 (c) for semiconverter and full converter.

6.3.2.3. Analysis of two-pulse bridge converter with continuous conduction. In this part of the section, steady state analysis of single-phase two-pulse converter of both the types is presented.

Semiconverter. During conduction period, the voltage equation for the circuit of Fig. 6.11 (a), is

$$v_0 = v_s = R\,i_0 + L\frac{di_0}{dt} + E \qquad\qquad ...(6.30)$$

for
$$\alpha < \omega t \leq \pi$$

During freewheeling period, the voltage equation is

$$0 = R\,i_0 + L\frac{di_0}{dt} + E \qquad\qquad ...(6.31)$$

for
$$\pi < \omega t \leq (\pi + \alpha)$$

Eqs. (6.30) and (6.31) can be solved in time domain if required. Under steady-state operating conditions, only average value of output voltage and current are required. Therefore, from these two equations, we get steady-state solution as

$$V_0 = RI_0 + E \qquad\qquad ...(6.32)$$

where $\qquad\qquad V_0 = \dfrac{V_m}{\pi}(1 + \cos\alpha)$ = average voltage applied to the load

$\qquad\qquad\qquad I_0$ = average load current ; E = Load circuit emf

In case load is a dc motor, then $E = K_m\,\omega_m$; $R = r_a$, armature-circuit resistance ; $I_0 = I_a$, armature current ; $T_e = K_m\,I_a$,

where $\qquad\qquad\qquad T_e$ = electromagnetic torque in Nm.

$\qquad\qquad\qquad K_m$ = torque constant in Nm/A, or emf constant in V–sec/rad

$\therefore\qquad\qquad\qquad V_0 = r_a I_a + K_m\omega_m$

or $\qquad\qquad\qquad \omega_m = \dfrac{V_0 - r_a I_a}{K_m}$

or $\qquad\qquad\qquad \omega_m = \dfrac{(V_m/\pi)(1 + \cos\alpha)}{K_m} - \dfrac{r_a}{K_m^2}\,T_e$ $\qquad$ ...(6.33)

Full-converter. The voltage equation, for the circuit of Fig. (6.10) (a), is

$$v_0 = v_s = Ri_0 + L\dfrac{di_0}{dt} + E$$

Its average value, as in a semi-converter, is

$\qquad\qquad\qquad V_0 = RI_0 + E$ $\qquad\qquad\qquad$ ...(6.32)

where $\qquad\qquad\qquad V_0 = \dfrac{2\,V_m}{\pi}\cos\alpha$

In case load is a dc motor, $V_0 = r_a I_a + K_m\omega_m$

or $\qquad\qquad\qquad \omega_m = \dfrac{(2\,V_m/\pi)\cos\alpha}{K_m} - \dfrac{r_a}{K_m^2}\,T_e$ $\qquad$ ...(6.34)

A comparison of the waveforms for i_{T1} and i_o in Fig. 6.10 (b), for a single-phase full-converter, reveals that $i_{T1} = \dfrac{1}{2}\,i_o$ for continuous load current.

$\therefore$ Average value of thyristor current $= \dfrac{1}{2}$ (average value of load current)

or $\qquad\qquad\qquad I_{TA} = \dfrac{1}{2}I_o$ $\qquad\qquad\qquad$ ...(6.35)

This, however, is not the case in single-phase semiconverter.

Example 6.7. *A single-phase full converter bridge is connected to RLE load. The source voltage is 230 V, 50 Hz. The average load current of 10 A is constant over the working range. For R = 0.4 Ω and L = 2m H, compute*

(a) firing angle delay for E = 120 V,

(b) firing angle delay for E = – 120 V.

Indicate which source is delivering power to load in parts (a) and (b). Sketch the time variations of output voltage and load current for both the parts.

(c) In case output current is assumed constant, find the input pf for both parts (a) and (b).

Solution. (a) For $E = 120$ V, the full converter is operating as a controlled rectifier.

$$\therefore \quad \frac{2\,V_m}{\pi}\cos\alpha = E + I_0 R$$

or

$$\frac{2\cdot\sqrt{2}\cdot 230}{\pi}\cos\alpha = 120 + 10\times 0.4 = 124 \text{ V}$$

or

$$\alpha = 53.208 \cong 53.21°$$

For $\alpha = 53.21°$, power flows from ac source to dc load.

(b) For $E = -120$ V, the full converter is operating as a line commutated inverter.

$$\therefore \quad \frac{2\cdot\sqrt{2}\cdot 230}{\pi}\cos\alpha = -120 + 10\times 0.4 = -116 \text{ V} \quad \text{or } \alpha = 124.075 \cong 124.1°$$

For $\alpha = 124.1°$, the power flows from dc source to ac load.

Output voltage and load current waveforms for $\alpha = 53.21°$ can be drawn by referring to Fig. 6.10 (b) and for $\alpha = 124.1°$ from Fig. 6.10 (c).

(c) For constant load current, rms value of load current I_{or} is

$$I_{or} = I_0 = 10 \text{ A}$$

$$\therefore \quad V_s \cdot I_{or} \cos\phi = EI_0 + I_{or}^2\, R$$

For $\alpha = 53.21°$, $\cos\phi = \dfrac{120\times 10 + 10^2\times 0.4}{230\times 10} = 0.5391 \text{ lag}$

For $\alpha = 124.1°$, $\cos\phi = \dfrac{120\times 10 - 40}{230\times 10} = 0.5043 \text{ lag.}$

Example 6.8. (a) *A single-phase full converter delivers power to a resistive load R. For ac source voltage V_s, show that average output voltage V_0 is given by*

$$V_0 = \frac{\sqrt{2}\,V_s}{\pi}(1 + \cos\alpha)$$

Sketch the time variations of source voltage, output voltage, output current and voltage across one pair of SCRs. Hence find therefrom the circuit turn-off time.

(b) *For the converter of part (a), show that rms value of output current is given by*

$$I_{or} = \frac{V_s}{R}\left[\frac{1}{\pi}\left\{(\pi - \alpha) + \frac{1}{2}\sin 2\alpha\right\}\right]^{1/2}$$

Solution. Time variations of source voltage, load voltage and load current are shown in Fig. 6.12. At $\omega t = \pi$, $v_0 = v_s = 0$ and for resistive load R,

$$i_s = \frac{v_s}{R} = 0 \text{ and } i_0 = \frac{v_0}{R} = 0,$$

soon after $\omega t = \pi$, supply voltage reverse biases T1, T2; this pair is therefore turned off. When T3, T4 is triggered at $\omega t = \pi + \alpha$, output voltage $v_0 = v_s$ up to $\omega t = 2\pi$. Note that no SCR conducts during 0 to α, π to $(\pi + \alpha)$ and so on, Fig. 6.12. For the output voltage waveform v_0, average output voltage V_0 is

$$V_0 = \frac{1}{\pi}\int_\alpha^\pi V_m \sin\omega t \cdot d(\omega t)$$

$$= \frac{\sqrt{2}V_s}{\pi}(1 + \cos\alpha)$$

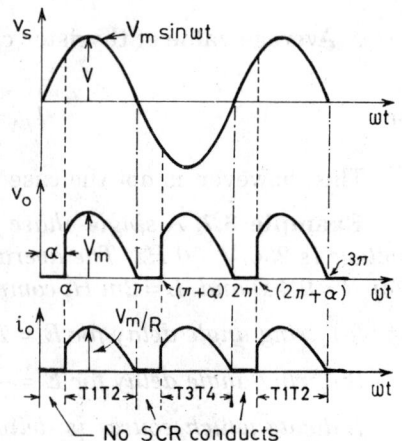

Fig. 6.12. Pertaining to Example 6.8.

The other waveforms can be drawn by referring to Fig. 6.10 (b).

(b) Rms value of the output current can be obtained from the waveform i_0 shown in Fig. 6.12.

$$I_{or} = \left[\frac{1}{\pi} \int_\alpha^\pi \left(\frac{V_m}{R} \sin \omega t \right)^2 d(\omega t) \right]^{1/2}$$

$$= \frac{V_s}{R} \left[\frac{1}{\pi} \int_\alpha^\pi (1 - \cos 2\omega t) \, d(\omega t) \right]^{1/2}$$

$$= \frac{V_s}{R} \left[\frac{1}{\pi} \left\{ (\pi - \alpha) + \frac{1}{2} \sin 2\alpha \right\} \right]^{1/2}$$

Example 6.9. (a) A single-phase controlled rectifier bridge consists of one SCR and three diodes as shown in Fig. 6.13 (a). Sketch output voltage waveform for a firing angle α for the SCR and hence obtain an expression for the average output voltage under the assumption of constant current. Show the conduction of various components as well.

(b) Draw waveforms of current through T1, D1, D2 and D3 assuming constant load current.

(c) For an ac source voltage of 230 V, 50 Hz and firing angle of 45°, find the average output current and power delivered to battery in case load consists of $R = 5 \, \Omega$, $L = 8 \, mH$ and $E = 100 \, V$.

Solution. (a) For the circuit of Fig. 6.13 (a), output voltage waveform v_0 is shown in Fig.

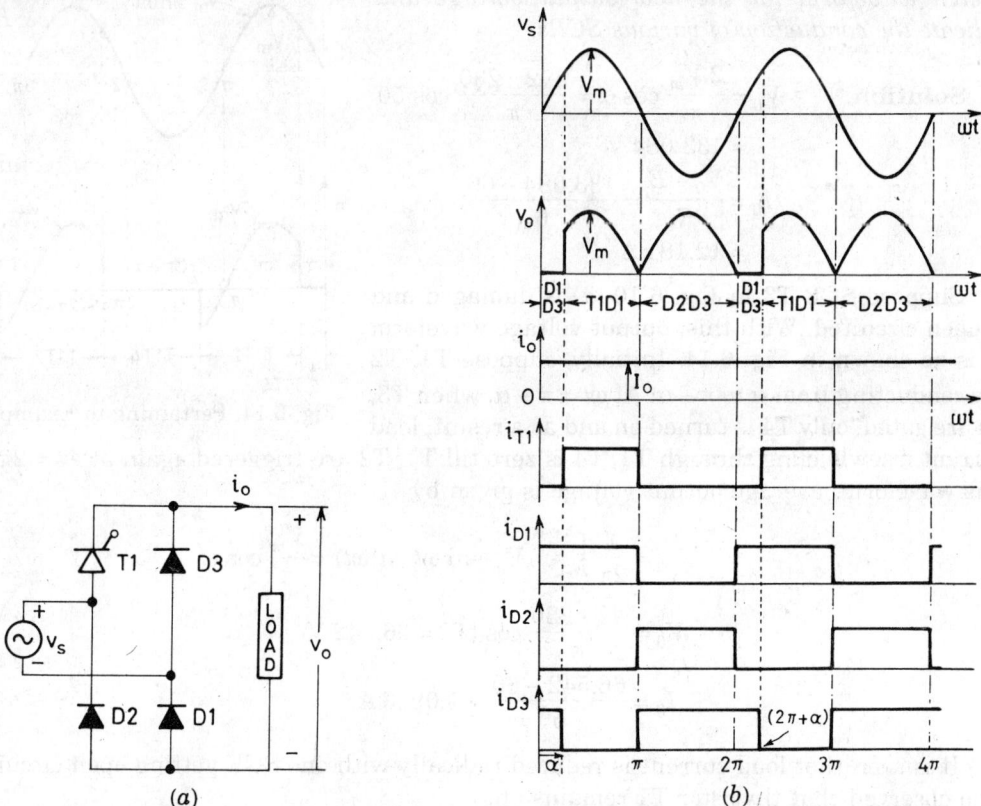

Fig. 6.13. (a) Circuit diagram for Example 6.9 (b) Various voltage and current waveforms.

6.13 (b). The conduction of various components is also indicated. It is seen that average value of v_0 is given by

$$V_0' = \frac{1}{2\pi}\left[\int_\alpha^\pi V_m \sin \omega t \cdot d(\omega t) - \int_\pi^{2\pi} V_m \sin \omega t \cdot d(\omega t)\right] = \frac{V_m}{2\pi}[3 + \cos \alpha]$$

(b) The conduction of various elements shown helps in drawing the waveforms for currents through T1, D1, D2 and D3. For example, D3 conducts from $\omega t = 0$ to α, from π to $2\pi + \alpha$, from 3π to $4\pi + \alpha$ and so on ; this is shown as i_{D3} in Fig. 6.13 (b).

(c)
$$V_0 = \frac{\sqrt{2} \cdot 230}{2 \cdot \pi}(3 + \cos 45°) = 191.88 \text{ V}$$

$$I_0 = \frac{191.88 - 100}{5} = 18.376 \text{ A}$$

Power delivered to battery $= E\,I_0 = \dfrac{100 \times 18.376}{1000} = 1.8376 \text{ kW.}$

Example 6.10. *A single-phase full converter feeds power to RLE load with $R = 6\,\Omega$, $L = 6$ mH and $E = 60$ V. The ac source voltage is 230 V, 50 Hz. For continuous conduction, find the average value of load current for a firing angle delay of 50°.*

In case one of the four SCRs gets open circuited due to a fault, find the new value of average load current taking the output current as continuous. Sketch waveform for the new output voltage and indicate the conduction of various SCRs.

Solution.
$$V_0 = \frac{2 V_m}{\pi}\cos \alpha = \frac{2\sqrt{2} \cdot 230}{\pi}\cos 50°$$

$$= 133.084 \text{ V}$$

$$I_0 = \frac{V_0 - E}{R} = \frac{133.084 - 60}{6}$$

$$= 12.181 \text{ A}$$

Suppose SCR T3 in Fig. 6.10 (a) is damaged and is open circuited. With this, output voltage waveform v_0 is as shown in Fig. 6.14. Initially, suppose T1, T2 are conducting from α to $\pi + \alpha$. At $\omega t = \pi + \alpha$, when T3, T4 are gated, only T4 is turned on and as a result, load

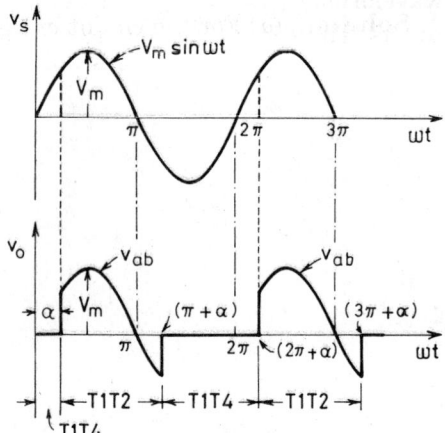

Fig. 6.14. Pertaining to Example 6.10.

current freewheeling through T1, T4 is zero till T1, T2 are triggered again at $\omega t = 2\pi + \alpha$. For this waveform, average output voltage is given by

$$V_0 = \frac{1}{2\pi}\int_\alpha^{\pi+\alpha} V_m \sin \omega t \cdot d(\omega t) = \frac{V_m}{\pi}\cos \alpha$$

$$V_0 = \frac{\sqrt{2} \cdot 230}{\pi}\cos 50° = 66.542 \text{ V}$$

$$I_0 = \frac{66.542 - 60}{6} = 1.0903 \text{ A.}$$

It is seen that load current is reduced radically with one SCR getting open circuited. It is also observed that thyristor T1 remains on.

6.4. SINGLE-PHASE TWO-PULSE CONVERTERS WITH DISCONTINUOUS LOAD CURRENT

So far, single-phase two-pulse converters have been studied on the assumption of continuous load current. In practice, the output current may become discontinuous at high values of firing angle or at low values of load current. The term *discontinuous* is applied to the condition when load current reaches zero during each half cycle before the next SCR in sequence is fired. The term *continuous* means that load current never ceases but continues to flow through SCR/diode or their combination. The load performance deteriorates if load current becomes discontinuous. It is therefore preferable to operate dc load in continuous current mode. This is promoted by having freewheeling action and using an external inductor in series with the load. In this article, working of both single-phase full converter and semiconverter is studied with their load current discontinuous.

6.4.1. Single-phase Full Converter with Discontinuous Current

Power circuit diagram for a single-phase full converter is shown in Fig. 6.10 (*a*). For this converter, when SCR pair T1 T2 is triggered at $\omega t = \alpha$, load current begins to build up from zero as shown. At some angle β, known as extinction angle, load current decays to zero. Here $\beta > \pi$. As T1 and T2 are reverse biased after $\omega t = \pi$, this pair is commutated at $\omega t = \beta$ when $i_0 = 0$. From α to β, output voltage v_0 follows source voltage v_s. From β to $(\pi + \alpha)$, no SCR conducts, the load voltage therefore jumps from $V_m \sin \beta$ to E as shown. At $\omega t = \pi + \alpha$, as pair T3 T4 is triggered, load current starts to build up again as before and load voltage v_0 follows v_s waveform as shown. At $\pi + \beta$, i_0 falls to zero, v_0 changes from $V_m \sin (\pi + \beta)$ to E as no SCR conducts. The source current i_s is also shown in Fig. 6.15 (*a*).

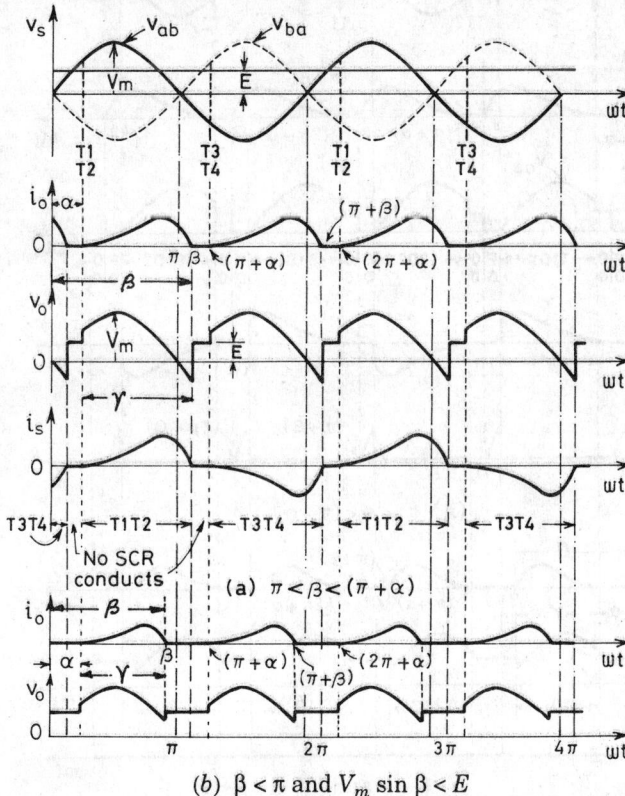

(a) $\pi < \beta < (\pi + \alpha)$

(b) $\beta < \pi$ and $V_m \sin \beta < E$

Fig. 6.15. Voltage and current waveforms for discontinuous load current for a single-phase full converter.

Under some conditions, load current may become zero at $\omega t = \beta$, where β is less than π. It is assumed here that $V_m \sin \beta < E$. At β, v_0 jumps from $V_m \sin \beta$ to E. The waveforms for load current i_0 and load voltage v_0 are shown in Fig. 6.15 (b). No SCR conducts from β to $(\pi + \alpha)$ and during this interval, therefore $v_0 = E$.

From above, the following observations can be made :

(i) Conduction period, $\alpha < \omega t < \beta$, T1, T2 conduct and $v_0 = v_s$. Also

$$(\pi + \alpha) < \omega t < (\pi + \beta), \text{ T3 T4 conduct and } v_o = v_s \text{ and so on.}$$

(ii) Idle period , $\beta < \omega t < (\pi + \alpha)$, no circuit element conducts and $v_0 = E$.

The output voltage during discontinuous current mode is less than given by Eq. (6.26). As stated before, load performance during discontinuous conduction is impaired.

6.4.2. Single-phase Semiconverter with Discontinuous Current

For this converter, power circuit diagram is given in Fig. 6.11 (a). For this controlled 2-pulse converter, when SCR T1 is triggered at $\omega t = \alpha$, load current builds up from zero, rises to a maximum and then decays to zero at $\beta > \pi$.

From α to π, T1 D1 conduct and $v_0 = v_s$. At $\omega t = \pi$, as v_s tends to become negative, FD is forward biased and starts conducting the load current. When FD conducts from π to β, $v_0 = 0$.

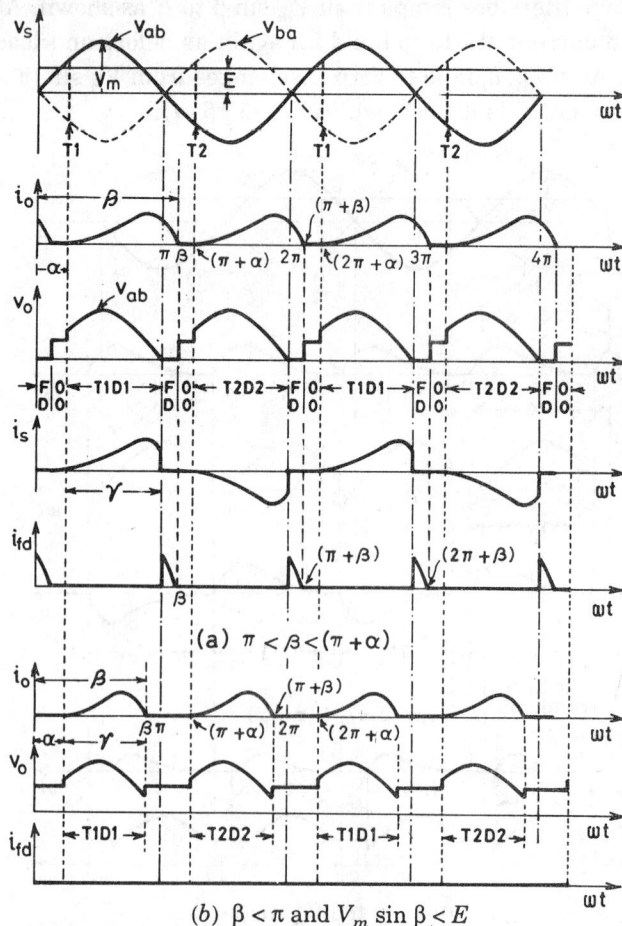

(a) $\pi < \beta < (\pi + \alpha)$

(b) $\beta < \pi$ and $V_m \sin \beta < E$

Fig. 6.16. Voltage and current waveforms for discontinuous conduction for a single-phase semiconverter.

From β to $\pi + \alpha$, no circuit component conducts, therefore $v_0 = E$ as shown in Fig. 6.16 (a). During β to $\pi + \alpha$, as load current is zero, this makes the load current discontinuous. When T2 is triggered at $\pi + \alpha$, i_0 builds as shown. At 2π, FD is forward biased and starts conducting till $\pi + \beta$. During the time FD conducts, $v_0 = 0$. From $\pi + \beta$ to $(2\pi + \alpha)$, no circuit component conducts, therefore $v_0 = E$. At $(2\pi + \alpha)$, T1 is triggered again and the above process repeats. Source current i_s is also shown in Fig. 6.16 (a).

In case load current becomes zero before π, *i.e.* for β less than π, then the current and voltage waveforms are shown in Fig. 6.16 (b). Here $V_m \sin \beta$ is assumed less than E. During β to $\pi + \alpha$, no circuit component conducts, therefore $v_0 = E$.

From the waveforms for single-phase semiconverter, the following observations are made:

(a) *When $\pi < \beta < \pi + \alpha$:*

(i) Conduction period, $\quad \alpha < \omega t < \pi$, T1D1 conduct and $v_0 = v_s$. Also

for $\pi + \alpha < \omega t < 2\pi$, T2D2 conduct and $v_0 = v_s$ and so on.

(ii) Freewheeling period, $\quad \pi < \omega t < \beta$, FD conducts, $i_{fd} = i_0$ and $v_0 = 0$. Also

for $2\pi < \omega t < \pi + \beta$, FD conducts, $i_{fd} = i_0$ and $v_0 = 0$ and so on.

(iii) Idle period, $\beta < \omega t < \pi + \alpha$, no circuit component conducts, $i_0 = 0$ and $v_0 = E$

(b) *When $\beta < \pi$ and $V_m \sin \beta < E$:*

(i) Conduction period, $\alpha < \omega t < \beta$, T1D1 conduct and $v_0 = v_s$. Also

for $\pi + \alpha < \omega t < \pi + \beta$, T2 D2 conduct and $v_0 = v_s$ and so on

(ii) Freewheeling period, absent and $i_{fd} = 0$.

(iii) Idle period, $\beta < \omega t < \pi + \alpha$ and $\pi + \beta < \omega t < 2\pi + \alpha$, no circuit element conducts, $i_0 = 0$ and $v_0 = E$.

The output voltage during discontinuous conduction is not given by Eq. (6.28). The load performance with discontinuous load current deteriorates as stated before.

Average output voltage and current. For single-phase *full converter,* for $\beta > \pi$ or $\beta < \pi$, the average load voltage $\bar{V}_o$, obtained from v_o waveform in Fig. 6.15 (a) or (b), is given by

$$V_o = \frac{1}{\pi} \left[\int_{\alpha}^{\beta} V_m \sin \omega t \, d\, (\omega t) + E\, (\pi + \alpha - \beta) \right]$$

$$= \frac{V_m}{\pi} (\cos \alpha - \cos \beta) + E \left(1 - \frac{\gamma}{\pi} \right)$$

where $\gamma = \beta - \alpha = $ conduction angle

Average load current, $I_o = \dfrac{V_o}{R} = \dfrac{V_m}{\pi R} (\cos \alpha - \cos \beta) + \dfrac{E}{R} \left(1 - \dfrac{\gamma}{\pi} \right)$. For single-phase *semiconverter* with $\beta > \pi$, the average output voltage V_o, obtained from v_o waveform in Fig. 6.16 (a), is given by

$$V_o = \frac{1}{\pi} \left[\int_{\alpha}^{\pi} V_m \sin \omega t \, . \, d\, (\omega t) + E\, (\pi + \alpha - \beta) \right]$$

$$= \frac{V_m}{\pi} (1 + \cos \alpha) + E \left(1 - \frac{\gamma}{\pi} \right)$$

Average load current, $\quad I_o = \dfrac{V_o}{R} = \dfrac{V_m}{\pi R} (1 + \cos \alpha) + \dfrac{E}{R} \left(1 + \dfrac{\gamma}{\pi} \right)$

6.5. PERFORMANCE PARAMETERS OF TWO-PULSE CONVERTERS

Here the performance parameters of single-phase full converter and single-phase semiconverter are derived from their input and output waveforms already obtained. The load, or output, current is assumed continuous during these derivations.

In general, the instantaneous input current to a converter can be expressed in Fourier series as under :

$$i(t) = a_o + \sum_{n = 1, 2, 3}^{\alpha} (a_n \cos n\omega t + b_n \sin n\omega t)$$

where

$$a_o = \frac{1}{2\pi} \int_o^{2\pi} i(t) \, d(\omega t)$$

$$a_n = \frac{1}{\pi} \int_0^{2\pi} i(t) \cos n\omega t \, d(\omega t) \text{ and } b_n = \frac{1}{\pi} \int_0^{2\pi} i(t). \sin n\omega t. \, d(\omega t)$$

The performance parameters are now obtained first for single-phase full-converter and then for single-phase semiconductor.

6.5.1. Single-phase full converter

In Fig. 6.10 (b), the variation of input current, or source current i_s, from α to $(\pi + \alpha)$, from $(\pi + \alpha)$ to $(2\pi + \alpha)$ is continuous but not constant. Here i_s is assumed to be ripple free with amplitude I_o during each half cycle, where I_o = constant load current.

$\therefore$

$$i_s(t) = I_o + \sum_{n = 1, 2, 3}^{\alpha} C_n \sin (n\omega t + \theta_n)$$

where

$$C_n = \sqrt{a_n^2 + b_n^2} \text{ and } \theta_n = \tan^{-1}\left(\frac{a_n}{b_n}\right)$$

Now

$$I_o = \frac{1}{2\pi} [\int_\alpha^{\pi + \alpha} I_o. \, d(\omega t) - \int_{\pi + \alpha}^{2\pi + \alpha} I_o. \, d(\omega t)] = 0$$

$$a_n = \frac{1}{\pi} [\int_0^{\pi + \alpha} I_o. \cos n\omega t. \, d(\omega t) - \int_{\pi + \alpha}^{2\pi + \alpha} I_o \cos n\omega t. \, d(\omega t)]$$

$$= \frac{I_o}{n\pi} [| \sin n\omega t |_\alpha^{\pi + \alpha} - | \sin n\omega t |_{\pi + \alpha}^{2\pi + \alpha}]$$

$$= -\frac{4 I_o}{n\pi} \sin n \alpha \dots\dots\dots \text{ for } n = 1, 3, 5,\dots$$

$$= 0 \dots\dots\dots \text{ for } n = 2, 4, 6,\dots\dots$$

$$b_n = \frac{1}{\pi} \left[\int_\alpha^{\pi + \alpha} I_o. \sin n\omega t. \, d(\omega t) - \int_{\pi + \alpha}^{2\pi + \alpha} I_o \sin n\omega t. \, d(\omega t) \right]$$

$$= \frac{I_o}{n\pi} [|- \cos n\omega t |_\alpha^{\pi + \alpha} - | - \cos n\omega t |_{\pi + \alpha}^{2\pi + \alpha}]$$

$$= \frac{4I_o}{n\pi} \cos n\alpha \dots\dots\dots\dots \text{ for } n = 1, 3, 5,\dots$$

$$= 0 \dots\dots\dots\dots\dots\dots \text{for } n = 2, 4, 6,\dots$$

$$C_n = \left[\left(-\frac{4 I_o}{n\pi} \sin n\alpha \right)^2 + \left(\frac{4 I_o}{n\pi} \cos n\alpha \right)^2 \right]^{1/2} = \frac{4 I_o}{n\pi}$$

$$\theta_n = \tan^{-1} [- \tan n\alpha] = - n\alpha$$

= displacement angle of nth harmonic current.

$$\therefore \quad i_s(t) = \sum_{n=1,3,5}^{\alpha} \frac{4 I_0}{n\pi} \sin(n\omega t - n\alpha) \quad \dots(6.36)$$

The rms value of nth harmonic input current, from Eq. (6.36), is

$$I_{sn} = \frac{4I_0}{\sqrt{2}.n\pi} = \frac{2\sqrt{2}.I_0}{n\pi} \quad \dots(6.36a)$$

Rms value of fundamental current, $I_{s1} = \dfrac{2\sqrt{2}.I_o}{\pi} = 0.90032\, I_o$

Rms value of total input current, $I_s = \left[\dfrac{I_o^2.\pi}{\pi} \right]^{1/2} = I_o$. Also $\theta_1 = -\alpha$

Negative sign for θ_1 indicates that fundamental current lags the source voltage. The various parameters are now obtained.

From Eq. (3.52), displacement factor, $DF = \cos\theta_1 = \cos(-\alpha) = \cos\alpha$ $\quad\dots(6.37)$

From Eq. (3.53), current distortion factor, $CDF = \dfrac{I_{s1}}{I_s} = \dfrac{2\sqrt{2}.I_o}{\pi}\gamma\dfrac{1}{I_o} = \dfrac{2\sqrt{2}}{\pi} = 0.90032$

Power factor, PF $= CDF \times DF = \dfrac{2\sqrt{2}}{\pi}\cos\alpha$ $\quad\dots(6.38)$

Harmonic factor HF or THD $= \left[\dfrac{1}{CDF^2} - 1 \right]^{1/2}$

$$= \left[\left(\frac{\pi}{2\sqrt{2}}\right)^2 - 1 \right]^{1/2} = 0.48343 \text{ or } 48.343\%$$

From Eq. (3.66 a), voltage ripple factor $= \left[\left(\dfrac{V_{or}}{V_o}\right)^2 - 1 \right]^{1/2}$

Substituting the values of V_{or} from Eq. (6.27) and V_o from Eq. (6.26), we get

$$\text{VRF} = \left[\frac{V_m^2}{2} \times \frac{\pi^2}{4 V_m^2.\cos^2\alpha} - 1 \right]^{1/2} = \left[\frac{\pi^2}{8\cos^2\alpha} - 1 \right]^{1/2} \quad \dots(6.39)$$

Active power input. Only the *rms* fundamental component of input current contributes to the active input power to the converter.

$\therefore$ Active power input, $\quad P_i = rms$ value of source voltage $\times$ *rms* fundamental component of source current $\times$ displacement factor

$$= V_s . I_{s1}. \cos\theta_1$$

$$= V_s\left(\frac{2\sqrt{2}\,I_o}{\pi}\right).\cos\alpha = \frac{2 V_m}{\pi} I_o \cos\alpha = V_o I_o \quad \dots(6.40)$$

Reactive power input, $\quad Q_i = V_s. I_{s1}. \sin\alpha$

$$= V_s\left(\frac{2\sqrt{2}\,I_o}{\pi}\right)\sin\alpha = \frac{2 V_m}{\pi} I_o \sin\alpha \quad \dots(6.41\,a)$$

$$= \frac{V_o}{\cos\alpha} I_o \sin\alpha = V_o I_o \tan\alpha \quad \dots(6.41\,b)$$

6.5.2. Single-phase semiconverter

For this converter, the variation of input current i_s, in Fig. 6.11 (b), is shown continuous from α to π, $(\pi + \alpha)$ to 2π and so on, but i_s is not constant. As before, i_s is now assumed ripple free also with its amplitude I_o.

Here
$$I_o = \frac{1}{2\pi} [\int_\alpha^\pi I_o . \, d\,(\omega t) - \int_{\pi+\alpha}^{2\pi} I_o \, d\,(\omega t)] = 0$$

$$a_n = \frac{1}{\pi} [\int_\alpha^\pi I_o \cos n\omega t \, d\,(\omega t) - \int_{\pi+\alpha}^{2\pi} I_o \cos n\omega t. \, d\,(\omega t)]$$

$$= \frac{I_o}{n\pi} [\mid \sin n\omega t \mid_\alpha^\pi - \mid \sin n\omega t \mid_{n+\alpha}^{2\pi}]$$

$$= -\frac{2 I_o}{n\pi} \sin n\,\alpha \dots\dots\dots \text{ for } n = 1, 3, 5,\dots$$

$$= 0 \dots\dots\dots\dots\dots\dots\dots \text{for } n = 2, 4, 6, \dots$$

$$b_n = \frac{1}{\pi} [\int_\alpha^\pi I_o \sin n\omega t . \, d\,(\omega t) - \int_{n+\alpha}^{2\pi} I_o . \sin n\omega t. \, d\,(\omega t)]$$

$$= \frac{2 I_o}{n\pi} (1 + \cos n\,\alpha) \dots\dots \text{ for } n = 1, 3, 5,\dots$$

$$= 0 \dots\dots\dots\dots\dots\dots\dots \text{for } n = 2, 4, 6,\dots$$

$$C_n = \left[\left(-\frac{2 I_o}{n\pi} \sin n\,\alpha \right)^2 + \left(\frac{2 I_o}{n\pi} (1 + \cos n\,\alpha) \right)^2 \right]^{1/2}$$

$$= \frac{2\sqrt{2}}{n\pi} (1 + \cos n\alpha)^{1/2}$$

It is known that $1 + \cos n\,\theta = 2 \cos^2 \dfrac{n\theta}{2}$

$\therefore$
$$C_n = \frac{2\sqrt{2}}{n\pi} \left[2\cos^2 \frac{n\,\alpha}{2} \right]^{1/2} = \frac{4 I_o}{n\pi} \cos \frac{n\alpha}{2}$$

$$\theta_n = \tan^{-1} \left(\frac{a_n}{b_n} \right) = \tan^{-1} \left[-\frac{\sin n\alpha}{1 + \cos n\alpha} \right]$$

$$= \tan^{-1} \left[-\frac{2\sin \dfrac{n\alpha}{2} \cos \dfrac{n\alpha}{2}}{2\cos^2 \dfrac{n\alpha}{2}} \right] = -\frac{n\alpha}{2}$$

$\therefore$
$$i_s (t) = \sum_{n = 1, 3, 5}^\infty \frac{4 I_o}{n\pi} \cos \frac{n\alpha}{2} . \sin \left(n\omega t - \frac{n\alpha}{2} \right) \qquad \dots(6.42)$$

Rms value of nth harmonic input current, from Eq. (6.42), is

$$I_{sn} = \frac{4 I_o}{\sqrt{2} . n\pi} \cos \frac{n\alpha}{2} = \frac{2\sqrt{2} I_o}{n\pi} . \cos \frac{n\alpha}{2} \qquad \dots(6.43)$$

Rms fundamental circuit, $I_{s1} = \dfrac{2\sqrt{2} . I_o}{\pi} . \cos \dfrac{\alpha}{2}$

Rms value of total input current, $I_s = \left[\dfrac{I_o^2\,(\pi - \alpha)}{\pi}\right]^{1/2} = I_o\left[\dfrac{\pi - \alpha}{\pi}\right]^{1/2}$

Also, $\qquad\qquad \theta_1 = -\dfrac{\alpha}{2}$

$\therefore \qquad\qquad DF = \cos\theta_1 = \cos\left(-\dfrac{\alpha}{2}\right) = \cos\dfrac{\alpha}{2}$

$$\text{CDF} = \frac{I_{s1}}{I_s} = \frac{2\sqrt{2}\,.\,I_o}{\pi}\cos\frac{\alpha}{2}\frac{\sqrt{\pi}}{I_o\sqrt{\pi - \alpha}} = \frac{2\sqrt{2}\cos\dfrac{\alpha}{2}}{\sqrt{\pi}\,(\pi - \alpha)} \qquad ...(6.44)$$

$$\text{HF or THD} = \left[\frac{1}{\text{CDF}^2} - 1\right]^{1/2} = \left[\frac{\pi(\pi - \alpha)}{8\cos^2\dfrac{\alpha}{2}} - 1\right]^{1/2}$$

$$= \left[\frac{\pi\,(\pi - \alpha)}{4\,(1 + \cos\alpha)} - 1\right]^{1/2} \qquad ...(6.45)$$

Power factor $= \text{CDF} \times DF = \dfrac{2\sqrt{2}\,.\,\cos\dfrac{\alpha}{2}}{\sqrt{\pi}\,(\pi - \alpha)} \times \cos\dfrac{\alpha}{2}$

$$= 2\left[\frac{2}{\pi\,(\pi - \alpha)}\right]^{1/2}\cos^2\frac{\alpha}{2} = \left[\frac{2}{\pi\,(\pi - \alpha)}\right]^{1/2}.\,(1 + \cos\alpha) \qquad ...(6.46)$$

$$VRF = \left[\left(\frac{V_{or}}{V_o}\right)^2 - 1\right]^{1/2}$$

$$= \left[\frac{V_m^2}{2\pi}\left[(\pi - \alpha) + \frac{\sin 2\alpha}{2}\right] \times \frac{\pi^2}{V_m^2\,(1 + \cos\alpha)^2} - 1\right]^{1/2}$$

$$= \left[\frac{\left[(\pi - \alpha) + \dfrac{1}{2}\sin 2\,\alpha\right]\pi}{2\,(1 + \cos\alpha)^2} - 1\right]^{1/2} \qquad ...(6.47)$$

Power input, $\quad P_i = V_s.\,I_{s1}.\cos\theta_1$

$$= V_s.\,\frac{2\sqrt{2}\,.\,I_o}{\pi}\cos\frac{\alpha}{2}.\cos\frac{\alpha}{2} = \frac{V_s\,.\,I_o\,.\,\sqrt{2}}{\pi}\,(1 + \cos\alpha)$$

$$= \frac{V_m}{\pi}\,(1 + \cos\alpha).\,I_o = V_o\,I_o \qquad ...(6.48)$$

Reactive power, $\quad Q_i = V_s.\,I_{s1}.\sin\theta_1$

$$= V_s.\,\frac{2\sqrt{2}\,I_o}{\pi}\cos\frac{\alpha}{2}\sin\frac{\alpha}{2} = \frac{V_m}{\pi}\,I_o.\sin\alpha \qquad ...(6.49\,a)$$

$$= \frac{1}{2}\,[\text{Reactive power required in 1-phase full converter for the same } I_0]$$

Also, $\qquad V_o = \dfrac{V_m}{\pi}\,(1 + \cos\alpha)$ or $\dfrac{V_m}{\pi} = \dfrac{V_o}{1 + \cos\alpha}$

$$\therefore \qquad Q_i = \frac{V_o\,I_o}{1 + \cos\alpha}.\sin\alpha = V_o\,I_o.\tan\frac{\alpha}{2} \qquad ...(6.49\,b)$$

Example 6.11. *A single-phase full converter, connected to 230 V, 50 Hz source, is feeding a load R = 10 Ω in series with a large inductance that makes the load current ripple free. For a firing angle of 45°, calculate the input and output performance parameters of this converter.*

Solution. From Eq. (6.26), $V_o = \dfrac{2V_m}{\pi} \cos \alpha = \dfrac{2 \times \sqrt{2} \times 230}{\pi} \cos 45° = 146.423$ V

$$I_o = \frac{V_o}{R} = \frac{146.423}{10} = 14.6423 \text{ A}$$

From Eq. (6.27), $V_{or} = \dfrac{V_m}{\sqrt{2}} = V_s = 230$ V

As load current is ripple free, rms value of load current, $I_{or} = I_o = 14.6423$ A.

From Eq. (3.60), $P_{dc} = V_o I_o = 146.423 \times 14.6423 = 2143.97$ W

Rms value of source current, $I_s = I_o = I_{or} = 14.6423$ A

Output ac power, $P_{ac} = V_{or} \times I_{or} = 230 \times 14.6423 = 3367.73$ W

From Eq. (3.61), rectification efficiency $= \dfrac{P_{dc}}{P_{ac}} = \dfrac{2143.97}{3367.73} = 0.6366$ or 63.66%

From Eq. (3.64), $FF = \dfrac{V_{or}}{V_o} = \dfrac{230}{146.423} = 1.5708$

From Eq. (3.66 a), $\text{VRF} = \sqrt{FF^2 - 1} = \sqrt{1.5708^2 - 1} = 1.2114$

As load current is ripple free, CRF = 0

From Eq. (6.36 a), $I_{s1} = \dfrac{2\sqrt{2}}{\pi} \times 14.6423 = 13.183$ A

$$\theta_1 = -\alpha = -45°$$

From Eq. (6.37), $\text{DF} = \cos \alpha = \cos (+45) = 0.707$

Rms value of total input current, $I_s = I_o = 14.6423$ A

$$CDF = 0.90032, PF = CDF \times DF = 0.90032 \times 0.707 = 0.63653 \text{ lag}$$

$$THD \text{ or } HF = \left[\frac{1}{CDF^2} - 1 \right]^{1/2} = \left[\frac{1}{0.90032^2} - 1 \right] = 0.48342$$

From Eq. (6.40), active power $= V_o I_o = 146.423 \times 14.6423 = 2143.97$ W

From Eq. (6.41 a), reactive power

$$= \frac{2 V_m}{\pi} I_o \sin \alpha = \frac{2 \sqrt{2} \times 230}{\pi} \times 14.6423 \times \sin 45° = 2143.963 \text{ VAr}$$

Also, from Eq. (6.41b), $Q_1 = V_0 I_0 \tan \alpha = 146.423 \times 14.6423 \times 1 = 2143.97$ VAr

Example 6.12. *Repeat Example 6.11 for a single-phase semiconductor.*

Solution. From Eq. (6.29), $V_o = \dfrac{V_m}{\pi} (1 + \cos \alpha) = \dfrac{\sqrt{2} \times 230}{\pi} (1 + \cos 45°) = 176.72$ V

$$I_o = \frac{V_o}{R} = \frac{176.72}{10} = 17.672 \text{ A}$$

From Eq. (6.30), $V_{or} = V_s \left[\dfrac{1}{\pi} \left\{ (\pi - \alpha) + \dfrac{\sin 2\alpha}{2} \right\} \right]$

$$= 230 \left[\frac{1}{\pi} \left\{ \left(\pi - \frac{\pi}{4} \right) + \frac{\sin 90}{2} \right\} \right] = 219.3 \text{ V}$$

$$I_{or} = I_o \; (\because \text{load current is ripple free}) = 17.672 \text{ A}$$

From Eq. (6.30), $\quad P_{dc} = V_o I_o = 176.72 \times 17.672 = 3122.996 \text{ W}$

Output *ac* power, $\quad P_{ac} = V_{or} I_{or} = 219.3 \times 17.672 = 3875.47 \text{ W}$

Rectification efficiency $\quad = \dfrac{P_{dc}}{P_{ac}} = \dfrac{3122.996}{3875.47} = 0.8058 \text{ or } 80.58\%$

$$FF = \frac{V_{or}}{V_o} = \frac{219.3}{176.72} = 1.241$$

$$VRF = \sqrt{FF^2 - 1} = \sqrt{1.241^2 - 1} = 0.735, \, CRF = 0$$

From Eq. (6.43), $\quad I_{s1} = \dfrac{2\sqrt{2}}{\pi} \times 17.672 \times \cos \dfrac{45°}{2} = 14.697 \text{ A}$

$$I_s = 17.672 \sqrt{\frac{3\pi}{4.\,\pi}} = 15.304 \text{ A}$$

$$DF = \cos \theta_1 = \cos \left(-\frac{\alpha}{2} \right) = \cos \frac{\alpha}{2} = \cos \frac{45}{2} = 0.9239$$

$$CDF = \frac{2\sqrt{2} \times \cos 22.5}{\left[\pi \left(\pi - \dfrac{\pi}{4} \right) \right]^{1/2}} = 0.9603$$

$\therefore \quad PF = CDF \times DF = 0.9603 \times 0.9239 = 0.8872 \text{ lag}$

$$THD = \left[\frac{1}{CDF^2} - 1 \right]^{1/2} = \left[\frac{1}{0.9603^2} - 1 \right]^{1/2} = 0.2905$$

Power input $\quad = V_o I_o = 176.72 \times 17.672 = 3122.996 \text{ W}$

Reactive power input, $\quad Q_i = \dfrac{V_m}{\pi} I_o \sin \alpha = \dfrac{\sqrt{2} \times 230}{\pi} \times 17.672 \times \cos 45° = 1293.79 \text{ VAr}$

6.6. SINGLE-PHASE SYMMETRICAL AND ASYMMETRICAL SEMICONVERTERS

A single-phase semi-converter topology employing two SCRs and three diodes is commonly used in industrial applications and therefore, the term "single-phase semiconverter" implies this converter only. Here single-phase semiconverters, or single-phase half-controlled converters, employing two SCRs and two diodes are also studied. Both of these semiconverters have two legs.

A single-phase semiconverter, employing one SCR and one diode in each leg, is called single-phase *symmetrical semiconverter*. The other configuration, using two SCRs in one leg and two diodes in the other leg, is called single-phase *asymmetrical semiconverter*. These are now studied briefly.

6.6.1. Single-phase symmetrical semiconverter

It is also called single-phase half-controlled symmetrical converter, or single-phase two-pulse symmetrical converter. As stated before, it has two legs, each leg is made up of one SCR and one diode in series as shown in Fig. 6.17 (*a*). The voltage and current waveforms are drawn as shown in Fig. 6.17 (*b*).

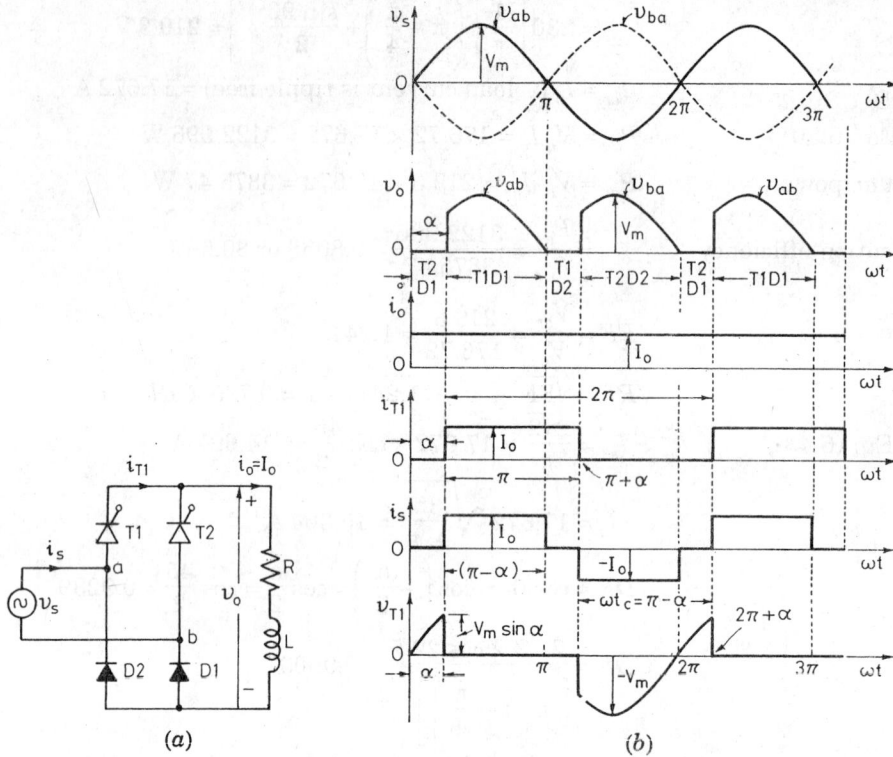

Fig. 6.17. Single-phase symmetrical semiconverter (a) circuit diagram and (b) waveforms.

From $\omega t = 0$ to $\omega t = \alpha$, let constant load current I_o free wheel T2 $D1$. Soon after $\omega t = 0$, T1 gets forward biased through T2. At a firing angle α, T1 is turned on. Load current shifts from T2 to T1; thyristor T2 is therefore turned off. At $\omega t = \alpha$, T2 is subjected to a reverse voltage of $V_m \sin \alpha$ which aids in the commutation of $T2$. Load current now flows through $T1, RL, D1, b$, source v_s, a and T1 and load voltage $v_0 = v_{ab}$. At $\omega t = \pi$, $v_s = 0$. Soon after $\omega t = \pi$, b becomes somewhat positive with respect to a, $D2$ gets forward biased through $D1$; as a consequence, load current now freewheels through $T1 D2$ from $\omega t = \pi$ to $(\pi + \alpha)$. At $\omega t = (\pi + \alpha)$, forward biased $T2$ is turned on, $T1$ is therefore, turned off. Current now flows through $T2 D2$ from $\omega t = (\pi + \alpha)$ to 2π and $v_o = v_{ba}$. Soon after $\omega t = 2\pi$, $D1$ is forward biased, therefore I_o now begins to freewheel through $T2 D1$ and so on. Various waveforms for voltages and currents are drawn in Fig. 6.17 (b). During freewheeling periods, $v_o = 0$ because devices are considered ideal. Waveform for v_{T1} shows that circuit turn-off time t_c is given by

$$t_c = \frac{\pi - \alpha}{\omega} \text{ sec}$$

6.6.2. Single-phase asymmetrical semiconverter

For this semiconverter, power circuit diagram is drawn in Fig. 6.18 (a) and the relevant waveforms in Fig. 6.18 (b). An examination of Figs. 6.17 (a) and 6.18 (a) reveals that in single-phase symmetrical semiconverter of Fig. 6.17 (a), the two cathodes of SCRs T1 and T2 are connected together, these cathodes are, therefore, at the same potential. As such, only one triggering circuit is sufficient for this semiconverter. When single gate pulse is applied to both the thyristors, the SCR which is forward-biased at that instant will get turned on. In asymmetrical semiconverter of Fig. 6.18 (a), however, two separate triggering circuits must be employed.

The thought-process, leading to the understanding of previous single-phase symmetrical semiconverter, can now be extended to this converter configuration also.

For both symmetrical and unsymmetrical topologies, output voltage V_o is given by Eq. (6.28) and its *rms* value V_{or} by Eq. (6.29). The other performance parameters can be evaluated as desired.

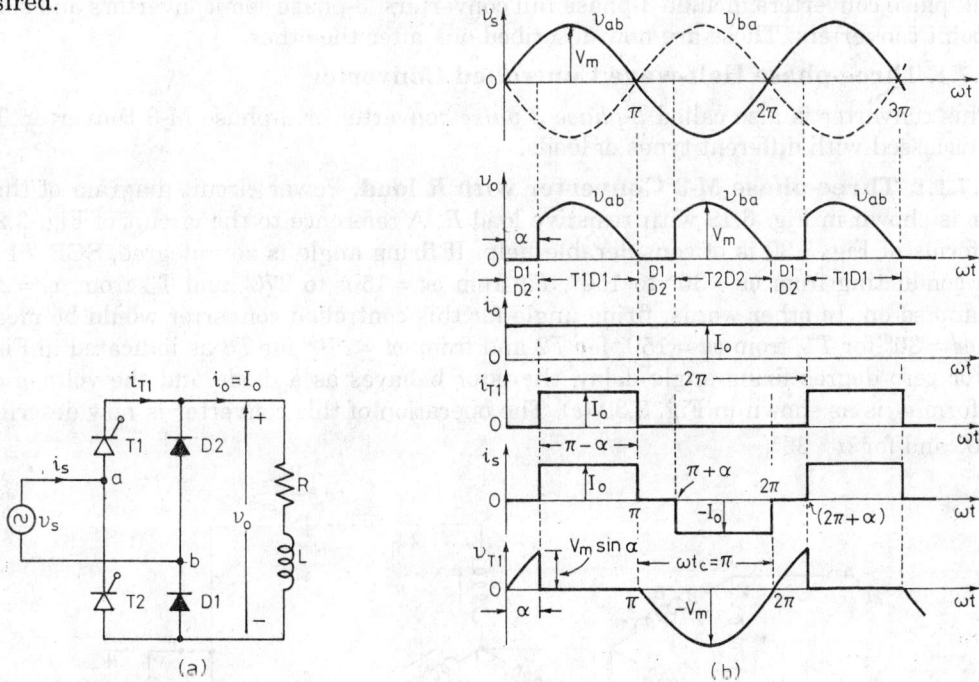

Fig. 6.18. Single-phase asymmetrical semiconverter (a) circuit diagram and (b) waveforms.

A comparison between single-phase two-pulse type semiconverters and full converters can now be made as under :

(i) Single-phase semiconverter requires two SCRs and three (or two) diodes but a single-phase full converter needs from SCRs. Single-phase semiconverter circuits are, therefore, cheaper.

(ii) Single-phase semiconverter offers one-quadrant operation, whereas single-phase full convertor can furnish two-quadrant operation.

(iii) Freewheeling action in semiconverter circuits render power factor better than its value in full converter circuits.

Performance parameters of semiconverter circuits are superior than their corresponding values in full converter circuits.

6.7. THREE-PHASE THYRISTOR CONVERTERS

The advantages of using three-phase controlled converters over single-phase controlled converters are the same as possessed by 3-phase diode rectifiers over 1-phase diode rectifiers enumerated in Art. 3.9. It is also discussed there why three-phase delta-star transformer is employed for delivering power to three-phase converters. All three-phase controlled converters use line-commutation for the turning-off of thyristors.

Three-phase thyristor converters may be classified as under :

(a) Three-pulse converters

(b) Six-pulse converters

(c) Twelve-pulse converters.

Six-pulse converters include 3-phase full converters, 3-phase semiconverters and six-pulse mid-point converters. These are now described one after the other.

6.7.1. Three-phase Half-wave Controlled Converter

This converter is also called *3-phase 3-pulse* converter or 3-phase M-3 converter. This is now discussed with different types of loads.

6.7.1.1. Three-phase M-3 Converter with R load. Power circuit diagram of this converter is shown in Fig. 6.19 with resistive load R. A reference to the circuit of Fig. 3.29 and waveforms of Fig. 3.30 is of considerable help. If firing angle is zero degree, SCR $T1$ would begin conducting from $\omega t = 30°$ to $150°$, $T2$ from $\omega t = 150°$ to $270°$ and $T3$ from $\omega t = 270°$ to $390°$ and so on. In other words, firing angle for this controlled converter would be measured from $\omega t = 30°$ for $T1$, from $\omega t = 150°$ for $T2$ and from $\omega t = 270°$ for $T3$ as indicated in Fig. 6.20 (a). For zero degree firing angle delay, thyristor behaves as a diode and the voltage output waveform v_o is as shown in Fig. 3.30 (c). The operation of this converter is now described for $\alpha < 30°$ and for $\alpha > 30°$.

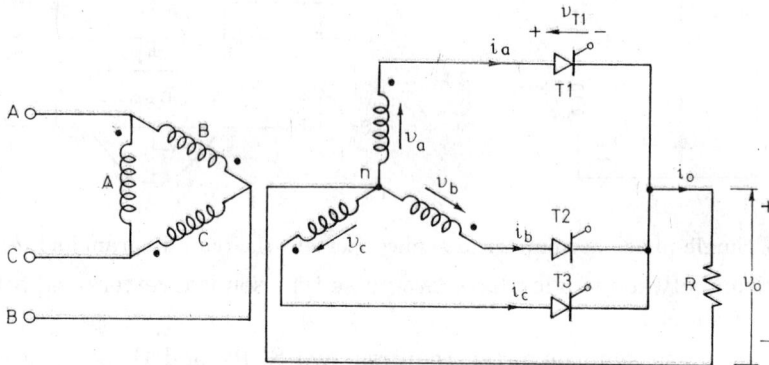

Fig. 6.19. Three-phase half-wave thyristor converter feeding R load.

Firing angle < 30°. The output voltage waveform v_o, for firing angle less than 30° (say around 15°) is sketched in Fig. 6.20 (b), where $T1$ conducts from $\omega t = 30° + \alpha$ to $\omega t = 150° + \alpha = \dfrac{5\pi}{6} + \alpha$, $T2$ from $150° + \alpha$ to $270° + \alpha$ and so on. Each SCR conducts for 120°. The waveform of load current i_o (not shown) would be identical with voltage waveform v_o.

Average value of output voltage, $V_o = \dfrac{3}{2\pi} \int\int_{\alpha + \frac{\pi}{6}}^{\alpha + \frac{5\pi}{6}} V_{mp} \sin \omega t \, d\,(\omega t)$

$$= \frac{3\sqrt{3}}{2\pi} V_{mp}. \cos \alpha = \frac{3 V_{ml}}{2\pi} \cos \alpha \qquad \qquad ...(6.50)$$

where V_{mp} = maximum value of phase (line to neutral) voltage

V_{ml} = maximum value of line voltage = $\sqrt{3}. V_{mp}$

α = firing-angle delay

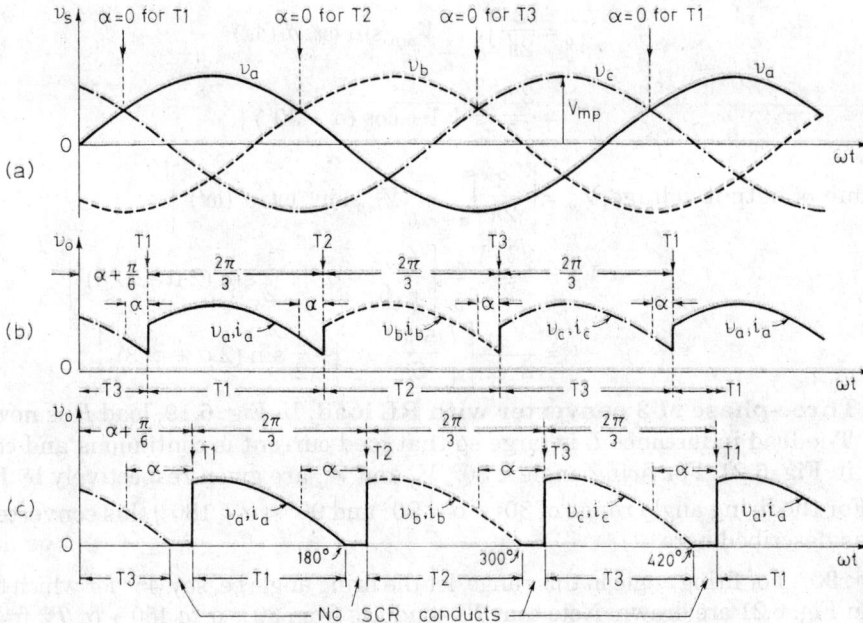

Fig. 6.20. Three-phase 3-pulse converter with R load (a) line to neutral source voltages.
Load voltage waveforms for (b) $0 < \alpha < 30°$ and (c) $\alpha > 30°$

Average load current, $I_o = \dfrac{V_o}{R} = \dfrac{3\,V_{ml}}{2\,\pi.\,R}\cos\alpha$ ...(6.50 a)

Rms value of output, or load voltage is

$$V_{or} = \left[\frac{3}{2\,\pi}\int_{\alpha+\pi/6}^{\alpha+5\,\pi/6} V_{mp}^2 \sin^2\omega t.\,d\,(\omega t)\right]^{1/2}$$

or

$$V_{or}^2 = \frac{3\,V_{mp}^2}{4\,\pi}\left[\;\mid\omega t\mid_{\alpha+\pi/6}^{\alpha+5\,\pi/6} - \mid\frac{\sin 2\,\omega t}{2}\mid_{\alpha+\pi/6}^{\alpha+5\,\pi/6}\;\right]$$

$$= \frac{3\,V_{mp}^2}{4\,\pi}\left[\frac{2\,\pi}{3} + \frac{\sqrt{3}}{2}\cos 2\,\alpha\right]$$

or

$$V_{or} = V_{mp}\left[\frac{1}{2} + \frac{3\sqrt{3}}{8\,\pi}\cos 2\,\alpha\right]^{1/2}$$

$$= \sqrt{3}\,V_{mp}\left[\frac{1}{6} + \frac{\sqrt{3}}{8\pi}\cos 2\,\alpha\right]^{1/2} = V_{ml}\left[\frac{1}{6} + \frac{\sqrt{3}}{8\,\pi}\cos 2\,\alpha\right]^{1/2} \quad ...(6.51)$$

Rms load current, $\quad I_{or} = \dfrac{V_{or}}{R} = \dfrac{V_{ml}}{R}\left[\dfrac{1}{6} + \dfrac{\sqrt{3}}{8\,\pi}\cos 2\,\alpha\right]^{1/2}$ ...(6.51 a)

Firing angle > 30°. When firing angle is more than 30°, $T1$ would conduct from $30° + \alpha$ to 180°, $T2$ from $150° + \alpha$ to 300° and so on as shown in Fig. 6.20 (b). For R load, when phase voltage v_a reaches zero at $\omega t = 180°$, current $i_o = 0$, T1 is therefore turned off. Thus, T1 would conduct from $30° + \alpha$ to 180°. Same is true for other SCRs. This shows that each SCR, for firing angle > 30°, conducts for $(150° - \alpha)$ only. This also implies that for R load, *maximum* possible value of *firing angle is 150°.* Waveform of i_o agrees with v_o waveform, Fig. 6.20 (c). Average value of load voltage,

$$V_o = \frac{3}{2\pi} \int_{\alpha + \frac{\pi}{6}}^{\pi} V_{mp} \sin \omega t . \, d(\omega t)$$

$$= \frac{3 \, V_{mp}}{2\pi} [\, 1 + \cos(\alpha + 30°) \,] \qquad\qquad ...(6.52)$$

Rms value of output voltage, $V_{or} = \left[\dfrac{3}{2\pi} \displaystyle\int_{\alpha + \pi/6}^{\pi} V_{mp}^2 \sin^2 \omega t . \, d(\omega t) \right]^{1/2}$

$$\therefore \qquad V_{or} = \frac{\sqrt{3} . \, V_{mp}}{2\sqrt{\pi}} \left[\left(\frac{5\pi}{6} - \alpha \right) + \frac{1}{2} \sin(2\alpha + \pi/3) \right]^{1/2} \qquad ...(6.53\ a)$$

$$= \frac{V_{ml}}{2\sqrt{\pi}} \left[\left(\frac{5\pi}{6} - \alpha \right) + \frac{1}{2} \sin(2\alpha + \pi/3) \right]^{1/2} \qquad ...(6.53\ b)$$

6.7.1.2. Three-phase M-3 converter with RL load. In Fig. 6.19, load R is now replaced by load RL. The load inductance L is large so that load current is continuous and constant at I_0 as shown in Fig. 6.21. For firing angle < 30°, V_o and V_{or} are given respectively by Eqs. (6.50) and (6.51). For the firing angle range of 30° < α < 90° and 90° < α < 180°, this converter behaves differently as described here.

30° < α < 90°. For firing angle in this range, let the firing angle be, say, 45° for which the various waveforms in Fig. 6.21 are drawn. Note that $T1$ conducts from 30 + α to 150 + α, $T2$ from 150 + α to 270° + α, $T3$ from 270° + α to 390° + α and so on. Thus, each SCR conducts for 120°.

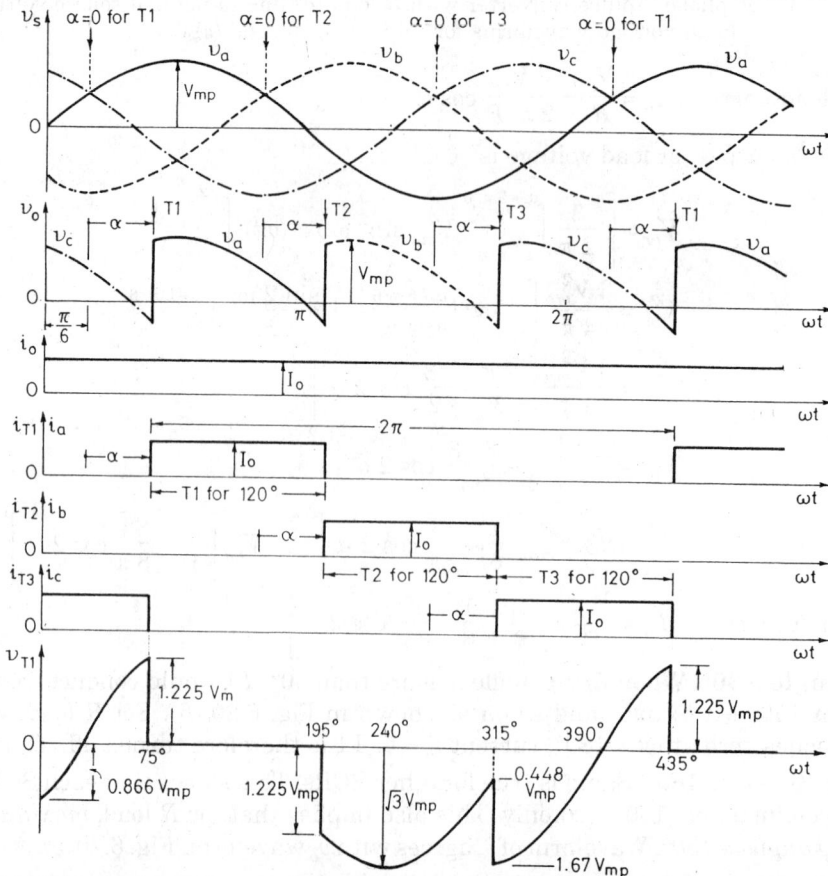

Fig. 6.21. Three-phase M-3 converter waveforms for 30° < α < 90° for ripple-free load current

Δt $\omega t = \pi$, phase voltage v_a is zero, but i_{T1} (or i_a) is not zero because of RL load. Therefore, $T1$ would continue conducting beyond $\omega t = \pi$. As such, $v_o = v_a$ goes negative beyond $\omega t = \pi$. When T_2 is turned on at $\omega t = 150° + \alpha$, load current shifts from $T1$ to $T2$ and a voltage $v_a - v_b = [V_m \sin(150 + \alpha) - V_m \sin(30 + \alpha)]$ appears as reverse bias across $T1$ to aid its commutation. SCR $T2$ conducts from $(150° + \alpha)$ to $(270° + \alpha)$ and so on. The waveform for i_{T1} or i_a, i_{T2} or i_b and i_{T3} or i_c are as shown in Fig. 6.21.

The waveform v_{T1} for voltage across $T1$, on the assumption of firing angle $45°$ can be drawn as under ;

When $T1$ is on, $v_{T1} = v_a - v_a = 0$ from $\omega t = 75°$ to $195°$, Fig. 6.21.

When T_2 is on, $v_{T1} = v_a - v_b$ from $\omega t = 195°$ to $315°$ and

When T_3 is on, $v_{T1} = v_a - v_c$ from $\omega t = 315°$ to $435°$ and so on.

When $T2$ is turned on at $\omega t = 195°$, $v_{T1} = v_a - v_b = -V_{mp} \sin 15 - V_{mp} \sin 75° = -1.225\, V_{mp}$; at $\omega t = 210°$, $v_{T1} = -1.5\, V_{mp}$; at $\omega t = 240°$, $v_{T1} = \sqrt{3}\, V_{mp}$; at $\omega t = 270°$, $v_{T1} = -1.5\, V_{mp}$; at $\omega t = 300°$, $v_{T1} = -V_{mp} \sin 60 - 0 = -0.866\, V_{mp}$. At $\omega t = 315°$, $v_{T1} = -V_{mp} \sin 45° + V_{mp} \sin 15° = -0.448\, V_{mp}$.

Also, at $\omega t = 315°$, $T2$ gets turned-off whereas $T3$ is turned on.

$\therefore$
$$v_{T1} = v_a - v_c = -V_{mp} \sin 45° + V_{mp} \sin 75° = -1.673\, V_{mp}.$$

This shows that at $\omega t = 315°$, v_{T1} at once changes from $-0.448\, V_{mp}$ to $-1.673\, V_{mp}$ as shown in Fig. 6.21.

At $\omega t = 330°$, $v_{T1} = -V_{mp} \sin 30 - V_{mp} = -1.5\, V_{mp}$

At $\omega t = 360°$, $v_{T1} = 0 - 0.866\, V_{mp} = -0.866\, V_{mp}$

At $\omega t = 390°$, $v_{T1} = 0.5\, V_{mp} - 0.5\, V_{mp} - 0$

At $\omega t = 420°$, $v_{T1} = 0.866\, V_{mp} - 0 = 0.866\, V_{mp}$

At $\omega t = 435°$, $v_{T1} = V_{mp} \sin 75° + V_{mp} \sin 15° = 1.225\, V_{mp}$

and also $v_{T1} = v_a - v_a = 0$ and so on.

Average and rms values of output voltage are the same as given by Eqs. (6.50) and (6.51) respectively.

$90° < \alpha < 180°$. For firing angle in this range, let α be, say, $165°$. Under the assumption of ripple free load current I_o, the various waveforms are shown in Fig. 6.22. As the output voltage waveform v_o is below the reference line, average value of v_o must be negative. It is also evident from $V_o = (3\, V_{ml}/2\pi) \cos \alpha$ that when firing angle α is more than $90°$, V_o is negative. For $\alpha > 90°$, three-phase 3-pulse converter operates as a line-commutated inverter which is possible only if the load circuit has a dc voltage source of reverse polarity, as in a single-phase full converter already discussed in Art. 6.3.2.1.

It is also seen from Figs. 6.21 and 6.22 that average value of thyristor current, I_{TA} = average value of source current, $I_{sA} = (I_o \times 120)/360 = I_o/3$.

Rms value of thyristor or source current, $I_{Tr} = I_{sr} = \left[\dfrac{I_o^2 \times 120}{360} \right]^{1/2} = \dfrac{I_o}{\sqrt{3}}$

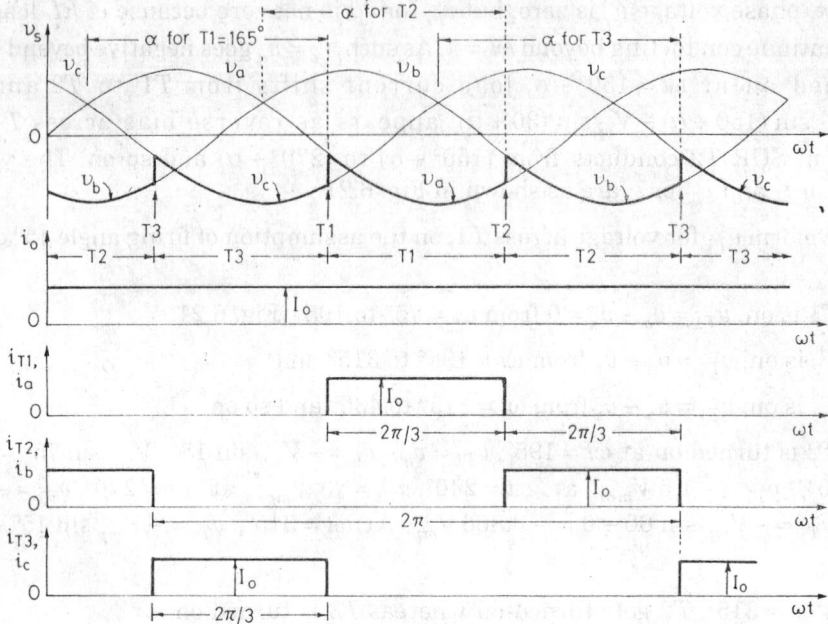

Fig. 6.22. Three-phase 3-pulse converter waveforms for 90° < α < 180° for constant load current.

Waveforms of i_a, i_b, i_c in Figs. 6.21 and 6.22 show that transformer windings have to carry *dc* current which is harmful to the transformer. The problem can, however, be solved by using delta-zigzag connection instead of delta-star connection as shown in Fig. 6.23.

In this figure, delta-zigzag transformer feeds RL load through a 3-phase 3-pulse converter. Load current I_o enters the neutral n of secondary zigzag, divides equally in the three half-windings, *i.e.* each half winding a, b, c shares a load current $I_o/3$. This current flows through other half windings $b1, c1, a1$, through SCRs $T1, T2$ $T3$ and load RL as shown in Fig. 6.23. Note that each secondary winding is separated into two equal halves which are appropriately connected to result in zigzag secondary.

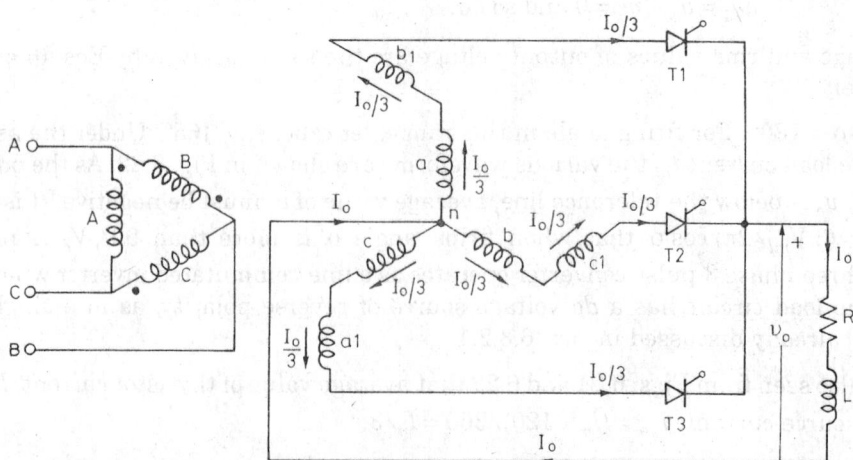

Fig. 6.23. A delta-zigzag transformer feeding a 3-phase 3-pulse thyristor converter.

'A careful observation of zigzag winding in Fig. 6.23 reveals that same phase winding, dividing into two halves as $a, a1$; carries current $I_o/3$ in both these halves, but in opposite directions. Same is true for phase b, c windings. Since each half, of the three secondary windings, carries direct current in opposite direction, their magnetic effects cancel each other. As a result, the core flux and therefore core loss and temperature rise remain unaffected. This shows that a 3-phase 3-pulse converter can be used for energizing a *dc* load provided a delta-zigzag transformer is employed on its input side.

Example 6.13. *A 3-phase M-3 converter is operated from 3-phase, 230 V, 50 Hz supply with load resistance R = 10 Ω. An average output voltage of 50% of the maximum possible output voltage is required. Determine (a) the firing angle (b) average and rms values of load current and (c) rectification efficiency.*

Solution. *(a)* For R load, voltage is continuous when $\alpha \leq 30°$ and average output voltage is given by

$$V_o = \frac{3\,V_{ml}}{2\,\pi} \cos \alpha$$

Its maximum possible value is, $V_{om} = \dfrac{3\,V_{ml}}{2\,\pi} = \dfrac{3\sqrt{2} \times 230}{2\,\pi} = 155.3$ V

Required average output voltage $= \dfrac{V_{om}}{2} = \dfrac{155.3}{2} = 77.65$ V

∴ $$V_o = \frac{3\,V_{ml}}{2\,\pi} \cos \alpha = V_{om} \cos \alpha$$

Now $$V_o = \frac{1}{2}\,V_{om} = V_{om} \cos \alpha, \qquad \therefore \ \alpha = 60°$$

This shows that actual value of $V_o = \dfrac{1}{2}\,V_{om} = 77.65$ V cannot be obtained for $\alpha < 30°$, but for $\alpha > 30°$. So by using Eq (6.52), we get

$$V_o = \frac{3\,V_{mp}}{2\pi}\,[1 + \cos\,(\alpha + 30°)] = \frac{3\,V_{ml}}{2\pi\,.\,\sqrt{3}}\,[1 + \cos\,(\alpha + 30°)]$$

∴ $\dfrac{1}{\sqrt{3}}\,[1 + \cos\,(\alpha + 30°)] = V_o \times \dfrac{2\pi}{3\,V_{ml}} = \dfrac{V_o}{V_{om}} = \dfrac{1}{2}$

∴ $\alpha = 67.7°$

(b) $$I_o = \frac{V_o}{R} = \frac{77.65}{10} = 7.765 \text{ A}$$

Rms voltage from Eq. (6.53 *b*) is

$$V_o = \frac{V_{ml}}{2\sqrt{\pi}}\left[\left(\frac{5\pi}{6} - \alpha\right) + \frac{1}{2}\sin\,(2\,\alpha + \pi/3)\right]$$

$$= \frac{\sqrt{2} \times 230}{2\sqrt{\pi}}\left[\left(\frac{5\,\pi}{6} - \frac{67.7 \times \pi}{180}\right) + \frac{1}{2}\sin\,(2 \times 67.7 + 60°)\right] = 104.765 \text{ V}$$

$$I_o = \frac{104.765}{10} = 10.477 \text{ A}$$

(c) Rectifier efficiency $= \dfrac{V_o\,I_o}{V_{or}\,I_{or}} = \dfrac{77.65 \times 7.765}{104.765 \times 10.477} = 0.5493$ or 54.93%

Example 6.14. *Derive expressions for the average and rms output voltages for a 3-phase 3-pulse controlled converter by using cosine function for the supply voltage. Assume continuous conduction.*

Solution. The variation of output voltage for 3-phase 3-pulse converter is shown in Fig. 6.24. For using the cosine function for the input voltage, the origin must be taken when instantaneous voltage is maximum. So here origin is taken at OO' as shown in Fig. 6.24. With OO' as the origin, $v_a = V_m \cos \omega t$ and integration must be made from instant 1 where $\omega t = -\left(\dfrac{\pi}{2} - \alpha\right)$ to instant 2 where $\omega t = \left(\dfrac{\pi}{3} + \alpha\right)$. Thus, the average output voltage V_o for a 3-phase M-3 converter is

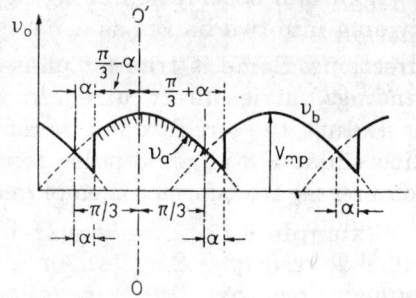

Fig. 6.24. Pertaining to Example 6.14

$$V_o = \frac{3}{2\pi} \int_{-\left(\frac{\pi}{3} - \alpha\right)}^{\frac{\pi}{3} + \alpha} V_{mp} \cos \omega t \, d(\omega t) = \frac{3 \, V_{mp}}{2\pi} \left| -\sin \omega t \right|_{-\left(\frac{\pi}{3} - \alpha\right)}^{\frac{\pi}{3} + \alpha}$$

$$= \frac{3\sqrt{3} \, V_{mp}}{2\pi} \cos \alpha = \frac{3 \, V_{ml}}{2\pi} \cos \alpha \qquad \qquad ...(6.50)$$

Now, *rms* value of output voltage V_{or} is

$$V_{or} = \frac{3}{2\pi} \int_{-\left(\frac{\pi}{3} - \alpha\right)}^{\frac{\pi}{3} + \alpha} V_{mp}^2 \cos^2 \omega t . \, d(\omega t)$$

$$V_{or}^2 = \frac{3 \, V_{mp}^2}{4\pi} \left| \omega t + \frac{\sin 2\omega t}{2} \right|_{-\left(\frac{\pi}{3} - \alpha\right)}^{\frac{\pi}{3} + \alpha}$$

or

$$V_{or} = \sqrt{3} \, V_{mp} \left[\frac{1}{6} + \frac{\sqrt{3}}{8\pi} \cos 2\alpha \right]^{1/2}$$

$$= V_{ml} \left[\frac{1}{6} + \frac{\sqrt{3}}{8\pi} \cos 2\alpha \right]^{1/2} \qquad \qquad ...(6.51)$$

Example 6.15. *A 3-phase half-wave controlled converter is fed from 3-phase, 400 V, 50 Hz source and is connected to load taking a constant current of 36 A. Thyristors have a voltage drop of 1.4 V. (a) Calculate average vbalue of load voltage for a firing angle of 30° and 60°. (b) Determine average and rms current ratings as well as PIV of thyristors. (c) Find the average power dissipated in each thyristor.*

Solution. Here, average output voltage,

$$V_o = \frac{3 \, V_{ml}}{2\pi} \cos \alpha - v_T; \ V_{ml} = \sqrt{2} \times 400 \text{ V and } v_T = 1.4 \text{ V}$$

For a firing angle of 30°, $\quad V_o = \dfrac{3\sqrt{2} \times 400}{2\pi} \cos 30° - 1.4 = 232.474 \text{ V}$

For $\alpha = 60°$, $\qquad\qquad V_o = \dfrac{3\sqrt{} \times 400}{2\pi} \cos 60° - 1.4 = 133.63 \text{ V}$

(b) Average current rating of SCR, $I_{TA} = \dfrac{I_o}{3} = \dfrac{36}{3} = 12 \text{ A}$

Rms current rating of SCR, $\qquad I_{Tr} = \dfrac{I_o}{\sqrt{3}} = \dfrac{36}{\sqrt{3}} = 20.785 \text{ A}$

PIV of SCR $\qquad\qquad\qquad = \sqrt{3} \, V_{mp} = V_{ml} = \sqrt{2} \times 400 = 565.6 \text{ V}$

(c) Average power dissipated in each SCR $= I_{TA} \, v_T = 12 \times 1.4 = 16.8 \text{ W}$

Example 6.16. *A 3-phase 3-pulse converter, fed from delta-star transformer, is connected to a load requiring ripple free current. A freewheeling diode is connected across the load. Sketch waveforms for source voltage, output voltage, load current, line current and freewheeling diode current. Obtain expressions for average and rms value of output voltage, thyristor current and freewheeling-diode current.*

Solution. A 3-phase 3-pulse converter feeding RL load and with freewheeling diode **across** RL is shown in Fig. 6.25 (a). For firing angle < 30°, freewheeling diode does not come into **play**. So here, firing angle is taken, say 60°, just to illustrate how freewheeling diode comes **into** play and to examine its effect on the performance of the converter.

At $\omega t = \pi$, as phase voltage v_a tends to go negative, freewheeling diode gets forward **biased** through T1. Therefore, freewheeling diode starts conducting from $\omega t = \pi$ till $T2$ is turned **on** at $\omega t = 150 + \alpha$. Similarly, when v_b and v_c tend to go negative, freewheeling diode comes **into** play, as shown in Fig. 6.25 (b). Note that each SCR conducts for $(150° - \alpha)$ and freewheeling diode for $(\alpha - 30°)$.

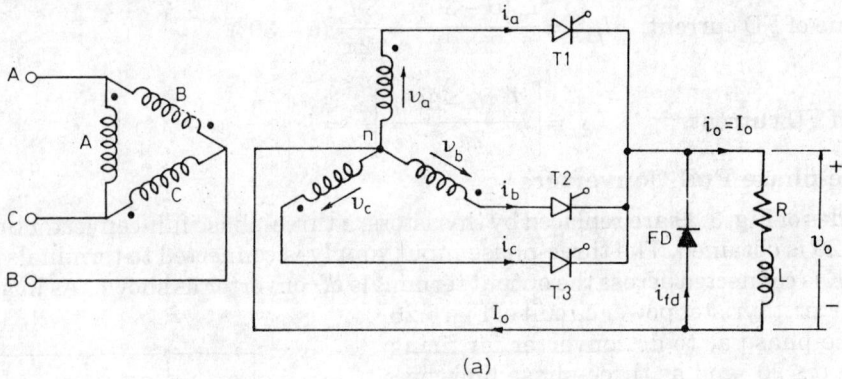

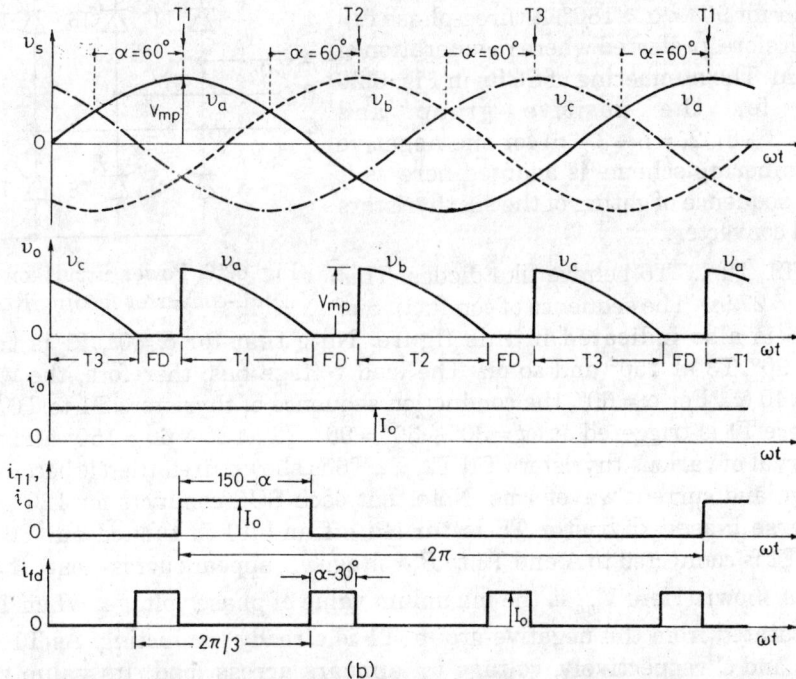

Fig. 6.25. Three-phase M-3 converter (*a*) circuit (*b*) waveform Example 6.16.

Average value of output voltage, $V_o = \dfrac{3}{2\pi} \displaystyle\int_{\alpha + \pi/6}^{\pi} V_{mp} \sin \omega t.\, d\,(\omega t)$

$$= \dfrac{3\,V_{mp}}{2\pi} [1 + \cos (\alpha + \pi/6)] \qquad \ldots(6.52)$$

Similarly, $V_{or} = \dfrac{V_{ml}}{2\sqrt{\pi}} \left[\left(\dfrac{5\pi}{6} - \alpha \right) + \dfrac{1}{2} \sin (2\alpha + \pi/3) \right]^{1/2} \qquad \ldots(6.53\ b)$

Average thyristor current, $I_{TA} = \dfrac{I_o \left(\dfrac{5\pi}{6} - \alpha \right)}{2\pi} = \dfrac{I_o}{2\pi} \left[\dfrac{5\pi}{6} - \alpha \right]$

Rms thyristor current, $I_{Tr} = \left[\dfrac{I_o^2}{2\pi} \left\{ \dfrac{5\pi}{6} - \alpha \right\} \right]^{1/2} = I_o \left[\dfrac{1}{2\pi} \left\{ \dfrac{5\pi}{6} - \alpha \right\} \right]^{1/2}$

Average value of FD current, $I_{fd.A} = \dfrac{I_o\,(\alpha - 30°)}{2\pi/3} = \dfrac{3\,I_o}{2\pi}\,(\alpha - 30°)$

Rms value of FD current, $I_{fd.r} = \left[\dfrac{I_o^2\,(\alpha - 30°)}{2\pi/3} \right]^{1/2} = I_o \left[\dfrac{3}{2\pi}\,(\alpha - 30°) \right]^{1/2}$

6.7.2. Three-phase Full Converters

If all the diodes of Fig. 3.39 are replaced by thyristors, a three-phase full-converter bridge as shown in Fig. 6.26 is obtained. The three-phase input supply is connected to terminals A, B, C and the load RLE is connected across the output terminals of converter as shown. As in a single-phase full-converter, thyristor power circuit of Fig. 6.26 works as a three-phase ac to dc converter for firing angle delay $0° < \alpha \le 90°$ and as three-phase line-commutated inverter for $90° < \alpha < 180°$. A three-phase full converter is, therefore, preferred where regeneration of power is required. The numbering of SCRs in Fig. 6.26 is 1, 3, 5 for the positive group and $4 (= 1 + 3)$, $6 (= 3 + 3)$, $2 (= 5 + 3 - 6)$ for the negative group. This numbering scheme is adopted here as it agrees with the sequence of gating of the six thyristors in a 3-phase full converter.

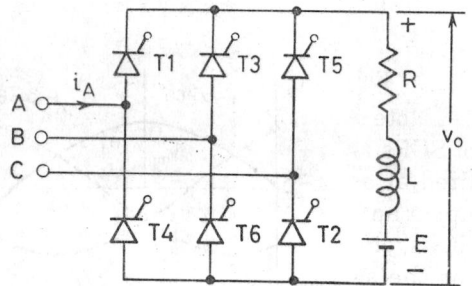

Fig. 6.26. Power circuit for a 3-phase full-converter feeding RLE load.

For $\alpha = 0°$; T1, T2,......T6 behave like diodes. This is shown in Fig. 6.27 (a). The sequence of conduction of SCRs T1 to T6 is also indicated in this figure. Note that for $\alpha = 0°$, T1 is triggered at $\omega t = \pi/6$, T2 at 90°, T3 at 150° and so on. The load voltage has, therefore, the waveform as shown in Fig. 3.40 (c). For $\alpha = 60°$, the conduction sequence of thyristors T1 to T6 is shown in Fig. 6.27 (b). Here T1 is triggered at $\omega t = 30° + 60° = 90°$, T2 at $90 + 60 = 150°$ and so on. If the conduction interval of various thyristors T1, T2, ... T6 is shown first, then it becomes easier to draw the voltage and current waveforms. Note that each SCR conducts for 120°, when T1 is triggered, reverse biased thyristor T5 is turned off and T1 is turned on. T6 is already conducting. As T1 is connected to A and T6 to B, voltage v_{ab} appears across load. It varies from $1.5\,V_m$ to zero as shown. Here V_{mp} is the maximum value of phase voltage. When T2 is turned on, T6 is commutated from the negative group. T1 is already conducting. As T1 and T2 are connected to A and C respectively, voltage v_{ac} appears across load. Its value varies from $1.5\,V_{mp}$ to zero as shown. This sequence of triggering is continued for other SCRs.

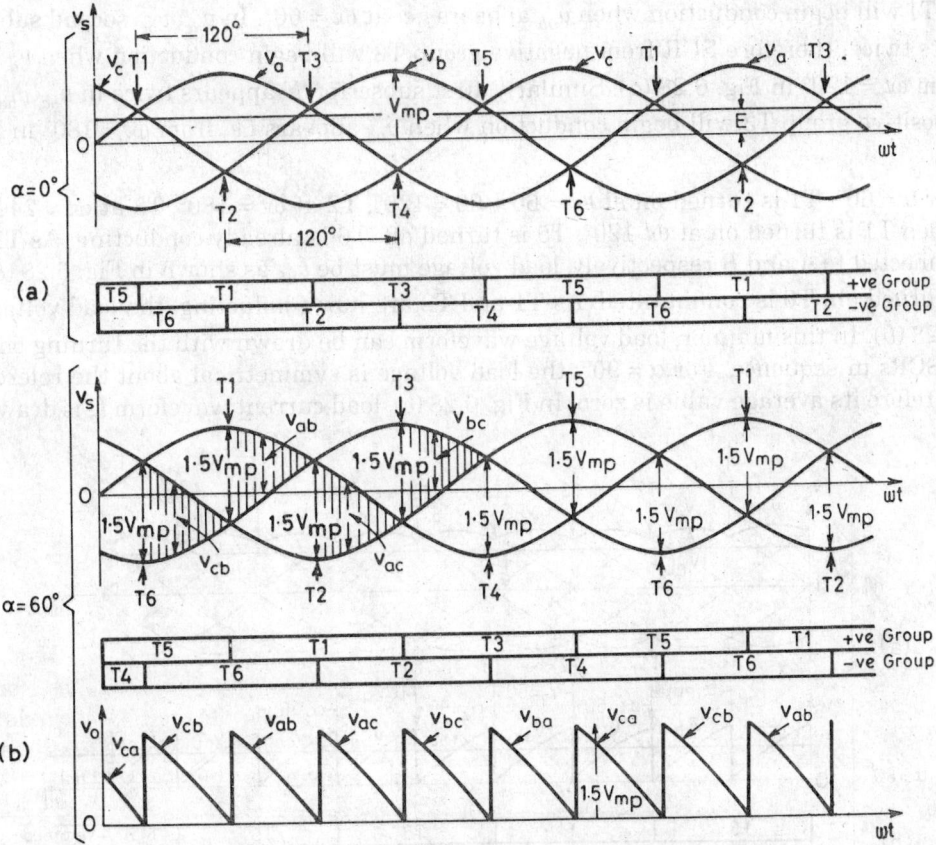

Fig. 6.27. Voltage waveforms and conduction of thyristors for a 3-phase full converter.

Note that positive group of SCRs are fired at an interval of 120°. Similarly, negative group of SCRs are fired with an interval of 120° amongst them. But SCRs from both the groups are fired at an interval of 60°. This means that commutation occurs every 60°, alternatively in upper and lower group of SCRs. Each SCR from both groups conducts for 120°. At any time, two SCRs, one from the positive group and the other from negative group, must conduct together for the source to energise the load. For *ABC* phase sequence of the three-phase supply, thyristors conduct in pairs ; T1 and T2, T2 and T3, T3 and T4 and so on.

The sequence of events in Fig. 6.27 can also be shown more conveniently if line voltages, instead of phase voltages, are considered. In Fig. 6.28 (a) are shown line voltages $v_{ab}, v_{ac}, v_{bc}, v_{ba}$ etc. For $\alpha = 0°$, SCRs T1, T2,....T6 behave as diodes and the output voltage waveform is as shown in Fig. 6.28 (a) by v_{ab}, v_{ac}, v_{bc} etc. In this figure, for $\alpha = 0$, T1 is turned on at $\omega t = 60°$, T2 at $\omega t = 120°$, T3 at $\omega t = 180°$ and so on. In Fig. 6.28 (a), therefore, firing angle is measured from $\omega t = 60°$ for T1, from $\omega t = 120°$ for T2, from $\omega t = 180°$ for T3 and so on.

The question may arise in the minds of the readers as to why T1, for $\alpha = 0$, conducts from $\omega t = 60°$ and not from $\omega t = 0°$. Here the use of subscripts ab, ac, bc, ba etc come to the rescue of readers. As observed, the subscripts in sequence appear twice. When *first* subscript appears twice, the SCR in the positive group pertaining to that line conducts for 120°. Likewise, when *second* subscript comes twice, the SCR in the negative group pertaining to that line conducts for 120°. For example, first subscript 'a' appears twice in v_{ab}, v_{ac} ; therefore SCR from positive group T1 will begin conduction when v_{ab} appears *i.e.* at $\omega t = 60°$. In v_{ac}, v_{bc}, second subscript 'C' appears twice, therefore SCR from negative group T2 will begin conduction when v_{ac} appears *i.e.* from $\omega t = 120°$ in Fig. 6.28 (*a*). Similarly, first subscript 'b' appears twice in v_{bc}, v_{ba}, so SCR from positive group T3 will begin conduction when v_{bc} appears *i.e.* from $\omega t = 180°$ in Fig. 6.28 (*a*).

For $\alpha = 60°$, T1 is turned on at $\omega t = 60 + 60 = 120°$, T2 at $\omega t = 180°$, T3 at $\omega t = 240°$ and so on. When T1 is turned on at ωt 120°, T5 is turned off. T6 is already conducting. As T1 and T6 are connected to A and B respectively, load voltage must be v_{ab} as shown in Fig. 6.28 (*b*). When T2 is turned on, T6 is commutated. As T1 and T2 are now conducting, the load voltage is v_{ac}, Fig. 6.28 (*b*). In this manner, load voltage waveform can be drawn with the turning on or off of other SCRs in sequence. For $\alpha = 90°$, the load voltage is symmetrical about the reference line ωt, therefore its average value is zero. In Fig. 6.28 (*c*), load current waveform i_o is drawn on the

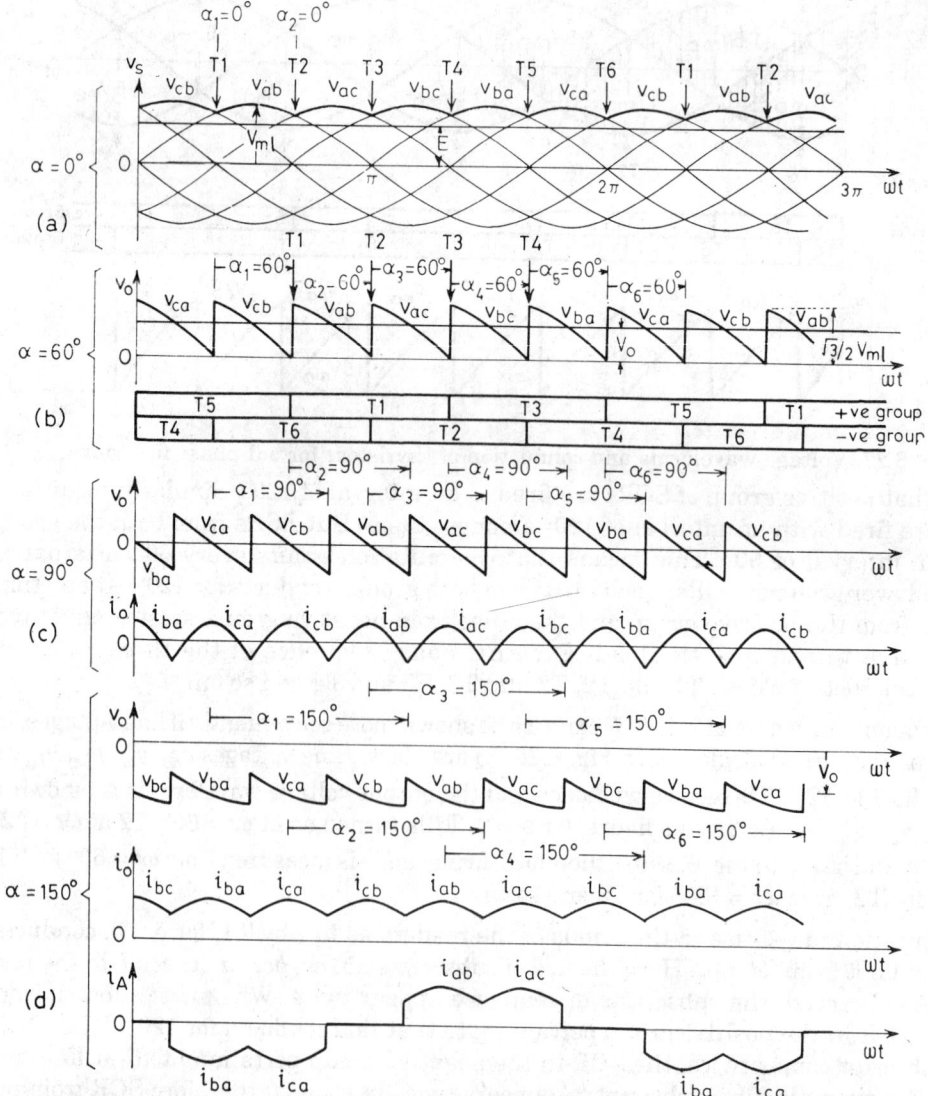

Fig. 6.28. Voltage and current waveforms for a 3-phase full-converter for different firing angles.

assumption that load is pure L. When v_{ca} is peak positive, slope di_o/dt is maximum positive so that $L. (di_o/dt)$ equals peak positive v_{ca}. Similarly, di_o/dt is maximum negative when v_{ca} is peak negative. When current i_o has peak value, di_o/dt is zero and likewise v_{ca} is zero as shown in Fig. 6.28 (c). For $\alpha = 150°$, T1 is triggered at $\omega t = 210°$, T2 at 270° and so on. The output voltage waveform is shown in Fig. 6.28 (d). It is seen from this figure that average voltage is reversed in polarity. This means that dc source is delivering power to ac source ; this is called line-commutated inverter operation of the 3-phase full converter bridge. It may be seen from above that for $\alpha = 0°$ to 90°, power circuit of Fig. 6.26 works as a 3-phase full converter delivering power from ac source to dc load and for $\alpha = 90°$ to 180°, it works as a line-commutated inverter delivering power from dc source to ac load. It can work in the inverter mode only if the load has a direct emf E due to a battery or a dc motor. It should be noted that direction of current for both converter and inverter operations remains fixed but the polarity of output voltage reverses.

Source current i_A in phase A is also drawn in Fig. 6.28 (d) for $\alpha = 150°$. For the arrow direction indicated in Fig. 6.26, i_A is treated as positive. Therefore i_A is positive when T1 is conducting, i.e. when first subscript for voltages or currents is 'a'. Likewise i_A is negative when T4 is conducting, i.e. when the second subscript for voltages or currents is 'a'. Source current waveforms for other two phases can also be drawn accordingly. For other firing angles, source currents can be drawn similarly.

Expression for the average output voltage V_0 can be obtained by referring to Fig. 6.29 where v_{ab}, v_{ac} etc. are sketched from Fig. 6.28 (a) for firing angle delay $\alpha < 30°$. Note that periodicity of output voltage is $\pi/3$ radians. Average value of output voltage is obtained by finding the dashed area $abcd$ over a periodic cycle, Fig. 6.29, and then dividing it by the periodic time. With OO' as the origin at the maximum value of v_{ab}, V_0 is given by

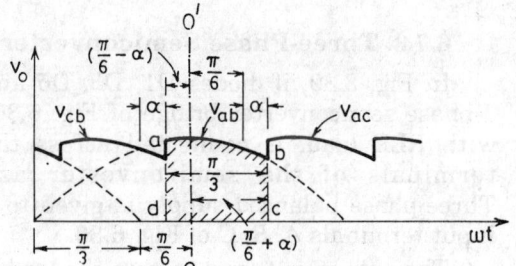

Fig. 6.29. Output voltage waveform for a 3-phase full converter.

$$V_0 = \frac{3}{\pi} \int_{-\left(\frac{\pi}{6} - \alpha\right)}^{\left(\frac{\pi}{6} + \alpha\right)} V_{ml} \cos \omega t \cdot d(\omega t)$$

$$= \frac{3V_{ml}}{\pi} \left[\sin\left(\alpha + \frac{\pi}{6}\right) - \sin\left(\alpha + \frac{\pi}{6}\right) \right] = \frac{3V_{ml}}{\pi} \cos \alpha \qquad ...(6.54)$$

Here V_{ml} is the maximum value of line voltage.

If sine function is used for the source voltage, then $v_{ab} = V_{ml} \sin \omega t$ because $v_{ab} = 0$ at $\omega t = 0$.

$$\therefore \qquad V_0 = \frac{3}{\pi} \int_{\frac{\pi}{3} + \alpha}^{\frac{2\pi}{3} + \alpha} V_{ml} \sin \omega t \cdot d(\omega t)$$

$$= -\frac{3V_{ml}}{\pi} \left[\cos\left(\frac{2\pi}{3} + \alpha\right) - \cos\left(\frac{\pi}{3} + \alpha\right) \right] = \frac{3V_{ml}}{\pi} \cos \alpha \qquad ...(6.54)$$

Rms value of output voltage V_{or} is

$$V_{or} = \left[\frac{3}{\pi} \int_{\frac{\pi}{3}+\alpha}^{\frac{2\pi}{3}+\alpha} V_{ml}^2 \sin^2 \omega t \, d\,(\omega t) \right]^{1/2}$$

$$V_{or}^2 = \frac{3\,V_{ml}^2}{2\pi} \int_{\frac{\pi}{3}+\alpha}^{\frac{2\pi}{3}+\alpha} (1 - \cos 2\omega t) \, d\,(\omega t)$$

or

$$V_{or} = V_{ml} \sqrt{\frac{3}{2\pi}} \left[\frac{\pi}{3} + \frac{\sqrt{3}}{2} \cos 2\alpha \right]^{1/2} \qquad \qquad ...(6.54a)$$

It is observed from Fig. 6.28 that source current for phase A, *i.e.* i_A (or for any other phase) flows for 120° for every 180°. Therefore, in case output current is assumed constant at I_0, the rms value of source current is

$$I_s = \sqrt{I_0^2 \frac{2\pi}{3} \times \frac{1}{\pi}} = I_0 \sqrt{\frac{2}{3}}$$

Each SCR conducts for 120° for every 360°. Therefore, the rms value of thyristor current is

$$I_{Th} = \sqrt{I_0^2 \frac{2\pi}{3} \times \frac{1}{2\pi}} = I_0 \sqrt{\frac{1}{3}}$$

6.7.3. Three-Phase Semiconverters

In Fig. 3.39, if diodes D1, D3, D5 are replaced by thyristors T1, T2, T3 respectively, a 3-phase semiconverter bridge of Fig. 6.30 is obtained. A freewheeling diode FD, in parallel with RLE load, is connected across the output terminals of the semiconverter as shown. Three-phase balanced supply is given to the three input terminals A, B, C of Fig. 6.30.

The output voltage v_0 across the load terminals is controlled by varying the firing angles of SCRs T1, T2 and T3. The diodes D1, D2 and D3 provide merely a return path for the current to the most negative line terminal.

The semiconverter bridge operation for different firing angles is shown in Fig. 6.31 in the form of voltage and current waveforms. The conduction angles for the SCRs, diodes or FD are also shown.

Fig. 6.30. Power circuit for a 3-phase semi-converter feeding RLE load.

For a firing angle delay of $\alpha = 0°$, thyristors T1, T2, T3 would behave as diodes and the output voltage of semiconverter would be symmetrical six-pulse per cycle as shown in Fig. 6.31 (a). The output voltage consisting of pulses $v_{cb}, v_{ab}, v_{ac}, v_{bc}$ etc. shown in this figure is similar to that shown in Fig. 3.40 (c). The output voltage consists of pulses $v_{ab}, v_{ac}, v_{bc}, v_{ba}$ etc. as in Fig. 3.40 (c). When the firing angle is delayed to $\alpha = 15°$ (say) as shown in Fig. 6.31 (b), the triggering of SCRs T1, T2, T3 is delayed but return diodes D1, D2, D3 remain unaffected so that only alternate pulses are altered. The load current is continuous and has little ripple. The FD does not come into play for $\alpha = 15°$. Each SCR and diode conduct for 120°.

In Fig. 6.31 (a), v_{cb} is the load voltage from $\omega t = 0°$ to 60°. As the first subscript indicates conducting element in the positive group, v_{cb} shows that T3 is already conducting through diode D2 of negative group. Voltages v_{ab}, v_{ac} indicate that, according to the first subscript, T1 conducts for 120° and it begins to conduct at $\omega t = 60°$ for $\alpha = 0°$ as shown in Fig. 6.31 (a).

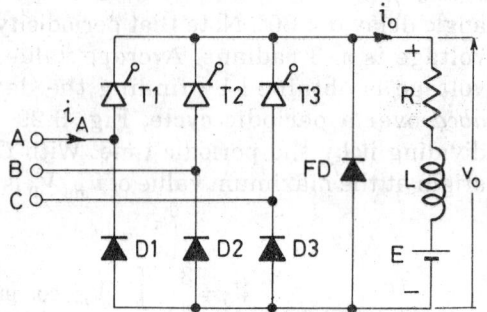

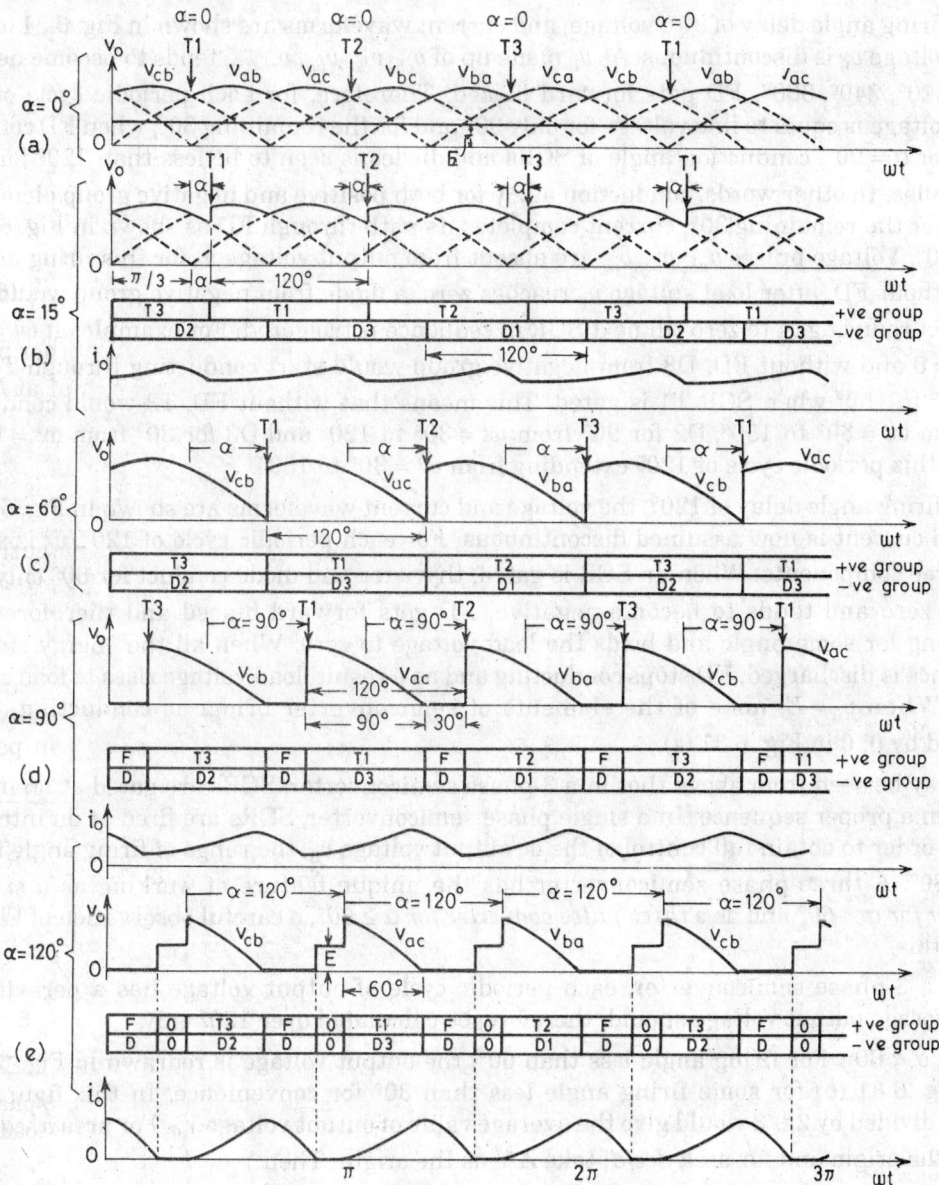

Fig. 6.31. Voltage and current waveforms for a 3-phase semiconverter for different firing angles.

Similarly, v_{bc}, v_{ba} indicate that T2 conducts for 120° and it begins to conduct at $\omega t = 180°$ for $\alpha = 0°$. An SCR with zero degree firing angle behaves like a simple diode. Thus, as per the definition of firing angle, it should be measured from $\omega t = 60°$ for T1, from $\omega t = 180°$ for T2, from $\omega t = 300°$ for T3 and so on.

For $\alpha = 60°$, Fig. 6.31 (c), the thyristors are fired so that current returns through one diode during each 120° conduction period. For voltage v_{ac}, T1 and D3 conduct simultaneously for 120° as shown. Similarly, other elements conduct. FD does not come into play even for $\alpha = 60°$. Further note that voltage pulses v_{ab}, v_{bc}, v_{ca} do not appear in the output voltage waveform for $\alpha = 60°$. It will be seen that for $\alpha \geq 60°$, voltage pulses v_{ab}, v_{bc}, v_{ca} are eliminated. The load current, assumed continuous for $\alpha = 60°$, is not shown in Fig. 6.31 (c).

For firing angle delay of 90°, voltage, and current waveforms are shown in Fig. 6.31 (d). The output voltage v_0 is discontinuous. As v_0 made up of $v_{cb}, v_{ac}, v_{ba}, v_{cb},...,$ tends to become negative at $\omega t = 120°, 240°, 360°$, FD gets forward biased. Therefore, for each periodic cycle of 120°, output voltage is equal to line voltage for only 90° and for the remaining 30°, when FD conducts, $v_0 = 0$. For $\alpha = 90°$, conduction angle of SCRs and diodes is seen to be less than 120° for every output pulse. In other words, conduction angle for both positive and negative group elements is 90° and for the remaining 30°, current completes its path through FD as shown in Fig. 6.31 (d) for $\alpha = 90°$. Voltage pulses v_{ab}, v_{bc}, v_{ca} are absent from output voltage v_0 for this firing angle as well. Without FD, after load voltage v_0 reaches zero, a diode from negative group would begin to conduct reducing v_0 to zero till next SCR in sequence is triggered. For example, at $\omega t = 120°$, $v_0 = v_{cb} = 0$ and without FD, D3 from negative group would start conducting through T3 from $\omega t = 120°$ to 150° when SCR T1 is gated. This means that without FD, T3 would conduct for 120° from $\omega t = 30°$ to 150°, D2 for 90° from $\omega t = 30°$ to 120° and D3 for 30° from $\omega t = 120°$ to 150° for this periodic cycle of 120° extending from $\omega t = 30°$ to 150°.

For firing angle delay of 120°, the voltage and current waveforms are shown in Fig. 6.31 (e). The load current is now assumed discontinuous. For each periodic cycle of 120°, v_0 is seen to have three components. When an SCR is gated, thyristor and diode conduct for 60° only. As v_0 reaches zero and tends to become negative, FD gets forward biased and therefore starts conducting for some angle and holds the load voltage to zero. When all the energy stored in inductance is discharged, FD stops conducting and as a result, load voltage rises to load counter emf E. When $v_0 = E$, none of the elements of semiconverter bridge is conducting, this is indicated by 0, 0 in Fig. 6.31 (e).

It may be seen from above that in a 3-phase semiconverter, SCRs are gated at an interval of 120° in a proper sequence. In a single phase semiconverter, SCRs are fired at an interval of 180°. In order to obtain full control of the dc output voltage v_0, the range of firing angle is from 0° to 180°. A three-phase semiconverter has the unique feature of working as a *six-pulse converter for $\alpha < 60°$* and as a *three-pulse converter for $\alpha \geq 60°$*, a careful observation of Fig. 6.31 reveals this.

For a 3-phase semiconverter, each periodic cycle of output voltage has a periodicity of 120°. Average output voltage should, therefore, be calculated over 120° only.

For $\alpha < 60°$. For firing angle less than 60°, the output voltage is redrawn in Fig. 6.32 (a) from Fig. 6.31 (b) for some firing angle less than 30° for convenience. In this figure, area *abcefda* divided by $2\pi/3$ would give the average value of output voltage V_0. For area *abcde*, take OO' as the origin and for area *dcefd*, take AA' as the origin. Then

$$V_0 = \frac{3}{2\pi} [\text{Area } abcda + \text{Area } dcefd]$$

$$= \frac{3}{2\pi} \left[\int_{-\left[\frac{\pi}{6} - \alpha\right]}^{\pi/6} V_{ml} \cos \omega t \cdot d(\omega t) + \int_{-\pi/6}^{\left(\frac{\pi}{6} + \alpha\right)} V_{ml} \cos \omega t \cdot d(\omega t) \right]$$

or

$$V_0 = \frac{3\, V_{ml}}{2\pi} (1 + \cos \alpha) \qquad \qquad ...(6.55)$$

With OO' as the origin, angle $\left(\frac{\pi}{6} - \alpha\right)$ is measured to the left of OO', therefore minus sign is put before $\left(\frac{\pi}{6} - \alpha\right)$. Similarly, minus sign is put before $\frac{\pi}{6}$.

Voltage $v_{ab} = 0$ at $\omega t = 0$ and $v_{ac} = 0$ at $\omega t = \dfrac{\pi}{3}$ in Fig. 6.32 (a). Therefore, V_0 can also be obtained as

$$V_0 = \frac{3}{2\pi}\left[\int_{\frac{\pi}{3}+\alpha}^{2\pi/3} V_{ml}\sin\omega t\, d(\omega t) + \int_{\pi/3}^{\frac{2\pi}{3}+\alpha} V_m\sin\omega t\, d(\omega t)\right]$$

$$= \frac{3\,V_{ml}}{2\pi}(1+\cos\alpha) \qquad\qquad ...(6.55)$$

Rms value of output voltage V_{or}, for $\alpha < 60°$, is given by

$$V_{or} = \left[\frac{3}{2\pi}\left\{\int_{-\left(\frac{\pi}{6}-\alpha\right)}^{\pi/6} V_{ml}^2\cos^2\omega t\, d(\omega t) + \int_{\pi/6}^{\frac{\pi}{6}+\alpha} V_{ml}^2\cos^2\omega t\, d(\omega t)\right\}\right]^{1/2}$$

$$V_{or}^2 = \frac{3\,V_{ml}^2}{4\pi}\left[\left|\;\omega t + \frac{\sin 2\omega t}{2}\;\right|_{-\left(\frac{\pi}{6}-\alpha\right)}^{\pi/6} + \left|\;\omega t + \frac{\sin 2\omega t}{2}\;\right|_{-\pi/6}^{\frac{\pi}{6}+\alpha}\right]$$

or

$$V_{or} = \frac{V_{ml}}{2}\sqrt{\frac{3}{\pi}}\left[\frac{2\pi}{3} + \frac{\sqrt3}{2}(1+\cos 2\alpha)\right]^{1/2} \qquad\qquad ...(6.56)$$

For $\alpha \ge 60°$. For $\alpha \ge 60°$, the output voltage waveform is drawn in Fig. 6.32 (b) for a firing angle $60° < \alpha < 90°$ for convenience. With OO' as the origin in this figure, the average output voltage V_0 is given by

$$V_0 = \frac{3}{2\pi}[\text{Area } abcda].$$

$$= \frac{3}{2\pi}\left[\int_{-\left[\frac{\pi}{2}-\alpha\right]}^{\pi/2} V_{ml}\cos\omega t\cdot d(\omega t) = \frac{3\,V_{ml}}{2\pi}(1+\cos\alpha)\right. \qquad ...(6.55)$$

Voltage $v_{ac} = 0$ at $\omega t = \dfrac{\pi}{3}$ and α is measured from $\omega t = \pi/3$ as shown in Fig. 6.32 (b). Therefore, V_0 can also be obtained as

$$V_0 = \frac{3}{2\pi}\int_{\alpha}^{\pi} V_{ml}\sin\omega t\cdot d(\omega t)$$

$$= \frac{3\,V_{ml}}{2\pi}(1+\cos\alpha) \qquad\qquad ...(6.55)$$

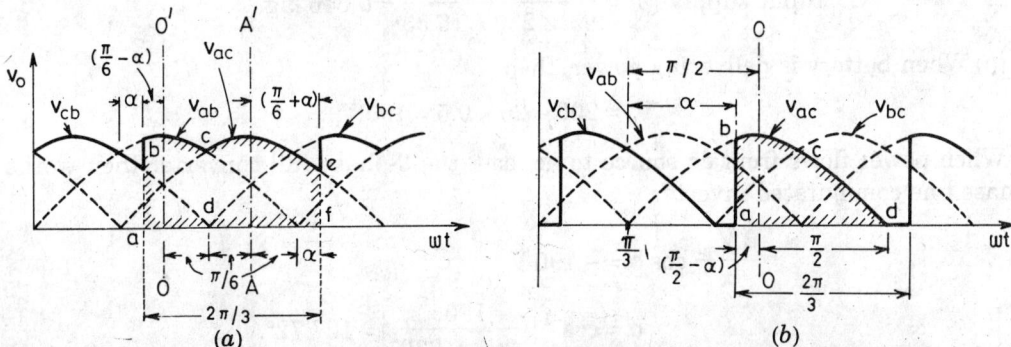

Fig. 6.32. Output voltage waveforms for a 3-phase semiconverter for (a) $\alpha < 60°$ and (b) $\alpha > 60°$.

It is seen from above that expression for average output voltage is the same for both six-pulse and three-pulse operating modes of a 3-phase semiconverter.

Rms value of output voltage V_{or}, for $\alpha > 60°$ is given by

$$V_{or} = \left[\frac{3}{2\pi} \int_{-\left(\frac{\pi}{2}-\alpha\right)}^{\pi/2} V_{ml}^2 \cos^2 \omega t. \, d(\omega t) \right]^{1/2}$$

$$= \frac{V_{ml}}{2} \cdot \sqrt{\frac{3}{\pi}} \left[(\pi - \alpha) + \frac{1}{2} \sin 2\alpha \right]^{1/2} \qquad ...(6.57)$$

Example 6.17. *(a) A 3-phase full converter charges a battery from a three-phase supply of 230 V, 50 Hz. The battery emf is 200 V and its internal resistance is 0.5 Ω. On account of inductance connected in series with the battery, charging current is constant at 20 A. Compute the firing angle delay and the supply power factor.*

(b) In case it is desired that power flows from dc source to ac load in part (a), find the firing angle delay for the same current.

Solution. (a) The battery terminal voltage V_0 is

$$V_0 = 200 + 20 \times 0.5 = 210 \text{ V}$$

But
$$V_0 = \frac{3V_{ml}}{\pi} \cos \alpha = 210 \text{ V}$$

∴
$$\alpha = \cos^{-1} \frac{210 \times \pi}{3\sqrt{2} \times 230} = 47.453°.$$

For constant load current of $I_0 = 20$ A, Fig. 6.28 (d) reveals that supply current i_A is of rectangular (or square) wave of amplitude 20 A. It is also seen from this figure that i_A flows for 120° (or $2\pi/3$ radians) over every half cycle of 180° or π radians.

∴ Rms value of the supply current I_s over π radians is

$$I_s = \left[\frac{1}{\pi} (20)^2 \frac{2\pi}{3} \right]^{1/2} = 20 \sqrt{\frac{2}{3}} = 16.33 \text{ A}$$

Rms value of output current, $I_{or} = 20$ A

Power delivered to load $= EI_0 + I_{or}^2 \cdot r = 200 \times 20 + (20)^2 \times 0.5 = 4200 \text{ W}$

Now $\sqrt{3} \, V_s I_s \cos \phi = 4200 \text{ W}$

∴   Input supply $pf = \dfrac{4200}{\sqrt{3} \times 230 \times 16.33} = 0.646 \text{ lag.}$

(b) When battery is delivering power, then

$$V_0 = 200 - 20 \times 0.5 = 190 \text{ V}$$

When power flows from dc source to ac load, the 3-phase full converter then works as a 3-phase line commutated inverter.

∴
$$\frac{3 \, V_{ml}}{\pi} \cos \alpha = -190 \text{ V}$$

or
$$\alpha = \cos^{-1} \left[\frac{-190 \times \pi}{3\sqrt{2} \times 230} \right] = 127.72°.$$

Example 6.18. *For a 3-phase full converter, sketch the time variations of input voltage and the voltage across one thyristor for one complete cycle for a firing angle delay of (a) 0° and (b) 30°.*

For both the angles, find the magnitude of reverse voltage across this SCR and its commutation time for a three-phase supply voltage of 230 V , 50 Hz.

Solution. (a) The three-phase input voltage waveform is shown as $v_{cb}, v_{ac}, v_{bc}, v_{ba}$ etc. in Fig. 6.33 (a). A conducting SCR has zero voltage across it. Let the variation of voltage across SCR T_1 belonging to positive group of Fig. 6.26 be plotted in Fig. 6.33 (a).

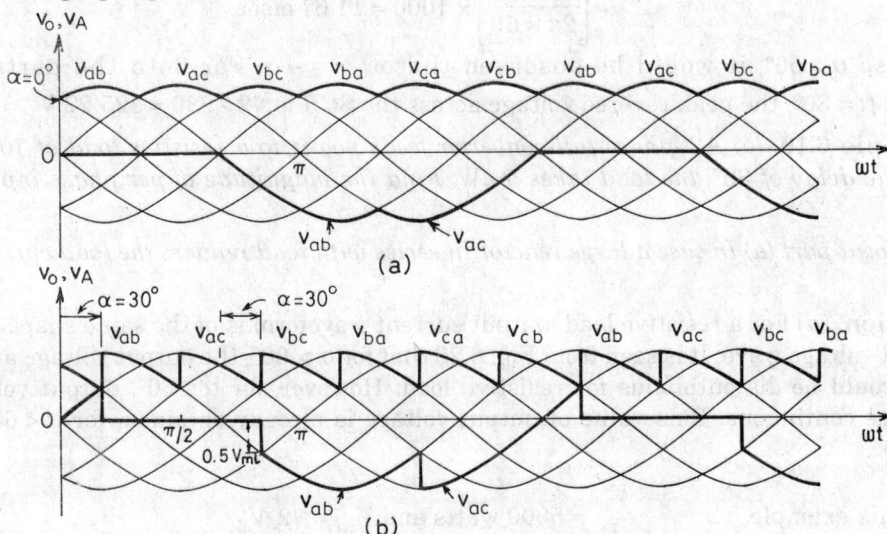

Fig. 6.33. Pertaining to Example 6.18.

For voltages v_{ub}, v_{uc}, SCR T1 conducts, therefore voltage across this SCR is $v_A = v_a - v_a = 0$ for a period of 120°, *i.e.* from $\omega t = 0$ to $\omega t = 120°$ as shown. After $\omega t = 120°$, SCR T3 conducts for 120° with v_{bc}, v_{ba} as the output voltages. Now cathode of T1 is connected to supply terminal B through T3 for a period of 120° and its anode to supply terminal A. Therefore, voltage across T1 from $\omega t = 120°$ to 240° is $v_A = v_a - v_b$. This voltage reverse biases T1 and it is shown as v_{ab} below the reference line in Fig. 6.33 (a).

After $\omega t = 240°$, for voltages v_{ca}, v_{cb}; SCR T5 conducts for 120° and therefore cathode of T1 is connected to terminal C through T5. Thus, voltage across T1 is $v_A = v_a - v_c = v_{ac}$ below the reference line for a period of 120° from $\omega t = 240°$. At $\omega t = 2\pi$, T1 is again gated and voltage $v_A = 0$ as shown.

For firing angle delay of zero degree, each SCR is reverse biased for $240° = \dfrac{4\pi}{3}$ rad.

∴ Commutation time available for SCR turn-off = circuit turn-off time, $t_c = \dfrac{4\pi}{3\omega}$ sec

$$= \frac{4\pi \times 1000}{3 \times 2\pi \times 50} = 13.33 \text{ msec.}$$

(b) For $\alpha = 30°$, the output voltage waveform and voltage v_A across SCR T1 are shown in Fig. 6.33 (b). At $\omega t = 150°$, when SCR T1 stops conducting, voltage across T1 must follow v_{ab} curve for 120° as discussed in part (a) above. Therefore, v_A jumps from zero to

$V_{ml} \sin 30 = 0.5\,V_{ml}$ as shown. As T3 remains on for 120°, v_A follows v_{ab} curve below the reference line for 120°. At $\omega t = 270°$, when T5 is gated, v_A follows v_{ac} for 120° till T1 is gated again, Fig. 6.33 (b). This figure reveals that each SCR is reverse biased for $(240° - \alpha)$.

$\therefore$ Commutation time available for SCR turn-off = Circuit turn-off time,

$$t_c = \frac{4\pi/3 - \alpha}{\omega} \text{ sec}$$

$$= \left[\frac{\dfrac{4\pi}{3} - \dfrac{\pi}{6}}{2\pi \times 50} \right] \times 1000 = 11.67 \text{ msec.}$$

In case $\alpha \ge 60°$, it would be observed that $\omega t_c = \pi - \alpha$. For both the parts, i.e. for $\alpha = 0°$ and $\alpha = 30°$, the peak reverse voltage across the SCR is $\sqrt{2} \cdot 230 = 325.22$ V.

Example 6.19. (a) A 3-phase full-converter feeds power to a resistive load of 10 Ω. For a firing angle delay of 30°, the load takes 5 kW. Find the magnitude of per phase input supply voltage.

(b) Repeat part (a) in case a large reactor in series with load renders the load current ripple free.

Solution. (a) For a resistive load, output current waveform is of the same shape as that of the output voltage wave. It is seen from Fig. 6.28 that for $\alpha > 60°$, the output voltage and output current would be discontinuous for resistive load. However, for $\alpha \le 60°$, output voltage and current are continuous. Rms value of output voltage is already obtained, for $\alpha < 60°$, in Eq. (6.54a).

For this example, $\quad \dfrac{V_{or}^2}{R} = 5000$ watts and $V_{ml} = \sqrt{2}\,V_s$

$\therefore \qquad 2\,V_s^2 \dfrac{3}{2\pi} \left[\dfrac{\pi}{3} + \dfrac{\sqrt{3}}{2} \cos 60 \right] = 5000 \times 10$

or $\qquad\qquad V_s = 188.08$ V

$\therefore$ Per phase voltage, $\quad V_{ph} = \dfrac{V_s}{\sqrt{3}} = 108.591$ V

(b) For a constant load current, average load current I_0 = rms load current,

$$I_{or} = \frac{V_0}{R} = \left(\frac{3V_{ml}}{\pi} \cos \alpha \right) \frac{1}{R}$$

$\therefore \qquad I_{or}^2 \times R = \left[\dfrac{3\,V_{ml}}{\pi} \cos \alpha \right]^2 \dfrac{1}{R} = 5000$ W

or $\qquad\qquad V_s = \sqrt{50000} \times \dfrac{\pi}{\sqrt{2} \times 3 \cos 30°} = 191.22$ V

$\therefore \qquad\qquad V_{ph} = 110.40$ V.

Example 6.20. (a) A 3-phase semiconverter feeds power to a resistive load of 10 Ω. For a firing angle delay of 30°, the load takes 5 kW. Find the magnitude of per phase input supply voltage.

(b) Repeat part (a) in case load current is made ripple free by connecting an inductor in series with the load.

Solution. It is seen from Fig. 6.31 that for $\alpha < 30°$, the output voltage is continuous. For a resistive load, output current is also continuous. Rms value of output voltage, for $\alpha < 30°$, is already obtained in Eq. (6.56).

$\therefore$ For $\quad \alpha = 30°, \dfrac{2 V_s^2}{4} \dfrac{3}{\pi}\left[\dfrac{2\pi}{3} + \dfrac{\sqrt{3}}{2}(1 + \cos 60)\right] = 5000 \times 10$

or $\quad V_s = 175.67$ V and $V_{ph} = 101.43$ V

(a) For constant load current, I_{or} = average load current, $I_0 = \dfrac{V_0}{R} = \dfrac{3V_{ml}}{2\pi}(1 + \cos \alpha)\dfrac{1}{R}$

$\therefore \quad I_{or}^2 \times R = \left[\dfrac{3 V_{ml}}{2\pi}(1 + \cos 30°)\right]^2 \dfrac{1}{10} = 5000$ W

or $\quad V_s = \sqrt{50000} \times \dfrac{\sqrt{2}\,\pi}{3 \times 1.866} = 177.44$ V and $V_{ph} = 102.45$ V

Example 6.21. *Repeat Example 6.15 in case firing angle delay is 90°.*

Solution. (a) Fig. 6.31 shows that for $\alpha = 90°$, the output voltage is discontinuous. For a resistive load, output current is also discontinuous, for $\alpha > 60°$, rms value of output voltage is already obtained in Eq. (6.57).

$\therefore$ For $\alpha = 90°, \quad \dfrac{2 V_s^2}{4} \cdot \dfrac{3}{\pi}\left[\left(\pi - \dfrac{\pi}{2}\right) + \dfrac{1}{2}\sin 180°\right] \times \dfrac{1}{10} = 5000$ W

or $\quad V_s = \sqrt{50000} \times \dfrac{4}{3} = 298.14$ V

and $\quad V_{ph} = 172.14$ V

(b) For constant load current, $I_{or} = I_0 = \dfrac{V_0}{R} = \left[\dfrac{3V_{ml}}{2\pi}(1 + \cos \alpha)\right] \times \dfrac{1}{R}$

$\therefore \quad I_{or}^2 \times R = \left[\dfrac{3 V_{ml}}{2\pi}(1 + \cos 90)\right]^2 \dfrac{1}{10} = 5000$ W

or $\quad V_s = \sqrt{50000} \times \dfrac{\sqrt{2}\,\pi}{3} = 331.153$ V and $V_{ph} = 191.2$ V

Example 6.22. *Repeat part (a) of Example 6.17 in case 3-phase full converter is replaced by a 3-phase semiconverter.*

Solution. The battery terminal voltage is
$$V_0 = 200 + 20 \times 0.5 = 210 \text{ V}$$

But $\quad V_0 = \dfrac{3 V_{ml}}{2\pi}(1 + \cos \alpha) = 210$ V

$\therefore \quad \alpha = \cos^{-1}\left[\dfrac{210 \times 2\pi}{3 \cdot \sqrt{2},\, 230} - 1\right] = 69.37°$

An examination of Fig. 6.31 reveals that for firing angle $\alpha > 60°$, each SCR conducts for $180 - \alpha$. So, in this example, each SCR conducts for $(180 - 69.37) = 110.63°$. For constant load current of $I_0 = 20$ A, supply current i_A is of square wave of amplitude 20 A. As i_A flows for $110.63°$ over every half cycle of $180°$, the rms value of supply current I_s is given by

$$I_{sr} = \left[\frac{1}{\pi}(20)^2 \frac{110.63 \times \pi}{180}\right]^{1/2} = 20\sqrt{\frac{110.63}{180}} = 15.68 \text{ A}$$

Power delivered to load

$$= V_0 I_0 = 210 \times 20 = 4200 \text{ W}$$

∴ Input supply

$$pf = \frac{4200}{\sqrt{3} \times 230 \times 15.68} = 0.6724 \text{ lag.}$$

Example 6.23. *A 3-phase full converter, fed from delta-star transformer, is connected to load RL requiring ripple free load current. For a firing angle delay of around 45°, sketch waveforms for (a) input voltage v_{ab}, v_{ac}, v_{bc} etc. (b) load voltage, load current, thyristor T1 and T4 currents and phase a, b, c currents.*

In case ac supply is 3-phase, 400 V, 50 Hz, level load current = 15 A and α = 45°, calculate rectification efficiency, TUF and input power factor.

Solution. The power circuit diagram of 3-phase full converter, fed from three-phase delta-star transformer, is shown in Fig. 6.34 (a).

(a) The input voltage waveform v_{ab}, v_{ac}, v_{bc} etc. is sketched in Fig. 6.34 (b) with v_{ab} zero at $\omega t = 0$. The maximum value of v_{ab}, v_{ac} or v_{bc} is V_{ml}, where V_{ml} = maximum value of line voltage $V_l = \sqrt{3} \, V_{mp}$. Note that v_{ab}, v_{ac}, v_{bc} etc. are displaced from each other by an angle of $60°$.

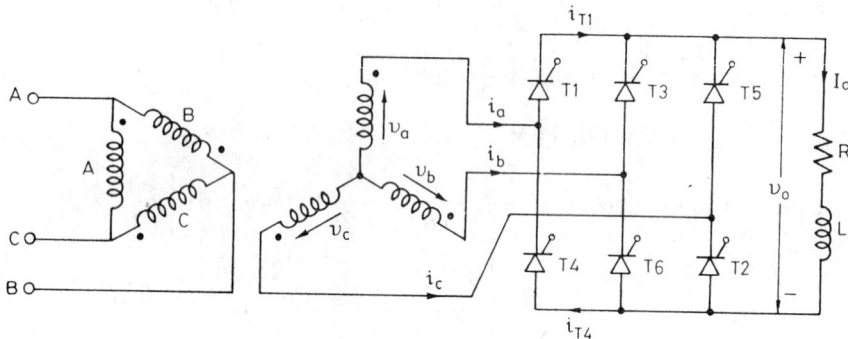

Fig. 6.34 (a) Pertaining to Example 6.23.

For a firing-angle delay of 45°, load voltage waveform v_o is drawn. Load current I_o is constant throughout, SCR $T1$ conducts when first subscript in load voltage is 'a'. So $T1$ conducts when load voltage pulses are v_{ab}, v_{ac}. Likewise, T4 conducts for load voltage pulses v_{ba}, v_{ca}. Source current i_a, as shown in Fig. 6.34 (b), is positive when $T1$ conducts and negative when T4 conducts. Current i_b is shown displaced by 120° from i_a. Similarly, i_c is drawn.

(b) Here $I_o = 15$ A, $I_{or} = 15$ A, because load current is ripple free. Average value of SCR current, from i_{T1} waveform, is

$$I_{TA} = I_o \times \frac{120}{360} = \frac{I_o}{3} = \frac{15}{3} = 5 \text{ A}$$

Rms value of thyristor current, $I_{Tr} = \sqrt{I_o^2 \times \frac{120}{360}} = \frac{I_o}{\sqrt{3}} = \frac{15}{\sqrt{3}} = 8.66 \text{ A}$

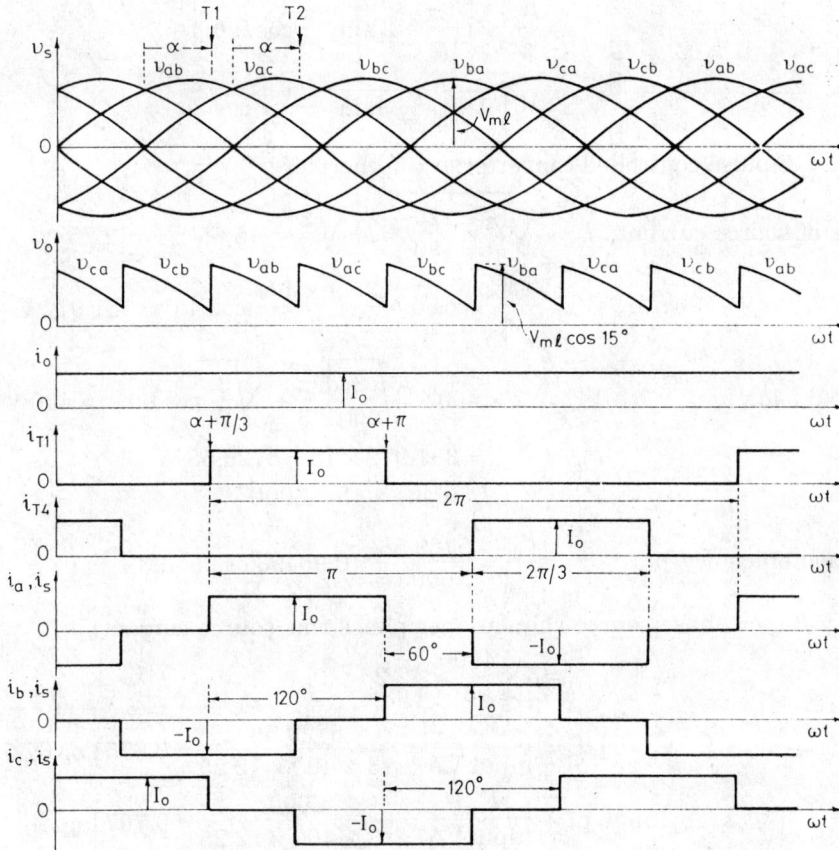

Fig. 6.34 (*b*) Waveform for a 3-phase full converter, Example 6.23.

Fig. 6.35 illustrates the output voltage waveform for a *p*-pulse controlled converter. A review of Art. 3.9.3 and Fig. 3.34 is also helpful. The origin is taken at OO', the peak value of the instantaneous output voltage v_o. Firing angle must be measured from the intersection of phase voltage waveforms as shown. Therefore, the limits of integration are from $-\left(\dfrac{\pi}{p}-\alpha\right)$ to $\left(\dfrac{\pi}{p}+\alpha\right)$ as illustrated in Fig. 6.35.

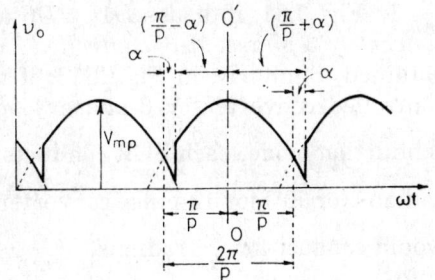

Fig. 6.35. Output voltage waveform for *p*-pulse controlled converter

∴ Average output voltage, V_o is

$$V_o = \frac{p}{2\pi} \int_{\left(\frac{\pi}{p}-\alpha\right)}^{\frac{\pi}{p}+\alpha} V_{mp} \cos \omega t . d(\omega t) = \frac{p . V_{mp}}{2\pi} \left| \sin \omega t \right|_{-\left(\frac{\pi}{p}-\alpha\right)}^{\frac{\pi}{p}+\alpha}$$

$$= V_{mp} . \left(\frac{p}{\pi}\right) \sin\left(\frac{\pi}{p}\right) . \cos \alpha \qquad ...(6.58)$$

Rms value of output voltage, V_{or} is

$$V_{or} = \left[\frac{p}{2\pi} \int_{-\left(\frac{\pi}{p}-\alpha\right)}^{\frac{\pi}{p}+\alpha} V_{mp}^2 \cos^2 \omega t \, d(\omega t) \right]^{1/2}$$

$$V_{or}^2 = \frac{p \cdot V_s^2}{2\pi}\left[\frac{2\pi}{p} + 2\sin\frac{2\pi}{p}\cos 2\alpha\right]$$

or
$$V_{or} = V_s\left[1 + \left(\frac{p}{2\pi}\right)\sin\left(\frac{2\pi}{p}\right)\cos 2\alpha\right]^{1/2} \qquad ...(6.59)$$

For 3-phase, 6-pulse controlled converter, $p = 6$, therefore,

rms value of source current, $I_{sr} = \sqrt{I_o^2 \times \frac{120}{180}} = I_o\sqrt{\frac{2}{3}} = 15.\sqrt{\frac{2}{3}} = 12.25$ A

Here,
$$V_o = \frac{3\,V_{ml}}{\pi}\cos\alpha = \frac{3\sqrt{2}\times 400}{\pi}\cos 45° = 381.972 \text{ V}$$

From Eq. (6.54a),
$$V_{or} = \sqrt{2}\times 400 \times \sqrt{\frac{3}{2\pi}\left[\frac{\pi}{3} + \sqrt{\frac{3}{2}}\cos 90°\right]} = 400 \text{ V}$$

$$P_{dc} = V_o\,I_o = 381.972 \times 15 = 5729.58 \text{ W}$$
$$P_{ac} = V_{or} \cdot I_{or} = 400 \times 15 = 6000 \text{ W}$$

Rectification efficiency
$$= \frac{P_{dc}}{P_{ac}} = \frac{5729.58}{6000} = 0.95493 \text{ or } 95.493\%$$

Input VA = 3 (per phase source voltage) (per phase rms source current)
$$= 3 \times \frac{400}{\sqrt{3}} \times 12.25$$

$$\therefore \qquad \text{TUF} = \frac{P_{dc}}{\text{input VA}} = \frac{5729.58}{\sqrt{3}\times 400 \times 12.25} = 0.6751 \text{ or } 67.51\%$$

$$\text{Input pf} = \frac{P_{ac}}{\text{input VA}} = \frac{6000}{\sqrt{3}\times 400 \times 12.25} = 0.707 \text{ lag}.$$

6.7.4. Multi-pulse Controlled Converters

In Fig. 3.31, if diodes $D1,.....D6$ are replaced by thyristors $T1,.....T6$, a *6-pulse controlled converter, 3-phase M-6 controlled converter* or *6-phase half-wave controlled converter* is obtained. Similarly, in Fig. 3.42, if twelve diodes are replaced by twelve SCRs, a 12-pulse controlled converter or 3-phase 12-pulse controlled converter is obtained. For continuous conduction mode, each SCR conducts for $\frac{2\pi}{6}$ radians (or 60°) for 6-pulse converter and for $\frac{2\pi}{12}$ radians (or 30°) for 12-pulse converter. In general, for a *p*-pulse controlled converter, each SCR would conduct for $\frac{2\pi}{p}$ radians.

$$V_o = V_{mp}\left(\frac{6}{\pi}\right)\sin\left(\frac{\pi}{6}\right)\cos\alpha = \frac{3\,V_{mp}}{\pi}\cos\alpha$$

$$V_{or} = V_s\left[1 + \left(\frac{6}{2\pi}\right)\sin\left(\frac{2\pi}{6}\right)\cos 2\alpha\right]^{1/2}$$

$$= V_s\left[1 + \frac{\sqrt{3}}{4\pi}\cos 2\alpha\right]^{1/2}$$

For a 3-phase 12-pulse converter, $p = 12$, therefore,

$$V_o = V_{mp} \cdot \left(\frac{12}{\pi}\right)\sin\left(\frac{\pi}{12}\right)\cos\alpha = 0.98816\,V_{mp}\cos\alpha$$

$$V_{or} = V_s \left[1 + \left(\frac{12}{2\pi} \right) \sin \left(\frac{2\pi}{12} \right) \cos 2\alpha \right]^{1/2}$$

$$= V_s \left[1 + \frac{3}{\pi} \cos 2\alpha \right]^{1/2}$$

In a 6-pulse converter, with resistive load, continuous conduction occurs for $0 < \alpha < 60°$ and discontinuous conduction for $60 < \alpha \leq 120°$. The maximum possible firing angle is therefore, 120°.

In a 12-pulse converter, with resistive load, continuous conduction occurs for $0 < \alpha < 75°$ and discontinuous conduction for $75 < \alpha \leq 105°$. The maximum possible firing angle is 105°.

6.8. PERFORMANCE PARAMETERS OF 3-PHASE FULL CONVERTERS

Here the performance parameters of 3-phase full converter are derived from the various waveforms sketched in Fig. 6.36. In a 3- phase system, power-factor angle is defined as the angle between phase voltage and phase current. This is the reason for showing the phase voltage v_a, v_b, v_c in Fig. 6.36. Line voltage v_{ab}, v_{ac}, v_{bc} etc. facilitate the sketching of output voltage waveform v_o, so these are also shown in Fig. 6.36. As before, load current is assumed ripple free.

The source current i_s is given by

$$t_s(t) = I_o + \sum_{n=1,2,3}^{\infty} C_n \sin (n\omega t + \theta_n)$$

Here, $I_o = 0$, because positive and negative half cycles of source current i_s, about the reference line ωt, are identical in Fig. 6.36. Further, waveform of source current i_s reveals that for phase 'a', positive pulse is from $\left(\alpha + \frac{\pi}{6} \right)$ to $\left(\alpha + \frac{5\pi}{6} \right)$ and negative pulse is from $\left(\alpha + \frac{7\pi}{6} \right)$ to $\left(\alpha + \frac{11.\pi}{6} \right)$.

$$\therefore \qquad a_n = \frac{1}{\pi} \left[\int_{\alpha + \frac{\pi}{6}}^{\alpha + \frac{5\pi}{6}} I_o . \cos n\omega t . d(\omega t) - \int_{\alpha + \frac{7.\pi}{6}}^{\alpha + \frac{11.\pi}{6}} I_o \cos n\omega t . d(\omega t) \right]$$

$$= \frac{I_o}{n.\pi} \left[\left| \sin n\omega t \right|_{\alpha + \pi/6}^{\alpha + 5\pi/6} - \left| \sin n\omega t \right|_{\alpha + 7\pi/6}^{\alpha + 11\pi/6} \right]$$

For $n = 1$, $\qquad a_1 = \frac{4 I_o}{\pi} \left[- \sin \alpha . \frac{\sqrt{3}}{2} \right]$

In general, $a_n = - \frac{4 I_o}{n\pi} . \sin n\alpha . \sin \frac{n\pi}{3}$ ..... for $n = 1, 3, 5.$

$$= 0 \text{ .... for } n = 1, 4. 6 \text{ ....}$$

$$b_n = \frac{1}{\pi} \left[\int_{\alpha + \frac{\pi}{6}}^{\alpha + \frac{5.\pi}{6}} I_o \sin n\omega t . d(\omega t) - \int_{\alpha + \frac{7.\pi}{6}}^{\alpha + \frac{11\pi}{6}} I_o . \sin n\omega t \, d(\omega t) \right]$$

$$= \frac{I_o}{n\pi} \left[\left| - \cos n\omega t \right|_{\alpha + \pi/6}^{\alpha + 5\pi/6} - \left| - \cos n\omega t \right|_{\alpha + 7\pi/6}^{\alpha + 11\pi/6} \right]$$

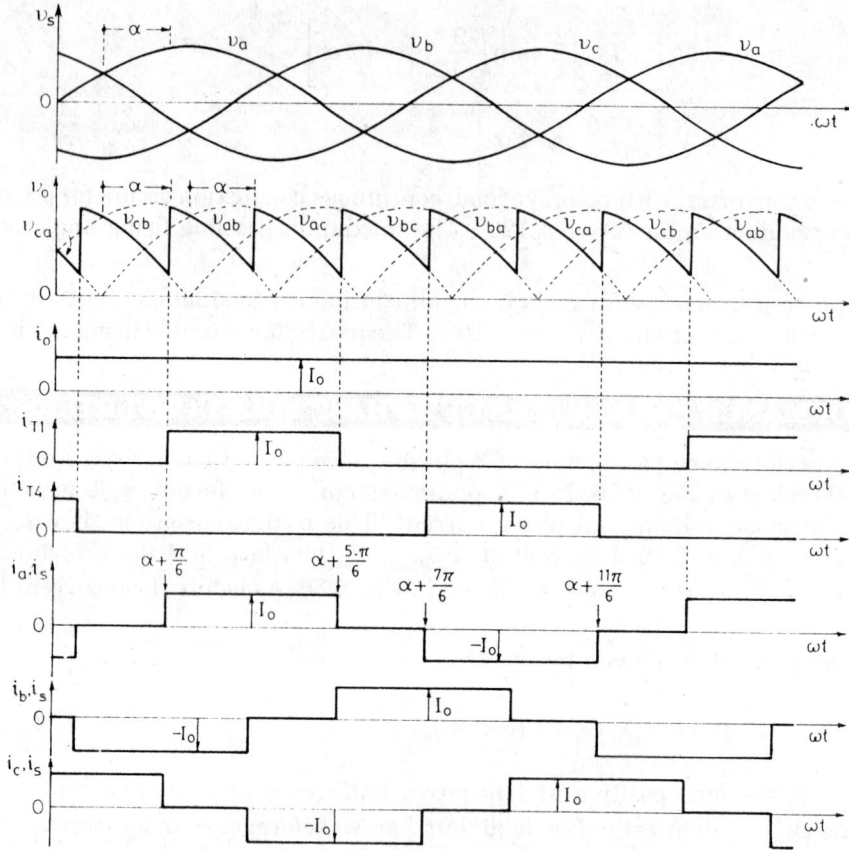

Fig. 6.36. Waveforms for 3-phase full converter with $v_a = 0$ when $\omega t = 0$.

$$= \frac{4 I_o}{n \pi} \cos n \alpha . \sin \frac{n\pi}{3} \quad ...... \text{ for } n = 1, 3, 5, ...$$

$$= 0 \quad\quad ...... \text{ for } n = 2, 4, 6, ...$$

$\therefore$

$$C_n = \left[\left(-\frac{4 I_o}{n\pi} . \sin n\alpha . \sin \frac{n\pi}{3} \right)^2 + \left(\frac{4 I_o}{n\pi} . \cos n\alpha . \sin \frac{n\pi}{3} \right)^2 \right]^{1/2}$$

$$= \frac{4 I_o}{n\pi} \sin \frac{n\pi}{3} \quad \text{and} \quad \theta_n = \tan^{-1} [-\tan n\alpha] = -n\alpha$$

$$\theta_n = \tan^{-1} \left(\frac{a_n}{b_n} \right) = \tan^{-1} (-\tan n\alpha) = -n\alpha$$

$\therefore$

$$i_s (t) = \sum_{n = 1, 3, 5,}^{\infty} \frac{4 I_o}{n\pi} . \sin \frac{n\pi}{3} \sin (n\omega t - n\alpha) \qquad ...(6.60)$$

Rms value of nth harmonic, $I_{sn} = \dfrac{4 I_o}{\sqrt{2} . n\pi} \sin \dfrac{n\pi}{3}$

Rms value of fundamental current, $I_{s1} = \dfrac{2\sqrt{2} I_o}{\pi} \sin 60° = \dfrac{\sqrt{6}}{\pi} I_o \qquad ...(6.61)$

Rms value of total source current, $I_s = \sqrt{I_s^2 \times \dfrac{2\pi}{3} \times \dfrac{1}{\pi}} = I_o\sqrt{\dfrac{2}{3}}$ ...(6.62)

$$DF = \cos\theta_1 = \cos(-\alpha) = \cos\alpha$$

$$CDF = \frac{I_{s1}}{I_s} = \frac{\sqrt{6}}{\pi}. I_o \times \frac{1}{I_o}. \sqrt{\frac{3}{2}} = \frac{3}{\pi} = 0.955$$

$$PF = CDF \times DF = \frac{3}{\pi}\cos\alpha = 0.995\cos\alpha \qquad ...(6.63)$$

$$HF = THD = \left[\frac{1}{CDF^2} - 1\right]^{1/2} = \left[\left(\frac{\pi}{3}\right)^2 - 1\right]^{1/2} = 0.31084 \text{ or } 31.084\%$$

From Eq. (6.55), $V_{or} = V_{ml}\sqrt{\dfrac{3}{2\pi}}\left[\dfrac{\pi}{3} + \dfrac{\sqrt{3}}{2}\cos 2\alpha\right]^{1/2}$

$\therefore$ $VRF = \left[\left(\dfrac{V_{or}}{V_o}\right)^2 - 1\right]^{1/2} = \left[\dfrac{3.V_{ml}^2}{2\pi}\left(\dfrac{\pi}{3} + \dfrac{\sqrt{3}}{2}\cos 2\alpha\right).\dfrac{\pi^2}{9.V_{ml}^2.\cos^2\alpha} - 1\right]^{1/2}$

$$= \left[\frac{\pi}{6\cos^2\alpha}\left(\frac{\pi}{3} + \frac{\sqrt{3}}{2}\cos 2\alpha\right) - 1\right]^{1/2} \qquad ...(6.64)$$

Active power input, $P_i = 3\,V_s\,I_{s1}.\cos\theta_1 = 3.\dfrac{V_l}{\sqrt{3}}.\dfrac{\sqrt{6}}{\pi}.I_o.\cos\alpha$

$$= \left(\frac{3V_{ml}}{\pi}.\cos\alpha\right)I_o = V_o I_o \qquad ...(6.65)$$

where $V_s = \dfrac{\text{line voltage, } V_l}{\sqrt{3}}$ = per-phase source voltage.

Reactive power input, $Q_i = 3\,V_s.\,I_{s1}.\sin\theta_1$

$$= 3.\frac{V_l}{\sqrt{3}}.\frac{\sqrt{6}}{\pi}.I_o.\sin\alpha = \frac{3V_{ml}}{\pi}I_o.\sin\alpha \qquad ...(6.66\ a)$$

But $V_o = \dfrac{3\,V_{ml}}{\pi}\cos\alpha$ or $\dfrac{3\,V_{ml}}{\pi} = \dfrac{V_o}{\cos\alpha}$

$\therefore$ $Q_i = \dfrac{V_o}{\cos\alpha}.I_o.\sin\alpha = V_o\,I_o\tan\alpha$ ...(6.66 b)

Example 6.24. *A 3-phase full converter delivers a ripple free load current of 10 A with a firing angle delay of 45°. The input voltage is 3-phase, 400 V, 50 Hz.*

(a) Express the source current in Fourier series.

(b) Find the DF, CDF, THD and PF.

(c) Calculate the active a reactive input powers.

Solution. It is seen from Eq. (6.60) that source current for a 3-phase full converter is given by

$$i_s(t) = \sum_{n=1,3,5}^{\infty} \frac{4\,I_o}{n\pi}\sin\frac{n\pi}{3}\sin(n\omega t - n\alpha)$$

Amplitudes of source current for different harmonics are as under:

For $n = 1$, $\dfrac{4\,I_o}{\pi}\sin\dfrac{\pi}{3} = \dfrac{4 \times 10}{\pi} \times \dfrac{\sqrt{3}}{2} = 11.03$ A and $\alpha = 45°$

For $n = 3$, $\dfrac{4 \times 10}{3\pi}\sin\pi = 0$

For $n = 5$, $\dfrac{4 \times 10}{5 \times \pi} \sin 300° = -2.205$ A

For $n = 7$, $= 4 \times \dfrac{10}{7 \times \pi} \sin 420° = 1.575$ A

$\therefore i_s(t) = 11.03 \sin(\omega t - 45) - 2.205 \sin 5(\omega t - 45) + 1.575 \sin 7(\omega t - 45°)$.

(b) DF $= \cos 45 = 0.707$

Here $\qquad I_{s1} = \dfrac{4 I_o}{\sqrt{2} . \pi} \sin \dfrac{\pi}{3} = \dfrac{\sqrt{6}}{\pi} I_o$

Rms value of source current, from Eq. (6.62), is $I_{sr} = I_o \sqrt{\dfrac{2}{3}}$

$\therefore \qquad CDF = \dfrac{I_{s1}}{I_s} = \dfrac{\sqrt{6} I_o}{\pi} \times \dfrac{1}{I_o} \sqrt{\dfrac{3}{2}} = \dfrac{3}{\pi} = 0.955$

$$THD = \left[\dfrac{1}{CDF^2} - 1\right]^{1/2} = \left[\left(\dfrac{\pi}{3}\right)^2 - 1\right]^{1/2} = 0.31084 \text{ or } 31.084\%$$

$$PF = CDF \times DF = \dfrac{3}{\pi} \times \cos 45° = 0.6751 \text{ lag}$$

(c) Active input power, from Eq.(6.65), is

$$P = \left(\dfrac{3 V_{ml}}{\pi} \cos \alpha\right) I_o = \left(\dfrac{3 \sqrt{2} \times 400}{\pi} \cos 45\right) \times 10 = 3819.72 \text{ W}$$

From Eq. (6.66), $\quad Q = \left(\dfrac{3 V_{ml}}{\pi} \sin \alpha\right). I_o = \left(\dfrac{3 \sqrt{2} \times 400}{\pi} \sin 45\right) \times 10 = 3819.72 \text{ VAr}$

Example 6.25. *A 3-phase full converter, fed from 3-phase, 400 V, 50 Hz source, is connected to load R = 10 Ω, E = 350 V and large inductance so that output current is ripple free. Calculate the power delivered to load and input pf for (a) firing angle of 30° and (b) firing advance angle of 60°.*

Solution. (a) For $\alpha = 30°$, $V_o = \dfrac{3 V_{ml}}{\pi} \cos \alpha = \dfrac{3 \sqrt{2} \times 400}{\pi} \cos 30° = 467.734$ V

But $\qquad V_o = E + I_o R$

$\therefore$ Load current $\quad I_o = \dfrac{467.734 - 350}{10} = 11.7734$ A

As load current is constant rms, value of load current $I_{or} = 11.7734$ A.

Power delivered to load $= EI_o + I_{or}^2 \times R$

$\qquad = 350 \times 11.7734 \times 11.7734^2 \times 10 = 5506.82$ W

or $\qquad = V_o I_o = 467.734 \times 11.7734 = 5506.82$ W

Rms value of source current, from Eq. (6.62), is $I_{sr} = I_o \sqrt{\dfrac{2}{3}} = 11.7734. \sqrt{\dfrac{2}{3}} = 9.613$ A

Input VA $= 3 V_s . I_{sr} = 3 \times \dfrac{400}{\sqrt{3}} \times 9.613$

$\therefore$ Power factor $\quad = \dfrac{\text{Power in load}}{\text{Input VA}} = \dfrac{5506.82}{\sqrt{3} \times 400 \times 9.613} = 0.82686$ lag

Also from Eq. (6.63), $pf = \dfrac{3}{\pi}. \cos 30° = 0.82697$ lag

(b) For firing advance angle of 60°, $\alpha = 180 - 60 = 120°$

$\therefore \qquad V_o = \dfrac{3 \sqrt{2} \times 400}{\pi} \cos 120° = -270.05$ V

As V_o is negative, this converter is operating as line-commutated inverter. The polarity of load emf E must therefore, be reversed.

Now
$$V_o = -E + I_o R$$
$$-270.05 = -350 + I_o R$$

∴ Load current, $I_o = \dfrac{350 - 270.05}{10} = 7.995$ A

Rms value of load current, $I_{or} = I_o = 7.995$ A

Power delivered by the battery to the *ac* source through the line-commutated inverter,
$$P = EI_o - I_{or}^2 R = V_o I_o = 270.05 \times 7.995 = 2159.05 \text{ W}$$

Rms value of source current, $I_{sr} = I_o . \sqrt{\dfrac{2}{3}} = 7.995 . \sqrt{\dfrac{2}{5}} = 6.528$ A

Supply VA $= 3 . V_s . I_{sr} = 3 \times \dfrac{400}{\sqrt{3}} \times 6.528$

∴ Power factor $= \dfrac{\text{Power delivered to ac source}}{\text{supply VA}} = \dfrac{2159.05}{\sqrt{3} \times 400 \times 6.528} = 0.4774$ lag

Also, from Eq. (6.63), $pf = \dfrac{3}{\pi} \cos \alpha = \dfrac{3}{\pi} \cos 120° = 0.4775$ lag.

6.9. EFFECT OF SOURCE IMPEDANCE ON THE PERFORMANCE OF CONVERTERS

For single-phase and three-phase full converters, derivation of the average output voltages, as given by Eqs. (6.26) and (6.54), has been obtained on the assumption that current transfers from the outgoing SCRs to the incoming SCRs instantaneously. This means that when incoming SCRs T1 and T2 are fired in a single-phase full converter, Fig. 6.10 (*a*), outgoing SCRs T3 and T4 get turned off due to the application of reverse voltage and the current shifts to SCRs T1 and T2 instantaneously. This is possible only if the voltage source has no internal impedance. Actually, the source does possess internal impedance. If the source impedance is resistive, then there will be a voltage drop across the resistance and the average voltage output of a converter gets reduced by an amount equal to $I_0 . r_s$ for a single-phase converter and by $2I_0 . r_s$ for a 3-phase converter. Here I_0 is the constant dc load current and r_s is the source resistance per phase. Since source resistance is usually low, it is assumed that duration of the commutation is very small and the current transfer takes place immediately after the incoming SCRs are fired. However, a voltage drop caused by source resistance must be taken into consideration as discussed above.

In the following lines, the source impedance is taken as purely inductive. The load inductance is assumed large so that output current is virtually constant. The source inductance causes the outgoing and incoming SCRs to conduct together. During the commutation period (when both incoming and outgoing SCRs are conducting together), the output voltage is equal to the average value of the conducting-phase voltages. For a single-phase converter, the load voltage will be zero and for a 3-phase converter, the load voltage is $(v_a + v_b)/2$ (average value of the conducting phases *a* and *b*). The commutation period in seconds, when outgoing and incoming SCRs are conducting together, is also known as the *overlap period*. The angular-period, during which both the incoming and outgoing SCRs are conducting, is known as *commutation angle* or *overlap angle* μ in degrees or radians. The effect of source inductance is investigated in this section for both single-phase and three-phase full converters. It would be observed that the effect of source inductance is (i) to lower the mean output voltage, (ii) to distort the output voltage and current waveforms and (iii) to modify the performance parameters of the converter.

6.9.1. Single-phase Full Converter

The commutation overlap is more predominant in full converters than in semiconverters.

In the single-phase full converter shown in Fig. 6.37 (a), L_s is the source inductance. The load current is assumed constant (analysis with pulsating load current is more involved). Fig. 6.37 (b) gives the equivalent circuit for Fig. 6.37 (a) for analytical purposes. When terminal 1 of source voltage v_s is positive in Fig. 6.37 (a), current i_1 flows through L_s, T1, load and T2; this is shown as v_1, L_s, T1 T2 and load in Fig. 6.37 (b). Similarly, when terminal 2 of v_s is positive, load current i_2 flows through T3, load, T4 ; this is shown as v_2, L_s, T3 T4 and load in Fig. 6.37 (b).

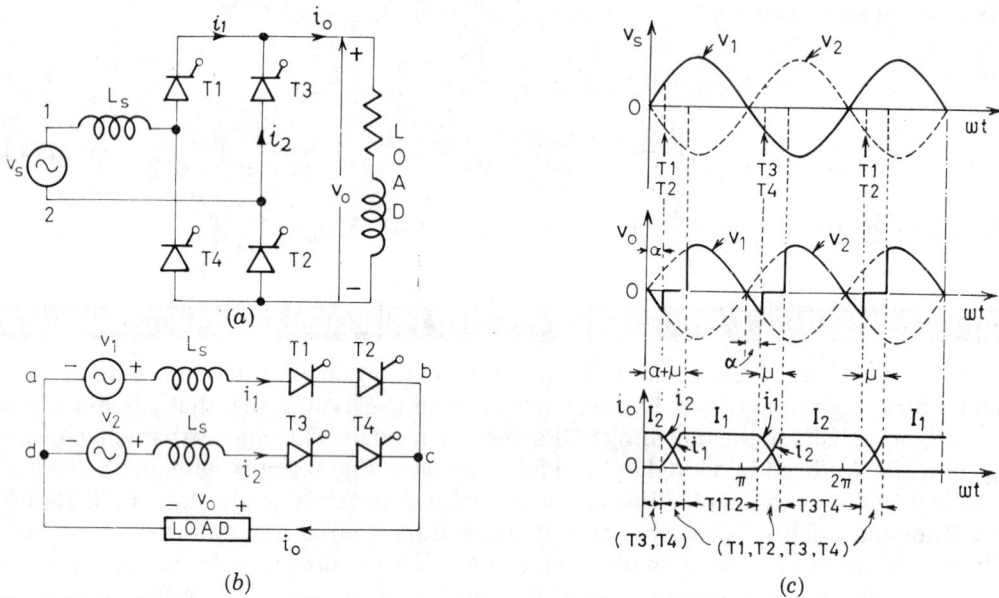

Fig. 6.37. (a) Single-phase full converter with source inductance L_s (b) its equivalent circuit and (c) typical current and voltage waveforms with L_s.

When T1, T2 are triggered at a firing angle α, the commutation of already conducting thyristors T3, T4 begins. Because of the presence of source inductance L_s, the current through outgoing devices T3, T4 decreases gradually to zero from its initial value of I_0 ; whereas in incoming thyristors T1, T2 ; the current builds up gradually from zero to full value of load current I_0. During the commutation of T1, T2 and T3, T4 ; i.e. during the overlap angle μ, KVL for the loop abcda of Fig. 6.37 (b) gives,

$$v_1 - L_s \cdot \frac{di_1}{dt} = v_2 - L_s \frac{di_2}{dt}$$

or

$$v_1 - v_2 = L_s \left(\frac{di_1}{dt} - \frac{di_2}{dt} \right)$$

It is seen from Fig. 6.37 (c) that if $v_1 = V_m \sin \omega t$, then $v_2 = -V_m \sin \omega t$.

$\therefore$

$$L_s \left(\frac{di_1}{dt} - \frac{di_2}{dt} \right) = 2 V_m \sin \omega t \qquad \qquad ...(6.67)$$

As the load current is assumed constant throughout, $i_1 + i_2 = I_0$. Differentiating this with respect to t, we get

$$\frac{di_1}{dt} + \frac{di_2}{dt} = 0 \qquad \qquad ...(6.68)$$

From Eq. (6.67), $\qquad \frac{di_1}{dt} - \frac{di_2}{dt} = \frac{2V_m}{L_s} \sin \omega t \qquad \qquad ...(6.69)$

Addition of Eqs. (6.68) and (6.69) gives

$$\frac{di_1}{dt} = \frac{V_m}{L_s} \sin \omega t \qquad \qquad ...(6.70)$$

Load current i_1 through thyristor pair T1, T2 builds up from zero to $I_1 = I_0$ during the overlap angle μ; i.e. at $\omega t = \alpha$, $i_1 = 0$ and at $\omega t = (\alpha + \mu)$, $i_1 = I_0$

$\therefore$ From Eq. (6.70), $\qquad \int_0^{I_0} di_1 = \frac{V_m}{L_s} \int_{\alpha/\omega}^{(\alpha + \mu)/\omega} \sin \omega t \cdot dt$

or $\qquad \qquad I_0 = \frac{V_m}{\omega L_s} [\cos \alpha - \cos (\alpha + \mu)] \qquad \qquad ...(6.71)$

It is seen from Fig. 6.37 (c) (middle figure) that output voltage v_0 is zero from α to $(\alpha + \mu)$. Thus the average output voltage V_{ox} is given by

$$V_{ox} = \frac{V_m}{\pi} \int_{(\alpha + \mu)}^{(\alpha + \pi)} \sin \omega t \cdot d(\omega t) = \frac{V_m}{\pi} [\cos (\alpha + \mu) - \cos (\alpha + \pi)]$$

$$= \frac{V_m}{\pi} [\cos \alpha + \cos (\alpha + \mu)] \qquad \qquad ...(6.72a)$$

Average value of output voltage at no load, $V_o = \dfrac{2V_m}{\pi} \cos \alpha$

Maximum mean output voltage, $V_{om} = \dfrac{2V_m}{\pi}$

Eq. (6.72a) can therefore be expressed as

$$V_{ox} = \frac{\text{Maximum mean output voltage at no load}}{2} [\cos \alpha + \cos (\alpha + \mu)]$$

$$= \frac{V_{om}}{2} [\cos \alpha + \cos (\alpha + \mu)] \qquad \qquad ...(6.72\ b)$$

From Eq. (6.71), $\cos (\alpha + \mu) = \cos \alpha - \dfrac{\omega L_s}{V_m} I_0$

Substituting this value of $\cos (\alpha + \mu)$ in Eq. (6.72a) gives

$$V_{ox} = \frac{2 V_m}{\pi} \cos \alpha - \frac{\omega L_s}{\pi} I_0 = \frac{2V_m}{\pi} \cos \alpha - 2fL_s \cdot I_0 \qquad \qquad ...(6.73)$$

Also, from Eq. (6.71), $\cos \alpha = \dfrac{\omega L_s}{V_m} I_0 + \cos (\alpha + \mu)$. Substituting this value of $\cos \alpha$ in Eq. (6.72a) gives

$$V_{ox} = \frac{2 V_m}{\pi} \cos (\alpha + \mu) + \frac{\omega L_s}{\pi} I_0 \qquad \qquad ...(6.74)$$

Voltage regulation due to source inductance $= \dfrac{\omega L_s \cdot I_o}{\pi} \times \dfrac{1}{V_o \text{ at no load}}$

$$= \frac{2fL_s \cdot I_o \cdot \pi}{2V_m \cdot \cos \alpha} = \frac{\pi f L_s \cdot I_o}{V_m \cos \alpha} \qquad \qquad ...(6.75\ a)$$

From Eq. (6.71), $\dfrac{\omega L_s \cdot I_o}{\pi} = \dfrac{V_m}{\pi}[\cos \alpha - \cos(\alpha + \mu)]$

For a full-wave diode rectifier, $\alpha = 0°$

$\therefore$ $\dfrac{\omega L_s I_o}{\pi} = \dfrac{V_m}{\pi}[1 - \cos \mu]$

$\therefore$ Inductive voltage regulation $= \dfrac{\dfrac{\omega L_s \cdot I_o}{\pi}}{2V_m/\pi} = \dfrac{\dfrac{V_m}{\pi}(1 - \cos \mu)}{2\,V_m/\pi}$

$$= \left[\frac{1 - \cos \mu}{2}\right] \qquad \qquad ...(6.75\ b)$$

With the help of Eq. (6.74), a dc equivalent circuit for a 2-pulse single-phase full converter can be drawn as shown in Fig. 6.38 (a). In this figure, diode D merely indicates that load current is unidirectional. This equivalent circuit reveals that the effect of source inductance L_s is to present an equivalent resistance of magnitude $\omega L_s/\pi$ ohms in series with internal voltage of the rectifier $(2V_m/\pi) \cos \alpha$. The voltage drop due to L_s is proportional to I_0 and L. Thus as the load current (or source inductance) increases, the commutation interval (or overlap angle) increases and as a consequence, the average output voltage decreases as illustrated in Fig. 6.33 (b).

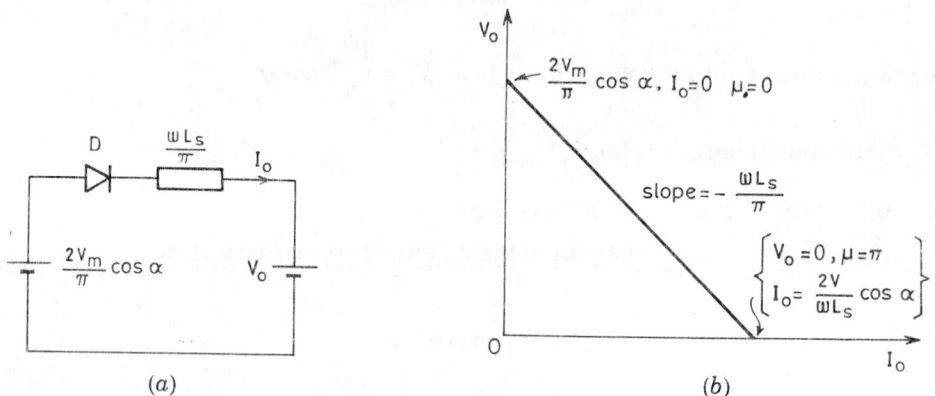

Fig. 6.38. (a) DC equivalent circuit of single-phase full converter.

In single-phase full converter, as long as $\mu < \pi$, the output voltage is given by Eq. (6.73). When $\mu = \pi$, the load will be permanently short circuited by SCRs and the output voltage will be zero because during the overlap angle, all SCRs will be conducting.

Another observation can be made from Fig. 6.37 (c) as follows. If $\alpha = 0$, then the mean output voltage can be controlled over $\mu < \alpha < 180°$. Also, the maximum value of firing angle can be $180 - \mu$. In practical circuits, thyristor takes some time to regain its forward blocking capability. Therefore, the maximum possible firing angle can be $180 - \mu - \delta = 180 - (\mu + \delta)$ where (δ/ω) is the thyristor turn-off time including the factor of safety. Here δ is called the *recovery angle*.

Example 6.26. *A single-phase full converter is made to deliver a constant load current. For zero degree firing angle, the overlap angle is 15°. Calculate the overlap angle when firing angle is (a) 30° (b) 45° and (c) 60°.*

Solution. The *dc* load current, from Eq. (6.71), is given by

$$I_o = \frac{V_m}{\omega L_s} [\cos \alpha - \cos (\alpha + \mu)]$$

Let μ_1 be the overlap angle for firing angle α_1.

$$\therefore \quad I_o = \frac{V_m}{\omega L_s} [\cos \alpha - \cos (\alpha + \mu)] = \frac{V_m}{\omega L_s} [\cos \alpha_1 - \cos (\alpha_1 + \mu_1)]$$

or $\qquad\qquad \cos \alpha_1 - \cos (\alpha_1 + \mu_1) = \cos \alpha - \cos (\alpha + \mu)$

It is given that for $\alpha = 0$, $\mu = 15°$

$\therefore \qquad\qquad \cos \alpha_1 - \cos (\alpha_1 + \mu_1) = \cos 0° - \cos (0 + 15°) = 1 - \cos 15° = 0.03407$

or $\qquad\qquad \cos (\alpha_1 + \mu_1) = \cos \alpha_1 - 0.03407$

(*a*) For firing angle $\alpha_1 = 30°$, $\cos (30 + \mu_1) = \cos 30 - 0.03407$ $\qquad \therefore \mu_1 = 3.7°$.

(*b*) For $\alpha_2 = 45°$, $\cos (45 + \mu_1) = \cos 45° - 0.03407$ $\qquad\qquad\quad \therefore \mu_1 = 2.7°$

(*c*) For $\alpha_3 = 60°$, $\cos (60 + \mu_1) = \cos 60° - 0.03407$ $\qquad\qquad\quad \therefore \mu_1 = 2.23°$

This example demonstrates that for constant load current, overlap angle decreases as the firing angle is increased.

6.9.2. Three-phase Full Converter Bridge

Fig. 6.39 shows a three-phase full-converter bridge with a source inductance L_s in each line. The load current is assumed constant as the analysis with pulsating current is quite complicated.

In Fig. 6.40 (*b*) is shown the conduction of various SCRs with firing angle $\alpha = 0$ and overlap angle $\mu = 0$. In this figure ; T5, T6 conduct upto $\omega t = 30°$. From $\omega t = 30°$ to 90° (*i.e.* for 60°), T1, T6 conduct. From $\omega t = 90°$ to 150°; T1, T2 conduct and so on. It is seen that only two SCRs conduct at a time, one from the positive group and the other from the negative group.

Fig. 6.40 (*c*) shows the effect of overlap. From $\omega t = 0°$ to 30° ; T5, T6 conduct. At $\omega t = 30°$, T5 is outgoing SCR and T1 is incoming SCR and both T5, T6 belong to the

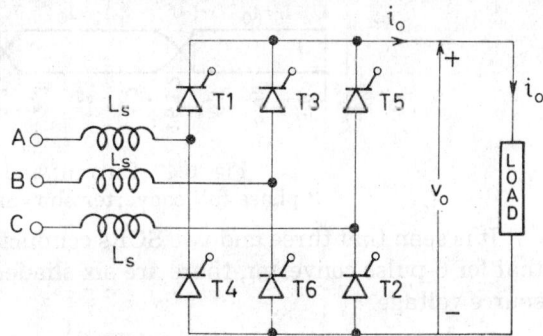

Fig. 6.39. Three-phase full converter with source inductance L_s in each line.

positive group. As T1 is triggered, current through T5 starts decaying while through T1 current begins to build up. At $\omega t = 30° + \mu$, I_5 is zero while $I_1 = I_0$. Therefore, from $\omega t = 30°$ to $30° + \mu$, three SCRs T5, T6, T1 conduct. After $\omega t = 30° + \mu$; T6, T1 conduct. At $\omega t = 90°$, as T2 is triggered, I_6 begins to decrease and I_2 starts to build up. Therefore, from $\omega t = 90°$ to $90° + \mu$, three SCRs T6, T1, T2 conduct. At $\omega t = 90° + \mu$, $I_6 = 0$ and $I_2 = I_0$. After $\omega t = 90° + \mu$, only two SCRs T1, T2 conduct. This sequence of operation repeats with other SCRs of the full converter. It may be observed from this that when positive group of SCRs are undergoing commutation, two SCRs from the positive group and one SCR from the negative group conduct. After the commutation of positive group is completed ; only two SCRs conduct, one from the positive group and the other from the negative group. Similarly, when negative group of SCRs are undergoing commutation, three SCRs conduct, two from the negative group and one from the positive group and these are followed by two SCRs, one from negative group and one from positive group and so on. Conduction of various thyristors as shown in Fig. 6.40 (c) is as follows :

5–6, 5–6–1, 6–1, 6–1–2, 1–2, 1–2–3, 2–3, 2–3–4, 3–4, 3–4–5, 4–5, 4–5–6, 5–6 and so on.

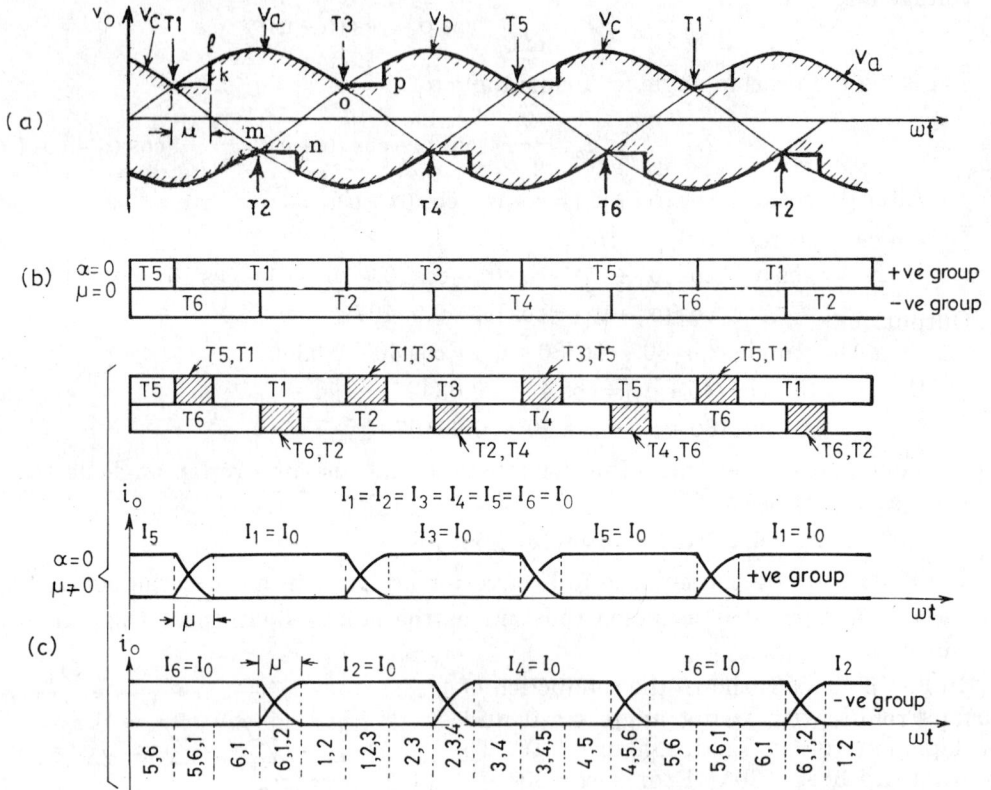

Fig. 6.40. Current and voltage waveforms for a
3-phase full converter showing commutation during overlap.

It is seen that three and two SCRs conduct alternately. It is also observed from Fig. 6.40 (c) that for 6-pulse converter, there are six shaded areas indicating six commutations per cycle of source voltage.

During commutation of T5 and T1 (transfer of current from T5 to T1), the output voltage is obtained by taking average of corresponding phase voltages v_c and v_a of the positive group. This means that voltage from $\omega t = 30°$ to $30° + \mu$ follows the curve $\dfrac{v_a + v_c}{2}$ from the positive group ; this is indicated by jk in Fig. 6.40 (a). During commutation of T6 and T2, the voltage waveform from the negative group is $\dfrac{v_b + v_c}{2}$ as indicated by $m\,n$. Similarly, during commutation of T1, T3; the voltage is $\dfrac{v_a + v_b}{2}$ as shown by curve op and so on. In Fig. 6.40, firing angle delay has been taken as zero just to highlight the effect of source inductance. The above treatment is, however, applicable for any firing angle delay provided overlap angle is less than 60°. In Fig. 6.40 (a) ; v_a, v_b, v_c are the phase voltages and the output voltage is in between the hatched portion as shown.

The effect of source inductance L_s is to reduce the average dc output voltage. This reduction is proportional to the triangular (almost) area $jk\,l$ shown in Fig. 6.40 (a). The average value of this fall in output voltage due to overlap is equal to the triangular area $j\,k\,l$ divided by the periodicity of this triangular area which is equal to $\pi/3$. Thus, average value of fall in output voltage due to overlap

$$= \frac{3}{\pi} \int_0^\mu v_L \, d(\omega t) = \frac{3}{\pi} \int_0^\mu L_s \frac{di}{dt} \, d(\omega t)$$

$$= \frac{3 L_s}{\pi} \int_0^{\mu/\omega} \omega \cdot \frac{di}{dt} \, dt = \frac{3 \omega L_s}{\pi} \int_0^{I_0} di = \frac{3 \omega L_s}{\pi} I_0$$

Alternatively, average value of fall in output voltage due to overlap

$$= \frac{3}{\pi/\omega} \int_0^t L_s \cdot \frac{di}{dt} \cdot dt = \frac{3 \omega L_s}{\pi} \int_0^{I_0} di = \frac{3 \omega L_s}{\pi} I_0$$

Output voltage with no overlap = internal voltage of the 3-phase full converter

$$= \frac{3 V_{ml}}{\pi} \cos \alpha = V_0$$

Output voltage with overlap, $V_{ox} = \dfrac{3 V_{ml}}{\pi} \cos \alpha - \dfrac{3 \omega L_s}{\pi} I_0$ ...(6.76)

In general, for p-pulse converter, fall in output voltage due to overlap

$$= \frac{p}{2\pi} \int_0^\mu L_s \left(\frac{di}{dt} \right) d(\omega t) = \frac{p \omega L_s}{2\pi} \int_0^{\mu/\omega} \left(\frac{di}{dt} \right) dt$$

$$= \frac{p \omega L_s}{2\pi} \int_0^{I_0} di = \frac{p \omega L_s \cdot I_0}{2\pi} = \left(\frac{p}{2\pi} \right) \omega L_s I_0$$

From Eq. (6.58), mean output voltage is a p-phase converter, with overlap, is given by

$$V_{ox} = V_{mp} \left(\frac{p}{\pi} \right) \sin \left(\frac{\pi}{p} \right) \cos \alpha - \left(\frac{p}{2\pi} \right) \omega L_s I_o$$

For a 2-pulse converter, $\quad V_{ox} = \dfrac{2 V_m}{\pi} \cos \alpha - \dfrac{\omega L_s \cdot I_o}{\pi}$ ...(6.73)

For a 3-pulse converter, $\quad V_{ox} = \dfrac{3 \sqrt{3} \cdot V_{mp}}{2\pi} \cos \alpha - \dfrac{3 \omega L_s I_o}{2\pi}$ ...(6.77)

For a 6-pulse converter, (here replace V_{mp} by V_{ml}),

$$V_{ox} = \frac{3 V_{ml}}{\pi} \cos \alpha - \frac{3 \omega L_s \cdot I_o}{\pi}$$ ...(6.76)

An expression, similar to Eq. (6.72a) for a single-phase full converter, can be written for a 3-phase full converter as

$$I_o = \frac{V_{ml}}{2\omega L_s} \left[\cos \alpha - \cos (\alpha + \mu) \right]$$ ...(6.78)

In Eq. (6.78), $2 \omega L_s$ is written in place of ωL_s just because two lines carry current I_o and each line has source inductance L_s.

Substituting the value of I_o from Eq. (6.78) in Eq. (6.76), we get

$$V_{ox} = \frac{3 V_{ml}}{\pi} \cos \alpha - \frac{3 \omega L_s}{\pi} \left[\frac{V_{ml}}{2 \omega L_s} (\cos \alpha - \cos (\alpha + \mu)) \right]$$

$$= \frac{3\,V_{ml}}{2\pi}\,[\cos\alpha + \cos(\alpha + \mu)] \qquad\qquad ...(6.79\,a)$$

$$= \frac{\text{Maximum mean output voltage at no load}}{2}\,[\cos\alpha + \cos(\alpha + \mu)]$$

$$= \frac{V_{om}}{2}\,[\cos\alpha + \cos(\alpha + \mu)] \qquad\qquad ...(6.79\,b)$$

where $V_{om} = \dfrac{3\,V_{ml}}{\pi}$ for a 3-phase full converter.

In case overlap angle μ is zero, Eq. (6.79 a) gives $V_{ox} = \dfrac{3\,V_{ml}}{\pi}\cos\alpha = V_o$. This is as expected.

Output voltage for a 3-phase full converter, similar to that given for a 1-phase full converter in Eq. (6.74), is given by

$$V_0 = \frac{3\,V_{ml}}{\pi}\cos(\alpha + \mu) + \frac{3\,\omega L_s}{\pi}\,I_0 \qquad\qquad ...(6.80)$$

In a 3-phase full converter, voltage regulation due to source inductance

$$= \frac{3\,\omega L_s\,I_s}{\pi} \times \frac{1}{V_o\ \text{at no load}}$$

$$= \frac{2\,\pi f. L_s. I_o}{V_{ml}.\cos\alpha} \qquad\qquad ...(6.81)$$

Example 6.27. *In a three-phase full converter thyristor bridge shown in Fig. 6.39, load consists of a resistor R. For this converter, do the following :*

(a) Sketch waveforms for 3-phase input voltages $v_{ab}, v_{ac}, v_{bc}, v_{ba}$ etc.

(b) From (a), sketch waveforms of the output voltage v_0 for a firing angle of zero degree and overlap angle of $\mu = 30°$. Indicate the conduction of various SCRs.

(c) Repeat part (b) if firing angle $\alpha = 30°$ and $\mu = 30°$.

(d) From (c), sketch the waveform of input current i_a for $\alpha = 30°$ and $\mu = 30°$. In case input voltage is 400 V and R = 200 Ω, indicate the peak magnitude of current i_a.

Solution. (*a*) The line voltage waveforms v_{ab}, v_{ac}, v_{bc} etc. for a 3-phase full converter are sketched in Fig. 6.41 (*a*).

(*b*) The effect of overlap angle of 30° with zero degree firing angle is shown in Fig. 6.40 (*a*) with phase voltages. In Fig. 6.41 (*a*), the output voltage waveform v_0 is sketched with line voltages, for an overlap angle of 30° and with firing angle $\alpha = 0°$. The conduction of SCRs T1 to T6 is also indicated. When T5 is outgoing SCR and T1 the incoming SCR, the average output voltage is $\dfrac{v_{cb} + v_{ab}}{2}$; this is shown accordingly in Fig. 6.41 (*a*). In v_{cb}, v_{ab} ; first subscripts c and a indicate the conduction of SCRs T5, T1 together from the positive group and second subscript b indicates the conduction of SCR T6 from negative group. When T1, T6 conduct, the output voltage is v_{ab} as shown. For T6 as the outgoing SCR and T2 the incoming SCR, the average output voltage is $\dfrac{v_{ab} + v_{ac}}{2}$. The second subscripts in v_{ab}, v_{ac} indicate the conduction of SCRs 6 and 2 together from negative group and first subscript a indicates the conduction of SCR T1 from the positive group. The shape of output voltage waveform and the conduction of SCRs for the remaining part of a cycle can be explained similarly.

(*c*) For firing angle of 30°, and overlap angle of 30°, the output voltage waveform v_0 is drawn in Fig. 6.41 (*b*). For obtaining v_0, it is preferrable to indicate first the conduction of various SCRs and then draw the output voltage waveform v_0. In Fig. 6.41 (*b*), the conduction of different

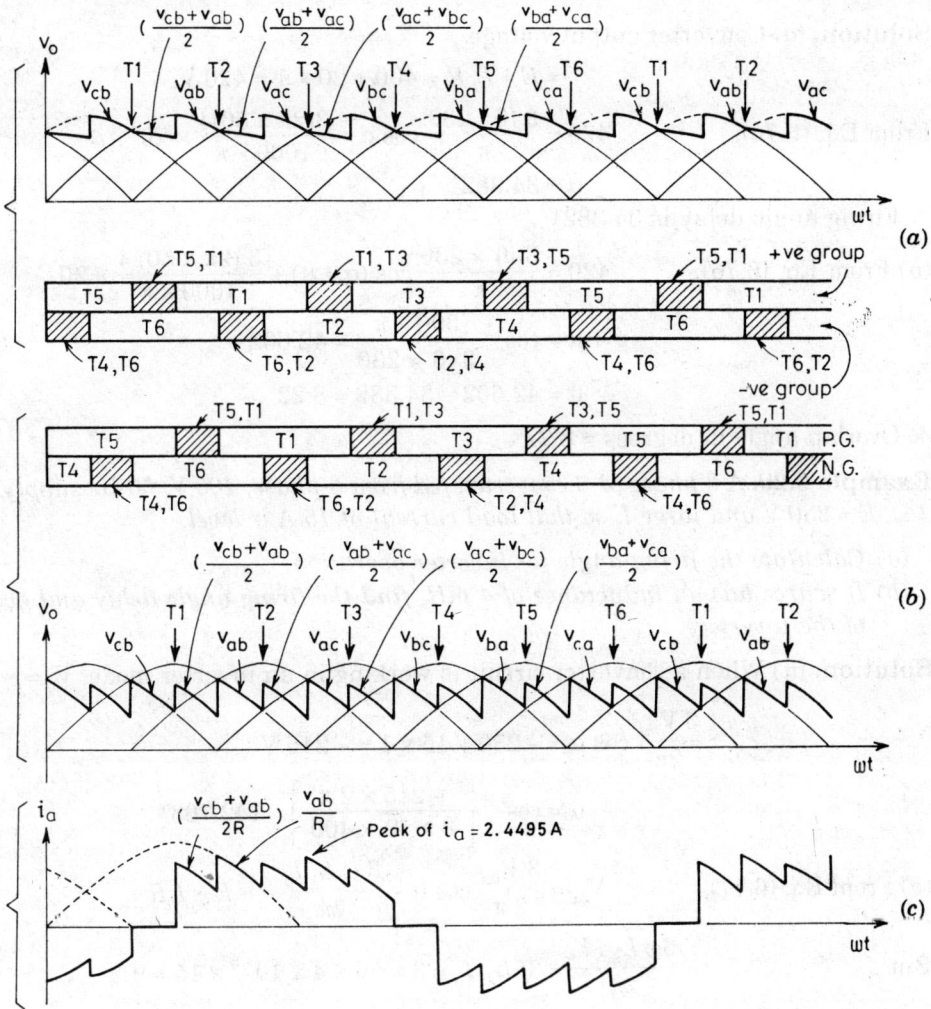

Fig. 6.41. Pertaining to Example 6.27.

thyristors is shifted to the right by $\alpha = 30°$ with respect to their position in Fig. 6.41 (a) where $\alpha = 0°$. Now the waveform for v_0, as shown in Fig. 6.41 (b), can be drawn easily.

(d) The waveform for phase a input current, $i.e.\ i_a$, is shown in Fig. 6.41 (c). As the load is resistive, the waveform of i_a is identical to the waveform of voltage v_0 of Fig. 6.41 (b). Current i_a is positive when T1 is conducting and negative when T4 is conducting. Note that i_a without overlap would have four ripples but with overlap, it is seen to have five ripples in both positive and negative half cycles in Fig. 6.41 (c).

Example 6.28. *A 3-phase full converter bridge is connected to supply voltage of 230 V per phase and a frequency of 50 Hz. The source inductance is 4 mH. The load current on dc side is constant at 20 A. If the load consists of a dc source of internal emf 400 V with internal resistance of 1 Ω, then calculate :*

(a) firing angle delay and (b) overlap angle in degrees.

Solution. (*a*) Converter output voltage

$$= E + I_0 R = 400 + 20 \times 1 = 420 \text{ V}.$$

From Eq. (6.76), $420 = \dfrac{3\sqrt{6} \cdot 230}{\pi} \cos \alpha - \dfrac{3(2\pi \times 50)4}{1000 \times \pi} \times 20$

or $\alpha = 34.382°$

∴ Firing angle delay is 34.382°

(*b*) From Eq. (6.76), $420 = \dfrac{3\sqrt{6} \times 230}{\pi} \cos (\alpha + \mu) + \dfrac{3 (2\pi \times 50) 4}{1000 \times \pi} \times 20$

or $\alpha + \mu = \cos^{-1} \dfrac{396 \times \pi}{3\sqrt{6} \times 230} = 42.602°$

∴ $\mu = 42.602 - 34.382 = 8.22°$

∴ Overlap angle in degrees = 8.22°.

Example 6.29. *A 3-phase M-3 converter, fed from 3-phase, 400 V, 50 Hz supply, has a load R = 1 Ω, E = 230 V and large L so that load current of 15 A is level.*

   (*a*) *Calculate the firing angle for inverter operation.*
   (*b*) *If source has an inductance of 4 mH, find the firing angle delay and overlap angle of the inverter.*

Solution. (*a*) When a converter circuit is working in an inverter mode, $V_o = - E + I_o R$

∴ $\dfrac{3 V_{ml}}{2\pi} \cos \alpha = - 230 + 15 \times 1 = - 215 \text{ V}$

∴ $\alpha = \cos^{-1} \left[\dfrac{- 215 \times 2 \pi}{3 \times \sqrt{2} \times 400} \right] = 142.763°$

(*b*) From Eq. (6.77), $V_{ox} = \dfrac{3 V_{ml}}{2 \pi} \cos \alpha - \dfrac{3 \omega L_s I_o}{2\pi} = - \overset{\cdot}{E} + I_o R$

But $\dfrac{3\omega L_s \cdot I_o}{2\pi} = 3f L_s \cdot I_o = 3 \times 50 \times 4 \times 10^{-3} \times 15 = 9 \text{ V}$

∴ $\dfrac{3 \sqrt{2} \times 400}{2 \pi} \cos \alpha - 9 = - 230 + 15 = - 215 \text{ V}$

∴ $\alpha = \cos^{-1} \left[\dfrac{- 206 \times 2\pi}{3 \sqrt{2} \times 400} \right] = 139.712°$

For a 3-phase converter, $\cos (\alpha + \mu) = \cos \alpha - \dfrac{3f \cdot L_s \cdot I_o}{V_{mp}}$

∴ $\cos (139.712° + \mu) = \cos 139.712° - \dfrac{9 \times \sqrt{3}}{\sqrt{2} \times 400} = \cos 142.22°$

∴   Overlap angle $\mu = 142.22° - 139.712° = 2.508°$

Example 6.30. *A 3-phase full converter feeds a load with ripple free direct current I_o. If source inductance is taken into consideration, show that load current I_o is given by*

$$I_o = \dfrac{V_{ml}}{2 \omega L_s} \left[\cos \alpha - \cos (\alpha + \mu) \right] \qquad \qquad \text{...(6.78)}$$

Solution. Waveform of output voltage v_o in Fig. 6.40 (*a*) is sketched with firing angle $\alpha = 0°$ and with overlap angle μ. This figure shows that when T1 is triggered at $\alpha = 0°$, current in T5 begins to decay from $I_5 (= I_o)$ and that in T1 begins to rise from zero, during the overlap angle μ.

Negative group SCR T6 is already conducting a current I_o. Thus, during overlap period, equivalent circuit of Fig. (6.42) can be obtained by referring to Fig. 6.39. This equivalent circuit is similar to that given in Fig. 6.37 (b) for a single-phase full converter.

As T1 and T5 from positive group are conducting together, these got short circuited as shown. KVL for the loop a $k\,l\,m\,n$, during the overlap period, gives

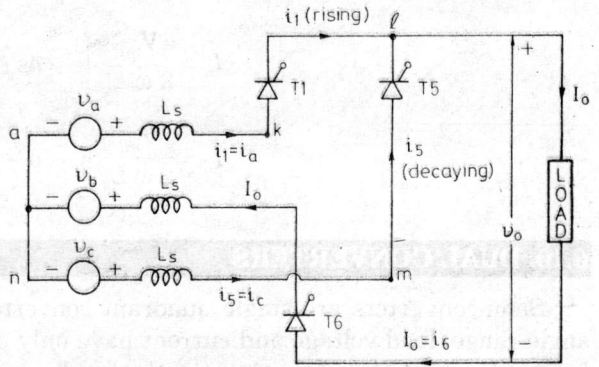

Fig. 6.42. Equivalent circuit of 3-phase full converter with conducting SCRs T1, T5, T6 during overlap, Example 6.29.

$$v_a - L_s \frac{di_1}{dt} = v_c - L_s \frac{di_5}{dt}$$

or $$v_a - v_c = L_s \left[\frac{di_1}{dt} - \frac{di_5}{dt} \right] \qquad \qquad \ldots(i)$$

Fig. 6.40 (a) shows that $v_a = V_{mp} \sin \omega t$. As v_c lags v_a by 240°, or v_c leads v_a by 120°, $v_c = V_{mp} \sin (\omega t + 120°)$.

$\therefore$
$$v_a - v_c = V_{mp} [\sin \omega t - \sin (\omega t - 120°)]$$
$$= \sqrt{3} \, V_{mp} \sin (\omega t - 30°) = V_{ml} . \sin (\omega t - 30°) \qquad \ldots(ii)$$

From Eq. (i), we get
$$\frac{di_1}{dt} - \frac{di_5}{dt} = \frac{V_{ml}}{L_s} \sin (\omega t - 30°) \qquad \qquad \ldots(iii)$$

Load current during overlap period is also assumed constant.

$\therefore$
$$i_1 + i_5 = I_o$$

or
$$\frac{di_1}{dt} + \frac{di_5}{dt} = 0 \qquad \qquad \ldots(iv)$$

Addition of Eqs. (iii) and (iv) gives

$$\frac{di_1}{dt} = \frac{V_{ml}}{L_s} \sin (\omega t - 30°) \qquad \qquad \ldots(v)$$

Fig. 6.40 (c), for zero degree firing angle, shows that current i_1 rises to I_o in overlap angle μ.

$\therefore$ When $\omega t = 30°$, $i_1 = 0$ and when $\omega t = 30° + \mu$, $i_1 = I_1 = I_o$

If firing angle is taken into consideration in Fig. 6.40 (c), then, when $\omega t = 30° + \alpha$, $i_1 = 0$ and when $\omega t = 30° + \alpha + \mu$, $i_1 = I_1 = I_o$. Therefore, in order to take into account the firing angle α, the limits of integration for Eq. (v) must be $(30° + \alpha) = (\pi/6 + \alpha)$ to $(30° + \alpha + \mu) = \left(\dfrac{\pi}{6} + \alpha + \mu \right)$

$\therefore$
$$\int_o^{I_o} di_i = \frac{V_{ml}}{2L_s} \int_{\left(\frac{\pi}{6} + \alpha \right)/\omega}^{\left(\frac{\pi}{6} + \alpha + \mu \right)/\omega} \sin (\omega t - \pi/6) \, dt$$

$$I_o = \frac{V_{ml}}{2\,\omega\,L_s} \left| -\cos\left(\omega t - \pi/6\right) \right| \begin{array}{l} \left(\frac{\pi}{6} + \alpha + \mu\right)/\omega \\ \left(\frac{\pi}{6} + \alpha\right)/\omega \end{array}$$

$$I_o = \frac{V_{ml}}{2\,\omega\,L_s} \left[\cos \alpha - \cos\left(\alpha + \mu\right)\right] \qquad \qquad \qquad ...(6.78)$$

6.10. DUAL CONVERTERS

Semi-converters are single quadrant converters. This means that over the entire firing angle range, load voltage and current have only one polarity as shown in Fig. 6.9 (a). In this figure, V_0 and I_0 represent, respectively, the average positive voltage and current of the semi-converter indicating rectification mode and power flow from ac source to the dc load. In full converters, direction of current cannot reverse because of the unidirectional properties of SCRs but polarity of output voltage can be reversed as shown in Fig. 6.9 (b). Thus, a full converter operates as a rectifier in first quadrant (both V_0, I_0 positive) from $\alpha = 0°$ to $90°$ and as an inverter (V_0 negative but I_0 positive) from $\alpha = 90°$ to $180°$ in the fourth quadrant. This shows that a full converter can operate as a two-quadrant converter, Fig. 6.9 (b). In the first quadrant, power flows from ac source to the dc load and in fourth quadrant, power flows from dc circuit to the ac source.

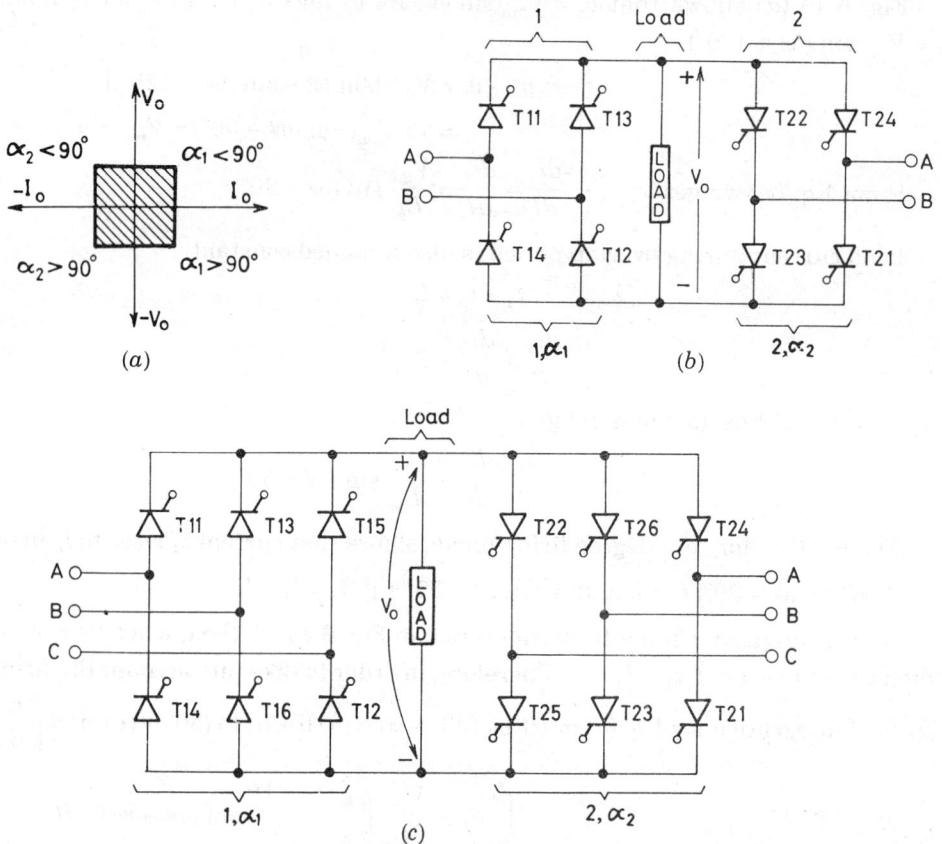

Fig. 6.43. (a) Four-quadrant diagram. Non-circulating type (b) single-phase dual converter and (c) three-phase dual converter.

In case four quadrant operation is required without any mechanical changeover switch, two full converters can be connected back to back to the load circuit. Such an arrangement using two full converters in antiparallel and connected to the same dc load is called a *dual converter*.

There are two functional modes of a dual converter, one is non-circulating-current mode and the other is circulating-current mode. Non-circulating types of dual converters using single-phase and three-phase configurations are shown in Fig. 6.43 (*b*) and (*c*) respectively. If full converter marked 1, to the left of load circuit in Fig. 6.43 (*b*) and (*c*), is working alone, operation in first and fourth quadrants can be obtained. With full converter marked 2 working alone in Fig. 6.43 (*b*) and (*c*), polarity of load voltage as well as direction of load current, with respect to converter 1, can be reversed. Hence, full converter marked 2 can operate in both second and third quadrants. Thus, a dual converter using two full converters can give four quadrant operation as shown in Fig. 6.43 (*a*).

6.10.1. Ideal Dual Converter

Assume that the dual converter consists of two ideal converters and that there is no ripple in their output voltages. Such a dual converter can be represented by an equivalent circuit shown in Fig. 6.44 (*a*). V_{01} and V_{02} are the magnitudes of average output voltages of converters 1 and 2 respectively. Diodes D1 and D2 shown here in series with the dc voltage sources V_{01} and V_{02} indicate the unidirectional flow of current. The current in load circuit can, however, flow in either direction.

The firing angles of both the converters are controlled in such a manner that their average output voltages are equal in magnitude and have the same polarity. This can happen only if one converter is operating as a rectifier and the other as an inverter.

The average output voltages for both single-phase and 3-phase converters are of the form.

$$V_{01} = V_{om} \cos \alpha_1 \qquad \qquad ...(6.82)$$

and

$$V_{02} = V_{om} \cos \alpha_2 \qquad \qquad ...(6.83)$$

where, for a single-phase full converter, $V_{om} = (2V_m/\pi)$

and for a three-phase full converter, $V_{om} = (3V_{ml}/\pi)$

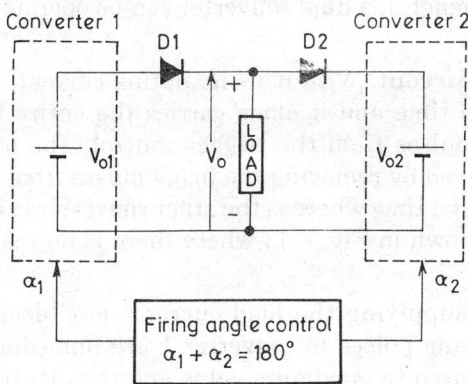

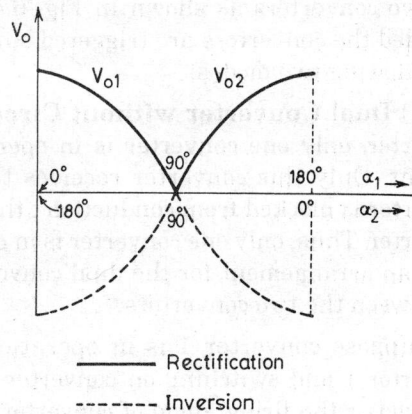

Fig. 6.44. (*a*) Equivalent circuit of an ideal dual converter (*b*) Variation of terminal voltage for an ideal dual converter with firing angle.

Under normal operation, output voltage V_{01} of converter 1 has upper positive and lower negative polarities and output voltage V_{02} of converter 2 has upper negative and lower positive polarities in Figs. 6.43 (*b*) and (*c*). It is assumed that the two converters have their average output voltages equal in magnitude. Their output voltages would have the same polarity only if polarity of V_{02} is reversed. In other words, their average output voltage V_0 can be expressed as

$$V_0 = V_{01} = -V_{02} \qquad \qquad ...(6.84)$$

Substitution of V_{01} and V_{02} from Eqs. (6.82) and (6.83) in Eq. (6.84) gives,

$$V_{om} \cos \alpha_1 = - V_{om} \cos \alpha_2$$

or $$\cos \alpha_1 = - \cos \alpha_2 = \cos (180 - \alpha_2)$$

or $$\alpha_1 + \alpha_2 = 180° \qquad \qquad ...(6.85)$$

Also $$\cos \alpha_1 = - \cos \alpha_2 = \cos (180 + \alpha_2)$$

or $$\alpha_1 = 180° + \alpha_2 \qquad \qquad ...(6.85\ a)$$

As per Eq. (6.85 a), for some value of firing angle α_2, α_1 is always greater than 180°. But α_1 can never be greater than 180°. Therefore, the solution as given by Eq. (6.85) is only possible.

From Eqs. (6.82), (6.83) and (6.85), the variation of output voltage with firing angle for the two converters is as shown in Fig. 6.44 (b). The firing control circuit changes the firing angle α_1 and α_2 in such a manner that Eq. (6.85) is always satisfied.

6.10.2. Practical Dual Converter

With the firing angles controlled in a manner that $\alpha_1 + \alpha_2 = 180°$ and with both the converters in operation, their average output voltages are equal and have the same polarity. One converter will be operating as a rectifier with firing angle α_1 and the other as an inverter with firing angle $(180° - \alpha_1)$. Though their average output voltages are equal, yet their instantaneous voltages v_{01} and v_{02} are out of phase in a practical dual converter. This results in a voltage difference when the two converters are interconnected and as a consequence, a large circulating current flows between the two converters but not through the load. In practical dual converters, this circulating current is limited to a tolerable value by inserting a reactor between the two converters as shown in Fig. 6.45. The circulating current can however, be avoided provided the converters are triggered suitably. In general, a dual converter can be operated in the following two modes.

(a) **Dual Converter without Circulating Current.** With non-circulating current dual converter, only one converter is in operation at a time and it alone carries the entire load current. Only this converter receives the firing pulses from the trigger control. The other converter is blocked from conduction ; this is achieved by removing the firing pulses from this converter. Thus, only one converter is in operation at a time whereas the other converter is idle. Such an arrangement for the dual converters is shown in Fig. 6.43 where there is no reactor in-between the two converters.

Suppose converter 1 is in operation and is supplying the load current. For blocking convertor 1 and switching on converter 2, first firing pulses to converter 1 are immediately removed or the firing angle of converter 1 is increased to maximum value and then its firing pulses are blocked. With this, load current would decay to zero and then only converter 2 is made to conduct by applying the firing pulses to it. Now the current in converter 2 would build up through the load in the reverse direction. So long as converter 2 is in operation, converter 1 is idle as firing pulses are withdrawn from it. It should be ensured that during changeover from one converter to the other, the load current must decay to zero. After the outgoing converter has stopped conducting, a delay time of 10 to 20 msec is introduced before the firing pulses are applied to switch on the incoming converter. This time delay ensures reliable commutation of

SCRs in the outgoing converter. If the incoming converter is triggered before the outgoing converter has been completely turned-off, a large circulating current would flow between the two converters.

With non-circulating current mode of dual converter, the load current may be continuous or discontinuous. The control circuitry for the dual converter is so designed as to give satisfactory operation during continuous as well as discontinuous load current.

(b) **Dual converter with circulating current.** In the circulating current mode of dual converter, a reactor is inserted in-between converters 1 and 2 as shown in Fig. 6.45. This reactor limits the magnitude of circulating current to a reasonable value.

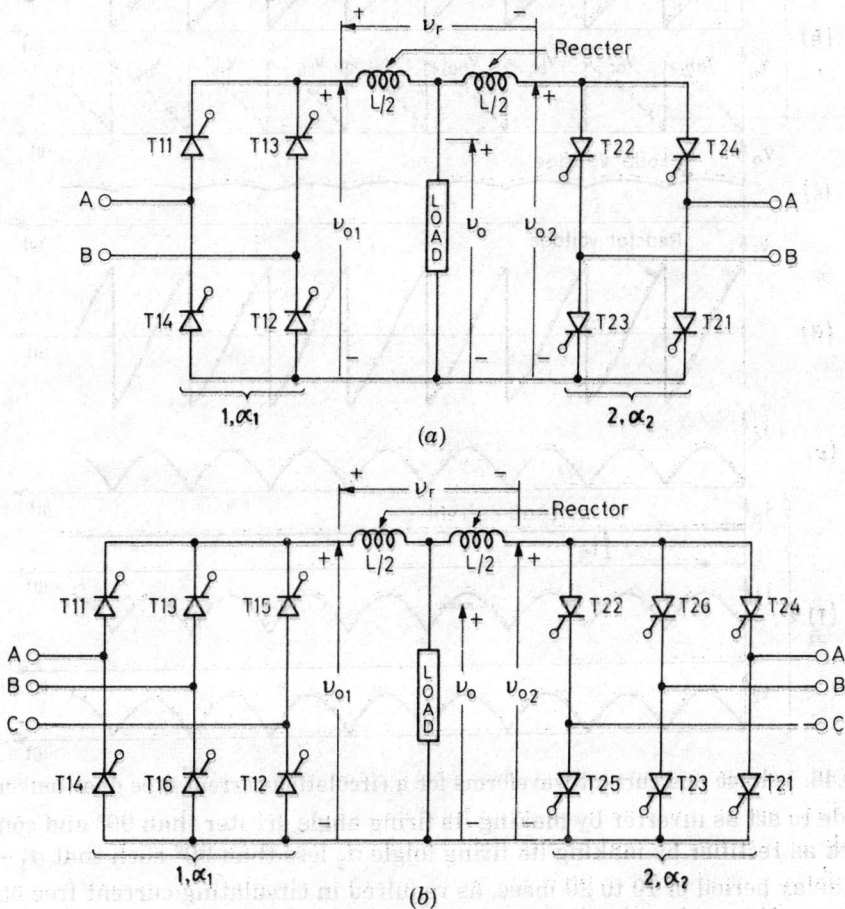

Fig. 6.45. Circulating current type dual converter for (a) single-phase supply and (b) three-phase supply.

The firing pulses of the two converters are so adjusted that $\alpha_1 + \alpha_2 = 180°$. As for example, if firing angle of converter 1 is 60°, then firing angle of converter 2 must be 120°. Therefore, for these firing angles, converter 1 is working as a rectifier and converter 2 as an inverter. Though the output voltage at the terminals of both converters 1 and 2 has the same average value and also has the same polarity, their instantaneous output voltage waveforms, however, are not similar as shown by v_{01} and v_{02} in Fig. 6.46 (b). As a consequence of it, circulating current flows between the two converters. This circulating current is limited by the reactor. If the load current is to be reversed, the role of two converters is interchanged. This means that converter

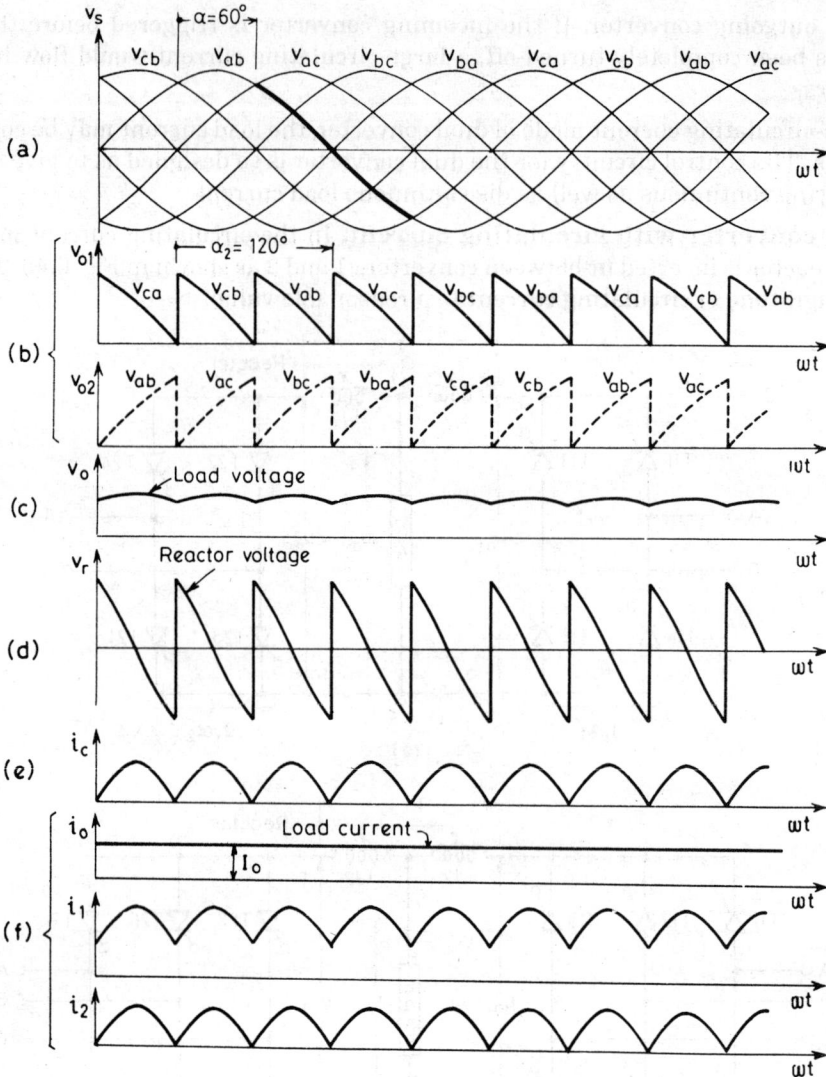

Fig. 6.46. Voltage and current waveforms for a circulating-current type dual converter.

1 is now made to act as inverter by making its firing angle greater than 90° and converter 2 is made to work as rectifier by making its firing angle α_2 less than 90° such that $\alpha_1 + \alpha_2 = 180°$. The normal delay period of 10 to 20 msec, as required in circulating-current free operation, is not needed here. This makes the dual converter with circulating current operation faster. The main disadvantages of this dual converter are as under :

(i) A reactor is required to limit the circulating current. The size and cost of this reactor may be quite significant at high power levels.

(ii) Circulating current gives rise to more losses in the converters, hence the efficiency and power factor are low.

(iii) As the converters have to handle load as well as circulating currents, the thyristors for the two converters are rated for higher currents.

In spite of these drawbacks, a dual converter with circulating current mode is preferred if load current is to be reversed quite frequently and a fast response is desired in the four-quadrant operation of the dual converter.

Dual converter operation with waveforms. The operation of the dual converter in the circulating-current mode is described here under the following assumptions :

(*i*) The reactor is lossless.

(*ii*) The firing angles of the two converters are so controlled that $\alpha_1 + \alpha_2 = 180°$.

In Fig. 6.46 (*a*), supply line voltages v_{ab}, v_{ac}, v_{bc} etc. are shown. As an illustrative example for describing the working of a dual converter with waveforms, let α_1 be equal to 60°. Then, for converter 2, $\alpha_2 = 180° - 60° = 120°$. With $\alpha_1 = 60°$, the output voltage v_{01} for converter 1 is indicated by thick line in Fig. 6.46 (*a*). This output voltage v_{01} is now shown as v_{ab} in Fig. 6.46 (*b*) from $\omega t = 120°$ to 180°. In this manner, v_{01} is drawn for other intervals of time in Fig. 6.46 (*b*). With $\alpha_2 = 120°$ for converter 2, the output voltage is negative as shown by thick line in Fig. 6.46 (*a*). As the average values of output voltages of both the converters are positive, the output voltage v_{02} of converter 2 must also be shown positive above the reference line ωt. This output voltage v_{ab} indicated by thick line in Fig. 6.46 (*a*) is now shown as v_{ba} dotted in Fig. 6.46 (*b*) from $\omega t = 180°$ to 240°. In this manner, v_{02} waveform is drawn positive as v_{ba}, v_{ca}, v_{cb} etc. in Fig. 6.46 (*b*).

The load voltage v_0 is equal to the average value of the instantaneous converter output voltages v_{01} and v_{02}, *i.e.*

$$v_0 = \frac{v_{01} + v_{02}}{2} \qquad \qquad ...(6.86)$$

At $\omega t = 0°$, $\qquad\qquad v_0 = \dfrac{V_{ml} \sin 60° + 0}{2} = 0.433 \, V_{ml}$

At $\omega t = 30°$, $\qquad\qquad v_0 = \dfrac{V_{ml} \sin 30° + V_{ml} \sin 30°}{2} = 0.5 \, V_{ml}$

At $\omega t = 60°$, $\qquad\qquad v_0 = \dfrac{0 + V_{ml} \sin 60°}{2} = 0.433 \, V_{ml}$ and so on.

This load voltage waveform v_0 is shown in Fig. 6.46 (*c*).

The reactor voltage v_r is equal to the difference of converter output voltages v_{01} and v_{02}, *i.e.*

$$v_r = v_{01} - v_{02} \qquad \qquad ...(6.87)$$

At $\omega t = 0°$, $\qquad\qquad v_r = V_{ml} \sin 60° - 0 = 0.866 \, V_{ml}$

At $\omega t = 30°$, $\qquad\qquad v_r = V_{ml} \sin 30 - V_{ml} \sin 30° = 0$

At $\omega t = 60°$, $\qquad\qquad v_r = 0 - V_{ml} \sin 60° = - \, 0.866 \, V_{ml}$

and so on. The variation of reactor voltage v_r is plotted in Fig. 6.46 (*d*)

Now $\qquad\qquad\qquad v_r = L \dfrac{di_c}{dt}$

where i_c is the circulating current through both the converters and reactor L. The waveform of i_c can be drawn from the waveform of v_r as under :

At $\omega t = 0°$, v_r is maximum and positive, therefore slope of i_c must be maximum and positive to satisfy the relation $v_r = L \dfrac{di_c}{dt}$. Thus, i_c is shown rising in Fig. 6.46 (*e*) with a maximum positive slope. At $\omega t = 30°$, $v_r = 0$, therefore slope of $i_c = 0$; this is possible only when i_c is maximum with $\dfrac{di_c}{dt} = 0$. After $\omega t = 30°$, v_r starts becoming negative, the value of i_c also starts falling so that di_c/dt is negative. At $\omega t = 60°$, v_r is maximum but negative, therefore slope of i_c must be negative but maximum. Based on this logic, waveform of i_c is drawn as shown in Fig. 6.46 (e).

No-load. At no load, both converters handle only circulating current i_c as shown in Fig. 6.46 (e).

On-load. Converter 1 (with $\alpha_1 < 90°$) works as a rectifier and carries load current as well as circulating current. Assuming load current I_0 as constant, the converter 1 current is $i_1 = I_0 + i_c$. Converter 2 (with $\alpha_2 > 90°$) works as an inverter and handles only the circulating current i_c. The waveforms for these currents i_1 and i_2 are shown in Fig. 6.46 (f).

The expression for the circulating current in the dual converter can be obtained from voltage v_r across the reactor.

It is seen from Fig. 6.46 (b) that for the time interval $(\pi/3 + \alpha_1) < \omega t < (\pi/3 + \alpha_1 + \pi/3)$, the converter output voltages are

$$v_{01} = v_{ab} \text{ and } v_{02} = v_{bc}$$

The reactor voltage v_r, from Eq. (6.87), is

$$v_r = v_{01} - v_{02} = v_{ab} - v_{bc} \qquad \qquad ...(6.88)$$

It is seen from Fig. 6.46 (a) that

$$v_{ab} = V_{ml} \sin \omega t. \text{ As } v_{bc} \text{ lags } v_{ab} \text{ by } 120°, \text{ it is given by}$$
$$v_{bc} = V_{ml} \sin (\omega t - 120°)$$
$$\therefore \qquad v_r = V_{ml} [\sin \omega t - \sin (\omega t - 120°)]$$
or
$$v_r = \sqrt{3} \, V_{ml} \sin (\omega t + \pi/6) \qquad \qquad ...(6.89)$$

The circulating current i_c is obtained from the time integral of reactor voltage v_r and is given by

$$i_c = \frac{1}{L} \int_{(\alpha_1 + \pi/3)/\omega}^{t} v_r \cdot dt = \frac{\sqrt{3} \cdot V_{ml}}{L} \int_{(\alpha_1 + \pi/3)/\omega}^{t} \sin (\omega t + \pi/6) \, dt$$

$$= \frac{\sqrt{3} \cdot V_{ml}}{\omega L} \int_{\alpha_1 + \pi/3}^{\omega t} \sin (\omega t + \pi/6) \cdot d(\omega t) \qquad \qquad ...(6.90)$$

or
$$i_c = \frac{\sqrt{3} \cdot V_{ml}}{\omega L} [- \sin \alpha_1 - \cos (\omega t + \pi/6)] \qquad \qquad ...(6.90)$$

It is seen from Eq. (6.90) that the **magnitude of circulating** current depends upon the firing angle α_1 and upon ωt. For any value of firing angle, the peak value of circulating current occurs when $\omega t = 5\pi/6$. This peak value i_{cp} is then given by

$$i_{cp} = \frac{\sqrt{3} \, V_{ml}}{\omega L} [1 - \sin \alpha_1] \qquad \qquad ...(6.91)$$

Eq. (6.91) shows that peak value of circulating current depends upon the firing angle α_1. For $\alpha_1 = 0$, the maximum value of i_{cp} is $\sqrt{3} \, V_{ml}/\omega L$ and for $\alpha_1 = 90°$, $i_{cp} = 0$.

In Eqs. (6.90) and (6.91), V_{ml} is the maximum value of line voltage.

Example 6.31. *A 3-phase dual converter, operating in the circulating-current mode, has the following data :*

Per phase supply voltage = 230 V, f = 50 Hz, $\alpha_1 = 60°$, current limiting reactor, L = 15 mH. Calculate the peak value of circulating current.

Solution. The peak value of circulating current, for firing angle $\alpha_1 = 60°$, is given by Eq. (6.91).

$$\therefore \qquad i_{cp} = \frac{\sqrt{3} \cdot \sqrt{6} \cdot 230}{2\pi \times 50 \times 15 \times 10^{-3}} [1 - \sin 60°] = 27.7425 \text{ A.}$$

6.11. SOME WORKED EXAMPLES

In this article, some typical problems on phase controlled rectifiers are solved.

Example 6.32. *A single-phase full converter is supplied from 230 V, 50 Hz source. The load consists of R = 10 Ω and a large inductance so as to render the load current constant. For a firing angle delay of 30°, determine (a) average output voltage (b) average output current (c) average and rms values of thyristor currents and (d) the power factor.*

Solution. The waveforms for source voltage v_s, load current i_0, load voltage v_0, thyristor current i_{T1} (or i_{T2}) and source current i_s (refer to Fig. 6.10) are drawn in Fig. 6.47.

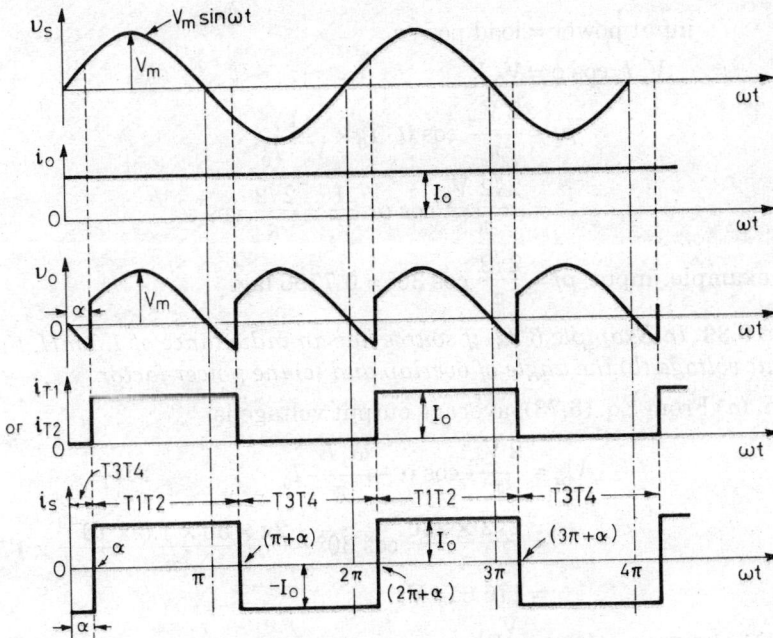

Fig. 6.47. Pertaining to Example 6.32.

(a) For a single-phase full converter, average output voltage V_0, Eq. (6.26), is given by

$$V_0 = \frac{2V_m}{\pi} \cos \alpha = \frac{2\sqrt{2} \times 230}{\pi} \cos 30° = 179.303 \text{ V}$$

(b) Average output current, $I_0 = \dfrac{V_0}{R} = \dfrac{179.303}{10} = 17.93 \text{ A}$

(c) It is seen from the waveform of thyristor current i_{T1} (or i_{T2}) that its average value is given by

$$I_{T.A} = I_0 \cdot \frac{\pi}{2\pi} = \frac{I_0}{2} = \frac{17.93}{2} = 8.965 \text{ A}$$

Rms value of thyristor current is

$$I_{T \cdot r} = \sqrt{I_0^2 \cdot \pi \times \frac{1}{2\pi}} = \frac{I_0}{\sqrt{2}} = \frac{17.93}{\sqrt{2}} = 12.68 \text{ A}$$

(d) Rms value of source current, $I_s = \sqrt{I_0^2 \cdot \frac{\pi}{\pi}} = I_0 = 17.93$ A

Load power $= V_0 I_0 = 179.3 \times 17.93$ W

Input power $= V_s I_s \cos \phi$

For no loss in the power converter,

$$V_s I_s \cos \phi = V_0 I_0$$

$\therefore$ Power factor, $\cos \phi = \dfrac{179.3 \times 17.93}{230 \times 17.93} = 0.7796$ lag

In general, for a 1-phase full converter with ripple free load current as in this example and with no device drops,

$$\text{input power} = \text{load power}$$

or $V_s I_s \cos \phi = V_0 I_0$

$\therefore$ Input $pf = \dfrac{2 V_m}{\pi} \cos \alpha \cdot I_0 \times \dfrac{1}{V_s \cdot I_0}$

$$= \frac{2\sqrt{2}\, V_s}{\pi} \cos \alpha \cdot \frac{1}{V_s} = \frac{2\sqrt{2}}{\pi} \cos \alpha$$

For this example, input $pf = \dfrac{2\sqrt{2}}{\pi} \cos 30° = 0.7796$ lag.

Example 6.33. *In Example 6.32, if source has an inductance of 1.5 mH, then determine (a) average output voltage (b) the angle of overlap and (c) the power factor.*

Solution. (a) From Eq. (6.73), average output voltage is

$$V_{ox} = \frac{2 V_m}{\pi} \cos \alpha - \frac{\omega_0 L_s}{\pi} I_0$$

$$= \frac{2\sqrt{2} \times 230}{\pi} \cos 30° - \frac{2\pi \times 50 \times 1.5 \times 10^{-3}}{\pi} \times 17.93$$

$$= 176.614 \text{ V}$$

(b) From Eq. (6.71), $I_0 = \dfrac{V_m}{\omega L_s} [\cos \alpha - \cos (\alpha + \mu)]$

or $17.93 = \dfrac{\sqrt{2} \times 230 \times 10^3}{2\pi \times 50 \times 1.5} [\cos 30 - \cos (30 + \mu)]$

Its simplification gives overlap angle,

$$\mu = 32.855 - 30 = 2.855°$$

(c) Power factor $= \dfrac{V_0 I_0}{V_s I_s} = \dfrac{176.614 \times 17.93}{230 \times 17.93} = 0.7679$ lag.

Example 6.34. *A 3-phase fully-controlled bridge converter with 415 V supply, 0.04 Ω resistance per phase and 0.25 Ω reactance per phase is operating in the inverting mode at a firing advance angle of 35°. Calculate the mean generator voltage when the current is level at 80 A. The thyristor voltage drop is 1.5 V.*

                                                           [I.A.S., 1994]

Solution. Power circuit diagram of a 3-phase full converter reveals that source resistance r_s will lead to a voltage drop of $2\,I_0\,r_s$. Two thyristors, one from positive group and another from negative group, conduct together, therefore there will be a constant thyristor voltage drop of $2\,V_T$. The source reactance leads to overlap and its effect is taken care of by Eq. (6.76). By taking into consideration these voltage drops, the average, or mean, output voltage V_0 in a 3-phase full converter is given by

$$V_{ox} = \frac{3\,V_{ml}}{\pi}\cos\alpha - 2\,I_0\,r_s - 2\,V_T - \frac{3\,\omega L_s}{\pi}\,I_0$$

In case 3-phase full converter is working in the inverting mode, then the load emf E or V_g (mean generator voltage in this example) can be obtained from the relation :

$$\frac{3\,V_{ml}}{\pi}\cos\alpha = -E + 2\,I_0\,r_s + 2V_T + \frac{3\omega L_s}{\pi}\,I_0$$

$$\therefore \qquad \frac{3\sqrt{2}\times 415}{\pi}\cos(180-35) = -E + 2\times 80\times 0.04 + 2\times 1.5 + \frac{3\times 0.25}{\pi}\times 80$$

or $\qquad\qquad E = 459.022 + 6.4 + 3 + 19.1 = 487.522$ V

$\therefore$ Mean generator voltage $\quad = E = 487.522$ V.

Example 6.35. *In Example 6.34, in case load consists of RLE, with $R = 0.2\;\Omega$, inductance large enough to make load current level at 80 A and emf E, then find the mean value of E for (i) firing angle of 35° and (ii) firing advance angle of 35°.*

Solution. (*i*) When firing angle is 35°, 3-phase full converter is in the rectifying mode. Therefore, from Example 6.34,

$$V_{ox} = E + I_0\,R = \frac{3\,V_{ml}}{\pi}\cos\alpha - 2\,I_0\,r_s - 2V_T - \frac{3\,\omega L_s}{\pi}\,I_0$$

or $\qquad \dfrac{3\,V_{ml}}{\pi}\cos\alpha = E + I_0 R + 2\,I_0\,r_s + 2V_T + \dfrac{3\,\omega L_s}{\pi}\,I_0$

$$\therefore \qquad \frac{3\sqrt{2}\times 415}{\pi}\cos 35° = E + 80\times 0.2 + 2\times 80\times 0.04 + 2\times 1.5 + \frac{3\times 0.25}{\pi}\times 80$$

or $\qquad\qquad E = 414.522$ V.

(*ii*) For firing advance angle of 35°, the full converter is in the inverting mode. From Example 6.34,

$$\frac{3\,V_{ml}}{\pi}\cos\alpha = -E + I_0 R + 2\,I_0\,r_s + 2V_T + \frac{3\,\omega L_s}{\pi}\,I_0$$

or $\qquad\qquad E = 459.22 + 16 + 6.4 + 3 + 19.1 = 503.522$ V.

Example 6.36. *Fig. 6.48 (a) shows a battery charging circuit using SCRs. The input voltage from neutral to any line is 230 V (rms) and firing angle for thyristors is 30°. Find the average current flowing through the battery.*

Derive the expression used.

Solution. For the parameters given in this example, the waveform of load current is drawn in Fig. 6.48 (*b*). When thyristor A1 is gated at $\alpha = 30°$, it begins conduction at $\omega t = 30 + \alpha = 60°$. After its turn-on, when $V_{mp}\sin\beta = 150$ V, thyristor A1 gets turned off at $\omega t = \beta$. Note that here β is more than 90° as is seen from Fig. 6.48 (*b*). Equation governing the conduction period in Fig. 6.48 is

$$V_m \sin\omega t - E = i_0 R$$

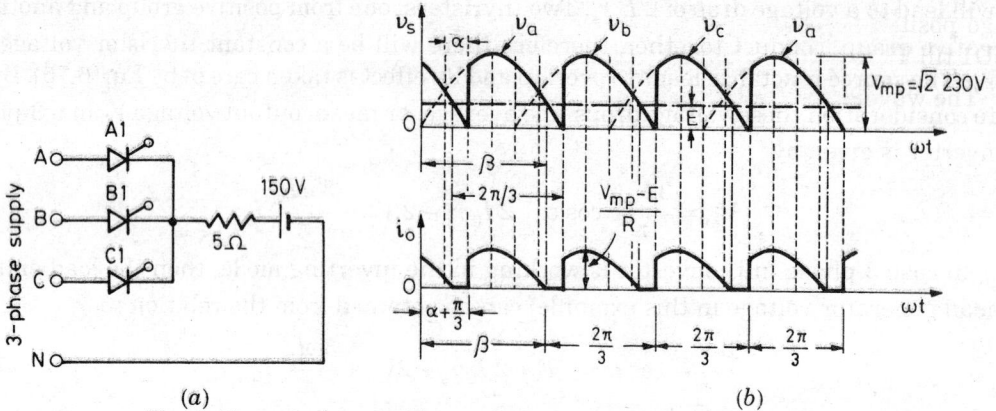

(a)   (b)

Fig. 6.48. (a) 3-phase half-wave battery charging circuit and
(b) its relevant waveforms, Example 6.36.

between the limits of $\left(\dfrac{\pi}{6} + \alpha\right)$ and $\beta > 90°$. Thus, average output current is given by

$$I_0 = \frac{3}{2\pi} \int_{\alpha + \frac{\pi}{6}}^{\beta} \frac{V_{mp} \sin \omega t - E}{R} \cdot d\,(\omega t)$$

$$= \frac{3}{2\pi R} \left[V_{mp} \left| \cos \omega t \right|_{\beta}^{\alpha + \pi/6} - E\,(\beta - \alpha - 30°) \right]$$

$$= \frac{3}{2\pi R} \left[V_{mp} \{\cos(\alpha + 30) - \cos \beta\} - E\,(\beta - \alpha - 30°) \right]$$

Here $\sqrt{2} \cdot 230 \sin \beta = 150$ V. This gives $\beta = 27.47°$ or $152.53°$. As $\beta > 90°$, therefore $\beta = 152.53°$. This gives the value of average battery current I_0 as under :

$$I_0 = \frac{3}{2\pi \times 5} \left[\sqrt{2} \cdot 230\,(\cos 60° - \cos 152.53°) - 150 \left(152.53 - 30 - 30) \times \frac{\pi}{180} \right) \right]$$

$$= \frac{3}{10\pi} \,[208.91928] = 19.95 \text{ A}.$$

Example 6.37. *A single-phase semiconverter, using two thyristors and two diodes as shown in Fig. 6.49 (a), is supplied from 230 V, 50 Hz source. The load consists of R = 10 Ω, E = 100 V and a large inductance so as to render the load current level. For a firing delay angle of 30°, determine (a) average output voltage (b) average output current (c) average and rms values of thyristor currents (d) average and rms values of diode currents (e) input power factor and (f) circuit turn-off time.*

Solution. The waveforms for voltages and currents are sketched in Fig. 6.49 (b).

When forward-biased thyristor T1 is triggered at firing angle α, T1D1 start conducting the constant current I_0. Soon after $\omega t = \pi$, as supply voltage tends to go negative, diode D2 gets forward biased through D1. Therefore, from $\omega t = \pi$, load current begins to freewheel through T1D2. Thyristor T2 gets forward biased after $\omega t = \pi$. At $\omega t = \pi + \alpha$, when T2 is turned

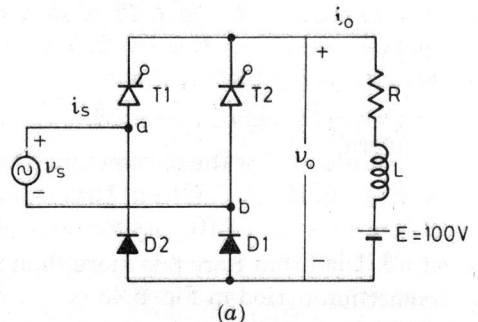

(a)

Fig. 6.49. Pertaining to Example 6.37.

on, current I_0 begins to flow through T2D2 as shown. Soon after $\omega t = 2\pi$, as supply voltage tends to go positive, diode D1 gets forward biased through D2. As a result, current flows through T2D1 till T1 is turned on at $\omega t = 2\pi + \alpha$ and so on.

The waveform of output voltage v_0 shows that average value of output voltage is given by

$$V_0 = \frac{V_m}{\pi}(1 + \cos \alpha)$$

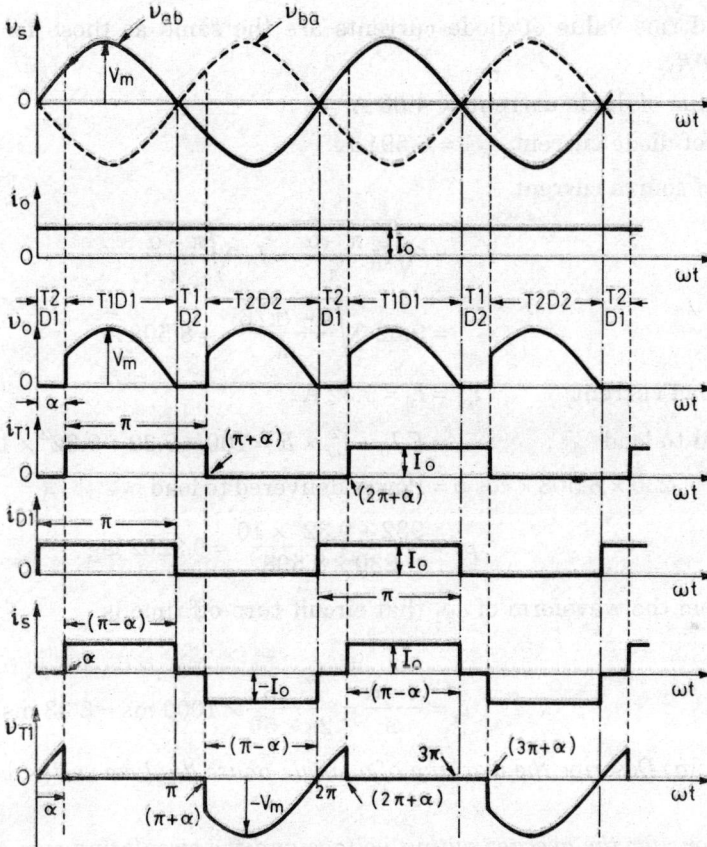

Fig. 6.49. (b) Voltage and current waveforms pertaining to Example 6.37.

(a) Average value of output voltage

$$V_0 = \frac{\sqrt{2} \cdot 230}{\pi}(1 + \cos 30°) = 193.172 \text{ V}$$

(b)
$$V_0 = E + I_0 R$$
$$193.172 = 100 + I_0 \times 10$$

Average value of output current

$$I_0 = \frac{93.172}{10} = 9.32 \text{ A}$$

(c) It is seen from the waveforms of thyristor current i_{T1} and diode current i_{D1} that both conduct for π radians for any value of firing delay angle. On account of this, the circuit of Fig. 6.49 (a) is sometimes called *symmetrical configuration* for a single-phase semiconverter.

Average value of thyristor current

$$I_{T \cdot A} = I_0 \frac{\pi}{2\pi} = \frac{I_0}{2} = \frac{9.32}{2} = 4.66 \text{ A}$$

Rms value of thyristor current

$$I_{T \cdot r} = \sqrt{I_0^2 \frac{\pi}{2\pi}} = \frac{I_0}{\sqrt{2}} = \frac{9.32}{\sqrt{2}} = 6.591 \text{ A}$$

(d) Average and rms value of diode currents are the same as those for a thyristor as discussed in (c) above.

∴   Average value of diode current = 4.66 A

Rms value of diode current   = 6.591 A

(e) Rms value of source current

$$I_{s \cdot r} = \sqrt{I_0^2 \frac{\pi - \alpha}{\pi}} = I_0 \sqrt{\frac{\pi - \alpha}{\pi}}$$

$$= 9.32 \sqrt{\frac{\pi - (\pi/6)}{\pi}} = 8.508 \text{ A}$$

Rms value of load current $I_{or} = I_0 = 9.32$ A.

Power delivered to load $= E I_0 + I_{or}^2 \times R = 100 \times 9.32 + 9.32^2 \times 10$

Also $230 \times 8.508 \times \cos \phi = $ Power delivered to load

∴ Input $pf = \dfrac{932 + 9.32^2 \times 10}{230 \times 8.508} = 0.9202$ lag

(f) It is seen from the waveform of v_{T1} that circuit turn-off time is

$$t_c = \frac{\pi - \alpha}{\omega} = \frac{\pi - \dfrac{\pi}{6}}{2\pi \times 50} \times 1000 \text{ ms} = 8.33 \text{ ms.}$$

Example 6.38. (a) Describe the working of a single-phase dual converter with appropriate waveforms.

Derive expressions for the average output voltage and the circulating current.

(b) A single-phase dual converter is fed from 230 V, 50 Hz source. The load is R = 30 Ω and the current-limiting reactor has L = 0.05 H. For $\alpha_1 = 30°$, calculate the peak value of circulating current and also the peak currents of both the converters.

Solution. (a) The circuit diagram of single-phase dual converter is shown in Fig. 6.45 (a). Single-phase voltage applied across terminals A, B is sketched in Fig. 6.50 (a) as v_s. Let the firing angle of converter 1 be α_1, say around 30° or so. Waveform of output voltage v_{01} across output terminals of converter 1 is shown in Fig. 6.50 (b). For converter 2, $\alpha_2 = 180 - \alpha_1 = 150°$ and waveforms of its output voltage v_{02} is shown in Fig. 6.50 (c). Since v_{02} is mostly below the reference line ωt, its average is negative. As per the polarity markings of output voltages v_{01} and v_{02} in Fig. 6.45 (a), the average values of output voltages of both the converters must be positive. Thus, the waveform of output voltage v_{02} of converter 2 must be shown positive above the reference line ωt, this is shown in Fig. 6.50 (d).

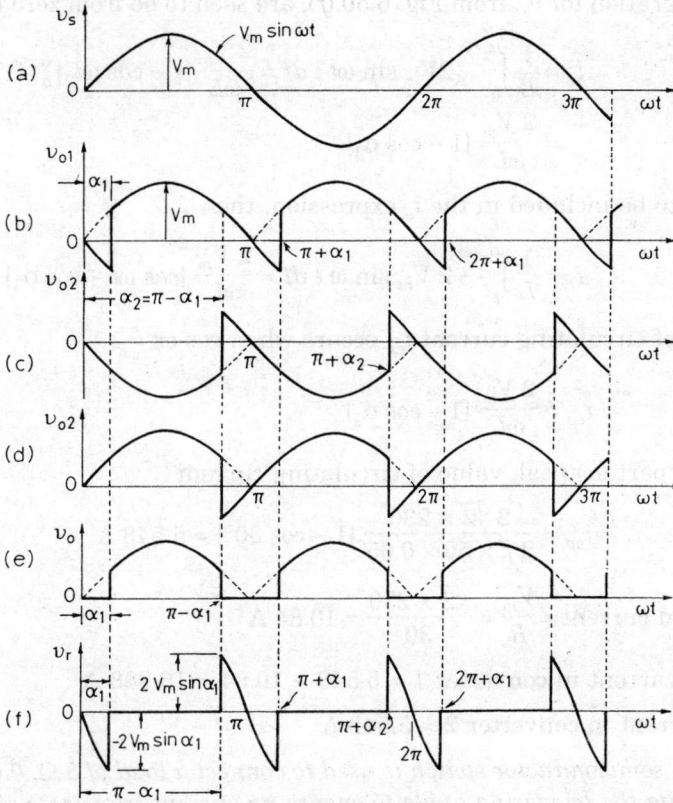

Fig. 6.50. Waveforms for single-phase dual converter, Example 6.38.

Now $v_{01} = V_m \sin \omega t$ and $v_{02} = V_m \sin \omega t$

$\therefore$ Load voltage, $v_o = \dfrac{v_{01} + v_{02}}{2} = V_m \sin \omega t$ as shown in Fig. 6.50 (e). Note that from $\omega t = 0$ to $\omega t = \alpha_1$ and from $\omega t = \pi - \alpha_1$ to $\omega t = \pi$, load voltage $v_o = 0$

$\therefore$ Average value of load voltage, $V_o = \dfrac{1}{\pi} \displaystyle\int_{\alpha_1}^{\pi - \alpha_1} V_m \sin \omega t \cdot d\,(\omega t) = \dfrac{2V_m}{\pi} \cos \alpha_1$ ...(i)

As per Eq. (6.88), voltage v_r across reactor is $v_r = v_{o1} - v_{oi}$. Waveforms of v_{o1} and v_{o2} of Fig. 6.50 (d) reveal that

from $\omega t = 0$ to $\omega t = \alpha_1$, $v_r = -2\,V_m \sin \omega t$

from $\omega t = \alpha_1$ to $\omega t = \pi - \alpha_1$, $v_r = 0$

and from $\omega t = \pi - \alpha_1$ to $\omega t = \pi$, $v_r = 2\,V_m \sin \omega t$ and so on. Waveshape of v_r is sketched in Fig. 6.50(f). Note that when $\omega t = \alpha_1$, $v_r = -2\,V_m \sin \alpha_1$ and at $\omega t = \pi - \alpha_1$, $v_r = 2\,V_m \sin \alpha_1$.

If i_c is the circulating current due to v_r, then $v_r = L\,\dfrac{di_c}{dt}$

or $\qquad\qquad\qquad\qquad i_c = \dfrac{1}{L} \displaystyle\int v_r\, dt$

The limits of integration for v_r, from Fig. 6.50 (f), are seen to be from zero to α_1.

$$\therefore \qquad i_c = \frac{1}{L} \int_o^{\alpha_1/\omega} 2V_m \sin \omega t \cdot dt = \frac{2V_m}{\omega L} \mid -\cos \omega t \mid_o^{\alpha_1/\omega}$$

$$= \frac{2V_m}{\omega L} [1 - \cos \alpha_1] \qquad \qquad ...(ii)$$

In case time t is to be included in the i_c expression, then

$$i_c = \frac{1}{L} \int_t^{\alpha_1/\omega} 2 V_m \sin \omega t \, dt = \frac{2 V_m}{\omega L} [\cos \omega t - \cos \alpha_1]$$

Maximum value of circulating current i_{cp} occurs when $\cos \omega t = 1$.

$$\therefore \qquad i_{cp} = \frac{2 V_m}{\omega L} [1 - \cos \alpha_1] \qquad \qquad ...(ii)$$

(b) From Eq. (ii), part (a) peak value of circulating current

$$i_{cp} = \frac{2\sqrt{2} \times 230}{2\pi \times 50 \times 0.05} [1 - \cos 30°] = 5.548 \text{ A}$$

Peak value of load current $= \dfrac{V_m}{R} = \dfrac{\sqrt{2} \times 230}{30} = 10.84$ A

$\therefore$ Peak value of current in converter 1 = 5.548 + 10.84 = 16.388 A

Peak value of current in converter 2 = 5.548 A

Example 6.39. *A semiconductor switch is used to connect a load of 5 Ω, 0.05 H to a 240 V, 50 Hz supply. Estimate the triggering angle to ensure no current transients. Also indicate the initial triggering angle for the worst transient.*

Solution. A semiconductor switch connecting a load RL to ac supply is shown in Fig. 6.2 (a). The solution for the load current i_o for this converter, given by Eq. (6.7) is repeated here.

$$i_o = \frac{V_m}{Z} \sin (\omega t - \varphi) - \frac{V_m}{Z} \sin (\alpha - \varphi) \exp. \left\{ -\frac{R}{\omega L} (\omega t - \alpha) \right\} = i_s + i_t$$

Here magnitude of transient component of load current i_o is

$$i_t = \frac{V_m}{Z} \sin (\alpha - \phi) \exp \left\{ -\frac{R}{\omega L} (\omega t - \alpha) \right\}$$

In order that no current transients occur, when SCR is turned on at a triggering angle α, i_t must be zero. This is possible only if $\sin (\alpha - \phi) = 0 = \sin 0°$ or $\sin 180°$.

$\therefore$ Triggering angle $\qquad \alpha = \phi = \tan^{-1} \dfrac{2\pi \times 50 \times 0.05}{5} = 72.34°$

Therefore, when triggering angle $\alpha = \phi = 72.34°$, no current transients would occur.

For worst transient current, $\sin (\alpha - \phi) = 1 = \sin 90°$

$\therefore$ Triggering angle $\qquad \alpha = 90 + \phi = 90 + 72.34° = 162.34°$

Therefore, when triggering angle $\alpha = 162.34°$, worst current transients would occur in the circuit.

6.1. (*a*) Give at least five applications of phase-controlled rectifiers.

'(*b*) What is an ideal thyristor switch ?

(*c*) Power flow from 1-phase source to load R can be controlled through the use of a thyristor. Discuss why this method of power flow control is called phase-controlled converter.

(*d*) Give at least two definitions of firing angle. Illustrate your answer with relevant circuit and waveforms.

6.2. (*a*) A single-phase half-wave SCR circuit feeds power to a resistive load. Draw waveforms for source voltage, load voltage, load current and voltage across the SCR for a given firing angle α. Hence obtain expressions for average and rms load voltages in terms of source voltage and firing angle.

(*b*) A resistive load of 10 Ω is connected through a half-wave SCR circuit to 220 V, 50 Hz, single-phase source. Calculate the power delivered to load for a firing angle of 60°. Find also the value of input power factor. [**Ans**. (*b*) 1946 . 887 W, 0.6342]

6.3. (*a*) An RL load is fed from single-phase supply through a thyristor. Derive an expression for load current in terms of supply voltage, frequency, R, L etc. Indicate the time limits during which this solution is applicable.

For this thyristor-load combination, draw waveforms for load voltage, load current, source current and voltage across the thyristor.

(*b*) An RL load, energised from single-phase, 230 V, 50 Hz source through a single thyristor, has $R = 10\ \Omega$ and $L = 0.08$ H. If thyristor is triggered in every positive half cycle at α = 75°, find current expression as a function of time.

[**Ans**. (*b*) $12.023 \sin (314\,t - 68.3°) - 2.3614\,e^{-125t}$]

6.4. A single-phase half-wave converter is operated from 230 V, 50 Hz source and the load resistance is $R = 12\ \Omega$. For a firing angle delay of 30°, determine (*a*) the rectification efficiency (*b*) form factor (*c*) voltage ripple factor (*d*) transformer utilization factor and (*e*) PIV of thyristor. [**Ans**. (*a*) 36.31% (*b*) 1.66 (*c*) 1.325 (*d*) 0.253 (*e*) 325.22 V]

6.5. (*a*) For a single-phase one-pulse controlled converter system, sketch waveforms for load voltage and load current for (*i*) RL load and (*ii*) RL load with freewheeling diode across RL. From a comparison of these waveforms, discuss the advantages of using a freewheeling diode.

(*b*) A battery is charged by a single-phase one-pulse thyristor controlled rectifier. The supply is 30 V, 50 Hz and battery emf is constant at 6 V. Find the resistance to be inserted in series with the battery to limit the charging current to 4 A on the assumption that SCR is triggered continuously. Take a voltage drop of 1 V across the SCR. Derive the expression used. [**Ans** (*b*) 2.5467 Ω]

6.6. (*a*) A dc battery is charged through a resistor R as shown in Fig. 6.5 (*a*). Derive an expression for the average value of charging current in terms of V_m, E, R, firing-angle delay α etc. Here α > θ₁ where θ₁ can be obtained from $V_m \sin \theta_1 = E$.

(*b*) For an ac source voltage of 230 V, 50 Hz, find the value of average charging current for $R = 10\ \Omega$, $E = 110$ V and for firing angle delay = 30°.

(*c*) Derive an expression for the rms value of charging current for the circuit described in part (*a*) above. Also, calculate the power delivered to battery and that dissipated in the resistor.

(*d*) Calculate the supply *pf*.

[**Ans**. (*a*) Eq. (6.18), (*b*) 5.3743 A

(*c*) $\left[\dfrac{1}{2\pi R^2} \left\{ (V_s^2 + E^2)\,(\pi - \theta_1 - \alpha) + \dfrac{V_s^2}{2}\,(\sin 2\alpha + \sin 2\theta_1) - 2\,V_m\,E\,(\cos \alpha + \cos \theta_1) \right\} \right]^{1/2}$

591.173 W, 926.175 W (*d*) 0.6855 lag]

6.7. A single-phase one-pulse SCR controlled converter feeds an RL load with a freewheeling diode across the load. Discuss how freewheeling diode comes into play when supply voltage is passing through zero and becoming negative. Sketch waveforms for supply and load voltages, load current, supply current, freewheeling diode current and voltage across the SCR.

Derive expressions for the load current as a function of time during conduction as well as freewheeling periods. Derive also an expression for average load current.

6.8. (a) Describe the working of a single-phase one-pulse SCR controlled converter with RLE load through the waveforms of supply voltage, load voltage, load current and voltage across the SCR. Hence derive expression for the load current in terms of supply voltage, load impedance, firing angle, load voltage E etc.

(b) A single-phase one-pulse converter with RLE load has the following data :
Supply voltage = 230 V at 50 Hz, $R = 2\,\Omega$, $L = 1$ mH, $E = 120$ V,
Extinction angle $\beta = 220°$, firing angle $\alpha = 25°$.

(i) Calculate the voltage across thyristor at the instant SCR is triggered.

(ii) Find the voltage that appears across SCR when current decays to zero.

(iii) Find the peak inverse voltage for the SCR.   **Ans.** [17.465 V, 329.079 V, 445.27 V]

6.9. Describe the operation of a single-phase two-pulse mid-point converter with relevant voltage and current waveforms. Discuss how each SCR is subjected to a reverse voltage equal to double the supply voltage in case turns ratio from primary to each secondary is unity.

Find the circuit turn-off time provided to each SCR by this converter configuration.

6.10. In a single-phase mid-point converter, turns ratio from primary to each secondary is 1.25. The source voltage is 230 V, 50 Hz. For a resistive load of $R = 2\,\Omega$, determine

(a) maximum value of average output voltage and load current and the corresponding firing and conduction angles,

(b) maximum average and rms thyristor currents,

(c) maximum possible values of positive and negative voltages across SCRs,

(d) the value of α for load voltage of 100 V,

(e) the value of voltage across SCR at the instant of commutation for α of part (d).

$$\left[\textbf{Hint.}\ (b)\ \text{Maximum average thyristor current} = \frac{1}{2\pi}\int_0^\pi \frac{V_m}{R}\sin \omega t \cdot d(\omega t)\ \text{etc.}\right]$$

[**Ans.** (a) 165.63 V, 82.82 A, $\alpha = 0°$, $\gamma = 180°$ (b) 41.41 A, 65.054 A
(c) 520.4 V, 520.4 V (d) 78.025 (e) 509.03 V]

6.11. (a) A single-phase full converter charges a battery which offers a constant value of E. A resistor R is inserted to limit the battery charging current. Derive an expression for the average charging current in terms of V_m, E, R etc. on the assumption that each pair of SCRs is fired continuously in each half cycle. Take V_r as the voltage drop in conducting SCRs.

(b) Find the value of R in case battery charging current is 6 A, supply voltage is 40 V, 50 Hz, $E = 12$ V and $V_r = 1$ volt.

(c) Find the power dissipated in R.

(d) Find the supply power factor.

[**Hint.** Refer to example 6.2. (b) $\theta_1 = \sin^{-1}\dfrac{13}{\sqrt{2}\cdot 40}$

(c) Use Eq. (3.39) with 2π replaced by π in the denominator].

[**Ans.** (a) $\dfrac{1}{\pi R}[2 V_m \cos\theta_1 - (E + V_r)(\pi - 2\theta_1)]$

(b) 3.994 Ω (c) 206.446 W (d) 0.9891 lag]

6.12. A single-phase semi-converter delivers power to RLE load with $R = 5\,\Omega$, $L = 10$ mH and $E = 80$ V. The ac source voltage is 230 V, 50 Hz. For a continuous conduction, find the average value of output current for a firing angle delay of 50°.

If main SCR T2 is damaged and open circuited, find the new value of average output current on the assumption of continuous conduction. Sketch the output voltage and current waveforms and indicate the conduction of various components.

[**Ans.** 18.013 A ; T1 D1, FD ; T1 D1, FD and so on ; 1.0063 A]

6.13. A single-phase full converter feeding RLE load has the following data. Source voltage = 230 V, 50 Hz; $R = 2.5\,\Omega$, $E = 100$ V, Firing angle = 30°.

If load inductance is large enough to make the load current virtually constant, then

(a) sketch the time variations of source voltage, source current, load voltage, load current, current through one SCR and voltage across it,

(b) compute the average value of load voltage and load current,

(c) compute the input *pf*.

[**Ans.** (a) Refer to Fig. 6.10 (b), (b) 179.30 V, 31.72 A (c) 0.7796 lag]

6.14. Repeat Prob. 6.13 in case a freewheeling diode is connected across RLE load.

$$\left[\textbf{Hint.}\ (c)\ \text{Rms value of source current} = \left[I_0^2 \frac{150}{180}\right]^{1/2}\right]$$ [**Ans.** (b) 193.17 V, 37.268 A (c) 0.92 lag.]

6.15. Describe the working of a single-phase full converter in the rectifier mode with RLE load. Discuss how one pair of SCRs is commutated by an incoming pair of SCRs. Illustrate your answer with waveforms for source voltage, E, output voltage and current, source current, current through and voltage across one thyristor. Assume continuous conduction.

Derive an expression for the average output voltage in terms of source voltage and firing angle. From the voltage differential equation of this converter, show that $V_0 = I_0 R + E$.

6.16. Describe the working of a single-phase full converter in the inverter mode with RLE load. Illustrate your answer with waveforms for source voltage, E, load voltage and current, source current, current through and voltage across one SCR. Assume continuous conduction.

Find also the circuit turn-off time. Should the average output voltage be more than E during inverter operation ? Discuss.

6.17. A single-phase semiconverter bridge feeds RLE load. Discuss how freewheeling diode comes into operation and holds the output voltage to almost zero for a given firing angle delay. Sketch the time variations of supply voltage, E, load voltage and current, freewheeling diode current and current through each pair consisting of SCR and diode. Find also the circuit turn-off time. Assume the load current continuous.

Also, derive an expression for the average output voltage in terms of source voltage and firing angle.

6.18. (a) Describe how a freewheeling diode improves power factor in a converter system.

(b) A separately-excited dc motor fed through a single-phase semiconverter runs at a speed of 1200 rpm when ac supply voltage is 230 V, 50 Hz and the motor counter emf is 140 V. The firing angle delay is 50°. Armature circuit resistance is 3 Ω. Compute the average armature current and motor torque.

[**Hint.** (b) Find $K_m = 1.114$ Nm/A and proceed]   [**Ans.** (b) [10.021 A, 11.164 Nm]

6.19. A single-phase full converter is connected to RLE load. For discontinuous load current, draw the source voltage, output voltage, load current and source current waveforms as a function of time when

(a) extinction angle $\beta > \pi$

(b) extinction angle $\beta < \pi$ with $V_m \sin\beta < E$.

Explain how the various waveforms are obtained and discuss their nature.

6.20. A single-phase semiconverter feeds power to RLE load. For discontinuous load current, draw the source voltage, output voltage, load current, source current and freewheeling diode current waveforms as a function of time when

(a) extinction angle $\beta > \pi$

(b) extinction angle $\beta < \pi$ with $V_m \sin \beta < E$.

Explain how various waveforms are obtained and discuss their nature.

6.21. A single-phase full-converter supplies power to RLE load. The source voltage is 230 V, 50 Hz and for load $R = 2\ \Omega$, $L = 10$ mH, $E = 100$ V. For a firing angle of 30°, find the average value of output current and output voltage in case the load current extinguishes at (a) 200° and (b) 170°.

Derive the expressions used.   **[Ans.** (a) 96.2424 A, 192.485 V (b) 106.911 A, 213.822 V]

6.22. A single-phase mid-point SCR converter supplies constant load current of 5 A when the triggering angle is maintained at 35°. The input voltage to the converter is 220 V at 50 Hz. The turns ratio from primary to each secondary is $\frac{1}{2}$. Determine the load voltage and input power factor.

[Hint : Primary AT_s = secondary AT_s, $\therefore$ $I_s N_1 = I_0 \cdot N_2$ or $I_s = 10$ A etc]

[Ans.   324.45 V, 0.7374 lag]

6.23. A single-phase full converter delivers a constant load current I_o. Express its source current in Fourier series and derive therefrom the expressions for the following performance parameters :

displacement factor, current distortion factor, power factor, total harmonic distortion, voltage ripple factor, active and reactive power inputs.

6.24. A 1-phase full converter delivers ripple free current to RL load with $R = 15\ \Omega$. The source voltage is 230 V, 50 Hz. For a firing angle of 30°, calculate rectification efficiency, voltage ripple factor, displacement factor, current distortion factor, power-factor, THD, active and reactive powers.

[Ans. 77.96%, 0.804, 0.866, 0.90032, 0.7797 lag, 0.4834, 2143.173 W, 1237.362 VAr]

6.25. A single-phase semi-converter delivers a constant load current I_o. Express its source current in Fourier series and derive therefrom the expressions for the following performance parameters :

displacement factor, current distortion factor, power factor, total harmonic distortion, voltage ripple factor, active and reactive power inputs.

6.26. Repeat Prob. 6.24 for a single-phase semiconverter.

[Ans. 85.22%, 0.6139, 0.966, 0.9525, 0.9201 lag, 0.31973, 2487.64 W, 666.57 VAr]

6.27. A single-phase asymmetrical semiconverter feeds an RL load with $R = 10\ \Omega$ and a large L so that load current is level.

The source voltage is 230 V, 50 Hz. For a firing angle delay of 30°, determine

(a) average value of output voltage and output current,

(b) average and rms values of diode, thyristor and source currents,

(c) input power factor and circuit turn-off time.

[Ans. (a) 193.172 V, 19.32 A (c) 11.27 A, 14.756 A; 8.05 A, 12.471 A; zero, 17.64 A (c) 0.92 lag, 10 ms]

6.28. For a 3-phase half-wave diode rectifier, derive an expression for the average output voltage V_0 in terms of maximum value of source voltage from line to neutral.

If this rectifier feeds RL load with $R = 5\ \Omega$ and $L = 3$ mH, find the average load current for 3-phase input voltage of 400 V, 50 Hz.   **[Ans.** 54.011 A]

6.29. (a)

For a three-phase half-wave SCR converter delivering continuous output current, derive expressions for the average output voltage for firing angle of (i) $0° < \alpha < 30°$ and (ii) $30° < \alpha < 150°$.

(b) A three-phase half-wave SCR converter delivers constant load current of 30 A over the firing angle range of $0°$ to $80°$. At these two firing angles, compute the power delivered to load for an ac input voltage of 400 V from a delta-star transformer.

$$\left[\textbf{Ans:} \quad (a) \text{ For both } (i) \text{ and } (ii), \quad V_0 = \frac{3\sqrt{3}\, V_{mp}}{2\pi} \cos \alpha \quad (b) \ 8.102 \text{ kW} \quad (c) \ 1.4068 \text{ kW} \right]$$

6.30. A delta-star transformer feeds power to a load $R = 10\ \Omega$ through a 3-phase M-3 converter. The input voltage to converter is 400 V, 50 Hz. Find the power delivered to load for a firing-angle delay of (a) $15°$ and (b) $60°$. Derive the expressions used. [**Ans.** (a) 7243.2 W (b) 4000 W]

6.31. An M-3 converter operates on a 400 V, 3-phase, 50 Hz mains and delivers power to the armature of a *dc* motor with negligible resistance and infinitely large reactor in the *dc* bus. The transformer has Dy 11 connection with unity phase turns ratio. Back emf is 300 V. Determine the trigger angle.

[**Hint.** Input line voltage to M-3 converter is $\sqrt{3}$. 400 V] [**Ans.** 50.113°]

6.32. A 3-phase, 3-pulse converter is connected to RLE load. The source voltage is 3-phase, 230 V, 50 Hz and the load current is level at 10 A. For $R = 0.5\ \Omega$ and $L = 2\ H$, determine (a) firing angle for $E = 134\ V$ and (b) firing-angle advance for $E = -134\ V$.

[**Ans.** (a) 26.472° (b) 33.824°]

6.33. A 3-phase full converter is connected to a resistive load. Show that the average output voltage is given by

$$V_0 = \frac{3\, V_{ml}}{\pi} \cos \alpha \qquad \text{for } 0 < \alpha < \frac{\pi}{3}$$

and $$V_0 = \frac{3\, V_{ml}}{\pi} \left[1 + \cos \left(\alpha + \frac{\pi}{3} \right) \right] \qquad \text{for } \frac{\pi}{3} < \alpha < \frac{2\pi}{3}$$

where V_{ml} = maximum value of line voltage.

6.34. For a 3-phase full converter, explain how output voltage wave, for a firing angle of $30°$, is obtained by using

(a) phase voltages and (b) line voltages.

6.35. (a) A resistive load of 10 Ω is connected to a 3-phase full converter. The load takes 5 kW for a firing angle delay of $70°$. Find the magnitude of per phase input supply voltage. Derive the expression required for the output voltage in terms of firing angle etc.

(b) Repeat part (a) in case an inductor connected in series with the load makes the load current constant.

(c) Repeat part (a) in case an inductor connected in series with the load makes the load current continuous.

[**Hint.** (a) First derive an expression for the rms value of output voltage,

$$V_{or} = V_{ml} \sqrt{\frac{3}{2\pi} \left[\left(\frac{2\pi}{3} - \alpha \right) + \frac{1}{4} \left(\sqrt{3} \cos 2\alpha - \sin 2\alpha \right) \right]}^{1/2}$$

[**Ans.** (a) 214.242 V (b) 279.55 V (c) 213.254 V]

6.36. For a 3-phase semiconverter, draw output voltage waveforms for a firing angle delay of $45°$ indicating the conduction of its various elements on the assumption of continuous output current. Discuss whether freewheeling diode comes into play or not. Hence obtain an expression for the average output voltage in terms of ac supply voltage, firing angle delay etc. by using both sine and cosine functions for the supply voltage.

6.37. Sketch output voltage waveform for a 3-phase semiconverter for a firing angle delay of $75°$. Indicate the conduction of various elements and discuss whether freewheeling diode comes into play on the assumption of continuous load current. Hence obtain an expression for the average output voltage by using both sine and cosine functions for the supply voltage.

6.38. A 3-phase semiconverter is connected to RLE load. For a firing angle delay of 120°, draw output voltage and load current waveforms in case load current is (a) continuous and (b) discontinuous. For both parts, indicate the conducting elements of the semiconverter during three periodic times of the output voltage wave. Discuss briefly the nature of waveforms obtained.

6.39. A separately-excited dc motor fed from 3-phase semiconverter develops a full load torque at 1500 rpm when firing angle is zero, the armature taking 50 A at 400 V dc and having an armature-circuit resistance of 0.5 Ω. Calculate the supply voltage per phase. Find also the range of firing angle required to give speeds between 1500 rpm and 750 rpm at full-load torque.

[**Hint.** For $\alpha = 0°$ and 1500 rpm, $\dfrac{3\,V_{ml}}{\pi} = 400$, $K = 0.25$ V/rpm etc.]

[**Ans.** 171.006 V, $\alpha = 0°$ to 86.42°]

6.40. A 3-phase full converter thyristor bridge feeds a resistive load R.
(a) Sketch input voltage waveforms for v_{ab}, v_{ac}, v_{bc} etc.
(b) From (a), sketch the waveform of the output current i_0 for a firing angle of 30°.
(c) From (b), sketch the waveform of input current i_a for phase A for $\alpha = 30°$. Show the duration of conducting thyristors. In case input voltage is 400 V and $R = 200$ Ω, indicate the peak magnitude of current i_a. [**Ans.** (c) Peak magnitude of $i_a = 2.828$ A]

6.41. For a 3-phase full converter, sketch the input voltage waveforms for v_{ab}, v_{ac}, v_{bc} etc and voltage variation across any one thyristor for one complete cycle for a firing angle delay of (a) 60° and (b) 120°.
Find the magnitude of reverse voltage across this SCR and its commutation time for both parts (a) and (b) for a supply voltage of 230 V, 50 Hz.

[**Ans.** (a) 325.22 V, 10 m sec (b) 281.69 V, 3.33 m sec]

6.42. A battery is charged from 3-phase supply mains of 230 V, 50 Hz through a 3-phase semiconverter. The battery emf is 190 V and its internal resistance is 0.5 Ω. An inductor connected in series with the battery renders the charging current of 20 A ripple free. Compute the firing angle delay and the supply power factor. [**Ans.** 73.263°, 0.652 lag]

6.43. (a) A 3-phase full converter is used for charging a battery with an emf of 110 V and an internal resistance of 0.2 Ω. For a constant charging current of 10 A, compute the firing angle delay for ac line voltage of 220 V. Find also the supply power factor.
(b) For the purpose of delivering energy from dc source to 3-phase system, the firing angle of the 3-phase converter has been increased to 150°. For the same value of dc source current of 10 A, compute the output ac line voltage.

[**Ans.** (a) 67.85°, 0.36 lag (b) 92.36 V]

6.44. A 3-phase full converter is delivering a constant load current of 50 A at 230 V dc when its input is 3-phase, 415 V, 50 Hz. If each thyristor has a voltage drop of 1.1 V when conducting, calculate (a) the firing angle delay of SCRs (b) the rms current of SCRs (c) rms source current (d) the mean power loss in each SCR and (e) input pf. (f) In case ac supply has an inductance of 3 mH per phase, find the new value of firing angle for the same dc power output as before.
[**Hint :** (e) Input power = output power + power lost in SCRs

$$\sqrt{3} \times 415 \times 40.825 \times \cos \phi = 230 \times 50 + 1.1 \times \frac{50}{3} \times 6 \text{ etc.}]$$

[**Ans.** (a) 65.52° (b) 28.87 (c) 40.825 A (d) 18.33 W (e) 0.3936 lag (f) 60.35°]

6.45. (a) Discuss the effect of source inductance on the performance of a single-phase full converter indicating clearly the conduction of various thyristors during one cycle.

Derive expressions for its output voltage in terms of (i) maximum voltage V_m, firing angle α and overlap angle and (ii) V_m, a, L_s and load current I_0. Here L_s is the source inductance. Show that the effect of source inductance is to present an equivalent resistance of $\dfrac{\omega L_s}{\pi}\,\Omega$ in series with the internal rectifier voltage.

(b) A single-phase full-converter fed from 220 V, 50 Hz supply gives an output voltage of 180 V at no load. When loaded with a constant output current of 10 A, the overlap angle is found to be 6°. Compute the value of source inductance in henries.

[**Ans.** (b) 4.8084 mH]

6.46. (a) Show that the performance of a single-phase full converter as effected by source inductance is given by the relation

$$\cos(\alpha + \mu) = \cos\alpha - \frac{\omega L_s I_0}{V_m}$$

where the symbols used have their usual meanings.

(b) A single-phase full converter is connected to ac supply of 330 sin 314 t volt. It operates with a firing angle $\alpha = \pi/4$ rad. The total load current is maintained constant at 5 A and the load voltage is 140 V. Calculate the source inductance, angle of overlap and the load resistance. [**Ans.** 17.113 mH, 6.267°, 28 Ω]

6.47. (a) Show that the performance of a three-phase full converter as influenced by source inductance is given by the relation

$$\cos(\alpha + \mu) = \cos\alpha - \frac{2\,\omega L_s}{V_{ml}} I_0$$

The symbols used have their usual meanings.

(b) A 3-phase fully controlled bridge converter is fed from a 3-phase 400 V, 50 Hz mains. For firing angle of 60°, output current is level at 25 A and output voltage is 250 V. Calculate the load resistance, source inductance and angle of overlap.

[**Ans.** 10 Ω, 2.667 mH, 4.8°]

6.48. Describe the effect of source inductance on the performance of a 3-phase full converter with the help of phase voltage waveforms. Indicate the sequence of conduction of various thyristors and sketch load current waveforms for both positive and negative group of thyristors. State the various assumptions made.

Derive an expression for its output voltage in terms of supply voltage, source inductance, load current etc.

6.49. Repeat Example 6.27 for a firing-angle delay of 45° and overlap angle of 45°.

6.50. A 3-phase full converter is fed from 3-phase, 230 V, 50 Hz supply having per-phase source inductance of 4 mH. The load current is 10 A ripple free.

(a) Calculate the voltage drop in dc output voltage due to source inductance.

(b) If dc output voltage is 210 V, calculate the firing angle and the overlap period.

(c) In case the bridge is made to operate as a line-commutated inverter with dc voltage of 210 V, calculate the firing angle for the same load current.

[**Ans.** (a) 12 V, (b) 44.37°, 0.3344 ms (c) 129.61°]

6.51. Explain how two 3-phase full converters can be connected back to back to form a circulating current type of dual converter. Discuss its operation with the help of voltage waveforms across (a) each converter (b) load and (c) reactor, Take $\alpha_1 = 0°$.

Describe how circulating current waveform can be obtained from reactor voltage waveform. If one of the two converters is loaded, sketch the waveforms of their load currents.

6.52. (a) For a 3-phase dual converter, derive an expression for the circulating-current in terms of supply voltage, reactor inductance, firing-angle delay etc. Relevant voltage and current waveforms, needed for this derivation, must be sketched.

(b) Two 3-phase full-converters are connected in antiparallel to form a dual converter of the circulating-current type. The input to the dual converter is 3-phase, 400 V, 50 Hz. If peak value of circulating current is limited to 20 A, find the value of inductance needed for the reactor for firing angles of (a) $\alpha_1 = 30°$ and (b) $\alpha_1 = 60°$

[**Ans.** (b) (i) 77.97 mH (ii) 20.892 mH]

6.53. A 3-phase full converter, fed from 3-phase, 400 V source, has an output voltage of 450 V dc for a firing-angle delay of 30°. Calculate the overlap angle and the voltage drop due to overlap.

[**Ans.** 6.84°, 17.82 V]

Chapter 7

Choppers

Many industrial applications require power from dc voltage sources. Several of these applications, however, perform better in case these are fed from variable dc voltage sources. Examples of such dc systems are subway cars, trolley buses, battery-operated vehicles, battery-charging etc.

From ac supply systems, variable dc output voltage can be obtained through the use of phase-controlled converters (discussed in Chapter 6) or motor-generator sets. The conversion of fixed dc voltage to an adjustable dc output voltage, through the use of semiconductor devices, can be carried out by the use of two types of dc to dc converters given below [5].

AC Link Chopper. In the ac link chopper, dc is first converted to ac by an inverter (dc to ac converter). AC is then stepped-up or stepped-down by a transformer which is then converted back to dc by a diode rectifier, Fig. 7.1 (a). As the conversion is in two stages, dc to ac and then ac to dc, ac link chopper is costly, bulky and less efficient.

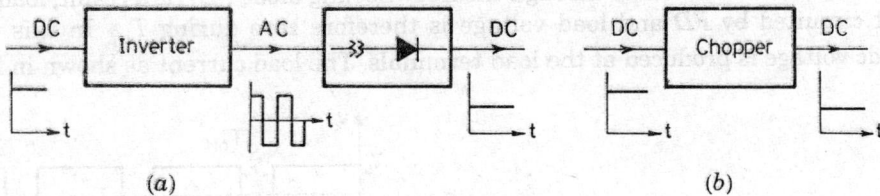

(a)   (b)

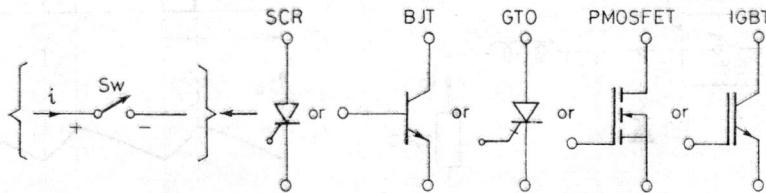

(c)

Fig. 7.1 (a) AC link chopper and (b) dc chopper (or chopper)
(c) Representation of a power semiconductor device.

DC Chopper. A chopper is a static device that converts fixed dc input voltage to a variable dc output voltage directly, Fig. 7.1 (b). A chopper may be thought of as dc equivalent of an ac transformer since they behave in an identical manner. As choppers involve one stage conversion, these are more efficient.

Choppers are now being used all over the world for rapid transit systems. These are also used in trolley cars, marine hoists, forklift trucks and mine haulers. The future electric automobiles are likely to use choppers for their speed control and braking. Chopper systems offer smooth control, high efficiency, fast response and regeneration.

The power semiconductor devices used for a chopper circuit can be force-commutated thyristor, power BJT, power MOSFET, GTO or IGBT. These devices, in general, can be represented by a switch SW with an arrow as shown in Fig. 7.1 (c). When the switch is off, no current can flow. When the switch is on, current flows in the direction of arrow only. The power semiconductor devices have on-state voltage drops of 0.5 V to 2.5 V across them. For the sake of simplicity, this voltage drop across these devices is neglected.

As stated above, a chopper is dc equivalent to an ac transformer having continuously variable turns ratio. Like a transformer, a chopper can be used to step down or step up the fixed dc input voltage. As step-down dc choppers are more common, the term dc chopper, or chopper, in this book would mean a step-down dc chopper unless stated otherwise.

The object of this chapter is to discuss the basic principles of chopper operation and the more common types of chopper configurations using ideal switches.

7.1. PRINCIPLE OF CHOPPER OPERATION

A chopper is a high speed on/off semiconductor switch. It connects source to load and disconnects the load from source at a fast speed. In this manner, a chopped load voltage as shown in Fig. 7.2 (b) is obtained from a constant dc supply of magnitude V_s. In Fig. 7.2 (a), chopper is represented by a switch SW inside a dotted rectangle, which may be turned-on or turned-off as desired. For the sake of highlighting the principle of choper operation, the circuitry used for controlling the on, off periods of this switch is not shown. During the period T_{on}, chopper is on and load voltage is equal to source voltage V_s. During the interval T_{off}, chopper is off, load current flows through the freewheeling diode FD. As a result, load terminals are short circuited by FD and load voltage is therefore zero during T_{off}. In this manner, a chopped dc voltage is produced at the load terminals. The load current as shown in Fig. 7.2 (b)

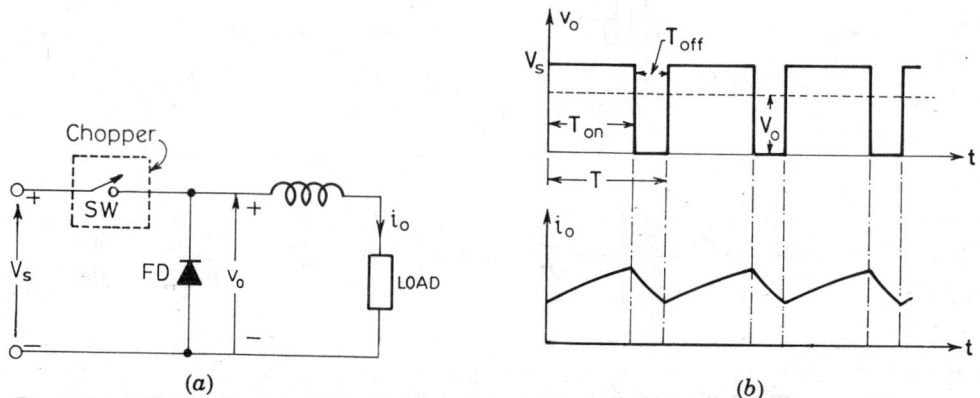

Fig. 7.2 (a) Elementry chopper circuit and (b) output voltage and current waveforms.

is continuous. During T_{on}, load current rises whereas during T_{off}, load current decays. From Fig. 7.2 (b), average load voltage V_0 is given by

$$V_0 = \frac{T_{on}}{T_{on} + T_{off}} V_s = \frac{T_{on}}{T} V = \alpha V_s \qquad \qquad ...(7.1)$$

where
$$T_{on} = \text{on–time} \; ; T_{off} = \text{off–time}$$
$$T = T_{on} + T_{off} = \text{chopping period}$$
$$\alpha = \frac{T_{on}}{T} = \text{duty cycle}$$

Thus load voltage can be controlled by varying duty cycle α. Eq. (7.1) shows that load voltage is independent of load current. Eq. (7.1) can also be written as

$$V_0 = f \cdot T_{on} \cdot V_s \qquad \qquad ...(7.2)$$

where
$$f = \frac{1}{T} = \text{chopping frequency}$$

7.2. CONTROL STRATEGIES

It is seen from Eq. (7.1) that average value of output voltage V_0 can be controlled through α by opening and closing the semiconductor switch periodically. The various control strategies for varying duty cycle α are as follows :

1. Time ratio control (TRC) and

2. Current-limit control

These are now described one after the other.

7.2.1. Time Ratio Control (TRC).

As the name suggests, in this control scheme, time ratio T_{on}/T (as duty ratio or duty cycle) is varied. This is realized in two different strategies called *constant frequency system* and *variable frequency system* as detailed below :

1. Constant Frequency System

In this scheme, the on-time T_{on} is varied but chopping frequency f (or chopping period T) is kept constant. Variation of T_{on} means adjustment of pulse width, as such this scheme is also called *pulse-width-modulation scheme*.

Fig. 7.3 illustrates the principle of pulse-width modulation. Here chopping period T is constant. In Fig. 7.3 (a), $T_{on} = \frac{1}{4} T$ so that $\alpha = 0.25$ or $\alpha = 25\%$. In Fig. 7.3 (b), $T_{on} = \frac{3}{4} T$ so that $\alpha = 0.75$ or 75%. Ideally α can be varied from zero to unity. Therefore output voltage V_0 can be varied between zero and source voltage V_s.

2. Variable Frequency System

In this scheme, the chopping frequency f (or chopping period T) is varied and either (i) on-time T_{on} is kept constant or (ii) off-time T_{off} is kept constant. This method of controlling α is also called frequency-modulation scheme.

Fig. 7.4 illustrates the principle of frequency modulation. In Fig. 7.4 (a), T_{on} is kept constant but T is varied. In the upper diagram of Fig. 7.4 (a), $T_{on} = \frac{1}{4} T$ so that $\alpha = 0.25$. In the lower diagram of Fig. 7.4 (a), $T_{on} = \frac{3}{4} T$ so that $\alpha = 0.75$. In Fig. 7.4 (b), T_{off} is kept constant and T is

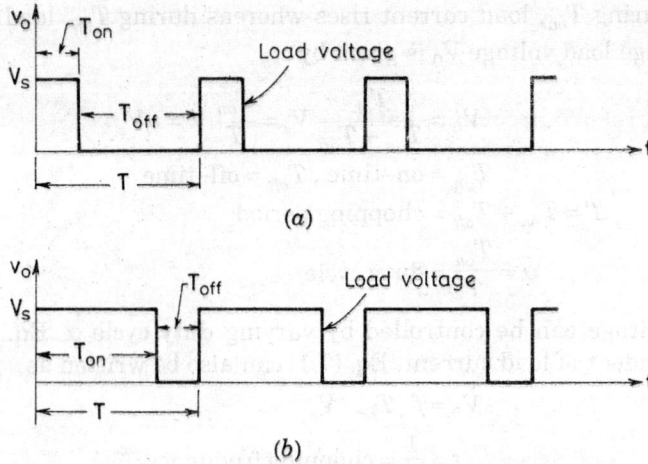

Fig. 7.3. Principle of pulse-width modulation (constant T).

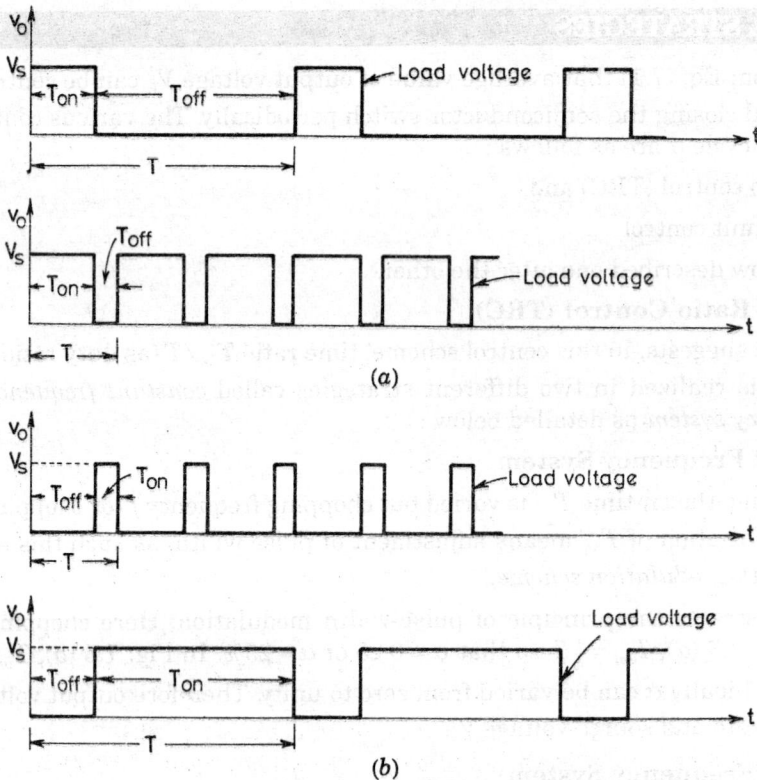

Fig. 7.4. Principle of frequency modulation.
(a) on-time T_{on} constant and (b) off-time T_{off} constant.

varied. In the upper diagram of this figure, $T_{on} = \frac{1}{4} T$ so that $\alpha = 0.25$ and in the lower diagram $T_{on} = \frac{3}{4} T$ so that $\alpha = 0.75$.

Frequency modulation scheme has some disadvantages as compared to pulse-width modulation scheme. These are as under :

(*i*) The chopping frequency has to be varied over a wide range for the control of output voltage in frequency modulation. Filter design for such wide frequency variation is, therefore, quite difficult.

(*ii*) For the control of α, frequency variation would be wide. As such, there is a possibility of interference with signalling and telephone lines in frequency modulation scheme.

(*iii*) The large off-time in frequency modulation scheme may make the load current discontinuous which is undesirable.

It is seen from above that constant frequency (*PWM*) scheme is better than variable frequency scheme. *PWM* technique has, however, a limitation. In this technique, T_{on} cannot be reduced to near zero for most of the commutation circuits used in choppers. As such, low range of α control is not possible in *PWM*. This can, however, be achieved by increasing the chopping period (or decreasing the copping frequency) of the chopper.

7.2.2. Current-limit Control

In this control strategy, the on and off of chopper circuit guided by the previous set value of load current. These two set values are maximum load current $I_{o.mx}$ and minimum load current $I_{o.mn}$

When load current reaches the upper limit $I_{o.mx}$, chopper is switched off. Now load current freewheels and begins to decay exponentially. When it falls to lower limit $I_{o.mn}$, chopper is switched on and load current begins to rise as shown in Fig. 7.5. Profile of load current shows that it fluctuates between $I_{o.mx}$ and $I_{o.mn}$, and therefore cannot be discontinuous.

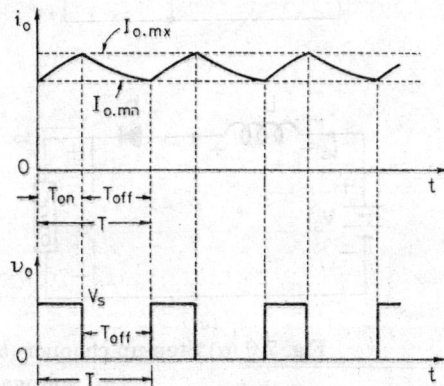

Fig. 7.5. Current-limit control for chopper.

Switching frequency of chopper can be controlled by setting $I_{o.mx}$ and $I_{o.min}$. Ripple current ($= I_{o.mx} - I_{o.mn}$) can be lowered and this in turn necessitates higher switching frequency and therefore more switching losses.

Current-limit control involves feedback loop, the trigger circuitry for the chopper is therefore more complex. PWM technique is, therefore, the commonly chosen control strategy for the power control in chopper circuits.

7.3. STEP-UP CHOPPERS

For the chopper configuration of Fig. 7.2 (*a*), average output voltage V_0 is less than the input voltage V_s, i.e. $V_0 < V_s$; this configuration is therefore called step-down chopper. Average output voltage V_0 greater than input voltage V_s can, however, be obtained by a chopper called *step-up chopper*. Fig. 7.6 (*a*) illustrates an elementary form of a step-up chopper. In this article, working principle of a step-up chopper is presented.

In this chopper, a large inductor L in series with source voltage V_s is essential as shown in Fig. 7.6 (*a*). When the chopper CH is on, the closed current path is as shown in Fig. 7.6 (*b*) and inductor stores energy during T_{on} period. When the chopper CH is off, as the inductor current cannot die down instantaneously, this current is forced to flow through the diode and load for a time T_{off}, Fig. 7.6 (*c*). As the current tends to decrease, polarity of the emf induced in L is

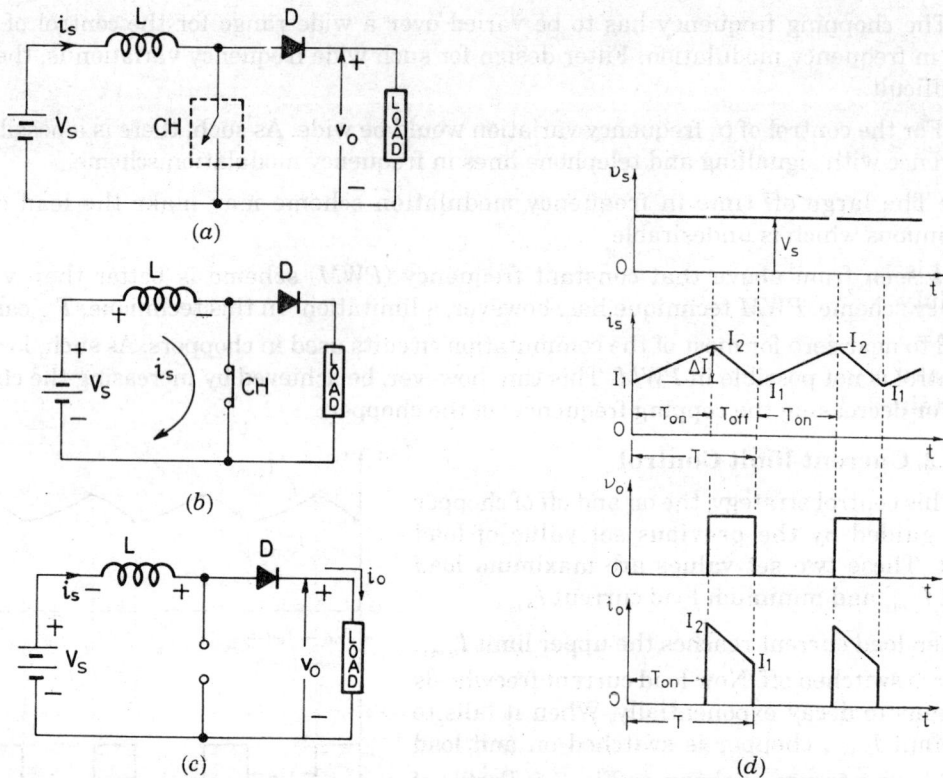

Fig. 7.6 (a) Step-up chopper (b) L stores energy (c) $L \cdot di/dt$ is added to V_s
(d) voltage and current waveforms.

reversed as shown in Fig. 7.6 (c). As a result, voltage across the load, given by $V_0 = V_s + L \, (di/dt)$, exceeds the source voltage V_s. In this manner, the circuit of Fig. 7.6 (a) acts as a step-up chopper and the energy stored in L is released to the load.

When CH is on, current through the inductance L would increase from I_1 to I_2 as shown in Fig. 7.6 (d). When CH is off, current would fall from I_2 to I_1. With CH on, source voltage is applied to L i.e. $v_L = V_s$. When CH is off, KVL for Fig. 7.6 (c) gives $v_L - V_0 + V_s = 0$, or $v_L = V_0 - V_s$. Here v_L = voltage across L. Variation of source voltage v_s, source curren i_s, load voltage v_o and load current i_o is sketched in Fig. 7.6 (d). Assuming linear variation of output current, the energy input to inductor from the source, during the period T_{on}, is

$$W_{in} = \text{(voltage across } L\text{) (average current through } L\text{)} \, T_{on}$$

$$= V_s \cdot \left(\frac{I_1 + I_2}{2}\right) T_{on} \qquad \qquad ...(7.3)$$

During the time T_{off}, when chopper is off, the energy released by inductor to the load is

$$W_{off} = \text{(voltage across } L\text{) (average current through } L\text{)} \, T_{off}$$

$$= (V_0 - V_s)\left(\frac{I_1 + I_2}{2}\right) \cdot T_{off} \qquad \qquad ...(7.4)$$

Considering the system to be lossless, these two energies given by Eqs. (7.3) and (7.4) will be equal.

$$\therefore \qquad V_s \left(\frac{I_1 + I_2}{2}\right) T_{on} = (V_0 - V_s)\left(\frac{I_1 + I_2}{2}\right) \cdot T_{off}$$

$$V_s \cdot T_{on} = V_o\, T_{off} - V_s \cdot T_{off}$$

$$V_o\, T_{off} = V_s\,(T_{on} + T_{off}) = V_s \cdot T$$

or
$$V_o = V_s \frac{T}{T_{off}} = V_s \frac{T}{T - T_{on}} = V_s \frac{1}{1 - \alpha} \qquad \qquad ...(7.5)$$

It is seen from Eqn. (7.5) that average voltage across the load can be stepped up by varying the duty cycle. If chopper of Fig. 7.6 (a) is always off, $\alpha = 0$ and $V_0 = V_s$. If this chopper is always on, $\alpha = 1$ and $V_0 = \infty$ (infinity) as shown in Fig. 7.7 (a). In practice, chopper is turned on and off so that α is variable and the required step-up average output voltage, more than source voltage, is obtained.

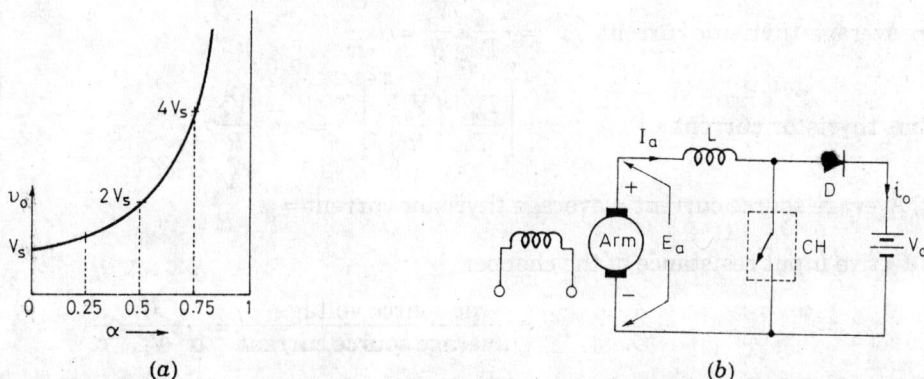

Fig. 7.7. (a) Variation of load voltage v_o with duty cycle (b) regenerative braking of dc motor.

The principle of step-up chopper can be employed for the regenerative braking of dc motors. This is illustrated in Fig. 7.7 (b) where motor armature voltage E_a represents V_s of Fig. 7.6 (a). Voltage V_o is the dc source voltage. When CH is on, L stores energy. When CH is off, L releases energy. In case $E_a/(1 - \alpha)$ exceeds V_o, dc machine begins to work as a dc generator and armature current I_a flows opposite to motoring mode. Power now flows from dc machine to source V_o causing regenerative braking of dc motor. Motor armature voltage E_a is directly proportional to field flux and motor speed. Therefore, even at decreasing motor speeds, regenerative braking can be made to take place provided duty cycle and field flux are so adjusted that $E_a/(1 - \alpha)$ is more than the fixed source voltage V_o.

Example 7.1. *For the basic dc to dc converter of Fig. 7.2 (a), express the following variables as functions of V_s, R and duty cycle α in case load is resistive :*

(a) Average output voltage and current

(b) Output current at the instant of commutation

(c) Average and rms freewheeling diode currents

(d) Rms value of the output voltage

(e) Rms and average thyristor currents

(f) Effective input resistance of the chopper.

Solution. The load voltage variation is shown in Fig. 7.2 (*b*). For a resistive load, output or load current waveform is similar to load voltage waveform.

(*a*) Average output voltage, $V_0 = \dfrac{T_{on}}{T} V_s = \alpha V_s$

Average output current, $I_0 = \dfrac{V_0}{R} = \dfrac{T_{on}}{T} \cdot \dfrac{V_s}{R} = \alpha \dfrac{V_s}{R}$

(*b*) The output current is commutated by the thyristor at the instant $t = T_{on}$. Therefore, output current at the instant of commutation is V_s/R.

(*c*) For a resistive load, freewheeling diode FD does not come into play. Therefore, average and rms values of freewheeling diode currents are zero.

(*d*) Rms value of output voltage $= \left[\dfrac{T_{on}}{T} \cdot V_s^2\right]^{1/2} = \sqrt{\alpha} \cdot V_s$

(*e*) Average thyristor current $= \dfrac{T_{on}}{T} \cdot \dfrac{V_s}{R} = \alpha \dfrac{V_s}{R}$

Rms thyristor current $= \left[\dfrac{T_{on}}{T} \cdot \left(\dfrac{V_s}{R}\right)^2\right]^{1/2} = \sqrt{\alpha} \cdot \dfrac{V_s}{R}$

(*f*) Average source current = average thyristor current $= \alpha \cdot \dfrac{V_s}{R}$

Effective input resistance of the chopper

$$= \frac{\text{dc source voltage}}{\text{average source current}} = \frac{V_s \cdot R}{\alpha \cdot V_s} = \frac{R}{\alpha}$$

Example 7.2. *For type-A chopper of Fig. 7.2 (a), dc source voltage = 230 V, load resistance = 10 Ω. Take a voltage drop of 2 V across chopper when it is on. For a duty cycle of 0.4, calculate*

(*a*) *average and rms values of output voltage and*

(*b*) *chopper efficiency.*

Solution. (*a*) When chopper is on, output voltage is $(V_s - 2)$ volts and during the time chopper is off, output voltage is zero.

∴ Average output voltage $= \dfrac{(V_s - 2) T_{on}}{T} = \alpha (V_s - 2)$

$= 0.4 (230 - 2) = 91.2$ V

Rms value of output voltage,

$$V_{or} = \left[(V_s - 2)^2 \cdot \frac{T_{on}}{T}\right]^{1/2} = \sqrt{\alpha} (V_s - 2)$$

$$= \sqrt{0.4} (230 - 2) = 144.2 \text{ V}$$

(*b*) Power output or power delivered to load,

$$P_0 = \frac{V_{or}^2}{R} = \frac{(144.2)^2}{10} = 2079.364 \text{ W}$$

Power input to chopper, $P_i = V_s \cdot I_0 = 230 \times \dfrac{91.2}{10} = 2097.6$ W

Chopper efficiency $= \dfrac{P_0}{P_i} = \dfrac{2079.364}{2097.6} \times 100 = 99.13\%.$

Example 7.3. *A step-up chopper has input voltage of 220 V and output voltage of 660 V. If the conducting time of thyristor-chopper is 100 μs, compute the pulse width of output voltage.*

In case output-voltage pulse width is halved for constant frequency operation, find the average value of new output voltage.

Solution. From Eq. (7.5), $660 = 220 \dfrac{1}{1 - \alpha}$

or $$\alpha = \frac{2}{3} = \frac{T_{on}}{T}$$

It is seen from Fig. 7.6 (d) that conducting time of chopper is $T_{on} = \dfrac{2}{3} T = 100$ μs. This gives

chopping period $T = 100 \times \dfrac{3}{2} = 150$ μ s.

∴ Pulse width of output voltage $= T_{off} = T - T_{on} = 150 - 100 = 50$ μs

When pulse width of output voltage is halved, $T_{off} = \dfrac{50}{2} = 25$ μs.

For constant frequency operation, $T = 150$ μs, $T_{on} = 150 - 25 = 125$ μs

∴ $$\alpha = \frac{T_{on}}{T} = \frac{125}{150} = \frac{5}{6}$$

∴ Average value of new output voltage, $V_o = 220 \dfrac{1}{1 - \dfrac{5}{6}} = 1320$ V

7.4. TYPES OF CHOPPER CIRCUITS

Power semiconductor devices used in chopper circuits are unidirectional devices ; polarities of output voltage V_0 and the direction of output current I_0 are, therefore, restricted. A chopper can, however, operate in any of the four quadrants by an appropriate arrangement of semiconductor devices. This characteristic of their operation in any of the four quadrants forms the basis of their classification as type-A chopper, type-B chopper etc. Some authors describe this chopper classification as class A, class B, ... in place of type-A, type-B .... respectively.

In the chopper-circuit configurations drawn henceforth, the current directions and voltage polarities marked in the power circuit would be treated as positive. In case current directions and voltage polarities turn out to be opposite to those shown in the circuit, these currents and voltages must be treated as negative.

In this section, the classification of various chopper configurations is discussed.

7.4.1. First-quadrant, or Type-A, Chopper

This type of chopper is shown in Fig. 7.8 (a). It is observed that chopper circuit of Fig. 7.2 (a) is also type-A chopper. In Fig. 7.8 (a), when chopper CH1 is on, $v_0 = V_s$ and current i_0 flows in the arrow direction shown. When CH1 is off , $v_0 = 0$ but i_0 in the load continues flowing in the same direction through freewheeling diode FD, Fig. 7.2 (b). It is thus seen that average values of both load voltage and current, *i.e.* V_0 and I_0 are always positive : this fact is shown by the hatched area in the first quadrant of $V_0 - I_0$ plane in Fig. 7.8 (b).

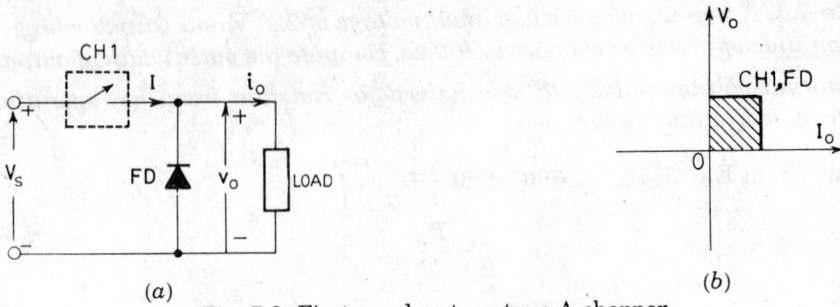

Fig. 7.8. First-quadrant, or type-A chopper.

The power flow in type-A chopper is always from source to load. This chopper is also called *step-down chopper* as average output voltage V_0 is always less than the input dc voltage V_s.

7.4.2. Second-quadrant, or Type-B, Chopper

Power circuit for this type of chopper is shown in Fig. 7.9 (a). Note that load must contain a dc source E, like a battery (or a dc motor) in this chopper.

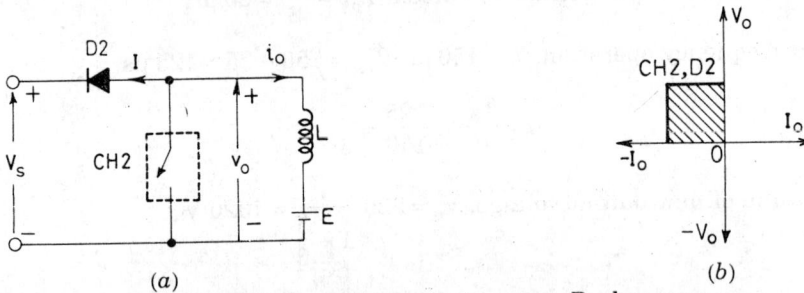

Fig. 7.9. Second-quadrant, or type-B, chopper.

When CH2 is on, $v_0 = 0$ but load voltage E drives current through L and CH2. Inductance L stores energy during T_{on} (= on period) of CH2. When CH2 is off, $v_0 = \left(E + L \dfrac{di}{dt} \right)$ exceeds source voltage V_s. As a result, diode D2 is forward biased and begins conduction, thus allowing power to flow to the source. Chopper CH2 may be on or off, current I_0 flows out of the load, current i_0 is therefore treated as negative. Since V_0 is always positive and I_0 is negative, power flow is always from load to source. As load voltage $V_0 = \left(E + L \dfrac{di}{dt} \right)$ is more than source voltage V_s, type-B chopper is also called *step-up chopper.*

Both type-A and type-B chopper configurations have a common negative terminal between their input and output circuits.

7.4.3. Two-quadrant type-A chopper, or Type-C Chopper

This type of chopper is obtained by connecting type-A and type-B choppers in parallel as shown in Fig. 7.10 (a). The output voltage V_0 is always positive because of the presence of freewheeling diode FD across the load. When chopper CH2 is on, or freewheeling diode FD conducts, output voltage $v_0 = 0$ and in case chopper CH1 is on or diode D2 conducts, output voltage $v_0 = V_s$. The load current i_0 can, however, reverse its direction. Current i_0 flows in the arrow direction marked in Fig. 7.10 (a), i.e. load current is positive when CH1 is on or FD conducts. Load current is negative if CH2 is on or D2 conducts. In other words, CH1 and FD operate together as type-A chopper in first quadrant. Likewise, CH2 and D2 operate together as type-B chopper in second quadrant.

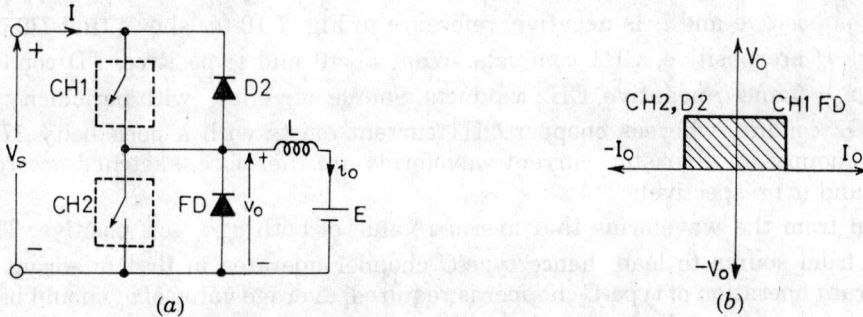

Fig. 7.10. Two-quadrant type-A chopper, or type-C chopper.

Average load voltage is always positive but average load current may be positive or negative as explained above. Therefore, power flow may be from source to load (first-quadrant operation) or from load to source (second-quadrant operation). Choppers CH1 and CH2 should not be on simultaneously as this would lead to a direct short circuit on the supply lines. This type of chopper configuration is used for motoring and regenerative braking of dc motors. The operating region of this type of chopper is shown in Fig. 7.10 (b) by hatched area in first and second quadrants.

Example 7.4. *Sketch output voltage, output current, source current and thyristor (or chopper) current waveforms for type-C chopper for its operation in first quadrant.*

Solution. For this example, refer to chopper circuit of Fig. 7.10 (a). When D2 conducts or CH1 is on, load voltage v_o is equal to V_s during T_{on}. When CH2 is on or FD conducts, $v_o = 0$ during T_{off}. The output voltage waveform v_o is sketched as V_s during T_{on} and zero during T_{off} in Fig. 7.11 (a). Load current i_o is assumed negative at $t = 0$. It changes from negative to positive value during T_{on} and from positive to negative value during T_{off} as shown in Fig. 7.11 (b).

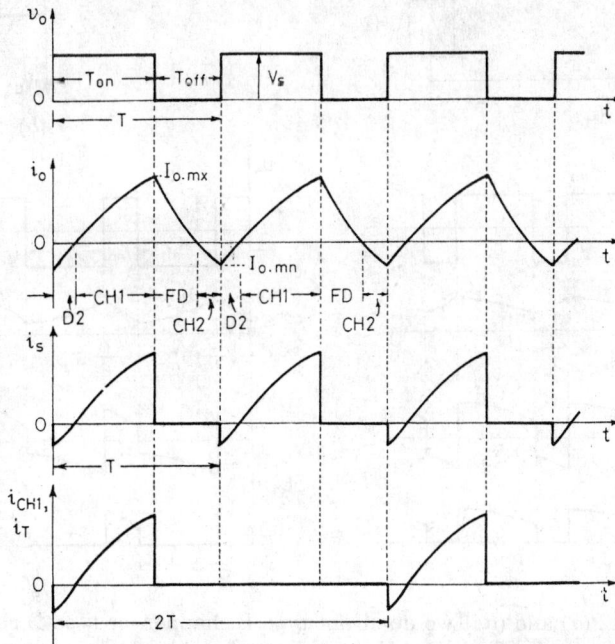

Fig. 7.11. Waveforms for type-C chopper, Example 7.4.

When v_o is positive and i_o is negative, reference to Fig. 7.10 (a) shows that D2 conducts; when both v_o, i_o are positive, CH1 conducts. What $v_o = 0$ and i_o positive, FD conducts and further with $v_o = 0$ and i_o negative, CH2 conducts. Source current i_s, with periodicity T, exists when CH1, D2 conduct whereas chopper CH1 current exists with a periodicity $2T$. Source current and chopper, or thyristor, current waveforms are therefore, sketched accordingly in Fig. 7.11 (c) and (d) respectively.

It is seen from the waveforms that average value of both v_o, i_o are positive. Therefore, power flows from source to load, hence type-C chopper operates in first quadrant. In case second quadrant operation of type-C chopper is required, average value of v_o should be positive but that of i_o must be negative.

In Fig. 7.10 (a), if load current is assumed positive and continuous, then CH1 would conduct during T_{on} and FD during T_{off}. This would result in first quadrant operation of type-C chopper. Chopper CH2 and D2 would, however, remain idle under the assumption of positive load current.

7.4.4. Two-quadrant Type-B Chopper, or Type-D Chopper

The power circuit diagram for two-quadrant type-B chopper, or type-D chopper, is shown in Fig. 7.12 (a). The output voltage $v_0 = V_s$ when both CH1 and CH2 are on and $v_0 = -V_s$ when both choppers are off but both diodes $D1$ and $D2$ conduct. Average output voltage V_0 is positive when choppers turn-on time T_{on} is more than their turn-off time T_{off} as shown in Fig. 7.12 (c).

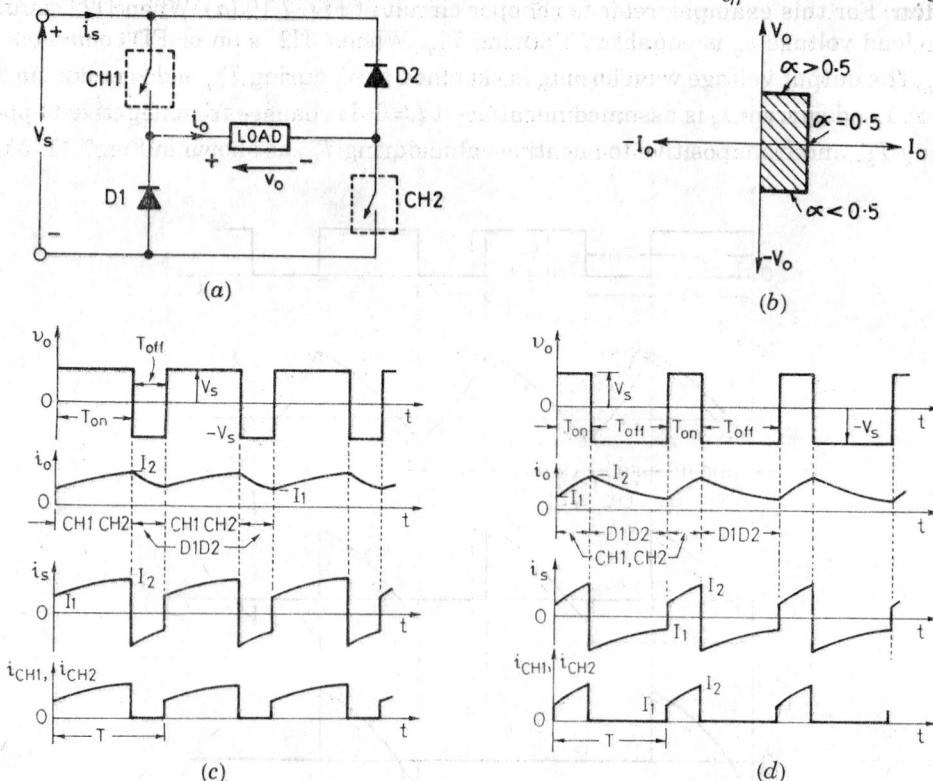

Fig. 7.12 (a) and (b) Two-quadrant type-B chopper, or type-D chopper
(c) waveforms for $T_{on} > T_{off}$, V_o is positive, first quadrant operation and
(d) waveforms for $T_{on} < T_{off}$, V_o is negative, fourth quadrant operation.

The direction of load current is always positive, Fig. 7.12 (c), because choppers and diodes can conduct current only in the direction of arrows shown in Fig. 7.12 (a). Waveform of source current i_s and chopper current i_{CH1} or i_{CH2} are also sketched in Fig. 7.12 (c).

As average values of both v_o, i_o are positive, chopper operation in first quadrant is obtained and power flows from source to load.

Various waveforms for $T_{on} < T_{off}$ are also sketched in Fig. 7.12 (d). It is seen that average value of v_o is negative, but that of i_o is positive. Thus, fourth quadrant operation of type-D chopper is obtained and power flows from load to source.

Average value of output voltage, from Fig. 7.12 (c) and (d) is

$$V_o = \frac{V_s T_{on} - V_s \cdot T_{off}}{T} = V_s \cdot \frac{T_{on} - T_{off}}{T}$$

(i) In case $T_{on} > T_{off}$, $\alpha > 0.5$, V_o is positive as in Fig. 7.12 (b) and (c).

(ii) In case $T_{on} < T_{off}$, $\alpha < 0.5$, V_o is negative as in Fig. 7.12 (b) and (d).

(iii) In case $T_{on} = T_{off}$, $\alpha = 0$, $V_o = 0$, Fig. 7.12 (b).

7.4.5. Four-quadrant Chopper, or Type-E Chopper

The power circuit diagram for a four-quadrant chopper is shown in Fig. 7.13 (a). It consists of four semiconductor switches CH1 to CH4 and four diodes D1 to D4 in antiparallel. Numbering of choppers CH1, ... , CH4 corresponds to their respective quadrant operation. For example; for first quadrant operation, only CH1 is operated; for second quadrant operation, only CH2 is operated and so on. Working of this chopper in the four quadrants is explained as under :

First quadrant : For first-quadrant operation of Fig. 7.13 (a), CH4 is kept on, CH3 is kept off and *CH1 is operated*. With CH1, CH4 on, load voltage $v_0 = V_s$ (source voltage) and load current i_0 begins to flow. Here both v_0 and i_0 are positive giving first quadrant operation. When CH1 is turned off, positive current freewheels through CH4, D2. In this manner, both V_0, I_0 can be controlled in the first quadrant.

Note down that type-E chopper operates as a step-down chopper in this quadrant.

Second quadrant : Here *CH2 is operated* and CH1, CH3 and CH4 are kept off. With CH2 on, reverse (or negative) current flows through L, CH2, D4 and E. Inductance L stores energy during the time CH2 is on. When CH2 is turned off, current is fed back to source through diodes D1, D4. Note that here $\left(E + L \dfrac{di}{dt}\right)$ is more than the source voltage V_s. As load voltage V_0 is positive and I_0 is negative, it is second quadrant operation of chopper. Also, power is fed back from load to source. For second quadrant operation, load must contain emf E as shown in Fig. 7.13 (a). In second quadrant, configuration operates as a step-up chopper.

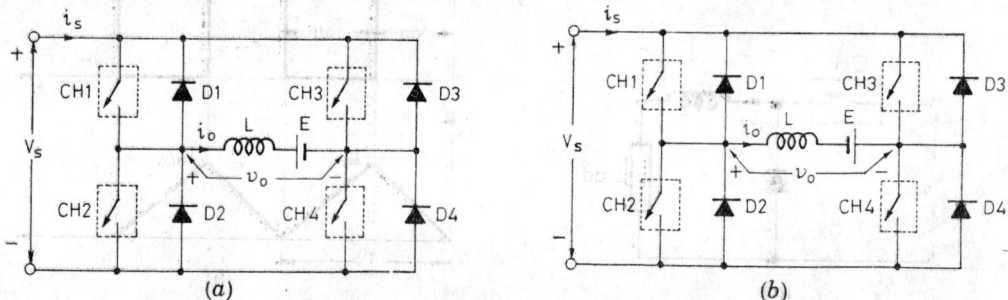

Fig. 7.13. Four-quadrant, or Type-E chopper circuit diagram with
(a) load emf E and (b) load emf E reversed.

Third quadrant. For third quadrant operation, CH1 is kept off, CH2 is kept on and *CH3 is operated*. Polarity of load emf E must be reversed for this quadrant working; this is shown in Fig. 7.13 (b). When CH3 is on, load gets connected to source V_s so that both v_o, i_o are negative leading to third-quadrant operation. When CH3 is turned-off, negative current freewheels through CH2, *D4*. In this manner, v_o and i_o can be controlled in the third quadrant. Here chopper operates as a step-down chopper operates as a step-down chopper.

Fourth quadrant. Here *CH4 is operated* and other devices are kept off. Load emf E has its polarity as shown in Fig. 7.13 (b) for its operation in the fourth quadrant. With CH4 on, positive current flows through CH4, D2, L and E. Inductance L stores energy during the time CH4 is on. When CH4 is turned off, current is fed back to source through diodes D2, D3. Here load voltage is negative, but load current is positive leading to the chopper operation in the fourth quadrant. Also power is fed back from load to source. Here chopper operates as a step-up chopper.

The devices conducting in the four quadrants are indicated in Fig. 7.13 (c).

1 chopper on *step-up* chopper CH2 operated CH2-D4 : L stores energy CH2 – off ; then D1–D4 conduct	2 choppers on *step-down* chopper CH1 operated CH1–CH4 on CH1 – off ; then CH4-D2 conduct
2 choppers on *step-down* chopper CH3 – operated CH3–CH2 : on CH3 – off ; then CH2-D4 conduct E reversed	1 chopper on *step-up* chopper CH4 operated CH4–D2 : L stores energy CH4 – off, then D2, D3 conduct E reversed

Fig. 7.13 (c) Type-E chopper; operation of various devices in the four quadrants.

Example 7.5. *Show that for a basic dc to dc converter, the critical inductance of the filter circuit is given by*

$$L = \frac{V_0^2 (V_s - V_0)}{2f V_s P_0}$$

where V_0, V_s, P_0 *and* f *are load voltage, source voltage, load power and chopping frequency respectively.*

Solution. The critical inductance L is that value of inductance for which the output current falls to zero at $t = T$ during the turn-off period of the chopper. A typical waveform of output current, with critical inductance in the load circuit, is shown in Fig. 7.14 (b). If current variation, from zero to I_{mx} during T_{on} and from I_{mx} to zero during T_{off}, is assumed linear, then average value of output current I_0 is given by

$$I_0 \cdot T = \frac{1}{2} I_{mx} T_{on} + \frac{1}{2} I_{mx} T_{off} = \frac{1}{2} I_{mx} \left(T_{on} + T_{off} \right) = \frac{1}{2} I_{mx} T$$

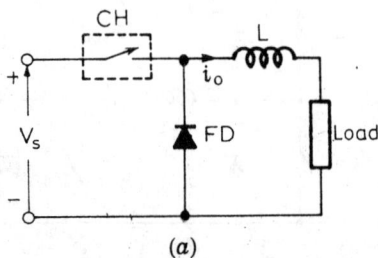

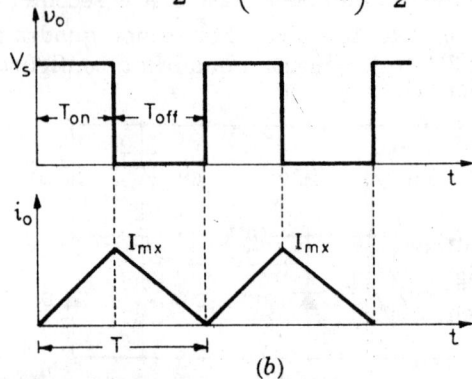

Fig. 7.14. Pertaining to Example 7.5.

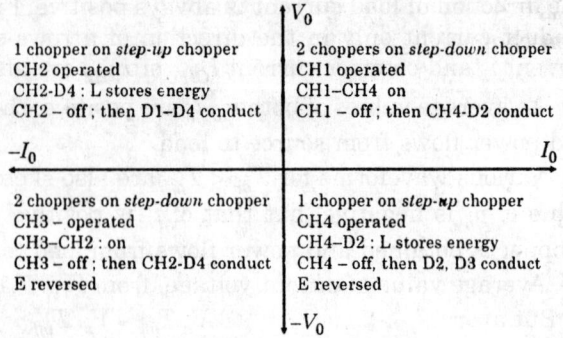

or $I_{mx} = 2\,I_0$ = maximum value of chopper current at $t = T_{on}$. It is seen from Fig. 7.14 (a) that when chopper CH is on,

$$V_0 + L\frac{di}{dt} = V_s \quad \text{or} \quad V_0 + L\frac{I_{mx}}{T_{on}} = V_s$$

or

$$L\frac{2\,I_0}{T_{on}} = V_s - V_0$$

∴

$$L = \frac{(V_s - V_0)\,T_{on}}{2\,I_0} \qquad \qquad ...(i)$$

But average value of output voltage $V_0 = f\,T_{on}\,V_s$ and output, or load, power $P_0 = V_0\,I_0$. This gives

$$T_{on} = \frac{V_0}{f\cdot V_s} \quad \text{and} \quad I_0 = \frac{P_0}{V_0}$$

Substituting these values of T_{on} and I_0 in Eq (i), we get

$$L = \frac{(V_s - V_0)\,V_0^2}{2f\,V_s\,P_0}$$

7.5. STEADY STATE TIME-DOMAIN ANALYSIS OF TYPE-A CHOPPER

For the type-A chopper of Fig. 7.8 (a) with RLE load, the waveforms for gate signal i_g, load current i_0 and load voltage v_0 are as shown in Fig. 7.15 (a) for continuous conduction and, in Fig. 7.15 (b) for discontinuous conduction. In Fig. 7.15 (b), periodic time T is more than that in Fig. 7.15 (a). The determination of load current expression is useful for knowing (i) the current profile over periodic time T, (ii) the current ripple and (iii) whether the current is continuous or discontinuous. The object of this article is to study the type-A chopper with RLE load for current variation over T, current ripple and also for the Fourier analysis of output voltage.

For RLE type load, E is the load voltage which may be a dc motor or a battery. When CH1 is on in Fig. 7.8 (a), the equivalent circuit is as shown in Fig. 7.15 (c). For this mode of operation, the differential equation governing its performance is

$$V_s = R\,i + L\frac{di}{dt} + E \qquad \qquad ...(7.6)$$

for

$$0 \le t \le T_{on}.$$

When CH1 is off, the load current continues flowing through the freewheeling diode and the equivalent circuit is as shown in Fig. 7.15 (d). For this circuit, the differential equation is

$$0 = R\,i + L\frac{di}{dt} + E \qquad \qquad ...(7.7)$$

for

$$T_{on} < t \le T.$$

Solution of Eqs. (7.6) and (7.7) may be obtained by the use of Laplace transform. It is seen from Fig. 7.15 (a) that initial value of current is I_{mn} for Eq. (7.6) and I_{mx} for Eq. (7.7). Therefore, Laplace transform of Eqs. (7.6) and (7.7) is

$$RI(s) + L[sI(s) - I_{mn}] = \frac{V_s - E}{s} \qquad \qquad ...(7.8)$$

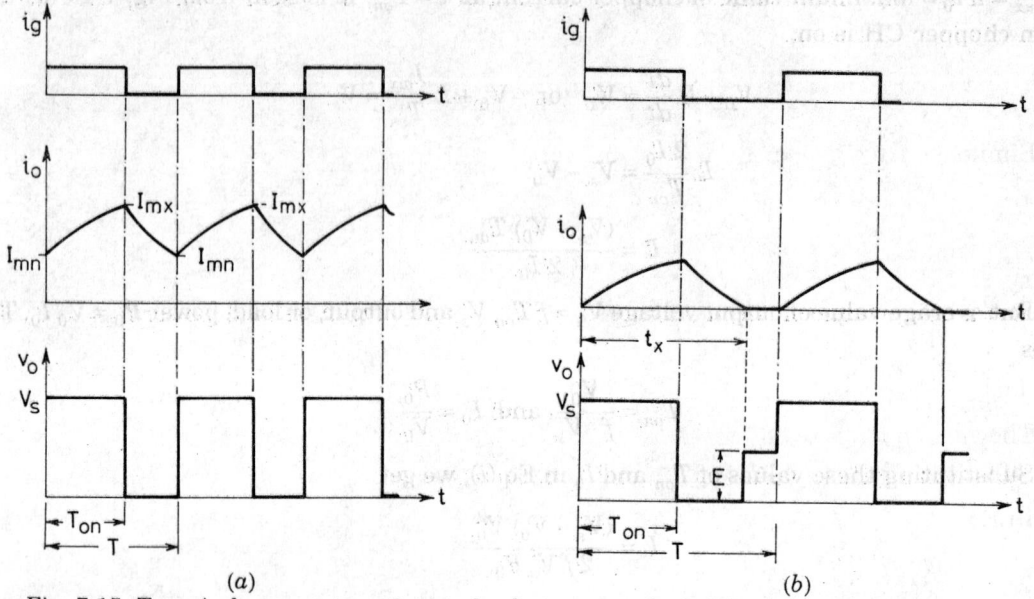

(a)   (b)

Fig. 7.15. Type-A chopper (a) continuous load current and (b) Discontinuous load current.

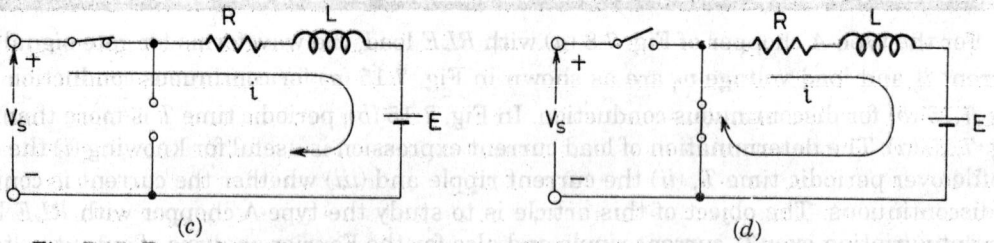

(c)   (d)

Fig. 7.15. Equivalent circuit for type-A chopper with (c) CH1 on and (d) CH1 off.

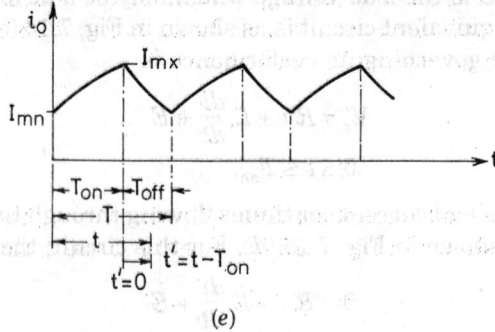

(e)

Fig. 7.15 (e) Pertaining to t'.

and $RI(s) + L[\cdot I(s) - I_{mx}] = -\dfrac{E}{s}$ ...(7.9)

From Eq. (7.8), $I(s) = \dfrac{V_s - E}{s(R + s \cdot L)} + \dfrac{L \cdot I_{mn}}{R + s \cdot L} = \dfrac{V_s - E}{s \cdot L\left(s + \dfrac{R}{L}\right)} + \dfrac{I_{mn}}{s + \dfrac{R}{L}}$

Laplace inverse of the above expression is

$$i(t) = \frac{V_s - E}{R}\left(1 - e^{-\frac{R}{L}t}\right) + I_{mn}\, e^{-\frac{R}{L}t} \qquad \qquad ...(7.10)$$

for $\qquad 0 \le t \le T_{on}.$

Similarly, the time-domain expression for current from Eq. (7.9) is

$$i(t') = -\frac{E}{R}\left(1 - e^{-\frac{R}{L}t'}\right) + I_{mx}\, e^{-\frac{R}{L}t'} \qquad \qquad ...(7.11)$$

for $\qquad T_{on} < t \le T$

where $t' = t - T_{on}$, see Fig. 7.15 (e), so that when

$$t = T_{on}, \quad t' = 0$$

and for $\qquad t = T, \qquad t' = T - T_{on} = T_{off}.$

The variation of current $i(t)$ from I_{mn} to I_{mx} for $0 \le t \le T_{on}$ can be plotted from Eq. (7.10) and that of $i(t')$ from I_{mx} to I_{mn} for $0 < t' \le T_{off}$ from Eq. (7.11).

In Eq. (7.10), at $t = T_{on}$, $i(t) = I_{mx}$.

$\therefore \qquad \qquad I_{mx} = \dfrac{V_s - E}{R}\left(1 - e^{-T_{on}/T_a}\right) + I_{mn}\, e^{-T_{on}/T_a} \qquad \qquad ...(7.12)$

In Eq. (7.11), at $t' = T_{off} = T - T_{on}$, $i(t') = I_{mn}$

$\therefore \qquad \qquad I_{mn} = -\dfrac{E}{R}\left(1 - e^{-(T - T_{on})/T_a}\right) + I_{mx} \cdot e^{-(T - T_{on})/T_a} \qquad \qquad ...(7.13)$

where $\qquad \qquad T_a = \dfrac{L}{R}.$

Eqs. (7.12) and (7.13) can be solved for I_{mx} and I_{mn} as under :

From Eq. (7.12), $\quad I_{mx} = \dfrac{V_s}{R}\left(1 - e^{-(T_{on}/T_a)}\right) - \dfrac{E}{R}\left(1 - e^{-T_{on}/T_a}\right) + I_{mn}\, e^{-T_{on}/T_a}$

Substitution of I_{mn} from Eq. (7.13) in the above expression gives

$$I_{mx} = \frac{V_s}{R}\left(1 - e^{-T_{on}/T_a}\right) - \frac{E}{R} + \frac{E}{R}\, e^{-T_{on}/T_a} - \frac{E}{R}\, e^{-T_{on}/T_a}$$

$$\qquad \qquad + \frac{E}{R} \cdot e^{-(T - T_{on})/Ta} \cdot e^{-T_{on}/T_a} + I_{mx}\, e^{-(T - T_{on})/T_a} \cdot e^{-T_{on}/T_a}$$

$$= \frac{V_s}{R}\left(1 - e^{-T_{on}/T_a}\right) - \frac{E}{R}\left(1 - e^{-T/T_a}\right) + I_{mx} \times e^{-T/T_a}$$

or $\qquad \qquad I_{mx}\left(1 - e^{-T/T_a}\right) = \dfrac{V_s}{R}\left(1 - e^{-T_{on}/T_a}\right) - \dfrac{E}{R}\left(1 - e^{-T/T_a}\right)$

or $\qquad \qquad I_{mx} = \dfrac{V_s}{R}\left[\dfrac{1 - e^{-T_{on}/T_a}}{1 - e^{-T/T_a}}\right] - \dfrac{E}{R} \qquad \qquad ...(7.14)$

Substitution of I_{mx} from Eq. (7.14) in (7.13) gives

$$I_{mx} = -\frac{E}{R} + \frac{E}{R} \cdot e^{-(T - T_{on})/T_a} + \frac{V_s}{R}\left[\frac{1 - e^{-T_{on}/T_a}}{1 - e^{-T/T_a}}\right]e^{-(T - T_{on})/T_a} - \frac{E}{R} \cdot e^{-(T - T_{on})/T_a}$$

$$= \frac{V_s}{R}\left[\frac{1-e^{-T_{on}/T_a}}{1-e^{-T/T_a}}\right]\frac{e^{T_{on}/T_a}}{e^{T/T_a}} - \frac{E}{R}$$

$$I_{mn} = \frac{V_s}{R}\left[\frac{e^{T_{on}/T_a}-1}{e^{T/T_a}-1}\right] - \frac{E}{R} \qquad \qquad ...(7.15)$$

In case CH1 conducts continuously, then $T_{on} = T$ and from Eqs. (7.14) and (7.15).

$$I_{mx} = I_{mn} = \frac{V_s - E}{R} \qquad \qquad ...(7.16)$$

The maximum I_{mx} and minimum I_{mn} values of load current can be obtained from Eqs. (7.14) and (7.15) respectively for given V_s, R, α, T_a and E.

For those who are not familiar with Laplace-transform technique, the following method may be adopted for solving Eqs. (7.6) and (7.7).

Eq. (7.6) can be re-written as

$$Ri + L\frac{di}{dt} = V_s - E$$

$$(R + Lp)\,i = V_s - E \qquad \qquad ...(i)$$

where

$$p \equiv \frac{d}{dt}$$

The solution of Eq. (i) consists of two parts, complementary function and the particular integral.

Complementary Function : It is obtained from force-free equation $(R + Lp)\,i = 0$. Its solution is of the type

$$i = Ae^{pt}.$$

Here p is the root of the auxiliary equation $R + Lp = 0$ and this root is $p = -\frac{R}{L}$.

$$\therefore \qquad i_{C.F.} = Ae^{-\frac{R}{L}t}$$

Particular Integral : It is obtained from Eq. (i) by putting $p = 0$.

$$\therefore \qquad Ri = V_s - E$$

or

$$i_{P.I.} = \frac{V_s - E}{R}$$

Therefore, the complete solution of Eq. (7.6) is

$$i = i_{P.I.} + i_{C.F.}$$

$$= \frac{V_s - E}{R} + Ae^{-\frac{R}{L}t}$$

Constant A can be obtained from the initial condition. It is seen from Fig. 7.15 (a) that at $t = 0$, the initial current during T_{on} is $i = I_{mn}$.

$$\therefore \qquad I_{mn} = \frac{V_s - E}{R} + A$$

or

$$A = \left(I_{mn} - \frac{V_s - E}{R}\right)$$

$$\therefore \qquad i = \frac{V_s - E}{R}\left(1 - e^{-\frac{R}{L}t}\right) + I_{mn} - e^{-\frac{R}{L}t} \qquad \qquad ...(7.10)$$

This agrees with the solution in Eq. (7.10). Solution of Eq. (7.7) can also be obtained similarly as in Eq. (7.11).

7.5.1. Steady State Ripple

It is seen from Fig. 7.15 (e) that current pulsates between I_{mx} and I_{mn}. The ripple current $(I_{mx} - I_{mn})$ can be obtained from Eqs. (7.14) and (7.15) as follows :

$$I_{mx} - I_{mn} = \frac{V_s}{R}\left[\frac{1 - e^{-T_{on}/T_a}}{1 - e^{-T/T_a}} - \frac{e^{T_{on}/T_a} - 1}{e^{T/T_a} - 1}\right]$$

$$= \frac{V_s}{R}\left[\frac{1 - e^{-T_{on}/T_a}}{1 - e^{-T/T_a}} - \frac{(1 - e^{-T_{on}/T_a})\, e^{T_{on}/T_a}}{(1 - e^{-T/T_a}) \cdot e^{T/T_a}}\right]$$

$$= \frac{V_s}{R}\left[\frac{(1 - e^{-T_{on}/T_a}) - (1 - e^{-T_{on}/T_a})\, e^{(T_{on} - T)/T_a}}{1 - e^{-T/T_a}}\right]$$

$$= \frac{V_s}{R}\left[\frac{(1 - e^{-T_{on}/T_a})\,(1 - e^{-(T - T_{on})/T_a})}{1 - e^{-T/T_a}}\right] \qquad ...(7.17)$$

The ripple current given by Eq. (7.17) is seen to be independent of load counter emf E. With $T_{on} = \alpha T$ and $T - T_{on} = (1 - \alpha)\, T$, Eq. (7.17) can be written as

$$I_{mx} - I_{mn} = \frac{V_s}{R}\left[\frac{(1 - e^{-\alpha T/T_a})\,(1 - e^{-(1 - \alpha) T/T_a})}{1 - e^{-T/T_a}}\right]$$

Per unit ripple current $= \dfrac{I_{mx} - I_{mn}}{V_s/R} = \dfrac{(1 - e^{-\alpha T/T_a})\,(1 - e^{-(1 - \alpha)T/T_a})}{1 - e^{-T/T_a}}$ $\qquad$ ...(7.18)

For $\alpha = 0.5$ and $T/T_a = 5$, p.u. ripple current $= 0.848$. For $\alpha = 0.5$ and $T/T_a = 25$, pu ripple current $= 1$. In this manner, the variation of pu ripple current as a function of duty cycle α and ratio T/T_a can be plotted as shown in Fig. 7.16. Its value is maximum when $\alpha = 0.5$. As L increases, $T_a\,(= L/R)$ increases and T/T_a reduces and pu ripple current decreases, Fig. 7.16.

The peak to peak ripple current has maximum value ΔI_{mx} when duty cycle $\alpha = 0.5$ in Eq. (7.18).

Putting $\dfrac{T}{T_a} = x$ for convenience, ΔI_{mx} from Eq. (7.18) is

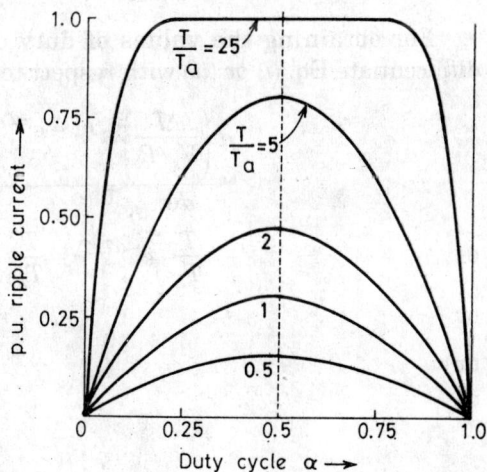

Fig. 7.16. Per unit ripple current as a function of α and T/T_a.

$$\Delta I_{mx} = \frac{V_s}{R}\left[\frac{(1 - e^{-0.5x})\,(1 - e^{-0.5x})}{1 - e^{-x}}\right]$$

$$= \frac{V_s}{R}\left[\frac{(1 - e^{-0.5x})\,(1 - e^{-0.5x})}{(1 + e^{-0.5x})\,(1 - e^{-0.5x})}\right]$$

$$= \frac{V_s}{R}\left[\frac{1 - e^{-0.5x}}{1 + e^{-0.5x}}\right] = \frac{V_s}{R}\tanh\frac{1}{4}x$$

$$= \frac{V_s}{R}\tanh\frac{T}{4T_a}$$

But $\qquad T = \dfrac{1}{f}$ and $T_a = \dfrac{L}{R}$

$$\therefore \qquad \Delta I_{mx} = \frac{V_s}{R} \tanh \frac{R}{4fL}.$$

In case $4fL \gg R$, then $\tanh \dfrac{R}{4fL} \cong \dfrac{R}{4fL}$. Under this condition, maximum value of ripple current is

$$\Delta I_{mx} = \frac{V_s}{R} \cdot \frac{R}{4fL} = \frac{V_s}{4fL}. \qquad \qquad ...(7.18a)$$

This shows that maximum value of ripple current is inversely proportional to chopping frequency and the circuit inductance.

Example 7.6. *In the continuous conduction mode of type-A chopper, show that per unit ripple in the load current is maximum when duty cycle is equal to 0.5.*

Solution. Eq. (7.18) can be used to prove that ripple in the output current, or per-unit ripple in load current is maximum when the duty cycle is equal to 0.5.

Therefore, from Eq. (7.18), ripple current ΔI is

$$\Delta I = \frac{V_s}{R} \left[\frac{(1 - e^{-\alpha T/T_a})(1 - e^{-(1-\alpha)T/T_a})}{1 - e^{-T/T_a}} \right] \qquad \qquad ...(i)$$

Also per-unit ripple in load current is

$$\frac{\Delta I}{V_s/R} = \frac{(1 - e^{-\alpha T/T_a} - e^{-(1-\alpha)T/T_a} + e^{-T/T_a})}{1 - e^{-T/T_a}} \qquad \qquad ...(ii)$$

For obtaining the values of duty cycle for which the ripple in current is maximum, differentiate Eq. (*i*) or (*ii*) with respect to α and equate to zero.

$$\therefore \qquad \frac{d\left(\dfrac{\Delta I}{V_s/R}\right)}{d\alpha} = \frac{(1 - e^{-T/T_a})\left[0 - e^{-\alpha T/T_a} \cdot \left(-\dfrac{T}{T_a}\right) - e^{-(1-\alpha)T/T_a} \cdot \dfrac{T}{T_a} + 0 \right]}{(1 - e^{-T/T_a})^2} = 0$$

or $$\frac{T}{T_a} \cdot e^{-\alpha T/T_a} - \frac{T}{T_a} \cdot e^{-(1-\alpha)T/T_a} = 0$$

or $$e^{-\alpha T/T_a} = e^{-(1-\alpha)T/T_a}$$

or $$\frac{\alpha T}{T_a} = (1 - \alpha) \frac{T}{T_a}$$

$$\therefore \qquad \alpha = (1 - \alpha) \quad \text{or} \quad \alpha = \frac{1}{2} = 0.5.$$

This shows that for duty cycle equal to 0.5, ripple in load current is maximum.

7.5.2. Limit of Continuous Conduction

In a chopper, if T_{on} is reduced, T_{off} increases for a constant chopping period T. At some low value of T_{on}, the value of T_{off} is large and the current i may fall to zero. Since the current in type-A chopper cannot reverse, it stays at zero. The limit of continuous conduction is reached when I_{mn} in Eq. (7.15) goes to zero. The value of duty cycle α at the limit of continuous conduction is obtained by equating I_{mn} in Eq. (7.15) to zero. Therefore,

$$I_{mn} = \frac{V_s}{R}\left[\frac{e^{T_{on}/T_a} - 1}{e^{T/T_a} - 1} \right] - \frac{E}{R} = 0$$

or
$$\frac{e^{T_{on}/T_a}-1}{e^{T/T_a}-1} = \frac{E}{V_s} = m$$

or
$$e^{T_{on}/T_a} = 1 + m\,(e^{T/T_a}-1)$$

or
$$\alpha' = \frac{T_{on}}{T} = \frac{T_a}{T}\ln\left[1 + m\left(e^{T/T_a}-1\right)\right] \qquad\qquad ...(7.19)$$

For given E, V_s, T and T_a ; if duty cycle is α' as given by Eq. (7.19), then the current is just continuous. If actual duty cycle is less than α' as given here, the load current would be discontinuous. For some value of T_a/T, say 1, value of α' is calculated for various values of m from 0 to 1. For example, for $m = 0.5$,

$$\alpha' = 1\ln\,[1 + 0.5\,(e^1 - 1)] = 0.6201.$$

For other values of m, α' is computed and plotted as curve OAB in Fig. 7.17 with T_a/T as one parameter. For this value of $T_a/T(= 1.0)$, the $OABCO$ represents continuous conduction region and the other area $OABDO$ as the region of discontinuous conduction. In Fig. 7.17, for $m = 0.5$ and $T_a/T = 1$, point A gives the limit of continuous conduction with $\alpha' = 0.6201$. For these values of $m(= 0.5)$ and $(T_a/T\,(= 1)$, if

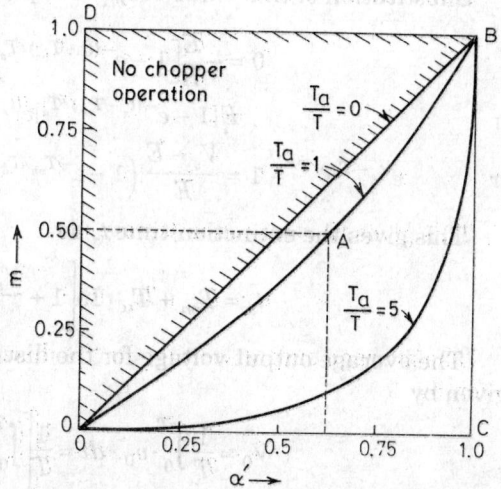

Fig. 7.17. Limit of continuous conduction for type-A chopper.

actual duty cycle α is equal to 0.7 (say) , then this point would lie in the region $OABCO$; this value of α would therefore give continuous conduction for the load current. Straight line OB corresponds to $T_a/T = 0$. Actually, the chopper operation for the area between (Fig. 7.17)

(i) the straight line OB and BOD is not possible as T_a/T can never be less than zero,

(ii) the straight line OB and curve OAB represents discontinuous current mode for $0 < \dfrac{T_a}{T} < 1.$

(iii) curve OAB and OCB represents continuous current mode for $\dfrac{T_a}{T} > 1.$

In case actual duty cycle α is equal to 0.6 (say), then as this point would lie in the region between the curve OAB and straight line OB, the load current would be discontinuous. Curves like OAB can also be drawn for other values of T_a/T. One such curve for $T_a/T = 5$ is shown in Fig. 7.17. For this value of $T_a/T = 5$, load current is discontinuous for an operating point between straight line OB and curve marked $(T_a/T) = 5$; load current is continuous for an operating point between the curve marked $(T_a/T) = 5$ and OCB.

7.5.3. Computation of Extinction Time t_x

The expressions for load current obtained in Eqs. (7.10) to (7.16) apply only when load current is continuous. For $T_{on} < t < T$, the load current may become discontinuous due to large T_{off}. The time t_x, called *extinction time* and measured from the instant $t = 0$, Fig. 7.15 (b), can be calculated as under :

As t_x is within the time limits of $T_{on} < t < T$, Eqs. (7.11) and (7.13) should only be used for computing t_x. The value of I_{mx} needed in these equations should, however, be obtained from Eq. (7.12). Thus, in Eq. (7.12), $I_{mn} = 0$ at $t = t_x$, i.e. within $T_{on} < t < T$. This equation gives I_{mx} as

$$I_{mx} = \frac{V_s - E}{R}\left(1 - e^{-T_{on}/T_a}\right) \qquad \qquad ...(7.20)$$

Substitution of this value of I_{mx} in Eq. (7.11) at $t' = t_x - T_{on}$ gives $i(t') = 0$ as

$$0 = -\frac{E}{R}\left[1 - e^{-(t_x - T_{on})/T_a}\right] + \frac{V_s - E}{R}\left[1 - e^{-T_{on}/T_a}\right]e^{-(t_x - T_{on})/T_a}$$

or $\qquad\qquad E[1 - e^{-(t_x - T_{on})/T_a}]e^{(t_x - T_{on})/T_a} = (V_s - E)(1 - e^{-T_{on}/T_a})$

or $\qquad\qquad e^{(t_x - T_{on})/T_a} - 1 = \dfrac{V_s - E}{E}\left(1 - e^{-T_{on}/T_a}\right)$

This gives the extinction time t_x as

$$t_x = T_{on} + T_\alpha \cdot \ln\left[1 + \frac{V_s - E}{E}\left(1 - e^{-T_{on}/T_a}\right)\right] \qquad ...(7.21)$$

The average output voltage for the discontinuous current mode as shown in Fig. 7.15 (b) is given by

$$V_0 = \frac{1}{T}\int_0^T v_0 \, dt = \frac{1}{T}\left[\int_0^{T_{on}} V_s \cdot dt + \int_{T_{on}}^{t_x} 0 \cdot dt + \int_{t_x}^T E \, dt\right]$$

$$= V_s \frac{T_{on}}{T} + \frac{E}{T}(T - t_x)$$

or $\qquad\qquad V_0 = \alpha\, V_s + \left(1 - \dfrac{t_x}{T}\right)E$ volts. $\qquad\qquad ...(7.22)$

7.5.4. Fourier Analysis of Output Voltage

For continuous load current, the load voltage waveform v_0 is as shown in Fig. 7.15 (a). This voltage waveform is periodic in nature and is independent of load circuit parameters. Voltage wave of Fig. 7.15 (a) can be resolved into Fourier series as

$$v_0 = V_0 + \sum_{n=1}^{\infty} v_n \qquad\qquad ...(7.23)$$

where v_n = value of nth harmonic voltage

$$= \frac{2\,V_s}{n\pi} \cdot \sin n\pi\, \alpha \cdot \sin(n\,\omega t + \theta_n) \qquad\qquad ...(7.24)$$

$$V_0 = \alpha V_s, \; \alpha = T_{on}/T \; \text{ and } \; \theta_n = \tan^{-1}\frac{\sin 2\pi n\, \alpha}{1 - \cos 2\pi n\, \alpha} = \tan^{-1}\left[\frac{\cos \pi\, n\, \alpha}{\sin \pi\, n\, \alpha}\right].$$

The average value of output voltage V_0 can be controlled by varying the duty cycle α. The amplitude of the harmonic voltages, i.e. $2\,V_s/n\pi \sin n\pi\, \alpha$, depends on n, the order of harmonic and also on the duty cycle α. The maximum value of nth harmonic occurs when $\sin n\pi\, \alpha = 1$ and its value is

$$\frac{2\,V_s}{\pi n} = \frac{0.6366 V_s}{n} \text{ volts} \qquad\qquad ...(7.25a)$$

and its rms value is

$$\frac{2 V_s}{\pi n \sqrt{2}} = \frac{0.45 V_s}{n} \text{ volts} \qquad \qquad ...(7.25b)$$

The harmonic current in the load is given by

$$i_n = \frac{v_n}{Z_n}$$

where Z_n is the load impedance at harmonic frequency $nf\,Hz$ and is given by

$$Z_n = \sqrt{R^2 + (n\omega L)^2}\;.$$

For negligible load resistance R, $i_n = \dfrac{v_n}{n\omega L}$ or $i_n \propto \dfrac{V_s}{n^2}$. This shows that harmonic current decreases as n, the order of harmonic, increases. Another term used for knowing the harmonic content of a waveform, without calculating its harmonic components, is the ac ripple voltage V_r. It is defined as

$$V_r = \sqrt{V_{rms}^2 - V_0^2} \qquad \qquad ...(7.26)$$

In the above equation, $V_{rms} = V_{or}$ and V_0 are respectively the rms and average values of the output voltage. It is seen from Example 7.1 that

$$V_{or} = \sqrt{\alpha}\, V_s \quad \text{from part } (d)$$

and

$$V_0 = \alpha V_s \quad \text{from part } (a)$$

$$\therefore \qquad V_r = \sqrt{\alpha V_s^2 - \alpha^2 V_s^2} = V_s \sqrt{\alpha - \alpha^2} \qquad \qquad ...(7.27)$$

Ripple factor, defined as the ratio of ac ripple voltage to average voltage is given by

$$\text{ripple factor} = \frac{V_r}{V_0}$$

or

$$RF = V_s \cdot \frac{\sqrt{\alpha - \alpha^2}}{V_s \cdot \alpha} = \sqrt{\frac{1 - \alpha}{\alpha}} = \sqrt{\frac{1}{\alpha} - 1} \qquad \qquad ...(7.28)$$

Example 7.7. *(a) For an ideal type-A chopper feeding RLE load, show that the average input (or thyristor) current is given by*

$$I_{TAV} = \frac{\alpha(V_s - E)}{R} - \frac{L}{RT}(I_{mx} - I_{mn}).$$

(b) For the chopper of part (a), derive an expression for the average current in the freewheeling diode for a continuous load current.

(c) From parts (a) and (b), prove that average value of load current I_{av} is given by

$$I_{av} = \frac{V_0 - E}{R}.$$

Solution. For type-A chopper, the output voltage waveform v_0 and the waveforms for load current i_0, input (or thyristor) current i_T and freewheeling diode current i_{fd} are as shown in Fig. 7.18.

(*a*) When chopper is on, the voltage equation for the chopper circuit of Fig. 7.8 (*a*) is

$$R i_T + L \frac{di_T}{dt} + E = V_s$$

or

$$R i_T \cdot dt + L \frac{di_T}{dt}\, dt = (V_s - E)\, dt.$$

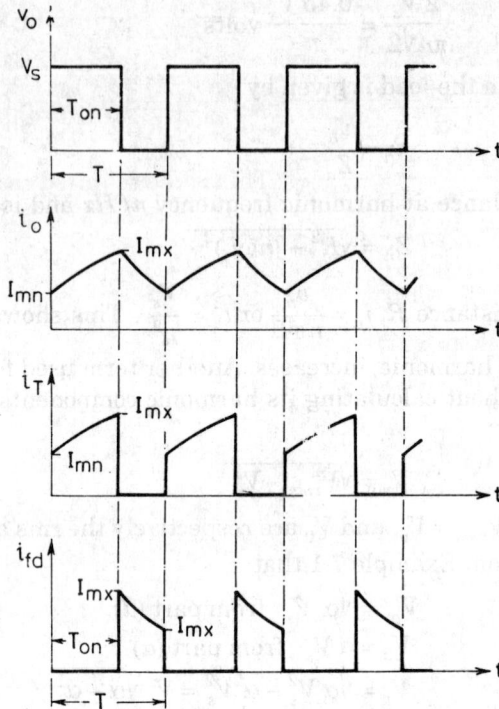

Fig. 7.18. Pertaining to Example 7.7.

Its average value, Fig. 7.18, is

$$R \cdot \frac{1}{T} \int_0^{T_{on}} i_T \cdot dt + \frac{1}{T} \int_0^{T_{on}} L \cdot \frac{di_T}{dt} dt = \frac{(V_s - E)}{T} \int_0^{T_{on}} dt$$

$$R \, I_{TAV} + \frac{1}{T} \int_{I_{mn}}^{I_{mx}} L \, di_T = (V_s - E) \frac{T_{on}}{T}$$

or

$$R \, I_{TAV} + \frac{L}{T} (I_{mx} - I_{mn}) = (V_s - E) \, \alpha$$

or

$$I_{TAV} = \frac{\alpha \, (V_s - E)}{R} - \frac{L}{RT} (I_{mx} - I_{mn}) \qquad ...(7.29)$$

(b) When the freewheeling diode is conducting, load voltage is zero. The voltage equation, when FD is conducting, is given by

$$R \, i_{fd} + L \frac{d \, i_{fd}}{dt} + E = 0$$

Its average value, Fig. 7.18, is

$$R \cdot \frac{1}{T} \int_{T_{on}}^{T} i_{fd} \cdot dt + L \frac{1}{T} \int_{T_{on}}^{T} \frac{di_{fd}}{dt} dt + E \frac{1}{T} \int_{T_{on}}^{T} dt = 0$$

or

$$R \cdot I_{fd} + \frac{L}{T} \int_{I_{mx}}^{i_{mn}} di_{fd} = - E \frac{T - T_{on}}{T} \quad \text{or} \quad R \, I_{fd} + \frac{L}{T} (I_{mn} - I_{mx}) = - E(1 - \alpha)$$

$$\therefore \qquad I_{fd} = \frac{L \, (I_{mx} - I_{mn})}{TR} - \frac{E \, (1 - \alpha)}{R} \qquad\qquad ...(7.30)$$

(c) Average load current over a complete cycle can be obtained by adding I_{TAV} and I_{fd} from Eqs. (7.29) and (7.30) respectively.

$$\therefore \qquad I_{av} = \frac{\alpha(V_s - E)}{R} - \frac{L}{RT}(I_{mx} - I_{mn}) + \frac{L}{RT}(I_{mx} - I_{mn}) - \frac{E(1-\alpha)}{R}$$

$$= \frac{\alpha V_s - E}{R} = \frac{V_0 - E}{R} \cdot \text{ This is the required result.}$$

Example 7.8. *For type-A chopper, feeding an RLE load, obtain maximum value of average current rating for the thyristor in case load current remains constant.*

Solution. For constant load current I_0, current waveform for thyristor current i_T is as shown in Fig. 7.19. Here

$$I_0 = \frac{V_0 - E}{R}$$

The average thyristor current I_T is given by

$$I_T = I_0 \frac{T_{on}}{T} = \frac{V_0 - E}{R} \alpha.$$

$$= \frac{\alpha V_s - E}{R} \cdot \alpha = \frac{\alpha^2 V_s - \alpha E}{R} \qquad ...(7.31)$$

This will give a maximum value when

$$\frac{dI_T}{d\alpha} = \frac{2\alpha V_s - E}{R} = 0$$

and from this, $\alpha = \dfrac{E}{2V_s}$ $\qquad\qquad ...(7.32)$

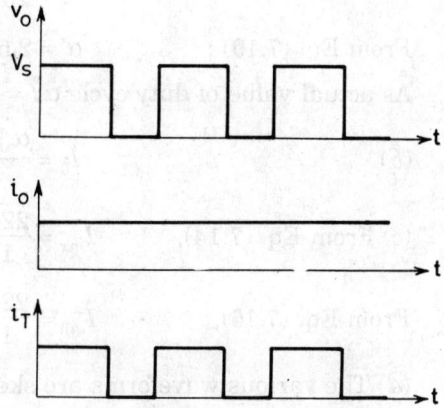

Fig. 7.19. Pertaining to Example 7.8

Therefore maximum value of average thyristor current is obtained by substituting the value of α from Eq. (7.32) in Eq. (7.31).

$$\therefore \qquad I_{T.mx} = \frac{E}{2V_s R} \cdot \left(\frac{E}{2V_s} \cdot V_s - E\right) = \frac{E^2}{4V_s R} \text{ Amps.}$$

Example 7.9. *For type-A chopper circuit, source voltage $V_s = 220$ V, chopping period $T = 2000$ μs, on-period $= 600$ μs, load circuit parameters : $R = 1$ Ω, $L = 5$ mH and $E = 24$ V.*

(a) Find whether load current is continuous or not.

(b) Calculate the value of average output current.

(c) Compute the maximum and minimum values of steady state output current.

(d) Sketch the time variations of gate signal i_g, load voltage v_0, load current i_0, thyristor current i_T, freewheeling diode current i_{fd} and voltage across thyristor v_T.

(e) Find rms values of the first, second and third harmonics of the load current.

(f) Compute the average value of supply current.

(g) Compute input power, the power absorbed by the load counter emf and the power loss in the resistor.

(h) Compute rms value of load current using the results of (b) and (g).

(i) Using results of (e), find the rms value of load current. Compare the result with that obtained in part (h).

Solution. (a)
$$T_a = \frac{L}{R} = 5 \times \frac{10^{-3}}{1} = 5 \times 10^{-3} \text{ sec}$$

$$\frac{T_a}{T} = \frac{5 \times 10^{-3}}{2000 \times 10^{-6}} = 2.5$$

$$\frac{T}{T_a} = 0.4 \; ; \; m = \frac{E}{V_s} = \frac{24}{220} = 0.11$$

$$\alpha = \frac{600}{2000} = 0.3, \; \frac{T_{on}}{T_a} = \frac{600}{5000} = 0.12.$$

From Eq. (7.19) ; $\alpha' = 2.5 \ln [1 + 0.11 \, (e^{0.4} - 1)] = 0.13172$

As actual value of duty cycle $\alpha \, (= 0.3)$ is more than α', load current is continuous.

(b)
$$I_0 = \frac{\alpha V_s - E}{R} = \frac{0.3 \times 220 - 24}{1} = 42 \text{ A}$$

(c) From Eq. (7.14), $I_{mx} = \frac{220}{1} \left[\frac{1 - e^{-0.12}}{1 - e^{-0.4}} \right] - \frac{14}{1} = 51.46 \text{ A}$

From Eq. (7.15), $I_{mn} = \frac{220}{1} \left[\frac{e^{0.12} - 1}{e^{0.4} - 1} \right] - 24 = 33.031 \text{ A}$

(d) The various waveforms are sketched in Fig. 7.20.

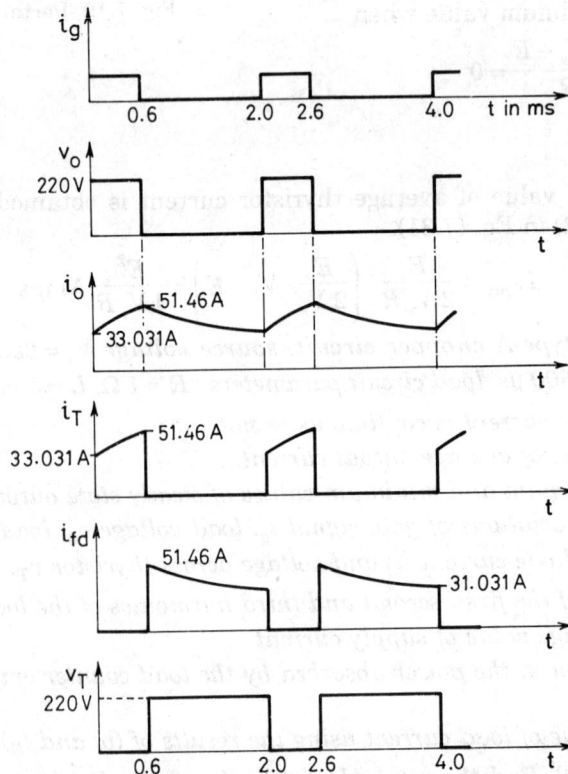

Fig. 7.20. Pertaining to Example 7.9.

(e) From Eq. (7.24), rms value of first harmonic voltage is

$$V_1 = \frac{2V_s}{\sqrt{2}\,\pi} \sin (\pi \times 0.3) = \frac{2 \times 220}{\sqrt{2}\,\pi} \sin 54° = 80.121 \text{ V}$$

Here chopping frequency, $f = \dfrac{1}{T} = \dfrac{10^6}{2000} = 500$ Hz

$\therefore$

$$Z_1 = \sqrt{R^2 + (\omega L)^2} = \sqrt{1^2 + (2\pi \times 500 \times 5 \times 10^{-3})^2} = 15.739762 \ \Omega$$

$\therefore$

$$I_1 = \frac{V_1}{Z_1} = \frac{80.121}{15.739762} = 5.0903 \text{ A}$$

Similarly,

$$I_2 = \frac{2 \times 220}{2 \cdot \sqrt{2} \cdot \pi} \, [\sin 108°] \, \frac{1}{\sqrt{1^2 + (2\pi \times 500 \times 2 \times 5 \times 10^{-3})^2}}$$

$$= 1.4983 \text{ A}$$

$$I_3 = \frac{2 \times 220}{3 \cdot \sqrt{2} \cdot \pi} \, [\sin 162°] \, \frac{1}{\sqrt{1^2 + (2\pi \times 1500 \times 5 \times 10^{-3})^2}}$$

$$= 0.21643 \text{ A}$$

(f) From Eq. (7.29), average supply current is

$$I_{TAV} = \frac{0.3 \, (220 - 24)}{1} - \frac{5 \times 10^{-3} \, (51.46 - 33.031)}{1 \times 2000 \times 10^{-6}}$$

$$= 58.8 - 46.0725 = 12.7275 \text{ A}$$

(g) Input power $= V_s \times$ average supply current

$$= 220 \times 12.7275 = 2800.05 \text{ watts}$$

Power absorbed by load emf $= E \times$ average load current

$$= 24 \times 42 = 1008 \text{ watts}$$

Power loss in resistor $R = 2800.05 - 1008 = 1792.05$ W

(h)

$$I_{or} = \sqrt{I_{av}^2 + I_1^2 + I_2^2 + I_3^2}$$

$$= \sqrt{42^2 + 5.0903^2 + 1.4983^2 + 0.21643^2} = 42.31 \text{ A}$$

(i) Power loss in resistor $= I^2 R = 1792.05$ W

$\therefore$

$$I_{or} = \sqrt{\frac{1792.05}{1}} = 42.333 \text{ A}$$

The results obtained in parts (h) and (i) are in complete agreement.

Example 7.10. *For type-A chopper, source voltage $V_s = 220$ V, chopping frequency $f = 500$ Hz, $T_{on} = 800$ μs, $R = 1\ \Omega, L = 1$ mH and $E = 72$ V.*

(a) Find whether load current is continuous or not.

(b) Calculate the values of average output voltage and average output current.

(c) Compute the maximum and minimum values of steady state output current.

(d) Sketch the time variations of gate signal i_g, load current i_0, load voltage v_0, thyristor current i_T. freewheeling diode current i_{fd} and voltage across thyristor v_T.

Solution. Here

$$T_a = \frac{L}{R} = 1 \times 10^{-3} \text{ sec,}$$

$$T = \frac{1}{f} = \frac{1}{500} = 2000 \text{ μsec}$$

$$\frac{T_a}{T} = fT_a = 500 \times 10^{-3} = 0.5, \quad \frac{T}{T_a} = \frac{1}{fT_a} = 2,$$

$$m = \frac{E}{V_s} = \frac{72}{220} = 0.327$$

$$\alpha = \frac{T_{on}}{T} = f \cdot T_{on} = 500 \times 800 \times 10^{-6} = 0.4,$$

$$\frac{T_{on}}{T_a} = 0.8$$

(a) From Eq. (7.19),

$$\alpha' = 0.5 \ln [1 + 0.327 (e^{-2} - 1)] = 0.564$$

As α is less than α', load current is discontinuous.

(b) From Eq. (7.21),

$$t_x = 800 \times 10^{-6} + 1 \times 10^{-3}$$
$$\ln \left[1 + \frac{220 - 72}{72} (1 - e^{-0.8}) \right]$$
$$= 1.55703 \times 10^{-3} \text{ sec}$$

From Eq. (7.22),

$$V_0 = 0.4 \times 220 + \left(\frac{1 - 1.55703 \times 10^{-3}}{2 \times 10^{-3}} \right) \cdot 72$$
$$= 103.95 \text{ volts.}$$

$$I_0 = \frac{103.75 - 72}{1} = 31.95 \text{ A.}$$

(c) As the load current is discontinuous, $I_{mn} = 0$. The maximum value of current from Eq. (7.20) is

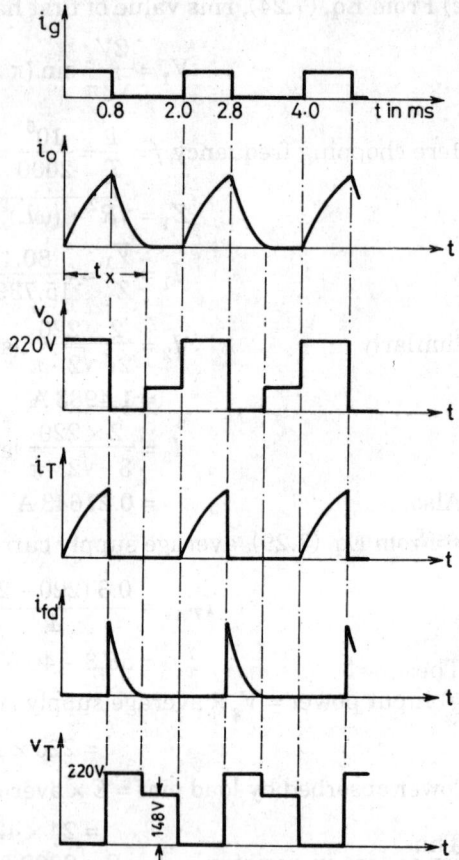

Fig. 7.21. Pertaining to Example 7.10.

$$I_{mx} = \frac{220 - 72}{1} (1 - e^{-0.8}) = 81.5 \text{ A}$$

(d) The time variations of various waveforms are sketched in Fig. 7.21.

Example 7.11. *An RLE load is operating in a chopper circuit from a 500-volt dc source. For the load, L = 0.06 H, R = 0 and constant E. For a duty cycle of 0.2, find the chopping frequency to limit the amplitude of load current excursion to 10 A.*

Solution. The average output, or load, voltage is given by

$$V_0 = \alpha V_s$$

As the average value of voltage drop across L is zero,

$$E = V_0 = \alpha V_s = 0.2 \times 500 = 100 \text{ Volts.}$$

During T_{on}, the difference in source voltage V_s and load emf E, *i.e*, $(V_s - E)$ appears across inductance L as shown in Fig. 7.22.

$\therefore$ During T_{on}, volt-time area applied to inductance

$$= (500 - 100) T_{on} = 400 T_{on} \text{ Volt-sec}$$

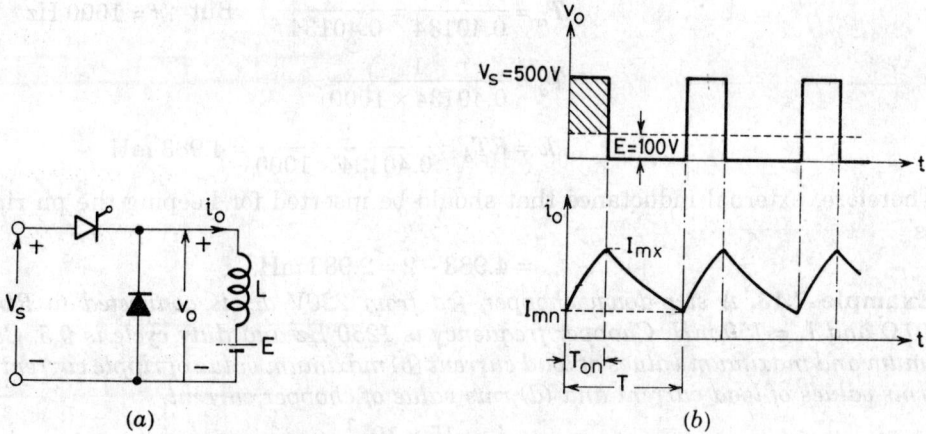

Fig. 7.22. Pertaining to Example 7.11.

Also, during T_{on}, the current through L rises from I_{mn} to I_{mx}. From this, volt-time area across L during this current change is given by

$$\int_0^{T_{on}} v_L \cdot dt = \int_0^{T_{on}} L\, \frac{di}{dt}\, dt = \int_{I_{mn}}^{I_{mx}} L \cdot di = L\,(I_{mx} - I_{mn}) = L \cdot \Delta I$$

These two volt-time areas during T_{on} must be equal.

$\therefore \qquad\qquad 400\, T_{on} = L \cdot \Delta I$

or $\qquad\qquad T_{on} = \dfrac{0.06 \times 10}{400} = 1.5\ \text{msec}$

Thus, chopping frequency, $f = \dfrac{1}{T} = \dfrac{\alpha}{T_{on}} = \dfrac{0.2}{1.5 \times 10^{-3}} = 133.33\ \text{Hz}.$

Example 7.12. *A series motor used for a rapid transit system is fed through a dc chopper. The series motor has total circuit resistance of 2 Ω and inductance of 2 mH. What external inductance should be inserted in series with the armature circuit in order to limit the per unit ripple in armature current to 10% for a duty cycle ratio of 0.5. The chopping frequency is 1 kHz.*

Solution. From Eq. (7.18), per unit ripple in current is given by

$$\frac{I_{mx} - I_{mn}}{V_s / R} = \left[\frac{(1 - e^{-\alpha T/T_a})(1 - e^{-(1-a)T/T_a})}{1 - e^{-T/T_a}} \right]$$

Let $\dfrac{T}{T_a} = x$. Here $\alpha = 0.5$ and pu ripple $= \dfrac{10}{100} = 0.1$. Substitution of these in the above expression gives

$$0.1 = \frac{(1 - e^{-0.5x})(1 - e^{-0.5x})}{1 - e^{-x}} = \frac{(1 - e^{-0.5x})(1 - e^{-0.5x})}{(1 - e^{-0.5x})(1 + e^{-0.5x})} = \frac{1 - e^{-0.5x}}{1 + e^{-0.5x}}$$

or $\qquad\qquad 0.1 = \dfrac{1 - y}{1 + y}\ $ where $y = e^{-0.5x}$

or $\qquad\qquad y = \dfrac{0.9}{1.10} = e^{-0.5x}$

or $\qquad\qquad e^{0.5x} = \dfrac{1.10}{0.9} = 1.22222\quad$ or $\quad x = \dfrac{T}{T_a} = 0.40134$

$$\therefore \qquad T_a = \frac{T}{0.40134} = \frac{1}{0.40134\,f} \qquad \text{But} \quad f = 1000 \text{ Hz}$$

$$\therefore \qquad T_a = \frac{1}{0.40134 \times 1000}$$

or

$$L = RT_a = \frac{2}{0.40134 \times 1000} = 4.983 \text{ mH}$$

Therefore, external inductance that should be inserted for keeping the pu ripple within limits

$$= 4.983 - 2 = 2.983 \text{ mH}.$$

Example 7.13. *A step-down chopper, fed from 220V dc, is connected to RL load with R = 10 Ω and L = 150 mH. Chopper frequency is 1250 Hz and duty cycle is 0.5. Calculate (a) minimum and maximum values of load current (b) maximum value of ripple current (c) average and rms values of load current and (d) rms value of chopper current.*

Solution. Here

$$T_a = \frac{L}{R} = \frac{15 \times 10^{-3}}{10} = 1.5 \times 10^{-3} \text{ s}$$

$$T = \frac{1}{f} = \frac{1}{1250} = 0.8 \times 10^{-3} \text{s}, \; T_{on} = \alpha T = 0.4 \times 10^{-3}\text{s}$$

$$\frac{T_{on}}{T_a} = \frac{0.4 \times 10^{-3}}{1.5 \times 10^{-3}} = 0.267, \frac{T}{T_a} = \frac{0.8 \times 10^{-3}}{1.5 \times 10^{-3}} = 0.533$$

From Eq. (7.14),

$$I_{mx} = I_2 = \frac{220}{10}\left[\frac{1 - e^{-0.267}}{1 - e^{-0.533}}\right] = 12.478 \text{ A}$$

From Eq. (7.15),

$$I_{mn} = I_1 = \frac{220}{10}\left[\frac{e^{0.267} - 1}{e^{0.533} - 1}\right] = 9.56 \text{ A}$$

(b) Maximum value of ripple current, $\Delta I = I_2 - I_1 = 12.478 - 9.56 = 2.918$ A

From Eq. (7.18 a), approximate value of maximum ripple $= \dfrac{V_s}{4\,fL}$

$$= \frac{220 \times 10^{-3}}{4 \times 1250 \times 15} = 2.933 \text{ A}$$

(c) Average output voltage, $V_o = \alpha\, V_s = 0.5 \times 220 = 110$ V

Average load current, $I_o = \dfrac{110}{10} = 11$ A

Also $I_0 = \dfrac{I_1 + I_2}{2} = \dfrac{12.478 + 9.56}{2} = 11.02$ A

Assuming linear variation of current from I_1 to I_2, the current during T_{on} can be expressed as

$$i_o = I_1 + \frac{I_2 - I_1}{T_{on}}\,t = I_1 + \frac{\Delta I}{T_{on}}\,t$$

$\therefore$ Rms value of load current, $I_{or}^2 = \dfrac{1}{T_{on}}\displaystyle\int_0^{T_{on}}\left(I_1 + \frac{\Delta I}{T_{on}} \cdot t\right)^2 . dt$

$$= \frac{1}{T_{on}}\int_0^{T_{on}}\left(I_1^2 + \left(\frac{\Delta I}{T_{on}}\,t\right)^2 + 2\,I_1 \cdot \frac{\Delta I}{T_{on}}\,t\right)dt$$

or
$$I_{or} = \left[I_1^2 + \frac{\Delta I^2}{3} + I_1 . \Delta I \right]^{1/2}$$

$$= \left[12.478^2 + \frac{2.918^2}{3} + 12.478 \times 2.918 \right]^{1/2} = 13.962 \text{ A}$$

Rms value of chopper current, $I_{ch.r}^2 = \frac{1}{T} \int_0^{T_{on}} \left(I_1 + \frac{\Delta I}{T_{on}} . t \right)^2 dt$

$$= \sqrt{\frac{T_{on}}{T}} \left[I_1^2 + \frac{\Delta I^2}{3} + I . \Delta I \right]^{1/2} = \sqrt{\alpha} . I_{or}$$

$$\therefore \qquad I_{chr} = \sqrt{0.5} \times 13.962 = 9.873 \text{ A}$$

Example 7.14. *Show that the critical inductance in the load circuit of a step-down chopper is proportional to x (1 – x) where x is the duty cycle.*

Solution. This example is just an extension of Example 7.5. From this example, from Eq. (*i*), the critical inductance L is given by

$$L = \frac{(V_s - V_o) T_{on}}{2 I_o}$$

Now $T_{on} = \alpha T$ and $V_o = \alpha V_s$, substituting these values, we get

$$L = \frac{V_s (1 - \alpha) - \alpha T}{2 I_o} = \frac{V_s T}{2 I_o} . \alpha (1 - \alpha)$$

Here $\alpha = x$ = duty cycle, $\qquad \therefore L = \frac{V_s . T}{2 I_o} . x (1 - x)$

$\therefore$ Critical inductance, $L \propto x (1 - x)$

7.6. THYRISTOR CHOPPER CIRCUITS

So far a chopper has been shown as a switch inside a dotted rectangle. Actually, a chopper consists of main power semiconductor device together with their turn-on and turn-off mechanisms. In low-power chopper circuits ; power transistors, GTOs etc. are being used widely. In high-power levels, however, thyristors are in common use. The object of this section is to study the thyristor chopper circuits along with their commutation circuitry.

The process of opening, or turning-off, a conducting thyristor is called commutation. In dc choppers, it is essential to provide a separate commutation circuitry to commutate the main power SCR. It may be recalled that a conducting thyristor can be turned off by reducing its anode current below holding current value and then applying a reverse voltage across the device to enable it to regain its forward blocking capability. There are several ways of turning-off of a thyristor. All these methods differ from one another in the manner in which commutation is achieved. In dc choppers, commutation circuitry has passed through numerous innovations. All these commutation circuits can, however, be broadly classified into two groups as under :

(*a*) **Forced Commutation.** In forced commutation, external elements L and C which do not carry the load current continuously, are used to turn-off a conducting thyristor. Forced commutation can be achieved in the following two ways :

(*i*) **Voltage commutation**. In this scheme, a conducting thyristor is commutated by the application of a pulse of large reverse voltage. This reverse voltage is usually applied by switching a previously charged c. pacitor. The sudden application of reverse voltage across the conducting thyristor reduces the anode current to zero rapidly. Then the presence of reverse voltage across the SCR aids in the completion of its turn-off process (*i.e.* aids in gaining the forward blocking capability of SCR).

(*ii*) **Current Commutation**. In this scheme, an external pulse of current greater than the load current is passed in the reversed direction through the conducting SCR. When the current pulse attains a value equal to the load current, net pulse current through thyristor becomes zero and the device is turned off. The current pulse is usually generated by an initially charged capacitor.

An important feature of current commutation is the connection of a diode in antiparallel with the main thyristor so that voltage drop across the diode reverse biases the main SCR. Since this voltage drop is of the order of 1 volt, the commutation time in current commutation is more as compared to that in voltage commutation.

In both voltage and current commutation schemes, commutation is initiated by gating an auxiliary SCR.

(*b*) **Load Commutation**. In load commutation, a conducting thyristor is turned off when load current flowing through a thyristor either

(*i*) becomes zero due to the nature of load circuit parameters or

(*ii*) is transferred to another device from the conducting thyristor.

Only three commutation principles listed above are described in what follows though there are numerous other commutation schemes. Chopper circuits with type-A configuration are only studied in this section. For understanding the performance of voltage and current commutated chopper circuits, Example 7.15 should be studied carefully.

Example 7.15. *In the circuit shown in Fig. 7.23 (a), capacitor C is initially charged to a voltage V_0. If the switch S is closed at $t = 0$, obtain an expression for current i (t). Sketch time variation of voltage across capacitor and inductor and of the current i (t). Indicate also the flow of energy handled by L and C.*

In case L = 1.6 mH and C = 4µF, find the time for which current i (t) flows in the circuit of Fig. 7.23 (a).

Solution. When the switch S is closed, KVL for the circuit is

$$L \frac{di\,(t)}{dt} + \frac{1}{C} \int i(t)\,dt = 0$$

Its Laplace transform is $sLI(s) + \dfrac{1}{C}\left[\dfrac{I(s)}{s} - \dfrac{CV_0}{s} \right] = 0$

Here negative sign before CV_0/s is used because current i (t) leaves the positive terminal of C and enter its negative terminal when switch S is closed.

$\therefore$ $$I(s) = \frac{V_0}{L} \frac{1}{s^2 + 1/LC} = \frac{V_o}{L} \cdot \frac{1/\sqrt{LC}}{1/\sqrt{LC}\ [s^2 + (1/\sqrt{LC})^2]}$$

or $$I(s) = \frac{V_0}{\omega_0 L} \frac{\omega_0}{[s^2 + \omega_0^2]} \text{ where } \omega_0 = \frac{1}{\sqrt{LC}}$$

Its Laplace inverse gives $i(t) = \dfrac{V_0}{\omega_0 L} \sin \omega_0\, t$ ...(7.33)

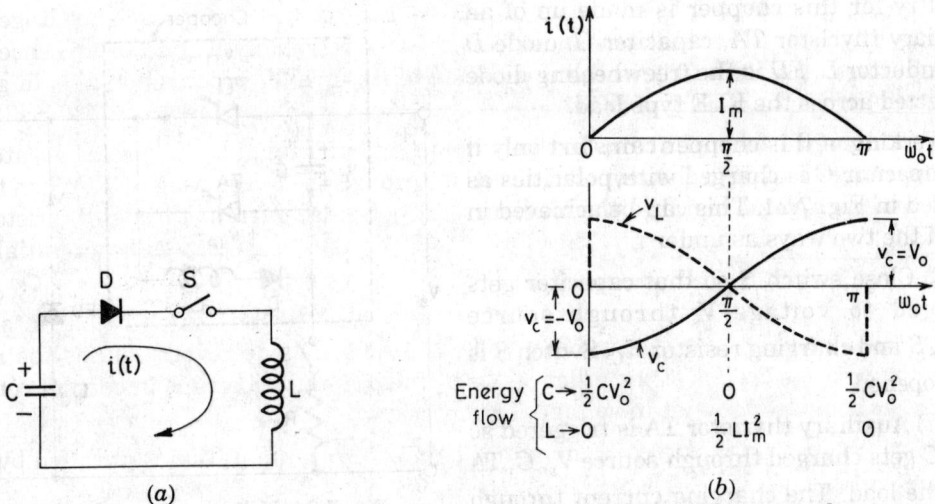

Fig. 7.23 (a) Circuit for Example 7.15 and (b) time variations of $i(t)$, v_c and v_L

Now
$$v_c = \frac{1}{c}\int_o^t i(t) \cdot dt - V_o$$

$$= \frac{1}{C} \cdot \frac{V_o}{\omega_o L}\int_o^t \sin \omega_o t \, dt - V_o$$

$$= \frac{V_o}{LC} \cdot \frac{1}{\omega_o^2} \mid - \cos \omega_o t \mid_o^t - V_o$$

$$= V_o (1 - \cos \omega_o t) - V_o = - V_o \cos (\omega_o t) \qquad ...(7.34)$$

where $\omega_o^2 LC = 1$ and $\omega_o = \dfrac{1}{\sqrt{LC}}$ is called the resonant frequency in rad/sec.

Also
$$v_L = L \frac{di(t)}{dt} = L \cdot \frac{V_0}{\omega_0 L}(\cos \omega_0 t) \cdot \omega_0 = V_0 \cos \omega_0 t \qquad ...(7.35)$$

The time variation of $i(t)$, v_c and v_L obtained from Eqs. (7.33) to (7.35) is shown in Fig. 7.23 (b). When $i(t) = I_m$, $v_c = v_L = 0$. When $i(t)$ becomes zero at $\omega_0 t = \pi$, further conduction stops because diode can't conduct in the reversed direction. Therefore, at $\omega_0 t = \pi$, capacitor is charged to a voltage V_0 but with reversed polarity.

When switch S is closed, voltage across L at once becomes V_0, its value is zero at $\omega_0 t = \pi/2$ and $- V_0$ at $\omega_0 t = \pi$. As current $i(t)$ reduces to zero at $\omega_0 t = \pi$, v_L finally reduces to zero as shown.

The flow of energy in C and L is also shown in Fig. 7.23 (b). At $\omega_0 t = 0$, capacitor has stored energy equal to $W_c = \frac{1}{2} CV_0^2$, at $\omega_0 t = \pi/2$, $W_c = 0$ and at $\omega_0 t = \pi$, $W_c = \frac{1}{2} CV_0^2$ again. The flow of energy in L is also indicated in Fig. 7.23 (b).

7.6.1. Voltage-Commutated Chopper

One of the earliest chopper circuits which has been in wide use is the voltage-commutated chopper. This chopper is generally used in high-power circuits where load fluctuation is not very large. This chopper is also known as *parallel-capacitor turn-off chopper*, *impulse-commutated chopper* or *classical chopper*. Fig. 7.24 gives the power circuit diagram for this type of chopper. In this diagram, thyristor T1 is the main power switch. Commutation circuitry for this chopper is made up of an auxiliary thyristor *TA*, capacitor *C*, diode *D* and inductor *L*. *FD* is the freewheeling diode connected across the RLE type load.

Working of this chopper can start only if the capacitor *C* is charged with polarities as marked in Fig. 7.24. This can be achieved in one of the two ways as under :

(*i*) Close switch *S* so that capacitor gets charged to voltage V_s through source V_s, C, S and charging resistor R_C. Switch *S* is then opened.

(*ii*) Auxiliary thyristor TA is triggered so that *C* gets charged through source V_s, *C*, TA and the load. The charging current through capacitor *C* decays and as it reaches zero, $v_c = V_s$ and TA is turned off.

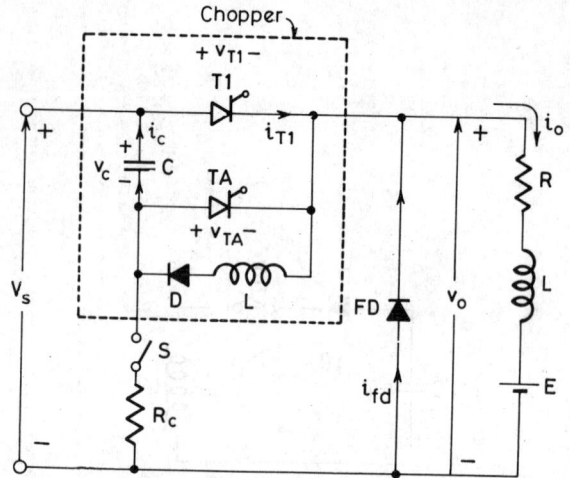

Fig. 7.24. Voltage-commutated chopper.

With capacitor *C* charged with the polarities as shown in Fig. 7.24, the chopper circuit is ready for operation. The current i_c, i_{T1}, i_{fd} and i_0 are taken as positive in the arrow directions marked. Similarly, the voltages v_c, v_{T1}, v_{TA} and v_0 across *C*, T1, TA and load are taken as positive with the polarities marked.

Simplifying assumptions for this chopper are (*i*) load current is constant and (*ii*) thyristors and diodes are ideal elements. The chopper operation, for convenience, is divided into certain modes and is explained as under :

Mode I. The main thyristor is triggered at $t = 0$ and RLE load gets connected across source V_s so that load voltage $v_0 = V_s$. During this mode, there are two current paths as shown in Fig. 7.25 (*a*). Load current I_0, assumed constant, constitute one path and commutation current i_c the other path. Load current I_0 flows through source V_s, main thyristor T1 and load whereas the current i_c flows through the oscillatory circuit formed by *C*, T1, *L* and *D*. The capacitor (or commutation) current first rises from zero to a maximum value when voltage across *C* is zero at $t = t_1/2$. As i_c decreases to zero, capacitor is charged to voltage $(- V_s)$ as shown at $t = t_1$ in Fig. 7.26, see Example 7.15. The capacitor current changes sinusoidally whereas the capacitor voltage cosinusoidally from $t = 0$ to $t = t_1$. This voltage is held constant at $(- V_s)$ by diode *D*. Voltage across TA is $(- V_s)$ at $t = 0$, zero at $t_1/2$ and V_s at t_1, this variation is shown as cosine wave in Fig. 7.26. The thyristor current i_{T1} has a peak at $t_1/2$, because $i_{T1} = i_c + I_0$ between $t = 0$ and $t = t_1$. At the end of mode *I*, i.e. at t_1 ; $i_c = 0, i_{T1} = I_0, v_c = - V_s, v_{TA} = V_s, v_0 = V_s$ as shown in Fig. 7.26.

Mode II. The conditions existing at t_1 continue during mode II. In other words, for $t_1 \le t \le t_2, i_c = 0, i_{T1} = I_0, v_c = - V_s, v_{TA} = V_s, v_0 = V_s, i_D = 0$ as shown. Note that during this mode, only main SCR T1 is conducting.

Mode III. When main thyristor T1 is to be turned off, auxiliary thyristor TA is triggered at the desired instant t_2. With the turning on of TA at t_2, capacitor voltage $(- V_s)$ appears across T1, it is therefore reverse biased and turned off. As the capacitor voltage does the required job of commutating the main thyristor T1, it is called *voltage commutated chopper*. Current i_{T1}

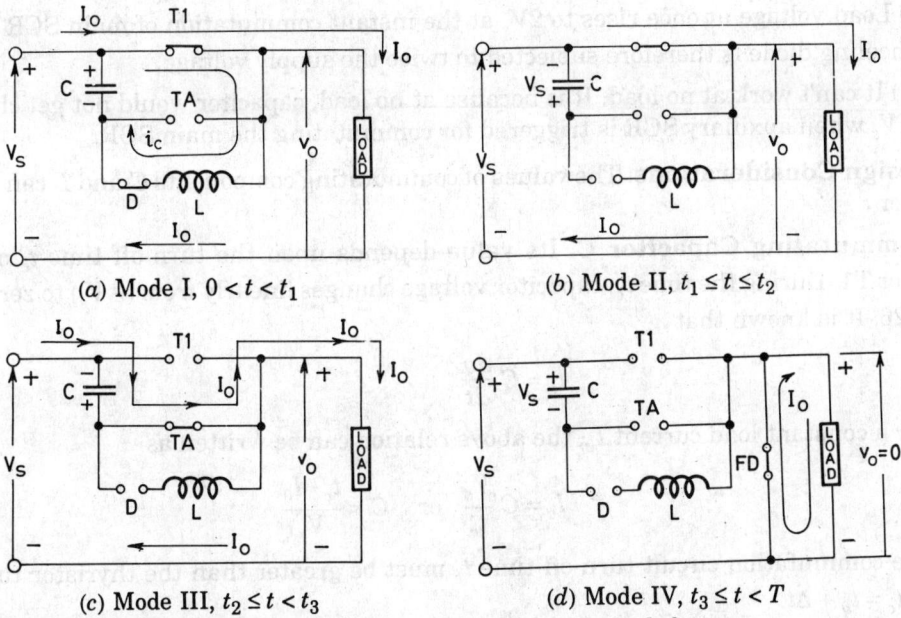

(a) Mode I, $0 < t < t_1$

(b) Mode II, $t_1 \leq t \leq t_2$

(c) Mode III, $t_2 \leq t < t_3$

(d) Mode IV, $t_3 \leq t < T$

Fig. 7.25. Different modes of voltage-commutated chopper.

becomes zero at t_2. After T1 is turned off, capacitor C and auxiliary SCR TA provide the path for load current I_0 through V_s, C, TA and the load, see Fig. 7.25 (c). The load voltage is the sum of source voltage and the voltage across capacitor. Therefore, at instant t_2, load voltage is $v_0 = V_s + V_s = 2V_s$ and it decreases linearly as the voltage across capacitor decreases. During this mode, $v_c = v_{T1}$, because capacitor is directly connected across T1 through TA. As the capacitor discharges through the load, v_c and v_{T1} change from $(-V_s)$ to zero at $(t_2 + t_c)$. Load voltage v_0 changes from $2V_s$ at t_2 to V_s at $(t_2 + t_c)$. After $(t_2 + t_c)$, v_c and v_{T1} start rising from zero towards V_s whereas v_0 starts falling towards zero. For mode III, $t_2 \leq t \leq t_3$. Note that v_c and v_{T1} change linearly* from $(-V_s)$ at t_2 to V_s at t_3, because load current I_0 is assumed constant. Similarly v_0 changes linearly from $2V_s$ at t_2 to zero at t_3. Note also that $i_c = -I_0$ and $i_{TA} = I_0$ during mode III.

Mode IV. For this mode, $t_3 \leq t < T$. At t_3, $v_c = v_{T1} = V_s$, $v_0 = 0$, i_c or i_{TA} becomes zero. As i_c (or i_{TA}) tends to go negative soon after t_3, thyristor TA is turned off naturally at t_z. As capacitor is slightly overcharged at t_3, freewheeling diode FD gets forward biased. The load current after t_3 freewheels through the load and FD, see Fig. 7.25 (d). Note that during freewheeling period from t_3 to T, v_{TA} is slightly negative as C is somewhat overcharged. During this mode, $i_c = 0$, $i_{T1} = 0$, $i_{fd} = I_0$, $v_{T1} = V_s$, $v_c = V_s + \Delta V$, $v_{TA} = -\Delta V$, $v_0 = 0$, $i_{TA} = 0$.

At $t = T$, the main thyristor T1 is triggered and the cycle as described from $t = 0$ to $t = T$ repeats.

Voltage commutated chopper is simple, it has therefore been used extensively. It, however, suffers from the following disadvantages.

(*i*) A starting circuit is required.

---

*In the relation $i = C(dv/dt)$, if i is constant, C is already constant, then dv/dt must be constant.

(ii) Load voltage at once rises to $2V_s$ at the instant commutation of main SCR is initiated. Freewheeling diode is therefore subjected to twice the supply voltage.

(iii) It can't work at no load. It is because at no load, capacitor would not get charged from $-V_s$ to V_s when auxiliary SCR is triggered for commutating the main SCR.

Design Considerations. The values of commutating components C and L can be obtained as under :

Commutating Capacitor C. Its value depends upon the turn-off time t_c of the main thyristor T1. During the time t_c, capacitor voltage changes linearly from $(-V_s)$ to zero, mode III, Fig. 7.26. It is known that

$$i_c = C \frac{dv}{dt}$$

For a constant load current I_0, the above relation can be written as

$$I_c = C \frac{V_s}{t_c} \quad \text{or} \quad C = \frac{t_c \cdot I_0}{V_s}$$

The commutation circuit turn-off time t_c must be greater than the thyristor turn-off time t_q. Let $t_c = t_q + \Delta t$.

$$\therefore \qquad C = \frac{(t_q + \Delta t) \cdot I_0}{V_s} \qquad \qquad \text{...(7.36)}$$

Commutating inductor L. It can be designed from a consideration of the oscillatory current established when main thyristor T1 is turned-on. The current i_c, when T1 is triggered, flows through the ringing circuit formed by C, T1, L, D and is given by

$$i_c = \frac{V_s}{\omega_0 L} \sin \omega_0 t \quad \text{where } \omega_0 = \frac{1}{\sqrt{LC}}$$

The peak capacitor current $\quad I_{cp} = \frac{V_s}{\omega_0 L} = V_s \cdot \sqrt{\frac{C}{L}}$

This current flows through T1 when it is turned-on. As T1 handles load current as well as I_{cp}, peak capacitor current should not be too large. It is usual to take I_{cp} less than, or equal to, load current I_0, i.e.

$$I_{cp} \le I_0 \quad \text{or} \quad V_s \sqrt{\frac{C}{L}} \le I_0$$

or $\qquad\qquad\qquad\qquad L \ge \left(\frac{V_s}{I_0}\right)^2 C \qquad\qquad\qquad\qquad \text{...(7.37)}$

The second method of designing the value of L is from a consideration of the circuit turn-off time for auxiliary thyristor TA. It is seen from time-variation of v_{TA} in Fig. 7.26 that turn-off time for TA is t_{c1} and it is given by

$$t_{c1} = \frac{t_1}{2}$$

But $\qquad\qquad\qquad\qquad \omega_0 t_1 = \pi \quad \text{or} \quad t_1 = \frac{\pi}{\omega_0} = \pi\sqrt{LC}$

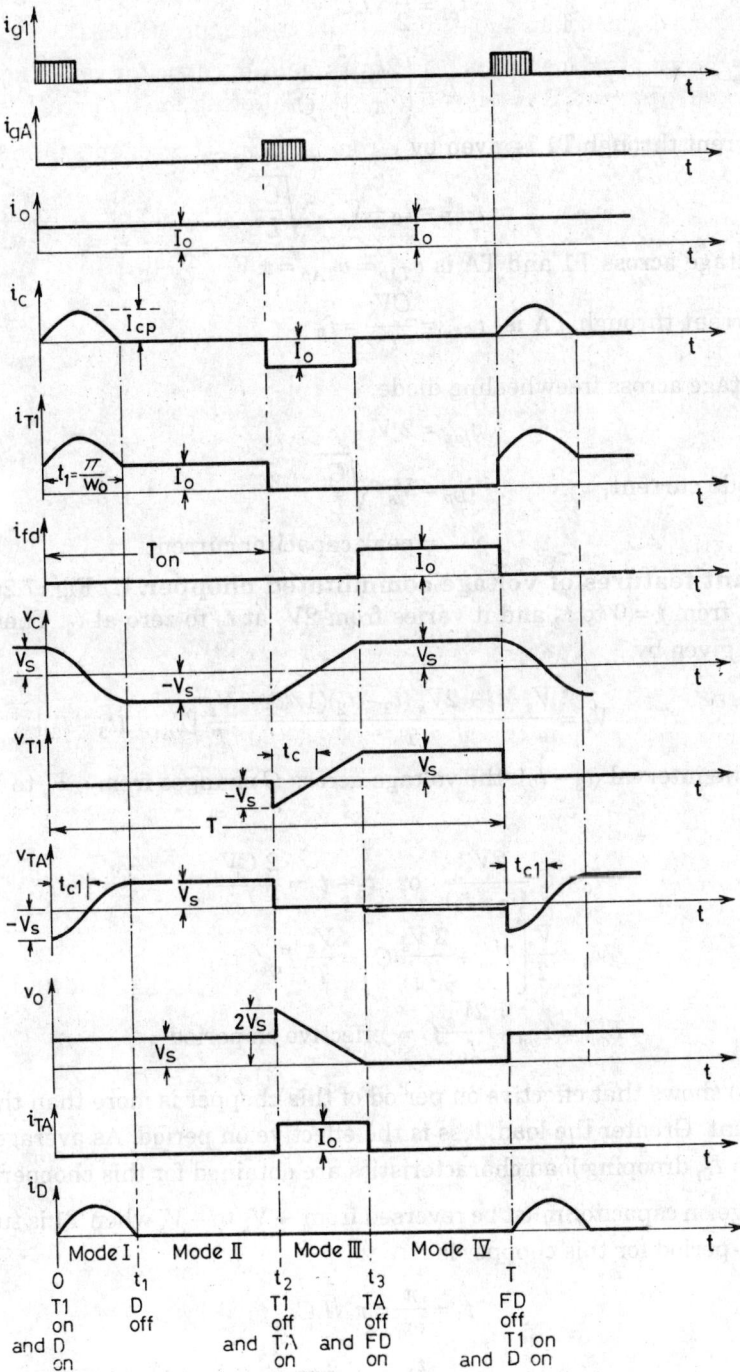

Fig. 7.26. Current and voltage waveforms for voltage-commutated chopper.

$$\therefore \qquad t_{c1} = \frac{\pi}{2} \sqrt{LC} \qquad\qquad\qquad ...(7.38)$$

or

$$L = \left(\frac{2t_{c1}}{\pi}\right)^2 \cdot \frac{1}{C} \qquad\qquad\qquad ...(7.39)$$

Peak current through T1 is given by

$$i_{T1P} = I_0 + V_s \cdot \sqrt{\frac{C}{L}} \qquad\qquad\qquad ...(7.40)$$

Peak voltage across T1 and TA is $v_{T1p} = v_{TAp} = \pm V_s$ \qquad ...(7.41)

Peak current through TA is $i_{TAp} = \dfrac{CV_s}{t_c} = I_0$ \qquad\qquad ...(7.42)

Peak voltage across freewheeling diode,

$$v_{fdp} = 2 V_s \qquad\qquad\qquad ...(7.43)$$

Peak diode current, \qquad $i_{Dp} = V_s \sqrt{\dfrac{C}{L}}$ \qquad\qquad ...(7.44)

$$= \text{peak capacitor current.}$$

Important features of voltage-commutated chopper. (*i*) Fig. 7.26 reveals that load voltage is V_s from $t = 0$ to t_2 and it varies from $2V_s$ at t_2 to zero at t_3. Therefore average load voltage V_0 is given by

$$V_0 = \frac{V_s \cdot t_2 + 2V_s (t_3 - t_2)(1/2)}{T} = \frac{V_s}{T} [T_{on} + (t_3 - t_2)] \qquad (\text{As } t_2 = T_{on})$$

During the interval $(t_3 - t_2)$, the voltage across C changes from $- V_s$ to V_s, *i.e.* total change is $2 V_s$.

$$\therefore \qquad I_0 = C \frac{2V_s}{(t_3 - t_2)} \quad \text{or} \quad t_3 - t_2 = \frac{2 CV_s}{I_0}$$

$$\therefore \qquad V_0 = \frac{V_s}{T}\left[T_{on} + \frac{2 V_s}{I_0} C \right] = \frac{V_s}{T} T_{on}' \qquad\qquad ...(7.45)$$

where \qquad $T_{on}' = T_{on} + \dfrac{2V_s}{I_0} C = $ effective on period \qquad ...(7.45a)

Eq. (7.45) shows that effective on period of this chopper is more than the period T_{on} and is load dependent. Greater the load, less is the effective on period. As average load voltage V_0 is dependent on I_0, drooping load characteristics are obtained for this chopper.

(*ii*) Charge on capacitor must be reversed from $+ V_s$ to $- V_s$ when T1 is turned on. Therefore minimum on-period for this chopper is

$$t_1 = \frac{\pi}{\omega_0} = \pi \sqrt{LC}$$

Minimum duty cycle, \qquad $\alpha_{mn} = \dfrac{t_1}{T} = \pi f \sqrt{LC}$ \qquad\qquad ...(7.46a)

Minimum load voltage, \qquad $V_{0.mn} = \alpha_{mn} \cdot V_s + \dfrac{2V_s \cdot 2t_c}{2T}$

$$= V_s (\alpha_{mn} + 2ft_c) = V_s [\pi f \sqrt{LC} + 2ft_c]$$

$$\therefore \qquad V_{0.mn} = f \cdot V_s [t_1 + 2t_c] \qquad\qquad ...(7.46b)$$

Maximum on period is $(T - 2t_c)$. This gives maximum value of duty cycle as

$$\alpha_{mx} = \frac{T - 2t_c}{T} = (1 - 2ft_c) \qquad\qquad ...(7.47)$$

Maximum load or output voltage V_{0mx} is given by

$$V_{0.mx} = \alpha_{mx} \cdot V_s + \frac{2V_s \cdot 2t_c}{2T} = V_s [\alpha_{mx} + 2ft_c]$$

Substituting the value of α_{mx} gives

$$V_{0.mx} = V_s [1 - 2ft_c + 2ft_c] = V_s$$

(*iii*) If main thyristor T1 fails to turn off when *TA* is triggered, *C* is completely discharged. The commutation, therefore, cannot be resumed in the next cycle and therefore control lost must be regained by turning off T1 by interrupting the supply.

(*iv*) The circuit cannot be operated at no load.

Example 7.16. *A voltage-commutated chopper feeds power to a battery-powered electric car. The battery voltage is 60 V, starting current is 60 A and thyristor turn-off time is 20 μsec. Calculate the values of the commutating capacitor C and the commutating inductor L.*

Solution. Let circuit turn-off time $t_c = t_q + \Delta t = 20 + 20 = 40$ μs for reliable turn-off of T1 and TA. In Eqs. (7.36) and (7.37), I_0 is the maximum load current that the commutation circuitry must be able to commutate.

$$\therefore \text{ From Eq. (7.36),} \qquad C = \frac{40 \times 60 \times 10^{-6}}{60} = 40 \text{ μF}$$

$$\text{From Eq. (7.37),} \qquad L \geq \left(\frac{60}{60}\right)^2 \times 40 \times 10^{-6} = 40 \text{ μH}$$

$$\text{From Eq. (7.39),} \qquad L = \left(\frac{2 \times 40 \times 10^{-6}}{\pi}\right)^2 \times \frac{1}{40 \times 10^{-6}} = 16.21 \text{ μH}$$

As per Eq. (7.37), value of L should be equal to, or more than 40 μH, but Eq. (7.39) gives $L = 16.21$ μH. Low value of L increases the peak value of capacitor current, see Eq. (7.44). So higher value of $L = 40$ μH should be chosen.

Example 7.17. *A voltage commutated chopper is shown in Fig. 7.27. With the main thyristor T1 conducting, the capacitor C is charged to a source voltage V_s with the polarities as marked.*

Now, after the auxiliary thyristor TA is turned on, the main SCR remains reverse biased for 100 microseconds. Calculate the value of the commutating component C.

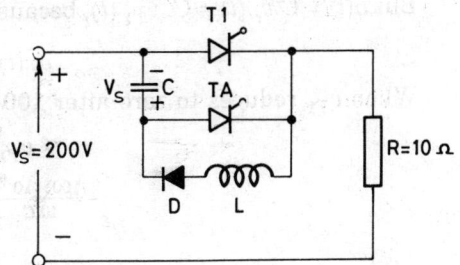

Fig. 7.27. Pertaining to Example 7.17.

Find also the value of commutating component L in case maximum permissible current through the main SCR is

(a) 2.5 times the load current and (b) 1.5 times the peak diode current.

Solution. When auxiliary SCR *TA* is turned on, capacitor voltage V_s at once appears across main thyristor *TI* and it is therefore turned off. After it, the load current completes its path through source V_s, *C*, *TA* and the load. The voltage equation for this circuit is

$$iR + \frac{1}{C}\int i\, dt = V_s \quad \text{or} \quad R\frac{dq}{dt} + \frac{q}{C} = V_s$$

The above equation can also be written as

$$\left(Rp + \frac{1}{C}\right)q = V_s \quad \text{where} \quad p \cong \frac{d}{dt}$$

Solution of this equation has two components : steady state and transient. Particular integral gives its steady state solution whereas complementary function provides its transient solution.

P.I. Its particular integral is obtained from $\frac{1}{C}q = V_s$

$\therefore$
$$q_{P.I.} = CV_s$$

C.F. Its complementary function is obtained from $\left(Rp + \frac{1}{C}\right)q = 0$.

$\therefore$
$$q_{C.F.} = Ae^{pt}$$

where p is obtained from $Rp + \frac{1}{C} = 0$. This gives $p = -\frac{1}{CR}$

$\therefore$
$$q_{C.F.} = Ae^{-t/RC}$$

Therefore,
$$q(t) = q_{P.I.} + q_{C.F.} = CV_s + Ae^{-t/RC}$$

At $t = 0$, $q = -CV_s$. Here minus sign is used before CV_s; it is because charging current leaves the positive terminal of *C* and enters its negative terminal. In other words; *C*, charged to V_s, now discharges, so negative sign is used before CV_s.

$\therefore$
$$-CV_s = CV_s + A \quad \text{or} \quad A = -2CV_s$$

This gives
$$q(t) = CV_s - 2CV_s\, e^{-t/RC}$$
$$= CV_s\, [1 - 2e^{-(t/RC)}]$$

But $q(t) = C.v_c(t) = C.\, v_{T1}(t)$, because $v_c = v_{T1}$.

$\therefore$
$$v_c(t) = v_{T1}(t) = V_s\left[1 - 2\, e^{-(t/RC)}\right] \qquad \qquad ...(7.48)$$

When v_{T1} reduces to zero after 100 μsec, thyristor T1 is turned off. Thus from Eq. (7.48),

$$v_{T1} = 0 = V_s\, [1 - 2e^{-(t/RC)}]$$

or
$$e^{-\frac{100 \times 10^{-6}}{10\, C}} = \frac{1}{2} \quad \text{or } C = 14.427\ \mu F$$

(*a*) Here load current,
$$I_0 = \frac{V_s}{R}$$

Maximum permissible current through the main SCR, from Eq. (7.40) is

$$i_{T1p} = \frac{V_s}{R} + V_s\sqrt{\frac{C}{L}} = 2.5\frac{V_s}{R} \quad \text{or} \quad V_s\sqrt{\frac{C}{L}} = 1.5\frac{V_s}{R}$$

or
$$L = \frac{4}{9}CR^2 = \frac{4}{9}(14.427)(10)^2 = 641.2\ \mu H.$$

(b) Peak diode current, from Eq. (7.44), is

$$i_{Dp} = V_s \sqrt{\frac{C}{L}}$$

$\therefore$

$$1.5 \, V_s \sqrt{\frac{C}{L}} = i_{T1p} = \frac{V_s}{R} + V_s \sqrt{\frac{C}{L}}$$

or

$$\frac{1}{2} V_s \sqrt{\frac{C}{L}} = \frac{V_s}{R}$$

or

$$L = \frac{CR^2}{4} = \frac{14.427 \times 10^{-6} \times 100}{4} = 360.68 \, \mu H.$$

Example 7.18. *A voltage-commutated chopper has the following parameters :*

$$V_s = 220 \, V, \, load \, circuit \, parameters = 0.5 \, \Omega, \, 2 \, mH, \, 40 \, V$$

Commutation circuit parameters :

$$L = 20 \, \mu H, \, and \, C = 50 \, \mu F$$
$$T_{on} = 800 \, \mu sec, \, T = 2000 \, \mu s.$$

For a constant load current of 80 A , compute the following :

(a) Effective on period.

(b) Peak currents through main thyristor T1 and auxiliary thyristor TA.

(c) Turn-off times for T1 and TA.

(d) Total commutation interval.

(e) Capacitor voltage 150 μs after TA is triggered.

(f) Time needed to recharge the capacitor to voltage V_s.

Solution. (a) From Eq. (7.45), effective-on period

$$= T_{on} + \frac{2V_s}{I_0} C = 800 \times 10^{-6} + \frac{2 \times 220}{80} \times 50 \times 10^{-6} = 1075 \, \mu s.$$

(b) From Eq. (7.40), $\quad i_{T1p} = 80 + 220 \sqrt{\frac{50}{20}} = 427.85 \, A$

Since load current is given as constant at 80 A, peak current through TA is 80 A.

(c) From Eq. (7.36), turn-off time for T1 is

$$t_c = \frac{CV_s}{I_0} = \frac{50 \times 10^{-6} \times 220}{80} = 137.5 \, \mu s.$$

Turn-off time for thyristor TA, from Eq. (7.38), is

$$t_{c1} = \frac{\pi}{2} \sqrt{LC} = \frac{\pi}{2} \sqrt{20 \times 50} \times 10^{-6}$$
$$= 49.673 \, \mu s$$

(d) Total commutation interval

$$= 2 \, t_c = 2 \times 137.5 = 275 \, \mu s$$

(e) Capacitor voltage before TA is triggered is equal to $(- V_s)$. After TA is triggered, capacitor begins to charge from $(- V_s)$. Therefore, capacitor voltage, after TA is triggered, is given by

$$v_c = \frac{I_0 \cdot t}{C} - V_s$$

where t = time measured from the instant TA is triggered

$$\therefore \qquad v_c = \frac{80 \times 150 \times 10^{-6}}{50 \times 10^{-6}} - 220 = 20 \text{ V}$$

(f) Time needed to recharge the capacitor from $(-V_s)$ to V_s is given by

$$\frac{(\text{Total change in voltage})\ (\text{Capacitance})}{\text{Load current}}$$

$$= \frac{[V_s - (-V_s)]\ C}{I_0} = \frac{2\ V_s \cdot C}{I_0} = \frac{2 \times 220 \times 50}{80} = 275 \text{ μs}$$

Example 7.19. *An impulse-commutated chopper feeds inductive load requiring a constant current of 260 A. The source voltage is 220 V dc and the chopping frequency is 400 Hz. Turn-off time for main thrystor is 18 μs. Peak current through main thyristor is limited to 1.8 times the constant load current. Taking a factor of safety 2 for the main thyristor, calculate the values of (a) commutating components C and L and (b) the minimum and maximum output voltage.*

Solution. (a) Load current $I_0 = 260$ A, $V_s = 220$ V.

For a factor of safety 2, the commutation circuit turn-off time for main thyristor

$$t_c = 2 \times 18 = 36 \text{ μs}.$$

From Eq. (7.36), $\qquad C = \dfrac{36 \times 10^{-6} \times 260}{220} = 42.545 \text{ μF}$

Peak current through main thyristor, from Eq. (7.40), is

$$1.8 I_0 = I_0 + V_s \sqrt{\frac{C}{L}} \quad \text{or} \quad 220 \sqrt{\frac{C}{L}} = 0.8 \times I_0$$

$$\therefore \qquad L = \left(\frac{220}{0.8 \times 260}\right)^2 \times 42.545 = 47.596 \text{ μH}.$$

(b) From Eq. (7.46a), minimum value of duty cycle is

$$\alpha_{mn} = \pi f \sqrt{LC} = \pi \times 400 \sqrt{42.545 \times 47.596} \times 10^{-6} = 0.0565$$

Minimum value of output voltage, from Eq. (7.46b), is

$$V_{0.mn} = V_s (\alpha_{mn} + 2 f t_c)$$

$$= 220\ (0.0565 + 2 \times 400 \times 36 \times 10^{-6}) = 18.766 \text{ V}$$

Maximum value of output voltage is $V_{0.mx} = 220$ V.

7.6.2. Current-Commutated Chopper

The power-circuit diagram for current-commutated chopper is shown in Fig. 7.28. In this diagram, T1 is the main thyristor. The other components, namely, auxiliary thyristor TA, capacitor C, inductor L, diodes D1 and D2 constitute commutation circuitry. FD is the freewheeling diode and R_c is the charging resistor.

Simplifying assumptions for the chopper are as follows :

   (i) Load current is constant.

  (ii) SCRs and diodes are ideal switches.

 (iii) Charging resistor R_c is so large that it can be treated as open circuit during the commutation interval.

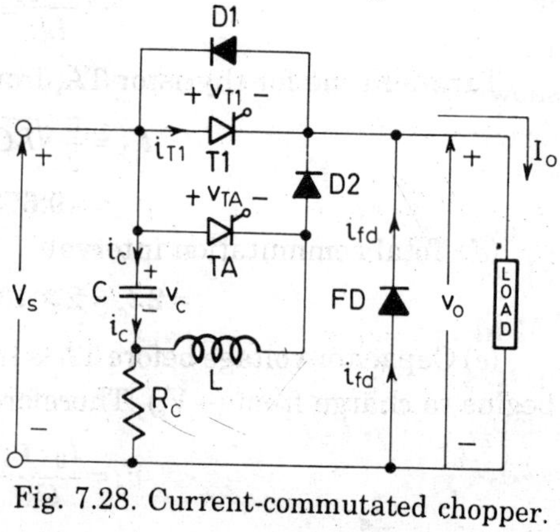

Fig. 7.28. Current-commutated chopper.

Currents i_c, i_{T1}, i_{fd} and i_0 are treated as positive when these are in the arrow directions marked. Similarly, voltages v_c, v_{T1}, v_{TA} and v_0 are taken as positive with the polarities as marked in Fig. 7.28.

Like voltage-commutated chopper, energy for current-commutation comes from the energy stored in a capacitor. Therefore, first of all, capacitor C is charged to a voltage V_s so that energy for commutation process is available. Capacitor C in Fig. 7.28 is charged through source V_s, capacitor C and the charging resistor R_c to a voltage V_s. After this, main thyristor T1 is fired at $t = 0$ so that load voltage $v_0 = V_s$ and load current $i_0 = I_0$ as shown in Fig. 7.30 up to $t = t_1$. With the turning on of T1, commutation circuitry remains inactive. Initiation of commutation process begins with the turning-on of thyristor TA. The commutation process for its easy grasp is divided into various modes as follows :

Mode I. At time $t = t_1$, auxiliary thyristor TA is triggered to commutate main thyristor T1. With the turning-on of TA, an oscillatory current $i_c = \dfrac{V_s}{\omega_0 L} \sin \omega_0 t$ [Eq. (7.33)] is set up in the circuit consisting of C, TA and L as shown in Fig. 7.29 (a). For the time interval $(t_2 - t_1)$, i_c and v_c vary sinusoidally through half cycle, Fig. 7.30. During this mode when v_c is zero, i_c is maximum though negative. At t_2, as i_c tends to reverse in the auxiliary thyristor TA, it gets naturally commutated. At t_2, $v_c = -V_s$ as shown in Fig. 7.30 ; i.e. in Fig. 7.28, lower plate is positive and upper plate is negative. During this mode T1 remains uneffected, therefore load current and load voltage remain I_0 and V_s respectively.

Mode II. As TA is turned off at t_2, oscillatory current i_c begins to flow through C, L, D2 and T1 as shown in Fig. 7.29 (b). Note that after t_2, i_c would flow through T1 and not through D1. It is because D1 is reverse biased by a small voltage drop across conducting thyristor T1. So after t_2, i_c would pass through T1 and not through D1.

In thyristor T1, i_c is in opposition to load current i_0 so that $i_{T1} = I_0 - i_c$. Note that i_{T1} is in the forward direction through T1. At t_3, i_c rises to I_0 so that $i_{T1} = 0$; as a result main SCR T1 is turned off at t_3. Since the oscillatory current through T1 turns it off, it is called *current-commutated chopper*.

During this mode, load voltage remains V_s through T1. For this mode, $t_2 < t < t_3$.

Mode III. As T1 is turned off at t_3, i_c becomes more than I_0. After t_3, i_c supplies load current I_0 and the excess current $i_{D1} = i_c - I_0$ is conducted through diode D1 as shown in Fig. 7.29 (c) and 7.30. The voltage drop in D1 due to $(i_c - I_0)$ keeps T1 reverse biased for $(t_4 - t_3) = t_c$; this is shown in the waveform for v_{T1}. At t_4, in case v_c exceeds V_s, FD comes into conduction, otherwise mode IV would follow. During mode III, when i_c is at its peak value of $I_{cp} \left(= \dfrac{V_s}{\omega_0 L} \right)$, $v_c = 0$. After this peak, capacitor voltage reverses and at t_4, upper plate is positive and lower plate is negative in Fig. 7.28.

Mode IV. At t_4, i_c reduces to I_0, as a result $i_{D1} = 0$ and diode D1 is therefore turned off. After t_4, a constant current equal to I_0 flows through source V_s, C, L, D2 and load and therefore capacitor C is charged linearly to source voltage V_s at t_5, Fig. 7.29 (d). So during the time $(t_5 - t_4)$, $i_c = I_0$.

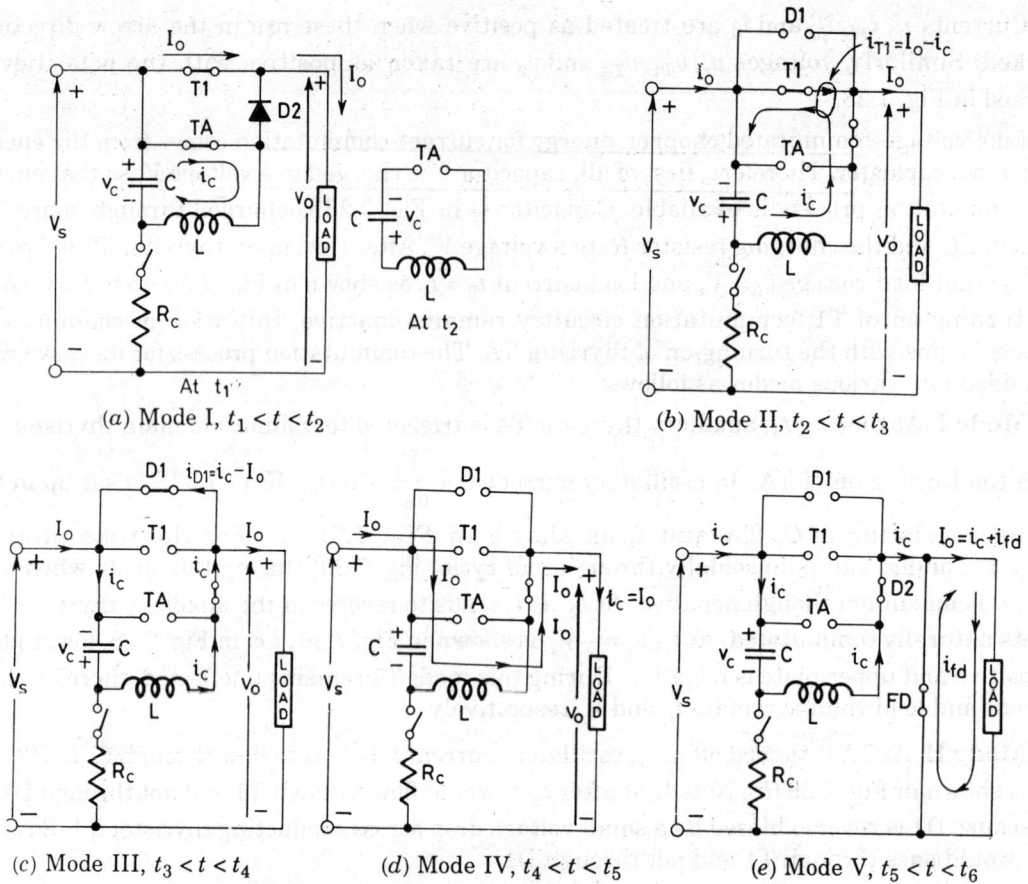

(a) Mode I, $t_1 < t < t_2$

(b) Mode II, $t_2 < t < t_3$

(c) Mode III, $t_3 < t < t_4$

(d) Mode IV, $t_4 < t < t_5$

(e) Mode V, $t_5 < t < t_6$

Fig. 7.29. Various modes of current-commutated chopper.

As D1 is turned off at t_4, $v_{T1} = v_{TA} = v_c$; this is shown as ab in Fig. 7.30 for v_c, v_{T1} and v_{TA}. Now the load voltage $v_0 = V_s - v_c = V_s -$ voltage ab at t_4. This is also marked in the waveform for v_0 at t_4. At t_5, $v_c = V_s$, therefore load voltage $v_0 = V_s - V_s = 0$ at t_5. During the interval $(t_5 - t_4)$, v_c increases linearly, therefore load voltage v_0 decreases to zero linearly during this interval.

Mode V. At t_5, capacitor C is actually overcharged to a voltage somewhat more than source voltage V_s. Therefore, FD gets forward biased and starts to conduct load current I_0 at t_5. Load voltage v_0 is reduced to zero at t_5 as stated in mode IV. As i_c is not zero at t_5, the capacitor C is still connected to load through source V_s, C, L and D2 ; as a consequence C is overcharged by the transfer of energy from L to C. At t_6, $i_c = 0$ and v_c becomes more than source voltage V_s. During $(t_6 - t_5)$, $i_c + i_{fd} = I_0$. With the build up of i_{fd}, i_c decays and finally at t_6, $i_c = 0$ and $i_{fd} = I_0$. Commutation process is completed at t_6. Total turn-off time, or commutation interval, is $(t_6 - t_1)$.

At t_6, v_c is shown as equal to xy. As D2 is turned off at t_6, $v_c = v_{TA} =$ voltage xy at t_6, Fig. 7.30.

From t_6 onwards, i_0 freewheels through FD. As i_c is zero and D2 is open circuited ; C now discharges through R_c for the freewheeling interval of the chopper. After t_5, v_{T1} remains constant at V_s, because V_s reaches T1 terminals through FD. At $t = T$, the main SCR T1 is again triggered and the cycle repeats.

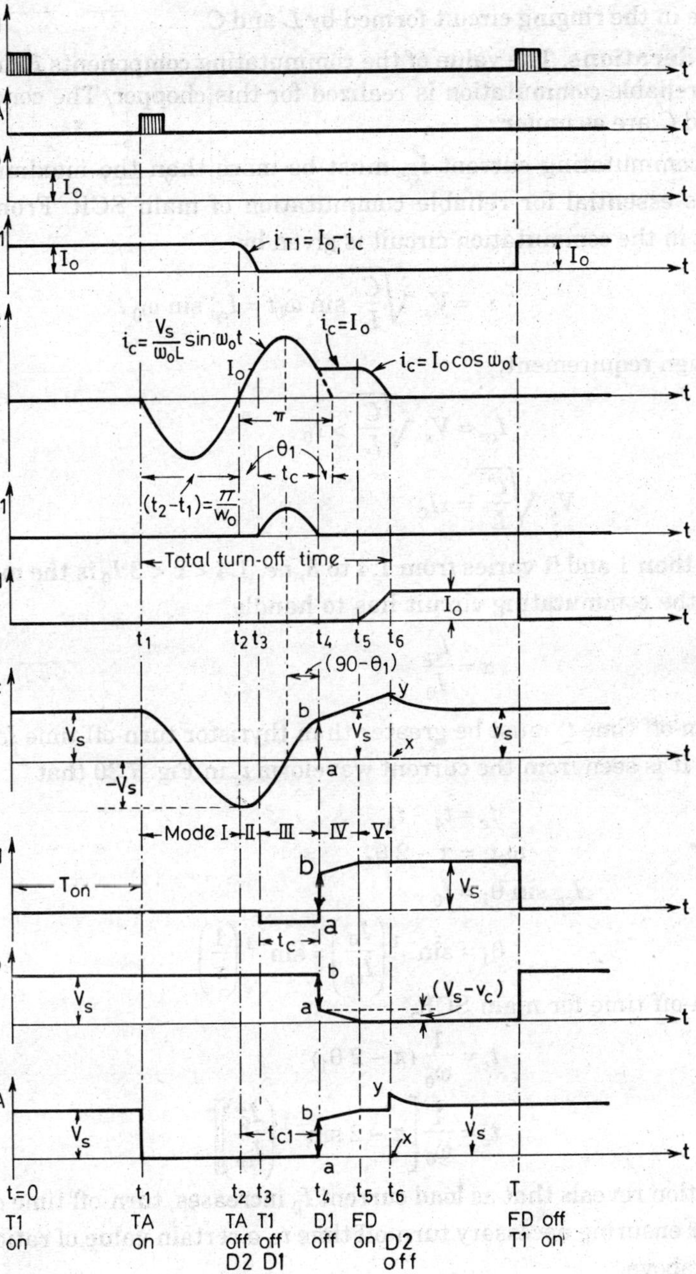

Fig. 7.30. Current and voltage waveforms for current-commutated chopper.

This chopper, developed by Hitachi Electric Co., Japan, is widely used in traction cars. The merits of this chopper are as under :

(i) Commutation is reliable so long as the load current is less than the peak commutating current I_{cp}.

(ii) Capacitor is always charged with the correct polarity.

(*iii*) Auxiliary thyristor TA is naturally commutated as its commutating current passes through zero value in the ringing circuit formed by L and C.

Design considerations. The value of the commutating components L and C should be so calculated that a reliable commutation is realized for this chopper. The conditions governing the design of L and C are as under :

(*i*) The peak commutating current I_{cp} must be more than the maximum possible load current I_0. This is essential for reliable commutation of main SCR. From Eq. (7.33), the oscillating current in the commutation circuit is given by

$$i_c = V_s \sqrt{\frac{C}{L}} \sin \omega_0 t = I_{cp} \sin \omega_0 t$$

As per the design requirement,

$$I_{cp} = V_s \sqrt{\frac{C}{L}} > I_0$$

or $$V_s \sqrt{\frac{C}{L}} = x I_0 \qquad \qquad ...(7.49)$$

where x is greater than 1 and it varies from 1.4 to 3, *i.e.* $1.4 < x < 3. I_0$ is the maximum possible load current that the commutating circuit has to handle.

$$\therefore \qquad \qquad x = \frac{I_{cp}}{I_0} .$$

(*ii*) Circuit turn-off time t_c must be greater than thyristor turn-off time for the main SCR. That is $t_c = t_q + \Delta t$. It is seen from the current waveform i_c in Fig. 7.30 that

$$t_c = t_4 - t_3$$
or $$\omega_0 t_c = \pi - 2 \theta_1$$
Also $$I_{cp} \sin \theta_1 = I_0$$
or $$\theta_1 = \sin^{-1}\left(\frac{I_0}{I_{cp}}\right) = \sin^{-1}\left(\frac{1}{x}\right)$$

$\therefore$ Circuit turn-off time for main SCR,

$$t_c = \frac{1}{\omega_0}(\pi - 2\theta_1) \qquad \qquad ...(7.50\ a)$$

$$t_c = \frac{1}{\omega_0}\left[\pi - 2\sin^{-1}\left(\frac{I_0}{I_{cp}}\right)\right] \qquad \qquad ...(7.50\ b)$$

The above relation reveals that as load current I_0 increases, turn-off time of main thyristor decreases. Thus for ensuring necessary turn-off time t_c, a certain value of ratio (I_0/I_{cp}) must be maintained. From above,

$$t_c = [\pi - 2\sin^{-1}(1/x)]\sqrt{LC} \qquad \qquad ...(7.51)$$

or $$\sqrt{C} = \frac{t_c}{[\pi - 2\sin^{-1}(1/x)]\sqrt{L}}$$

Substitution of this value of $\sqrt{C}$ in Eq. (7.49) gives

$$\frac{V_s}{L} \cdot \frac{t_c}{[\pi - 2\sin^{-1}(1/x)]} = x\, I_0$$

or
$$L = \frac{V_s \cdot t_0}{x\, I_0\, [\pi - 2 \sin^{-1}(1/x)]} \qquad \text{...(7.52)}$$

From Eq. (7.51),
$$\frac{1}{\sqrt{L}} = \frac{1}{t_c}\, [\pi - 2 \sin^{-1}(1/x)]\, \sqrt{C}$$

Substituting this value of $\sqrt{L}$ in the Eq. (7.49) gives

$$x\, I_0 = \frac{V_s}{t_c}\, [\pi - 2 \sin^{-1}(1/x)]\, C$$

or
$$C = \frac{x\, I_0 \cdot t_c}{V_s [\pi - 2 \sin^{-1}(1/x)]} \qquad \text{...(7.53)}$$

Total commutation interval. The total turn-off time, or the commutation interval, is $(t_6 - t_1)$, Fig. 7.30. It can be expressed as a sum of the following components :

$$t_6 - t_1 = (t_2 - t_1) + (t_4 - t_2) + (t_5 - t_4) + (t_6 - t_5).$$

The above four time-components of $(t_6 - t_1)$ can be obtained as under :

$(t_2 - t_1)$: During $(t_2 - t_1)$, waveform of current i_c , oscillating at frequency ω_0, completes one negative half cycle of π radians, Fig. 7.30.

∴ $(t_2 - t_1)$ = time period of half–cycle of oscillating current

$$= \frac{\pi}{\omega_0} = \pi\sqrt{LC}. \qquad \text{...(7.54)}$$

$(t_4 - t_2)$: For $(t_4 - t_2)$, sine current waveform of i_c is examined. At t_2, $i_c = 0$ and at t_4, i_c attains a value I_0 after passing through its peak, Fig. 7.30. Angle covered by i_c from t_2 to t_4 is equal to $(\pi - \theta_1)$ radians and as ω_0 is the angular frequency for i_c, $(t_4 - t_2)$ is given by

$$(t_4 - t_2) = \frac{\pi - \theta_1}{\omega_0} = (\pi - \theta_1)\sqrt{LC} \qquad \text{...(7.55)}$$

$(t_5 - t_4)$: The time $(t_5 - t_4)$ can be obtained from voltage considerations across C at t_4 and at t_5.

Voltage across C at $t_4 = ab = v_c$ at t_4. As this voltage at the instant t_4 is $(90 - \theta_1)^\circ$ away from zero crossing of the v_c sine wave, $ab = V_s \sin(90 - \theta_1)$.

Voltage across C at $t_5 = V_s$

∴ Increase in voltage across C during $(t_5 - t_4) = V_s - V_s \sin(90 - \theta_1)$

We know that $i = C \dfrac{dv}{dt}$. As I_0 is constant,

$$I_0 = C\, \frac{V_s - V_s \sin(90 - \theta_1)}{(t_5 - t_4)}$$

or
$$(t_5 - t_4) = CV_s\, \frac{1 - \sin(90 - \theta_1)}{I_0} = CV_s\, \frac{1 - \cos\theta_1}{I_0} \qquad \text{...(7.56)}$$

$(t_6 - t_5)$: During $(t_6 - t_5)$, current i_c is assumed to be $I_0 \cos \omega_0 t$. As $(t_6 - t_5)$ is equal to one quarter cycle $(\pi/2 \text{ rad.})$ of a sine wave,

$$(t_6 - t_5) = \frac{1}{2}\, \frac{\pi}{\omega_0} = \frac{\pi}{2}\, \sqrt{LC} \qquad \text{...(7.57)}$$

Addition of Eqs. (7.54) to (7.57) gives

$$(t_6 - t_1) = \text{total commutation interval}$$

$$= \left(\frac{5\pi}{2} - \theta_1\right)\sqrt{LC} + CV_s \frac{1 - \cos\vartheta_1}{I_0} \qquad ...(7.58a)$$

$$= \left(\frac{5\pi}{2} - \theta_1\right)\sqrt{LC} + 2\,CV_s \frac{\sin^2\theta_1/2}{I_0} \qquad ...(7.58b)$$

Turn-off times. For main SCR, turn-off time from Eq. (7.51) is

$$t_4 - t_3 = t_c = (\pi - 2\,\theta_1)\sqrt{LC}$$

$$= [\pi - 2\sin^{-1}(1/x)]\sqrt{LC}.$$

Turn-off time for auxiliary thyristor, from Eq. (7.55), is

$$t_4 - t_2 = t_{c1} = (\pi - \theta_1)\sqrt{LC} = [\pi - \sin^{-1}(1/x)]\sqrt{LC}$$

Peak capacitor voltage. Waveform of v_c in Fig. 7.30 reveals that maximum capacitor voltage xy is reached at t_6.

∴ Voltage at $t_6 = V_{cp} = $ voltage at $t_5 + $ voltage rise due to the energy transferred from L to C during $(t_6 - t_5)$.

At t_5, energy in L is $\frac{1}{2}LI_0^2$ and at t_6, this entire energy is transferred to C. Thus the voltage rise of C due to this transfer of energy is

$$\frac{1}{2}CV_c^2 = \frac{1}{2}LI_0^2$$

or

$$V_c = I_0\sqrt{\frac{L}{C}}$$

∴

$$V_{cp} = V_s + I_0\sqrt{\frac{L}{C}}. \qquad ...(7.59)$$

The values of I_{cp} and V_{cp} are also the peak ratings of the thyristors T1 and TA.

The value of charging resistor R_c is usually taken such that the periodic time $T \geq 3CR_c$.

Example 7.20. (a) *For a current commutated chopper, peak commutating current is twice the maximum possible load current. The source voltage is 230 V dc and main SCR turn-off time is 30 μ sec. For a maximum load current of 200 A, calculate*

(i) the values of the commutating inductor and capacitor,

(ii) maximum capacitor voltage and

(iii) the peak commutating current.

(b) Repeat part (a), in case peak commutating current is thrice the maximum possible load current.

Compare the results obtained in parts (a) and (b).

Solution. (a) Here

$$x = 2, \quad t_q = 30 \text{ μsec}$$

$$t_c = t_q + \Delta t.$$

Taking

$$\Delta t = 30 \text{ μsec}, \ t_c = (30 + 30) \text{ μsec} = 60 \text{ μsec}$$

(i) From Eq. (7.52),

$$L = \frac{230 \times 60 \times 10^{-6}}{2 \times 200\left[\pi - 2\sin^{-1}\left(\frac{1}{2}\right)\right]} = 16.473 \text{ μH}$$

From Eq. (7.53),

$$C = \frac{2 \times 200 \times 60 \times 10^{-6}}{230\left[\pi - 2\sin^{-1}\left(\frac{1}{2}\right)\right]} = 49.822 \text{ μF}$$

(ii) From Eq. (7.59), peak capacitor voltage is

$$V_{cp} = 230 + 200 \sqrt{\frac{16.473}{49.822}} = 345 \text{ volts.}$$

(iii) Peak commutating current,

$$I_{cp} = x I_0 = 2 \times 200 = 400 \text{ A.}$$

(b) (i) Here $x = 3$,

∴

$$L = \frac{230 \times 60 \times 10^{-6}}{3 \times 200 \left[\pi - 2 \sin^{-1}\left(\frac{1}{3}\right)\right]} = 9.342 \text{ μH}$$

$$C = \frac{3 \times 200 \times 60 \times 10^{-6}}{230 \left[\pi - 2 \sin^{-1}\left(\frac{1}{3}\right)\right]} = 63.577 \text{ μF}$$

(ii) Peak capacitor voltage,

$$V_{cp} = 230 + 200 \sqrt{\frac{9.342}{63.577}} = 306.67 \text{ V.}$$

(iii) Peak commutating current,

$$I_{cp} = 3 \times 200 = 600 \text{ A.}$$

It is seen from above that with the increase of x, L is reduced while C is increased, peak capacitor voltage is reduced but peak commutating current is increased.

Example 7.21. *A current commutated chopper is fed from a dc source of 230 V. Its commutating components are L = 20 μH and C = 50 μF. If load current of 200 A is assumed constant during the commutation process, then compute the following :*

(a) Turn-off time of main thyristor.

(b) Total commutation interval.

(c) Turn-off time of auxiliary thyristor.

Solution. (a) Peak commutating current from Eq. (7.49) is

$$I_{cp} = V_s \sqrt{\frac{C}{L}} = 230 \sqrt{\frac{50}{20}} = 363.66 \text{ A}$$

∴

$$x = \frac{I_{cp}}{I_0} = \frac{363.66}{200} = 1.8183$$

Turn-off time of main SCR, from Eq. (7.51), is

$$t_c = \left[\pi - 2 \sin^{-1}(1/1.8183)\right] \sqrt{20 \times 50 \times 10^{-12}} = 62.52 \text{ μsec}$$

(b)

$$\theta_1 = \sin^{-1}\left(\frac{I_0}{I_{cp}}\right) = \sin^{-1}\left(\frac{200}{363.66}\right) = 33.365°$$

Total commutation interval, from Eq. (7.58), is

$$\left(\frac{5\pi}{2} - \frac{33.365 \times \pi}{180}\right) \sqrt{1000} \times 10^{-6} + 50 \times 10^{-6} \times 230 \frac{1 - \cos 33.365°}{200}$$

$$= 229.95 \times 10^{-6} + 9.477 \times 10^{-6} = 239.427 \text{ μsec.}$$

(c) Turn-off time of auxiliary thyristor, from Eq. (7.55), is

$$\left(\pi - \frac{33.365 \times \pi}{180}\right) \sqrt{1000} \times 10^{-6} = 80.931 \text{ μsec.}$$

7.6.3. Load-Commutated Chopper

A load-commutated chopper is shown in Fig. 7.31. It consists of four thyristors T1 – T4 and one commutating capacitor C. The thyristors T1, T2 act together as one pair and thyristors T3, T4 act together as the second pair for conducting the load current alternately. When T1, T2 are conducting, these act as main thyristors and T3, T4 and C as the commutating components. Likewise, with the conduction of T3, T4 ; these become main thyristors and T1, T2 and C as the commutating components. FD is the freewheeling diode across the load.

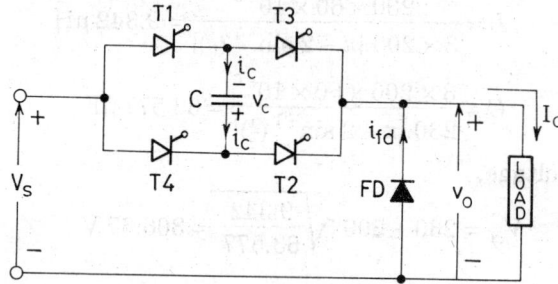

Fig. 7.31. Load commutated chopper.

Initially, the capacitor is charged to a voltage V_s with upper plate negative and lower plate positive as shown in Fig. 7.31. Assumptions made in current-commutated chopper also apply here. The working of this chopper can be explained in various modes as under :

Mode I. With the capacitor C charged with lower plate positive, the load commutated chopper is ready for operation. When thyristor pair T1, T2 is triggered at $t = 0$, circuit consisting of V_s, T1, C, T2 and load shows that load voltage at once shoots to $v_0 = V_s + v_c = 2\,V_s$. Load current now flows from source to load as shown in Fig. 7.32 (a).

The capacitor C is charged linearly by constant load current I_0 from V_s at $t = 0$ to $(-V_s)$ at t_1. When the capacitor voltage becomes $(-V_s)$, the load voltage falls from $2\,V_s$ to $v_0 = V_s - V_s = 0$ at t_1, Fig. 7.33. At $t = 0$, when T1, T2 are turned on, T3, T4 are reverse biased by capacitor voltage, i.e. at $t = 0$, $v_{T3} = v_{T4} = -V_s$. At t_1, $v_{T3} = v_{T4} = V_s$, i.e. T3, T4 are forward biased at t_1.

Mode II. At t_1, capacitor C is slightly overcharged, as a result freewheeling diode gets forward biased and load current is transferred from T1, T2 to FD. From t_1 onwards, load current freewheels through FD, Fig. 7.32 (b). During $(t_2 - t_1)$, $v_c = -V_s$, $v_0 = 0$, $i_c = 0$, $i_{fd} = I_0$, $i_{T1} = i_{T2} = 0$, $v_{T3} = v_{T4} = V_s$ and $v_{T1} = v_{T2} = -\Delta V_s$ as capacitor is overcharged by a small voltage ΔV_s.

Mode III. At t_2, thyristor pair T3, T4 is triggered, load voltage at once becomes $v_0 = V_s + v_c = 2\,V_s$. Thyristor pair T1, T2 is reverse biased by v_c, this pair is therefore turned off at t_2. The load current, now flowing through V_s, T4, C, T3 and load charges capacitor linearly from $(-V_s)$ at t_2 to V_s at t_3, Fig. 7.32 (c). Load voltage accordingly falls from $2\,V_s$ at t_2 to zero at t_3. During $(t_3 - t_2)$; $i_c = -I_0$, $i_{T3} = i_{T4} = I_0$ but $v_{T1} = v_{T2} = -V_s$ at t_2 and V_s at t_3 ; i.e. thyristor pair T1, T2 gets forward biased at t_3, Fig. 7.33.

At t_3, capacitor C is somewhat overcharged, FD gets forward biased and therefore after t_3, load current freewheels through FD and load. This is not shown in Fig. 7.32.

When T1, T2 are turned on at t_4, mode I repeats.

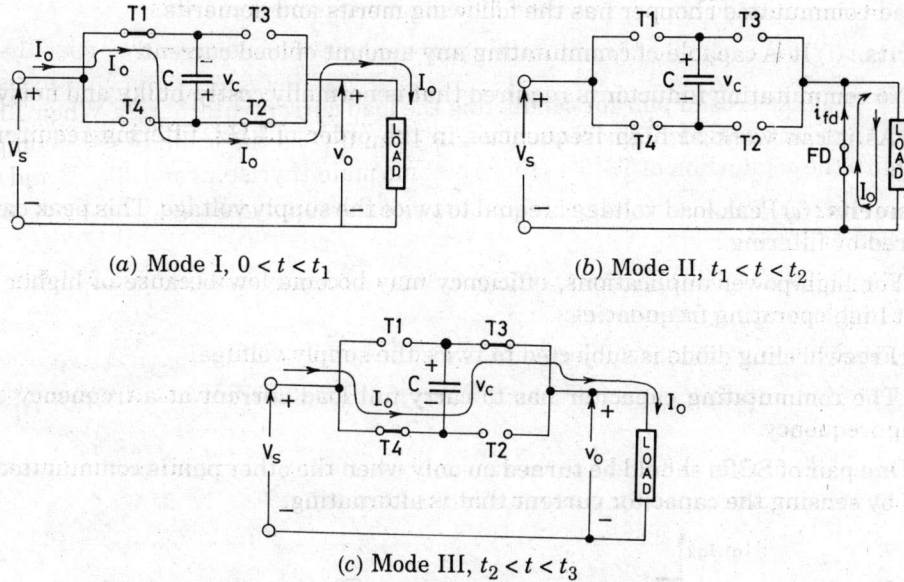

(a) Mode I, $0 < t < t_1$

(b) Mode II, $t_1 < t < t_2$

(c) Mode III, $t_2 < t < t_3$

Fig. 7.32. Different operating modes of load-commutated chopper.

Design of commutating capacitance. For constant load current I_0, capacitor voltage changes from $-V_s$ to V_s in time T_{on}, i.e. total change in voltage is $2V_s$ in time T_{on}, Fig. 7.33.

$$\therefore \qquad I_0 = C \, \frac{2V_s}{T_{on}}$$

or

$$C = \frac{I_0 \cdot T_{on}}{2V_s} \qquad \qquad ...(7.60)$$

Output voltage,

$$V_0 - \frac{1}{2}\,(2V_s)\,T_{on} \cdot \frac{1}{T} = V_s \cdot T_{on}\,f \qquad ...(7.61)$$

Substitution of $T_{on} = \dfrac{2V_s \cdot C}{I_0}$ in Eq. (7.61) gives average output voltage as

$$V_0 = V_s \cdot f \cdot \frac{2CV_s}{I_0} = \frac{2 \cdot V_s^2 \cdot C \cdot f}{I_0} \qquad ...(7.62)$$

Minimum chopping period, $T_{min} = T_{on}$

$\therefore$ Maximum chopping frequency,

$$f_{max} = \frac{1}{T_{min}} = \frac{1}{T_{on}}$$

From Eq. (7.60),

$$C = \frac{I_0}{2V_s} \cdot \frac{1}{f_{max}} \qquad \qquad ...(7.63)$$

It is seen from the waveform of v_{T1}, v_{T2}, v_{T3}, and v_{T4} in Fig. 7.33 that circuit turn-off time for each thyristor is

$$t_c = \frac{1}{2}\,T_{on} = \frac{1}{2}\,C\,\frac{2V_s}{I_0} = \frac{CV_s}{I_0} \qquad ...(7.64)$$

Total commutation interval $= T_{on} = \dfrac{2CV_s}{I_0}$.

A load-commutated chopper has the following merits and demerits.

Merits : (*i*) It is capable of commutating any amount of load current.

(*ii*) No commutating inductor is required that is normally costly, bulky and noisy.

(*iii*) As it can work at high frequencies in the order of kHz, filtering requirements are minimal.

Demerits : (*i*) Peak load voltage is equal to twice the supply voltage. This peak can however be reduced by filtering.

(*ii*) For high-power applications, efficiency may become low because of higher switching losses at high operating frequencies.

(*iii*) Freewheeling diode is subjected to twice the supply voltage.

(*iv*) The commutating capacitor has to carry full load current at a frequency of half the chopping frequency.

(*v*) One pair of SCRs should be turned on only when the other pair is commutated. This can be done by sensing the capacitor current that is alternating.

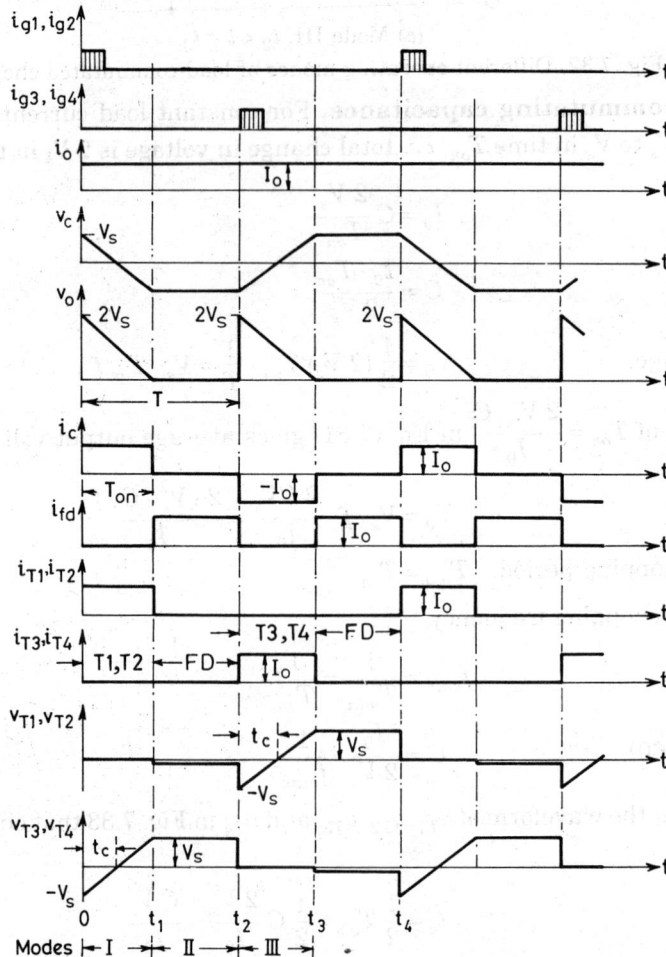

Fig. 7.33. Voltage and current waveforms for a load-commutated chopper.

7.7. MULTIPHASE CHOPPERS

A multiphase chopper is one that consists of two or more choppers connected in parallel. The two-chopper configuration shown in Fig. 7.34 is called a two-phase chopper. Similarly, three choppers connected in parallel will constitute a 3-phase chopper.

A multiphase chopper may be operated in two modes, *viz.* in-phase operation mode and the phase-shifted operation mode. In the in-phase operation mode, all the parallel connected choppers are on and off *at the same instant*. In the phase-shifted operation mode, different choppers are on and off at different instants of time.

In the two-phase chopper configuration shown in Fig. 7.34, inductance L in series with each chopper is assumed to be sufficiently large in that each chopper operates independent of each other. Let the load current be I_0 and ripple free. For a duty cycle of $\alpha = 30\%$, Fig. 7.35 (a) shows the in-phase operation of this chopper when both the choppers are on and off at the same instant. The input

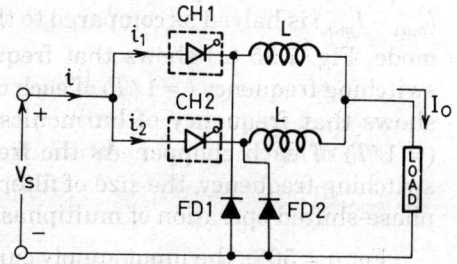

Fig. 7.34. Two-phase chopper.

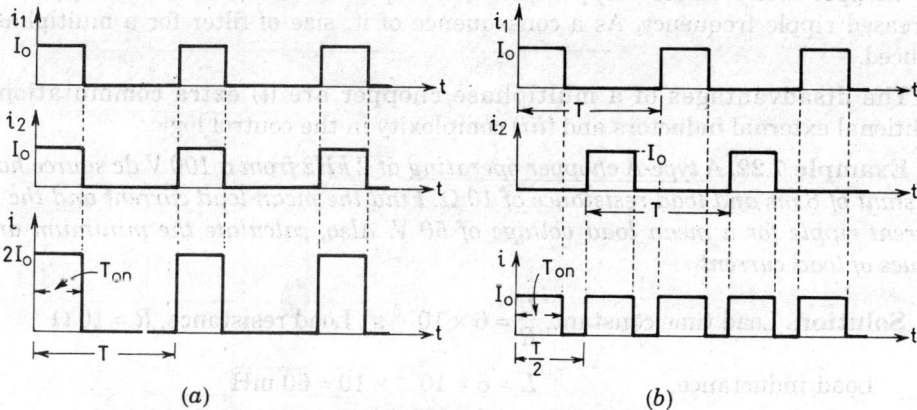

(a) (b)

Fig. 7.35. Input current waveforms for duty cycle $\alpha = 0.30$ for
(a) in-phase operation and (b) phase-shifted operation.

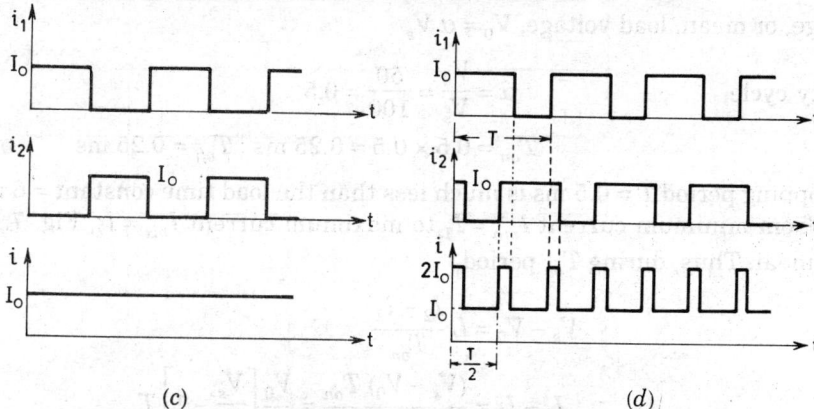

(c) (d)

Fig. 7.35. Current waveform for phase-shifted operation for
(c) $\alpha = 0.50$ and (d) $\alpha = 0.60$.

current i, obtained by the addition of i_1 and i_2, is seen to be doubled as shown. The in-phase operation of multiphase chopper is equivalent to a single-chopper operation.

Fig. 7.35 (b) shows the phase shifted operation for $\alpha = 30\%$. Chopper CH1 is on for 0.3 T from $t = 0$. Chopper CH2 is made on such that input current obtained from $i_1 + i_2$ is periodic in nature. A comparison of Figs. 7.35 (a) and (b) reveals that for phase-shifted operation, the frequency of input current is doubled and its ripple current amplitude (proportional to $I_{max} - I_{min}$) is halved as compared to the inphase-operation of chopper. In the in-phase operating mode, Fig. 7.35 (a) shows that frequency of harmonics in the input current is equal to the switching frequency ($= 1/T$) of each chopper. But in phase-shifted operating mode, Fig. 7.35 (b) shows that frequency of harmonics in the input current is twice the switching frequency ($= 1/T$) of each chopper. As the frequency of harmonics in the input current is twice the switching frequency, the size of filter is reduced in the phase-shifted chopper. This shows that phase-shifted operation of multiphase choppers is usually preferred.

For $\alpha = 50\%$, the input supply current of phase-shifted operation is continuous and without any ripples, Fig. 7.35 (c). For $\alpha = 60\%$, the supply current is continuous but with a pedestal of half the load current, Fig. 7.35 (d).

A multiphase chopper is used where large load current is required. The main advantage of this chopper over a single chopper is that its input current has reduced ripple amplitude and increased ripple frequency. As a consequence of it, size of filter for a multiphase chopper is reduced.

The disadvantages of a multiphase chopper are (i) extra commutation circuits (ii) additional external inductors and (iii) complexity in the control logic.

Example 7.22. *A type-A chopper operating at 2 kHz from a 100 V dc source has a load time constant of 6 ms and load resistance of 10 Ω. Find the mean load current and the magnitude of current ripple for a mean load voltage of 50 V. Also, calculate the minimum and maximum values of load current.*

Solution. Load time constant, $\dfrac{L}{R} = 6 \times 10^{-3}$ s ; Load resistance, $R = 10\ \Omega$

∴ Load inductance, $\qquad L = 6 \times 10^{-3} \times 10 = 60$ mH

Chopping period, $\qquad T = \dfrac{1}{f} = \dfrac{1}{2000} \times 1000 = 0.5$ ms

Average, or mean, load voltage, $V_0 = \alpha\, V_s$

∴ Duty cycle, $\qquad \alpha = \dfrac{V_0}{V_s} = \dfrac{50}{100} = 0.5$

$$T_{on} = 0.5 \times 0.5 = 0.25 \text{ ms} \; ; \; T_{off} = 0.25 \text{ ms}.$$

As chopping period $T = 0.5$ ms is much less than the load time constant $= 6$ ms, the current variation from minimum current $I_{mn} = I_1$ to maximum current $I_{mx} = I_2$, Fig. 7.15 (a), must be taken as linear. Thus, during T_{on} period,

$$V_s - V_0 = L\, \frac{I_2 - I_1}{T_{on}}$$

or, $\qquad I_2 - I_1 = \dfrac{(V_s - V_0)\, T_{on}}{L} = \dfrac{V_0}{L}\left[\dfrac{V_s}{V_0} - 1\right] T_{on}$

$$= \frac{V_0}{L}\left[\frac{1}{\alpha} - 1\right]T_{on} = \frac{V_0}{L}\left[\frac{T}{T_{on}} - 1\right]T_{on}$$

$$\therefore \qquad I_2 - I_1 = \frac{V_0}{L} T_{off} = \Delta I, \text{ current ripple} \qquad\qquad ...(i)$$

$\therefore$ Magnitude of current ripple,

$$\Delta I = I_2 - I_1 = \frac{V_0}{L} T_{off} = \frac{50}{60 \times 10^{-3}} \times 0.25 \times 10^{-3}$$

$$= 0.2083 \text{ A}.$$

Also, average or mean value of load current,

$$I_0 = \frac{I_1 + I_2}{2} = \frac{V_0}{R}$$

$$= \frac{50}{10} = 5 \text{ A}$$

An examination of Fig. 7.15 (a) shows that maximum value of load current I_2 is given by

$$I_2 = I_0 + \frac{\Delta I}{2} = I_0 + \frac{V_0}{2L} \cdot T_{off} = 5 + \frac{50 \times 0.25 \times 10^{-3}}{2 \times 60 \times 10^{-3}} = 5.104 \text{ A}$$

Minimum value of load current,

$$I_1 = I_0 - \frac{\Delta I}{2} = 5 - 0.104 = 4.896 \text{ A}.$$

Example 7.23. *For the circuit shown in Fig. 7.36, show that*

$$I_{max} = \frac{V_0}{R} + \frac{V_0}{2fL}\left[1 - \frac{V_0}{V_s}\right]$$

and $$I_{min} = \frac{V_0}{R} - \frac{V_0}{2fL}\left[1 - \frac{V_0}{V_s}\right]$$

where f = operating frequency of chopper switch S.

Assume L to be large enough to ensure linear growth and decay of the current through it and have continuous current.

Solution. The function of capacitor C across load resistance R is to make the output voltage continuous. In this figure, LC is a filter and FD is a flywheel diode. Inductor stores energy during T_{on} and delivers it during T_{off} of a cycle.

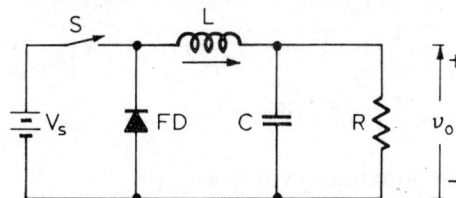

Fig. 7.36. Pertaining to Example 7.23.

The solution up to Eq. (i) in Example 7.22 is the same for this example also. Therefore, Eq. (i) in Example 7.22 gives current ripple as

$$\Delta I = I_2 - I_1 = I_{max} - I_{min} = \frac{V_0}{L} T_{off}$$

Average load voltage, $V_0 = \alpha V_s = \dfrac{T_{on}}{T} V_s$

or

$$T_{on} = \dfrac{V_0}{V_s} T = \dfrac{V_0}{fV_s}$$

Also,

$$T_{off} = T - T_{on} = \dfrac{1}{f} - \dfrac{V_0}{fV_s} = \dfrac{1}{f}\left(1 - \dfrac{V_0}{V_s}\right)$$

Average load current, $I_0 = \dfrac{V_0}{R}$

It is seen from Fig. 7.15 (a) that

$$I_{max} = I_0 + \dfrac{\Delta I}{2} = \dfrac{V_0}{R} + \dfrac{V_0}{2L} T_{off}$$

$$= \dfrac{V_0}{R} + \dfrac{V_0}{2fL}\left(1 - \dfrac{V_0}{V_s}\right)$$

Similarly,

$$I_{min} = \dfrac{V_0}{R} - \dfrac{V_0}{2fL}\left(1 - \dfrac{V_0}{V_s}\right)$$

Example 7.24. *For type-A chopper feeding an RLE load, show that maximum value of rms current rating for the freewheeling diode, in case load current is ripple free, is given by 0.3849* $\dfrac{V_s}{R}\left(1 - \dfrac{E}{V_s}\right)^{3/2}$

Solution. For chopper with RLE load, average load current I_0 is

$$I_0 = \dfrac{V_0 - E}{R} = \dfrac{\alpha V_s - E}{R}$$

Freewheeling-diode current flows during the period T_{off}. Therefore, rms value of freewheeling-diode current, when I_0 is ripple free, is given by

$$I_{fdr} = \sqrt{\dfrac{T_{off}}{T}} \cdot I_0 = \left[\dfrac{T - T_{on}}{T}\right]^{1/2}\left[\dfrac{\alpha V_s - E}{R}\right]$$

$$= \dfrac{1}{R}(\alpha V_s - E)(\sqrt{1 - \alpha}) \qquad \qquad \dots(i)$$

$$= \dfrac{1}{R}\left[\alpha \sqrt{1 - \alpha} \cdot V_s - \sqrt{1 - \alpha} \cdot E\right]$$

or

$$I_{fd.r} = \dfrac{1}{R}\left[\sqrt{\alpha^2 - \alpha^3} \cdot V_s - \sqrt{1 - \alpha} \cdot E\right]$$

This current $I_{fd.r}$ will have maximum value when

$$\dfrac{d\, I_{fd.r}}{d\,\alpha} = \dfrac{1}{R}\left[\dfrac{1}{2}\dfrac{(2\alpha - 3\alpha^2)\, V_s}{\sqrt{\alpha^2 - \alpha^3}} + \dfrac{1}{2}\dfrac{E}{\sqrt{1 - \alpha}}\right] = 0$$

or

$$\dfrac{(2\alpha - 3\alpha^2)\, V_s}{\sqrt{\alpha^2 - \alpha^3}} = -\dfrac{E}{\sqrt{1 - \alpha}}$$

or

$$\dfrac{(2\alpha - 3\alpha^2)\sqrt{1 - \alpha}}{\alpha \sqrt{1 - \alpha}} = -\dfrac{E}{V_s}$$

or
$$3\alpha - 2 = \frac{E}{V_s} \quad \text{or} \quad \alpha = \frac{1}{3}\left(2 + \frac{E}{V_s}\right)$$

Substituting this value of α in Eq. (i), we get maximum value of rms current rating $I_{fd.rm}$ of freewheeling diode as under :

$$
\begin{aligned}
I_{fd.rm} &= \frac{1}{R}\left[\frac{1}{3}\left(2 + \frac{E}{V_s}\right)V_s - E\right]\left[1 - \frac{1}{3}\left(2 + \frac{E}{V_s}\right)\right]^{1/2} \\
&= \frac{1}{R}\left[\frac{1}{3}\left(\frac{2V_s + E}{V_s}\right)V_s - E\right]\left[1 - \frac{2V_s + E}{3V_s}\right]^{1/2} \\
&= \frac{1}{R}\left[\frac{2V_s + E - 3E}{3}\right]\left[\frac{3V_s - 2V_s - E}{3V_s}\right]^{1/2} \\
&= \frac{1}{R}\left[\frac{2V_s - 2E}{3}\right]\left[\frac{V_s - E}{3V_s}\right]^{1/2} = \frac{1}{R} \cdot \frac{2}{3}\left[V_s - E\right]^{3/2} \cdot \frac{1}{\sqrt{3V_s}} \\
&= \frac{1}{R} \cdot \frac{2}{3} \frac{V_s \cdot \sqrt{V_s}}{\sqrt{3V_s}}\left[1 - \frac{E}{V_s}\right]^{3/2} \\
&= \frac{2}{3\sqrt{3}} \cdot \frac{V_s}{R}\left[1 - \frac{E}{V_s}\right]^{3/2} \\
&= 0.3849 \frac{V_s}{R}\left[1 - \frac{E}{V_s}\right]^{3/2}
\end{aligned}
$$

Example 7.25. *The speed of a separately excited dc motor is controlled below base speed by type-A chopper. The supply voltage is 220 V dc. The armature circuit has $r_a = 0.5~\Omega$ and $L_a = 10~mH$. The motor constant is $k = 0.1~V/rpm$. The motor drives a constant torque load requiring an average armature current of 30 A. On the assumption of continuous armature current, calculate (a) the range of speed control and (b) the range of duty cycle.*

Solution. For a motor, $V_t = V_0 = E_a + I_a r_a$

The minimum possible speed of dc motor is zero. This gives motor counter emf

$$E_a = 0$$

$\therefore$
$$\alpha V_s = V_0 = 0 + I_a r_a$$

or
$$\alpha \times 220 = 0 + 30 \times 0.5 = 15~V$$

or
$$\alpha = \frac{15}{220} = \frac{3}{44}$$

Maximum possible value of duty cycle is 1

$\therefore$
$$\alpha V_s = E_a + I_a r_a$$

or
$$1 \times 220 = K.N + 30 \times 0.5$$

or
$$N = \frac{220 - 15}{0.10} = 2050~rpm$$

Therefore (a) range of speed control is $0 < N < 2050$ rpm and

(b) the range of duty cycle is $\frac{3}{44} < \alpha < 1$.

Example 7.26. *In a battery-powered dc drive scheme, a chopper-controlled motor rated at 72 V, 200 A and 2500 rpm is separately-excited at a flux corresponding to its full rating. The* *current pulsation during acceleration is maintained between 180 A and 230 A. The motor resistance is 0.045 Ω, while the inductance is 7 mH. The battery resistance is 0.065 Ω. Neglecting the semiconductor losses, determine the chopping frequency and the duty cycle ratio when the speed is 1000 rpm. Draw diagrams to show the circuit arrangement and performance during one chopping cycle.*   *(I.A.S., 1988)*

Solution. For a separately-excited dc motor,

$$V_t = V_0 = E_a + I_a\, ra$$

or $$72 = K.N + 200 \times 0.045$$

∴ $$K = \frac{72 - 9}{2500} = \frac{63}{2500} \text{ V/rpm}$$

At 1000 rpm, counter emf of motor

$$E_a = K \times 1000 = \frac{63}{2500} \times 1000 = 25.2 \text{ V}$$

From Eq. (7.12), $$I_{mx} = \frac{V_s - E}{R}\left(1 - e^{-T_{on}/T_a}\right) + I_{mn}\, e^{-T_{on}/T_a}$$

Here circuit time constant, $T_a = \dfrac{L}{R} = \dfrac{7 \times 10^{-3}}{0.045 + 0.065} = 0.064 \text{ s}, R = 0.11\,\Omega$

∴ $$I_{mx} = 230 = \frac{72 - 25.2}{0.11}\left[1 - e^{-T_{on}/0.064}\right] + 180\, e^{-T_{on}/0.064}$$

or $$245.45\, e^{-T_{on}/0.064} = 195.45$$

or $$T_{on} = 0.01458 \text{ s}$$

From Eq. (7.13), $$I_{mn} = -\frac{E}{R}\left(1 - e^{-T_{off}/T_a}\right) + I_{mx}\, e^{-T_{off}/T_a}$$

During the freewheeling period,

$$T_a = \frac{L}{R} = \frac{7 \times 10^{-3}}{0.045} = 0.1556 \text{ s} \quad \text{and} \quad R = 0.045\,\Omega$$

∴ $$180 = -\frac{25.2}{0.045}(1 - e^{-T_{off}/0.1556}) + 230\, e^{-T_{off}/0.1556}$$

or $$T_{off} = 0.1556 \ln \frac{790}{740} = 0.010174 \text{ s}$$

Chopping period, $$T = T_{on} + T_{off} = 0.01458 + 0.010174 = 0.024754 \text{ s}$$

Chopping frequency, $$f = \frac{1}{T} = \frac{1}{0.024754} = 40.4 \text{ Hz}$$

Duty cycle ratio, $$\alpha = \frac{T_{on}}{T} = \frac{0.01458}{0.024754} = 0.588995$$
$$= 0.589 \text{ or } 58.9\%$$

Example 7.27. *Show by diagram with necessary explanation a battery-powered dc drive scheme using a chopper controlled motor. Sketch the voltage and current waveforms. Define chopping frequency and duty cycle ratio.*

In the drive scheme stated above, the maximum possible value of accelerating current is 425 A, the lower limit of the current pulsation is 180 A, length of ON period is 14 ms and that of OFF period is 11 ms, the time constant being 63.5 ms. Determine the higher limit of current pulsation, the chopping frequency and the duty cycle ratio.   *[I.A.S., 1989]*

Solution. Maximum value of current, $I_{mx} = 425$ A

Lower limit of current pulsation = 180 A

∴ Minimum value of current, $I_{mn} = 425 - 180 = 245$ A

Here $T_{on} = 14$ ms, $T_{off} = 11$ ms, $T = 25$ ms, $T_a = 63.5$ ms, $\alpha = \dfrac{14}{25} = 0.56$

From Eq. (7.12),
$$I_{mx} = \frac{V_s - E}{R}\left(1 - e^{-T_{on}/T_a}\right) + I_{mn} \cdot e^{-T_{on}/T_a}$$

or
$$425 = \frac{V_s - E}{R}\left(1 - e^{-14/63.5}\right) + 245\, e^{-14/63.5}$$

or
$$\frac{V_s - E}{R} = 1175.023\ \text{A}$$

For higher current pulsation, T_{on} is kept 14 ms. For duty cycle ratio α less than 0.56, the current pulsation will be more. For $\alpha = 0.5$, the current pulsation is the highest, So with $\alpha = 0.5$, $T_{on} = 14$ ms, from Eq. (7.12), we get

$$425 = 1175.023\,(1 - e^{-14/63.5}) + I_{mn}\, e^{-14/63.5}$$

or
$$I_{mn} = 239 \cdot 9956 \cong 240\ \text{A}$$

Higher limit of current pulsation $= 425 - 240 = 185$ A

Chopping period,
$$T = \frac{T_{on}}{\alpha} = \frac{14}{0.5} = 28\ \text{ms}$$

Chopping frequency,
$$f = \frac{1}{T} = \frac{1}{28 \times 10^{-3}} = 35.714\ \text{Hz}$$

Duty cycle ratio,
$$\alpha = 0.50$$

Example 7.28. *Fig. 7.37 (a) shows a chopper circuit operating at 100 Hz with $K_d = 0.5$. The load current at steady state is continuous but varies between 3 amps and 10 amps. Sketch the waveshape of*

(a) *the load current i_L*

(b) *the current i_f through the freewheeling diode DF*

(c) *the current i_c through the commutating capacitor.*   [GATE, 1994 ; I.A.S., 1997]

Solution. The circuit of Fig. 7.37 (a) is redrawn in Fig. 7.37 (b) where the positive directions for i_c, v_c, i_L, v_L, i_{TH1} and i_f are indicated. Source voltage V_s charges commutating

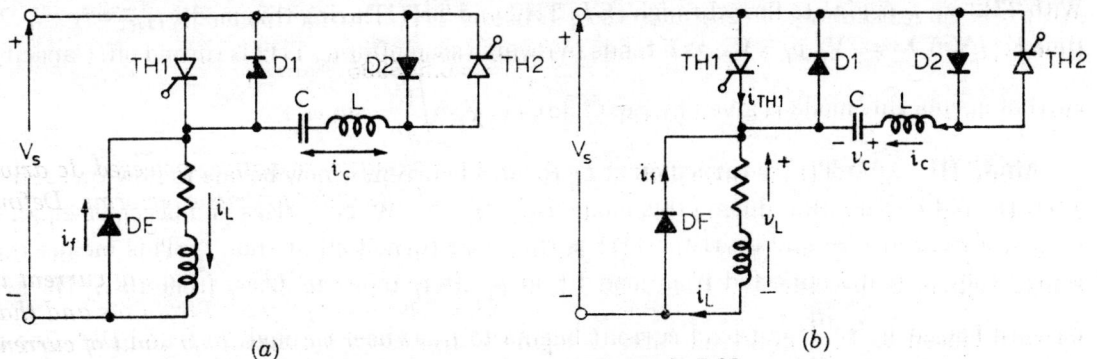

Fig. 7.37. Pertaining to Example 7.28.

capacitor C to voltage $+ V_s$ through D2, L, C, load parameters R, L and source voltage V_s. The working of this chopper is divided into certain modes as under :

Mode I : Main thyristor TH1 is turned on at $t = 0$, load voltage $v_L = V_s$ and load current i_L begins to build up from 3 A. Capacitor voltage remains dormant. The waveforms for v_C, v_L, i_L and $i_{TH1}\,(= i_L)$ are shown in Fig. 7.39. The circuit operation during this mode is shown in Fig. 7.38 (a).

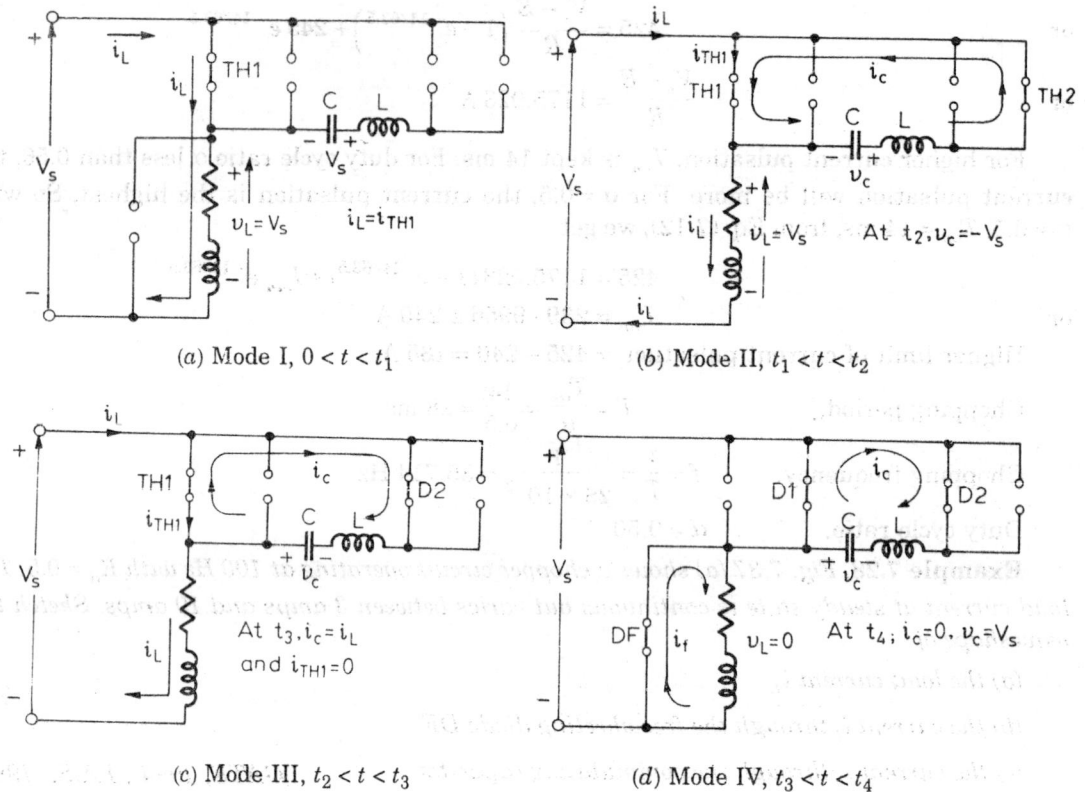

(a) Mode I, $0 < t < t_1$

(b) Mode II, $t_1 < t < t_2$

(c) Mode III, $t_2 < t < t_3$

(d) Mode IV, $t_3 < t < t_4$

Fig. 7.38

Mode II : Auxiliary thyristor TH2 is turned on at t_1 to commutate main thyristor TH1. With TH2 on, i_c begins to flow through C, L, TH2 and TH1. During this mode, $i_{TH1} = i_L + i_C$. At time t_2 ; $i_c = 0$, $v_c = - V_s$, $v_L = V_s$. As i_c tends to reverse soon after t_2, TH2 is turned off. Capacitor current during this mode is given by Eq. (7.33), i.e. $V_s \sqrt{\dfrac{C}{L}} \cdot \sin \omega_0 t$.

Mode III : After TH2 is turned off at t_2, reversed current i_c now begins to flow through C, TH1, D2 and L. Note that during this mode, $i_{TH1} = i_L - i_C$. When i_c rises and becomes equal to i_L, i_{TH1} decays to zero and thyristor TH1 is therefore turned off at time t_3. This means that source voltage is disconnected from load RL at t_3. As i_L tends to decay from 10 A, DF gets forward biased by $L \dfrac{di}{dt}$ and load current begins to freewheel through R, L and DF after t_3. Eventually, i_L decays to 3 A at t_5. At t_3, capacitor voltage v_c is still negative as shown in Fig. 7.39.

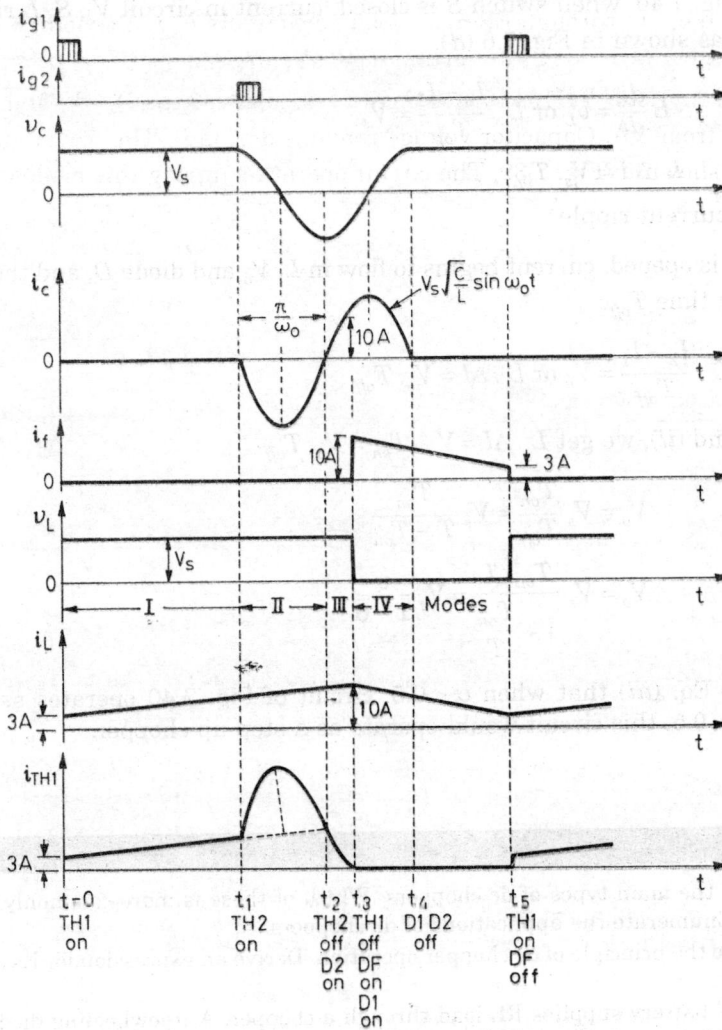

Fig. 7.39. Pertaining to chopper circuit of Example 7.28.

Mode IV : After TH1 is turned off at t_3, i_c rises through the oscillating circuit formed by D1, D2, L and C. Finally, at t_4, $i_c = 0$ and $v_c = V_s$. Soon after t_4, as i_c tends to reverse, diodes D1, D2 get turned off. Capacitor C charged to proper polarity and voltage is now suitable for next repeat cycle initiated by triggering TH1 on at t_5. During the interval $(t_5 - t_3)$, load current freewheels through DF, R and L and decays from 10 A at t_3 to 3 A at t_5 as shown in Fig. 7.39.

Example 7.29. *In the circuit shown in Fig. 7.40, S is an ideal semiconductor switch, V_s is the dc source voltage and V_o is the load emf. Show that this circuit configuration can be made to operate as a step-up or step-down chopper.*

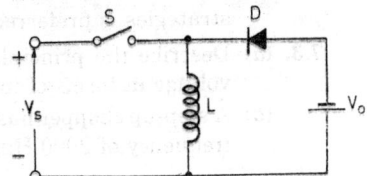

Fig. 7.40. Pertaining to Example 7.29.

Solution. In Fig. 7.40, when switch S is closed, current in circuit V_s, S, L rises from I_1 to I_2 during time T_{on} as shown in Fig. 7.6 (d).

$$\therefore \qquad L\frac{di}{dt} = v_L \text{ or } L \cdot \frac{I_2 - I_1}{T_{on}} = V_s$$

or $\qquad\qquad L \cdot \Delta I = V_s \cdot T_{on}$ ...(i)

where $I_2 - I_1 = \Delta I$, current ripple.

When switch S is opened, current begins to flow in L, V_o and diode D, and the current falls from I_2 to I_1 during time T_{off}.

$$\therefore \qquad L \cdot \frac{I_2 - I_1}{T_{off}} = V_o \text{ or } L \cdot \Delta I = V_o \cdot T_{off}$$ ...(ii)

From Eqs. (i) and (ii), we get $L \cdot \Delta I = V_s \cdot T_{on} = V_o \cdot T_{off}$.

or $\qquad\qquad V_o = V_s \dfrac{T_{on}}{T_{off}} = V_s \dfrac{T_{on}}{T - T_{on}}$

$$\therefore \qquad V_o = V_s \frac{T_{on}/T}{1 - \dfrac{T_{on}}{T}} = V_s \frac{\alpha}{1 - \alpha}$$ ...(iii)

It is seen from Eq. (iii) that when $\alpha < 0.5$, circuit of Fig. 7.40 operates as a step-down chopper. In case $\alpha > 0.5$, this circuit would operate as a step-up chopper.

PROBLEMS

7.1. (a) Discuss the main types of dc choppers. Which of these is more commonly employed and why ? Enumerate the applications of dc choppers.

(b) Describe the principle of dc chopper operation. Derive an expression for its average output voltage.

(c) A 120 V battery supplies RL load through a chopper. A freewheeling diode is connected across RL load having $R = 5\,\Omega$ and $L = 60$ mH. Load current varies between 7A and 9A. Calculate time ratio T_{on}/T_{off} for this chopper.

[**Hint.** (c) $I_o = 8\,A$, $\alpha = 1/3$ etc.]   [**Ans.** (c) 0.5]

7.2. (a) What is time ratio control in dc choppers ? Explain the use of TRC for controlling the output voltage in choppers.

(b) What is current limit control ? How does it differ from TRC ? Which of these control strategies is preferred over the other and why ?

7.3. (a) Describe the principle of step-up chopper. Derive an expression for the average output voltage in terms of input voltage and duty cycle. State the assumptions made.

(b) A step-up chopper has output voltage of two to four times the input voltage. For a chopping frequency of 2000 Hz, determine the range of off-periods for the gate signal.

[**Ans.** (b) 250 μs to 125 μs]

7.4. (a) What is meant by step-up chopper ? Explain its operation. Sketch the input voltage, input current, output voltage and output current waveforms. State the various assumption made.

How can a step-up chopper be used for the regenerative braking of dc motors ? Discuss.

(b) A step-up chopper with a pulse-width of 100 μs is operating from 230 V dc supply. Compute the average value of load voltage for a chopping frequency of 2000 Hz.

[**Ans.** (b) 287.5 V]

7.5. (a) What is a dc chopper ? Describe the working of type-B chopper. Does it operate as a step down or step-up chopper ? Explain.

(b) A dc battery is to be charged from a constant dc source of 220 V. The dc battery is to be charged from its internal emf of 90 V to 122 v. The battery has internal resistance of 1 Ω. For a constant charging current of 10 A, computer the range of duty cycle.

[**Ans.** (b) 0.4545 to 0.6]

7.6. For type-A chopper, express the following variables in terms of V_s, R, I_0 and duty cycle α in case load inductance causes the load current I_0 to remain constant at a value $I_0 = V_0/R$. Here V_s is the source voltage.

(a) Average output voltage and current.

(b) Output current at the instant of commutation.

(c) Average and rms values of freewheeling diode current.

(d) Rms value of the output voltage.

(e) Average and rms values of thyristor current.

Sketch the time variations of gate signal i_g, output voltage v_0, output current i_0, thyristor current i_T and freewheeling diode current i_{fd}.

[**Ans.** (a) αV_s, V_0/R (b) V_0/R (c) $(1-\alpha)I_0$, $\sqrt{(1-\alpha)}\,I_0$ (d) $\sqrt{\alpha}\,V_s$ (e) αI_0, $\sqrt{\alpha}\,I_0$]

7.7. Draw the power circuit diagram for a type-A chopper. Show load voltage waveforms for (i) $\alpha = 0.3$ and (ii) $\alpha = 0.8$. For both these duty cycles, calculate

(a) the average and rms values of output voltage in terms of source voltage V_s,

(b) the output power in case of resistive load R and

(c) the ripple factors.

[**Ans.** (a) (i) 0.3 V_s, 0.5477 V_s (ii) 0.8 V_s, 0.8944 V_s

(b) 0.3 V_s^2/R, 0.8 V_s^2/R (c) 1.5275, 0.5.]

7.8. (a) A chopper controls power given to an R–L load. For $T/T_a = Q$, derive an expression for the value of duty cycle α below which the per unit value of minimum load current falls below x per unit of V_s/R

(b) A chopper has the following data :

$T = 1000$ μs, $R = 2\Omega$, $L = 5$ mH

Find the duty cycle α so that per unit value of minimum load current does not fall below (i) 0.1 and (ii) 0.3 of V_s/R.

$$\left[\text{Hint. } (a) \text{ Use Eq. (7.15) with } E = 0. \text{ Here } x = \frac{I_{mn}}{V_s/R}\right]$$

$$\left[\textbf{Ans. } (a)\ \alpha = \frac{1}{Q}\ln\{1 + x\,(e^Q - 1)\}\ (b)\ 0.12,\ 0.344\right]$$

7.9. (a) A chopper, fed from a 220-V dc source, is working at a frequency of 50 Hz and is connected to an R–L load of $R = 5\ \Omega$ and $L = 40$ mH. Determine the value of duty cycle at which the minimum load current will be (i) 5 A (ii) 10 A (iii) 20 A and (iv) 30 A.

(b) For the values of α obtained in (a), calculate the corresponding values of maximum currents and the ripple factors. [**Ans.** (a) 0.328, 0.50582, 0.7222, 0.86184

(b) 26.823 A, 1.4313 ; 34.3996 A, 0.9884 ; 40.055 A, 0.6202 ; 42.3767 A, 0.4004]

7.10. (a) A dc chopper feeds power to an RLE load with $R = 2\ \Omega$, $L = 10$ mH and $E = 6$ V. If this chopper is operating at a chopping frequency of 1 kHz and with duty cycle of 10% from a 220-V dc source, compute the maximum and minimum currents taken by the load.

(b) A dc chopper is used to control the speed of a separately excited dc motor. The dc supply voltage is 220 V, armature resistance $r_a = 0.2\ \Omega$ and motor constant $K_a\ \phi = 0.08$ V/rpm.

The motor drives a constant torque load requiring an average armature current of 25 A. Determine (i) the range of speed control (ii) the range of duty cycle. Assume the motor current to be continuous. [*I.A.S., 1990*]

[**Ans.** (a) 9.016 A, 7.0367 A (b) Duty cycle : $1/44 < \alpha < 1$; Speed: $0 < N < 268.5$ rpm]

7.11. Sketch output voltage, output current, source current and thyristor current waveforms for type-C chopper for its operation in second quadrant. Indicate the conduction of various devices. For ripple free load current, discuss the operation of type-C chopper.

7.12. Describe the working of type-D chopper with appropriate waveforms to demonstrate its operation in first as well as fourth quadrants. Indicate clearly the range of duty cycle for which it operates in first and fourth quadrants.

7.13. Describe the working of type-E chopper with relevant circuit diagrams and its operation in all the four quadrants. Record in each quadrant, the brief account of its working as step-up or step-down chopper.

7.14. Write voltage equations governing the performance of type-A chopper during T_{on} and T_{off} periods for RLE type load. Hence obtain therefrom expressions for the maximum and minimum currents taken by the load on the assumption of continuous output current.

7.15. A step-down chopper is fed from 230 V dc and its duty cycle is 0.5. Calculate rms value of output voltage for fundamental, second and third harmonic components. Hence, express the output voltage as a function of Fourier series.

[**Ans.** 103.552 V, zero, −34.52 V. $v_o = 115 + 146.423 \sin \omega t - 48.81 \sin 3\ \omega t$]

7.16. For type-A chopper connected to RLE load, write the basic voltage equations and derive the expressions for the maximum and minimum values of load current in terms of source voltage V_s, R, E etc.

Hence show that the expression for per unit ripple in the load current is given by

$$\frac{(1 - e^{-\alpha T/T_a})\ (1 - e^{-(1-\alpha) T/T_a})}{(1 - e^{-T/T_a})}$$

where T = chopping period, α = duty cycle and $T_a = L/R$

7.17. A type-A chopper feeds RLE load. For low value of T_{on}, limit of continuous conduction is reached when load current during $T_{on} < t < T$ falls to zero. Derive an expression for this load current from basic voltage equations and hence obtain therefrom that the duty cycle α' at the limit of continuous conduction is given by

$$\alpha' = \frac{T_a}{T} \ln\left[1 + \frac{E}{V_s}\ (e^{T/T_a} - 1)\right]$$

where V_s = source voltage, $T_a = L/R$ and T = chopping period.

7.18. For type-A chopper feeding an RLE load, find an expression for the duty cycle for which the average current rating of freewheeling diode would be maximum. Assume constant load current. $\left[\text{**Ans.** } \dfrac{1}{2}\left(1 + \dfrac{E}{V_s}\right), \alpha > \dfrac{1}{2}\right]$

7.19. A battery with its terminal voltage of 200 V is supplied with power from type-A chopper circuit. The output voltage of the chopper consists of rectangular pulses of 2 ms duration in an overall cycle time of 5 ms. Internal resistance of the battery is negligible. Calculate :

(a) ripple factor
(b) average and rms values of output voltage
(c) rms value of the fundamental component of output voltage
(d) ac ripple voltage [**Ans.** (a) 1.2247 (b) 80 V 126.49 V (c) 85.63 V (d) 97.98 V]

7.20. A type-A chopper feeds power to RLE load with $R = 1.5\,\Omega$, $L = 6$ mH and $E = 44$ V. Other data for this chopper is as under :

Source voltage = 220 V dc, chopping frequency = 1 kHz,

Output voltage pulse duration = 400 μsec.

Repeat parts (a) to (i) of Example 7.9.   **[Ans.** (a) $\alpha' - 0.221$, current is continuous

(b) 29.33 A (c) 33.764 A, 24.975 A (e) 2.496 A, 0.7719 A, – 0.1716 A,

(f) 11.781 A, (g) 2591.82 W, 1290.52 W, 1300.5 W, (h) 29.445 A (i) 29.445 A**]**.

7.21. A 500 kW dc series motor used in a high-speed train is controlled by a chopper circuit. The inductance of the armature and series field windings is augmented by an external inductance. The dc source voltage for the train is 1000 V. The duty cycle varies from 0.15 to 0.9. Find the range of total inductance (its maximum and minimum values) in the armature circuit in terms of chopping period in case the amplitude of armature current excursion is limited to 25 A.

[Ans. 3.6 T to 10 T henries**]**

7.22. Describe a voltage-commutated chopper with relevant current and voltage waveforms as a function of time. The chopper operation may be divided into certain well-defined modes. Enumerate the various simplifying assumptions made.

Show that effective on-period for this chopper is load dependent. Find also the minimum permissible on-period in terms of commutating parameters.

7.23. A voltage-commutated chopper delivers power to RLE load for which $R = 0$ and $L = 8$ mH. For a chopping frequency of 200 Hz and dc source voltage of 400 V, find the chopper duty cycle so as to limit the load current excursion to 40 A.   **[Ans.** 0.8 or 0.2**]**

7.24. (a) Derive expressions for computing the magnitude of commutating components C and L for a voltage-commutated chopper. Relevant voltage and current waveforms may be drawn to assist these derivations. Discuss the considerations on which design value of these components depend.

(b) Describe the various important features of a voltage-commutated chopper, such as effective on period, minimum on-period etc.

7.25. Draw the power circuit diagram for a current commutated chopper. Explain the working of this chopper by dividing its commutation-process interval into some well-defined modes. Show distinctly the total turn-off time, turn-off times for main and auxiliary thyristors in the relevant waveforms drawn.

7.26. (a) Discuss the conditions governing the design of commutating components L and C for a current-commutated chopper and hence obtain expressions for these parameters in terms of source voltage, load current, circuit turn-off time etc. The current and voltage waveforms needed for obtaining the expressions of L and C may be sketched.

(b) A current-commutated chopper has the following data : Source voltage = 220 V dc ; Peak commutating current = 1.8 times the load current ; Main SCR t_q = 20 μs ; Factor of safety =2 ; Load current = 180 A.

Determine the values of the commutating inductor and capacitor, maximum capacitor voltage and the peak commutating current.   **[Ans.** 13.832 μH, 30.002 μF, 342.22 V, 324 A**]**

7.27. The commutating components for a current-commutated chopper are $C = 40$ μF and $L = 18$ μH. DC source voltage is 220 V and load current is constant at a value of 180 A during the commutation interval. For this chopper, calculate

(a) circuit turn-off time for main thyristor,

(b) circuit turn-off time for auxiliary thyristor, and

(c) total commutation interval.

Derive the expressions used in parts of (a), (b) and (c).

[Ans. (a) 53.12 μs (b) 68.71 μs (c) 203.176 μs**]**

7.28. (a) Discuss the working of a load-commutated chopper with relevant voltage and current waveforms. Show voltage variation across each pair of SCRs as a function of time.

Derive an expression from which the value of commutating capacitor of this chopper can be computed.

(b) A load-commutated chopper, fed from 230 V dc source, has a constant load current of 50 A. For a duty cycle of 0.4 and a chopping frequency of 2 kHz, compute

    (i) the average output voltage,

    (ii) the value of commutating capacitance,

    (iii) circuit turn-off time for one thyristor pair and

    (iv) total commutation interval.　　　　　**[Ans.** (b) 92 V ; 21.739 µF ; 100 µs ; 200 µs]

7.29. What is a multiphase chopper ? Bring out clearly, with appropriate waveforms, the difference between the in-phase operation and phase-shifted operation of a multiphase chopper. Hence show why phase-shifted operation is always preferred.

Enumerate the merits and demerits of multiphase choppers.

7.30. Describe a three-phase chopper with appropriate circuit diagram. Draw the input current waveforms for both in-phase and phase-shifted operations of this chopper on the assumption of ripple free output current. Suitable values of duty cycle may be chosen to illustrate your answer.

7.31. (a) A chopper circuit drives an inductive load from a 200 V dc supply. Given the load resistance of 40 Ω, the average load current of 30 A and the operating chopper frequency of 400 Hz, compute the ON and OFF periods of the chopper.

(b) A separately-excited dc motor is supplied from a 60 V dc source through a fixed frequency chopper. The rated speed is 900 rpm and the rated current is 30 A. Armature circuit resistance is 0.25 Ω. Find the duty cycle ratio of the chopper at rated motor torque for a speed of 300 rpm ignoring current pulsations.　**[Ans.** (a) 2.083 ms, 0.417 ms (b) 0.4167]

7.32. (a) A chopper circuit as shown in Fig. 7.41 (a) is inserted between a battery, V_s = 150 V and a load resistance R = 10 Ω. The circuit turn-off time for the main thyristor T1 is 110 µs and the maximum permissible current through it is 30 A. Calculate the values of the commutating components L and C.　　　　　　　　　　　　　　　*[I.A.S., 1995]*

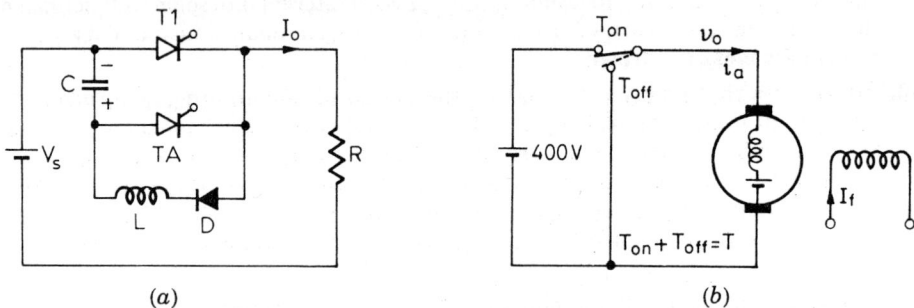

(a)　　　　　　　　　　　　　　　　　　　(b)

Fig. 7.41. Pertaining to Problem 7.32.

(b) Fig. 7.41 (b) shows the circuit schematic of a chopper driven separately-excited dc motor. The single-pole double-throw switch operates with a switching period T of 1 ms. The duty ratio of the switch (T_{on}/T) is 0.2. The motor may be assumed lossless, with an armature inductance of 10 mH. The motor draws an average current of 20 A at a constant back emf of 80 V, under steady state.

    (i) Sketch and label the voltage waveform $v_0(t)$ of the chopper for one switching period

    (ii) Sketch and label the motor current $i_a(t)$ for one switching period T

    (iii) Evaluate the peak-to-peak current ripple of the motor.

                         [Ans. (a) L = 1.587 mH, C = 15.87 µF

(b) Minimum current = 16.8 A, maximum current = 23.2 A,

peak-to-peak current = 6.4 A]

7.33. In Fig. 7.42. (a), the ideal switch S is switched on and off with a switching frequency $f = 10$ kHz. The circuit is operated in steady state at the boundary of continuous and discontinuous conduction, so that the inductor current i is as shown in Fig. 7.42 (b). Find (b) on-time T_{on} of the switch and (b) the value of the peak current I_p. [*GATE, 2002*]

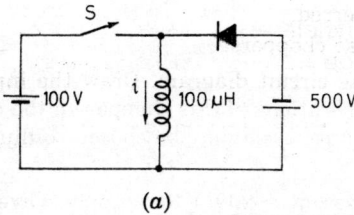

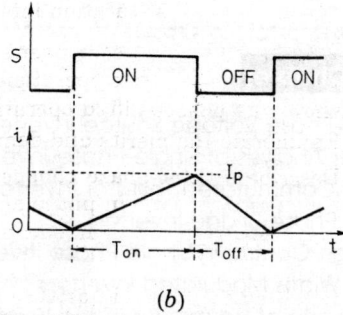

Fig. 7.42. Pertaining to Problem 7.33.

[**Hint.** $L \cdot I_p = 100 \cdot T_{on}$ during T_{on} etc. See Example 7.29]   [**Ans.** (a) 83.33 μs (b) 83.33 A]

Chapter 8

Inverters

In this Chapter

- Single-Phase Voltage Source Inverters : Operating Principle
- Fourier Analysis of Single-Phase Inverter Output Voltage
- Force-Commutated Thyristor Inverters
- Three Phase Bridge Inverters
- Voltage Control In Single-Phase Inverters
- Pulse-Width Modulated Inverters
- Reduction of Harmonics in the Inverter Output Voltage
- Current Source Inverters
- Series Inverters
- Single-Phase Parallel Inverter
- Good Inverter

A device that converts dc power into ac power at desired output voltage and frequency is called an inverter. Some industrial applications of inverters are for adjustable-speed ac drives, induction heating, stand by air-craft power supplies, UPS (uninterruptible power supplies) for computers, hvdc transmission lines etc. Phase-controlled converters, when operated in the inverter mode, are called *line-commutated inverters*, Chapter 6. But line-commutated inverters require at the output terminals an existing ac supply which is used for their commutation. This means that line-commutated inverters can't function as isolated ac voltage sources or as variable frequency generators with dc power at the input. Therefore, voltage level, frequency and waveform on the ac side of line-commutated inverters cannot be changed. On the other hand, force commutated inverters provide an independent ac output voltage of adjustable voltage and adjustable frequency and have therefore much wider applications. In this chapter, force-commutated and load commutated inverters are described.

The dc power input to the inverter is obtained from an existing power supply network or from a rotating alternator through a rectifier or a battery, fuel cell, photovoltaic array or magneto hydrodynamic (MHD) generator. The configuration of ac to dc converter and dc to ac inverter is called a dc-link converter. The rectification is carried out by standard diodes or thyristor converter circuits discussed in Chapter 6. The inversion is performed by the methods discussed in this chapter.

Inverters can be broadly classified into two types ; voltage source inverters and current source inverters. A voltage-fed inverter (VFI), or voltage-source inverter (VSI), is one in which the dc source has small or negligible impedance. In other words, a voltage source inverter has stiff dc voltage source at its input terminals. A current-fed inverter (CFI) or current-source inverter (CSI) is fed with adjustable current from a dc source of high impedance, *i.e.* from a stiff dc current source. In a CSI fed with stiff current source, output current waves are not affected by the load.

In VSIs using thyristors, some type of forced commutation is usually required; however, load commutation is possible only if the load is underdamped. In VSIs using GTOs, switching-off is achieved by applying a negative gate-current pulse. VSIs using transistors, like BJTs, PMOSFETs, IGBTs or SITs, can be turned off by the control of their base current. Switching-off of the devices with the help of their gate or base currents is called *self-commutation.*. So the self-commutated inverters using GTOs and transistors do not require additional commutation circuitry as needed in thyristor-based inverters. This reduces the complexity and cost of the self-commutated inverter circuits and at the same time, enhances the reliability of their operation.

From the viewpoint of connections of semiconductor devices, inverters are classified as under :

    1. Bridge inverters   2. Series inverters   3. Parallel inverters

The object of this chapter is to describe the operating principles of both single-phase and three-phase inverters and to present their elementary analysis. As before, switching devices are assumed to possess ideal characteristics. Since bridge inverters are more popular, more emphasis is given to their description.

8.1. SINGLE-PHASE VOLTAGE SOURCE INVERTERS : OPERATING PRINCIPLE

In this section, operating principle of single-phase voltage source inverters is discussed.

8.1.1. Single-phase Bridge Inverters

Single-phase bridge inverters are of two types, namely (*i*) single-phase half-bridge inverters and (*ii*) single-phase full-bridge inverters. Basic principles of operation of these two types are presented here.

Power circuit diagrams of the two configurations of single-phase bridge inverter, as stated above, are shown in Fig. 8.1 (*a*) for half-bridge inverter and in Fig. 8.2 (*a*) for full-bridge inverter. In these diagrams, the circuitry for turning-on or turning-off of the thyristors is not shown for simplicity. The gating signals for the thyristors and the resulting output voltage waveforms are shown in Figs. 8.1 (*b*) and 8.2 (*b*) for half-bridge and full-bridge inverters respectively. These voltage waveforms are drawn on the assumption that each thyristor conducts for the duration its gate pulse is present and is commutated as soon as this pulse is removed. In Figs. 8.1 (*b*) and 8.2 (*b*), $i_{g1} - i_{g4}$ are gate signals applied respectively to thyristors T1-T4.

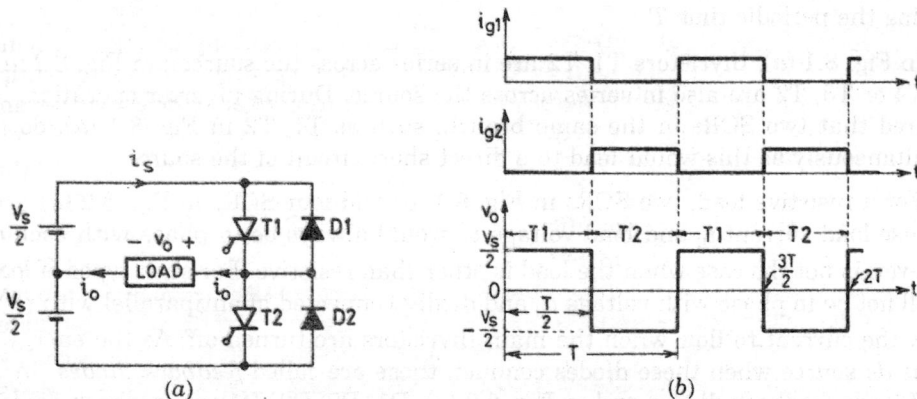

Fig. 8.1. Single-phase half-bridge inverter.

Single-phase half bridge inverter, as shown in Fig. 8.1 (a), consists of two SCRs, two diodes and three-wire supply. It is seen from Fig. 8.1 (b) that for $0 < t \leq T/2$, thyristor T1 conducts and the load is subjected to a voltage $V_s/2$ due to the upper voltage source $V_s/2$. At $t = T/2$, thyristor T1 is commutated and T2 is gated on. During the period $T/2 < t \leq T$, thyristor T2 conducts and the load is subjected to a voltage $(- V_s/2)$ due to the lower voltage source $V_s/2$. It is seen from Fig. 8.1 (b) that load voltage is an alternating voltage waveform of amplitude $V_s/2$ and of frequency $1/T$ Hz. Frequency of the inverter output voltage can be changed by controlling T.

The main drawback of half-bridge inverter is that it requires 3-wire dc supply. This difficulty can, however, be overcome by the use of a full-bridge inverter shown in Fig. 8.2 (a). It consists of four SCRs and four diodes. In this inverter, number of thyristors and diodes is twice of that in a half bridge inverter. This, however, does not go against full inverter because the amplitude of output voltage is doubled whereas output power is four times in this inverter as compared to their corresponding values in the half-bridge inverter. This is evident from Figs. 8.1 (b) and 8.2 (b).

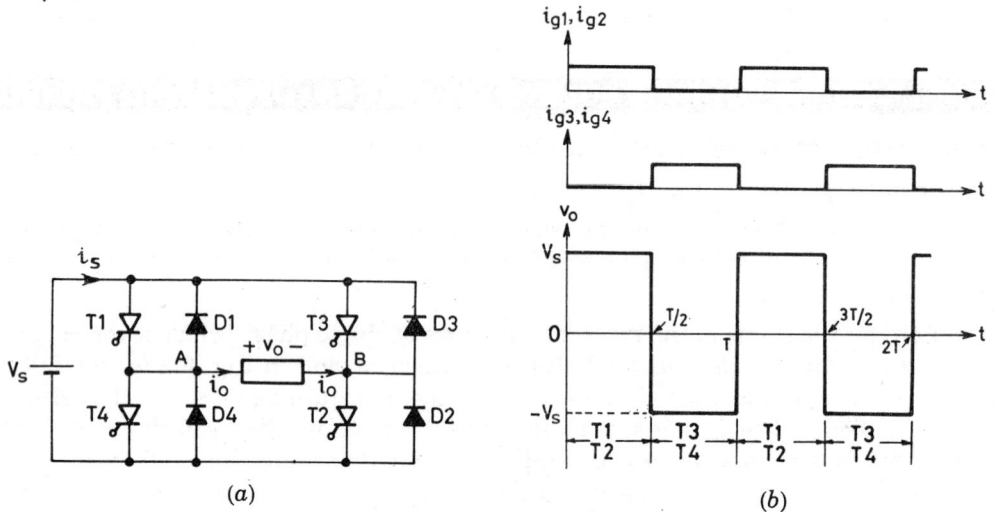

Fig. 8.2. Single-phase full-bridge inverter.

For full-bridge inverter, when T1, T2 conduct, load voltage is V_s and when T3, T4 conduct load voltage is $- V_s$ as shown in Fig. 8.2 (b). Frequency of output voltage can be controlled by varying the periodic time T.

In Fig. 8.1 (a), thyristors T1, T2 are in series across the source ; in Fig. 8.2 (a) thyristors T1, T4 or T3, T2 are also in series across the source. During inverter operation, it should be ensured that two SCRs in the same branch, such as T1, T2 in Fig. 8.1 (a), do not conduct simultaneously as this would lead to a direct short circuit of the source.

For a resistive load, two SCRs in Fig. 8.1 (a) and four SCRs in Fig. 8.2 (a) would suffice, because load current i_0 and load voltage v_0 would always be in phase with each other. This, however, is not the case when the load is other than resistive. For such types of loads, current i_0 will not be in phase with voltage v_0 and diodes connected in antiparallel with thyristors will allow the current to flow when the main thyristors are turned off. As the energy is fed back to the dc source when these diodes conduct, these are called *feedback diodes*. In Fig. 8.1 (a), D1, D2 are feedback diodes and in Fig. 8.2 (a), D1, D2, D3, D4 are feedback diodes.

8.1.2. Steady-state Analysis of Single-phase Inverter

Figs. 8.1 (*b*) and 8.2 (*b*) reveal that load voltage waveform does not depend on the nature of load. The load voltage is given by

for half-bridge inverter, $v_0 = \dfrac{V_s}{2}$ .... $0 < t << T/2$

$$= -\dfrac{V_s}{2} \text{ .... } T/2 < t < T$$

and for full-bridge inverter, $v_0 = V_s$ .... $0 < t < T/2$

$$= -V_s \text{ .... } T/2 < t < T$$

The load current is, however, dependent upon the nature of load. Let the load, in general, consist of RLC in series. The circuit model of single-phase half-bridge or full-bridge inverter is as shown in Fig. 8.3 (*a*). In this circuit, load current would finally settle down to steady state conditions and would vary periodically as shown in Figs. 8.3 (*c*) to (*f*). It is seen from these waveforms that

$$i_0 = \pm I_0 \text{ .... at } t = 0, T, 2T, 3T, ...$$

and

$$i_0 = \pm I_0 \text{ .... at } t = T/2, 3T/2, 5T/2, ...$$

The voltage equation for the circuit model of Fig. 8.3 (*a*) for half-bridge inverter and for $0 < t < T/2$ is given by

$$\frac{V_s}{2} = R\, i_0 + L\frac{d\,i_0}{dt} + \frac{1}{C}\int i_0 dt + V_{c1} \qquad \qquad ...(8.1)$$

For full-bridge inverter, replace $V_s/2$ by V_s in Eq. (8.1). In this equation, V_{c1} is the initial voltage across capacitor at $t = 0$.

For $T/2 < t < T$, or $0 < t' < T/2$, the voltage equation for half-bridge inverter is

$$-\frac{V_s}{2} = R\, i_0 + L\frac{di}{dt'} + \frac{1}{C}\int i dt' + V_{c2} \qquad \qquad ...(8.2)$$

and for a full-bridge inverter, replace $(-V_s/2)$ by $(-V_s)$ in Eq. (8.2). In this equation, V_{c2} is the initial voltage across capacitor at $t' = 0$.

Differentiation of Eqs. (8.1) and (8.2) gives

$$\frac{d^2 i_0}{dt^2} + \frac{R}{L}\frac{d\,i_0}{dt} + \frac{1}{LC}\,i_0 = 0$$

and

$$\frac{d^2 i_0}{dt'^2} + \frac{R}{L}\frac{d\,i_0}{dt'} + \frac{1}{LC}\,i_0 = 0$$

The solution of these second order differential equations can be obtained by using initial conditions as specified above. Components constituting the load decide the nature of load current waveforms.

For a full inverter, the rectangular output voltage waveform is shown in Fig. 8.3 (*b*). For this inverter, various current waveforms for different load characteristics are drawn in Fig. 8.3 (*c*) to (*f*). The nature of these current waveforms is briefly discussed in what follows :

R load. For a resistive load R, load current waveform i_0 is identical with load voltage waveform v_0 and diodes D1-D4 do not come into conduction, Fig. 8.3 (*c*).

RL and RLC overdamped loads. The load current waveforms for RL and RLC overdamped loads are shown in Figs. 8.3 (*d*) and (*e*) respectively. Before $t = 0$, thyristors T3,

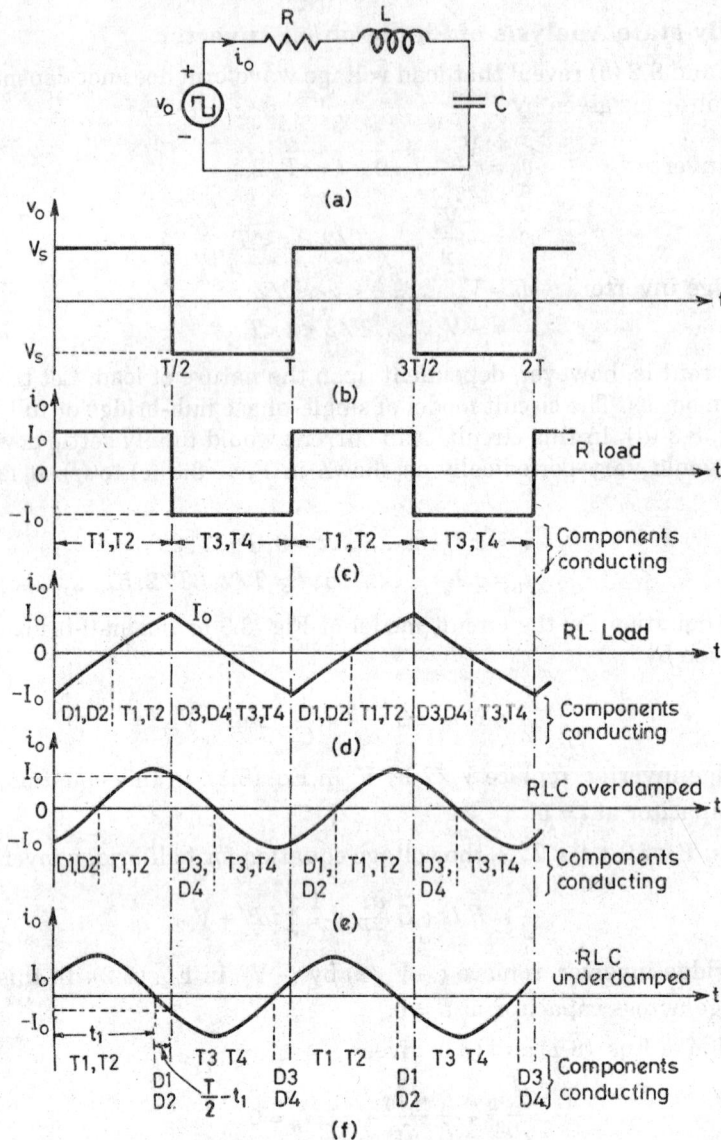

Fig. 8.3. Load voltage and current waveforms for single-phase bridge inverter.

T4 are conducting and load current i_0 is flowing from B to A, *i.e.* in the reversed direction, Fig. 8.2 (a). This current is shown as $-I_0$ at $t = 0$ in Figs. 8.3 (d) and (e). After T3, T4 are turned off at $t = 0$, current i_0 cannot change its direction immediately because of the nature of load. As a result, diodes D1, D2 start conducting after $t = 0$ and allow i_0 to flow against the supply voltage V_s. As soon as D1, D2 begin to conduct, load is subjected to V_s as shown. Though T1, T2 are gated at $t = 0$, these SCRs will not turn on as these are reverse biased by voltage drops across diodes D1 and D2. When load current through $D1$, $D2$ falls to zero, T1 and T2 become forward biased by source voltage V_s, T1 and T2 therefore get turned on as these are gated for a period $T/2$ sec. Now load current i_0 flows in the positive direction from A to B. At $t = T/2$; T1, T2 are turned off by forced commutation and as load current cannot reverse immediately, diodes D3, D4 come into conduction to allow the flow of current i_0 after $T/2$.

Thyristors T3, T4, though gated, will not turn on as these are reverse biased by the voltage drop in diodes D3, D4. When current in diodes D3, D4 drops to zero ; T3, T4 are turned on as these are already gated. The conduction of various components of the full-bridge inverter is shown in Figs. 8.3 (*d*) and (*e*).

RLC underdamped load. The load current i_0 for RLC underdamped load is shown in Fig. 8.3 (*f*). After $t = 0$; T1, T2 are conducting the load current. As i_0 through T1, T2 reduces to zero at t_1, these SCRs are turned off before T3, T4 are gated. As T1, T2 stop conducting, current through the load reverses and is now carried by diodes D1, D2 as T3, T4 are not yet gated. The diodes D1, D2 are connected in antiparallel to T1, T2 ; the voltage drop in these diodes appears as a reverse bias across T1, T2. If duration of this reverse bias is more than the SCR turn-off time t_q, *i.e.* If $(T/2 - t_1) > t_q$; T1, T2 will get commutated naturally and therefore no commutation circuitry will be needed. This method of commutation, knows as *load commutation*, is in fact used in high frequency inverters used for induction heating.

In single-phase bridge inverters shown in Figs. 8.1 (*a*) and 8.2 (*a*), thyristors are shown as switching devices. Note that basic inverter operation is not dependent on the particular semiconductor device used. It means that if *npn* transistors (or GTOs, IGBTs etc) are used as switching devices in place of thyristors as shown in Fig. 8.4, normal inverter operation is obtained. The operating principle of an inverter using transistors, Fig. 8.4, can be described merely replacing T (for thyristor) by TR (for transistor) in Figs. 8.1 (*b*), 8.2 (*b*) and 8.3 (*c* to *f*).

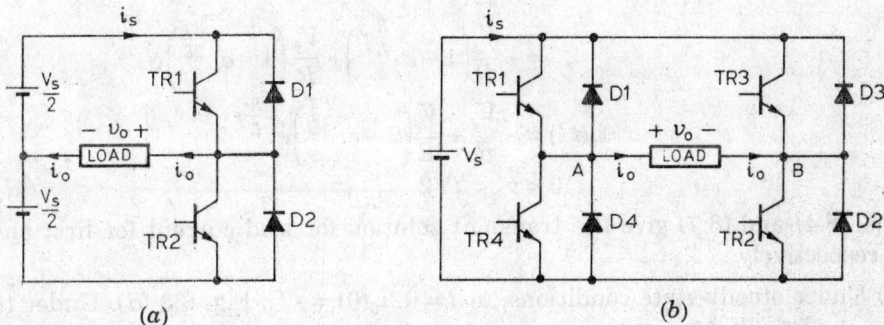

Fig. 8.4. Single-phase (*a*) half-bridge and (*b*) full-bridge inverters using transistors.

Example 8.1. (*a*) *A single-phase full bridge inverter is connected to an RL load. For a dc source voltage of V_s and output frequency $f = 1/T$, obtain expressions for load current as a function of time for the first two half cycles of the output voltage.*

(*b*) *Derive also the expressions for steady-state current for the first two half cycles.*

(*c*) *For $R = 20\ \Omega$ and $L = 0.1\ H$, obtain current expressions for parts (a) and (b) in case source voltage is 240 V dc and frequency of output voltage is 50 Hz.*

Solution. (*a*) For the first half cycle, Fig. 8.3 (*b*), *i.e.* for $0 < t < T/2$, the voltage equation for *RL* load is

$$V_s = Ri_0 + L\frac{di_0}{dt} \qquad \qquad ...(8.3)$$

Its Laplace transform, with zero initial conditions, is

$$\frac{V_s}{s} = R\,I(s) + Ls \cdot I(s) = I(s)\,[R + Ls]$$

Its time solution is,

$$i_0(t) = \frac{V_s}{R}\left(1 - e^{-\frac{R}{L}t}\right) \qquad \qquad ...(8.4)$$

for $\qquad \qquad 0 < t < T/2.$

This is the expression of current as a function of time for the first half cycle from the instant of switching in with $i_0(t) = 0$ at $t = 0$.

At $t = T/2$, current $i_0(t)$ of Eq. (8.4) becomes the initial value for second half cycle.

$$\therefore \qquad i_0(T/2) = \frac{V_s}{R}\left(1 - e^{-\frac{RT}{2L}}\right) \qquad \qquad ...(8.5)$$

For second half cycle, time limit is from $T/2$ to T or $0 < t' < T/2$ where $t' = t - T/2$. The voltage equation for RL load during second half cycle is

$$-V_s = R\,i_0 + L\,\frac{di_0}{dt'} \qquad \qquad ...(8.6)$$

Its Laplace transform, with initial current $i_0\,(T/2)$ given by Eq. (8.5), is

$$-\frac{V_s}{s} = I(s)\,[R + Ls] - i_0(T/2)\cdot L$$

or $$\qquad I(s) = -\frac{V_s}{s(R + Ls)} + \frac{L\cdot i_0(T/2)}{R + Ls}$$

Its time solution is $$\qquad i_0(t') = -\frac{V_s}{R}\left(1 - e^{-\frac{R}{L}t'}\right) + i_0\,(T/2)e^{-\frac{R}{L}t'}$$

$$= -\frac{V_s}{R}\left(1 - e^{-\frac{R}{L}t'}\right) + \frac{V_s}{R}\left(1 - e^{-\frac{RT}{2L}}\right)e^{-\frac{R}{L}t'}$$

or $$\qquad i_0(t') = -\frac{V_s}{R} + \frac{V_s}{R}\left(2 - e^{-\frac{RT}{2L}}\right)e^{-\frac{R}{L}t'} \qquad \qquad ...(8.7)$$

for $$\qquad 0 < t' < T/2.$$

Eqs. (8.4) and (8.7) give the transient solution for load current for first and second half cycles respectively.

(b) Under steady-state conditions, at $t = 0$, $i_0(0) = -I_0$, Fig. 8.3 (d). Under this condition, Laplace transform of Eq. (8.3), is

$$\frac{V_s}{s} = I(s)\,[R + Ls] + L\,I_0$$

Its time solution is $$\qquad i_0(t) = \frac{V_s}{R}\left(1 - e^{-\frac{R}{L}t}\right) - I_0\,e^{-\frac{R}{L}t} \qquad \qquad ...(8.8)$$

At $t = T/2$, $i_0\,(t) = I_0$ Fig. 8.3 (d), therefore from Eq. (8.8)

$$i_0(T/2) = I_0 = \frac{V_s}{R}\left(1 - e^{-\frac{RT}{2L}}\right) - I_0\cdot e^{-\frac{RT}{2L}}$$

or $$\qquad I_0 = \frac{V_s}{R}\cdot\frac{1 - e^{-\frac{RT}{2L}}}{1 + e^{-\frac{RT}{2L}}} \qquad \qquad ...(8.9)$$

Substituting this value of I_0 in Eq. (8.8), gives

$$i_0(t) = \frac{V_s}{R}\left(1 - e^{-\frac{R}{L}t}\right) - \frac{V_s}{R}\frac{1 - e^{-\frac{RT}{2L}}}{1 + e^{-\frac{Rt}{2L}}}e^{-\frac{R}{L}t} \qquad \qquad ...(8.10)$$

Eq. (8.10), gives the steady-state solution during the first half cycle, i.e. for $0 < t < T/2$.

For second half cycle, at $t = T/2$, $i_0(T/2) = I_0$, Fig. 8.3 (d). Under this initial condition, Laplace transform of Eq. (8.6), is

$$-\frac{V_s}{R} = I(s)\,[R + Ls] - L\,I_0$$

Its time solution is
$$i_0(t) = -\frac{V_s}{R}\left(1 - e^{-\frac{R}{L}t'}\right) + I_0\,e^{-\frac{R}{L}t'}$$

$$= -\frac{V_s}{R}\left(1 - e^{-\frac{R}{L}t'}\right) + \frac{V_s}{R}\cdot\frac{1 - e^{-\frac{RT}{2L}}}{1 + e^{-\frac{RT}{2L}}}\cdot e^{-\frac{R}{L}t'} \qquad \text{...(8.11)}$$

Eq. (8.11) gives the steady-state solution during the second half cycle, i.e. for $0 < t' < T/2$ where $t' = t - T/2$.

(c) Here $\dfrac{R}{L} = 200$, $T = \dfrac{1}{f} = \dfrac{1}{50} = 0.02$ sec., $\dfrac{RT}{2L} = 2$.

Expression for transient current during the first half cycle from Eq. (8.4), is

$$i_0(t) = \frac{240}{20}\left(1 - e^{-200\,t}\right) = 12\left(1 - e^{-200\,t}\right)$$

Expression for transient current for second half cycle, from Eq. (8.7), is

$$i_0(t') = -\frac{240}{20} + \frac{240}{20}\left(2 - e^{-2}\right)e^{-200\,t'}$$

$$= -12 + 22.376\,e^{-200\,t'}$$

From Eq. (8.9),
$$I_0 = \frac{240}{20}\cdot\frac{1 - e^{-2}}{1 + e^{-2}} = 9.139 \text{ Amps.}$$

Steady-state current for the first half cycle, i.e. for $0 < t < T/2$, is obtained from Eq. (8.10) as

$$i_0(t) = 12\,(1 - e^{-200\,t}) - 9.139\,e^{-200\,t}$$

$$= 12 - 21.139\,e^{-200\,t} \text{ for } 0 < t' < T/2$$

For the second half cycle, steady state current from Eq. (8.11), is

$$i_0(t') = -12\,(1 - e^{-200\,t'}) + 9.139\,e^{-200\,t'}$$

$$= -12 + 21.139\,e^{-200\,t'} \text{ for } 0 < t' < T/2.$$

Example 8.2. (a) *Repeat parts (a) and (b) of Example 8.1 in case load consists of resistor R and capacitor C in series.*

(b) *For $R = 20\,\Omega$ and $C = 50\,\mu F$, obtain current expressions for part (a) for input voltage of 240 V and output frequency of 500 Hz.*

Solution. (a) For the first half cycle, i.e. for $0 < t < T/2$, voltage equation for RC load is

$$V_s = R\,i_0 + \frac{1}{C}\int i_0\,dt$$

or
$$R\frac{dq}{dt} + \frac{q}{C} = V_s \qquad \text{...(8.12)}$$

Its Laplace transform, with zero initial conditions, is

$$R[s\,Q(s) - q(0)] + \frac{Q(s)}{C} = \frac{V_s}{s}$$

or
$$Q(s) = \frac{V_s}{R}\cdot\frac{1}{s(s + 1/RC)}$$

Its time solution is,
$$q(t) = CV_s\,(1 - e^{-t/RC})$$

or
$$v_c(t) = \frac{q(t)}{C} = V_s\left(1 - e^{-t/RC}\right) \qquad \text{...(8.13)}$$

Also,
$$i_0(t) = C\frac{dv_c(t)}{dt} = \frac{V_s}{R}e^{-t/RC} \qquad \text{...(8.14)}$$

Eqs. (8.13) and (8.14) give the transient solution for $v_c(t)$ across C and $i_0(t)$ through RC load **during** the first half cycle.

At $t = T/2$, $v_c(t)$ of Eq. (8.13) becomes the initial value of the second half cycle,

$$\therefore \qquad v_c(T/2) = V_s\left[1 - e^{-\frac{T}{2RC}}\right] \qquad \text{...(8.15)}$$

Voltage equation for RC load for second half cycle is

$$-V_s = R\,i_0 + \frac{1}{C}\int i_0\,dt'$$

or
$$R\frac{dq}{dt'} + \frac{q}{C} = -V_s \qquad \text{...(8.16)}$$

Its Laplace transform, with initial voltage $v_c(T/2)$ given in Eq. (8.15), is

$$R[s\,Q(s) - C\cdot v_c(T/2)] + \frac{Q(s)}{C} = -\frac{V_s}{s}$$

or
$$Q(s) = -\frac{C\cdot V_s}{s(RCs + 1)} + \frac{C\,v_c(T/2)}{\left(s + \dfrac{1}{RC}\right)}$$

Its time solution is
$$q(t') = -C\,V_s(1 - e^{-t'/RC}) + C\,v_c(T/2)\,e^{-t'/RC}$$

or
$$v_c(t') = -V_s\left(1 - e^{-\frac{t'}{RC}}\right) + V_s\left(1 - e^{-\frac{T}{2RC}}\right)e^{-\frac{t'}{RC}}$$

$$= -V_s + V_s\left(2 - e^{-\frac{T}{2RC}}\right)e^{-\frac{t'}{RC}} \qquad \text{...(8.17)}$$

Also
$$i_0(t') = C\frac{dv_c(t')}{dt} = -\frac{V_s}{R}\left(2 - e^{-\frac{T}{2RC}}\right)e^{-\frac{t'}{RC}} \qquad \text{...(8.18)}$$

Voltage and current solutions in the first half cycle are given by Eqs. (8.13) and (8.14) and in the second half cycle by Eqs. (8.17) and (8.18).

(b) Under steady state working, the waveform for **voltage** v_0 across capacitor and load current i_0 through RC are as shown in Fig. 8.5.

At $t = 0$, $v_c(0) = -V_0$. Therefore, Laplace transform of **Eq.** (8.12) under this condition is

$$R\left[s\,Q(s) + CV_0\right] + \frac{Q(s)}{C} = \frac{V_s}{s}$$

or
$$Q(s)\left[Rs + \frac{1}{C}\right] = \frac{V_s}{s} - CR\,V_0$$

or
$$Q(s) = \frac{V_s}{R}\cdot\frac{1}{s\left(s + \dfrac{1}{RC}\right)} - \frac{CV_0}{\left(s + \dfrac{1}{RC}\right)}$$

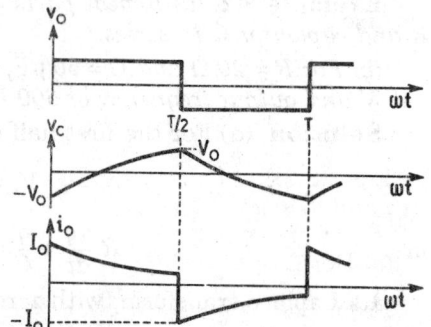

Fig. 8.5. Pertaining to Example 8.2.

Its time solution is $\quad q(t) = CV_s\,(1 - e^{-t/RC}) - C\,V_0\,e^{-t/RC}$

or $\qquad\qquad\qquad\quad v_c(t) = V_s\,(1 - e^{-t/RC}) - V_0\,e^{-t/RC}$ ...(8.19)

At $t = T/2$, $v_c(T/2) = V_0$, Fig. 8.5, therefore, from Eq. (8.19),

$$v_c(T/2) = V_0 = V_s\left(1 - e^{-\frac{T}{2RC}}\right) - V_0\,e^{-\frac{T}{2RC}}$$

or $\qquad\qquad\qquad\qquad V_0 = V_s\,\dfrac{1 - e^{-\frac{T}{2RC}}}{1 + e^{-\frac{T}{2RC}}}$ ...(8.20)

From Eq. (8.19), $\quad v_c(t) = V_s\left(1 - e^{-\frac{T}{RC}}\right) - V_s\,\dfrac{1 - e^{-\frac{T}{2RC}}}{1 + e^{-\frac{T}{2RC}}}\cdot e^{-\frac{t}{RC}}$ ...(8.21)

From above equation, $\quad i_0(t) = C\,\dfrac{dv_c(t)}{dt} = \dfrac{2\,V_s}{R}\cdot\dfrac{e^{-\frac{t}{RC}}}{1 + e^{-\frac{T}{2RC}}}$ ...(8.22)

Eqs. (8.21) and (8.22) give the steady-state solution for $v_c(t)$ and load current $i_0(t)$ for the first half cycle, i.e. for $0 < t < T/2$.

For second half cycle, at $t = T/2$, $v_c(T/2) = V_0$. Under this condition, Laplace transform of Eq. (8.16) is

$$R[s\,Q(s) - CV_0] + \frac{Q(s)}{C} = -\frac{V_s}{s}$$

or $\qquad\qquad\qquad Q(s) = -\dfrac{V_s}{R}\cdot\dfrac{1}{s\left(s + \dfrac{1}{RC}\right)} + \dfrac{CV_0}{s + \dfrac{1}{RC}}$

Its time solution is $\quad q(t') = -CV_s\,(1 - e^{-t'/RC}) + CV_0\,e^{-t'/RC}$

or $\qquad\qquad\qquad v_c(t') = V_s\,(1 - e^{-t'/RC}) + V_0\,e^{-t'/RC}$

$$= -V_s\left(1 - e^{-\frac{t'}{RC}}\right) + V_s\,\dfrac{1 - e^{-\frac{T}{2RC}}}{1 + e^{-\frac{T}{2RC}}}\,e^{-\frac{t'}{RC}}$$ ...(8.23)

Also $\qquad\qquad\qquad i_0(t') = C\,\dfrac{d\,v_c(t')}{dt}$

$$= -\dfrac{2V_s}{R}\,\dfrac{e^{-\frac{t'}{RC}}}{1 + e^{-\frac{T}{RC}}}$$ ...(8.24)

Eqs. (8.23) and (8.24) give respectively the steady-state solution for capacitor voltage $v_c(t')$ and load current $i_0(t')$ for second half cycle, i.e. for $0 < t' < T/2$.

(c) Here $R = 20\ \Omega$. $C = 50\ \mu F$. $\dfrac{1}{RC} = 1000$, $T = \dfrac{1}{f} = 0.002$

From Eqs. (8.13) and (8.14), the transient expressions for the first half cycle are

$$v_c(t) = 240\,(1 - e^{-1000\,t})$$

and

$$i_0(t) = 12\,e^{-1000\,t})$$

From Eqs. (8.17) and (8.18), the transient expressions for the second half cycle are

$$v_c(t') = -240 + 240\,(2 - e^{-2})\,e^{-1000\,t'}$$

$$= -240 + 512\cdot 48\,e^{-1000\,t'}$$

and $\qquad i_0(t') = -12 (2 - e^{-2}) e^{-1000 t'} = -22.376\, e^{-1000 t'}$

From Eq. (8.20), $\qquad V_0 = 240\, \dfrac{1 - e^{-1}}{1 + e^{-1}} = 110.91$ volts.

For the first half cycle, steady state voltage and current, from Eqs. (8.21) and (8.22) are

$$v_c(t) = 240 (1 - e^{-1000 t}) - 110.91\, e^{-1000 t} = 240 - 350.91\, e^{-1000 t}$$

and $\qquad i_0(t) = \dfrac{2 \times 240}{20} \cdot \dfrac{e^{-1000 t}}{1 + e^{-1}} = 17.545\, e^{-1000 t}$

For the second half cycle, steady state voltage and current, from Eqs. (8.23) and (8.24) are

$$v_c(t') = -240 (1 - e^{-1000 t'}) + 110.91\, e^{-1000 t'} = -240 + 350.91\, e^{-1000 t'}$$

and $\qquad i_0(t') = -17.545\, e^{-2000 t'}$

Example 8.3. *A single-phase bridge inverter delivers power to a series connected RLC load with R = 2 Ω and ωL = 10 Ω. The periodic time T = 0.1 msec. What value of C should the load have in order to obtain load commutation for the SCRs. The thyristor turn-off time is 10 μsec. Take circuit turn off time as 1.5 t_q. Assume that load current contains only fundamental component.*

Solution. The value of C should be such that RLC load is underdamped. Moreover, when load voltage passes through zero, the load current must pass through zero before the voltage wave, *i.e.* the load current must lead the load voltage by an angle θ as shown in Fig. 8.6. Recall the phasor diagram for RLC series circuit. From this phasor diagram,

$$\tan \theta = \frac{X_C - X_L}{R}$$

Here $X_C > X_L$ as the current is leading the voltage. Now (θ/ω) must be at least equal to circuit turn-off time, *i.e.* $1.5 \times 10 = 15$ μsec.

Fig. 8.6. Pertaining to Example 8.3.

$$\therefore \quad \frac{\theta}{\omega} = 15 \times 10^{-6} \text{ sec}$$

Now $\qquad f = \dfrac{10^3}{0.1} = 10^4 \text{ Hz}$

$\therefore \qquad \theta = 2\pi \times 10^4 \times 15 \times 10^{-6} = 0.9424778 \text{ rad} = 54°$

$\therefore \qquad \tan 54° = \dfrac{X_C - 10}{2}$

or $\qquad X_C = 12.752764 = \dfrac{1}{2\pi \times 10^4 \times C}$

or $\qquad C = 1.248 \text{ μF.}$

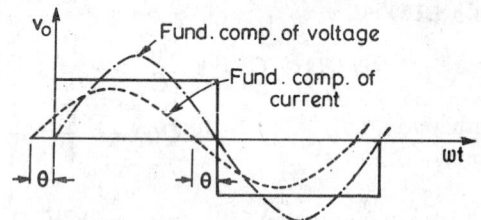

8.2. FOURIER ANALYSIS OF SINGLE-PHASE INVERTER OUTPUT VOLTAGE

The output voltage v_0 is shown in Fig. 8.1 (b) for a single-phase half-bridge inverter and in Fig. 8.2 (b) for a single-phase full-bridge inverter. These output or load voltage waveforms do not depend on the nature of load. Voltage waveshapes of Figs. 8.1 (b) and 8.2 (b) can be resolved into Fourier series as under :

$$v_0 = \sum_{n = 1, 3, 5, \dots}^{\infty} \frac{2V_s}{n\pi} \sin n\omega t \text{ volts} \qquad \dots(8.25)$$

for single-phase half-bridge inverter and

$$v_0 = \sum_{n = 1, 3, 5, \ldots}^{\infty} \frac{4\,V_s}{n\pi} \sin n\omega t \text{ volts} \qquad \ldots(8.26)$$

for single-phase full-bridge inverter.

Here n is the order of the harmonic and $\omega = 2\pi f$ is the frequency of the output voltage in rad/s.

The load current i_0 can, therefore, be expressed as

$$i_0 = \sum_{n = 1, 3, 5, \ldots}^{\infty} \frac{4\,V_s}{n\pi \cdot Z_n} \sin(n\omega t - \phi_n) \text{ Amps} \qquad \ldots(8.27)$$

where Z_n = load impedance at frequency $n.f$

$$= \left[R^2 + \left(n\omega L - \frac{1}{n\omega C} \right)^2 \right]^{1/2} \qquad \ldots(8.28)$$

and phase angle ϕ_n is

$$\phi_n = \tan^{-1} \frac{\left[n\omega L - \dfrac{1}{n\omega C} \right]}{R} \text{ rad} \qquad \ldots(8.29)$$

The output, or load, current at the instant of commutation is obtained from Eq. (8.27) by putting $\omega t = \pi$. Its value is

$$i_0 = I_0 \text{ at } \omega t = \pi \text{ rad}$$

In case $I_0 > 0$, forced commutation is essential. If $I_0 < 0$, no forced commutation is required and load commutation, as described for RLC underdamped load in Art. 8.1.2, can be relied upon.

If I_{01} = rms value of the fundamental component of load current, then the fundamental load power P_{01} is given by

$$P_{01} = I_{01}^2 R = V_{01} I_{01} \cos \phi_1$$

where V_{01} = rms value of fundamental output voltage.

The fundamental output power P_{01} does the useful work in most of the applications (*e.g.* electric motor drives). The output power associated with harmonic current does no useful work and is dissipated as heat leading to rise in load temperature.

Example 8.4. *A single-phase half-bridge inverter has load $R = 2\Omega$. and dc source voltage*
$$\frac{V_s}{2} = 115\ V$$

(a) Sketch the waveforms for v_0, load current i_{01}, currents through thyristor 1 and diode 1 and voltage across thyristor T1. Harmonics other than fundamental component are neglected. Indicate the devices that conduct during different intervals of one cycle.

(b) Find the power delivered to load due to fundamental current.

(c) Check whether forced commutation is required.

Solution. (a) The fundamental component of output voltage, from Eq. (8.25), is

$$v_{01} = \frac{2\,V_s}{\pi} \sin \omega t$$

The rms value of this voltage, $V_{01} = \dfrac{2 \times 230}{\pi \cdot \sqrt{2}} = 103.552\ V$

and the load current,

$$I_{01} = \frac{V_{01}}{R} = \frac{103.552}{2} = 51.776 \text{ A}$$

The fundamental frequency component of load current is

$$i_{01} = 51.776 \sqrt{2} \sin \omega t$$

The waveforms for the various voltages and currents are shown in Fig. 8.7. For resistive load, diodes do not come into conduction, therefore i_{D1} is zero. When T1 conducts, $v_{T1} = 0$. When T2 conducts, $v_{T1} = V_s$ as shown.

(b) Power delivered to load

$$= I_{01}^2 R = (51.776)^2 \times 2 = 5361.5 \text{ watts}$$

When T1 is conducting, power to load is delivered by upper source $\frac{V_s}{2}$ and when T2 is on, lower source delivers power to load.

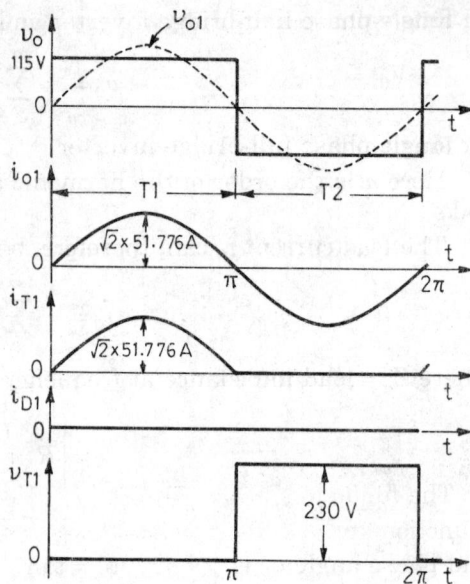

Fig. 8.7. Pertaining to Example 8.4.

Power delivered by each source $\qquad = \frac{V_s}{2} \cdot I_s$

Here I_s = average value of fundamental component of source current over one cycle.

$$= \frac{1}{2\pi} \int_0^\pi \sqrt{2} \, I_{01} \sin \omega t \cdot d\,(\omega t) = \frac{\sqrt{2} \, I_{01}}{\pi}$$

$$= \frac{\sqrt{2} \times 51.776}{\pi} = 23.304 \text{ A}$$

Power delivered by each source $= 115 \times 23.304 = 2679.96$ watts

Power delivered by both the sources $= 2 \times 2679.96 \cong 5360$ W

Power delivered by both the sources is equal to that consumed by the load.

(c) As the diodes do not conduct, forced commutation is essential.

Example 8.5. *For a single-phase full-bridge inverter, $V_s = 230$ V dc, $T = 1$ ms. The load consists of RLC in series with $R = 1\ \Omega$, $\omega L = 6\ \Omega$ and $\frac{1}{\omega C} = 7\ \Omega$.*

(a) Sketch the waveforms for load voltage v_0, fundamental component of load current i_{01}, source current i_s and voltage across thyristor 1. Indicate the devices under conduction during different intervals of one cycle.

(b) Find the power delivered to load due to fundamental component.

(c) Check whether forced commutation is required or not. Take thyristor turn-off time as 100 μs.

Solution. (a) The load voltage waveform v_0 and its fundamental component v_{01} are shown in Fig. 8.8.

Rms value of load voltage, from Eq. (8.26), is

$$V_{01} = \frac{4\,V_s}{\pi\,\sqrt{2}} = \frac{4 \times 230}{\pi\,\sqrt{2}} = 207.1 \text{ V}$$

Rms value of current, $I_{01} = \dfrac{V_{01}}{Z_1}$

$$= \frac{V_{01}}{\left[R^2 + \left(\omega L - \dfrac{1}{\omega C}\right)^2\right]^{1/2}}$$

$$= \frac{207.1}{\left[1^2 + (-1)^2\right]^{1/2}} = \frac{207.1}{\sqrt{2}} = 146.46 \text{ A}$$

$$\phi_1 = \tan^{-1}\frac{X_L - X_C}{R} = \tan^{-1}(-1) = -45°$$

The fundamental component of current i_{01} as a function of time is

$$i_{01} = \sqrt{2}\,I_{01} \sin(\omega t - \phi_1)$$

$$= \sqrt{2}\,\frac{207.1}{\sqrt{2}} \sin(\omega t + 45°)$$

$$= 207.1 \sin(\omega t + 45°)$$

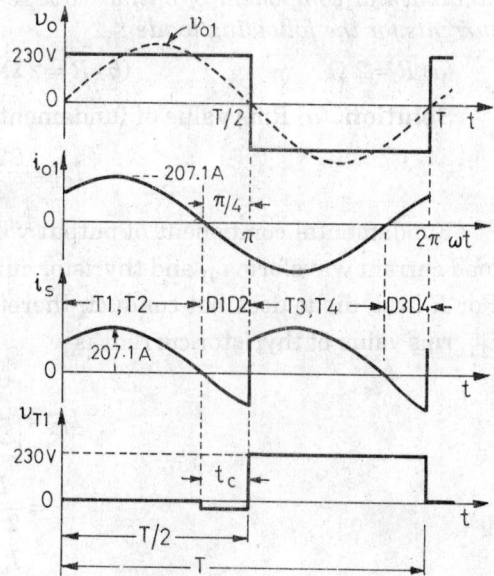

Fig. 8.8. Pertaining to Example 8.5.

Load current i_{01} and source current i_s are plotted in Fig. 8.8 and the conducting components are also indicated.

(b) Power delivered to load $= I_{01}^2\,R = \left(\dfrac{207.1}{\sqrt{2}}\right)^2 \times 1 = 21.445 \text{ kW}$

This must be equal to the power P_s delivered by the source.

$$\therefore \qquad P_s = V_s\,I_s \text{ watts}$$

where I_s = average value of the fundamental component of source current

$$= \frac{1}{\pi} \int_0^{\pi} \sqrt{2}\,I_{01} \sin(\omega t + 45°)\,d(\omega t)$$

$$= \frac{207.1}{\pi} \left[-\cos(\omega t + 45°)\right]_0^{\pi} = \frac{207.1}{\pi} [2 \cos 45°] = 93.23 \text{ A}$$

$$\therefore \qquad P_s = 230 \times 93.23 = 21.443 \text{ kW}$$

(c) Fig. 8.8 reveals that v_{T1} is negative for some time before T3, T4 are triggered. Thus circuit turn-off time can be obtained from

or

$$\omega\,t_c = \frac{\pi}{4}$$

$$t_c = \frac{1}{4} \cdot \frac{T}{2} = 0.125 \text{ ms} = 125 \text{ μs}$$

As voltage drop in diodes D1, D2 reverse biases T1, T2 for 125 μs, which is more than the thyristor turn-off time of 100 μs, no forced commutation is required.

Example 8.6. *A single-phase full-bridge inverter is fed from a dc source such that fundamental component of output voltage is 230 V. Find the rms value of thyristor and diode currents for the following loads :*

(a) $R = 2\,\Omega$ (b) $R = 2\,\Omega,\ X_L = 8\,\Omega,\ X_C = 6\,\Omega.$

Solution. (a) Rms value of fundamental component of load current

$$I_{01} = \frac{230}{2} = 115\,\text{A}$$

Fundamental component of output voltage v_{01} is shown in Fig. 8.9 (a). For resistive load, load current waveform i_{01} and thyristor current i_{T1} handled by T1 are also shown in Fig. 8.9 (a). For R load, diode does not conduct, therefore diode current i_{D1} is zero. From the waveform of i_{T1}, rms value of thyristor current is

$$I_{T1} = \left[\frac{1}{2\pi} \int_0^\pi (I_m \sin \omega t)^2 \cdot d\,(\omega t) \right]^{1/2}$$

$$= \frac{I_m}{2\sqrt{\pi}} \left[\int_0^\pi (1 - \cos 2\,\omega t) \cdot d\,(\omega t) \right]^{1/2}$$

$$= \frac{I_m}{2} = \frac{115\sqrt{2}}{2} = 81.33\,\text{A}$$

Rms value of diode current, $I_{D1} = 0$

(b) Rms value of load current, $I_{01} = \dfrac{230}{[2^2 + (8-6)^2]^{1/2}} = 81.317\,\text{A}$

Phase angle, $\phi_1 = \tan^{-1} \dfrac{X_L - X_C}{R} = 45°$

For $R = 2\,\Omega, X_L = 8\,\Omega$ and $X_C = 6\,\Omega$, the fundamental component of load current lags the output voltage by 45°, this is shown in Fig. 8.9 (b). The thyristor-current waveform i_{T1} and

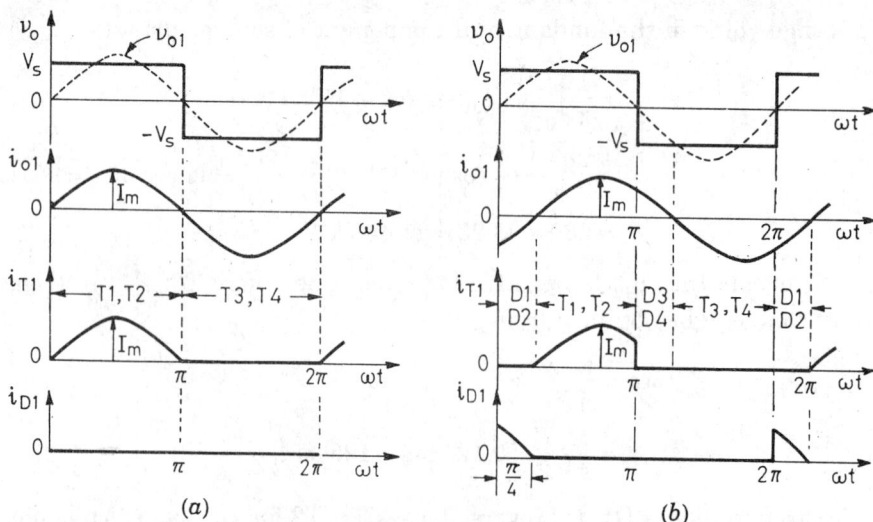

(a)   (b)

Fig. 8.9. Pertaining to Example 8.6.

diode-current waveform i_{D1} are also shown in Fig. 8.9 (b). It is observed from this figure that rms value of thyristor current is

$$I_{T1} = \left[\frac{1}{2\pi} \int_0^{3\pi/4} (I_m \sin \omega t)^2 \cdot d\,(\omega t) \right]^{1/2}$$

$$= \frac{I_m}{2\sqrt{\pi}} \left[\left| \omega t - \frac{\sin 2\,\omega t}{2} \right|_0^{\frac{3\pi}{4}} \right]^{1/2} = 0.47675\,I_m$$

$$= 0.47675 \times \sqrt{2} \times 81.317 = 54.818 \text{ A}$$

Rms value of diode current, $I_{D1} = \left[\frac{1}{2\pi} \int_0^{\pi/4} (I_m \sin \omega t)^2 \cdot d\,(\omega t) \right]^{1/2}$

$$= 0.1507025\,I_m = 0.1507025 \times \sqrt{2} \times 81.317 = 17.328 \text{ A}$$

As the load current i_0 does not change from positive to negative at an angle $\omega t < \pi$, no time is available for SCR to turn off ; forced commutation is therefore essential.

Example 8.7. *A single-phase full-bridge inverter has RLC load of R = 4 Ω, L = 35 mH and C = 155 μF. The dc input voltage is 230 V and the output frequency is 50 Hz.*

(a) Find an expression for load current up to fifth harmonic. Also, calculate

(b) rms value of fundamental load current,

(c) the power absorbed by load and the fundamental power,

(d) the rms and peak currents of each thyristor,

(e) conduction time of thyristors and diodes if only fundamental component is considered.

Solution. (a) From Eq. (8.26), an expression for output voltage is

$$v_0 = \frac{4V_s}{\pi} \sin \omega t + \frac{4V_s}{3\pi} \sin 3\,\omega t + \frac{4V_s}{5\pi} \sin 5\,\omega t$$

$$= \frac{4 \times 230}{\pi} \left[\sin \omega t + \frac{1}{3} \sin 3\omega t + \frac{1}{5} \sin 5\,\omega t \right]$$

$$= 292.85 \sin 314t + 97.62 \sin (3 \times 314t) + 58.57 \sin (5 \times 314t)$$

Load impedance at frequency $n.f$ is

$$Z_n = 4 + j \left(2\pi \times 50 \times 35 \times 10^{-3} \times n - \frac{10^6}{2\pi \times 50 \times 155 \times n} \right)$$

$$= 4 + j \left(11n - \frac{20.54}{n} \right) \Omega$$

$$\therefore \qquad Z_1 = \sqrt{4^2 + (11 - 20.54)^2} = 10.345 \ \Omega$$

and $\qquad \phi_1 = \tan^{-1} \left(\frac{11 - 20.54}{4} \right) = -67.25°$

$$Z_3 = \sqrt{4^2 + \left(11 \times 3 - \frac{20.54}{3} \right)^2} = 26.46 \ \Omega$$

and $\qquad \phi_3 = \tan^{-1} \left[\frac{33 - 20.54/3}{4} \right] = 81.3°$

Similarly, $\qquad Z_5 = 51.05 \ \Omega$ and $\phi_5 = 85.5°$

Load current, from Eq. (8.27), is given by

$$i_0 = \frac{292.85}{10.345} \sin(\omega t + 67.25°) + \frac{97.62}{26.46} \sin(3\omega t - 81.3°)$$

$$+ \frac{58.57}{51.05} \sin(5\omega t - 85.5°)$$

$$= 28.31 \sin(314t + 67.25°) + 3.689 \sin(3 \times 314t - 81.3°)$$

$$+ 1.1473 \sin(5 \times 314t - 85.5°)$$

(b) Rms value of fundamental load current,

$$I_{01} = \frac{I_{m1}}{\sqrt{2}} = \frac{28.31}{\sqrt{2}} = 20.02 \text{ A}$$

(c) Peak load current

$$I_m = \sqrt{28.31^2 + 3.689^2 + 1.1473^2} = 28.572 \text{ A}$$

Rms load current $= \dfrac{28.572}{\sqrt{2}} = 20.207$ A

Load power $= (20.207)^2 \times 4 = 1633.3$ W

Fundamental load power, $P_{01} = I_{01}^2 \times R = (20.02)^2 \times 4 = 1603.2$ W

(d) Peak thyristor current $= I_m = 28.572$ A

Rms value of thyristor current

$$= \frac{28.572}{2} = 14.286 \text{ A}$$

(e) Fundamental component of current is

$$i_{01} = 28.31 \sin(314t + 67.25°)$$

This current leads the fundamental voltage component by 67.25°. This means that diode conducts for 67.25° and thyristor for $180 - 67.25 = 112.75°$

∴ Conduction time for thyristors

$$= \frac{112.75 \times \pi}{180 \times 314} = 6.267 \text{ ms}$$

Conduction time for diodes $= \dfrac{67.25 \times \pi}{180 \times 314} = 3.738$ ms,

In case SCR turn-off time is less than 3.738 ms, load commutation will occur and no forced commutation will be required under the assumption of no harmonics.

8.3. FORCE-COMMUTATED THYRISTOR INVERTERS

For low- and medium-power applications, inverters using transistors, GTOs and IGBTs are becoming increasingly popular. However, for high-voltage and high-current applications, thyristors are more suitable.

In voltage fed inverters, thyristors remain forward biased by the dc supply voltage. This entails the use of forced commutation for inverter circuits using thyristors. As stated earlier, forced commutation requires a precharged capacitor of correct polarity to turn-off an already conducting thyristor. A large variety of forced commutation circuits have been described in the technical literature. Here popularly used McMurray technique will be described leading to the discussion of two types of force-commutated inverters, *viz* modified McMurray inverter and McMurray-Bedford inverter. These are now described in what follows :

8.3.1. Modified McMurray Half-bridge Inverter

Load commutated voltage-source inverter has been discussed in the latter part of the Section 8.1.2. It is shown there that for obtaining load commutation in VSI, the load circuit must be underdamped, *i.e.* capacitive reactance of the load must be more than its inductive reactance.

The object of this section is to describe modified McMurray half-bridge inverter which is a current-commutated VSI.

Fig. 8.10 shows a single-phase modified McMurray half-bridge inverter. It consists of main thyristors T1, T2 and main diodes D1, D2. The commutation circuit consists of auxiliary thyristors TA1, TA2, auxiliary diodes DA1, DA2 ; damping resistor R_d, inductor L and capacitor C. Three-wire dc supply is required and ac load is connected between terminals A and B as shown in Fig. 8.10. The function of capacitor is to provide the energy required for commutating the main thyristors. Inductance L_1 is for limiting di/dt to a safe value in main and auxiliary thyristors. As an auxiliary thyristor is used for commutating the main thyristor, this inverter is also known as *auxiliary-commutated inverter.*

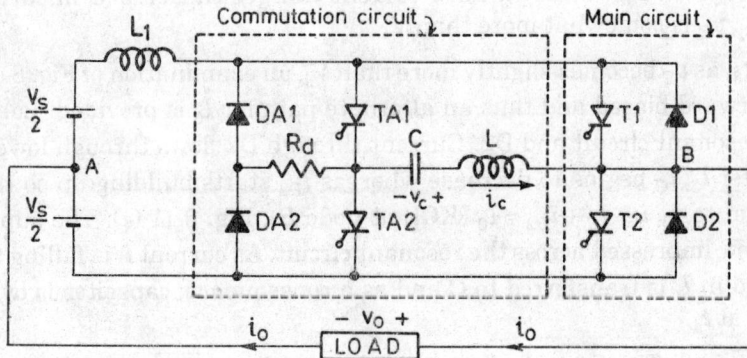

Fig. 8.10. Power circuit diagram for a single-phase modified McMurray half-bridge inverter.

In the original inverter circuit given by McMurray, elements DA1, DA2 and R_d were not present, hence the circuit of Fig. 8.10 is now commonly known as the modified McMurray inverter. As three-wire dc supply is required, as in Fig. 8.1, the term "half-bridge" is added to this inverter of Fig. 8.10.

The following simplifying assumptions are made for this inverter :

(*i*) Load current remains constant during the commutation interval.

(*ii*) SCRs and diodes are ideal switches.

(*iii*) Inductor L and capacitor C are ideal in that they have no resistance.

The operation of this inverter of Fig. 8.10 may be subdivided into various modes as follows:

Mode I : Thyristor T1 is conducting a constant load current I_0, *i.e.* $i_{T1} = I_0$. Capacitor C is already charged to a voltage V_s with right hand plate positive because of the commutation of previously conducting thyristor T2. In this mode, the equivalent circuit of this inverter is as shown in Fig. 8.11 (*a*). With T1 conducting, commutation circuit is passive in this mode.

Mode II. Auxiliary thyristor TA1 is triggered at $t = 0$ to turn off the main thyristor T1. As TA1 is fired, capacitor current i_c starts building up through resonant circuit consisting of L, T1, TA1 and C. Voltage drop across T1 reverse biases D1, current i_c can therefore flow only through T1 and not through D1. As I_0 is constant, an increase in i_c causes a corresponding decrease in i_{T1} so that $i_{T1} = I_0 - i_c$ (KCL at node B). Waveforms of i_c, i_{T1}, I_0 and v_c are shown in Fig. 8.12 and equivalent circuit is as shown in Fig. 8.11 (b). At t_1, i_c rises to I_0 and therefore $i_{T1} = 0$. As a result, main thyristor T1 is turned off at t_1.

Mode III. After t_1, as resonant current i_c exceeds I_0, the excessive current $i_c - I_0 = i_{D1}$ circulates through feedback diode D1, Fig. 8.11 (c). The voltage drop across D1 reverse biases T1 to bring it to forward blocking capability. When capacitor voltage v_c discharges to zero, resonant current i_c rises to peak value I_{cp} as shown in Fig. 8.12. After attaining I_{cp}, i_c begins to decrease and in so doing, C begins to get charged in the reverse direction. At t_2, i_c falls to I_0. In case v_C is somewhat more than source voltage V_s at t_2, diode D2 gets forward biased and starts conducting. In such a case, mode IV is absent ; otherwise mode IV follows.

Mode IV. After t_2, as i_C tends to fall below I_0, diode current i_{D1} becomes zero and D1, therefore, stops conducting. Constant load current I_0 continues flowing through $V_s/2$, TA1, C, L and load as shown in Fig. 8.11 (d). Load current charges capacitor C linearly with reverse polarity and at t_3, v_C is somewhat more than V_s.

Mode V. At t_3, as v_C becomes slightly more than V_s, an examination of Fig. 8.10 reveals that diode D2 gets forward biased and thus an alternate path for I_0 is provided. Load current I_0 is now shared by resonant circuit and D2. Current through D2 flows through lower source $V_s/2$, D2 and load. After t_3, i_C begins to decrease whereas i_{D2} starts building up so that the sum of i_C and i_{D2} is equal to I_0, i.e. $i_C + i_{D2} = I_0$ (KCL at node B), Fig. 8.11 (e). The supply voltage V_s, through D2, is now impressed across the resonant circuit. As current i_c is falling from I_0 to zero, the energy stored in L is transferred to C and as a consequence, capacitor is overcharged to a peak voltage V_m at t_4.

The main thyristor T2 is usually given the trigger pulse $\pi \sqrt{LC}$ seconds after thyristor TA1 is fired, i.e. between the interval t_3 and t_4. But T2 will not turn on because of the reverse bias applied to it by the voltage drop in D2.

Mode VI. At t_4, i_{D2} rises to I_0 and at the same time i_C falls to zero. As i_C tends to reverse, TA1 is turned off at t_4. Now $v_c > V_s$, capacitor C therefore discharges through R_d, DA1, source voltage V_s, D2, L and C, Fig. 8.11 (f). Note that current i_C is now negative as it is flowing opposite to its positive direction. For a constant load current I_0, KCL at node B gives $i_{D2} = i_C + I_0$. During this mode, i_{D2} is more than I_0.

Here the question arises : how can diode current i_{D2} in D2 be established and deliver load current I_0 when source voltage V_s opposes i_{D2} ? It is the overcharging of capacitor C to a voltage higher than V_s that causes the establishment of diode current i_{D2}.

The circuit traced by i_C is usually critically damped so that v_C gradually reduces to V_s. At t_5, i_C becomes zero, $v_c = -V_s$ and $i_{D2} = I_0$. This is the end of mode VI and also the end of commutation interval.

The voltage drop across R_d and DA1 applies a reverse bias across TA1 and completes its commutation process.

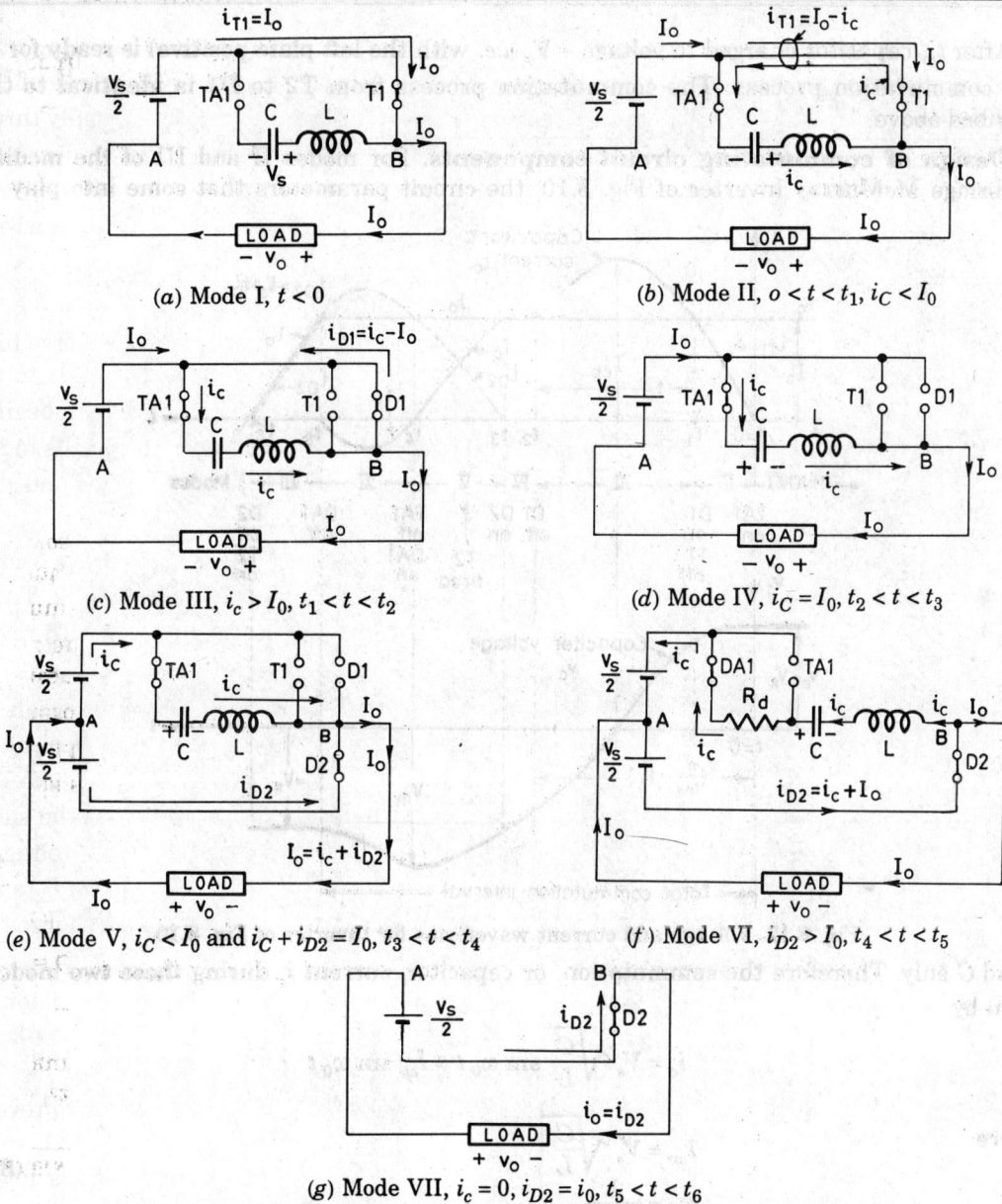

(a) Mode I, $t < 0$

(b) Mode II, $o < t < t_1$, $i_C < I_0$

(c) Mode III, $i_c > I_0$, $t_1 < t < t_2$

(d) Mode IV, $i_C = I_0$, $t_2 < t < t_3$

(e) Mode V, $i_C < I_0$ and $i_C + i_{D2} = I_0$, $t_3 < t < t_4$

(f) Mode VI, $i_{D2} > I_0$, $t_4 < t < t_5$

(g) Mode VII, $i_c = 0$, $i_{D2} = i_0$, $t_5 < t < t_6$

Fig. 8.11. Operating modes of modified McMurray inverter of Fig. 8.10.

Mode VII. After t_5, the circuit model is as shown in Fig. 8.11 (g). The decreasing load current $i_0 = i_{D2}$ becomes zero at t_6. Main thyristor T2 is already gated on during $t_4 - t_3$ interval, i.e. $\pi \sqrt{LC}$ seconds after TA1 is fired. But it will not get turned on at this moment because of the reverse bias applied to it by voltage drop in D2 due to current i_{D2}. At t_6, $i_{D2} = 0$ and T2 is no longer reverse biased. Therefore, after t_6, source voltage applies a forward bias across T2 and the trigger pulse already applied to it turns it on. Load is now subjected to negative current as desired. Note that load was already subjected to negative voltage through D2 at t_3 at the commencement of its conduction.

After t_6, capacitor charged to voltage $-V_s$ (i.e. with the left plate positive) is ready for the next commutation process. The commutation process from T2 to D1 is identical to that described above.

Design of commutating circuit components. For modes II and III of the modified half-bridge McMurray inverter of Fig. 8.10, the circuit parameters that come into play are

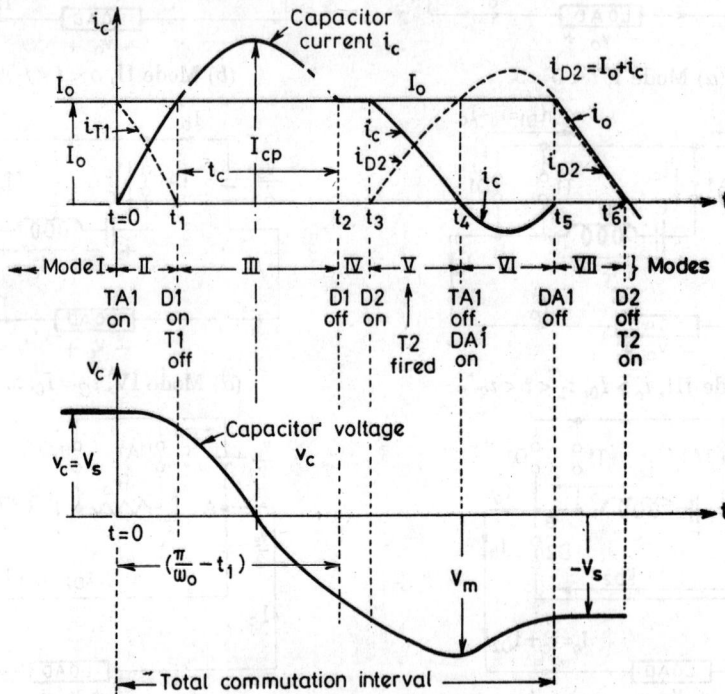

Fig. 8.12. Voltage and current waveforms for inverter of Fig. 8.10.

L and C only. Therefore the commutation, or capacitor, current i_c during these two modes is given by

$$i_c = V_s \sqrt{\frac{C}{L}} \, \sin \omega_0 t = I_{cp} \sin \omega_0 t$$

where

$$\left. \begin{array}{l} I_{cp} = V_s \sqrt{\dfrac{C}{L}} \\[2ex] \omega_0 = \dfrac{1}{\sqrt{LC}} \end{array} \right\} \qquad \qquad \text{...(8.30)}$$

and

From Fig. 8.12, at $t_1, i_c = I_0 = I_{cp} \sin \omega_0 t_1$

or

$$t_1 = \frac{1}{\omega_0} \sin^{-1}\left(\frac{I_0}{I_{cp}}\right) \qquad \qquad \text{...(8.31)}$$

Fig. 8.12 also reveals that

$$\omega_0 t_2 = \pi - \omega_0 t_1 = \pi - \sin^{-1}(I_0/I_{cp})$$

or

$$t_2 = 1/\omega_0 \, [\pi - \sin^{-1}(I_0/I_{cp})]$$

Therefore, circuit turn-off time for main thyristor T1 is

$$t_c = t_2 - t_1 = \frac{1}{\omega_0} [\pi - 2 \sin^{-1}(I_0/I_{cp})] \qquad \qquad \text{...(8.32)}$$

This time t_c must be greater than the thyristor turn-off time t_q. In practice, this condition can be realized by several different combinations of C and L as shown in Fig. 8.13 (a). The current commutation pulse i_c that gives the required turn-off time with the minimum amount of capacitor energy $\frac{1}{2} CV^2$, is the optimum pulse. This can be achieved from Eqs. (8.31) and (8.32) as under.

From Eq. (8.32),
$$\frac{\omega_0 t_c}{2} = \frac{\pi}{2} - \sin^{-1}\left(\frac{I_0}{I_{cp}}\right)$$

or
$$\frac{I_0}{I_{cp}} = \sin\left(\frac{\pi}{2} - \frac{\omega_0 t_c}{2}\right) = \cos\frac{\omega_0 t_c}{2}$$

Let $\frac{I_{cp}}{I_0} = x$ so that
$$\cos\frac{\omega_0 t_c}{2} = \frac{1}{x} \qquad \qquad ...(8.33)$$

or
$$\frac{t_c}{\sqrt{LC}} = 2\cos^{-1}(1/x) = g(x) \qquad \qquad ...(8.34)$$

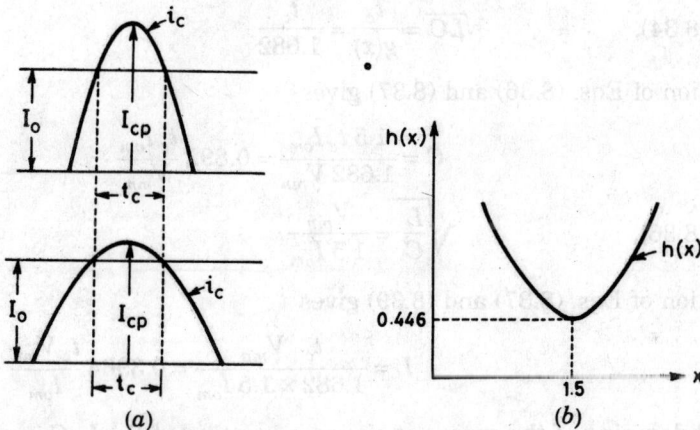

Fig. 8.13. Pertaining to the choice of L and C for the inverter circuit of Fig. 8.10.

The energy W that the commutating capacitor should provide for commutating the main thyristor is

$$W = \frac{1}{2} CV_s^2 = \frac{1}{2} LI_0^2$$

Substituting the value of $V_s = I_{cp}\sqrt{\dfrac{L}{C}}$ from Eq. (8.30) in the above relation, we get

$$W = \frac{1}{2} CV_s \cdot I_{cp}\sqrt{\frac{L}{C}} = \frac{1}{2}\sqrt{LC} \cdot V_s I_{cp}$$

From Eq. (8.34), $\sqrt{LC} = \dfrac{t_c}{2\cos^{-1}(1/x)}$. Substituting this value of $\sqrt{LC}$ in the above relation, we get

$$W = \frac{1}{2} \frac{t_c \cdot V_s \cdot I_{cp}}{2\cos^{-1}(1/x)} = \frac{t_c \cdot V_s \cdot x I_0}{4\cos^{-1}(1/x)}$$

Here the product $V_s I_0 t_c$ has the dimensions of energy. In normalized form, the above relation is

$$\frac{W}{V_s I_0 t_c} = \frac{x}{4 \cos^{-1}(1/x)} = h(x) \qquad \qquad \text{...(8.35)}$$

On plotting $h(x)$ against x from Eq. (8.35), it is found that normalized commutation energy $h(x)$ has a minimum value of 0.446 when $x = 1.5$, Fig. 8.13 (b)

From Eq. (8.34), $g(x) = 2 \cos^{-1}(1/1.5) = 1.682$

The design is carried out on the basis of worst operating conditions which consist of minimum supply voltage V_{mn} and maximum load current I_{om}.

∴ From Eq. (8.30), $V_{mn} \sqrt{\dfrac{C}{L}} = I_{cp} = x I_{om} = 1.5 I_{om}$

or $\sqrt{\dfrac{C}{L}} = \dfrac{1.5 I_{om}}{V_{mn}}$ $\qquad \qquad \text{...(8.36)}$

From Eq. (8.34), $\sqrt{LC} = \dfrac{t_c}{g(x)} = \dfrac{t_c}{1.682}$ $\qquad \qquad \text{...(8.37)}$

Multiplication of Eqs. (8.36) and (8.37) gives

$$C = \frac{1.5 \, t_c \, I_{om}}{1.682 \, V_{mn}} = 0.892 \frac{t_c \, I_{om}}{V_{mn}} \qquad \qquad \text{...(8.38)}$$

From Eq. (8.36), $\sqrt{\dfrac{L}{C}} = \dfrac{V_{mn}}{1.5 I_{om}}$ $\qquad \qquad \text{...(8.39)}$

Multiplication of Eqs. (8.37) and (8.39) gives

$$L = \frac{t_c \cdot V_{mn}}{1.682 \times 1.5 \, I_{om}} = 0.3964 \frac{t_c \, V_{mn}}{I_{om}} \qquad \qquad \text{...(8.40)}$$

For critical damping of the resonant circuit consisting of R_d, L, C in series in Fig. 8.10,

$$\sqrt{\frac{1}{LC} - \left(\frac{R_d}{2L}\right)^2} = 0$$

From above, the value of resistance that gives critical damping is

$$R_d = 2 \sqrt{\frac{L}{C}} \qquad \qquad \text{...(8.41)}$$

8.3.2. Modified McMurray full-bridge inverter

A single-phase modified McMurray full-bridge inverter is shown in Fig. 8.14. The number of thyristors, diodes and other components in full-bridge inverter is double of those in half-bridge inverter of Fig. 8.10. The operation of this inverter during commutation process is similar to that described in the previous section for a single-phase half-bridge inverter. For example, for mode I, thyristors T1, T2 are conducting and the load current completes its path through source V_s, T1, load, T2 and back to source. For mode II, TA1 and TA2 are triggered together for commutating main SCRs T1 and T2. For mode III, commutating current i_c in both the circuits goes beyond load current I_0 so that T1 and T2 are turned off and so on.

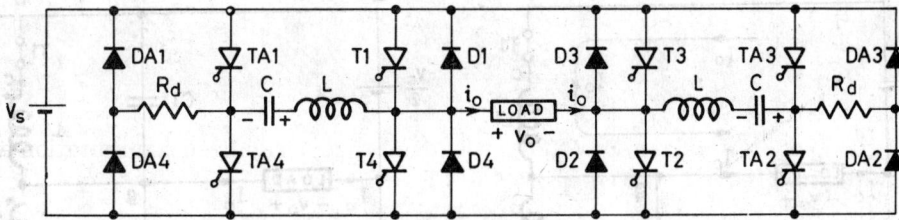

Fig. 8.14. Single-phase modified McMurray full-bridge inverter.

8.3.3. McMurray-Bedford half-bridge inverter

Power circuit diagram of a McMurray-Bedford half-bridge inverter is shown in Fig. 8.15. It uses less number of thyristors and diodes as compared to modified McMurray half-bridge inverter. The number of capacitors and inductors is, however, large. This inverter, Fig. 8.15, consists of main thyristors T1, T2 and feedback diodes D1, D2. Commutation circuitry consists of two capacitors C1, C2 and magnetically coupled inductors L1, L2. Actually L1 and L2 constitute one inductor with a centre tap so that L1 = L2 = L. The inductance of this centre-tapped inductor is usually of the order of about 50 μH. The inductor is wound on a core with an air gap so as to avoid saturation. The value of capacitance for the two capacitors is the same, i.e. C1 = C2 = C. The inverter of Fig. 8.15 is a voltage-commutated VSI.

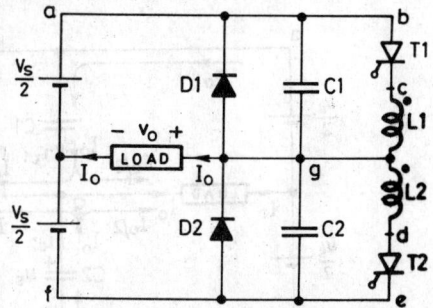

Fig. 8.15. Single-phase McMurray-Bedfored half-bridge inverter with L1 = L2 = L and C1 = C2 = C.

In a branch consisting of two tightly coupled inductors in series with two thyristors, if one thyristor is turned on, the other conducting thyristor gets turned off. This type of commutation is called *complementary commutation*. As a consequence, inverter shown in Fig. 8.15 is also known as *complementary-commutated inverter*.

The simplifying assumptions are the same as for the inverter discussed in previous section.

The working of this inverter can be explained in different modes as follows :

Mode I. In this mode, thyristor T1 is conducting and upper dc source supplies load current I_0 to the load, Fig. 8.16 (a). As the load current is almost constant, voltage drop across commutating inductance L1 $\left(\text{proportional to L1.}\dfrac{di}{dt}\right)$ is negligible. With zero voltage drop across L1 and T1, voltage across C1 is zero and voltage across C2 is V_s because point g is now effectively connected to point a through L1 and T1 and lower plate of C2 is connected to point f. The equivalent circuit for this mode is as shown in Fig. 8.16 (a). In this figure, voltage of node g with respect to point f is V_s. The potential of points d and c is the same as that of point g with respect to f. In other words, the potential of all the three nodes g, d, c with respect to point f is V_s; this is shown in Fig. 8.17.

Mode II. At $t = 0 -$, thyristor T2 is triggered to initiate the commutation of T1. With the turning-on of T2, point d gets connected to e or f, i.e. to the negative supply terminal. Voltage across C1 and C2 cannot change instantaneously, therefore a voltage V_s appears across L2. As L1 and L2 are magnetically coupled, an equal voltage is induced across L1 with terminal c

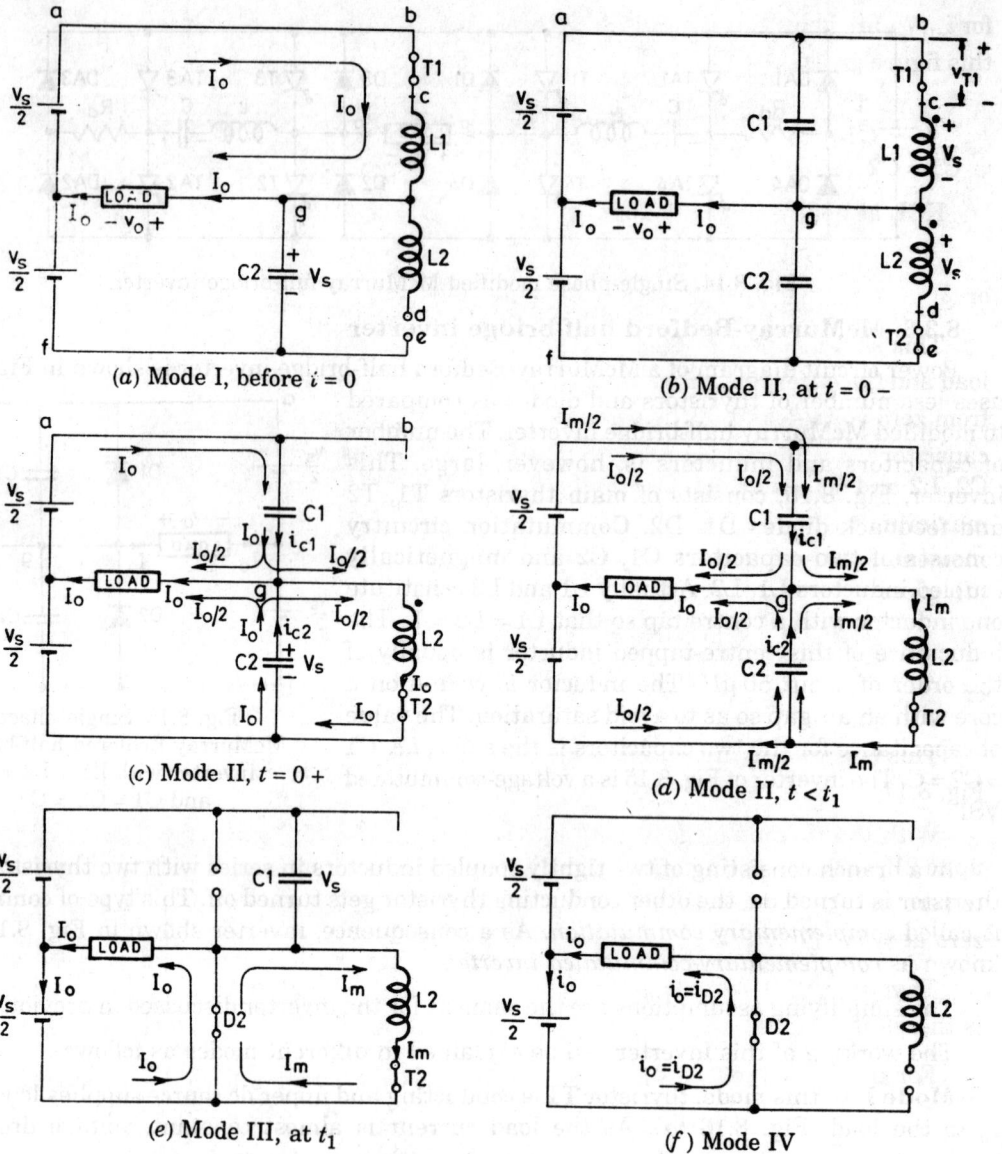

(a) Mode I, before $i = 0$

(b) Mode II, at $t = 0 -$

(c) Mode II, $t = 0 +$

(d) Mode II, $t < t_1$

(e) Mode III, at t_1

(f) Mode IV

Fig. 8.16. Different modes for the inverter of Fig. 8.15.

positive. Voltage v_{T1} across terminals of thyristor T1 can be found by applying KVL for the loop $c\ b\ a\ f\ e\ d\ c$ in Fig. 8.16 (b), therefore, $v_{T1} - \dfrac{V_s}{2} - \dfrac{V_s}{2} + V_s + V_s = 0$

$$\therefore \qquad v_{T1} = - V_s$$

This shows that point c is positive with respect to b by V_s volts, *i.e.* T1 is subjected to a reverse voltage of $- V_s$; it is therefore turned off at $t = 0 +$. Load current I_0 flowing through T1 and L1 is at once transferred to L2 and T2 so as to maintain constant *mmf* (proportional to L1, I_0) in the centre-tapped inductor as per the constant flux linkage theorem. Current directions for i_{c1}, i_{c2} are shown in Fig. 8.16 (c). KVL for the loop consisting of C1, C2 and the source V_s for this figure gives

$$-\frac{1}{C1} \int i_{c1} \, dt + \frac{1}{C2} \int i_{c2} \, dt - V_s \text{ (voltage across C2)} + V_s \text{ (source voltage)} = 0 \text{ or } i_{c1} = i_{c2},$$

$\because$ C1 = C2.

KCL at node g in Fig. 8.16 (c) gives

$$i_{c1} + i_{c2} = I_0 + I_0$$

or

$$i_{c1} = i_{c2} = I_0$$

This shows capacitor C1 and C2 both carry current I_0 at $t = 0+$. Half of i_{c1} flows through load and the other half through L2 ; same is true for i_{c2}, Fig. 8.16 (c). Now C1 is getting charged from zero voltage and at the same time C2 is getting discharged from V_s at the same rate. As capacitor C2 is placed across L2, an oscillating current is set up in the closed loop formed by C2, L2 and T2. After one-fourth of a cycle, this oscillating current rises (initial value I_0 to a maximum value of I_m in L2 and T2 and at the same time v_{c2} across C2 falls to zero. At this moment, *i.e.* one-fourth of a cycle after the instant T2 is triggered, KCL at node g, Fig. 8.16 (d), gives

$$i_{c1} + i_{c2} = I_0 \text{ (load current)} + I_m \text{ (current in L2 and T2)}$$

or

$$i_{c1} = i_{c2} = \frac{I_0 + I_m}{2}$$

This is shown in Fig. 8.16 (d). The variation of i_{c1}, i_{c2}, i_{T2} and i_0 from $t = 0$ to t_1 is shown in Fig. 8.17.

With the turning on of T2, voltage of node d drops to zero whereas that of node c shoots up to $2V_s$ with respect to node f at $t = 0+$, Fig. 8.16 (b). From $t = 0$ to t_1, voltage of node g drops to zero in one-fourth of a cycle of $V_s \cos \omega_0 t$. Similarly, voltage of node c reduces from $2V_s$ to zero at t_1 in one-fourth of a cycle, Fig. 8.17.

Thyristor T1 is reverse biased until voltage of node c falls to V_s, commutation time for T1 is therefore t_c as shown in Fig. 8.17.

For the circuit consisting of C2, L2 and T2, ringing frequency is

$$\omega_0 = \frac{1}{\sqrt{LC}}$$

Periodic time

$$T_0 = \frac{1}{f_0} = \frac{2\pi}{\omega_0} = 2\pi \sqrt{LC}$$

$\therefore$

$$t_1 = \frac{T_0}{4} = \frac{\pi}{2} \sqrt{LC}$$

Here L is the inductance of L2 and C is the capacitance of C2.

Circuit turn-off time $t_c <$ one quarter of a cycle

or

$$t_c < t_1$$

or

$$t_c < \frac{\pi}{2} \sqrt{LC}$$

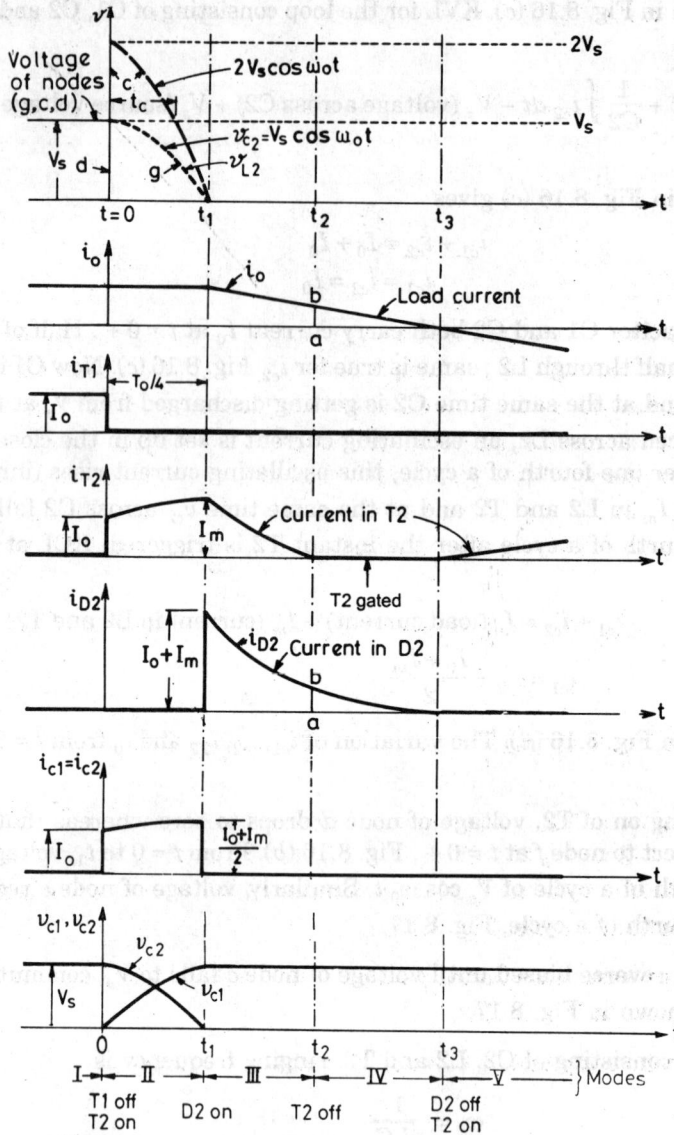

Fig. 8.17. Voltage and current wave-forms for the inverter of Fig. 8.15.

Mode III. At t_1, capacitor C1 is charged to supply voltage V_s and therefore no current can flow through C1, i.e. $i_{c1} = 0$. After one-fourth of a cycle from $t = 0$, i.e. at t_1, $v_{c2} = 0$. Just after t_1, $(I_0 + I_m)/2$ through C2 tends to charge it with bottom plate positive. As a result, diode D2 gets forward biased at t_1. Thus, now entire current $(I_0 + I_m)$ is transferred to D2 so that both $i_{c1} = i_{c2} = 0$ just after t_1 but $i_{D2} = I_0 + I_m$; this is shown in the equivalent circuit of Fig. 8.16 (e). Diode current $i_{D2} = I_0 + I_m$ is also shown in Fig. 8.17.

The energy stored in inductor L2 at t_1 is dissipated in the closed circuit made up of L2, T2 and D2. At time t_2, this energy is entirely dissipated, therefore current i_{T2} decays to zero and as a result, T2 is turned off at t_2, Fig. 8.17. Sometimes, a small resistance is included in series with the diode to speed up the dissipation of stored energy in L2. As i_{T2} decays from I_m at t_1 to zero at t_2, i_{D2} also decays from $I_0 + I_m$ at t_1 to $i_{D2} = i_0 = ab$ at t_2. Load current also decreases from I_0 at t_1 to $i_0 = i_{D2} = ab$ at t_2.

Mode IV. When the current i_{T2} through L2 and T2 has decayed to zero, the equivalent circuit is as shown in Fig. 8.16 (f). A load current $i_0 = i_{D2}$ still continues flowing through the diode D2 as i_{D2} during $(t_3 - t_2)$ interval, Fig. 8.16 (f).

Mode V. Finally, load current i_0 through the diode D2 and load decays to zero at t_3 and is then reversed. As soon as i_0, equal to i_{D2}, tends to reverse, D2 is blocked. The reverse bias across T2, due to voltage drop in D2 no longer exists. Therefore, thyristor T2 already gated during the interval $(t_3 - t_2)$ gets turned on to carry the load current in the reversed direction as shown in Fig. 8.17. The capacitor C1, now charged to the source voltage V_s, Fig. 8.16 (e), is ready for commutating the main thyristor T2.

The magnitude of commutating circuit parameters L and C, for minimum trapped energy, is given by

$$L = 2.35 \frac{V_s t_q}{I_{om}} \qquad \qquad ...(8.42)$$

and

$$C = 2.35 \frac{I_{om} t_q}{V_s} \qquad \qquad ...(8.43)$$

where t_q = thyristor turn-off time.

and I_{om} = maximum load current to be commutated.

8.3.4. McMurray-Bedford Full-bridge Inverter

A single-phase McMurray-Bedford full-bridge inverter can be realized by connecting two half-bridge inverters as shown in Fig. 8.18. The various components required are double of those in the half-bridge inverter of Fig. 8.15. The working of this inverter is similar to that described for half-bridge inverter in the previous section. For example, for mode I, thyristors T1, T2 are conducting and load current flows through V_s, T1, L1, load, L2, T2. Voltage across C1, C2 is zero but C3, C4 are charged to voltage V_s. For initiating commutation of T1, T2 ; thyristors T3, T4 are triggered. This reverse biases T1, T2 by voltage $- V_s$, these thyristors are therefore turned off and so on.

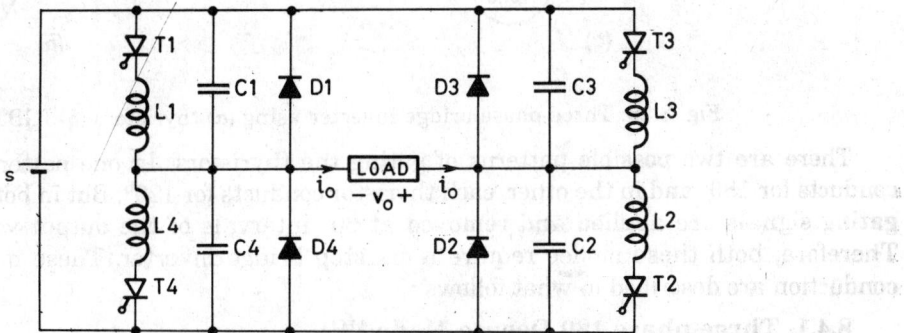

Fig. 8.18. Single-phase McMurray-Bedford full-bridge inverter.

8.4. THREE PHASE BRIDGE INVERTERS

For providing adjustable-frequency power to industrial applications, three-phase inverters are more common than single-phase inverters. Three-phase inverters, like single-phase inverters, take their dc supply from a battery or more usually from a rectifier.

A basic three-phase inverter is a six-step bridge inverter. It uses a minimum of six thyristors. In inverter terminology, a step is defined as a change in the firing from one thyristor to the next thyristor in proper sequence. For one cycle of 360°, each step would be of 60° interval for a six-step inverter. This means that thyristors would be gated at regular intervals of 60° in proper sequence so that a 3-phase ac voltage is synthesized at the output terminals of a six-step inverter.

Fig. 8.19 (a) shows the power circuit diagram of a three-phase bridge inverter using six thyristors and six diodes. As stated earlier, the transistor family of devices is now very widely used in inverter circuits. Presently, the use of IGBTs in single-phase and three-phase inverters is on the rise. The basic circuit configuration of inverter, however, remains unaltered as shown in Fig. 8.19 (b) for a three-phase bridge inverter using IGBTs in place of thyristors. A large capacitor connected at the input terminals tends to make the input dc voltage constant. This capacitor also suppresses the harmonics fed back to the dc source.

In Fig. 8.19 (a) inverter using six thyristors, commutation and snubber circuits are omitted for simplicity. It may be seen from Figs. 8.1 and 8.19 that a three-phase bridge inverter consists of three half-bridge inverters arranged side by side. The three-phase load is assumed to be star connected. In Fig. 8.19 (a), the thyristors are numbered in the sequence in which they are triggered to obtain voltages v_{ab}, v_{bc}, v_{ca} at the output terminals a, b, c of the inverter.

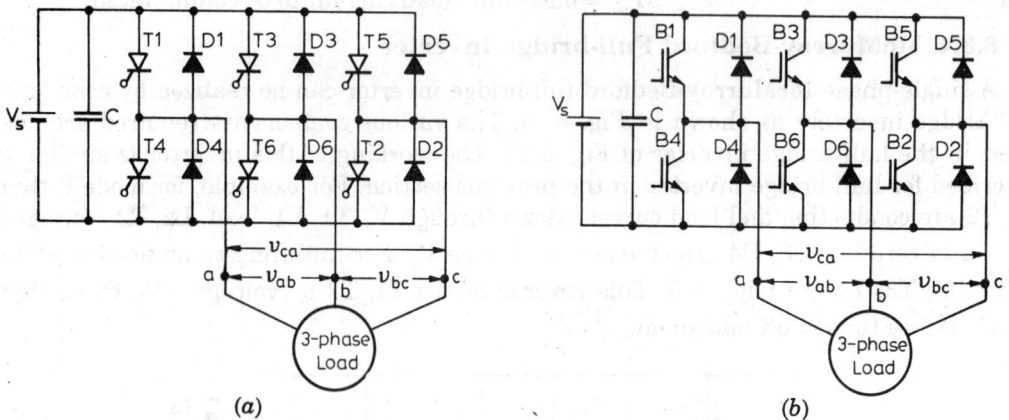

Fig. 8.19. Three-phase bridge inverter using (a) thyristors (b) IGBTs.

There are two possible patterns of gating the thyristors. In one pattern, each thyristor conducts for 180° and in the other, each thyristor conducts for 120°. But in both these patterns, gating signals are applied and removed at 60° intervals of the output voltage waveform. Therefore, both these modes require a six step bridge inverter. These modes of thyristor conduction are described in what follows :

8.4.1. Three-phase 180 Degree Mode VSI

In the three-phase inverter of Fig. 8.19, each SCR conducts for 180° of a cycle. Thyristor pair in each arm, *i.e.* T1, T4 ; T3, T6 and T5, T2 are turned on with a time interval of 180°.

It means that T1 conducts for 180° and T4 for the next 180° of a cycle. Thyristors in the upper group, *i.e.* T1, T3, T5 conduct at an interval of 120°. It implies that if T1 is fired at $\omega t = 0°$, then T3 must be fired at $\omega t = 120°$ and T5 at $\omega t = 240°$. Same is true for lower group of SCRs. On the basis of this firing scheme, a table is prepared as shown at the top of Fig. 8.20. In this table, first row shows that T1 from upper group conducts for 180°, T4 for the next 180° and

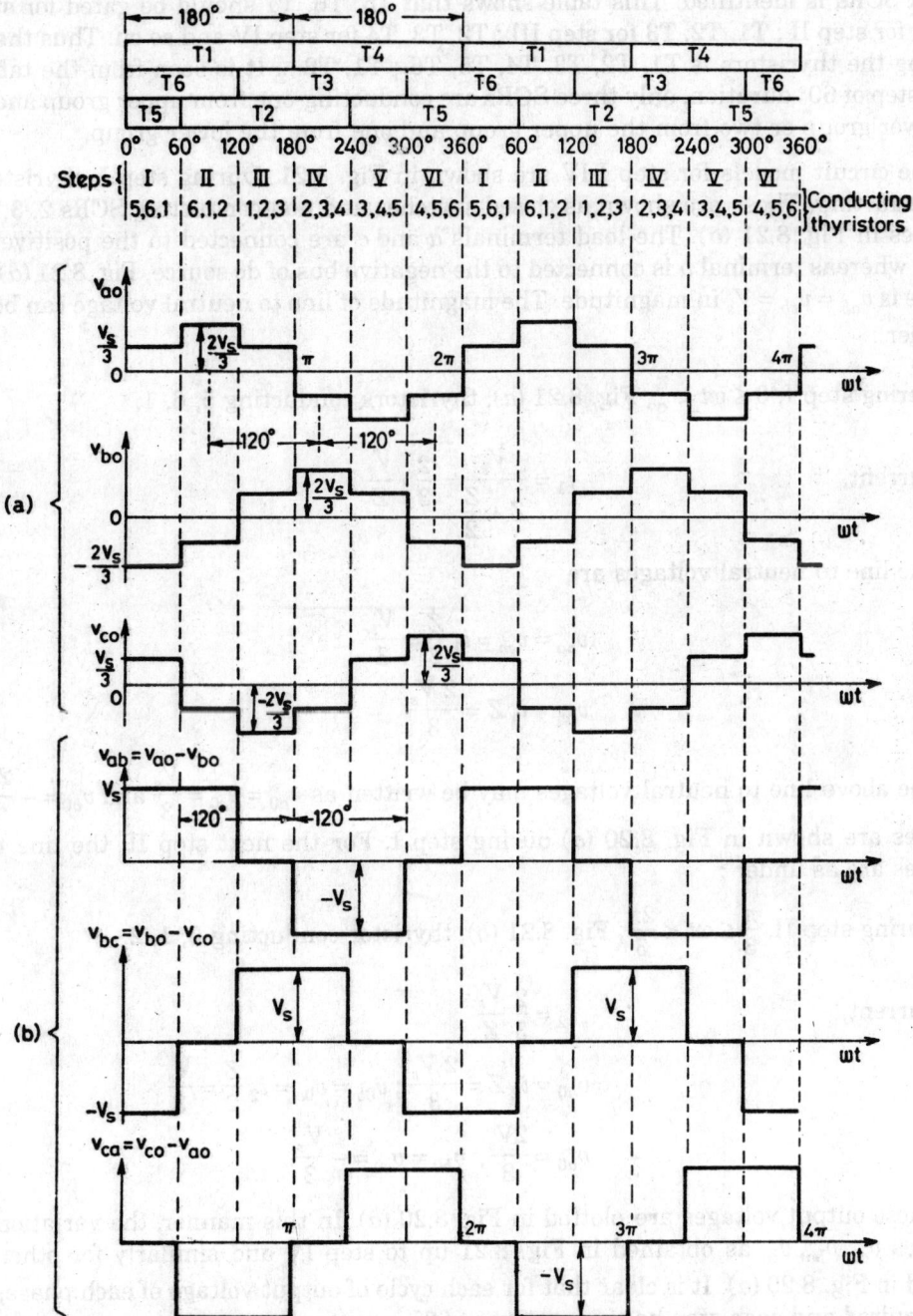

Fig. 8.20. Voltage waveforms for 180° mode 3-phase VSI.

then again T1 for 180° and so on. In the second row, T3 from the upper group is shown to start conducting 120° after T1 starts conducting. After T3 conduction for 180°, T6 conducts for the next 180° and again T3 for the next 180° and so on. Further, in the third row, T5 from the upper group starts conducting 120° after T3 or 240° after T1. After T5 conduction for 180°, T2 conducts for the next 180°, T5 for the next 180° and so on. In this manner, the pattern of firing the six SCRs is identified. This table shows that T5, T6, T1 should be gated for step I ; T6, T1, T2 for step II ; T1, T2, T3 for step III ; T2, T3, T4 for step IV and so on. Thus the sequence of firing the thyristors is T1, T2, T3, T4, T5, T6 ; T1, T2.... It is seen from the table that in every step of 60° duration, only three SCRs are conducting-one from upper group and two from the lower group or two from the upper group and one from the lower group.

The circuit models for step I-IV are shown in Fig. 8.21. During step I, thyristors 5, 6, 1 are conducting. These are shown as closed switches and non-conducting SCRs 2, 3, 4 as open switches in Fig. 8.21 (a). The load terminals a and c are connected to the positive bus of dc source whereas terminal b is connected to the negative bus of dc source, Fig. 8.21 (a). The load voltage is $v_{ab} = v_{cb} = V_s$ in magnitude. The magnitude of line to neutral voltage can be obtained as under :

During step I, $0 \le \omega t < \dfrac{\pi}{3}$, Fig. 8.21 (a), thyristors conducting 5, 6, 1.

Current,
$$i_1 = \frac{V_s}{Z + \dfrac{Z}{2}} = \frac{2}{3} \cdot \frac{V_s}{Z}$$

The line to neutral voltages are

$$v_{ao} = v_{co} = i_1 \frac{Z}{2} = \frac{V_s}{3}$$

and
$$v_{0b} = i_1 Z = \frac{2 V_s}{3}$$

The above line to neutral voltages may be written as $v_{a0} = v_{c0} = \dfrac{V_s}{3}$ and $v_{b0} = -\dfrac{2V_s}{3}$. These voltages are shown in Fig. 8.20 (a) during step I. For the next step II, the line to neutral voltages are as under :

During step II, $\dfrac{\pi}{3} \le \omega t < \dfrac{2\pi}{3}$, Fig. 8.21 (b), thyristor conducting 6, 1, 2.

Current,
$$i_2 = \frac{2}{3} \frac{V_s}{Z}$$

∴
$$v_{a0} = i_2 Z = \frac{2 V_s}{3} \; ; \; v_{0b} = v_{0c} = i_2 \frac{Z}{2} = \frac{V_s}{3}$$

or
$$v_{a0} = \frac{2V_s}{3}, \; v_{b0} = v_{c0} = -\frac{V_s}{3}.$$

These output voltages are plotted in Fig. 8.20 (a). In this manner, the variation of phase voltages v_{ao}, v_{bo}, v_{co} as obtained in Fig. 8.21 up to step IV and similarly for other steps, is plotted in Fig. 8.20 (a). It is clear that for each cycle of output voltage of each phase, six steps are required and each step has a duration of 60°.

Step I

$$v_{ao} = v_{co} = V_s/3$$
$$v_{bo} = -v_{ob} = -2V_s/3$$

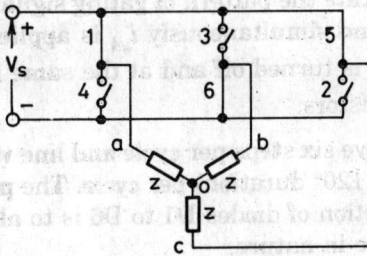

(a) 0—60° ; 5, 6, 1 closed.

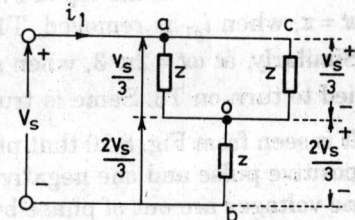

Step II

$$v_{ao} = 2V_s/3$$
$$v_{bo} = v_{co} = -V_s/3$$

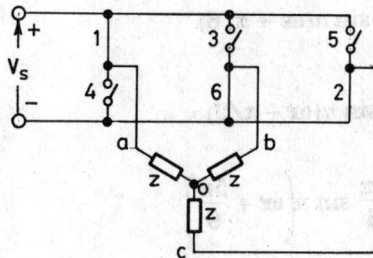

(b) 60—120° ; 6, 1,2 closed.

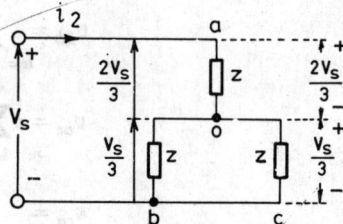

Step III

$$v_{ao} = v_{bo} = V_s/3$$
$$v_{co} = -2V_s/3$$

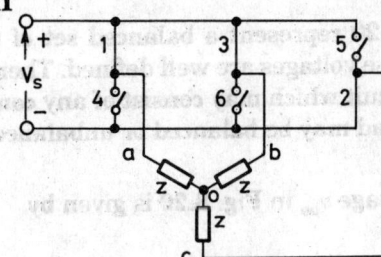

(c) 120—180° ; 1, 2, 3 closed.

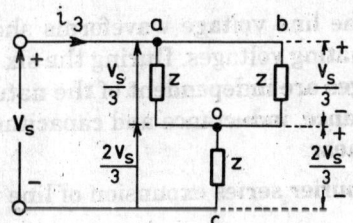

Step IV

$$v_{bo} = 2V_s/3$$
$$v_{ao} = v_{co} = -V_s/3$$

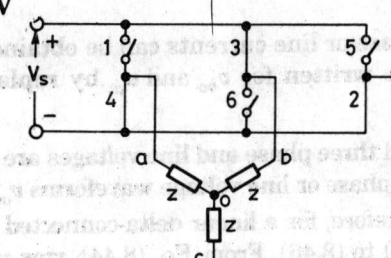

(d) 180—240° ; 2, 3, 4 closed.

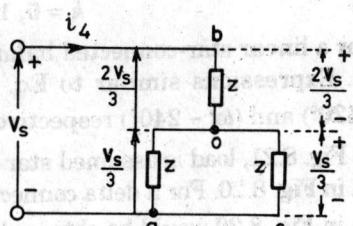

Fig. 8.21. Equivalent circuits for a 3-phase six-step 180° mode inverter with a balanced star-connected load.

The line voltage $v_{ab} = v_{ao} + v_{ob}$ or $v_{ab} = v_{ao} - v_{bo}$ is obtained by reversing v_{bo} and adding it to v_{ao} as shown in Fig. 8.20 (b). Similarly, line voltages $v_{bc} = v_{bo} - v_{co}$ and $v_{ca} = v_{co} - v_{ao}$ are plotted in Fig. 8.20 (b).

The three rows in the top of Fig. 8.20 also indicate the pattern of gating signal waveforms. At $\omega t = \pi$, when i_{g1} is removed, T1 is turned off and simultaneously i_{g4} is applied to turn on T4. Similarly, at $\omega t = 2\pi/3$, when i_{g6} is cut off, T6 is turned off and at the same instant i_{g3} is applied to turn on T3. Same is true for other thyristors.

It is seen from Fig. 8.20 that phase voltages have six steps per cycle and line voltages have one positive pulse and one negative pulse (each of 120° duration) per cycle. The phase as well as line voltages are out of phase by 120°. The function of diodes D1 to D6 is to allow the flow of currents through them when the load is reactive in nature.

The three line output voltages can be described by the Fourier series as follows :

$$v_{ab} = \sum_{n = 1, 3, 5,}^{\infty} \frac{4V_s}{n\pi} \cos \frac{n\pi}{6} \sin n(\omega t + \pi/6) \qquad ..(8.44)$$

$$v_{bc} = \sum_{n = 1, 3, 5,}^{\infty} \frac{4V_s}{n\pi} \cos \frac{n\pi}{6} \sin n(\omega t - \pi/2) \qquad ...(8.45)$$

$$v_{ca} = \sum_{n = 1, 3, 5, -}^{\infty} \frac{4V_s}{n\pi} \cos \frac{n\pi}{6} \sin n\left(\omega t + \frac{5\pi}{6} \right) \qquad ...(8.46)$$

For $n = 3$, $\cos \dfrac{3\pi}{6} = 0$. Thus, all triplen harmonics are absent from the line voltages as given by Eqs. (8.44) to (8.46).

The line voltage waveforms shown in Fig. 8.20 represent a balanced set of three-phase alternating voltages. During the six intervals, these voltages are well defined. Therefore, these voltages are independent of the nature of load circuit which may consist of any combination of resistance, inductance and capacitance and the load may be balanced or unbalanced, linear or nonlinear.

Fourier series expansion of line to neutral voltage v_{ao} in Fig. 8.20 is given by

$$v_{ao} = \sum_{n = 6k \pm 1}^{\infty} \frac{2 V_s}{n\pi} \sin n\omega t \qquad ...(8.47)$$

where $\qquad k = 0, 1, 2,...$

For a linear star-connected balanced load, phase or line currents can be obtained from Eq. (8.47). Expressions similar to Eq. (8.47) can be written for v_{bo} and v_{co} by replacing ωt by ($\omega t - 120°$) and ($\omega t - 240°$) respectively.

In Fig. 8.21, load is assumed star connected and three phase and line voltages are obtained as shown in Fig. 8.20. For a delta connected load also, phase or line voltage waveforms v_{ab}, v_{bc}, v_{ca} as shown in Fig. 8.20 would be obtained directly. Therefore, for a linear delta-connected load, phase and line currents can be obtained from Eqs. (8.44) to (8.46). From Eq. (8.44), rms value of nth component of line voltage is

$$V_{Ln} = \frac{4 V_s}{\sqrt{2} \; n\pi} \cos \frac{n\pi}{6} \qquad ...(8.48)$$

Rms value of fundamental line voltage,

$$V_{L1} = \frac{4 V_s}{\sqrt{2} \cdot \pi} \cos \frac{\pi}{6} = 0.7797 \, V_s \qquad ...(8.49)$$

It is seen from line voltage waveform v_{ab} in Fig. 8.20 (a) that line voltage is V_s from 0° to 120°. Therefore, rms value of line voltage V_L is

$$V_L = \left[\frac{1}{\pi}\int_0^{2\pi/3} V_s^2\, d\,(\omega t)\right]^{1/2} = \sqrt{\frac{2}{3}}\, V_s = 0.8165\, V_s \qquad\qquad ...(8.50)$$

Rms value of phase voltage V_p is

$$V_p = \frac{V_L}{\sqrt{3}} = \frac{\sqrt{2}}{3}\, V_s = 0.4714\, V_s \qquad\qquad ...(8.51)$$

Rms value of fundamental phase voltage, from Eq. (8.47), is

$$V_{p1} = \frac{2V_s}{\sqrt{2}\,\pi} = 0.4502\, V_s = \frac{V_{L1}}{\sqrt{3}} \qquad\qquad ...(8.52)$$

8.4.2. Three-phase 120 Degree Mode VSI

The power circuit diagram of this inverter is the same as that shown in Fig. 8.19. For the 120-degree mode VSI, each thyristor conducts for 120° of a cycle. Like 180° mode, 120° mode inverter also requires six steps, each of 60° duration, for completing one cycle of the output ac voltage.

For this inverter too, a table giving the sequence of firing the six thyristors is prepared as shown in the top of Fig. 8.22. In this table, first row shows that T1 conducts for 120° and for the next 60°, neither T1 nor T4 conducts. Now T4 is turned on at $\omega t = 180°$ and it further conducts for 120°, *i.e.* from $\omega t = 180°$ to $\omega t = 300°$. This means that for 60° interval from $\omega t = 120°$ to $\omega t = 180°$, series connected SCRs T1, T4 do not conduct. At $\omega t = 300°$, T4 is turned off, then 60° interval elapses before T1 is turned on again at $\omega t = 360°$. In the second row, T3 is turned on at $\omega t = 120°$ as in 180° mode inverter. Now T3 conducts for 120°, then 60° interval elapses during which neither T3 nor T6 conducts. At $\omega t = 300°$, T6 is turned on, it conducts for 120° and then 60° interval elapses after which T3 is turned on again. The third row is also completed similarly. This table shows that T6, T1 should be gated for step I ; T1, T2 for step II ; T2, T3 for step III and so on. The sequence of firing the six thyristors is the same as for the 180° mode inverter. During each step, only two thyristors conduct for this inverter — one from the upper group and one from the lower group ; but in 180° mode inverter, three thyristors conduct in each step.

The circuit models for steps I-IV are shown in Fig. 8.23, where load is assumed to be resistive and star connected. During step I, thyristors 6, 1 are conducting and as such load terminal a is connected to the positive bus of dc source whereas terminal b is connected to negative bus of dc source, Fig. 8.23 (a). Load terminal c is not connected to dc bus. The line to neutral voltages, from Fig. 8.23 (a) are

$$v_{ao} = \frac{V_s}{2}, \quad v_{ob} = \frac{V_s}{2}$$

or

$$v_{bo} = -\frac{V_s}{2}$$

and

$$v_{co} = 0$$

These voltages are shown in Fig. 8.22 (a) during step I of 0° – 60°. For step II, thyristors 1, 2 conduct and load voltages are $v_{ao} = V_s/2$, $v_{co} = -V_s/2$ and $v_{bo} = 0$, Fig. 8.23 (b) ; these voltages are plotted in Fig. 8.22 (a). This procedure is followed for obtaining load voltages for the remaining steps and these phase voltages are then plotted in Fig. 8.22 (a).

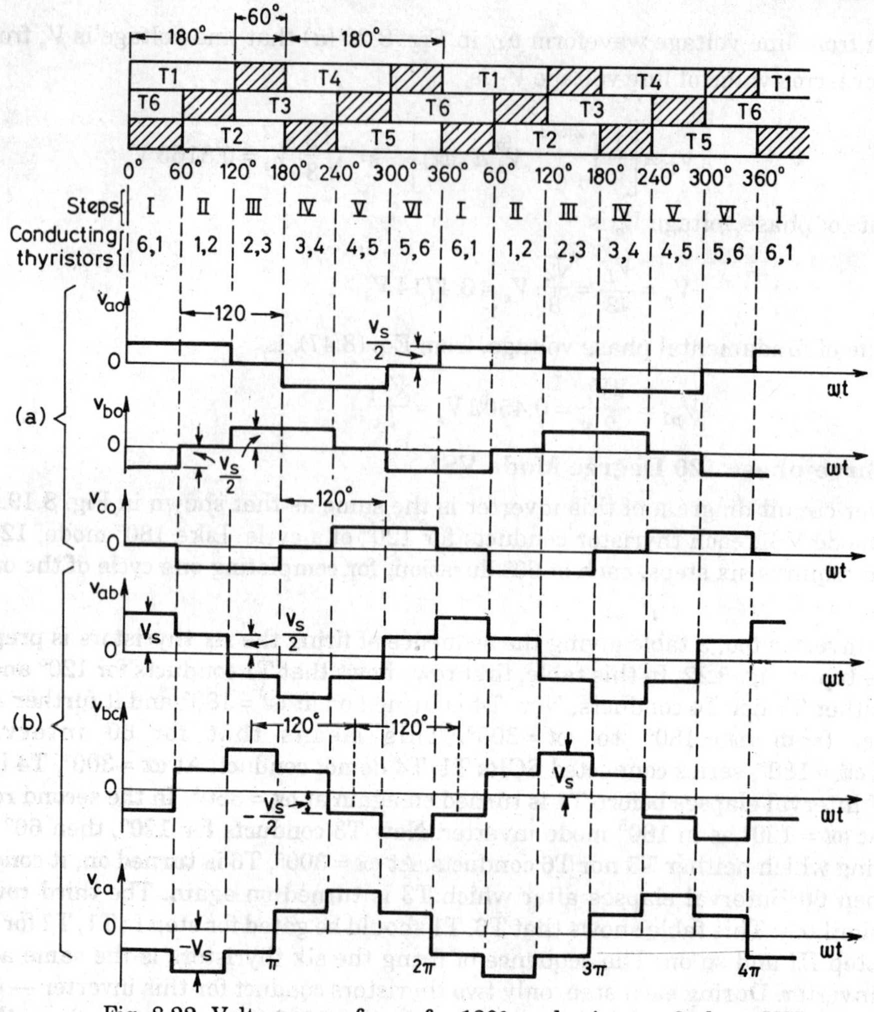

Fig. 8.22. Voltage waveforms for 120° mode six-step 3-phase VSI.

The line voltages

$$v_{ab} = v_{ao} - v_{bo}$$
$$v_{bc} = v_{bo} - v_{co}$$

and

$$v_{ca} = v_{co} - v_{ao}$$

are also plotted in Fig. 8.22 (b).

It is seen from Fig. 8.22 that phase voltages have one positive pulse and one negative pulse (each of 120° duration) for one cycle of output alternating voltage. The line voltages, however, have six steps per cycle of output alternating voltage.

As stated before, the three rows in the top of Fig. 8.22 indicate the pattern of gating signal waveforms.

The merits and demerits of 120-degree mode inverter over 180-degree mode inverter are as follows :

(i) In the 180° mode inverter, when gate signal i_{g1} is cut-off to turn off T1 at $\omega t = 180°$, gating signal i_{g4} is simultaneously applied to turn on T4 in the same leg. In practice, a commutation interval must exist between the removal of i_{g1} and application of i_{g4}, because otherwise dc source would experience a direct short-circuit through SCRs T1 and T4 in the same leg.

Step I

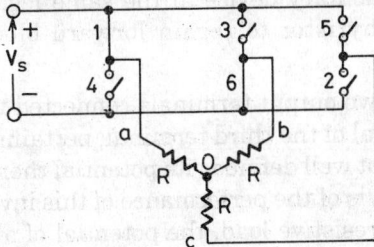

(a) 0—60° ; 6, 1 closed

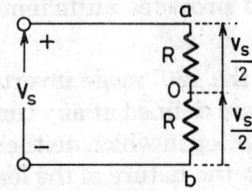

$$v_{ao} = V_s/2$$
$$v_{bo} = -V_s/2 \text{ and } v_{co} = 0$$

Step II

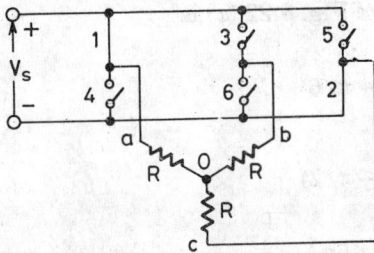

(b) 60—120° ; 1, 2 closed

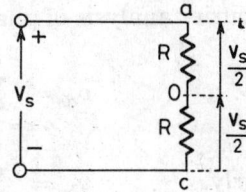

$$v_{ao} = V_s/2$$
$$v_{co} = -V_s/2 \text{ and } v_{bo} = 0$$

Step III

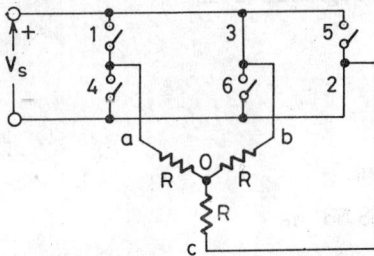

(c) 120—180° ; 2, 3 closed

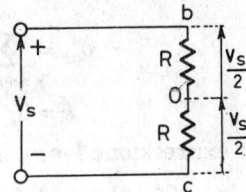

$$v_{bo} = V_s/2$$
$$v_{co} = -V_s/2 \text{ and } v_{ao} = 0$$

Step IV

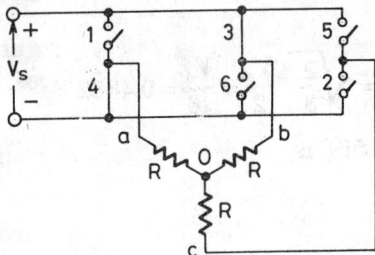

(d) 180—240° ; 3, 4 closed

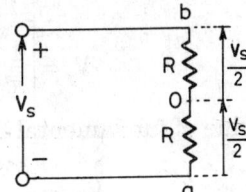

$$v_{bo} = V_s/2$$
$$v_{ao} = -V_s/2 \text{ and } v_{co} = 0$$

Fig. 8.23. Equivalent circuits for a 3-phase six-step 120° mode inverter with balanced star-connected resistive load.

This difficulty is overcome considerably in 120-degree mode inverter. In this inverter, there is a 60° interval between the turning off of T1 and turning on of T4. During this 60° interval, T1 can be commutated safely. In general, this angular interval of 60° exists between the turning-off of one device and turning-on of the complementary device in the same leg. This 60° period provides sufficient time for the outgoing thyristor to regain forward blocking capability.

(*ii*) In the 120° mode inverter, the potentials of only two output terminals connected to the dc source are defined at any time of the cycle. The potential of the third terminal, pertaining to a particular leg in which neither device is conducting, is not well defined ; its potential therefore depends on the nature of the load circuit. Thus, the analysis of the performance of this inverter is complicated for a general load circuit. For a balanced resistive load, the potential of all the three terminals is, however, well defined. This is the reason load is assumed resistive in Fig. 8.23. For a balanced delta-connected resistive load, the line voltages as shown in Fig. 8.22 (*b*) are obtained directly.

The Fourier analysis of phase voltage waveform v_{ao} of Fig. 8.22 (*a*) is

$$v_{ao} = \sum_{n = 1, 3, 5}^{\infty} \frac{2 V_s}{n\pi} \cos \frac{n\pi}{6} \sin n \, (\omega t + \pi/6) \qquad \qquad ...(8.53)$$

Similarly,
$$v_{bo} = \sum_{n = 1, 3, 5, ...}^{\infty} \frac{2V_s}{n\pi} \cos \frac{n\pi}{6} \sin n \, (\omega t - \pi/2) \qquad \qquad ...(8.54)$$

and
$$v_{co} = \sum_{n = 1, 3, 5, ...}^{\infty} \frac{2V_s}{n\pi} \cos \frac{n\pi}{6} \sin n \left(\omega t + \frac{5\pi}{6} \right) \qquad \qquad ...(8.55)$$

The Fourier analysis of line voltage waveform v_{ab} of Fig. 8.22 (*b*) is

$$v_{ab} = \sum_{n = 6k \pm 1}^{\infty} \frac{3 \cdot V_s}{n\pi} \sin n \left(\omega t + \frac{\pi}{3} \right) \qquad \qquad ...(8.56)$$

where $\quad k = 0, 1, 2, 3...$

Similar expressions for v_{bc} and v_{ca} can also be written.

Rms value of fundamental phase voltage, from Eq. (8.53), is

$$V_{p1} = \frac{2V_s}{\sqrt{2} \cdot \pi} \cdot \cos \frac{\pi}{6} = 0.3898 \, V_s \qquad \qquad ...(8.57)$$

Rms value of phase voltage,

$$V_p = \left[\frac{1}{\pi} \int_0^{2\pi/3} \left(\frac{V_s}{2} \right)^2 \cdot d\,(\omega t) \right]^{1/2} = \sqrt{\frac{2}{3}} \cdot \frac{V_s}{2} = \frac{V_s}{\sqrt{6}} = 0.4082 \, V_s \qquad ...(8.58)$$

Rms value of fundamental line voltage, from Eq. (8.56), is

$$V_{L1} = \frac{3V_s}{\sqrt{2} \, \pi} = 0.6752 \, V_s = \sqrt{3} \, V_{p1} \qquad \qquad ...(8.59)$$

Rms value of line voltage,

$$V_L = \sqrt{3} \, V_p = \frac{V_s}{\sqrt{2}} = 0.7071 \, V_s \qquad \qquad ...(8.60)$$

Example 8.8. *A three-phase bridge inverter delivers power to a resistive load from a 450 V dc source. For a star-connected load of 10 Ω per phase, determine for both (a) 180° mode and (b) 120° mode,*

    (i) rms value of load current
    (ii) rms value of thyristor current
    (iii) load power.

Solution. For a resistive load, the waveform of load current is the same as that of the applied voltage. In view of this, waveforms of phase-load current and thyristor current are as shown in Fig. 8.24 (a) for 180° mode operation and in Fig. 8.24 (b) for 120° mode operation.

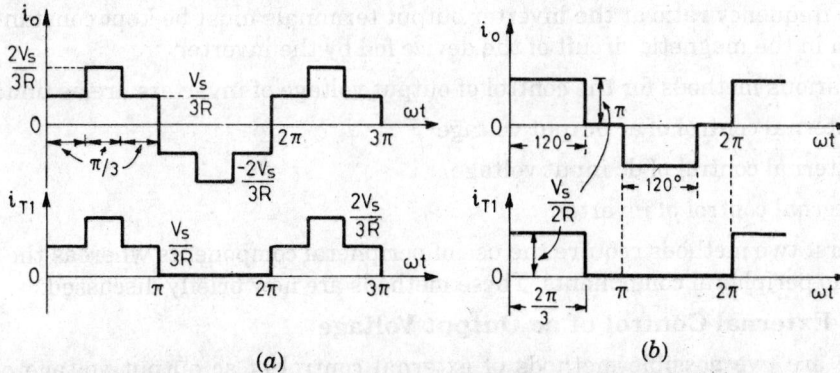

(a)   (b)

Fig. 8.24. Pertaining to example 8.8 (a) 180° mode (b) 120° mode.

(a) **180° mode** : Upper waveform of Fig. 8.24 (a) shows that rms value of per-phase load current I_{or} is given by

$$I_{or} = \left[\frac{1}{\pi}\left\{\left(\frac{V_s}{3R}\right)^2 \frac{\pi}{3} + \left(\frac{2V_s}{3R}\right)^2 \times \frac{\pi}{3} + \left(\frac{V_s}{3R}\right)^2 \frac{\pi}{3}\right\}\right]^{1/2}$$

$$= \left[\left(\frac{450}{3 \times 10}\right)^2 \times \frac{2}{3} + \left(\frac{2 \times 450}{3 \times 10}\right)^2 \times \frac{1}{3}\right]^{1/2} = \sqrt{450} = 21.213 \text{ A}$$

Rms value of thyristor current is

$$I_{T1} = \left[\frac{1}{2\pi}\left\{\left(\frac{450}{3 \times 10}\right)^2 \times \frac{2\pi}{3} + \left(\frac{2 \times 450}{3 \times 10}\right)^2 \times \frac{\pi}{3}\right\}\right]^{1/2}$$

$$= \sqrt{225} = 15 \text{ A}$$

Power delivered to load

$$= 3 I_{or}^2 R = 3\left(\sqrt{450}\right)^2 \times 10 = 13.5 \text{ kW}$$

(b) **120° mode** : Upper waveform in Fig. 8.24 (b) gives rms value of per-phase load current I_{or} as under :

$$I_{or} = \left[\frac{1}{\pi}\left(\frac{450}{2 \times 10}\right)^2 \times \frac{2\pi}{3}\right]^{1/2} = \sqrt{337.5} = 18.371 \text{ A}$$

Rms value of thyristor current,

$$I_{T1} = \left[\frac{1}{2\pi}\left(\frac{450}{2 \times 10}\right)^2 \times \frac{2\pi}{3}\right]^{1/2} = 12.99 \text{ A}$$

Load power $= 3 I_{or}^2 R = 3\left(\sqrt{337.5}\right)^2 \times 10 = 10.125 \text{ kW}$

8.5. VOLTAGE CONTROL IN SINGLE-PHASE INVERTERS

AC loads may require constant or adjustable voltage at their input terminals. When such loads are fed by inverters, it is essential that output voltage of the inverters is so controlled as to fulfil the requirement of ac loads. Examples of such requirements are as under :

(*i*) An ac load may require a constant input voltage though at different levels. For such a load, any variations in the dc input voltage must be suitably compensated in order to maintain a constant voltage at the ac load terminals at a desired level.

(*ii*) In case inverter supplies power to a magnetic circuit, such as an induction motor, the voltage to frequency ratio at the inverter output terminals must be kept constant. This avoids saturation in the magnetic circuit of the device fed by the inverter.

The various methods for the control of output voltage of inverters are as under :

(*a*) External control of ac output voltage

(*b*) External control of dc input voltage

(*c*) Internal control of inverter.

The first two methods require the use of peripheral components whereas the third method requires no peripheral components. These methods are now briefly discussed.

8.5.1. External Control of ac Output Voltage

There are two possible methods of external control of ac output voltage obtained from inverter output terminals. These methods are :

(*a*) AC voltage control

(*b*) Series-inverter control

These are now discussed briefly.

(*a*) **AC voltage control** : In this method, an ac voltage controller is inserted between the output terminals of inverter and the load terminals as shown in Fig. 8.25. The voltage input to the ac load is regulated through the firing angle control of ac voltage controller. This method gives rise to higher harmonic content in the output voltage ; particularly when the output voltage from the ac voltage controller is at low level. This method is, therefore, rarely employed except for low power applications.

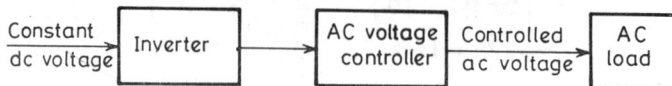

Fig. 8.25. External control of ac output voltage.

(*b*) **Series-inverter control** : This method of voltage control involves the use of two or more inverters in series. Fig. 8.26 (*a*) illustrates how the output voltage of two inverters can be summed up with the help of transformers to obtain an adjustable output voltage. In this figure, the inverter output is fed to two transformers whose secondaries are connected in series. Phasor sum of the two fundamental voltages V_{01}, V_{02} gives the resultant fundamental voltage V_0 as shown in Fig. 8.26 (*b*). Here V_0 is given by

$$V_0 = \left[V_{01}^2 + V_{02}^2 + 2\, V_{01} \cdot V_{02} \cos \theta \right]^{1/2}$$

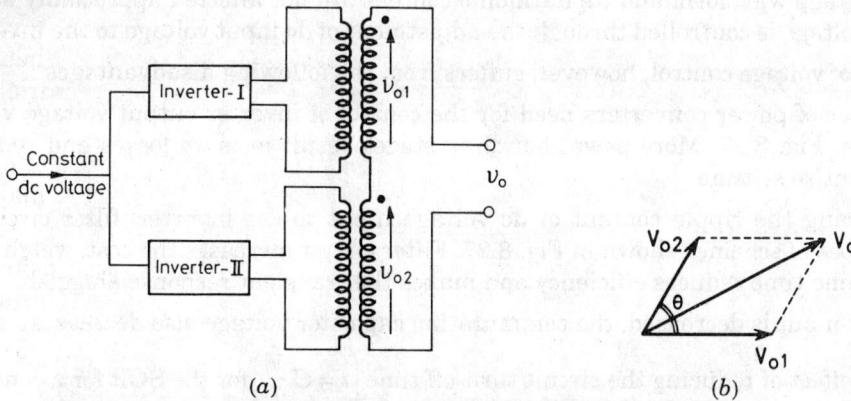

Fig. 8.26. Series inverter control of two inverters.

It is essential that the frequency of output voltages V_{01}, V_{02} from the two inverters is the same. When θ is zero, $V_0 = V_{01} + V_{02}$ and for $\theta = \pi$, $V_0 = 0$ in case $V_{01} = V_{02}$. The angle θ can be varied by the firing angle control of two inverters. The series connection of inverters, called *multiple converter control*, does not augment the harmonic content even at low output voltage levels.

8.5.2. External Control of dc Input Voltage

In case the available voltage source is ac, then dc voltage input to the inverter is controlled through a fully-controlled rectifier, Fig. 8.27 (a) ; through an uncontrolled rectifier and a chopper, Fig. 8.27 (b) ; or through an ac voltage controller and an uncontrolled rectifier, Fig. 8.27 (c). If available voltage is dc, then dc voltage input to the inverter is controlled by means of a chopper as shown in Fig. 8.27 (d).

Input voltage-control techniques shown in Fig. 8.27, in which dc voltage input to inverter is controlled by means of components external to the inverter, has the following main advantage.

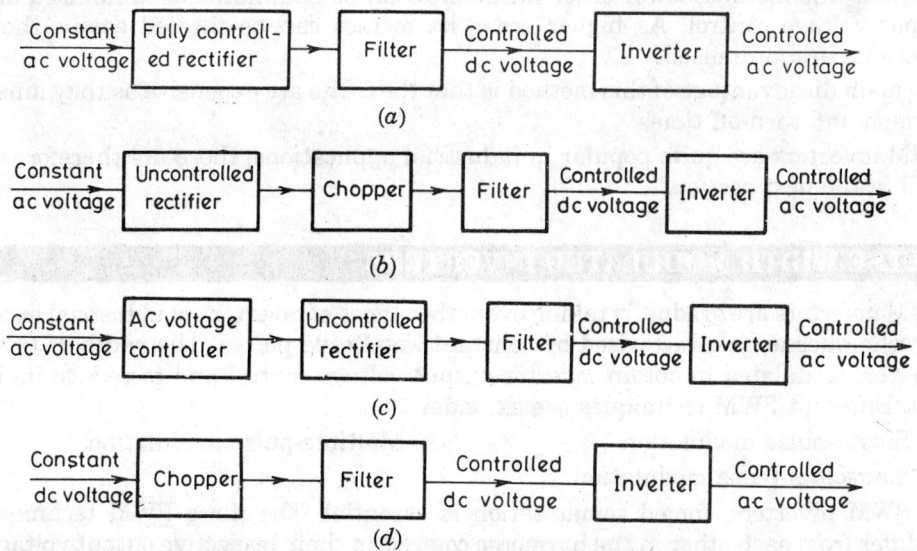

Fig. 8.27. External control of dc input voltage to inverter ; (a), (b) and (c) with ac source on the input (d) with dc source on the input.

(*i*) Output voltage waveform and its harmonic content are not affected appreciably as the inverter output voltage is controlled through the adjustment of dc input voltage to the inverter.

This method of voltage control, however, suffers from the following disadvantages :

(*i*) The number of power converters used for the control of inverter output voltage varies from two to three, Fig. 8.27. More power-handling stages result in more losses and reduced efficiency of the entire scheme.

(*ii*) For reducing the ripple content of dc voltage input to the inverter, filter circuit is required in all types of schemes shown in Fig. 8.27. Filter circuit increases the cost, weight and size and at the same time reduces efficiency and makes the transient response sluggish.

(*iii*) As the dc input is decreased, the commutating capacitor voltage also decreases.

This has the effect of reducing the circuit turn-off time $\left(t = C\,\dfrac{V}{I} \right)$ for the SCR for a constant load current. Therefore, for a large variation of output voltage for a constant load current, control of dc input voltage is not conducive. This difficulty can, however, be overcome by a separate fixed dc source for charging the commutating capacitor, but this makes the scheme costly and complicated.

8.5.3. Internal Control of Inverter

Output voltage from an inverter can also be adjusted by exercising a control within the inverter itself. The most efficient method of doing this is by pulse-width modulation control used within an inverter. This is discussed briefly in what follows :

Pulse width modulation control. In this method, a fixed dc input voltage is given to the inverter and a controlled ac output voltage is obtained by adjusting the on and off periods of the inverter components. This is the most popular method of controlling the output voltage and this method is termed as *pulse–width modulation* (PWM) control.

The advantages possessed by PWM technique are as under :

(*i*) The output voltage control with this method can be obtained without any additional components.

(*ii*) With this method, lower order harmonics can be eliminated or minimised along with its output voltage control. As higher order harmonics can be filtered easily, the filtering requirements are minimised.

The main disadvantage of this method is that the SCRs are expensive as they must possess low turn-on and turn-off times.

PWM inverters are quite popular in industrial applications, these are therefore discussed in detail in the next section.

8.6. PULSE-WIDTH MODULATED INVERTERS

PWM inverters are gradually taking over other types of inverters in industrial applications. PWM techniques are characterised by constant amplitude pulses. The width of these pulses is, however, modulated to obtain inverter output voltage control and to reduce its harmonic content. Different PWM techniques are as under :

(*a*) Single-pulse modulation   (*b*) Multiple-pulse modulation

(*c*) Sinusoidal-pulse modulation.

In PWM inverters, forced commutation is essential. The three PWM techniques listed above differ from each other in the harmonic content in their respective output voltages. Thus, choice of a particular PWM technique depends upon the permissible harmonic content in the inverter output voltage.

In industrial applications, PWM inverter is supplied from a diode bridge rectifier and an *LC* filter. The inverter topology remains the same as in Fig. 8.2 (*a*) for a single-phase inverter and in Fig. 8.19 for a three-phase inverter. But now the devices are switched on and off several times within each half cycle to control the output voltage which has low harmonic content.

In the following lines, the basic principles of PWM techniques for single-phase inverters are illustrated and then the methods of obtaining such output voltages are considered.

8.6.1. Single-pulse Modulation

The output voltage from single-phase full-bridge inverter is shown in Fig. 8.28 (*a*). When this waveform is modulated, the output voltage is of the form shown in Fig. 8.28 (*b*). It consists of a pulse of width $2d$ located symmetrically about $\pi/2$ and another pulse located symmetrically about $3\pi/2$. The range of pulse width $2d$ varies from 0 to π ; *i.e.* $0 < 2d < \pi$. The output voltage is controlled by varying the pulse-width $2d$. This shape of the output voltage wave shown in Fig. 8.28 (*b*) is called *quasi-square wave*.

Fourier analysis of Fig. 8.28 (*b*) is as under :

$$b_n = \frac{2}{\pi} \int_{(\pi/2 - d)}^{(\pi/2 + d)} V_s \sin n\omega t \cdot d\,(\omega t) = \frac{4V_s}{n\pi}\left[\sin \frac{n\pi}{2} \sin nd\right] \qquad ...(8.61)$$

Positive and negative half cycles of v_0 in Fig. 8.28 (*b*) are symmetrical about $\pi/2$ and $3\pi/2$ respectively. In addition, these half cycles are also identical. As a result, coefficient $a_n = 0$. Thus the waveform of Fig. 8.28 (*b*) can be described by Fourier series as

$$v_0 = \sum_{n=1,3,5}^{\infty} \frac{4V_s}{n\pi} \sin \frac{n\pi}{2} \sin nd \sin n\omega t \qquad ...(8.62)$$

or

$$v_0 = \frac{4V_s}{\pi}\left[\sin d \sin \omega t - \frac{1}{3} \sin 3\,d \sin 3\,\omega t + \frac{1}{5} \sin 5\,d \sin 5\,\omega t......\right] \quad ...(8.63)$$

When pulse width $2d$ is equal to its maximum value of π radians, then the fundamental component of output voltage, from Eq. (8.63), has a peak value of

$$v_{01m} = \frac{4V_s}{\pi} \qquad ...(8.64)$$

For pulse width other than $2\,d = \pi$ radians, the peak value of fundamental component, from Eq. (8.63), is $\dfrac{4V_s}{\pi} \sin d$.

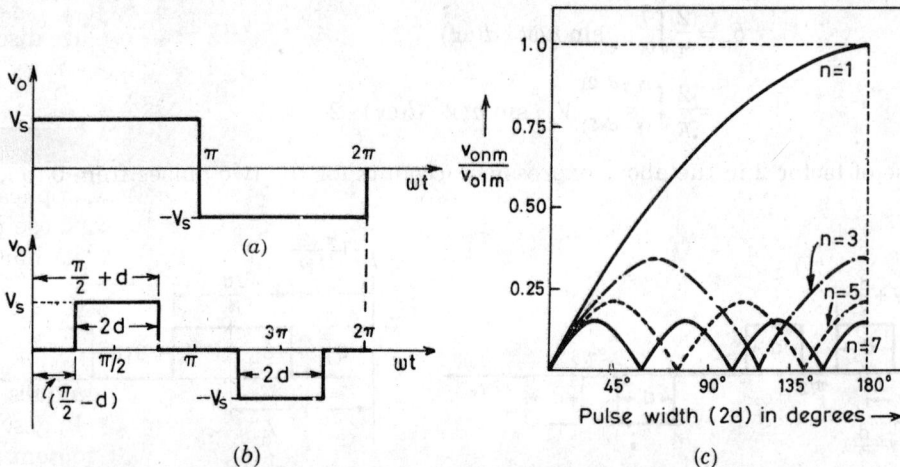

Fig. 8.28. (*a*), (*b*) Single-pulse modulation (SPM) (*c*) Harmonic content in SPM.

If nd is made equal to π or $d = \dfrac{\pi}{n}$ or if pulse width is made equal to $2d = \dfrac{2\pi}{n}$, Eq. (8.62) shows that nth harmonic is eliminated from the inverter output voltage. For example, for eliminating third harmonic, pulse width of $2d$ must be equal to $\dfrac{2\pi}{3} = 120°$.

The peak value of nth harmonic, from Eq. (8.62), is

$$v_{onm} = \frac{4 V_s}{n\pi} \sin nd \qquad \qquad ...(8.65)$$

From Eqs. (8.64) and (8.65),

$$\frac{v_{onm}}{v_{01m}} = \frac{\sin nd}{n} \qquad \qquad ...(8.66)$$

In Eq. (8.66), note that v_{01m} is the peak value of the fundamental component of square voltage wavefrom of width $2d = \pi$. The ratio as given by Eq. (8.66) is plotted in Fig. 8.28 (c) for $n = 1$ (plot of $\sin d$), $n = 3$ (plot of $\sin 3d/3$), $n = 5, 7$ for different pulse widths. It is seen from these curves that when fundamental component is reduced to 0.5 for $2d = 60°$, the amplitude of third harmonic is $\frac{1}{3} \sin 90 = 0.33$. When fundamental component is reduced to about 0.143, all the three harmonics (3, 5, 7) become almost comparable to the fundamental. This shows that in this method of voltage control, a great deal of harmonic content is introduced in the output voltage, particularly at low output voltage levels.

The rms value of output voltage, from Fig. 8.28 (b), is

$$V_{or} = \left[\frac{V_s^2 \cdot 2d}{\pi} \right]^{1/2} = V_s \left[\frac{2d}{\pi} \right]^{1/2} \qquad \qquad ...(8.67)$$

8.6.2. Multiple-pulse Modulation

This method of pulse modulation is an extension of single-pulse modulation. In multiple-pulse modulation (MPM), several equidistant pulses per half cycle are used. For simplicity, the effect of using two symmetrically spaced pulses per half cycle, Fig. 8.29 (a), is investigated here. In this figure, pulse width is taken half of that in Fig. 8.28 (b), but their amplitudes are the same. This means that rms values of pulses in Figs. 8.28 (b) and 8.29 (a) are equal to that given in Eq. (8.67). For the waveform of Fig. 8.29 (a), Fourier constants are as under :

$$b_n = \frac{2}{\pi} \int_0^\pi v_0 \sin n\omega t \cdot d(\omega t)$$

$$= \frac{2}{\pi} \int_{(\gamma - d/2)}^{(\gamma + d/2)} V_s \cdot \sin n\omega t \cdot d(\omega t) \cdot 2$$

The use of factor 2 in the above expression accounts for the two pulses from 0 to π in Fig. 8.29 (a)

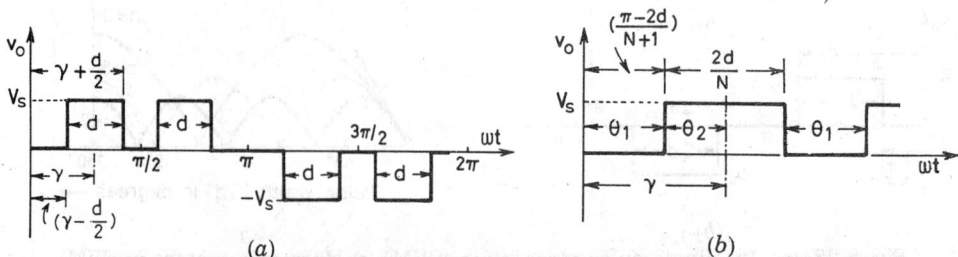

Fig. 8.29. Symmetrical two-pulse modulation pertaining to MPM.

$$\therefore \qquad b_n = \frac{4\,V_s}{n\,\pi}\left|\cos n\,\omega t\right|_{\gamma+d/2}^{\gamma-d/2} = \frac{8V_s}{n\pi}\sin n\gamma \sin \frac{nd}{2} \qquad ...(8.68)$$

As in Fig. 8.28 (b), $a_n = 0$ in Fig. 8.29 (a) also.

Therefore, the waveform of Fig. 8.29 (a) can be described by Fourier series as

$$v_0 = \sum_{n=1,3,5}^{\infty} \frac{8\,V_s}{n\pi}\sin n\gamma \sin \frac{nd}{2} \sin n\omega t \qquad ...(8.69)$$

or $\quad v_0 = \dfrac{8\,V_s}{\pi}\left[\sin \gamma \sin \dfrac{d}{2}\sin \omega t + \dfrac{1}{3}\sin 3\gamma \cdot \sin \dfrac{3d}{2}\sin 3\,\omega t + \dfrac{1}{5}\sin 5\gamma \sin \dfrac{5d}{2}\sin 5\omega t + ...\right]$

$$...(8.70)$$

The amplitude of the nth harmonic of the two-pulse waveform of Fig. 8.29 (a), from Eq. (8.69), is

$$v_n = \frac{8\,V_s}{n\pi}\sin n\gamma \cdot \sin \frac{nd}{2} \qquad ...(8.71)$$

Eq. (8.71) shows that magnitude of v_n depends upon γ and d. This expression also shows that when $\gamma = \dfrac{\pi}{n}$ or $d = \dfrac{2\pi}{n}$, nth harmonic can be eliminated from the output voltage. But this has the effect of reducing the fundamental component of output voltage. For example, take pulse width $2d = 72°$ for single-pulse modulation of Fig. 8.28 (b). Then, from Eq. (8.65), the peak value of fundamental voltage component is

$$v_{01m} = \frac{4\,V_s}{\pi}\sin 36° = 0.7484\,V_s.$$

For two-pulse modulation and pulse width $d = 36°$, γ in Fig. 8.29 (a), is

$$\gamma = \frac{180 - 72}{3} + \frac{72}{4} = 54°$$

or in general. $\qquad \gamma = \dfrac{\pi - 2d}{N+1} + \dfrac{2d}{2N} = \dfrac{\pi - 2d}{N+1} + \dfrac{d}{N} \qquad ...(8.72)$

Eq. (8.72) is valid in case pulses of equal width are symmetrically spaced. Here N is the number of pulses per half cycle.

Eq. (8.72) can also be obtained by referring to Fig. 8.29 (b). For N pulses per half cycle, there are $(N+1)$ intervening equidistant spaces, each of width θ_1 as shown in Fig. 8.29 (b). Note that for these equidistant spaces, $v_0 = 0$. Total width of these $(N+1)$ equidistant spaces $= (N+1)\,\theta_1 = (\pi - \text{width of } N \text{ pulses}) = (\pi - 2d)$

or $\qquad \theta_1 = \dfrac{\pi - 2d}{N+1}$

Fig. 8.29 (b) shows that $\theta_2 = $ half of the pulse width $= \dfrac{d}{N}$. This figure also reveals that

$$\gamma = \theta_1 + \theta_2$$

or $\qquad \gamma = \dfrac{\pi - 2d}{N+1} + \dfrac{d}{N} \qquad ...(8.72)$

Peak value of fundamental voltage component, from Eq. (8.71), is

$$v_{01m} = \frac{8\,V_s}{\pi}\sin 54 \sin 18° = 0.637\,V_s.$$

It is seen from above that fundamental component of output voltage is lower ($0.637\ V_s$) for two-pulse modulation than it is for single-pulse modulation ($0.7484\ V_s$). It can be shown that for more number of pulses per half cycle, the amplitudes of lower order harmonics are reduced but those of some higher harmonics are increased significantly. But this is no disadvantage as higher order harmonics can be filtered out easily.

The symmetrical modulated wave shown in Fig. 8.29 (a) can be generated by comparing an adjustable square voltage wave V_r of frequency ω with a triangular carrier wave V_c of frequency ω_c as shown in Fig. 8.30 (b). This comparison is done in a comparator, Fig. 8.30 (a). In Fig. 8.29 (a), there are only two pulses per half-cycle but in Fig. 8.30 (b), there are four

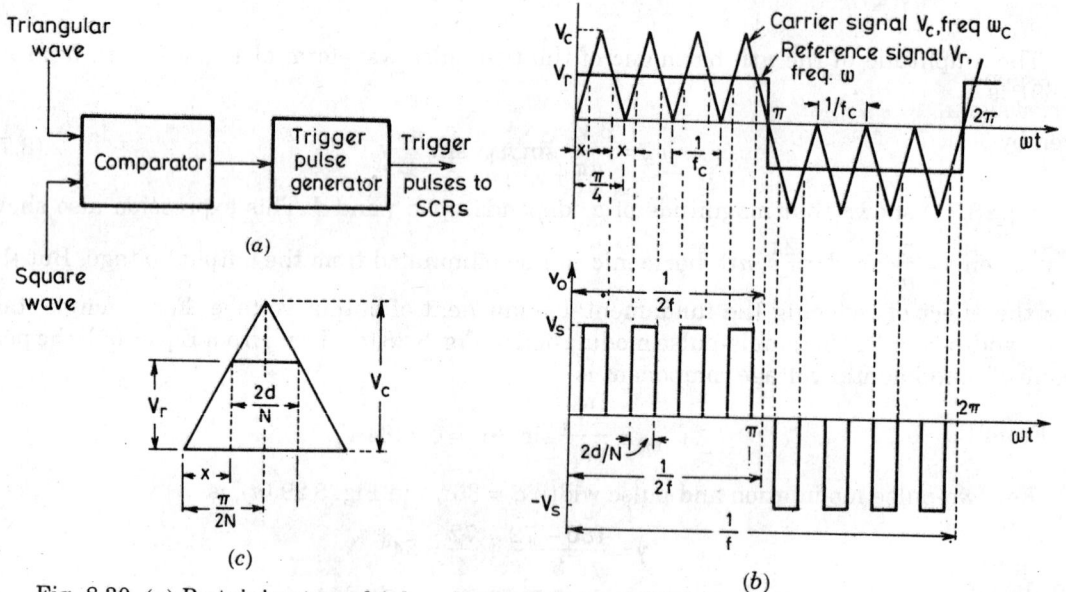

Fig. 8.30. (a) Pertaining to multiple-pulse modulation (MPM) (b) Output voltage waveform with MPM (c) V_c and V_r shown on a larger scale.

pulses per half cycle. The triggering pulses for thyristors are generated at the points of intersection of the carrier and reference signal waves. The firing pulses so generated turn-on the SCRs so that output voltage v_0 is available during the interval triangular voltage wave exceeds the square modulating wave shown in Fig. 8.30 (b). In this figure, f_c and f are the frequencies in Hz for the carrier signal and reference signal respectively. This figure reveals that $\dfrac{1}{f_c} = \dfrac{\pi}{4}$ and $\dfrac{1}{2f} = \pi$ and the number of trigger pulses is $\dfrac{4}{\pi} \times \dfrac{\pi}{1} = 4$. In general, the number of pulses generated per half cycle can be determined from Fig. 8.30 (b) as under :

For triangular carrier wave, pulse width $= \dfrac{1}{f_c}$.

For square reference wave, width of half-cycle $= \dfrac{1}{2f}$.

∴ Number of pulses per half-cycle,

$$N = \text{Number of hill-tops per half-cycle,}$$

$$N = \frac{\text{Length of half-cycle of square reference wave}}{\text{Width of one cycle of triangular carrier wave}}$$

or
$$N = \frac{1/2f}{1/f_c} = \frac{f_c}{2f} = \frac{\omega_c}{2\omega} \qquad \qquad ...(8.73)$$

Note that N in Eq. (8.73) must be an integer. The pulse height of the reference, or modulating, signal can be controlled within the range $0 < V_r < V_c$ and pulse width $\frac{2d}{N}$ varied in the range $0 < \frac{2d}{N} < \frac{\pi}{N}$ by adjusting the magnitude V_r of the reference square wave. The pulse width is $2d/N$ on the assumption of same rms voltage as in single-pulse modulation.

In Fig. 8.30 (b), pulse width $2d/N$ is given by

$$\frac{2d}{N} = \left(\frac{\pi}{4} - 2x \right)$$

A general expression for the pulse width can be obtained by sketching the first cycle of carrier signal on a larger scale as in Fig. 8.30 (c). From this figure, pulse width, in general, is given by

$$\frac{2d}{N} = \left(\frac{\pi}{4} - 2x \right) \qquad \qquad ...(8.74)$$

where x, defined in Fig. 8.30 (c), is

$$\frac{V_c}{\pi/2N} = \frac{V_r}{x}$$

or
$$x = \frac{\pi}{2N} \cdot \frac{V_r}{V_c}$$

From Eq. (8.74), the pulse width is

$$\frac{2d}{N} = \left(\frac{\pi}{N} - \frac{\pi}{N} \cdot \frac{V_r}{V_c} \right) = \left(1 - \frac{V_r}{V_c} \right) \frac{\pi}{N} \qquad \qquad ...(8.75)$$

In MPM method, lower order harmonics can be eliminated by a proper choice of $2d$ and γ. But the rms voltage in Figs. 8.28 to 8.30 is the same, *i.e.*

$$V_{or} = V_s \left[\frac{2d}{\pi} \right]^{1/2}$$

This means that if lower order harmonics are eliminated, the magnitude of higher order harmonics would go up. But this is not a disadvantage, as higher order harmonics can be filtered out by the use of filters at the output terminals of the inverters.

8.6.3. Sinusoidal-pulse Modulation (SPWM)

In this method of modulation, several pulses per half cycle are used as in the case of multiple-pulse modulation (MPM). In MPM, the pulse width is equal for all the pulses. But in SPWM, the pulse width is a sinusoidal function of the angular position of the pulse in a cycle as shown in Fig. 8.31.

For realizing SPWM, a high-frequency triangular carrier wave v_c is compared with a sinusoidal reference wave v_r of the desired frequency. The intersection of v_c and v_r waves determines the switching instants and commutation of the modulated pulse. In Fig. 8.31, V_C is the peak value of triangular carrier wave and V_r that of the reference, or modulating, signal.

The carrier and reference waves are mixed in a comparator as in Fig. 8.30 (a). When sinusoidal wave has magnitude higher than the triangular wave, the comparator output is

(a)

(b)

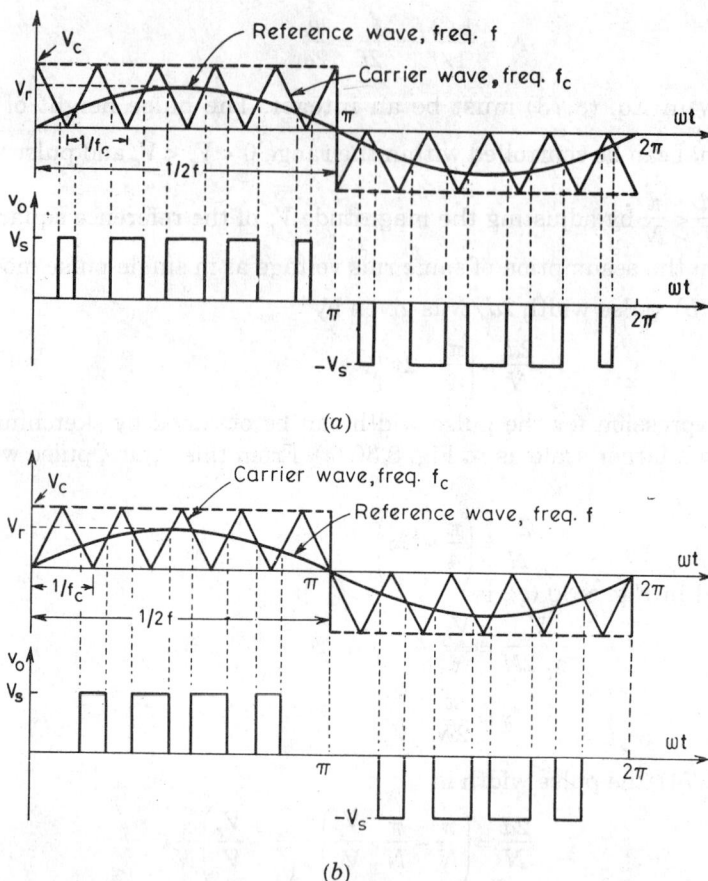

Fig. 8.31. Output voltage waveforms with sinusoidal pulse modulation.

high, otherwise it is low. The comparator output is processed in a trigger pulse generator in such a manner that the output voltage wave of the inverter has a pulse width in agreement with the comparator output pulse width.

When triangular carrier wave has its peak coincident with zero of the reference sinusoid, there are $N = \dfrac{f_c}{2f}$ pulses per half cycle ; Fig. 8.31 (a) has five pulses. In case zero of the triangular wave coincides with zero of the reference sinusoid, there are $(N-1)$ pulses per half cycle ; Fig. 8.31 (b) has $\left(\dfrac{f_c}{2f} - 1\right)$, i.e. four, pulses per half cycle.

The ratio of V_r/V_c is called the *modulation index* (MI) and it controls the harmonic content of the output voltage waveform. The magnitude of fundamental component of output voltage is proportional to MI, but MI can never be more than unity. Thus the output voltage is controlled by varying MI.

Harmonic analysis of the output modulated voltage wave reveals that SPWM has the following important features :

(i) For MI less than one, largest harmonic amplitudes in the output voltage are associated with harmonics of order $f_c/f \pm 1$ or $2N \pm 1$, where N is the number of pulses per half cycle. Thus, by increasing the number of pulses per half cycle, the order of dominant harmonic frequency can be raised, which can then be filtered out easily. In Fig. 8.31 (a), N = 5, therefore harmonics of order 9 and 11 become significant in the output voltage. It may be noted that the highest order of significant harmonic of a modulated voltage wave is centred around the carrier frequency f_c [in Fig. 8.31 (a), $f_c = 10$].

It is observed from above that as N is increased, the order of significant harmonic increases and the filtering requirements are accordingly minimised. But higher value of N entails higher switching frequency of thyristors. This amounts to more switching losses and therefore an impaired inverter efficiency. Thus a compromise between the filtering requirements and inverter efficiency should be made.

(*ii*) For *MI* greater than one, lower order harmonics appear, since for *MI* > 1, pulse width is no longer a sinusoidal function of the angular position of the pulse.

In addition to the three PWM techniques discussed above, there is another PWM technique called multiple-pulse modulation with selective reduction (MPMSR). In this technique, the number of *M* pulse positions in each quarter cycle are so selected as to reduce or eliminate *M* harmonics from the output voltage waveform [6]. This PWM technique will, however, not be discussed here.

8.6.4. Realization of PWM in Single-phase Bridge Inverters

The output voltage waveforms shown in Figs. 8.28 to 8.31 reveal that output voltage from an inverter is V_s, zero or $- V_s$. Such waveforms can be realized in single-phase inverters as under :

(*a*) **Single-phase full-bridge inverter.** In the inverter of Fig. 8.2 (*a*), when + V_s is to be obtained in the positive half cycle, thyristors T1, T2 are turned on. For obtaining – V_s in the negative half cycle, thyristors T3, T4 should be turned on. For zero output voltage, *i.e.* if the load is to be short-circuited ; then T1, D3 or T3, D1 from positive group ; or T4, D2 or T2, D4 from negative group should conduct depending upon the direction of load current. This means that for obtaining zero output voltage at the end of each pulse, one of the two conducting SCRs should only be turned off. Under this strategy, only one thyristor need be turned on for obtaining the next voltage pulse. Switching on and commutation of thyristors should be so arranged as to utilize the thyristors symmetrically. Let us illustrate this with an example.

Suppose output voltage of pulse width $2\pi/3$ radians is to be obtained in each half cycle. This pulse width is symmetrically placed as shown in Fig. 8.32. The waveform of load current i_0 is

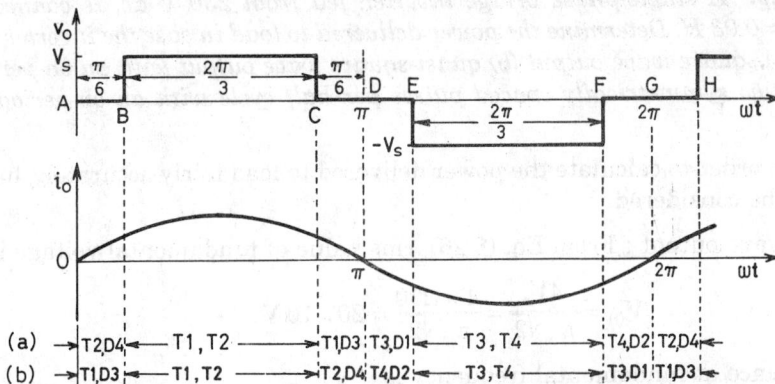

Fig. 8.32. Conduction of various components for single-phase bridge inverter of Fig. 8.2 (*a*).

assumed as sketched in Fig. 8.32. It is obvious from these two waveforms that from B to C ; T1, T2 should conduct and from E to F ; T3, T4 should be on.

From C to D, $v_0 = 0$ but current i_0 is positive. Therefore, from C to D ; either T1, D3 or T2, D4 should conduct, this is shown in Fig. 8.32.

From D to E, $v_0 = 0$ but i_0 is negative. Negative current with zero output voltage can exist only if T3 or T4 together with one diode are on. When T3 is on, then T3, D1 should conduct and with T4, D2 should conduct, Fig. 8.32.

From F to G, $v_0 = 0$, i_0 is negative. For this ; T4, D2 or T3, D1 should conduct. From D to E, if T3, D1 conduct, then now T2, D4 must conduct in order to utilize the thyristors symmetrically. In case T4, D2 conduct from D to E, then T3, D1 should conduct from F to G.

From G to H ; T2, D4 or T1, D3 conduct as shown. The conduction from G to H is similar to that from A to B.

It may be observed from Fig. 8.32 that, during one cycle, each thyristor conducts for 150° and each diode for 30°.

(b) **Single-phase half-bridge inverter.** In single-phase half-bridge inverter of Fig. 8.1 (a), zero value of output voltage cannot be obtained. The output voltage can either be $V_s/2$ or $-V_s/2$. In Fig. 8.33, $V_s/2$ from A to B is obtained with T1 on, from B to C with T2 on, from C to D with T1 on and so on. For obtaining a symmetrical waveform for output voltage in Fig. 8.33, interval AB = interval DE ; interval BC = interval EF and so on. The output voltage can be controlled through the adjustment of width $2d$.

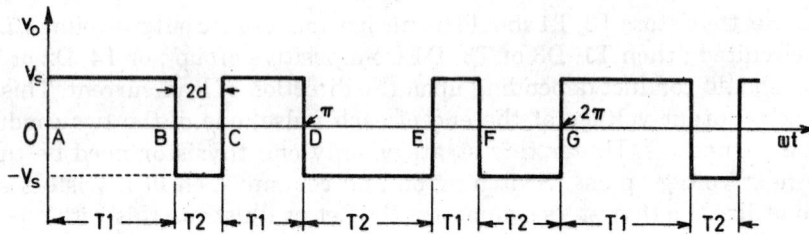

Fig. 8.33. Output voltage waveform obtained through PWM in half-bridge inverter.

Example 8.9. *A single-phase bridge inverter, fed from 230 V dc, is connected to load $R = 10\ \Omega$ and $L = 0.03$ H. Determine the power delivered to load in case the inverter is operating at 50 Hz with (a) square wave output (b) quasi-square wave output with an on-period of 0.5 of a cycle and (c) two symmetrically spaced pulses per half cycle with an on-period of 0.5 of a cycle.*

Solution. In order to calculate the power delivered to load fairly accurately, harmonics up to seventh may be considered.

(a) Square-wave output : From Eq. (8.26), rms value of fundamental voltage is

$$V_{01} = \frac{4V_s}{\pi \cdot \sqrt{2}} = \frac{4 \times 230}{\pi \cdot \sqrt{2}} = 207.10\ \text{V}$$

Load impedance at fundamental frequency is

$$Z_1 = [10^2 + (2\pi \times 50 \times 0.03)^2]^{1/2} = 13.7414\ \Omega$$

$$I_{01} = \frac{207.10}{13.7414} = 15.0712 \text{ A}$$

$$V_{03} = \frac{4 \times 230}{3 \times \pi \times \sqrt{2}} = 69.035 \text{ V}$$

and

$$Z_3 = \sqrt{10^2 + (2\pi \times 50 \times 3 \times 0.03)^2} = 29.9906 \ \Omega$$

$$I_{03} = \frac{69.035}{29.9906} = 2.302 \text{ A}$$

Similarly,

$$I_{o5} = \frac{920}{5 \times \pi \times \sqrt{2}} \times \frac{1}{\sqrt{10^2 + (2\pi \times 50 \times 5 \times 0.03)^2}} = 0.8598 \text{ A}$$

$$I_{07} = \frac{920}{7 \times \pi \times \sqrt{2}} \times \frac{1}{\sqrt{10^2 + (2\pi \times 50 \times 7 \times 0.03)^2}} = 0.4434 \text{ A}$$

Rms value of resultant load current,

$$I_0 = \left[I_{01}^2 + I_{03}^2 + I_{05}^2 + I_{07}^2 \right]^{1/2}$$

Power delivered to load $= I_0^2 R$

$$= [15.0712^2 + 2.302^2 + 0.8598^2 + 0.4434^2] \times 10$$

$$= 2333.76 \text{ W}$$

(b) **Quasi-square wave output** : For quasi-square wave or single-pulse modulated wave, use Eq. (8.62), where pulse width, $2d = 0.5 \times 180° = 90°$ or $d = 45°$. From this equation, rms value of fundamental voltage is

$$V_{01} = \frac{4V_s}{\pi \cdot \sqrt{2}} \sin d = \frac{4 \times 230}{\pi \cdot \sqrt{2}} \sin 45° = 146.423 \text{ V}$$

$$I_{01} = \frac{146.423}{13.7414} = 10.6556 \text{ A}$$

$$V_{03} = \frac{4 \times 230}{3 \times \pi \times \sqrt{2}} \sin 3 \times 45° = 48.8075 \text{ V}$$

$$I_{03} = \frac{48.8075}{29.9906} = 1.6274 \text{ A}$$

Similarly,

$$I_{05} = \frac{4 \times 230}{5 \times \pi \times \sqrt{2}} \sin (5 \times 45°) \times \frac{1}{48.17324} = -0.6079 \text{ A}$$

$$I_{07} = \frac{4 \times 230}{7 \times \pi \times \sqrt{2}} \sin (7 \times 45°) \times \frac{1}{66.727} = -0.3135 \text{ A}$$

Power delivered to load $= (10.6556^2 + 1.6274^2 + 0.6079^2 + 0.3135^2) \times 10$

$$= 1166.58 \text{ W}$$

(c) For two symmetrically spaced pulses per half cycle, use Eq. (8.69). For this equation, $2d = 0.5 \times 180 = 90°$ or $d = 45°$ and from Eq. (8.72), $\gamma = \frac{180 - 90}{3} + \frac{45}{2} = 52.5°$. From Eq. (8.69), rms value of fundamental voltage is

$$V_{01} = \frac{8V_s}{\pi \cdot \sqrt{2}} \sin \gamma \sin \frac{d}{2} = \frac{8 \times 230}{\pi \cdot \sqrt{2}} \sin 52.5° \sin \frac{45}{2} = 125.755 \text{ V}$$

$$I_{01} = \frac{125.755}{13.7414} = 9.1515 \text{ A}$$

$$V_{03} = \frac{8 \times 230}{3 \times \pi \times \sqrt{2}} \cdot \sin (52.5 \times 3) \cdot \sin \left(\frac{45}{2} \times 3 \right) = 48.815 \text{ V}$$

$$I_{03} = \frac{48.815}{29.9906} = 1.6277 \text{ A}$$

$$I_{05} = \frac{8 \times 230}{5 \times \pi \sqrt{2}} \sin(52.5 \times 5) \sin(22.5 \times 5) \times \frac{1}{48.17324} = 1.575 \text{ A}$$

$$I_{07} = \frac{8 \times 230}{7 \times \pi \times \sqrt{2}} \sin(52.5 \times 7) \sin(22.5 \times 7) \times \frac{1}{66.727} = 0.0443 \text{ A}$$

Power delivered to load $\quad = (9.1515^2 + 1.6277^2 + 1.575^2 + 0.0443^2) \times 10$

$$= 888.82 \text{ W}.$$

8.7. REDUCTION OF HARMONICS IN THE INVERTER OUTPUT VOLTAGE

There are several industrial applications which may allow a harmonic content of 5% of its fundamental component of input voltage when inverters are used. Actually, the inverter output voltage may have harmonic content much higher than 5% of its fundamental component. In order to bring this harmonic content to a reasonable limit of 5%, one method is to insert filters between the load and inverter. If the inverter output voltage contains high frequency harmonics, these can be reduced by a low-size filter. For the attenuation of low-frequency harmonics, however, the size of filter components increases. This makes the filter circuit costly, bulky and weighty and in addition, the transient response of the system becomes sluggish. This shows that lower order harmonics from the inverter output voltage should be reduced by some means other than the filter. Subsequent to this, high frequency component from this voltage can easily be attenuated by a low-size, low-cost filter. The object of this section is to study these methods of reducing low-order harmonics from the output voltage of an inverter.

8.7.1. Harmonic Reduction by PWM

It has already been discussed that when there are several pulses per half cycle, lower-order harmonics are eliminated. Fig. 8.34 illustrates output voltage waveform that can be obtained from a single-phase full-bridge inverter. This waveform can also be obtained from a single-phase half-bridge inverter, but then the amplitude of voltage wave would be $V_s/2$. The waveform of Fig. 8.34 needs ten commutations per cycle (= 360°) instead of two in an unmodulated wave. The voltage waveform of Fig. 8.34 is symmetrical about π as well as $\pi/2$.

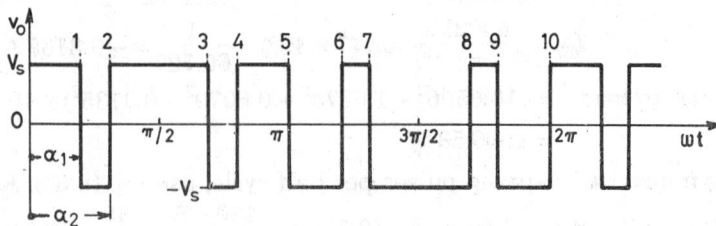

Fig. 8.34. Harmonic reduction by PWM in single-phase inverter.

As this voltage waveform has quarter-wave symmetry, $a_n = 0$.

$$\therefore \qquad b_n = \frac{4}{\pi} V_s \left[\int_0^{\alpha_1} \sin n\omega t \cdot d(\omega t) - \int_{\alpha_1}^{\alpha_2} \sin n\omega t \cdot d(\omega t) + \int_{\alpha_2}^{\pi/2} \sin n\omega t \cdot d(\omega t) \right] \qquad ...(8.76)$$

$$= \frac{4V_s}{\pi} \left[\frac{1 - 2\cos n\alpha_1 + 2\cos n\alpha_2}{n} \right\} \qquad ...(8.76)$$

If third and fifth harmonics are to be eliminated, then from Eq. (8.76),

$$b_3 = \frac{4V_s}{\pi} \left[\frac{1 - 2 \cos 3\,\alpha_1 + 2 \cos 3\,\alpha_2}{3} \right] = 0$$

and

$$b_5 = \frac{4V_s}{\pi} \left[\frac{1 - 2 \cos 5\alpha_1 + 2 \cos 5\alpha_2}{5} \right] = 0$$

or

$$1 - 2 \cos 3\alpha_1 + 2 \cos 3\alpha_2 = 0$$

and

$$1 - 2 \cos 5\alpha_1 + 2 \cos 5\alpha_2 = 0$$

The above two simultaneous equations can be solved numerically to calculate α_1 and α_2 under the condition that $0 < \alpha_1 < 90°$ and $\alpha_1 < \alpha_2 < 90°$. This gives $\alpha_1 = 23.62°$ and $\alpha_2 = 33.304°$.

With these values of α_1, α_2, the amplitudes of 7th, 9th and 11th harmonics, from Eq. (8.76), are as under :

$$b_7 = \frac{4V_s}{7\pi} [1 - 2 \cos 7 \times 23.62 + 2 \cos 7 \times 33.304] = 0.31555\,V_s$$

$$b_9 = \frac{4V_s}{9\pi} [1 - 2 \cos 9 \times 23.62 + 2 \cos 9 \times 33.304] = 0.5202\,V_s$$

and

$$b_{11} = \frac{4V_s}{11\pi} [1 - 2 \cos 11 \times 23.62 + 2 \cos 11 \times 33.304] = 0.3867\,V_s$$

The amplitude of the fundamental component for these values of α_1 and α_2 is

$$b_1 = \frac{4V_s}{\pi} [1 - 2 \cos 23.62 + 2 \cos 33.304] = 1.0684\,V_s$$

The amplitude of the fundamental component of unmodulated output voltage wave is

$$b_1{}_n = \frac{4V_s}{\pi} = 1.27324\,V_o$$

In terms of the fundamental component of unmodulated voltage wave, the amplitude of 7th, 9th and 11th harmonics are respectively 24.78% (= $0.31555 \times 100/1.27324$), 40.86% and 30.37% but third and fifth harmonics are eliminated from the inverter output voltage wave. The amplitude of the fundamental voltage is 83.91% or 0.8391 times the amplitude of fundamental component of unmodulated voltage wave. Thus, with this method of harmonic reduction, inverter is derated by (100-83.91) 16.09%. Another disadvantage of this method is that there are additional eight commutations per cycle and this leads to more switching losses in the thyristors.

8.7.2. Harmonic Reduction by Transformer Connections

Output voltage from two or more inverters can be combined by means of transformers to get a net output voltage with reduced harmonic content. The essential condition of this scheme is that the output voltage waveforms from the inverters must be similar but phase-shifted from each other. Fig. 8.35 (a) illustrates two transformers in series. Their output voltages, v_{01} from inverter 1 and v_{02} from inverter 2, are shown in Fig. 8.34 (b). Here v_{02} waveform is taken to have a phase shift of $\pi/3$ radians with respect to v_{01} waveform as shown. The resultant output voltage v_0 is obtained by adding the vertical ordinates of v_{01} and v_{02}. It is seen that v_0 has an amplitude of $2\,V_s$ from $\frac{\pi}{3}$ to π, $\frac{4\pi}{3}$ to 2π and so on. Note that shape of the output voltage wave v_0 is a quasi-square wave.

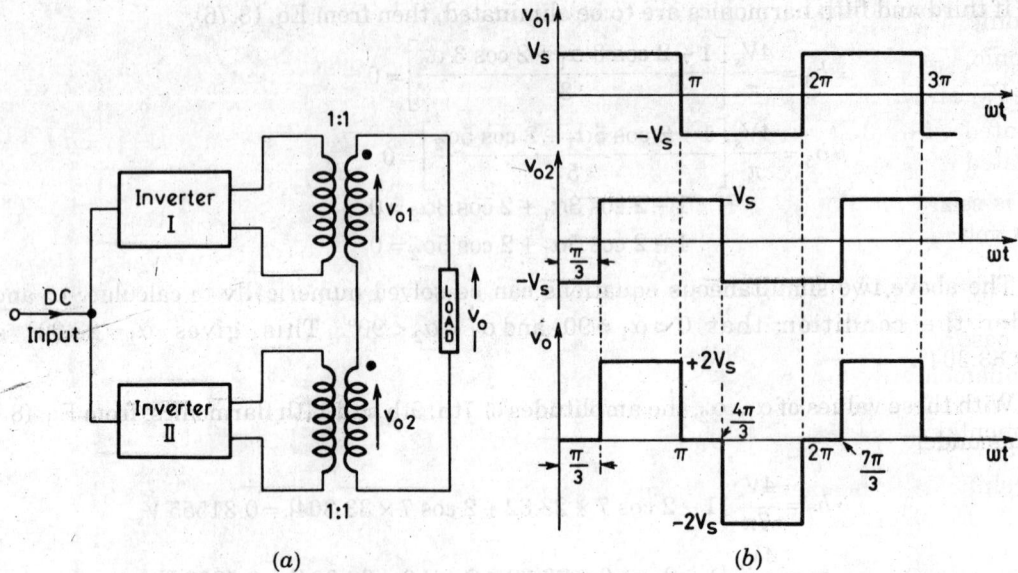

Fig. 8.35. (a) Harmonic reduction by transformer connections (b) Elimination of third and other triplen harmonics.

The Fourier analysis of waveforms v_{01} and v_{02} gives

$$v_{01} = \frac{4V_s}{\pi}\left[\sin\omega t + \frac{1}{3}\sin 3\omega t + \frac{1}{5}\sin 5\omega t + \frac{1}{7}\sin 7\omega t + \dots\right]$$

$$v_{02} = \frac{4V_s}{\pi}\left[\sin\left(\omega t - \frac{\pi}{3}\right) + \frac{1}{3}\sin 3\left(\omega t - \frac{\pi}{3}\right) + \frac{1}{5}\sin 5\left(\omega t - \frac{\pi}{3}\right)\right.$$
$$\left. + \frac{1}{7}\sin 7\left(\omega t - \frac{\pi}{3}\right) + \dots\right]$$

The resultant voltage v_0 is $= v_{01} + v_{02}$

$$= \frac{4V_s}{\pi}\sqrt{3}\left[\sin\left(\omega t - \frac{\pi}{6}\right) + \frac{1}{5}\sin\left(5\omega t + \frac{\pi}{6}\right) + \frac{1}{7}\sin\left(7\omega t - \frac{\pi}{6}\right) + \dots\right] \quad \dots(8.77)$$

The expression for resultant voltage v_0 as given above can be obtained from v_{01} and v_{02} analytically or graphically. The summation by graphical method is carried out as under :

An examination of the expressions for v_{01} and v_{02} reveals that for the fundamental frequency ; V_{02} lags V_{01} by 60°, this is shown in Fig. 8.36 (a). The resultant of V_{01} and V_{02} must be $\sqrt{3}$ times V_{01} (or V_{02}) and at the same time, the resultant lags V_{01} by 30°. Net value of fundamental frequency voltage must, therefore, be associated with $\sqrt{3}\sin(\omega t - \pi/6)$. For third

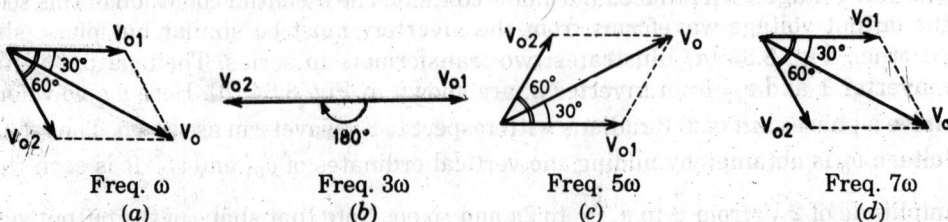

Fig. 8.36. Pertaining to the summation of first, third, fifth and seventh harmonic voltages.

harmonic, V_{02} lags V_{01} by 180°, Fig. 8.36 (b), their resultant is therefore zero. For fifth harmonic, V_{02} lags V_{01} by 300° or V_{02} leads V_{01} by 60°, see Fig. 8.36 (c) ; its resultant is $\sqrt{3}$ times V_{01} or V_{02} and it leads V_{01} by 30°. Thus, the resultant of fifth harmonic voltage must be associated with $\sqrt{3} \sin(\omega t + \pi/6)$. Similarly, resultant of seventh harmonic voltage must be associated with $\sqrt{3} \sin(\omega t - \pi/6)$, Fig. 8.36 (d).

It is seen from Eq. (8.77) that third and other triplen harmonics are eliminated from net output voltage wave. The amplitude of fundamental component of v_0 is

$$V_{01m} = \frac{4V_s}{\pi} \sqrt{3}$$

In case output voltages v_{01}, v_{02} from inverters 1 and 2 has no phase shift, then amplitude of the fundamental voltage wave is $8 V_s/\pi$. This shows that with phase shift, the amplitude of the fundamental voltage is $\dfrac{4V_s}{\pi} \sqrt{3} \times \dfrac{\pi}{8V_s} = \dfrac{\sqrt{3}}{2}$ times the amplitude of fundamental voltage with no phase shift. With this method of harmonic reduction, there is thus a derating of

$$\left[\frac{\dfrac{8V_s}{\pi} I - \dfrac{4}{\pi} V_s \cdot \sqrt{3}\, I}{\dfrac{8V_s}{\pi} I} \right] 100 = \left(1 - \frac{\sqrt{3}}{2} \right) 100 = 13.4\%$$

in their net output power so far as fundamental component is concerned. The degree of derating with this method is, however, less than that obtained in PWM harmonic reduction method.

The disadvantage of this method of harmonic reduction is the need for more number of inverters and transformers of similar ratings.

8.7.3. Harmonic Reduction by Stepped-wave Inverters

In this method, pulses of different widths and heights are superimposed to produce a resultant stepped wave with reduced harmonic content. Fig. 8.37 illustrates two stepped-wave inverters fed from a common dc supply. The two transformers used have different turns ratio from primary to secondary. In this figure, the turns ratio from primary to secondary is assumed three for transformer 1 and unity for transformer 2.

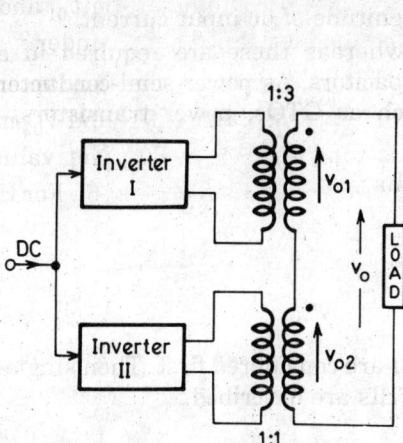

Fig. 8.37. Harmonic reduction by stepped
wave inverters.

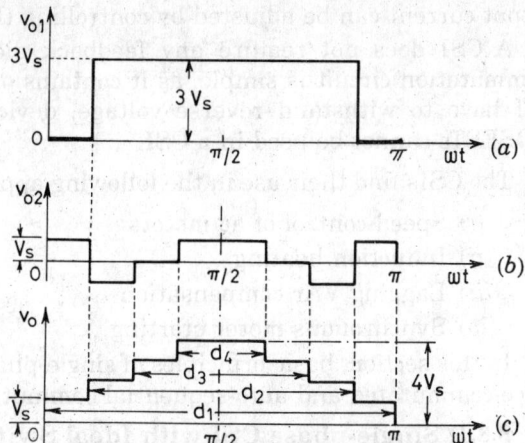

Fig. 8.38. Waveforms for stepped-wave
inverters.

The inverter I is so gated that its output voltage is v_{01} as shown in Fig. 8.38 (a). During the first-half cycle, output voltage level is either zero or positive. During second half cycle (not shown in the figure), the output voltage would be either zero or negative. This output voltage waveform is given the name *two-level modulation*.

For inverter II, the triggering is so arranged as to give output voltage v_{02} as shown in Fig. 8.38 (b). It is seen from v_{02} waveform that the level of output voltage is positive, negative or zero during the first half cycle, this inverter has therefore *three-level modulation*. The resultant output voltage from a series combination of inverters I and II is obtained by superimposing the waveforms of Figs. 8.38 (a) and (b). This summation depicted in Fig. 8.38 (c) shows that the amplitude of output voltage is $4\ V_s$ and waveform has four steps. Fourier analysis of Fig. 8.38 (c) would give harmonics whose amplitudes would depend upon the values of d_1, d_2, d_3, d_4 and amplitude of v_0. By a proper choice of these parameters ; third, fifth and seventh harmonics can be eliminated or attenuated considerably and the fundamental component optimised.

Note that it is the three-level modulation of second inverter that helps in achieving the required wave-stepping of the resultant output voltage waveform. It is seen from the waveform of Fig. 8.38 (c) that this waveshape is more nearer to sinusoidal wave.

8.8. CURRENT SOURCE INVERTERS

So far, voltage-source inverters have been discussed. In these inverters, input voltage is maintained constant and the amplitude of output voltage does not depend on the load. However, the waveform of load current as well as its magnitude depends upon the nature of the load impedance.

In the current-source inverters (CSIs), input current is constant but adjustable. The amplitude of output current from CSI is independent of the load. However, the magnitude of output voltage and its waveform output from CSI is dependent upon the nature of load impedance. The dc input to CSI is obtained from a fixed voltage ac source through a controlled rectifier bridge, or through a diode bridge and a chopper. In order that current input to CSI is almost ripple free, L-filter is used before CSI.

A CSI converts the input dc current to an ac current at its output terminals. The output frequency of ac current depends upon the rate of triggering the SCRs. The amplitude of ac output current can be adjusted by controlling the magnitude of dc input current.

A CSI does not require any feedback diodes, whereas these are required in a VSI. Commutation circuit is simple, as it contains only capacitors. As power semi-conductors in a CSI have to withstand reverse voltage, devices such as GTOs, power transistors, power MOSFETs cannot be used in a CSI.

The CSIs find their use in the following applications :

    (i) Speed control of ac motors

    (ii) Induction heating

    (iii) Lagging VAr compensation

    (iv) Synchronous motor starting.

In this section, basic principles of single-phase CSI are considered first. Then single-phase force-commutated and auto-sequential commutated CSIs are described.

8.8.1. Single-phase CSI with Ideal Switches

A single-phase CSI with ideal thyristors is shown in Fig. 8.39 (a). Here a thyristor is assumed an ideal switch with zero commutation time. Positive directions for load voltage v_0 and

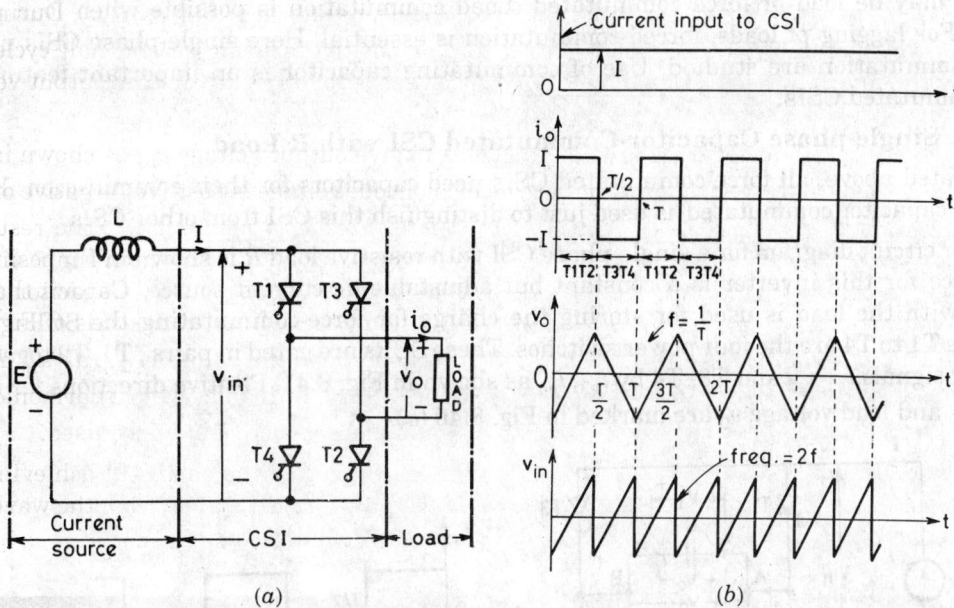

Fig. 8.39. (a) Power circuit diagram and (b) waveforms for an ideal single-phase CSI.

load current i_0 are indicated in Fig. 8.39 (a). The source consists of a voltage source E and a large inductance L in series with it. The function of high-impedance reactor in series with voltage source is to maintain a constant current at the input terminals of CSI. In other words, dc input current I to CSI is constant as shown in Fig. 8.39 (b).

In Fig. 8.39 (a), when T1, T2 are on, load current i_0 is positive and equal to I. When T3, T4 are on, load current i_0 is negative and equal to $-I$ as shown in Fig. 8.39 (b). The output frequency of i_0 can be varied by controlling the frequency of triggering the thyristor pairs T1, T2 and T3, T4. It is seen from Fig. 8.39 (b) that output current i_0 is a square wave of amplitude equal to the dc input current I.

Assume that load consists of a capacitor C. It is known for a capacitor that

$$i_0 = C \frac{dv_0}{dt}$$

As i_0 is constant, slope $\frac{dv_0}{dt}$ must be constant over every half cycle. This slope is positive from zero to $T/2$ and negative from $T/2$ to T. On this basis, waveform of load voltage v_0 is drawn in Fig. 8.39 (b). The input voltage to the CSI, i.e. $v_{in} = v_0$ when T1, T2 conduct and $v_{in} = -v_0$ when T3, T4 conduct. Note that waveshape of v_{in} can be drawn by referring to the waveform of v_0. For $f = 1/T$ as the frequency of output voltage v_0 or current, input voltage v_{in} has a frequency of $2f$, this is revealed by an examination of Fig. 8.39 (b).

The dc current I, input to CSI, is always unidirectional. If average value of v_{in} is positive, power flows from source to load. In case average value of v_{in} is negative, power flows from load to source, i.e. regeneration of power takes place.

In a practical inverter, the load current waveform is not a square wave owing to the fact that rise and fall of current can't be instantaneous as shown in Fig. 8.39 (b). On account of finite commutation time, a practical inverter has finite times for the rise and fall of current.

CSIs may be load or force commutated. Load commutation is possible when load pf is leading. For lagging pf loads, forced-commutation is essential. Here single-phase CSIs using forced commutation are studied. Use of commutating capacitor is an important feature of force-commutated CSIs.

8.8.2. Single-phase Capacitor-Commutated CSI with R Load

As stated above, all force-commutated CSIs need capacitors for their commutation. Here the term 'capacitor commutated' is used just to distinguish this CSI from other CSIs.

Power circuit diagram for a single-phase CSI with resistive load R is shown in Fig. 8.40 (a). The source for this inverter is a constant but adjustable dc current source. Capacitor C in parallel with the load is used for storing the charge for force-commutating the SCRs. The thyristors T1 to T4 are the four power switches. These SCRs are gated in pairs ; T1, T2 together by gating signals i_{g1}, i_{g2} and T3, T4 by i_{g3}, i_{g4} as shown in Fig. 8.41. Positive directions for load current i_0 and load voltage v_0 are marked in Fig. 8.40 (a).

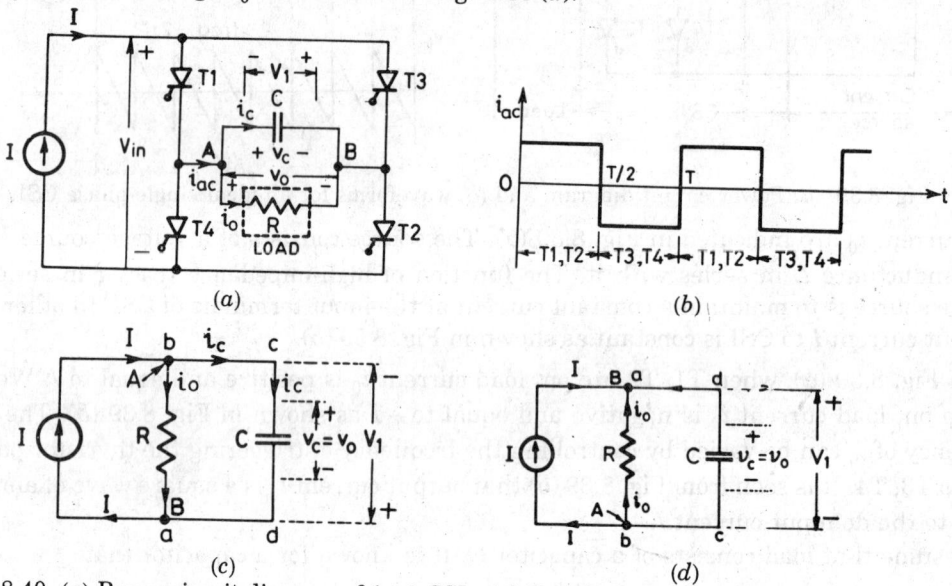

Fig. 8.40. (a) Power circuit diagram of 1—ϕ CSI with R load (b) AC output current waveform (c) Equivalent circuit of Fig. (a) for $0 < t < T/2$ and (d) Equivalent circuit of Fig. (a) for $T/2 < t < T$.

Before $t = 0$, let the capacitor voltage be $v_c = -V_1$, i.e. capacitor has left plate negative and right plate positive in Fig. 8.40 (a). When T1, T2 are gated at $t = 0$, the capacitor voltage v_c reverse biases conducting thyristors T3, T4 ; these are therefore commutated immediately. The source current I now flows through T1, parallel combination of R and C and through T2. From zero to $T/2$, $i_{T1} = i_{T2} = I$, output current $i_{ac} = I$; capacitor voltage v_c changes from $-V_1$ to V_1 through the charging of C by current i_c. Note that here load voltage $v_0 = v_c$. Thus, the waveform of $i_0 = \dfrac{v_0}{R} = \dfrac{v_c}{R}$ has the same nature as that of v_c, see Fig. 8.41. When T3, T4 are gated at $t = T/2$, $v_c = V_1$ reverse biases T1, T2 ; these are therefore turned-off immediately. The source current now flows through T3, parallel combination of R, C and T4. From $T/2$ to T, $i_{T3} = i_{T4} = +I$ but $i_{ac} = -I$. The variation of ac current i_{ac} is shown in Fig. 8.40 (b).

Under steady state operation of the CSI, various current and voltage waveforms are sketched in Fig. 8.41. At $t = 0$, when the capacitor is charged with voltage $v_c = -V_1$, then

$v_0 = v_c = -V_1$ and load current $i_0 = -\dfrac{V_1}{R_1} = -I_1$. From $t = 0$ to $T/2$, capacitor charges from

$-V_1$ to V_1. Therefore, at $t = \dfrac{T}{2}$, $i_0 = \dfrac{v_c}{R} = \dfrac{v_0}{R} = \dfrac{V_1}{R} = I_1$. This is shown in Fig. 8.41. Input voltage

$v_{in} = v_0$ from $t = 0$ to $T/2$ whereas $v_{in} = -v_0$ from $T/2$ to T.

It is seen from Fig. 8.40 (a) that when T1, T2 are conducting for $0 < t < T/2$, currents i_c, i_0 are leaving the node A and current I is entering the node A. Therefore, equivalent circuit for Fig. 8.40 (a) for $0 < t < \dfrac{T}{2}$ may be drawn as shown in Fig. 8.40 (c). KCL at node A in Fig. 8.40 (a) or at node b in Fig. 8.40 (c) gives

$$i_0 + i_c = I \quad \text{or} \quad i_c = I - i_0$$

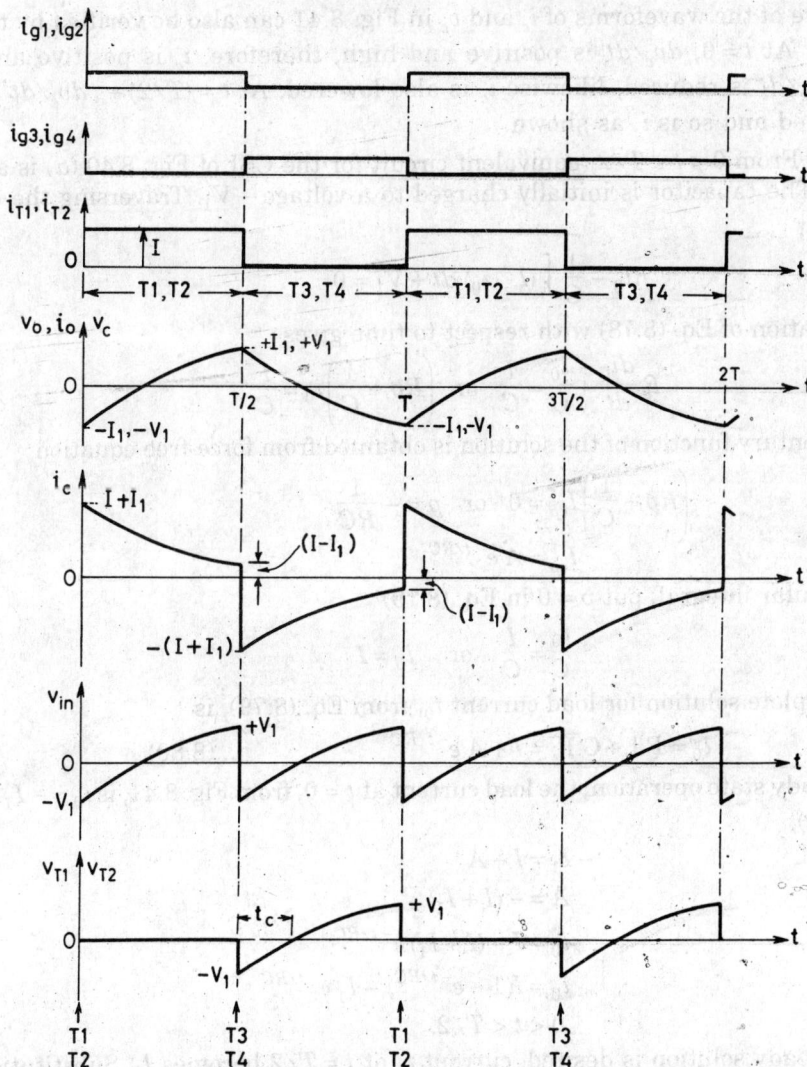

Fig. 8.41. Current and voltage waveforms for single-phase CSI with R load.

At $t = 0$, $i_0 = -I_1$, therefore, $i_c = I + I_1$. Just before $T/2$, $i_0 = I_1$, therefore $i_c = I - I_1$. This is shown in Fig. 8.41.

Just after $T/2$, when T1, T2 are off and T3, T4 are conducting, currents i_0, i_c will continue flowing in the same direction. As such, all the three currents i_0, i_c, I enter the node B in Fig. 8.40 (a). This gives the equivalent circuit of Fig. 8.40 (d). KCL at node B of this figure gives

$$i_0 + i_c + I = 0$$

or
$$i_c = -I - i_0$$

At $t = (T/2) +$, $i_0 = +I_1$, therefore, $i_c = -I - I_1 = -(I + I_1)$. This is shown in Fig. 8.41.

At $t = T -$, $i_0 = -I_1$, therefore, $i_c = -I + I_1 = -(I - I_1)$ and so on.

Voltage across thyristor T1 is zero when T1, T2 are on. At $T/2$, when T3, T4 are turned on, $v_{T1} = v_{T2} = -v_c = -v_0 = v_{in}$ from $T/2$ to T. These waveforms are sketched in Fig. 8.41.

The nature of the waveforms of i_c and v_c in Fig. 8.41 can also be verified by the relation $i_c = C \cdot dv_c/dt$. At $t = 0$, dv_c/dt is positive and high, therefore, i_c is positive and high. At $t = (T/2) -$, dv_c/dt is reduced, likewise i_c is also lowered. At $t = (T/2) +$, dv_c/dt is negative and pronounced and so is i_c as shown.

Analysis. From $0 < t < T/2$, equivalent circuit for the CSI of Fig. 8.40 (a) is as shown in Fig. 8.40 (c). The capacitor is initially charged to a voltage $-V_1$. Traversing the closed path $a\,bcd\,a$, we get

$$Ri_0 - \frac{1}{C} \int (I - i_0)\, dt + V_1 = 0 \qquad \qquad ...(8.78)$$

Differentiation of Eq. (8.78) with respect to time gives

$$R\frac{di_0}{dt} + \frac{i_0}{C} = \frac{I}{C} \quad \text{or} \quad \left(Rp + \frac{1}{C}\right)i_0 = \frac{I}{C} \qquad ...(8.79)$$

Complementary function of the solution is obtained from force-free equation

$$\left(Rp + \frac{1}{C}\right)I_{cp} = 0 \quad \text{or} \quad p = -\frac{1}{RC}$$

$$\therefore \qquad \qquad I_{cp} = A\, e^{-t/RC}$$

For particular integral, put $p = 0$ in Eq. (8.79)

$$\therefore \qquad \qquad \frac{i_0}{C} = \frac{I}{C} \quad \text{or} \quad i_0 = I$$

Thus, complete solution for load current i_0, from Eq. (8.79), is

$$i_0 = \text{P.I.} + \text{C.F.} = I + A\, e^{-t/RC} \qquad ...(8.80)$$

Under steady state operation, the load current at $t = 0$, from Fig. 8.41, is $i_0 = -I_1$. Therefore, from Eq. (8.80),

$$-I_1 = I + A$$

or
$$A = -(I + I_1)$$

$$\therefore \qquad \qquad i_0 = I - (I + I_1)\, e^{-t/RC}$$

or
$$i_0 = I(1 - e^{-t/RC}) - I_1\, e^{-t/RC} \qquad ...(8.81)$$

fcr
$$0 < t < T/2.$$

As only steady solution is desired, current i_0 at $t = T/2$ becomes I_1. Substitution of these values in Eq. (8.81), gives

$$I_1 = I(1 - e^{-T/2\,RC}) - I_1\, e^{T/2\,RC}$$

or
$$I_1 = I\left[\frac{1 - e^{-T/2RC}}{1 + e^{-T/2RC}}\right] \qquad ...(8.82)$$

$$= I \text{ if } \frac{T}{2RC} \gg 1 \quad \text{or} \quad T \gg RC$$

Substitution of I_1 from Eq. (8.82) in Eq. (8.81), gives

$$i_0 = I(1 - e^{-t/RC}) - I\left\{\frac{1 - e^{-T/2RC}}{1 + e^{-T/2RC}}\right\} \cdot e^{-t/RC}$$

or
$$i_0 = I\left[1 - 2\frac{e^{-t/RC}}{1 + \exp(-T/2RC)}\right] \qquad ...(8.83)$$

The output voltage v_0, or capacitor voltage v_c is given by

$$v_0 = v_c = R\,i_0 = RI\left[1 - 2\frac{e^{-t/RC}}{1 + \exp(-T/2RC)}\right]$$

The turn-off time t_c provided by the circuit to each SCR is obtained from the condition that when $t = t_c$, $v_0 = v_c = i_0 R = 0$. Therefore, from the above equation,

$$v_0 = v_c = i_0 R = RI\left[1 - 2\frac{e^{-t_c/RC}}{1 + e^{-T/RC}}\right] = 0$$

or
$$\exp(-t_c/RC) = \frac{1}{2}[1 + \exp(-T/2RC)]$$

or
$$t_c = RC \ln\left[\frac{2}{1 + \exp(-T/2RC)}\right] \qquad ...(8.84)$$

The average value of the input voltage V_{in} is obtained from the equation

$$V_{in} = \frac{1}{T/2}\int_0^{T/2} i_0 R\, dt$$

$$= \frac{2}{T} IR \int_0^{T/2}\left[1 - 2\frac{e^{-\frac{t}{RC}}}{1 + e^{-\frac{T}{2RC}}}\right] dt$$

or
$$V_{in} = IR\left[1 - \frac{4RC}{T}\left(\frac{1 - \exp.(-T/2RC)}{1 + \exp.(-T/2RC)}\right)\right] \qquad ...(8.85)$$

When input power $V_{in} \cdot I$ is positive, power is delivered to the load.

Design considerations. (i) It is seen from Eq. (8.84) that as T is reduced (e.g. with $T = 0$, $t_c = 0$), or inverter frequency is increased, t_c reduces. But the circuit commutating time t_c should not be less than the SCR turn off time t_q. This means that there is an upper limit to the inverter frequency beyond which inverter SCRs will fail to commutate.

(ii) When T is large or inverter frequency ($= 1/T$) is low, the plot of i_0, or v_0, versus time t, from Eq. (8.83), becomes flatter as shown by dotted curve in Fig. 8.42. As this curve shape is nearer to a square wave, it can be inferred that for low inverter

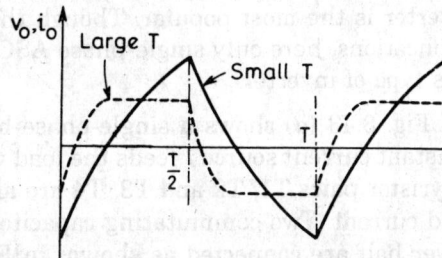

Fig. 8.42. Waveforms for 1-phase CSI with R load.

frequencies, inverter has square wave output for load current i_0 or load voltage v_0.

When T is small, or inverter frequency is high, waveform of v_0, or i_0 is shown by full-line curve in Fig. 8.42. As this full-line curve is closer to a sine wave, it can be said that for high inverter frequency, CSI has sinusoidal waveshape for output load current or load voltage.

(iii) *Square–wave current.* It has been found that for obtaining square wave of the load current, $\dfrac{T}{2RC} > 5.00$.

If t_q is the turn-off time for the SCRs used in the CSI, then from Eq. (8.84),

$$t_q = RC \ \text{In} \ \frac{2}{1+e^{-5}} \cong RC \ln 2 = 0.69 \, RC$$

or

$$C = \frac{t_q}{0.69 \, R} \qquad\qquad ...(8.86)$$

For $\dfrac{T}{2RC} = 5$ or for $T = 10 \, RC$, maximum frequency,

$$f_{max} = \frac{1}{T} = \frac{1}{10RC}$$

Substituting the value of C from Eq. (8.86),

$$f_{max} = \frac{1}{10 \, R} \cdot \frac{0.69 \, R}{t_c} = \frac{0.069}{t_q}$$

(iv) *Sinusoidal wave output.* For obtaining sinusoidal waveshape for load current, frequency domain analysis shows that capacitive reactance X_c at 3 times the minimum frequency f_{min} should be less than $R/2$, i.e.

$$X_c \text{ at } 3 f_{min} \leq \frac{R}{2}$$

or

$$\frac{1}{3 \cdot 2\pi f_{min} \cdot C} \leq \frac{R}{2}$$

or

$$\frac{R}{2} \cdot \frac{3 \cdot 2\pi f_{min} \cdot C}{1} \geq 1$$

or

$$C \geq \frac{0.106}{R \cdot f_{min}} \qquad\qquad ...(8.87)$$

8.8.3. Single-phase Auto-sequential Commutated Inverter (1-phase ASCI)

Out of the force-commutated current source inverters, auto-sequential commutated inverter is the most popular. Though three-phase ASCI is the universal choice in industrial applications, here only single-phase ASCI is presented so as to highlight the basic concepts of this type of inverter.

Fig. 8.43 (a) shows a single-phase bridge inverter with auto-sequential commutation. A constant current source I feeds the load which is assumed here an inductance L for simplicity. Thyristor pairs T1, T2 and T3, T4 are alternatively switched to obtain a nearly square wave load current. Two commutating capacitors, one C1 in the upper half and the other C2 in the lower half are connected as shown. In Fig. 8.43 (a), diodes D1 to D4 are connected in series with each SCR to prevent the commutation capacitors from discharging into the load. The

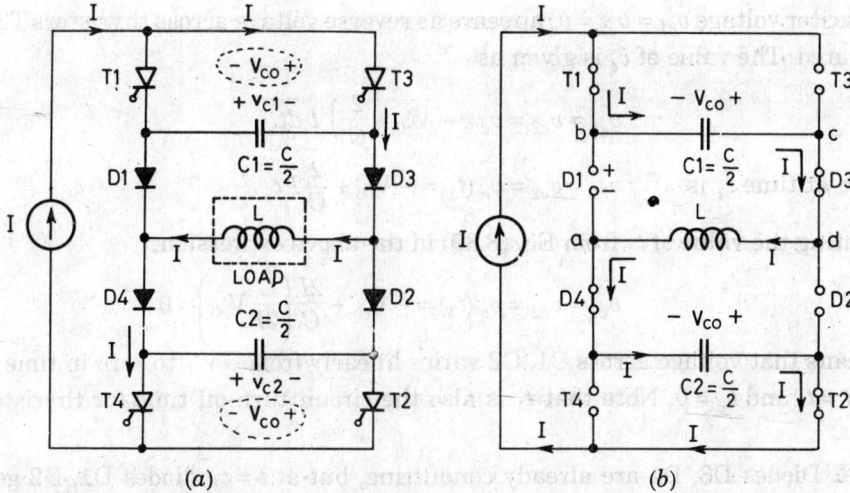

Fig. 8.43. (a) Single-phase bridge auto-sequential commutated inverter with L load (b) Mode I.

inverter output frequency is controlled by adjusting the period T through the triggering circuits of thyristors. The operation of this inverter can be explained in two modes as under.

Mode I. In the beginning, *i.e.* before $t = 0$, assume that T3, T4 are conducting and a steady current I flows through the path T3, D3, L, D4, T4 and source I as shown in Fig. 8.43 (a). The commutating capacitors are assumed to be initially charged equally with the polarity as shown in Fig. 8.43 (a), *i.e.* $v_{c1} = v_{c2} = -V_{c0}$. This means that both capacitors C1 and C2 have right hand plate positive and left hand plate negative.

At time $t = 0$, thyristors T1, T2 are gated. The thyristor pair T3, T4 is turned-off by the application of reverse capacitor voltage V_{c0}. Now pair T1, T2 conducts current I. The path for this current I is through T1, C1, D3, L, D4, C2 and T2 as shown in Fig. 8.43 (b). Both the capacitors will now begin charging linearly from $-V_{c0}$ by the constant current I. The diodes D1, D2 remain reverse biased by V_{c0} initially. The voltage v_{D1} across D1, when it is forward biased, can be obtained by traversing-closed path abcda as

$$v_{D1} + V_{c0} - \frac{1}{C/2} \int I \, dt = 0$$

Note that voltage across L is zero because of constant current I.

$$\therefore \qquad v_{D1} = -V_{c0} + \frac{2}{C} \int I \, dt \qquad \qquad \qquad ...(8.88)$$

As the capacitor charges, voltages v_{D1} across D1 rises linearly. Eventually, at some time t_1, reverse bias across D1 vanishes, v_{D1} becomes zero and diode D1 starts conducting. An equation, identical to Eq. (8.88), holds good for diode D2 also. Actually, D1 and D2 start conducting at the same instant t_1. The time for which D1, D2 remain reverse biased is obtained from Eq. (8.88) by equating $v_{D1} = 0$.

$$\therefore \qquad 0 = -V_{c0} + \frac{2}{C} I \, t_1$$

or

$$t_1 = \frac{C}{2I} V_{c0} \qquad \qquad \qquad ...(8.89)$$

The capacitor voltage $v_{c1} = v_{c2} = v_c$ appears as reverse voltage across thyristors T3, T4 when T1, T2 are gated. The value of v_c is given as

$$v_{c1} = v_{c2} = v_c = -V_{c0} + \frac{2}{C}\int I\, dt$$

Its value at time t_1 is $$v_{c1} = v_{c2} = v_c\,(t_1) = -V_{c0} + \frac{2}{C} I t_1$$

Substituting the value of t_1 from Eq. (8.89) in the above expression,

$$v_{c1} = v_{c2} = v_c\,(t_1) = -V_{c0} + \frac{2I}{C}\left(\frac{C}{2I} V_{c0}\right) = 0$$

This means that voltage across C1, C2 varies linearly from $-V_{c0}$ to zero in time t_1. Mode I ends when $t = t_1$ and $v_c = 0$. Note that t_1 is also the circuit turn-off time for thyristors of Fig. 8.43.

Mode II. Diodes D3, D4 are already conducting, but at $t = t_1$, diodes D1, D2 get forward biased and start conducting. Thus, at the end of time t_1, all four diodes D1, D2, D3 and D4 conduct. As a result, commutating capacitors now get connected in parallel with the load as shown in Fig. 8.44 (a). For simplicity in analysis, Fig. 8.44 (a) is redrawn as shown in Fig. 8.44(b) where two parallel capacitors C1 and C2 are combined into one as $C\ (= C/2 + C/2)$. In this figure, KCL gives

$$I + i_0 = i_c\ (= i_{c1} + i_{c2})$$

As $$i_{c1} = i_{c2},\quad i_{c1} = i_{c2} = \frac{1}{2} i_c$$

For this figure, KVL gives, $$L\frac{di_0}{dt} + \frac{1}{C}\int i_c \cdot dt = 0$$

or $$L\frac{di_0}{dt} + \frac{1}{C}\int (I + i_0)\, dt = 0$$

or $$L\frac{d^2 i_0}{dt^2} + \frac{i_0}{C} = -\frac{I}{C} \qquad\qquad ...(8.90)$$

For solving Eq. (8.90), the initial conditions at $t = 0$ are

$$i_0 = I \quad\text{and}\quad \frac{di_0}{dt} = 0$$

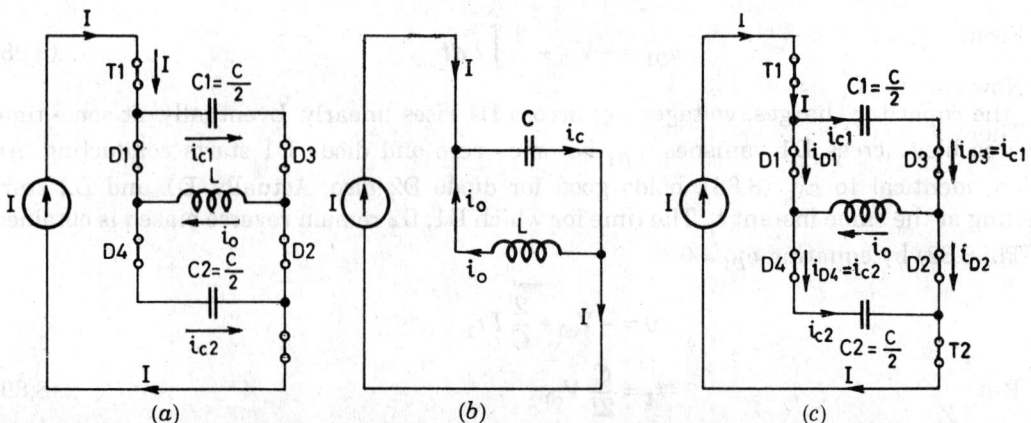

(a)   (b)   (c)

Fig. 8.44. Equivalent circuits for 1-phase ASCI with load L.

It should be noted that in Eq. (8.90), time t is measured from the instant D1, D2 begin conduction, *i.e.* time t for mode II is measured from the instant mode I is over.

In Eq. (8.90), for particular integral

$$\frac{i_{os}}{C} = -\frac{I}{C}$$

or

$$i_{os} = -I$$

and for complementary function,

$$\left(Lp^2 + \frac{1}{C}\right)i_0 = 0$$

or

$$Lp^2 + \frac{1}{C} = 0$$

This gives

$$p^2 = -\frac{1}{LC} = -\omega_0^2 = j^2\omega_0^2$$

or

$$p = \pm j\omega_0$$

where

$$\omega_0 = \frac{1}{\sqrt{LC}}$$

$$\therefore \qquad i_{ot} = Ae^{j\omega_0 t} + Be^{-j\omega_0 t}$$

The total solution for the current is obtained by adding i_{os} and i_{ot}

$$\therefore \qquad i_0(t) = i_{os} + i_{ot}$$

$$= -I + Ae^{j\omega_0 t} + Be^{-j\omega_0 t} \qquad \text{...(8.91)}$$

At

$$t = 0, \quad i_0 = I$$

$$\therefore \qquad I = -I + A + B$$

or

$$A + B = 2I \qquad \text{...(8.92)}$$

At $t = 0$, $\dfrac{di_0}{dt} = 0$, from Eq. (8.91),

$$\frac{di_0}{dt} = j\omega_0 Ae^{j\omega_0 t} - j\omega_0 Be^{-j\omega_0 t} = 0$$

or

$$j\omega_0(A - B) = 0$$

or

$$(A - B) = 0 \qquad \text{...(8.93)}$$

From Eqs. (8.92) and (8.93), $A = B = I$

From Eq. (8.91), $\quad i_c(t) = -I + 2I\left[\dfrac{e^{j\omega_0 t} + e^{-j\omega_0 t}}{2}\right] = I[2\cos\omega_0 t - 1] \qquad \text{...(8.94)}$

Now capacitor current, $\quad i_c = I + i_0 = 2I\cos\omega_0 t \qquad \text{.. (8.95)}$

The voltage across capacitor is

$$v_c = \frac{1}{C}\int i_c \cdot dt = \frac{2I}{\omega_0 C}\sin\omega_0 t \qquad \text{...(8.96)}$$

This expression for v_c can also be obtained as

$$v_c = v_L = L\frac{di_0}{dt} = L\omega_0 \cdot 2I \cdot \sin\omega_0 t$$

But

$$\omega_0^2 = \frac{1}{LC} \quad \text{or} \quad L\omega_0 = \frac{1}{\omega_0 C}$$

$$\therefore \qquad v_c = \frac{2I}{\omega_0 C} \sin \omega_0 t \qquad \qquad \qquad ...(8.96)$$

Also
$$v_c = \frac{2I}{C} \cdot \sqrt{LC} \cdot \sin \omega_0 t = 2I \cdot \sqrt{L/C} \sin \omega_0 t \qquad ...(8.97)$$

From Eq. (8.95),
$$i_{c1} = i_{c2} = \frac{1}{2} i_c = I \cos \omega_0 t \qquad \qquad ...(8.98)$$

The existence of various currents i_{C1}, i_{C2}, i_{D1} etc. is marked in Fig. 8.44 (c). Diode currents are given by

$$\left. \begin{array}{l} i_{D3} = i_{C1} = I \cos \omega_0 t \\ \text{and} \quad i_{D1} = I - i_{C1} = I\,(1 - \cos \omega_0 t) \end{array} \right\} \text{ for } 0 < t < t_2$$

As current i_{C1} tends to reverse, diode D3 prevents its reversal. Similarly, diode D4 prevents the reversal of current i_{C2}.

From the initiation of mode II, a time t_2 must elapse for the current i_{C1} to become zero. This time t_2 can be obtained by equating i_{C1} to zero.

$$\therefore \qquad i_{C1} = I \cos \omega_0 t_2 = 0$$

or
$$\cos \omega_0 t_2 = \cos \frac{\pi}{2}$$

or
$$t_2 = \frac{\pi}{2\omega_0} \qquad \qquad ...(8.99)$$

The capacitor voltage v_c at the end of mode II, i.e. at $t = t_2 = \dfrac{\pi}{2\omega_0}$, can be obtained from Eq. (8.96).

$$\therefore \qquad v_c = \frac{2I}{\omega_0 C}$$

Load current at $t = t_2$ is $i_0 = 2I \cos \dfrac{\pi}{2} - I = -I$. This shows that load current has reversed from $+I$ to $-I$ during mode II of t_2 duration.

The waveforms for v_c, i_0, i_{C1}, i_{D1} etc. are plotted in Fig. 8.45. It is seen from the waveform for v_c that capacitor voltage changes by $2\,V_{co}$ during each commutation interval.

The total commutation interval t_c from Eqs. (8.89) and (8.99) is given as

$$t_c = t_1 + t_2 = \frac{C}{2I}\,V_{co} + \frac{\pi}{2\omega_0} \qquad \qquad ...(8.100)$$

It is seen from Eq. (8.96) that v_c is maximum and equal to V_{co} when $\omega_0 t = \pi/2$

$$\therefore \qquad V_{co} = \frac{2I}{\omega_0 C}$$

From Eq. (8.89),
$$t_1 = \frac{C}{2I}\,V_{co} = \frac{C}{2I} \cdot \frac{2I}{\omega_0 C} = \frac{1}{\omega_0} = \sqrt{LC}$$

Therefore, commutation interval from Eq. (8.100) is

$$t_c = \left(\frac{1}{\omega_0} + \frac{\pi}{2\omega_0} \right) = \left(1 + \frac{\pi}{2} \right) \sqrt{LC} \qquad \qquad ...(8.101)$$

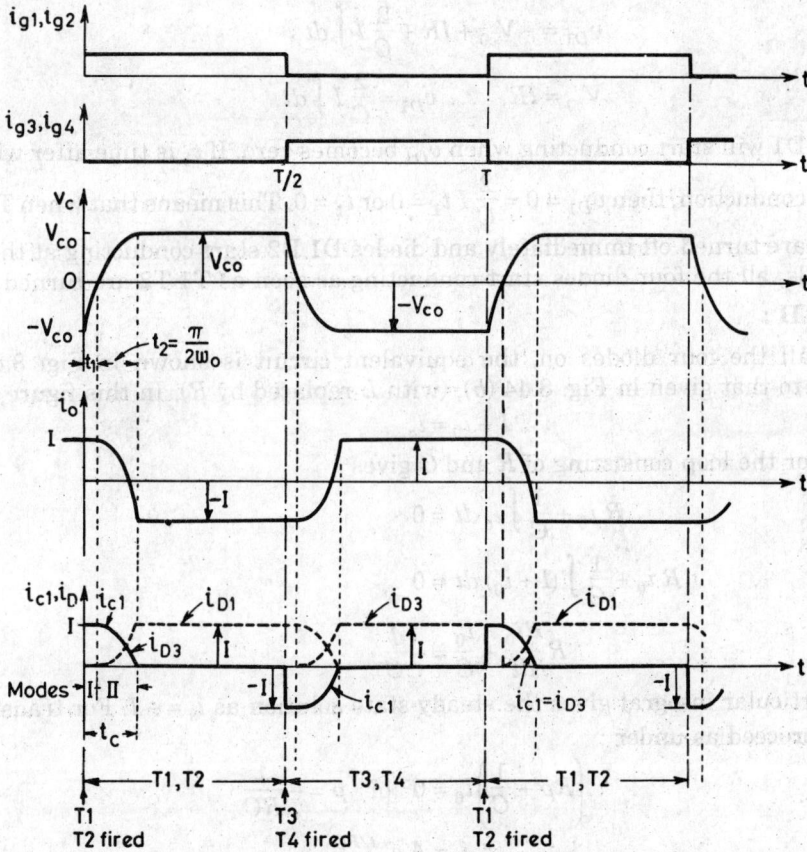

Fig. 8.45. Voltage and current waveforms for 1-phase ASCI.

At the end of total commutation interval $(t_1 + t_2)$, the steady input current I flows through T1, D1, load L, D2 and T2. This constant current continues to flow till the next commutation process is initiated by gating SCRs, T3 and T4.

Example 8.10. *A single-phase autosequential commutated current source inverter feeds a load R. Describe its working with appropriate circuit and waveforms. Find also the circuit turn-off time for the thyristors.*

Solution. For the circuit diagram, replace L by R in Fig. 8.43 (a). In other words, Figs. 8.43 and 8.44 are being used here to describe the working of this inverter with R as load in place of L. Working of this inverter is explained in two modes as before.

Mode I : Initially, it is assumed that T3, T4 are conducting. A constant current I is flowing through the path T3 D3 R D4 T4 and the source I. Capacitors are initially charged so that $v_{c1} = v_{c2} = -V_{c0}$. Voltage across load is $IR = V_{c0}$.

When T1 T2 are turned on at $t = 0$, T3 T4 are turned off by the reverse voltage V_{c0} appearing across them. Current I now starts flowing through T1 C1 D3 R D4 T2. Capacitor voltages v_{c1} and v_{c2} start rising from $-V_{c0}$ towards zero. If v_{D1} is the forward voltage drop across D1, then KVL for the loop $abcd$ in Fig. 8.43 (b), (with L replaced by R), gives

$$v_{D1} + V_{c0} - \frac{2}{C} \int I \, dt - IR = 0$$

or
$$v_{D1} = -V_{c0} + IR + \frac{2}{C} I \int dt$$

But
$$V_{c0} = IR, \quad \therefore \quad v_{D1} = \frac{2}{C} I \int dt.$$

Diode D1 will start conducting when v_{D1} becomes zero. If t_1 is time after which $v_{D1} = 0$ and D1 begins conduction, then $v_{D1} = 0 = \frac{2}{C} I t_1 = 0$ or $t_1 = 0$. This means that when T1 T2 are turned on, T3 T4 are turned off immediately and diodes D1 D2 start conducting at the same time. In other words, all the four diodes start conducting as soon as T1 T2 are turned on.

Mode II :

With all the four diodes on, the equivalent circuit is shown in Fig. 8.44 (a) which is simplified to that given in Fig. 8.44 (b), (with L replaced by R). In this figure, KCL gives
$$I + i_0 = i_c$$

KVL for the loop consisting of R and C gives
$$R i_0 + \frac{1}{C} \int i_c \, dt = 0$$

$$R i_0 + \frac{1}{C} \int (I + i_0) \, dt = 0$$

or
$$R \frac{di_0}{dt} + \frac{i_0}{C} = -\frac{I}{C}$$

Its particular integral gives the steady-state solution as $i_s = -I$. For transient part of the solution, proceed as under :

$$\left(Rp + \frac{1}{C} \right) i_0 = 0 \quad \text{or} \quad p = -\frac{1}{RC}$$

$\therefore$
$$i_t = A e^{-t/RC}$$

Total solution for load current is $i_0 = i_s + i_t = -I + A e^{-t/RC}$

When $t = 0$, initial current through R is $i_0 = I$, $\therefore A = 2I$

$\therefore$ Load current,
$$i_0 = -I + 2 I e^{-t/RC} = I (2 e^{-t/RC} - 1) \qquad \qquad ...(i)$$

Capacitor current,
$$i_c = I + i_0 = 2 I e^{-t/RC}$$

$\therefore$
$$i_{c1} = i_{c2} = \frac{1}{2} i_c = I e^{-t/RC} \qquad \qquad ...(ii)$$

Diode D1 current,
$$i_{D1} = I - i_{c1} = I (1 - e^{-t/RC}) \qquad \qquad ...(iii)$$

Capacitor voltage
$$v_{c1} = \frac{1}{C/2} \int i_{c1} \cdot dt = \frac{2}{C} \int I e^{-t/RC} \cdot dt$$

$$= -\frac{2I}{C} e^{-t/RC} \cdot (-RC) + k$$

or
$$v_{c1} = -2IR e^{-t/RC} + k = -2 V_{c0} e^{-t/RC} + k$$

When $t = 0$, initial voltage across capacitor is $v_{c1} = -V_{c0}$, this gives $k = V_{c0}$.

$\therefore$
$$v_{c1} = V_{c0} [1 - 2 e^{-t/RC}] \qquad \qquad ...(iv)$$

When T1 T2 are on ; v_{c1}, v_{c2} appear as reverse bias across T3, T4 respectively. Therefore, circuit turn-off time t_c for T3, T4 (or for any thyristor in Fig. 8.43 with R as the load) can be obtained from Eq. (iv) by putting $v_{c1} = 0$.

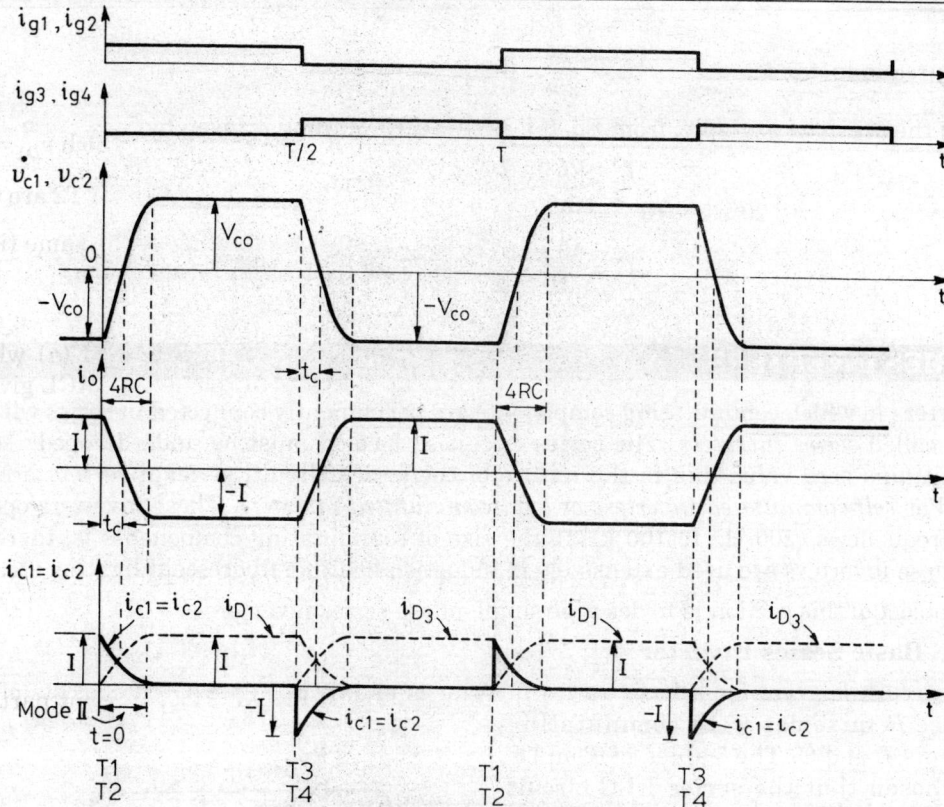

Fig. 8.46. Pertaining to Example 8.10.

$$\therefore \qquad 0 = V_{c0}\,[1 - 2\,e^{-t_c/RC}]$$

or
$$t_c = RC \ln 2 \qquad\qquad\qquad ...(v)$$

From above, waveforms for v_{c1} from Eq. (iv), i_0 from Eq. (i), $i_{c1} = i_{c2}$ from Eq. (ii) and i_{D1} from Eq. (iii) are plotted in Fig. 8.46. It is seen from Eq. (iv) that v_{c1} will change from $- V_{c0}$ at $t = 0$ to $+ V_{c0}$ after an infinite time. But it is usual to take that in time 4 RC, v_{c1} varies from $- V_{c0}$ to $+ V_{c0}$ as shown in Fig. 8.46.

Example 8.11. *A single-phase auto-sequential commutated CSI is fed from 220 V dc source. The load is R = 10 Ω. Thyristors have turn-off time of 20 μs and inverter output frequency is 50 Hz. Take a factor of safety of 2.*

Determine suitable value of source inductance assuming a maximum current change of 0.5 A in one cycle. Neglect all losses. Find also the values of commutating capacitors.

Solution. Time of one cycle, $\quad T = \dfrac{1}{f} = \dfrac{1}{50}$ sec

$\therefore$ Rate of change of current, $\dfrac{di}{dt} = \dfrac{0.5A}{T} = 0.5 \times 50 = 25$ A/sec

A short circuit at the load terminals of the inverter puts the most severe conditions on the source. So the value of source inductance must be obtained from these considerations.

$$\therefore \qquad\qquad V_s = L\,\frac{di}{dt}$$

∴ Source inductance, $L = \dfrac{220}{25} = 8.8$ H

From the previous example, from Eq. (v), circuit-turn-off time is given by

$$t_c = RC \ln 2$$

or $20 \times 2 \times 10^{-6} = 10 \cdot C \ln 2$

or $C = \dfrac{40 \times 10^{-6}}{10 \ln 2} = 5.77$ μF.

8.9. SERIES INVERTERS

Inverters in which commutating components are permanently connected in series with the load are called *series inverters*. The series circuit so formed must be underdamped. As the current attains zero value due to the nature of the series circuit, series inverters are also classified as *self-commutated inverters* or *load-commutated inverters*. These inverters operate at high frequencies (200 Hz to 100 kHz), the size of commutating components is, therefore, small. These inverters are used extensively in induction heating, fluorescent lighting etc.

The object of this section is to describe single-phase series inverters.

8.9.1. Basic Series Inverter

The circuit diagram for a basic series inverter is shown in Fig. 8.47. It consists of load resistance R in series with commutating components L and C. The values of L and C are so chosen that the series RLC circuit forms an underdamped circuit. Two thyristors T1 and T2 are turned on appropriately so that output voltage of desired frequency can be obtained.

When thyristor T1 is turned on, with T2 off, current i starts building up in the RLC circuit, Fig. 8.48. As the circuit is underdamped ; the load current, after reaching some peak value, decays to zero at point a, Fig. 8.48. At point a, as the load current tends to reverse, SCR T1 is turned off.

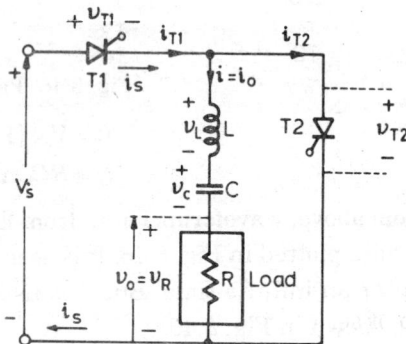

Fig. 8.47. Basic series inverter.

After instant a, some minimum time $t_{q \cdot min}$ must elapse for T1 to regain its forward blocking capability. This minimum time is given by

$$t_{q.min} = \frac{\pi}{\omega} - \frac{\pi}{\omega_r} = \frac{1}{2}\left(\frac{1}{f} - \frac{1}{f_r}\right) \qquad \qquad ...(8.102)$$

where   ω = output frequency in rad/sec
and ω_r = circuit ringing frequency in rad/sec.

In Fig. 8.48, time interval between the instant T1 is turned off and the instant T2 is turned on is indicated by $T_{off} = ab$, where $T_{off} > t_{q \cdot min}$. After thyristor T1 has commutated, upper plate of capacitor attains positive polarity. Now when T2 is turned on at instant b, capacitor begins to discharge and load current in the reversed direction builds up to some peak negative value and then decays to zero at instant c. After this, time $T_{off} = cd$ must elapse for T2 to recover. At d, T1 is again turned on and the process repeats. In this manner, dc is converted to ac with the

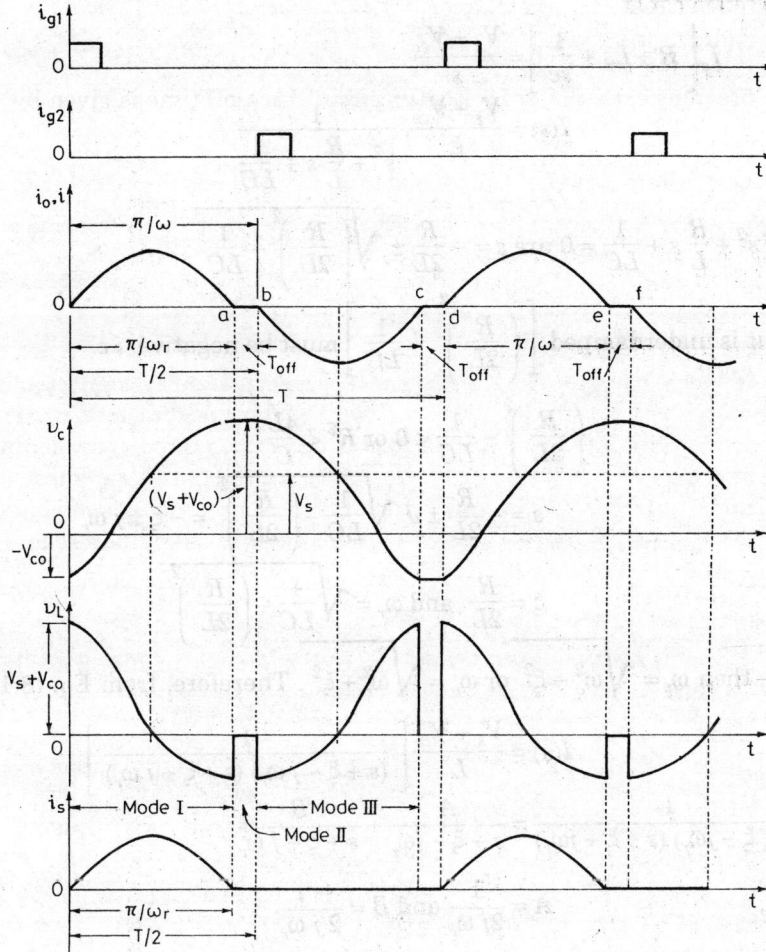

Fig. 8.48. Current and voltage waveforms for basic series inverter of Fig. 8.47.

help of series inverter. In Fig. 8.48, $T_{off} = ab$, or cd, is called *circuit turn-off time* or *dead-zone time*.

The capacitor stores charge during one half cycle and releases the same amount of charge during the next half cycle. As a consequence, the positive half cycle of current is identical with the negative half cycle of load current. In a practical series inverter, positive and negative half cycle may not be sine waves.

8.9.2. Analysis of Basic Series Inverter

The operation of this inverter is described in three different modes as below.

Mode I. With $T2$ off, the equivalent circuit of basic series inverter of Fig. 8.47 is as shown in Fig. 8.49 (a). In this figure, the capacitor C is assumed to be initially charged to voltage V_{co} with lower plate positive. Mode I begins with the turning on of thyristor $T1$. Thus, with $T1$ on, KVL for the closed circuit of Fig. 8.49 (a) can be written as

$$Ri + L\frac{di}{dt} + \frac{1}{C}\int i\,dt = V_s + V_{co} \qquad \qquad ...(8.103)$$

Its Laplace transform is

$$I_s \left[R + Ls + \frac{1}{sc} \right] = \frac{V_s + V_{co}}{s}$$

or

$$I(s) = \frac{V_s + V_{co}}{L} \cdot \frac{1}{s^2 + \frac{R}{L}s + \frac{1}{LC}} \qquad \qquad ...(8.104)$$

The roots of $s^2 + \frac{R}{L}s + \frac{1}{LC} = 0$ are $s = -\frac{R}{2L} \pm \sqrt{\left(\frac{R}{2L}\right)^2 - \frac{1}{LC}}$

As the circuit is underdamped, $\left[\left(\frac{R}{2L}\right)^2 - \frac{1}{LC} \right]$ must be negative, i.e.

$$\left(\frac{R}{2L}\right)^2 - \frac{1}{LC} < 0 \text{ or } R^2 < \frac{4L}{C}$$

$$\therefore \qquad s = -\frac{R}{2L} \pm J \sqrt{\frac{1}{LC} - \left(\frac{R}{2L}\right)^2} = -\xi \pm j\,\omega_r$$

where

$$\xi = \frac{R}{2L} \text{ and } \omega_r = \sqrt{\frac{1}{LC} - \left(\frac{R}{2L}\right)^2}$$

If $\omega_o = \frac{1}{\sqrt{LC}}$, then $\omega_r = \sqrt{\omega_o^2 - \xi^2}$ or $\omega_o = \sqrt{\omega_r^2 + \xi^2}$. Therefore, from Eq. (8.104),

$$I(s) = \frac{V_s + V_{co}}{L} \left[\frac{1}{(s + \xi - j\,\omega_r)(s + \xi + j\,\omega_r)} \right]$$

Let

$$\frac{1}{(s + \xi - j\omega_r)(s + \xi + j\omega_r)} = \frac{A}{s + \xi - j\omega_r} + \frac{B}{s + \xi + j\,\omega_r}$$

From above,

$$A = \frac{1}{2j\,\omega_r} \text{ and } B = \frac{-1}{2j\,\omega_r}$$

$$\therefore \qquad I(s) = \frac{V_s + V_{co}}{L} \cdot \frac{1}{2j\,\omega_r} \left[\frac{1}{s + \xi - j\,\omega_r} - \frac{1}{s + \xi + j\omega_r} \right]$$

$$= \frac{V_s + V_{co}}{L} \cdot \frac{1}{\omega_r} \left(\frac{\omega_r}{(s + \xi)^2 + \omega_r^2} \right)$$

Its Laplace inverse is

$$i(t) = \frac{V_s + V_{co}}{\omega_r \cdot L} - e^{-\xi t} \sin \omega_r t \qquad \qquad ...(8.105)$$

Here $\xi = \frac{R}{2L}$ is called *damping factor*, $\omega_o = \frac{1}{\sqrt{LC}}$ is resonant frequency in rad/sec, ω_r is known as *circuit ringing frequency* in rad/sec and ω = operating, or output, frequency in rad/sec = $2\,\pi f$.

The plot of this current of Eq. (8.105) shows that load current $i(t)$ will be zero when $\omega_r\,t = \pi$ or $t = \frac{\pi}{\omega}$ sec. This is shown as $oa = \pi/\omega_r$ in Fig. 8.48.

Time period of oscillation, $oa = \frac{\pi}{\omega_r} = \frac{\pi}{\sqrt{1/LC - (R/2L)^2}} \qquad \qquad ...(8.106)$

Output frequency,

$$f = \frac{1}{2(oa + T_{off})}$$

or $\qquad f = \dfrac{1}{\dfrac{2\,\pi}{\sqrt{1/LC - (R/2L)^2}} + 2\,T_{off}}$ ...(8.107)

It is seen from Eq. (8.107) that output frequency can be controlled by varying (i) T_{off}, it depends upon power semiconductor device used, (ii) load resistance R, (iii) L or (iv) C. In any case, output frequency f cannot be more than the circuit ringing frequency

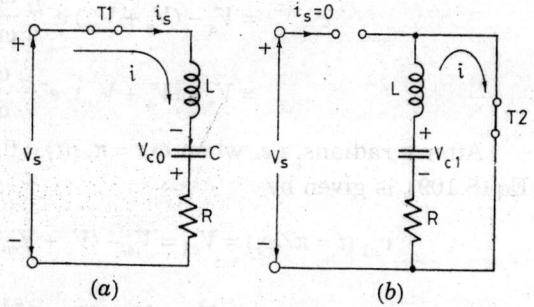

Fig. 8.49. Equivalent circuit of Fig. 8.47 with (a) T1 on, T2 off, (b) T1 off, T2 on.

$$f_r = \frac{1}{2\pi} \cdot \frac{1}{\sqrt{1/LC - (R/2L)^2}} \text{ Hz}, \text{ i.e. } f < f_r$$

The voltage across inductance L, from Eq. (8.105), is

$$v_L = L\frac{di}{dt} = L \cdot \frac{V_s + V_{co}}{L} \cdot \frac{1}{\omega_r} [e^{-\xi t} \cdot \omega_r \cdot \cos \omega_r t - \xi \cdot e^{-\xi t} \sin \omega_r t]$$

$$= \frac{V_s + V_{co}}{\omega_r} \cdot e^{-\xi t} [\omega_r \cos \omega_r t - \xi \sin \omega_r t]$$

$$= \frac{V_s + V_{co}}{\omega_r} e^{-\xi t} \sqrt{\omega_r^2 + \xi^2} \left[\frac{\omega_r}{\sqrt{\omega_r^2 + \xi^2}} \cos \omega_r t - \frac{\xi}{\sqrt{\omega_r^2 - \xi^2}} \sin \omega_r t \right]$$

$$= \frac{V_s + V_{co}}{\omega_r} - e^{-\xi t} \sqrt{\omega_r^2 + \xi^2} [\cos \psi \cos \omega_r t - \sin \psi \sin \omega_r t]$$

$$= (V_s + V_{co}) \cdot \frac{\omega_o}{\omega_r} \cdot e^{-\xi t} \cos (\omega_r t + \psi) \qquad ...(8.108)$$

where ω_o = resonant frequency = $\sqrt{\omega_r^2 + \xi^2}$ and $\psi = \tan^{-1}\left(\dfrac{\xi}{\omega_r}\right)$, see Fig. 8.50.

The voltage across capacitor C is

$$v_c = V_s - v_R - v_L$$

$$= V_s - \frac{V_s + V_{co}}{\omega_r} \cdot R_L e^{-\xi t} - \sin \omega_r t$$

$$- (V_s + V_{co}) \frac{\omega_o}{\omega_r} \cdot e^{-\xi t} \cdot \cos (\omega_r t + \psi)$$

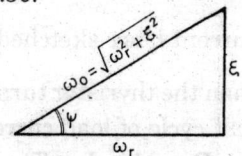

Fig. 8.50. Pertaining to ω_r, ξ and ω_o.

$$= V_s - (V_s + V_{co}) \cdot e^{-\xi t} \left[\frac{1}{\omega_r} \cdot \frac{R}{L} \cdot \sin \omega_r t + \frac{\omega_o}{\omega_r} \cos (\omega_r t + \psi) \right]$$

But $\qquad \dfrac{R}{L} = 2\,\xi$

$\therefore \qquad v_c = V_s - (V_s + V_{co}) e^{-\xi t} \left[\dfrac{2\,\xi}{\omega_r} \cdot \dfrac{\omega_o}{\omega_o} \sin \omega_r t + \dfrac{\omega_o}{\omega_r} \cos (\omega_r t + \psi) \right]$

$\qquad = V_s - (V_s + V_{co}) e^{-\xi t} \dfrac{\omega_o}{\omega_r} \left[\dfrac{2\,\xi}{\omega_o} \sin \omega_r t + \cos (\omega_r t + \psi) \right]$

$$= V_s - (V_s + V_{co}) \, e^{-\xi t} \, \frac{\omega_o}{\omega_r} \, [2 \sin \psi \sin \omega_r t + \cos \omega_r t \cos \psi - \sin \omega_r t \sin \psi]$$

$$= V_s - (V_s + V_{co}). \, e^{-\xi t} \frac{\omega_o}{\omega_r} \cos (\omega_r t - \psi) \qquad \qquad ...(8.109)$$

After π radians, *i.e.* when $\omega_r t = \pi$, $i(t) = 0$, therefore capacitor voltage v_{c1} at $t = \pi/\omega_r$, from Eq (8.109), is given by

$$v_{c1} \, (t = \pi/\omega_r) = V_{c1} = V_s - (V_s + V_{co}) \, e^{-\xi.\pi/\omega_r}. \, \frac{\omega_o}{\omega_r} \cos (\pi - \psi)$$

$$= V_s + (V_s + V_{co}) \, e^{-\pi \xi/\omega_r}. \, \frac{\omega_o}{\omega_r} . \frac{\omega_r}{\omega_o}$$

$$\therefore \qquad \qquad V_{c1} = V_s + (V_s + V_{co}) \exp [-\pi \, \xi/\omega_r] \qquad \qquad ...(8.110)$$

Mode II. During this mode, both SCRs $T1$ and $T2$ are off. Therefore, $v_c = V v_{c1}$, $i = 0$, $v_L = 0$ during mode II.

Mode III. This mode begins with the turning on of thyristor $T2$. Equivalent circuit for this mode is shown in Fig. 8.49(*b*). Time origin for this mode is taken when $T2$ is switched on. Therefore, KVL for the circuit of Fig. 8.49 (*b*) is

$$R_i + L \frac{di}{dt} + \frac{1}{C} \int i. \, dt = V_{c1} \qquad \qquad ...(8.111)$$

Initial voltage V_{c1} causes the capacitor to have upper plate positive. Therefore, with the turning on of $T2$, the current in load R is reversed as desired. It is seen that Eq. (8.111) is identical with Eq.(8.103). Their solutions must, therefore, be identical. Thus, solution of Eq. (8.111), from Eq. (8.105), is given by

$$i \, (t) = \frac{V_{c1}}{\omega_r . L}. \, e^{-\xi +} \sin \omega_r t \qquad \qquad ...(8.112)$$

Peak value of positive current, as given by Eq. (8.105) would be equal to the peak value of the negative current given by Eq. (8.112) only when $V_{c1} = V_s + V_{co}$.

The waveforms for load current i_o or i, voltage, v_c across capacitor, v_L across L and source current i_s are sketched in Fig. 8.48. As stated before, dead zone $ab \left(= \frac{\pi}{\omega} - \frac{\pi}{\omega_r} \right)$ must be greater than the thyristor turn-off time t_q. It is seen that source current i_s flows only during the positive half cycle of load current.

Drawbacks. The basic series inverter shown in Fig. 8.47 is very simple. It, however, suffers from the following drawbacks.

1. For a load power, the load current is taken from the supply during the positive half cycle only. This has the effect of increasing the peak current rating of the *dc* source.
2. Since the source current flows during the positive half cycle only, its harmonic content is much pronounced.
3. The maximum operating frequency of the inverter is limited because this frequency ω has to be less than the circuit ringing frequency.
4. For output frequency much lower than the circuit ringing frequency, the load voltage waveform gets distorted considerably due to the increased duration of the dead zone.
5. Amplitude and duration of load current flow in each half cycle depends on the load circuit parameters. Therefore, this inverter suffers from poor output regulation.
6. In this inverter, commutating components have to carry the load current continuously, therefore high rating of these components is necessary.

Out of the drawbacks enumerated above; 4, 5 and 6 are inherent in all types of series inverters. However, the drawbacks listed at 1, 2 and 3 can be mitigated by modifying the configuration of the basic series inverter. This is discussed in what follows.

8.9.3. Modified Series Inverter

Fig. 8.51 shows a modified series inverter in which the drawback of limited operating frequency is overcome. It consists of two thyristors $T1$, $T2$, dc source voltage V_s, load R, capacitor C and mid-tapped inductor. Actually, inductor is one but it has a centre tapping so that inductance of both halves are equal, *i.e.* $L_1 = L_2$.

Let $T1$ be switched on so that load current i builds up through R and then decreases. When current i is nearing zero value, v_c across C is less than $V_s + V_{co}$, Fig. 8.48 and at this instant, turn on SCR $T2$. Since current i is decaying, the polarity of emf induced in L_1 would be as indicated in Fig. 8.51. Since L_1, L_2 are mutually coupled, voltage across L_2 has induced voltage with polarity as shown. Now KVL for the loop consisting of $T1$, V_s, R, C, L_1 and $T1$ gives $v_{T1} - V_s + iR$ drop $+ V_s + V_{co}$ − emf in L_1 (say $0.5\ V_{co}$) = 0

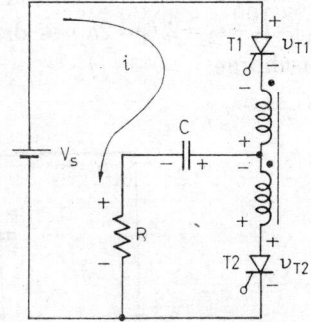

Fig. 8.51. Modified series inverter with coupled inductors.

$$\therefore \qquad v_{T1} = (-0.5\ V_{co} - iR \text{ drop})$$

Since v_{T1} is negative, $T1$ is reverse biased, as a result $T1$ is turned off by v_{T1}. This shows that when $T2$ is turned on, even before load current has reached zero, $T2$ gets turned off. At the same time, when $T2$ is turned on, it gets forward biased by a voltage v_{T2} = emf in $L_2 + (V_s + V_{co}) + iR$ drop.

In basic series inverter of Fig. 8.47, if $T2$ is turned on before current in $T1$ vanishes, there would be a dead short circuit of source voltage V_s; this will eventually damage the inverter. However, in modified series inverter of Fig. 8.51, there is no danger of any short circuit. This important feature permits the modified series inverter of Fig. 8.51 to operate at frequency higher than the circuit ringing frequency.

8.9.4. Half-bridge Series Inverter

In the circuit of Fig. 8.51, the load draws power from dc source in positive half cycle only, even though its operating frequency limit is surmounted. This drawback of intermittent power flow can be overcome by *half-bridge series inverter* shown in Fig. 8.52. This inverter consists of two thyristors, mutually-coupled inductors $L_1 = L_2$, two capacitors $C_1 = C_2 = C$, dc source V_s and load R. In this inverter, power is drawn from dc source during both the half cycles as explained below.

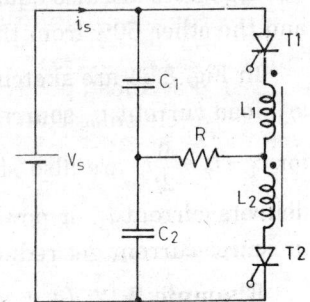

Fig. 8.52. Half-bridge series inverter.

Assume that capacitor C_2 is initially charged to voltage V_{co} with the lower plate positive, Fig. 8.53 (a). Since C_1 and C_2 are in series across the battery, (voltage across C_1 + voltage across C_2) must be equal to battery voltage V_s. Capacitor C_1 would, therefore, be charged to voltage $V_{c1} = V_s + V_{co}$ with upper plate positive.

Let the thyristor $T1$ be turned on at $t = 0$. With $T1$ on, two current paths i_1 and i_2 are established in Fig. 8.53 (a). One path is for i_1 (shown dotted), it consists of driving voltage $V_s + V_{co}$ (net voltage in the loop of i_1), T_1, L_1, R and $C_2 = C$. The second path is for i_2 (shown full line), it consists of driving voltage $V_s + V_{co}$ (net voltage across C_1), T_1, L_1, R and $C_1 = C$. Since these two paths for i_1 and i_2 are identical, currents i_1 and i_2 must be equal, i.e. $i_1 = i_2$. Since current i_1 is driven by dc source and i_2 by capacitor C_1; 50% of the load current $i_o (= i_1 + i_2 = 2\,i_1 = 2i_2)$ is drawn from dc source and the remaining 50% from the capacitor discharge.

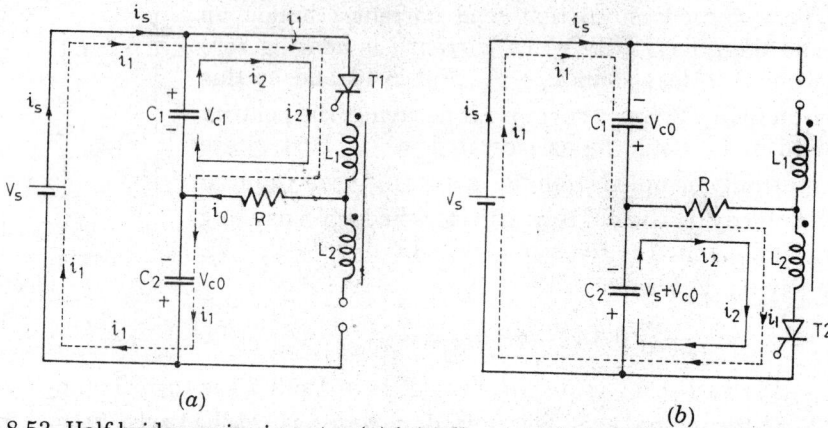

Fig. 8.53. Half-bridge series inverter (a) initially $v_{c1} = V_{co}$ and (b) initially $v_{c2} = V_{co}$.

At the end of positive half cycle at a, Fig. 8.54, load current i_o becomes zero, T_1 is therefore, turned off. Now the voltage across C_1 is V_{co} and that across C_2 is $V_s + V_{co}$ with the polarities as indicated in Fig. 8.53 (b).

When T_2 is turned on at instant b, Fig. 8.54, currents i_1 and i_2 are established as shown. Load current $i_o (= i_1 + i_2)$ now get reversed as desired in an inverter. Current i_1 (shown dotted) is driven by voltage $V_s + V_{co}$ and the path it traverses is C_1, R, L_2, T_2, V_s and $C_1 = C$. Current i_2 (shown full) is driven by voltage $V_s + V_{co}$ and the local loop at traverses is made up of C_2, R, L_2, T_2 and $C_2 = C$. Since the paths negotiated by currents i_1 and i_2 are identical and the driving forces are also equal, $i_1 = i_2$. As before, 50% of the load power comes from the dc source and the other 50% from the capacitor discharge.

In Fig. 8.54 are sketched the waveforms for voltage across C_1 as v_{c1}, voltage across C_2 as v_{c2}, load current i_o, source current i_s or i_1 and voltage across thyristor T_1 as v_{T1}. Waveforms for $i_1 = i_2 = \dfrac{1}{2}\,i_o$ are also shown in this figure. It is seen from the waveforms that dc source delivers current i_s, or power, to load during both the half cycles, as a consequence (i) ripples in source current get reduced and (ii) peak current rating of dc source is also curtailed.

Example 8.12. *In a self-commutated SCR circuit, the load consists of $R = 10\ \Omega$ in series with commutating components of $L = 10$ mH and $C = 10$ µF. Check whether the circuit will commutate by itself when triggered from zero voltage condition on the capacitor. What will be the voltage across the capacitor and inductor at the time of commutation ?*

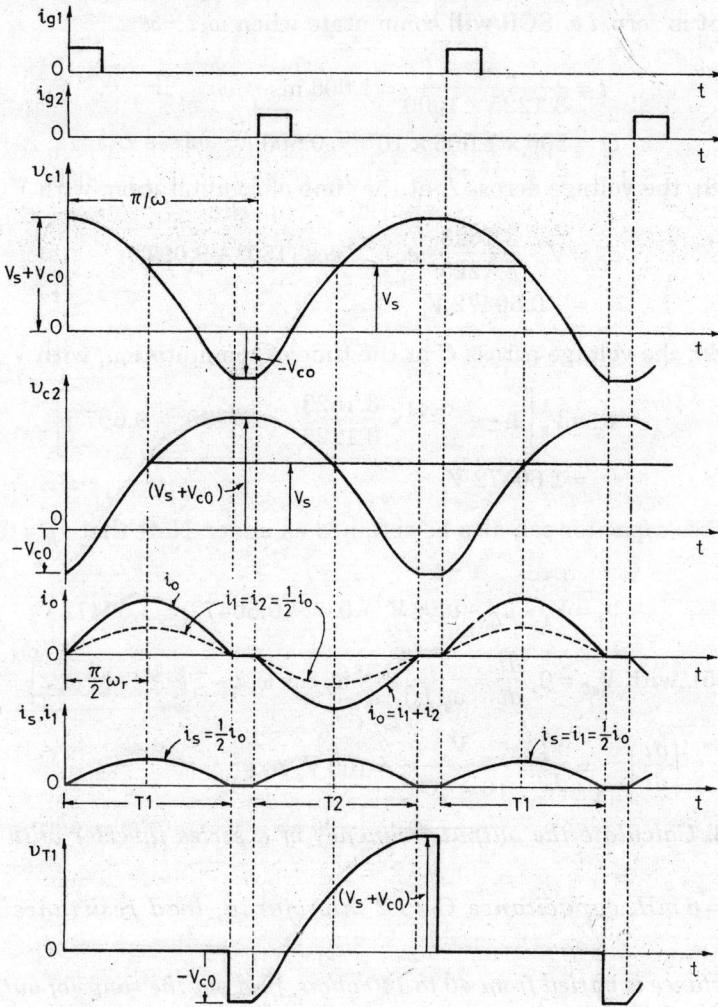

Fig. 8.54. Waveforms for half-bridge series inverter of Fig. 8.52.

Find also $\dfrac{di}{dt}$ *at* $t = 0$

Solution. Here $\quad R^2 = 10 \times 10 = 100$; $\dfrac{4L}{C} = \dfrac{4 \times 10 \times 10^{-3}}{10 \times 10^{-6}} = 4000$

As $R^2 < \dfrac{4L}{C}$, the circuit is underdamped. Therefore, the series circuit will commutate on its own when triggered from zero voltage condition on the capacitor.

Also $\qquad \xi = \dfrac{R}{2L} = \dfrac{10 \times 1000}{2 \times 10} = 500$

$$\omega_0 = \dfrac{1}{\sqrt{LC}} = \sqrt{10^7} = 3.1623 \times 10^3 \text{ rad/sec}$$

$$\omega_r = \sqrt{\omega_0^2 - \xi^2} = \left[10^7 - 2.5 \times 10^5\right]^{1/2} = 3.1225 \times 10^3 \text{ rad/sec}$$

$$\psi = \tan^{-1}\left(\dfrac{\xi}{\omega_r}\right) = \tan^{-1}\left(\dfrac{500}{3.1225 \times 10^3}\right) = 9.097°$$

The load current is zero, *i.e.* SCR will commutate when $\omega_r\, t = \pi$

or
$$t = \frac{\pi}{3.1225 \times 1000} = 1.006 \text{ ms}$$

$$\xi t = 500 \times 1.006 \times 10^{-3} = 0.503$$

From Eq. (8.108), the voltage across L at the time of commutation, with $V_{c0} = 0$, is

$$v_L = V_s \cdot \frac{3.1623}{3.1225} \cdot e^{-0.503} \cos\,(180° + 9.097°)$$

$$= -0.60472\, V_s$$

From Eq. (8.109), the voltage across C at the time of commutation, with $V_{c0} = 0$, is

$$v_c = V_s \left[1 - e^{-0.503} \times \frac{3.1623}{3.1225} \cdot \cos\,(180° - 9.097°) \right]$$

$$= 1.60472\, V_s$$

The voltage across capacitor can also be obtained as under. Note that $v_R = 0$ at the time of commutation.

$$v_c = V_s - v_R - v_L = V_s - 0 - (-0.60472) = 1.60472\, V_s$$

From Eq. (8.105), with $V_{c0} = 0$, $\dfrac{di}{dt} = \dfrac{V_s}{\omega_r\, L} \left[e^{-\xi t}\, \omega_r\, \cos\,\omega_r\, t - \xi\, e^{-\xi t} \cdot \sin\,\omega_r\, t \right]$

$$\therefore \qquad \left(\frac{di}{dt} \right)_{t=0} = \frac{V_s}{L} = \frac{V_s}{10 \times 10^{-3}} = 100\, V_s \text{ A/s}$$

Example 8.13. *Calculate the output frequency of a series inverter with the following parameters :*

Inductance $L = 6\, mH$, capacitance $C = 1.2$ microfarad, load resistance $R = 100$ ohms, $T_{off} = 0.2\, ms$.

If the load resistance is varied from 40 to 140 ohms, find out the range of output frequency.

[I.A.S , 1986]

Solution. From Eq. (8.106), the time period of oscillation is given by

$$\frac{\pi}{\sqrt{1/LC - (R/2L)^2}} = \frac{\pi}{\left[\dfrac{10^3 \times 10^6}{6 \times 1.2} - \dfrac{100^2 \times 10^6}{4 \times 36} \right]^{1/2}} = \frac{\pi}{1000\,(8.333)} = 0.377 \text{ ms}$$

From Eq. (8.107), the output frequency,

$$f = \frac{10^3}{0.377 \times 2 + 2 \times 0.2} = 866.55 \text{ Hz.}$$

When $R = 40\ \Omega$, output frequency

$$= \frac{1}{\dfrac{2\pi}{\left[\dfrac{10^3 \times 10^6}{6 \times 1.2} - \dfrac{1600 \times 10^6}{4 \times 36} \right]^{1/2}} + 0.4 \times 10^{-3}} = 1046.2 \text{ Hz.}$$

When $R = 140\ \Omega$, output frequency

$$= \cfrac{1}{\cfrac{2\pi}{\left[\dfrac{10^3 \times 10^6}{6 \times 1.2} - \dfrac{140^2 \times 10^6}{4 \times 36}\right]^{1/2}} + 0.4 \times 10^{-3}} = 239.8\ \text{Hz}$$

$\therefore$ Range of output frequency = 239.8 Hg to 1046.2 Hz

Example 8.14. *In a single-phase series inverter, the operating frequency is 50 kHz and the thyristor turn-off time $t_q = 10\ \mu s$. Circuit parameter are : $R = 3\ \Omega$, $L = 60\ \mu H$, $C = 7.5\ \mu F$ and $V_s = 220\ V$ dc.*

Determine (a) the circuit turn-off time and (b) maximum possible operating frequency, assuming a factor of safety = 1.5.

Solution. (*a*) Operating frequency, $\omega = 2\pi \times 5000$ rad/s

$$\xi = \frac{R}{2L} = \frac{3 \times 10^6}{2 \times 60} = 0.025 \times 10^6$$

$$\omega_o = \frac{1}{\sqrt{LC}} = \frac{10^6}{\sqrt{60 \times 7.5}} = 0.0471 \times 10^6$$

Circuit ringing frequency, $\omega_r = \sqrt{\omega_o^2 - \xi^2} = 10^6 \sqrt{0.0471^2 - 0.025^2}$

$$= 0.03992 \times 10^6 \simeq 0.04 \times 10^6\ \text{rad/s}$$

From Eq. (8.102), the circuit turn-off time (= time of dead zone) is

$$t_c = \frac{\pi}{\omega} - \frac{1}{\omega_r} = \frac{\pi}{2\pi \times 500} - \frac{\pi}{0.04 \times 10^6} = 21.46\ \mu s$$

As circuit turn-off time (= 21.46 μs) is more than the SCR turn-off time, the series inverter would operate satisfactorily.

(*b*) when the dead zone is just equal to $t_q \times$ factor of safety = $10 \times 1.5 = 15\ \mu s$, the maximum possible operating frequency would be obtained. From Eq. (8.102), we get

$$t_q \times 1.5 = \frac{\pi}{\omega} - \frac{\pi}{\omega_r}$$

$$15 \times 10^{-6} = \frac{\pi}{2\pi \cdot f_{\max}} - \frac{\pi}{0.04 \times 10^6}$$

or $\qquad\qquad f_{\max} = 5345.5\ \text{Hz}.$

Example 8.15. *In Example 8.14, initial voltage across capacitor is 80 V and output current is assumed sinusoidal. Calculate the power delivered to load, and average and rms values of thyristor current.*

Solution. It is seen from Fig. 8.48 or 8.54 that maximum value of load current across

when $\omega_r\, t_1 = \dfrac{\pi}{2}$ or when $t_1 = \dfrac{\pi}{2\,\omega_r}$.

$\therefore \qquad\qquad t_1 = \dfrac{\pi}{2 \times 0.04 \times 10^6} = 32.27\ \mu s$

From Eq. (8.105), the maximum value of load current $I_{o.mx}$ would occur when $\sin \omega_r t = 1$.

$\therefore \qquad\qquad I_{o.mx} = \dfrac{V_s + V_{co}}{\omega_r.L}\, e^{-\xi t_1}$

$$= \frac{220 + 80}{0.04 \times 10^6 \times 60 \times 10^{-6}} \exp[-0.025 \times 10^6 \times 32.27 \times 10^{-6}]$$

$$= 55.79 \text{ A} = \text{peak value of thyristor current}$$

$\therefore$ Rms value of load current $\quad = \dfrac{I_{o.mx}}{\sqrt{2}} = \dfrac{55.79}{\sqrt{2}} = 39.46 \text{ A}$

Load power $\qquad\qquad\qquad = (39.46)^2 \times 3 = 4671.27 \text{ W}$

Since thyristor current has a periodicity of 2π radians, *rms* value of thyristor current

$$= \frac{I_{o.mx}}{2} = \frac{55.79}{2} = 27.895 \text{ A}$$

Assuming lossless converter system, average value of source current,

$$I_{sA} = \frac{4671.27}{220} = 21.233 \text{ A}$$

Average thyristor current, $\quad I_{TA} = \dfrac{I_{SA}}{2} = \dfrac{21.233}{2} = 10.617 \text{ A}$

8.10. SINGLE-PHASE PARALLEL INVERTER

The basic inverter circuit for a single-phase parallel inverter, utilizing capacitor for its commutation, is shown in Fig. 8.55. It consists of two thyristors T1 and T2, an inductor L, an output transformer and a commutating capacitor C. The transformer turns ratio from each primary half to secondary winding is assumed unity. The output voltage and current are v_0 and i_0 respectively. The function of L is to make the source current constant at I_0. Positive directions for voltages and currents are marked in Fig. 8.55. During the working of this inverter, capacitor C comes in parallel with the load *via* the transformer. That is why it is called a *parallel inverter*.

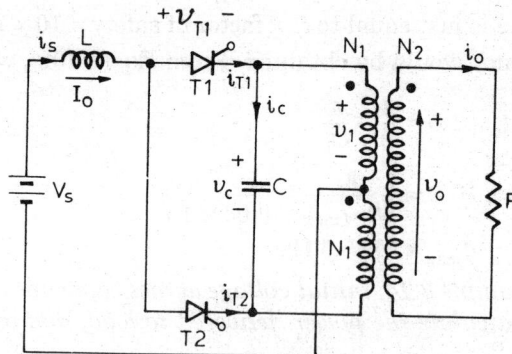

Fig. 8.55. Single-phase capacitor commutated parallel inverter with centre-tapped transformer.

The operation of this inverter can be explained in some well-defined modes as under :

Mode I : In this mode, thyristor T1 is conducting and a current flows in the upper half of primary winding. Thyristor T2 is off. This current establishes magnetic flux that links both the halves of primary winding. As a result, an emf V_s is induced across upper as well as lower half of primary winding. In other words, total voltage across primary winding is $2V_s$. This voltage charges the commutating capacitor C to a voltage of $2V_s$ with upper plate positive as

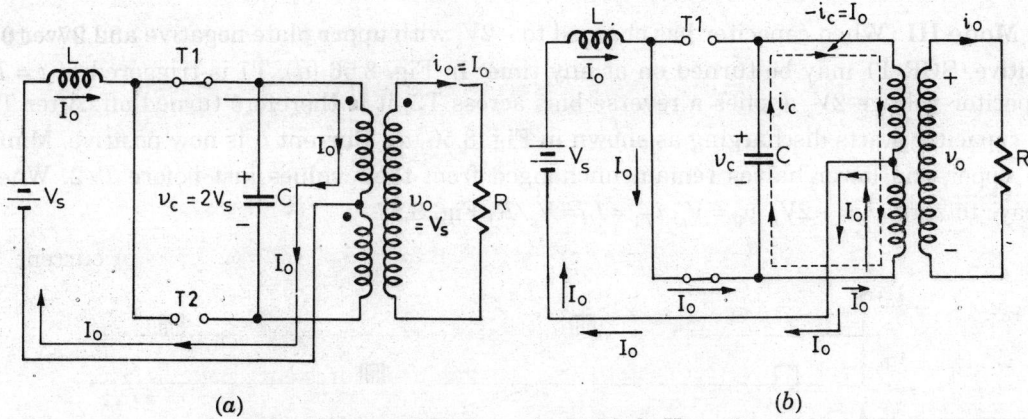

Fig. 8.56. (a) Mode I, $t < 0$; (b) Mode II, $t = 0 +$

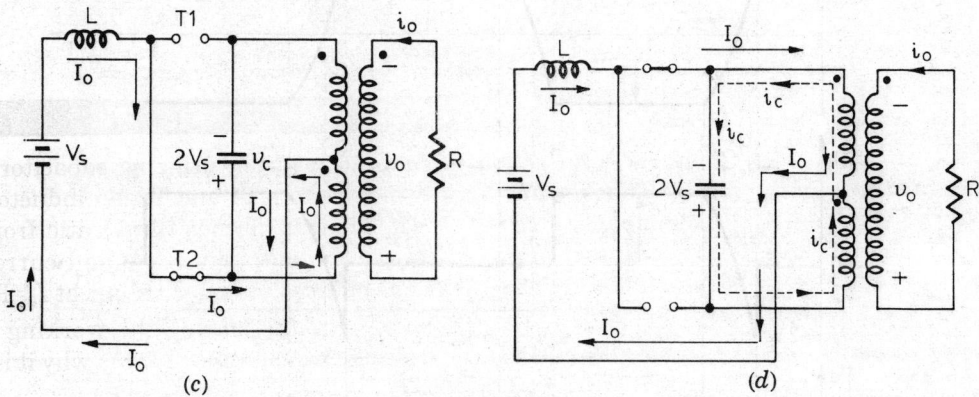

Fig. 8.56. (c) Mode II, $t_1 \leq t < T/2$ (d) Mode III, just after $t = T/2$.

shown in Fig. 8.56 (a). Thyristor T2 is forward biased through T1 by the capacitor voltage $2V_s$. Eventually, a steady state current I_0 flows through V_s, L, T1 and upper half of primary winding. During this mode, $v_0 = V_s$, $v_c = 2V_s$, $i_0 = I_0$, $v_{T1} = 0$, $i_c = 0$, $i_{T1} = I_0$ as shown in mode I in Fig. 8.56(a).

Mode II : At time $t = 0$, thyristor T2 is turned on by applying a triggering pulse to its gate. At this time $t = 0$, capacitor voltage $2V_s$ appears as a reverse bias across T1, it is therefore turned off. A current I_0 begins to flow through T2, lower half of primary winding, V_s and L as shown in Fig. 8.56 (b). At the same time, capacitor voltage $2V_s$ is applied across the total transformer primary and a capacitor current $- i_c$ is established. Negative sign before i_c means that current i_c flows opposite to its positive direction assumed in Fig. 8.55. Before T2 is on, i.e. at $t = 0 -$, mmf in upper primary winding is $I_0 N_1$ and zero in the lower primary winding. Soon after T2 is on, i.e. at $t = 0 +$, mmfs linking both upper and lower halves cannot change suddenly. Therefore, at $t = 0 +$, $- i_c = I_0$ such that mmf in the lower half remains zero and mmf in the upper half is equal to mmf at $t = 0 -$. After $t = 0 +$, capacitor C discharges and current i_c is such that it supplies the load current i_0 and balances the primary and secondary ampere turns of the transformer. Capacitor current continues flowing till capacitor has charged from $+ 2V_s$ to $- 2V_s$ at time $t = t_1$. Load voltage also changes from V_s at $t = 0$ to $- V_s$ at $t = t_1$, Fig. 8.56 (c) and Fig. 8.57.

Mode III : When capacitor has charged to $-2V_s$ with upper plate negative and lower plate positive, SCR T1 may be turned on at any time. In Fig. 8.56 (d), T1 is triggered at $t = T/2$. Capacitor voltage $2V_s$ applies a reverse bias across T2, it is therefore turned off. After T2 is off, capacitor starts discharging as shown in Fig. 8.56 (d). Current i_c is now positive. Mmfs in the upper and lower halves remain unchanged from their values just before $T/2$. When i_c decays to zero, $v_c = +2V_s$, $v_0 = V_s$, $i_{T1} = I_0 = V_s/R$, Fig. 8.57.

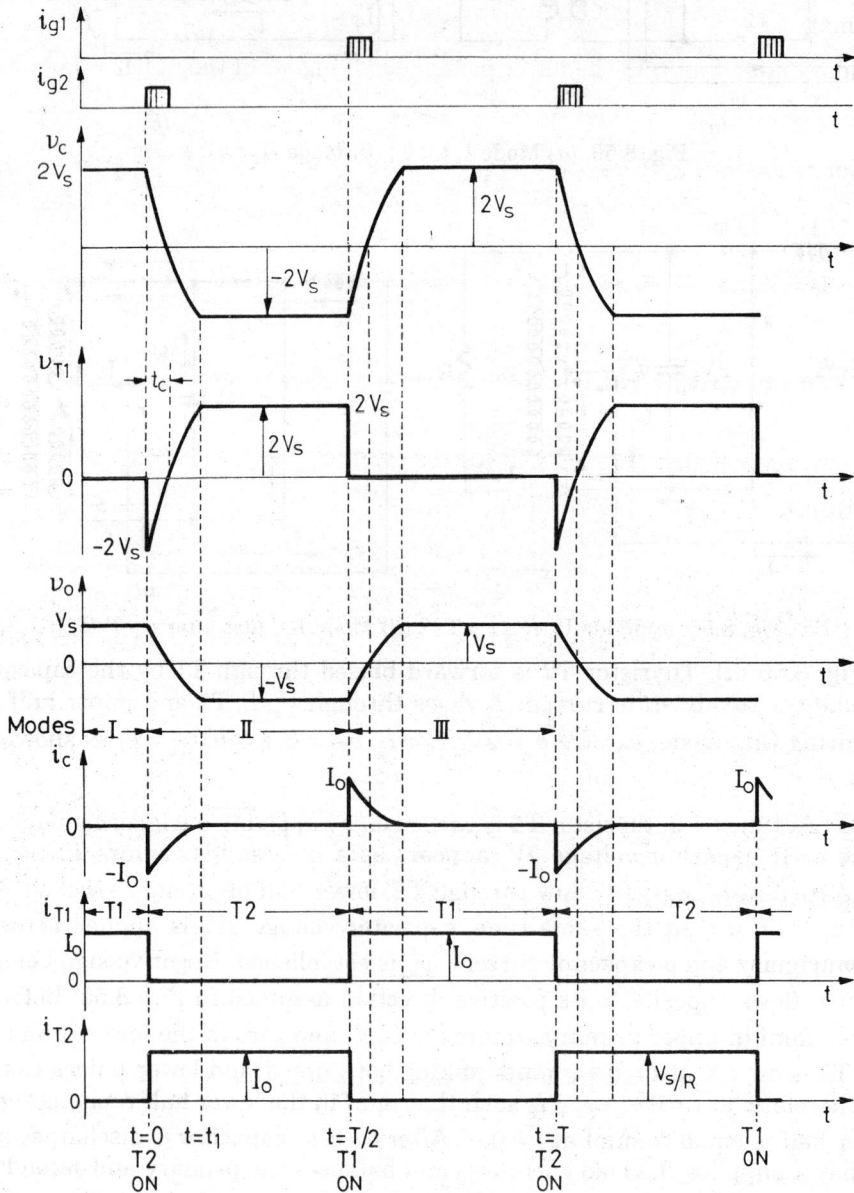

Fig. 8.57. Waveforms for currents and voltages pertaining to single-phase parallel inverter.

The waveforms for v_c, v_{T1}, v_0 etc are shown in Fig. 8.57. At $t = 0 +$, when T2 is turned on; $v_{T1} = -2V_s$, $i_c = -I_0$, $i_{T1} = 0$ and $i_{T2} = I_0$ as shown. As the turns ratio from whole primary to secondary winding is 2, the load voltage has half the amplitude of capacitor voltage. However, the load voltage has the same waveform as the capacitor voltage, see Fig. 8.57.

8.10.1. Analysis of Parallel Inverter

In Fig. 8.56 (b), the secondary mmf must be balanced by the primary mmf at any time t. It is seen from this figure that net current in the lower primary half is $(I_0 - i_c)$ upward and in the upper primary half is i_c downward.

Secondary $mmf = mmf$ in the lower primary half + mmf in the upper primary half.

$$\therefore \qquad i_0 N_2 = (I_0 - i_c) N_1 - i_c N_1$$

Load or secondary current, $\quad i_0 = (I_0 - 2 i_c) \dfrac{N_1}{N_2}$

But $$i_c = C \frac{dv_c}{dt}$$

$$\therefore \qquad i_0 = \left(I_0 - 2 C \frac{dv_c}{dt} \right) \frac{N_1}{N_2} \qquad\qquad \text{...(8.113)}$$

Also, load voltage $v_0 = i_0 R$. This voltage when referred to whole primary is

$$v_c = \frac{i_0 R}{N_2} (2 N_1) \qquad\qquad \text{...(8.114)}$$

Substituting the value of i_0 from Eq. (8.113) in Eq. (8.114), we get

$$v_c = \left(I_0 - 2 C \frac{dv_c}{dt} \right) \frac{N_1}{N_2} \cdot 2 R \frac{N_1}{N_2}$$

$$= \left(I_0 - 2C \frac{dv_c}{dt} \right) 2R \left(\frac{N_1}{N_2} \right)^2$$

If $\quad \dfrac{N_2}{N_1} = n$, then $\dfrac{n^2 \cdot v_c}{2R} = I_0 - 2C \dfrac{dv_c}{dt}$

or $$\frac{dv_c}{dt} + \frac{n^2 \cdot v_c}{4 RC} = \frac{I_0}{2C}$$

Its particular integral is, $\quad V_{P.I.} = \dfrac{I_0}{2C} \times \dfrac{4RC}{n^2} = \dfrac{2 I_0 R}{n^2}$

Its complementary function is,

$$V_{C.F.} = A \, e^{-\frac{n^2}{4RC} t}$$

$$\therefore \qquad v_c = \frac{2 I_0 R}{n^2} + A \, e^{-\frac{n^2}{4RC} t} \qquad\qquad \text{...(8.115)}$$

Here $\dfrac{R}{n^2} = R \left(\dfrac{N_1}{N_2} \right)^2$ is the load resistance referred to one half of the primary winding. I_0 is

the load current on the primary side when $\dfrac{dv_c}{dt} = 0$.

$\therefore 2 I_0$ (resistance referred to one half of primary winding) $= 2V_s$

or
$$\frac{2 I_0 R}{n^2} = 2V_s$$

With this, Eq. (8.112) becomes

$$v_c = 2V_s + A\, e^{-\frac{n^2}{4RC} t}$$

At $t = 0$, $v_c = 2V_s$, $\therefore 2V_s = 2V_s + A$. But this gives redundant solution.

At $t = T/2$, voltage across capacitor becomes opposite to what it is at $t = 0$. This fact can be expressed as

$$[v_c \text{ at } t = 0] = -[v_c \text{ at } t = T/2]$$

or
$$2V_s + A = -2V_s - A\, e^{-\frac{n^2 T}{8RC}}$$

or
$$A = \frac{-4V_s}{1 + e^{-n^2 T/8RC}}$$

$\therefore$
$$v_c = 2V_s\left[1 - \frac{2\, e^{-n^2 t/4RC}}{1 + \exp\left(-\dfrac{n^2 T}{8RC}\right)}\right] \qquad \ldots(8.116)$$

For practical circuits, $\exp\left[-\dfrac{n^2 T}{8RC}\right] \ll 1$.

$\therefore$
$$v_c = 2V_s\left[1 - 2\exp\left(-\frac{n^2 t}{4RC}\right)\right] \qquad \ldots(8.117)$$

In Fig. 8.57, v_c at $t = 0$ is $2V_s$. But in Eq. (8.117); at $t = 0$, $v_c = -2V_s$. In order that Eq. (8.117) is compatible with the waveform for v_c in Fig. 8.57, we re-write Eq. (8.117) as

$$v_c = 2V_s\left[2\exp\left(-\frac{n^2 t}{4RC}\right) - 1\right] \qquad \ldots(8.118)$$

It is seen from the waveform of v_c and v_{T1} in Fig. 8.57 during the interval $t = 0$ to $t = t_1$ that $v_{T1} = -v_c = 2V_s\left[1 - \exp\left(-\dfrac{n^2 t}{4RC}\right)\right]$. Circuit turn-off time t_c is the time taken by v_{T1} to become zero from its initial value of $-2V_s$. Therefore,

$$v_{T1} = 2V_s\left[1 - 2e^{-\frac{n^2 t_c}{4RC}}\right] = 0$$

or
$$\frac{n^2 t_c}{4RC} = \ln 2$$

or
$$t_c = \frac{4RC \cdot \ln 2}{n^2} \qquad \ldots(8.119)$$

Also, commutating capacitance,

$$C = \frac{n^2 \cdot t_c}{4R \cdot \ln 2} \qquad \ldots(8.120)$$

Eq. (8.119) reveals that for a given value of load resistance, the value of capacitor should be so selected as to give adequate circuit turn-off time t_c for the off-going thyristor. In Eq. (8.119), $t_c > t_{off}$, where t_{off} is the thyristor turn-off time. It is seen from Eq. (8.120) that if load resistance is small, size of capacitor required is large.

Example 8.16. *In a single-phase capacitor commutated parallel inverter using two thyristors and a center-tapped transformer, the source voltage is 220 V dc. The centre-tapped transformer has a turns ratio from each half primary winding to secondary winding of 3 : 1. For a load resistance of 20 Ω, find the value of capacitor C to obtain 20 μs turn-off time on the thyristor. Assume the inductor L large and transformer ideal. Take factor of safety as 2.*

Solution. From Eq. (8.118), $v_c = 2V_s \left[2 \exp\left(-\dfrac{n^2 t}{4RC} \right) - 1 \right]$

Here $n = \dfrac{1}{3}$. Circuit turn-off time $t_c = 2 \times 20 = 40$ μs

The circuit turn-off time is obtained when v_c reduces to zero from $2V_s$. This gives

$$C = \frac{n^2 t_c}{4R \ln 2}$$

$$= \frac{1}{9} \cdot \frac{40 \times 10^{-6}}{4 \times 20 \ln 2} = 0.0815 \ \mu\text{F}.$$

8.11. GOOD INVERTER

Here, some of the requirements of a good inverter are enumerated.
1. Its output voltage waveform should be sinusoidal.
2. Its gain (ac output voltage/dc input voltage) should be high.
3. Its output voltage and frequency should be controllable in the desired usage. For example, inverter must be capable of keeping V/f constant for some applications.
4. The power required by its controlling circuit should be minimum.
5. The semiconductor devices used in the inverter should have minimum switching and conduction losses
6. Its overall cost must be minimum without sacrificing its reliability.
7. Its working life must be long.
8. It should produce minimum electromagnetic interference (EMI).
9. The size of the filters required should be small.

Many of the points listed above, apply equally well to other types of power-electronics converters.

PROBLEMS

8.1. (*a*) What is an inverter ? List a few industrial applications of inverters.
  (*b*) What are line-commutated inverters ? How do they operate ? Explain the difference between line-commutated and force-commutated inverters.
  (*c*) What are the two main types of inverters ? Distinguish between them explicitly.
8.2. (*a*) Describe the working of a single-phase half-bridge inverter. What is its main drawback ? Explain how this drawback is overcome.
  Discuss how output power in single-phase full-bridge inverter becomes four times the power handled by a single-phase half-bridge inverter.

(b) What is the purpose of connecting diodes in antiparallel with thyristors in inverter circuits ? Explain how these diodes come into play.

8.3. A single-phase full-bridge inverter may be connected to a load consisting of (a) R (b) RL or RLC overdamped (c) RLC underdamped. For all these loads, draw the load voltage and load current waveforms under steady operating conditions. Discuss the nature of these waveforms. Also indicate the conduction of the various elements of the inverter circuit.

Is it possible for this inverter to have load commutation ? Explain.

8.4. For a single-phase full bridge inverter, $V_s = 230$ V dc, $T = 1$ ms. The load consists of RLC in series with $R = 1.2\ \Omega$, $\omega L = 8\ \Omega$, $\dfrac{1}{\omega C} = 7\ \Omega$.

(a) Sketch the waveforms for load voltage v_0, fundamental component of output current i_{01}, source current i_s and voltage across thyristor 1. Indicate the devices that conduct during different intervals of one cycle. Find also the rms value of fundamental component of load current.

(b) Find the power delivered to load due to fundamental component.

(c) Check whether forced commutation is required or not.

[**Ans.** (a) 132.586 A (b) 21094.8 W (c) It is required.]

8.5. (a) Write Fourier series expression for the output voltages and currents obtained from single-phase half-bridge and full-bridge inverters.

(b) A single-phase bridge inverter is fed from 230 V dc. In the output voltage wave, only fundamental component of voltage is considered. Determine the rms current ratings of an SCR and a diode of the bridge for the following types of loads :

(i) $R = 2\ \Omega$ (ii) $\omega L = 2\ \Omega$

Find also the repetitive peak voltage that may appear across a thyristor in parts (i) and (ii).

[**Ans.** (i) 73.211 A, zero A, 230 V (ii) 51.776 A, 51.776 A, 230 V]

8.6. (a) A single-phase full bridge inverter is connected to a dc source of V_s. Resolve the output voltage waveshape into Fourier series.

(b) A single-phase full-bridge inverter delivers power to RLC load with $R = 3\ \Omega$ and $X_L = 12\ \Omega$. The bridge operates with a periodicity of 0.2 ms. Calculate the value of C so that load commutation is achieved for the thyristors. Turn-off time for thyristors is 12 μs. Factor of safety is 2. Assume the load current to contain only the fundamental component. [**Ans.** (b) $C = 2.148$ μF.]

8.7. A single-phase full-bridge inverter feeds power at 50 Hz to RLC load with $R = 5\ \Omega$, $L = 0.3$ H and $C = 50$ μF. The dc input voltage is 220 V dc.

(a) Find an expression for load current up to fifth harmonic. Also, calculate :

(b) power absorbed by the load and the fundamental power,

(c) the rms and peak currents of each thyristor,

(d) conduction time of thyristors and diodes if only fundamental component were considered.

[**Ans.** (a) $9.036 \sin(\omega t - 80.72°) + 0.357 \sin(3\omega t - 88.90°) + 0.122 \sin(5\omega t - 89.38°)$
(b) 204.54 W, 204.12 W (c) 9.044 A, 4.522 A (d) 5.513 ms, 4.487 ms]

8.8. Describe modified McMurray half-bridge inverter with appropriate voltage and current waveforms. The total commutation interval may be subdivided into certain well-defined modes for the purpose of explaining its operation.

8.9. (a) For a single-phase modified McMurray half-bridge inverter, find an expression that gives the circuit turn-off time for the main thyristor in terms of load current, peak capacitor current etc.

Discuss how commutating circuit components can be designed on the basis of minimum commutation energy. The relevant voltage and current waveforms must be drawn.

(b) A single-phase modified McMurray inverter is fed by a dc source of 230 V. The dc source voltage may fluctuate by ± 25%. The load current during commutation may vary from 20 to 80 A. If thyristor turn-off time is 20 μsec, calculate the values of C and L. Use a factor of safety cf 2 .

Also, obtain the value of resistance that gives critical damping.

[**Ans.** (b) 16.547 μF, 34.1895 μH, 2.875 Ω]

8.10. Describe McMurray-Bedford half-bridge single-phase inverter with relevant voltage and current waveforms. The working of this inverter may be explained in certain well defined modes. Enumerate the simplifying assumptions made.

8.11. Discuss the principle of working of a three-phase bridge inverter with an appropriate circuit diagram. Draw phase and line voltage waveforms on the assumption that each thyristor conducts for 180° and the resistive load is star-connected. The sequence of firing of various SCRs should also be indicated in the diagram.

8.12. Repeat Problem 8.11 in case each thyristor conducts for 120°.

8.13. Repeat Problem 8.11 in case load is delta-connected.

8.14. A star-connected load of 15 Ω per phase is fed from 420 V dc source through a 3-phase bridge inverter. For both (a) 180° mode and (b) 120° mode, determine

(i) rms value of load current

(ii) rms value of thyristor current

(iii) load power.   [**Ans.** (a) 13.2 A, 9.333 A, 7840.8 W (b) 11.43 A, 8.083 A, 5879.02 W]

8.15. (a) What is the need for controlling the voltage at the output terminals of an inverter ? Describe briefly and compare the various methods employed for the control of output voltage of inverters.

(b) For the series-inverter control of voltage, two single-phase inverters are connected in series. Each inverter has output voltage of 400 V and each transformer has primary to secondary turns ratio of 1/2. Calculate the resultant output voltage from this scheme in case firing angles for the two inverters differ by 30°   [**Ans :** (b) 1545.48 V]

8.16. (a) What is pulse width modulation ? List the various PWM techniques. How do these differ from each other ?

(b) For a single-pulse modulation used in inverters, show that output voltage can be expressed as

$$v_0 = \sum_{n=1,3,5}^{\infty} \frac{4V_s}{n\pi} \cdot \sin\frac{n\pi}{2} \sin nd \sin n\omega t$$

where $2d$ is the pulse width.

Sketch the variation of $\dfrac{v_{0nm}}{v_{01m}}$ as function of pulse width $2d$. Here v_{onm} is the peak value of nth harmonic component and v_{01m} that of the fundamental component.

(c) A single-phase full bridge inverter has rms value of the fundamental component of output voltage, with single-pulse modulation, equal to 110 V. Compute the pulse width required and the rms value of output voltage in case dc source voltage is 220 V.

[**Ans :** (c) 67.471° 134.693 V]

8.17. For a single-phase bridge inverter, source voltage is 230 V dc and the load is series RLC with $R = 1\,\Omega$, $\omega L = 2\Omega$ and $\dfrac{1}{\omega C} = 1.5\,\Omega$. The output voltage is controlled by single pulse modulation and the pulse width is 120°. Determine the magnitude of rms values of fundamental, third, fifth and seventh harmonic components of the output current.

Also, find the power delivered to load.

[**Ans :** 160.43 A, zero, 3.6786 A, 1.8537 A, 25754.7 W]

8.18. (a) For the symmetrical two-pulse modulation shown in Fig. 8.29, prove that

(i) the magnitude of nth harmonic voltage is

$$\frac{8 V_s}{n\pi} \sin n\gamma \sin \frac{nd}{2}$$

(ii)
$$\gamma = \frac{\pi - 2d}{n+1} + \frac{d}{N}$$

where N = number of pulses per half cycle.

(b) Describe how multiple-pulse modulated wave can be generated from carrier and reference waves. Hence show that

(i) number of pulse per half-cycle, $N = \frac{\omega_c}{2\omega}$.

(ii) pulse width, $\frac{2d}{N} = \left(1 - \frac{V_r}{V_c}\right) \cdot \frac{\pi}{N}$

where V_c and V_r are the amplitudes of carrier and reference signals respectively.

8.19. Explain sinusoidal-pulse modulation as used in PWM inverters. Discuss the conditions under which the number of pulses generated per half cycle are $\frac{f_c}{2f}$ or $\left(\frac{f_c}{2f} - 1\right)$. Here f_c and f are the frequencies of carrier and reference signals respectively.

8.20. A single-phase bridge inverter feeds power to a load of $R = 12\ \Omega$ and $L = 0.04$ H from a 400 V dc source. If the inverter operates at a frequency of 50 Hz, determine the power delivered to load for (a) square wave operation (b) quasi-square wave operation with an on-period of 0.6 of a cycle (c) two symmetrically spaced pulse per half cycle with an on-period of 0.6 of a cycle.

[**Ans.** (a) 5285.56 W (b) 3400.96 W (c) 2706.34 W]

8.21. (a) For harmonic reduction in single-phase inverters, two identical transformers are used in series. If their rectangular output voltage waveforms are shifted from each other by 120°, then sketch these voltage waveforms and their resultant waveform on the assumption that transformer secondary voltages oppose each other. Find also an expression for the net output voltage as a function of time. Hence find the percentage derating of the inverter so far as its fundamental power component is concerned.

(b) A single-phase full bridge inverter has waveforms for its output voltage v_0 and output current i_0 as shown in Fig. 8.58. Explain and indicate the devices that conduct in the various intervals throughout the cycle.

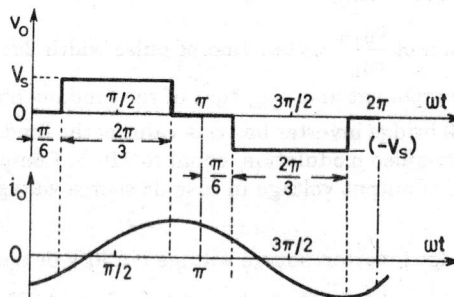

Fig. 8.58. Pertaining to Problem 8.21 (b).

$$\left[\textbf{Ans.} \quad (a)\ v_0 = \frac{4V_s}{\pi}\sqrt{3}\left[\sin\left(\omega t + \frac{\pi}{6}\right) + \frac{1}{5}\sin\left(5\omega t - \frac{\pi}{6}\right) + \frac{1}{7}\sin\left(7\omega t + \frac{\pi}{6}\right)\cdots\right],\ 13.4\%\right]$$

(b) T4D2, D1D2, T1T2, T1D3, T2D4, D3D4, T3T4, T3D1, T4D2 etc

or T3D1, D1D2, T1T2, T1D3, T2D4, D3D4, T3T4, T4D2, T3D1 etc.

8.22. The stepped wave output voltage waveforms of Figs. 8.59 (*a*) and (*b*) are to be obtained by the series cascading of two stepped-wave inverters. Describe how these wave-forms can be realized. Illustrate your answer with appropriate waveforms through the use of two-level or three-level modulations for Fig. 8.59 (*a*) ; and only two-level modulations for Fig. 8.59 (*b*).

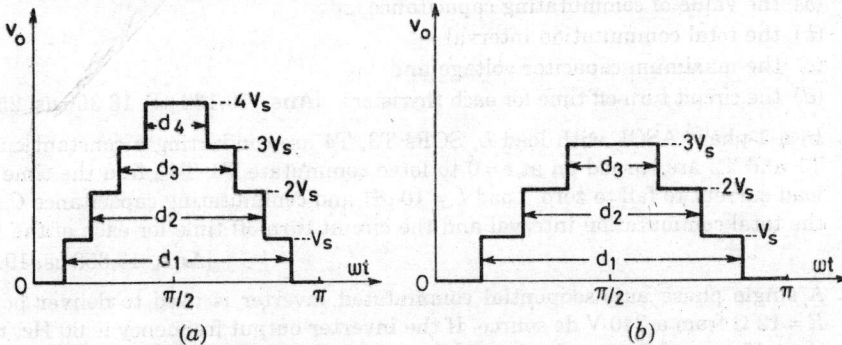

Fig. 8.59. Pertaining to Problem 8.22.

8.23. (*a*) A single-phase CSI is fitted with ideal SCRs. Describe its working when its load is a capacitor *C*. Show that the frequency of input voltage to CSI is twice the frequency of triggering the thyristors.

(*b*) A single-phase CSI (with ideal switches) has the following data :

$$I = 30 \text{ A}, \ f = 500 \text{ Hz}, \text{ Load capacitance} = 20 \text{ μF}$$

For this inverter, calculate

(*i*) the circuit turn-off time.

(*ii*) the peak value of reverse voltage that appears across thyristors.

[**Ans.** (*b*) 500 μs, 750 V]

8.24. (*a*) Describe a single-phase capacitor-commutated CSI connected to load *R* with the help of its power circuit diagram and waveforms for gating signals, load current, capacitor voltage and current, input voltage and voltage across one thyristor.

From the equations governing its performance, show that load current is given by

$$i_0 = I \left[1 - 2 \, \frac{\exp. \, (- \, t/RC)}{1 + \exp. \, (- \, T/2RC)} \right]$$

(*b*) A single-phase capacitor-commutated CSI connected to load *R* has the following data :

$$R = 40 \, \Omega, \ C = 50 \text{ μF}, \ f = 500 \text{ Hz}, \text{ Source current} = 40 \text{ A}$$

For this CSI.

(*i*) obtain an expression for the output current as a function of time and find its value at *t* = 0 and *t* = *T*/2,

(*ii*) find the circuit turn-off time,

(*iii*) compute the average value of input voltage and the power delivered to load.

[**Ans.** (*b*) 40 [1 – 1.245 exp. (– 500 *t*)], – 9.8 A, 9.8 A ; 438.14 μs ; 32 52 V, 1300.8 W]

8.25. Describe a single-phase auto-sequential commutated CSI with *L* load. Write appropriate expressions governing its performance and prove therefrom that total circuit turn-off time for this inverter is given by

$$t_c = \left(1 + \frac{\pi}{2} \right) \sqrt{LC}$$

Waveforms for gating signals, capacitor voltage and current and load current should also be sketched. Find also the circuit turn-off time for each thyristor.

[**Ans.** t_c for each thyristor = $\sqrt{LC}$]

8.26. A single-phase ASCI has the following data :

Load inductance, $L = 4\,\mu H,$ Source current = 20 A

Time during which four diodes conduct = 10 μs

For this CSI, determine

(a) the value of commutating capacitance,

(b) the total commutation interval

(c) the maximum capacitor voltage and

(d) the circuit turn-off time for each thyristor.   [**Ans.** 10.132 μF, 16.366 μs, 25.133 V, 6.366 μs]

8.27. In a 1-phase ASCI, with load L, SCRs T3, T4 are conducting a constant current $I = 10$ A. If T1 and T2 are turned on at $t = 0$ to force commutate T3, T4 ; find the time required for the load current to fall to zero. Load $L = 10\,\mu H$ and commutating capacitance $C = 6\,\mu F$. Find also the total commutation interval and the circuit turn-off time for each of the SCRs.

[**Ans.** 15.858 μs, 19.913 μs, 7.746 μs]

8.28. A single phase auto-sequential commutated inverter is used to deliver power to a load of $R = 12\,\Omega$ from a 240 V dc source. If the inverter output frequency is 60 Hz, thyristor turn-off time 15 μs and factor of safety 2, then determine the suitable values for source inductance and the commutating capacitors. Neglect all losses and assume a maximum current change of 0.4 in one cycle.

Derive the formula used for determining C.   [**Ans.** 10 H, 3.6067 μF]

8.29. (a) Describe the working of a single-phase series inverter with appropriate circuit and waveforms.

(b) For this inverter, derive an expression for the output frequency in terms of circuit parameters and T_{off}.

8.30. In a self-commutated SCR circuit, the load consists of an inductance of 12 mH, a capacitance of 8 μF and a resistance of 15 Ω all connected in series. Check that the circuit will commutate by itself when triggered from zero charge conditions on the capacitor.

Calculate the voltage across the inductor and capacitor at the time of commutation. Find also $\left(\dfrac{di}{dt}\right)$ at $t = 0$.

[**Ans.** It will commutate, $-0.538\,V_s$, $1.538\,V_s$, 83.33 A/s]

8.31. (a) A single-phase quasi-square wave inverter operates on a 24-V battery and it is required to feed power to a 230-V, 50 Hz load. PMOSFETs are used as switches.

Draw the power circuit diagram and explain its operation

(b) For a heater load of 1 A at 230 V, find the *rms* value of fundamental component of inverter output current in case inverter of part (a) operates with (i) square-wave output and (ii) quasi-square output with 120° pulse width. Treat the transformer ideal.

[**Ans.** (b) (i) 10.643 A (ii) 12.29 A]

8.32. Draw the circuit diagram of a series inverter and indicate the need for an optimum time margin. Also, indicate the merits and demerits of this inverter.

8.33. Enumerate the various limitations in a basic series inverter. Explain how the limitation of operating frequency is surmounted in this inverter.

8.34. Discuss the method of overcoming the intermittent power flow in a basic series inverter. Illustrate your answer with relevant circuits and waveforms.

8.35. A single-phase modified series inverter with coupled inductors has the following data :

$V_s = 220\ Vdc,\ R = 2\,\Omega,\ L_1 = L_2 = 40\,\mu H,\ C = 8\,\mu F$, operating frequency = 5 kHz, $t_q = 12$ μs.

Determine (a) circuit turn-off time and (b) the maximum possible operating frequency, taking a factor of safety of 2.   [**Ans.** (a) 37.17 μs (b) 5758.2 Hz]

8.36. In Prob. 8.35, initial voltage across capacitor is 60 V and output current is assumed sinusoidal. Calculate the load power, and average and *rms* values of thyristor current.

[**Ans.** 4075.6 W ; 18.525 A, 31.916 A]

8.37. (a) Calculate the maximum possible frequency of a series inverter with the following data :
$R = 100\,\Omega$, $L = 0.05\,H$, $C = 10\,\mu F$

(b) Calculate the output frequency of the series inverter if the parameters are :
$L = 5\,mH$, $C = 0.2\,\mu F$, $R = 200\,\Omega$, $T_{off} = 0.2\,ms$.

[**Ans.** (a) 159.15 Hz (b) 2442.36 Hz]

8.38. Describe the working of a single-phase parallel inverter with relevant circuit and waveforms.

8.39. A single-phase parallel inverter delivers power to a resistive load through a centre-tapped ideal transformer. Derive expression for the capacitor voltage on the assumption of constant source current. Hence obtain therefrom an expression for the circuit turn-off time.

8.40. Discuss the requirements of a good inverter.

Chapter 9

AC Voltage Controllers

In this Chapter

- Principle of Phase Control
- Principle of Integral Cycle Control
- Single-Phase Voltage Controllers
- Sequence Control of AC Voltage Controllers (Transformer Tap Changers)

AC voltage controllers are thyristor based devices which convert fixed alternating voltage directly to variable alternating voltage without a change in the frequency. Some of the main applications of ac voltage controllers are for domestic and industrial heating, transformer tap changing, lighting control, speed control of single-phase and three-phase ac drives and starting of induction motors. Earlier, the devices used for these applications were auto-transformers, tap-changing transformers, magnetic amplifiers, saturable reactors etc. But these devices are now replaced by thyristor-and triac-based ac voltage controllers because of their high efficiency, flexibility in control, compact size and less maintenance. AC voltage controllers are also adaptable for closed-loop control systems. Since the ac voltage controllers are phase-controlled devices, thyristors and triacs are line commutated and as such no complex commutation circuitry is required in these controllers. The main disadvantage of ac voltage controllers is the introduction of objectionable harmonics in the supply current and load voltage waveforms, particularly at reduced output voltage levels.

The object of this chapter is to study single-phase ac voltage controllers so far as their principle of working and gating signal requirements are concerned. Their use in transformer tap changers is also considered.

For regulating the power flow in *ac* voltage controllers, control strategies are of two types :

1. Phase control   2. Integral cycle control

These are now described in what follows.

9.1. PRINCIPLE OF PHASE CONTROL

Principal of phase control with **respect** to phase-controlled rectifiers has already been described in Art. 6.1. The basic philosophy of phase control in single-phase voltage controllers is the same. Here, switching device is so operated that load gets connected to *ac* source for a part of each half-cycle (or a part of each cycle) of the input voltage.

For illustrating the principle of phase control in single-phase voltage controllers, consider the power circuit diagram of Fig. 9.1 (*a*). It consists of one thyristor in antiparallel with one diode. This circuit configuration is called *single-phase half-wave voltage controller*. When SCR is forward biased during positive half cycle, it is turned on firing angle α. Load voltage at once

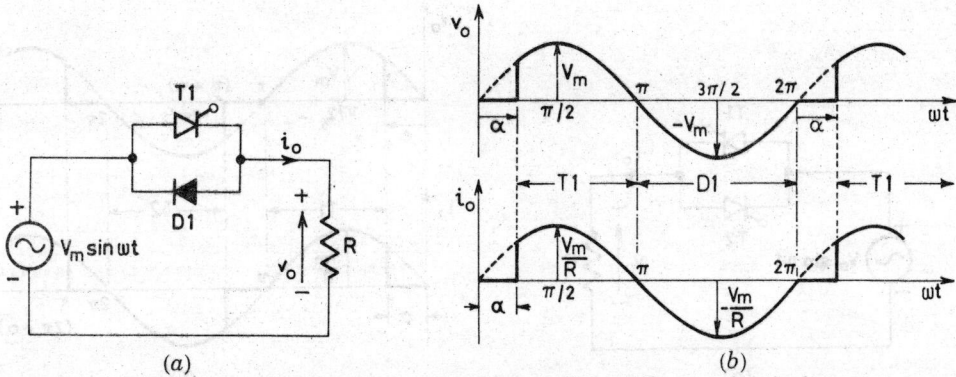

Fig. 9.1. Single-phase half-wave ac voltage controller (a) Power-circuit diagram
and (b) voltage and current waveforms.

jumps to $V_m \sin \alpha$, likewise load current becomes $\dfrac{V_m}{R} \sin \alpha$. The waveforms of load voltage v_o
and load current i_o are drawn in Fig. 9.1 (b). Thyristor get turned off at $\omega t = \pi$ for R load. After
$\omega t = \pi$, negative half cycle forward biases diode $D1$, therefore $D1$ conducts from $\omega t = \pi$ to 2π.
Note that only positive half cycle can be controlled, negative half cycle cannot be controlled. As
such, single-phase half wave voltage controller is also called *single-phase unidirectional voltage
controller*. Fig. 9.1 (b) reveals that positive half cycle is not identical with negative half-cycle for
both voltage and current waveforms. As a result, *dc* component is introduced in the supply and
load circuits which is undesirable.

From Fig 9.1 (b), *rms* value of output voltage is given by

$$V_{or} = \left[\frac{1}{2\pi} \int_{\alpha}^{2\pi} V_m^2 \sin^2 \omega t \, d\,(\omega t) \right]^{1/2}$$

$$V_{or}^2 = \frac{V_m^2}{4\pi} \int_{\alpha}^{2\pi} (1 - \cos 2 \omega t) \, d\,(\omega t) = \frac{V_m^2}{4\pi} \left[(2\pi - \alpha) + \frac{\sin 2\alpha}{2} \right]$$

$$\therefore \quad V_{or} = \frac{V_m}{2} \left[\frac{1}{\pi} \left\{ (2\pi - \alpha) + \frac{\sin 2\alpha}{2} \right\} \right]^{1/2} \qquad \ldots(9.1)$$

Rms load current, $\quad I_{or} = \dfrac{V_{or}}{R}$

Average value of output voltage is, $V_o = \dfrac{1}{2\pi} \displaystyle\int_{\alpha}^{2\pi} V_m \sin \omega t \, . \, d\,(\omega t)$

or $$V_o = \frac{V_m}{2\pi} \left| - \cos \omega t \right|_{\alpha}^{2\pi}$$

$$= \frac{V_m}{2\pi} (\cos \alpha - 1) \qquad \ldots(9.2)$$

Power circuit diagram of *single-phase full-wave voltage controller* is shown in Fig. 9.2 (a).
It consists of two SCRs connected in antiparallel. During positive halfcycle, T1 is triggered at
firing angle α, it conducts from $\omega t = \alpha$ to π for R load. During negative half cycle, T2 is triggered
at $\omega t = \pi + \alpha$, it conducts from $\omega t = \pi + \alpha$ to 2π. Voltage and current waveforms are sketched in
Fig. 9.2 (b). This figure reveals that positive half-cycle is identical with negative half cycle for
both voltage and current waveforms. The power circuit of Fig. 9.2 (a), therefore, introduces no
direct component in the supply and load circuit. This circuit is thus more suited to practical

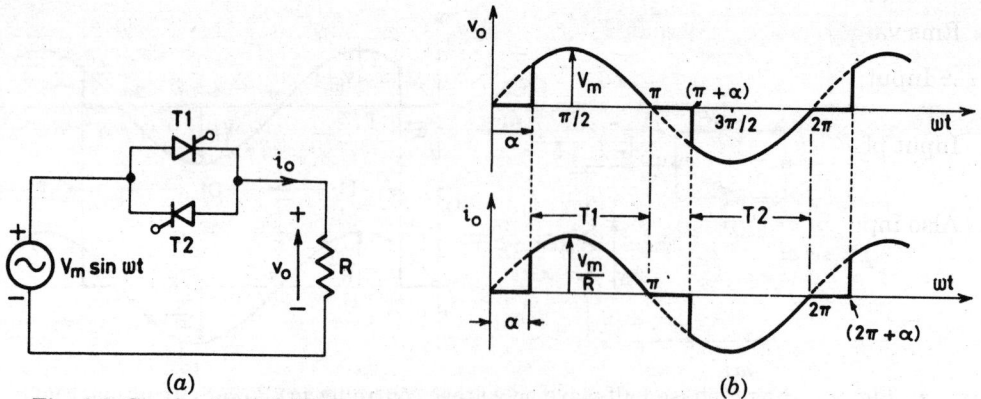

Fig. 9.2. Single-phase full-wave ac voltage controller (a) Power-circuit diagram
and (b) voltage and current waveforms.

circuits than single-phase half-wave circuit. In this chapter, therefore, only full-wave *ac* voltage controllers are described. Single phase full-wave *ac* voltage controller is also called *single phase bidirectional voltage-controller.*

Fig. 9.2 (b) shows that *rms* value of output voltage is

$$V_{or} = \left[\frac{1}{\pi} \int_\alpha^\pi V_m^2 \sin^2 \omega t \cdot d(\omega t) \right]^{1/2}$$

$$V_{or}^2 = \frac{V_m^2}{2\pi} \left| \omega t - \frac{\sin 2\omega t}{2} \right|_\alpha^\pi$$

or

$$V_{or} = \frac{V_m}{\sqrt{2}} \left[\frac{1}{\pi} \left\{ (\pi - \alpha) + \frac{\sin 2\alpha}{2} \right\} \right]^{1/2} \qquad \qquad ...(9.3)$$

Average value of output voltage would be zero.

It has been stated above that *ac* voltage controllers are phase-controlled converters. The principle of phase control has been illustrated in Figs. 9.1 and 9.2. In these figures, the *phase* relationship between the start of load current and the supply voltage is *controlled* by varying the firing angle. Recall a similar statement made in Art. 6.1 for phase-controlled rectifiers. As the controlled output in Fig. 9.1 and 9.2 is *ac*, these are called phase-controlled *ac* voltage controllers or *ac* voltage controllers.

Example 9.1. *A single-phase half-wave ac voltage controller feeds a load of R = 20 Ω with an input voltage of 230 V, 50 Hz. Firing angle of thyristor is 45°. Determine (a) rms value of output voltage (b) power delivered to load and input pf and (c) average input current.*

Solution. Here $V_s = 230$ V, $V_m = \sqrt{2} \times 230$ V, $\alpha = 45° = \frac{\pi}{4}, R = 20$ Ω

(a) From Eq. (9.1), *rms* value of load voltage is

$$V_{or} = \frac{V_m}{2} \left[\frac{1}{\pi} \left\{ (2\pi - \alpha) + \frac{\sin 2\alpha}{2} \right\} \right]^{1/2}$$

$$= \frac{\sqrt{2} \times 230}{2} \left[\left\{ \frac{1}{\pi} \left(2\pi - \frac{\pi}{4} \right) + \frac{\sin 90}{2} \right\} \right]^{1/2} = 224.682 \text{ V}$$

(b) Rms value of load current, $I_{on} = \frac{V_{or}}{R} = \frac{224.682}{20} = 11.2341$ A

Load power, $P_o = I_{or}^2 \times R = 11.2341^2 \times 20 = 2524.1$ W

Rms value of source current, $I_s = rms$ value of load current, $I_{or} = 11.2341$ A

$\therefore$ Input $\qquad VA = V_s \cdot I_s = 230 \times 11.2341$ VA

Input pf $\qquad = \dfrac{P_o}{\text{Input } VA} = \dfrac{2524.1}{230 \times 11.2341} = 0.9769$ (lagging)

Also input $pf = \dfrac{P_o}{VA} = \dfrac{V_{or} \cdot I_{or}}{V_r \cdot I_s} = \dfrac{V_{or} \cdot I_{or}}{V_s \cdot I_{or}} = \dfrac{V_{or}}{V_s}$. From Eq. (9.1),

$$\frac{V_{or}}{V_s} = \frac{1}{\sqrt{2}}\left[\frac{1}{\pi}\left\{(2\pi - \alpha) + \frac{\sin 2\alpha}{2}\right\}\right]^{1/2} = \text{input pf.}$$

$\therefore$ Input $pf = \dfrac{1}{\sqrt{2}}\left[\dfrac{1}{\pi}\left\{\left(2\pi - \dfrac{\pi}{4}\right) + \dfrac{\sin 90}{2}\right\}\right]^{1/2} = 0.9770$ (lagging)

(c) From Eq. (9.2), average output voltage

$$V_o = \frac{\sqrt{2} \times 230}{2\pi}(\cos 45 - 1) = -15.17 \text{ V}$$

Average input current, $\qquad I_{ON} = \dfrac{V_o}{R} = -\dfrac{15.17}{20} = -0.7585$ A

Negative value of average input current is due to the fact that current during positive half cycle is less than that during the negative half-cycle.

Example 9.2. *A single-phase full-wave ac voltage controller feeds a load of $R = 20\ \Omega$ with an input voltage of 230 V, 50 Hz. Firing angle for both the thyristors is 45°. Calculate (a) rms value of output voltage (b) load power and input pf (c) average and rms current of thyristors.*

Solution. Here $V_s = 230$ V, $V_m = \sqrt{2} \times 230$ V, $\alpha = 45° = \dfrac{\pi}{4}$, $R = 20\ \Omega$

(a) From Eq. (9.3), *rms* value of output voltage is

$$V_{or} = 230\left[\frac{1}{\pi}\left\{\left(\pi - \frac{\pi}{4}\right) + \frac{\sin 90°}{2}\right\}\right]^{1/2} = 219.304 \text{ V}$$

(b) Rms value of load current, $I_{or} = \dfrac{V_{or}}{R} = \dfrac{219.304}{20} = 10.9652$ A

Load power, $\qquad P_o = I_{or}^2 \times R = 10.9652^2 \times 20 = 2404.71$ W

Rms value of source current, $I_s = rms$ value of load current, $I_{or} = 10.9652$ A

$\therefore$ Input pf $\qquad = \dfrac{2404.71}{230 \times 10.9652} = 0.9535$ lagging.

Also, $\qquad pf = \dfrac{\text{load power, } P_o}{\text{input } VA} = \dfrac{V_{or} - I_{or}}{V_s \cdot I_{or}} = \dfrac{V_{or}}{V_s}$

From Eq. (9.3), $\qquad \dfrac{V_{or}}{V_s} = \text{input pf} = \left[\dfrac{1}{\pi}\left\{\left(\pi - \dfrac{\pi}{4}\right) + \dfrac{\sin 90}{2}\right\}\right]^{1/2} = 0.9535$

(c) Average thyristor current, $I_{TA} = \dfrac{1}{2\pi R}\displaystyle\int_{\alpha}^{\pi} V_m \sin \omega t \cdot d\,(\omega t)$

$$= \frac{V_m}{2\pi R}(1 + \cos \alpha)$$

$$= \frac{\sqrt{2} \times 230}{2\pi \times 20}(1 + \cos 45°) = 4.418 \text{ A}$$

Rms value of thyristor current, $I_{Tr} = \left[\dfrac{1}{2\pi R^2} \displaystyle\int_\alpha^\pi V_m^2 \sin^2 \omega t \cdot d(\omega t) \right]^{1/2}$

$$= \frac{V_m}{2R} \left[\frac{1}{\pi} \left\{ (\pi - \alpha) + \frac{\sin 2\alpha}{2} \right\} \right]^{1/2}$$

$$= \frac{\sqrt{2} \times 230}{2 \times 20} \left[\frac{1}{\pi} \left\{ \left(\pi - \frac{\pi}{4} \right) + \frac{\sin 90°}{2} \right\} \right]^{1/2} = 7.7524 \text{ A}$$

It is seen from the expression of current I_{Tr} that it can be re-written as

$$I_{Tr} = \frac{V_m}{2R} \left[\frac{V_{or}}{V_s} \right] = \frac{V_{or}}{\sqrt{2}.R} = \frac{219.304}{\sqrt{2} \times 20} = 7.736 \text{ A}$$

9.2. PRINCIPLE OF INTEGRAL CYCLE CONTROL

It is stated above that ac voltage controllers are phase-controlled devices. The principle of phase control is illustrated in Figs. 9.1 and 9.2. In these figures, the *phase* relationship between the start of load current and the supply voltage is *controlled* by varying the firing angle. As the controlled output is ac, these are called phase-controlled ac voltage controllers or ac voltage controllers.

In industry, there are several applications in which mechanical time constant or thermal time constant is of the order of several seconds. For example, mechanical time constant for many of the speed-control drives, or thermal time constant for most of the heating loads is usually quite high. For such applications, almost no variation in speed or temperature will be noticed if control is achieved by connecting the load to source for some on-cycles and then disconnecting the load for some off-cycles. This form of power control is called integral cycle control. So *integral cycle control* consists of switching on the supply to load for an integral number of cycles and then switching off the supply for a further number of integral cycles, Fig. 9.3.

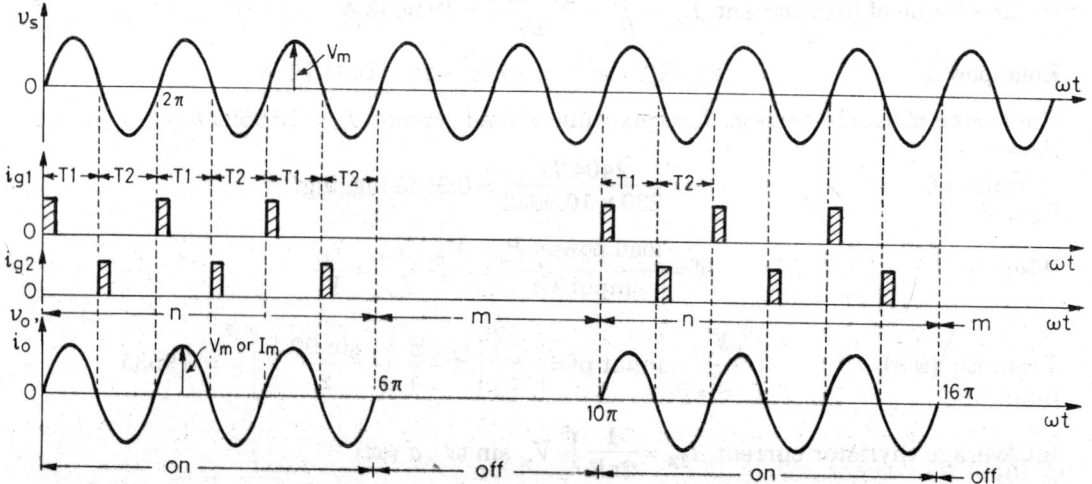

Fig. 9.3. Waveforms pertaining to integral cycle control.

The principle of integral cycle control can be explained by referring to Fig. 9.2 for a single-phase voltage controller with resistive load. Gate pulses i_{g1}, i_{g2} turn on the thyristors T1, T2 respectively at zero-voltage crossing of the supply voltage. The source energises the load for n (= 3) cycles. When gate pulses are withdrawn, load remains off for m (= 2) cycles. In this manner, process of turn-on and turn-off is repeated for the control of load power. By varying the number of n and m cycles, power delivered to load can be regulated as desired. The waveforms for source voltage v_s, gate pulses and output voltage v_0 are shown in Fig. 9.3 for $n = 3$ and $m = 2$. Power is delivered to load for n cycles. No power is delivered to load for m cycles. It is the average power in the load that is controlled.

In literature, integral cycle control is also known as *on-off control, burst firing, zero-voltage switching, cycle selection* or *cycle syncopation.*

For sinusoidal supply voltage, the rms value of output voltage V_{or} can be obtained as under :

$$V_{or}^2 = \frac{1}{\text{periodicity}}\left[\int_0^{2\pi} V_m^2 \sin^2 \omega t\, d\,(\omega t), \text{ for first on-cycle}\right.$$

$$+ \int_0^{2\pi} V_m^2 \sin^2 \omega t \cdot d\,(\omega t), \text{ for second on-cycle} + \dots\dots +$$

$$\left.\int_0^{2\pi} V_m^2 \sin^2 \omega t \cdot d\,(\omega t), \text{ for } n\text{th on-cycle}\right]$$

For n on-cycles and m off-cycles, the periodicity = $(n + m)\, 2\pi$ radians, see Fig. 9.3.

$$\therefore \qquad V_{or} = \left[\frac{n}{2\pi\,(n+m)} \int_0^{2\pi} V_m^2 \sin^2 \omega t \cdot d\,(\omega t)\right]^{1/2}$$

$$= \left[\frac{n\,V_m^2}{4\pi\,(n+m)} \int_0^{2\pi} (1 - \cos 2\omega t)\, d\,(\omega t)\right]^{1/2}$$

or $\qquad V_{or} = \frac{V_m}{\sqrt{2}} \cdot \sqrt{\frac{n}{n+m}} = V_s \sqrt{\frac{n}{n+m}} = V_s \sqrt{k}$ \qquad ...(9.4)

where $\qquad V_s$ = rms value of source voltage

and $\qquad k = \dfrac{n}{n+m}$ is the duty cycle of ac voltage controller.

Rms load current, $\qquad I_{or} = \dfrac{V_{or}}{R}$

Power delivered to load $\quad = \dfrac{V_{or}^2}{R} = \dfrac{V_s^2}{R}\left(\dfrac{n}{n+m}\right) = \dfrac{k \cdot V_s^2}{R}$ \qquad ...(9.5)

Rms value of input current, I_s = rms value of load current, I_{or}

Input $VA = V_s$ (rms value of source current)

$$= V_s \cdot I_s = V_s \cdot I_{or} = V_s \cdot \frac{V_{or}}{R}$$

Input VA × pf = power delivered to load

$\therefore$ Input pf $\quad = \dfrac{V_{or}^2}{R} \cdot \dfrac{R}{V_s \cdot V_{or}} = \dfrac{V_{or}}{V_s} = \sqrt{\dfrac{n}{n+m}} = \sqrt{k}$ \qquad ...(9.6)

As each thyristor conducts for π radians during each cycle of n on-cycles, the average value of thyristor current is given by

$$I_{TA} = \frac{1}{2\pi} \int_0^\pi I_m \sin \omega t \cdot d\,(\omega t), \text{ for first on-cycle } +$$

$$\frac{1}{2\pi} \int_0^\pi I_m \cdot \sin \omega t \cdot d\,(\omega t), \text{ for second on-cycle } + \ldots\ldots +$$

$$\frac{1}{2\pi} \int_0^\pi I_m \cdot \sin \omega t \cdot d\,(\omega t), \text{ for } n\text{th on-cycle]}$$

$$= \frac{n}{2\pi\,(n+m)} \int_0^\pi I_m \cdot \sin \omega t \cdot d\,(\omega t)$$

$$= \frac{I_m}{\pi} \cdot \frac{n}{m+n} = \frac{k \cdot I_m}{\pi} \qquad \ldots(9.7)$$

Similarly, rms value of thyristor current is

$$I_{TR} = \left[\frac{n}{2\pi\,(n+m)} \int_0^\pi I_m^2 \sin^2 \omega t \cdot d\,(\omega t) \right]^{1/2}$$

$$= \frac{I_m}{2} \sqrt{\frac{n}{n+m}} = \frac{I_m \sqrt{k}}{2} \qquad \ldots(9.8)$$

Integral cycle control introduces less harmonics into the supply system, the supply undertakings therefore insist upon the consumers to use integral-cycle method for heating loads and for motor-control drives.

AC voltage controllers with on-off control have specific applications as discussed above. Phase-controlled ac voltage controllers are, however, more common. As such, phase-controlled ac voltage controllers will only be discussed and analysed in what follows :

Example 9.3. *A single-phase voltage controller has input voltage of 230 V, 50 Hz and a load of R = 15 Ω. For 6 cycles on and 4 cycles off, determine (a) rms output voltage, (b) input pf and (c) average and rms thyristor currents.*

Solution. (*a*) From Eq. (9.4), rms value of output voltage is

$$V_{or} = V_s \sqrt{\frac{n}{n+m}} = 230 \sqrt{\frac{6}{6+4}} = 178.157 \text{ V}$$

(*b*) From Eq. (9.6), input pf $= \sqrt{k} = \sqrt{\frac{n}{n+m}} = \sqrt{\frac{6}{6+4}} = 0.7746 \text{ lag.}$

Also power delivered to load $= I_{or}^2 R = \frac{V_{or}^2}{R} = \frac{178.157^2}{15} = 2116 \text{ W}$

Input VA $= 230 \times \frac{230 \sqrt{6}}{15} = 2731.74 \text{ VA}$

$\therefore$ Input pf $= \frac{2116}{2731.74} = 0.7746 \text{ lag}$

(*c*) Peak thyristor current, $I_m = \frac{230 \sqrt{2}}{15} = 21.681 \text{ A}$

From Eq. (9.7), average value of thyristor current,

$$I_{TA} = \frac{k\,I_m}{\pi} = \frac{0.6 \times 21.681}{\pi} = 4.1407 \text{ A}$$

From Eq. (9.8), rms value of thyristor current,

$$I_{TR} = \frac{I_m \cdot \sqrt{k}}{2} = \frac{21.681 \times \sqrt{0.6}}{2} = 8.397 \text{ A}$$

9.3. SINGLE-PHASE VOLTAGE CONTROLLERS

Fig. 9.4 shows four possible configurations of single-phase ac voltage controllers. Fig. 9.4 (a), similar to Fig. 9.2 (a), uses two thyristors connected in antiparallel. The trigger sources for the two thyristors must be isolated from one another because otherwise the two cathodes would be connected together and the two thyristors would be out of circuit as shown in Fig. 9.4 (b). Thus, no control of the output voltage would be possible.

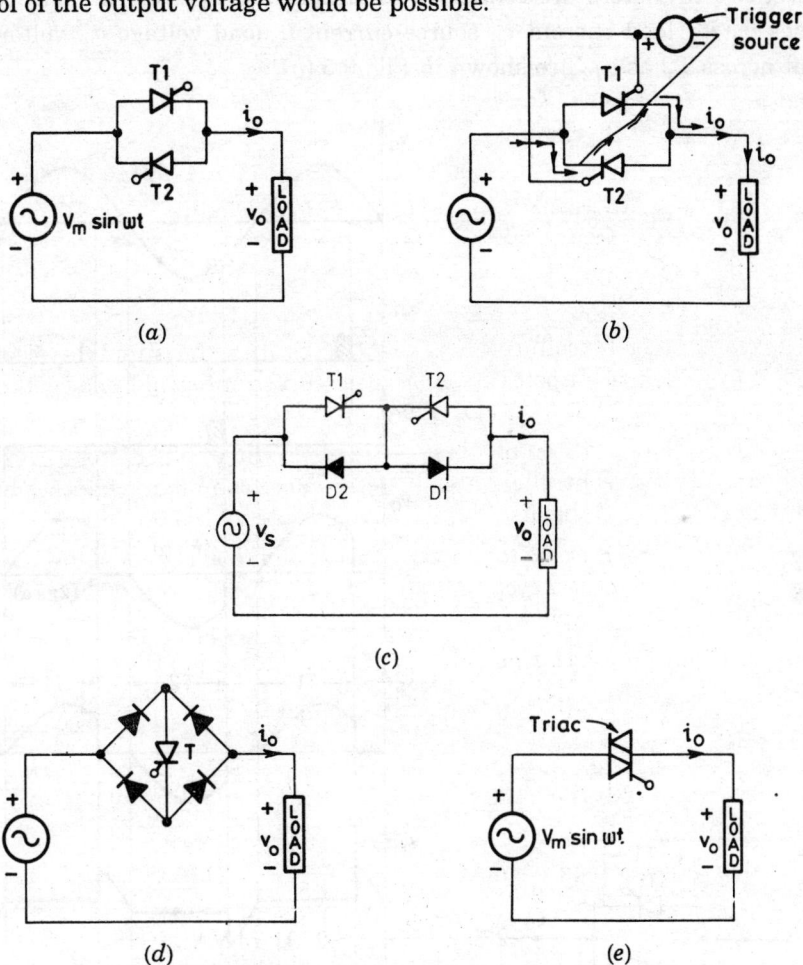

Fig. 9.4. Single-phase ac voltage controllers.

Single-phase full-wave voltage controller can also be realized by using two thyristors $T1$, $T2$ and two diodes $D1$, $D2$ as shown in Fig. 9.4 (c). Since cathodes of $T1$, $T2$ are connected together, only one triggering circuit can serve the purpose. $T1$, $D1$ conduct during positive half cycle and $T2$, $D2$ during negative half cycle. As two devices conduct at the same time, there are more conduction losses and, therefore, reduced efficiency of the circuit arrangement of Fig. 9.4 (c).

Scheme shown in Fig. 9.4 (d) employs four diodes and one thyristor. For this circuit, there is no need for any isolation between control and power circuits. This scheme, therefore offers a cheap ac voltage controller. The voltage drop in the three conducting devices (two diodes and one thyristor) will, however, be more than in Fig. 9.4 (a) and consequently its efficiency would be low.

The circuit shown in Fig. 9.4 (e) uses one triac. This configuration is suitable for low-power applications where the load is resistive or has only a small inductance. The triggering circuit for the triac need not be isolated from the power circuit.

9.3.1. Single-phase Voltage Controller with R Load

Fig. 9.5 (a) shows a single-phase voltage controller feeding power to a resistive load R. As stated before, two thyristors are connected in antiparallel. Waveforms for source voltage v_s, gating pulses i_{g1}, i_{g2}, load current i_0, source current i_s, load voltage v_0, voltage across T1 as v_{T1} and that across T2 as v_{T2} are shown in Fig. 9.5 (b).

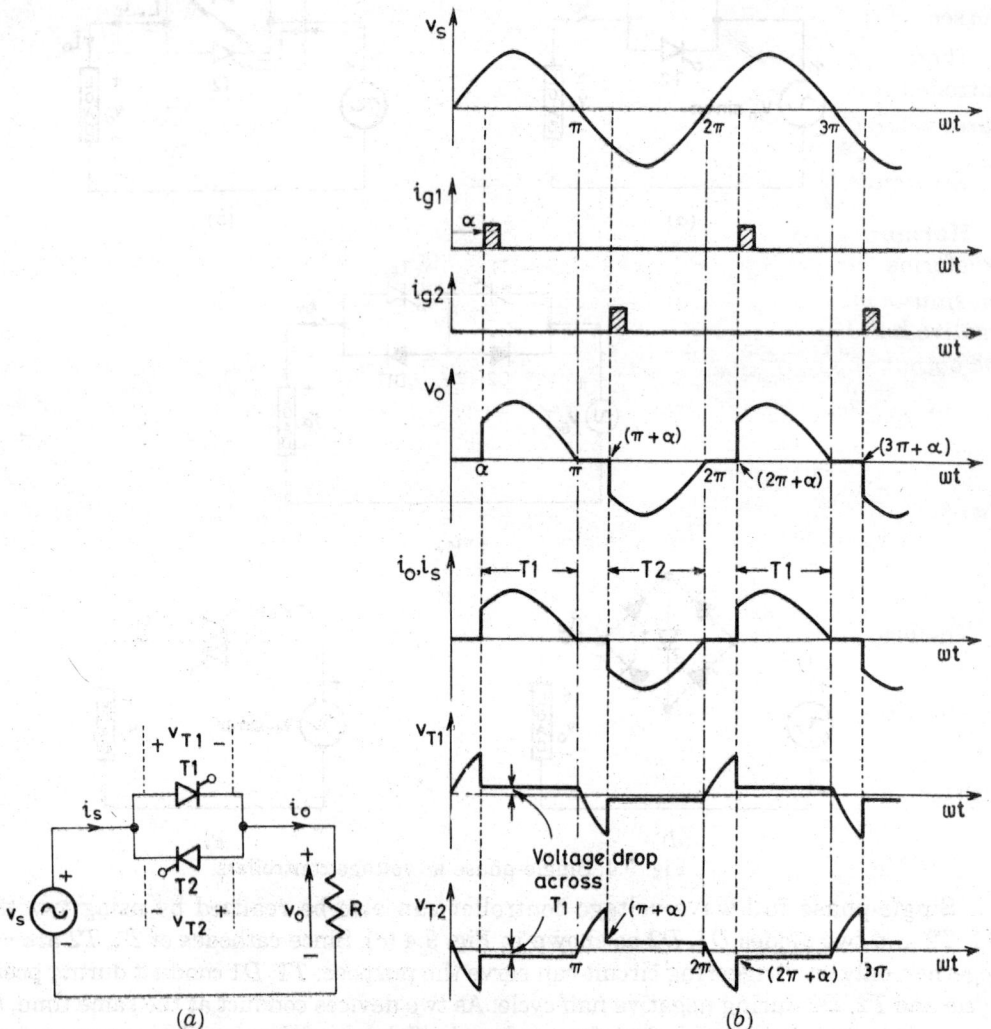

Fig. 9.5. (a) Single-phase ac voltage controller with R load
(b) Voltage and current waveforms for figure (a).

Thyristors T1 and T2 are forward biased during positive and negative half cycles respectively. During positive half cycle, T1 is triggered at a firing angle α. T1 starts conducting and source voltage is applied to load from α to π. At π, both v_0, i_0 fall to zero. Just after π, T1 is subjected to reverse bias, it is therefore turned off. During negative half cycle, T2 is triggered at $(\pi + \alpha)$. T2 conducts from $\pi + \alpha$ to 2π. Soon after 2π, T2 is subjected to a reverse bias, it is therefore commutated. Load and source currents have the same waveform.

From zero to α, T1 is forward biased, $v_{T1} = v_s$ as shown. From α, T1 conducts, v_{T1} is therefore about 1 V. After π, T1 is reverse biased by source voltage, therefore $v_{T1} = v_s$ from π to $\pi + \alpha$. From $\pi + \alpha$ to 2π, T2 conducts ; T1 is therefore reverse biased by voltage drop across T2 which is about 1 to 1.5 V. The voltage variation v_{T1} across SCR T1 is shown in Fig. 9.5(b). Similarly, the variation of voltage v_{T2} across T2 can be drawn. In Fig. 9.5 (b), voltage drop across thyristors T1 and T2 is purposely shown just to highlight the duration of reverse bias across T1 and T2. Examination of this figure reveals that for any value of α, each thyristor is reverse biased for π/ω sec.

There is thus no restriction on the value of firing angle α. Firing angle can, therefore, be controlled from zero to π and rms output voltage from V_s to zero. Here V_s is the rms value of source voltage.

$\therefore$ Circuit turn-off time, $\quad t_c = \dfrac{\pi}{\omega}$ sec.

Harmonics of output quantities and input current. It is seen from Fig. 9.5 (b) that waveforms for output quantities (voltage v_0 and current i_0) and input current i_s are non-sinusoidal. These waveforms can be described by Fourier series. As the positive and negative half cycles are identical, dc component and even harmonics are absent.

The output voltage v_0 can be represented by Fourier series as under :

$$v_o = \sum_{n = 1, 3, 5, \ldots}^{\infty} a_n \cos n\omega t + b_n \sin n\omega t \qquad \ldots(9.9)$$

where $$a_n = \frac{2}{\pi} \int_o^\pi v_o\,(\omega t).\cos n\omega t.\,d\,(\omega t) \qquad \ldots(9.10)$$

and $$b_n = \frac{2}{\pi} \int_o^\pi v_o\,(\omega t).\sin n\omega t.\,d\,(\omega t) \qquad \ldots(9.11)$$

The load voltage v_o during the first half-cycle is

$$v_o = V_m \sin \omega t \ldots\ldots \alpha < \omega t < \pi$$

Substitution of $v_o = V_m \sin \omega t$ in Eqs. (9.10) and (9.11) gives

$$a_n = \frac{2V_m}{\pi} \int_\alpha^\pi \sin \omega t.\cos n\omega t.\,d\,(\omega t)$$

$$= \frac{V_m}{\pi} \int_\alpha^\pi [\sin (n + 1)\,\omega t - \sin (n - 1)\,\omega t].\,d\,(\omega t)$$

$$= \frac{V_m}{\pi} \left[\frac{\cos (n + 1)\,\alpha - 1}{n + 1} - \frac{\cos (n - 1)\,\alpha - 1}{n - 1} \right] \qquad \ldots(9.12)$$

and $$b_n = 2\frac{V_m}{\pi} \int_\alpha^\pi \sin \omega t.\sin n\omega t.\,d\,(\omega t)$$

$$= \frac{V_m}{\pi} \int_\alpha^\pi [\cos (n - 1)\,\omega t - \cos (n + 1)\,\omega t].\,d\,(\omega t)$$

$$= \frac{V_m}{\pi} \left[\frac{\sin(n+1)\alpha}{n+1} - \frac{\sin(n-1)\alpha}{n-1} \right] \qquad \ldots(9.13)$$

where $V_m = \sqrt{2}\, V_s$ and V_s = rms value of source voltage.

For obtaining Eq. (9.12), note that for $n = 1, 3, 5\ldots \cos(n+1)\pi = 1$ and $\cos(n-1)\pi = 1$.

The amplitude of the nth harmonic output voltage V_{nm} and its phase ϕ_n are given by

$$V_{nm} = \sqrt{a_n^2 + b_n^2} \text{ and } \phi_n = \tan^{-1}\frac{a_n}{b_n} \qquad \ldots(9.14a)$$

and $$I_{nm} = \frac{V_{nm}}{R} = n\text{th harmonic load current} \qquad \ldots(9.14b)$$

For fundamental frequency, i.e. for $n = 1$, V_{1m} and ϕ_1 cannot be obtained from Eqs. (9.12) to (9.14), because these become indeterminate for $n = 1$. This difficulty can, however, be overcome by putting $n = 1$ in Eqs. (9.10) and (9.11) and substituting the value of $v_0 = V_m \sin \omega t$.

and $$a_1 = \frac{2}{\pi}\int_\alpha^\pi V_m \sin \omega t \cdot \cos \omega t \cdot d(\omega t) = \frac{V_m}{\pi}\left[\frac{\cos 2\alpha - 1}{2}\right] \qquad \ldots(9.15)$$

$\therefore$ $$b_1 = \frac{2}{\pi}\int_\alpha^\pi V_m \sin^2 \omega t \cdot d(\omega t) = \frac{V_m}{\pi}\left[\frac{\sin 2\alpha}{2} + (\pi - \alpha)\right] \qquad \ldots(9.16)$$

From the coefficients a_1 and b_1, the peak value of the fundamental frequency voltage V_{1m} and its phase ϕ_1 are given by

$$V_{1m} = [a_1^2 + b_1^2]^{1/2}$$

$$= \frac{V_m}{\pi}\left[\left\{\frac{\cos 2\alpha - 1}{2}\right\}^2 + \left\{\frac{\sin 2\alpha}{2} + (\pi - \alpha)\right\}^2\right]^{1/2} \qquad \ldots(9.17a)$$

$$I_{1m} = \frac{V_{1m}}{R} = \text{amplitude of fundamental component of load or source current} \qquad \ldots(9.17b)$$

and $$\phi_1 = \tan^{-1}\frac{a_1}{b_1} = \tan^{-1}\left[\frac{\cos 2\alpha - 1}{\sin 2\alpha \pm 2(\pi - \alpha)}\right] \qquad \ldots(9.18)$$

When ac voltage controller is used for the speed control of a single-phase induction motor, only fundamental component is useful in producing the torque. The harmonics in the motor current merely increase the losses and therefore heating of the induction motor and reduced efficiency. Harmonics also cause more noise and vibrations in the motor.

For heating and lighting loads, however, both fundamental and harmonics are useful in producing the ac controlled power. In such applications, rms value V_{or} of the output voltage should be known. It can be obtained from $v_0 = V_m \sin \omega t$ as follows :

$$V_{or} = \left[\frac{1}{\pi}\int_\alpha^\pi V_m^2 \sin^2 \omega t \cdot d(\omega t)\right]^{1/2}$$

$$= \frac{V_m}{\sqrt{2}}\left[\frac{1}{\pi}\left\{(\pi - \alpha) + \frac{1}{2}\sin 2\alpha\right\}\right]^{1/2} \qquad \ldots(9.19a)$$

and $$I_{or} = \frac{V_{or}}{R} = \text{rms value of load, or source current} \qquad \ldots(9.19b)$$

The average power P delivered to load of resistance R is

$$P = I_{or}^2 R = \frac{V_{or}^2}{R} = \frac{V_m^2}{2\pi R}\left[(\pi - \alpha) + \frac{1}{2}\sin 2\alpha\right]$$

$$= \frac{V_s^2}{\pi R}\left[(\pi - \alpha) + \frac{1}{2}\sin 2\alpha\right] \qquad ...(9.20)$$

Maximum power P_{max} is delivered to load when $\alpha = 0$.

$$\therefore \qquad P_{max} = \frac{V_s^2}{R}$$

This gives $\dfrac{P}{P_{max}} = \dfrac{1}{\pi}\left[(\pi - \alpha) + \dfrac{1}{2}\sin 2\alpha\right]$ = per unit power.

In terms of harmonic components,

$$P = R\,(I_{01}^2 + I_{03}^2 + I_{05}^2 + ...)$$

$$= \frac{1}{R}\,(V_{01}^2 + V_{03}^2 + V_{05}^2 + ...)$$

Fig. 9.5 (b) shows that source current waveform is identical with load current waveform. This means that expressions for both load and source currents for the appropriate harmonics are the same.

Power Factor. Assuming that source voltage remains sinusoidal even though non-sinusoidal current is drawn from it, the power factor is given by

$$pf = \frac{\text{Real power}}{\text{Apparent power}} = \frac{V_s\, I_1 \cos\phi_1}{V_s \cdot I_{rms}} = \frac{I_1 \cdot \cos\phi_1}{I_{rms}} \qquad ...(9.21)$$

where $\qquad I_1 = \dfrac{I_{1m}}{\sqrt{2}}$ = rms value of fundamental component of load or source current

given by Eq. (9.17 b)

$I_{rms} = I_{or}$ = rms value of load or source current, Eq. (9.19 b),

ϕ_1 = phase angle between V_s and I_1, Eq. (9.18)

Another expression for pf can be obtained as follows :

Real power delivered to load $= \dfrac{V_{or}^2}{R}$

Apparent power delivered to load $= V_s \cdot I_{rms}$

$$= V_s \cdot I_{or} = V_s \cdot \frac{V_{or}}{R}$$

$$\therefore \qquad pf = \frac{V_{or}^2/R}{V_s \cdot V_{or}/R} = \frac{V_{or}}{V_s}$$

From Eq. (9.19a), $\quad pf = \dfrac{V_{or}}{V_s} = \left[\dfrac{1}{\pi}\left\{(\pi - \alpha) + \dfrac{1}{2}\sin 2\alpha\right\}\right]^{1/2}$ = [per unit power]$^{1/2}$ $\qquad ...(9.22)$

The maximum value of rms output voltage and current occurs at $\alpha = 0$ and are given by V_s and V_s/R respectively, Eq. (9.19). For $\alpha = 0$, harmonics are absent, these are therefore also the maximum values of fundamental rms voltage and current.

Example 9.4. *A single-phase voltage controller feeds power to a resistive load of 3 Ω from 230 V, 50 Hz source. Calculate :*

(a) the maximum values of average and rms thyristor currents for any firing angle α,

(b) the minimum circuit turn-off time for any firing angle α,

(c) the ratio of third-harmonic voltage to fundamental voltage for $\alpha = \dfrac{\pi}{3}$,

(d) the maximum value of di/dt occurring in the thyristors,

(e) the angle α at which the greatest forward or reverse voltage is applied to either of the thyristors and the magnitude of these voltages.

Solution. (*a*) It is seen from Fig. 9.5 (*b*) that current through thyristor flows from α to π for the first cycle of 2π radians. Therefore, average thyristor current from Example 9.2 is

$$I_{TA} = \frac{V_m}{2\pi R}(1 + \cos \alpha)$$

Its maximum value occurs when α = 0. Therefore, maximum value of average thyristor current is

$$I_{TAM} = \frac{V_m}{\pi R} = \frac{\sqrt{2} \times 230}{\pi \times 3} = 34.512 \text{ A}$$

Rms thyristor current, from Example 9.2, is

$$I_{TR} = \left[\frac{1}{2\pi} \int_{\alpha}^{\pi} \left(\frac{V_m}{R} \sin \omega t \right)^2 \cdot d(\omega t) \right]^{1/2}$$

$$= \frac{V_m}{2R \sqrt{\pi}} \left[(\pi - \alpha) + \frac{1}{2} \sin 2\alpha \right]^{1/2}$$

Its maximum value occurs at α = 0.

$$\therefore \qquad I_{TRM} = \frac{V_m}{2R} = \frac{\sqrt{2} \times 230}{2 \times 3} = 54.211 \text{ A}$$

Also

$$\frac{I_{TRM}}{I_{TAVM}} = \frac{V_m}{2R} \cdot \frac{\pi R}{V_m} = \frac{\pi}{2}$$

$$\therefore \qquad I_{TRM} = \frac{\pi}{2} \times 34.512 = 54.211 \text{ A}$$

(*b*) Waveforms for v_{T1}, v_{T2} in Fig. 9.5 (*b*) show that for any value of firing angle α, the circuit turn-off time is always proportional to π radians.

$$\therefore \text{ Circuit turn-off time} = \frac{\pi}{\omega} = \frac{\pi}{2\pi f} = \frac{1}{2f} = \frac{1}{2 \times 50} = 0.01 \text{ sec} = 10 \text{ m-sec.}$$

(*c*) For third harmonic, from Eq. (9.12),

$$a_3 = \frac{V_m}{\pi} \left[\frac{\cos 240° - 1}{4} - \frac{\cos 120° - 1}{2} \right] = 0.375 \frac{V_m}{\pi} \text{ and from Eq. (9.13),}$$

$$b_3 = \frac{V_m}{\pi} \left[\frac{\sin 240°}{4} - \frac{\sin 120°}{2} \right] = 0.64952 \frac{V_m}{\pi}$$

The amplitude of third harmonic voltage, from Eq. (9.14), is

$$V_{3m} = [a_3^2 + b_3^2]^{1/2} = 0.75 \frac{V_m}{\pi}$$

The amplitude of fundamental frequency voltage from Eq. (9.17a) is

$$V_{1m} = \frac{V_m}{\pi}\left[\left\{\frac{\sin 120}{2} + \left(\pi - \frac{\pi}{3}\right)\right\}^2 + \left\{\frac{\cos 120° - 1}{2}\right\}^2\right]^{1/2} = 2.63634 \frac{V_m}{\pi}$$

$$\therefore \qquad \frac{V_{3m}}{V_{1m}} = \frac{0.75}{2.63634} = 0.2845$$

(d) As there is a sudden rise of current from zero to $V_m/R \sin \alpha$ at firing angle α, di/dt is infinity.

(e) The waveforms of v_{T1}, v_{T2} in Fig. 9.5 (b) reveal that the greatest forward or reverse voltage would appear across either of the thyristors when $\alpha = \frac{\pi}{2}$ or $\alpha > \frac{\pi}{2}$. The magnitude of these voltages is $V_m = \sqrt{2} \, V_s$.

9.3.2. Single-phase Voltage Controller with RL Load

Fig. 9.6 (a) shows a single-phase voltage controller with RL load. In Fig. 9.6 (b) are shown waveforms for source voltage v_s, gate currents i_{g1} and i_{g2}, load and source currents i_0 and i_s, load voltage v_0, voltage v_{T1} across SCR T1 and voltage v_{T2} across thyristor T2.

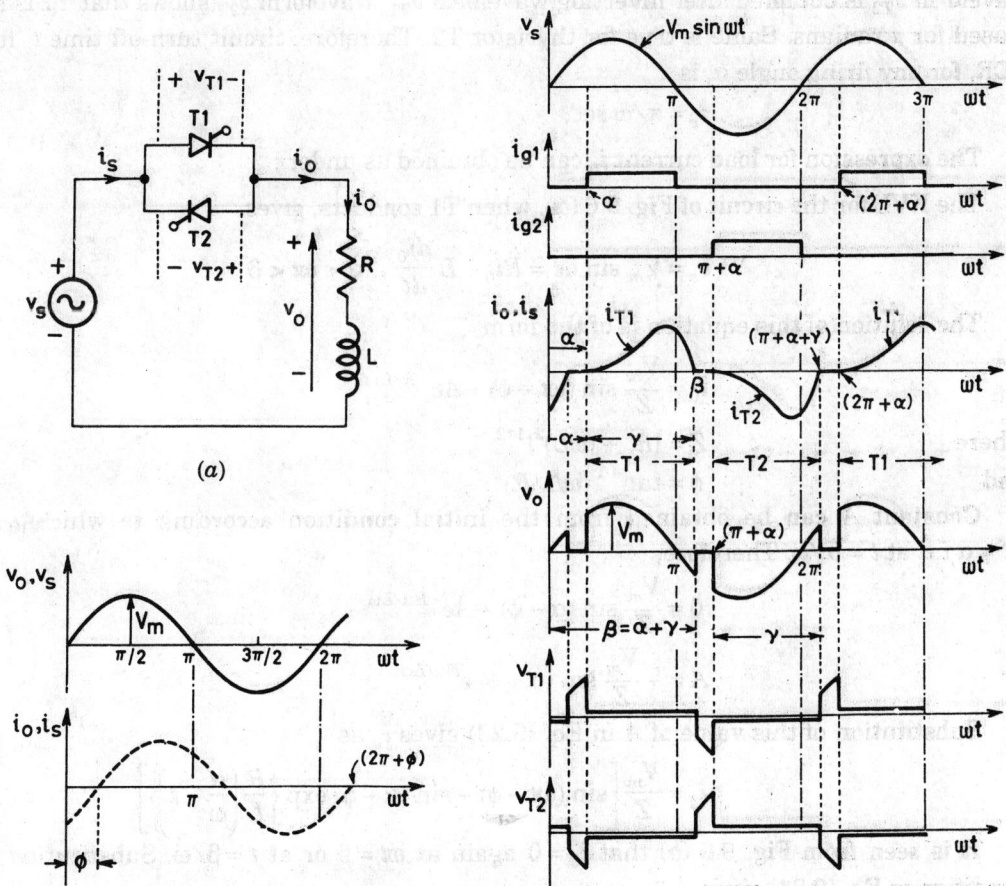

(a)

(c) $\alpha \le \phi$ (b) $\alpha > \phi$

Fig. 9.6. Single-phase voltage controller with RL load.

During zero to π, T1 is forward biased. At $\omega t = \alpha$, T1 is triggered and $i_0 = i_{T1}$ starts building up through the load. At π, load and source voltages are zero but the current is not zero because of the presence of inductance in the load circuit. At $\beta > \pi$, load current reduces to zero. Angle β is called the extinction angle. After π, T1 is reverse biased but does not turn off because i_0 is not zero. At β only, when i_0 is zero, T1 is turned off as it is already reverse biased. After the commutation of T1 at β, a voltage of magnitude $V_m \sin \beta$ at once appears as a reverse bias across T1 and as a forward bias across T2, Fig. 9.6 (b). From β to $\pi + \alpha$, no current exists in the power circuit, therefore, $v_0 = 0$, $v_{T1} = -v_s$ and $v_{T2} = v_s$. Thyristor T2 is turned on at $(\pi + \alpha) > \beta$. Current $i_0 = i_{T2}$ starts building up in the reversed direction through the load. At 2π, v_s and v_0 are zero but $i_{T2} = i_0$ is not zero. At $(\pi + \alpha + \gamma)$, $i_{T2} = 0$ and T2 is turned off because it is already reverse biased. At $(\pi + \alpha + \gamma)$, $V_m \sin (\pi + \alpha + \gamma)$ appears as a forward bias across T1 and as a reverse bias across T2, Fig. 9.6 (b). From $(\pi + \alpha + \gamma)$ to $(2\pi + \alpha)$, no current exists in the power circuit. As before, $v_0 = 0$, $v_{T1} = v_s$ and $v_{T2} = -v_s$. At $(2\pi + \alpha)$, T1 is turned on and current starts building up as before.

When T1 conducts, voltage drop across it appears as a reverse bias across T2. Similarly, when T2 conducts, v_{T2} appears as a reverse bias across T1. It is seen from Fig. (9.6 b) that waveform v_{T2} is obtained after inverting waveform v_{T1}. Waveform v_{T1} shows that T1 is reverse biased for π radians. Same is true for thyristor T2. Therefore, circuit turn-off time t_c for each SCR, for any firing angle α, is

$$t_c = \pi/\omega \text{ sec.}$$

The expression for load current i_0 can be obtained as under :

The KVL for the circuit of Fig. 9.6 (a), when T1 conducts, gives

$$v_s = V_m \sin \omega t = R i_0 + L \frac{d i_0}{dt} \quad ...\alpha < \omega t < \beta$$

The solution of this equation is of the form

$$i_0 = \frac{V_m}{Z} \sin (\omega t - \phi) + A e^{-(R/L) t} \qquad ...(9.23)$$

where $$Z = [R^2 + (\omega L)^2]^{1/2}$$
and $$\phi = \tan^{-1} (\omega L/R)$$

Constant A can be obtained from the initial condition according to which $i_0 = 0$ at $\omega t = \alpha$ i.e. at $t = \alpha/\omega$. Therefore,

$$0 = \frac{V_m}{Z} \sin (\alpha - \phi) + A e^{-R\alpha/L\omega}$$

or $$A = -\frac{V_m}{Z} \sin (\alpha - \phi) e^{R\alpha/L\omega}$$

Substitution of this value of A in Eq. (9.23) gives i_0 as

$$i_0 = \frac{V_m}{Z} \left[\sin (\omega t - \phi) - \sin (\alpha - \phi) \exp \left\{ \frac{R}{L} \left(\frac{\alpha}{\omega} - t \right) \right\} \right] \qquad ...(9.24)$$

It is seen from Fig. 9.6 (b) that $i_0 = 0$ again at $\omega t = \beta$ or at $t = \beta/\omega$. Substitution of this condition in Eq. (9.24) gives

$$\sin (\beta - \phi) = \sin (\alpha - \phi) \cdot \exp \left[\frac{R}{L} \left(\frac{\alpha - \beta}{\omega} \right) \right] \qquad ...(9.25)$$

Extinction angle β can be obtained from Eq. (9.25).

The conduction angle γ during which current flows from angle α to angle β is given by

$$\gamma = \beta - \alpha \qquad \qquad \qquad ...(9.26)$$

For a given value of load phase angle α, angle β is determined for various values of α from Eq. (9.25) and thus a relationship between γ and α can be realized from Eq. (9.26). For the various values of γ and α, curves shown in Fig. 9.7 are obtained for different values of ϕ. Note that phase angle ϕ cannot exceed 90°.

It is seen from Figs. 9.6 (b) and 9.7 that as α is decreased, the conduction angle γ increases. The waveform of current i_0 in Fig. 9.6 (b) reveals that for $\gamma < \pi$, i_{T1} through T1 flows from α to $(\alpha + \gamma) = \beta$. T1 remains off from $\alpha + \gamma$ upto $(\pi + \alpha)$. At $(\pi + \alpha)$, i_{T2} through T2 flows from $\pi + \alpha$ to $\pi + \alpha + \gamma$. T2 remains off from $\pi + \alpha + \gamma$ to $2\pi + \alpha$. At $2\pi + \alpha$, T1 is turned on. With progressive decrease in α, γ may become equal to π. Under this condition, when γ is just equal to π, T1 will be on from α to $\pi + \alpha$ and i_{T1} flows from α to $\pi + \alpha$. Further, T2 will be on from $\pi + \alpha$ to $2\pi + \alpha$ and current i_{T2} flows from $\pi + \alpha$ to $2\pi + \alpha$. Thus,

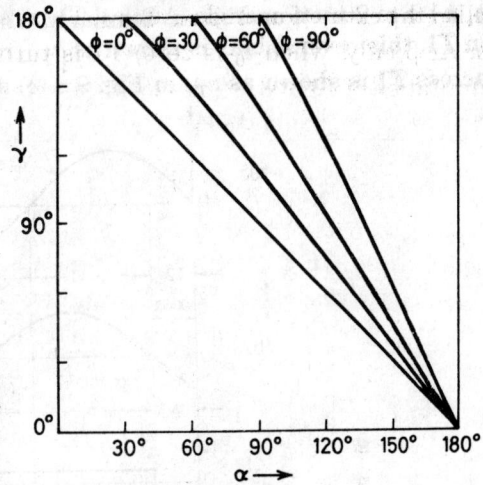

Fig. 9.7. γ versus α curves for ac voltage controller of Fig. 9.6 (a).

when $\gamma = \pi$,
$$\left.\begin{array}{l} 0 \text{ to } \alpha \text{ —T2 conducts} \\ \alpha \text{ to } (\pi + \alpha) \text{ —T1 conducts} \\ (\pi + \alpha) \text{ to } (2\pi + \alpha) \text{ —T2 conducts and so on} \end{array}\right\} \qquad ...(A)$$

This shows that load current will never become zero for any segment of time and therefore for all the time, load is connected to source. Thus, for $\gamma - \pi$, the load voltage is equal to sinusoidal source voltage provided the voltage drop in thyristors is neglected. Under these conditions, load behaves as if it is being fed directly by the ac source.

What is the value of α for which $\gamma = \pi$ and load is directly connected to ac source ? For this, consider that RL load, with load phase angle ϕ, is connected directly to ac source. Under steady state, the load current will be a sine wave and lag behind the voltage wave by an angle ϕ as shown in Fig. 9.6 (c). The current is positive from ϕ to $\pi + \phi$ and negative from $\pi + \phi$ to $2\pi + \phi$, Fig. 9.6 (c). If it is required to obtain the current waveform of Fig. 9.6 (c) through the operation of power circuit of Fig. 9.6 (a), then

from
$$\left.\begin{array}{l} 0 \text{ to } \phi \text{ —T2 conducts} \\ \phi \text{ to } (\pi + \phi) \text{ —T1 conducts} \\ (\pi + \phi) \text{ to } (2\pi + \phi) \text{ —T2 conducts and so on} \end{array}\right\} \qquad ...(B)$$

A comparison of expressions (A) and (B) reveals that when $\alpha = \phi$, $\gamma = \pi$. This can be verified by referring to Eqs. (9.25) and (9.26). When $\alpha = \phi$, Eq. (9.25) gives

$$\sin (\beta - \alpha) = 0 = \sin \pi$$
$$(\beta - \alpha) = \pi$$

or

From Eq. (9.26)
$$\gamma = \beta - \alpha = \pi.$$

This shows that for 1-phase ac voltage controller, waveforms of Fig. 9.6 (b) are applicable only when $\alpha > \phi$ and that of Fig. 9.6 (c) for $\alpha \le \phi$.

Operation with $\alpha \le \phi$. Assume that ac voltage controller is working under steady state with $\alpha = \phi$. From zero to ϕ, T2 conducts and from ϕ to $(\pi + \phi)$, T1 conducts ; from $\pi + \phi$ to $2\pi + \phi$, T2 conducts and so on as shown in Fig. 9.8 (b). When $T2$ conducts, T1 is reverse biased by voltage drop v_{T2} in T2. This reverse bias across T_1 is shown as v_{T2} from 0° to ϕ, $\pi + \phi$ to $2\pi + \phi$ and so on. Similarly, when T1 conducts, T2 is reverse biased by voltage drop in T1, this reverse bias is shown as v_{T1} from ϕ to $\pi + \phi$, $2\pi + \phi$ to $3\pi + \phi$ and so on. Voltage drop across T1 is shown as v_{T1} in Fig. 9.8 (c) and that across T2 in Fig. 9.8 (d).

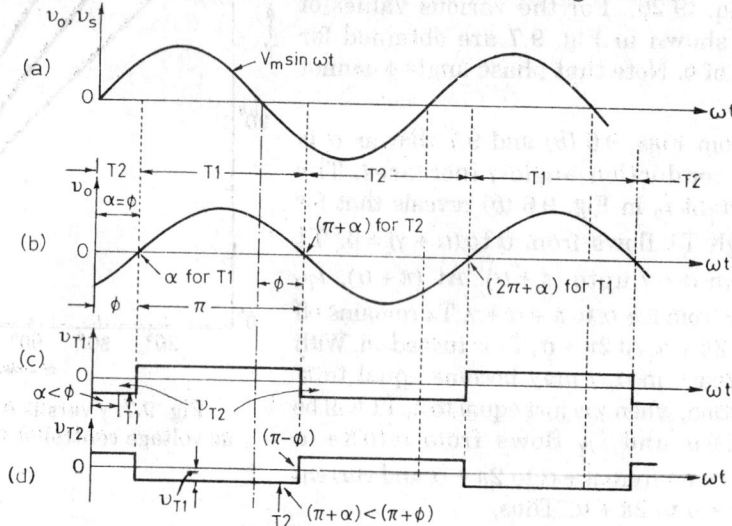

Fig. 9.8. Single-phase voltage controller with RL load and $\alpha \le \phi$.

Not let firing angle α be made less than ϕ. When $T1$ is triggered at $\alpha < \phi$, Fig. 9.8 (c), T1 will not get turned on because it is reverse biased by voltage drop v_{T2} in T2 which is conducting current i_{T2}. T1 will get turned on only at ϕ when $i_{T2} = 0$ and reverse bias due to voltage drop v_{T2} in T2 vanishes. Now T1 will conduct from ϕ to $\pi + \phi$. T2 will be triggered at an angle $(\pi + \alpha) < (\pi + \phi)$ as shown in Fig. 9.8 (d). As T1 is conducting, a voltage drop v_{T1} in T1 will apply a reverse bias across T2, as a result T2 will not get turned on at $\pi + \alpha$; but only at $\pi + \phi$, when $i_{T1} = 0$. Now T2 will conduct from $\pi + \phi$ to $2\pi + \phi$ and so on. This shows that load voltage and load current waveforms will not change from what these are at $\alpha = \phi$. Thus the reduction of α below ϕ is not able to control the load voltage and load current. The ac output power can be controlled only for $\alpha > \phi$. Note that for $\alpha \le \phi$, r remains equal to π. Thus the control range of firing angle in phase-controlled ac voltage controllers is $\phi < \alpha < 180°$.

Gating signal requirements. For R load as in Fig. 9.5, thyristor T1 stops conducting at π, T2 is now forward biased after π. When T2 is triggered at $\pi + \alpha$, it gets turned on as it is already forward biased by source voltage. Thus pulse gating is suitable for R load as shown in Fig. 9.5.

Pulse gating is, however, not suitable for RL loads. The reason for this can be explained by referring to Fig. 9.9(a). At α, T1 is fired and the current grows as shown. At $\pi + \alpha$, T2 is fired. As T1 is still conducting, voltage drop across T1 reverse biases T2 at $\pi + \alpha$, T2 is therefore not turned on at $\pi + \alpha$. At $(\alpha + \gamma)$, i_{T1} decays to zero and T1 stops conducting, as a result T2 gets forward biased but the gate pulse i_{g2} applied to T2 at $\pi + \alpha$ is already zero and therefore T2 does not get turned on. At $(2\pi + \alpha)$, gate pulse is applied to T1, it gets turned on because it is already forward biased by source voltage. At $(3\pi + \alpha)$, when T2 is pulse gated, it will not turn on as explained earlier. Thus, the a.c. voltage controller gives asymmetrical output voltage

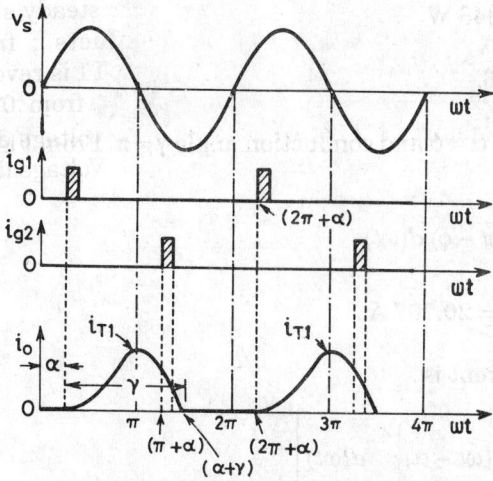

Fig. 9.9. (a) Single-phase voltage controller with RL load with pulse gating.

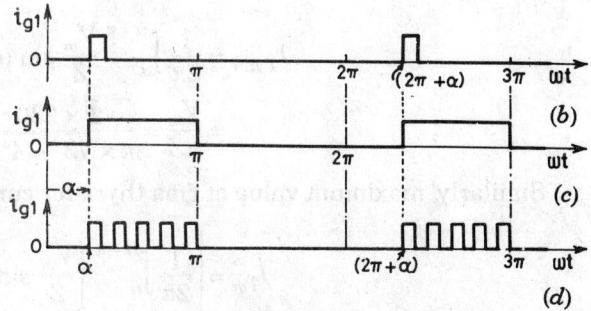

Fig. 9.9. Types of gating signals (b) pulse gating (c) continuous gating (d) high-frequency carrier gating.

waveform due to the conduction of T1 alone. This half-wave rectifier operation of the ac voltage controller is undesirable. This difficulty can, however, be overcome by applying a *continuous gate* single to the SCRs T1 and T2 so that when i_{T1} becomes zero at $(\alpha + \gamma)$, T2 gets turned on due to the presence of continuous signal. A continuous gate signal is shown in Fig. 9.9 (c). The duration of a continuous gate signal should last for a period of $(\pi - \alpha)/\omega$ seconds. Strictly speaking, sustained gate pulse may not last from α to π as shown. For $\alpha = \phi$ or $\alpha > \phi$, a gate pulse of narrow width is sufficient to trigger the SCR. In case $\alpha < \phi$, minimum width of gate pulse should be equal to ϕ plus the angle required for the current to reach latching current value.

In practice continuous gating is undesirable as it leads to more heating of the SCR gate and at the same time, it increases the size of the pulse transformer. The technique that mitigates the above disadvantages of continuous gate signal and ensures thyristor turn on is to use a train of firing pulses from α to π as shown in Fig. 9.9 (d). This type of signal is also termed as *high-frequency carrier gating*.

Example 9.5. *A single-phase voltage controller is employed for controlling the power flow from 230 V, 50 Hz source into a load circuit consisting of R = 3 Ω and ωL = 4 Ω. Calculate*

(a) the control range of firing angle,
(b) the maximum value of rms load current,
(c) the maximum power and power factor,
(d) the maximum values of average and rms thyristor currents,
(e) the maximum possible value of di/dt that may occur in the thyristor and
(f) the conduction angle for α = 0° and α = 120° assuming a gate pulse of duration π radian.

Solution. (a) For controlling the load, the minimum value of firing angle α = load phase angle, $\phi = \tan^{-1} \dfrac{\omega L}{R} = \tan^{-1} \dfrac{4}{3} = 53.13°$. The maximum possible value of α is 180°.

∴ Firing angle control range is $53.13° \le \alpha \le 180°$.

(b) The maximum value of rms load current I_{or} occurs when $\alpha = \phi = 53.13°$. But at this value of firing angle, the power circuit of ac voltage controller behaves as if load is directly connected to ac source. Therefore, maximum value of rms load current is

$$I_{0r} = \frac{230}{\sqrt{R^2 + (\omega L)^2}} = \frac{230}{\sqrt{3^2 + 4^2}} = 46 \text{ A.}$$

(c) Maximum power $= I_{0r}^2 \, R = 46^2 \times 3 = 6348$ W

$$\text{Power factor} = \frac{I_{0r}^2 \, R}{V_s \, I_0} = \frac{46 \times 3}{230} = 0.6.$$

(d) Average thyristor current is maximum when $\alpha = \phi$ and conduction angle $\gamma = \pi$. From Fig. 9.6 (c),

$$I_{TAM} = \frac{1}{2\pi} \int_{\alpha}^{\alpha + \pi} \frac{V_m}{Z} \sin (\omega t - \phi) \, d(\omega t)$$

$$= \frac{V_m}{\pi Z} = \frac{\sqrt{2} \times 230}{\pi \times \sqrt{3^2 + 4^2}} = 20.707 \text{ A.}$$

Similarly, maximum value of rms thyristor current is

$$I_{Tm} = \left[\frac{1}{2\pi} \int_{\alpha}^{\alpha + \pi} \left\{ \frac{V_m}{Z} \sin (\omega t - \alpha) \right\}^2 d(\omega t) \right]^{1/2}$$

$$= \frac{V_m}{2Z} = \frac{\sqrt{2} \times 230}{2 \times 5} = 32.527 \text{ A.}$$

(e) Maximum value of $\dfrac{di_0}{di}$ occurs when $\alpha = \phi$. From Eq. (9.23),

$$\frac{di_0}{dt} = \frac{\omega \cdot V_m}{Z} \cos (\omega t - \phi) - 0.$$

Its value is maximum when $\cos (\omega t - \phi) = 1$

$$\therefore \qquad \left(\frac{di_0}{dt} \right)_{max} = \frac{\sqrt{2} . 230 \cdot 2\pi \times 50}{5} = 2.0437 \times 10^4 \text{ A/sec.}$$

(f) For $\alpha = 0°$, Fig. 9.7 shown that conduction angle γ is 180°. For $\alpha = 120°$ and $\phi = 53.13°$, Fig. 9.7 gives a conduction angle of about 95°.

Example 9.6. *For the circuit shows in Fig. 9.10 (a), sketch the waveforms of output voltage and current for the following values of firing angles.*

(a) Only T2 is triggered at $\omega t = 0, 2\pi, 4\pi$ etc.

(b) Only T1 is triggered at $\omega t = 0, 2\pi, 4\pi$ etc.

(c) T2 is triggered at $\omega t = 0, 2\pi, 4\pi$ etc. But T1 is triggered at $\omega t = \alpha, 2\pi + \alpha, 4\pi + \alpha$ and so on. Take α around 40°.

Solution. (a) At $\omega = 0°$, supply voltage is passing through zero and becoming positive. Therefore, when thyristor T2 is triggered at $\omega t = 0, 2\pi, 4\pi$ etc. it gets turned on and load voltage v_0 equals source voltage v_s. At $\pi, 3\pi$ etc, as source voltage tends to become negative, T2 is turned off as load current i_0 is zero. Load voltage $v_0 = 0$ from π to 2π etc. as shown in Fig. 9.10 (b-i). For R load, current waveform is identical with voltage waveform.

(b) This part is similar to part (a), except that the voltage amplitude is now $\sqrt{2} \cdot 230$ volts, Fig. 9.10 (b-ii).

(c) At $\omega t = 0, 2\pi, 4\pi$ etc. when T2 is triggered, it gets turned on and $v_0 = \sqrt{2} . 115 \sin \omega t$. At $\omega t = \alpha, 2\pi + \alpha, 4\pi + \alpha$ etc. when forward biased thyristor T1 is triggered, it gets turned on. But when T1 gets on, a voltage equal to $\sqrt{2} \cdot 115 \sin \alpha$ appears as reverse bias across T2, it is therefore turned off at $\omega t = \alpha, 2\pi + \alpha$ etc. Thus, from $\omega t = 0$ to α, v_0 follows the curve

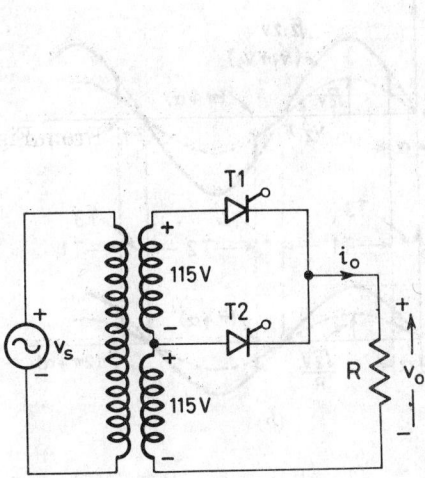

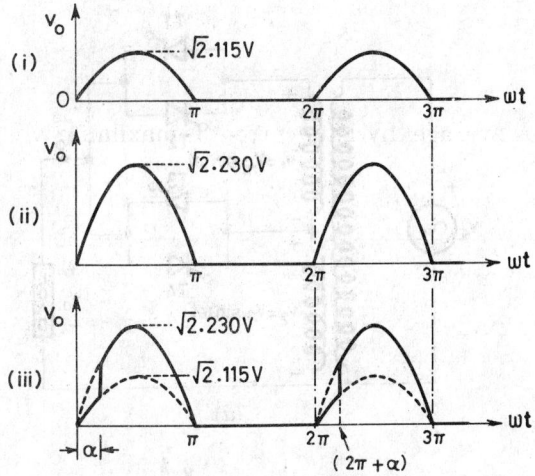

Fig. 9.10. Pertaining to Example 9.4.

$\sqrt{2} \cdot 115 \sin \omega t$ but from $\omega t = \alpha$ to π, v_0 follows the curve $\sqrt{2} \cdot 230 \sin \omega t$ as shown in Fig. 9.10 (*b-iii*). From π to 2π, $v_0 = 0$. From 2π to $2\pi + \alpha$, $v_0 = \sqrt{2}.115 \sin \omega t$ and from $2\pi + \alpha$ to 3π, $v_0 = \sqrt{2} \cdot 230 \sin \omega t$ and so on. As the load is resistive, load current waveform is identical with load voltage waveform.

9.4. SEQUENCE CONTROL OF AC VOLTAGE CONTROLLERS (TRANSFORMER TAP CHANGERS)

Sequence control of ac voltage controllers is employed for the improvement of system pf and for the reduction of harmonics in the input current and output voltage. Sequence control of ac voltage controllers means the use of two or more stages of voltage controllers in parallel for the regulation of output voltage. The term 'sequence control' means that the stages of voltage controllers in parallel are triggered in a proper sequence one after the other so as to obtain a variable output voltage with low harmonic content.

The object of this section is to describe two-stage as well as multistage sequence control of voltage controllers. A single-phase sinusoidal voltage controller is also explained.

9.4.1. Two-stage Sequence Control of Voltage Controllers

A two-stage sequence control of ac voltage regulators employs two stages in parallel as shown in Fig. 9.11 (*a*). The turns ratio from primary to each secondary is taken as unity for convenience. This means that for source voltage $v_s = V_m \sin \omega t$, $v_1 = v_2 = V_m \sin \omega t$ and sum of two secondary voltages is $2V_m \sin \omega t$.

The load may be R or RL. For both types of loads, for obtaining output voltage control from zero to rms value V, use only thyristor pair T3, T4 in Fig. 9.11 and keep T1, T2 off. For zero output voltage, α is 180° for T3, T4 and for V, α is zero. For output voltage control from V to $2V$, α for thyristor pair T3, T4 is always zero and for thyristor pair T1, T2 ; α is varied from zero to 180°.

The main advantage of two-stage sequence control of ac voltage controller over single-phase full-wave ac voltage controller is the reduction of harmonics in the load and supply currents.

(a)

(b)

(c)

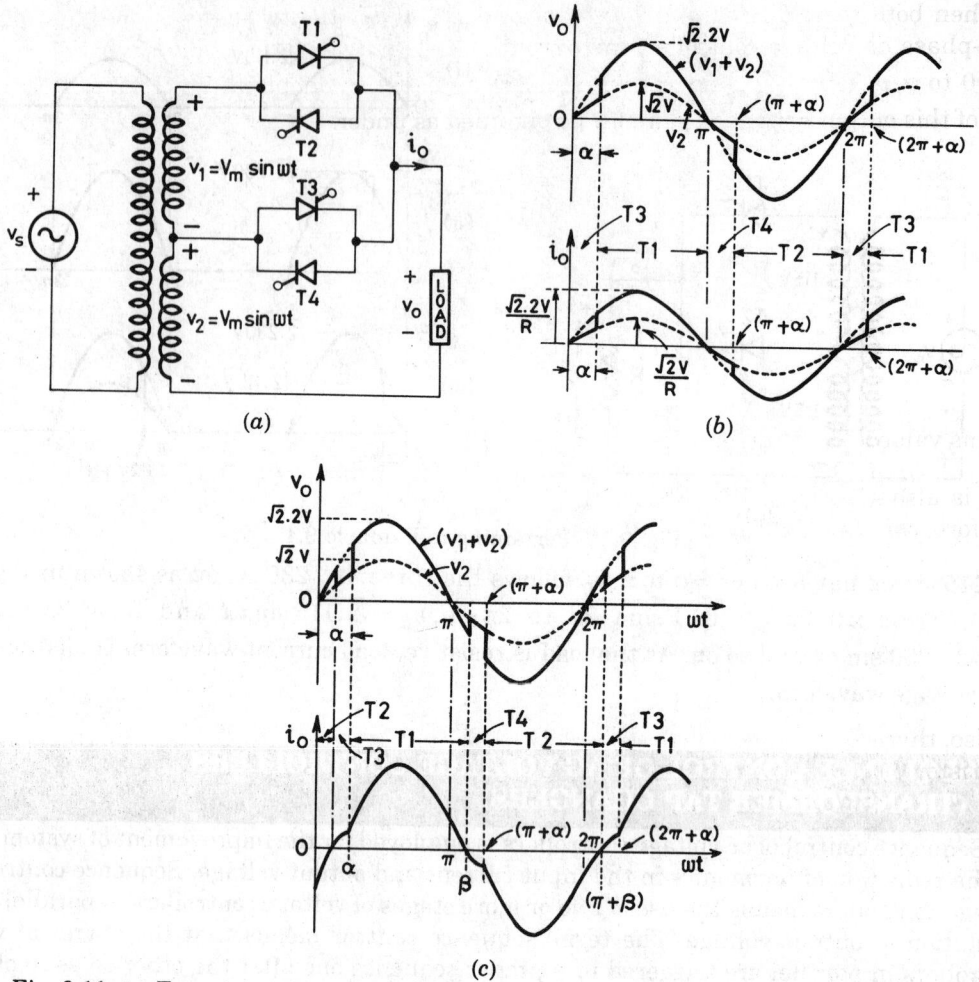

Fig. 9.11. (a) Two-stage sequence controlled ac voltage controller (b) R load (c) RL load.

(a) **Resistance load**. For resistance load, the load current waveform is identical with output voltage waveform. When thyristor pair T3, T4 is in operation with T1, T2 off, then the output voltage and current waveforms are as shown in Fig. 9.5 (b). The rms value of output voltage under this operation is given by Eq. (9.19 a). When both pairs T1, T2 and T3, T4 are in operation, then firing angle for T3, T4 is always zero whereas firing angle α for pair T1, T2 is varied from 180° to zero for obtaining output voltage from V to $2V$.

The output voltage, when thyristor T3 is triggered at $\omega t = 0$, follows $v_2 = V_m \sin \omega t$ curve. When SCR T1 is triggered at $\omega t = \alpha$, voltage v_1 reverse biases T3, it is therefore turned off. After this, T1 begins conduction and the output voltage jumps from v_2 to $(v_1 + v_2)$ and follows $2 V_m \sin \omega t$ curve. At $\omega t = \pi$, output voltage and current are zero. At this instant, T4 is triggered and output voltage follows $V_m \sin \omega t$ curve. At $\omega t = \pi + \alpha$, when forward biased SCR T2 is triggered, T4 is reverse biased by $V_m \sin \alpha$, it is therefore turned off. When T2 begins conduction, output voltage follows $2 V_m \sin \omega t$ curve as shown by the negative half cycle in Fig. 9.11 (b). In this figure, output current waveform i_0 is shown identical with output voltage waveform v_0.

When both pairs $T1$, $T2$ and $T3$, $T4$ are in use, or for the two-stage sequence control of single-phase ac voltage controller, the output voltage v_o waveform is shown in Fig. 9.11 (b). From 0 to α, v_o follows $V_m \sin \omega t$ curve and from α to π, v_o adopts to $2 V_m \sin \omega t$ curve. Rms value of this output voltage v_o can now be obtained as under.

$$V_{or} = \left[\frac{1}{\pi} \left\{ \int_o^\alpha V_m^2 \sin^2 \omega t \cdot d(\omega t) + \int_\alpha^\pi 4 V_m^2 \sin^2 \omega t \cdot d(\omega t) \right\} \right]^{1/2}$$

$$V_{or}^2 = \frac{1}{\pi} \left\{ \frac{V_m^2}{2} \int_o^\alpha (1 - \cos 2\omega t) \, d(\omega t) + 2 V_m^2 (1 - \cos 2\omega t). \, d(\omega t) \right\}$$

or

$$V_{or} = \left[\frac{V_m^2}{2\pi} \left(\alpha - \frac{\sin 2\alpha}{2} \right) + \frac{2 V_m^2}{\pi} \left(\pi - \alpha + \frac{\sin 2\alpha}{2} \right) \right]^{1/2} \qquad \qquad ...(9.27)$$

Rms value of load current, $I_{or} = \dfrac{V_{or}}{R}$

It is also seen from Fig. 9.11 (b) that $T3$ conducts for α degrees from $\omega t = 0°$ to $\omega t = \alpha$. Therefore, rms value of current for thyristor $T3$ (or $T4$) is

$$I_{T3r} = \left[\frac{1}{2\pi R^2} \int_o^\alpha V_m^2 \sin^2 \omega t. \, d(\omega t) \right]^{1/2}$$

$$= \frac{V_m}{2R} \left[\frac{1}{\pi} \left(\alpha - \frac{\sin 2\alpha}{2} \right) \right]^{1/2} \qquad \qquad ...(9.28)$$

Also, thyristor $T1$ conducts for $(\pi - \alpha)$ radians from $\omega t = \alpha$ to $\omega t = \pi$. Therefore, rms current of thyristor $T1$ (or T_2) is given by

$$I_{T1R} = \left[\frac{1}{2\pi R^2} \int_\alpha^\pi 4 V_m^2 \sin^2 \omega t \cdot d(\omega t) \right]^{1/2}$$

$$= \frac{V_m}{R} \left[\frac{1}{\pi} \left(\pi - \alpha + \frac{\sin 2\alpha}{2} \right) \right]^{1/2} \qquad \qquad ...(9.29)$$

When thyristors $T1$, $T2$, $T3$, $T4$ are in use, upper secondary winding handles both cycles of top SCR pair; whereas lower secondary winding has to handle both cycle of top and bottom pairs.

$\therefore$ Rms current rating of upper secondary,

$$I_1 = \sqrt{2} \times I_{T1r} \qquad \qquad ...(9.30)$$

and rms current rating of lower secondary,

$$I_3 = [(\sqrt{2} \times I_{T1r})^2 + (\sqrt{2} \cdot I_{T3\,r})^2]^{1/2} \qquad \qquad ...(9.31)$$

If voltage rating of upper and lower secondary windings are equal to V_s, then transformer rating $= V_s \, (I_1 + I_3)$ VA.

Example 9.7. *A two-stage sequence controlled single-phase ac voltage controller is feeding a load of R = 20 Ω. The source voltage is 230 V, 50 Hz and turns ratio from primary to each transformer secondary is unity. For two-stage sequence control, the firing angle of upper thyristors is 60°. Calculate (a) rms value of output voltage (b) rms value of current for upper thyristors (c) rms value of current for lower thyristors (d) transformer VA rating and (e) input power factor.*

Solution. Here $V_s = 230$ V, $V_m = \sqrt{2} \times 230$ V, $\alpha = 60°$, $R = 20$ Ω

(a) From Eq. (9.27), *rms* value of output voltage is

$$V_{or} = \left[\frac{2 \times 230^2}{2\pi} \left(\frac{\pi}{3} - \frac{\sin 120}{2} \right) + \frac{2 \times 2 \times 230^2}{\pi} \left(\pi - \frac{\pi}{3} + \frac{\sin 120°}{2} \right) \right]^{1/2} = 424.94 \text{ V}$$

(b) From Eq. (9.29) rms value of current for upper thyristors is

$$I_{T1r} = \frac{\sqrt{2} \times 230}{20} \left[\frac{1}{\pi} \left(\pi - \frac{\pi}{3} + \frac{\sin 120°}{2} \right) \right]^{1/2} = 14.585 \text{ A}$$

(c) From Eq. (9.28), rms value of current for lower thyristors is

$$I_{T3r} = \frac{\sqrt{2} \times 230}{2 \times 20} \left[\frac{1}{\pi} \left(\frac{\pi}{3} - \frac{\sin 120°}{2} \right) \right]^{1/2} = 3.595 \text{ A}$$

(d) Rms current rating of upper secondary,

$$I_1 = \sqrt{2} \times 14.585 = 20.623 \text{ A}$$

Rms current rating of lower secondary, from Eq. (9.31) is

$$I_3 = [(\sqrt{2} \times 14.585)^2 + (\sqrt{2} \times 3.595)^2]^{1/2} = 21.244 \text{ A}$$

Transformer rating $= V_s (I_1 + I_3)$

$$= 230 (20.623 + 21.244) = 9629.41 \text{ VA}$$

(e) Load Power, $P_o = \dfrac{V_{or}^2}{R} = \dfrac{424.94^2}{20} = 9028.7 \text{ W}$

$\therefore$ Input pf $\quad = \dfrac{P_o}{VA} = \dfrac{9028.7}{9629.41} = 0.9376$ (lagging)

RL load. When pair T3, T4 alone is in operation, then the waveforms of output voltage and current are as shown in Fig. 9.6 (b) for $\alpha > \phi$ and in Fig. 9.6 (c) for $\alpha \le \phi$.

For obtaining output voltage from V to $2 V$, firing angle for T3, T4 is always zero whereas firing angle for T1, T2 is varied from 180° to zero. It is assumed that during positive half cycle, firing pulses for T1, T3 last from $\omega t = 0$ to $\omega t = \pi$ and during negative half cycle, firing pulses for T2, T4 extend from $\omega t = \pi$ to 2π.

During positive half cycle, T3 is conducting and a voltage $v_2 = \sqrt{2} V \sin \omega t$ is applied to the load. At $\omega t = \alpha$, when T1 is triggered, T3 is turned off by reverse voltage v_1 and output voltage jumps to $v_1 + v_2 = 2 V_m \sin \omega t$ as shown. At $\omega t = \pi$, $(v_1 + v_2)$ reaches zero but output current i_0 is not zero because of the presence of L in the load. Thus, T1 continues conducting until $\omega t = \beta$, where i_0 decays to zero and T1, already reverse biased by $(v_1 + v_2)$, is turned-off. Thyristor T4, already gated at $\omega t = \pi$, starts conducting lowering the voltage to v_2 as shown. At $\omega t = \pi + \alpha$, T2 is triggered, v_1 turns off T4 and output voltage in the negative half cycle jumps to $(v_1 + v_2)$ as shown. At $\omega t = 2\pi$, $(v_1 + v_2)$ reaches zero but i_0 is not zero because of L. At $\omega t = \pi + \beta$, i_0 reaches zero. T2, already reverse biased by $(v_1 + v_2)$, is turned off and T3 already gated at $\omega t = 2\pi$ is turned on lowering the voltage to v_2 at $\omega t = \pi + \beta$. At $\omega t = 2\pi + \alpha$, already gated T1 turns on and T3 turns off and output voltage jumps from v_2 to $(v_1 + v_2)$ in the positive half cycle. The output voltage and output current waveforms are shown in Fig. 9.11 (c).

9.4.2. Multistage Sequence Control of Voltage Controllers

Multistage sequence control of ac voltage controllers is employed when it is desired to have harmonic content lower than that in a two-stage sequence control. Fig. 9.12 shows the power circuit for n-stage sequence control of voltage controllers. In this figure, the transformer has n secondary windings. Each secondary is rated for v_s/n where v_s is the source voltage. Voltage of terminal a with respect to 0 is v_s. Voltage of terminal b is $(n-1) v_s/n$ and so on. If voltage control from $v_{d0} = (n-3) v_s/n$ to $v_{co} = (n-2) v_s/n$ is required, then thyristor pair 4 is fired at $\alpha = 0°$ and the firing angle of thyristor pair 3 is controlled from $\alpha = 0°$ to $180°$ whereas all other thyristor pairs are kept off. Similarly, for controlling the voltage from $v_{bo} = (n-1) v_s/n$ to $v_{ao} = v_s$, thyristor pair 2 is triggered at $\alpha = 0°$ whereas for pair 1, α is varied from $0°$ to $180°$ by keeping the remaining $(n-2)$ SCR pairs off. Thus, the load voltage can be controlled from v_s/n to v_s by an appropriate control of triggering the adjacent thyristor pairs.

The presence of harmonics in the output voltage depends upon the magnitude of voltage variation. If this voltage variation is a small fraction of the total output voltage, the harmonic content in the output voltage is small. For example, for voltage control from $(n-2)$ v_s/n to $(n-1) v_s/n$, if voltage variation is $v_s/n \ll (n-1) v_s/n$, then the harmonic content in the output voltage would be small.

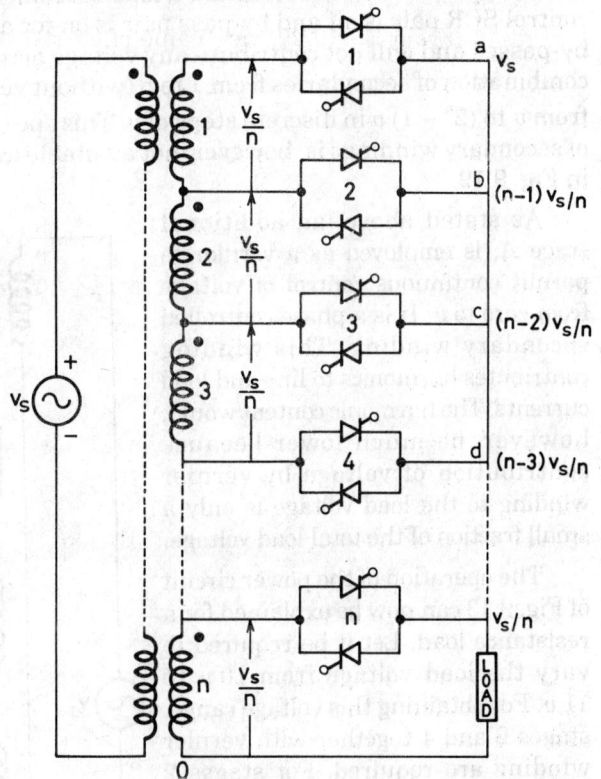

Fig. 9.12. Multistage sequence control of ac voltage controllers.

9.4.3. Single-phase Sinusoidal Voltage Controller

For obtaining continuous voltage control over a wide range with low harmonic content and improved pf, a multistage sequence voltage controller must have quite a large number of stages. This is, however, an expensive proposition. An alternative to this, with less number of stages and called single-phase sinusoidal voltage controller is usually employed ; this is shown in Fig. 9.13.

This voltage controller has one primary winding and $(n + 1)$ secondary windings, *i.e.* single-phase sinusoidal voltage controller employs $(n + 1)$ stages ; 0, 1, 2, 3...n. The top secondary winding, numbered A, is called *vernier winding*. Its rating is v volts. The voltage ratings of the remaining n windings are in geometric progression with a ratio of 2. Thus, if v is the voltage rating of secondary numbered 1, then voltage rating of secondary numbered 2 is $2v$, that of numbered 3 is $4v$ ($= 2^{3-1} \cdot v$), that of numbered 4 is $8v$ ($2^{4-1} \cdot v$) and that of numbered n is $2^{n-1} \cdot v$ volts. The power circuit of Fig. 9.13 uses two sets of thyristors for n windings. These are TC1, TC2,..... called *control thyristors* and TB1, TB2..... called *by-pass thyristors*. Vernier winding has two pairs of thyristors marked TCA and TBA. Thyristors pertaining to n stages, *i.e.* TC1, TC2,...,

TCn and TB1, TB2,..., TBn are either on or off throughout a cycle. This means that control and by-pass thyristors are made to act as switches which remain either on or off during a cycle. When control SCR pair for any stage is on and its by-pass thyristor pair is off, then voltage of that stage would appear across the load and a load current would flow accordingly. On the other hand, if control SCR pair is off and by-pass pair is on for any stage, then this particular stage would be by-passed and will not contribute any voltage across the load. Thus, with an appropriate series combination of secondaries from 1 to n (without vernier winding), the load voltage can be varied from v to $(2^n - 1) v$ in discrete steps of v. This special feature of choosing any series combination of secondary windings is, however, not available in the multistage sequence control of thyristors in Fig. 9.12.

As stated above, an additional stage A, is employed as a vernier to permit continuous control of voltage from zero to v. It is a phase-controlled secondary winding. This winding contributes harmonics to line and load currents. The harmonic content would, however, be much lower because contribution of voltage by vernier winding to the load voltage is only a small fraction of the total load voltage.

The operation of the power circuit of Fig. 9.13 can now be explained for a resistance load. Let it be required to vary the load voltage from $10\,v$ to $11\,v$. For obtaining this voltage range, stages 2 and 4 together with vernier winding are required. For stages 2 and 4, by-pass SCRs are kept off but their control SCRs are kept on all the time. For the remaining stages from 1 to n, all by-pass SCRs are kept on while their control SCRs are kept off all the time. With this, the circuit of Fig. 9.13 reduces to that shown in Fig. 9.14 (a).

During the positive half cycle, thyristor T3 is turned on at $\omega t = 0°$ and as a result, a voltage $10\,v$ is applied to the load. At $\omega t = \alpha$, T1 is turned on and this turns off T3, the voltage now jumps from $10v$ to $11v$, i.e. from $10\,V_m \sin \alpha$ to $11 V_m \sin \alpha$,

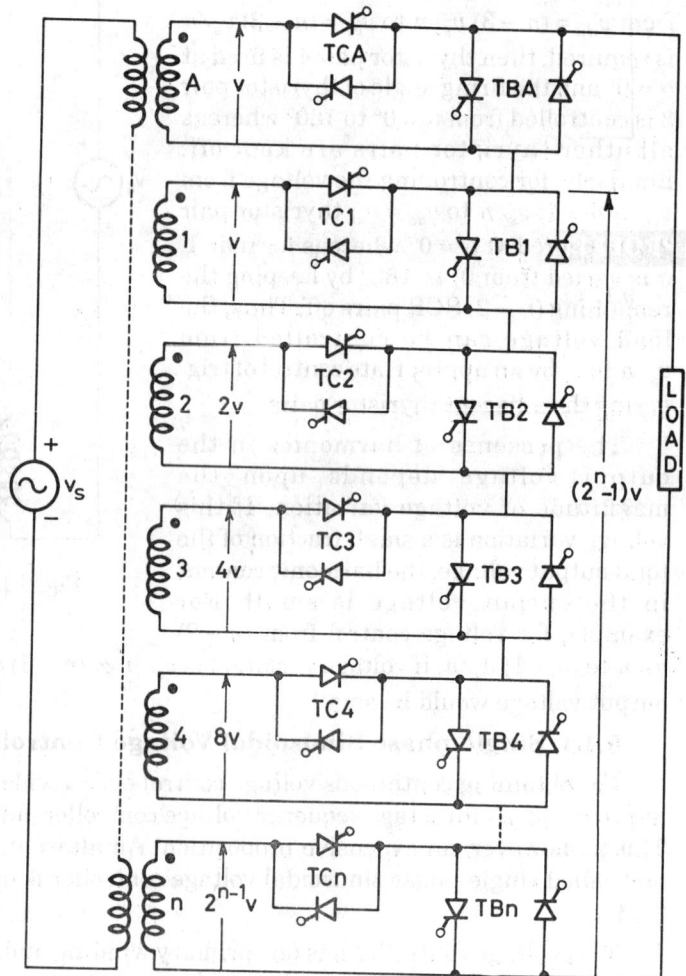

Fig. 9.13. Single-phase sinusoidal voltage controller.

Fig. 9.14 (b). After $\omega t = \alpha$, output voltage follows $11\,V_m \sin \omega t$ curve. At the end of positive half cycle, T4 is turned on at $\omega = \pi$, the load voltage is now $10\,v$. At $\omega t = \pi + \alpha$, T2 is turned on and with this T4 is turned off and load voltage jumps from $10\,v$ ($10\,V_m \sin \alpha$) to $11v$ ($11\,V_m \sin \alpha$) in the negative half cycle. In this manner, load voltage can be continuously controlled from zero to $2^n\,v$ by a suitable choice of output voltage range.

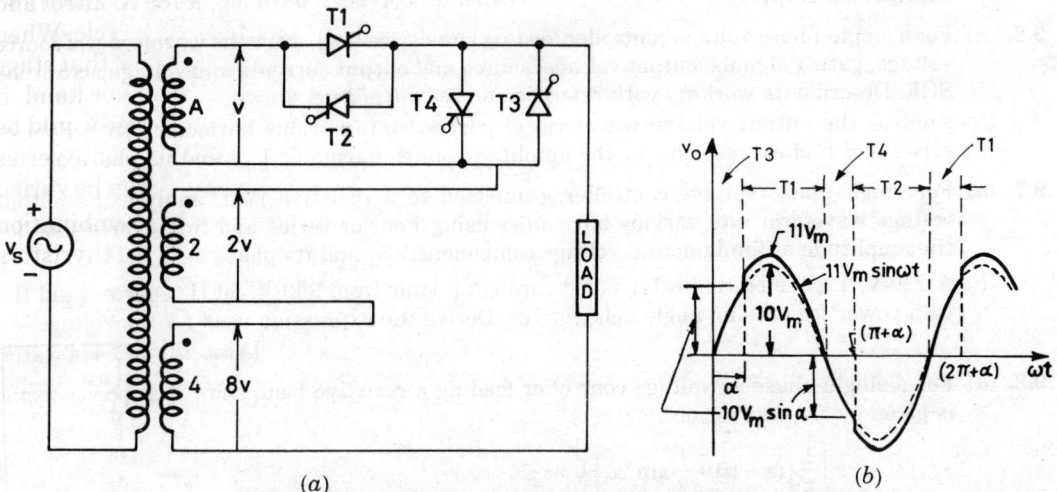

(a)

(b)

Fig. 9.14. Operation of voltage controller of Fig. 9.13.

PROBLEMS

9.1. (a) What is an *ac* voltage controller ? List some of its industrial applications. Enumerate its merits and demerits.

(b) What are the control strategies for the regulation of output voltage in ac voltage controllers ?

Discuss the merits and demerits of the control strategies listed above.

(c) What are the advantages and disadvantages of unidirectional as well as bidirectional controllers ? Which one of these is preferred and why ?

9.2. (a) Draw the possible configurations of a single-phase voltage controller and compare them.

A single-phase voltage controller using two SCRs in antiparallel must have its trigger sources isolated from each other. Why ? Explain with a suitable diagram.

(b) Describe the principle of phase control in single-phase half-wave *ac* voltage-controller. Derive expressions for the average and rms value of output voltage for this voltage controller.

9.3. A single-phase unidirectional voltage controller is connected to a load of $R = 10\ \Omega$. Input voltage is 230 V, 50 Hz. Firing-angle delay is 30°. Determine (a) rms value of output voltage (b) average and rms values of thyristor current (c) average and rms values of diode current and (d) input power factor.

[**Ans.** (a) 228.3 V (b) 9.6585 A, 16.146 A (c) 10.352 A, 16.261 A (d) 0.9926 lag]

9.4. (a) Discuss the principle of phase control in single-phase full-wave *ac* voltage controller. Derive expression for the rms value of its output voltage.

(b) A single-phase full-wave ac voltage controller has a load of $R = 5\ \Omega$ and the input voltage is 230 V, 50 Hz. If load power is 5 kW, find (a) firing-angle delay of thyristors and (b) input power factor.

[**Hint :** $\alpha - \dfrac{\sin 2\alpha}{2} = 1.657$. By trial and error $\alpha = 92.5°$]   [**Ans :** (a) 92.5° (b) 0.6874 lag]

9.5. (a) Describe the principle of burst firing for a single-phase *ac* voltage controller. Derive an expression for the rms value of output voltage.

(b) A single-phase *ac* voltage controller uses burst-firing control for heating a load of $R = 5\ \Omega$ with an input voltage of 230 V, 50 Hz. For a load power of 5 kW, determine (*i*) the duty cycle (*ii*) input power factor (*iii*) average and *rms* thyristor currents. Derive the expressions used.   **[Ans.** (*b*) 0.4726, 0.6875 lag, 9.785 A, 22.3575 A]

9.8. (*a*) For a single-phase voltage controller feeding a resistive load, draw the waveforms of source voltage, gating signals, output voltage, source and output currents and voltage across one SCR. Describe its working with reference to the waveforms drawn.

(*b*) Analyse the output voltage waveform of part (*a*) into various harmonics with Fourier series and find expressions for the amplitude of *n*th harmonic V_{nm} and its phase ϕ_n.

9.7. (*a*) For single-phase voltage controller, connected to a resistive load, analyse the output voltage waveform into various harmonics using Fourier series and find expressions for the amplitude of fundamental voltage component V_{1m} and its phase ϕ_1.

(*b*) A 230-V, 1 kW electric heater is fed through a triac from 230 V, 50 Hz source. Find the load power for a firing angle delay of 70°. Derive the expression used for the voltage.

                                                      [Ans. (*b*) 713.414 watts]

9.8. (*a*) For a single-phase ac voltage controller feeding a resistive load, show that power factor is given by the expression :

$$\left[\frac{1}{\pi} \left\{ (\pi - \alpha) + \frac{1}{2} \sin 2\alpha \right\} \right]^{1/2}$$

(*b*) A single-phase half-wave ac voltage controller, using one SCR in antiparallel with a diode, feeds 1 kW, 230 V heater. For a supply voltage of 230 V, 50 Hz, find the load power for a firing-angle delay of (*i*) 0° (*ii*) 180° (*iii*) 70°.   **[Ans.** (*b*) 1000 W, 500 W, 856.707 W]

9.9. (*a*) Compare the merits of controlling the heater power by a triac using integral cycle control over the phase-angle control.

(*b*) A heater load is controlled through a triac from a single-phase source. Determine the firing angle delay when the power is at (*i*) 50%, (*ii*) 70% of its maximum power. Derive the expression used.

(*c*) A heater load is controlled by means of single-phase ac voltage controller. Determine the firing angle delay, when the controlled power is at (*i*) 50%, (*ii*) 70% of its maximum power. Derive the expression used.   **[Ans.** (*b*) 90°, 71.5° (*c*) 180°, 99°]

9.10. (*a*) Define the term 'power factor'. Derive its expression for a single-phase voltage controller feeding a resistive load circuit and show that

Power factor = [per unit power]$^{1/2}$

(*b*) For the circuit shown in Fig. 9.15, do the following :

    (*i*) Sketch the waveforms for two cycles of supply voltage, supply current, load voltage and load current for a firing angle of about 45° for the two thyristors.

    (*ii*) For 230 V, 50 Hz as the supply voltage, find the power consumed by load in case $\alpha = 60°$ and $R = 10\ \Omega$. Derive the expression used for power.

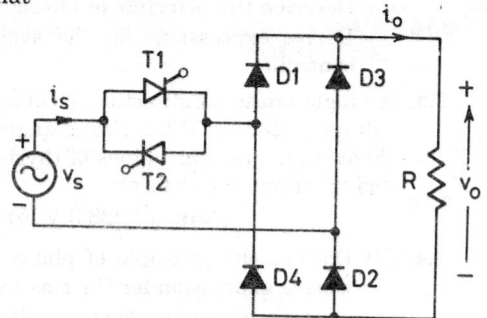

Fig. 9.15. Pertaining to Prob. 9.10 (*b*).

    (*iii*) In case diode D1 gets open circuited, draw the load current waveform and calculate the power delivered to load.   **[Ans.** (*b*) (*ii*) 4255.8 W (*iii*) 2127.9 W]

9.11. A single-phase voltage controller with resistance load has the following data :

Supply mains : 230 V, 50 Hz, $R = 4\ \Omega$. Calculate :

(*a*) the firing angle α at which the greatest forward or reverse voltage is applied to either of the thyristors and the magnitude of these voltages ;

(b) the greatest forward or reverse voltage that appears across either of the thyristors for firing angles of 60° and 120° ;

(c) the rms value of fifth harmonic current and its phase for α = π/4.

[Ans. (a) $\alpha = \dfrac{\pi}{2}$ or at $\alpha > \dfrac{\pi}{2}$, 325.27 V (b) 281.691 V, 325.27 V (c) 27.284 V, – 63.435°**]**

9.12. (a) A single-phase voltage controller, with two thyristors arranged in antiparallel, is connected to *RL* load. Discuss its working when firing angle is more than the load pf angle. Illustrate your answer with waveforms of source voltage, gate singals, load and source currents, output voltage and voltage across both the thyristors.

Hence derive an expression for the output current in terms of source voltage, load impedance, firing angle etc.

(b) A single-phase voltage controller has the following data :

Source voltage = 230 V at 50 Hz, Load = 0 + *j* 4 Ω. Calculate :

　(*i*) the control range of firing angle,

　(*ii*) the maximum value of rms load current,

　(*iii*) the maximum value of average and rms thyristor currents,

　(*iv*) the maximum value of *di/dt* that may occur in the thyristors,

　(*v*) the value of conduction angle for α = 90° assuming gate pulse width of π radians.

[Ans. (b) 90° ≤ α ≤ 180° ; 57.5 A ; 25.88 A, 40.66 A ; 2.5546×10^4 A/sec ; 180°**]**

9.13. (a) For a single-phase voltage controller, develop a relation between conduction angle γ and firing angle α and plot their variation as a function of load phase angle φ. Under what conditions conduction angle γ becomes equal to π ?

(b) Discuss the operation of a single-phase voltage controller with *RL* load when firing angle α is less than, or equal to, load phase angle φ. Hence show that for α less than φ, output voltage of the ac voltage controller cannot be regulated.

(c) For a single-phase voltage controller, discuss how pulse gating is suitable for *R* load and not for *RL* load. Hence show that high-frequency carrier gating is essential for *RL* loads.

9.14. (a) Describe the working of a two-stage sequence control of voltage controllers for *R* load. What is the advantage of this controller over single-phase full-wave voltage controller?

(b) Derive an expression for the *rms* value of output voltage when both stages are in operation.

9.15. A two-stage sequence controlled single-phase *ac* voltage controller is connected to a load *R* = 10 Ω and input voltage is 200 V, 500 Hz. Turns ratio from primary to each transformer secondary is unity. The firing angle of the upper group of thyristors is 30° for the two-stage control. Calculate (a) the *rms* value of output voltage (b) the *rms* value of current for upper thyristors (c) *rms* value of current for lower thyristors (d) transformer VA rating and (e) input power factor.

9.16. (a) Describe the working of a two-stage sequence control of voltage controllers for RL load.

(b) Distinguish between two-stage and multistage sequence control of voltage controllers. What are the advantages of multistage over two-stage sequence control ? Describe multistage sequence control of voltage controllers.

9.17. Describe a single-phase sinusoidal voltage controller with vernier winding. What are the functions of controlled and by-pass thyristors ? Discuss how output voltage waveform from 7 *v* to 8 *v* can be obtained from this voltage controller.

9.18. A single-phase bidirectional *ac* voltage controller delivers a load power of 50 kW with an efficiency and power factor of 0.9 and 0.85 respectively. The input voltage is 230 V, 50 Hz. Determine the maximum possible ratings of (a) average and *rms* thyristor currents and (b) repetitive voltage of the thyristors. **[Ans.** (a) 127.92 A, 200.93 A (c) 325.22 V**]**

Chapter 10

Cycloconverters

A device which converts input power at one frequency to output power at a different frequency with one-stage conversion is called a cycloconverter. A cycloconverter is thus a one-stage frequency changer. Basically, cycloconverters are of two types, namely :

(i) step-down cycloconverters and

(ii) step-up cycloconverters.

In step-down cycloconverters, the output frequency f_0 is lower than the supply frequency f_s, i.e. $f_0 < f_s$. In step-up cycloconverters, $f_0 > f_s$.

The operating principles of step-down cycloconverters were developed as far back as 1930. At that time, mercury-arc rectifier was used as a cycloconverter for converting three-phase 50 Hz supply to single-phase $16\frac{2}{3}$ Hz supply for use in ac traction system in Germany. A single-phase series motor, when operated at a lower frequency, gives better operating characteristics. In the United States, a cycloconverter comprising 18 thyratrons was employed to drive a 400-HP synchronous motor for several years in Logan power station [6]. The cycloconverter systems at that time did not find widespread use only because early systems were not technically attractive and economically viable.

With the advent of high-power thyristors, cycloconverters are again becoming popular. At present, the applications of cycloconverters include the following :

(i) Speed control of high-power ac drives

(ii) Induction heating

(iii) Static VAr Compensation

(iv) For converting variable-speed alternator voltage to constant frequency output voltage for use as power supply in aircraft or shipboards.

The general use of cycloconverter is to provide either a variable frequency power from a fixed input frequency power (as in ac motor speed control) or a fixed frequency power from a variable input frequency power (as in aircraft or shipboard power supplies or wind generators).

The object of this chapter is to present both single-phase and three-phase cycloconverters at an introductory level.

10.1. PRINCIPLE OF CYCLOCONVERTER OPERATION

In this section, basic principle of operation of step-up as well as step-down cycloconverter is presented. Single-phase to single-phase cycloconverter, though seldom used in practice, is considered here for describing the principle of operation of both the types of cycloconverters. A single-phase to single-phase device of the mid-point type is shown in Fig. 10.1 (*a*) and of the bridge type in Fig. 10.1 (*b*). With the help of this figure, the basic principles of both types of cycloconverters are described here.

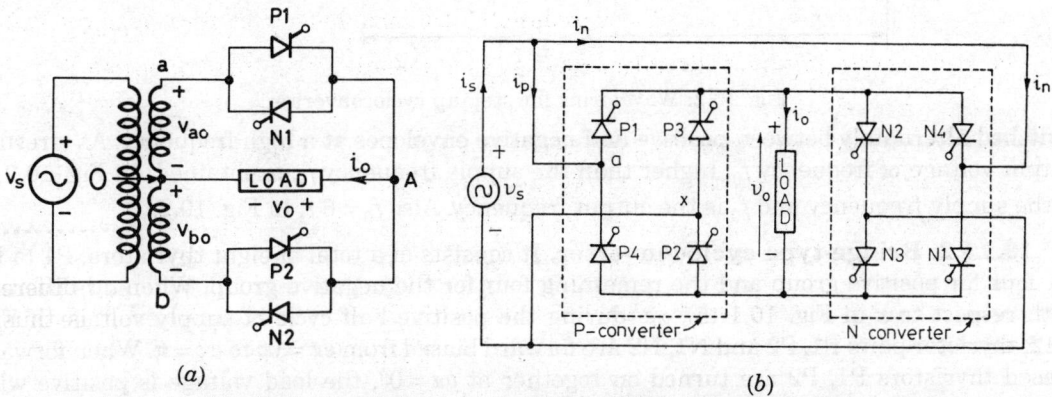

(*a*)   (*b*)

Fig. 10.1. Single-phase to single-phase cycloconverter circuit
(*a*) mid-point type and (*b*) bridge type.

10.1.1. Single-phase to Single-phase Circuit–Step-up Cycloconverter

For understanding the operating principle of step-up device, the load is assumed to be resistive for simplicity. It should be noted that a step-up cycloconverter requires forced commutation. The basic principle of step-up device is described here first for mid-point and then for bridge-type cycloconverters.

10.1.1.1. Mid-point cycloconverter.
It consists of a single-phase transformer with mid-tap on the secondary winding and four thyristors. Two of these thyristors P1, P2 are for positive group and the other two N1, N2 are for the negative group. Load is connected between secondary winding mid-point 0 and terminal *A* as shown in Fig. 10.1 (*a*). Positive directions for output voltage v_0 and output current i_0 are marked in Fig. 10.1.

In Fig. 10.1, during the positive half cycle of supply voltage of Fig. 10.2, terminal *a* is positive with respect to terminal *b*. Therefore, in this positive half cycle, both SCRs P1 and N2 are forward biased from $\omega t = 0$ to $\omega t = \pi$. As such SCR P1 is turned on at $\omega t = 0°$ so that load voltage is positive with terminal *A* positive and 0 negative. The load voltage now follows the positive envelope of the supply voltage, Fig. 10.2. At instant ωt_1, P1 is force commutated and forward-biased thyristor N2 is turned on so that load voltage is negative with terminal 0 positive and *A* negative. The load, or output, voltage now traces the negative envelope of the supply voltage, Fig. 10.2. At ωt_2, N2 is force commutated and P1 is turned on, the load voltage is now positive and follows the positive envelope of supply voltage, Fig. 10.2. After $\omega t = \pi$, terminal *b* is positive with respect to terminal *a* ; both SCRs P2 and N1 are therefore forward biased from $\omega t = \pi$ to 2π. At $\omega t = \pi$, N2 is force commutated and forward biased SCR P2 is turned on. At $\omega t = \dfrac{1}{2_{fs}} + \dfrac{1}{2_{fo}}$, P2 is force commutated and forward biased SCR N1 is turned on. In this manner, thyristors P1, N2 for first half cycle ; P2, N1 in the second half cycle and so on are

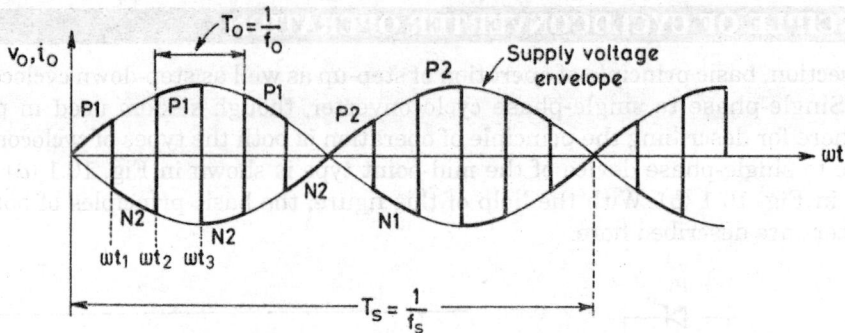

Fig. 10.2. Waveforms for step-up cycloconverter.

switched alternately between positive and negative envelopes at a high frequency. As a result, output voltage of frequency f_0, higher than the supply frequency f_s, is obtained. In Fig. 10.2, f_s is the supply frequency and f_0 is the output frequency. Also $f_0 = 6\,f_s$ in Fig. 10.2.

10.1.1.2. Bridge-type cycloconverter. It consists of a total of eight thyristors, P1 to P4 *i.e.* four for positive group and the remaining four for the negative group. When a is positive with respect to x in Fig. 10.1 (*b*), *i.e.* during the positive half cycle of supply voltage of Fig. 10.2, thyristor pairs P1, P2 and N1, N2 are forward biased from $\omega t = 0°$ to $\omega t = \pi$. When forward biased thyristors P1, P2 are turned on together at $\omega t = 0°$, the load voltage is positive with respect to x in Fig. 10.1 (*b*), forward-biased thyristors P1, P2 are turned on together at $\omega t = 0°$ so that load voltage is positive with terminal A positive with respect to 0. Load voltage now traverses the positive envelope of supply voltage, Fig. 10.2. At ωt_1, pair P1, P2 is force commutated and forward biased pair N1, N2 is turned on. With this, load voltage is negative with terminal 0 positive with respect to A. Load voltage now follows the negative envelope of source voltage, Fig. 10.2. At ωt_2 ; N1, N2 are force commutated and P1, P2 are turned on. The load voltage is now positive and follows the positive envelope of source voltage. After $\omega t = \pi$, thyristor pairs P3, P4 and N3, N4 are forward biased, these can therefore be turned on and force commutated from $\omega t = \pi$ to $\omega t = 2\pi$. In this manner, a high-frequency turning-on and force commutation of pairs P1 P2, N1 N2 and pairs P3 P4, N3 N4 gives a carrier-frequency modulated output voltage across load terminals.

In Fig. 10.2 conduction of thyristors P1, P2 and N1, N2 for mid-point cycloconverter of Fig. 10.1 (*a*) is only shown. It is fairly easy to indicate the conduction of thyristors P1 to P4 and N1 to N4 in Fig. 10.2.

10.1.2. Single-phase to Single-phase Circuit–Step-down Cycloconverter

A step-down cycloconverter does not require forced commutation. It requires phase-controlled converters connected as shown in Fig. 10.1. These converters need only line, or natural, commutation which is provided by ac supply. Both mid-point and bridge-type cycloconverters are described in what follows :

10.1.2.1. Mid-point cycloconverter. This type of cycloconverter will be described both for discontinuous as well as continuous load current. The load is now assumed to consist of R and L in series.

(*a*) **Discontinuous load current.** When a is positive with respect to 0 in Fig. 10.1 (*a*), forward biased SCR P1 is triggered at $\omega t = \alpha$. With this, load current i_0 starts building up in the positive direction from A to O. Load current i_0 becomes zero at $\omega t = \beta > \pi$ but less than $(\pi + \alpha)$, Fig. 10.3 (*c*). Thyristor P1 is thus naturally commutated at $\omega t = \beta$ which is already reverse

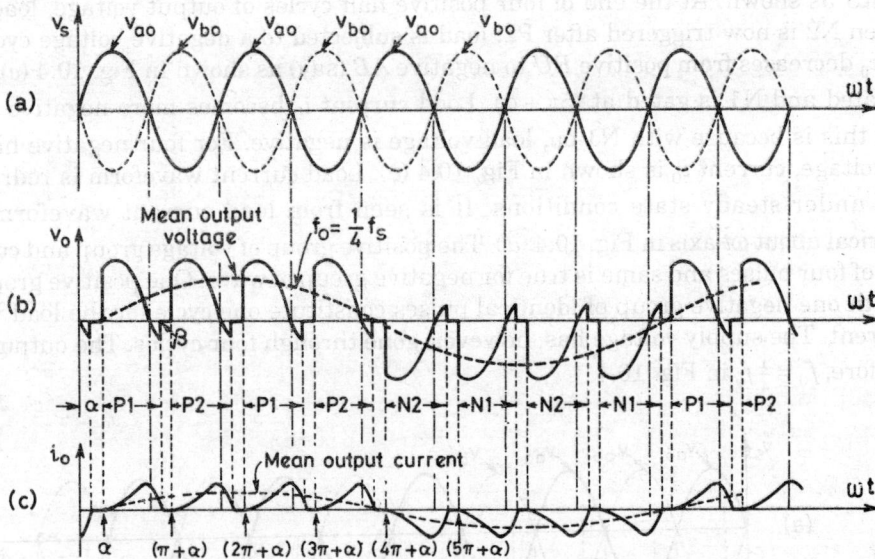

Fig. 10.3. Voltage and current waveforms for
step-down cycloconverter with discontinuous load current.

biased after π. After half a cycle, b is positive with respect to 0. Now forward biased thyristor
P2 is triggered at $\omega t = \pi + \alpha$. Load current is again positive from A to O and builds up from zero
as shown in Fig. 10.3 (c). At $\omega t = \pi + \beta$, i_0 decays to zero and P2 is naturally commutated. At
$2\pi + \alpha$, P1 is again turned on. Load current in Fig. 10.3 (c) is seen to be discontinuous. After four
positive half cycles of load voltage and load current, thyristor N2 (after P2, N2 should be fired)
is gated at $(4\pi + \alpha)$ when 0 is positive with respect to b. As N2 is forward biased, it starts
conducting but load current direction is reversed, $i.e.$ it is now from 0 to A. After N2 is triggered,
load current builds up in the negative direction as shown in Fig. 10.3 (c). In the next half-cycle,
0 is positive with respect to a but before N1 is fired, i_0 decays to zero and N2 is naturally
commutated. Now when N1 is gated at $(5\pi + \alpha)$, i_0 again builds up but it decays to zero before
thyristor N2 in sequence is again gated. In this manner, four negative half cycles of load voltage
and load current, equal to the number of four positive half cycles, are generated. Now P1 is
again triggered to fabricate further four positive half cycles of load voltage and so on. For
discontinuous load current, natural commutation is achieved, $i.e.$ P1 goes to blocking state
before P2 is gated and so on.

In Fig. 10.3, mean output voltage and current waves are also shown. It is seen from this
figure that frequency of output voltage and current is $f_0 = \dfrac{1}{4} f_s$.

(b) **Continuous load current.** When a is positive with respect to 0 in Fig. 10.1 (a), P1 is
triggered at $\omega t = \alpha$, positive output voltage appears across load and load current starts building
up, Fig. 10.4 (c). At $\omega t = \pi$, supply and load voltages are zero. After $\omega t = \pi$, P1 is reverse biased.
As load current is continuous, P1 is not turned off at $\omega t = \pi$. When P2 is triggered in sequence
at $\pi + \alpha$, a reverse voltage appears across P1, it is therefore turned off by natural commutation.
When P1 is commutated, load current has built up to a value equal to RR, Fig. 10.4 (c). With
the turning on of P2 at $(\pi + \alpha)$, output voltage is again positive as it was with P1 on. As a
consequence, load current builds up further than RR as shown in Fig. 10.4 (c). At $(2\pi + \alpha)$, when
P1 is again turned on, P2 is naturally commutated and load current through P1 builds up

beyond *RS* as shown. At the end of four positive half cycles of output voltage, load current is *RU*. When N2 is now triggered after P2, load is subjected to a negative voltage cycle and load current i_0 decreases from positive *RU* to negative *AB* (say) as shown in Fig. 10.4 (c). Now N2 is commutated and N1 is gated at $(5\pi + \alpha)$. Load current i_0 becomes more negative than *AB* at $(6\pi + \alpha)$, this is because with N1 on, load voltage is negative. For four negative half cycles of output voltage, current i_0 is shown in Fig. 10.4 (c). Load current waveform is redrawn in Fig. 10.4 (d) under steady state conditions. It is seen from load current waveform that i_0 is symmetrical about ωt axis in Fig. 10.4 (d). The positive group of voltage group and current wave consists of four pulses and same is true for negative group of wave. One positive group of pulses along with one negative group of identical pulses constitute one cycle for the load voltage and load current. The supply voltage has, however, gone through four cycles. The output frequency is, therefore, $f_0 = \frac{1}{4} f_s$ in Fig. 10.4.

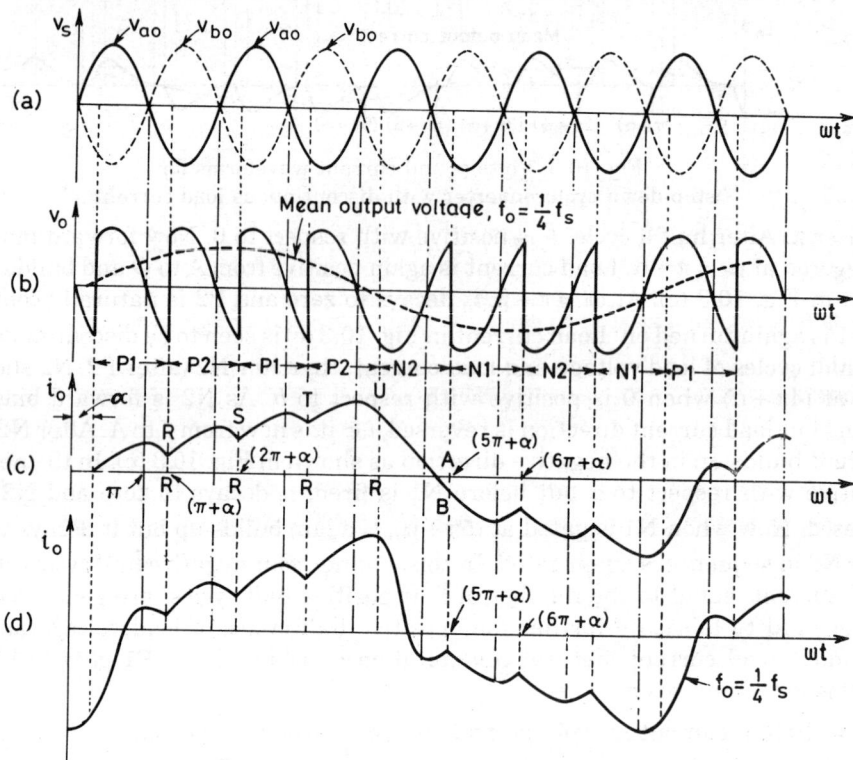

Fig. 10.4. Voltage and current waveforms for
step-down cycloconverter with continuous load current.

10.1.2.2. Bridge-type cycloconverter. The operation of bridge type cycloconverter shown in Fig. 10.1 (b) can be easily explained for both discontinuous and continuous load currents. The voltage and current waveforms would again be as shown in Fig. 10.3 for discontinuous load current and as in Fig. 10.4 for continuous load current. The explanation of bridge-type cycloconverter is left as an exercise to the reader.

Example 10.1. (a) *A single-phase bridge type cycloconverter feeds a load R. For an output frequency equal to one-third of the input frequency, sketch output voltage waveform for a firing angle of about 30°*

(b) *Derive an expression for the rms output voltage in part (a).*

Solution. (a) For this example, refer to Fig. 10.1 (b) which is the circuit diagram for single-phase bridge-type cycloconverter. The source voltage waveform v_s and load voltage waveform v_o are sketched in Fig. 10.5. The half cycles of source voltage are numbered 1, 2, 3, ... for the sake of convenience.

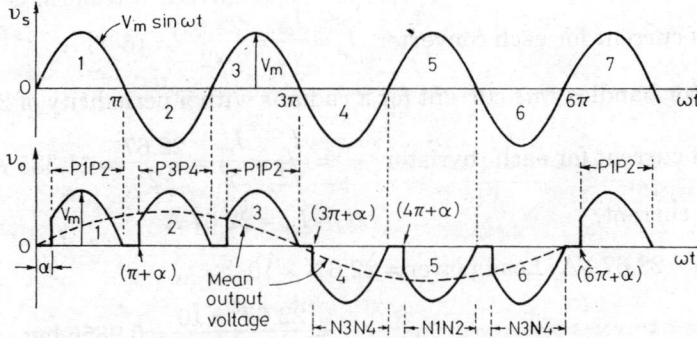

Fig. 10.5. Voltage waveform pertaining to Example 10.1.

During positive half cycle 1, thyristors P_1, P_2 are forward biased, these are therefore triggered at α. P_1, P_2 conduct from α to π because of resistive load. During negative half cycle 2 of source voltage ; $P3, P4$ are forward biased, these are therefore turned on at $\pi + \alpha$ and so on. During negative half cycle 4; $N3, N4$ are forward biased, these are, therefore, turned on at $\omega t = 3\pi + \alpha$ and so on. For output frequency $f_o = \frac{1}{3} f_s$, three positive half cycles are obtained first and then three negative half cycles are secured as shown in Fig. 10.5. Waveform of mean output voltage v_o is also shown in Fig. 10.5. For one half-cycle of low-frequency f_o, there are three half cycle of source frequency f_s, we have $f_o = \frac{1}{3} f_s$.

It is seen from output voltage waveform, consisting of three positive and three negative half cycles, that *rms* value of output voltage is given by

$$V_{or} = \left[\frac{1}{\pi} \int_\alpha^\pi V_m^2 \sin^2 \omega t \, d\,(\omega t) \right]^{1/2}$$

This expression is similar to that leading to Eq. (9.3). Therefore, from Eq. (9.3), rms value V_{or} is

$$\therefore \qquad V_{or} = \frac{V_m}{\sqrt{2}} \left[\frac{1}{\pi} \left(\pi - \alpha + \frac{\sin 2\alpha}{2} \right) \right]^{1/2} \qquad \qquad ...(10.1)$$

Example 10.2. *A single-phase bridge-type cycloconverter has input voltage of 230 V, 50 Hz and load of R = 10 Ω. Output frequency is one-third of input frequency. For a firing angle delay of 30°, calculate (a) rms value of output voltage (b) rms current of each converter (c) rms current of each thyristor and (d) input power factor.*

Solution. Here $V_s = 230$ V, $V_m = \sqrt{2} \times 230$ V, $R = 10\ \Omega$, $\alpha = 30° = \frac{\pi}{6}$.

(a) From Eq. (10.1), we get

$$V_{or} = \frac{\sqrt{2} \times 230}{\sqrt{2}} \left[\frac{1}{\pi} \left(\pi - \frac{\pi}{6} + \frac{\sin 60°}{2} \right) \right]^{1/2} = 226.66\ V$$

Rms value of load current, $I_{or} = \dfrac{V_{or}}{R} = \dfrac{226.66}{10} = 22.67$ A

(b) For an output frequency of f_o, each converter conducts for $\dfrac{1}{2f_o} = \pi$ radians, with a periodicity of 2π radians.

$\therefore$ Rms value of current for each converter. $I_p = \dfrac{I_{or}}{\sqrt{2}} = \dfrac{22.67}{\sqrt{2}} = 16.03$ A

(c) Each thyristor handles *rms* current for π radians with a periodicity of 2π rad.

$\therefore$ Rms value of current for each thyristor $= \dfrac{I_p}{\sqrt{2}} = \dfrac{I_{or}}{2} = \dfrac{22.67}{2} = 11.335$ A

(d) Rms source current, $I_s = I_{or} = 22.67$ A

Input $VA = 230 \times 22.67$ VA, Load power $= 22.67^2 \times 10$

$\therefore$ Input pf $= \dfrac{22.67^2 \times 10}{230 \times 22.67} = 0.9856$ lag.

10.2. THREE-PHASE HALF-WAVE CYCLOCONVERTERS

The object of this section is to consider how single-phase low-frequency output voltage is fabricated from the segments of 3-phase input voltage waveform. Then three-phase to three-phase cyclcconverters are described.

10.2.1. Three-phase to Single-phase Cycloconverters

For converting three-phase supply at one frequency to single-phase supply at a lower frequency, the basic principle is to vary progressively the firing angle of the three thyristors of a 3-phase half-wave circuit. In Fig. 10.6, firing angle at A is 90°, at B firing angle is somewhat less than 90°, at C the firing angle is still further reduced than it is at B and so on. In this manner, a small delay in firing angle is introduced at C, D, E, F and G. At G, the firing angle is zero and the mean output voltage, given by $V_0 = V_{do} \cos \alpha$, is maximum at G. At A, the mean output voltage is zero as $\alpha = 90°$. After point G, a small delay in firing angle is further introduced progressively at points H, I, J, K, L and M. At M, the firing angle is again 90° and the value of mean output voltage is zero. The gating circuitry is suitably designed to introduce progressive firing angle delay as discussed here. In Fig. 10.6, the single-phase output voltage, fabricated from 3-phase input voltage, is shown by thick curve. Mean output voltage wave is obtained by joining points pertaining to average voltage values. For example, at A, $\alpha = 90°$, $V_0 = 0$; at G, $\alpha = 0°$, therefore V_0 has maximum mean output voltage and so on. It is

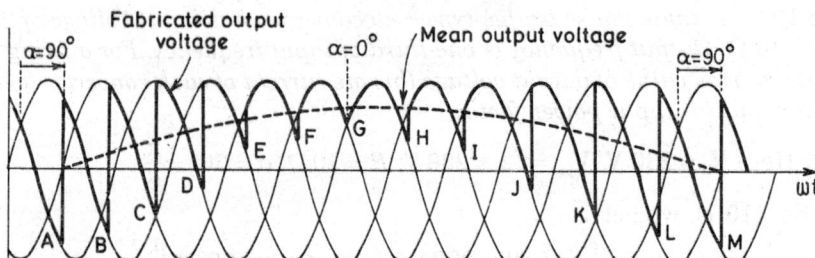

Fig. 10.6. Fabricated and mean output voltage waveforms for a single-phase cycloconverter.

seen from Fig. 10.6 that fabricated output voltage given by thick curve can be resolved into fundamental frequency output voltage plus several other harmonic components. The load inductance can, however, filter out the high-frequency unwanted harmonics. Fig. 10.6 reveals that for one half-cycle of fundamental frequency output voltage (marked mean output voltage in this figure), there are eight half cycles of supply frequency voltage. This shows that output frequency $f_0 = \frac{1}{8} f_s$ where f_s is the supply frequency.

It is obvious from Fig. 10.6 that for obtaining positive half cycle of low-frequency output voltage, firing angle is varied from 90° to zero degree and then to 90°. For obtaining one cycle (consisting of one positive half cycle and one negative half cycle) of low frequency output voltage, the firing angle should be varied from 90° to zero degree to 90° for positive half cycle and from 90° to 180° and back to 90° for negative half cycle. This is illustrated in Fig. 10.7.

A careful inspection of Fig. 10.6 reveals that the magnitude of progressive change in firing angle is given by

$$\text{(reduction factor in frequency)} \times 120° \qquad \qquad ...(10.2)$$

For example, in Fig. 10.6, the reduction factor, given by (output frequency/input frequency) is 1/8.

$\therefore$ Progressive step variation in firing angle $= \frac{1}{8} \times 120° = 15°$

Thus, in Fig. 10.6, $\alpha = 90°$ at A, $\alpha = 90 - 15 = 75°$ at β, $\alpha = 60°$ at C, $\alpha = 45°$ at D, ..., $\alpha = 0°$ at G, $\alpha = 15°$ at H and so on till $\alpha = 90°$ at M. From A to M, there is one half cycle of low frequency output voltage and eight half cycle of supply frequency, indicating $f_o = \frac{1}{8} f_s$ as stated before.

It is thus seen from above that a complete cycle of low-frequency output voltage can be fabricated from the segments of 3-phase input voltage waveform by the use of phase-controlled converters. The cycloconverter can be made to deliver any pf load. In Fig. 10.7, the device is shown to deliver a lagging pf load. In a thyristor converter circuit, current can only flow in one direction. For allowing the flow of current in both the directions during one complete cycle of

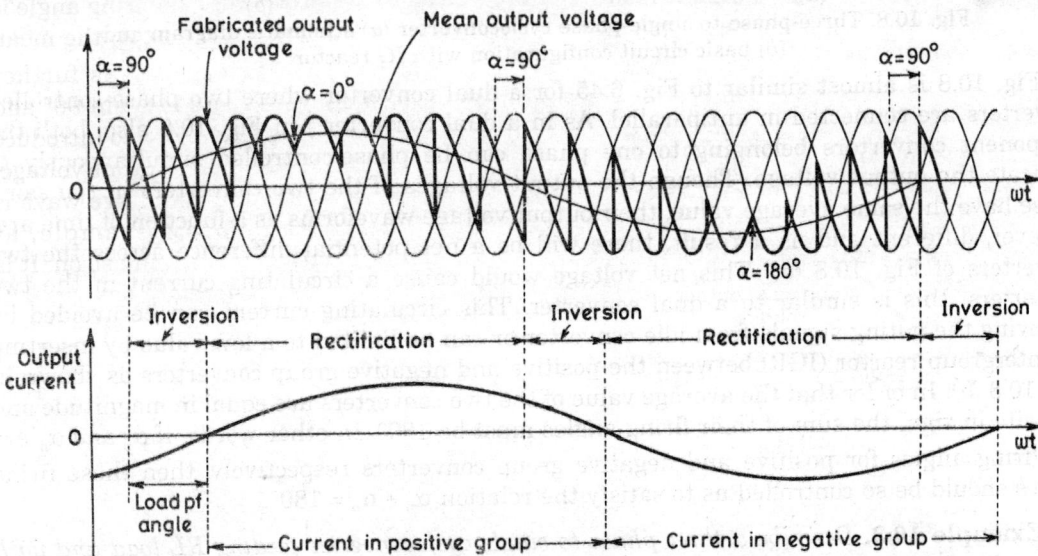

Fig. 10.7. Voltage and current waveforms for a 3-phase half-wave cycloconverter.

load curren*, two three-phase half-wave converters must be connected in antiparallel as shown in Fig. 10.8. The converter circuit that permits the flow of current during positive half cycle of low-frequency output current is called *positive converter group*. The other group permitting the flow of current during the negative half cycle of output current is called *negative converter group*. For a three-phase to single phase cycloconverter, schematic diagram is shown in Fig. 10.8 (a) and basic circuit configuration in Fig. 10.8 (b). This figure uses two 3-phase half wave converters in anti-parallel, the positive group for the conduction of positive load current and the negative group for the flow of negative load current.

Examination of Fig. 10.7 reveals that when output current is positive (above the reference line ωt), positive converter conducts. Under this condition, positive converter acts as a rectifier when output voltage is positive and as an inverter when output voltage is negative. When output current is negative, the negative converter conducts ; under this condition, negative converter acts as a rectifier when output voltage is negative and as an inverter when output voltage is positive. It can thus be inferred, in general, that one of two component converters in Fig. 10.8 would operate as rectifier if the output voltage and current have the same polarity and as an inverter if these are of opposite polarity.

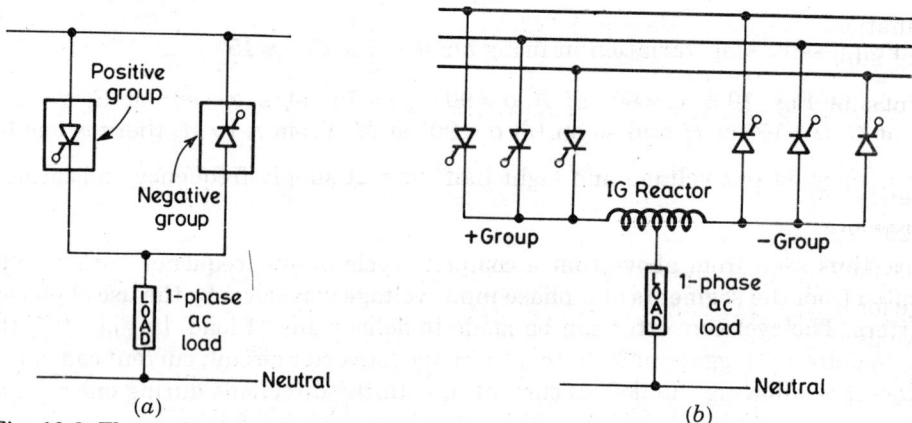

Fig. 10.8. Three-phase to single-phase cycloconverter (a) schematic diagram and (b) basic circuit configuration with *IG* reactor.

Fig. 10.8 is almost similar to Fig. 6.45 for a dual converter where two phase-controlled converters are connected in antiparallel. As in a dual converter ; in Fig. 10.8 also, both the component converters belonging to one phase can be phase-controlled simultaneously to fabricate the output voltage. Though the output voltages of the two converters in the same phase have the same average value, their output voltage waveforms as a function of time are, however, different and as a result, there will be a net potential difference across the two converters of Fig. 10.8 (a). This net voltage would cause a circulating current in the two converters, this is similar to a dual converter. This circulating current can be avoided by removing the gating signals from idle converter or can be limited to a low value by inserting an intergroup reactor (IGR) between the positive and negative group converters as shown in Fig. 10.8 (b). In order that the average value of the two converters are equal in magnitude and opposite in sign, the sum of their firing angles must be 180°. In other words, if α_p and α_n are the firing angles for positive and negative group converters respectively, then these firing angles should be so controlled as to satisfy the relation $\alpha_p + \alpha_n = 180°$.

Example 10.3. *Describe a three-phase to single cycloconverter feeding RL load and with output frequency one-fourth of input frequency. Illustrate your answer by showing one cycle of low-frequency output voltage. Indicate clearly the triggering of various thyristors.*

Solution. It is seen from Eq. (10.2) that progressive variation in triggering angle required is given by $\frac{1}{4} \times 120 = 30°$.

Power circuit diagram for a 3-phase three-pulse converter is given in Fig. 10.9 (a). Firing angle are α_1 for thyristor T_1, α_2 for T_2 and α_3 for T_3.

In Fig. 10.9 (a), thyristor T_1 is triggered at $\alpha_1 = 90°$ at instant x as shown in Fig. 10.9 (b), where 3-phase voltage waveforms v_a, v_b, v_c are drawn. Now T_2 is triggered at $\alpha_2 = 90 - 30 = 60°$, T_3 at $\alpha_3 = 60 - 30°$, $T1$ at $\alpha_1 = 0°$, T_2 at $\alpha_2 = 30°$ and so on till $\alpha_1 = 90°$ for T_1 is obtained at instant y. Mean output voltage of low frequency is sketched from instant $x\,(\alpha_1 = 90°)$ to instant $y\,(\alpha_1 = 90°)$. Examination of waveforms indicate that for one half cycle of supply output voltage, there are four half cycles of supply frequency, i.e. $f_o = \frac{1}{4} f_s$ is obtained.

For sketching negative half cycle of low frequency output voltage, T_2 is triggered at $\alpha_2 = 120°$, T_3 at $\alpha_3 = 150°$, T_1 at $\alpha_1 = 180°$, till $\alpha_1 = 90°$ for $T1$ is obtained at instant z. Mean output voltage for low frequency is again sketched. Fig. 10.9 (b) reveals that there are four half cycles of supply voltage for one half cycle of low frequency negative voltage. It is thus seen that output frequency

$$f_o = \frac{1}{4} \times \text{supply frequency } f_s.$$

10.2.2. Three-phase to Three-phase Cycloconverters

When 3-phase low-frequency output is required, then three sets of phase-controlled 3-phase to single-phase circuits are inter connected as shown in Fig. 10.10 (a). Each phase of the 3-phase output must have a phase displacement of 120°. Fig. 10.10 (b) shows the circuit arrangement of 3-phase to 3-phase cycloconverter using three sets of three-phase half-wave circuits employing a total of 18 thyristors. The device of Fig. 10.10 is also called three-pulse 3-phase to 3-phase cycloconverter.

Three-phase bridge circuit. Out of the several configurations of 3-phase to 3-phase cycloconverters ; here only one important scheme, used for large industrial drives, is presented.

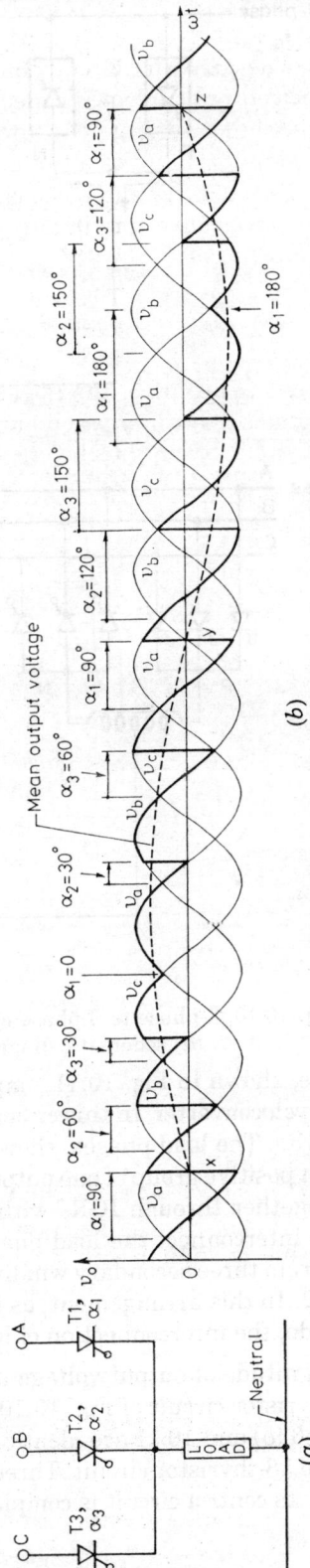

Fig. 10.9. (a). Three-phase three-pulse converter and (b) voltage waveform for cycloconverter. Example 10.3.

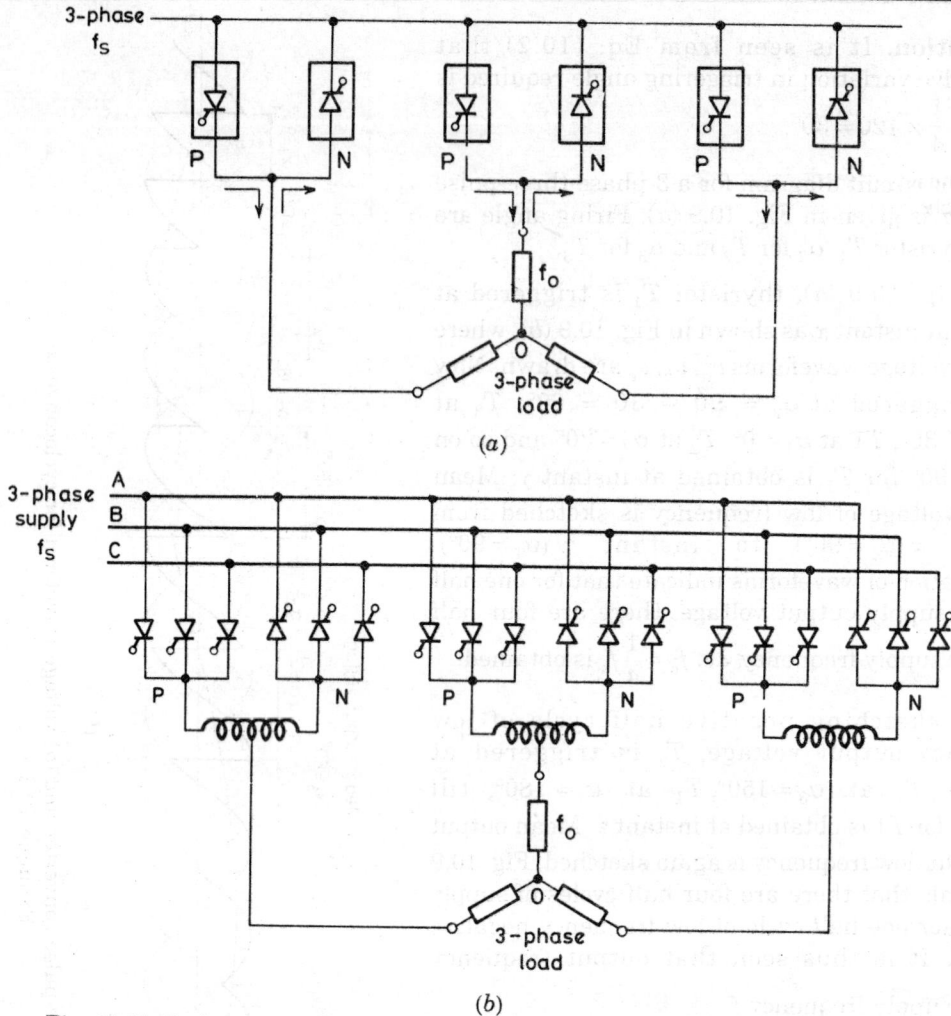

Fig. 10.10. 3-phase to 3-phase cycloconverter employing 3-phase half-wave circuits
(a) schematic diagram and (b) basic circuit arrangement.

This scheme, shown in Fig. 10.11, employs thirty-six thyristors and is called 6-pulse, 3-phase to 3-phase cycloconverter. In this circuit, each phase group consists of a 3-phase dual-converter with two IGRs. The load phases, shown in star in Fig. 10.11, must not be interconnected. If it is done, then positive group of one output phase and negative group of other output phase would be joined together through IGRs without load impedance which is undesirable. In case it is essential to interconnect the load phases in star or delta, then each phase group is supplied separately from three secondary windings S1, S2 and S3 of a three-phase transformer as shown in Fig. 10.12. In this arrangement, as individual phase groups are isolated from each other on the input side, the interconnection of load phases in star or delta is permissible.

The magnitude of output voltage in a 3-phase bridge circuit of Fig. 10.11 is double of that in the 18-thyristor circuit of Fig. 10.10 (b). In case voltage and current ratings of all the SCRs in Figs. 10.10 (b) and 10.11 are identical, then total VA rating of bridge circuit would be double of that of the 18-thyristor circuit. Three-phase bridge circuit gives a smooth variation of output voltage, but its control circuit is complex and expensive.

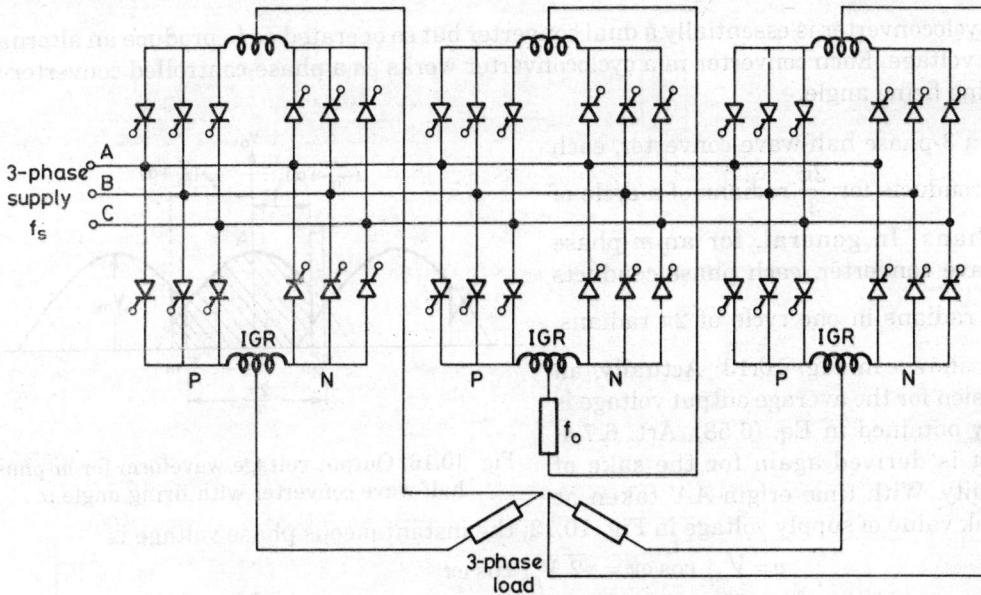

Fig. 10.11. Six-pulse, 3-phase to 3-phase cycloconverter using 36 thyristors
and with isolated load phases.

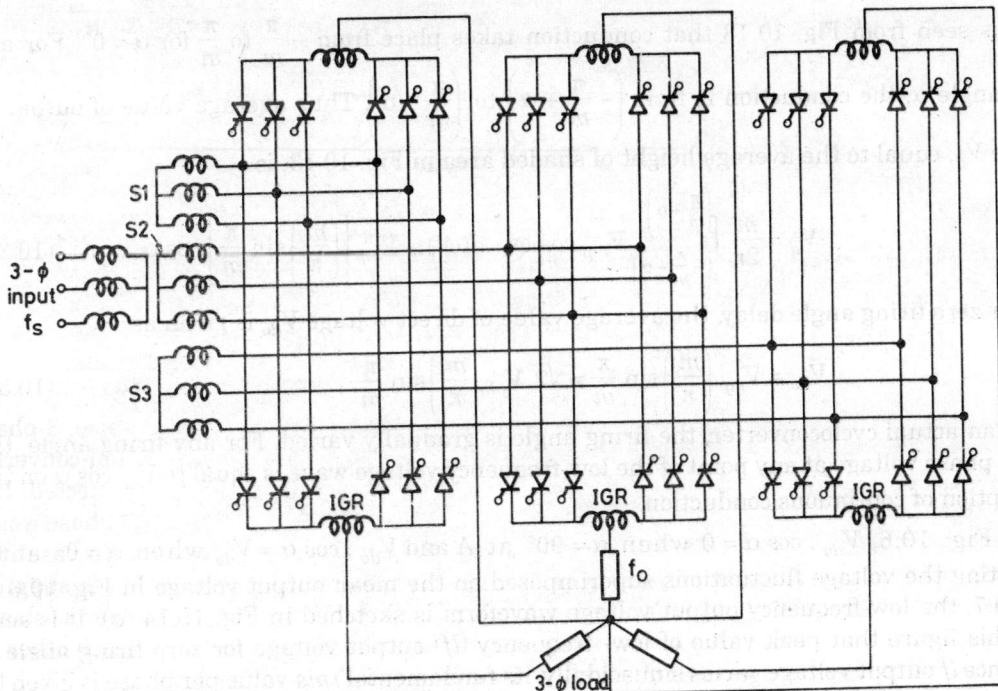

Fig. 10.12. Three-phase bridge cycloconverter using 36-thyristors
and with non-isolated load phases.

10.3. OUTPUT VOLTAGE EQUATION FOR A CYCLOCONVERTER

In this section, emf expressions for the line-commutated phase-controlled cycloconverters
are discussed.

A cycloconverter is essentially a dual converter but so operated as to produce an alternating output voltage. Each converter in a cycloconverter works as a phase-controlled converter with a varying firing angle.

In a 3-phase half-wave converter, each phase conducts for $\dfrac{2\pi}{3}$ radians of a cycle of 2π radians. In general, for an m-phase half-wave converter, each phase conducts for $\dfrac{2\pi}{m}$ radians in one cycle of 2π radians, this is shown in Fig. 10.13. Actually, an expression for the average output voltage is already obtained in Eq. (6.58), Art. 6.7.4. Here it is derived again for the sake of continuity. With time origin AA' taken at

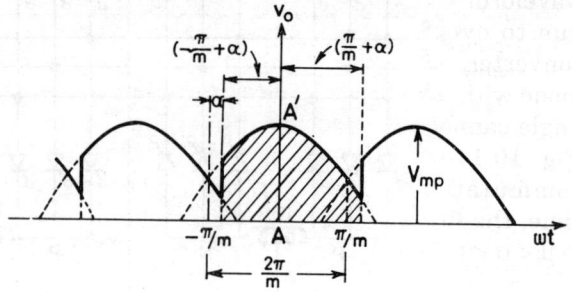

Fig. 10.13. Output voltage waveform for m-phase half-wave converter with firing angle α.

the peak value of supply voltage in Fig. 10.13, the instantaneous phase voltage is

$$v = V_{mp} \cos \omega t = \sqrt{2}\, V_{ph} \cos \omega t$$

where $\qquad V_{ph} = \dfrac{V_{mp}}{\sqrt{2}}$ = rms value of per-phase supply voltage.

It is seen from Fig. 10.13 that conduction takes place from $-\dfrac{\pi}{m}$ to $\dfrac{\pi}{m}$ for $\alpha = 0°$. For any firing angle α, the conduction is from $\left(-\dfrac{\pi}{m}+\alpha\right)$ to $\left(\dfrac{\pi}{m}+\alpha\right)$. Thus average value of output dc voltage V_d, equal to the average height of shaded area in Fig. 10.13, is

$$V_d = \frac{m}{2\pi} \int_{\left(-\frac{\pi}{m}+\alpha\right)}^{\left(\frac{\pi}{m}+\alpha\right)} V_{mp} \cos \omega t \cdot d(\omega t) = V_{mp}\left[\left(\frac{m}{\pi}\right) \sin \frac{\pi}{m}\right] \cos \alpha \qquad \ldots(10.3a)$$

For zero firing angle delay, the average value of direct voltage V_{do} is given as

$$V_{do} = V_{mp}\left(\frac{m}{\pi}\right) \sin \frac{\pi}{m} = \sqrt{2}\, V_{ph}\left(\frac{m}{\pi}\right) \sin \frac{\pi}{m} \qquad \ldots(10.3b)$$

In an actual cycloconverter, the firing angle is gradually varied. For any firing angle, the output phase voltage at any point of the low-frequency voltage wave is equal to $V_{do} \cos \alpha$ on the assumption of continuous conduction.

In Fig. 10.6, $V_{do} \cdot \cos \alpha = 0$ when $\alpha = 90°$ at A and $V_{do} \cdot \cos \alpha = V_{do}$ when $\alpha = 0°$ at G. Neglecting the voltage fluctuations superimposed on the mean output voltage in Fig. 10.6 or Fig. 10.7, the low-frequency output voltage waveform is sketched in Fig. 10.14 (a). It is seen from this figure that peak value of low- frequency (lf) output voltage for zero firing angle is V_{do}. Since lf output voltage varies sinusoidally, its fundamental rms value per phase is given by

$$V_{or} = \frac{V_{do}}{\sqrt{2}} = \frac{V_{mp}}{\sqrt{2}} \cdot \frac{m}{\pi} \cdot \sin \frac{\pi}{m}$$

or $\qquad V_{or} = V_{ph} \cdot \left(\frac{m}{\pi}\right) \cdot \sin\left(\frac{\pi}{m}\right) \qquad\qquad \ldots(10.4)$

The effect of source inductance leads to commutation overlap as discussed in Art. 6.9.1. If firing angle is zero, the output voltage waveform would be as shown in Fig. 10.14 (b) due to overlap angle in a single-phase full converter. At the same time, in the inversion mode with $\alpha > 90°$, the maximum value of firing angle cannot be more than $180 - \mu$ as shown in Fig. 10.14 (c). This shows that on account of commutation overlap and thyristor turn-off time, the firing angle range in a cycloconverter is $\mu < \alpha < (180 - \mu)$.

This implies that, in practice, the firing angle α_p of the positive converter cannot be zero, see Fig. 10.14 (b) and at the same time, maximum value of firing angle α_n of negative

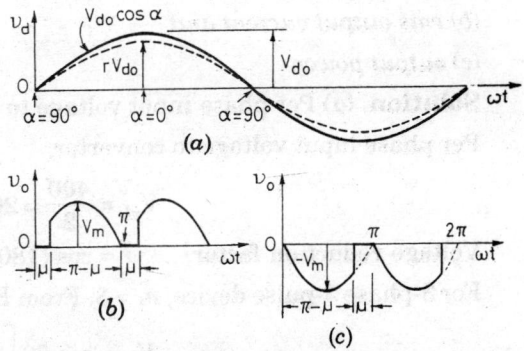

Fig. 10.14 (a) Low-frequency output voltage wave form of a cycloconverter. Single-phase full-converter output voltage waveform when overlap angle is μ and (a) $\alpha = 0$, rectifying mode and (b) $\alpha = 180 - \mu$, inverting mode.

converter cannot exceed $180 - \mu$, see fig. 10.14 (c). This shows that positive converter must have some minimum value of firing angle, let this value be $\alpha_p = \alpha_{min}$. For this firing angle, amplitude of lf output voltage per phase is

$$V_{d.\,mx} = V_{do}.\cos \alpha_{mn} = r.\,V_{do} \qquad \qquad ...(10.5)$$

where $r = \cos \alpha_{mn}$ and is called *voltage reduction factor* (VRF). The amplitude $r.\,V_{do}$ is shown in Fig. 10.14 (a).

Now the expression for the fundamental *rms* phase value of the *lf* output voltage of a cycloconverter, from Eq. (10.5), is given by

$$V_{or} = \frac{V_{d.mx}}{\sqrt{2}} = \frac{r.\,V_{do}}{\sqrt{2}} = r.\frac{V_{mp}}{\sqrt{2}}\frac{m}{\pi}.\sin\frac{\pi}{m}$$

$$= r\left[V_{ph}.\left(\frac{m}{\pi}\right)\sin\left(\frac{\pi}{m}\right)\right] \qquad \qquad ...(10.6)$$

As α_{mn} is always greater than zero, the voltage reduction factor, r, is always less than unity.

Eq. (10.6) gives the *rms* value of the per-phase output voltage for a 3-phase to 3-phase or 3-phase to single-phase cycloconverter employing m-phase half-wave circuits shown in Fig. 10.8 or 10.10. Note that these converter circuits consist of two converters, one 3-pulse positive group converter and the other 3-pulse negative group converter.

Eq. (10.6) is also applicable for 3-phase to 3-phase or 3-phase to single-phase cycloconverter employing 6-pulse bridge converter of Fig. 10.11 or 10.12, but then m is equal to the number of pulses, *i.e.* $m = 6$. Since bridge converters of Fig. 10.11 or 10.12 employ 3-phase full converters, V_{ph} in Eqs. (10.4) and (10.6) must be replaced by line to line voltage V_l as done in 3-phase full converters.

Example 10.4. *A 3-phase to single-phase cycloconverter employs 3-pulse positive and negative group converters. Each converter is supplied from delta/star transformer with per phase turns ratio of 2 : 1. The supply voltage is 400 V, 50 Hz. The RL load has R = 2 Ω and at low output frequency, $\omega_0 L = 1.5$ Ω. In order to account for commutation overlap and thyristor turn-off time, the firing angle in the inversion mode should not exceed 160°. Compute*

(a) the value of the fundamental rms output voltage.

(b) rms output current and

(c) output power.

Solution. (a) Per phase input voltage to transformer = 400 V.

Per phase input voltage to converter,

$$V_{ph} = \frac{400}{2} = 200 \text{ V}$$

Voltage reduction factor, $r = \cos(180 - 160) = \cos 20°$

For 3-phase 3-pulse device, $m = 3$. From Eq. (10.6), the rms value of fundamental voltage is

$$V_{or} = \cos 20 \left[200 \left(\frac{3}{\pi} \right) \cdot \sin \frac{\pi}{3} \right] = 155.424 \text{ V}$$

(b) Rms output current

$$= \frac{155.424}{\sqrt{2^2 + 1.5^2}} \left| - \tan^{-1} \frac{1.5}{20} \right.$$

$$I_{or} = 62.17 \left| - 36.87° \right. \text{ Amps.}$$

(c) Output power $= I_{or}^2 \cdot R = (62.17)^2 \times 2 = 7730.22 \text{ W.}$

Example 10.5. *Repeat Example 10.4 in case 3-phase to 1-phase cycloconverter employs 6-pulse bridge converter.*

Solution. (a) Per phase input voltage to converter = 200 V

Line voltage input to bridge converter = $200 \sqrt{3}$ V

Voltage reduction factor, $r = \cos 20°$

For 3-phase, 6-pulse device, $m = 6$. From Eq. (10.6), the rms value of output voltage is

$$V_{or} = \cos 20 \left[200 \sqrt{3} \left(\frac{6}{\pi} \right) \sin \frac{\pi}{6} \right] = 310.84 \text{ V}$$

This example demonstrates that output voltage in a 6-pulse bridge converter employing 36 thyristors is double of that in a 3-pulse half-wave converter using 18 thyristors.

(b) Rms output current

$$= \frac{310.84}{\sqrt{2^2 + 1.5^2}} \left| - \tan^{-1} \frac{1.5}{2} \right.$$

$$= 124.34 \left| - 36.87° \right. \text{ Amps}$$

(c) Rms output power $= (124.34)^2 \times 2 = 30920.88 \text{ W.}$

This example shows that output power handled by a 6-pulse bridge converter is four times the power handled by a 3-pulse converter.

Example 10.6. *In a single-phase cycloconverter arrangement for changing the frequency of the supply voltage to a load, as shown in Fig. 10.15 (a), the load terminal L should be connected to (a) A (b) B (c) C (d) D or E.*

Solution. A careful examination of Fig. 10.15 (a) reveals that this figure is the same as that of 1-phase to 1-phase mid-point cycloconverter of Fig. 10.1 (a). The thyristor are marked P1, P2, N1, N2 in Fig. 10.1 (a) and on the same pattern, these are marked as P1, P2, N1, N2 in Fig. 10.15 (b). In Fig. 10.1 (a), load terminal A is connected directly to P1, P2, N1, N2. But, in Fig. 10.15 (b), load terminal L must be connected to mid-point B of the intergroup reactor. Therefore, here the answer is (b).

In Fig. 10.15 (a), if IG reactor is not used, then load terminal L may be connected to A, B or C as shown in Fig. 10.15 (c). Fig. 10.15 (c) is exactly identical to Fig. 10.1 (a).

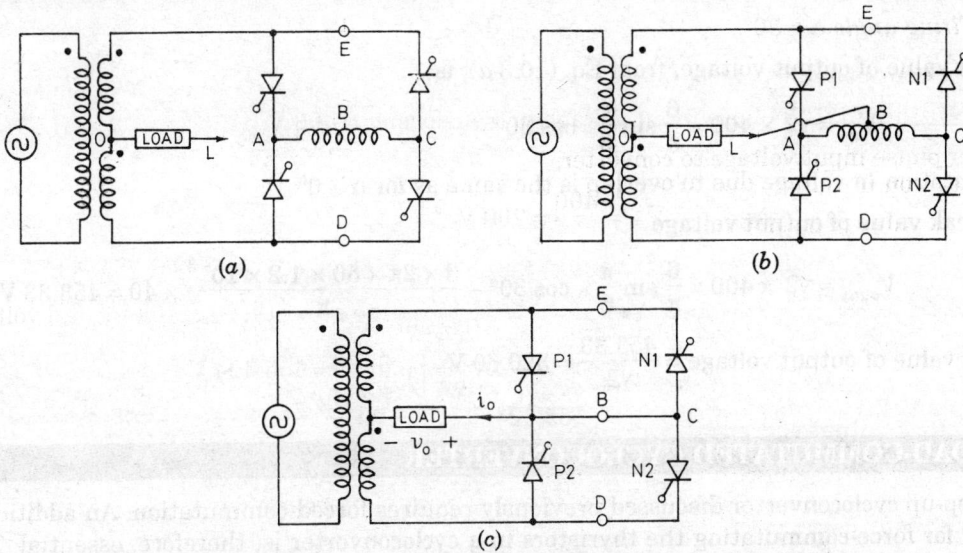

Fig. 10.15 (a) Pertaining to Example 10.6 (b) Load terminal L connected to B
(c) 1-phase to 1-phase mid-point cycloconverter redrawn.

Basic principle of working of 1-phase to 1-phase mid-point cycloconverter can also be explained with the help of Fig. 10.15 (b) if reactor is used and with Fig. 10.15 (c) if reactor is not used.

Example 10.7. *A six-pulse cycloconverter, fed from 3-phase, 400 V, 50 Hz source, is delivering a load current of 40 A to a 1- phase resistive load. The source has an inductance of 1.2 mH per phase. Calculate the rms value of load voltage for firing angle delays of (a) 0° and (b) 30°.*

Solution. The object here is to first calculate V_{do} of Fig. 10.14 (a) and r. V_{do} and then its rms value.

(a) *Firing angle* $\alpha = 0°$.

Peak value of output voltage, for a 3-pulse converter, from Eq. (10.3 b), is

$$V_{do} = V_{mp} \left(\frac{m}{\pi} \right) \sin \left(\frac{\pi}{m} \right)$$

For a 6-pulse converter, V_{mp} must be replaced by $V_{ml} = \sqrt{2} \cdot V_l$.

$$\therefore \qquad V_{do} = \sqrt{2} \times 400 \times \frac{6}{\pi} \sin \frac{\pi}{6}$$

Reduction in voltage due to source inductance, from Eq. (6.76), for a 6-pulse converter is

$$\frac{3 \, \omega L_s}{\pi} I_o$$

∴ Peak value of output voltage,

$$\dot{V}_{o.mx} = \sqrt{2} \times 400 \times \frac{6}{\pi} \sin 30° - \frac{3 \times 2\pi \times 50 \times 1.2 \times 10^{-3}}{\pi} \times 40 = 525.71 \text{ V}$$

∴ Rms value of load voltage $= \dfrac{525.71}{\sqrt{2}} = 371.79$ V   ●

(b) *Firing angle* $\alpha = 30°$.

Peak value of output voltage, from Eq. (10.3 a), is

$$V_{do} = \sqrt{2} \times 400 \times \frac{6}{\pi} \sin\frac{\pi}{6} \cdot \cos 30°$$

Reducation in voltage due to overlap is the same as for $\alpha = 0°$

∴ Peak value of output voltage

$$V_{o.mx} = \sqrt{2} \times 400 \times \frac{6}{\pi} \sin\frac{\pi}{6} \times \cos 30° - \frac{3 \times 2\pi \times 50 \times 1.2 \times 10^{-3}}{\pi} \times 40 = 453.33 \text{ V}$$

Rms value of output voltage $= \dfrac{453.33}{\sqrt{2}} = 320.60$ V

10.4. LOAD-COMMUTATED CYCLOCONVERTER

A step-up cycloconverter discussed previously requires forced-commutation. An additional circuitry for force-commutating the thyristors in a cycloconverter is, therefore, essential. The process of forced commutation, however, does not depend upon the source or load voltage.

In the step-down cycloconverter discussed in the previous sections, the phase-controlled cycloconverter rely on natural commutation for their operation. The natural, or line, commutation is provided by the supply voltage. In these cycloconverters, the output frequency f_0 is less than the input supply frequency f_s.

A load-commutated cycloconverter differs from the force-commutated and line-commutated cycloconverters discussed so far. In load-commutated circuit, the thyristors are commutated by the reversal of the load voltage. This implies that the load circuit must have a generated emf that should be independent of the source voltage. The most usual example for such a load is wound-field or permanent-magnet synchronous machine. For such loads, the load frequency may be equal to, or greater than, the source frequency and for both these cases, thyristors will be naturally commutated by the reversal of the load circuit emf.

PROBLEMS

10.1. (a) What is a cycloconverter ? Enumerate some of its industrial applications.

(b) Describe the operating principle of single-phase to single-phase step-up cycloconverter with the help of mid-point and bridge-type configurations. Illustrate your answer with appropriate circuit and waveforms. The conduction of various thyristors must also be indicated on the waveforms.

10.2. Describe the basic principle of working of single-phase to single-phase step-down cycloconverter for both continuous and discontinuous conductions for a bridge type cycloconverter. Mark the conduction of various thyristors also.

10.3. A single-phase to single-phase mid-point cycloconverter is delivering power to a resistive load. The supply transformer has turns ratio of 1 : 1 : 1. The frequency ratio is $f_o/f_s = 1/5$. The firing angle delay for all the four SCRs are the same. Sketch the time variations of the following waveforms for $\alpha = 0°$ and $\alpha = 30°$.

(a) Supply voltage (b) Output current and (c) Supply current. Indicate the conduction of various thyristors also.

10.4. A single-phase to single-phase bridge-type cycloconverter is fed from 230 V, 50 Hz source and has a load $R = 20$ Ω. Output frequency is one-fifth of input frequency. For a firing angle delay of 45°, calculate (a) rms value of output voltage (b) rms current for each converter (c) rms current of each thyristor and (d) input power factor.

Derive the expression for rms output voltage for this cycloconverter.

     [**Ans.** (a) 219.3 V (b) 7.775 A (c) 5.483 A (d) 0.9535 lag]

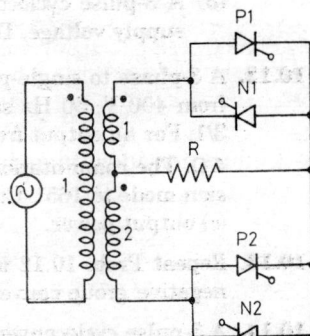

10.5. A single-phase to single-phase cycloconverter of Fig. 10.16 is used for obtaining an output frequency of $\frac{1}{3}$ times the input frequency. Turns ratio from primary to upper secondary is 1/1 and to lower secondary is 1/2. The thyristors are so triggered as to obtain a symmetrical output voltage waveform. Sketch the output voltage and output current waveforms for a resistive load for firing angle $\alpha = 0°$ and $\alpha = 30°$. Indicate the conduction of various thyristors.

Fig. 10.16. Pertaining to Example 10.5.

10.6. For the circuit of Fig. 10.16, source voltage is $V_m \sin \omega t$. Derive expression for the *rms* value of output voltage (a) for firing angle $\alpha = 0$ and (b) for firing angle α.

$$\left[\textbf{Hint: } (a) \; V_{or} = \left[\frac{1}{3\pi} \int_0^\pi 6V_m^2 \sin^2 \omega t \cdot d\,(\omega t) \right]^{1/2} \text{ etc.} \right.$$

$$\left. \textbf{Ans.} \; (a) \; V_m \qquad (b) \; V_m \left[\frac{1}{\pi} \left(\pi - \alpha + \frac{\sin 2\alpha}{2} \right) \right]^{1/2} \right]$$

10.7. In Prob. 10.6, source voltage is 230 V, 50 Hz and load is $R = 20\ \Omega$. Find the power delivered to load for (a) $\alpha = 0°$ and (b) $\alpha = 30°$.   [**Ans.** (a) 5290 W (b) 5136.01 W]

10.8. Cycloconverter circuit of Fig. 10.16 is used for obtaining on output frequency $f_o = \frac{1}{3} \times$ source frequency f_s. Turns ratio from primary to upper secondary is 1/1 and to lower secondary is 1/n. Derive expression for the rms value of output voltage for firing angle α.

For. $V_s = 230$ V, 50 Hz; $n = 3$; $R = 20\ \Omega$ and $\alpha - 45°$, find the load power.

$$\left[\textbf{Ans.} \; V_m \left[\frac{n^2 + 2}{6\pi} \left(\pi - \alpha + \frac{\sin 2\alpha}{2} \right) \right]^{1/2}, \; 8814.625 \text{ W} \right]$$

10.9. Describe how single-phase low-frequency output voltage can be fabricated from the segments of 3-phase input-voltage waveform through the use of a 3-phase half-wave circuit ? Show a complete cycle of low-frequency outputs voltage. In case load current lags the low-frequency output voltage, discuss the operation of positive and negative group phase-controlled converters.

10.10. (a) Discuss why 3-phase to 1-phase cycloconverter requires positive and negative group phase-controlled converters. Under what conditions, the groups work as inverters or rectifiers ? How should the firing angles of the two converters be controlled ?

     (b) Describe 3-phase to 3-phase cycloconverter with relevant circuit arrangements using 18 SCRs and 36 SCRs.

What are the advantages of 3-phase bridge circuit cycloconverter over 18-thyristor device ?

10.11. (a) Show that the fundamental rms value of per-phase output voltage of low-frequency for an m-pulse cycloconverter is given by

$$V_{or} = V_{ph} \left(\frac{m}{\pi} \right) \sin \left(\frac{\pi}{m} \right)$$

Hence express V_{or} in terms of voltage reduction factor.

(b) A 3-pulse cycloconverter feeds a single-phase load at 200 V. Estimate the value of the supply voltage. Derive the formula used.   [**Ans.** (b) V_{ph} = 241.85 V]

10.12. A 3-phase to single-phase cycloconverter employs a 6-pulse bridge circuit. This device is fed from 400 V, 50 Hz supply through a delta/star transformer whose per-phase turns ratio is 3/1. For an output frequency of 2 Hz, the load resistance is $\omega_0 L = 3 \Omega$. The load resistance is 4 Ω. The commutation overlap and thyristor turn-off time limit the firing angle in the inversion mode to 165°. Compute (a) peak value of rms output voltage (b) rms output current and (c) output power.   [**Ans.** 301.215 V (b) 42.605 ∠ – 36.87° A (c) 7260.74 W]

10.13. Repeat Prob. 10.12 in case 3-phase to 1-phase cycloconverter employs 3-pulse positive and negative group converters.   [**Ans.** (a) 150.602 V (b) 21.301 ∠ – 36.87° (c) 1815 W]

10.14. A 3-pulse cycloconverter, fed from 3-phase, 400 V, 50 Hz supply is delivering a load current of 30 A to a 1-phase resistive load. The supply has an inductance of 1 mH per phase. Calculate the rms value of load voltage for a firing angle of (a) 0° and (b) 45°.

[**Ans.** (a) 187.83 V (b) 152.76 V]

Chapter 11
Some Applications

Several applications of power electronics are listed in Art. 1.2. Study of all these applications will be a voluminous task. Even then, some of these applications described in this chapter will be of interest to the reader.

11.1. SWITCHED MODE POWER SUPPLY (SMPS)

With advances in electronics, need for dc power supplies for use in integrated circuits (ICs) and digital circuits has increased manifold. For such electronic circuits, NASA was the first to develop a light-weight and compact switched mode power supply in the 1960s for use in its space vehicles. Subsequently, this power supply became popular and presently, annual production of SMPSs may be as high as 70 to 80% of the total number of power supplies produced.

At this juncture, a question may arise that controlled dc supply can also be obtained from phase-controlled rectifiers. Then why go in for SMPS ? An ac to dc rectifier operates at supply frequency of 50 (or 60) Hz. In order to obtain almost negligible ripple in the dc output voltage, physical size of filter circuits required is quite large. This makes the dc power supply inefficient, bulky and weighty. On the other hand, SMPS works like a dc chopper. By operating the on/off switch very rapidly, ac ripple frequency rises which can be easily filtered by L and C filter circuits which are small in size and less weighty. It may therefore be inferred that it is the requirement of small physical size and weight that has led to the wide spread use of SMPSs.

As stated above, SMPS is based on the chopper principle. The output dc voltage is controlled by varying the duty cycle of chopper by PWM or FM techniques. The circuit configurations used for SMPS can be classified into four broad categories; namely flyback, pushpull, half bridge and full-bridge.

In SMPS circuits discussed here, PWM technique is used for the inverter. The output of the inverter is then converted to dc by a diode rectifier. As the inverter is made to operate at very high frequency, the ripples on the dc output voltage can be filtered out easily by using small filter components. If the switching devices are power transistors, the chopping frequency is limited to 40 kHz. For power MOSFETs, the chopping frequency is of the order of 200 kHz;

as a result, size of the filter circuit and transformer decreases leading to considerable savings. At such high frequency, ferrite core is used in transformers.

The four categories of SMPSs listed above are now discussed briefly.

11.1.1. Flyback Converter

The circuit configuration for flyback converter is shown in Fig. 11.1. It consists of a power MOSFET M1, transformer for isolation purposes, diode D, capacitor C and load. An uncontrolled rectifier converts ac to dc output which is fed to flyback SMPS as shown in Fig. 11.1.

When power MOSFET is turned on, supply voltage V_s is applied to the transformer primary, i.e. $v_1 = V_s$. A corresponding voltage v_2, with the polarity as shown in Fig. 11.2(a), is induced in the transformer secondary, i.e. $v_2 = \dfrac{V_s}{N_1} N_2$. As v_2 reverse biases diode D, equivalent circuit of Fig. 11.2(a) is obtained. Filter

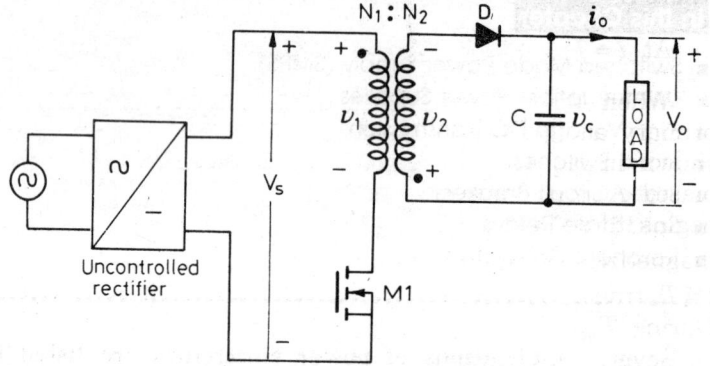

Fig. 11.1. Flyback SMPS.

capacitance C is assumed large enough so that capacitor voltage $v_c(t)$ = load or output voltage V_0 is taken as almost constant. When M1 is turned off, a voltage of opposite polarity is induced in primary and secondary windings as shown in Fig. 11.2(b). Voltage across transformer secondary is $v_2 = -V_0 = -\dfrac{V_s}{N_1} N_2$. Diode D is forward biased and starts conducting a current i_D. As a result, energy stored in the transformer core is delivered partly to load and partly to charge the capacitor C.

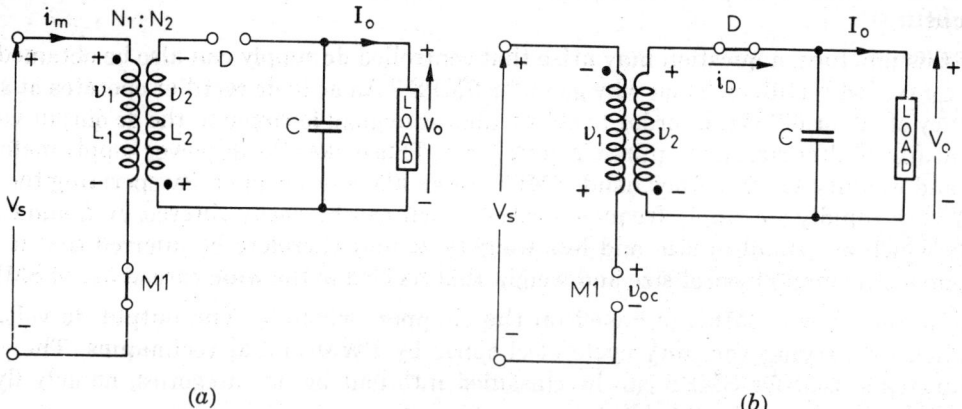

(a)   (b)

Fig. 11.2. Flyback SMPS equivalent circuit during (a) T_{on} and (b) T_{off}.

Waveforms for v_1, v_2, transformer magnetizing current i_m and diode current i_D are shown in Fig. 11.3. During the time M1 is on, $v_1 = V_s$, $v_2 = \dfrac{V_s}{N_1} \cdot N_2$. For magnetizing current, it is assumed that transformer core is not demagnetized completely at the end of periodic time

$T = T_{on} + T_{off}$. In other words, it means that transformer magnetizing current at $t = 0$ is not zero but has some positive value I_{mo}. Therefore, during T_{on}, magnetizing current rises linearly from its initial value I_{mo} to I_{m1} at $t = T_{on}$. With the rise of i_m during T_{on}, magnetic energy gets stored in the transformer core. The variation of i_m as shown in Fig. 11.3 can be expressed as under :

$$i_m(t) = I_{mo} + \frac{V_s}{L} t \quad \ldots\ldots \quad 0 < t < T_{on} \tag{11.1}$$

where L = transformer magnetizing inductance, H

At $t = T_{on}$, $\quad i_m(T_{on}) = I_{m1} = I_{mo} + \frac{V_s}{L} \cdot T_{on}$ ...(11.2)

When M1 is turned off, the emfs induced in primary and secondary windings are reversed as shown in Fig. 11.2(b). Diode D is now forward biased. A current in transformer secondary winding begins to flow through D. As this current i_D or magnetizing current i_m reduces from I_{m1} to I_{mo} at $t = T$, transformer core energy is delivered to load. During T_{off}, M1 is off and $v_2 = -V_0$. This voltage when referred to primary is $v_1 = -\frac{V_0}{N_2} N_1$. The fall of current i_m during T_{off} can be expressed as under:

$$i_m(t) = I_{m1} - \frac{V_0}{N_2} N_1 \cdot \frac{1}{L} (t - T_{on}) \quad \ldots\ldots T_{on} < t < T$$

...(11.3)

At $t = T$,

$$i_m(T) = I_{m1} - \frac{V_0}{N_2} \cdot N_1 \cdot \frac{1}{L} (T - T_{on})$$

Substituting the value of I_{m1} from Eq. (11.2) in the above expression, we get

$$i_m(T) = I_{mo} + \frac{V_s}{L} \cdot T_{on} - \frac{V_0}{N_2} \cdot N_1 \cdot \frac{1}{L} (T - T_{on})$$

...(11.4)

Since the net energy stored in core over periodic time T is zero,

$$i_m(O) = i_m(T)$$

or [Eq. (11.1) at $t = 0$] = [Eq. (11.4) at $t = T$]

$$I_{mo} = I_{mo} + \frac{V_s}{L} \cdot T_{on} - \frac{V_0}{N_2} \cdot N_1 \cdot \frac{1}{L} (T - T_{on})$$

or $\quad V_s \cdot T_{on} = \frac{V_0}{a} (T - T_{on})$

∴ Load voltage, $\quad V_0 = \frac{a \cdot V_s \cdot T_{on}}{T - T_{on}} = \frac{a \cdot V_s \cdot k}{1 - k}$ ...(11.5)

where $a = \frac{N_2}{N_1}$, transformer turns ratio from secondary to primary

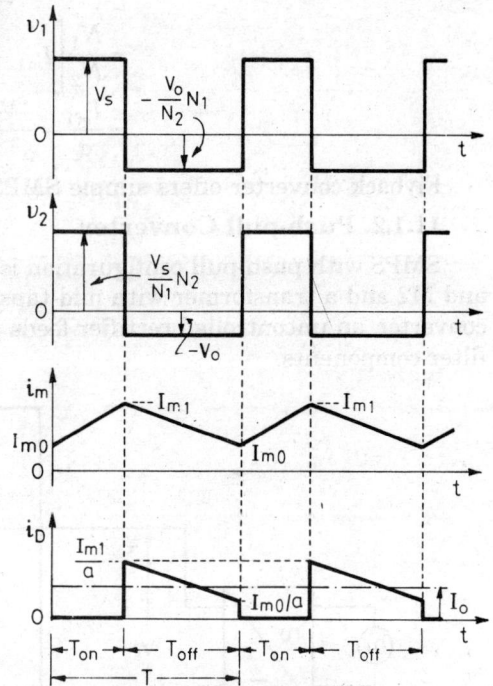

Fig. 11.3. Waveforms for flyback converter SMPS.

and $k = \dfrac{T_{on}}{T}$, duty cycle of flyback converter.

It is seen from Fig. 11.2 (b) that open circuit voltage across M1 is

$$V_{oc} = v_1 + V_s = \frac{V_0}{N_2} \cdot N_1 + V_s = \frac{V_0}{a} + V_s$$

From Eq. (11.5), $V_{oc} = \dfrac{Vs \cdot k}{1-k} + V_s = \dfrac{V_s}{1-k}$ ...(11.6)

Eq. (11.3) gives current on primary side of the transformer. This current, when referred to secondary side, is equal to diode current i_D.

$$\therefore \qquad i_D(t) = i_m(t) \cdot \frac{N_1}{N_2}$$

$$= \frac{N_1}{N_2}\left[I_{m1} - \frac{V_0}{N_2} \cdot N_1 \cdot \frac{1}{L}(t - T_{on}) \right]$$

$$= \frac{I_{m1}}{a} - \frac{V_0}{a^2 \cdot L}(t - T_{on}) \qquad\qquad ...(11.7)$$

Flyback converter offers simple SMPS and is useful for applications below about 500 W.

11.1.2. Push-pull Converter

SMPS with push-pull configuration is shown in Fig. 11.4. It uses two power MOSFETs M1 and M2 and a transformer with mid-taps on both primary and secondary sides. As in flyback converter, an uncontrolled rectifier feeds push-pull SMPS. Inductor L and capacitor C are the filter components.

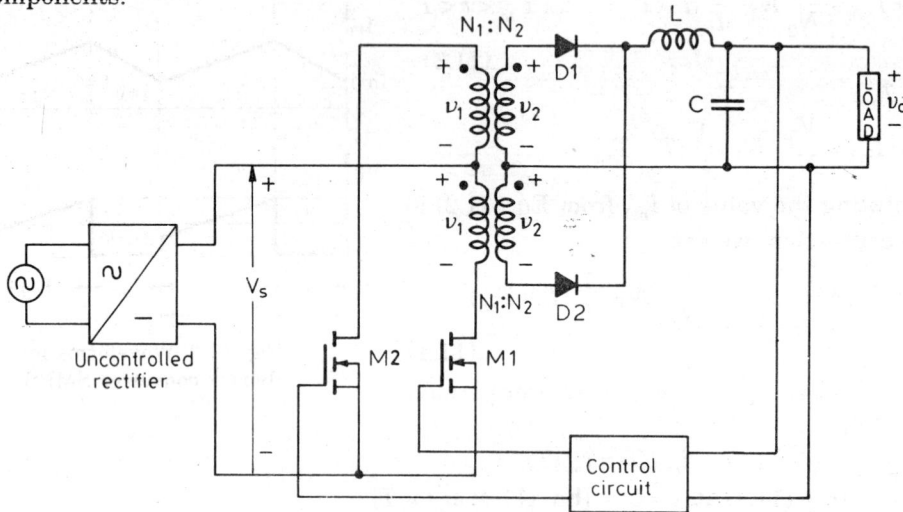

Fig. 11.4. Push-pull SMPS.

When M1 is turned on, V_s is applied to lower half of transformer primary, i.e. $v_1 = V_s$. As a result, voltage $v_2 = \dfrac{V_s}{N_1} N_2$ is induced in both the secondary windings. Voltage v_2 in the upper half secondary forward biases diode D1, therefore load voltage V_0 is given by

$$V_0 = \frac{V_s}{N_1} N_2 = aV_s.$$

When M2 is turned on, $v_1 = -V_s$ is applied to upper half of primary winding. Consequently, $v_2 = -\dfrac{V_s}{N_1} N_2$ is induced in both the transformer secondaries. As v_2 is negative, diode D2 gets forward biased and $V_0 = a\, V_s$ as before.

This shows that voltage on primary swings from $+V_s$ with M1 on to $-V_s$ with M2 on. Power MOSFETs M1 and M2 operate with duty cycle of 0.5. When M1 is off, the voltage across M1 terminals is $V_{oc} = 2V_s$. As both M1 and M2 are subjected to open-circuit voltage of $2V_s$, this configuration is suitable for low-voltage applications only.

11.1.3. Half-bridge Converter

The circuit for half-bridge SMPS configuration is shown in Fig. 11.5. It consists of an uncontrolled rectifier, two capacitors C1 and C2, two power MOSFETs M1 and M2, one transformer with mid-tap on the secondary side, two diodes D1 and D2 and filter components L and C.

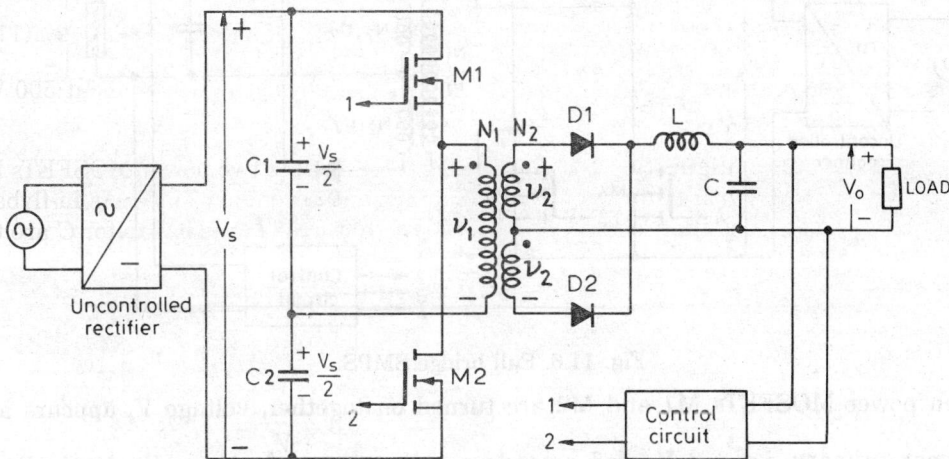

Fig. 11.5. Half-bridge SMPS.

Two capacitors C1 and C2 have equal capacitance, therefore voltage across each of the two is $\dfrac{V_s}{2}$. When M1 is turned on, voltage of C1 appears across transformer primary, i.e. $v_1 = \dfrac{V_s}{2}$ and voltage induced in secondary is $v_2 = \dfrac{V_s}{2N_1} \cdot N_2$ and diode D1 gets forward biased. When M2 is turned on, a reverse voltage of $\dfrac{V_s}{2}$ appears across transformer primary from C2, i.e. $v_1 = -\dfrac{V_s}{2}$ and voltage induced in secondary winding is $v_2 = -\dfrac{V_s}{2N_1} N_2$, therefore diode D2 gets forward biased. This means that transformer primary voltage swings from $-\dfrac{V_s}{2}$ to $+\dfrac{V_s}{2}$. Average output voltage, however, is

$$V_0 = \frac{V_s}{2N_1} \cdot N_2 = 0.5\, a\, V_s$$

When M1 is off, open circuit voltage across M1 terminals is $V_{oc} = V_s$. When M2 is off, as before $V_{oc} = V_s$. For *h.v. dc* applications, half-bridge converter is, therefore, preferred over push-pull converters. For *l.v. dc* applications, push-pull SMPS is preferred due to low MOSFET currents.

11.1.4. Full-bridge Converter

The circuit arrangement for a full-bridge SMPS is shown in Fig. 11.6. It consists of an uncontrolled rectifier, four power MOSFETs, transformer with mid-tap secondary, two diodes and *LC* filter circuit. As in all the previous circuits, the function of control circuit is to sense the output load voltage and to decide about the duty ratio of MOSFETs.

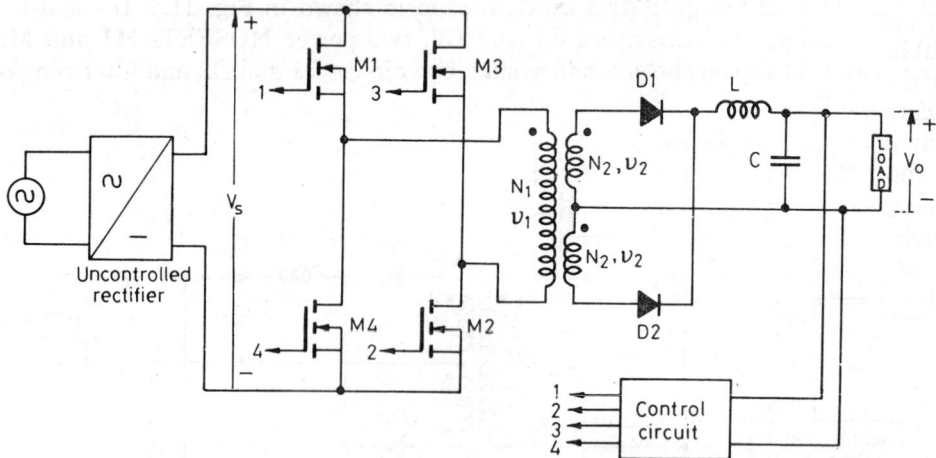

Fig. 11.6. Full bridge SMPS.

When power MOSFETs M1 and M2 are turned on together, voltage V_s appears across transformer primary, *i.e.* $v_1 = V_s$ and secondary voltage $v_2 = \dfrac{V_s}{N_1} \cdot N_2 = a\,V_s$. Diode D1 gets forward biased and $V_0 = a\,V_s$. When M3 and M4 are turned on together, the primary voltage is reversed, *i.e.* $v_1 = -\,V_s$ and $v_2 = -\dfrac{V_s}{N_1} N_2 = -\,a\,V_s$. Therefore, diode D2 now begins to conduct and the output voltage is again $V_0 = a\,V_s$.

The open circuit voltage across each MOSFET is $V_{oc} = V_s$. Of all the four configurations of SMPSs, full-bridge converter operates with minimum voltage and current stress on the power MOSFET. It is therefore very popular for high power applications above 750 W.

The overall size of SMPSs is dependent on its operating frequency. Use of power transistors is limited to approximately 40 to 50 kHz. Above this operating frequency, power MOSFETs are used up to about 200 kHz.

The main advantages of SMPSs over conventional linear power supplies are as under :

   (*i*) For the same power rating, SMPS is of smaller size, lighter in weight and possesses higher efficiency because of its high-frequency operation.

   (*ii*) SMPS is less sensitive to input voltage variations.

The disadvantages of SMPS are as under :

(i) SMPS has higher output ripple and its regulation is worse

(ii) SMPS is a source of both electromagnetic and radio interference due to high frequency switching.

(iii) Control of radio frequency noise requires the use of filters on both input and output of SMPS.

The advantages possessed by SMPSs far outweigh their shortcomings. This is the reason for their wide-spread popularity and growth.

11.2. UNINTERRUPTIBLE POWER SUPPLIES

There are several applications where even a temporary power failure can cause a great deal of public inconvenience leading to large economic losses. Examples of such applications are major computer installations, process control in chemical plants, safety monitors, general communication systems, hospital intensive care units etc. For such critical loads, it is of paramount importance to provide an uninterruptible power supply (UPS) system so as to maintain the continuity of supply in case of power outages.

Earlier UPS systems were based on an arrangement shown in Fig. 11.7. This scheme is usually called rotating-type UPS. This arrangement consists of DC motor-driven alternator, the shaft of which is also coupled to a diesel engine. The three-phase mains supply, after rectification, charges a dc battery-bank and feeds the dc motor as well. The uninterruptible

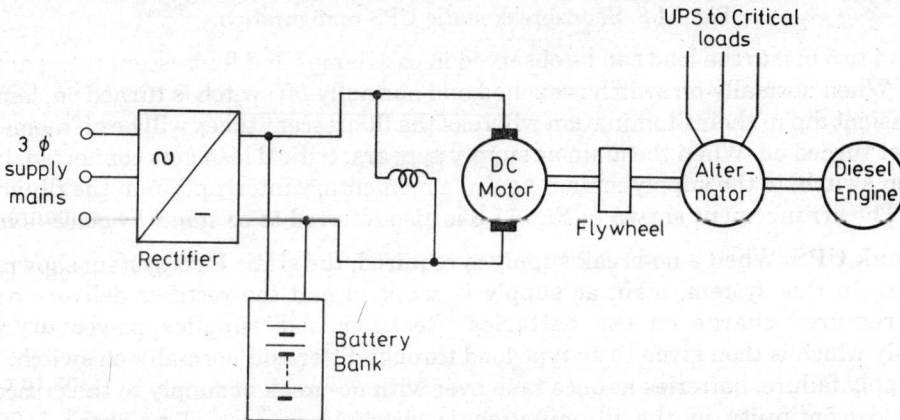

Fig. 11.7. Rotating-type UPS system based on dc motor/alternator set.

power supply needed is taken from the alternator output terminals. When mains supply fails, the diesel engine is run to take over the load. Starting of the diesel engine takes 10 to 15 seconds. During this period, the battery-bank is able to maintain the alternator speed through the dc motor and the flywheel, thus giving a no-break supply to the critical load. At present, however, static UPS systems are becoming popular up to a few kVA ratings.

Static UPS systems are of two types ; namely short-break UPS and no-break UPS. In short-break UPS, the load gets disconnected from the power source for a short duration of the order of 4 to 5 ms. In no-break UPS, load gets continuous uninterrupted supply from the power source. These are now discussed briefly.

Short-break UPS. In situations where short interruption (4 to 5 ms) in supply can be tolerated, the short-break UPS shown in Fig. 11.8 is used. In this system, main ac supply is rectified to dc. This dc output from the rectifier charges the batteries and is also converted to ac by an inverter, Fig. 11.8. After passing through the filter, ac can be delivered to load in case normally-off contacts are closed. Under normal circumstances, normally-on contacts are closed and normally-off contacts are open and the main supply delivers ac power to the load. At the same time, the rectifier supplies continuous trickle charge to batteries to keep them fully charged. In the event of power outage, normally-off switch is turned-on and the batteries deliver ac power to critical load through the inverter and filter. A momentary interruption in

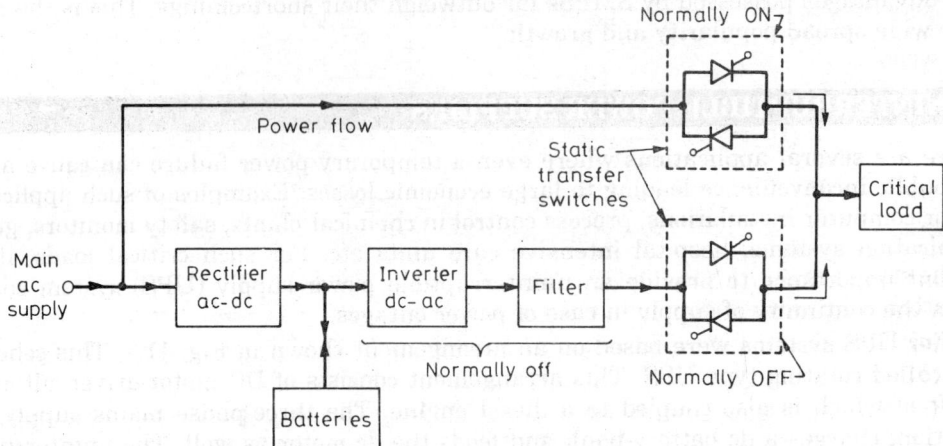

Fig. 11.8. Short-break static UPS configuraticn.

the supply (4 to 5 ms) to the load can be observed in case lamps and fluorescent tubes are a part of the load. When normally-on switch is opened and normally-off switch is turned on, lamps will have a transient dip in their illumination whereas the fluorescent tubes will be off momentarily and then get turned on. When the main ac supply appears, critical load gets connected, through normally-on switch, to the supply mains. Again, a momentary interruption in the illumination is noticed. The arrangement shown in Fig. 11.8 is also referred to as *stand-by power supply*.

No-break UPS. When a no-break supply is required, the static UPS system shown in Fig. 11.9 is used. In this system, main ac supply is rectified and the rectifier delivers power to maintain required charge on the batteries. Rectifier also supplies power to inverter continuously which is then given to ac-type load through filter and normally-on switch. In case of main-supply failure, batteries at once take over with no-break of supply to the critical load. No dip or discontinuity in the illumination is observed in case of no-break UPS. This configuration of Fig. 11.9 has the following additional advantages :

(*i*) The inverter can be used to condition the supply delivered to load.

(*ii*) Load gets protected from transients in the main ac supply.

(*iii*) Inverter output frequency can be maintained at the desired value.

In case inverter failure is detected, the load is switched on to the main ac supply directly by turning on the normally-off static switch and opening the normally-on static switch. The transfer of load from inverter to main ac supply takes 4 to 5 ms by static transfer switch as compared to 40 to 50 ms for a mechanical contactor. After inverter fault is cleared, uninterruptible power supply is again restored to the load through the normally on switch. The batteries are now recharged from the main supply by adjusting the charger at maximum charge rate so that batteries are charged to their full capacity in the shortest possible time.

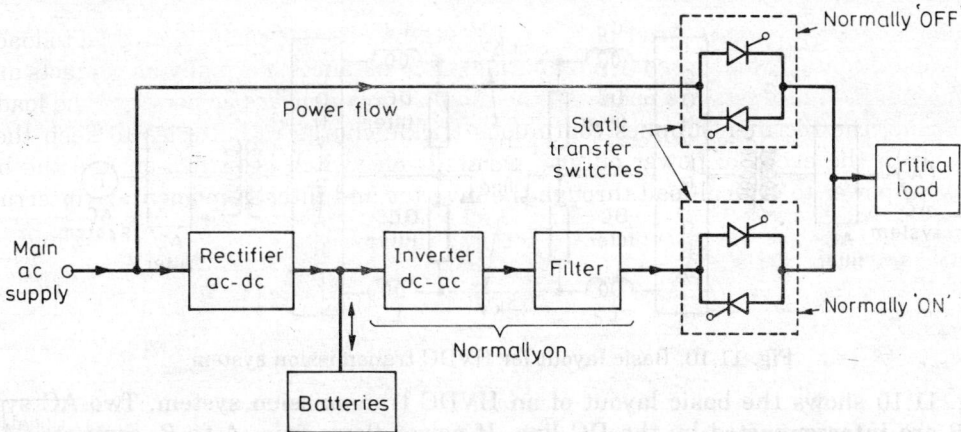

Fig. 11.9. No-break UPS configuration.

The standby batteries in the UPS system are either nickel-cadmium (NC) or lead-acid type. NC batteries have the following advantages :

(a) Their electrolyte is non-corrosive.

(b) Their electrolyte does not emit an explosive gas when charging.

(c) NC batteries cannot be damaged by overcharging or discharging, these have therefore longer life.

Cost of NC batteries is, however, two or three times that of lead-acid batteries.

The time period for which a battery or a battery-bank can deliver power to load through inverter at the required voltage level depends upon (i) the size of the batteries and (ii) nature of the load.

11.3. HIGH VOLTAGE DC TRANSMISSION

It is well known that electric power generated in power plants is transmitted to the load centre on three-phase ac transmission lines. However, for bulk power transmission over long distances, high voltage dc (HVDC) transmission lines are preferred. HVDC transmission possesses the following advantages over AC transmission system :

(i) In HVDC transmission system, one or two conductors and smaller towers are required as against three conductors and tall towers in AC transmission system. HVDC transmission, therefore, costs less.

(ii) Fault clearance in HVDC is faster, therefore DC transmission system possesses improved transient stability.

(iii) Size of conductors in DC transmission can be reduced as there is no skin effect.

(iv) Two AC systems at different frequencies can be interconnected through HVDC transmission lines.

(v) For power transmission through cables, HVDC is preferred as it requires no charging current and the reactive power.

The additional cost of converting and inverting equipments makes HVDC transmission uneconomical for low-power supply over short distances. However, for large-power transmission over long distances, HVDC turns out to be economical. As a result, HVDC links are being used worldwide at power levels of several gigawatts with the use of thyristor valve[*].

---

[*] The term "thyristor valve", used on HVDC systems, denotes a number of thyristors connected in series and parallel to get the required voltage and current ratings.

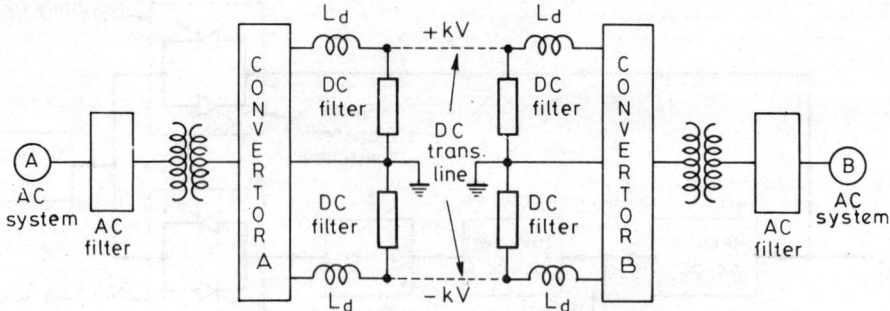

Fig. 11.10. Basic layout for HVDC transmission system.

Fig. 11.10 shows the basic layout of an HVDC transmission system. Two AC systems A and B are interconnected by the DC line. If power flows from A to B, converter A then operates as a rectifier and B as an inverter. Reverse power flow from B to A is also possible with B acting as a rectifier and A as an inverter. AC filters reduce the current harmonics generated by the converters from entering into ac systems. DC filters and smoothing inductors L_d reduce the ripple in the dc voltage. Both converters A and B have 12-pulse configuration. The centre-point of converters A and B is earthed with one line, or pole, at +kV and the other line, or pole, at –kV with respect to earth for a ± kV system. With both the ends earthed, the power flow can be maintained with +kV line and the ground or with –kV line and the ground.

11.3.1. Types of HVDC Link

There are two basic types of HVDC transmission systems. These are *monopolar link* and *bipolar link*.

Monopolar or unipolar link shown in Fig. 11.11 (*a*) offers the simplest arrangement. It uses a single conductor which has either positive, or negative, polarity. It is preferred to have

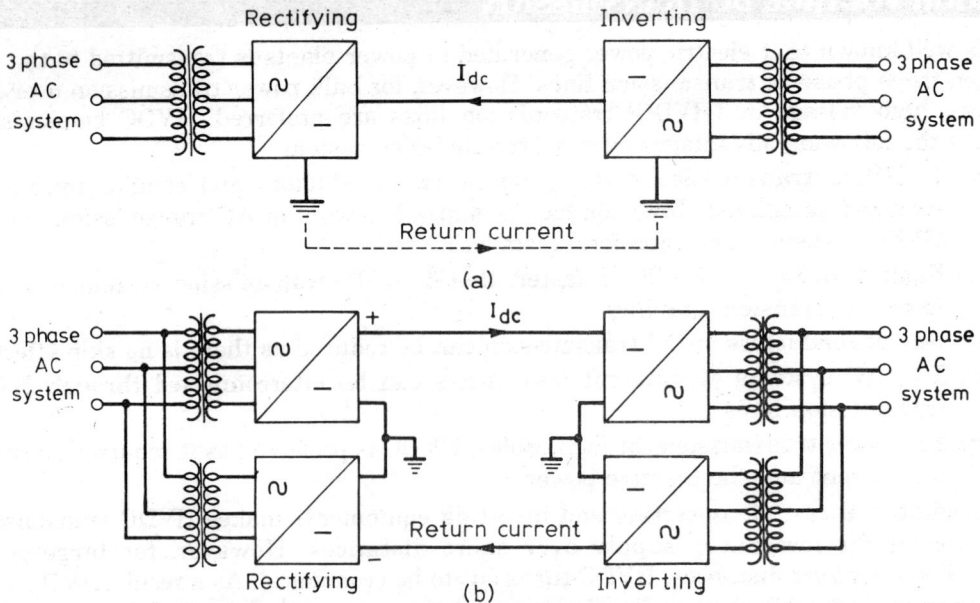

Fig. 11.11. Types of HVDC link (*a*) monopolar link and (*b*) bipolar link.

negative polarity for the single conductor as it produces less radio interference. The return path is provided by ground or sea. The return current through ground or sea leads to higher conduction losses, electrolytic action and large potential gradients.

In bipolar HVDC transmission, two conductors are used, one is positive and other is negative with respect to the ground as shown in Fig. 11.11 (b). As stated before, the neutral points are grounded at both the ends. As the positive and negative conductors carry equal currents, there is no earth current. In case one line is opened due to fault, the other conductor and the ground will form unipolar link and half the rated power can be transmitted untill the fault is cleared. It is obvious from above that bipolar system of HVDC is more reliable than the unipolar or monopolar system. As such, HVDC bipolar link is more commonly employed. A typical bipolar HVDC arrangement is described in what follows.

11.3.2. Bipolar HVDC System

Three-phase 6-pulse phase-controlled converters were discussed in Chapter 6. Current harmonics generated on the ac side and the voltage ripple produced on the dc side of the

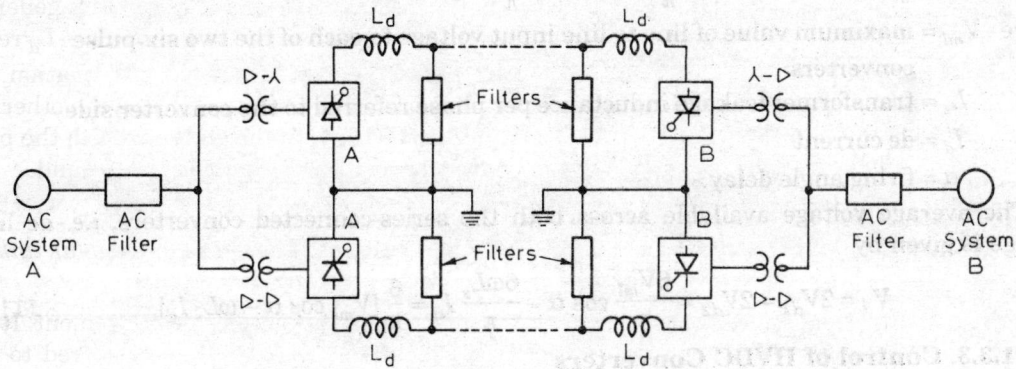

(a)

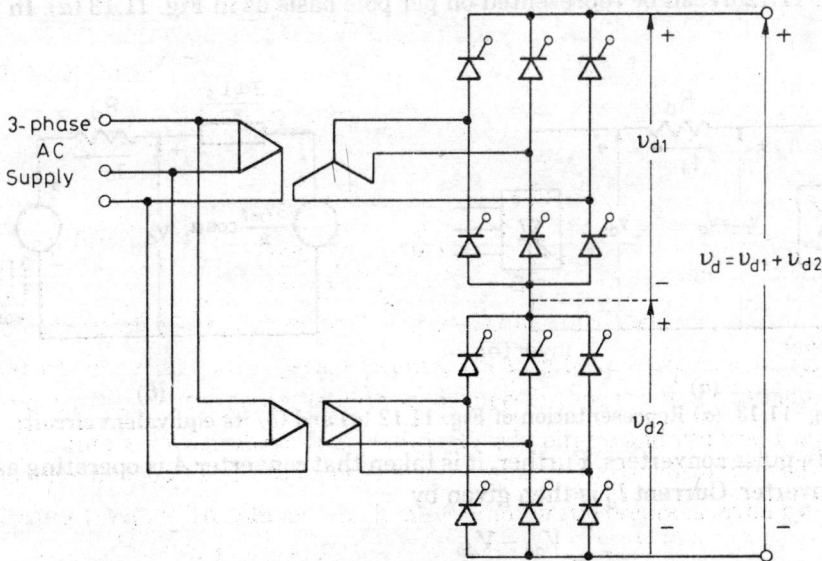

(b)

Fig. 11.12. (a) Schematic diagram for bipolar HVDC system
(b) **Twelve-pulse converter obtained by connecting two six-pulse converters.**

converter used in HVDC system must be reduced. This is achieved by using twelve-pulse converter operation which requires the use of two six-pulse converters fed from delta-delta and delta-star transformers as shown in the schematic diagram of Fig. 11.12 (a). The details of the two converter connections are shown in Fig. 11.12 (b). The delta-star bridge gives six-pulse output voltage v_{d1} whereas the other delta-delta bridge also delivers six-pulse output voltage v_{d2}. The two secondaries, one in star and the other in delta, cause a displacement of 30° in the two six-pulse output voltages such that a twelve-pulse output voltage is obtained from Fig. 11.12. It is seen from this figure that the two six-pulse converters are connected in series on the dc side and in parallel on the ac side. The series connection on the dc side helps to meet the high-voltage requirement in HVDC systems.

When the power flow is from system A to system B, converters A operate as rectifiers. The average value of output voltage of converters A is as under :

Average value of v_{d1} = average value of v_{d2}

or
$$V_{d1} = V_{d2} = \frac{3V_{ml}}{\pi} \cos \alpha - \frac{3\omega L_s}{\pi} I_d$$

where V_{ml} = maximum value of line to line input voltage to each of the two six-pulse converters

L_s = transformer leakage inductance per phase referred to the converter side

I_d = dc current

α = firing angle delay.

The average voltage available across both the series-connected converters, *i.e.* dc link voltage is given by

$$V_d = 2V_{d1} = 2V_{d2} = \frac{6V_{ml}}{\pi} \cos \alpha - \frac{6\omega L_s}{\pi} I_d = \frac{6}{\pi} [V_{ml} \cos \alpha - \omega L_s I_d] \qquad ...(11.8)$$

11.3.3. Control of HVDC Converters

Positive and negative poles are operated under identical conditions. Therefore, HVDC system of Fig. 11.12(a) can be represented on per pole basis as in Fig. 11.13 (a). In this figure,

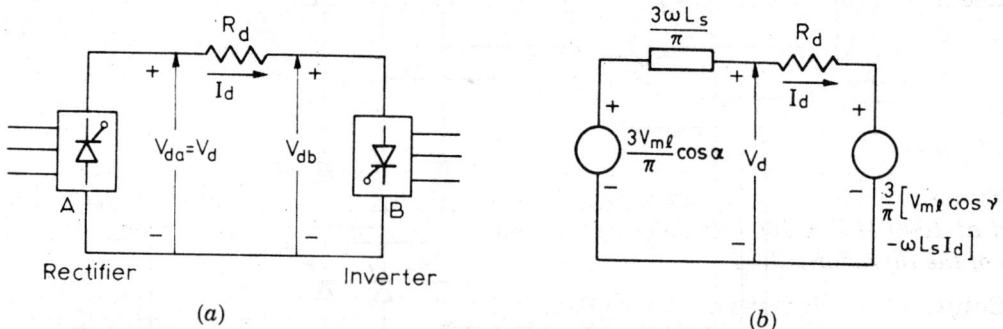

Fig. 11.13. (a) Representation of Fig. 11.12 (a) and (b) its equivalent circuit.

A and B are six-pulse converters. Further, it is taken that converter A is operating as a rectifier and B as an inverter. Current I_d is then given by

$$I_d = \frac{V_{da} - V_{db}}{R_d}$$

where R_d = resistance of one transmission line conductor.

In practice, one converter is made to control the transmission line voltage and the other to control the current I_d. The inverter is usually made to operate at a constant margin angle $\gamma = \pi - \alpha - \mu$ from commutation considerations. It, therefore, follows that inverter is assigned the job of controlling V_d. The current I_d and therefore power level is controlled by the rectifier.

At constant extinction angle γ, the inverter dc voltage is given by

$$V_{db} = \left[\frac{3V_{ml}}{\pi}\cos\gamma - \frac{3\omega L_s}{\pi}I_d\right] = \frac{3}{\pi}\left[V_{ml}\cos\gamma - \omega L_s I_d\right]$$

and the rectifier output voltage V_d is

$$V_d = V_{da} = V_{db} + I_d R_d = \frac{3}{\pi}\left[V_{ml}\cos\gamma - \omega L_s I_d\right] + I_d R_d \qquad ...(11.9a)$$

Also, $$V_d = \frac{3}{\pi}\left[V_{ml}\cos\alpha - \omega L_s I_d\right] \qquad ...(11.9b)$$

Eq. (11.9) leads to the equivalent circuit of HVDC system as shown in Fig. 11.13(b).

Example 11.1. *Two six-pulse converters are used in bipolar HVDC transmission system. The ac systems are 3-phase, 11000V, 50Hz. The input transformers have a leakage inductance of 10 mH per phase. The current in dc line is 300A. The inverter marginal angle is 20°. Resistance of each dc transmission line is 1 Ω. Calculate firing angle of the rectifier, its output voltage and dc link voltage.*

Solution. It is seen from the equivalent circuit of Fig. 11.13 (b) that

$$\frac{3V_{ml}}{\pi}\cos\alpha - \frac{3\omega L_s \cdot I_d}{\pi} = \frac{3V_{ml}}{\pi}\cos\gamma - \frac{3\omega L_s \cdot I_d}{\pi} + I_d \cdot R_d$$

or $$\cos\alpha = \cos\gamma + \frac{\pi}{3V_{ml}}\cdot I_d \cdot R_d$$

$$= \cos 20° + \frac{\pi}{3\sqrt{2}\times 11000}\times 300 \times 1$$

or $$\alpha = 16.283°$$

Rectifier output voltage, $$V_d = \frac{3V_{ml}}{\pi}\cos\alpha - \frac{3\omega L_s \cdot I_d}{\pi}$$

$$= \frac{3}{\pi}[\sqrt{2}\times 11000 \times \cos 16.283° - 2\pi \times 50 \times 10 \times 10^{-3}\times 300]$$

$$= 13357.2\ V$$

DC link voltage $$= 2V_d = 2 \times 13357.2 = 26714.4\ V = 26.714\ kV.$$

Example 11.2. *Two six-pulse converters, used for bipolar HVDC transmission system, are rated at 1000 MW, ± 200 kV. Calculate the rms current and peak reverse voltage ratings for each of the thyristor valves.*

Solution. The dc transmission voltage

$$= 200 + 200 = 400\ kV$$

Direct current in the transmission lines,

$$I_d = \frac{1000 \times 10^3}{400} = 2500\ A$$

It is seen from the working of a 3 phase full converter that each thyristor conducts for 120° for a periodicity of 360°.

$\therefore$ Rms current rating of thyristor

$$= I_d \sqrt{\frac{120}{360}} = 2500 \sqrt{\frac{1}{3}} = 1443.38 \text{ A}$$

Also, $\quad \dfrac{3V_{ml}}{\pi} \cos \alpha = 200 \text{ kV}$

For extreme cases, $\alpha = 0°$ and $\dfrac{3V_{ml}}{\pi} = 200 \text{ kV}$

or $\qquad\qquad V_{ml} = \dfrac{200 \times \pi}{3} = 209.44 \text{ kV}$

Since there are two SCRs conducting simultaneously in a six-pulse converter, the peak reverse voltage across each thyristor valve $= \dfrac{209.44}{2} = 104.72 \text{ kV}$.

11.4. STATIC SWITCHES

A switch having no moving parts is called a static switch. Power semiconductor devices which can be turned on and off within a few microseconds can be used as fast-acting static switches. For high-power applications, thyristors are being used as static switches whereas for low-power applications, power transistors are preferred. Static switches are now replacing mechanical and electromechanical switches because of several advantages listed below :

(i) On time of a static switch (SS) is of the order of 3 µs, it has therefore very high switching speed.

(ii) SS has no moving parts, its maintenance is therefore very low.

(iii) SS has no bouncing at the time of turning on.

(iv) SS has long operational life.

In static switches, however, attention must be paid to leakage current during their off periods.

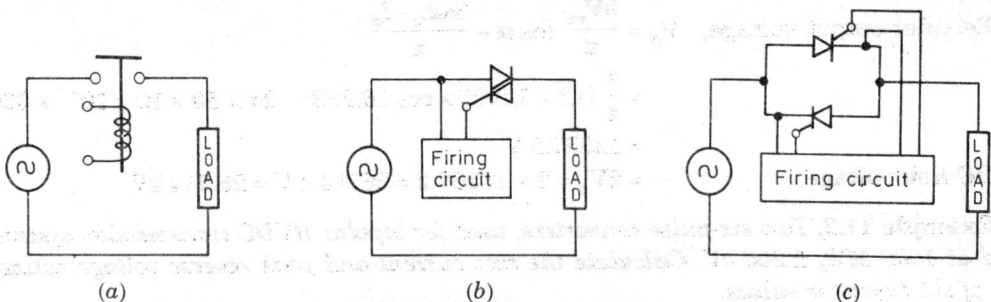

(a)   (b)   (c)

Fig. 11.14. Single-phase switches
(a) electromechanical (b) triac and (c) two-thyristors in anti-parallel.

An electromechanical switch shown in Fig. 11.14(a) is actuated by magnetic coil or plunger. The static switches using triac and two thyristors in anti-parallel are shown in Fig. 11.14 (b) and (c) respectively. Static switches are now being used for relays, circuit breakers, fuses, flashers, UPS, automobile blinkers etc.

It may be observed that circuit of Fig. 11.14(c) is the same as that of Fig. 9.4(a) for single-phase ac voltage controller. But these two circuits are operated differently. Note that static means changeless. In other words, this implies that static switch merely connects a load to the supply or disconnects a load from the supply. Static switch does not change or control the power delivered to load as it is done in a single-phase voltage controller. In static switches, the semiconductor switches are turned on at zero-crossing of load current, whereas it is not so in single-phase voltage controllers.

Static switches can also be used for latching, current and voltage detection, time-delay circuits, transducers etc.

Static switches are of two types : (*i*) ac switches and (*ii*) dc switches. If the input is ac, ac SSs are used and for dc input, dc SSs are used. Switching speed for ac switches is governed by the supply frequency and turn-off time of thyristors. For dc static switches, the switching speed depends on the commutation circuitry and turn-off time of fast thyristors. AC switches may be single-phase or three-phase. Static switches are discussed in what follows:

11.4.1. Single-phase AC Switches

The circuit diagram of a single-phase ac switch is shown in Fig. 11.15 (*a*). Here two thyristors are connected in anti-parallel. For resistive load in Fig. 11.15 (*a*), the waveforms for source voltage v_s, triggering pulse i_{g1} for T1 and pulse i_{g2} for T2, load voltage v_0 and load current i_0 are shown in Fig. 11.15(*c*). Note that T1 is triggered at $\omega t = 0°$, $\omega t = 2\pi$, ... and T2 is triggered at $\omega t = \pi$, 3π ..... when the load current waveform is passing through zero. For *RL* load, output or load current i_0 lags v_0 by load power-factor angle $\phi = \tan^{-1}\dfrac{\omega L}{R}$. For *RL* load, T1 must be triggered at $\omega t = \phi$, $2\pi + \phi$,.... and T2 at $\pi + \phi$, $3\pi + \phi$ and so on, Fig. 11.15(*d*). A triac *TR* can replace two anti-parallel thyristors as shown in Fig. 11.15(*b*). For triac, only one pulse $i_g = i_{g1} = i_{g2}$ in each half cycle will be required.

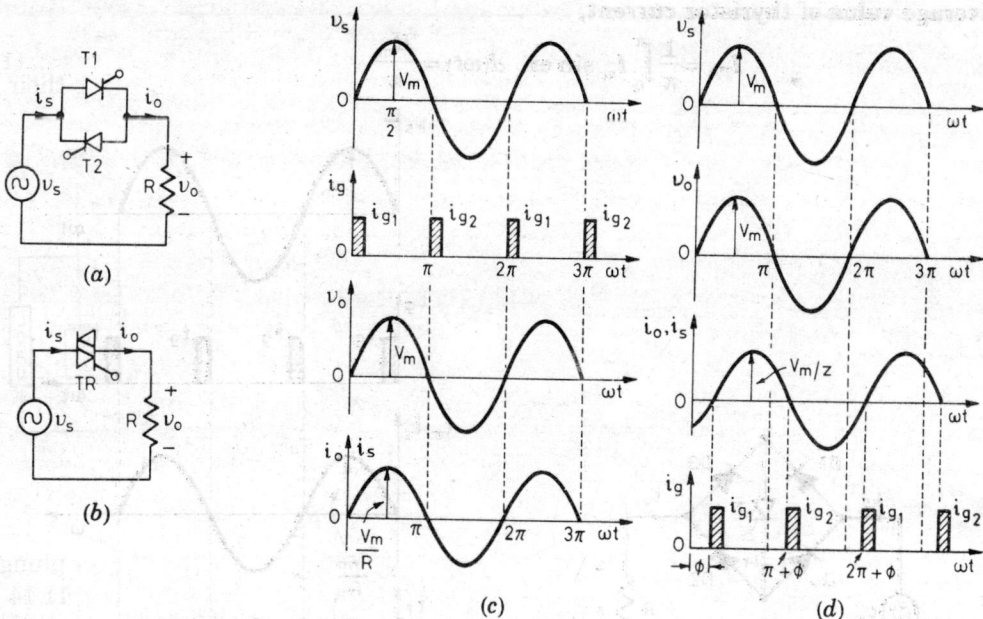

Fig. 11.15. Circuit diagram for single-phase *ac* switch (*a*) using two thyristors (*b*) using one triac (*c*) waveforms for *R* load (*d*) waveforms for *RL* load.

As load current waveform i_0 is a sine wave, it can be expressed as

$$i_0 = I_m \sin \omega t$$

Rms value of load current i_0, or source current i_s, in Fig. 11.15 (c) and (d) is

$$I_{or} = I_{sr} = \left[\frac{1}{\pi} \int_0^\pi I_m^2 \sin^2 \omega t \cdot d(\omega t) \right]^{1/2} = \frac{I_m}{\sqrt{2}} \qquad ...(11.10)$$

Note that each thyristor carries current for 180° for each cycle of load current. Therefore, average value of thyristor current is

$$I_{TA} = \frac{1}{2\pi} \int_0^\pi I_m \sin \omega t \cdot d(\omega t) = \frac{I_m}{\pi} \qquad ...(11.11)$$

and rms value of thyristor current,

$$I_{Tr} = \left[\frac{1}{2\pi} \int_0^\pi I_m^2 \sin^2 \omega t \cdot d(\omega t) \right]^{1/2} = \frac{I_m}{2} \qquad ...(11.12)$$

A bi-directional switch, using four diodes and one thyristor (or a triac or a GTO) is shown in Fig. 11.16(a). This circuit, in performance, is similar to that of Fig. 11.15(a). The waveforms for source voltage v_s, triggering pulse i_g for thyristor T, load current i_0 or source current i_s and thyristor current i_T are shown in Fig. 11.16(b). It is seen from this figure that *rms* value of thyristor current,

$$I_{Tr} = \left[\frac{1}{\pi} \int_0^\pi I_m^2 \sin^2 \omega t \cdot d(\omega t) \right]^{1/2} = \frac{I_m}{\sqrt{2}} \qquad ...(11.13)$$

and average value of thyristor current,

$$I_{TA} = \frac{1}{\pi} \int_0^\pi I_m \sin \omega t \cdot d(\omega t) = \frac{2 I_m}{\pi} \qquad ...(11.14)$$

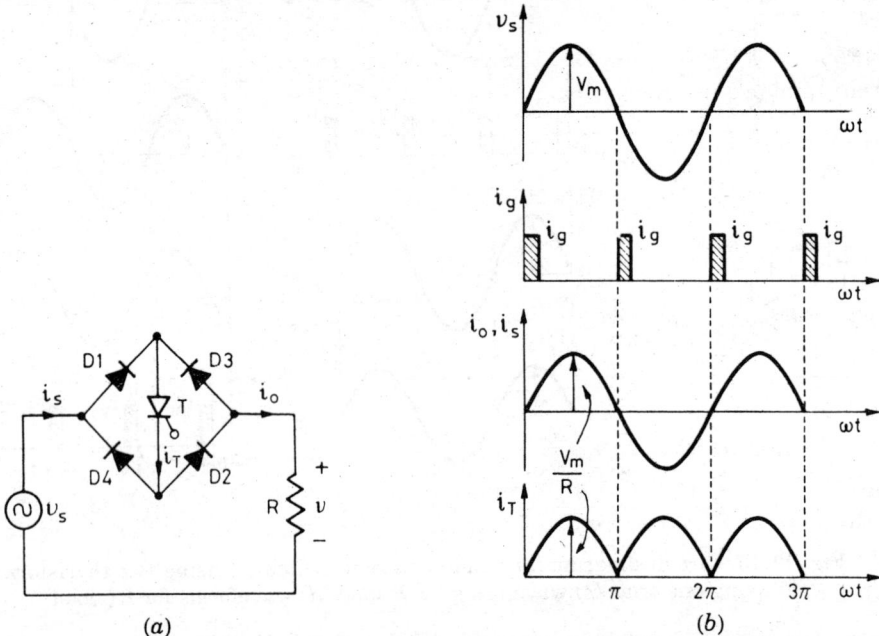

Fig. 11.16. (a) Single-phase ac bi-directional switch (b) its waveforms.

11.4.2. DC Switches

As stated before, in dc switches the input voltage is dc. Power semiconductor devices used in a dc switch may be transistors, thyristors or GTOs.

When thyristor is used, it must have forced commutation circuitry as an integral part of dc switch. One such circuit giving the principle of operation of a dc switch is shown in Fig. 11.17(*a*). Here T1 is the main thyristor and *TA* is the auxiliary thyristor. Capacitor *C* is charged to source voltage V_s with lower plate positive. When T1 is on, normal load current I_0 flows from source to load through T1. For breaking the dc circuit, auxiliary thyristor *TA* is turned on. Capacitor *C* at once applies a reverse voltage across T1 turning it off at once. After this, load current flows as shown in Fig. 11.17(*b*). Capacitor now gets charged from $+V_s$ to $-V_s$ and current through *TA* falls below its holding current to turn it off. Subsequently, freewheeling diode takes over and load current eventually decays to zero. For further details, refer to Figs. 7.24 to 7.26.

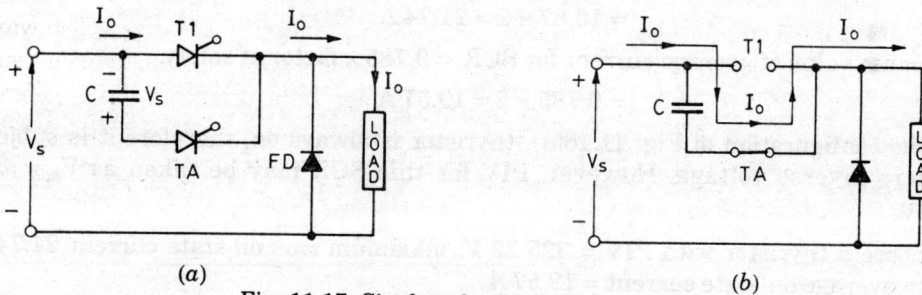

Fig. 11.17. Single-pole thyristor dc switch.

Single-pole dc switches using a transistor and a GTO are shown in Fig. 11.18(*a*) and (*b*) respectively. In Fig. 11.18(*a*), when forward base current is applied to transistor, *TR* gets turned on and dc voltage V_s appears across load. When base current is removed, *TR* gets turned off and load current falls to zero. A freewheeling diode *FD* is necessary for the inductive load.

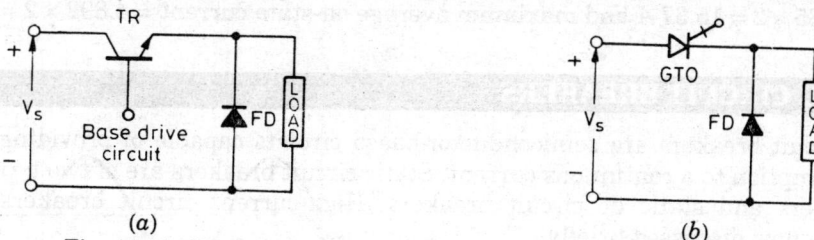

Fig. 11.18. Single-pole (*a*) transistor dc switch (*b*) GTO dc switch.

In Fig. 11.18 (*b*), gate turn-off thyristor is turned on by a short positive gate pulse as in ordinary thyristors. When required, GTO can be turned off by a short negative pulse applied to its gate terminals. Note that a GTO requires no forced-commutation circuitry.

11.4.3. Design of Static Switches

The design of static switches involves the determination of voltage and current ratings of power semiconductor devices employed. Their design is illustrated with an example.

Example 11.3. *A single-phase ac switch of Fig. 11.16 (a) is used in between a 230–V, 50 Hz source and a load of 2 kW at pf of 0.8 lagging. Determine the voltage and current ratings of (a) the thyristor and (b) diodes of the bridge. Take a factor of safety of two.*

Solution. (*a*) Peak value of load current,

$$I_m = \frac{2000\sqrt{2}}{230 \times 0.8} = 15.37 \text{ A}$$

Rms value of thyristor current, from Eq. (11.13) is

$$I_{Tr} = \frac{I_m}{\sqrt{2}} = \frac{15.37}{\sqrt{2}} = 10.87 \text{ A}$$

Average value of thyristor current, from Eq. (11.14), is

$$I_{TA} = \frac{2I_m}{\pi} = \frac{2 \times 15.37}{\pi} = 9.785 \text{ A}$$

Maximum value of rms current for SCR = 10.87 × factor of safety.

$$= 10.87 \times 2 = 21.74 \text{ A}$$

Maximum value of average current for SCR = 9.785 × factor of safety.

$$= 9.785 \times 2 = 19.57 \text{ A}$$

For the configuration of Fig. 11.16(*a*), thyristor is always on, therefore it is subjected to almost zero inverse voltage. However, PIV for this SCR may be taken as $V_m = \sqrt{2} \times 230$ = 325.22 V.

So choose a thyristor with PIV = 325.22 V, maximum rms on-state current 21.74 A and maximum average on-state current = 19.57 A.

(*b*) Each diode conducts for 180° for a periodicity of 360°. This gives maximum value of diode current $= \dfrac{I_m}{2} = \dfrac{15.37}{2} = 7.685 \text{ A}$ and maximum value of average diode current $= \dfrac{I_m}{\pi}$ $= \dfrac{15.37}{\pi} = 4.892 \text{ A}.$

Diode also experiences zero inverse voltage under ideal conditions. So diode PIV may be selected as for the thyristor. So choose a diode with PIV = 325.22 V, maximum rms on-state current = 7.685 × 2 = 15.37 A and maximum average on-state current = 4.892 × 2 = 9.784 A.

11.5. STATIC CIRCUIT BREAKERS

Static circuit breakers are semiconductor-based circuits capable of providing a fast and reliable interruption to a continuous current. Static circuit breakers are of two types; static ac circuit breakers and static dc circuit breakers. High-current circuit breakers employing thyristors are now discussed briefly.

11.5.1. Static AC Circuit Breakers

Static ac switch can be made to operate as a static ac circuit breaker. In Fig. 11.19 (*a*) is shown a simplified circuit configuration for static ac circuit breaker and Fig. 11.19 (*b*) gives relevant voltage and current waveforms. As in static switches, thyristors 1 and 2 in Fig. 11.19(*a*) are turned on at the instant load current is passing through zero. For breaking the circuit, the triggering pulse is withdrawn. For example, at $\omega t = 4\pi + \phi$, if triggering pulse i_{g1} is not applied to T1, it will not get turned on. T2 is already off just before $\omega t = 4\pi + \phi$. Therefore,

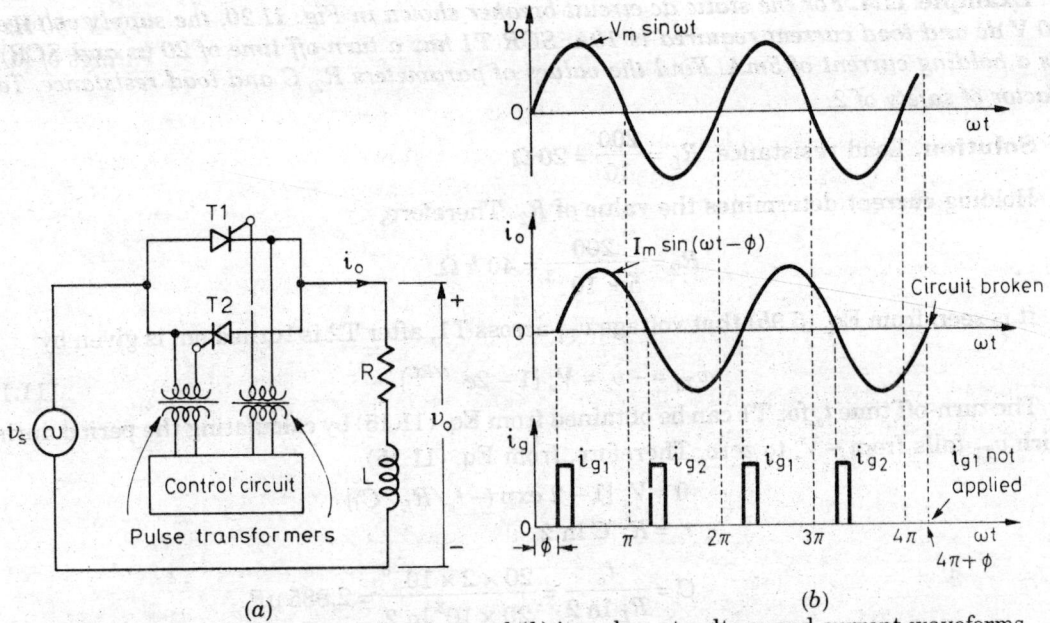

Fig. 11.19. (a) Static ac circuit breaker and (b) its relevant voltage and current waveforms.

the continuity of the circuit is broken. So when turn-off command is received by the control circuit due to some system fault, the gating pulse is withdrawn from T1 or T2 and eventually the circuit is broken. In case turn-off command is received just after $3\pi + \phi$, load current will be broken only at $4\pi + \phi$, i.e. a delay of π radians or half-cycle is a must. If turn-off command is received at the instant $(3\pi + \phi) < \omega t < (4\pi + \phi)$, even then the circuit is broken at the instant $\omega t = 4\pi + \phi$ only. This shows that maximum time delay for breaking the circuit is one half-cycle

i.e. $\dfrac{\pi}{\omega}$ seconds after turn-off command is accepted by the control circuit due to some exigencies in the system.

11.5.2. Static DC Circuit Breakers

A simple arrangement of static dc circuit breaker is shown in Fig. 11.20. This circuit is similar to that shown in Fig. 5.4(a), pertaining to class-C commutation. As stated before, when input voltage to a circuit consisting of thyristors is dc, forced commutation is essential for turning off a thyristor. For complete analysis, refer to section 5.3. Here only brief discussion is given.

When main thyristor T1 is turned on, load voltage becomes equal to source voltage V_s and capacitor C begins to charge through the circuit V_s, R_2, C and T1. Eventually capacitor C gets charged with right hand plate positive. For breaking the circuit, auxiliary thyristor T2 is turned on. Capacitor voltage v_c at once applies a reverse voltage V_s across SCR T1 and turns it off. After T1 is force commutated, capacitor will charge from $+V_s$ to $-V_s$ through the circuit V_s, load, C and T2. When C is fully charged to

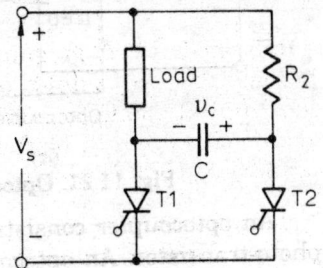

Fig. 11.20.
Static dc circuit breaker.

$-V_s$ (left hand plate positive), current through load will be zero and at the same time current through R_2 will be less than the holding current of SCR T2. As a result, T2 will get turned off naturally. From this, the value of R_2 can be determined.

Example 11.4. *For the static dc circuit breaker shown in Fig. 11.20, the supply voltage is 200 V dc and load current required is 10A. SCR T1 has a turn-off time of 20 μs and SCR T2 has a holding current of 5mA. Find the values of parameters R_2, C and load resistance. Take a factor of safety of 2.*

Solution. Load resistance, $R_L = \dfrac{200}{10} = 20\ \Omega$

Holding current determines the value of R_2. Therefore,

$$R_2 = \frac{200}{5 \times 10^{-3}} = 40\ k\ \Omega.$$

It is seen from Eq. (5.9b) that voltage v_{T1} across T1, after T2 is turned on, is given by

$$v_{T1} = -v_c = V_s\,[1 - 2e^{-t/RC}] \qquad\qquad ...(11.15)$$

The turn-off time t_c for T1 can be obtained from Eq. (11.15) by calculating the period during which v_{T1} falls from $-V_s$ to zero. Therefore, from Eq. (11.15),

$$0 = V_s\,[1 - 2\exp(-t_c/R_L \cdot C)]$$

or
$$t_c = R_L\,C \ln 2$$

∴
$$C = \frac{t_c}{R_L \ln 2} = \frac{20 \times 2 \times 10^{-6}}{20 \times 10^3 \ln 2} = 2.885\ \mu F.$$

11.6. SOLID STATE RELAYS

AC and dc static switches can be used as solid state relays (SSRs) in ac and dc circuits respectively. In ac circuits, thyristors or triacs are used whereas in dc circuits, transistors are preferred. Solid state relays have no contacts or moving parts. These are now being used extensively and are replacing the conventional contact-type electromagnetic relays in applications like control of motor drives, resistance heating etc. SSRs need electrical isolation between control circuit and the load circuit by means of optocouplers or pulse transformers.

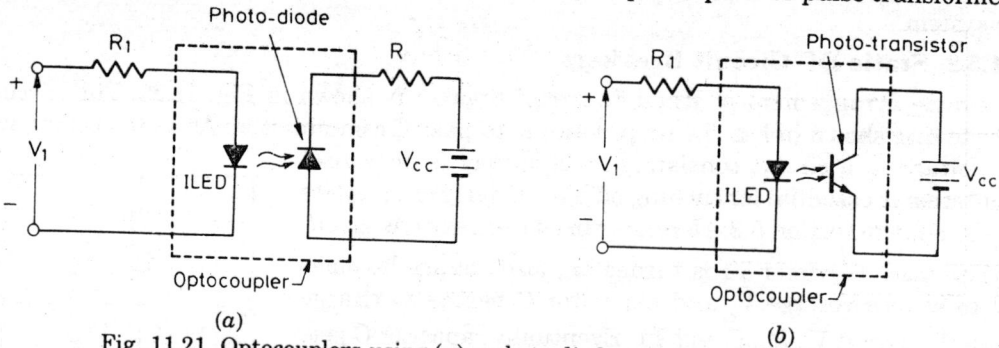

Fig. 11.21. Optocouplers using (a) a photo-diode and (b) a photo-transistor.

An optocoupler consists of infra-red light emitting diode (ILED) and a photo-diode or a photo-transistor. An optocoupler having ILED and photo-diode is shown in Fig. 11.21 (a). A short pulse V_1 applied to ILED will cause it to emit light on to photo-diode which will then begin to conduct in the reverse direction as shown. An optocoupler using photo-transistor is shown in Fig. 11.21 (b). As before, a short pulse V_1 applied to ILED will throw light on the base of photo-transistor and turn it on. As photo-transistor is more sensitive than a photo-diode, optocouplers based on opto-transistors are more common.

11.6.1. DC Solid State Relays

A dc solid state relay using opto-coupler for isolation purposes is shown in Fig. 11.22. When control pulse V_C is applied to ILED, it emits light and turns on the photo-transistor. The current output from the photo-transistor acts as the base current for transistor TR. Consequently TR is turned on and source voltage V_s is applied to load.

When control pulse V_C is absent, TR gets turned off and load voltage is zero.

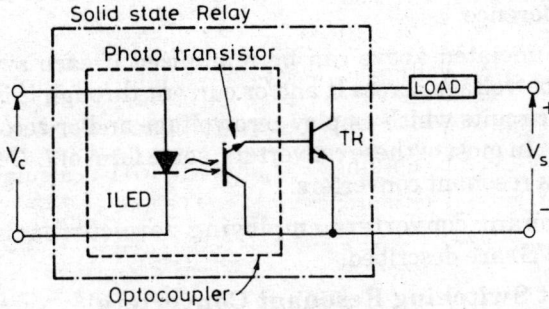

Fig. 11.22. DC solid-state relay using an optocoupler.

11.6.2. AC Solid State Relays

Fig. 11.23 shows two basic circuits for ac solid-state relays. Fig. 11.23 (a) uses a pulse transformer for isolation purposes and in Fig. 11.23 (b), isolation is provided by an optocoupler. When control signal appears across the primary of pulse transformer, its secondary applies a triggering pulse to turn on the triac. As a result, circuit is completed through v_s, load and triac and therefore, source voltage is applied to the load.

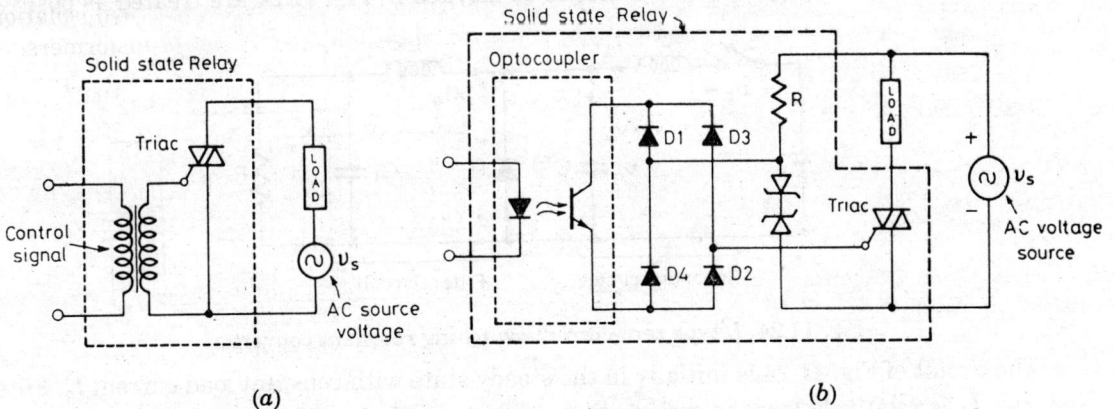

Fig. 11.23. AC solid-state relays using (a) a pulse transformer and (b) an optocoupler.

In Fig. 11.23 (b), control signal turns on the photo-transistor. If the ac supply has upper terminal positive as shown in Fig. 11.23(b), the current will flow through R, D1, photo-transistor, D2, triac gate and source. This current will turn on the triac and load gets energised by source voltage v_s. The function of R is to limit the flow of gate current of triac. If lower terminal of ac supply is positive, the current will flow through triac gate, D3, photo-transistor, D4, R and source v_s. Triac gets turned on and source voltage is applied to load.

11.7. RESONANT CONVERTERS

In SMPSs discussed in Art. 11.1 and in the PWM inverters described in Chapter 8, the switching devices are made to turn-on and turn-off the entire load current at high di/dt. The devices handling high di/dt also experience high-voltage stresses across them; due to these two effects, there are increased power losses in the switching devices. In case size and weight of the converter components is to be reduced, switching frequencies are increased. At these high frequencies, switching losses and high-voltage stresses are further aggravated. Another major drawback of high di/dt and high dv/dt caused by rapid on and off of the switching devices is the electromagnetic interference.

The shortcomings enunciated above can be minimised if each switch in a converter is turned on and off when the voltage across it and/or current through it is zero at the instant of switching. The converter circuits which employ zero-voltage and/or zero-current switching are called *resonant converters*. In most of these converters, some form of L-C resonance is used, that is why these are known as resonant converters.

In this section, resonant converters employing zero-current switching (ZCS) and zero-voltage switching (ZVS) are described.

11.7.1. Zero-Current Switching Resonant Converters

There are two types of ZCS resonant converters, L-type and M-type. Both of these circuit topologies use L and C as a series resonant circuit; in addition L also limits di/dt of the switching current. Here first L-type and then M-type ZCS resonant converters are presented.

11.7.1.1. L-type ZCS Resonant Converters.
An L-type ZCS resonant converter is shown in Fig. 11.24. The switching device S in the figure can be a GTO, thyristor, BJT, power MOSFET or IGBT. At low kilohertz range; GTO, thyristor, transistor or IGBT is used whereas for megahertz range, power MOSFETs are preferred. Inductor L and capacitor C near the dc source V_s form a resonant circuit whereas L_1, C_1 near the load constitute a filter circuit. Direction of currents and plarities of voltages as marked in Fig. 11.24 are treated as positive.

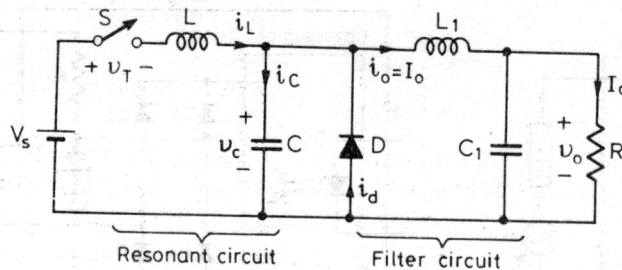

Fig. 11.24. L-type zero-current-switching resonant converter.

The circuit of Fig. 11.24 is initially in the steady state with constant load current I_0. Filter inductor L_1 is relatively large to assume that current i_0 in L_1 is almost constant at I_0. Initially, switch S is open; resonant circuit parameters have $i_L = 0$ in L and $v_c = 0$ across C and the load current I_0 freewheels through the diode D.

For the sake of convinience, working of this converter is divided into five modes as under. For all these modes, time t is taken as zero at the beginning of each mode.

Mode I. $(0 \le t \le t_1)$. At $t = 0$, switch S is turned on. As I_0 is freewheeling through diode D, voltage across ideal diode $v_D = 0$ and also $v_C = 0$, Fig. 11.25(a). It implies that source voltage

V_s gets applied across L and the switch current i_L begins to flow through V_s, switch S, L and diode D, Fig. 11.25(a). Therefore, $V_s = L \, di/dt$. It gives $i_L = \dfrac{V_s}{L} t$. It shows that inductor or switch current i_L rises linearly from its zero initial value. The diode current i_D is given by

$$i_D = I_0 - i_L = I_0 - \frac{V_s}{L} t \qquad \qquad \text{...(11.15)}$$

At $t = t_1$, $\qquad \qquad i_L = \dfrac{V_s}{L} \cdot t_1 = I_0$. This gives

$$t_1 = \frac{I_0 \cdot L}{V_s}$$

Also, at $t = t_1$, $i_D = I_0 - I_0 = 0$. Soon after t_1, as i_D tends to reverse, diode D gets turned off. As a result of this, short circuit across C is removed.

Mode II ($0 \leq t \leq t_2$). Switch S remains on. As D turns off at $t = 0$, current I_0 flows through V_s, L, L_1 and R. In Figs. 11.25(a) and (b), constant current through L_1 and R is represented by current source I_0. Also, a current i_C begins to build up through resonant circuit consisting of V_s, L and C in series. The inductor current i_L is, therefore, given by

$$i_L = I_0 + i_C = I_0 + I_m \sin \omega_0 t \qquad \qquad \text{...(11.16)}$$

where $I_m = V_s \sqrt{\dfrac{C}{L}} = \dfrac{V_s}{Z_0}$ and $\omega_0 = \dfrac{1}{\sqrt{LC}}$. Here $Z_0 = \sqrt{\dfrac{L}{C}}$ is the characteristic impedance of the resonant circuit.

The capacitor current is $i_c = I_m \sin \omega_0 t$ and capacitor voltage v_c, from Eq. 3.10 (a), is given by

$$v_c(t) = V_s (1 - \cos \omega_0 t) \qquad \qquad \text{...(11.17)}$$

The peak value of current i_L is $I_p = I_0 + I_m$ and it occurs at $t = \dfrac{\pi}{2\omega_0} = \dfrac{\pi}{2} \sqrt{LC}$. At this instant,

$$v_c = V_s \left[1 - \cos \frac{\pi}{2} \right] = V_s \text{ and } i_c = I_m.$$

When $t = t_2 = \dfrac{\pi}{\omega_0} = \pi \sqrt{LC}$, capacitor voltage reaches peak value $V_{cp} = V_s [1 - \cos \pi] = 2V_s$ and $i_c = 0$.

Also, at $t = t_2$, $i_L = I_0$, i.e. switch current drops from peak value $(I_0 + I_m)$ to I_0.

Mode III ($0 \leq t \leq t_3$). Switch S remains on. At $t = 0$, capacitor voltage is $2V_s$. As i_c tends to reverse at $t = 0$, capacitor begins to discharge and force a current $i_c = V_s \sqrt{\dfrac{C}{L}} \sin \omega_0 t$ opposite to i_L, Fig. 11.25(c), so that inductor or device current i_L is given by

$$i_L = I_0 - i_c = I_0 - I_m \sin \omega_0 t \qquad \qquad \text{...(11.18)}$$

and capacitor voltage $v_c = 2V_s \cos \omega_0 t$. Current i_L falls to zero when $t = t_3$,

i.e. $\qquad \qquad i_L = 0 = I_0 - I_m \sin \omega_0 t_3$

or $\qquad \qquad t_3 = \sqrt{LC} \, \sin^{-1} (I_0/I_m)$

At $t = t_3$, $\qquad \qquad v_c = 2V_s \cos \omega_0 t_3 = 2V_s \left[\dfrac{\sqrt{I_m^2 - I_0^2}}{I_m} \right] = V_{c3}$

During this mode, $i_c = I_m \sin \omega_0 t$ and as i_L falls to zero at t_3, switching device S gets turned off. Note that current i_c in this mode flows opposite to its positive direction, it is therefore shown negative in Fig. 11.25(c) and in Fig. 11.26. At $t = t_3$, the value of $i_c = -I_0$.

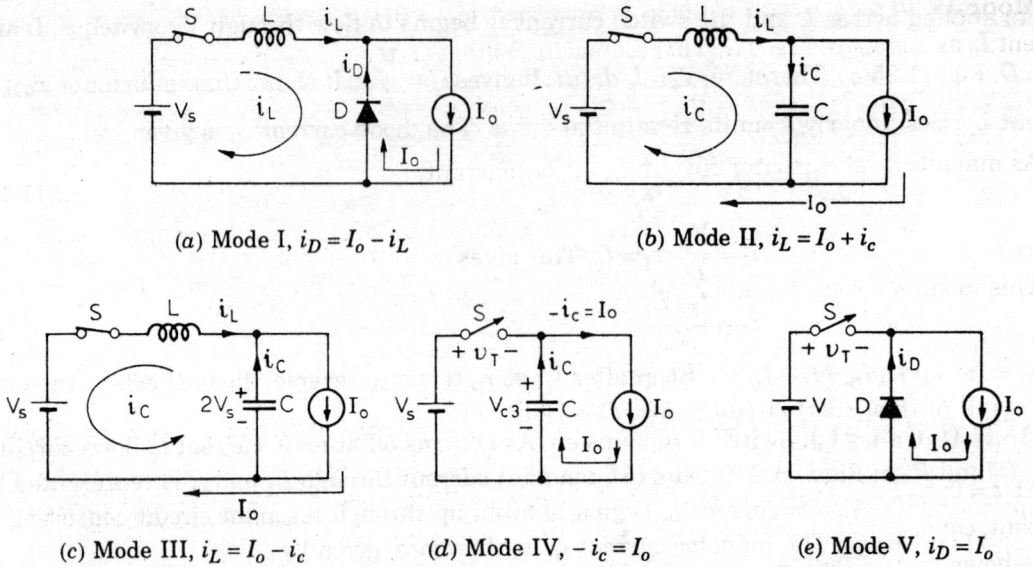

(a) Mode I, $i_D = I_o - i_L$ (b) Mode II, $i_L = I_o + i_c$

(c) Mode III, $i_L = I_o - i_c$ (d) Mode IV, $-i_c = I_o$ (e) Mode V, $i_D = I_o$

Fig. 11.25. Equivalent circuits for the operating modes of L-type ZCS resonant converter.

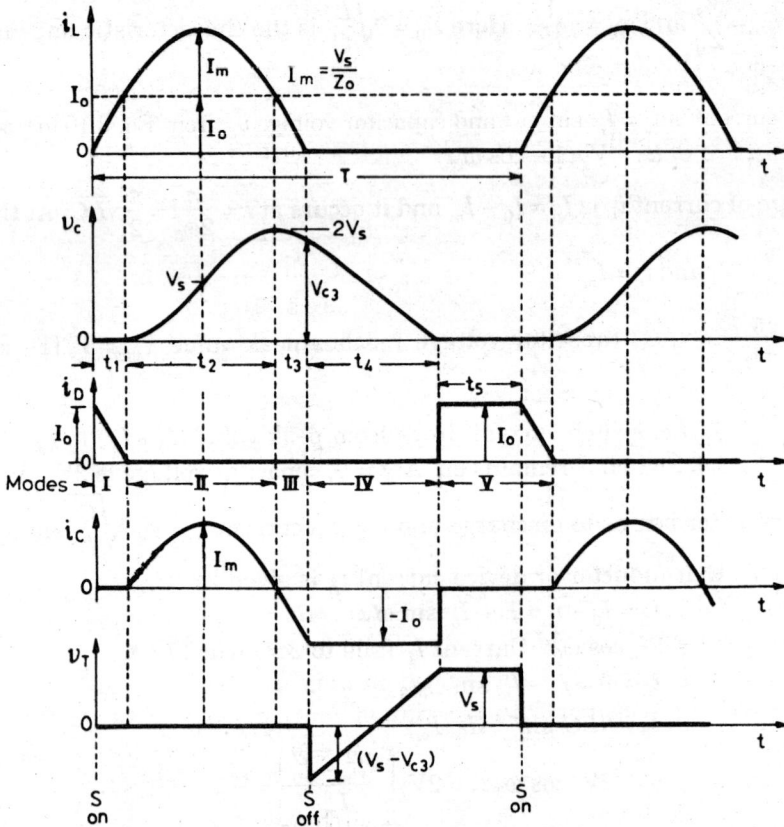

Fig. 11.26. Waveforms for L-type ZCS resonant converter.

Mode IV. ($0 \leq t \leq t_4$). As switch S is turned off at $t = 0$, capacitor begins to supply the load current I_0 as shown in Fig. 11.25 (d). Capacitor voltage at any time t is given by

$$v_c = V_{c3} - \frac{1}{C} \int i_c \cdot dt$$

As magnitude of capacitor current $i_c = I_0$ is constant,

$$v_c = V_{c3} - \frac{I_0}{C} t \qquad \qquad ...(11.19)$$

This mode comes to an end when v_c falls to zero at $t = t_4$.

or

$$0 = V_{C3} - \frac{I_0}{C} t_4$$

or

$$t_4 = \frac{C \cdot V_{C3}}{I_0} \qquad \qquad ...(11.20)$$

At $t = 0$, $v_T = V_s - V_{C3}$ and at $t = t_4$, $v_T = V_s - 0 = V_s$ as shown in Fig. 11.25(d). As I_0 is constant, capacitor discharges linearly from V_{C3} to zero and v_T varies linearly from $(V_s - V_{C3})$ to V_s as shown in Fig. 11.26.

Mode V ($0 \leq t \leq t_5$). At the end of mode IV or in the beginning of mode V, capacitor voltage v_c is zero as shown in Fig. 11.26. As v_c tends to reverse at $t = 0$, diode D gets forward biased and starts conducting, Fig. 11.25(e). The load current I_0 flows through the diode D so that $i_D = I_0$ during this mode.

This mode comes to an end when switch S is again turned on at $t = t_5$. The cycle is now repeated as before. Here $t_5 = T - (t_1 + t_2 + t_3 + t_4)$.

The waveforms for switch or inductor current i_L, capacitor voltage v_L, diode current i_D, capacitor current i_C and voltage across switch S as v_T are shown in Fig. 11.26. It is seen that at turn-on at $t = 0$ ($0 \leq t \leq t_1$), switch current $i_L = 0$, therefore switching loss $v_T i_L = 0$. Similarly, at turn-off at t_3 ($0 \leq t \leq t_3$), $i_L = 0$ and therefore $v_T i_L = 0$. It shows that the switching loss during turn-on and turn-off processes is almost zero. The peak resonant current $I_m = \frac{V_s}{Z_0}$ must be more than the load current I_0, otherwise switch current i_L will not fall to zero and switch S will not get turned off.

The load voltage v_0 can be regulated by varying the period t_5. It is obvious that longer the period t_5, lower is the load voltage.

11.7.1.2. M-type ZCS Resonant Converter. An M-type ZCS resonant converter is shown in Fig. 11.27. As before, L and C form the resonant circuit and L_1, C_1 the filter circuit. Capacitor C is connected across the series combination of switch S and L; but in L-type converter, C is connected across diode D. Working of this converter can be divided into five modes as for the L-type resonant converter. The time origin $t = 0$ is redefined at the beginning of each mode.

Mode I ($0 \leq t \leq t_1$). Prior to mode I, *i.e.* before switch S is turned on at $t = 0$, load current I_0 freewheels through diode D. Also, voltage $v_c = V_s$ before S is closed. At $t = 0$, switch S is turned on. Now current i_L begins to develop through V_s, L and D as shown in the equivalent circuit of

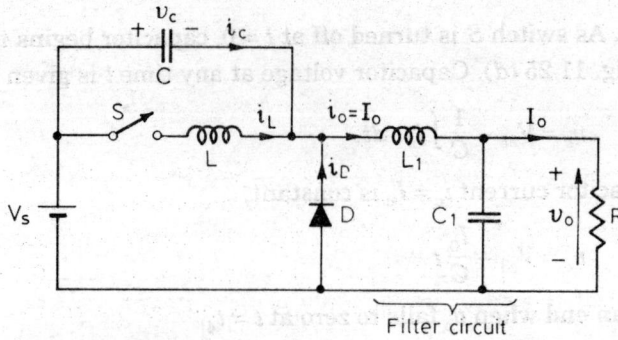

Fig. 11.27. *M*-type zero-current-switching resonant converter.

this mode, Fig. 11.28 (a). It is seen that $i_L = \dfrac{V_s}{L} t$ rises linearly as in *L*-type converter. Also, diode current $i_D = I_0 - i_L$

At $t = t_1$,
$$i_L = I_0 = \frac{V_s}{L} t_1$$

or
$$t_1 = \frac{I_0 \cdot L}{V_s} \qquad\qquad ...(11.21)$$

At $t = t_1, i_L = I_0$ and $i_D = I_0 - I_0 = 0$, diode *D* is therefore turned off. Voltage v_c across capacitor stays at V_s through *D*.

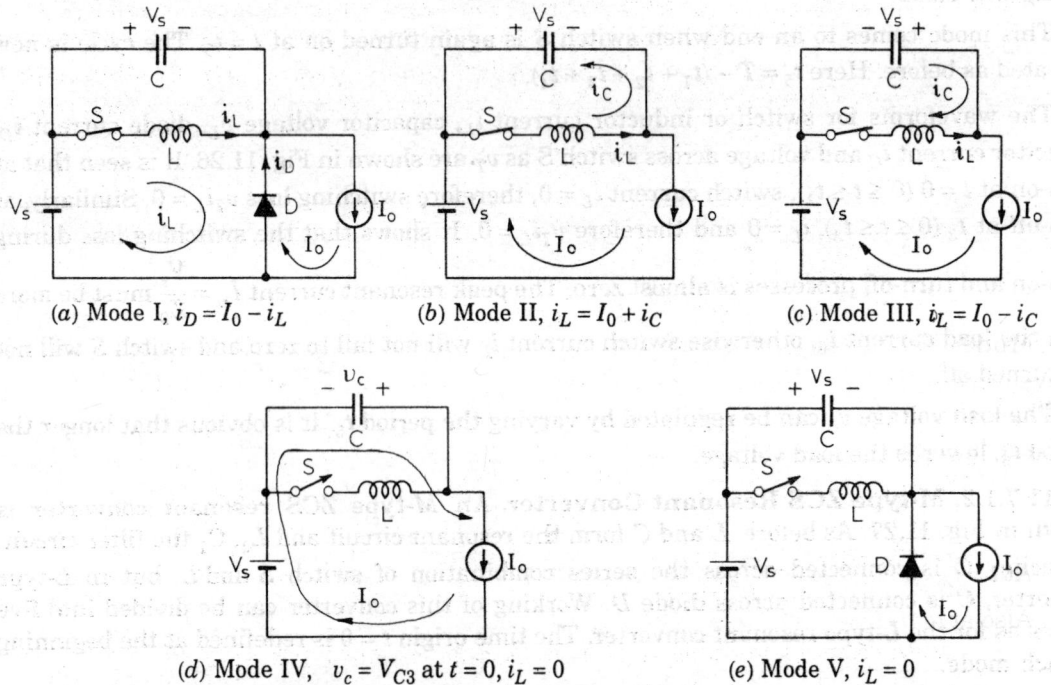

(a) Mode I, $i_D = I_0 - i_L$ (b) Mode II, $i_L = I_0 + i_C$ (c) Mode III, $i_L = I_0 - i_C$

(d) Mode IV, $v_c = V_{C3}$ at $t = 0$, $i_L = 0$ (e) Mode V, $i_L = 0$

Fig. 11.28. Equivalent circuits for *M*-type ZCS resonant converter.

Mode II ($0 \le t \le t_2$). After *D* turns off, load current I_0 flows through V_s and *L* as shown in Fig. 11.28(b). Also, *C* and *L* form a resonant circuit where the current i_c is given by

$$i_c = V_s \sqrt{\frac{C}{L}} \sin \omega_0 t = I_m \sin \omega_0 t$$

and capacitor voltage v_c is given by $v_c = V_s \cos \omega_0 t$

At $t = 0$, $v_c = V_s$, $i_c = 0$, $i_L = I_0$. When $\omega_0 t_2 = \pi$ or $t_2 = \dfrac{\pi}{\omega_0} = \pi \sqrt{LC}$, capacitor voltage $v_c = - V_s$, $i_c = 0$ and $i_L = I_0$.

When $t = \dfrac{\pi}{2\omega_0} = \dfrac{t_2}{2}$, $i_c = I_m$, $i_L = I_0 + I_m$ and $v_c = 0$.

During this mode, $i_L = I_0 + i_C = I_0 + I_m \sin \omega_0 t$ and capacitor gets charged from V_s to $- V_s$ as shown in Fig. 11.29.

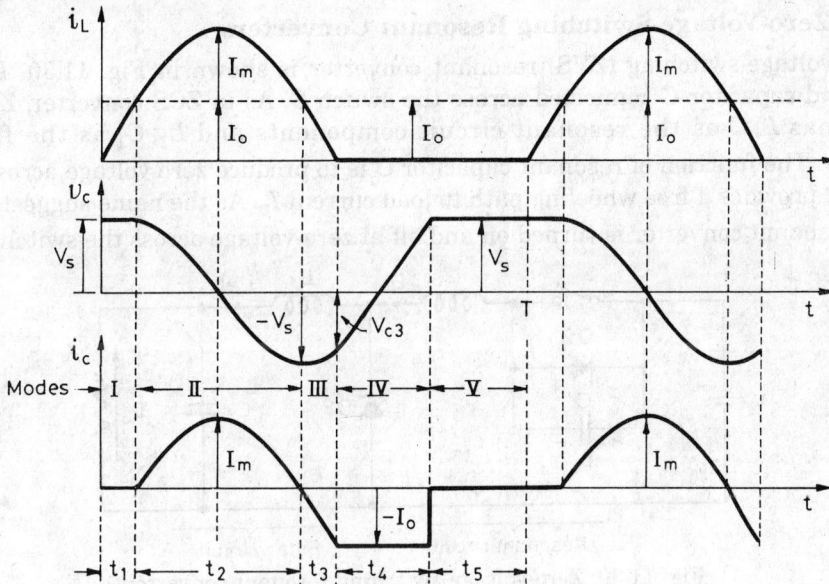

Fig. 11.29. Waveforms for M-type ZCS resonant converter.

Mode III ($0 \leq t \leq t_3$). Equivalent circuit for this mode is given in Fig. 11.28(c). Various waveforms for M-type ZCS resonant converter are shown in Fig. 11.29. It is seen that capacitor voltage at the end of second mode is negative, i.e. $v_c = - V_s$. During mode III, the capacitor voltage is given by

$$v_c = - V_s \cos \omega_0 t, \quad i_c = I_m \sin \omega_0 t$$

and
$$i_L = I_0 - i_c = I_0 - I_m \sin \omega_0 t$$

At $t = t_3$, switch current i_L falls to zero as in L-type converter.

Also at $t = t_3$, $v_c = - V_s \cos \omega_0 t_3 = - V_{C3}$ and $i_c = I_m \sin \omega_0 t_3 = I_0$ so that $i_L = 0$. This gives

$$t_3 = \frac{1}{\omega_0} \sin^{-1} \left(\frac{I_0}{I_m} \right) = \sqrt{LC} \sin^{-1} \left(\frac{I_0}{I_m} \right) \qquad \dots(11.22)$$

Mode IV ($0 \leq t \leq t_4$). In the previous mode, as i_L falls to zero and tends to reverse, switch S is naturally turned off. In this mode, therefore, S remains off and the equivalent circuit of

Fig. 11.28(d) applies. Load current I_0 flows through V_s and C. At $t = 0$, $v_c = -V_{C3}$. Current I_0 charges C from $-V_{C3}$ at $t = 0$ to V_s at $t = t_4$, Fig. 11.29. Therefore,

$$I_0 = C\frac{dv}{dt} = I\frac{V_s + V_{C3}}{t_4}$$

or
$$t_4 = C\frac{V_s + V_{C3}}{I_0} \qquad \qquad \dots(11.23)$$

At t_4, $v_c = V_s$. Actually, at t_4, C is somewhat overcharged with left hand plate positive and consequently diode D gets forward biased.

Mode V ($0 \le t \le t_5$). At $t = 0$, diode D starts conducting and I_0 freewheels through D as shown in Fig. 11.28 (e). Switch S is open and voltage v_c stays at V_s through D. Switch current i_L remains zero as S is open. At time $t = T$, switch S is again turned on and the cycle repeats.

11.7.2. Zero-Voltage-Switching Resonant Converters

A zero-voltage-switching (ZVS) resonant converter is shown in Fig. 11.30. It consists of diode D1 and capacitor C connected across the switch S. As in ZCS converter, ZVS resonant converter has L, C as the resonant circuit components and L_1, C_1 as the filter circuit components. The function of resonant capacitor C is to produce zero voltage across the switch S. Diode D2 provides a free wheeling path to load current I_0. As the name suggests, the switch S in ZVS resonant converter is turned on and off at zero-voltage across the switch.

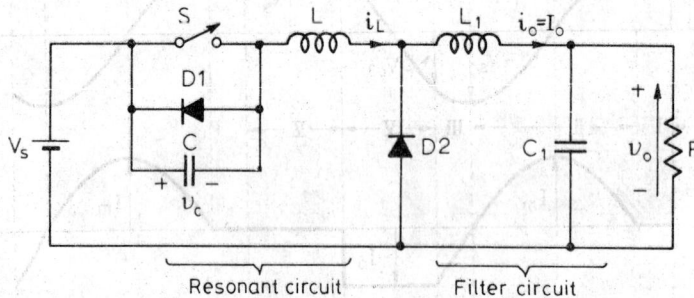

Fig. 11.30. Zero-voltage-switching resonant converter.

The working of this converter can be divided into five modes with equivalent circuits as shown in Fig. 11.31. As before, the time origin $t = 0$ is redefined at the beginning of each mode. Load current I_0 is assumed constant and filter inductor current i_0 is also taken to remain level at I_0 as filter inductor is relatively large. Initially, switch S is on and conducting I_0. Therefore, inductor current $i_L = I_0$ and initial voltage across capacitor $V_{C0} = 0$.

Mode I ($0 \le t \le t_1$). At $t = 0$, switch S is turned off. From the equivalent circuit of mode I, Fig. 11.31(a), it is seen that constant current I_0 flows through V_s, C and L. As a result, voltage across switch S or C builds up linearly from zero to V_s at time $t = t_1$. Diode D2 is off. As the capacitor is charged from zero to V_s, capacitor voltage v_c is given by

$$I_0 = C\frac{dv}{dt}$$

or
$$v_c = \frac{I_0}{C}t$$

At time $t = t_1$,
$$v_c = \frac{I_0}{C}t_1 = V_s \quad \text{or} \quad t_1 = \frac{CV_s}{I_0} \qquad \qquad \dots(11.24)$$

Note that voltage across diode D2 is $v_{D2} = V_s$ at $t = 0$ and $v_{D2} = 0$ at $t = t_1$.

Also, at $t = 0$, $v_c = 0$; therefore switch S is turned off at zero voltage as required.

Mode II $(0 \leq t \leq t_2)$. At $t = 0$, actually capacitor is somewhat overcharged, i.e. $v_c > V_s$; therefore diode D2 becomes forward biased. Now a resonant current i_L is set up in series circuit V_s, C, L and D2, Fig. 11.31(b), where i_L is given by $i_L = I_0 \cos \omega_0 t$.

The capacitor voltage v_c is given by

$$v_c = V_s + V_m \sin \omega_0 t \qquad \qquad ...(11.25)$$

where $V_m = I_0 \sqrt{\dfrac{L}{C}} = I_0 Z_0$ and $Z_0 = \sqrt{\dfrac{L}{C}}$ is the characteristic impedance of the circuit in

ohms. The peak switch or capacitor voltage V_{pk} occurs when $\omega_0 t = \pi/2$ or $t = \dfrac{\pi}{2}\sqrt{LC}$ and its value is

$$V_{pk} = V_s + V_m = V_s + I_0 Z_0$$

At $t = t_2$, $i_L = -I_0$ where $\omega_0 t_2 = \pi$ or $t_2 = \pi \sqrt{LC}$ and capacitor voltage is $v_c = V_s$.

Diode D2 current is given by $i_{D2} = I_0 - I_0 \cos \omega_0 t$.

At $t = 0$, $i_{D2} = 0$, at $t = \dfrac{\pi}{2}\sqrt{LC}$, $i_{D2} = I_0$ and at $t = t_2$, $i_{D2} = 2I_0$.

It may be observed from the waveforms that a ZVS resonant converter is the dual of ZCS resonant converter.

Mode III $(0 \leq t \leq t_3)$. Initially, i.e. at $t = 0$, $v_c = V_s$ and $i_L = -I_0$. With time t reckoned zero from the beginning of this mode, capacitor voltage is given by

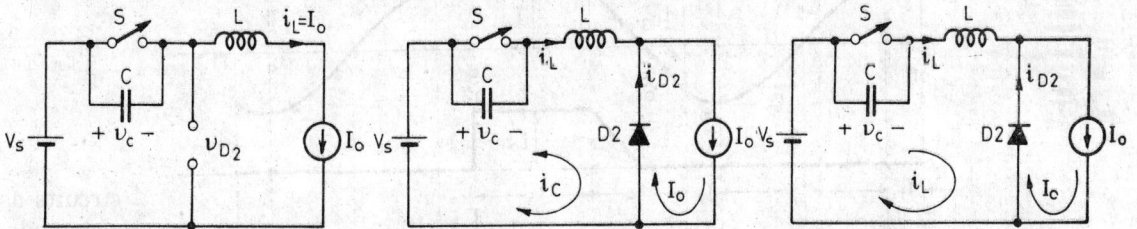

(a) Mode I, $i_L = I_0$ (b) Mode II, $i_L = I_0 \sin \omega_0 t$ (c) Mode III, $v_c = V_s$ at $t = 0$

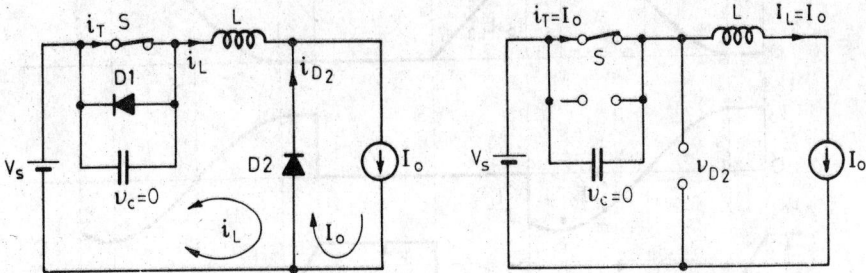

(d) Mode IV, $i_L = -I_{L3} + \dfrac{V_s}{L} t$ (e) Mode V, $i_T = i_L = I_0$

Fig. 11.31. Equivalent circuit for ZVS resonant converter.

$$v_c = V_s - V_m \sin \omega_0 t$$

and

$$i_L = - I_0 \cos \omega_0 t$$

so that

$$i_{D2} = I_0 - I_0 \cos \omega_0 t$$

At time $t = t_3$, $v_c = 0$, $i_L = I_{L3}$ and $i_{D2} = I_0 - I_{L3}$. This gives

$$0 = V_s - V_m \sin \omega_0 t_3$$

or

$$t_3 = \sqrt{LC} \sin^{-1} \left(\frac{V_s}{I_0} \right) \sqrt{\frac{C}{L}} \qquad \qquad ...(11.26)$$

At the end of this mode, *i.e.* at $t = t_3$, $v_c = 0$; as a result reverse bias across D1 vanishes and i_L begins to flow through D1.

Mode IV $(0 \le t \le t_4)$. During this mode, capacitor voltage is clamped to zero by diode D1 conducting negative current i_L. As soon as antiparallel diode D1 begins to conduct at $t = 0$, gate drive is applied to switch S. The inductor current i_L rises linearly from $- I_{L3}$ to zero. At this instant, reverse bias of D1 vanishes and already gated switch S turns on. This shows that switch S turns on at zero voltage and zero current. After this, current rises linearly to I_0 in the circuit formed by V_s, S, L and D2. The linear variation of current from I_{L3} is given by

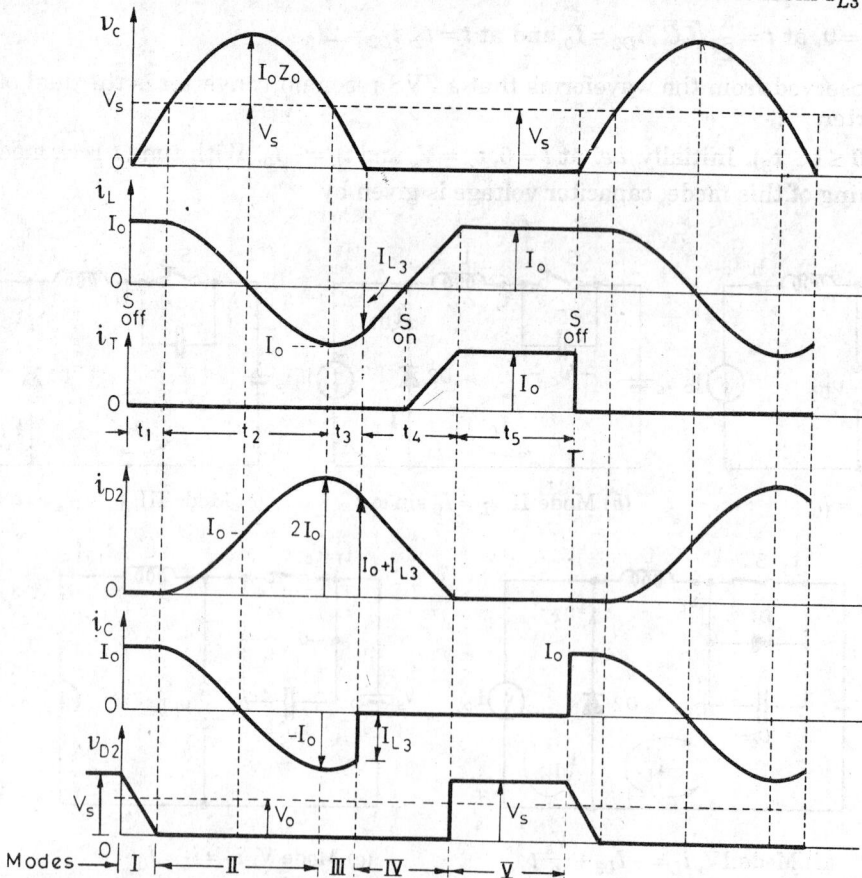

Fig. 11.32. Waveforms for ZVS resonant converter.

$$i_L = -I_{L3} + \frac{V_s}{L} t$$

At $t = t_4$, $i_L = I_0 = -I_{L3} + \frac{V_s}{L} t_4$. This gives

$$t_4 = (I_0 + I_{L3}) \left(\frac{L}{V_s} \right) \qquad \qquad ...(11.27)$$

Diode current $i_{D2} = I_0 + i_L$. At $t = 0$, $i_{D2} = I_0 + I_{L3}$ and time $t = t_4$, $i_{D2} = 0$. During modes II, III and IV, diode D2 is in conduction, therefore $v_{D2} = 0$, as shown in the waveforms of Fig. 11.32.

Mode V ($0 \le t \le t_5$). At the end of mode IV, or in the beginning of mode V at $t = 0$, i_L reaches I_0 and therefore diode D2 turns off. Switch S continues conducting I_0 as shown in Fig. 11.31(e). Note that voltage $v_{D2} = V_s$ during this mode. Mode V ends at $t = t_5$ when switch S is turned off again at zero voltage. The cycle now repeats as before.

The various waveforms for these five modes are now sketched in Fig. 11.32. It is seen from these waveforms that for a ZVS resonant converter :

(*i*) switch, or inductor, current is limited to I_0

(*ii*) average value of output voltage V_0 can be controlled by controlling the interval t_5.

This shows that average power delivered to load can be controlled by regulating the output voltage V_0 for a given load current I_0.

11.7.3. Comparison between ZCS and ZVS Converters

In ZCS, the switch is required to handle a peak current of $I_0 + \frac{V_s}{Z_0}$. For natural turn-off, $\frac{V_s}{Z_0}$ must be more than I_0. There is, therefore, an upper limit to the value of load current in ZCS converters.

In ZVS, the switch is required to withstand a peak voltage of $V_s + I_0 Z_0$. This shows that peak switch voltage is dependent on the load current I_0. A wide variation of load current would need large voltage across the switch. As peak voltage across the switch is a dominating factor, ZVS converters are used only for constant load applications.

In general, ZVS is preferred over ZCS at high switching frequencies, primarily due to internal capacitances associated with the switch.

PROBLEMS

11.1. (a) What is SMPS ? Give its operating principle and industrial applications.

(b) List the various types of SMPSs. Describe SMPS with a pushpull configuration.

11.2. Describe flyback SMPS with relevant equivalent circuits and waveforms. Derive the various expressions for voltages and currents involved.

11.3. A flyback SMPS supplies a load of 40A at 5V. The source voltage is 240V dc and the transformer initial magnetizing current is 0.4 A. The power MOSFET is operating at a frequency of 50 kHz with a duty cycle of 0.4. Determine the transformer turns ratio from primary to secondary and its inductance. Assume ideal components and no ripple in load voltage.

Find also the open-circuit voltage across the semiconductor device.

[**Ans.** 32, $L = 0.5714$ mH, 400 V]

11.4. Describe SMPSs using half-bridge and full-bridge configurations. Enumerate the advantages and disadvantages possessed by SMPSs.

11.5. What is an UPS ? Give its industrial applications. Describe rotating-type, short-break static and no-break static UPS configurations.

Why are nickel-cadmium batteries preferred over lead-acid type batteries in UPSs ?

11.6. (a) Give the merits and demerits of HVDC transmission system over ac transmission system.

(b) Describe both types of HVDC links with relevant circuits.

Derive the equivalent circuit of an HVDC system.

11.7. (a) Two six-pulse converters are used in bipolar HVDC transmission system. The ac systems are 3 phase, 11 kV, 50 Hz. The input transformers have a leakage inductance of 8 mH per phase. Resistance of each transmission line is 0.8 Ω. The inverter marginal angle is 18° and rectifier firing angle is 15°. Calculate current in dc line, rectifier output voltage and dc link voltage.

(b) An HVDC transmission system, using two six-pulse converters for bipolar transmission, is rated at 1000 MW, ± 250 kV. Determine the rms current and peak reverse voltage ratings for each of the thyristor valves.

[Ans. (a) 276.07 A, 13684.3 V, 27.368 kV (b) 1154.7 A, 130.9 kV]

11.8. (a) What is a static switch ? List the merits of static switches over mechanical switches.

(b) The circuit of single-phase ac voltage controller is the same as that used for single-phase ac switch. Discuss how these two differ from each other.

(c) Describe single-phase ac switches using (i) one triac and (ii) bidirectional switches. Derive average and rms values of currents for the semiconductor devices used.

11.9. (a) Describe single-pole dc switches based on (i) a thyristor (ii) a transistor and (iii) a GTO.

(b) A single-phase ac switch, using two thyristors in antiparallel, is inserted between 230 V, 50 Hz source and a load of 10 kW at a pf of 0.8 lagging. Determine (i) the voltage and current ratings of thyristors and (ii) the firing angles of thyristors. Take a factor of safety of 2.   [**Ans.** (b) (i) 650.44 V, 76.85 A, 48.924 A (ii) 36.87° and 216.87°]

11.10. (a) What is a static circuit breaker ? Describe static ac as well as static dc circuit breakers.

(b) A static dc circuit breaker of Fig. 11.20 has input voltage of 220 V dc and load current of 5 A. Thyristor T1 has turn-off time of 15 μs and thyristor 2 has holding current of 6 mA. Find the values of parameters R_2, C and load resistance. Take a factor of safety of 2.5.

[Ans. (b) 36.67 k Ω, 1.23 μF, 44 Ω]

11.11. (a) What are solid state relays ? How is electrical isolation obtained in these relays ?

(b) Describe dc solid state and ac solid state relays with relevant circuit diagrams.

11.12. (a) What are resonant converters ? Give their advantages over PWM controlled converters.

(b) Describe M-type ZCS resonant converter with relevant circuits and waveforms.

11.13. (a) Give the advantages and limitations of ZCS resonant converters.

(b) Describe L-type ZCS resonant converter with relevant circuits and waveforms.

11.14. (a) Give the principle of ZVS resonant converter.

(b) Describe a ZVS resonant converter with appropriate circuits and waveforms.

11.15. (a) Give the advantages and limitations of ZVS resonant converters.

(b) What is the difference between L-type and M-type ZCS converters ?

(c) Compare ZCS and ZVS converters.

Chapter 12

Electric Drives

In this chapter, first the concept of electric drive is given and then dc and ac drives are described. The object of this chapter is not to discuss electric drives exhaustively but at an introductory level.

12.1. CONCEPT OF ELECTRIC DRIVE

In many of the industrial applications, an electric motor is the most important component. A complete production unit consists primarily of three basic components; an electric motor, an energy-transmitting device and the working (or driven) machine.

An electric motor is the source of motive power. An energy transmitting device delivers power from electric motor to the driven machine (or the load); it usually consists of shaft, belt, chain, rope etc. A working machine is the driven machine that performs the required production process. Examples of working machines are lathes, centrifugal pumps, drilling machines, lifts, conveyer belts, food-mixers etc. An electric motor together with its control equipment and energy-transmitting device forms an *electric drive* (10). An electric drive together with its working machine constitutes an *electric-drive system* (10). A ceiling-fan motor with its speed regulator but without blades is an example of electric drive. Other examples of electric drives are : a food-mixer without food to be processed, a motor and conveyer-belt without any material on its belt. Some examples of electric-drive systems are : a ceiling-fan motor with regulator and also with blades, a food-mixer with food to be processed, a motor and conveyer-belt with material on its belt and so on.

Fig. 12.1 shows an electric drive system. The electric drive, consisting of electric motor, its power controller and energy-transmitting shaft is also indicated in Fig. 12.1. A modern electric drive system using a feedback loop is illustrated in Fig. 12.2. In this chapter, electric drives controlled through power-electronic converters are only described.

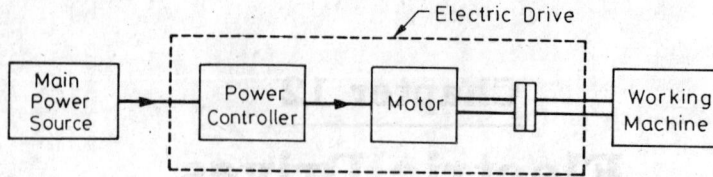

Fig. 12.1. An electric-drive system.

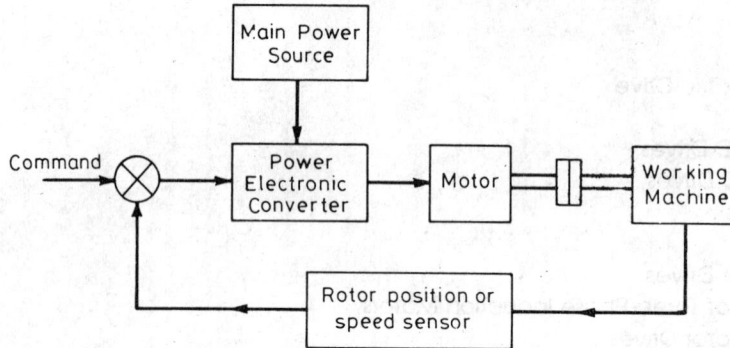

Fig. 12.2. Block diagram for a modern electric drive system using power electronic converter.

Electric drives are mainly of two types : dc drives and ac drives. The two types differ from each other in that the motive power in dc and ac drives is provided by dc motors and ac motors respectively.

12.2. DC DRIVES

DC motors are used extensively in adjustable-speed drives and position control applications. Their speeds below base speed can be controlled by armature-voltage control. Speeds above base speed are obtained by field-flux control. As speed control methods for dc motors are simpler and less expensive than those for ac motors, dc motors are preferred where wide-speed control range is required.

Phase-controlled converters provide an adjustable dc output voltage from a fixed ac input voltage. DC choppers also provide dc output voltage from a fixed dc input voltage. The use of phase-controlled rectifiers and dc choppers for the speed control of dc motors have revolutionized the modern industrial controlled systems.

The dc motors used in conjunction with power-electronic converters are *dc separately excited motors* or *dc series motors*. These motors will, therefore, be studied here. Depending upon the type of ac source or the method of voltage control, dc drives are classified as under:

1. Single-phase dc drives

2. Three-phase dc drives

3. Chopper drives.

First the basic operating characteristics of dc motors are presented and then three speed control strategies as mentioned above are described.

12.2.1. Basic Performance Equations of DC Motors

Equivalent circuit and basic performance equations for a separately-excited dc motor and a dc series motor are presented in what follows.

(a) **Separately-excited dc motor.** The equivalent circuit for a separately-excited dc motor coupled with a load is shown in Fig. 12.3 (a) under steady-state conditions. The load torque T_L opposes the electromagnetic torque T_e. For field circuit, $V_f = I_f \cdot r_f$

For armature circuit, $V_t = E_a + I_a r_a$ ...(12.1)

Motor back e.m.f. or motor armature e.m.f.,

$$E_a = K_a \phi \omega_m = K_m \omega_m$$
$$T_e = K_a \phi I_a = K_m I_a$$

Also, $T_e = D \omega_m + T_L$

where V_t = motor terminal voltage, V

I_a = armature current, A

ϕ = field flux per pole, Wb

$K_m = K_a \phi$ = torque constant, Nm/A or, emf constant, V-sec/rad

r_a = armature circuit resistance, Ω

ω_m = angular speed of motor, rad/sec

r_f = field circuit resistance, Ω

D = viscous friction constant, Nm-sec/rad.

Electromagnetic power, $P = \omega_m \cdot T_e$ watts

From Eq. (12.1), $E_a = K_m \omega_m = V_t - I_a r_a$

or $$\omega_m = \frac{V_t - I_a r_a}{K_m} = \frac{V_t - I_a r_a}{K_a \phi}$$ ...(12.2)

It is seen from Eq. (12.2) that speed can be controlled by varying (i) armature terminal voltage V_t, known as the *armature-voltage control* and (ii) the field flux ϕ, known as the *field-flux control*.

Base speed is defined as the speed at which motor runs under rated armature voltage, rated field current and rated armature current. Speeds below base speed are obtained by armature-voltage control. During this control, armature current and field flux (or field current) are kept constant so as to meet the torque demand. So the armature voltage control method is also termed as *constant-torque drive method* because motor torque $T_e = K_a \phi I_a$ remains almost constant.

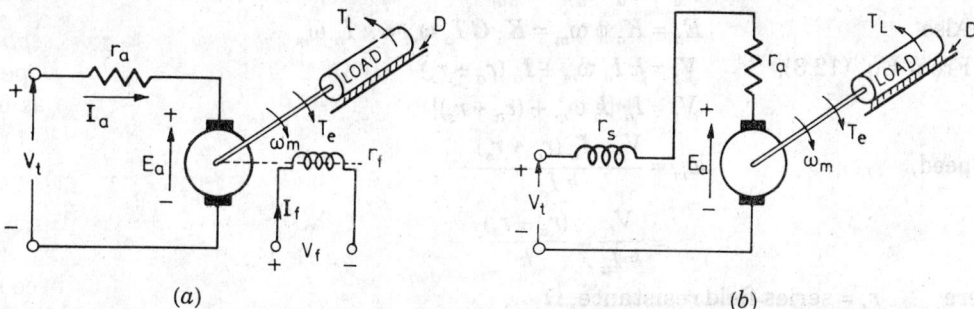

Fig. 12.3. Equivalent circuit of a (a) separately-excited dc motor and (b) dc series motor.

Speeds above base speed are obtained by varying the field current or field flux and by keeping V_t and I_a constant at their rated values. As flux decreases, speed increases so that motor e.m.f. E_a remains almost constant. Consequently, field-flux control method is also called *constant-power drive method* as power $P = E_a I_a$ remains substantially constant. The variations of T_e, P, I_a, I_f, ϕ and V_t against speed are shown in Fig. 12.4 for a separately-excited dc motor.

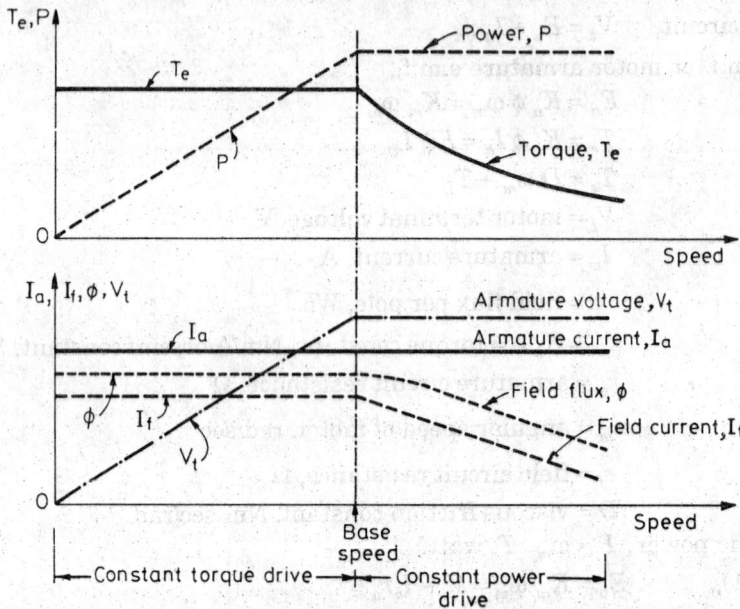

Fig. 12.4. Characteristics of a separately-excited dc motor.

(b) **DC series motor.** For a series motor, field winding in series with the armature circuit is designed to carry the rated armature current. Fig. 12.3 (b) gives the equivalent circuit of a dc series motor driving a load torque T_L.

For the armature circuit in Fig. 12.3 (b),

$$V_t = E_a + I_a (r_a + r_s) \qquad \qquad \qquad ...(12.3)$$
$$T_e = K_a \phi I_a$$

For no saturation in the magnetic circuit, $\phi = C I_a$

$$\therefore \qquad \qquad T_e = K_a C I_a^2 = k\, I_a^2$$

Also $\qquad \qquad E_a = K_a \phi \omega_m = K_a C I_a \omega_m = k\, I_a \omega_m$

From Eq. (12.3), $\qquad V_t = k I_a \omega_m + I_a (r_a + r_s)$

$$V_t = I_a [k \omega_m + (r_a + r_s)]$$

or speed, $\qquad \qquad \omega_m = \dfrac{V_t - I_a (r_a + r_s)}{k I_a}$

$$= \dfrac{V_t}{k I_a} - \dfrac{(r_a + r_s)}{k} \qquad \qquad ....(12.4)$$

where $\qquad r_s$ = series-field resistance, Ω

$\qquad \qquad k = K_a C$ = a constant in Nm/A^2 or in V-s/A. rad.

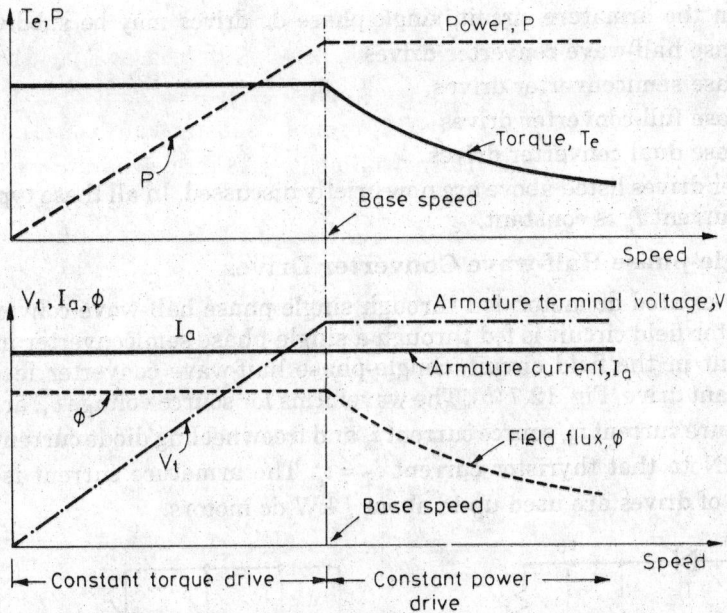

Fig. 12.5. Characteristics of a dc series motor.

For speed control up to base speed, armature terminal voltage V_t is varied with I_a kept constant. Therefore, P $(= V_t I_a)$ varies linearly and torque $T_e = k I_a^2$ remains constant. For speeds above base speed, series field flux is decreased by the use of diverter or tapped-field control and I_a is kept constant. Therefore, torque $T_e = K_a \phi I_a$ decreases but power $P = E_a I_a$ remains substantially constant.

Speed control of dc motors, when fed through single-phase or three-phase converters, is now studied in what follows.

12.3. SINGLE-PHASE DC DRIVES

Fig. 12.6 illustrates the general circuit arrangement for the speed control of a separately-excited dc motor from a single-phase source. The firing angle control of converter 1 regulates the armature voltage applied to dc motor armature. Thus, the variation of delay angle α_1 of converter 1 gives speed control below base speed. The variation of the firing angle α_2 of converter 2 installed in the field circuit gives speeds above base speed. At low values of α_1 for converter 1, armature current may become discontinuous. The discontinuous armature current causes (i) more losses in the armature and (ii) poor speed regulation. It is usual to insert an inductor L in series with the armature circuit to reduce the ripple in the armature current and to make the armature current continuous for low values of motor speeds. Depending upon the type of power-electronic

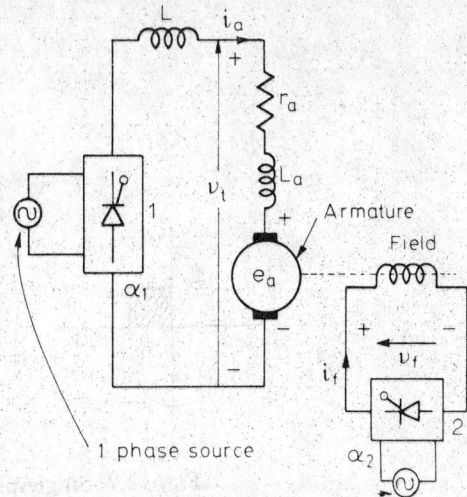

Fig. 12.6. General circuit arrangement for single-phase dc drives.

converter used in the armature circuit, single-phase dc drives may be subdivided as under:

1. Single-phase half-wave converter drives
2. Single-phase semiconverter drives
3. Single-phase full-converter drives
4. Single-phase dual converter drives.

The converter drives listed above are now briefly discussed. In all these types, it is assumed that armature current I_a is constant.

12.3.1. Single-phase Half-wave Converter Drives

A separately-excited dc motor, fed through single-phase half-wave converter, is shown in Fig. 12.7 (a). Motor field circuit is fed through a single-phase semiconverter in order to reduce the ripple content in the field circuit. Single-phase half-wave converter feeding a dc motor offers one-quadrant drive, Fig. 12.7 (b). The waveforms for source voltage v_s, armature terminal voltage v_t, armature current i_a, source current i_s and freewheeling diode current i_{fd} are sketched in Fig. 12.7 (c). Note that thyristor current $i_T = i_s$. The armature current is assumed ripple free. Such types of drives are used up to about $\frac{1}{2}$ kW dc motors.

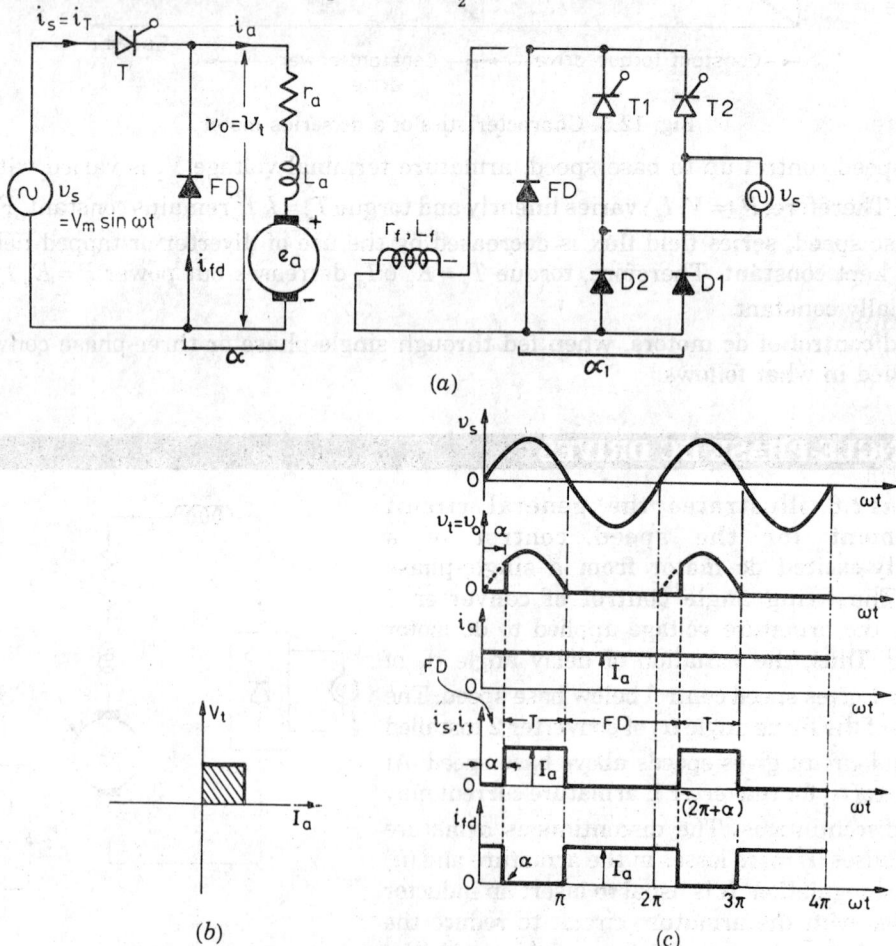

Fig. 12.7. Single-phase half-wave converter drive
(a) circuit diagram (b) quadrant diagram and (c) waveforms.

For single-phase half-wave converter, average output voltage of converter, V_0 = armature terminal voltage, V_t is given by Eq. (6.1) as

$$V_0 = V_t = \frac{V_m}{2\pi}(1 + \cos \alpha) \quad \text{for } 0 < \alpha < \pi \quad \text{...(12.5)}$$

where V_m = maximum value of source voltage.

For single-phase semiconverter in the field circuit, the average output voltage is given by Eq. (6.28) as

$$V_f = \frac{V_m}{\pi}(1 + \cos \alpha_1) \quad \text{for } 0 < \alpha_1 < \pi \quad \text{...(12.6)}$$

It is seen from the waveforms of Fig. 12.7 (c) that
rms value of armature current, $I_{ar} = I_a$
rms value of source or thyristor current,

$$I_{sr} = \sqrt{I_a^2 \frac{\pi - \alpha}{2\pi}} = I_a \left(\frac{\pi - \alpha}{2\pi}\right)^{1/2} \quad \text{...(12.7)}$$

rms value of freewheeling-diode current,

$$I_{fdr} = \sqrt{I_a^2 \frac{\pi + \alpha}{2\pi}} = I_a \left(\frac{\pi + \alpha}{2\pi}\right)^{1/2} \quad \text{...(12.8)}$$

Apparent input power = (rms source voltage) (rms source current)
$$= V_s \cdot I_{sr}$$

Power delivered to motor = $E_a I_a + I_a^2 \cdot r_a = (E_a + I_a r_a) I_a = V_t \cdot I_a$

$$\text{Input supply} \quad \text{pf} = \frac{E_a I_a + I_a^2 r_a}{V_s \cdot I_{sr}} = \frac{V_t \cdot I_a}{V_s \cdot I_{sr}} \quad \text{...(12.9)}$$

Example 12.1. *A separately-excited dc motor is supplied from 230 V, 50 Hz source through a single-phase half-wave controlled converter. Its field is fed through 1-phase semiconverter with zero degree firing-angle delay. Motor resistance $r_a = 0.7\,\Omega$ and motor constant = 0.5 V-sec/rad. For rated load torque of 15 Nm at 1000 rpm and for continuous ripple free currents, determine*
(a) firing-angle delay of the armature converter
(b) rms value of thyristor and freewheeling diode currents
(c) input power factor of the armature converter.
Solution. (a) Motor constant = 0.5 V-sec/rad = 0.5 Nm/A = K_m
But motor torque, $T_e = K_m I_a$

$$\therefore \text{ Armature current } = \frac{15}{0.5} = 30 \text{ A}$$

Motor emf, $\quad E_a = K_m \cdot \omega_m = 0.5 \times \frac{2\pi \times 1000}{60} = 52.36 \text{ V}$

For 1-phase half-wave converter feeding a dc motor,

$$V_t = \frac{V_m}{2\pi}(1 + \cos \alpha) = E_a + I_a r_a$$

or

$$V_t = \frac{\sqrt{2} \times 230}{2\pi}(1 + \cos \alpha) = 52.36 + 30 \times 0.7 = 73.36 \text{ V}$$

$$\therefore \qquad \alpha = \cos^{-1}\left[\frac{73.36 \times 2\pi}{\sqrt{2} \times 230} - 1\right] = 65.336°$$

Thus, firing-angle delay of converter 1 is 65.336°

(b) Rms value of thyristor current, from Eq. (12.7), is

$$I_{Tr} = I_a \left(\frac{\pi - \alpha}{2\pi} \right)^{1/2} = 30 \left(\frac{180 - 65.336}{360} \right)^{1/2} = 16.931 \text{ A} = I_{sr}$$

Rms value of free wheeling-diode current, from Eq. (12.8), is

$$I_{fd \cdot r} = I_a \left(\frac{\pi + \alpha}{2\pi} \right)^{1/2} = 30 \left(\frac{180 + 65.336}{360} \right)^{1/2} = 24.766 \text{ A}$$

(c) From Eq. (12.9), input power factor of armature converter

$$= \frac{V_t \cdot I_a}{V_s \cdot I_{sr}} = \frac{73.36 \times 30}{230 \times 16.931} = 0.5651 \text{ lag.}$$

12.3.2. Single-phase Semiconverter Drives

A separately-excited dc motor, fed through two single-phase semiconverters, one for the armature circuit and the other for the field circuit, is shown in Fig. 12.8 (a). Both converters 1 and 2 are connected to the same single-phase source. This converter also offers one-quadrant

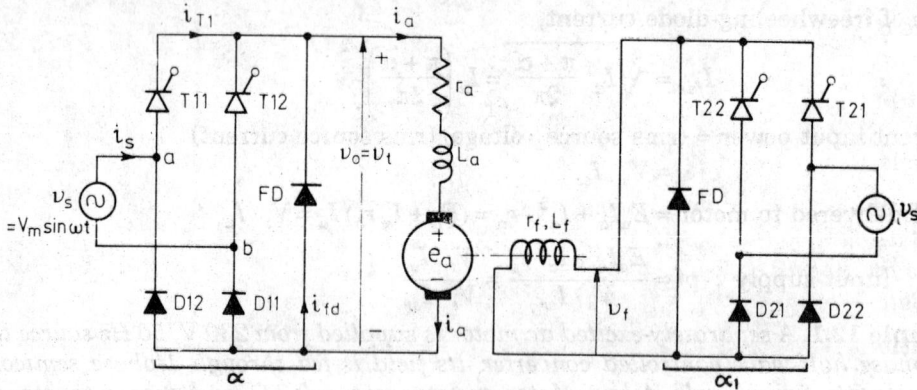

(a)

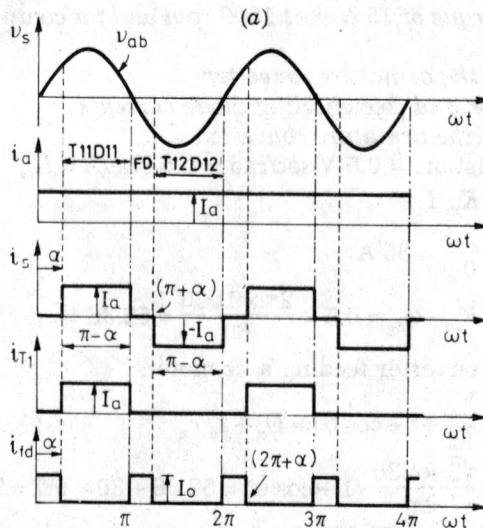

(b)

Fig. 12.8. Single-phase semiconverter drive (a) circuit diagram and (b) waveforms.

drive and is used up to about 15 kW dc drives. The waveforms for currents and voltages are sketched in Fig. 12.8 (b) on the assumption of ripple free armature current. Load voltage waveform for $v_0 = v_t$ is the same as shown in Fig. 6.11 (b).

For a single-phase semiconverter, average output voltage, from Eq. (6.28), is given by

$$V_0 = V_t = \frac{V_m}{\pi} (1 + \cos \alpha) \qquad \qquad ...(12.10 \, a)$$

For field circuit, $\qquad \qquad V_f = \frac{V_m}{\pi} (1 + \cos \alpha_1) \qquad \qquad ...(12.10 \, b)$

It is seen from the waveforms in Fig. 12.8 (b) that

rms value of source current, $\qquad I_{sr} = I_a \left[\frac{\pi - \alpha}{\pi} \right]^{1/2} \qquad \qquad ...(12.11)$

rms value of freewheeling-diode current, $I_{fdr} = I_a \left[\frac{\alpha}{\pi} \right]^{1/2} \qquad \qquad ...(12.12)$

rms value of thyristor current, $\; I_{Tr} = I_a \left(\frac{\pi - \alpha}{2\pi} \right)^{1/2} \qquad \qquad ...(12.13)$

Input pf $\qquad \qquad \qquad = \frac{V_t \cdot I_a}{V_s \cdot I_{sr}}$

A single-phase semiconverter is also called single-phase half-controlled bridge converter.

Example 12.2. *A separately-excited dc motor, operating from a single-phase half-controlled bridge at a speed of 1400 rpm, has an input voltage of 330 sin 314t and a back emf 80 V. The SCRs are fired symmetrically at α = 30° in every half cycle and the armature has a resistance of 4 Ω. Calculate the average armature current and the motor torque.*

Solution. For a single-phase semiconverter feeding a separately-excited motor,

$$V_0 = V_t = \frac{V_m}{\pi} (1 + \cos \alpha) = E_a + I_a r_a$$

$$\frac{330}{\pi} (1 + \cos 30°) = 80 + I_a \cdot 4$$

$$196.01 = 80 + I_a \cdot 4$$

∴ Average armature current,

$$I_a = \frac{196.01 - 80}{4} = 29.003 \text{ A}$$

Motor emf, $\qquad \qquad E_a = K_m \, \omega_m = K_m \frac{2\pi \times 1400}{60}$

or $\qquad \qquad \qquad K_m = \frac{80 \times 60}{2\pi \times 1400} = 0.546 \text{ V-s/rad or } 0.546 \text{ Nm/A}.$

∴ Motor torque, $\qquad \qquad T_e = K_m I_a = 0.546 \times 29.003 = 15.836 \text{ Nm}.$

Example 12.3. *The speed of a 15 hp, 220 V, 1000 rpm dc series motor is controlled using a 1-phase half-controlled bridge converter. The combined armature and field resistance is 0.2 Ω. Assuming continuous and ripple free motor current and speed of 1000 rpm and k = 0.03 Nm/amp², determine (a) motor current (b) motor torque for a firing angle α = 30°. AC voltage is 250 V. Derive any formula used.* (I.A.S., 1991)

Solution.

Refer to Fig. 12.3 (b) for a dc series motor and Fig. 12.8 for a single-phase semiconverter. From Eq. (12.3) for a dc series motor,

$$V_t = E_a + I_a(r_a + r_s)$$

Motor torque, $T_e = K_a \phi I_a$. For no saturation, $\phi = CI_a$

$$\therefore \qquad T_e = K_a C I_a^2 = k I_a^2$$

where k is a constant in Nm/amp^2.

Also $\qquad E_a = K_a \phi \omega_m = K_a C I_a \omega_m = k I_a \omega_m$

Constant k in the expressions for T_e and E_a is the same.

(a) From above, $\qquad V_t = V_0 = E_a + I_a(r_a + r_s)$

or

$$V_t = V_0 = \frac{V_m}{\pi}(1 + \cos\alpha) = E_a + I_a(r_a + r_s) = kI_a\omega_m + I_a(r_a + r_s)$$

$$\therefore \qquad \frac{\sqrt{2} \times 250}{\pi}(1 + \cos 30°) = 0.03\, I_a \times \frac{2\pi \times 1000}{60} + 0.2\, I_a$$

$$209.97 = 3.3416\, I_a$$

$$\therefore \text{ Motor armature current, } I_a = \frac{209.97}{3.3416} = 62.84 \text{ A}$$

(b) Motor torque, $\qquad T_e = k I_a^2 = 0.03\,(62.84)^2 = 118.466 \text{ Nm}.$

12.3.3. Single-phase Full Converter Drives

Two full converters, one feeding the armature circuit and other feeding the field circuit of a separately-excited dc motor, are shown in Fig. 12.9 (a). This scheme offers two-quadrant drive, Fig. 12.9 (b) and its use is limited to about 15 kW. For regenerative braking of the motor, the power must flow from motor to the ac source and this is feasible only if motor counter emf is reversed because then $e_a i_a$ would be negative. Note that direction of current cannot be reversed as SCRs are unidirectional devices. So, for regenerative breaking, the polarity of e_a must be reversed which is possible by reversing the direction of motor field current by making delay angle of full converter 2 more than 90°. In order that current in field winding can be reversed, the field winding must be energised through single-phase full converter as in Fig. 12.9 (a).

For the armature converter 1, $V_0 = V_t = \dfrac{2V_m}{\pi}\cos\alpha \quad$ for $0 < \alpha < \pi$ $\qquad$ ...(12.14 a)

For the field converter 2, $\qquad V_f = \dfrac{2V_m}{\pi}\cos\alpha_1 \quad$ for $0 < \alpha_1 < \pi$ $\qquad$ ...(12.14 b)

From the waveforms in Fig. 12.9 (c), it is seen that

rms value of source current, $\qquad I_{sr} = \sqrt{I_a^2 \cdot \dfrac{\pi}{\pi}} = I_a$

rms value of thyristor current, $\qquad I_{Tr} = \left[I_a^2 \cdot \dfrac{\pi}{2\pi}\right]^{1/2} = \dfrac{I_a}{\sqrt{2}}$ $\qquad$ ...(12.15)

From Eq. (12.9), input supply $\text{pf} = \dfrac{V_t \cdot I_a}{V_s \cdot I_{sr}} = \dfrac{2V_m}{\pi}\cos\alpha \cdot \dfrac{I_a \cdot \sqrt{2}}{V_m \cdot I_a}$

$$= \frac{2\sqrt{2}}{\pi}\cos\alpha \qquad \qquad \qquad \text{...(12.16)}$$

It is seen from Eq. (12.16) that input pf depends on the firing angle α only under the assumptions of constant armature current.

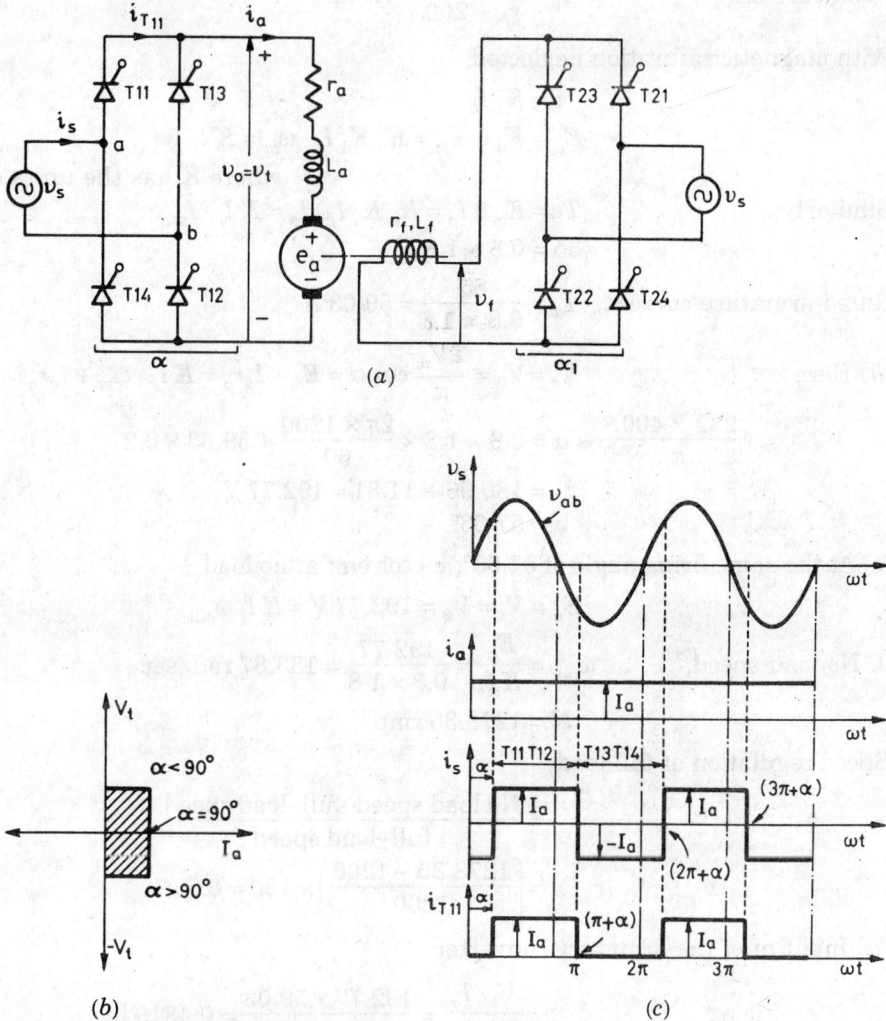

Fig. 12.9. Single-phase full converter drive
(a) circuit diagram (b) two-quadrant diagram and (c) waveforms.

Example 12.4. *A separately-excited dc motor drives a rated load torque of 85 Nm at 1200 rpm. The field circuit resistance is 200 Ω and armature circuit resistance is 0.2 Ω. The field winding, connected to 1-phase, 400 V source, is fed through 1-phase full converter with zero degree firing angle. The armature circuit is also fed through another full converter from the same 1-phase, 400 V source. With magnetic saturation neglected, the motor constant is 0.8 V-sec/A-rad. For ripple free armature and field currents, determine*

  (a) rated armature current

  (b) firing-angle delay of armature converter at rated load

  (c) speed regulation at full load

  (d) input pf of the armature converter and the drive at rated load.

Solution. (a) For field converter, firing-angle delay = 0°

∴ Field voltage, $V_f = \dfrac{2V_m}{\pi} = \dfrac{2\sqrt{2} \times 400}{\pi} = 360\text{V}$

Field current, $\qquad I_f = \dfrac{V_f}{r_f} = \dfrac{360}{200} = 1.8\text{ A}$

With magnetic saturation neglected,

$$\phi = K_1\, I_f$$

$\therefore \qquad E_a = K_a\, \phi\, \omega_m = K_a\, K_1\, I_f \cdot \omega_m = K\, I_f \cdot \omega_m,$

$$\text{where } K \text{ has the units of V-sec/A-rad.}$$

Similarly, $\qquad Te = K_a\, \phi\, I_a = K_a\, K_1\, I_f \cdot I_a = K\, I_f \cdot I_a$

$\therefore \qquad 85 = 0.8 \times 1.8\, I_a$

Rated armature current, $\quad I_a = \dfrac{85}{0.8 \times 1.8} = 59.03\text{ A}$

(b) Here $\qquad V_t = V_0 = \dfrac{2V_m}{\pi}\cos\alpha = E_a + I_a r_a = K\, I_f \cdot \omega_m + I_a r_a$

or $\qquad \dfrac{2\sqrt{2} \times 400}{\pi}\cos\alpha = 0.8 \times 1.8 \times \dfrac{2\pi \times 1200}{60} + 59.03 \times 0.2$

$$= 180.96 + 11.81 = 192.77\text{ V}$$

or $\qquad \alpha = 57.63°$

(c) At the same firing angle of 57.63°, motor emf at no load,

$$E_a = V_t = V_0 = 192.77\text{ V} = K\, I_f\, \omega_{m0}$$

$\therefore$ No load speed, $\qquad \omega_{m0} = \dfrac{E_a}{K\, I_f} = \dfrac{192.77}{0.8 \times 1.8} = 133.87\text{ rad/sec}$

or $\qquad N = 1278.35\text{ rmp}$

Speed regulation at full load

$$= \dfrac{\text{No load speed–full–load speed}}{\text{full–load speed}}$$

$$= \left(\dfrac{1278.35 - 1200}{1200}\right) \times 100 = 6.53\%.$$

(d) Input pf of the armature converter

$$= \dfrac{V_t \cdot I_a}{V_s \cdot I_{ar}} = \dfrac{192.77 \times 59.03}{400 \times 59.03} = 0.4819\text{ lag.}$$

Also, from Eq. (12.16), input pf of the armature converter

$$= \dfrac{2\sqrt{2}}{\pi}\cos\alpha = \dfrac{2\sqrt{2}}{\pi}\cos 57.63° = 0.4819\text{ lag}$$

Rms value of current in armature converter,

$$I_{ar} = I_a = 59.03\text{ A}$$

Rms value of current in field circuit,

$$I_{fr} = I_f = 1.8\text{ A}$$

Total rms current taken from the source,

$$I_{sr} = \sqrt{I_{ar}^2 + I_{fr}^2} = \sqrt{59.03^2 + 1.8^2} = 59.06\text{ A}$$

Input $\qquad VA = V_s \cdot I_{sr} = 400 \times 59.06$

With no loss in the converters, total power input to motor and field

$$= V_t \cdot I_a + V_f \cdot I_f$$
$$= 192.77 \times 59.03 + 360 \times 1.8 = 12027.2 \text{ watts}$$

Input *pf* of the drive $= \dfrac{\text{Power input in W}}{\text{Input in VA}} = \dfrac{12027.2}{400 \times 59.06} = 0.5091 \text{ lag.}$

Example 12.5. *In Example 12.4, the polarity of the counter emf is reversed by reversing the field excitation to its maximum value. Calculate (a) delay angle of the field converter (b) delay angle of the armature converter at 1200 rpm to maintain the armature current constant at 50 A and (c) the power fed back to the supply during regenerative braking of the motor.*

Solution. (a) The field voltage is reversed to its maximum value of 360 V.

$\therefore \qquad V_f = \dfrac{2V_m}{\pi} \cos \alpha_1 = -360 \text{ V}$

or $\qquad \alpha_1 = 180°$

(b) With field current reversed, motor emf E_a is also reversed.

$\therefore \qquad V_0 = V_t = -E_a + I_a r_a$

$\dfrac{2V_m}{\pi} \cos \alpha = -180.96 + 50 \times 0.1 = -175.96 \text{ V}$

or $\qquad \alpha = \cos^{-1}\left[\dfrac{-175.96 \times \pi}{2\sqrt{2} \times 400}\right] = 119.254°$

(c) Power fed back to the ac supply

$$= V_t \cdot I_a = 175.96 \times 50 = 8798 \text{ watts}$$

Example 12.6. *A 220 V, 1500 rpm, 10 A separately-excited dc motor has an armature resistance of 1 ohm. It is fed from a single-phase fully-controlled bridge rectifier with an ac source voltage of 230 V, 50 Hz. Assuming continuous load current, compute*

(a) motor speed at the firing angle of 30° and torque of 5 Nm

(b) developed torque at the firing angle of 45° and speed of 1000 rpm. (GATE, 1996)

Solution. Under rated operating conditions of the separately-excited dc motor,

$$V_t = E_a + I_a r_a = K_m \omega_m + I_a \cdot r_a$$

or $\qquad 220 = K_m \cdot \dfrac{2\pi \times 1500}{60} + 10 \times 1 = 50 \cdot \pi \cdot K_m + 10$

$\therefore$ Motor constant, $\quad K_m = \dfrac{220 - 10}{50\pi} = 1.337 \text{ V-s/rad or } 1.337 \text{ Nm/A.}$

(a) For a torque of 5 Nm, motor armature current,

$$I_a = \dfrac{5}{1.337} = 3.74 \text{ A}$$

The equation giving the operation of converter-motor is

$$V_0 = V_t = E_a + I_a r_a$$

$$\dfrac{2V_m}{\pi} \cos \alpha = K_m \cdot \omega_m + I_a r_a$$

$$\dfrac{2\sqrt{2} \times 230}{\pi} \cos 30° = 1.337 \omega_m + 3.74 \times 1$$

or $\qquad \omega_m = \dfrac{179.3 - 3.74}{1.337} = 131.31 \text{ rad/sec}$

or $\qquad \dfrac{2\pi \cdot N}{60} = 131.31 \text{ rad/sec}$

$$\therefore \qquad \text{Motor speed} = \frac{131.31 \times 60}{2\pi} = \textbf{1253.92 rpm}$$

(b) For $\alpha = 45°$,

$$\frac{2\sqrt{2} \times 230}{\pi} \cos 45° = 1.337 \times \frac{2\pi \times 1000}{60} + I_a \times 1$$

$$146.4 = 140.01 + I_a \times 1$$

or
$$I_a = \frac{6.39}{1} = 6.39 \text{ A}$$

Motor developed torque, $T_e = K_m I_a = 1.337 \times 6.39 = \textbf{8.543 Nm}$

Example 12.7. *A 220V, 1000 rpm, 60A separately-excited dc motor has an armature resistance of 0.1 Ω. It is fed from a single-phase full converter with an ac source voltage of 230V, 50Hz. Assuming continuous conduction, compute*

(a) *firing angle for rated motor torque at 600 rpm*

(b) *firing angle for rated motor torque at (–500) rpm*

(c) *motor speed for* $\alpha = 150°$ *and half rated-torque.*

Solution. Under rated operating conditions of the motor,

$$V_t = E_a + I_a \, r_a = K_m \, \omega_m + I_a \, r_a$$

or
$$220 = K_m \frac{2\pi \times 1000}{60} + 60 \times 0.1$$

$$K_m = \left[\frac{220 - 6}{2\pi \times 1000}\right] \times 60 = 2.044 \text{ V-s/rad or } 2.044 \text{ Nm/A.}$$

(a) For rated motor torque, armature current = 60 A

$$\therefore \qquad V_0 = V_t = K_m \, \omega_m + I_a \, r_a$$

$$\frac{2\sqrt{2} \times 230}{\pi} \cos \alpha = 2.044 \frac{2\pi \times 600}{60} + 60 \times 0.1 = 134.43 \text{ V}$$

or
$$\alpha = \cos^{-1}\left[\frac{134.43 \times \pi}{2\sqrt{2} \times 230}\right] = 49.512°$$

(b) At (–500) rpm,

$$\frac{2\sqrt{2} \times 230}{\pi} \cos \alpha = 2.044 \frac{2\pi \, (-500)}{60} + 60 \times 0.1$$

$$= -107.024 + 6 = -101.024 \text{ V}$$

$$\alpha = \cos^{-1}\left[\frac{-101.024 \times \pi}{2\sqrt{2} \times 230}\right] = 119.274°$$

(c) At half-rated torque, motor armature current

$$= \frac{1}{2} \times \text{rated current} = \frac{1}{2} \times 60 = 30 \text{ A}$$

$$\therefore \qquad \frac{2\sqrt{2} \times 230}{\pi} \cos (150°) = 2.044 \times \omega_m + 30 \times 0.1$$

$$-179.30 = 2.044 \, \omega_m + 3$$

$$\omega_m = \frac{-182.3°}{2.044} = -89.188 \text{ rad/sec}$$

$\therefore$ Speed,
$$N = -\frac{89.188 \times 60}{2\pi} = -851.683 \text{ rpm.}$$

12.3.4. Single-phase Dual Converter Drives

A single-phase dual converter, obtained by connecting two full-converters in anti-parallel, is shown feeding a separately-excited dc motor in Fig. 12.10 (a). Its use is limited to about 15 kW dc drives. It offers four-quadrant operation, Fig. 12.10 (b). For working in first and fourth quadrants, converter 1 is in operation. For operation in second and third quadrants, converter 2 is energised. Four-quadrant operation demands that field winding of the motor is energised from a single-phase, or three-phase, full converter.

For converter 1 in operation, $V_t = \dfrac{2V_m}{\pi} \cos \alpha_1$ for $0 \le \alpha_1 \le \pi$

For converter 2 in operation, $V_t = \dfrac{2V_m}{\pi} \cos \alpha_2$ for $0 \le \alpha_2 \le \pi$

where $\alpha_1 + \alpha_2 = \pi$

For field converter, $V_f = \dfrac{2V_m}{\pi} \cos \alpha_3$ for $0 \le \alpha_3 \le \pi$

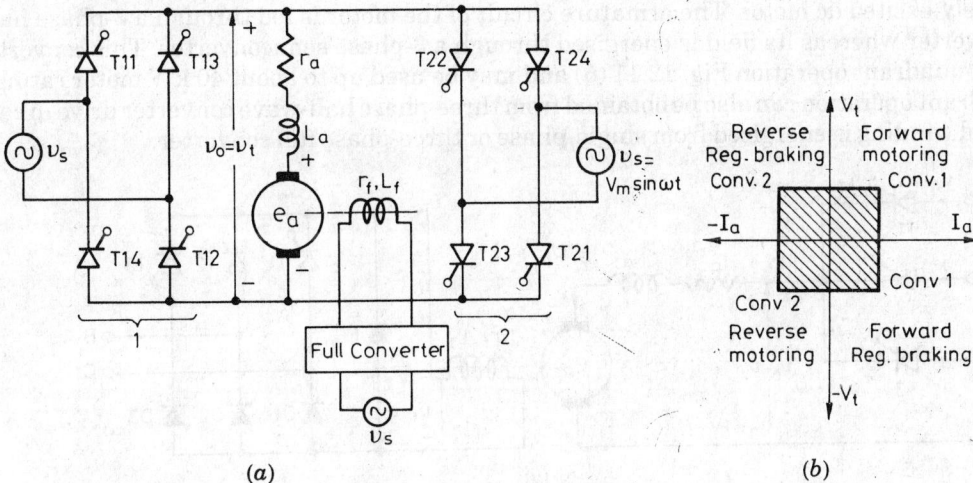

Fig. 12.10. (a) Single-phase dual converter feeding a separately-excited dc motor
(b) four-quadrant diagram.

Note that in Fig. 12.10,

(i) Converter 1 with $\alpha_1 < 90°$ operates the motor in forward motoring mode in quadrant 1.

(ii) Converter 1 with $\alpha_1 > 90°$ and with field excitation reversed operates the motor in forward regenerative braking mode in quadrant 4.

(iii) Converter 2 with $\alpha_2 < 90°$ operates the motor in reverse motoring mode in quadrant 3.

(iv) Converter 2 with $\alpha_2 > 90°$ and with field excitation reversed operates the motor in reverse regenerative braking mode in quadrant 2.

12.4. THREE-PHASE DC DRIVES

Large dc motor drives are always fed through three-phase converters for their speed control. A three-phase controlled converter feeds power to the armature circuit for obtaining speeds below base speed. Another three-phase controlled converter is inserted in the field circuit for getting speeds above base speed.

The output frequency of three-phase converters is higher than those of single-phase converters. Therefore, for reducing the armature current ripple, the inductance required in a three-phase dc drive is of lower value than that in a single-phase dc drive. As the armature current is mostly continuous, the motor performance in 3-phase dc drives is superior to those in single-phase dc drives.

The three-phase dc drives, as in single-phase dc drives, may be subdivided as under :

1. Three-phase half-wave converter drives
2. Three-phase semiconverter drives
3. Three-phase full-converter drives
4. Three-phase dual-converter drives

These converter controlled dc drives are now described one after the other. Armature current is assumed ripple free for convenience.

12.4.1. Three-phase half-wave converter drives.

Fig. 12.11 (a) illustrates a 3-phase half-wave converter drive consisting of two converters and a separately-excited dc motor. The armature circuit of the motor is fed through a 3-phase half-wave converter whereas its field is energised through a 3-phase semiconverter. This converter offers one-quadrant operation Fig. 12.11 (b) and may be used up to about 40 kW motor ratings. Two-quadrant operation can also be obtained from three-phase half-wave converter drive in case motor field winding is energised from single-phase or three-phase full converter.

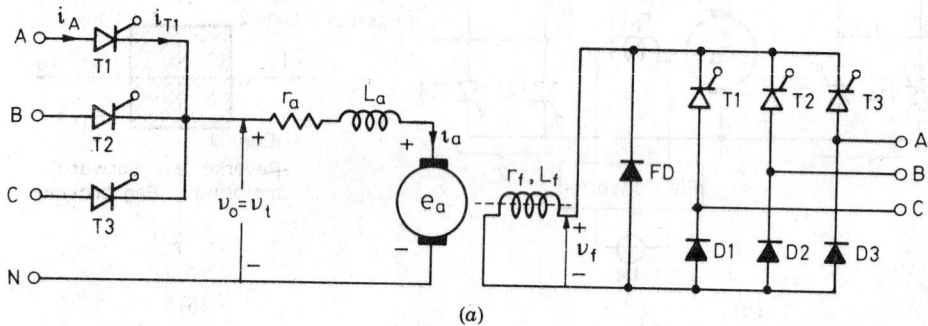

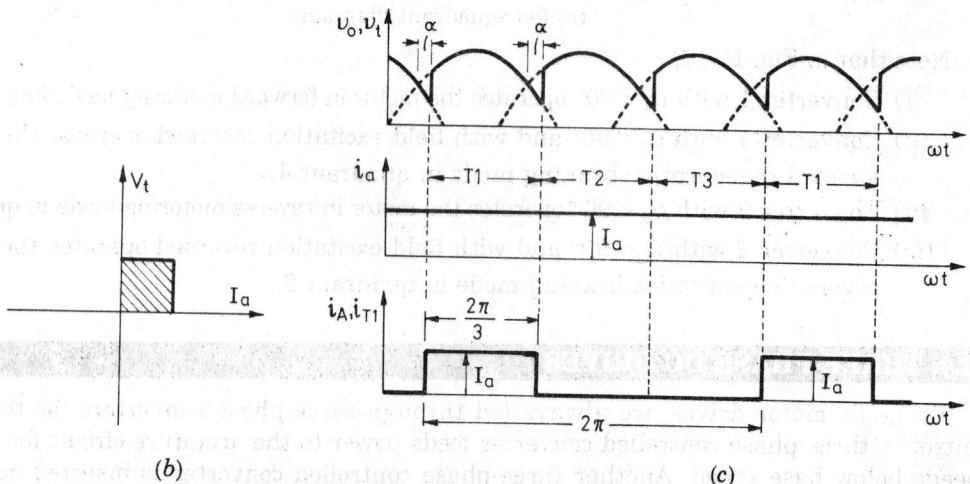

(a)

(b)   (c)

Fig. 12.11. Three-phase half-wave converter drive
(a) circuit diagram (b) quadrant diagram (c) waveforms.

For a 3-phase half-wave converter, average value of output voltage or armature terminal voltage, from Eq. (6.50), is

$$V_0 = V_t = \frac{3V_{ml}}{2\pi} \cos \alpha \quad \text{for } 0 \le \alpha < \pi \qquad \qquad ...(12.17)$$

where V_{ml} = maximum value of line voltage and α is the firing angle for converter 1. The voltage expression of Eq. (12.17) is valid only for continuous armature current. For three-phase semiconverter, the average value of field voltage, from Eq. (6.55), is given by

$$V_f = \frac{3V_{ml}}{2\pi} (1 + \cos \alpha_1) \quad \text{for } 0 \le \alpha_1 \le \pi \qquad \qquad ...(12.18)$$

A three-phase half-wave converter drive is not normally used in industrial applications as it introduces dc component in the ac supply line.

It is seen from the waveforms of Fig. 12.11 (c) that

rms value of armature current, $I_{ar} = I_a$

rms value of phase or line current,

$$I_{sr} = \sqrt{I_a^2 \frac{2\pi}{3} \cdot \frac{1}{2\pi}} = I_a \sqrt{\frac{1}{3}} \qquad \qquad ...(12.19)$$

average thyristor current, $\quad I_{TA} = I_a \cdot \frac{2\pi}{3} \cdot \frac{1}{2\pi} = \frac{1}{3} I_a \qquad \qquad ...(12.20)$

rms thyristor current, $\qquad I_{Tr} = I_{sr} = \sqrt{\frac{1}{3}} \, I_a \qquad \qquad ...(12.19)$

12.4.2. Three-phase Semiconverter Drives

The circuit diagram for a 3-phase semiconverter feeding a separately-excited dc motor is shown in Fig. 12.12. The field winding of the motor is also connected to three-phase semiconverter. This drive offers one quadrant operation and is used up to about 115 kW ratings.

On the assumption of continuous and ripple free armature current, waveforms for line current i_A and thyristor current i_{T1} are sketched in Fig. 12.13 for firing angle $\alpha = 30°$ and also for $\alpha = 90°$. An examination of these waveforms would reveal that

(*i*) for firing angle $\alpha \le 60°$, each thyristor conducts for 120° and

(*ii*) for $60° < \alpha < 180°$, each thyristor conducts for $(180° - \alpha)$.

Note that freewheeling diode comes into conduction only when firing angle of 3-phase semiconverter is more than 60° *i.e.* when $\alpha > 60°$. The waveforms of Fig. 12.13 reveal that freewheeling diode conducts for $(\alpha - 60°)$ in case load, or armature, current is continuous. Also, conduction angle of thyristor + conduction angle of freewheeling diode = 120°, when armature current is continuous.

As armature current is ripple free, rms value of armature current, $I_{ar} = I_a$. It is also seen from Fig. 12.13 as under :

For $\alpha < 60°$, rms value of supply line current, i_A is given by,

$$I_{sr} = \sqrt{I_a^2 \frac{2\pi}{3} \cdot \frac{1}{\pi}}$$

$$= I_a \sqrt{\frac{2}{3}} \qquad \qquad ...(12.21)$$

and rms value of thyristor current i_{T1} is given by

$$I_{Tr} = \left[I_a^2 \frac{2\pi}{3} \cdot \frac{1}{2\pi} \right]^{1/2} = I_a \sqrt{\frac{1}{3}} \qquad \qquad ...(12.22)$$

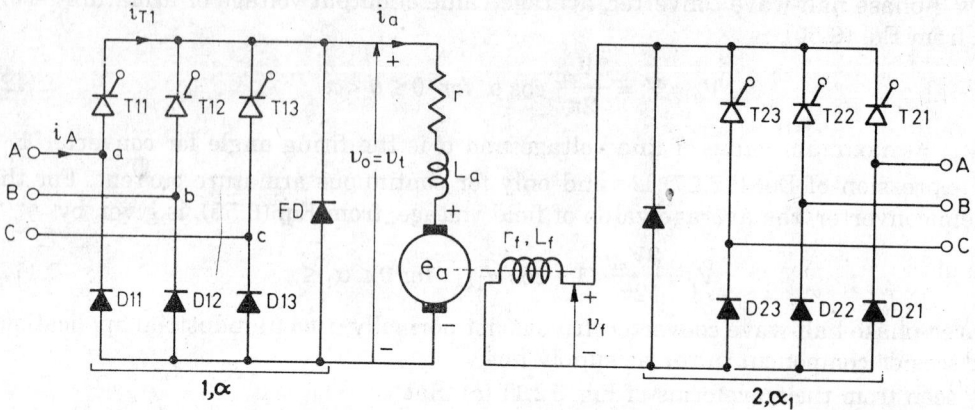

Fig. 12.12. Three-phase semiconverters feeding a separately-excited dc motor.

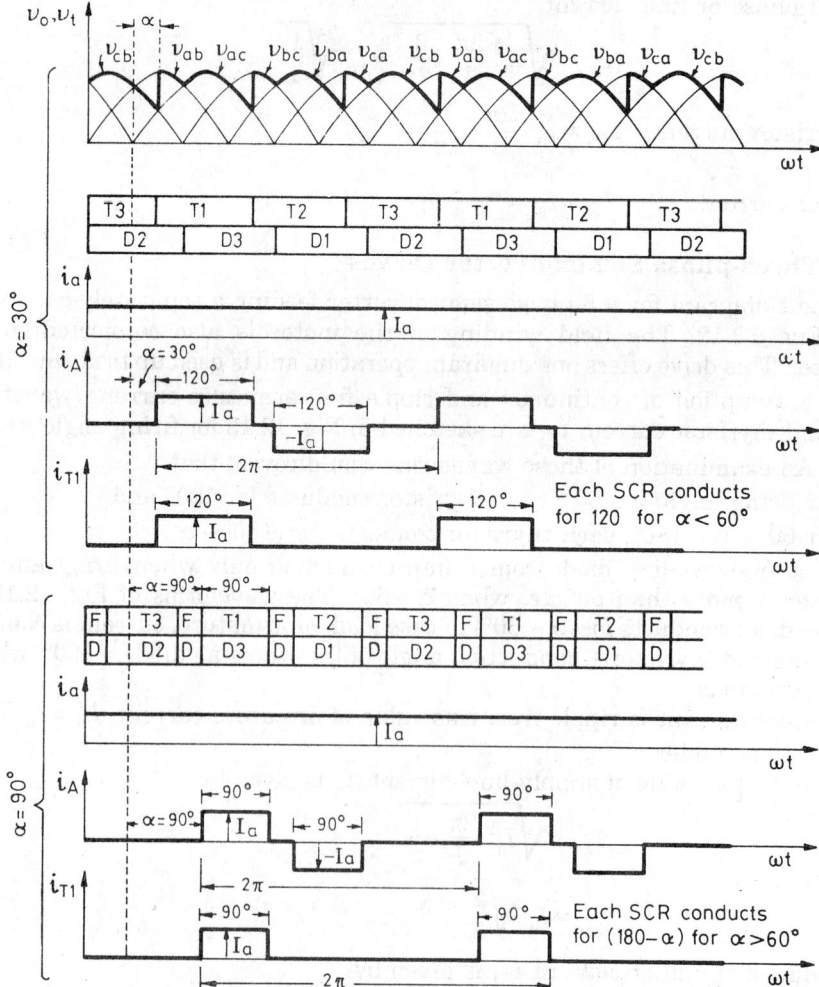

Fig. 12.13. Voltage and current waveforms for a
three-phase semiconverter drive of Fig. 12.12.

For $60° < \alpha < 180°$, rms value of supply line current i_A is given by

$$\bar{i}_{sr} = \left[I_a^2 \left(\frac{180 - \alpha}{180} \right) \right]^{1/2} = I_a \sqrt{\frac{180 - \alpha}{180}} \qquad \text{...(12.23)}$$

and rms value of thyristor current i_{T1} is given by

$$I_{Tr} = I_a \left(\frac{180 - \alpha}{360} \right)^{1/2} \qquad \text{...(12.24)}$$

From above, it is obvious that average thyristor current is $\frac{1}{3} I_a$ for $\alpha < 60°$ and $\left(\frac{180 - \alpha}{360} \right) I_a$ for $60° < \alpha < 180°$.

For $60° < \alpha < 180°$, freewheeling diode has rms value of $I_a \sqrt{\frac{\alpha - 60}{120}}$ and average value of $I_a \cdot \frac{\alpha - 60}{120}$.

For converter 1, $\qquad V_0 = V_t = \dfrac{3V_{ml}}{2\pi} (1 + \cos \alpha)$ for $0 < \alpha < \pi$ $\qquad$ ...(12.25 a)

For converter 2, $\qquad V_f = \dfrac{3V_{ml}}{2\pi} (1 + \cos \alpha_1)$ for $0 < \alpha_1 < \pi$ $\qquad$ ...(12.25 b)

Example 12.8. *The speed of a separately-excited dc motor is controlled by means of a 3-phase semiconverter from a 3-phase, 415V, 50 Hz supply. The motor constants are : inductance 10 mH, resistance 0.9 ohm and armature constant 1.5 V/rad/s (Nm/A). Calculate the speed of this motor at a torque of 50 Nm when the converter is fired at 45°. Neglect losses in the converter.*

Solution. Armature constant, $K_m = 1.5$ V/rad/s or 1.5 Nm/A.

Motor torque, $\qquad\qquad T_e = K_m I_a = 50$ Nm

∴ Motor armature current, $I_a = \dfrac{50}{1.5} = \dfrac{100}{3}$ A

The equation for the converter-motor combination is

$$\frac{3V_{ml}}{2\pi} (1 + \cos \alpha) = E_a + I_a r_a = K_m \omega_m + I_a r_a$$

$$\frac{3\sqrt{2} \times 415}{2\pi} (1 + \cos 45°) = 1.5 \times \omega_m + \frac{100}{3} \times 0.9$$

$$478.3 = 1.5\, \omega_m + 30$$

or $\qquad\qquad \omega_m = \dfrac{478.3 - 30}{1.5} = 298.867$ rad/s

or $\qquad\qquad \dfrac{2\pi N}{60} = \omega_m = 298.867$ rad/s

∴ Motor speed, $\qquad N = \dfrac{298.867 \times 60}{2\pi} = 2853.97$ rpm

Example 12.9. *A 600V, 1500 rpm, 80A separately-excited dc motor is fed through a three-phase semiconverter from 3-phase 400V supply. Motor armature resistance is 1 Ω. Armature current is assumed constant.*

(a) For a firing angle of 45° at 1200 rpm, compute the rms values of source and thyristor currents, average value of thyristor current and the input supply power factor.

(b) Repeat part (a) for a firing angle of 90° at 700 rpm.

Solution. Under rated operating conditions,

$$V_t = E_a + I_a \, r_a = K_m \, \omega_m + I_a r_a$$

$$600 = K_m \frac{2\pi \times 1500}{60} + 80 \times 1$$

or Motor constant $\qquad K_m = \dfrac{520 \times 60}{2\pi \times 1500} = 3.31$ V–s/rad (or Nm/A).

(a) For the converter-motor combination,

$$V_t = \frac{3V_{ml}}{2\pi}(1 + \cos\alpha) = E_a + I_a \, r_a = K_m \, \omega_m + I_a \, r_a$$

$$\frac{3\sqrt{2} \times 400}{2\pi}(1 + \cos 45°) = 3.31 \times \frac{2\pi \times 1200}{60} + I_a \times 1$$

$$461.01 = 415.95 + I_a$$

∴ Armature current, $\qquad I_a = 45.06$ A

From Eq. (12.21), rms value of source current,

$$I_{sr} = I_a \sqrt{\frac{2}{3}} = 45.06 \sqrt{\frac{2}{3}} = 36.791 \text{ A}$$

From Eq. (12.22), rms value of thyristor current,

$$= I_a \frac{1}{\sqrt{3}} = 45.06 \frac{1}{\sqrt{3}} = 26.015 \text{ A}$$

Average value of thyristor current

$$= \frac{1}{3} \times 45.06 = 15.02 \text{ A}$$

Input supply power factor $\qquad = \dfrac{V_t \cdot I_a}{\sqrt{3} \cdot V_s \cdot I_{sr}} = \dfrac{461.01 \times 45.06}{\sqrt{3} \times 400 \times 36.791} = 0.815$ lag

(b) $V_t = \dfrac{3\sqrt{2} \times 400}{2\pi}(1 + \cos 90°) = 3.31 \times \dfrac{2\pi \times 700}{60} + I_a \times 1$

$$270.05 = 242.64 + I_a$$

∴ Armature current, $\qquad I_a = 27.41$ A

Rms value of source current, $\qquad I_{sr} = 27.41 \cdot \sqrt{\frac{2}{3}} = 22.38$ A

Rms value of thyristor current, $I_{Tr} = 27.41 \dfrac{1}{\sqrt{3}} = 15.825$ A

Average value of thyristor current $= \dfrac{27.41}{3} = 9.137$ A

Input supply power factor $\qquad = \dfrac{270.05 \times 27.41}{\sqrt{3} \times 400 \times 22.38} = 0.4774$ lag.

12.4.3. Three-phase Full-converter Drives

The circuit diagram, consisting of one three-phase full converter in the armature circuit and another 3-phase (or 1-phase) full converter in the field circuit, is as shown in Fig. 12.14. It offers two-quadrant drive and is used up to about 1500 kW drives. For regenerative purposes, the polarity of counter emf is reversed by reversing the field excitation by making the firing-angle delay of converter 2 more than 90°.

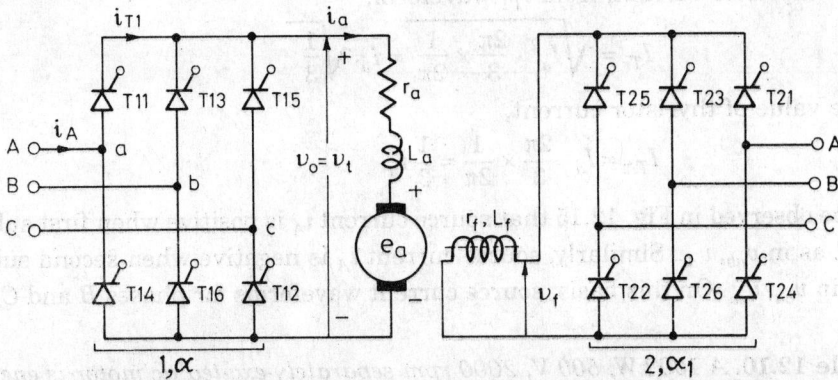

Fig. 12.14. Three-phase full converters feeding a separately-excited dc motor.

For converter 1 in the armature circuit, the average output voltage, from Eq. (6.54), is given by

$$V_0 = V_t = \frac{3 V_{ml}}{\pi} \cos \alpha \text{ for } 0 \le \alpha \le \pi \qquad \text{...(12.26 }a\text{)}$$

For converter 2 in the field circuit,

$$V_f = \frac{3 V_{ml}}{\pi} \cos \alpha_1 \text{ for } 0 \le \alpha_1 \le \pi \qquad \text{...(12.26 }b\text{)}$$

where V_{ml} = maximum value of line voltage.

Voltage and current waveforms for $\alpha = 30°$ and for constant armature current are sketched in Fig. 12.15. It is seen from this figure that each thyristor conducts for 120° for continuous armature current. This gives rms value of armature current, $I_{ar} = I_a$.

rms value of source current, from i_A waveform,

$$I_{sr} = \sqrt{I_a^2 \cdot \frac{2\pi}{3} \times \frac{1}{\pi}} = I_a \sqrt{\frac{2}{3}} \qquad \text{...(12.27)}$$

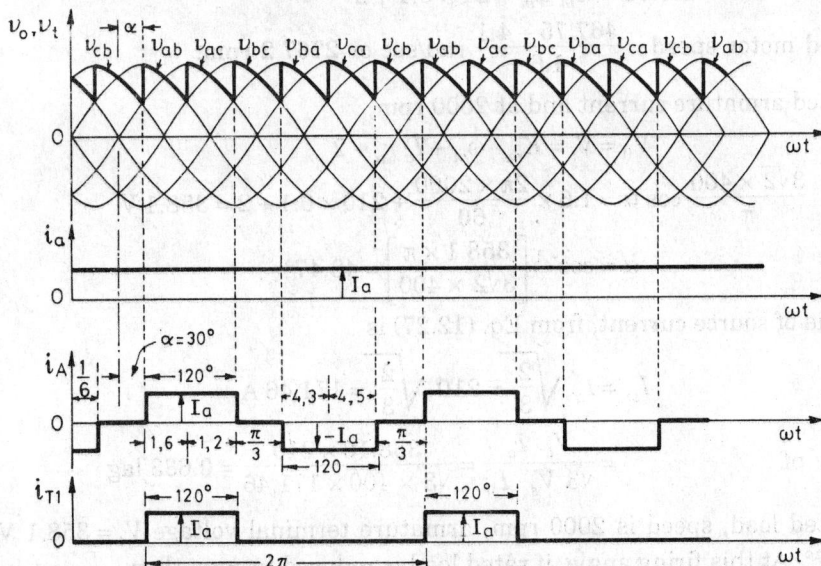

Fig. 12.15. Voltage and current waveforms for firing angle of 30°
for a three-phase full-converter drive of Fig. 12.14.

rms value of thyristor current, from i_{T1} waveform,

$$I_{Tr} = \sqrt{I_a^2 \cdot \frac{2\pi}{3} \times \frac{1}{2\pi}} = I_a \sqrt{\frac{1}{3}} \qquad \qquad ...(12.28)$$

and average value of thyristor current,

$$I_{TA} = I_a \cdot \frac{2\pi}{3} \times \frac{1}{2\pi} = \frac{1}{3} I_a \qquad \qquad ...(12.29)$$

It may be observed in Fig. 12.15 that source current i_A is positive when first subscript with voltage is a, as in v_{ab}, v_{ac}. Similarly, source current i_A is negative when second subscript is a, just as it is in v_{ba}, v_{ca}. On this basis, source current waveforms for phases B and C can also be sketched.

Example 12.10. *A 100 kW, 500 V, 2000 rpm separately-excited dc motor is energised from 400 V, 50 Hz, 3-phase source through a 3-phase full converter. The voltage drop in conducting thyristors is 2V. The dc motor parameters are as under :*

$$r_a = 0.1 \ \Omega, \ K_m = 1.6 \ V\text{-}s/rad, \ L_a = 8 \ mH.$$

Rated armature current = 210 A. No-load armature current = 10% of rated current. Armature current is continuous and ripple free.

(a) Find the no-load speed at firing angle of 30°.

(b) Find the firing angle for a speed of 2000 rpm at rated armature current. Determine also the supply power factor.

(c) Find the speed regulation for the firing angle obtained in part (b).

Solution. (a) The motor terminal voltage,

$$V_0 = V_t = \frac{3\sqrt{2} \times 400}{\pi} \cos 30° = 467.75 \ V$$

Also $\qquad \qquad V_t = E_a + I_a r_a + 2$

or $\qquad \qquad 467.75 = K_m \ \omega_m + 21 \times 0.1 + 2$

$\therefore$ No-load motor speed $= \dfrac{467.75 - 4.1}{1.6}$ rad/sec or **2767.2 rpm.**

(b) At rated armature current and at 2000 rpm,

$$V_0 = V_t = K_m \cdot \omega_m + I_a \ r_a + 2$$

or $\qquad \dfrac{3\sqrt{2} \times 400}{\pi} \cos \alpha = 1.6 \times \dfrac{2\pi \times 2000}{60} + 210 \times 0.1 + 2 = 358.1 \ V$

or $\qquad \alpha = \cos^{-1}\left[\dfrac{358.1 \times \pi}{3\sqrt{2} \times 400} \right] = 48.47°$

Rms value of source current, from Eq. (12.27) is

$$I_{sr} = I_a \sqrt{\frac{2}{3}} = 210 \sqrt{\frac{2}{3}} = 171.46 \ A$$

$\therefore$ Supply pf $\qquad = \dfrac{V_t \cdot I_a}{\sqrt{3} \ V_s \cdot I_{sr}} = \dfrac{358.10 \times 210}{\sqrt{3} \times 400 \times 171.46} = 0.633$ lag

(c) At rated load, speed is 2000 rpm, armature terminal voltage $V_t = 358.1$ V and firing angle is 48.47°. At this firing angle, if rated load is reduced to zero, then

$$V_0 = V_t = 358.1 = K_m \cdot \omega_m + 21 \times 0.1 + 2$$

or $\qquad\qquad \omega_m = \dfrac{358.1 - 4.1}{1.6}$ rad/sec or 2112.8 rpm

∴ Speed regulation $\qquad = \dfrac{2112.8 - 2000}{2000} \times 100 = 5.64\%$

Example 12.11. *A 230V, 1500 rpm, 20 A separately-excited dc motor is fed from 3-phase full converter. Motor armature resistance is 0.6 Ω. Full converter is connected to 400 V, 50 Hz source through a delta-star transformer. Motor terminal voltage is rated when converter firing angle is zero.*

(a) Calculate the transformer phase turns-ratio from primary to secondary.

(b) Calculate the firing angle delay of the converter when (i) the motor is running at 1000 rpm at rated torque and (ii) the motor is running at (– 900) rpm and at half the rated torque.

Solution. (a) For zero degree firing angle, motor terminal voltage is rated *i.e.* 230 V. Therefore,

$$\frac{3\sqrt{2}\ V_l}{\pi} \cos 0° = V_t = 230 \text{ V}$$

or $\qquad\qquad V_l = \dfrac{230 \times \pi}{3\ \sqrt{2} \times 1} = 170.34 \text{ V}$

Here V_l is the line voltage. Per-phase voltage on transformer star side is

$$V_{ph} = \frac{170.34}{\sqrt{3}} = 98.35 \text{ V}$$

Per-phase voltage input to transformer delta = 400 V

∴ Transformer phase turns ratio from primary to secondary

$$= \frac{400}{98.35} = 4.067.$$

(b) (i) At 1500 rpm, $\qquad E_a = V_t - I_a\,r_a = 230 - 20 \times 0.6 = 218$ V

At 1000 rpm, motor emf $\quad = \dfrac{218}{1500} \times 1000 = 145.33$ V

For this motor emf, armature terminal voltage at rated torque is

$$V_t = E_a + I_a\,r_a = 145.33 + 20 \times 0.6 = 157.33 \text{ V}$$

But $\qquad\qquad \dfrac{3V_{ml}}{\pi} \cos \alpha = V_0 = V_t = 157.33$ V

or $\qquad\qquad \alpha = \cos^{-1}\left[\dfrac{157.33 \times \pi}{3\ \sqrt{2} \times 170.34}\right] = 46.84°$

(ii) At half the rated torque, armature current $I_a = \dfrac{1}{2} \times$ rated current

$$= \frac{1}{2} \times 20 = 10 \text{ A}$$

∴ $\qquad \dfrac{3\ \sqrt{2} \times 170.34}{\pi} \cos \alpha = -\dfrac{900}{1500} \times 218 + 10 \times 0.6 = -124.8$ V

or $\qquad\qquad \alpha = \cos^{-1}\left[\dfrac{-124.8 \times \pi}{3\ \sqrt{2} \times 170.34}\right] = 122.861°$

Example 12.12. *A 230 V, 10 kW, 1000 rpm separately-excited dc motor has its armature resistance of 0.3 Ω and field resistance of 300 Ω. The speed of this motor is controlled by two 3-phase full converters, one in the armature circuit and the other in the field circuit and both are fed from 400 V, 50 Hz source. The motor constant is 1.1 V-s/A.rad. Armature and field currents are ripple free.*

(a) With field converter setting to maximum field current, calculate firing angle for the armature converter for load torque of 60 Nm at rated speed.

(b) With the load torque as in part (a) and zero degree firing angle for armature converter, speed is to be raised to 3000 rpm. Determine the firing angle of the field converter.

Solution. (a) For maximum field current, firing angle of field converter is zero. Therefore, field voltage,

$$V_f = \frac{3V_{ml}}{\pi} = \frac{3\sqrt{2} \times 400}{\pi} = 540.1 \text{ V}.$$

Field current, $I_f = \dfrac{540.1}{300} = 1.8 \text{ A}$

Motor emf, $E_a = K_a \, \phi \, \omega_m$

With no saturation, $\phi = KI_f,$ $\therefore E_a = K_a \, K I_f \cdot \omega_m = k \, I_f \cdot \omega_m$

where k is a constant in V-s/A.rad

Motor torque, $T_e = K_a \, \phi \, I_a = K_a \, K I_f \cdot I_a = k \, I_f \cdot I_a$

or $60 = 1.1 \times 1.8 \times I_a$

$\therefore$ Motor current, $I_a = \dfrac{60}{1.1 \times 1.8} = 30.30 \text{ A}$

For the motor converter, $V_t = V_0 = \dfrac{3V_{ml}}{\pi} \cos \alpha = E_a + I_a \, r_a = k \, I_f \cdot \omega_m + I_a \, r_a$

$$\frac{3\sqrt{2} \times 400}{\pi} \cos \alpha = 1.1 \times 1.8 \times \frac{2\pi \times 1000}{60} + 30.30 \times 0.3 = 216.435 \text{ V}$$

$\therefore$ Firing angle of armature converter,

$$\alpha = \cos^{-1}\left[\frac{216.435 \times \pi}{3\sqrt{2} \times 400}\right] = 66.376°$$

(b) With zero degree firing angle of the armature converter,

$$\frac{3\sqrt{2} \times 400}{\pi} \cos 0° = 1.1 \times I_f \times \frac{2\pi \times 3000}{60} + 30.30 \times 0.3$$

or $I_f = \dfrac{540.1 - 9.09}{345.58} = 1.5366 \text{ A}$

$\therefore$ Field voltage, $V_f = I_f \cdot r_f = 1.5366 \times 300 = \dfrac{3\sqrt{2} \times 400}{\pi} \cos \alpha_1$

$\therefore$ Firing angle of field converter,

$$\alpha_1 = \cos^{-1}\left[\frac{300 \times 1.5366 \times \pi}{3\sqrt{} \times 400}\right] = 31.406°$$

Example 12.13. *In a speed controlled dc drive, the load torque is 40 Nm. At time t = 0, the operation is under steady state and the speed is 500 rpm. Under this condition at t = 0+, the generated torque is instantly increased to 100 Nm. The inertia of the drive is 0.01 Nm · sec²/rad. The friction is negligible.*

(a) *Write down the differential equation governing the speed of the drive for t > 0.*

(b) *Evaluate the time taken for the speed to reach 1000 rpm.* [GATE, 1998]

Solution. (a) At $t = 0$, steady state exists and therefore, generated torque, $T_e = T_L$, load torque

In general, the dynamic equation for the motor-load combination is generated (or motor) torque = inertia torque + friction torque + load torque

or

$$T_e = J \frac{d\omega_m}{dt} + D\omega_m + T_L$$

As friction torque is zero, $D\omega_m = 0$. This gives the differential equation, governing the speed of the drive at $t > 0$, as

$$T_e = J \frac{d\omega_m}{dt} + T_L$$

$$100 = 0.01 \frac{d\omega_m}{dt} + 40 \qquad \qquad ...(i)$$

(b) From Eq. (i), $\qquad \dfrac{d\omega_m}{dt} = \dfrac{60}{0.01} = 6000$

or $\qquad \qquad dt = \dfrac{d\omega_m}{6000}$

Its integration gives, $\qquad t = \dfrac{1}{6000} \cdot \omega_m + A \qquad \qquad ...(ii)$

Initial speed at $t = 0 +$ remains 500 rpm. Therefore

$$\omega_{mo} = \frac{2\pi \times 500}{60} = \frac{100\,\pi}{6} \text{ rad/sec}$$

From Eq. (ii), $\qquad 0 = \dfrac{1}{6000} \times \dfrac{100\,\pi}{6} + A \quad$ or $\quad A = \dfrac{-\pi}{360}$

$\therefore \qquad \qquad t = \dfrac{\omega_m}{6000} - \dfrac{\pi}{360}$

Final speed $\qquad \omega_m = \dfrac{2\pi \times 1000}{60} = \dfrac{200\,\pi}{6} \text{ rad/sec}$

$\therefore \qquad \qquad t = \dfrac{200\,\pi}{6000 \times 6} - \dfrac{\pi}{360} = \dfrac{\pi}{360} \text{ sec} = 0.0873 \text{ sec}$

$\therefore$ Time taken for the speed to reach 1000 rpm = 0.0873 sec.

Example 12.14. *A dc motor driven from a fully-controlled 3-phase converter shown in Fig. 12.16 draws a dc current of 100 A with negligible ripple.*

(a) *Sketch the ac line current i_A for one cycle.*

(b) *Determine the 3rd and 5th harmonic components of the line current as a percentage of the fundamental current.* [GATE, 1998]

Solution. (a) The ac line current i_A for one cycle is sketched in Fig. 12.15 for a firing angle α under the assumption of negligible ripple in the armature current $I_a = 100$ A.

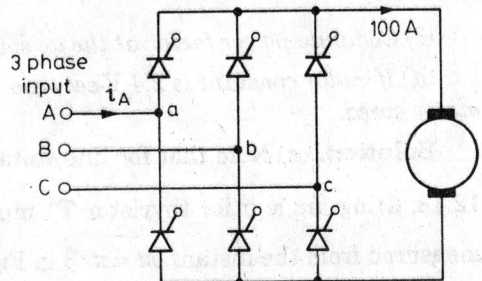

Fig. 12.16. Pertaining to Example 12.14.

(b) The line current i_A shown in Fig. 12.15 can be expressed in Fourier series as

$$i_{sn} = i_{An} = \sum_{n=1,3,5,-}^{\infty} \frac{4I_a}{n\pi} \cos \frac{n\pi}{6} \sin(n\omega t - n\alpha)$$

Rms value of the nth harmonic line current is given by

$$I_{sn} = \frac{4I_a}{\sqrt{2} \cdot n\pi} \cos \frac{n\pi}{6} = \frac{2\sqrt{2} \cdot I_a}{n\pi} \cdot \cos \frac{n\pi}{6}$$

Rms value of fundamental current,

$$I_{s1} = \frac{2\sqrt{2}\,I_a}{\pi} \cos 30° = \frac{\sqrt{6}}{\pi} I_a$$

Rms value of third-harmonic current,

$$I_{s3} = \frac{2\sqrt{2}\,I_a}{3\pi} \cos 90° = 0$$

Rms value of fifth-harmonic current,

$$I_{s5} = \frac{2\sqrt{2}\,I_a}{5\pi} \cos 150° = -\frac{\sqrt{6}}{5\pi} I_a$$

From above, third harmonic current as a percentage of fundamental current = 0% and fifth harmonic current as a percentage of fundamental current

$$= \frac{I_{s5}}{I_{s1}} \times 100 = -\frac{\sqrt{6}\,I_a}{5\pi} \times \frac{\pi}{\sqrt{6} \cdot I_a} \times 100 = -20\%.$$

Example 12.15. *A dc motor driven from a 3-phase full converter shown in Fig. 12.17 draws a dc line current of 60 A with negligible ripple.*

(a) Sketch the line voltage v_{ab} taking it zero-crossing and becoming positive at $\omega t = 0$. Also, sketch the line current i_A for one cycle for $\alpha = 150°$. Indicate also the conduction of devices. Thyristor current i_T should also be sketched.

(b) Calculate average and rms values of thyristor current.

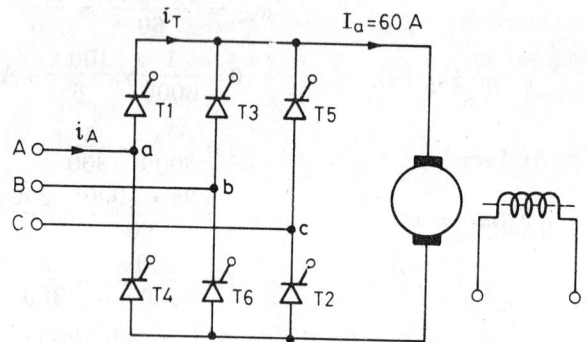

Fig. 12.17. Three-phase full converter feeding a dc motor, Example 12.15.

(c) Compute power factor at the ac source.

(d) If motor constant is 2.4 V-sec/rad and armature circuit resistance is 0.5 Ω, calculate the motor speed.

Solution. (a) Note that for line voltages $v_{ab}, v_{ac}, v_{bc}, v_{ba}$ etc. and with v_{ab} as shown in Fig. 12.18, firing angle α for thyristor T1 must be measured from $\omega t = \frac{\pi}{3}$. Accordingly, $\alpha = 150°$ is measured from the instant $\omega t = \pi/3$ in Fig. 12.18.

Motor current $I_a = 60$ A is shown constant in Fig. 12.18. At $\alpha = 150°$, T1 is turned on. So voltage v_{ab} will send constant current I_a through T1, T6. Thyristor T1 will conduct for 120°; for

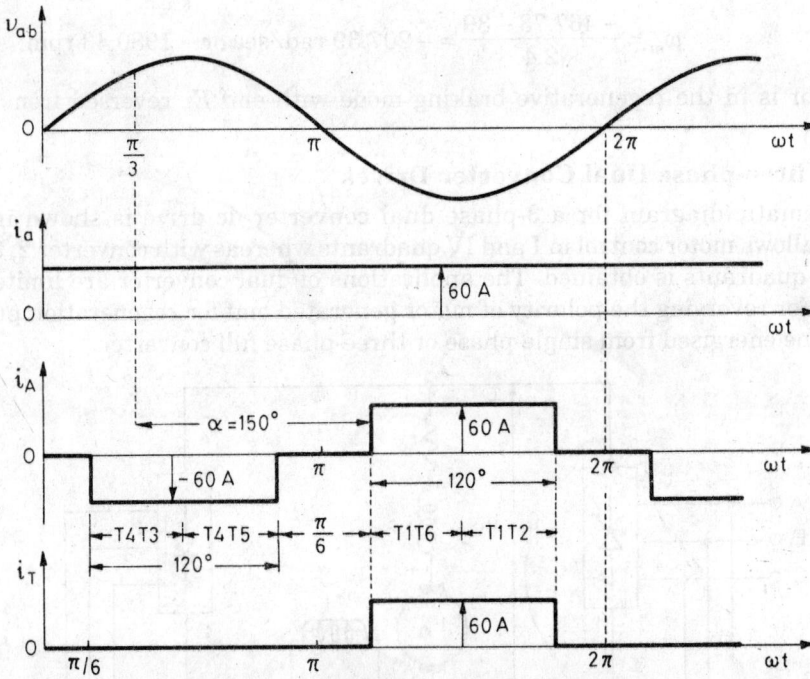

Fig. 12.18. Waveforms for Example 12.15.

the first 60°, T1, T6 conduct together. For the next 60°; T1, T2 conduct together as shown in Fig. 12.18. Voltage $v_{ba} = -v_{ab}$ will cause T3, T4 to conduct for 60° and v_{ca} will force T5, T4 to conduct for the next 60° as shown. Note that voltages v_{ba}, v_{ca} will cause line current i_A to be negative whereas for v_{ab}, v_{ac}, line current i_A is positive. Thyristor current i_T through T1 will flow only when i_A is positive.

(b) Average thyristor current, $I_{TA} = \dfrac{I_a}{3} = \dfrac{60}{3} = 20$ A

Rms thyristor current, $I_{Tr} = \dfrac{I_a}{\sqrt{3}} = \dfrac{60}{\sqrt{3}} = 34.642$ A

(c) Rms value of source current, $I_{sr} = I_a \sqrt{\dfrac{2}{3}} = 60 \cdot \sqrt{\dfrac{2}{3}}$ A

Power delivered to motor $= V_t \cdot I_a$

Power factor at ac source
$$= \dfrac{V_t \cdot I_a}{\sqrt{3} \cdot V_s \cdot I_{sr}}$$

$$= \left[\dfrac{3\sqrt{2}\, V_s}{\pi} \cos 150° \right] \times I_a \dfrac{\sqrt{3}}{\sqrt{3}\, V_s \cdot I_a \sqrt{2}}$$

$$= \dfrac{3}{\pi} \cos 150 = -0.827.$$

Minus sign for the power factor merely indicates the system to be in the inversion mode.

(d)
$$V_0 = V_t = \dfrac{3\sqrt{2} \times 400}{\pi} \cos 150° = K_m\, \omega_m + 60 \times 0.5$$

$$\therefore \qquad \omega_m = \frac{-467.73 - 30}{2.4} = -207.39 \text{ rad/sec or } -1980.43 \text{ rpm.}$$

The motor is in the regenerative braking mode with emf E_a reversed from its motoring mode polarity.

12.4.4. Three-phase Dual Converter Drives

The schematic diagram for a 3-phase dual converter dc drive is shown in Fig. 12.19. Converter 1 allows motor control in I and IV quadrants whereas with converter 2, the operation in II and III quadrants is obtained. The applications of dual converter are limited to about 2 MW-drives. For reversing the polarity of motor generated emf for regeneration purposes, field circuit must be energised from single-phase or three-phase full converter.

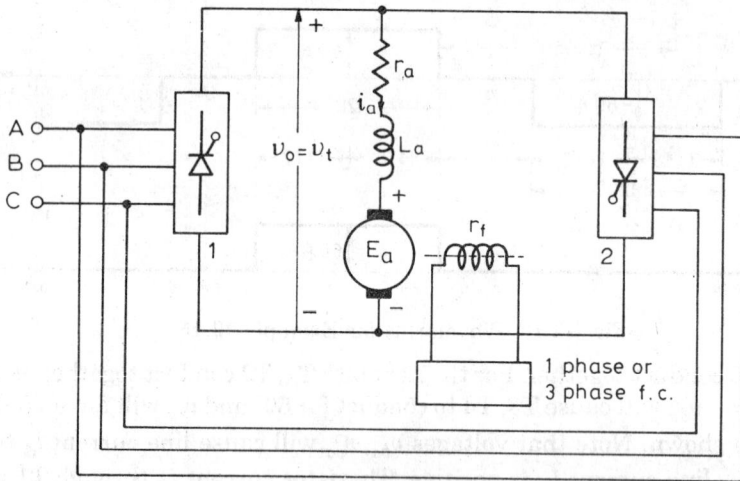

Fig. 12.19. Three-phase dual converter controlled separately-excited dc motor.

When converter 1, or 2, is in operation, average output voltage is

$$V_0 = V_t = \frac{3V_{ml}}{\pi} \cos \alpha_1 \quad \text{for } 0 \le \alpha_1 \le \pi \qquad \text{...(12.30)}$$

With a 3-phase full converter in the field circuit,

$$V_f = \frac{3V_{ml}}{\pi} \cos \alpha_f \qquad \text{for } 0 \le \alpha_f \le \pi \qquad \text{...(12.31)}$$

In case circulating current-type dual converter of Fig. 6.45 is used, then as per Eq. (6.85),

$$\alpha_1 + \alpha_2 = 180°$$

12.5. CHOPPER DRIVES

When variable dc voltage is to be obtained from fixed dc voltage, dc chopper is the ideal choice. Use of chopper in traction systems is now accepted all over the world. A chopper is inserted in between a fixed voltage dc source and the dc motor armature for its speed control below base speed. In addition, chopper is easily adaptable for regenerative braking of dc motors and thus kinetic energy of the drive can be returned to the dc source. This results in overall energy saving which is the most welcome feature in transportation systems requiring frequent stops, as for example in rapid transit systems. Chopper drives are also used in battery-operated vehicles where energy saving is of prime importance.

Though choppers can be used for dynamic braking and for combined regenerative and dynamic control of dc drives, only the following two control modes are described in what follows.

1. Power control or motoring control.

2. Regenerative-braking control.

Both the chopper control methods are now described. In addition, two-quadrant and four-quadrant chopper drives are also discussed.

12.5.1. Power Control or Motoring Control

Fig. 12.20 (*a*) shows the basic arrangement of a dc chopper feeding power to a dc series motor. The chopper is shown to consist of a force-commutated thyristor, it could equally well be a transistor switch. It offers one-quadrant drive, Fig. 12.20 (*b*). Armature current is assumed continuous and ripple free. The waveforms for the source voltage V_s, armature terminal voltage $v_t = v_0$, armature current i_a, dc source current i_c and freewheeling-diode current i_{fd} are sketched in Fig. 12.20 (*c*). From these waveforms, the following relations can be obtained :

$$\text{Average motor voltage, } V_0 = V_t = \frac{T_{on}}{T} \cdot V_s = \alpha V_s = f\,T_{on} \cdot V_s \qquad \qquad \qquad ...(12.32)$$

where

$$\alpha = \text{duty cycle} = \frac{T_{on}}{T}$$

and

$$f = \text{chopping frequency} = \frac{1}{T}$$

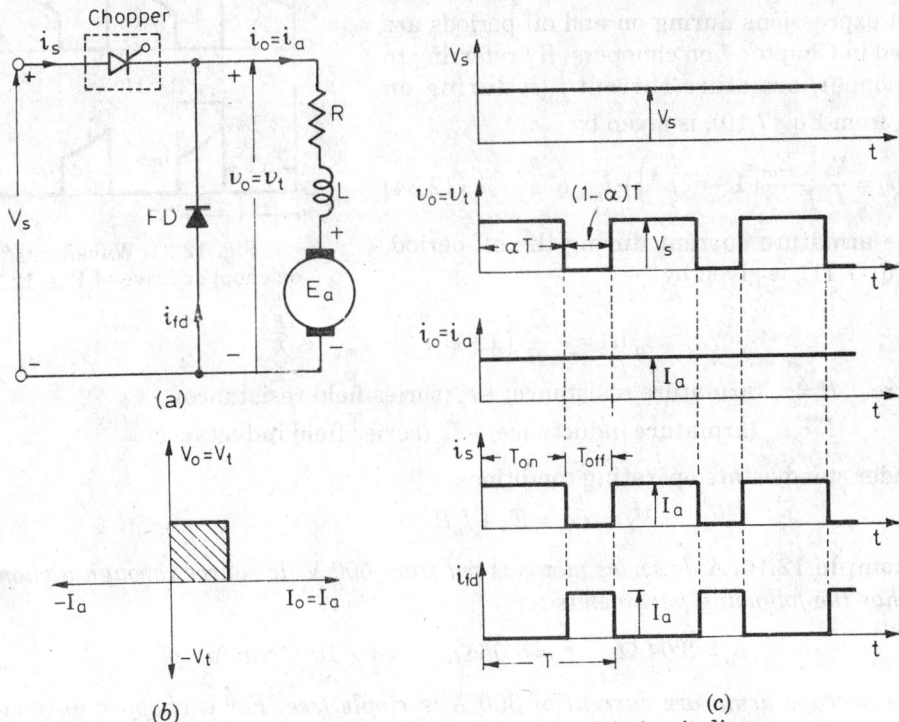

Fig. 12.20 D.C. Chopper for series motor drive (*a*) circuit diagram
(*b*) quadrant diagram and (*c*) waveforms.

Power delivered to motor = (Average motor voltage) (average motor current)

$$= V_t \cdot I_a = \alpha \cdot V_s \cdot I_a$$

Average source current $= \dfrac{T_{on}}{T} \cdot I_a = \alpha \cdot I_a$

Input power to chopper = (average input voltage) (average source current)

$$= V_s \cdot \alpha I_a$$

For the motor armature circuit,

$$V_t = \alpha V_s = E_a + I_a (r_a + r_s) = K_m \cdot \omega_m + I_a (r_a + r_s)$$

or $$\omega_m = \frac{\alpha V_s - I_a (r_a + r_s)}{K_m} \qquad \ldots(12.33)$$

It is seen from Eq. (12.33) that by varying the duty cycle α of the chopper, armature terminal voltage can be controlled and thus speed of the dc motor can be regulated.

So far, armature current i_a has been assumed ripple free and accordingly, waveforms in Fig. 12.20 are sketched. Actually, the motor armature current will rise during chopper on period and fall during off period as shown in Fig. 12.21. The current expressions during on and off periods are obtained in Chapter 7 on choppers. By referring to this chopper, armature current $i_a(t)$ during on period, from Eq. (7.10), is given by

$$i_a (t) = \frac{V_s - E_a}{R} \left(1 - e^{-\frac{R}{L} t} \right) + I_{mn} \, e^{-\frac{R}{L} t} \quad \ldots(12.34)$$

The armature current during the off-period, from Eq. (7.11), is given by

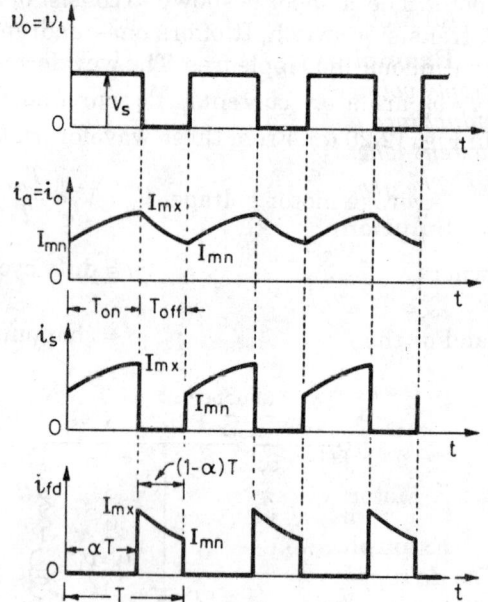

Fig. 12.21. Waveforms for dc chopper drive of Fig. 12.20 (a).

$$i_a (t) = -\frac{E_a}{R} \left(1 - e^{-\frac{R}{L} t} \right) + I_{mx} \cdot e^{-\frac{R}{L} t} \qquad \ldots(12.35)$$

Here $R = r_a$ (armature resistance) $+ r_s$ (series-field resistance)

$L = L_a$ (armature inductance) $+ L_s$ (series-field inductance)

Under steady-state operating conditions,

$$V_t = \alpha V_s = E_a + I_a R.$$

Example 12.16. *A dc series motor is fed from 600 V dc source through a chopper. The dc motor has the following parameters :*

$$r_a = 0.04 \, \Omega, \quad r_s = 0.06 \, \Omega, \quad k = 4 \times 10^{-3} \, Nm/amp^2$$

The average armature current of 300 A is ripple free. For a chopper duty cycle of 60%, determine :

(a) input power from the source

(b) motor speed and (c) motor torque.

Solution. (*a*) Power input to motor

$$= V_t \cdot I_a = \alpha \, V_s \cdot I_a$$
$$= 0.6 \times 600 \times 300 = 108 \text{ kW}.$$

(*b*) For a dc series motor,

$$\alpha V_s = E_a + I_a R = k \, I_a \, \omega_m + I_a R$$
$$0.6 \times 600 = 4 \times 10^{-3} \times 300 \times \omega_m + 300 \, (0.04 + 0.06)$$
$$\omega_m = \frac{360 - 30}{1.2} = 275 \text{ rad/sec or } 2626.1 \text{ rpm}$$

(*c*) Motor torque, $\quad T_e = k \, I_a^2 = 4 \times 10^{-3} \times 300^2 = 360 \text{ Nm}.$

Example 12.17. *The chopper used for on-off control of a dc separately-excited motor has supply voltage of 230V dc, an on-time of 10 m sec and off-time of 15 m sec. Neglecting armature inductance and assuming continuous conduction of motor current, calculate the average load current when the motor speed is 1500 rpm and has a voltage constant of $K_v = 0.5$ V/rad per sec. The armature resistance is 3 Ω.* [*I.A.S., 1985*]

Solution. Chopper duty cycle

$$\alpha = \frac{T_{on}}{T_{on} + T_{off}} = \frac{10}{10 + 15} = 0.4$$

For the motor armature circuit,

$$V_t = \alpha V_s = E_a + I_a \, r_a = K_m \cdot \omega_m + I_a r_a$$
$$0.4 \times 230 = 0.5 \times \frac{2\pi \times 1500}{60} + I_a \times 3$$

$\therefore$ Motor load current, $\quad I_a = \dfrac{92 - 25 \times \pi}{3} = 4.487 \text{ A}$

Example 12.18. *A dc chopper is used to control the speed of a separately-excited dc motor. The dc supply voltage is 220 V, armature resistance $r_a = 0.2$ Ω and motor constant $K_a \phi = 0.08$ V/rpm.*

This motor drives a constant torque load requiring an average armature current of 25 A. Determine (a) the range of speed control (b) the range of duty cycle α. Assumed the motor current to be continuous. [*I.A.S., 1990*]

Solution. For the motor armature circuit,

$$V_t = \alpha V_s = E_a + I_a r_a$$

As motor drives a constant torque load, motor torque T_e is constant and therefore armature current remains constant at 25 A.

Minimum possible motor speed is $N = 0$. Therefore,

$$\alpha \times 220 = 0.08 \times 0 + 25 \times 2.0 = 5$$
$$\alpha = \frac{5}{220} = \frac{1}{44}$$

Maximum possible motor speed corresponds to $\alpha = 1$, *i.e.* when 220 V dc is directly applied and no chopping is done.

$\therefore \qquad 1 \times 220 = 0.08 \times N + 25 \times 0.2$

or $\qquad N = \dfrac{220 - 5}{0.08} = 2687.5 \text{ rpm}$

$\therefore$ Range of speed control : $0 < N < 2687.5$ rpm and corresponding range of duty cycle :
$\dfrac{1}{44} < \alpha < 1.$

Example 12.19. *A separately-excited dc motor is fed from 220 V dc source through a chopper operating at 400 Hz. The load torque is 30 Nm at a speed of 1000 rpm. The motor has $r_a = 0$, $L_a = 2$ mH and $K_m = 1.5$ V-sec/rad. Neglecting all motor and chopper losses, calculate*

(a) the minimum and maximum values of armature current and the armature current excursion,

(b) the armature current expressions during on and off periods.

Solution. As the armature resistance is neglected, armature current varies linearly between its minimum and maximum values.

(a) Average armature current, $I_a = \dfrac{T_e}{K_m} = \dfrac{30}{1.5} = 20$ A

Motor emf, $E_a = K_m \cdot \omega_m = 1.5 \times \dfrac{2\pi \times 1000}{60} = 157.08$ V

Motor input voltage, $\alpha V_s = V_t = E_a + I_a r_s = 157.08 + 0$

$\therefore$ $\alpha = \dfrac{157.08}{220} = 0.714$

Periodic time, $T = \dfrac{1}{f} = \dfrac{1}{400} = 2.5$ ms

On-period, $T_{on} = \alpha\, T = 0.714 \times 2.5 = 1.785$ ms

Off-period, $T_{off} = (1 - \alpha)\, T = 0.715$ ms

During on-period T_{on}, armature current will rise which is governed by the equation,

$$0 + L\,\frac{di_a}{dt} + E_a = V_s$$

or

$$\frac{di_a}{dt} = \frac{V_s - E_a}{L} = \frac{220 - 157.08}{0.02} = 3146 \text{ A/s}$$

During off period, $\dfrac{di_a}{dt} = -\dfrac{E_a}{L} = \dfrac{-157.08}{0.02} = -7854$ A/s

With current rising linearly, it is seen from Fig. 12.21 that

$$I_{mx} = I_{mn} + \left(\frac{di_a}{dt} \text{ during } T_{on}\right) \times T_{on}$$

$$= I_{mn} + 3146 \times 1.785 \times 10^{-3}$$

or $I_{mx} = I_{mn} + 5.616$ ...(i)

For linear variation between I_{mn} and I_{mx}, average value of armature current

$$I_a = \frac{I_{mx} + I_{mn}}{2} = 20 \text{ A}$$

or $I_{mx} = 40 - I_{mn}$ ...(ii)

Solving Eqs. (i) and (ii), we get $I_{mx} = 22.808$ A

and $I_{mn} = 17.912$ A.

$\therefore$ Armature current excursion $= I_{mx} - I_{mn} = 22.808 - 17.912 = 5.616$ A

(b) Armature current expression during turn-on,

$$i_a(t) = I_{mn} + \left(\frac{di_a}{dt} \text{ during } T_{on}\right) \times t$$

$$= 17.192 + 3146\,t \quad \text{for } 0 \le t \le T_{on}$$

Armature current expression during turn-off,

$$i_a(t) = I_{mx} + \left(\frac{di_a}{dt} \text{ during } T_{off}\right) \times t$$

$$= 22.808 - 7854\, t \quad \text{for } 0 \le t \le T_{off}$$

Example 12.20. *Repeat Example 12.19, in case motor has a resistance of 0.2 Ω for its armature circuit.*

Solution. (a) From Example 12.19, armature current, $I_a = 20$ A and motor emf, $E_a = 157.08$ V; source voltage, $V_s = 220$ V.

For armature circuit, $\alpha V_s = V_0 = V_t = E_a + I_a r_a = 157.08 + 20 \times 0.2 = 161.08$ V

$$\therefore \qquad \alpha = \frac{161.08}{220} = 0.7322$$

$$T_{on} = \alpha T = 0.7322 \times 2.5 = 1.831 \text{ ms}$$

$$T_{off} = T - T_{on} = 0.669 \text{ ms}, \quad \frac{R}{L} = \frac{0.2}{0.02} = 10$$

During T_{on}, from Eq. (12.34), armature current is

$$i_a(t) = \frac{220 - 157.08}{0.2}(1 - e^{-10t}) + I_{mn} \cdot e^{-10t}$$

$$= 314.6(1 - e^{-10t}) + I_{mn} \cdot e^{-10t}$$

At $t = T_{on} = 1.831$ ms, current become I_{mx}. This gives

$$i_a(t) = I_{mx} = 5.7079 + 0.98187\, I_{mn} \qquad \text{...(i)}$$

During T_{off}, from Eq. (12.35), armature current is

$$i_a(t) = \frac{-157.08}{0.2}(1 - e^{-10t}) + I_{mx} \cdot e^{-10t}$$

$$= -785.4(1 - e^{-10t}) + I_{mx} \cdot e^{-10t}$$

At $t = 0.669$ ms, $i_a(t) = I_{mn}$. This gives

$$i_a(t) = I_{mn} = -5.237 + 0.9933\, I_{mx} \qquad \text{...(ii)}$$

Solving Eqs. (i) and (ii), we get

$$I_{mx} = 5.7079 + 0.98187(-5.237 + 0.9933\, I_{mx})$$

$$= 0.5658 + 0.9753\, I_{mx}$$

or

$$I_{mx} = \frac{0.5658}{0.0247} = 22.907 \text{ A}$$

$$I_{mn} = -5.237 + 0.9933 \times 22.907 = 17.516 \text{ A}$$

∴ Armature current excursion

$$= I_{mx} - I_{mn} = 22.907 - 17.516 = 5.39 \text{ A}$$

(b) Armature current expression during turn-on period is

$$i_a(t) = 314.6(1 - e^{-10t}) + 17.516\, e^{-10t}$$

Armature current expression during turn-off period is

$$i_a(t) = -785.4(1 - e^{-10t}) + 22.907\, e^{-10t}$$

12.5.2. Regenerative-Braking Control

In regenerative-braking control, the motor acts as a generator and the kinetic energy of the motor and connected load is returned to the supply.

During motoring mode, armature current $I_a = \dfrac{V_t - E_a}{r_a}$, i.e. armature current is positive and the motor consumes power. In case load drives the motor at a speed such that average value of motor counter emf E_a ($= K_m \cdot \omega_m$) exceeds V_t, I_a is reversed and power is delivered to the dc bus. The motor is then working as a generator in the regenerative braking mode.

The principle of regenerative braking mode is explained with the help of Fig. 12.22 (a), where a separately-excited dc motor and a chopper are shown. For active loads, such as a train going down the hill or a descending hoist, let it be assumed that motor counter emf E_a is more than the source voltage V_s. When chopper CH is on, current through armature inductance L_a rises as the armature terminals get short circuited through CH. Also, $v_t = 0$ during T_{on}. When chopper is turned off, E_a being more than source voltage V_s, diode D conducts and the energy stored in armature inductance is transferred to the source. During T_{off}, $v_t = V_s$. On the assumption of continuous and ripple free armature current, the relevant voltage and current waveforms are shown in Fig. 12.22 (b).

With respect to first quadrant operation as offered by motoring control of Fig. 12.20 (a), regenerative braking control offers second quadrant operation as armature terminal voltage has the same polarity but the direction of armature current is reversed, Figs. 12.22 (a) and (c). From the waveforms of Fig. 12.22 (b), the following relations can be derived :

The average voltage across chopper (or armature terminals) is

$$V_t = \frac{T_{off}}{T} \cdot V_s = (1 - \alpha) V_s \qquad \qquad ...(12.36)$$

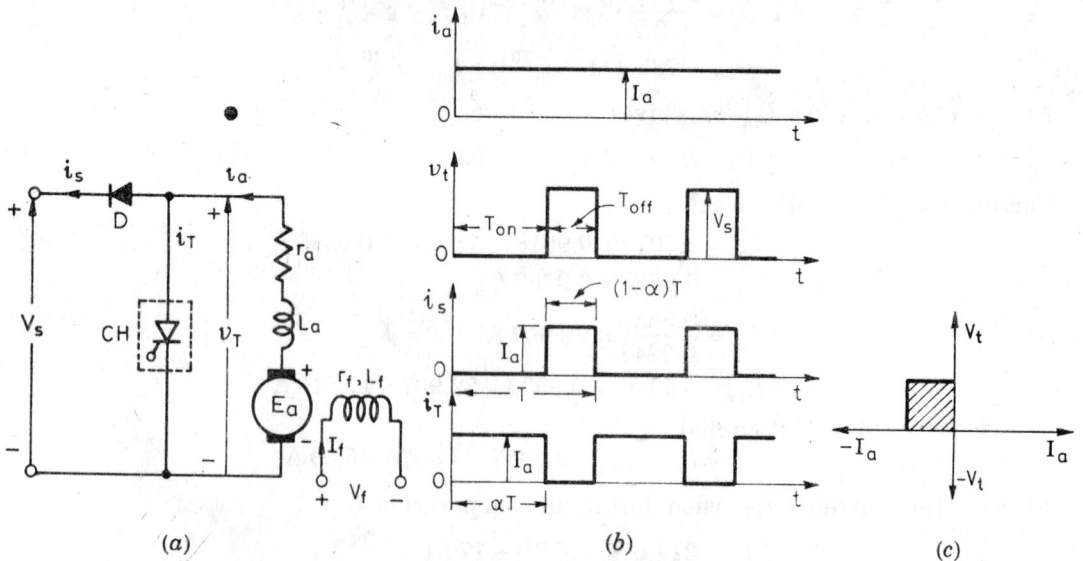

Fig. 12.22. Regenerative braking of a separately-excited dc motor
(a) circuit diagram (b) waveforms (c) quadrant diagram.

Power generated by the motor

$$= V_t \cdot I_a = (1 - \alpha) V_s \cdot I_a$$

Motor emf generated, $E_a = K_m \omega_m = V_t + I_a r_a$

$$= (1 - \alpha) V_s + I_a r_a \qquad \qquad ...(12.37)$$

Motor speed during regenerative braking,

$$\omega_m = \frac{(1 - \alpha) V_s + I_a r_a}{K_m}$$

When chopper is on, $E_a - I_a r_a - L_a \dfrac{di_a}{dt} = 0$

or

$$(E_a - I_a r_a) = L_a \cdot \frac{di_a}{dt}$$

With chopper on, L_a must store energy and current must rise, i.e. $\dfrac{di_a}{dt}$ must be positive or

$$(E_a - I_a r_a) \geq 0 \qquad \qquad ...(12.38)$$

When chopper is off, $E_a - I_a r_a - L_a \cdot \dfrac{di_a}{dt} = V_s$

or

$$V_s - (E_a - I_a r_a) = -L_a \cdot \frac{di_a}{dt}$$

With chopper off, $(E_a - I_a r_a)$ must be more than V_s for regeneration purposes and therefore $[V_s - (E_a - I_a r_a)]$ must be negative. This is possible only if current decreases during off period, i.e. $\dfrac{di_a}{dt}$ in the above expression must be negative.

$$\therefore \qquad [V_s - (E_a - I_a r_a)] \leq 0$$
$$- (E_a - I_a r_a) \leq (- V_s)$$

or

$$(E_a - I_a r_a) \geq V_s \qquad \qquad ...(12.39)$$

Eqs. (12.38) and (12.39) can be combined to give the conditions for controlling the power during regenerative braking as

$$0 \leq (E_a - I_a r_a) \geq V_s \qquad \qquad ...(12.40)$$

Eq. (12.40) gives the conditions for the two voltages and their polarity for the regenerative braking control of dc separately-excited motor.

Minimum braking speed is obtained when $E_a - I_a r_a = 0$

or

$$K_m \omega_{mn} = I_a r_a$$

$\therefore$ Minimum braking speed $\omega_{mn} = \dfrac{I_a r_a}{K_m}$ $\qquad \qquad ...(12.41)$

Maximum possible braking speed is obtained when

$$E_a - I_a r_a = V_s$$

$\therefore$ Maximum braking speed, $\omega_{mx} = \dfrac{V_s + I_a r_a}{K_m}$ $\qquad \qquad ...(12.42)$

Thus regenerative braking control is effective only when motor speed is less than ω_{mx} and more than ω_{mn}. This can be expressed as

$$\omega_{mn} < \omega_m < \omega_{mx}$$

$$\frac{I_a\,r_a}{K_m} < \omega_m < \frac{V_s + I_a\,r_a}{K_m}$$

Therefore, the speed range for regenerative braking is $\dfrac{V_s + I_a r_a}{K_m} : \dfrac{I_a\,r_a}{K_m}$ or $(V_s + I_a\,r_a) : I_a r_a$.

Regenerative braking of chopper-fed separately-excited dc motor is stable, it is therefore discussed here. DC series motors, however, offer unstable operating characteristics during regenerative braking. As such, regenerative braking of chopper-controlled series motors is difficult.

Example 12.21. *A dc chopper is used for regenerative braking of a separately-excited dc motor. The dc supply voltage is 400V. The motor has $r_a = 0.2\ \Omega$, $K_m = 1.2$ V·s/rad. The average armature current during regenerative braking is kept constant at 300 A with negligible ripple.*

For a duty cycle of 60% for a chopper, determine

(a) power returned to the dc supply

(b) minimum and maximum permissible braking speeds and

(c) speed during regenerative braking.

Solution. (a) Average armature terminal voltage,

$$V_t = (1 - \alpha)\,V_s = (1 - 0.6) \times 400 = 160 \text{ V}.$$

Power returned to the dc supply $= V_t I_a = 160 \times 300\ W = 48 \text{ kW}$

(b) From Eq. (12.41), minimum braking speed is

$$\omega_{mn} = \frac{I_a \cdot r_a}{K_m} = \frac{300 \times 0.2}{1.2} = 50 \text{ rad/s or } 477.46 \text{ rpm}$$

From Eq. (12.42), maximum braking speed is

$$\omega_{mx} = \frac{V_s + I_a \cdot r_a}{K_m} = \frac{400 + 300 \times 0.2}{1.2}$$

$$= 383.33 \text{ rad/s or } 3660.6 \text{ rpm}$$

(c) When working as a generator during regenerative braking, the generated emf is
$E_a = K_m\,\omega_m = V_t + I_a r_a = 160 + 300 \times 0.2 = 220 \text{ V}$

$\therefore$ Motor speed, $\omega_m = \dfrac{220}{1.2}$ rad/s or 1750.7 rpm

12.5.3. Two-quadrant Chopper Drives

Motoring control circuit for chopper drives offer only first-quadrant drive, because armature voltage and armature current remain positive over the entire range of speed control. In regenerative braking, second-quadrant drive is obtained as armature terminal voltage remains positive but direction of armature current is reversed.

In two-quadrant dc motor drive, both motoring mode as well as regenerative braking mode are carried out by one chopper configuration. One such circuit is shown in Fig. 12.23 (a) which consists of two choppers CH1, CH2 and two diodes D1, D2 and a separately-excited dc motor.

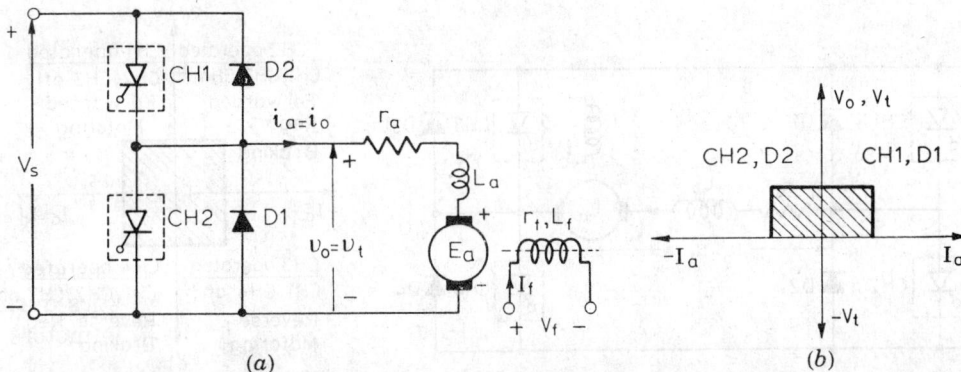

Fig. 12.23. Two-quadrant dc chopper drive (a) circuit diagram and (b) two-quadrant diagram.

Motoring mode. When chopper CH1 is on, the supply voltage V_s gets connected to armature terminals and therefore armature current i_a rises. When CH1 is turned off, i_a free wheels through D1 and therefore i_a decays. This shows that with CH1 and D1, motor control in first quadrant is obtained.

Regenerative mode. When CH2 is turned on, the motor acts as a generator and the armature current i_a rises and therefore energy is stored in armature inductance L_a. When CH2 is turned off, D2 gets turned on and therefore direction of i_a is reversed. Now the energy stored in L_a is returned to dc source and second quadrant operation is obtained, Fig. 12.23 (b). In this figure, first-quadrant operation of dc motor is sometimes called *forward-motoring mode* and second-quadrant operation as *forward regenerative-braking mode*.

12.5.4. Four-quadrant Chopper Drives

In four-quadrant dc chopper drives, a motor can be made to work in forward-motoring mode (first quadrant), forward regenerative braking mode (second quadrant), reverse motoring mode (third quadrant) and reverse regenerative-braking mode (fourth quadrant). The circuit shown in Fig. 12.24 (a) offers four-quadrant operation of a separately-excited dc motor. This circuit consists of four choppers, four diodes and a separately-excited dc motor. Its operation in the four quadrants can be explained as under :

Forward motoring mode. During this mode or first-quadrant operation, choppers CH2, CH3 are kept off, CH4 is kept on whereas *CH1 is operated*. When CH1, CH4 are on, motor voltage is positive and positive armature current rises. When CH1 is turned off, positive armature current free-wheels and decreases as it flows through CH4, D2. In this manner, controlled motor operation in first quadrant is obtained.

Forward regenerative-braking mode. A dc motor can work in the regenerative-braking mode only if motor generated emf is made to exceed the dc source voltage. For obtaining this mode, CH1, CH3 and CH4 are kept off whereas *CH2 is operated*. When CH2 is turned on, negative armature current rises through CH2, D4, E_a, L_a, r_a. When CH2 is turned off, diodes D1, D2 are turned on and the motor acting as a generator returns energy to the dc source. This results in forward regenerative-braking mode in the second-quadrant.

Reverse motoring mode. This operating mode is opposite to forward motoring mode. Choppers CH1, CH4 are kept off, CH2 is kept on whereas *CH3 is operated*. When CH3 and CH2 are on, armature gets connected to source voltage V_s so that both armature voltage V_t and armature current i_a are negative. As armature current is reversed, motor torque is reversed and consequently motoring mode in third quadrant is obtained. When CH3 is turned

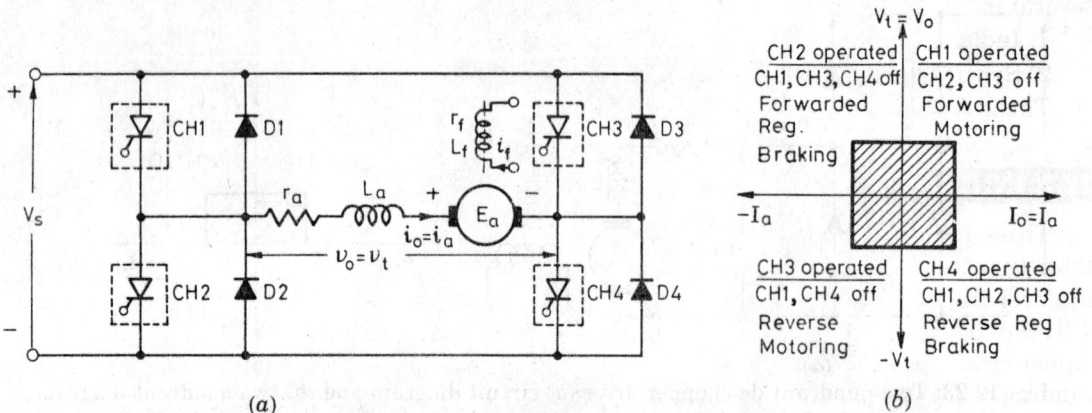

Fig. 12.24. Four-quadrant dc chopper drive (a) circuit diagram and (b) four-quadrant diagram.

off, negative armature current freewheels through CH2, D4, E_a, L_a, r_a; armature current decreases and thus speed control is obtained in third quadrant. Note that during this mode, polarity of E_a must be opposite to that shown in Fig. 12.24 (a).

Reverse Regenerative-braking mode. As in forward braking mode, reverse regenerative-braking mode is feasible only if motor generated emf is made to exceed the dc source voltage. For this operating mode, CH1, CH2 and CH3 are kept off whereas *CH4 is operated*. When CH4 is turned on, positive armature current i_a rises through CH4, D2, r_a, L_a, E_a. Note that in this mode also, polarity of motor emf E_a must be opposite to that shown in Fig. 12.24 (a). When CH4 is turned off, diodes D2, D3 begin to conduct and motor acting as a generator returns energy to the dc source. This leads to reverse regenerative-braking operation of the dc separately-excited motor in fourth quadrant.

Note that in Fig. 12.24 (a), the numbering of choppers is done to agree with the quadrants in which these are operated. For example, CH1 is operated for first quadrant, ..., CH4 for fourth quadrant etc.

12.6. A.C. DRIVES

Primarily, electric drives can be divided into two groups, dc drives and ac drives. DC drives have already been discussed in this chapter. Now ac drives are described at their introductory level only. Advantages and disadvantages of ac drives with respect to dc drives are as under :

Advantages of ac drives

(*i*) For the same rating, ac motors are lighter in weight as compared to dc motors.

(*ii*) AC motors require low maintenance as compared to dc motors.

(*iii*) AC motors are less expensive as compared to equivalent dc motors.

(*iv*) AC motors can work in hazardous areas like chemical, petrochemical etc. whereas dc motors are unsuitable for such environments because of commutator sparking.

Disadvantages of ac drives

(*i*) Power converters for the control of ac motors are more complex.

(*ii*) Power converters for ac drives are more expensive.

(*iii*) Power converters for ac drives generate harmonics in the supply system and load circuit. As a result, ac motors get derated.

The advantages of ac drives outweigh their disadvantages. As such, ac drives are used for several industrial applications. In general, there are two types of ac drives :

1. Induction motor drives
2. Synchronous motor drives.

These are now described in what follows.

12.7. INDUCTION-MOTOR DRIVES

Three-phase induction motors are more commonly employed in adjustable-speed drives than three-phase synchroncus motors. Three-phase induction motors are of two types, squirrel-cage induction motors (SCIMs) and slip-ring (or wound-rotor) induction motors (SRIMs). Stator windings of both types carry three-phase windings. Rotor of SCIM is made of copper or aluminium bars short-circuited by two end rings. Rotor of SRIM carries three-phase winding connected to three slip rings on the rotor shaft.

When 3-phase supply is connected to three-phase stator winding, rotating magnetic field is produced. The speed of this rotating field, called synchronous speed, is given by

$$N_s = \frac{120 f_1}{P} \text{ rpm} \quad \text{or} \quad n_s = \frac{2f_1}{P} \text{ rps} \qquad \qquad ...(12.43)$$

Also,

$$\omega_s = \frac{4\pi f_1}{P} = \frac{2\omega_1}{P} \text{ rad/sec} \qquad \qquad ...(12.44)$$

where
f_1 = supply frequency in Hz,

ω_1 = supply frequency in rad/s

P = number of stator poles.

Rotor cannot attain synchronous speed. It must run at a speed N_r less than N_s, where

$$N_r = N_s (1 - s)$$
$$\omega_m = \omega_s (1 - s)$$

where
N_r = rotor speed in rpm

ω_m = rotor speed in rad/s

$$s = \text{slip} = \frac{N_s - N_r}{N_s} = \frac{\omega_s - \omega_m}{\omega_s}$$

Analysis and performance. Per-phase equivalent circuit of a three-phase induction motor is shown in Fig. 12.25. In this circuit, r_2 = rotor resistance referred to stator, x_2 = rotor leakage reactance referred to stator. $r_1 + jx_1$ is the stator leakage impedance and X_m = magnetizing reactance. In this

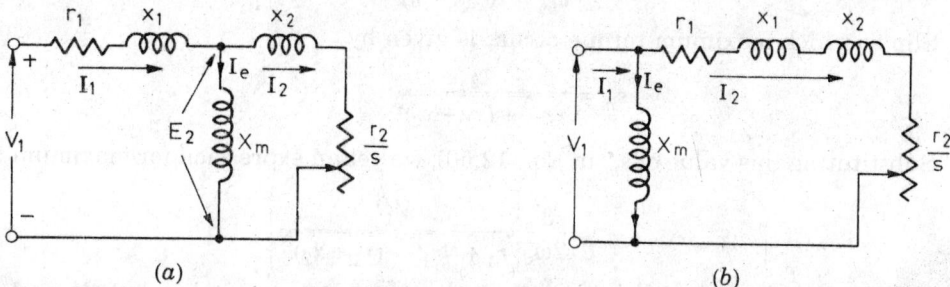

Fig. 12.25. Per-phase equivalent circuit of a three-phase induction motor referred to stator.

figure, core-loss resistance R_c is not shown. But, for determining the shaft power or shaft torque, the core loss must be taken into consideration. In case stator impedance drop is assumed negligible as compared to terminal voltage V_1, X_m can be moved to stator terminals to get the simplified equivalent circuit of Fig. 12.25 (b). From this per-phase circuit model, stator current I_1 and rotor current I_2 can be calculated. Once I_1, I_2 are known, the performance parameters of a 3- phase induction motor can be determined as follows :

From Fig. 12.25 (b),
$$I_2 = \frac{V_1}{\left(r_1 + \dfrac{r_2}{s}\right) + j\,(x_1 + x_2)} \qquad \qquad \therefore(12.45)$$

Air-gap power (power transferred from stator to rotor through the air gap),

$$P_g = 3\,I_2^2\,\frac{r_2}{s} \qquad \qquad \qquad ...(12.46)$$

Rotor ohmic loss $= 3I_2^2 r_2$

Developed power in rotor, $P_m = P_g -$ rotor ohmic loss

$$= 3I_2^2\,r_2\left(\frac{1-s}{s}\right) \qquad \qquad ...(12.47)$$

Developed rotor torque, $T_e = \dfrac{P_m}{\omega_m} = 3I_2^2\,r_2\left(\dfrac{1-s}{s}\right)\cdot\dfrac{1}{\omega_s\,(1-s)}$

$$= \frac{3}{\omega_s}\cdot I_2^2\,\frac{r_2}{s} \qquad \qquad \qquad ...(12.48)$$

Also, $T_e = \dfrac{P_g}{\omega_s}$ $\qquad \qquad \qquad ...(12.49)$

Substituting the value of I_2 from Eq. (12.45), we get

$$T_e = \frac{3}{\omega_s}\cdot\frac{V_1^2}{\left(r_1 + \dfrac{r_2}{s}\right)^2 + (x_1 + x_2)^2}\cdot\frac{r_2}{s} \qquad ...(12.50)$$

Motor power input, $P = P_g +$ stator core loss + stator copper loss

Output, or shaft, power, $P_{sh} = P_m -$ fixed loss (friction and windage loss)

Motor efficiency, $\eta = \dfrac{P_{sh}}{P}$

Output, or shaft, torque, $T_{sh} = \dfrac{P_{sh}}{\omega_m} = \dfrac{P_{sh}}{\omega_s\,(1-s)}$ $\qquad \qquad ...(12.51)$

Slip at which maximum torque occurs is given by

$$s_m = \frac{r_2}{\sqrt{r_1^2 + (x_1 + x_2)^2}} \qquad \qquad ...(12.52)$$

Substituting this value of s_m in Eq. (12.50), we get an expression for maximum torque as

$$T_{e.m} = \frac{3}{2\omega_s}\left[\frac{V_1^2}{r_1 + \sqrt{r_1^2 + (x_1 + x_2)^2}}\right] \qquad ...(12.53)$$

The maximum torque, also called *pull-out* or *breakdown* torque, is independent of rotor resistance. However, s_m is directly proportional to rotor resistance.

It is seen from Eq. (12.50) that if three-phase induction motor is energised from fixed voltage at a constant frequency, motor torque is a function of the slip. For different values of slip, the speed-torque characteristic of a three-phase induction motor is plotted in Fig. 12.26. In motoring mode under normal running ($s < s_m$), as stator current is not high, the air-gap flux remains substantially constant and torque increases with increase in slip from zero to s_m. After maximum torque T_{em} at slip s_m, as the slip increases, stator current rises much more than the rated current, air-gap flux decreases and therefore torque decreases with an increase in slip.

In case rotor resistance r_1 is neglected, which is usually true in large induction motors, the expressions for slip and torque are as under :

$$s_m = \frac{r_2}{x_1 + x_2} \qquad ...(12.54)$$

Fig. 12.26. Speed-torque characteristics of a three-phase induction motor.

$$T_e = \frac{3}{\omega_s} \cdot \frac{V_1^2}{\left(\dfrac{r_2}{s}\right)^2 + (x_1 + x_2)^2} \cdot \frac{r_2}{s} \qquad ...(12.55)$$

$$T_{em} = \frac{3}{\omega_s} \cdot \frac{V_1^2}{2(x_1 + x_2)} \qquad ...(12.56)$$

From Eqs. (12.55) and (12.56), it can be obtained that

$$\frac{T_e}{T_{e.m}} = \frac{2}{\dfrac{s}{s_m} + \dfrac{s_m}{s}} \qquad ...(12.57)$$

If $s < s_m$, then

$$\frac{T_e}{T_{e.m.}} = \frac{2 \cdot s}{s_m} = \frac{2(\omega_s - \omega_m)}{s_m \cdot \omega_s} \qquad ...(12.58)$$

or

$$\frac{\omega_s \cdot s_m \cdot T_e}{2 T_{em}} = (\omega_s - \omega_m)$$

or Motor speed,

$$\omega_m = \omega_s \left[1 - \frac{s_m \cdot T_e}{2 T_{em}} \right] \qquad ...(12.59)$$

It is seen from Eqs. (12.58) and (12.59) that for small values of slips, motor torque is proportional to slip Eq. (12.58), and motor speed decreases as torque $T_e =$ load torque T_L, increases, Eq. (12.59).

From Eqs. (12.54) and (12.59),

$$\omega_m = \omega_s \left[1 - \frac{r_2}{x_1 + x_2} \cdot \frac{T_e}{2 T_{em}} \right] \qquad ...(12.60)$$

This expression shows that drop in speed from no load to full load depends on rotor resistance.

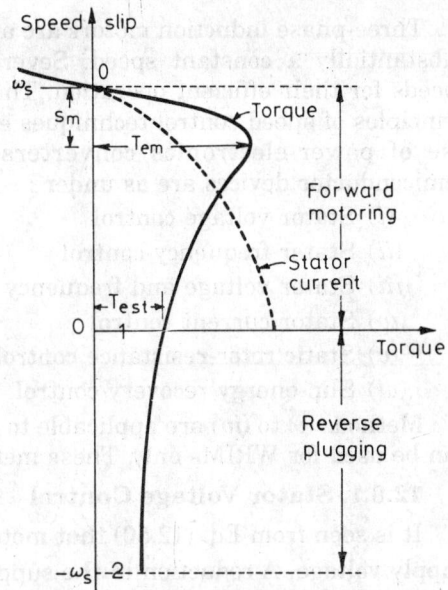

12.8. SPEED CONTROL OF THREE-PHASE INDUCTION MOTORS

Three-phase induction motors are admirably suited to fulfil the demand of loads requiring substantially a constant speed. Several industrial applications, however, need adjustable speeds for their efficient operation. The object of the present section is to describe the basic principles of speed control techniques employed to three-phase induction motors through the use of power-electronics converters. The various methods of speed control through semiconductor devices are as under :

    (*i*) Stator voltage control

    (*ii*) Stator frequency control

    (*iii*) Stator voltage and frequency control

    (*iv*) Stator current control

    (*v*) Static rotor-resistance control

    (*vi*) Slip-energy recovery control.

Methods (*i*) to (*iv*) are applicable to both SCIMs and WRIMs whereas methods (*v*) and (*vi*) can be used for WRIMs only. These methods are now described in what follows.

12.8.1. Stator Voltage Control

It is seen from Eq. (12.50) that motor torque T_e is proportional to the square of the stator supply voltage. A reduction in the supply voltage will reduce the motor torque and therefore the speed of the drive. If the motor terminal voltage is reduced to KV_1 where $K < 1$, then the motor torque is given by

$$T_e = \frac{3}{\omega_s} \cdot \frac{(KV_1)^2}{\left(r_1 + \dfrac{r_2}{s}\right)^2 + (x_1 + x_2)^2} \cdot \frac{r_2}{s} \qquad \qquad ...(12.61)$$

For the purpose of varying the voltage applied to a 3-phase induction motor so as to achieve a speed control, a 3-phase ac voltage controller is usually employed. Fig. 12.27 (*a*) shows a three-phase ac voltage controller feeding a three-phase induction motor. By controlling the firing angle of the thyristors connected in antiparallel in each phase, the rms value of the stator voltage can be regulated. As a consequence, motor torque and thus speed of the drive is controlled. In Fig. 12.27 (*b*), for load torque T_L, *a* is the operating point at rated voltage and

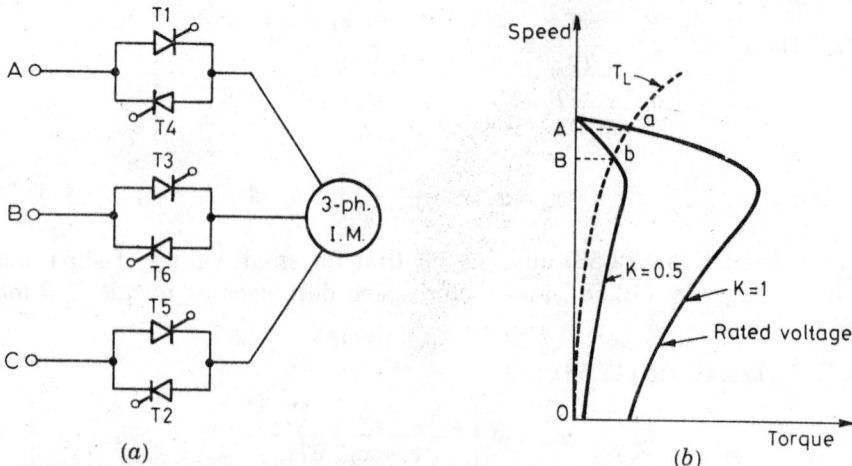

             (*a*)   (*b*)

Fig. 12.27. (*a*) Three-phase ac voltage controller feeding a 3-phase induction motor
(*b*) Speed-torque characteristics as effected by stator voltage control.

OA is the motor speed. For reduced stator voltage ($K = 0.5$), b is the operating point and *OB* is the reduced motor speed for load torque T_L. This method is suitable for motors having large value of s_m. For low-slip motors, the range of speed control is very narrow.

Stator-voltage-control method offers limited speed range. It is usual to use 3-phase voltage controllers. Their use, however, introduces pronounced harmonic contents and input supply power factor for the voltage controller is quite low. These are, therefore, used for low-power drives like fans, blowers and centrifugal pumps requiring low starting torque. For these types of loads, the load torque is proportional to speed squared and input current is maximum when slip $s = 1/3$, this is proved in Example 12.22.

Example 12.22. (*a*) *A three-phase SCIM drives a blower-type load. No-load rotational losses are negligible. Show that rotor current is maximum when motor runs at a slip s = 1/3. Find also an expression for maximum rotor current.*

(*b*) *If three-phase SCIM runs at speed of (i) 1455 rpm and (ii) 1350 rpm, determine the maximum current in terms of rated current at these speeds. The IM drives a fan and no-load rotational losses are ignored.*

Solution. (*a*) The torque required by a blower-type load is proportional to speed squared.

$$\therefore \qquad T_L = k\omega_m^2$$

Mechanical power developed in motor, $P_m = (1 - s)\,P_g$

As no-load rotational losses are negligible, P_m = power required by load.

or $\qquad (1-s)\,P_g = T_L \cdot \omega_m$

$$3I_2^2\,\frac{r_2}{s}\,(1 - s) = T_L \cdot \omega_m$$

or
$$I_2 = \left[\frac{\omega_m \cdot T_L \cdot s}{3r_2\,(1 - s)}\right]^{1/2} \qquad \qquad \ldots(i)$$

But $\omega_m = \omega_s\,(1 - s)$ and $T_L = k\,\omega_m^2$. Substituting these in (*i*), we get,

$$I_2 = \left[\frac{\omega_s\,(1 - s) \cdot T_L \cdot s}{3r_2\,(1 - s)}\right]^{1/2} = \left[\frac{\omega_s \cdot k \cdot \omega_m^2 \cdot s}{3r_2}\right]^{1/2}$$

$$= \omega_m\left[\frac{s \cdot k \cdot \omega_s}{3r_2}\right]^{1/2} = (1 - s)\,\omega_s\left[\frac{s \cdot k \cdot \omega_s}{3r_2}\right]^{1/2}$$

or
$$I_2 = \sqrt{s}\,(1 - s)\left[\frac{k \cdot \omega_s^3}{3r_2}\right]^{1/2} \qquad \qquad \ldots(ii)$$

The slip at which rotor current I_2 becomes maximum can be found by obtaining $\dfrac{dI_2}{ds}$ and equating it to zero.

$$\therefore \qquad \frac{dI_2}{ds} = \frac{1}{2} \cdot \frac{1}{\sqrt{s}}\,(1 - s)\left[\frac{k\omega_s^3}{3r_2}\right]^{1/2} + \sqrt{s}\,(- 1)\left[\frac{k\omega_s^3}{3r_2}\right]^{1/2} = 0$$

or $\qquad \dfrac{1 - s}{2} = s \qquad$ or $\qquad s = \dfrac{1}{3} \qquad \qquad \ldots(iii)$

This shows that I_2 is maximum at a slip of $\frac{1}{3}$. The maximum value of I_2 is obtained by putting $s = \frac{1}{3}$ in Eq. (ii).

$$\therefore \qquad I_{2 \cdot mx} = \sqrt{\frac{1}{3} \cdot \frac{2}{3} \left[\frac{k\omega_s^3}{3r_2} \right]^{1/2}} = \frac{2}{9} \cdot \omega_s \left[\frac{k \, \omega_s}{r_2} \right]^{1/2} \qquad \dots (iv)$$

(b) From Eq. (ii), $\qquad I_2 = \sqrt{s} \, (1 - s) \left[\frac{k \cdot \omega_s^3}{3r_2} \right]^{1/2}$

(i) For 1455, full-load slip

$$s_1 = \frac{1500 - 1455}{1500} = 0.03$$

From above, $\qquad \dfrac{I_{2 \cdot mx}}{I_{2 \cdot r}} = \dfrac{\sqrt{\frac{1}{3}} \cdot \frac{2}{3}}{\sqrt{s_1} \, (1 - s_1)} = \dfrac{\sqrt{\frac{1}{3}} \cdot \frac{2}{3}}{\sqrt{0.03} \, (1 - 0.03)} = 2.291$

Here I_{2r} is the rated, or full-load, rotor current.

(ii) For 1350 rpm, full-load slip

$$s_1 = \frac{1500 - 1350}{1500} = 0.1$$

$$\therefore \qquad \dfrac{I_{2 \cdot mx}}{I_{2 \cdot r}} = \dfrac{\sqrt{\frac{1}{3}} \left(\frac{2}{3} \right)}{\sqrt{0.1} \, (1 - 0.1)} = 1.352$$

12.8.2. Stator Frequency Control

By changing the supply frequency, motor synchronous speed can be altered and thus torque and speed of a 3-phase induction motor can be controlled. For a three-phase induction motor, per-phase supply voltage is $V_1 = \sqrt{2} \, \pi f_1 N_1 \phi \, k_{w1}$. This expression shows that under rated voltage and frequency operation, flux will be rated. In case supply frequency is reduced with constant V_1, the air-gap flux increases and the induction motor magnetic circuit gets saturated. The motor parameters will change leading to inaccurate speed-torque characteristics. Further, at low frequencies, reactances will be low leading to high motor currents, more losses and reduced efficiency. In view of this, induction motor (I.M.) speed control with constant supply voltage and reduced supply frequency is rarely used in practice.

With constant supply voltage, if the supply frequency is increased, the synchronous speed and therefore motor speed rises. But, with increase in frequency, flux and torque also get reduced. IM performance at constant voltage and increased frequency can be obtained by neglecting X_m and r_1 from the equivalent circuit of Fig. 12.25 (a). This assumption is not going to introduce any noticeable error as magnetizing current at high frequency is quite small. Thus, rotor current under this assumption is given by

$$I_2 = \frac{V_1}{\left[\left(\dfrac{r_2}{s} \right)^2 + (x_1 + x_2)^2 \right]^{1/2}} \qquad \dots (12.62)$$

Synchronous speed, $\omega_s = \dfrac{4\pi f_1}{P} = \dfrac{2\omega_1}{P}$ rad/s

Motor torque, $\qquad T_e = \dfrac{3}{\omega_s} \cdot I_2^2 \dfrac{r_2}{s}$

$$= \dfrac{3P}{2\omega_1} \cdot \dfrac{V_1^2}{\left(\dfrac{r_2}{s}\right)^2 + (x_1 + x_2)^2} \cdot \dfrac{r_2}{s} \qquad\qquad ...(12.63)$$

Slip, $\qquad\qquad s = \dfrac{f_2}{f_1} = \dfrac{\omega_2}{\omega_1} \quad$ or $\quad \omega_2 = s\,\omega_1$

Here f_2 and ω_2 are the rotor frequencies in Hz and rad/s respectively. Substituting the value

of slip $s = \dfrac{\omega_2}{\omega_1}$ in Eq. (12.63), we get

$$T_e = \dfrac{3P}{2\omega_1} \cdot \dfrac{V_1^2 \cdot \omega_1}{\dfrac{r_2 \cdot \omega_1^2}{\omega_2^2} + \omega_1^2\,(l_1 + l_2)^2} \cdot \dfrac{r_2}{\omega_2}$$

$$= \dfrac{3P}{2\omega_1^2} \cdot \dfrac{V_1^2 \cdot \omega_2}{r_2^2 + \omega_2^2\,(l_1 + l_2)^2} \cdot r_2 \qquad\qquad ...(12.64)$$

Slip at which maximum torque occurs is given in Eq. (12.54) as

$$s_{mT} = \dfrac{r_2}{x_1 + x_2} \qquad\qquad ...(12.65)$$

Rotor frequency in rad/s at which maximum torque occurs is given by

$$\omega_{2m} = s_{mT} \cdot \omega_1 = \dfrac{\omega_1 \cdot r_2}{\omega_1\,(l_1 + l_2)} = \dfrac{r_2}{l_1 + l_2}$$

Note that ω_{2m} does not depend on the supply frequency ω_1. Substituting $r_2 = \omega_{2m} \cdot (l_1 + l_2)$ in Eq. (12.64) gives maximum torque $T_{e.m}$ as

$$T_{e.m} = \dfrac{3P}{2\omega_1^2} \cdot \dfrac{V_1^2 \cdot \omega_{2m}^2 \cdot (l_1 + l_2)}{\omega_{2m}^2\,(l_1 + l_2)^2 + \omega_{2m}^2\,(l_1 + l_2)^2}$$

$$= \dfrac{3P}{4\omega_1^2} \cdot \dfrac{V_1^2}{l_1 + l_2} \qquad\qquad ...(12.66)$$

Eq. (12.66) indicates that T_{em} is inversely proportional to supply-frequency squared. Also,

$$T_{em} \cdot \omega_1^2 = \dfrac{3P}{4} \cdot \dfrac{V_1^2}{l_1 + l_2}$$

At given source voltage V_1, $\dfrac{3P}{4} \cdot \dfrac{V_1^2}{l_1 + l_2}$ is constant, therefore, $T_{em} \cdot \omega_1^2$ is also constant. As the

operating frequency ω_1 is increased, $T_{em} \cdot \omega_1^2$ remains constant but maximum torque at increased frequency ω_1 gets reduced as shown in Fig. 12.28. Supposing rated frequency for a motor is 50 Hz and $T_{e \cdot m} = 100$ Nm. If the motor is now operated at 100 Hz, then $100\,(2\,\pi \times 50)^2 = $ (new maximum torque) $(2\pi \times 100)^2$ or the maximum torque at increased frequency of 100 Hz is 25 Nm. Such type of IM behaviour is similar to the working of dc series motors. With constant voltage and increased-frequency operation, air-gap flux gets reduced; therefore, during this control, IM is said to be working in *field-weakening mode*. Constant voltage and variable frequency control of Fig. 12.28 can be obtained by feeding 3-phase IM through three-phase inverters discussed in Chapter 8.

Example 12.23. *A 3-phase, 400V, 15kW, 1440 rpm, 50 Hz, star-connected induction motor has rotor leakage impedance of $0.4 + j 1.6 \, \Omega$. Stator leakage impedance and rotational losses are assumed negligible.*

If this motor is energised from 120 Hz, 400V, 3-phase source, then calculate

    (a) *the motor speed at rated load*

    (b) *the slip at which maximum torque occurs and*

    (c) *the maximum torque.*

Fig. 12.28. Speed torque characteristics of a 3-phase IM with stator frequency control with constant supply voltage.

Solution. Full-load torque, $T_{e \cdot fl} = \dfrac{P}{\omega_m} = \dfrac{60}{2\pi \times 1440} \times 15000 = 99.472 \, \text{Nm}$

(a) At 120 Hz, synchronous speed,

$$\omega_s = \frac{4\pi f_1}{P} = \frac{4\pi \times 120}{4} = 120 \, \pi \, \text{rad/s.}$$

Rotor impedance at 120 Hz $= 0.4 + j \, 1.6 \times \dfrac{120}{50} = 0.4 + j \, 3.84 \cong j \, 3.84 \, \Omega$

$\therefore$ $\quad T_{e \cdot fl} = 99.472 = \dfrac{3}{120 \, \pi} \cdot \dfrac{(400/\sqrt{3})^2}{(3.84)^2} \times \dfrac{0.4}{s}$

$\therefore$ Full-load slip, $\quad s = \dfrac{1}{40\pi} \times \dfrac{0.4}{99.472} \times \left(\dfrac{400}{\sqrt{3} \times 3.84}\right)^2 = 0.1157$

Motor speed at rated load $\quad = \dfrac{120 \times 120}{4} (1 - 0.1157) = 3183.48 \, \text{rpm}$

(b) Slip at which maximum torque occurs is given by

$$s_m = \frac{r_2}{x_2} = \frac{0.4}{3.84} = 0.1042$$

(c) From Eq. (12.56), with $x_1 = 0$, maximum torque,

$$T_{e \cdot m} = \frac{3}{\omega_s} \cdot \frac{V_1^2}{2x_2} = \frac{3}{120 \, \pi} \cdot \frac{(400/\sqrt{3})^2}{2 \times 3.84} = 55.262 \, \text{Nm.}$$

12.8.3. Stator Voltage and Frequency Control

For a 3-phase IM, stator voltage per phase is given by

$$V_1 = \sqrt{2} \, \pi f_1 \cdot N_{ph1} \cdot \phi \cdot k_{\omega 1} \qquad \qquad ...(12.66)$$

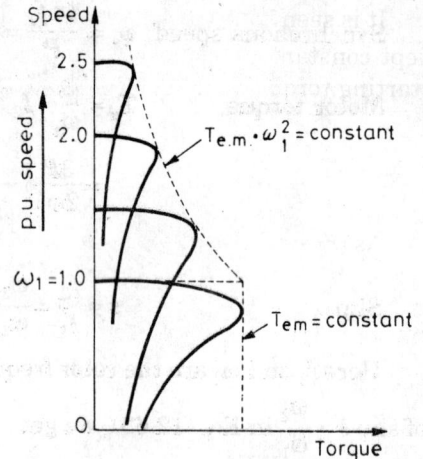

It is seen from above equation that if the ratio of supply voltage V_1 to supply frequency f_1 is kept constant, the air-gap flux ϕ remains constant. From Fig. 12.25 (b) and Eq. (12.50), the starting torque is given by

$$T_{e.st} = \frac{3}{\omega_s} \cdot \frac{V_1^2}{(r_1 + r_2)^2 + (x_1 + x_2)^2} \cdot r_2$$

As $(r_1 + r_2) << (x_1 + x_2)$ and $\omega_s = \dfrac{2\omega_1}{P}$, we get

$$T_{e \cdot st} = \frac{3P}{2\omega_1} \cdot \frac{V_1^2 \cdot r_2}{\omega_1^2 (l_1 + l_2)^2}$$

$$= \frac{3P}{2\omega_1} \cdot \left(\frac{V_1}{\omega_1}\right)^2 \cdot \frac{r_2}{(l_1 + l_2)^2} \qquad \qquad ...(12.67)$$

From Eq. (12.56), maximum torque is given by

$$T_{e.m} = \frac{3}{\omega_s} \cdot \frac{V_1^2}{2 (x_1 + x_2)}$$

$$= \frac{3P}{2\omega_1} \cdot \frac{V_1^2}{2 \cdot \omega_1 (l_1 + l_2)}$$

$$= \frac{3P}{4} \cdot \left(\frac{V_1}{\omega_1}\right)^2 \frac{1}{l_1 + l_2} \qquad \qquad ...(12.68)$$

Eq. (12.68) shows that if V_1/ω_1, or air-gap flux ϕ, is kept constant, the maximum torque remains unaltered. Eq. (12.67) indicates that starting torque is inversely proportional to supply frequency ω_1, even if air-gap flux is kept constant. At low values of frequencies, the effect of resistances cannot be neglected as compared to the reactances. This has the effect of reducing the magnitude of maximum torque at lower frequencies as shown in Fig. 12.29. In practice, at low frequencies, the supply voltage is increased to maintain the level of maximum torque. This method of speed control is also called *volts/hertz* control.

If stator resistance is neglected, then from Fig. 12.25 (b), the slip at which maximum torque occurs is given by

$$s_m = \frac{r_2}{x_1 + x_2}$$

$$= \frac{r_2}{\omega_1 (l_1 + l_2)} \qquad \qquad ...(12.69)$$

As the supply frequency ω_1 is reduced, the slip at maximum torque increases.

In Fig. 12.29, load torque T_L for a certain load is also shown. It is seen from this figure that as both voltage and frequency are varied (usually below their rated values), speed of the drive can be controlled. The control of both voltage and frequency can be carried out (so as to keep $\dfrac{V}{f}$ constant) through the use of three-phase inverters or

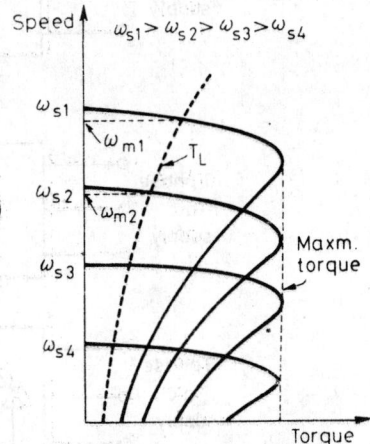

Fig. 12.29. Speed-torque characteristics of a 3-phase IM with volts/hertz control.

cycloconverters. Inverters are used in low and medium power drives whereas cycloconverters are suitable for high-power drives like cement mills, locomotives etc.

Variable voltage and variable frequency can be obtained from voltage-source inverters. Four such circuit configurations are shown in Fig. 12.30. In Fig. 12.30 (a), three-phase ac is converted to constant dc by diode rectifier. Voltage and frequency are both varied by PWM inverter. The circuitry between the rectifier and the inverter consists of an inductor L and capacitor C, called filter circuit. The function of filter circuit is to smooth dc input voltage to the inverter. This circuitry in between rectifier and inverter is called *dc link*. In Fig. 12.30 (a), regeneration is not possible because of diode rectifier. Also, inverter would inject harmonics into the 3-phase ac supply.

In Fig. 12.30 (b), three-phase ac is converted to dc by diode rectifier. Chopper varies the dc input voltage to the inverter and frequency is controlled by the inverter. Use of chopper reduces the harmonic injection into the ac supply. Regeneration is not feasible in the scheme of Fig. 12.30 (b).

Fig. 12.30 (c) uses a 3-phase controlled rectifier, dc link consisting of L and C and a force-commutated VSI. Voltage is regulated by controlled rectifier and frequency is varied within the inverter. Here regeneration is possible if three-phase full converter is used. Regeneration is also feasible in the scheme shown in Fig. 12.30 (d). It uses a 3-phase dual converter, L-C filter and inverter. Level of dc input voltage to the inverter is regulated in dual converter whereas frequency is varied within the VSI inverter.

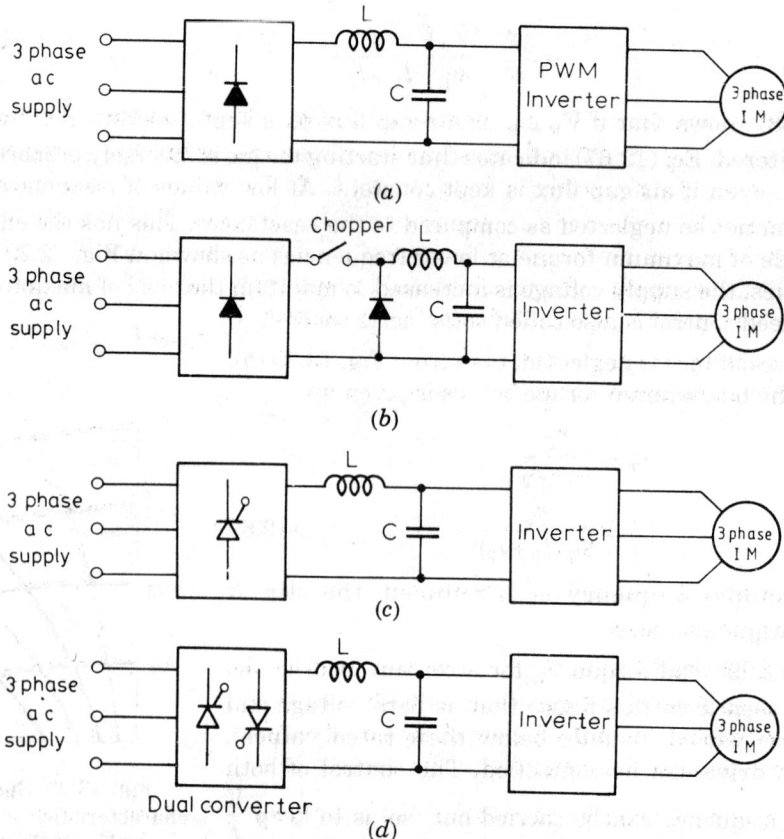

Fig. 12.30. Three-phase induction motor speed control through voltage source inverters.

It may be observed from above that volts/hertz control offers speed control from standstill upto rated speed of IM. This method is similar to the armature-voltage control method used for the speed control of a dc motor.

Example 12.24. *A 3-phase, 20 kW, 4-pole, 50 Hz, 400 V delta connected induction motor has the following per phase parameters referred to stator :*

$$r_1 = 0.6 \ \Omega, \ r_2 = 0.4 \ \Omega, \ x_1 = x_2 = 1.6 \ \Omega$$

Its magnetizing reactance is neglected. If this motor is operated at 200V, 25 Hz with DOL starting, calculate

(a) current and pf at the instant of starting and under maximum torque conditions; compare the results with normal values,

(b) starting and maximum torques and compare with normal values.

Solution. The subscripts n and d are used here for normal operation and for reduced-voltage reduced-frequency operation respectively.

(a) Starting current :

$$I_n = \frac{V_1}{\left[\left(r_1 + \dfrac{r_2}{s}\right)^2 + (x_1 + x_2)^2\right]^{1/2}} = \frac{400}{\left[\left(0.6 + \dfrac{0.4}{1}\right)^2 + (3.2)^2\right]^{1/2}} = 119.31 \ A$$

$$I_d = \frac{400/2}{\left[\left(0.6 + \dfrac{0.4}{1}\right)^2 + \left(\dfrac{3.2}{2}\right)^2\right]^{1/2}} = 106 \ A$$

$$(\cos\theta)_n = \frac{1.0}{[1^2 + 3.2^2]^{1/2}} = 0.2983$$

$$(\cos\theta)_d = \frac{1.0}{\left[1^2 + \left(\dfrac{3.2}{2}\right)^2\right]^{1/2}} = 0.53$$

(b) Starting torque :

$$T_{en} = \frac{3}{\omega_s} \cdot I_n{}^2 \frac{r_2}{s}$$

$$= \frac{3}{50\pi} (119.31)^2 \left(\frac{0.4}{1}\right) = 108.75 \ Nm$$

$$T_{ed} = \frac{3}{25\pi} (106)^2 \times \frac{0.4}{1} = 171.673 \ Nm$$

It is seen from above that at reduced-voltage reduced-frequency operation

(i) starting current decreases, however, starting torque becomes more,

(ii) power factor at starting is improved, therefore power input to motor is more.

It can be inferred from above that with reduced-voltage and reduced-frequency (keeping V/f or flux constant) operation, the performance of IM at starting is improved.

Maximum torque. The slip at which maximum torque occurs is given by

$$\frac{0.4}{s_{m \cdot n}} = \sqrt{0.6^2 + 3.2^2}$$

$$\therefore \qquad s_{m \cdot n} = \frac{0.4}{\sqrt{0.6^2 + 3.2^2}} = 0.123$$

$$I_n = \frac{400}{\left[\left(0.6 + \dfrac{0.4}{0.123}\right)^2 + (3.2)^2\right]^{1/2}} = 79.875 \text{ A}$$

$$T_{em \cdot n} = \frac{3}{50 \, \pi} \, (79.875)^2 \times \frac{40}{0.123} = 396.26 \text{ Nm}$$

$$(\cos \theta)_n = \frac{0.6 + \dfrac{0.4}{0.123}}{\left[\left(0.6 + \dfrac{0.4}{0.123}\right)^2 + (3.2)^2\right]^{1/2}} = 0.7692 \text{ lag}$$

At reduced voltage and reduced frequency, the calculations are as under :

$$s_{m \cdot d} = \frac{0.4}{\left[0.6^2 + \left(\dfrac{3.2}{2}\right)^2\right]^{1/2}} = 0.234$$

$$I_d = \frac{200}{\left[\left(0.6 + \dfrac{0.4}{0.234}\right)^2 + 1.6^2\right]^{1/2}} = 71.187 \text{ A}$$

$$T_{em \cdot d} = \frac{3}{25 \cdot \pi} \cdot (71.187)^2 \times \frac{0.4}{0.237} = 330.885 \text{ Nm}$$

$$(\cos \theta)_d = \frac{0.6 + \dfrac{0.4}{0.234}}{\left[\left(0.6 + \dfrac{0.4}{0.234}\right)^2 + 1.6^2\right]^{1/2}} = 0.822 \text{ lag}$$

It is seen from above that at reduced voltage and reduced frequency

(*i*) current gets reduced whereas *pf* is improved under maximum torque conditions,

(*ii*) the maximum torque, however, gets reduced.

and (*iii*) maximum torque occurs at higher value of slip.

12.8.4. Stator Current Control

The developed torque and therefore the speed of a 3-phase IM can also be controlled by stator-current control instead of stator-voltage control. The behaviour of a motor with stator-current control is different from that obtained with voltage-source inverter.

Consider a constant current I_1 fed into the stator windings of a three-phase IM, Fig. 12.31 (*a*). So far as rotor current I_2 is concerned, stator leakage impedance $(r_1 + jx_1)$ plays no role. Therefore, the effect of $(r_1 + jx_1)$ can be omitted when studying constant-current mode of motor operation. However, $(r_1 + jx_1)$ does influence the magnitude of applied stator voltage V_1. For stator current I_1, the rotor current I_2, from Fig. 12.31 (*a*), is given by

$$I_2 = I_1 \frac{jX_m}{\dfrac{r_2}{s} + j \, (x_2 + X_m)} \qquad \qquad ...(12.70)$$

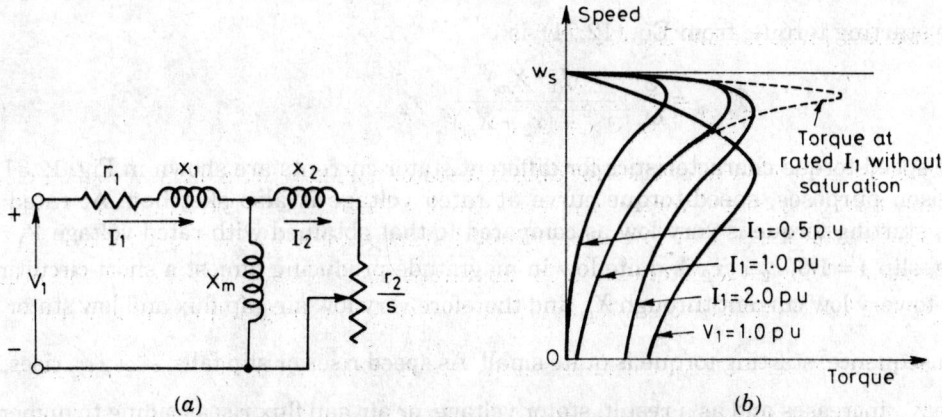

Fig. 12.31. (a) 3-phase I.M. per-phase equivalent circuit
(b) speed-torque curves with stator-current control.

The internal, or developed, electromagnetic torque T_e is

$$T_e = \frac{3}{\omega_s} \cdot I_2^2 \frac{r_2}{s}$$

$$= \frac{3}{\omega_s} \cdot \frac{(I_1 X_m)^2}{\left(\dfrac{r_2}{s}\right)^2 + (x_2 + X_m)^2} \cdot \frac{r_2}{s} \qquad \qquad ...(12.71)$$

Invoking the principle of maximum power transfer in the equivalent circuit of Fig. 12.31 (a) with $r_1 + jx_1 = 0$ and with constant current I_1, we get

$$\frac{r_2}{s_m} = x_2 + X_m$$

or

$$s_m = \frac{r_2}{x_2 + X_m} \qquad \qquad ...(12.72)$$

Eq. (12.72) gives the slip at which maximum torque occurs when 3-phase IM is fed from current-controlled source. Maximum torque T_{em} is obtained by substituting $\dfrac{r_2}{s_m} = x_2 + X_m$ in Eq. (12.71).

$$\therefore \qquad T_{e \cdot m} = \frac{3}{\omega_s} \cdot \frac{(I_1 \cdot X_m)^2}{2 (x_2 + X_m)^2} \cdot (x_2 + X_m)$$

$$= \frac{3}{\omega_s} \cdot \frac{(I_1 X_m)^2}{2 (x_2 + X_m)} \qquad \qquad ...(12.73)$$

Also,

$$T_{e \cdot m} = \frac{3P}{4\pi f_1} \cdot \frac{(I_1 \cdot 2\pi f_1 \cdot L_m)^2}{2 \cdot 2\pi f_1 \cdot (l_2 + L_m)}$$

$$= \frac{3P}{4} \cdot \frac{L_m^2}{l_2 + L_m} \cdot I_1^2 \qquad \qquad ...(12.74)$$

It is seen from Eq. (12.74) that maximum torque is (i) proportional to stator-current squared, (ii) independent of supply frequency f_1 and (iii) independent of rotor resistance.

The starting torque, from Eq. (12.71), is

$$T_{e \cdot st} = \frac{3}{\omega_s} \cdot \frac{(I_1 \cdot X_m)^2}{r_2^2 + (x_2 + X_m)^2} \cdot r_2 \qquad \qquad ...(12.75)$$

The speed-torque characteristics for different stator currents are shown in Fig. 12.31 (*b*). For comparison purposes, speed-torque curve at rated voltage is also sketched. At rated current $I_1 = 1.0$, starting torque is very low as compared to that obtained with rated voltage $V_1 = 1.0$. At starting, slip $s = 1.0$, $r_2 + jx_2$ is quite low in magnitude producing almost a short-circuiting effect leading to very low current through X_m and therefore very low air-gap flux and low stator voltage. As a consequence, starting torque is quite small. As speed rises or slip falls, $\frac{r_2}{s} + j x_2$ rises, current through X_m increases and as a result, stator voltage, or air-gap flux rises leading to higher torque. With no magnetic saturation, torque rises to quite a high value as shown by dotted line. In practice, the saturation will limit the peaking in maximum developed torque as shown by solid curve for $I_1 = 1.0$. The speed-torque curves for $I_1 = 0.5$ and 2.0 are also shown in Fig. 12.31 (*b*).

A constant current for the 3-phase IM can be obtained from a 3-phase current source inverter (CSI). The advantages of torque and speed control by CSI fed IM are (*i*) fault current level control and (*ii*) current input is almost unaffected by motor parameter variations. The disadvantages of current-fed drives are (*i*) generation of unwanted harmonics in the system and (*ii*) torque pulsations.

Fig. 12.32 shows two circuit configurations for current-fed inverter IM drives. In Fig. 12.32 (*a*), 3-phase controlled rectifier gives out controlled dc voltage. Inductor L converts this voltage to constant current. CSI regulates the output frequency and therefore the torque and speed of 3-phase IM. In Fig. 12.32 (*b*), uncontrolled dc voltage is regulated by chopper which is then converted to current source by inductor L. As before, CSI then controls the torque and speed of three-phase I.M.

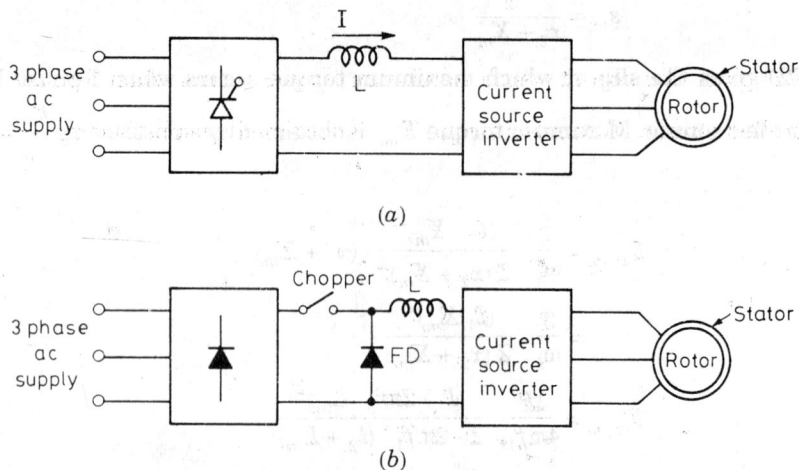

(*a*)

(*b*)

Fig. 12.32. (*a*) 3-phase IM speed control through current-source inverter.

Example 12.25. *A 400 V, 4-pole, 50 Hz, 3-phase, star-connected induction motor has* $r_1 = 0$, $x_1 = x_2 = 1\ \Omega$, $r_2 = 0.4\ \Omega$, $X_m = 50\ \Omega$, *all referred to stator. This induction motor is fed from* (*i*) *a constant-voltage source of 231V per phase and* (*ii*) *a constant-current source of 28A.*

For both parts (i) and (ii), calculate

(a) the slip for maximum torque

(b) the starting and maximum torques

(c) the supply voltage required to sustain the constant current at the maximum torque.

Solution. The equivalent circuit for this motor is shown in Fig. 12.33 (a). Its Thevenin's equivalent circuit is drawn in Fig. 12.33 (b) where X_e and V_e are given by

$$X_e = \frac{1 \times 50}{51} = 0.9804 \ \Omega$$

and

$$V_e = \frac{231 \times 50}{51} = 226.5 \ V$$

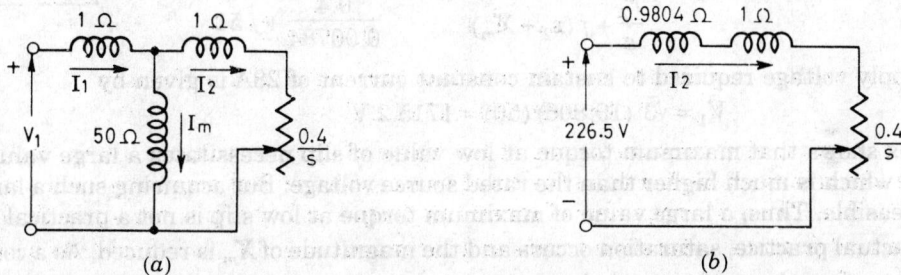

(a)   (b)

Fig. 12.33. Induction motor per-phase equivalent circuit, Example 12.25.

(a) (i) It is seen from the equivalent circuit of Fig. 12.33 (b) that the slip at which maximum torque occurs is given by

$$s_m = \frac{0.4}{1.9804} = 0.202$$

(ii) For constant-current operation of IM, slip s_m is given by Eq. 12.72.

$$\therefore \qquad s_{mT} = \frac{0.4}{1 + 50} = 0.00784$$

(b) (i) Synchronous speed, $\omega_s = \dfrac{4\pi f_1}{P} = \dfrac{4\pi \times 50}{4} = 50\pi \ \text{rad/s}$

For constant voltage input, $T_{e \cdot st}$ is given by

$$T_{e \cdot st} = \frac{3}{\omega_s} \cdot I_{2st}^{\ 2} \cdot r_2$$

$$= \frac{3}{50\pi} \cdot \frac{226.5^2}{0.4^2 + 1.9804^2} \times 0.4 = 96.012 \ \text{Nm}$$

$$T_{e \cdot m} = \frac{3}{\omega_s} \cdot \frac{V_e^{\ 2}}{2 \ (x_2 + X_e)}$$

$$= \frac{3}{50\pi} \cdot \frac{(226.5)^2}{2 \ (1.9804)} = 247.37 \ \text{Nm}$$

(ii) For constant current input, $T_{e \cdot st}$ from Eq. (12.75) is

$$T_{e \cdot st} = \frac{3}{50 \ \pi} \cdot \frac{(28 \times 50)^2}{0.4^2 + 51^2} \times 0.4 = 5.756 \ \text{Nm}$$

Maximum torque, from Eq. (12.73) is

$$T_{e \cdot m} = \frac{3}{50\pi} \cdot \frac{(28 \times 50)^2}{2 \times 51} = 366.993 \text{ Nm}$$

It is seen from above that for constant-current mode, $T_{e.st}$ is much smaller (5.756 Nm) as compared to $T_{e \cdot st}$ for constant-voltage mode (96.012 Nm). But maximum torque for constant-current mode is much more (366.993 Nm) than its value for constant-voltage mode (247.37 Nm). Also the slip at which maximum torque occurs is very low (0.00784) for constant-current mode as compared to its value for constant-voltage mode (0.202).

(c) At maximum torque, $s_{mT} = 0.00784$. From the equivalent circuit of Fig. 12.33 (a), the magnetizing current I_m is given by

$$I_m = I_1 \frac{\dfrac{r_2}{s} + j\, x_2}{\dfrac{r_2}{s} + j\,(x_2 + X_m)} = 28.0 \frac{\dfrac{0.4}{0.00784} + j\,1}{\dfrac{0.4}{0.00784} + j\,51} = 19.806 \underline{/43.865°}$$

Supply voltage required to sustain constant current of 28A is given by

$$V_1 = \sqrt{3}\,(19.806)\,(50) = 1715.2 \text{ V}$$

This shows that maximum torque at low value of slip necessitates a large value of source voltage which is much higher than the rated source voltage. But acquiring such a large voltage is not feasible. Thus, a large value of maximum torque at low slip is not a practical possibility.

In actual practice, saturation occurs and the magnitude of X_m is reduced. As a consequence, the value of maximum torque under constant-current mode has a much lower value than that computed here.

Summary of Characteristics of Adjustable-frequency Induction-motor Drives. Speed-torque characteristics of a three-phase induction-motor drives depend upon the methods of control techniques employed. For different stator frequencies, a family of speed-torque characteristics as shown in Fig. 12.34 can be obtained. These characteristics can be subdivided into three regions; constant-torque region, constant-power region and high- speed series-motor region. A summary of these characteristics is reviewed here.

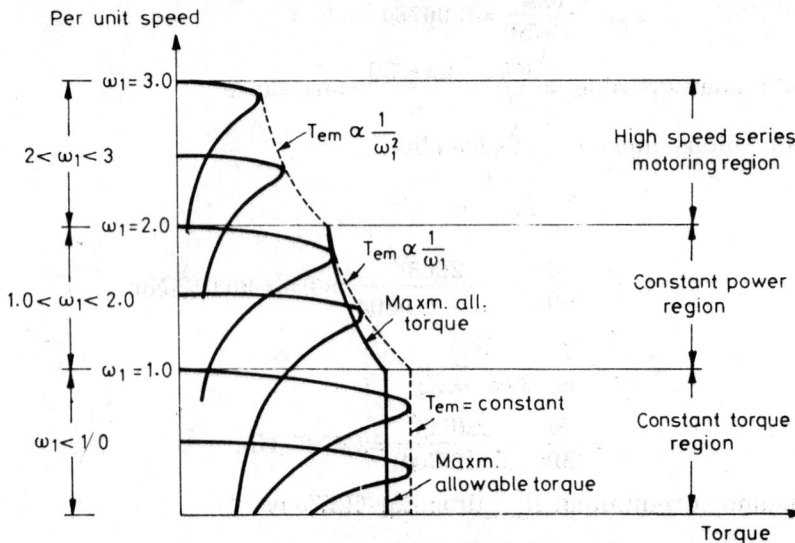

Fig. 12.34. Typical speed-torque characteristics of a
3-phase induction motor with variable-voltage and variable-frequency power supply.

Constant-torque region. As explained before, this region of constant torque can be obtained by volts/hertz control as shown in Fig. 12.29. In the low-frequency range of speed, the effect of stator resistance is compensated by a boost in the stator voltage as shown in Fig. 12.35. In this region, stator current is kept constant at its rated value. Power, equal to the product of constant torque and speed, varies linearly with speed as shown. Slip frequency remains constant during this region.

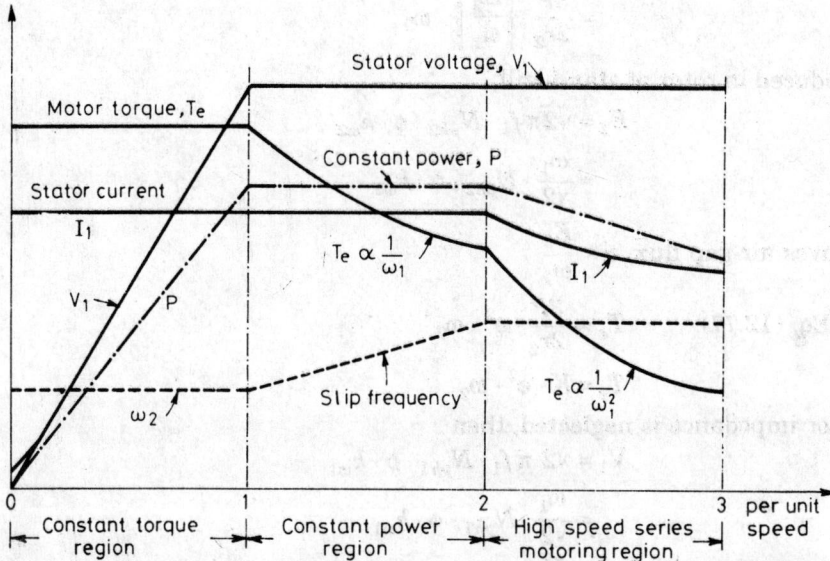

Fig. 12.35. Stator voltage, current, slip-frequency, torque and power variation
with speed for speed-torque characteristics of Fig. 12.34.

Constant-power region. When maximum speed, called base speed, is attained in the constant-torque region, stator voltage reaches its rated value. Motor speeds beyond base (or rated) speed are obtained by keeping stator voltage constant and lowering the stator frequency.

Torque in a 3-phase induction motor is given by

$$T_e = \frac{3}{\omega_s} \text{ (Power input to rotor)}$$

$$= \frac{3}{\omega_s} \cdot E_2 I_2 \cos \theta_2$$

At low rotor frequency, $\cos \theta_2 \left(= \dfrac{r_2}{\sqrt{r_2{}^2 + (2\pi f_2 l_2 s)^2}} \right)$ is almost nearer to unity.

$$\therefore \quad T_e = \frac{3}{\omega_s} \cdot E_2 I_2 \qquad\qquad\qquad ...(12.76)$$

But rotor current, $I_2 = \dfrac{E_2}{\dfrac{r_2}{s} + j\, x_2}$

At small slips, $\dfrac{r_2}{s} \gg x_2$. This gives $I_2 = \dfrac{sE_2}{r_2}$

Substituting this value of I_2 in Eq. (12.76), we get

$$T_e = \frac{3}{\omega_s} \cdot \frac{s\,E_2^{\,2}}{r_2}$$

$$= \frac{3}{\omega_s} \cdot \frac{\omega_2}{\omega_1} \cdot \frac{E_2^{\,2}}{r_2} = \frac{3P}{2\omega_1} \cdot \frac{\omega_2}{\omega_1} \cdot \frac{E_2^{\,2}}{r_2}$$

$$= \frac{3P}{2r_2} \cdot \left(\frac{E_2}{\omega_1}\right)^2 \cdot \omega_2 \qquad\qquad\qquad ...(12.77)$$

Emf induced in rotor at stand-still,

$$E_2 = \sqrt{2}\,\pi\,f_1 \cdot N_{ph2} \cdot \phi \cdot k_{\omega 2}$$

$$= \frac{\omega_1}{\sqrt{2}} \cdot N_{ph2} \cdot \phi \cdot k_{\omega 2}$$

This gives air-gap flux, $\phi \propto \dfrac{E_2}{\omega_1}$

From Eq. (12.77), $T_e \propto \dfrac{3P}{2r_2} \cdot \phi^2 \cdot \omega_2$

or $T_e = K \cdot \phi^2 \cdot \omega_2 \qquad\qquad\qquad ...(12.78)$

If stator impedance is neglected, then

$$V_1 = \sqrt{2}\,\pi\,f_1 \cdot N_{ph1} \cdot \phi \cdot k_{\omega 1}$$

$$= \frac{\omega_1}{\sqrt{2}} \cdot N_{ph1} \cdot \phi \cdot k_{\omega 1}$$

$\therefore$ Air-gap flux, $\phi \propto \dfrac{V_1}{\omega_1}$

From Eq. (12.78), $T_e \propto K \left(\dfrac{V_1}{\omega_1}\right)^2 \cdot \omega_2$

or $T_e = K_1 \dfrac{V_1^2}{\omega_1^2} \cdot \omega_2 = K_1 \cdot \dfrac{V_1^2}{\omega_1}\left(\dfrac{\omega_2}{\omega_1}\right) \qquad ...(12.79)$

Eq. (12.79) gives approximate value of motor torque for a 3-phase IM working in the low-slip range. It shows that if IM is operated at constant stator voltage V_1 and constant rotor frequency ω_2, the motor torque is inversely proportional to supply-frequency squared. In case slip (or rotor) frequency ω_2 is increased linearly as ω_1 is increased to obtain high-operating speed $\left(= \dfrac{2\omega_1}{P}\text{ rad/s}\right)$, then $\dfrac{\omega_2}{\omega_1}$ is constant and therefore $T_e \cdot \omega_1 = K \cdot V_1^2 \left(\dfrac{\omega_2}{\omega_1}\right)$ remains constant giving a constant power characteristic. During constant power operation, stator voltage is kept constant but stator frequency is raised. As a result, stator flux decreases or air-gap flux is weakened. In view of this, constant-power mode of a 3-phase IM is also called *field-weakening mode* of an IM.

The upper limit of constant-power mode is reached when maximum working value of rotor frequency is reached.

High-speed series-motoring region. From Eq. (12.79),

$$T_e = K_1 \cdot \frac{V_1^{\,2} \cdot \omega_2}{\omega_1^{\,2}} \qquad\qquad\qquad ...(12.79)$$

After the constant-power region, high-speed series-motoring region is obtained. In this region, stator voltage V_1 and rotor frequency ω_2 are maintained constant at their maximum values. It is seen from Eq. (12.79) that under constant V_1 and ω_2 operation of a 3-phase IM, output torque T_e is inversely proportional to supply frequency squared or T_e varies inversely as speed squared. In other words, $T_e \cdot \omega_1^2$ remains constant and series-type characteristics are thus obtained. As this region corresponds to high-speed series-type characteristics, operation under constant V_1, ω_2 and variable ω_1 is usually referred to as *high-speed series-motoring region*.

In Fig. 12.34, speed-torque characteristics at different stator frequencies in the constant-torque, constant-power and high-speed series-motoring regions are shown. The maximum torque, indicated by dashed line, is constant below base speed and decreases inversely with speed above base speed and upto $\omega_1 = 2.0$. In the constant-torque and constant-power regions, maximum allowable torque is shown somewhat lower than the maximum or breakdown torque T_{em} just as a matter of precaution because inverter current carrying capability is limited. Maximum allowable torque is indicated by solid line in the constant-torque region and by solid-curve in the constant-power region.

Example 12.26. *Is it possible to obtain Ward-Leonard type of characteristics from a 3-phase induction motor ? Discuss.*

Solution. Word-Leonard system of speed control for a separately-excited dc motor gives torque-speed and power-speed characteristics as shown in Fig. 12.36. A comparison of Figs. 12.35 and 12.36 shows that Ward-Leonard type of characteristics as obtained in a dc motor can be obtained from a 3-phase induction motor.

From base speed down to zero speed, the speed control is obtained by (*i*) armature voltage control in a dc motor and (*ii*) stator voltage and frequency control in a 3-phase induction motor. In both, the flux is kept constant by keeping (*i*) field current constant in a dc motor and (*ii*) V/f constant in a 3-phase IM. Armature current in a dc motor and stator current in IM are kept constant at their rated values. In both, constant torque and variable power characteristics are obtained from zero to base speed.

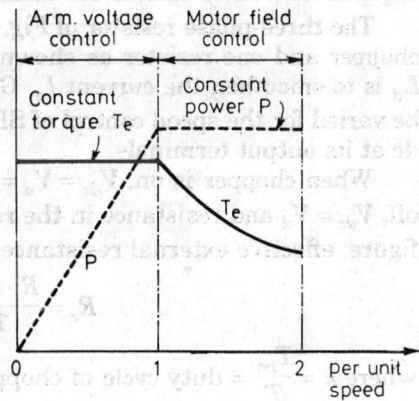

Fig. 12.36. Torque-speed and power-speed characteristics for Ward-Leonard system of speed control.

Above base speed, the speed control is obtained by field-weakening method in both dc and ac motors. Constant power and variable torque characteristics are obtained in both types of motors as shown. Armature current in dc motor and stator current in 3-phase IM are kept constant. In dc motor, armature voltage is kept constant whereas field flux is weakened by decreasing the field current. In IM, stator voltage is kept constant and air-gap flux is weakened by increasing stator frequency but by keeping $\dfrac{\omega_2}{\omega_1}$ constant.

12.8.5. Static Rotor-resistance Control

In a slip-ring induction motor (SRIM), a 3-phase variable resistor R_2 can be inserted in the rotor circuit as shown in Fig. 12.37 (*a*). By varying the rotor circuit resistance R_2, the motor torque can be controlled as shown in Fig. 12.37 (*b*). The starting torque and starting current can also be varied by controlling the rotor circuit resistance, Fig. 12.37 (*b*) and (*c*). The disadvantages of this method of speed control are : (*i*) reduced efficiency at low speeds (*ii*) speed

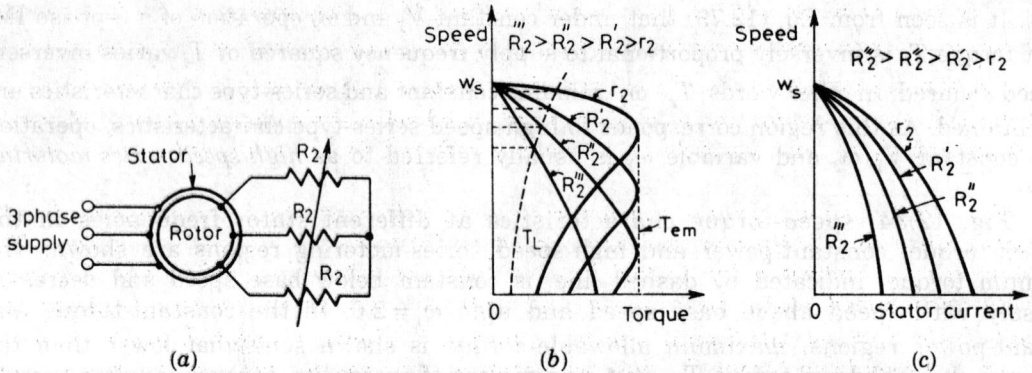

Fig. 12.37. Three-phase IM speed control by rotor resistance (a) circuit arrangement
(b) effect on developed torque (c) effect on stator current.

changes very widely with load variation (iii) unbalances in voltages and currents if rotor circuit resistances are not equal. In spite of these, this method of speed control is used when speed drop is required for a short time, as for example in overhead cranes, in load equalization etc.

The three-phase resistor of Fig. 12.37 (a) may be replaced by a three-phase diode rectifier, chopper and one resistor as shown in Fig. 12.38 (a). In this figure, the function of inductor L_d is to smoothen the current I_d. GTO chopper allows the effective rotor circuit resistance to be varied for the speed control of SRIM. Diode rectifier converts slip-frequency input power to dc at its output terminals.

When chopper is on, $V_{dc} = V_d = 0$ and resistance R gets short-circuited. When chopper is off, $V_{dc} = V_d$ and resistance in the rotor circuit is R. This is shown in Fig. 12.38 (b). From this figure, effective external resistance R_e is

$$R_e = \frac{R \cdot T_{off}}{T} = \frac{R\,(T - T_{on})}{T} = R\,(1 - k) \qquad \qquad ...(12.80)$$

where $k = \dfrac{T_{on}}{T}$ = duty cycle of chopper.

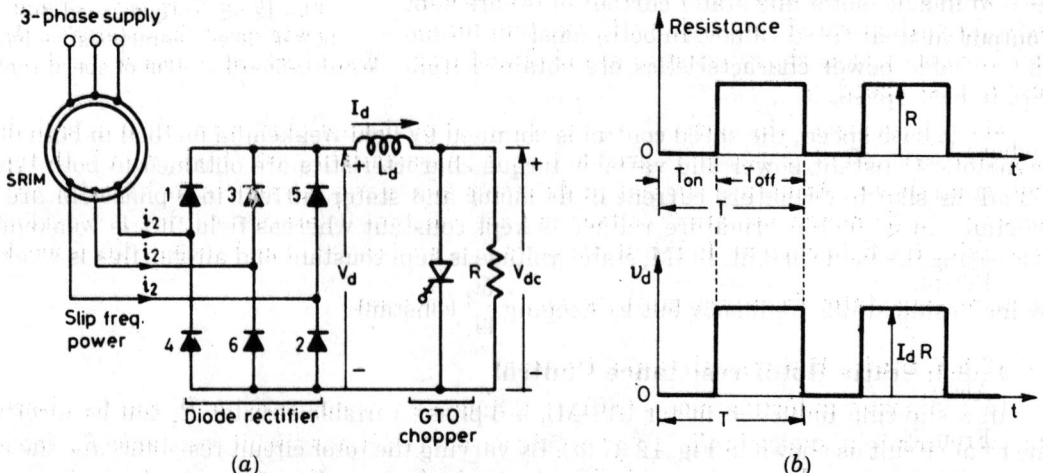

Fig. 12.38. (a) SRIM control by static variation of external rotor resistance
(b) waveforms pertaining to Fig. (a).

Analysis of induction motor with chopper control. The equivalent circuit for 3-phase IM, diode rectifier and chopper circuit of Fig. 12.38 (a) is as shown in Fig. 12.39 (a).

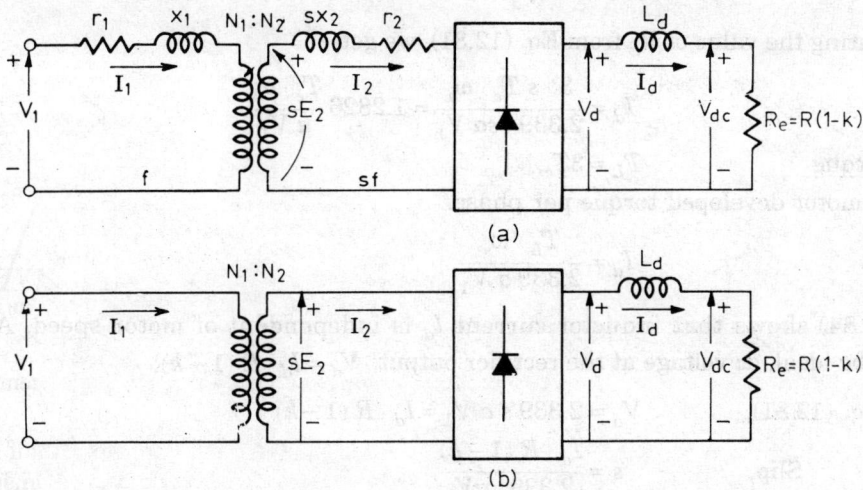

(a)

(b)

Fig. 12.39 (a). Equivalent circuit for Fig. 12.38 (a), (b) its approximate equivalent circuit.

If stator and rotor leakage impedances are neglected as compared to inductor L_d, equivalent circuit of Fig. 12.39 (b) is obtained. Stator voltage V_1 when referred to rotor circuit gives slip-frequency voltage as

$$s \cdot \frac{V_1}{N_1} \cdot N_2 = s \, a \, V_1 = s \, E_2$$

where E_2 = rotor induced emf per phase at stand-still

V_1 = stator voltage per phase

$a = \dfrac{\text{rotor effective turns per phase, } N_2}{\text{stator effective turns per phase, } N_1}$ = per phase turns ratio from rotor to stator.

Voltage $s E_2 = s a V_1$, after rectification by 3-phase diode bridge appears as V_d (rectifier output voltage).

$$\therefore \qquad V_d = \frac{3V_{ml}}{\pi} = \frac{3 \cdot \sqrt{3} \, V_{mp}}{\pi} = \frac{3\sqrt{3}}{\pi} \cdot \sqrt{2} \cdot sa \, V_1 = 2.339 \, sa V_1 \qquad \qquad ...(12.81)$$

where V_{mp} = maximum value of phase voltage = $\sqrt{2} \, sa \, V_1$

Total slip power = $3 \, s \, P_g$. For no losses in the rectifier, this must be equal to $V_d I_d$.

$$\therefore \qquad 3s \, P_g = V_d \, I_d$$

Per-phase developed power, $P_m = (1 - s) \, P_g$

$$= (1 - s) \frac{V_d \, I_d}{3s} \qquad \qquad ...(12.82)$$

Also, $\qquad \qquad P_m = T_e \cdot \omega_m = T_e \cdot \omega_s \, (1 - s) \qquad \qquad ...(12.83)$

From Eqs. (12.82) and (12.83), we get

$$V_d I_d \frac{(1 - s)}{3s} = T_e \cdot \omega_s \, (1 - s)$$

or
$$I_d = \frac{T_e \cdot \omega_s \cdot 3s}{V_d}$$

Substituting the value of V_d from Eq. (12.81), we get

$$I_d = \frac{3 \cdot s\, T_e \cdot \omega_s}{2.339 \cdot sa\, V_1} = 1.2826\, \frac{T_e \cdot \omega_s}{a\, V_1} \qquad \qquad ...(12.84\,a)$$

Load torque $\qquad\qquad T_L = 3T_e,$

where T_e = motor developed torque per phase.

$$\therefore \qquad\qquad I_d = \frac{T_L \cdot \omega_s}{2.339\, a\, V_1} \qquad\qquad ...(12.84\,b)$$

Eq. (12.84) shows that inductor current I_d is independent of motor speed. Assuming inductor to be ideal, dc voltage at the rectifier output, $V_d = I_d \cdot R\,(1-k)$.

From Eq. (12.81), $\qquad V_d = 2.339\, s\, a\, V_1 = I_d \cdot R\,(1-k)$

$$\therefore \qquad \text{Slip,} \qquad s = \frac{I_d \cdot R\,(1-k)}{2.339 \cdot aV_1} \qquad\qquad ...(12.85)$$

Motor speed, $\qquad \omega_m = \omega_s\,(1-s)$

$$= \omega_s\left[1 - \frac{I_d \cdot R\,(1-k)}{2.339 \cdot a\, V_1}\right] \qquad\qquad ...(12.86\,a)$$

Also $\qquad\qquad N_r = N_s\left[1 - \frac{I_d \cdot R\,(1-k)}{2.339 \cdot a\, V_1}\right]$

Substituting the value of I_d from Eq. (12.84 b) in Eq. (12.86 a), we get

$$\omega_m = \omega_s\left[1 - \frac{T_L \cdot \omega_s \cdot R\,(1-k)}{(2.339 \cdot a\, V_1)^2}\right] \qquad\qquad ...(12.86\,b)$$

For fixed value of duty cycle, speed falls as load torque T_L is increased.

Each diode in Fig. 12.38 (a) conducts for 120°. The waveform of rotor current i_2 is shown in Fig. 12.40. For a ripple-free output current I_d, it is seen from Fig. 12.40 that

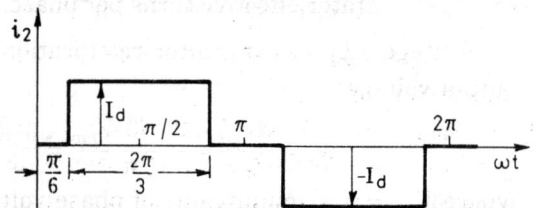

rms value of rotor current,

$$I_2 = \sqrt{I_d^2 \cdot \frac{2\pi}{3} \cdot \frac{1}{\pi}} = \sqrt{\frac{2}{3}} \cdot I_d \qquad ...(12.87)$$

Fig. 12.40. Rotor current waveform for ripple-free I_d.

Rotor current referred to stator,

$$I_1 = \frac{N_2}{N_1} I_2 = aI_2 = a \cdot I_d \cdot \sqrt{\frac{2}{3}} \qquad\qquad ...(12.88)$$

Fourier analysis of the waveform in Fig. 12.40 can be obtained from Eq. (8.61).

$$\therefore \qquad b_1 = \frac{2}{\pi}\int_{\pi/6}^{5\pi/6} I_d \sin \omega t \cdot d\,(\omega t) = \frac{2}{\pi} I_d \left|- \cos \omega t \right|_{\pi/6}^{5\pi/6} = \frac{2\sqrt{3}}{\pi} I_d$$

This gives **fundamental component of rotor current** as

$$I_{21} = \frac{b_1}{\sqrt{2}} = \frac{2\sqrt{3}}{\sqrt{2}\cdot\pi} = \frac{\sqrt{6}}{\pi} I_d \qquad\qquad ...(12.89)$$

Fundamental component of rotor current referred to stator

$$I_{11} = \frac{N_2}{N_1} I_{21} = a \cdot I_{21} = \frac{\sqrt{6}}{\pi} \cdot a \cdot I_d \qquad\qquad ...(12.90)$$

Example 12.27. *A 3-phase, 420V, 4-pole, 50Hz, star-connected SRIM has its speed controlled by means of GTO chopper in its rotor circuit. The effective phase turns ratio from rotor to stator is 0.8. The filter inductor makes the inductor current ripple free. Losses in the rectifier, inductor, GTO chopper and no-load losses of the motor are neglected. Load torque, proportional to speed squared, is 450 Nm at 1440 rpm.*

 (a) For a minimum motor speed of 1000 rpm, calculate the value of chopper resistance R.

 For the value of R obtained in part (a), if the speed is to be raised to 1320 rpm, calculate.

 (b) inductor current (c) duty cycle of the chopper (d) rectified output voltage (e) efficiency in case per-phase resistances for stator and rotor are 0.015 Ω and 0.02 Ω respectively.

Solution. Per-phase stator voltage,

$$V_1 = \frac{420}{\sqrt{3}} = 242.5 \text{ V}$$

Load torque at 1000 rpm, $T_L = 450\left(\dfrac{1000}{1440}\right)^2 = 217.01 \text{ Nm}$

Synchronous speed, $\omega_s = \dfrac{2\pi \times 1500}{60} = 50\,\pi \text{ rad/s}$

Minimum motor speed, $\omega_{m\cdot mn} = \dfrac{2\pi \times 1000}{60} = 104.72 \text{ rad/s}$

(a) From Eq. (12.84 b), the inductor current,

$$I_d = \frac{217.01 \times 50\,\pi}{2.339 \times 0.8 \times 242.5} = 75.122 \text{ A}$$

For minimum motor speed, duty cycle $k = 0$ in Eq. (12.86 a).

$$\therefore \qquad \omega_{m\cdot mn} = \omega_s\left[1 - \frac{I_d \cdot R\,(1-0)}{2.339\,aV_1}\right]$$

$$104.72 = 50\,\pi\left[1 - \frac{75.122 \times R}{2.339 \times 0.8 \times 242.5}\right]$$

This gives $R = 2.0134\,\Omega$

(b) New load torque at 1320 rpm, $T_L = 450\left(\dfrac{1320}{1440}\right)^2 = 378.125 \text{ Nm}$

Inductor current from Eq. (12.84 b) is

$$I_d = \frac{378.125 \times 50\,\pi}{2.339 \times 0.8 \times 242.5} = 130.895 \text{ A}$$

(c) $\omega_m = \dfrac{2\pi \times 1320}{60} = 138.23 \text{ rad/s.}$

From Eq. (12.86 a), $138.23 = 50\pi\left[1 - \dfrac{130.895 \times 2.0134 \times (1 - k)}{2.339 \times 0.8 \times 242.5}\right]$

$$= 50\,\pi\,[1 - 0.5808\,(1 - k)]$$

$\therefore$ Duty cycle of chopper, $k = 0.7934$

(d) Slip, $s = \dfrac{1500 - 1320}{1500} = 0.12$

From Eq. (12.81), $V_d = 2.339 \times 0.12 \times 0.8 \times 242.5 = 54.452$ V

(e) Power loss in chopper resistance

$$= V_d I_d = 54.452 \times 130.895 = 7127.5 \text{ W}$$

Also, power loss in chopper resistance

$$= I_d^2 \cdot R(1 - k)$$

$$= (130.895)^2 \times 2.0134 \,(1 - 0.7934) = 7127 \text{ W}$$

From Eq. (12.87), inductor current referred to rotor,

$$I_2 = \sqrt{\frac{2}{3}} \cdot I_d = \sqrt{\frac{2}{3}} \times 130.895 = 106.88 \text{ A}$$

Total rotor ohmic loss $= 3 I_2^2 \, r_2 = 3 \times 106.88^2 \times 0.02 = 685.4$ W

From Eq. (12.88), stator current,

$$I_1 = a I_2 = 0.8 \times 106.88 = 85.504 \text{ A}$$

Total stator ohmic loss $= 3 I_1^2 r_1 = 3 \times 85.504^2 \times 0.015 = 329$ W

Power output $= T_L \cdot \omega_m = 378.125 \times \dfrac{2\pi \times 1320}{60} = 52268.25$ W

Power input $= 52268.25 + 7127.5 + 685.4 + 329 = 60410.15$ W

Efficiency $= \dfrac{52268.25}{60410.15} \times 100 = 86.52\%$

Example 12.28. *A 3-phase, 400 V, 50 Hz, 960 rpm, star- connected SRIM has the following per-phase parameters referred to stator :*

$$r_1 = 0.1 \,\Omega, \; r_2 = 0.08 \,\Omega, \; x_1 = x_2 = 0.3 \,\Omega, X_m = 0$$

Per-phase turns ratio from rotor to stator = 0.7

Speed of this motor is controlled by a GTO chopper in its rotor circuit. For a speed of 800 rpm, the inductor current is 110 A and the chopper resistance is 2 Ω. Calculate

(a) the value of chopper duty cycle

(b) efficiency for a power output of 20 kW and for negligible no-load losses.

(c) the input power factor.

Solution. (a) Per-phase voltage $= \dfrac{400}{\sqrt{3}} = 230.9$ V

Synchronous speed, $N_s = 1000$ rpm

From Eq. (12.86 a), $800 = 1000 \left[1 - \dfrac{110 \times 2 \,(1 - k)}{2.339 \times 0.7 \times 230.9} \right]$

This gives chopper duty cycle, $k = 0.656$

(b) Power loss in chopper $= I_d^2 \cdot R\,(1 - k) = 110^2 \times 2 \,(1 - 0.656) = 8324.8$ W

Rms value of rotor current referred to stator, from Eq. (12.88), is

$$I_1 = a \cdot I_d \cdot \sqrt{\frac{2}{3}} = 0.7 \times 110 \times \sqrt{\frac{2}{3}} = 62.87 \text{ A}$$

Power loss in stator and rotor resistances

$$= 3 \,(62.87)^2 \times (0.1 + 0.08) = 2134.4 \text{ W}$$

Power input $= 20,000 + 8324.8 + 2134.4 = 30459.2$ W

Efficiency $= \dfrac{20,000}{30459.2} \times 100 = 65.66\%$

(c) From Eq. (12.90), fundamental component of rotor current referred to stator is

$$I_{11} = \frac{\sqrt{6}}{\pi} \cdot a \cdot I_d = \frac{\sqrt{6}}{\pi} \cdot (0.7) \times 110 = 60.04 \text{ A}$$

Power input $= \sqrt{3} \times 400 \times 60.04 \times \cos\theta = 30459.2 \text{ W}$

$\therefore$ Input $pf = \dfrac{30459.2}{\sqrt{3} \times 400 \times 60.04} = 0.7323 \text{ lag}$

12.8.6. Slip-Power Recovery Schemes

In chopper method of speed control for SRIM, the slip power is dissipated in the external resistance and it leads to poor efficiency of the drive. However, instead of wasting the slip power in the rotor circuit resistance, it can be conveniently converted by various schemes for the speed control of SRIM. Two important slip-power recovery schemes are static Kramer drive and static Scherbius drive. These are now discussed in what follows.

12.8.6.1. Static Kramer Drive. The circuit configuration for static Kramer drive is shown in Fig. 12.41. The slip-frequency power from the rotor circuit is converted to dc voltage which is then converted to line frequency and pumped back to the ac source. As the slip power can flow only in one direction, static Kramer drive offers speed control below synchronous speed only.

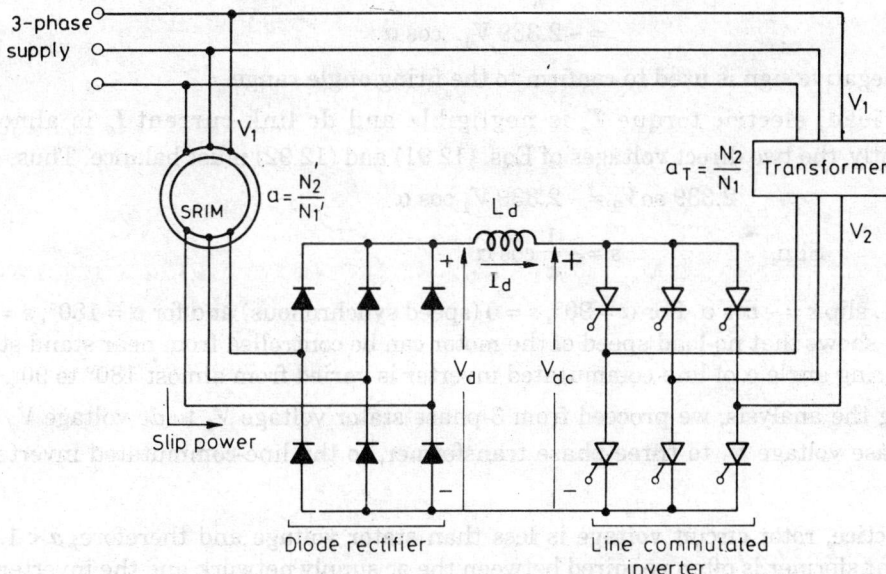

Fig. 12.41. Static Kramer drive.

The slip power from rotor is rectified to dc voltage by diode bridge. Inductor L_d smoothens the ripples in the rectified voltage V_d. This voltage V_d is then converted to ac voltage at line frequency by line-commutated inverter and fed back to 3-phase supply. As the power flow is from rotor circuit to supply, static Kramer drive offers *constant-torque drive*. As stated before, this scheme offers speed control below synchronous speed only. Simplified torque and speed expressions for this drive can be derived as follows.

Rotor voltage per phase $= sE_2$

where E_2 = per phase rotor e.m.f. at standstill

and s = slip

Voltage sE_2 is rectified to V_d by diode bridge. Uncontrolled output voltage of diode rectifier, from Eq. (6.54), with $\alpha = 0$, is

$$V_d = \frac{3 \times \text{maximum value of input line voltage}}{\pi}$$

$$= \frac{3\sqrt{2}\ (\sqrt{3}\ sE_2)}{\pi} = \frac{3\sqrt{6}}{\pi}\ s\ E_2$$

But
$$\frac{E_2}{N_2} = \frac{V_1}{N_1} \quad \text{or} \quad E_2 = V_1 \frac{N_2}{N_1} = a\ V_1$$

where
$$a = \frac{\text{effective rotor turns per phase, } N_2}{\text{effective stator turns per phase, } N_1}$$

$$V_1 = \text{supply voltage per phase.}$$

$\therefore$
$$V_d = \frac{3\sqrt{6}}{\pi} \cdot saV_1 = 2.339\ saV_1 \qquad \qquad \qquad ...(12.91)$$

For three-phase line-commutated inverter, average dc output voltage (with no transformer) is

$$V_{dc} = -\frac{3\sqrt{6}}{\pi}\ V_1 \cdot \cos \alpha \qquad 90° \leq \alpha \leq 180°$$

$$= -2.339\ V_1 \cdot \cos \alpha \qquad \qquad \qquad ...(12.92)$$

Here negative sign is used to confirm to the firing angle range.

At no load, electric torque T_e is negligible and dc link current I_d is almost zero. Consequently, the two direct voltages of Eqs. (12.91) and (12.92) must balance. Thus,

$$2.339\ saV_1 = -2.339\ V_1 \cos \alpha$$

or   Slip,
$$s = -\frac{1}{a} \cos \alpha \qquad \qquad \qquad ...(12.93)$$

If $a = 1$, slip $s = -\cos \alpha$. For $\alpha = 90°$, $s = 0$ (speed synchronous) and for $\alpha = 180°$, $s = 1$ (speed zero). This shows that no-load speed of the motor can be controlled from near stand-still to full speed as firing angle α of line-commutated inverter is varied from almost 180° to 90°.

During the analysis, we proceed from 3-phase stator voltage V_1 to dc voltage V_d and also from 3-phase voltage V_1 to three-phase transformer, to the line-commutated inverter and to V_{dc}.

In practice, rotor circuit voltage is less than stator voltage and therefore, $a < 1$. Thus, a 3-phase transformer is often required between the ac supply network and the inverter in order to step down the supply voltage to a level that is appropriate to V_d. Let the transformer turn-ratio be a_T where

$$a_T = \frac{\text{per phase output voltage of transformer as input voltage to inverter, } V_2}{\text{per phase supply voltage, } V_1}$$

Note that turns ratio a_T is usually less than unity. AC voltage across inverter terminals, $V_2 = a_T \cdot V_1$. The inverter dc voltage V_{dc}, from Eq. (12.92), is given by

$$V_{dc} = -\frac{3\sqrt{6}}{\pi} \cdot V_2 \cdot \cos \alpha = -\frac{3\sqrt{6}}{\pi} \cdot a_T \cdot V_1 \cos \alpha$$

$$= -2.339 \cdot a_T \cdot V_1 \cos \alpha \qquad \qquad \qquad ...(12.94)$$

With the use of transformer, from Eqs. (12.91) and (12.94), we get

$$2.339 \, sa V_1 = -2.339 \, a_T \cdot V_1 \cos \alpha$$

or

$$\text{slip, } s = -\frac{a_T}{a} \cos \alpha \qquad \qquad ...(12.95)$$

In order to develop motor torque, a rotor current I_2 is required and the rectifier rotor voltage V_d must force a current I_d against the inverter dc voltage V_{dc}. When the motor is loaded, speed falls and the increased rotor voltage sE_2 can overcome the voltage drops in the rotor windings, in the dc link circuit and the converters.

If resistance of the rotor circuit and inductor L_d are neglected, then

total slip power, $\qquad 3sP_g = V_{dc} \cdot I_d$

$$3 \cdot \omega_s \cdot T_e = V_{dc} \cdot I_d$$

or

$$T_e = \frac{V_{dc} \cdot I_d}{3s \cdot \omega_s} \qquad \qquad ...(12.96)$$

Substituting in Eq. (12.96), the values of s from Eq. (12.93) and V_{dc} from Eq. (12.92), we get

$$T_e = \frac{2.339 \, V_1 \cdot \cos \alpha \cdot I_d}{3 \cdot \dfrac{1}{a} \cdot \cos \alpha \cdot \omega_s} = \frac{2.339}{3} \cdot \frac{a \, V_1 \cdot I_d}{\omega_s} \qquad \qquad ...(12.97)$$

When a transformer is used in Fig. 12.41, then substitute in Eq. (12.96), the values of s from Eq. (12.95) and V_{dc} from Eq. (12.94) and we get

$$T_e = \frac{2.339 \cdot a_T \cdot V_1 \cos \alpha \cdot I_d}{3 \cdot \dfrac{a_T}{a} \cdot \cos \alpha \cdot \omega_s} = \frac{2.339}{3} \cdot \frac{a \, V_1 \, I_d}{\omega_s} \qquad \qquad ...(12.98)$$

An examination of Eqs. (12.97) and (12.98) reveal that these equations are valid whether a transformer is used in the static Kramer drive circuit or not. It is observed from Eq. (12.97) or Eq. (12.98) that steady state torque is

    (i) proportional to dc link current, I_d

    (ii) proportional to stator supply voltage, V_1

    (iii) proportional to effective rotor to stator turns ratio, a

and (iv) inversely proportional to synchronous speed, ω_s.

The dc link current I_d is given by

$$I_d = \frac{V_d - V_{dc}}{\text{resistance of dc link inductor, } R_d}$$

or

$$V_d = V_{dc} + I_d \cdot R_d = 2.339 \, sa V_1$$

$\therefore \qquad$ Slip, $\qquad$

$$s = \frac{V_{dc} + I_d \cdot R_d}{2.339 \, a \, V_1}$$

$$= \frac{-2.339 \cdot a_T \cdot V_1 \cos \alpha}{2.339 \, a V_1} + \frac{I_d R_d}{2.339 \, a V_1}$$

$$= -\frac{a_T}{a} \cos \alpha + \frac{I_d \cdot R_d}{2.339 \, a V_1}$$

Motor speed ω_m is given by

$$\omega_m = \omega_s (1 - s)$$

$$= \omega_s \left[1 + \frac{a_T}{a} \cos \alpha - \frac{I_d \cdot R_d}{2.339 \cdot aV_1} \right] \qquad \qquad ...(12.99)$$

Under steady state, from Eq. (12.98), total torque $3T_e$ is given by

$$3T_e = T_L = 2.339 \, \frac{aV_1 \cdot I_d}{\omega_s}$$

or

$$I_d = \frac{\omega_s \cdot T_L}{2.339 \, a \, V_1} \qquad \qquad ...(12.100)$$

Substituting this value of I_d in Eq. (12.99), we get

$$\omega_m = \omega_s \left[1 + \frac{a_T}{a} \cos \alpha - \frac{\omega_s \cdot R_d}{(2.339 \, aV_1)^2} \cdot T_L \right] \qquad \qquad ...(12.101)$$

$$= \omega_s \left[1 + \frac{a_T}{a} \cos \alpha - K \, T_L \right] \qquad \qquad ...(12.102)$$

where

$$K = \frac{\omega_s \cdot R_d}{(2.339. \, aV_1)^2}$$

From Eq. (12.101) or (12.102), the no-load speed of the drive is given by

$$\omega_m = \omega_s \left[1 + \frac{a_T}{a} \cos \alpha \right] \quad ...(12.103)$$

From Eq. (12.101), the speed-torque characteristics of static Kramer drive shown in Fig. 12.41 are plotted, for different firing angles, in Fig. 12.42. It is seen that these characteristics are similar to a separately excited dc motor with armature voltage control.

Static Kramer drive systems are used in large power-pumps and compressor type loads where speed control is within narrow range and below synchronous speed.

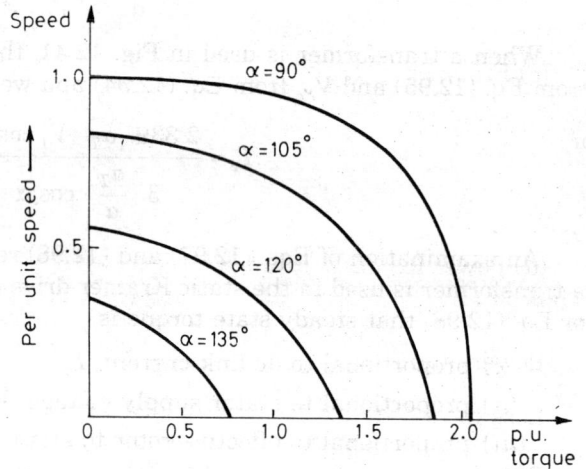

Fig. 12.42. Speed-torque characteristics of static Kramer drive for open loop system.

Example 12.29. *A 3-phase, 420 V, 4-pole, 50 Hz, star-connected SRIM has its speed controlled by means of static Kramer drive. The effective phase turns ratio from rotor to stator is 0.8 and transformer has phase turns ratio from l.v. to h.v. as 0.4. The inductor current is ripple free. Losses in diode rectifier, inductor, inverter and transformer are neglected. The load torque is proportional to speed squared and its value at 1200 rpm is 450 Nm. For a motor operating speed of 1000 rpm, calculate.*

(a) rotor rectified voltage (b) inductor current (c) delay angle of the inverter (d) efficiency, in case inductor resistance is 0.01 Ω and per-phase resistances for stator and rotor are 0.015 Ω and 0.02 Ω respectively.

(e) *For the firing angle obtained in part (c), the load torque is increased to 500 Nm, find the motor speed.*

Solution. (a) Per-phase stator voltage,

$$V_1 = \frac{420}{\sqrt{3}} = 242.5 \text{ V}$$

∴ Slip, $$s = \frac{1500 - 1000}{1500} = \frac{1}{3}$$

Turns ratio from rotor to stator, $a = 0.8$

From Eq. (12.91), rectified voltage,

$$V_d = 2.339 \, \varepsilon a V_1$$

$$= 2.339 \times \frac{1}{3} \times 0.8 \times 242.5 = 151.26 \text{ V}$$

(b) Load torque at 1000 rpm, $$T_L = 450 \times \left(\frac{1000}{1200}\right)^2 = 3T_e = 312.5 \text{ Nm}$$

Synchronous speed, $$\omega_s = \frac{4\pi f_1}{P} = \frac{4\pi \times 50}{4} = 50\,\pi \text{ rad/s}$$

From Eq. (12.100), inductor current is

$$I_d = \frac{\omega_s \cdot T_L}{2.339 \, a V_1} = \frac{50\pi \times 312.5}{2.339 \times 0.8 \times 242.5} = 108.18 \text{ A}$$

(c) From Eq. (12.95), $$s = -\frac{a_T}{a} \cdot \cos \alpha$$

or $$\frac{1}{3} = -\frac{0.4}{0.8} \cos \alpha$$

∴ $$\alpha = \cos^{-1}\left(-\frac{2}{3}\right) = 131.81°$$

(d) Power fed back to supply

$$= V_d I_d = 151.26 \times 108.18 = 16363.31 \text{ W}$$

Power output $$= T_L \cdot \omega_m = 312.5 \times \frac{2\pi \times 1000}{60} = 32724.92 \text{ W}$$

Loss in inductor $$= I_d^2 \cdot R_d = (108.18)^2 \times 0.01 = 117.03 \text{ W}$$

Rotor current, $$I_2 = \sqrt{\frac{2}{3}} \, I_d = \sqrt{\frac{2}{3}} \times 108.18 = 88.33 \text{ A}$$

Rotor ohmic loss $$= 3I_2^2 \, r_2 = 3\,(88.33)^2 \times 0.02 = 468.13 \text{ W}$$

Stator current, $$I_1 = a \, I_2 = 0.8 \times 88.33 = 70.664 \text{ A}$$

Stator ohmic loss $$= 3\,(70.664)^2 \times 0.015 = 224.70 \text{ W}$$

Power input $$= 32724.92 + 468.13 + 224.70 + 117.03 = 33534.78 \text{ W}$$

Efficiency $$= \frac{32724.92}{33534.78} \times 100 = 97.58\%$$

(e) From Eq. (12.101), we get motor speed as under :

$$\omega_m = 50\pi \left[1 + \frac{0.4}{0.8} \cos 131.81° - \frac{50\pi \times 0.01 \times 500}{(2.339 \times 0.8 \times 242.5)^2}\right]$$

$$= 104.121 \text{ rad/s} \quad \text{or} \quad 994.3 \text{ rpm.}$$

Example 12.30. *A static Kramer drive is used for the speed control of a 4-pole SRIM fed from 3-phase, 415 V, 50 Hz supply. The inverter is connected directly to the supply. If the motor is required to operate at 1200 rpm, find the firing advance angle of the inverter. Voltage across the open-circuited slip rings at stand-still is 700 V. Allow a voltage drop of 0.7 V and 1.5 V across each of the diodes and thyristors respectively. Inductor drop is neglected.*

Solution. Rotor induced emf at stand-still = 700 V (line)

$$\therefore \qquad E_2 = \frac{700}{\sqrt{3}} \text{ V}$$

Slip,
$$s = \frac{1500 - 1200}{1500} = 0.2$$

DC voltage across the diode rectifier is

$$V_d = \frac{3\sqrt{6}\ sE_2}{\pi} - 2 \times 0.7 = \frac{3\sqrt{2} \times 0.2 \times 700}{\pi} - 1.4$$

Inverter dc voltage is given by

$$V_{dc} = -\left[\frac{3\sqrt{2} \times 415}{\pi} \cos \alpha - 2 \times 1.5\right]$$

With no voltage drop in inductor, $V_{dc} = V_d$

$$-\frac{3\sqrt{2} \times 415}{\pi} \cos \alpha + 3 = \frac{3\sqrt{2} \times 0.2 \times 700}{\pi} - 1.4$$

or $\qquad \alpha = \cos^{-1}\left[\frac{-184.6379 \times \pi}{3\sqrt{2} \times 415}\right] = 109.24°$

$\therefore$ Firing advance angle of inverter = $180 - \alpha = \mathbf{180° - 109.24° = 70.76°}$

Example 12.31. *Repeat Example 12.30 in case there is an overlap angle of 18° in the rectifier and 4° in the inverter.*

Solution. Average output voltage for a 3-phase full converter, with overlap, is given by Eq. (6.79 a), which is reproduced here, for convenience.

$$V_{ox} = \frac{3V_{ml}}{2\pi} [\cos \alpha + \cos (\alpha + \mu)]$$

For uncontrolled rectifier bridge, dc output voltage is

$$V_d = \frac{3\sqrt{6}\ sE_2}{2\pi} [\cos 0° + \cos (0° + 18°)] - 2 \times 0.7$$

$$= \frac{3\sqrt{6} \times 0.2 \times 700}{2\pi \times \sqrt{3}} [1 + \cos 18°] - 1.4 = 183.012 \text{ V}$$

Similarly, inverter dc voltage is

$$V_{dc} = -\left[\left\{\frac{3V_{ml}}{2\pi} (\cos \alpha + \cos (\alpha + \mu)\right\} - 2 \times 1.5\right]$$

$$= -\left[\left\{\frac{3\sqrt{2} \times 415}{2\pi} (\cos \alpha + \cos (\alpha + 4°)\right\} - 3\right]$$

$$= -280.18 [\cos \alpha + \cos (\alpha + 4°)] + 3$$

With no-voltage drop in inductor,
$$V_{dc} = V_d$$

$\therefore$
$$-280.18 [\cos \alpha + \cos (\alpha + 4°)] + 3 = 183.012 \text{ V}$$
$$\cos \alpha + \cos (\alpha + 4°) = -0.6425$$
$$\cos \alpha + \cos \alpha \cdot \cos 4° - \sin \alpha \sin 4° = -0.6425$$
$$1.9976 \cos \alpha - 0.07 \sin \alpha = -0.6425 \qquad \qquad ...(i)$$

Note that $\quad A \cos \alpha - B \sin \alpha = \sqrt{A^2 + B^2} \cos(\alpha + \theta)$

where $\quad\quad\quad\quad\quad\quad \theta = \tan^{-1} \dfrac{B}{A}$

Eq. (i) gives $1.9976 \cos \alpha - 0.07 \sin \alpha = \sqrt{1.9976^2 + 0.07^2} \left[\cos \left(\alpha + \tan^{-1} \dfrac{0.07}{1.9976} \right) \right]$

$$= -0.6425$$

or $\quad\quad\quad\quad\quad\quad 1.999 \cos(\alpha + 2°) = -0.6425$

or $\quad\quad\quad\quad\quad\quad\quad\quad\quad\quad \alpha = 106.75°$

Firing angle of advance of inverter $= 180 - 106.75° = \textbf{73.25}°$.

Example 12.32. *Using the data of Example 12.30, find the voltage ratio of the transformer to be interposed between supply and the inverter for a minimum speed of 1200 rpm.*

Solution. Here $V_d = \dfrac{3\sqrt{6}\, sE_2}{\pi}$ and $V_{dc} = -\dfrac{3\sqrt{6}}{\pi} \cdot a_T \cdot V_1 \cdot \cos \alpha$

For an ideal inductor, $\quad V_d = V_{dc}$

$\therefore \quad\quad \dfrac{3\sqrt{6}\, sE_2}{\pi} = -\dfrac{3\sqrt{6}}{\pi} \cdot a_T \cdot V_1 \cdot \cos \alpha \quad$ or $\quad s = -\dfrac{a_T \cdot V_1}{E_2} \cdot \cos \alpha$

Minimum speed means maximum slip. Here s would be maximum in magnitude when firing angle of the inverter $\alpha = 180°$.

$\therefore \quad\quad\quad s = \dfrac{a_T \cdot V_1}{E_2} \quad$ or $\quad a_T = \dfrac{s\, E_2}{V_1} = \dfrac{0.2 \times 700}{415} = \dfrac{140}{415}$

$\therefore$ Transformer voltage ratio per phase from h.v. to l.v. is 415/140 or $a_T = 0.3373$.

12.8.6.2. Static Scherbius Drive. In static Kramer drive, speed of SRIM can be controlled below synchronous speed only. For the speed control both below and above synchronous speed, static Scherbius drive scheme is used. There are two possible configurations to obtain such a drive; these are (i) DC link static Scherbius drive and (b) cycloconverter static Scherbius drive. These are discussed briefly as under.

DC link Scherbius Drive. In subsynchronous speed control of WRIM, slip power is removed from the rotor circuit and is pumped back into the ac supply. In supersynchronous speed control, the additional power is fed into the rotor circuit at slip frequency. The circuit diagram of Fig. 12.43 allows both subsynchronous and supersynchronous speed control. It consists of one WRIM, two phase-controlled bridges, smoothing inductor and a transformer as shown.

For subsynchronous speed control, bridge 1 has firing angle less than 90° whereas bridge 2 has firing angle more than 90°. In other words, bridge 1 works as rectifier and bridge 2 as line-commentated inverter for subsynchronous motor control. The slip power flows from rotor circuit to bridge 1, bridge 2, transformer and to the supply.

For supersynchronous motor control, bridge 1 is made to work as line-commutated inverter with firing angle more than 90° and bridge 2 as a rectifier with firing angle less than 90°. The power flow is now from the supply to transformer, bridge 2, bridge 1 and to the rotor circuit.

Near synchronous speed, slip frequency emfs are insufficient for natural commutation of thyristors. This difficulty can, however, be overcome by using forced commutation. Thus, the provision of both subsynchronous and supersynchronous speed operation complicates the static converter system and nullifies the advantages of simplicity and economy which are inherent in a purely subsynchronous drive. In addition, static Scherbius drive is expensive than static Kramer drive because six diodes are replaced by six thyristors and their controlled circuitry. It offers constant torque-drive scheme.

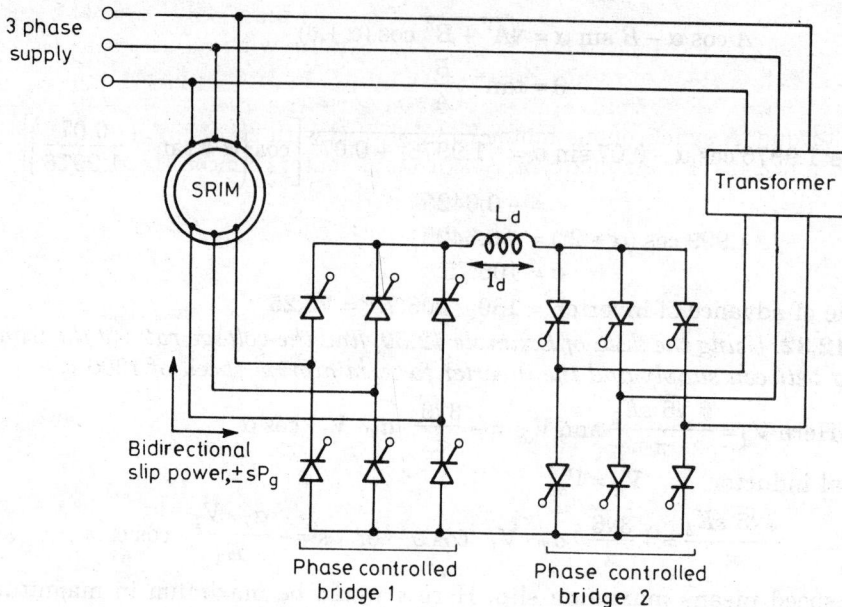

Fig. 12.43. DC link static Scherbius drive.

Cycloconverter Scherbius Drive. The dual controlled converter system used in dc link Scherbius drive is now replaced by one phase-controlled line-commutated cycloconverter as shown in Fig. 12.44. Such schemes are used for very high-power pumps and blower-type drives. Cycloconverter permits the slip-power flow in either direction and the machine can, therefore, be controlled in both subsynchronous and supersynchronous ranges with motoring and regeneration features. As the slip power is either returned to, or taken from, the supply mains, cycloconverter static Scherbius drive offers constant-torque drive scheme.

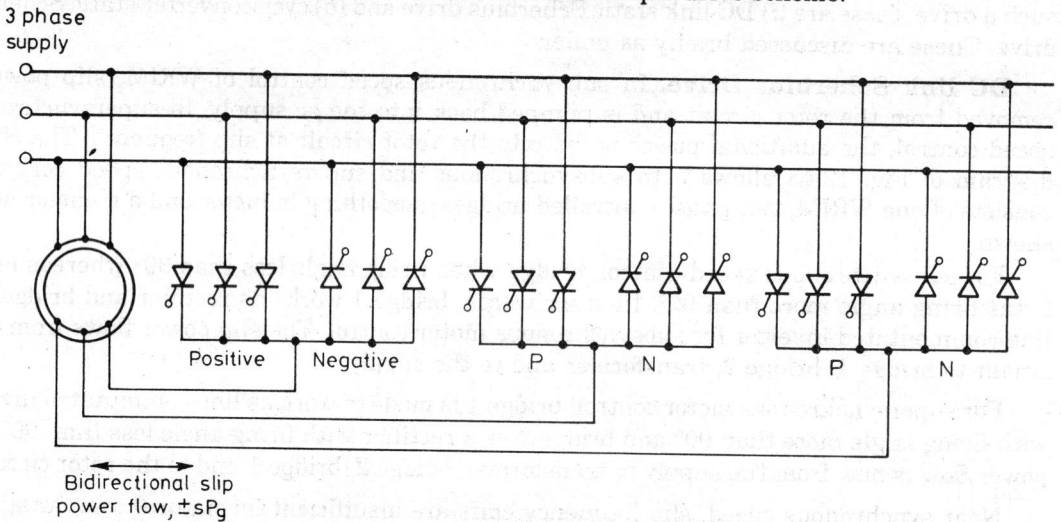

Fig. 12.44. Cycloconverter static Scherbius drive.

12.9. SYNCHRONOUS MOTOR DRIVES

Synchronous motors have two windings, one on the stator is three-phase armature winding and the other on the rotor is the field winding. The three-phase winding on its stator is similar to the 3-phase winding on the stator of a 3-phase IM. Field winding is excited with dc and it produces its own mmf called field mmf. Three-phase stator winding carrying three-phase balanced currents creates its own rotating armature mmf. The two mmfs combine together to produce resultant mmf. The field mmf interacts with the resultant mmf to produce electromagnetic torque. A synchronous motor runs always with zero slip, *i.e.* at synchronous speed given by Eq. (12.43). Power factor of synchronous motors can be controlled by varying its field current.

For the speed control of synchronous motors, both inverter and cyclonverters are employed. The various types of synchronous motors are :

(*i*) Cylindrical rotor motors

(*ii*) Salient-pole motors

(*iii*) Reluctance motors

(*vi*) Permanent-magnet motors.

These are now described briefly in the following lines.

12.9.1. Cylindrical Rotor Motors

These motors have uniform air gap. The per-phase equivalent circuit for a cylindrical-rotor synchronous motor is given in Fig. 12.45 (*a*). In this circuit, E_f = excitation voltage $= \sqrt{2} \cdot \pi \cdot f \cdot \phi_f \cdot N_{ph} \cdot k_\omega$, V_t = armature terminal voltage, r_a = armature resistance, X_s = synchronous reactance and $Z_s = \sqrt{r_a^2 + X_s^2}$ is called the synchronous impedance.

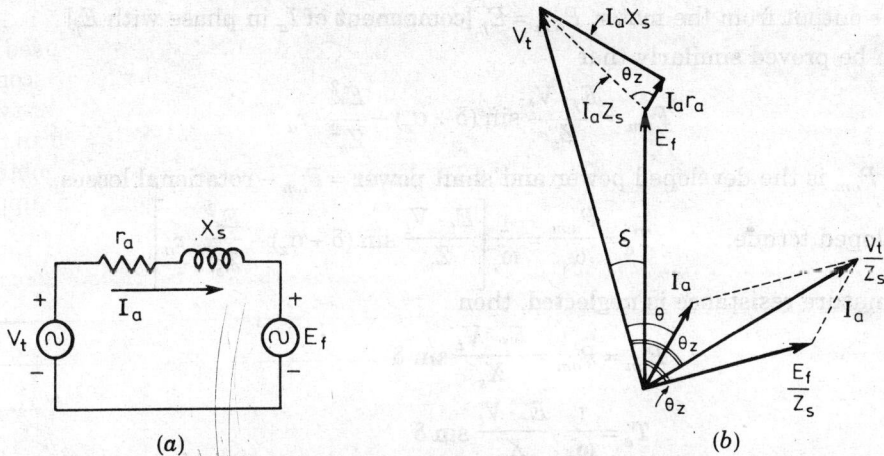

Fig. 12.45. Cylindrical-rotor synchronous motor
(*a*) equivalent circuit and (*b*) its phasor diagram at a lagging pf load.

It is seen from the equivalent circuit that

$$\overline{V}_t = \overline{E}_f + \overline{I}_a \cdot \overline{Z}_s$$

or

$$I_a = \frac{\overline{V}_t - \overline{E}_f}{\overline{Z}_s} = \frac{\overline{V}_t}{\overline{Z}_s} - \frac{\overline{E}_f}{\overline{Z}_s} \qquad \qquad ...(12.104)$$

Eq. (12.104) shows that armature current I_a is the difference of two currents $\dfrac{\overline{V}_t}{\overline{Z}_s}$ and $\dfrac{\overline{E}_f}{\overline{Z}_s}$, lagging behind their respective voltages by impedance angle θ_z as shown in Fig. 12.45 (*b*). Here impedance angle θ_z is given by

$$\theta_z = \tan^{-1} \left(\frac{X_s}{r_a} \right)$$

Power input to the motor is given by

$$P_{im} = V_t \text{ [Component of } I_a \text{ in phase with } V_t]$$

In the above expression, subcripts i and m denote input and motor respectively.

It is seen from Eq. (12.104) and Fig. 12.45 (b) that the component of I_a in phase with V_t is $\left[\dfrac{V_t}{Z_s} \cos \theta_z - \dfrac{E_f}{Z_s} \cos (\delta + \theta_z)\right]$.

$$\therefore \qquad P_{im} = V_t \left[\frac{V_t}{Z_s} \cos \theta_z - \frac{E_f}{Z_s} \cos (\delta + \theta_z)\right]$$

$$= \frac{V_t^2}{Z_s} \cos \theta_z - \frac{V_t \cdot E_f}{Z_s} \cos (\delta + \theta_z) \qquad \qquad ...(12.105)$$

Now $\cos \theta_Z = \dfrac{r_a}{Z_s}$ and $\theta_z = (90 - \alpha_z)$. Substituting these in Eq. (12.105), we get

$$P_{im} = \frac{V_r^2}{Z_s^2} \cdot r_a - \frac{E_f \cdot V_t}{Z_s} \cos \left\{(\delta - \alpha_z) + 90°\right\}$$

$$= \frac{E_f \cdot V_t}{Z_s} \sin (\delta - \alpha_z) + \frac{V_t^2}{Z_s^2} \cdot r_a \qquad \qquad ...(12.106)$$

Power output from the motor, $P_{o.m} = E_f$ [component of I_a in phase with E_f]

It can be proved similarly that

$$P_{om} = \frac{E_f \cdot V_t}{Z_s} \sin (\delta + \alpha_z) - \frac{E_f^2}{Z_s^2} \cdot r_a \qquad \qquad ...(12.107)$$

Here P_{om} is the developed power and shaft power = P_{um} – rotational losses.

Developed torque, $\qquad T_e = \dfrac{P_{o.m}}{\omega_s} = \dfrac{1}{\omega_s}\left[\dfrac{E_f \cdot V_t}{Z_s} \sin (\delta + \alpha_z) - \dfrac{E_f^2}{Z_s^2} \cdot r_a\right]$

If armature resistance is neglected, then

$$P_{im} = P_{om} = \frac{E_f \cdot V_t}{X_s} \sin \delta \qquad \qquad ...(12.108)$$

and $\qquad\qquad T_e = \dfrac{1}{\omega_s} \cdot \dfrac{E_f \cdot V_t}{X_s} \sin \delta \qquad \qquad ...(12.109)$

where $\qquad\qquad \omega_s = \dfrac{4\pi f}{P}$ = synchronous speed in **rad/s.**

and $\qquad\qquad\qquad\qquad \delta$ = load, or power, angle

The torque versus load angle characteristic for a cylindrical-rotor synchronous motor is shown in Fig. 12.46 (a). For stable operation of synchronous motor, the load angle δ should never exceed 90°.

Power factor of a synchronous motor depends on the field current. The variation of armature current with respect to field current for different loads on the synchronous motor is shown in Fig. 12.46 (b). As the shape of these curves resemble the letter V, these are called *V-curves* of a synchronous motor. Note that V-curves are obtained for constant shaft load and for constant terminal voltage. Unity pf curve is shown dotted. For low values of field current (under-excitation), synchronous motor operates at a lagging pf. As the field current is increased, it would start operating at unity pf (normal excitation). If the field current is still increased beyond the unity pf point (over-excitation), synchronous motor begins to operate at a leading pf.

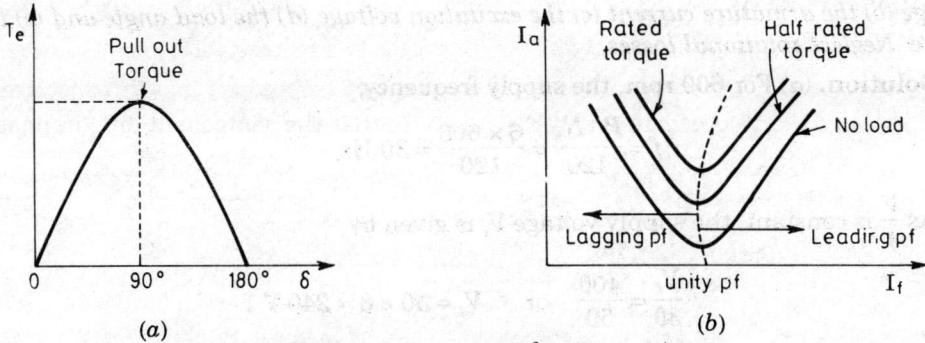

Fig. 12.46. Cylindrical-rotor synchronous motor
(a) torque- angle characteristic and (b) its V-curves.

The power given by Eqs. (12.107) and (12.108) and torque of Eq. (12.109) have per-phase values. For $\delta = 90°$, pull-out power P_{mx} and pull-out (or maximum) torque $T_{e \cdot m}$ are obtained.

$$\therefore \qquad P_{mx} = \frac{E_f \cdot V_t}{X_s} \qquad \qquad ...(12.110)$$

and

$$T_{e \cdot m} = \frac{1}{\omega_s} \cdot \frac{E_f \cdot V_t}{X_s} \qquad \qquad ...(12.111)$$

For fixed field excitation, the excitation emf E_f is directly proportional to supply frequency. The synchronous reactance X_s is also directly proportional to frequency. This shows that $\frac{E_f}{X_s}$ in Eqs. (12.110) and (12.111) is independent of frequency variation. If supply voltage V_t is varied in proportion to frequency so that V_t/f or $\frac{V_t}{\omega_s}$ is constant, then pull-out torque, Eq. (12.111) remains constant. Pull-out power $P_{mx} = T_{e.m} \times \omega_s$, however, rises linearly with speed as shown in Fig. 12.47.

At base speed ($\omega_s = 1.0$), rated voltage and rated frequency are reached. Beyond base speed, rated voltage is kept constant, but inverter frequency can be increased to obtain higher operating speeds of synchronous motor.

With constant field excitation, increase in frequency keeps $\frac{E_f}{X_s}$ constant as stated before.

With supply voltage remaining constant above base speed, pull-out power, Eq. (12.110), remains constant, but pull-out torque $= \frac{1}{\omega_s} \cdot P_{mx}$ falls inversely with rise in speed as shown in Fig. 12.47.

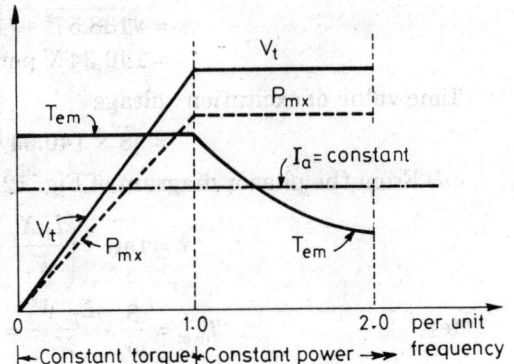

Fig. 12.47. Variation of pull-out torque, pull-out power, terminal voltage and armature current with frequency for a synchronous motor.

Example 12.33. *A 3-phase, 400 V, 50 Hz, 6 pole, star-connected round-rotor synchronous motor has $Z_s = 0 + j2 \ \Omega$. Load torque, proportional to speed squared, is 340 Nm at rated synchronous speed. The speed of the motor is lowered by keeping V/f constant and maintaining unity pf by field control of the motor. For the motor operation at 600 rpm, calculate (a) supply* *voltage (b) the armature current (c) the excitation voltage (d) the load angle and (e) the pull-out torque. Neglect rotational losses.*

Solution. (*a*) For 600 rpm, the supply frequency,

$$f = \frac{P \cdot N_s}{120} = \frac{6 \times 600}{120} = 30 \text{ Hz}.$$

As $\dfrac{V}{f}$ is constant, the supply voltage V_t is given by

$$\frac{V_t}{30} = \frac{400}{50} \quad \text{or} \quad V_t = 30 \times 8 = 240 \text{ V}$$

(*b*) Load torque at 600 rpm,

$$T_L = 340 \left(\frac{600}{1000}\right)^2 = 122.4 \text{ Nm}$$

Power output, $\qquad P = T_e \cdot \omega_s = T_L \cdot \omega_s = 122.4 \times \dfrac{2\pi \times 600}{60} = 7690.62 \text{ W}$

As rotational losses are neglected, $\sqrt{3} \; V_t I_a \cos \theta = P$

$\therefore$ Armature current, $I_a = \dfrac{7690.62}{\sqrt{3} \times 240 \times 1} = 18.50 \text{ A}$

(*c*) $\qquad\qquad X_s = \dfrac{30}{50} \times 2 = 1.2 \; \Omega$

Per-phase supply voltage, $V_t = \dfrac{240}{\sqrt{3}} = 138.57 \text{ V}$

Per-phase armature current, $I_a = 18.50 \text{ A}$

From the phasor diagram, excitation voltage is

$$E_f = \sqrt{V_t^2 + (I_a X_s)^2}$$
$$= \sqrt{138.57^2 + (18.5 \times 1.2)^2}$$
$$= 140.34 \text{ V per phase}$$

Line value of excitation voltage

$$= \sqrt{3} \times 140.34 = 243.07 \text{ V}$$

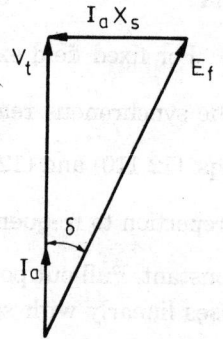

Fig. 12.48. Phasor diagram pertaining to Example 12.33.

(*d*) From the phasor diagram of Fig. 12.48, we get load angle δ as

$$\delta = \tan^{-1}\left(\frac{I_1 X_s}{V_t}\right) = \tan^{-1}\left(\frac{18.5 \times 1.2}{138.57}\right) = 9.10°$$

(*e*) $\qquad\qquad T_{e.m} = \dfrac{3}{\omega_s} \cdot \dfrac{E_f \cdot V_t}{X_s} = \dfrac{3}{20\pi} \cdot \dfrac{140.34 \times 138.57}{1.2} = 773.77 \text{ Nm}.$

12.9.2. Salient-pole Motors

The armature winding on the stator of a salient-pole motor is similar to that of a cylindrical-rotor motor. Field winding on the rotor is a concentrated winding on the salient poles. In this type of motor, the air gap is not uniform. For analysis purposes, its armature current is resolved into two components, called direct-axis current I_d and quadrature-axis current I_q. Likewise, there are two reactances, $X_d = d$-axis synchronous reactance and $X_q = q$-axis synchronous reactance.

The phasor diagram of a salient-pole synchronous motor with negligible armature resistance is shown in Fig. 12.49 (a). It is seen from this phasor diagram that

$$\overline{V}_t = \overline{E}_f + j\overline{I}_d X_d + j\overline{I}_q X_q \qquad ...(12.112)$$

$$I_d = I_a \sin(\theta - \delta), I_q = I_a \cos(\theta - \delta)$$

Also, $\qquad V_t \sin\delta = I_q \cdot X_q = X_q \cdot I_a \cos(\theta - \delta)$

$$= X_q I_a [\cos\theta \cdot \cos\delta + \sin\theta \sin\delta]$$

$$\sin\delta [V_t - I_a X_q \sin\theta] = X_q \cdot I_a \cos\theta \cdot \cos\delta$$

$$\therefore \qquad \tan\delta = \frac{I_a X_q \cos\theta}{V_t - I_a \cdot X_q \sin\theta} \qquad ...(12.113)$$

Power P is given by $P = V_t \cos\delta$ (current along q-axis) $+ V_t \cdot \sin\delta$ (current along d-axis)

$$= V_t \cdot \cos\delta \cdot I_q - V_t \cdot \sin\delta \cdot I_d \qquad ...(12.114)$$

But $\qquad I_d X_d = V_t \cos\delta - E_f$

or $\qquad I_d = \dfrac{V_t \cos\delta - E_f}{X_d}$

$$I_q X_q = V_t \sin\delta \quad \text{or} \quad I_q = \frac{V_t \sin\delta}{X_q}$$

Substituting these values of I_d, I_q in Eq. (12.114), we get

$$P = \frac{E_f \cdot V_t}{X_d} \sin\delta + \frac{V_t^2}{2}\left(\frac{1}{X_q} - \frac{1}{X_d}\right)\sin 2\delta \qquad ...(12.115)$$

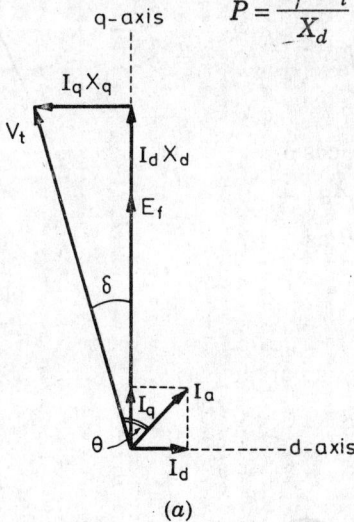

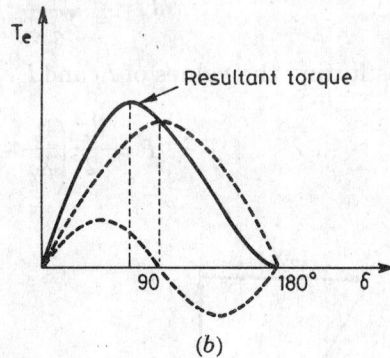

(a)

(b)

Fig. 12.49. Salient-pole synchronous motor.
(a) its phasor diagram and (b) its torque versus load angle characteristics.

The power in Eq. (12.115) has two components. The first component $\dfrac{E_f \cdot V_t}{X_d} \sin\delta$, similar to that of round-rotor motor, is called electromagnetic power. The other component $\dfrac{V_t^2}{2}\left(\dfrac{1}{X_q} - \dfrac{1}{X_d}\right)\sin 2\delta$ is called the reluctance power, as it is present due to different reluctances along direct axis and quadrature axis.

Developed torque T_e is given by

$$T_e = \frac{P}{\omega_s} = \frac{1}{\omega_s}\left[\frac{E_f \cdot V_t}{X_d}\sin\delta + \frac{V_t^2}{2}\left(\frac{1}{X_q} - \frac{1}{X_d}\right)\sin 2\delta\right] \quad ...(12.116)$$

As for power, the first and second components in Eq. (12.116) are called respectively the electromagnetic torque and the reluctance torque.

The torque T_e versus load angle δ characteristics are drawn in Fig. 12.49 (b) with the help of Eq. (12.116). It is seen that torque is maximum at a load angle less than 90°.

12.9.3. Reluctance Motors

A salient-pole synchronous motor connected to a voltage source runs at synchronous speed. If its field current is switched off, it continues running at synchronous speed as a reluctance motor. Thus, a machine designed to operate as a reluctance motor is similar to a salient-pole motor with no field winding on the rotor. Three-phase armature winding produces rotating magnetic field in the air gap. This rotating flux induces a field in the rotor which tends to align itself with the armature field, thus producing a reluctance torque at synchronous speed.

Reluctance motors are used for low power drives where constant-speed operation is required and where more than one motor is needed for the job so that number of motors can run in synchronism.

With zero field current, $E_f = 0$ and the phasor diagram for a reluctance motor can be drawn from Fig. 12.49 (a) by making $E_f = 0$, this is shown in Fig. 12.50 (a). From this figure, power P is given by

$$P = V_t \cdot \cos\delta \cdot I_q - V_t \cdot \sin\delta \, I_d$$

It is seen from Fig. 12.50 (a) that

$$I_q = \frac{V_t \sin\delta}{X_q} \quad \text{and} \quad I_d = \frac{V_t \cdot \cos\delta}{X_d}$$

Substituting the values of I_d and I_q, we get power P as

$$P = \frac{V_t^2}{2}\left(\frac{1}{X_q} - \frac{1}{X_d}\right)\sin 2\delta \quad ...(12.117)$$

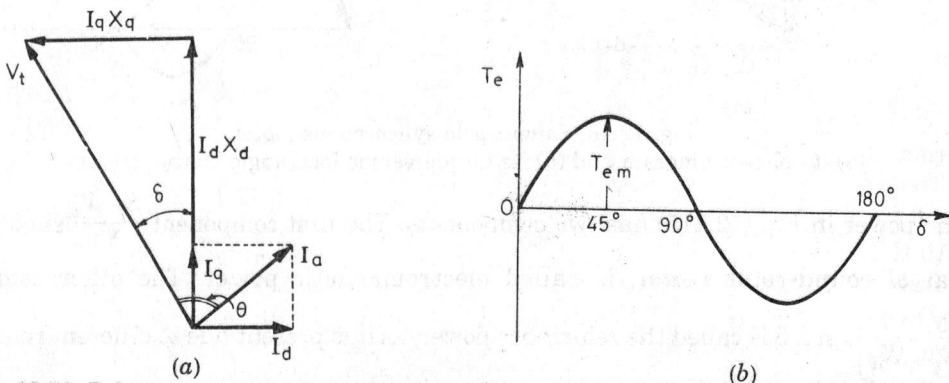

Fig. 12.50. Reluctance motor (a) its phasor diagram and (b) its torque-angle characteristics.

Phasor diagram of Fig. 12.50 (a) reveals that

$$V_t \cdot \sin \delta = I_q X_q = X_q \cdot I_a \cos(\theta - \delta).$$

This gives

$$\delta = \tan^{-1} \frac{I_a X_q \cos \theta}{V_t - I_a X_q \sin \theta} \qquad \qquad ...(12.118)$$

Reluctance torque, from Eq. (12.117) is

$$T_e = \frac{1}{\omega_s} \cdot \frac{V_t^2}{2} \left(\frac{1}{X_q} - \frac{1}{X_d} \right) \sin 2\delta \qquad \qquad ...(12.119)$$

Pull-out torque T_{em} is obtained when $\delta = 45°$

$$\therefore \qquad T_{e \cdot m} = \frac{V_t^2}{2\omega_s} \left(\frac{1}{X_q} - \frac{1}{X_d} \right) \qquad \qquad ...(12.120)$$

Variation of reluctance torque with load angle is shown in Fig. 12.50 (b).

Example 12.34. *A 3-phase, 400 V, 50 Hz, 4 pole, star-connected reluctance motor, with negligible armature resistance, has $X_d = 8\ \Omega$ and $X_q = 2\ \Omega$. For a load torque of 80 Nm, calculate (a) the load angle (b) the line current and (c) the input power factor. Neglect rotational losses.*

Solution. (a) Synchronous speed, $\omega_s = \dfrac{4\pi f}{P} = \dfrac{4\pi \times 50}{4} = 50\pi$ rad/s

From Eq. (12.117) $\quad P = T_e \cdot \omega_s = 80 \times 50\pi = \dfrac{3}{2} \left(\dfrac{400}{\sqrt{3}} \right)^2 \left[\dfrac{1}{2} - \dfrac{1}{8} \right] \sin 2\delta$

$$\delta = \sin^{-1} \left[\frac{80 \times 50\pi \times 8}{80000 \times 3} \right] = 12.383°$$

(b) Per phase voltage, $V_t = \dfrac{400}{\sqrt{3}} = 230.95$ V

$$I_d = \frac{V_t \cos \delta}{X_d} = \frac{230.95 \times \cos 12.382°}{8} = 28.197 \text{ A}$$

$$I_q = \frac{V_t \sin \delta}{X_q} = \frac{230.95 \times \sin 12.383°}{2} = 24.761 \text{ A}$$

$\therefore$ Armature current, $\ I_a = \sqrt{I_d^2 + I_q^2} = \sqrt{28.197^2 + 24.761^2} = 37.53$ A

(c) $\qquad \sqrt{3}\ V_t I_a \cos \theta = T_e \cdot \omega_s$

$$\sqrt{3} \times 400 \times 37.53 \times \cos \theta = 80 \times 50\pi \text{ W}$$

$\therefore$ Input pf $\qquad = \dfrac{4000 \times \pi}{\sqrt{3} \times 400 \times 37.53} = 0.4833$ lagging.

12.9.4. Permanent-magnet Motors

A permanent-magnet synchronous motor (PMSM) is similar to a salient-pole synchronous motor without the field winding on the poles. In PMSM, the required field flux is produced by permanent magnets mounted on the rotor. In these motors, the excitation emf E_f cannot be varied. All the equations governing the performance of a salient-pole synchronous motor are also applicable to PMSM with excitation emf E_f taken as constant. The absence of field winding, dc supply to field winding and two slip rings leads to reduction in motor losses. For the same frame size, PMSM has higher pull-out torque and more efficiency as compared to salient-pole motor.

These motors are used in robots and machine tools. A PMSM can be fed from rectangular current source or sinusoidal current source. A rectangular current-fed motor has concentrated winding on the stator and is used in low-power drives. A sinusoidal current-fed motor has distributed winding on the stator and is used in high-power drives.

12.10. SOME WORKED EXAMPLES

In this article, some typical problems pertaining to electric drives are solved.

Example 12.35. *A separately-excited dc motor is fed from 230 V, 50 Hz source through a 1-phase semiconverter. The motor armature resistance is 2 Ω and its torque constant is 1.2 Nm/A. The thyristors are fired at an angle of 100° and the discontinuous armature current extinguishes at 45° beyond voltage zero. Calculate the motor speed if load on the motor is 3 Nm.*

Solution. Circuit diagram for single-phase semiconverter feeding a separately-excited *dc* motor is shown in Fig. 12.8 (*a*).

The waveforms of discontinuous armature current and armature terminal voltage are shown in Fig. 12.51. When thyristor $T11$ is turned on at $\alpha = 100°$, armature current i_a builds up, reaches some maximum value and then decays to zero at $\omega t = 225°$; 45° beyond voltage zero, as specified. For $\omega t = 100°$ to 180°, $T11\, D11$ conduct and from $\omega t = 180°$ to 225°, FD conducts and so on.

When FD stops conducting at $\omega t = 225$, motor terminal voltage jumps from zero to E_a and stays there till $T12$ is triggered at $\alpha = 100°$ beyond $\omega t = \pi$. In other words, armature generated voltage E_a stays at constant value for 55° as shown.

From the waveform of voltage v_o, or v_t, mean value of armature terminal voltage is

$$V_t = \frac{1}{\pi}\left[E_a \times \frac{55 \times \pi}{180} + \int_{100}^{180°} V_m \sin \omega t \cdot d\,(\omega t) \right]$$
$$= 0.3056\, E_a + 0.263 \times \sqrt{2} \times 230 = [0.3056\, E_a + 85.53]\ \text{V}$$

Armature current $I_a = \dfrac{T_e}{K_m} = \dfrac{3}{1.2} = 2.5$ A

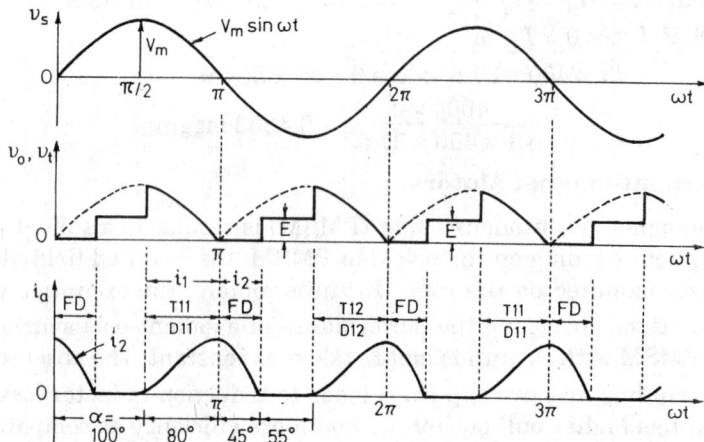

Fig. 12.51. Waveforms pertaining to Example 12.35.

Also $\qquad V_t = E_a + I_a\, ra = 0.3056\, E_a + 85.53$

or $\qquad E_a + 2.5 \times 2 = 0.3056\, E_a + 85.53$

$\therefore \qquad E_a = \dfrac{80.53}{0.6944} = 115.971\,\text{V} = K_m \cdot \omega_r$

$\therefore$ Motor speed, $\qquad \omega_m = \dfrac{115.971}{1.2} = 96.042\,\text{rad/s} = \dfrac{2\,\pi\,N}{60}$

$\therefore$ Motor speed, $\qquad N = 922.86\,\text{rpm}$

Example 12.36. *A separately-excited dc motor, with $r_a = 4\,\Omega$, $L_a = 0.04\,H$ and $K_m = 1\,Nm/A$, is fed from 230 V, 50 Hz supply via a 1-phase semiconverter. For a firing-angle delay of 75°, the motor runs at a speed of 1360 rpm.*

(a) Derive expression for motor armature current.

(b) Sketch the waveforms of source voltage, load voltage and motor armature current

(c) Calculate the average motor torque.

Solution. (a) At firing angle delay of 75°, $T11$ in Fig. 12.8 (a) is turned on and $T11\,D11$ begin to conduct. Under this condition, equivalent circuit of Fig. 12.52 (a) applies, for which the KVL gives

$$r_a \cdot i_1 + L_a \cdot \frac{di_1}{dt} + E_a = V_m \sin \omega t.$$

The solution of armature current i_1 has the following three components :

$i_1 = i_{ac}$, steady-state *ac* component $+ i_{dc}$, steady-state *dc* component + exponentially decaying component such that $i_1 = 0$ at $\omega t = \alpha = 75°$.

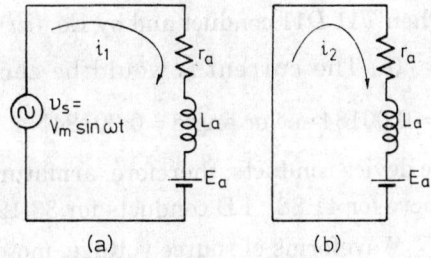

Fig. 12.52. Equivalent circuit of Fig. 12.8 (a) for example 12.36.

Current $\qquad i_{ac} = \dfrac{V_m}{Z} \sin \omega t$

Impedance $\qquad Z = 4 + j\, 2\pi\, 50 \times 0.04 = 4 + j\, 12.57 = 13.19\, \angle 72.35°\,\Omega$

$\therefore \qquad i_{ac} = \dfrac{\sqrt{2} \times 230}{13.19\, \angle\, 72.35°} \sin \omega t = 24.66 \sin (\omega t - 72.35°)$

Motor speed, $\qquad \omega_m = \dfrac{2\pi \times 1360}{60} = 142.42\,\text{rad/sec}$

Motor emf, $\qquad E_a = K_m - \omega_m = 1 \times 142.42 = 142.42\,\text{V}$

$\therefore \qquad i_{dc} = -\dfrac{E_a}{r_a} = -\dfrac{142.42}{4} = -35.60\,\text{A}$

$\qquad \tau = \dfrac{L}{R} = \dfrac{0.04}{4} = 0.01,\ \dfrac{1}{\tau} = 100$

$\therefore \qquad i_1 = i_{ac} + i_{dc} + A\, e^{-t/\tau}$

$\qquad = 24.66 \sin (\omega t - 72.35°) - 35.60 + A\, e^{-100\, t}$ ...(i)

When $\qquad \omega t = \alpha = 75°$ or $t = \dfrac{75 \times \pi}{180 \times 2\pi \times 50} = 0.0042\,\text{sec},\ i_1 = 0$

From Eq. (i), $\qquad 0 = 24.66 \sin (75 - 72.35) - 35.60 + A\, e^{-100 \times 0.0042}$

or $A = 52.45$

∴ $i_1 = 24.66 \sin (\omega t - 72.35°) - 35.60 + 52.45\, e^{-100t}$ ...(ii)

At $\omega t = \pi$, when voltage is zero, current is given by

$$i_1 = 24.66 \sin (180 - 72.35°) - 35.60 + 52.45\, e^{-100 \times \frac{\pi}{100\pi}} = 7.194 \text{ A}$$

Soon after $\omega t = \pi$, freewheeling diode FD begins to conduct; now the equivalent circuit of Fig. 12.52 (b) is applicable, for which KVL gives

$$ri_2 + L \frac{di_2}{dt} + E_a = 0$$

Here time t is measured from the instant FD begins to conduct. Its solution is

$$i_2 = -\frac{E_a}{r_a} + B \cdot e^{-t/\tau} = -35.6 + B\, e^{-100t}$$

At $t = 0$, $i_2 = 7.194$ A, ∴ $7.194 = -35.6 + B\, e^{-0}$ or $B = 42.794$

∴ $i_2 = -35.6 + 42.794\, e^{-100t}$ ...(iii)

Expression for motor armature current i_a $(= i_1 + i_2)$ is given by Eq. (i) from $\omega t = \alpha$ to $180°$ when $T11$ $D11$ conduct and by Eq. (iii) from $\omega t = 180°$ to $225°$ when FD conducts and so on.

(b) The current i_2 would be zero (after $\omega t = \pi$), when $i_2 = 0 = -35.6 + 42.794\, e^{-100t}$ or $t = 0.00184$ sec or angle $= 0.00184 \times 2\,\pi \times 50 \times \dfrac{180}{\pi} = 33.12° < \alpha = 75°$. For $75 - 33.12° = 41.88°$, no device conducts, therefore, armature terminal voltage jumps to E_a at $\omega t = 213.12°$ and stays there for $41.88°$. FD conducts for $33.12°$ from $\omega t = 180°$ to $213.12°$.

Waveforms of source voltage, motor terminal voltage and discontinuous armature current are shown in Fig. 12.53. Waveform of voltage v_t gives its average value as

$$V_t = \frac{1}{\pi}\left[\int_{75}^{180} \sqrt{2} \times 230 \sin\theta \cdot d\theta + 142.42 \times \frac{(75 - 33.12) \times \pi}{180} \right] = 163.45 \text{ V}$$

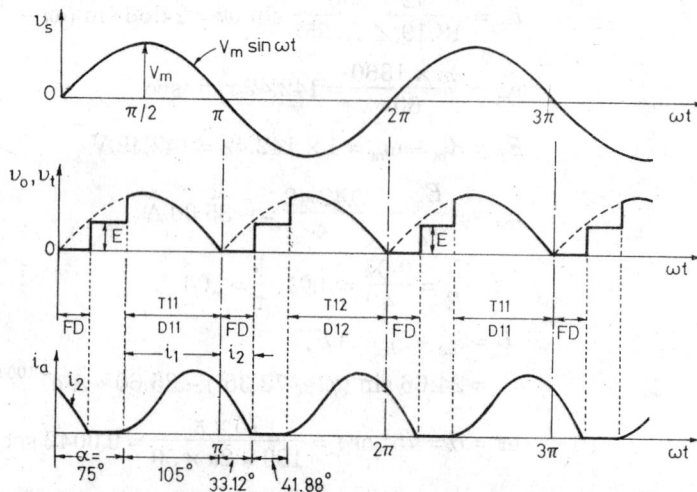

Fig. 12.53. Waveforms Pertaining to Example 12.36.

Motor emf, $E_a = 142.42 = V_t - I_a r_a$

$$\therefore \qquad I_a = \frac{V_t - E_a}{r_a} = \frac{163.45 - 142.42}{4} = 5.2575 \text{ A}$$

$\therefore$ Motor torque, $\quad T_e = K_m I_a = 1 \times 5.2575 = 5.2575 \text{ Nm}$

Example 12.37. *Repeat Example 12.36 with the same magnitude of firing-angle delay, but with load torque reduced so that motor speed rises to 2100 rpm.*

Solution. (*a*) Motor armature emf, $E_a = K_m \omega_m = 1 \times \dfrac{2\pi \times 2100}{60} = 219.91 \text{ V}$

$$\frac{E_a}{r_a} = \frac{219.91}{4} = 54.9775 \simeq 55 \text{ A}$$

The expression for the armature current, from the previous example, is given by

$$i_1 = 24.66 \sin(\omega t - 72.35°) - 55 + A\, e^{-100 t}$$

At $\qquad \omega t = \alpha = 75°$ or $t = \dfrac{75 \times 0.01}{180} = 0.0042 \text{ sec}, i_1 = 0$

$$\therefore \qquad 0 = 24.66 \sin(75 - 72.35°) - 55 + A\, e^{-100 \times 0.0042}$$

or $\qquad A = 81.98$

$$\therefore \qquad i_1 = 24.66 \sin(\omega t - 72.35°) - 55 + 81.98\, e^{-100 t} \qquad \qquad ...(i)$$

At $\omega t = \pi$, when the voltage is zero,

$$i_1 = 24.66 \sin(180 - 72.35°) - 55 + 81.98\, e^{-100 \times 0.01} = -1.342 \text{ A}$$

This shows that i_1 decays to zero before $\omega t = \pi$. The angle ($< 180°$) at which i_1 becomes zero can be obtained from Eq. (*i*), by hit and trial.

When $\quad \omega t = 175°, i_1 = 24.66 \sin(175 - 72.35) - 55 + 81.98\, e^{-100 \times \frac{175 \times 0.01}{180}} \simeq 0.0696$

So when $\omega t = 175°$, or when $t = \dfrac{175 \times 0.01}{180} = 0.00972 \text{ sec}$, the current i_1 decays to almost zero value.

(*b*) The waveforms for v_s, v_o and i_a are sketched in Fig. 12.54. It is seen that FD of the 1-phase semiconverter does not come into play. The current flows from $\omega t = \alpha = 75°$ to

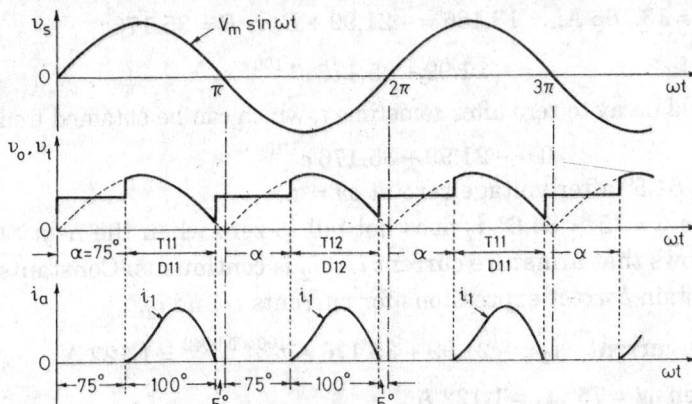

Fig. 12.54. Waveforms pertaining to Example 12.37.

$\omega t = 175°$. Armature voltage then jumps to E_a and stays there till $\omega t = 255°$, *i.e.* E_a remains constant for $255 - 175 = 80°$ as shown.

(c) Mean output voltage, $V_t = \dfrac{1}{\pi}\left[\displaystyle\int_{75}^{175°} V_m \sin\theta . d\theta + \dfrac{E_a . 80 \times \pi}{180}\right]$

$$= \dfrac{1}{\pi}\left[\int_{75}^{175°} \sqrt{2} \times 230 \sin\theta \, d\theta + \dfrac{219.91 \times 80 \times \pi}{180}\right] = 227.66 \text{ V}$$

But $V_t = E_a + I_a r_a$, $\therefore$ $I_a = \dfrac{227.66 - 219.91}{4} = 1.9375 \text{ A}$

$\therefore$ $T_e = K_m I_a = 1 \times 1.9375 = 1.9375 \text{ Nm}$

Example 12.38 : *Repeat Example 12.36 with the same firing-angle delay. Now the load torque is increased so that the motor speed reduces to 840 rpm.*

Solution. (a) Motor armature emf, $E_a = K_m . \omega_m = 1 \times \dfrac{2\pi \times 840}{60} = 87.96 \text{ V}$

$$\dfrac{E_a}{r_a} = \dfrac{87.96}{4} = 21.99 \text{ A}$$

From Example 12.36, armature current i_1 is given by

$$i_1 = 24.66 \sin(\omega t - 72.35°) - 21.99 + A . e^{-100 t} \qquad ...(i)$$

Assuming the current to be discontinuous when $\omega t = \alpha = 75°$, or when $t = \dfrac{75 \times 0.01}{180} = 0.0042$ sec, current $i_1 = 0$. From Eq. (*i*)

$$0 = 24.66 \sin(75 - 72.35°) - 21.99 + A \, e^{-100 \times 0.0042}$$

$\therefore$ $A = 31.74$

$\therefore$ $i_1 = 24.66 \sin(\omega t - 72.35°) - 21.99 + 31.74 \, e^{-100 t}$

When $\omega t = \pi = 180°$ cr $t = 0.01$ sec, then

$$i_1 = 24.66 \sin(180 - 72.35°) - 21.99 + 31.74 \, e^{-1} = 13.186 \text{ A}$$

After $\omega t = \pi$, FD begins to conduct. Now when $t = 0$, $i_2 = 13.186$ A, Current i_2 is given by

$$i_2 = -21.99 + B \, e^{-100t}$$

When $t = 0$, $i_2 = 13.186$ A, $\therefore$ $13.186 = -21.99 + B$, $\therefore$ $B = 35.176$

$\therefore$ $i_2 = -21.99 + 35.176 \, e^{-100 t}$ \qquad ...(ii)

Current i_2 would decay to zero after sometime t, which can be obtained from Eq (*ii*) as under :

$$0 = -21.99 + 35.176 \, e^{-100 t}$$

or $t = 0.0047 \, s$ or $84.6°$ after voltage zero at $\omega t = \pi$.

As firing angle $\alpha = 75° < 84.6°$, i_2 does not fall to zero when the next SCR in sequence is triggered. This shows that armature current $i_a = i_2$ is continuous. Constants A and B must be re-calculated to obtain correct expressions for currents i_1 and i_2.

When $\omega t = 75°$, current $i_2 = -21.99 + 35.176 \, e^{-100 \times 0.0042} = 1.122 \text{ A}$

Therefore, when $\omega t = 75°$, $i_1 = 1.122$ A

or $1.122 = 24.66 \sin(75 - 72.35) - 21.99 + A \, e^{-100 \times 0.0042}$

or $A = 33.443$

$\therefore$ $i_1 = 24.66 \sin(\omega t - 72.35) - 21.99 + 33.443\, e^{-100\,t}$

When $\omega t = \pi$, $i_1 = 24.66 \sin(180 - 72.35) - 21.99 + 33.443\, e^{-1}$

                                        $= 13.812\,A = i_2$ at $t = 0$, *i.e.* when voltage zero occurs.

For i_2, when $t = 0$, $i_2 = 13.812\,A$

or $13.812 = -21.99 + B \times 1$ or $B = 35.802$

$\therefore$ $i_2 = -21.99 + 35.802\, e^{-100\,t}$

(*b*) Relevant waveforms for voltage and armature current are sketched in Fig. 12.55.

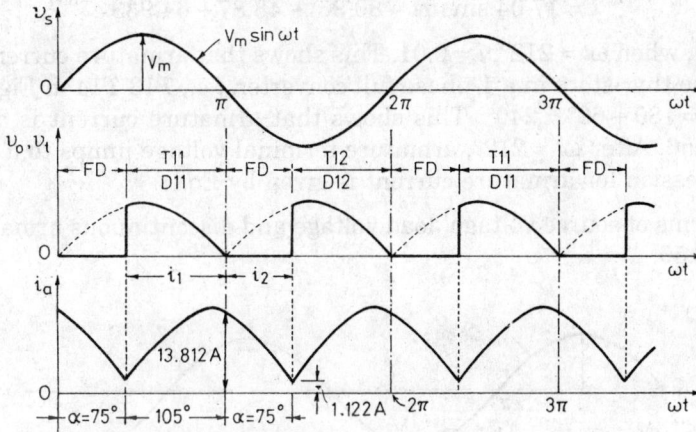

Fig. 12.55. Waveforms pertaining to Example 12.38.

(*c*) Mean voltage applied to armature, $V_t = \dfrac{V_m}{\pi}(1 + \cos \alpha)$

$$= \frac{\sqrt{2} \times 230}{\pi}(1 + \cos 75°) = 130.31\,V$$

But $V_t = E_a + I_a r_a$ $\therefore$ $130.31 = 87.96 + I_a \times 4$

$\therefore$ Armature current, $I_a = 10.5875\,A$

Torque, $T_e = K_m I_a = 1 \times 10.5875 = 10.5875\,N_m$

Example 12.39. *A single-phase full converter feeds a separately-excited dc motor having* $r_a = 3\,\Omega$, $L_a = 0.06\,H$ *and torque constant 1 Nm/A. For a source voltage of 230 V, 50 Hz, the motor runs at a speed of 1400 rpm for a firing-angle delay of 60°.*

(*a*) *Derive an expression for the armature current*

(*b*) *Sketch the waveforms of source voltage, load voltage and motor armature current*

(*c*) *Calculate the average motor torque.*

Solution. (*a*) Impedance $Z = [3^2 + (2\pi \times 50 \times 0.06)^2]^{1/2} = 19.087\,\Omega$

$$\overline{Z} = 19.087\,\angle \tan^{-1} \frac{18.85}{3} = 19.087\,\angle 80.96°$$

Motor counter emf, $E_a = K_m \cdot \omega_m = 1 \times \dfrac{2\pi \times 1400}{60} = 146.61\,V$

$\dfrac{E_a}{r_a} = \dfrac{146.61}{3} = 48.87\,A$, $\dfrac{r_a}{L_a} = \dfrac{3}{0.06} = 50$

$$\therefore \qquad\qquad i = \frac{\sqrt{2} \times 230}{19.087} \sin (\omega t - 80.96°) - 48.87 + A\, e^{-50\,t}$$

$$i = 17.04 \sin (\omega t - 80.96°) - 48.87 + A\, e^{-50\,t}$$

When $\omega t = 60°$, or $t = \dfrac{60 \times \pi}{180 \times 2\pi \times 50} = \dfrac{1}{300}$ sec, let the armature current be discontinuous so that $i = 0$.

$$\therefore \qquad\qquad 0 = 17.04 \sin (60 - 80.96) - 48.87 + Ae^{-\frac{50}{300}}$$

or
$$A = 64.933$$

$$i = 17.04 \sin (\omega t - 80.96) - 48.87 + 64.933\, e^{-50\,t} \qquad\qquad ...(i)$$

By hit and trial, when $\omega t = 212°$, $i = 0.01$. This shows that armature current decays to zero at $\omega t = 212°$. But the thyristors in a 1-phase full converter, *i.e.*, $T13\ T14$ in Fig. 12.9 (*a*), would be triggered at $\omega t = 180 + 60° = 240°$. This shows that armature current is discontinuous as depicted in Fig. 12.56. After $\omega t = 212°$, armature terminal voltage jumps to E and stays there till $\omega t = 240°$. Expression for armature current is given by Eq (*i*).

(*b*) The waveforms of source voltage, load voltage and discontinuous armature current are sketched in Fig. 12.56.

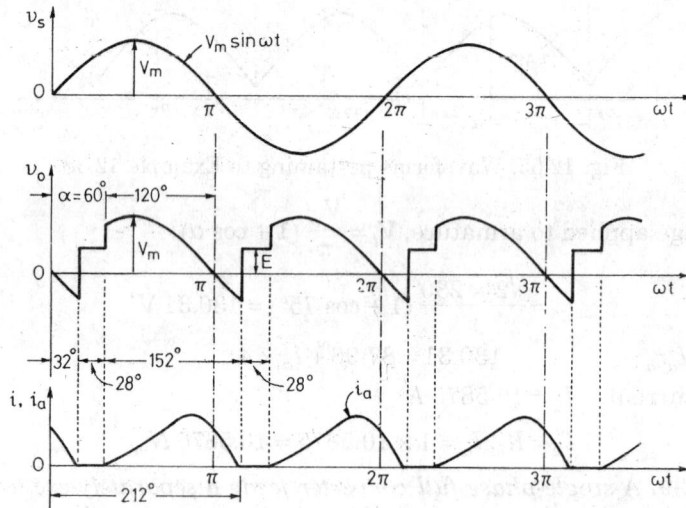

Fig. 12.56. Waveform pertaining to Example 12.39.

(*c*) Average value of output voltage,

$$V_o = \frac{1}{\pi} \left[\int_{60}^{212} V_m \sin \omega t\, d\, (\omega t) + \frac{E_a \times 28° \times \pi}{180} \right]$$

$$= \frac{\sqrt{2} \times 230}{\pi} \,[\cos 60 - \cos 212] + \frac{146.61 \times 28 \times \pi}{180} = 162.357 \text{ V}$$

But
$$V_o = V_t = E_a + r_a I_a$$

$\therefore$ Armature current, $I_a = \dfrac{162.357 - 146.61}{3} = 5.249$ A

Motor torque, $\qquad T_e = K_m I_a = 1 \times 5.249 = 5.249$ Nm

Example 12.40. *In Example 12.39, load torque on the motor is increased and this causes the speed to drop to 600 rpm with the same firing-angle delay. Calculate the load torque and expression for the motor armature current.*

Solution. Here $E_a = 1 \times \dfrac{2\pi \times 600}{60} = 62.831$ V, $\dfrac{E_a}{r_a} = 20.944$ A.

From Example 12.39, $i = 17.04 \sin (\omega t - 80.96) - 20.944 + A\, e^{-50t}$

Let the armature current be discontinuous, therefore when $\omega t = 60°$ or when $t = \dfrac{1}{300}$ sec, $i = 0$.

$\therefore \qquad 0 = 17.04 \sin (-20.96) - 20.944 + Ae^{-1/6}$ or $A = 31.943$

Thyristors $T13\ T14$ are triggered at $\omega t = 240°$, or at $t = \dfrac{4}{300} = \dfrac{1}{75}$ sec.

Armature current at this instant is given by

$$i = 17.04 \sin (240 - 80.96) - 20.944 + 31.943\, e^{-50/75} = 1.5515\ \text{A}$$

This shows that armature current is continuous. So calculate A again.

When $\omega t = 60°$, or $t = \dfrac{1}{300}$ sec, $i = 1.5515$ A.

$\therefore \qquad 1.5515 = 17.04 \sin (-20.96) - 20.944 + A\, e^{-1/6}$
$\therefore \qquad\qquad A = 33.776$

$\therefore$ Expression for armature current is

$$i = i_a = 17.04 \sin (\omega t - 80.96) - 20.944 + 33.776\, e^{-50t}$$

Since the armature current is continuous,

$$V_o = \frac{2V_m}{\pi} \cos \alpha = \frac{\sqrt{2} \times 2 \times 230}{\pi} \cos 60°$$

$$V_o = 103.52\ \text{V} = 62.831 + I_a\, r_a$$

$\therefore$ Armature current, $I_a = \dfrac{103.52 - 62.831}{3} = 13.563$ A

Motor torque, $\qquad T_e = K_m I_a = 1 \times 13.563 = 13.563$ Nm

Example 12.41. *An inverter feeds a three-phase induction motor whose parameters are given in Example 12.4. Calculate the source current and torque at a full-load slip of 0.04 when inverter output is 400 V, 50 Hz.*

If the inverter output is suddenly reduced to 360 V, 40 Hz ; determine, at that moment, the new value of source current and the torque.

Solution. Impedance seen by the voltage source at a slip of 0.04,

$$\overline{Z} = \left(r_1 + \frac{r_2}{s} \right) + j\,(x_1 + x_2)$$

$$= \left[0.6 + \frac{0.4}{0.04} \right] + j\,[1.6 + 1.6] = 11.073\ \angle\ 16.8°$$

Source current, $\qquad \overline{I_1} = \overline{I_2} = \dfrac{400}{11.073\ \angle\ 16.8} = 36.124\ \angle -16.8°$ A

Synchronous speed $\qquad = 1500$ rpm, $\omega s = 50\,\pi$ rad/sec

Torque,
$$T_e = \frac{3}{\omega_s} \cdot I_2^2 \frac{r_2}{s}$$

$$= \frac{3}{50\pi} \cdot (36.124)^2 \times \frac{0.4}{0.04} = 249.23 \ N_m = AB$$

Rotor speed, $N_r = 1500 \ (1 - 0.04) = 1440 \ \text{rpm} = 0 \ B$

New synchronous speed, $N_{s1} = \dfrac{120 \times 45}{4} = 1350 \ \text{rpm} = 0 \ D.$

$$\omega_{s1} = 45\pi \ \text{rad/sec.}$$

When the inverter output is suddenly reduced to 45 Hz, the motor speed will remain substantially constant for some time due to drive inertia.

New slip, $s_1 = \dfrac{N_{s1} - N_r}{N_{s1}} = \dfrac{1350 - 1440}{1350} = -\dfrac{1}{15}$

Impedance,

$$Z_1 = \left[0.6 - \frac{0.4}{1} \times 5 \right] + j \left[3.2 \times \frac{45}{50} \right]$$
$$= 6.12 \ \angle \ 151.96°$$

Fig. 12.57. Pertaining to Example 12.41.

Source current,

$$\overline{I}'_1 = \overline{I}'_2 = \frac{360}{6.12 \ \angle \ 151.93} = 58.823 \ \angle -151.93° \ A$$

Torque, $T_e = \dfrac{3}{\omega_{s1}} (I_2')^2 \dfrac{r_2}{s} = \dfrac{3}{45 \cdot \pi} (58.823)^2 \ (-0.4 \times 15) = -440.56 \ \text{Nm} = BC$

Since induction machine torque is negative, it is now working as induction generator in regenerative braking mode. The motor operating point A (BA = motoring torque = 249.33 Nm) at once shifts to C so that $BC = 440.56 \ Nm$ = regenerative braking torque. The induction generator now operates at power factor cos (-151.93) = 0.8824 or 0.8824 leading.

Now the operating point travels from C to D. Before point D is reached, if V/f is further reduced, regenerative braking action can be obtained uptill low values of operating speeds.

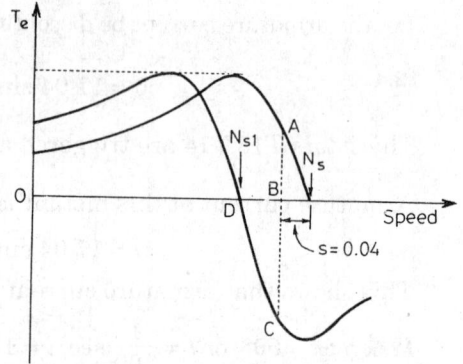

PROBLEMS

12.1. (a) Give the concept of electric drive. Illustrate your answer with examples.

   (b) Give two methods of speed control normally employed for dc motors. Hence, sketch the characteristics of a separately-excited dc motor based on these two methods. Indicate clearly constant-torque drive and constant-power drive regions.

   (c) Write down the basic performance equations for a dc series motor. Sketch also the characteristics of this motor indicating the two regions of constant-torque mode and constant-power mode.

12.2. (a) Give the general circuit layout for single-phase dc drives. Enumerate the various 1-phase dc drives used.

   (b) Describe single-phase half-wave converter feeding a separately-excited dc motor. Illustrate your answer with waveforms and appropriate expressions.

   (c) The speed of a separately-excited dc motor is controlled through single-phase half-wave controlled converter fed from 230-V mains. The motor armature resistance is 0.5 Ω and motor constant is $K = 0.4$ V-s/rad. For load torque of 20 Nm at 1500 rpm and for constant armature current, calculate (*i*) firing-angle delay of the converter (*ii*) rms value of thyristor current and (*iii*) input pf of the motor. [**Ans.** (*c*) 45.821°, 30.52 A, 0.6255 lag]

12.3. (*a*) Describe the working of a single-phase semiconverter fed dc separately-excited motor with relevant waveforms and expressions. State the assumptions made.

(*b*) A single-phase semiconverter feeds a separately-excited dc motor. If armature current is ripple free, then shown that input supply power factor is given by

$$(1 + \cos \alpha) \sqrt{\frac{2}{\pi (\pi - \alpha)}}$$

where $\alpha =$ firing-angle delay of semiconverter feeding the armature circuit of the motor.

12.4. A separately-excited dc motor has its armature circuit connected to one semiconverter and field winding to another semiconverter. The supply for both the converters is single-phase, 230 V, 50 Hz. Resistance for the field circuit is 100 Ω and that for the armature circuit is 0.2 Ω. Rated load torque is 80 Nm at 1000 rpm. The motor constant is 0.8 V-s/A-rad and magnetic saturation is neglected. For ripple free armature and field currents and with zero degree firing angle for field converter, determine (*a*) rated armature current (*b*) firing-angle delay of armature converter at rated load (*c*) speed regulation at full load (*d*) input pf of the armature converter and the drive at rated load. [**Ans.** (*a*) 48.31 A (*b*) 39.78° (*c*) 5.571% (*d*) 0.769 lag, 0.8338 lag]

12.5. A separately-excited dc motor is fed from two single-phase semiconverters, one in the armature circuit and the other in the field circuit. Field current is constant at 2A. Motor armature resistance is 0.8 Ω and motor constant is $k = 0.5$ V-s/A-rad. AC voltage is 230V, 50Hz. For a ripple-free armature current and speed of 1500 rpm, calculate.

(*a*) motor current and torque for a firing angle of 30° and

(*b*) input supply power factor. [**Ans.** (*a*) 45.12 A, 45.12 Nm (*b*) 0.92 lag]

12.6. (*a*) Describe, with appropriate voltage and current waveforms, the working of a single-phase full-converter fed dc drive.

Derive also an expression for its input pf. State the assumptions made.

(*b*) A 200 V, 1000 rpm, 10A separately-excited dc motor is fed from a single-phase full converter with ac source voltage of 230 V, 50 Hz. Armature circuit resistance is 1 Ω. Armature current is continuous. Calculate firing angle for

(*i*) rated motor torque at 500 rpm.

(*ii*) half the rated motor torque at (– 500) rpm. [**Ans.** (*b*) (*i*) 59.526° (*ii*) 115.766°]

12.7. The speed of a separately-excited dc motor is controlled by two single-phase full converters, one in the armature circuit and the other in the field circuit. Both converters are fed from the same single-phase, 230 V, 50 Hz source. Armature resistance is 0.5 Ω and field circuit resistance is 200 Ω. Firing angle for field converter is zero and motor constant is 0.8 V- s/A-rad. Armature and field currents are continuous and ripple free. If armature current is 30 A for a firing angle of 45°, then calculate

(*a*) motor speed

(*b*) motor torque

(*c*) input pf of the armature converter and

(*d*) input pf of the drive. [**Ans.** (*a*) 1515.1 rpm (*b*) 24.845 Nm (*c*) 0.6365 lag (*d*) 0.6672 lag]

12.8. Describe the use of a three-phase semiconverter for the speed control of a dc series motor. Illustrate your answer with appropriate waveforms.

Derive expressions for the rms values of source and thyristor currents and the average value of SCR current for (*a*) firing angle < 60° and (*b*) firing angle > 60°.

12.9. The speed of a dc series motor is controlled by a 3-phase semiconverter connected to 3-phase, 400V, 50Hz source. The motor constant is 0.4 V-s/A. rad. Total field and armature resistance is 1 Ω. Assuming continuous and ripple free armature current at a firing angle of 40° and speed of 1000 rpm, determine

(a) motor current and motor torque

(b) power delivered to motor

(c) reactive power drawn from the supply in VAr.

[**Ans.** (a) 11.12 A, 49.462 Nm (b) 5303.46 W (c) 3383.085 VAr]

12.10. (a) Describe how the speed of a separately-excited dc motor is controlled through the use of two 3-phase full converters. Discuss how two-quadrant drive can be obtained from this scheme.

Derive expressions for rms values of source and thyristor currents. State the assumptions made.

(b) The speed of a 50 kW, 500V, 120 A, 1500 rpm separately excited dc motor is controlled by a three-phase full converter fed from 400V, 50Hz supply. Motor armature resistance is 0.1 Ω. Find the range of firing angle required to obtain speeds between 1000 rpm and (– 1000) rpm at rated torque.   [**Ans.** (b) 51.35°, 125.46°]

12.11. The speed of a separately-excited dc motor is controlled by means of two 3-phase full converters, one in the armature circuit and the other in the field circuit and both are fed from 3-phase, 400 V, 50 Hz supply. Resistance of armature and field circuits is 0.2 Ω and 320 Ω respectively. The motor constant is 0.5 V-s/A. rad. Field converter has zero degree firing-angle delay. Armature and field currents have negligible ripple. For rated load torque of 60 Nm at 2000 rpm, calculate (a) rated armature current (b) firing angle delay of the armature converter (c) speed regulation at rated load and (d) input pf of the armature converter and the drive at rated load.   [**Ans.** (a) 71.10 A (b) 69.291° (c) 8.045% (d) 0.3376 lag, 0.3601 lag]

12.12. A dc motor driven from a 3-phase full converter shown in Fig. 12.17 draws a dc line current of 90 A with negligible ripple.

(a) Sketch the line voltage v_{ab} taking it zero-crossing and becoming positive at $\omega t = 0°$. Also, sketch line current i_A (for one cycle) and thyristor current i_T for a firing angle of 30°. Conduction of SCRs must also be indicated.

(b) Calculate average and rms values of thyristor current.

(c) Compute power factor at ac source.

(d) For motor constant of 2.5 V-s/rad and armature circuit resistance of 0.4 Ω, calculate the motor speed.

12.13. (a) Describe how the speed of a dc series motor can be controlled by means of a dc chopper.

(b) A dc series motor, fed from 400V dc source through a chopper, has the following parameters :

$$r_a = 0.05\ \Omega,\ r_s = 0.07\ \Omega,\ k = 5 \times 10^{-3}\ \text{Nm/amp}^2.$$

The average armature current of 200 A is ripple free. For a chopper duty cycle of 50%, determine

(i) input power from the source (ii) motor speed and (iii) motor torque.

[**Ans.** (b) (i) 40 kW (ii) 1680.68 rpm (iii) 200 Nm]

12.14. A 230 V dc source is connected to a separately-excited dc motor through a chopper operating at 500 Hz. The load torque at 1200 rpm is 32.5 Nm. The motor has $r_a = 0$, $L_a = 2mH$ and K_m = 1.3 V-s/rad. Motor and chopper losses are neglected.

(a) Calculate the minimum and maximum values of armature current and the armature-current excursion.

(b) Obtain the expressions for armature current during on and off periods of a chopper cycle.

[**Ans.** (a) 22.634 A, 27.366 A, 4.732 A (b) 22.634 + 3332 t, 27.366 - 8168 t]

12.15. Repeat Prob 12.14 in case armature circuit of the dc motor has a resistance of 0.3 Ω

[**Ans.** (*a*) 14.438 A, 18.95 A, 4.512 A

(*b*) 222.13 $(1 - e^{-15t}) + 14.43\, e^{-15t}$; $- 544.53\, (1 - e^{-15t}) + 18.95\, e^{-15t}$]

12.16. What is regenerative braking ? Describe the regenerative braking of a chopper-fed separately-excited dc motor. Illustrate your answer with circuit diagram and relevant waveforms.

Derive expressions for the minimum and maximum braking speeds for obtaining regenerative braking of the dc motor.

Show that the speed range for regenerative braking is $(V_s + I_a r_a) : I_a r_a$.

12.17. A 220 V, 60 A dc series motor, having combined resistance of armature and field of 0.15 Ω, is controlled in regenerative braking mode. The dc source voltage is 220 V. Motor constant is 0.05 V-s/A.rad. The average motor armature current is rated and ripple free. For a duty cycle of 50%, determine

(*a*) the power returned to the supply,

(*b*) minimum and maximum permissible braking speeds and

(*c*) speed during regenerative braking.

[**Ans.** (*a*) 6.6 kW (*b*) 28.65 rpm, 728.9 rpm (*c*) 378.82 rpm]

12.18. (*a*) Distinguish between two-quadrant and four-quadrant drives.

(*b*) Describe how a four-quadrant drive can be obtained from a chopper-fed separately-excited dc motor.

12.19. (*a*) What are ac drives ? Give the merits and demerits of ac drives with respect to dc drives.

(*b*) From the approximate equivalent circuit of a 3-phase induction motor, derive the following expressions :

Torque at any slip, slip at maximum torque, maximum torque, maximum torque and slip at which it occurs in case stator resistance is neglected.

12.20. (*a*) Enumerate the various methods of speed control of a 3-phase induction motor when fed through semiconductor devices.

(*b*) Describe stator-voltage-control technique for the speed control of a 3-phase induction motor.

12.21. (*a*) For fan-type loads, show that rotor current in a 3-phase induction motor is maximum when slip $s = 1/3$. State the assumptions made.

(*b*) A 400 V, 50 Hz, 3-phase SCIM develops full-load torque at 1470 rpm. If supply voltage reduces to 340 V, with load torque remaining constant, calculate the motor speed. Assume speed-torque characteristics of the motor to be linear in the stable region. Neglect stator resistance. [**Ans.** (*b*) 1458.5 rpm]

12.22. (*a*) Induction motor speed control with constant-supply voltage and reduced-supply frequency is rarely used in practice. Justify this statement.

(*b*) Describe stator frequency control for the speed control of a 3-phase induction motor. Derive expressions for motor torque, maximum torque and the slip at which it occurs. State the various assumptions made.

Discuss, why during this method of speed control, an induction motor is said to be working in field-weakening mode.

12.23. A 3-phase, 400V, 20 kW, 970 rpm, 50 Hz, delta-connected induction motor has rotor leakage impedance of 0.5 + j 2.00 Ω. Stator leakage impedance and rotational losses are assumed negligible.

If this motor is energised from a source of 3-phase, 400 V, 90Hz, then compute

(*a*) the motor speed at rated torque

(*b*) the slip at which maximum torque occurs and

(*c*) the maximum torque. [**Ans.** (*a*) 1819.8 rpm (*b*) 0.138 (*c*) 235.785 Nm]

12.24. Explain volts/hertz control for a 3-phase induction motor for its speed control. Enumerate its advantages.

Describe at least two inverter circuits used for volts/hertz control.

12.25. (a) Discuss how volts/hertz control for a 3-phase induction motor is similar to armature-voltage control of a dc motor.

(b) In stator-frequency control of a 3-phase induction motor, explain why

(i) ratio V/f is maintained constant for speeds below base speed

(ii) terminal voltage is maintained constant for speeds above base speed.

12.26. A 3-phase, 15kW, 420V, 4-pole, 50Hz, delta-connected induction motor has the following per-phase parameters referred to stator :

$$r_1 = 0.5 \ \Omega, r_2 = 0.4 \ \Omega, x_1 = x_2 = 1.5 \ \Omega, X_m = 0$$

If this motor is operated at 210V, 25Hz with DOL starting, calculate

(a) current and pf at the instant of starting and under maximum torque conditions ; compare the results with normal values,

(b) starting and maximum torques and compare with normal values.

[**Ans.** (a) 120.05 A, 0.5145 lag, 81.862 A, 0.8112 lag

At normal : 134.1 A, 0.2873 lag, 90.486 A, 0.763 lag

(b) 220.2 Nm, 404.702 Nm

At normal : 137.38 Nm, 475.662 Nm]

12.27. Describe stator-current-control method for the speed control of a 3-phase induction motor. Derive expressions for maximum torque, slip at maximum torque etc. by using approximate equivalent circuit.

Discuss the effect of saturation on the speed-torque characteristics obtained by this method of speed control.

12.28. A 420 V, 6-pole, 50Hz, 3-phase, star-connected IM has $r_1 = 0$, $x_1 = x_2 = 1.2 \ \Omega$, $r_2 = 0.5 \ \Omega$ and $X_m = 50 \ \Omega$ as its per-phase parameters referred to stator. This IM is fed from (i) constant-voltage source of 242.5 V per phase and (ii) constant-current source of 30A.

For both types of sources (i) and (ii), calculate

(a) the slip for maximum torque,

(b) the starting and maximum torques and

(c) the supply voltage required to sustain the constant current at the maximum torque.

[**Ans.** (a) 0.2108, 0.00976 (b) 136.71 Nm, 12.293 Nm ; 338.676 Nm, 629.47 Nm (c) 1838.1 V]

12.29. Sketch speed-torque characteristics of a 3-phase induction motor as influenced by different control techniques employed for its speed control by using semiconductor devices. Explain the three regions into which these characteristics can be subdivided.

Discuss the various methods employed for obtaining the three regions.

12.30. Discuss, in detail, how Ward-Leonard type of characteristics can be obtained from a 3-phase induction motor.

Explain also how high-speed series-motoring region is obtained in a 3-phase induction motor.

12.31. (a) Describe static rotor-resistance control method for the speed control of a 3-phase induction motor.

Derive expressions for inductor current and motor speed in terms of load torque, supply voltage, motor turns ratio, synchronous speed etc.

(b) A 3-phase SRIM uses rotor ON-OFF control by means of chopper for its speed control. The effective rotor resistance is increased to 10 times during off period. If the motor develops 0.4 pu torque at a slip of 0.02 for normal operation, calculate the average torque developed at the same slip for 20%, 50% and 80% duty cycles of the static rotor chopper.

[**Hint.** (b) At low values of slips,

$$T_e = K \frac{V^2}{r_2} s$$

$\therefore$ $T_{e1} = 0.4 = K\dfrac{V^2}{r_2} \times 0.02$ or $K\dfrac{V^2}{r_2} = 20$

$$T_{e2} = \frac{KV^2}{10r_2}s = \frac{20 \times 0.02}{10} = 0.04\ pu$$

Net torque, $T_{net} = \dfrac{T_{on} \cdot T_{e1} + T_{off} \cdot T_{e2}}{T}$ etc.]   [**Ans.** (b) 0.112 pu, 0.22 pu, 0.328 pu]

12.32. A 3-phase, 400V, 6-pole, 50 Hz, star-connected SRIM uses a chopper in its rotor circuit for its speed control. The effective turns ratio from rotor to stator is 0.6. Inductor current is ripple free. Losses in the rectifier, inductor, chopper and no-load losses of the motor are neglected. Load torque, proportional to speed squared, is 360 Nm at 970 rpm.

(a) For a minimum motor speed of 600 rpm, calculate the value of chopper resistance R.

For the value of R obtained in part (a) if the speed is to be raised to 800 rpm, calculate

(b) inductor current (c) duty cycle of the chopper (d) rectified output voltage (e) efficiency in case per-phase resistances for stator and rotor are 0.015 Ω and 0.02 Ω respectively.
                [**Ans.** (a) 3.496 Ω (b) 65.93 A (c) 0.7187 (d) 64.823 V (e) 79.285%]

12.33. (a) In static rotor-resistance control of a 3-phase SRIM, each diode in the rotor circuit conducts for 120°. Assuming ripple free rotor current, derive expressions for rms value of rotor current referred to stator, fundamental component of rotor current and its value referred to stator.

(b) A 3-phase, 415V, 50 Hz, 1470 rpm, star-connected SRIM has the following per-phase parameters referred to stator :
$$r_1 = 0.12\ \Omega,\ r_2 = 0.1\ \Omega,\ x_1 = x_2 = 0.4\ \Omega,\ X_m = 0$$

Effective per-phase turns ratio from rotor to stator = 0.8

Speed of this motor is controlled by rotor ON-OFF control. For a speed of 1200 rpm, the inductor current is 100 A and chopper resistance is 1.8 Ω. Calculate

(i) the value of chopper duty cycle

(ii) efficiency for a power output of 25 kW and for negligible no-load losses

(iii) the input power factor.   [**Ans.** (b)(i) 0.498 (ii) 67.84% (iii) 0.822 lag]

12.34. Describe static Kramer drive for the speed control of a 3-phase SRIM and show that steady-state torque is not influenced by whether a transformer is used or not.

Derive appropriate expressions to obtain speed-torque characteristics of static Kramer drive.

12.35. Speed of a 400V, 6-pole, 50Hz, star-connected SRIM is controlled by static Kramer drive. The effective phase turns ratio from rotor to stator is 0.6 and transformer has phase turns ratio from l.v. to h.v. as 0.4. The inductor current is ripple free. Losses in diode rectifier, inductor, inverter and transformer are neglected. The load torque is proportional to speed squared and its value is 250 Nm at 800 rpm. For a motor operating speed of 700 rpm, calculate

(a) rotor rectified voltage (b) inductor current (c) delay angle of the inverter (d) efficiency in case inductor resistance is 0.01 Ω and per-phase resistances for stator and rotor are 0.015 Ω and 0.02 Ω respectively (e) For the firing angle obtained in part (c), the load torque is increased to 350 Nm, find the motor speed.
        [**Ans.** (a) 97.23V (b) 61.84 A (c) 116.743° (d) 98.37% (e) 696.53 rpm]

12.36. (a) Describe a static Kramer drive and show that the slip s at which it operates is given by

$$s = -\frac{a_T}{a}\cos \alpha$$

where a and a_T pertain to per-phase turns ratio for SRIM and transformer respectively.

(b) Speed of a 6-pole SRIM fed from 400 V, 50Hz source is controlled by static Kramer drive. The inverter is directly connected to the supply. If the motor is required to operate at 800 rpm, determine the firing advance angle of the inverter. Voltage across the open-circuited slip rings at stand-still is 600V. There is a voltage drop of 0.7V and 1.5V across each of the diodes and thyristors respectively. Inductor drop is neglected.

In case transformer is to be interposed between supply and the inverter for obtaining a minimum speed of 600 rpm, determine the voltage ratio of the transformer from l.v. to h.v. **[Ans. (b) 106.97°, 0.6]**

12.37. Repeat Problem 12.36(b) in case there is an overlap angle of 12° in the rectifier and 5° in the inverter. **[Ans. 76.305°]**

12.38. Explain, with relevant circuit diagrams, both types of static Scherbius drives for obtaining speeds below as well as above synchronous speed.

12.39. (a) Enumerate the various types of synchrouous motors. Derive an expression for power developed in a cylindrical-rotor synchronous motor with negligible armature resistance.

(b) A 415V, 50Hz, 4-pole, star-connected synchronous motor has $X_s = 1.5 \ \Omega$. Load torque, proportional to speed, is 300 Nm at synchronous speed. The speed of the motor is lowered by keeping $\dfrac{V}{f}$ constant and maintaining 0.8 pf leading by field control. For the motor operation at 840 rpm, calculate (a) supply voltage (b) the armature current (c) the excitation voltage (d) load angle and (e) the pull-out torque. Neglect rotational losses.
[Ans. (a) 232.4 V (b) 44.96 A (c) 276.635 V (d) 10.904° (e) 870.15 Nm]

12.40. (a) Derive the expression for power developed in a salient-pole synchronous motor in terms of excitation voltage, load angle etc. Neglect armature resistance.

(b) Explain the working and uses of a permanent-magnet synchronous motor. How does the input-current waveforms effect the constructional features of PMSM ?

Chapter 13

Power Factor Improvement

There are several techniques available for the improvement of system power factor. Out of all these, the one employing thyristor controlled reactor (TCR) is now the most sought-after method. It is because TCR offers a continuous and a very fast control of reactive-power flow in a system for regulating its power factor. Primarily, popularity of this method is due to the use of thyristors which are fast, reliable, efficient and precise in their operation. In this chapter, power-factor improvement techniques in ac systems, with more emphasis to TCR, are discussed.

13.1. EFFECT OF POOR POWER-FACTOR

First of all, the effect of poor load power-factor on the system performance would be illustrated through an example.

Example 13.1. *An alternator, with fixed source voltage of 250 V, delivers power to a load. The transmission line reactance is 5 Ω and load current is 20 A. Calculate the load voltage, voltage regulation, system utilization and the energy consumed for a load power factor of (a) unity and (b) 0.5 lagging.*

Solution. (*a*) **Unity pf load.** The phasor diagram for unity load is shown in Fig. 13.1. (*b*), where V_L = load voltage and E = source voltage, 250 V. Fig. 13.1 (*a*) represents the circuit diagram where E, V_L, load and reactance are shown.

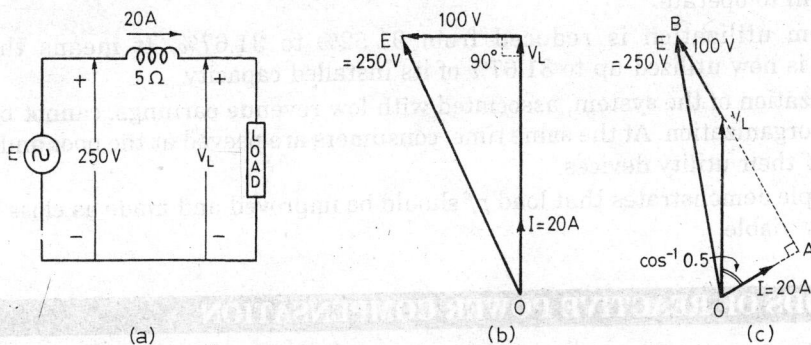

Fig. 13.1. Pertaining to Example 13.1 (*a*) circuit diagram and phasor diagrams at (*b*) unity *pf* and (*c*) *pf* = 0.5 lagging.

From Fig. 13.1 (b), $250^2 = V_L^2 + 100^2$

or   Load voltage, $V_L = 229.13$ V

$$\text{Voltage regulation} = \frac{250 - 229.13}{250} \times 100 = 8.35\%$$

Load power $= V_L I \cos\theta = 229.13 \times 20 \times 1 = 4582.6$ W

Maximum possible system rating $= 250 \times 20 \times 1 = 5000$ W

System utilization factor $= \dfrac{4582.6}{5000} \times 100 = 91.652\%$

Load energy delivered per hour $= \dfrac{4582.6}{1000} \simeq 4.6$ units

Assuming Rs. 5 per unit, revenue earned $= 4.6 \times 5 =$ Rs. 23 per hour.

(b) **Load power factor = 0.5 lagging.** For this power factor, the phasor diagram is as shown in Fig. 13.1 (c). From this figure,

$$OA^2 + AB^2 = E^2$$

or $(0.5\,V_L)^2 + (0.866\,V_L + 100)^2 = 250^2$ or $V_L = 158.35$ V

$$\text{Voltage regulation} = \frac{250 - 158.35}{250} \times 100 = 36.66\%$$

Load power $= V_L I \cos\theta = 158.35 \times 20 \times 0.5 = 1583.5$ W

System utilization factor $= \dfrac{1583.5}{5000} \times 100 = 31.67\%$

Energy delivered to load per hour $= \dfrac{1583.5}{1000} \simeq 1.58$ units

∴   Revenue earned $= 1.58 \times 5 =$ Rs. 7.90 per hour

It is seen from the above example that for the same source voltage and load (or line) current, a load at a poor *pf* effects the system performance as under :

(i) Load power is reduced by $\dfrac{4582.6 - 1583.5}{4582.6} \times 100 = 65.45\%$. As a consequence, revenue earned falls from Rs. 23 per hour to Rs. 7.90 per hour.

(ii) Load voltage falls to 158.35 V. As a result ; utility devices, like fluorescent tubes, lamps, refrigerators, washing machines etc., designed to operate at 230 V, would operate erratically or may even fail to operate.

(iii) System utilization is reduced from 91.52% to 31.67%. It means that system infrastructure is now utilized up to 31.67% of its installed capacity.

Underutilization of the system, associated with low revenue earnings, cannot be tolerated by an efficient organization. At the same time, consumers are peeved at the poor and unreliable performance of their utility devices.

This example demonstrates that load *pf* should be improved and made as close to unity as is economically viable.

13.2. METHODS OF REACTIVE POWER COMPENSATION

Industrial loads, which normally operate at poor *pf* are induction motors, arc and induction furnaces. Fluorescent tubes, fans etc also operate at low value of power factor. All these loads, working at low *pf*, need large amount of reactive power which results in reduced voltage level at the load terminals. A low voltage at the consumer terminals is undesirable as it leads to impaired performance of their utility devices.

The various methods of power-factor improvement are as under :

    (*i*) Use of capacitor banks

    (*ii*) Use of synchronous condensers

    (*iii*) Use of static VAr compensators.

These are now discussed in what follows.

13.2.1. Capacitor Banks

A bank of capacitors is connected across the load. Since the capacitor takes leading reactive power, overall reactive power taken from source decreases, consequently system power factor improves. Example 13.2 illustrates how capacitor bank renders the improvement in system power factor.

Example 13.2. *A single-phase induction motor, when running from 230 V, 50 Hz supply, gave the following data :*

No-load :	*2A, pf = 0.3 lag.*
Half-full load :	*5A, pf = 0.5 lag.*
Full-load :	*10A, pf = 0.7 lag.*

Calculate the capacitance required in parallel with the induction motor so that power-factor of the motor-capacitor combination (or the supply power factor) is raised to unity under all the three operating modes listed above.

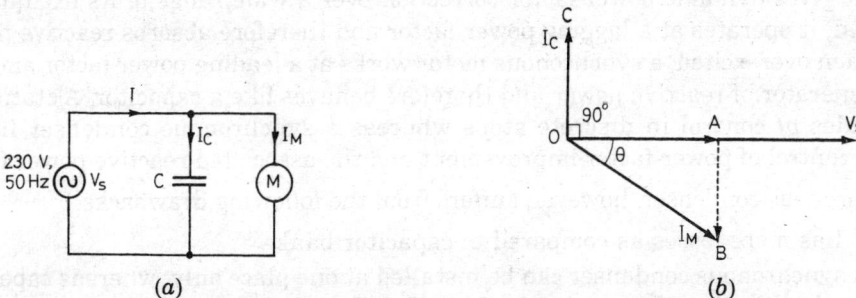

Fig. 13.2. (*a*) Circuit diagram and its (*b*) its phasor diagram. Pertaining to Example 13.2.

Solution. A capacitor, connected in parallel with 1-phase induction motor, is shown in **Fig.** 13.2 (*a*). Its phasor diagram is drawn in Fig. 13.2 (*b*), where OB = motor current I_m lags the source voltages V_s by motor operating *pf* cos θ and OC = capacitor current I_c leads V_s by 90°. For unity power factor of the combination, OC must be equal to $AB = I_m \sin θ$ for all the three operating modes.

    (*a*) At no load, $OC = AB$

or $I_c = I_m \sin θ = 2 \sin [\cos^{-1} 0.3] = 1.908$ A

But $I_c = \dfrac{V_s}{X_c} = 2\pi f C.V_s = 1.908$ A

∴ $C = \dfrac{1.908 \times 10^6}{2\pi \times 50 \times 230} = 26.406 \ \mu F$

(b) At half-full load, $OC = AB$

or
$$I_c = 5 \sin [\cos^{-1} 0.5] = 4.33 \text{ A} = \frac{V_s}{X_c} = 2\pi f C.V_s$$

$$\therefore \qquad C = \frac{4.33 \times 10^6}{2\pi \times 50 \times 230} = 59.925 \text{ } \mu\text{F}$$

(c) At full load, $OC = AB$

or
$$I_c = 10 \sin [\cos^{-1} 0.7] = 7.1413 \text{ A} = 2\pi f C.V_s$$

$$\therefore \qquad C = \frac{7.1413 \times 10^6}{2\pi \times 50 \times 230} = 98.834 \text{ } \mu\text{F}$$

This example illustrates that for keeping the supply power factor unity, the value of capacitance across the motor terminals must be varied as the load on the induction motor alters. This is called *dynamic VAr compensation* or *dynamic pf control* ; that is, reactive power compensation is carried out through switching-in or out of the capacitors so as to achieve a desired *pf* at all load conditions. A continuous control of the *pf* would entail the need of a large number of capacitors of small ratings. The switching-in or out is carried out by means of relays and circuit breakers. But this all is quite cumbersome and expensive. The mechanical switches and relays are sluggish, unreliable, require frequent maintenance and introduce switching transients. However, with the replacement of mechanical switches by thyristors, it has been made possible to continuously (i) regulate the reactive power flow and (ii) control the power factor and voltage profile by rapid switching-in or out of the static capacitors.

13.2.2. Synchronous Condensers

A 3-phase synchronous motor, when overexcited, works as a synchronous condenser, or a capacitor. It gives dynamic power-factor correction over a wide range of its excitation. When underexcited, it operates at a lagging power factor and therefore absorbs reactive power from the bus. When overexcited, a synchronous motor works at a leading power factor and therefore acts as a generator of reactive power and therefore behaves like a capacitor. A static capacitor bank provides *pf* control in discrete steps whereas a synchronous condenser furnishes a continuous control of power-factor improvement and the associated reactive power flow.

A synchronous condenser, however, suffers from the following drawbacks.

(i) It has more losses as compared to capacitor bank.

(ii) A synchronous condenser can be installed at one place only, whereas capacitor bank can be distributed at many places. A distributed capacitor bank is more effective in controlling the reactive-power flow and voltage profile.

(iii) A synchronous condenser is slow in response due to large time constant of its field circuit, whereas a capacitor bank offers faster response.

13.2.3. Thyristor Controlled Reactors (TCRs)

Thyristor controlled reactor is a major component of static VAr compensator. In this section, only TCR is described. Static VAr compensator is explained in the next article 13.3.

Static thyristor controlled reactors are connected in parallel with the load for the control of reactive power flow. With increase in the size of industrial connected loads, fast reactive power compensation has become necessary. For such loads, thyristor controlled reactors (TCRs) are now becoming increasingly popular. TCR is also called *thyristor controlled inductor* (TCI).

Fig. 13.3. (a) shows a linear reactor (or inductor) L connected to ac source v_s through two thyristors connected in antiparallel. This circuit configuration is also called *ac voltage controller*, Chapter 9. It can, therefore, be said that in Fig. 13.3 (a), a linear inductor L is

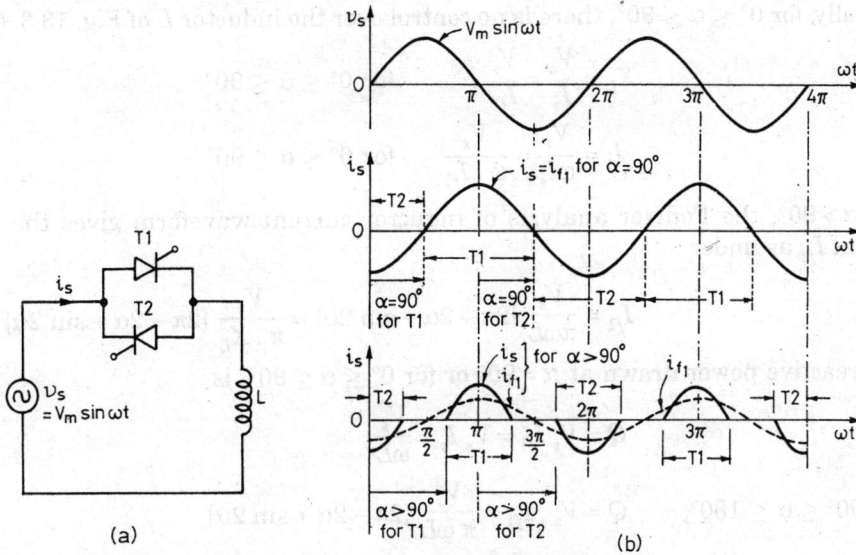

Fig. 13.3. Thyristor controlled reactor (a) circuit diagram and
(b) its voltage and current waveforms.

connected to 1-phase voltage controller. A reference to Art. 9.3 is therefore of considerable benefit.

During positive half cycle of source voltage, T1 is turned on and during the negative half cycle, T2 is turned on. For firing angle $\alpha = 90°$, the source current i_s is continuous as shown in Fig. 13.3 (b). The circuit behaves as if inductance L is directly connected to as source without SCRs. For $\alpha = 90°$, i_s is a sine wave, its fundamental component i_{f1} is the same as i_s and is therefore maximum. As a result, inductive reactance offered by reactor, $X_L = \dfrac{V_s}{I_{f1}}$ is minimum when $\alpha = 90°$. Here V_s = rms value of source voltage and I_{f1} is the rms value of fundamental component of source current, which for $\alpha = 90°$ is given by $I_{f1} = I_s$ (rms value of source current).

For firing angle $\alpha > 90°$, current i_s is discontinuous, Fig. 13.3. (b), but its fundamental component i_{f1} again lags V_s by 90°. With $\alpha > 90°$, as rms value of fundamental component I_{f1} has decreased, the inductive reactance offered by reactor ($= V_s/I_{f1}$) has become more. If α is further increased, fundamental component of i_s would be further reduced and therefore, reactance offered by the reactor would be more pronounced. For firing angle, $\alpha = 180°$, $i_s = 0$, $i_{f1} = 0$ and theoretically, the inductive reactance offered by the reactor would be infinite. This shows that with firing angle control from $\alpha = 90°$ to 180° ; the effective reactance of the reactor, as seen by the source, can be regulated from its actual value $X_L = 2\pi fL$ when $\alpha = 90°$, to an infinite value when $\alpha = 180°$.

As the fundamental component of source current lags the source voltage by 90°, the reactor (or inductor) consumes no power. It draws only the reactive power.

When $\alpha = 90°$, $I_{f1} = I_s = \dfrac{V_s}{\omega L}$, $\therefore$ $X_L = \dfrac{V_s}{I_s}\ \Omega$

where V_s = rms value of source voltage
        I_s = rms value of source current
        I_{f1} = rms value of fundamental component of source current.

Actually, for $0° \leq \alpha \leq 90°$, there is no control over the inductor L of Fig. 13.3. (a), therefore,

$$X_L = \frac{V_s}{I_s} = \frac{V_s}{I_{f1}} \qquad \text{for } 0° \leq \alpha \leq 90° \qquad \qquad ...(13.1a)$$

or

$$L = \frac{V_s}{\omega I_s} = \frac{V_s}{\omega . I_{f1}} \qquad \text{for } 0° \leq \alpha \leq 90° \qquad \qquad ...(13.1b)$$

For $\alpha > 90°$, the Fourier analysis of inductor current waveform gives the fundamental component I_{f1} as under :

$$I_{f1} = \frac{V_s}{\pi\omega L} [2\pi - 2\alpha + \sin 2\alpha] = \frac{V_s}{\pi . X_L} [2\pi - 2\alpha + \sin 2\alpha] \qquad ...(13.2)$$

The reactive power drawn at $\alpha = 90°$ or for $0° \leq \alpha \leq 90°$, is

$$Q = V_s I_{f1} = V_s I_s = \frac{V_s^2}{\omega L} \qquad \qquad ...(13.3)$$

For $90° \leq \alpha \leq 180°$, $\qquad Q = V_s . I_{f1} = \frac{V_s^2}{\pi \omega L} [2\pi - 2\alpha + \sin 2\alpha] \qquad \qquad ...(13.4)$

At $\alpha = 90°$, reactive power $\left(= \frac{V_s^2}{\omega L} \right)$ drawn is maximum. When $\alpha = 180°$, reactive power Q (from Eq. (13.4)) is zero.

13.3. STATIC VAr COMPENSATOR (SVC)

A static VAr compensator is also called *static VAr compensating system (SVS)* or *thyristor controlled compensator (TCC)*. It consists of a thyristor controlled reactor (TCR) in parallel with a fixed capacitor C. As stated before, TCR (shown in the dotted rectangle) is made up of two thyristors $T1$, $T2$ connected in antiparallel and a series-connected linear inductor L. An SVC, equal to TCR + fixed capacitor C, is shown in dash-dot rectangle in Fig. 13.4. Load is connected in parallel with static VAr compensator in Fig. 13.4.

Capacitance has a constant value C ; it, therefore, supplies constant leading-reactive power equal to $\omega C V_s^2$. When both SCRs are fired at $\alpha = 90°$, L seen by source is minimum, therefore TCR takes maximum lagging-reactive power equal to $V_s^2/\omega L$. The values of L and C are so selected that maximum value $V_s^2/\omega L$ is somewhat more than $\omega C V_s^2$; this means that SVC takes lagging-reactive power from the source when $\alpha = 90°$. As a result, *pf* of the combination (load plus SVC) gets impaired a little when $\alpha = 90°$.

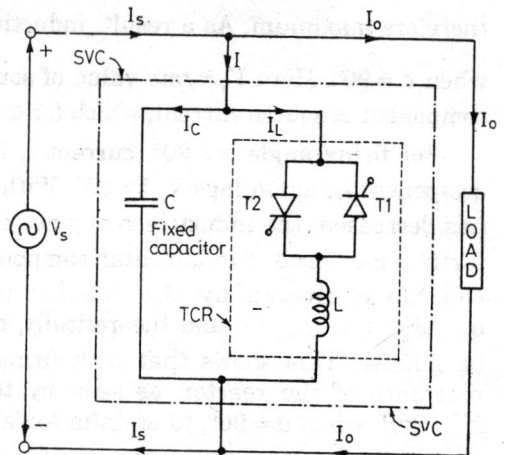

Fig. 13.4. Static VAr Compensator (SVC).

For firing angle $\alpha = 90°$, TCR current I_L is maximum. The current I_c through capacitor C is less than the magnitude of I_L for $\alpha = 90°$ as stated above. For a load *pf* $\cos \theta_0$, the phasor diagram indicating load current I_0, TCR current I_L, capacitor current I_c and source current I_s (= phasor sum of I_0, I_L, I_c) is as shown in Fig. 13.5. (a) for $\alpha = 90°$. The net reactive current

OA ($= I_L - I_c$) lags V_s by 90°, the *pf* of the combination, therefore, deteriorates from $\cos\theta_0$ to $\cos\theta$. The reactive power from the supply has increased from $V_s I_0 \sin\theta_0$ to $V_s I_s \sin\theta$. In other words, lagging-reactive power taken from the source has increased by $V_s(I_L - I_c)$. Note that fundamental component I_{f1} of TCR current I_L for $\alpha = 90°$ is the same as I_L, see Fig. 13.3 (b). In other words, $I_{f1} = I_L$ when $\alpha = 90°$.

For firing angle $\alpha = \alpha_1$, where $\alpha_1 > 90°$, the inductive reactance ωL offered by reactor is more, likewise I_{f1} (fundamental component of I_L) gets reduced. This is shown in Fig. 13.5(b). Now, net reactive current, leading V_s by 90°, is OB ($= I_c - I_{f1}$) and *pf* gets improved from $\cos\theta_0$ to $\cos\theta_1$. The net reactive power taken from the supply is now $V_s I_s \sin\theta_1$, less than the load reactive power $V_s I_o \cdot \sin\theta_0$.

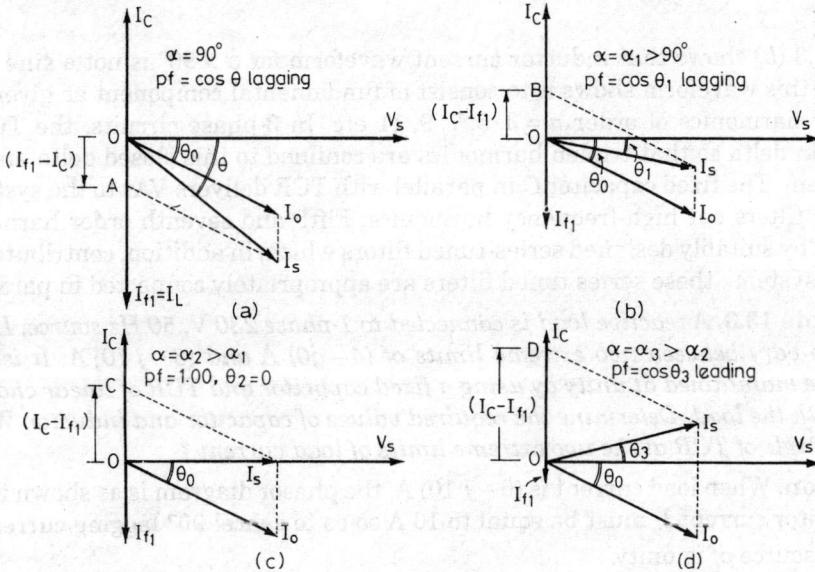

Fig. 13.5. Phasor diagrams illustrating the change in the firing angle of TCR
(a) $\alpha = 90°$ (b) $\alpha = \alpha_1 > 90°$ (c) $\alpha = \alpha_2 > \alpha_1$ (d) $\alpha = \alpha_3 > \alpha_2$.

For firing angle $\alpha = \alpha_2$, where $\alpha_2 > \alpha_1$, the inductive reactance offered is still more and I_L and therefore I_{f1} are further reduced. As a result, net reactive current leading V_s is now $OC > OB$ of Fig. 13.5 (b) and power factor is shown to become unity in Fig. 13.5 (c). The reactive power taken from the supply is now zero.

For firing angle $\alpha = \alpha_3$, where $\alpha_3 > \alpha_2$, the inductive reactance is more, I_L or I_{f1} is less. As a result, net reactive current OD ($= I_c - I_{f1}$), leading V_s by 90°, is more than OC of Fig. 13.5 (c). Power factor now becomes leading as shown in Fig. 13.5 (d). The reactive power, equal to $V_s I_s \sin\theta_3$ is now returned to the supply. For $\alpha = 180°$, I_L or $I_{f1} = 0$ and *pf* is still further improved (not shown).

Examination of Fig. 13.4 reveals that fundamental component of SVC current I is given by $I = I_c +$ fundamental component I_{f1} of TCR current given by Eq. (13.2). As I_c and I_{f1} oppose each other (see phasor diagrams in Fig. 13.5), $I = \dfrac{V_s}{X_c} - I_{f1}$

From Eq. (13.2), $\qquad I = \dfrac{V_s}{X_c} - \dfrac{V_s}{\pi\omega L}\ (2\pi - 2\alpha + \sin 2\alpha)$

or
$$I = V_s \left[\omega C - \frac{1}{\pi \omega L} (2\pi - 2\alpha + \sin 2\alpha) \right] \dots \text{ for } \frac{\pi}{2} < \alpha < \pi \qquad \dots(13.5)$$

If SVC net current I from Eq. (13.5) turns out to be positive, SVC delivers reactive power to load to improve (i) system pf and also (ii) the load voltage profile. In case net current I is negative, SVC absorbs reactive power from source and impairs the system pf as well as the voltage profile.

It is seen from above that by appropriately controlling the firing angle of TCR from 90° to 180°, the reactive power taken from the supply can be regulated continuously and power-factor of the combined static VAr compensator and the load can be improved. In other words, source pf gets improved from lagging to leading as the firing angle of TCR is progressively altered from 90° to 180°. Since the circuit response is fast, dynamic stability of the system voltage also improves.

Fig. 13.3 (b) shows that inductor current waveform for $\alpha > 90°$ is not a sine wave. Fourier analysis of this waveform shows it to consist of fundamental component as given by Eq. (13.2) and higher harmonics of order $n = 3, 5, 7, 9, 11$ etc. In 3-phase circuits, the TCI (or TCR) is connected in delta so that triplen harmonics are confined to this closed delta and do not enter the ac system. The fixed capacitor C in parallel with TCR delivers VAr to the system and at the same time, filters out high-frequency harmonics. Fifth and seventh order harmonics are also filtered out by suitably designed series-tuned filters which, in addition, contribute some leading VAr to the system ; these series tuned filters are appropriately connected in parallel with SVC.

Example 13.3. *A reactive load is connected to 1-phase 230 V, 50 Hz source. Load current is observed to vary between two extreme limits of $(4 - j0)$ A and $(6 - j 10)$ A. It is required that supply pf be maintained at unity by using a fixed capacitor and TCR of linear characteristics in parallel with the load. Determine the required values of capacitor and inductor. What should be the firing angle of TCR at the two extreme limits of load current ?*

Solution. When load current is $(6 - j 10)$ A, the phasor diagram is as shown in Fig. 13.6 (a). Now capacitor current I_c must be equal to 10 A so as to cancel 90° lagging current $I_L = 10$ A in order that source pf is unity.

$$\therefore \qquad I_c = 10 = \frac{V_s}{X_c} = 2\pi f C.V_s$$

$$\therefore \qquad C = \frac{10 \times 10^6}{2\pi \times 50 \times 230} = 138.396 \, \mu F$$

For load current of $(6 - j 0)$ A, TCR should not take any current, therefore its firing angle should be 180°.

When load current is $(4 - j0)$ A, the current I_c in the fixed capacitor must be cancelled by inductive current.

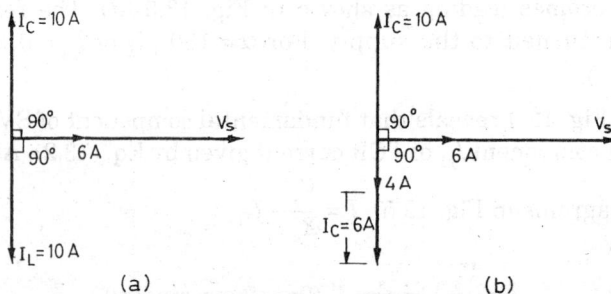

Fig. 13.6. Phasor diagram pertaining to Example (a) 13.3 and (b) 13.4.

As $I_c = 10$ A, inductive current I_L has to be 10 A.

$$\therefore \qquad I_L = \frac{V_s}{X_L} = \frac{V_s}{2\pi \times 50 \times L} = 10 \text{ A}$$

or
$$L = \frac{230 \times 10^3}{2\pi \times 50 \times 10} = 73.211 \text{ mH}.$$

For load current $(4 - j0)$, $\alpha = 90°$

$\therefore$ Firing angle of TCR for the two extreme limits is 90° and 180°.

Example 13.4. *In Example 13.3, if load current attains a value of $(6 - j4)$ A, find the firing angle of the TCR.*

Solution. Load current of $(6 - j4)$ A is shown in Fig. 13.6 (b). Here $I_c = 10$ A, load current = 4A lagging V_s by 90°. Therefore, fundamental component of TCI (or TCR) current I_{f1} must be $10 - 4 = 6$ A. In Example 13.3, $X_L = \dfrac{V_s}{I_L} = \dfrac{230}{10} = 23 \ \Omega$.

From Eq. (13.2),
$$I_{f1} = \frac{V_s}{\pi \cdot X_L} [2\pi - 2\,\alpha + \sin 2\,\alpha]$$

or
$$6 = \frac{230}{\pi \times 23} [2\pi - 2\alpha + \sin 2\alpha]$$

or
$$2\alpha - \sin 2\alpha = 4.39823$$

By hit and trial, $\qquad \alpha = 108.65°$

Example 13.5. *A TCI is fed from 230 V, 50 Hz and has an inductance of 10 mH. Calculate the effective inductance seen by the source for firing angles of 90°, 120°, 150°, 170°, 175° and 180°.*

Solution. From Eq. (13.1b), effective load inductance as seen by the source,

$$L_{eff} = \frac{V_s}{\omega I_{f1}}$$

From Eq. (13.2),
$$I_{f1} = \frac{V_s}{\pi X_L} [2\pi - 2\alpha + \sin 2\alpha]$$

$$\therefore \qquad L_{eff} = \frac{V_s}{\omega} \cdot \frac{\pi \cdot X_L}{V_s} \left[\frac{1}{2\pi - 2\alpha + \sin 2\alpha} \right] = \frac{\pi L}{[2\pi - 2\alpha + \sin 2\alpha]}$$

when $\alpha = 90°$,
$$L_{eff} = \frac{\pi L}{[2\pi - \pi + \sin 180°]} = L = 10 \text{ mH}$$

For $\alpha = 120°$,
$$L_{eff} = \frac{\pi \times 10 \times 10^{-3}}{[2\pi - \dfrac{4\pi}{3} + \sin 240°]} = 25.575 \text{ mH}$$

For $\alpha = 150°$,
$$L_{eff} = \frac{\pi \times 10 \times 10^{-3}}{\left[(360° - 300) \dfrac{\pi}{180} + \sin 300° \right]} = 173.40 \text{ mH}$$

For $\alpha = 170°$,
$$L_{eff} = \frac{\pi \times 10 \times 10^{-3}}{\left[(360 - 340) \dfrac{\pi}{180} + \sin 340° \right]} = 4.459 \text{ H}$$

For $\alpha = 175°$, $L_{eff} = \dfrac{\pi \times 10 \times 10^{-3}}{\left[(360 - 350) \dfrac{\pi}{180} + \sin 350° \right]} = 35.51$ H

For $\alpha = 180°$, $L_{eff} = \infty$.

This example demonstrates that effective inductance seen by the source rises as the firing angle of both the thyristors is increased from 90° to 180°.

Example 13.6. *A single-phase inductive load of 100 kW + j 50 kVAr is supplied from 11 kV, 50 Hz source. The static VAr compensator (SVC) has fixed capacitors of rating 100 kVAr whereas TCI can draw a maximum of 100 kVAr. For raising the system power factor to unity, find the firing-angle delay of TCI and its effective inductance.*

Solution. Maximum value of TCI reactive power, $Q = \dfrac{V_s^2}{X_L} = 100$ kVAr

$\therefore$ Inductance of TCI, $L = \dfrac{V_s^2}{2\pi f . Q} = \dfrac{11000^2}{2\pi \times 50 \times 100{,}000} = 3.8515$ H

It is seen from Fig. 13.5 that for unity *pf* operation, the reactive power of fixed capacitors = reactive power of TCI + reactive power of load.

$$100{,}000 = \dfrac{V_s^2}{\pi . X_L} [2\pi - 2\alpha + \sin 2\alpha] + 50{,}000$$

But $\dfrac{V_s^2}{X_L} = 100$ kVAr $= 100{,}000$ VAr

$\therefore$ $100{,}000 = \dfrac{100{,}000}{\pi} [2\pi - 2\alpha + \sin 2\alpha] + 50{,}000$

or $2\alpha - \sin 2\,\alpha = 1.5 \times \pi$

By hit and trial, $\alpha = \mathbf{113.83°}$

Effective value of TCI inductance at this firing-angle delay is given by (from Example 13.5)

$$L_{eff} = \dfrac{\pi \times 3.8915}{[2\pi - 2 \times 113.83 - \sin (2 \times 113.83)]} = \mathbf{7.7039}\ \mathbf{H}$$

13.4. SOME WORKED EXAMPLES

In this article, some typical examples on some topics in power electronics are presented.

Example 13.7. *Define the following parameters pertaining to inverters :*
harmonic factor of nth harmonic, total harmonic distortion.
How do these parameters help in evaluating the quality of inverters.

Solution. The output voltage obtained from inverters is not sine wave. It consists of fundamental component plus certain harmonics. Lower the harmonic content in the output voltage wave, better is the quality of an inverter. More important performance parameters of inverters are harmonic factor for any (*n*th) harmonic and total harmonic distortion (THD). These are explained below.

1. Harmonic factor of *n*th harmonic (ρ_n). It is defined as the ratio of rms value of *n*th harmonic voltage component, to the rms value of the fundamental voltage component.

$$\rho_n = \left| \dfrac{V_{on}}{V_{01}} \right|$$ ...(13.6)

where V_{on} = rms value of the nth harmonic component of output voltage

and V_{01} = rms value of fundamental component of output voltage.

For example, harmonic factor of 5th harmonic is

$$\rho_5 = \left| \frac{V_{05}}{V_{01}} \right|$$

where V_{05} = rms value of 5th harmonic component of output voltage.

Harmonic factor (HF) is a measure of the contribution of any individual harmonic to the inverter output voltage.

Higher the value of HF for any one harmonic component, greater is the contribution of that particular harmonic component.

2. Total harmonic distortion (THD). It is defined as the ratio of rms value of all the harmonic components, to the rms value of fundamental component.

$$\therefore \qquad THD = \frac{V_{oh}}{V_{01}} \qquad \qquad ...(13.7a)$$

where $V_{oh} = \sqrt{V_{or}^2 - V_{01}^2}$ $\qquad\qquad\qquad$ ...(13.8)

$\qquad$ = rms value of all harmonic components present in the inverter output voltage

and V_{or} = rms value of inverter output voltage, including fundamental plus all the harmonics.

$$\therefore \qquad THD = \frac{\sqrt{V_{or}^2 - V_{01}^2}}{V_{01}} = \left[\left(\frac{V_{or}}{V_{01}} \right)^2 - 1 \right]^{1/2} \qquad ...(13.7b)$$

THD is a measure of the waveform distortion. Lower the value of THD, closer is the waveform to sine wave.

Example 13.8. *A single-phase full-bridge inverter, using transistors and diodes, is feeding a load of $R = 3\Omega$ with input dc voltage of 60V. Calculate (a) rms value of (i) output voltage and (ii) fundamental-component of output voltage (b) output power (c) fundamental-frequency output power (d) average and peak currents of each transistor (a) harmonic factor for third harmonic and (g) THD.*

Solution. Here $V_s = 60$ V, $R = 3$ Ω

(a) Rms value of output voltage, $V_{or} = \sqrt{V_s^2 \times \dfrac{\pi}{\pi}} = V_s = 60$ V

From Eq. (8.26), rms value of fundamental component of output voltage,

$$V_{01} = \frac{4 \times 60}{\sqrt{2} \times \pi} = 54.02 \text{ V}$$

(b) Output power, $\qquad P_0 = \dfrac{V_{or}^2}{R} = \dfrac{60^2}{3} = 1200$ W

(c) Fundamental-frequency output power, $P_{01} = \dfrac{V_{01}^2}{R} = \dfrac{54.02^2}{3} = 972.72$ W

(d) Peak current of each transistor $= \dfrac{60}{3} = 20$ A. Each transistor conducts for π radians for every cycle of 2π radians.

$\therefore$ Average current of each transistor $= 20 \times \dfrac{\pi}{2\pi} = 10$ A

(e) Peak reverse blocking voltage of each transistor $= V_s = 60$ V.

(f) Here $\qquad V_{03} = \dfrac{4V_s}{3 \cdot \sqrt{2} \times \pi}, \quad V_{01} = \dfrac{4V_s}{\sqrt{2} \times \pi}$

$\therefore \qquad\qquad p_3 = \dfrac{4V_s}{3 \cdot \sqrt{2} \times \pi} \times \dfrac{\sqrt{2} \times \pi}{4V_s} = \dfrac{1}{3} = 0.3333$ or 33.33%

(g) From Eq. (13.8), rms value of all harmonic voltages is

$$V_{oh} = \sqrt{60^2 - 54.02^2} = 26.112 \text{ V}$$

From Eq. (13.7a), $\quad THD = \dfrac{V_{oh}}{V_{01}} = \dfrac{26.112}{54.02} = 0.4834$ or 48.34%.

Example 13.9. *A single-phase full bridge inverter, employing transistors, is fed from 220 V dc and output frequency is 50 Hz. Load is RLC with R = 6 Ω, L = 30 mH and C = 180 μF. (a) Calculate THD of the output voltage (b) Obtain an expression for load current in Fourier series. Also, compute (c) THD of the load current (d) load power and average dc source current.*

Considering only the fundamental component of load current, calculate (e) conduction time of each transistor and diode and (f) peak and rms current of each transistor.

Solution. Here $V_s = 220$ V, $f = 50$ Hz, $R = 6$ Ω, $L = 30$ mH and $C = 180$ μF.

$$X_L = 2\pi \times 50 \times 30 \times 10^{-3} = 9.425 \text{ Ω and } X_c = \dfrac{10^6}{2\pi \times 50 \times 180} = 17.684 \text{ Ω}.$$

$$Z_n = \left[6^2 + \left(9.425\, n - \dfrac{17.684}{n} \right)^2 \right]^{1/2} \text{ and } \phi_n = \tan^{-1} \left[\dfrac{9.425\, n - \dfrac{17.684}{n}}{6} \right]$$

(a) Rms value of output voltage, $V_{or} = V_s = 220$ V

Rms value of fundamental component of output voltage, $V_{01} = \dfrac{4 \times 220}{\sqrt{2} \times \pi} = 198.071$ V

Rms value of all harmonic voltages, $V_{oh} = \sqrt{V_{or}^2 - V_{01}^2} = \sqrt{220^2 - 197.071^2} = 95.751$ V

$\therefore \qquad\qquad THD = \dfrac{V_{oh}}{V_{01}} = \dfrac{95.751}{198.071} = 0.4834$ or 48.34%

(b) Fundamental component of load current, $I_{01} = \dfrac{V_{o1}}{Z_1}$

$\therefore \qquad\qquad I_{01} = \dfrac{198.071}{[6^2 + (9.425 - 17.684)^2]^{1/2}} = 19.403$ A

and $\qquad\qquad \phi_1 = \tan^{-1} \left[\dfrac{9.425 - 17.684}{6} \right] = -54°$

$$I_{03} = \dfrac{V_{03}}{Z_3} = \dfrac{4 \times 220}{3\pi \times \sqrt{2}} \times \dfrac{1}{\left[6^2 + \left(9.425 \times 3 - \dfrac{17.684}{3} \right)^2 \right]^{1/2}} = 2.849 \text{ A}$$

and $\qquad\qquad \phi_5 = \tan^{-1} \left[\dfrac{9.425 \times 3 - \dfrac{17.684}{3}}{6} \right] = 75°$

$$I_{05} = \frac{V_{05}}{Z_5} = \frac{4 \times 220}{5\pi \times \sqrt{2}} \times \frac{1}{\left[6^2 + \left(9.425 \times 5 - \frac{17.684}{5}\right)^2\right]^{1/2}} = 0.9 \text{ A}$$

and

$$\phi_5 = \tan^{-1}\left[\frac{9.425 \times 5 - \frac{17.684}{5}}{6}\right] = 82.16°$$

Similarly, $I_{07} = 0.444$ A and $\phi_7 = 84.6°$

Therefore, load current expression in Fourier series is

$i_o (t) = 19.403 \sin (\omega t + 54°) + 2.849 \sin (3\omega t - 75°) + 0.9 \sin$

$(5\omega t - 82.16°) + 0.444 \sin (7\omega t - 84.6°) + \dots$

(c) Peak load current, $I_m = [19.403^2 + 2.849^2 + 0.9^2 + 0.444^2]^{1/2} = 19.637$ A

Rms value of harmonic load current, $I_{oh} = \left[\frac{I_m^2 - I_{m1}^2}{2}\right]^{1/2}$

$$= \left[\frac{19.637^2 - 19.403^2}{2}\right]^{1/2} = 2.1372 \text{ A}$$

$\therefore \qquad THD = \frac{I_{0h}}{I_{01}} = \frac{2.1372 \times \sqrt{2}}{19.403} = 0.155773$ or 15.5773%

(d) Rms load current, $I_{0r} = \frac{I_m}{\sqrt{2}} = \frac{19.637}{\sqrt{2}} = 13.8855$ A

Load power, $P_0 = I_{or}^2 \times R = 13.8855^2 \times 6 = 1156.843$ W

Average value of source current $= \frac{P_0}{V_s} = \frac{1156.843}{220} = 5.2584$ A

(e) Expression for fundamental component of load current, $i_{01} = 19.403 \sin (\omega t + 54°)$. It shows that current leads the voltage by 54°.

$\therefore$ Conduction time of each transistor $= (180 - 54) \times \frac{\pi}{180} \times \frac{1}{2\pi \times 50} = 7.0$ ms

Conduction time of each diode $= \frac{1}{2f} - 7.0 \times 10^{-3} = \frac{1}{100} - 7 \times 10^{-3} = 3$ ms

(f) Peak transistor current, $I_p = I_{01m} = 19.403$ A

Since each transistor conducts for 126° for every 360° of output cycle, rms value of transistor current, from Example 8.6 is

$$I_{T1} = \frac{I_{01m}}{2\sqrt{\pi}}\left[126° \times \frac{\pi}{180} - \frac{\sin 252}{2}\right]^{1/2} = 0.46135 \, I_{01m}$$

$$= 0.46135 \times 19.403 = 8.952 \text{ A}$$

Example 13.10. *A 3-phase 180°-mode bridge inverter has star-connected load of R = 4 Ω and L = 20 mH. The inverter is fed from 220 V dc and its output frequency is 50 Hz. Obtain Fourier series expressions for line voltage v_{ab} (t), phase voltage v_a (t) and line current i_a (t). Also, calculate (a) rms values of phase and line voltages (b) rms value of fundamental component of phase and line voltages (c) THD for voltages (d) load power and average source current and (e) average value of thyristor current.*

Solution. Here $V_s = 220$ V, $R = 4$ Ω, $L = 20$ mH, $\omega L = 2\pi \times 50 \times 20 \times 10^{-3} = 6.283$ Ω

From Eq. (8.44), Fourier series expression for line voltage v_{ab} is

$$v_{ab}(t) = \frac{4V_s}{\pi}\left[\cos\frac{\pi}{6}\sin(\omega t + 30) + \frac{1}{3}\cos\frac{\pi}{2}\sin(3\omega t + 90°) + \frac{1}{5}\cos\frac{5\pi}{6}\sin(5\omega t + 150°)\right.$$

$$\left. + \frac{1}{7}\cos\frac{7\pi}{6}\sin(7\omega t + 210°) + \frac{1}{11}\frac{11\cdot\pi}{6}\sin(11\,\omega t + 330°)\right]$$

$$= \frac{4\times 220}{\pi}\left[\frac{\sqrt{3}}{2}\sin(\omega t + 30°) + 0 - \frac{1}{5}\cdot\frac{\sqrt{3}}{2}\sin(5\omega t + 150°) - \frac{1}{7}\cdot\frac{\sqrt{3}}{2}\cos(7\omega t + 210°)\right.$$

$$\left. + \frac{1}{\pi}\times\frac{\sqrt{3}}{2}\sin(11\omega t + 330°)\right]$$

$$= 242.585\sin(\omega t + 30°) - 48.517\sin(5\omega t + 150°) - 34.655\sin(7\omega t + 210°)$$
$$+ 22.053\sin(11\omega t + 330°)$$

From Eq. (8.47), phase voltage $v_a(t)$ in Fourier series is

$$v_a(t) = \frac{2V_s}{\pi}\left[\sin\omega t + \frac{1}{5}\sin 5\omega t + \frac{1}{7}\sin 7\omega t) + \frac{1}{11}\sin 11\omega t + ...\right]$$

Here $\quad \dfrac{2V_s}{\pi} = \dfrac{2\times 220}{\pi} = 140.06$ V

∴ $\quad v_a(t) = 140.06\sin\omega t + 28.011\sin 5\omega t + 20.008\sin 7\omega t + 12.732\sin 11\omega t + ...$

$$Z_n = [R^2 + (n\omega L)^2]^{1/2} = [4^2 + (6.283\,n)^2]^{1/2} \text{ and } \phi_n \doteq \tan^{-1}\left[\frac{6.283\,n}{4}\right]$$

Here $\quad I_{01m} = \dfrac{140.06}{[4^2 + 6.283^2]^{1/2}}\bigg/ -\tan^{-1}\dfrac{6.283}{4} = 18.804\angle -57.52°$

$\quad I_{05m} = \dfrac{28.011}{(4^2 + (6.283\times 5)^2)^{1/2}}\bigg/ -\tan^{-1}\dfrac{6.283\times 5}{4} = 0.8845\angle -82.744°$

Similarly $\quad I_{07m} = 0.455\angle -84.803°$ and $I_{0.11m} = 0.184\angle -86.68°$

∴ $\quad i_a(t) = 18.804\sin(\omega t - 57.52°) + 0.8845\sin(\omega t - 82.744°)$
$$+ 0.455\sin(7\omega t - 84.803°) + 0.184\sin(11\omega t - 86.68°)$$

(a) From Eq. (8.51), rms value of phasor voltage,

$$V_p = \frac{\sqrt{2}}{3}V_s = \frac{\sqrt{2}\times 220}{3} = 103.71 \text{ V}$$

From Eq. (8.50), rms value of line voltage,

$$V_L = \sqrt{3}\,V_p = \sqrt{\frac{2}{3}}\times 220 = 179.63 \text{ V}$$

(b) From Eq. (8.52), fundamental component of phase voltage is

$$V_{p1} = \frac{\sqrt{2}\cdot V_s}{\pi} = \frac{\sqrt{2}\times 220}{\pi} = 99.035 \text{ V}$$

Fundamental component of line voltage,

$$V_{L1} = \sqrt{3}\cdot V_{p1} = \sqrt{3}\times 99.035 = 171.534 \text{ V}$$

(c) Rms value of all harmonic voltages,

$$V_{oh} = [V_L^2 - V_{L1}^2]^{1/2} = [179.63^2 - 171.534^2]^{1/2} = 53.32 \text{ V}$$

∴ $\quad THD = \dfrac{V_{oh}}{V_{L1}} = \dfrac{53.32}{171.534} = 0.3108421$ or 31.08421%

(*d*) Rms value of phase, or line, current

$$I_{or} = \left[\frac{18.804^2 + 0.8845^2 + 0.455^2 + 0.184^2}{2} \right]^{1/2} = 18.831 \text{ A}$$

Load power $= 3 \times I_{or}^2 \cdot R = 3 \times 18.831^2 \times 4 = 4255.3 \text{ W}$

Average value of source current, $I_s = \dfrac{4255.3}{220} = 19.342 \text{ A}$

(*e*) For calculating the average thyristor current, Fig. 13.7(*a*) is drawn for step-I from **Fig.** 8.21 (*a*) for a 3-phase bridge inverter. For average source current I_s, current in SCR1 = current in SCR5 = $I_s/2$, as these two SCRs 1 and 5 are in parallel and current I_s is equally shared. For SCR1, current i_{T1} is shown $I_s/2$ for $\pi/3$ in Fig. 13.7 (*b*) for step I which is of $\pi/3$ radians duration. Reference to Fig. 8.21 shows that $i_{T1} = I_s$ for step II and $i_{T1} = I_s/2$ for step III ; **current** i_{T1} is sketched accordingly in Fig. 13.7 (*b*). Periodicity of i_{T1} is 2π radians.

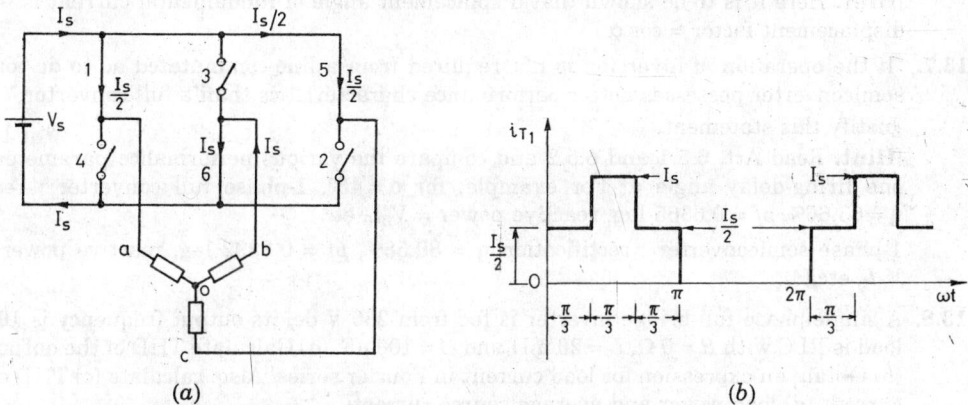

Fig. 13.7. Pertaining to three-phase bridge inverter ; Example 13.10.

From Fig. 13.7 (*b*), average value of thyristor current is given by

$$I_{TA} = \frac{1}{2\pi} \left[\left(\frac{I_s}{2} \right) \frac{\pi}{3} + I_s \cdot \frac{\pi}{3} + \left(\frac{I_s}{2} \right) \cdot \frac{\pi}{3} \right] = \frac{I_s}{3} = \frac{19.342}{3} = 6.4473 \text{ A}$$

PROBLEMS

13.1. What is thyristor controlled inductor ? Explain, how the inductance seen by the source can be altered in TCI. Illustrate your answer with appropriate waveforms and phasor diagrams. Hence, for any firing-angle delay, derive an expression for the reactive power handled by TCI.

13.2. What is SVC ? Describe how SVC regulates the reactive power flow and improves the **system** power factor. Illustrate your answer with relevant phasor diagrams.

Derive an expression for the resultant current drawn by an SVC and comment on the **flow of** reactive power.

13.3. (*a*) Derive an expression for the effective inductance in case of TCI.

(*b*) A TCI, fed from 6.6 kV, 50 Hz source, has an inductance of 1H. Calculate reactive **power** handled by TCI for firing angles of 90°, 120°, 150°, 175° and 180°.

[**Ans.** (*b*) 138.656 kVAr, 54.215 kVAr, 7.996 kVAr, 0.039 kVAr, **zero**]

13.4. A single-phase load of 20 kW+ j 12 kVAr is fed from 400 V, 50 Hz source. TCI in SVC has an inductance of 0.8 H. For a firing- angle delay of 120°, it is found that the system operates at unity *pf*. Find the value of capacitance of fixed capacitor and kVAr delivered by it.

[**Ans.** 243.68452 µF, 12248.92 VAr]

13.5. Which device will be the optimum choice for the following applications ? Give reasons for your choice.

(a) Single-stage, 50 kVA, self-commutated *ac* to *dc* converter with variable output voltage.

(b) Single-stage, 50 kVA, line-commutated *ac* to *dc* converter with variable output voltage.

(c) 200 VA self-commutated *dc* to *ac* inverter having 100 kHz as device switching frequency.

(d) 5 kVA self-commutated dc to ac inverter having 5 kHz as device switching frequency.

(e) 2 MVA self-commutated dc to ac inverter having 400 Hz as device switching frequency.

(f) 300 kVA self-commutated dc to ac inverter with 100 kHz as the device switching frequency.   [**Ans.** (a) GTO (b) SCR (c) PMOSFET (d) BJT (e) GTO (f) SITH.]

13.6. Show that for a 3-phase full converter, if the firing-angle delay is α, the fundamental component of the input phase current lags the respective phase voltage by angle α.

[**Hint.** Here it is to be shown that displacement angle of fundamental current is $-\alpha$ so that displacement factor = $\cos \alpha$.]

13.7. "If the operation of inversion is not required from a line-commutated ac to dc converter, a semiconverter possesses better performance characteristics than a full converter."

Justify this statement.

[**Hint.** Read Art. 6.5.1 and 6.5.2 and compare the various performance parameters for any one firing-delay angle α. For example, for $\alpha = 45°$, 1-phase full converter : rectification $\eta = 63.66\%, pf = 0.6365$ lag, reactive power = $V_0 I_0$ etc.

1-phase semiconverter : rectification $\eta = 80.58\%$, *pf* = 0.8892 lag, reactive power = 0.4142 $V_0 I_0$ etc.]

13.8. A single-phase full-bridge inverter is fed from 230 V dc, its output frequency is 100 Hz and load is RLC with $R = 6 \ \Omega$, $L = 20$ mH and $C = 100$ µF. (a) Calculate THD of the output voltage. (b) Obtain an expression for load current in Fourier series. Also, calculate (c) THD of the load current (d) load power and average source current.

Considering only the fundamental component of load current, calculate (e) conduction time of each thyristor and diode and (f) peak and rms current of each thyristor.

[**Ans.** (a) 48.346% (b) 30.124 sin (ωt + 29.205°) + 2.095 sin ($3\omega t$ − 79.506)

+ 0.691 sin ($5\omega t$ − 84.256°) + 0.3444 sin ($7\omega t$ − 88°). (c) 7.428%

(d) 2737.5 W, 11.9022 A (e) 4.189 ms ; 0.811 ms (f) 30.124 A ; 15.861 A]

13.9. A 3-phase 120°-mode bridge inverter feeds a star-connected load of $R = 5 \ \Omega$. DC source voltage is 230 V and output frequency is 50 Hz. Obtain Fourier series expression for line voltage $v_{ab} (t)$, phase voltage $v_a (t)$ and line current $i_a (t)$. Also, calculate (a) rms value of phase and line voltages (b) rms value of fundamental component of phase and line voltages (c) THD for voltages (d) load power and average source current and (e) average and rms value of thyristor current.

[**Ans.** $v_{ab} (t) = 219.634$ sin (ωt + 60°) + 43.93 sin ($5\omega t$ + 300°) + 31.376 sin ($7\omega t$ + 60°)

+ 19.967 sin ($11\omega t$ + 300°) ; $v_a (t) = 126.802$ sin (ωt + 30°) − 25.36 sin ($5\omega t$ + 150°)

− 18.115 sin ($7\omega t$ + 210°) + 11.53 sin ($11\omega t$ + 330°) ; $i_a (t) = 25.3604$ sin (ωt + 30°)

− 5.072 sin ($5\omega t$ + 150°) − 3.623 sin ($7\omega t$ + 210°) + 2.306 sin ($11\omega t$ + 330°).

(a) 93.897 V, 162.635 V (b) 89.665 V, 155.304 V (c) 31.0864%

(d) 10312.33 W, 44.836 A (e) 7.667 A, 13.28 A]

Appendix : A

Fourier Analysis

In power electronics, voltage and current waveforms are usually non-sinusoidal. These waveforms are, however, periodic in nature. A periodic function is one which repeats itself after regular intervals of time. A function $f(t)$ is said to be periodic function of time if

$$f(t) = f(t + T) \qquad \qquad ...(A-1)$$

where T = period, or periodic time of one cycle of $f(t)$. In power electronics, let $f(t)$ be the output voltage function of a power converter. Therefore, Eq. $(A.1)$ is written as

$$v_o(t) = v_o(t + T) \qquad \qquad ...(A.1)$$

If f is the frequency of output voltage in hertz, then $f = \dfrac{1}{T}$ and angular frequency ω is given by

$$\omega = 2\pi f = \frac{2\pi}{T} \qquad \qquad ...(A.2)$$

Eq. $(A.1)$ can now be written as

$$v_o(\omega t) = v_o(\omega t \pm 2\pi) \qquad \qquad ...(A.3)$$

A periodic function $v_o(t)$ with period T can be expanded into fourier series as under :

$$v_o(t) = a_0 + a_1 \cos \omega t + a_2 \cos \omega t + ... + a_n \cos n\omega t$$
$$+ b_1 \sin \omega t + b_2 \sin 2\omega t + ... + b_n \sin n\omega t$$
$$= a_0 + \sum_{n-1,2,3}^{\infty} (a_n \cos n\omega t + b_n \sin n\omega t) \qquad \qquad ...(A.4)$$

where a_o is the average value, or dc component, of the periodic wave over a complete cycle. In Eq. $(A.4)$, coefficients a_o, a_n and b_n can be determined from the following expressions :

a_o = average value, or dc value, of $v_o(t)$ over a period.

$$= \frac{1}{T} \int_0^T v_o(t)\, dt = \frac{1}{2\pi} \int_0^{2\pi} v_o(\omega t) . d(\omega t) \qquad \qquad ...(A.5)$$

a_n = amplitude of cos $n\omega t$ component of $v_o(t)$

$$= \frac{2}{T} \int_0^T v_o(t). \cos n\omega t . dt = \frac{1}{\pi} \int_0^{2\pi} v_o(\omega t). \cos n\omega t. d(\omega t) \qquad \qquad ...(A.6)$$

b_n = amplitude of sin $n\omega t$ component of $v_o(t)$

$$= \frac{2}{T} \int_0^T v_o(t) \sin n\omega t . dt) = \frac{1}{\pi} \int_0^{2\pi} v_o(\omega t) . \sin n\omega t . d(\omega t) \qquad \qquad ...(A.7)$$

Sine and cosine terms of Eq. $(A.4)$ can be combined into one term as under :

$$v_{on}(t) = a_n \cos n\,\omega t + b_n \sin n\omega t$$
$$= \sqrt{a_n^2 + b_n^2} \left[\frac{a_n}{\sqrt{a_n^2 + b_n^2}} \cos n\omega t + \frac{b_n}{\sqrt{a_n^2 + b_n^2}} \sin n\omega t \right]$$

Let us define θ_n as shown in Fig. *A-1.* $\therefore \theta_n = \tan^{-1}\left(\dfrac{a_n}{b_n}\right)$

$\therefore \qquad v_{on}(t) = C_n \left[\sin\theta_n \cos n\omega t + \cos\theta_n \sin n\omega t\right]$

$\qquad\qquad\qquad = C_n \sin(n\omega t + \theta_n)$

$\therefore \qquad v_o(t) = a_o + \sum\limits_{n=1,2,3.}^{\infty} C_n \sin(n\omega t + \theta_n)$ ...(A-8)

Fig. A-1. Pertaining to angle θ_n.

where $\qquad\qquad C_n = \sqrt{a_n^2 + b_n^2}$ and $\theta_n = \tan^{-1}\left(\dfrac{a_n}{b_n}\right)$ ...(A-9)

Here C_n and θ_n are respectively, the amplitude and the delay angle of the nth harmonic component of output voltage $v_o(t)$.

A.1 Half-wave or Mirror Symmetry. A waveform possesses half-wave symmetry if

$$\left.\begin{array}{l} v_o(t) = -v_o(t + T/2) \\ \text{or } v_o(\omega t) = -v_o(\omega t + \pi) \end{array}\right\} \qquad \text{...(A.10)}$$

Eq. (A.10) implies that for every positive value of $v_o(t)$ in a period, there is a corresponding negative value of $v_o(t)$, of the same magnitude at a distance $\dfrac{T}{2}$ or π. Waveforms shown in Fig. A.2 possess half-wave symmetry. The distance $\dfrac{T}{2}$, or π, can be taken anywhere, positive value of $v_o(t)$ must be equal to negative value of $v_o(t)$ for half-wave symmetry.

Also, a wave possesses half-wave symmetry if negative half-cycle is the mirror image of the positive half-cycle displaced horizontally by a distance $T/2$ or π.

Since areas of positive and negative half cycles are identical for waveforms having half-wave symmetry, average value of waveform $a_o = 0$. Also, a waveform with half-wave symmetry contains only odd harmonics (*i.e.* $n = 1, 3, 5...$).

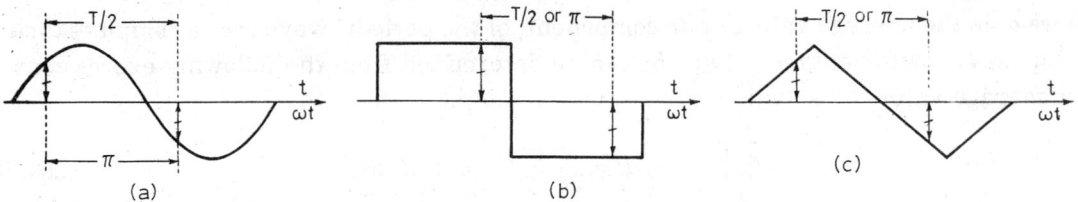

Fig. A.2. Pertaining to half-wave symmetry.

Summarizing, a waveform with half-wave symmetry has the following attributes :

(*i*) $a_o = 0$

(*ii*) From Eq. (A-6), for $n = 1, 3, 5, \ldots\ldots$

$$a_n = \frac{2}{T}\int_0^T v_o(t)\cos n\omega t\, dt = \frac{1}{\pi}\int_0^{2\pi} v_o(\omega t)\cdot \cos n\omega t\cdot d(\omega t) \qquad \text{...(A.11)}$$

(*iii*) From Eq. (A.7), for $n = 1, 3, 5,\ldots$

$$b_n = \frac{2}{T}\int_0^T v_o(t)\sin n\omega t\, dt = \frac{1}{\pi}\int_0^{2\pi} v_o(\omega t)\sin n\omega t\cdot d(\omega t) \qquad \text{...(A.12)}$$

$$\therefore \qquad v_o(t) = \sum\limits_{n=1,3,5}^{\infty} C_n \sin(n\,\omega t + \theta_n) \qquad \text{...(A.13)}$$

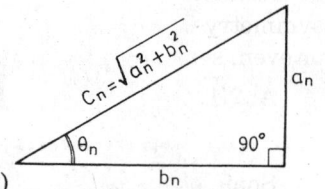

A.2. Odd and Even functions. It is seen from above that a waveform having half-wave symmetry is free from even harmonics and its average value is zero. When a waveform is **odd** or **even**, some more terms from Fourier series vanish.

A. 2.1. Odd function. A function is said to be an odd function if

$$v_o(t) = - v_o(-t)$$

Some of the odd functions are shown in Fig. A.3. For a point P, there is a corresponding point P' at an equal distance from the origin $t = 0$, or $\omega t = 0$. Note that the magnitude of $v_o(t)$ at $P = -$ magnitude of $v_o(t)$ at P'.

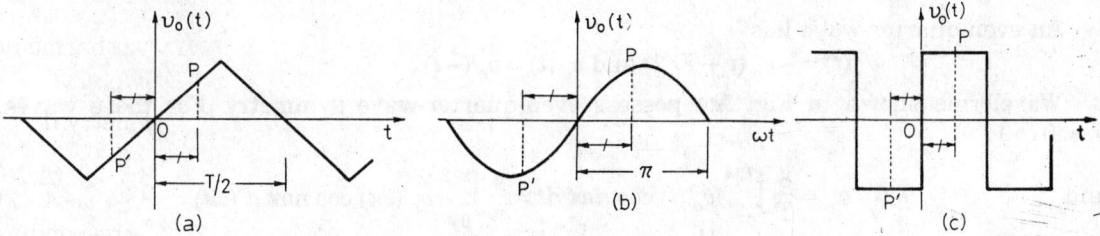

Fig. A.3. Waveforms for odd functions

If the positive cycle is folded about the horizontal reference axis and then again folded **about** the zero vertical axis, positive cycle will overlap the negative half cycle.

This shows that odd function has (a) symmetry about time origin and (b) symmetry about the zero-vertical axis. Thus, odd functions (i) possess half-wave symmetry (ii) **are** symmetry about horizontal reference line, or t-axis, or ωt-axis, (iii) are symmetrical about zero vertical axis (iv) have $a_o = 0$ and $a_n = 0$ (v) contain sine terms only and (vi)

$$b_n = \frac{4}{T} \int_0^{T/2} v_o(t) \sin n\, \omega t\, . \, dt = \frac{2}{\pi} \int_0^{\pi} v_o(\omega t) \sin n\omega t\, d(\omega t) \qquad \text{...(A.14)}$$

A.2.2. Even Function. A function is said to be an even function if $v_o(t) = v_o(-t)$.

Some even functions are shown in Fig. A.4. It is observed that even function exhibits **mirror** symmetry about the vertical axis at origin $t = 0$, or $\omega t = 0$, and also with respect to all **vertical** axis at a distance $nT/2$, or $n\pi$ from the origin.

Thus, even functions (i) possess mirror symmetry about vertical axis only

(ii) have $b_n = 0$

(iii) Contains cosine terms only and

(iv) $a_n = \frac{4}{T} \int_0^{T/2} v_o(t) \cos n\omega t = \frac{2}{\pi} \int_0^{\pi} v_o(\omega t). \cos n\omega t.\, d(\omega t) \qquad \text{...(A.15)}$

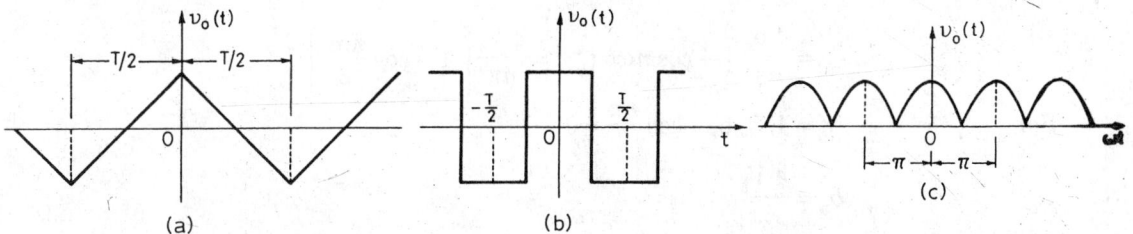

Fig. A.4. Waveforms having even symmetry.

A.2.3. Quarter-wave Symmetry. A combination of half-wave symmetry and odd symmetry gives a function odd-quarter-wave symmetry. Therefore, an odd-quarter wave should have

$$v_o(t) = -v_o\left(t + \frac{T}{2}\right) \text{ and } v_o(t) = -v_o(-t)$$

The waveforms shown in Fig. A.3 possess odd quarter-wave symmetry. For these waves; $a_0 = 0$, $a_n = 0$ and

$$b_n = \frac{8}{T}\int_0^{T/4} v_o(t) \sin n\omega t \,.\, dt = \frac{4}{\pi}\int_0^{\pi/2} v_o(\omega t)\,.\, \sin n\omega t\,.\, d(\omega t) \qquad ...(A.16)$$

An even-quarter-wave has

$$v_o(t) = -v_o(t + T/2) \text{ and } v_o(t) = v_o(-t)$$

Waveforms shown in Fig. A.4 possess even-quarter-wave symmetry. For these waves, $b_n = 0$

and

$$a_n = \frac{8}{T}\int_0^{T/4} v_o(t) \cos n\omega t\, dt = \frac{4}{\pi}\int_0^{\pi/2} v_o(\omega t) \cos n\omega t\, d(\omega t) \qquad ...(A.17)$$

A.3. Applications. Now some **waveforms** encountered in power electronic circuits are resolved into Fourier series.

A.3.1. Square wave. The square-wave voltage shown in Fig. A.5 (a) is obtained as the **output** voltage wave from a single-phase **full-bridge** inverter. It is a pure *ac* voltage wave.

Symmetry conditions :- (*i*) An areas of positive and negative half cycles are equal, $a_o = 0$

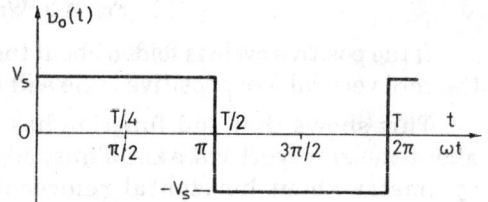

Fig. A.5 (a) Square-wave voltage.

(*ii*) $v_o(t = T/4) = V_s$ and $v_o\left(t = \frac{T}{4} + \frac{T}{2}\right) = -V_s$. As $v(t) = -v_o\left(t + \frac{T}{2}\right)$, the wave has half-wave symmetry, therefore only add harmonics are present in the Fourier series expansion.

(*iii*) $v_o\left(t = \frac{T}{4}\right) = V_s$ and $v_o(t = -T/4) = -V_s$. As $v_o(t) = -v_o(-t)$, it is odd function. It is observed from (*ii*) and (*iii*) above, that square wave of Fig. A-5 (a) possesses odd-quarter wave symmetry. Thus, $a_o = 0$ and $a_n = 0$

$$b_n = \frac{8}{T}\int_0^{T/4} V_s \,.\, \sin n\omega t \,.\, dt$$

or

$$b_n = \frac{8}{2\pi}\int_0^{\pi/2} V_s \sin n\omega t \,.\, d(\omega t)$$

$$= \frac{4\,v_s}{n\pi}\,|-\cos n\omega t\,|_0^{\pi/2} = \frac{4\,V_s}{n\pi}\left[1 - \cos\frac{\pi.n}{2}\right]$$

For

$$n = 1, 3, 5,\dots \cos\frac{n\pi}{2} = 0$$

$$\therefore \qquad b_n = \frac{4\,V_s}{n\pi}$$

$$\therefore \qquad v_o(t) = a_o + \sum_{n=1,3,5}^{\infty} a_n \cos n\omega t + b_n \sin n\omega t$$

$$= \sum_{n=1,3,5,}^{\infty} \frac{4 V_s}{n\pi} \sin n\omega t \qquad ...(A.18)$$

$$= \frac{4 V_s}{\pi}\left[\sin \omega t + \frac{1}{3} \sin 3\omega t + \frac{1}{5} \sin 5\omega t + \frac{1}{7} \sin 7\omega t - ... \right]$$

$$= v_{01} + v_{03} + v_{05} + v_{07} + ...$$

Rms value of nth harmonic component of square wave

$$V_{on} = \frac{4 V_s}{\sqrt{2} . n\pi} = \frac{1}{n} [0.900 V_s]$$

Rms value of fundamental component,

$$V_{01} = 0.900 v_s$$

Rms value of third-harmonic component,

$$V_{03} = 0.300 V_s \text{ and so on.}$$

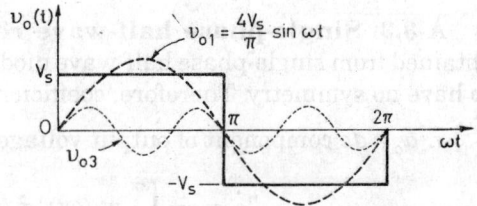

Fig. A.5. (b) Summation of fundamental v_{01}, third harmonic v_{03}, v_{05} etc gives square wave.

The sum of the ordinates of the various harmonic components of Fourier series *i.e.* $v_{01} + v_{03} + v_{05} + v_{07} + ...$ gives the actual square wave. In Fig. A.5 (b), only v_{01} and v_{03} are shown.

Note that $\qquad \dfrac{V_{on}}{V_{01}} = \dfrac{1}{n}$

or $\qquad V_{on} = \dfrac{1}{n} . V_{01}.$

A.3.2. Quasi-square Wave. This type of waveform shown in Fig. A-6 is obtained as the output voltage from a single-phase full-bridge inverter with single-pulse modulation. This waveform is also specified by the conditions of square wave.

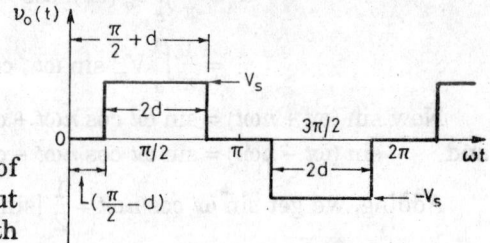

Fig. A.6. Quasi-square wave.

Therefore, $a_o = 0$ and $a_n = 0$.

$$\therefore \quad b_n = \frac{4}{\pi} \int_0^{\pi/2} v_o (\omega t) . \sin n\omega t . d (\omega t)$$

$$= \frac{4}{\pi} \int_{\left(\frac{\pi}{2}-d\right)}^{\pi/2} V_s . \sin n\omega t . d (\omega t) = \frac{4 V_s}{n . \pi} \left| - \cos n\omega t \right|_{\frac{\pi}{2}-d}^{\frac{\pi}{2}}$$

$$= \frac{4 V_s}{n\pi}\left[\cos \frac{n\pi}{2} \cos nd + \sin \frac{n\pi}{2} \sin nd + \cos \frac{n\pi}{2} \right]$$

For all values of odd n, $\cos \dfrac{n\pi}{2}$ is zero.

$$\therefore \quad b_n = \frac{4 V_s}{n\pi} . \sin \frac{n\pi}{2} \sin nd$$

$$\therefore \quad v_o (t) = \sum_{n=1,3,5}^{\infty} \frac{4 V_s}{n\pi} . \sin \frac{n\pi}{2} \sin nd . \sin n\omega t \qquad ...(A.19)$$

or $$v_o(t) = \frac{4 V_s}{\pi}\left[\sin d \sin \omega t - \frac{1}{3}\sin 3d \sin 3\omega t + \frac{1}{5}\sin 5d \sin 5\omega t - \frac{1}{7}\dots \right]$$

Here $$V_{on} = \frac{4 V_s}{\sqrt{2}\,.\,n\pi}.\sin\frac{n\pi}{2}\sin nd, \; V_{01} = \frac{4 V_s}{\sqrt{2}\,.\,\pi}\sin d$$

$$\therefore \quad \frac{V_{on}}{V_{01}} = \frac{1}{n}\frac{\sin nd}{\sin d}.\sin\frac{n\pi}{2} \text{ or } V_{on} = \left[\frac{1}{n}\frac{\sin nd}{\sin d}.\sin\frac{n\pi}{2}\right].V_{01}$$

A-3.3. Single-phase half-wave rectified voltage. Rectified output voltage wave obtained from single-phase half-wave diode rectifier is shown in Fig. A.7. This waveform is seen to have no symmetry. Therefore, coefficient a_o, a_n and b_n would have to be determined.

$\therefore a_o = dc$ component of output voltage, $v_o(t)$

$$= \frac{1}{2\pi}\int_0^{2\pi} v_o(\omega t).\,d(\omega t)$$

$$= \frac{1}{2\pi}\left[\int_0^{\pi} V_m \sin\omega t.\,d(\omega t) + 0\right]$$

$$= \frac{V_m}{\pi} \qquad\qquad \dots(i)$$

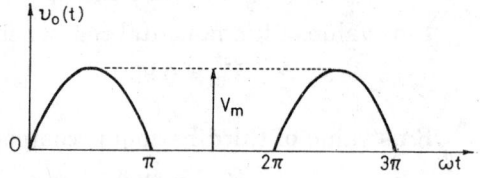

Fig. A.7. Single-phase half-wave rectified voltage wave.

$$a_n = \frac{1}{\pi}\int_o^{\pi} v_o(\omega t).\cos n\omega t.\,d(\omega t)$$

$$= \frac{1}{\pi}\int_o^{\pi} V_m \sin\omega t.\cos n\omega t.\,d(\omega t)$$

Now $\sin(\omega t + n\omega t) = \sin\omega t\cos n\omega t + \cos\omega t.\sin n\omega t$

and $\sin(\omega t - n\omega t) = \sin\omega t\cos n\omega t - \cos\omega t.\sin n\omega t$

Adding, we get $\sin\omega t\cos n\omega t = \frac{1}{2}[\sin(1+n)\omega t + \sin(1-n)\omega t]$

$$\therefore \quad a_n = \frac{V_m}{2\pi}\int_o^{\pi}\{\sin(1+n)\omega t + \sin(1-n)\omega t\}.\,d(\omega t)$$

$$= \frac{V_m}{2\pi}\left[\frac{1-\cos(1+n)\pi}{1+n} + \frac{1-\cos(1-n)\pi}{1-n}\right]$$

$$= \frac{V_m}{2\pi}\left[\frac{1-n+1+n}{1-n^2} - \frac{\cos\pi\cos n\pi - n\cos\pi\cos n\pi + \cos\pi\cos n\pi + n\cos\pi\cos n\pi}{1-n^2}\right]$$

$$= \frac{V_m}{2\pi}\left[\frac{2}{1-n^2} + \frac{2\cos n\pi}{1-n^2}\right]$$

$$= \frac{V_m}{\pi}\left[\frac{1+\cos n\pi}{1-n}\right] = \frac{V_m}{\pi}\left[\frac{1+(-1)^n}{1-n^2}\right], \text{ as } \cos n\pi = (-1)^n$$

For $n = 1$, a_1 is indeterminate.

For n odd, $a_n = 0$ and for n even, $a_n = -\dfrac{2 V_m}{\pi(n^2 - 1)}$

$$a_1 = \frac{1}{\pi}\int_0^{\pi} V_m \sin\omega t\cos\omega t.\,d(\omega t)$$

$$= \frac{V_m}{2\pi}\int_0^{\pi}\sin 2\omega t\,d(\omega t) = 0.$$

$$b_n = \frac{1}{\pi} \int_0^\pi V_m \sin \omega t \cdot \sin n\omega t \cdot d\,(\omega t)$$

From trigonometry, $\sin \omega t \sin n\omega t = \dfrac{1}{2} [\cos (1-n)\,\omega t - \cos (1+n)\,\omega t]$

$$\therefore \qquad b_n = \frac{V_m}{2\pi} \int_0^\pi \{\cos (1-n)\,\omega t - \cos (1+n)\,\omega t\}\, d(\omega t)$$

$$= \frac{V_m}{2\pi} \left[\left| \frac{\sin (1-n)\,\omega t}{1-n} - \frac{\sin (1+n)\,\omega t}{1+n} \right|_0^\pi \right] = 0$$

The above expression is indeterminate for $n = 1$ for b_1.

$$\therefore \qquad b_1 = \frac{1}{\pi} \int_0^\pi V_m \sin^2 \omega t\, d\,(\omega t) = \frac{V_m}{2}$$

$$\therefore \qquad v_o\,(t) = a_o + \sum_{n=1}^{1} b_n \sin n\omega t + \sum_{n=2,4,6}^{\infty} a_n \cos n\omega t$$

$$= \frac{V_m}{\pi} + \frac{V_m}{2} \sin \omega t - \frac{2\,V_m}{\pi} \sum_{n=2,4,6}^{\infty} \frac{1}{n^2-1} \cos n\omega t \qquad (A.20\ a)$$

$$= \frac{V_m}{\pi} + \frac{V_m}{2} \sin \omega t - \frac{2\,V_m}{3\,\pi} \cos 2\omega t - \frac{2\,V_m}{15\,\pi} \cos 4\omega t - \frac{2\,V_m}{35\,\pi} \cos 6\omega t \qquad ...(A.20\ b)$$

A.3.4. Single-phase Full-wave Rectified voltage. Rectified voltage obtained from a single-phase full-wave diode rectifier is shown in Fig. A.8. Such waveforms are encountered in chapter 3.

Waveform of Fig. A-8 may be treated as the sum of two waveforms shown in Fig. A.9. Waveform v_{02} is shown to lag v_{01} waveform by π radians but $v_0 = v_{01} + v_{02}$.

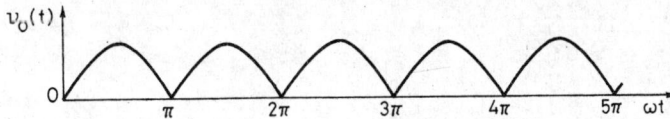

Fig. A.8. Single-phase two-pulse rectified voltage wave.

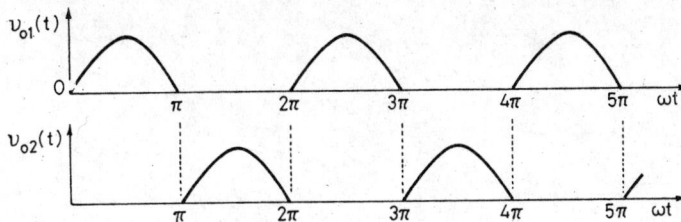

Fig. A.9. Sum of $v_{01}\,(t)$ and $v_{02}\,(t)$ gives $v_o\,(t)$ of Fig. A.8.

Voltage $v_{01}\,(t)$, identical with $v_o\,(t)$ and given by Eq. (A.20 a), is expressed as

$$v_{01}\,(t) = \frac{V_m}{\pi} + \frac{V_m}{2} \sin \omega t - \frac{2\,V_m}{\pi} \sum_{n=2,4,6}^{\infty} \frac{1}{n^2-1} \cos n\omega t \qquad ...(A.21\ a)$$

Waveform v_{02} lags waveform v_{01} by π radians. Therefore, Fourier series expression for v_{02} can be obtained from Eq. (A.21 a) by putting $(\omega t - \pi)$ for ωt.

$$\therefore \quad v_{02}(t) = \frac{V_m}{\pi} + \frac{V_m}{2} \sin(\omega t - \pi) - \frac{2V_m}{\pi} \sum_{n=2,4,6}^{\infty} \frac{1}{n^2 - 1} \cos n(\omega t - \pi)$$

$$= \frac{V_m}{\pi} - \frac{V_m}{2} \sin \omega t - \frac{2V_m}{\pi} \sum_{n=2,4,6}^{\infty} \frac{\cos n\omega t}{n^2 - 1} \qquad \qquad ...(A.21b)$$

By adding Eqs. (A.21 a) and (A.21 b), we get

$$v_o(t) = \frac{2V_m}{\pi} - \frac{4V_m}{\pi} \sum_{n=2,4,6}^{\infty} \frac{\cos n\omega t}{n^2 - 1} \qquad \qquad ...(A.22\ a)$$

or

$$v_o(t) = \frac{2V_m}{\pi} - \frac{4V_m}{3\pi} \cos 2\omega t - \frac{4V_m}{15\pi} \cos 4\omega t - \frac{4V_m}{35\pi} \cos 6\omega t \qquad \qquad ...(A.22\ b)$$

Appendix : B

Laplace Transforms

Time domain, $f(t)$: time domain	Laplace transform, $F(s)$: s-domain
1. Impulse function : $\delta(t)$	1. 1
2. Unit step function : $\acute{U}(t)$	2. $\dfrac{1}{s}$
3. t	3. $\dfrac{1}{s^2}$
4. e^{-at}	4. $\dfrac{1}{s+a}$
5. $\sin \omega t$	5. $\dfrac{\omega}{s^2+\omega^2}$
6. $\cos \omega t$	6. $\dfrac{s}{s^2+\omega^2}$
7. $t.e^{-at}$	7. $\dfrac{1}{(s+a)^2}$
8. $e^{-at}.\sin \omega t$	8. $\dfrac{\omega}{(s+a)^2+\omega^2}$
9. $e^{-at}.\cos \omega t$	9. $\dfrac{s+a}{(s+a)^2+\omega^2}$
10. $\dfrac{e^{-at}-e^{-bt}}{b-a}$	10. $\dfrac{1}{(s+a)(s+b)}$
11. $1-e^{-t/T}$	11. $\dfrac{1}{s(1+sT)}$
12. $1-\cos \omega t$	12. $\dfrac{\omega}{s(s^2+\omega^2)}$

Appendix : C

Objective Type Questions

In this appendix, multiple choice questions pertaining to Chapters 2 to 13 are given. The questions are so framed that four alternatives are provided, out of which one correct answer is to be tick-marked.

POWER SEMICONDUCTOR DIODES AND TRANSISTORS

1. For a diode, reverse recovery time is defined as the time between the instant diode current becomes zero and the instant reverse recovery current decays to
   (a) zero
   (b) 10% of reverse peak current I_{RM}
   (c) 25% of I_{RM}
   (d) 15% of I_{RM}

2. In a diode, the cut-in voltage and forward-voltage drop are respectively
   (a) 0.7 V, 0.7 V
   (b) 0.7 V, 1 V
   (c) 0.7 V, 0.6 V
   (d) 1 V, 0.7 V

3. The softness factor for soft-recovery and fast-recovery diodes are respectively
   (a) 1, > 1
   (b) < 1, 1
   (c) 1, 1
   (d) 1, < 1

4. Reverse recovery current in a diode depends upon
   (a) forward field current
   (b) storage charge
   (c) temperature
   (d) PIV

5. In a BJT,
   (a) $\beta = \dfrac{\alpha}{\alpha + 1}$
   (b) $\beta = \dfrac{\alpha}{\alpha - 1}$
   (c) $\alpha = \dfrac{\beta}{\beta - 1}$
   (d) $\alpha = \dfrac{\beta + 1}{\beta}$

6. A power MOSFET has three terminals called
   (a) collector, emitter and base
   (b) drain, source and base
   (c) drain, source and gate
   (d) collector, emitter and gate

7. As compared to power MOSFET, a BJT has
   (a) lower switching losses but higher conduction loss
   (b) higher switching losses and higher conduction loss
   (c) higher switching losses but lower conduction loss
   (d) lower switching losses and lower conduction loss

8. Choose the correct statement :
   (a) MOSFET has positive temperature coefficient (TC) whereas BJT has negative TC
   (b) Both MOSFET and BJT have positive TC
   (c) Both MOSFET and BJT have negative TC
   (d) MOSFET has negative TC whereas BJT has positive TC

9. Choose the correct statement :
   (a) Both MOSFET and BJT are voltage controlled devices (CDs)
   (b) Both MOSFET and BJT are current CDs
   (c) MOSFET is a voltage CD whereas BJT is a current CD
   (d) MOSFET is a current CD whereas BJT is a voltage CD

10. Secondary breakdown occurs in
    (a) MOSFET but not in BJT
    (b) both MOSFET and BJT
    (c) BJT but not in MOSFET
    (d) none of these

11. At present, the state of the art devices are available as under :

MOSFET	BJT
(a) 1200 V, 800 A	500 V, 140 A
(b) 500 V, 140 A	1200 V, 800 A
(c) 800 V, 1000 A	1000 V, 1200 A
(d) 200 V, 140 A	1500 V, 800 A

12. An IGBT has three terminals called
    (a) collector, emitter and base
    (b) drain, source and base
    (c) drain, source and gate
    (d) collector, emitter and gate

13. An MCT has three terminals called
    (a) anode, cathode and gate   (b) collector, emitter and gate
    (c) drain, source and base   (d) drain, source and gate

14. For the switching waveform shown in Fig. C.1 for a power transistor, the peak instantaneous power loss is
    (a) $\dfrac{V_s I_s}{4}$ (b) $\dfrac{V_s I_s}{6}$ (c) $\dfrac{V_s I_s}{3}$ (d) $\dfrac{V_s I_s}{8}$

15. In Fig. C.1, if $V_s = 100$ V, $I_s = 10\,A$, then peak instantaneous power loss in watts is
    (a) 500   (b) 166.67
    (c) 333.33   (d) 250

16. For the switching waveform shown in Fig. C.1 for a power transistor, the average value of switch-on power loss at a switching frequency f is ($f = 1/T$).
    (a) $\dfrac{V_s \cdot I_s}{4} \cdot t_{on} \cdot f$ (b) $\dfrac{V_s I_s}{6} \cdot t_{on} \cdot f$

    (c) $\dfrac{V_s I_s}{3} \cdot t_{on} \cdot f$ (d) $\dfrac{V_s \cdot I_s}{8} \cdot t_{on} \cdot f.$

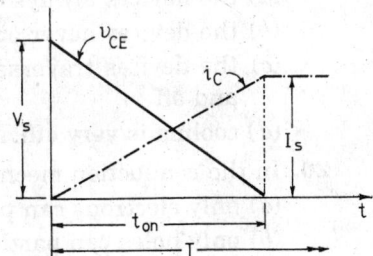

Fig. C.1

17. In Fig. C.1, if $V_s = 100$ V, $I_s = 10\,A$, $t_{on} = 1.2$ µs and $f = 10$ kHz, then average value of switch-on power loss is
    (a) 1 W   (b) 4 W
    (c) 2 W   (d) 3 W

18. Match the devices on the left hand side with the circuit symbols on right hand side and give the correct answer from the codes given below :

(A) BJT

(B) MOSFET

(C) IGBT

(D) MCT

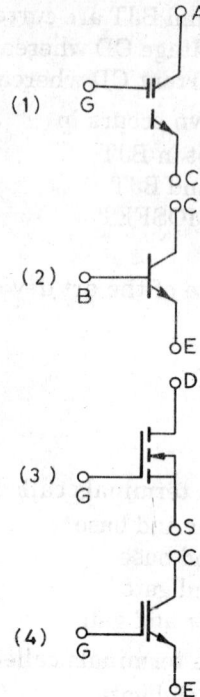

(1)

(2)

(3)

(4)

Codes

	A	B	C	D		A	B	C	D
(a)	2	3	4	1	(b)	2	3	1	4
(c)	3	2	1	4	(d)	3	2	4	1

19. Power-electronic equipment has very high efficiency, because
   (a) the devices always operate in active region
   (b) the devices never operate in active region
   (c) the devices traverse active region at high speed and stay at the two states, on and off
   (d) cooling is very efficient

20. In the conduction mechanism of Schottky diode
   (a) only electrons can participate
   (b) only holes can participate
   (c) both holes and electrons participate
   (d) none of the above.

21. Common emitter current gain h_{FE} of a BJT is
   (a) dependent on collector current I_c
   (b) dependent on collector-emitter voltage, V_{CE}
   (c) dependent on base-emitter voltage, V_{BE}
   (d) always constant

22. The semiconductor device which is suitable for induction hardening in radio frequency range is

(a) MCT   (b) BJT   (c) IGBT   (d) MOSFET

23. High-frequency operation of a circuit is limited by
    (a) On-state loss in the device   (b) off-state loss in the device
    (c) switching losses in the device   (d) all of the above.

24. Read the following statements carefully:
    1. PMOSFET is a majority carrier device
    2. IGBT is a bipolar device
    3. BJT is a majority carrier device
    4. MCT is unipolar device
    From above, the correct statements are
    (a) 1, 3   (b) 2, 4
    (c) 1, 4   (d) 1, 2

25. A bipolar junction transistor (BJT) is used as a power control switch by biasing it in the cut-off region (off state) or in the saturation region (ON state). In the ON state, for the BJT
    (a) both the base-emitter and base-collector junctions are reverse biased
    (b) the base-emitter junction is reverse biased, and the base-collector junction is forward biased
    (c) the base-emitter junction is forward-biased, and the base-collector junction is reverse biased
    (d) both the base-emitter and base-collector junctions are forward-biased.

ANSWERS

1. (c)	2. (b)	3. (d)	4. (a)	5. (b)	6. (c)
7. (c)	8. (a)	9. (c)	10. (c)	11. (b)	12. (d)
13. (a)	14. (a)	15. (d)	16. (b)	17. (c)	18. (a)
19. (c)	20. (a)	21. (a)	22. (d)	23. (c)	24. (d)
25. (d)					

DIODE CIRCUITS AND RECTIFIERS

1. In Fig. C.2, capacitor C is charged to $V_0 = 50$ V with upper plate positive. Switch S is closed at $t = 0$. Current through the circuit at $t = 0$ and final voltage across C are respectively,
   (a) 15 A, 200 V   (b) 20 A, 200 V
   (c) 25 A, 250 V   (d) 15 A, 150 V

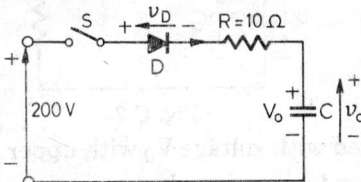

Fig. C.2

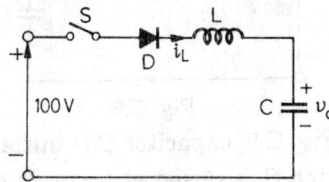

Fig. C.3

2. In Fig. C.2, suppose capacitor C is charged to 50 V with lower plate positive. Switch S is closed at $t = 0$. Current through the diode at $t = 0$ and final voltage across C are respectively
   (a) 25 A, 250 V   (b) 25 A, 200 V   (c) 20 A, 200 V   (d) 15 A, 150 V

3. In the circuit of Fig. C.3, switch S is closed at $t = 0$ with $i_L(0) = 0$ and $v_c(0) = 0$. In steady-state, v_c equals
   (a) 200 V   (b) 100 V
   (c) zero   (d) – 100 V

4. In Fig. C.4 ; V_1, V_2 and V_3 are zero centre PMMC voltmeters. The circuit is initially relaxed. Switch S is closed at $t = 0$. In steady state, readings of voltmeters V_1, V_2 and V_3 are respectively
   (a) 100 V, 100 V, – 100 V   (b) 100 V, 0, 200 V
   (c) – 100 V, 0, 200 V   (d) 100 V, 0, 100 V

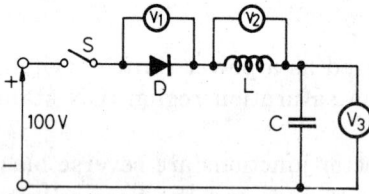   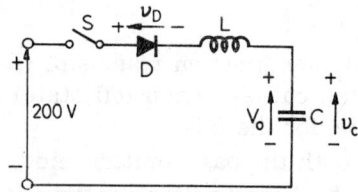

Fig. C.4   Fig. C.5

5. In Fig. C.5, initial voltage across capacitor is $V_0 = 50$ V with the polarity as shown. Switch S is closed at $t = 0$. In steady state, v_C and v_D are respectively given by
   (a) 400 V, – 200 V   (b) 350 V, – 150 V
   (c) 200 V, 0   (d) 450 V, – 250 V

6. In Fig. C.5, if $C = 8$ μF and $L = 0.2$ mH, the peak current handled by diode is
   (a) 40 A   (b) 50 A   (c) 10 A   (d) 30 A

7. In Fig. C.6, initial voltage across capacitor is $V_0 = 50$ V with the polarity as shown. Switch S is closed at $t = 0$. In steady-state, v_C and v_D are respectively given by
   (a) 400 V, – 200 V   (b) 350, – 150 V
   (c) 200 V, 0   (d) 450 V, – 250 V

8. In Fig. C.6, if $C = 8$ μF and $L = 0.2$ mH, the peak current through diode is given by
   (a) 40 A   (b) 50 A
   (c) 10 A   (d) 30 A

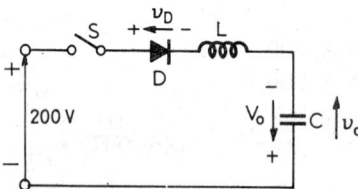   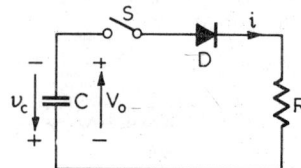

Fig. C.6   Fig. C.7

9. In Fig. C.7, capacitor C is initially charged with voltage V_0 with upper plate positive. Switch S is closed at $t = 0$. At $t = 0+$, v_c and i are given by
   (a) $0, \dfrac{V_0}{R}$ (b) $-V_0, \dfrac{V_0}{R}$
   (c) $-V_0, -\dfrac{V_0}{R}$ (d) $V_0, \dfrac{V_0}{R}$

10. In Fig. C.8, capacitor C is initially charged with voltage V_0. Switch S is closed at $t = 0$. In steady state, v_C and v_D are respectively given by

    (a) $V_0, -V_0$
    (c) $-V_0, 0$
    (b) $0, -V_0$
    (d) V_0, V_0

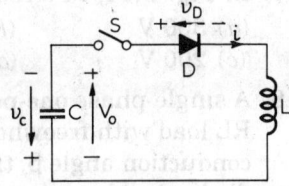

Fig. C.8

11. Each diode in Fig. C.9 can be described by a cut-in voltage and zero resistance. If the cut-in voltage of diode D1 is 0.2 V and of diode D2 is 0.6 V, the magnitude of current I_1 through D1 is ...... mA and magnitude of current through D2 is ...... mA.

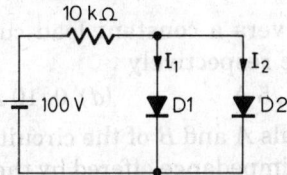

Fig. C.9

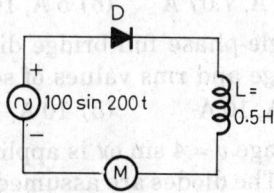

Fig. C.10

12. In Fig. C.10, ideal PMMC ammeter M will read

    (a) zero
    (c) 0.707 A
    (b) 1.414 A
    (d) 1 A

13. In Fig. C.10, if ammeter M is an ideal MI ammeter, then it will read

    (a) 0.707 A
    (c) 1.225 A
    (b) 1.414 A
    (d) 1 A

14. In Fig. C.11, ideal moving iron voltmeters M1 and M2 will respectively read

    (a) 141.4 V, 141.4 V
    (c) 0, 200 V
    (b) 0, 141.4 V
    (d) 141.4 V, 0

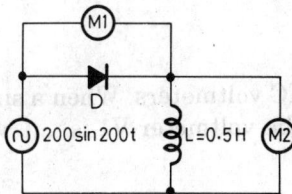

Fig. C.11

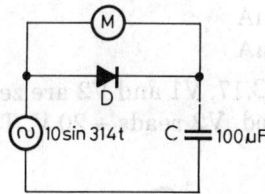

Fig. C.12

15. In Fig. C.12, an ideal moving iron voltmeter M will read

    (a) 7.07 V
    (c) 14.14 V
    (b) 12.25 V
    (d) 20.0 V

16. In Fig. C.13, zero-centre and ideal PMMC voltmeters M1 and M2 will read

    (a) – 10 V, 10 V   (b) 0, 10 V
    (c) – 10 V, 7.07 V   (d) 10 V, 7.07 V

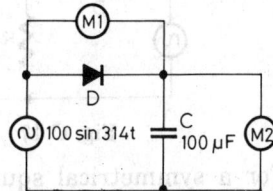

Fig. C.13

17. In Fig. C.14, PIV required for the diode is
   (*a*) 300 V   (*b*) 100 V
   (*c*) 200 V   (*d*) 400 V

18. A single-phase one-pulse diode rectifier is feeding an
   RL load with freewheeling diode across the load. For
   conduction angle β, the main diode and freewheeling
   diode would, respectively, conduct for
   (*a*) π, π − β   (*b*) π, β − π
   (*c*) β, π   (*d*) β − π, π

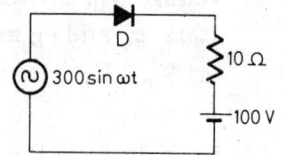

Fig. C.14

19. A single-phase full-bridge diode rectifier delivers a load current of 10 A, which is
   ripple free. Average and rms values of diode currents are respectively
   (*a*) 10 A, 7.07 A   (*b*) 5 A, 10 A   (*c*) 5 A, 7.07 A   (*d*) 7.07 A, 5 A

20. A single-phase full-bridge diode rectifier delivers a constant load current of 10 A.
   Average and rms values of source current are respectively
   (*a*) 5 A, 10 A   (*b*) 10 A, 10 A   (*c*) 5 A, 5 A   (*d*) 0, 10 A

21. A voltage $v = 4 \sin \omega t$ is applied to the terminals *A* and *B* of the circuit shown in Fig.
   C.15. The diodes are assumed to be ideal. The impedance offered by the circuit across
   the terminals *A* and *B* in kilo-ohms is
   (*a*) 5   (*b*) 20   (*c*) 10   (*d*) none of these

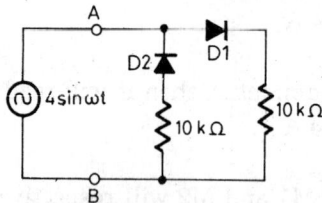

Fig. C.15

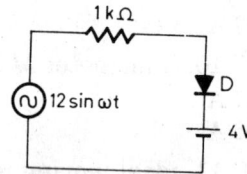

Fig. C.16

22. The peak current through the resistor of circuit of Fig. C.16, assuming the diodes to
   be ideal, is
   (*a*) 12 mA   (*b*) 4 mA
   (*c*) 16 mA   (*d*) 8 mA

23. In Fig. C.17, V1 and V2 are zero-centre PMMC voltmeters. When a sinusoidal signal
   is applied, V2 reads + 20 V. The reading of the voltmeter V1 is ...... volts.

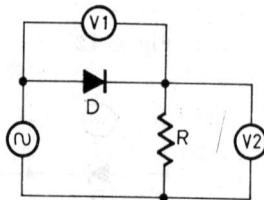

Fig. C.17

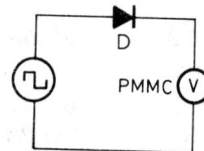

Fig. C.18

24. For a symmetrical square wave of 800 V peak to peak and for ideal diode, the
   voltmeter in Fig. C.18 will read
   (*a*) 200 V   (*b*) 400 V   (*c*) 800 V   (*d*) zero

25. The circuit in Fig. C.19 shows zener-regulated dc power supply. The zener-diode is ideal. The minimum value of R_L down to which the output voltage remains constant is

    (a) 27 Ω   (b) 45 Ω   (c) 15 Ω   (d) 24 Ω

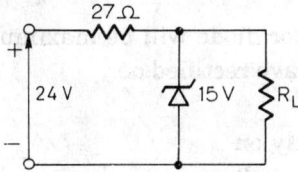 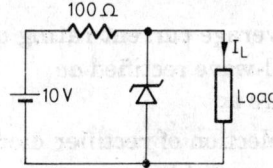

         Fig. C.19   Fig. C.20

26. In the circuit of Fig. C.20, the 5V zener diode requires a minimum current of 10 mA. For obtaining a regulated output of 5 V, the maximum permissible load current I_L is ...... mA and the minimum power rating of zener diode is ...... W.

27. Fig. C.21 shows an electronic voltage regulator. The zener diode may be assumed to require a minimum current of 25 mA for satisfactory operation. The value of R required for satisfactory operation is ...... ohms.

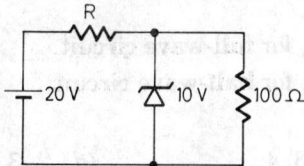 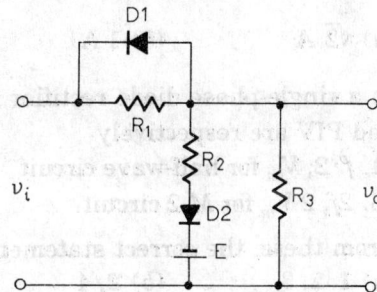

         Fig. C.21   Fig. C.22

28. In the circuit of Fig. C.22, the diode states at the extremely large negative value of the input voltage v_i are

    (a) D1 off, D2 off   (b) D1 on, D2 off
    (c) D1 off, D2 on   (d) D1 on, D2 on

29. A single-phase half-wave diode rectifier, connected to $V_m \sin \omega t$, feeds a load $R = 4.5\ \Omega$. The diode has forward resistance of $5\ \Omega$ and the remaining parameters are same as those of an ideal diode. The dc component of the source current is

    (a) $\dfrac{V_m}{50\ \pi}$ (b) $\dfrac{V_m}{50\ \pi \sqrt{2}}$ (c) $\dfrac{V_m}{100\ \pi \sqrt{2}}$ (d) $\dfrac{2\ V_m}{50\ \pi}$

30. In a single-phase diode bridge rectifier, the source voltage is $200 \sin \omega t$ with $\omega = 2\pi \times 50$ rad/sec and the load is $R = 50\ \Omega$. The power dissipated in the load resistor R is

    (a) $\dfrac{3200}{\pi}$ W   (b) $\dfrac{400}{\pi}$ W   (c) 400 W   (d) 800 W

31. The centre-tap full-wave single phase rectifier circuit uses two diodes. The transformer turns ratio from primary to each secondary is 2. In case transformer input voltage is 200 V at 50 Hz, then *rms* voltage across each diode is

    (a) 565.6 V   (b) 282.8 V   (c) 70.7 V   (d) 141.4 V

32. A single-phase two-pulse diode bridge has input supply of 200 sin ωt with load $R = 50\ \Omega$. Rms voltage across each diode is

    (a) 100 V   (b) 141.4 V   (c) 200 V   (d) $\dfrac{200}{\pi}$ V

33. The average current rating of a semiconductor diode will be maximum for
    (a) full-wave rectified ac   (b) half-wave rectified ac
    (c) pure ac   (d) pure dc

34. The selection of rectifier diode depends mostly on
    (a) forward voltage   (b) reverse voltage
    (c) fault current   (d) average load current

35. A single-phase bridge rectifier
    (a) can operate without an isolating transformer
    (b) cannot operate without an isolating transformer
    (c) cannot operate with isolating transformer
    (d) none of these

36. The *rms* value of half-wave rectified symmetrical square-wave current of 2 A is

    (a) $\sqrt{2}$ A   (b) 1 A   (c) $\dfrac{1}{\sqrt{2}}$ A   (d) $\sqrt{3}$ A

37. In a single-phase diode rectifier fed from $V_m \sin(2\pi ft)$, the lowest ripple frequency and PIV are respectively
    1. $f/2$, V_m for half-wave circuit.   2. f, V_m for full-wave circuit.
    3. $2f$, $2V_m$ for M-2 circuit.   4. f, V_m for half-wave circuit.

    From these, the correct statements are
    (a) 1, 2, 3   (b) 2, 4   (c) 3, 4   (d) 2, 3, 4

38. Current-ripple factor for a rectifier circuit is defined as

    (a) $\sqrt{\left(\dfrac{I_{or}}{I_o}\right)^2 - 1}$ (b) $\sqrt{\dfrac{I_{or}}{I_o} - 1}$ (c) $\sqrt{\left(\dfrac{I_o}{I_{or}}\right)^2 - 1}$ (d) $\sqrt{\dfrac{I_o}{I_{or}} - 1}$

    where I_{or} = rms value of output current
    and I_o = dc value of output current

39. A single-phase diode bridge rectifier has load R and filter capacitor across it. If one of the diodes is defective, then
    1. the dc load voltage would be lower than its expected value
    2. ripple frequency would be lower than the expected value
    3. the capacitor surge current would increase manifold.
    Of these statements,
    (a) 1 and 2 are correct   (b) 1 and 3 are correct
    (c) 2 and 3 are correct   (d) 1, 2 and 3 are correct.

40. Total harmonic distortion (THD) is defined as

    (a) $\sqrt{\dfrac{I_{s1}}{I_s} - 1}$ (b) $\sqrt{\left(\dfrac{I_s}{I_{s1}}\right)^2 - 1}$ (c) $\sqrt{\left(\dfrac{I_{or}}{I_s}\right)^2 - 1}$ (d) $\dfrac{I_s}{I_{s1}}$

    where I_s = rms value of supply phase current including fundamental and harmonics
    and I_{s1} = rms value of fundamental component of supply current.

41. Reactive power requirement of a rectifier system depends upon
   1. displacement factor   2. input power factor
   3. current distortion factor   4. crest factor

From these, the correct statements are

(a) 2 alone   (b) 1, 3, 4   (c) 2, 3, 4   (d) 1, 2, 3

42. In a 1-phase full-wave rectifier with R load and parallel capacitor filter C, with the increase of ωCR, the average value of output voltage
   (a) increases
   (b) decreases
   (c) remains unaltered
   (d) increases, reaches maximum value and then decreases.

43. Capacitor filter is ideal for currents which are
   (a) small   (b) medium   (c) large   (d) very large

44. An inductor filter at the output of a rectifier results in ripple which
   (a) increases with load resistance R
   (b) decreases with R
   (c) remains unaltered with increase of R
   (d) remains unaltered with decrease of R.

45. A capacitor filter at the output of a rectifier results in ripple which
   (a) increases with load resistance R
   (b) decreases with R
   (c) remains unaltered with increase of R
   (d) remains unaltered with decrease of R.

46. The function of a filter in a rectifier is to
   (a) limit the total current in the rectifier
   (b) limit the peak voltage of the rectifier
   (c) limit the dc current
   (d) reduce the ripple voltage in the output.

47. The disadvantage of half-wave diode rectifier circuit is that the
   (a) diode must have high PIV rating
   (b) diode must have high power rating
   (c) output voltage is difficult to filter
   (d) diode must have high current rating.

48. The function of centre-tapping on the secondary in a full-wave rectifier is to
   (a) step-up the voltage
   (b) step-down the voltage
   (c) isolate the load form ground
   (d) cause the diodes to conduct alternately.

49. A 3-phase M-6 (or six-phase half-wave) diode rectifier is fed from 400/1000 V, delta-star transformer, PIV of each diode is
   (a) 1414 V   (b) 707 V   (c) 1732 V   (d) 866 V

50. A delta-star transformer, with output line voltage of 1000 sin 100 πt volts, feeds 3-phase diode rectifier circuits. Their PIV is
   1. 1000 V for 3-pulse rectifier   2. 1155 V for M-6 rectifier

3. 1414 for 3-pulse rectifier   4. 1000 V for B-6 rectifier
5. 1414 for B-6 rectifier.

From these, the correct statements are

(a) 1, 2, 5   (b) 1, 2, 4   (c) 2, 3, 4   (d) 2, 3, 5

51. A 3-phase half-wave diode rectifier feeds a load of $R = 100\ \Omega$. For an input supply of 400 V, 50 Hz, the power delivered to load is

(a) 753.73 W   (b) 974.23 W   (c) 376.98 W   (d) 487.26 W

52. Read the following statements :
1. If load resistance R is low, ripple factor for L- filter is high
2. For C-filter, ripple factor gets reduced if R is increased
3. L-filter is suited to low-current loads
4. for C-filter, better-waveform in obtained if R is increased.
From these, the correct statement are

(a) 1, 4   (b) 2, 4   (c) 1, 3   (d) 2, 3

53. In rectifier circuits, if supply frequency is increased, then
1. ripple voltage increases with C-filter
2. ripple voltage decreases with L-filter
3. current ripple decreases with L-filter
4. output voltage increases with C-filter
From these, the correct statements are

(a) 1, 2, 3, 4   (b) 2, 4   (c) 3, 4   (d) 1, 4

54. A single-phase diode bridge rectifier is feeding a parallel combination of resistance load and a capacitor of high value. The nature of the input current drawn by the rectifier from the ac source will be
(a) a square wave of 180° duration
(b) a square wave of 120° duration
(c) peaky, with peaks at both positive and negative half cycle of the input voltage
(d) peaky, with peaks only at positive half cycle of the input voltage.

55. The current through the zener diode in Fig. C.23 is
(a) 33 mA   (b) 3.3 mA
(c) 2 mA   (d) 0 mA   [GATE, 2004]

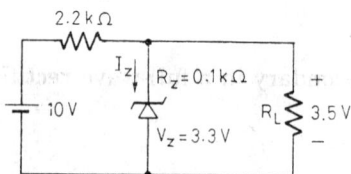

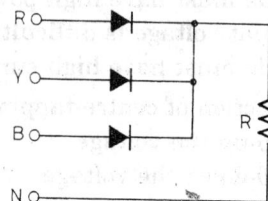

Fig. C.23   Fig. C.24

56. The circuit in Fig. C.24 showns a 3-phase half-wave rectifier. The source is a symmetrical 3-phase four wire system. The line to line voltage of the source is 100 V. The supply frequency is 400 Hz. The ripple frequency at the output is
(a) 400 Hz   (b) 800 Hz
(c) 1200 Hz   (d) 2400 Hz.   [GATE, 2004]

57. Assuming that the diodes are ideal in Fig. C.25, the current in diode $D1$ is

(a) 8 mA   (b) 5 mA   (c) 0 mA   (d) – 3 mA

[GATE, 2004]

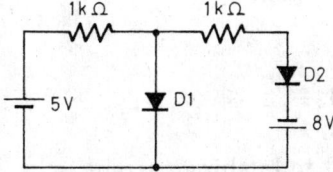

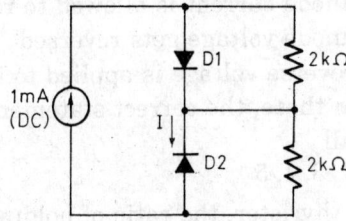

Fig. C.25   Fig. C.26

58. Assume that $D1$ and $D2$ in Fig. C.26 are ideal diodes. The value of the current I is

(a) 0 mA   (b) 0.5 mA   (c) 1 mA   (d) 2 mA

ANSWERS

1. (a)		**2.** (b)		**3.** (a)		**4.** (c)		**5.** (b)		**6.** (d)
7. (d)		**8.** (b)		**9.** (b)		**10.** (a)		**11.** 10 mA, 0	**12.** (d)
13. (c)		**14.** (b)		**15.** (b)		**16.** (a)		**17.** (d)		**18.** (b)
19. (c)		**20.** (d)		**21.** (c)		**22.** (d)		**23.** – 20V		**24.** (a)
25. (b)		**26.** 40 mA and 0.05 W				**27.** 80 Ω		**28.** (b)		**29.** (a)
30. (c)		**31.** (d)		**32.** (a)		**33.** (d)		**34.** (b)		**35.** (a)
36. (a)		**37.** (c)		**38.** (a)		**39.** (a)		**40.** (b)		**41.** (d)
42. (a)		**43.** (a)		**44.** (a)		**45.** (b)		**46.** (d)		**47.** (c)
48. (d)		**49.** (a)		**50.** (b)		**51.** (a)		**52.** (b)		**53.** (c)
54. (c)		**55.** (c)		**56.** (c)		**57.** (b)		**58.** (a)		

THYRISTORS

1. The number of p-n junctions in a thyristor is

(a) 1   (b) 2   (c) 3   (d) 4

2. When a thyristor is forward biased, the number of blocked p-n junctions is

(a) 1   (b) 2   (c) 3   (d) 4

3. When a thyristor is reverse biased, the number of blocked p-n junctions is

(a) 1   (b) 2   (c) 3   (d) 4

4. In a thyristor, anode current is made up of

(a) electronic only   (b) electrons or holes

(c) electrons and holes   (d) holes only.

5. A thyristor, when triggered, will charge from forward-blocking state to conduction state if its anode to cathode voltage is equal to

(a) peak repetitive off-state forward voltage

(b) peak working off-state forward voltage

(c) peak working off-state reverse voltage

(d) peak non-repetitive off-state forward voltage

6. Commutation or turn-off of a thyristor requires that
   1. anode current is reduced below holding current
   2. anode voltage is reduced to zero
   3. anode current is allowed to reverse
   4. anode voltage gets reversed
   5. reverse voltage is applied to it.
   From these, the correct statements are
   (a) all   (b) 1, 3, 4
   (c) 1, 3, 4, 5   (d) 1, 2, 4

7. In a thyristor, the ratio of holding current to latching current is
   (a) 0.4   (b) 1.0
   (c) 2.5   (d) 4.00

8. When a thyristor gets turned on, the gate drive
   (a) should not be removed as it will turn-off the SCR
   (b) may or may not be removed
   (c) should be removed
   (d) should be removed to avoid increased losses and higher junction temperature.

9. For normal SCRs, turn-on time is
   (a) less then turn-off time, t_q (b) more then t_q
   (c) equal to t_q (d) about half of t_q.

10. The forward voltage drop during SCR on-state is 1.5 V. This voltage drop
    (a) remains constant and is independent of load current
    (b) increase slightly with load current
    (c) decreases slightly with load current
    (d) varieus linearly with load current.

11. On-state voltage drop across a phase-controlled thyristor is as under:
    1. 0.5 V to 1.0 V for 250 V devices
    2. 1.5 V for 1 kV devices
    3. 1.5 V for 6 kV devices
    4. 3 V for 6 kV devices
    5. 1.5 V for 250 devices
    From these, the correct statements are
    (a) 1, 2, 3, 5   (b) 1, 2, 4   (c) 2, 3, 4, 5   (d) 2, 4, 5

12. The average current rating of a thyristor, as supplied by the manufacturers, corresponds to
    (a) resistive current   (b) inductive current
    (c) capacitive current   (d) none of the above

13. For reliable gate triggering of thyristors, it is advisable to employ
    (a) slight overtriggering   (b) very soft triggering
    (c) very hard triggering   (d) none of the above

14. It is easier to manufacture
    (a) fast SCR of high PIV   (b) fast SCR of low PIV
    (c) fast SCR of low losses   (d) none of the abcve.

15. The most efficient gate-triggering signal for SCR is
    (a) a steady *dc* level   (b) a short duration pulse
    (c) a high-frequency pulse train   (d) a low-frequency pulse train.

16. A driver circuit is required between the controller and the power circuit mainly for
    (a) isolation   (b) voltage level change
    (c) polarity change   (d) necessary drive power.

17. During forward blocking state, a thyristor is associated with
    (a) large current, low voltage   (b) low current, large voltage
    (c) medium current, large voltage   (d) low current, medium voltage

18. Once SCR starts conducting a forward current, its gate loses control over
    (a) anode circuit voltage only
    (b) anode circuit current only
    (c) anode circuit voltage and current
    (d) anode circuit voltage, current and time.

19. In a thyristor
    (a) latching current I_L is associated with turn-off process and holding current I_H with turn-on process
    (b) both I_L and I_H are associated with turn-off process
    (c) I_H is associated with turn-off process and I_L with turn-on process
    (d) both I_L and I_H are associated with turn-on process.

20. The SCR ratings, di/dt in $A/\mu sec$ and dv/dt in $V/\mu sec$, may vary, respectively, between
    (a) 20 to 500, 10 to 100   (b) both 20 to 500
    (c) both 10 to 100   (d) 50 to 300, 20 to 500.

21. A thyristor can be termed as
    (a) DC switch   (b) AC switch
    (c) either (a) or (b)   (d) square-wave switch.

22. Turn-on time of an SCR can be reduced by using a
    (a) rectangular pulse of high amplitude and narrow width
    (b) rectangular pulse of low amplitude and wide width
    (c) triangular pulse
    (d) trapezoidal pulse.

23. Turn-on time of an SCR in series with RL circuit can be reduced by
    (a) increasing circuit resistance R
    (b) decreasing R
    (c) increasing circuit inductance L
    (d) decreasing L.

24. For an SCR with turn-on time of 5 microsecond, an ideal trigger pulse should have
    (a) short rise time with pulse width = 3 µsec
    (b) long rise time with pulse width = 6 µsec
    (c) short rise time with pulse width = 6 µsec
    (d) long rise time with pulse width = 3 µsec.

25. A forward voltage can be applied to an SCR after its
    (a) anode current reduces to zero   (b) gate recovery time
    (c) reverse recovery time   (d) anode voltage reduces to zero.

26. Turn-off time of an SCR is measured from the instant
    (a) anode current becomes zero
    (b) anode voltage becomes zero
    (c) anode voltage and anode current become zero at the same time
    (d) gate current becomes zero.

27. Turn-on time for an SCR is 10 μsec. If an inductance is inserted in the anode circuit, then the turn-on time will be
    (a) 10 μsec   (b) less than 10 μsec
    (c) more than 10 μsec   (d) about 10 μsec.

28. In an SCR, anode current flows over a narrow region near the gate during
    (a) delay time t_d (b) rise time t_r and spread time t_p
    (c) t_d and t_p (d) t_d and t_r.

29. Gate characteristic of a thyristor
    (a) is a straight line passing through origin
    (b) is of the type $V_g = a + b\, I_g$
    (c) is a curve between V_g and I_g
    (d) has a spread between two curves of $V_g - I_g$.

30. The average on-state current for an SCR is 20 A for a conduction angle of 120°. Its average on-state current for 60° conduction angle would be
    (a) 20 A   (b) 10 A   (c) less than 20 A   (d) 40 A.

31. The average on-state current for an SCR is 20 A for a resistive load. If an inductance of 5 mH is included in the load, then average on-state current would be
    (a) more than 20 A   (b) less than 20 A
    (c) 15 A   (d) 20 A.

32. Specification sheet for an SCR gives its maximum rms on-state current as 35 A. This rms rating for a conduction angle of 120° would be
    (a) more than 35 A   (b) less than 35 A
    (c) 35 A   (d) 52.5 A.

33. Surge current rating of an SCR specifies the maximum
    (a) repetitive current with sine wave
    (b) non-repetitive current with rectangular wave
    (c) non-repetitive current with sine wave
    (d) repetitive current with rectangular wave.

34. The di/dt rating of an SCR is specified for its
    (a) decaying anode current   (b) decaying gate current
    (c) rising gate current   (d) rising anode current.

35. For an SCR , dv/dt protection is achieved through the use of
    (a) RL in series with SCR   (b) RC across SCR
    (c) L in series with SCR   (d) RC in series with SCR.

36. For an SCR, di/dt protection is achieved through the use of
    (a) R in series with SCR   (b) RL in series with SCR
    (c) L in series with SCR   (d) L across SCR.

37. The function of snubber circuit connected across an SCR is to
    (a) suppress dv/dt

(b) increase dv/dt

(c) decrease dv/dt

(d) keep transient overvoltage at a constant value.

38. The object of connecting resistance and capacitance across gate circuit is to protect the SCR gate against

(a) overvoltages   (b) dv/dt (c) noise signals   (d) overcurrents.

39. During the turn-off process in a thyristor, the current flow does not stop at the instant current reaches zero but continues to flow to a peak value in the reverse direction. This is due to

(a) commutation failure

(b) hole-storage effect

(c) presence of reverse voltage across the thyristor

(d) protective inductance in series with the thyristor.

40. The maximum di/dt in a thyristor circuit is

1. directly proportional to maximum value of supply voltage V_m

2. inversely proportional to V_m

3. inversely proportional to circuit inductance L

4. directly proportional to L

From above, the correct statements are

(a) 1, 3   (b) 2, 4   (c) 2, 3   (d) 1, 4

41. Thermal resistance in SCRs has the units of

(a) °C/W and heat sinks (HS) are made from aluminium

(b) W/°C and HS are made from steel

(c) °C/W and HS are made from copper

(d) °C/W and HS are made from copper alloy.

42. A power semiconductor device of thermal resistance 0.6°C/W has its heat sink at 90° C. In case the junction drop is 1.5 V for a load current of 30 A dc, the junction temperature would be

(a) 63°C   (b) 107°C   (c) 117°C   (d) 127°C

43. A thyristor converter of 415 V, 100 A is operating at rated load. Details of thyristor used are as follows :

Thermal resistance : junction to case = 0.01°C/W

                  case to sink = 0.08°C/W

       sink to atmosphere = 0.09°C/W

For an ambient temperature of 35°C, the junction temperature of 100% load would be

(a) 48.5°C   (b) 54.5°C   (c) 60°C   (d) 62°C

44. A thyristor can withstand a maximum junction temperature of 120°C with an ambient temperature of 75°C. If this SCR has thermal resistance from junction to ambient as 1.5°C/W, the maximum internal power dissipation allowed is

(a) 30 W   (b) 60 W   (c) 80 W   (d) 50 W

45. Practical way of obtaining static voltage equalization in series-connected SCRs is by the use of

(a) one resistor across the string

(b) resistors of different values across each SCR

(c) resistors of the same value across each SCR

(d) one resistor in series with the string.

46. For series connected SCRs, dynamic equalizing circuit consists of

(a) resistor R and capacitor C in series but with a diode D across C

(b) series R and D circuit but with C across R

(c) series R and C circuit but with D across R

(d) series C and D circuit but with R across C.

47. For dynamic equalizing circuit used for series connected SCRs, the choice of C is based on

(a) reverse recovery characteristic

(b) turn-on characteristics

(c) turn-off characteristics

(d) rise-time characteristics.

48. In an UJT, with V_{BB} as the voltage across two base terminals, the emitter potential at peak point is given by

(a) ηV_{BB} (b) ηV_D

(c) $\eta V_{BB} + V_D$ (d) $\eta V_D + V_{BB}$.

49. An UJT exhibits negative resistance region

(a) before the peak point   (b) between peak and valley points

(c) after the valley point   (d) both (a) and (c).

50. In an UJT, maximum value of charging resistance is associated with

(a) peak point

(b) valley point

(c) any point between peak and valley points

(d) after the valley point.

51. When an UJT is used for triggering an SCR, the waveshape of the voltage obtained from UJT circuit is a

(a) sine wave   (b) saw-tooth wave

(c) trapezoidal wave   (d) square wave.

52. For an UJT employed for the triggering of an SCR, stand-off ratio $\eta = 0.64$ and dc source voltage V_{BB} is 20 V. The UJT would trigger when the emitter voltage is

(a) 12.8 V   (b) 13.5 V   (c) 10 V   (d) 5 V.

53. UJTs are used for oscillators for the existence of

(a) peak-point potential

(b) valley-point potential

(c) positive resistance part of VA characteristics

(d) negative resistance part of VA characteristics.

54. An UJT is employed to fabricate a relaxation oscillator. When energised, it fails to oscillate. This may be due to

1. high base-terminal voltage V_{BB}

2. too large a capacitor

3. low value of charging resistor

4. large interbase resistance.

From these, the correct statements are

(a) 1, 3   (b) 1, 2, 3   (c) all   (d) 2, 4

55. A power semiconductor may undergo damage due to
   (a) high di/dt (b) low di/dt
   (c) high dv/dt (d) low dv/dt

56. A triace is equivalent to
   (a) two diodes in antiparallel   (b) one thyristor and one diode in parallel
   (c) two thyristors in parallel   (d) two thyristors in antiparallel

57. For a triac and SCR,
   (a) both are unidirectional devices
   (b) triac requires more current for turn-on than SCR at a particular voltage
   (c) a triac has less time for turn-off than SCR
   (d) both are available with comparable voltage and current ratings.

58. Consider the following statements :
   1. The triac is a five layer device
   2. The triac may be considered to consist of two parallel sections $p_1 n_1 p_2 n_2$ and $p_2 n_1 p_1 n_4$
   3. An additional lateral region serves as the control gate
   4. The triac is a double ended SCR.
   From above, the correct statements are
   (a) all   (b) 1, 2, 3
   (c) 1 only   (d) 1, 4

59. In a conventional reverse blocking thyristor
   (a) external layers are lightly doped and internal layers are heavily doped
   (b) external layers are heavily doped and internal layers are dightly doped
   (c) the p-layers are heavily doped and the n-layers are lightly doped
   (d) the p-layers are lightly doped and the n-layers are heavily doped.

60. If the amplitude of the gate pulse to thyristor is increased, then
   (a) both delay time and rise time would increase
   (b) the delay time would increase but the rise time would decrease
   (c) the delay time would decrease but the rise time would increase
   (d) the delay time would decrease while the rise time remains unaffected.

61. Use of a reverse conducting thyristor in place of antiparallel combination of thyristor and feedback diode in an inverter
   (a) effectively minimises the peak commutating current
   (b) decreases the operating frequency of operation
   (c) minimises effects of lead inductances on the commutation performance
   (d) causes deterioration in the commutation performance

62. Triacs are most suitable when the supply voltage is
   (a) dc   (b) low-frequency ac
   (c) high-frequency ac   (d) full-wave rectified ac

63. Which one of the following statements is correct ? A triac is a
   (a) 2 terminal switch
   (b) 2 terminal bilateral switch
   (c) 3 terminal unilateral switch
   (d) 3 terminal bidirectional switch

64. In the circuit of relaxation oscillator shown in Fig.
C. 27, what will be the change in voltage waveform
across capacitor, if the voltage V_1 is doubled.

  (*a*) The amplitude as well as the frequency of the
     waveform will get doubled.

  (*b*) The amplitude will get doubled, but the fre-
     quency will reduce to half its value.

  (*c*) The amplitude will get doubled, but the fre-
     quency will remain unchanged.

  (*d*) The amplitude will remain unchanged, but the frequency will get doubled ?

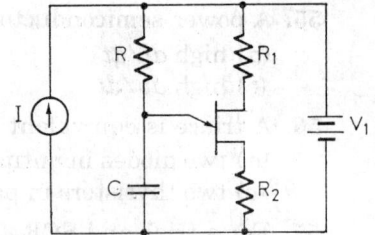

Fig. C.27

65. Which of the following characteristic of a silicon *p*- *n* junction diode makes it suitable
for use as an ideal diode ?

  1. it has very low saturation current

  2. It has a high value of forward cut-in voltage

  3. It can withstand large reverse voltage

  4. When compared with germanium diodes, silicon diodes show a lower degree of
     temperature dependence under reverse bias conditions.

Select the correct answer using the codes given below :

Codes :

  (*a*) 1 and 2   (*b*) 1, 2, 3 and 4

  (*c*) 2, 3 and 4   (*d*) 1 and 3

66. Which one of the following statements regarding the two-transistor model of the
p-n-p-n four-layer device is correct ?

  (*a*) It explains only the turn-on portion of the device characteristics.

  (*b*) It explains only the turn-off portion of the device characteristics

  (*c*) It explains only the negative-region portion of device characteristics

  (*d*) It explains all the regions of the device characteristics.

67. Thyristors can be turned off by

  1. reducing the current below the holding current

  2. applying a negative voltage to the anode of the device

  3. reducing the gate current.

Of these statements,

  (*a*) 1 and 3 are correct   (*b*) 2 and 3 are correct

  (*c*) 1, 2 and 3 are correct   (*d*) 2 and 3 are correct.

68. The turn-on time for an SCR is 30 µs. The pulse train at the gate has a frequency
of 2.5 kHz with a mark/space ratio of 0.1. This SCR will

  (*a*) turn-on

  (*b*) not turn-on

  (*c*) turn-on if pulse-frequency is increased

  (*d*) turn-on if pulse-frequency is decreased   [**Hint :** Here $\delta = 1/11$]

69. A thyristor is triggered by a pulse train of 5 kHz. The duty ratio is 0.4. If the
allowable average power is 100 W, the maximum allowable gate-drive power is

  (*a*) 100 $\sqrt{2}$ W   (*b*) 50 W   (*c*) 150 W   (*d*) 250 W

70. During turn-on process of a thyristor, maximum power losses occur during

  (*a*) t_p (*b*) t_r (*c*) t_d (*d*) equal in all

71. A pulse transformer is used in a driver circuit
  (a) to prevent a dc triggering
  (b) to shape the trigger signal
  (c) to generate high-frequency pulses
  (d) to provide isolation

72. Current unbalance in the parallel-connected SCRs is due to the non-uniformity in the
  (a) forward characteristics   (b) reverse characteristics
  (c) di/dt withstand capability   (d) dv/dt withstand capability.

73. In synchronized UJT triggering of an SCR, voltage v_c across capacitor reaches UJT thresh-hold voltage thrice in each half cycle so that there are three firing pulses during each half cycle. The firing angle of the SCR can be controlled
  (a) once in each half cycle   (b) thrice in each half cycle
  (c) twice in each half cycle   (d) four times in each half cycle.

74. The function of connecting a zener diode in an UJT circuit, used for the triggering of SCRs, is to
  (a) expedite the generation of triggering pulses
  (b) delay the generation of triggering pulses
  (c) provide a constant voltage to UJT to prevent erratic firing
  (d) provide a variable voltage to UJT as the source voltage changes.

75. A metal oxide varistor (MOV) is used for protecting
  (a) gate circuit against overcurrents
  (b) gate circuit against overvoltages
  (c) anode circuit against overcurrents
  (d) anode circuit against overvoltages.

76. The functions of connecting a resistor in series with gate-cathode circuit and a zener-diode across gate-cathode circuit are, respectively, to protect the gate circuit of a thysistor from
  (a) overvoltages, overcurrents   (b) overcurrents, overvoltages
  (c) overcurrents, noise signals   (d) noise signals, overvoltages.

77. In a GTO, anode current begins to fall when gate current
  (a) is negative peak at time $t = 0$
  (b) is negative peak at t = storage period t_s
  (c) just begins to become negative at $t = 0$
  (d) is negative peak at $t = (t_s + $ fall time$)$.

78. For a pulse transformer, the material used for its core and the possible turn-ratio from primary to secondary are, respectively,
  (a) ferrite ; 20 : 1   (b) laminated iron ; 1 : 1
  (c) ferrite ; 1 : 1   (d) powdered iron ; 1 : 1.

79. Protection against di/dt stress in a device is necessary because
  (a) it interferes with control electronics
  (b) it introduces voltage surges on supply lines
  (c) it destroys the device
  (d) none of the above are valid.

80. Overcurrent protection of thyristors is provided by
    (a) use of saturable di/dt coils   (b) use of circuit breaker and a fuse
    (c) use of snubber circuit   (d) liberal heat sinking.

81. The capacitance of a reverse biased junction of a thyristor is 20 picofarad. The charging current of this thyristor is 4 mA. The limiting value of dv/dt in $V/\mu s$ is
    (a) 50   (b) 100   (c) 200   (d) 500

82. SCR can be turned on by
    1. applying anode voltage at a sufficiently fast rate
    2. applying sufficiently large anode voltage
    3. increasing SCR temperature to a sufficiently large value
    4. applying adequately large gate current.
    From these, the correct statement are
    (a) all   (b) 4 only   (c) 2, 4   (d) 1, 2, 4

83. During forward blocking of two series-connected SCRs, a thyristor with
    1. high leakage impedance shares lower voltage
    2. high leakage impedance shares higher voltage
    3. low leakage impedance shares higher voltage
    4. low leakage impedance shares lower voltage.
    From these, the correct statements are
    (a) 1, 3   (b) 1, 4   (c) 2, 3   (d) 2, 4

84. Thyristor A has rated gate current of 2A and thyristor B a rated gate current of 100 mA.
    1. A is a GTO and B is a conventional SCR
    2. B is a GTO and A is a conventional SCR
    3. A may operate as a transistor
    4. B may operate as a thyristor.
    From the above the correct statements are
    (a) 1, 4   (b) 1, 3   (c) 2, 3   (d) 2, 4

85. A resistor connected across the gate-cathode terminals of a thyristor increases its
    1. dv/dt rating   2. holding current
    3. noise immunity   4. turn-off time.
    From these, the correct statements are
    (a) all   (b) 2, 3   (c) 1, 2, 3   (d) 2, 3, 4

86. To generate gate-triggering signals IC 555 is often used in
    (a) monostable mode   (b) bistable mode
    (c) astable mode   (d) none of the above.

87. A GTO with anode fingers has
    (a) no reverse blocking capability   (b) reduced tail current
    (c) high turn-off time   (d) reduced turn-off gain

88. A gold-doped GTO has
    (a) high reverse blocking capability
    (b) increased tail current
    (c) low on-state voltage drop
    (d) low turn-off time

From these, the correct statements are

(a) 1, 4   (b) 1, 3, 4   (c) all   (d) 1, 2, 4

89. The correct sequence of the given devices in the decreasing order of their speed of operation is

(a) power BJT, PMOSFET, IGBT, SCR
(b) IGBT, PMOSFET, power BJT, SCR
(c) SCR, PBJT, IGBT, PMOSFET
(d) PMOSFET, IGBT, PBJT, SCR

90. Consider the following statements :

1. BJT has lower power losses than PMOSFET
2. PMOSFET has lower power losses than IGBT
3. SCRs have lower power losses than PMOSFET and IGBT.

Which of these statements are correct ?

(a) 1, 2 and 3   (b) 1 and 2   (c) 2 and 3   (d) 1 and 3

91. Consider the following statements about SITH:

1. It is a $p^+ n n^+$ diode
2. It is normally-on device
3. It has low reverse blocking capability
4. It has p^+ gate electrodes buried in n layer
5. Its turn-off gain varies from 4 to 6.

From these, the correct statements are

(a) 1, 2, 4   (b) 1, 2, 3, 5   (c) 2, 3, 4, 5   (d) 1, 3, 4, 5

92. Consider the following statements about GTO and SITH.

1. Anode-shorting in GTO reduces its reverse-voltage blocking capability to zero
2. Anode-shorting in SITH reduces its reverse-voltage blocking capability to zero
3. Anode-fingers in GTO increase its turn-off gain
4. Both GTO and SITH have n^+ type fingers in the anode.

From above, the correct statements are

(a) all   (b) 2, 3, 4
(c) 1, 2, 3   (d) 2, 4

93. Cosine triggering control is used to linearise the relation between

(a) E_c and α (b) α and V_o
(c) V_o and I_o (d) E_c and V_o

where V_o = output voltage of converter, I_o = output current, E_c = control voltage to triggering scheme and α = trigger angle.

94. The triggering circuit of a thyristor is shown in Fig. C.28. The thyristor requires a gate current of 10 mA, for guaranteed turn-on. The value of R required for the thyristor to turn on reliably under all conditions of V_b variation is

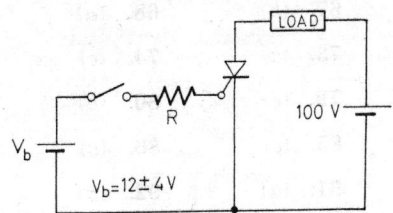

Fig. C.28

(a) 10,000 Ω (b) 1600 Ω (c) 1200 Ω (d) 800 Ω

[GATE, 2004]

95. An electronic switch S is required to block voltages of either polarity during its OFF state as shown in Fig. C.29 (a). The switch is required to conduct in only one direction during its ON state as shown in Fig. C.29 (b)

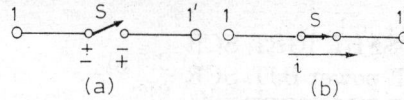

Fig. C.29

Which of the following are valid realization of the switch S ?

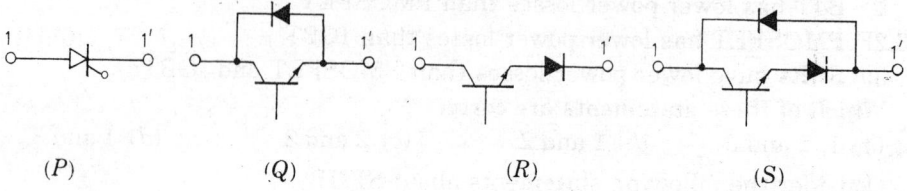

(P)	(Q)	(R)	(S)

(a) Only P (b) P and Q (c) P and R (d) R and S

[GATE, 2005]

ANSWERS

1. (c)	**2.** (a)	**3.** (b)	**4.** (c)	**5.** (b)	**6.** (c)
7. (a)	**8.** (d)	**9.** (a)	**10.** (b)	**11.** (b)	**12.** (a)
13. (c)	**14.** (a)	**15.** (c)	**16.** (d)	**17.** (b)	**18.** (c)
19. (c)	**20.** (b)	**21.** (a)	**22.** (a)	**23.** (d)	**24.** (c)
25. (b)	**26.** (a)	**27.** (c)	**28.** (d)	**29.** (d)	**30.** (c)
31. (a)	**32.** (c)	**33.** (c)	**34.** (d)	**35.** (b)	**36.** (c)
37. (a)	**38.** (c)	**39.** (b)	**40.** (a)	**41.** (a)	**42.** (c)
43. (d)	**44.** (a)	**45.** (c)	**46.** (c)	**47.** (a)	**48.** (c)
49. (b)	**50.** (a)	**51.** (b)	**52.** (b)	**53.** (d)	**54.** (a)
55. (a)	**56.** (d)	**57.** (b)	**58.** (a)	**59.** (b)	**60.** (d)
61. (c)	**62.** (b)	**63.** (d)	**64.** (b)	**65.** (d)	**66.** (a)
67. (b)	**68.** (a)	**69.** (d)	**70.** (b)	**71.** (d)	**72.** (a)
73. (a)	**74.** (c)	**75.** (d)	**76.** (b)	**77.** (b)	**78.** (c)
79. (c)	**80.** (b)	**81.** (c)	**82.** (a)	**83.** (d)	**84.** (b)
85. (c)	**86.** (c)	**87.** (c)	**88.** (c)	**89.** (d)	**90.** (d)
91. (a)	**92.** (b)	**93.** (d)	**94.** (d)	**95.** (c)	

THYRISTOR COMMUTATION TECHNIQUES

1. For the circuit shown in Fig. C.30, the conduction time for thyristor in microseconds is

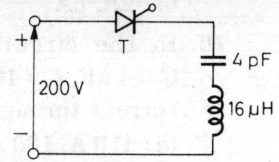

Fig. C.30

   (a) 0.393   (b) 2.546

   (c) 25.133   (d) 8.0

2. For the circuit in Fig. C.30, the capacitor voltage after SCR gets self-commutated is

   (a) 200 V   (b) 400 V   (c) 300 V   (d) 100 V

3. For the circuit shown in Fig. C.30, the voltage across thyristor, after it is self-commutated is

   (a) zero   (b) -1.5 V   (c) -200 V   (d) -400 V

4. In the circuit of Fig. C.30, the peak thyristor current is

   (a) 100 A   (b) 50 A   (c) 400 A   (d) 800 A

5. In the circuit of Fig. C.31, the maximum value of current through thyristors T1 and TA can respectively be

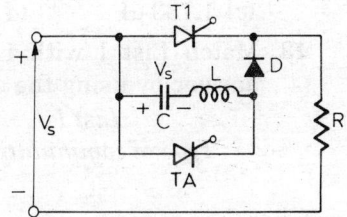

Fig. C.31

   (a) $\dfrac{V_s}{R}, \dfrac{V_s}{R} + V_s \sqrt{C/L}$

   (b) $\dfrac{V_s}{R} + V_s \sqrt{C/L}, \; V_s \sqrt{C/L}$

   (c) $V_s \sqrt{C/L}, \dfrac{V_s}{R}$

   (d) $\dfrac{V_s}{R}, \; V_s \sqrt{C/L}$

6. For the circuit shown in Fig. C.31, the peak value of resonant current is twice the load current. In case $V_s = 200$ V, the magnitude of voltage across the main thyristor, when it gets turned off, is equal to

   (a) 86.6 V   (b) 100 V   (c) 173.2 V   (d) 200 V

7. For the circuit in Fig. C.31, the peak value of current through auxiliary SCR is twice that through the main SCR. In case $V_s = 100$ V, $C = 10$ μF and constant load current $= 40$ A, the circuit turn-off time for main SCR, in microseconds is

   (a) 12.5   (b) 21.65   (c) 25   (d) 10

8. Read the following statements with regard to Fig. C.32, where capacitor C is charged to voltage V_s with polarity as shown :

   1. In order to turn-off T1, turn on T2.

   2. In order to turn-off T2, turn on T1.

   3. At the time of turn-on of SCR, initial thyristor current is $V_s \left[\dfrac{2}{R_1} + \dfrac{1}{R_2} \right]$

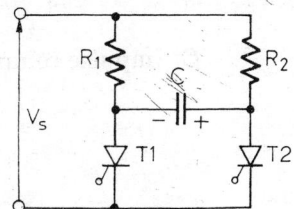

Fig. C.32

   4. At the time of turn-on of SCR, initial thyristor current is $V_s \left[\dfrac{1}{R_1} + \dfrac{2}{R_2} \right]$.

From above, the correct statements are

   (a) 2, 4   (b) 1, 3   (c) 1, 4   (d) 2, 3

9. In the circuit shown in Fig. C.32, $R_1 = 50\ \Omega$, $R_2 = 100\ \Omega$ and $V_s = 100$ V. The possible peak values of current through SCRs T1 and T2 are respectively
  (a) 2A, 5A   (b) 1A, 2A   (c) 4A, 2A   (d) 4A, 5A

10. In the circuit shown in Fig. C.33, $V_s = 200$ V, $C = 4\ \mu F$, $L = 16\ \mu H$ and $R = 20\ \Omega$. The peak value of current through T1 and D can respectively be
  (a) 110 A, 100 A   (b) 10 A, 110 A
  (c) 110 A, 10 A   (d) 100A, 110 A

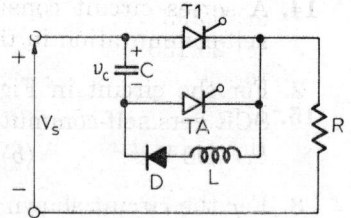

Fig. C.33

11. In the circuit of Fig. C.33 and for the parameters given in Prob. 10, the circuit turn-off time for main and auxiliary SCRs in microseconds are respectively
  (a) 8, 12.566   (b) 40, 1.2566
  (c) 80, 12.566   (d) 80, 25.132

12. In the circuit configuration shown in Fig. C.34, the circuit turn-off time for main thyristor is 34.657 μs. The value of capacitor C required, in this circuit, is
  (a) 5 μF   (b) 3.466 μF
  (c) 1.733 μF   (d) 10 μF

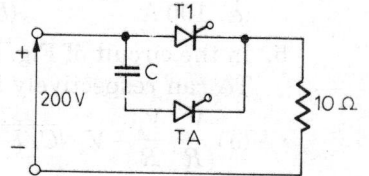

Fig. C.34

13. Match List I with List II and select the correct answer by using the codes given below the lists :

List I   *List II*
Type of commutation   *Power Circuit Diagram*

A. Self-commutation   (1)

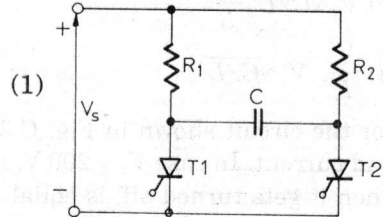

B. Complementary commutation   (2)

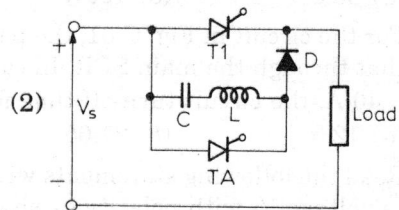

C. Impulse commutation   (3)

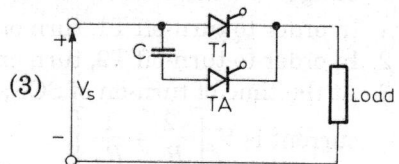

D. Resonant-pulse commutation (4)

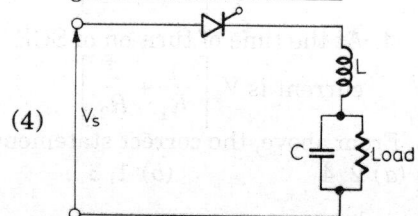

Codes :

	A	B	C	D		A	B	C	D
(a)	4	1	2	3	(b)	1	4	2	3
(c)	1	4	3	2	(d)	4	1	3	2

14. A series circuit consists of $R = 2.4\,\Omega$, $L = 25\,\mu H$, C and a thyristor. For obtaining self-commutation in the circuit, the value of C should be equal to

    (a) $50\,\mu F$ (b) $30\,\mu F$ (c) $20\,\mu F$ (d) $10\,\mu F$

15. Match the type of commutation in List I with their alternative names in List II and tick the correct answer from the codes given below :

List I
Type of commutation
A. Class A
B. Class B
C. Class C
D. Class D

List II
Alternative title
1. Voltage commutation
2. Parallel-capacitor commutation
3. Complementary-impulse commutation
4. Self-commutation
5. Natural commutation
6. Current commutation

Codes :

	A	B	C	D		A	B	C	D
(a)	4	6	3	1	(b)	5	1	4	6
(c)	4	6	3	2	(d)	4	6	2	4

16. Match the type of commutation in List I with those in List II and give the correct answer by using the codes given below the lists :

List I
Type of commutation
A. Load commutation
B. Impulse commutation
C. Line commutation
D. Resonant-pulse commutation

List II
Alternative title
1. Voltage commutation
2. Natural commutation
3. Resonant commutation
4. Parallel-capacitor commutation
5. Current commutation

Codes :

	A	B	C	D		A	B	C	D
(a)	5	1	2	3	(b)	2	4	3	5
(c)	3	4	2	5	(d)	3	1	2	4

17. In a commutation circuit employed to turn-off an SCR, satisfactory turn-off is obtained when

    (a) circuit turn-off time < device turn-off time
    (b) circuit turn-off time > device turn-off time
    (c) circuit time constant > device turn-off time
    (d) circuit time constant < device turn-off time

ANSWERS

1. (c)   2. (b)   3. (c)   4. (a)   5. (d)   6. (c)
7. (b)   8. (b)   9. (d)   10. (a)   11. (c)   12. (a)
13. (d)   14. (d)   15. (a)   16. (c)   17. (b)

PHASE CONTROLLED RECTIFIERS

1. A single-phase half-wave controlled rectifier has 400 sin 314 t as the input voltage and R as the load. For a firing angle of 60° for the SCR, the average output voltage is
   (a) $400/\pi$ (b) $300/\pi$ (c) $240/\pi$ (d) $200/\pi$.

2. A single-phase one-pulse controlled circuit has resistance and counter emf load and 400 sin 314 t as the source voltage. For a load counter emf of 200 V, the range of firing angle control is
   (a) 30° to 150°   (b) 30° to 180°   (c) 60° to 120°   (d) 60° to 180°

3. In a single-phase half-wave circuit with RL load, and a freewheeling diode across the load, extinction angle β is more than π. For a firing angle α, the SCR and freewheeling diode would conduct, respectively, for
   (a) $\pi - \alpha, \beta$ (b) $\beta - \alpha, \pi - \alpha$ (c) $\pi - \alpha, \beta - \pi$ (d) $\pi - \alpha, \pi - \beta$.

4. In a single-phase one-pulse circuit with RL load and a freewheeling diode, extinction angle β is less than π. For a firing angle α, the SCR and freewheeling diode would, respectively, conduct for
   (a) $\beta - \alpha, 0°$ (b) $\pi - \alpha, \pi - \beta$ (c) $\alpha, \beta - \alpha$ (d) $\beta - \alpha, \alpha$.

5. A single-phase full-wave mid-point thyristor converter uses a 230/200 V transformer with center tap on the secondary side. The P.I.V. per thyristor is
   (a) 100 V   (b) 141.4 V   (c) 200 V   (d) 282.8 V.

6. A single-phase two-pulse bridge converter has an average output voltage and power output of 500 V and 10 kW respectively. The SCRs used in the two-pulse bridge converter are now re-employed to form a single-phase two-pulse mid-point converter. This new controlled converter would give, respectively, an average output voltage and power output of
   (a) 500 V, 10 kW   (b) 250 V, 5 kW
   (c) 250 V, 10 kW   (d) 500 V, 5 kW.

7. In a single-phase full converter bridge, the average output voltage is given by
   (a) $\dfrac{1}{\pi} \displaystyle\int_{\alpha}^{\pi + \alpha} V_m \cos \theta \, d\theta$ (b) $\dfrac{1}{\pi} \displaystyle\int_{0}^{\alpha + \pi} V_m \cos \theta \cdot d\theta$
   (c) $\dfrac{1}{\pi} \displaystyle\int_{\alpha - (\pi/2)}^{\alpha + (\pi/2)} V_m \cos \theta \cdot d\theta$ (d) $\dfrac{1}{\pi} \displaystyle\int_{(\pi/2) - \alpha}^{(\pi/2) + \alpha} V_m \cos \theta \cdot d\theta$

8. In a single-phase semiconverter, the average output voltage is given by
   (a) $\dfrac{1}{\pi} \displaystyle\int_{\alpha}^{\pi} V_m \cos \theta \cdot d\theta$ (b) $\dfrac{1}{\pi} \displaystyle\int_{(\pi/2) - \alpha}^{(\pi/2)} V_m \cos \theta \cdot d\theta$
   (c) $\dfrac{1}{\pi} \displaystyle\int_{\alpha - (\pi/2)}^{\pi/2} V_m \cos \theta \cdot d\theta$ (d) $\dfrac{1}{\pi} \displaystyle\int_{\alpha - (\pi/2)}^{\pi} V_m \cos \theta \cdot d\theta$

9. In a single-phase full converter, for continuous conduction, each pair of SCRs conduct for
   (a) $\pi - \alpha$ (b) π (c) α (d) $\pi + \alpha$.

10. In a single-phase full converter, for discontinuous load current and extinction angle $\beta > \pi$, each SCR conducts for
    (a) α (b) $\beta - \alpha$ (c) β (d) $\alpha + \beta$

11. In a single-phase semi-converter, for continuous conduction, each SCR conducts for
    (a) α   (b) π   (c) α + π   (d) π − α

12. In a single-phase semiconverter, for discontinuous conduction and extinction angle β > π, each SCR conducts for
    (a) π − α   (b) β − α   (c) α   (d) β.

13. In a single-phase semiconverter, for discontinuous conduction and extinction angle β < π, each SCR conducts for
    (a) π − α   (b) β − α   (c) α   (d) β

14. In a single-phase semiconverter, for continuous conduction, freewheeling diode conducts for
    (a) α   (b) π − α   (c) π   (d) π + α

15. In a single-phase semiconverter, with discontinuous conduction and extinction-angle β > π, freewheeling diode conducts for
    (a) α   (b) β − π   (c) π + α   (d) β.

16. In a single-phase semiconverter, with discontinuous conduction and extinction angle β < π, freewheeling diode conducts for
    (a) α   (b) π − β   (c) β − π   (d) zero degree.

17. In a single-phase full converter, if α and β are firing and extinction angles respectively, then the load current is
    (a) discontinuous if (β − α) < π   (b) discontinuous if (β − α) > π
    (c) discontinuous if (β − α) = π   (d) continuous if (β − α) < π.

18. In a single-phase semiconverter with resistive load and for a firing angle α, each SCR and freewheeling diode conduct, respectively, for
    (a) α, 0°   (b) π − α, α   (c) π + α, α   (d) π − α, 0°.

19. In controlled rectifiers, the nature of load current, i.e. whether load current is continuous or discontinuous
    (a) does not depend on type of load and firing angle delay
    (b) depends both on the type of load and firing angle delay
    (c) depends only on the type of load
    (d) depends only on the firing angle delay.

20. In a single-phase full converter with resistive load and for a firing angle delay α, the load current is
    1. zero at α, π + α, 2π + α ...   2. remains zero for duration α.
    3. $\dfrac{V_m}{R} \sin α$ at α, π + α, 2π + α ...   4. remains zero for duration π − α
    5. $\dfrac{V_m}{R} \sin α$ at α only.

    From these, the correct statements are
    (a) 1, 2, 3   (b) 2, 3, 5   (c) 2, 3   (d) 4, 5

21. In a single-phase full converter, if output voltage has peak and average values of 325 V and 133 V respectively, then the firing angle is
    (a) 40°   (b) 140°   (c) 50°   (d) 130°

22. In a single-phase semiconverter, if output voltage has peak and average values of 325 and 133 V respectively, the firing angle is
    (a) 40°   (b) 73.40°   (c) 80°   (d) 140°

23. A single SCR is inserted in between voltage source 200 sin 314 t and a load $R = 10\ \Omega$. If the gate trigger voltage lags the ac supply voltage by 120°, then average load current is

(a) $-\dfrac{15}{\pi}$ A   (b) $\dfrac{15}{\pi}$ A   (c) $-\dfrac{5}{\pi}$ A   (d) $\dfrac{5}{\pi}$ A

24. The average value of dc voltage of a single-phase semiconverter, under continuous conduction, is

(a) $\dfrac{\sqrt{2}\cdot V_s\,(1 + \cos\alpha)}{\pi}$ (b) $\dfrac{\sqrt{2}\cdot V_s\,(1 - \cos\alpha)}{\pi}$

(c) $\dfrac{\sqrt{2}\,(1 + \cos\alpha)}{2\pi}$ (d) $\dfrac{\sqrt{2}\cdot V_s\,(1 - \cos\alpha)}{2\pi}$

25. A freewheeling diode is placed across the dc load
  1. to prevent reversal of load voltage
  2. to transfer the load current away from the source
  3. to transfer the load current away from conducting thyristor.
  The correct statements are

(a) 1, 3   (b) 2, 3
(c) 1, 2   (d) 1, 2, 3

26. A freewheeling diode across inductive load will provide

(a) quick turn-on   (b) slow turn-off
(c) reduced utilization factor   (d) improved power factor

27. When a flywheeling diode is connected across a full converter supplying ripple-free current at controlled output voltage
  1. dc voltage increases at higher value of α
  2. converter pf is improved
  3. SCR heating is reduced.
  From these, the correct statements are

(a) 2 only   (b) 1, 2   (c) 2, 3   (d) 1, 2, 3

28. Consider the following statements :
  Phase controlled converters
  1. do not provide smooth variation of output voltage
  2. inject harmonics into the power system
  3. draw non-unity pf current for finite triggering angle.
  Which of these statements are correct ?

(a) 1, 2 and 3   (b) 1 and 2   (c) 2 and 3   (d) 1 and 3

29. The purpose of commutating diode in a thyristor controlled ac to dc converter is to
  (a) reduce the current of its associated SCR to zero so that commutation can take place
  (b) share the load current of its associated SCR
  (c) conduct the load current when its associated SCR is turned off
  (d) maintain voltage across load at constant value.

30. Consider the following statements :
  The overlap angle of a phase-controlled converter will increase
  1. as the firing angle increases
  2. as the frequency of supply increases
  3. as the supply voltage decreases

Of these statements,

(a) 1, 2 and 3 are correct   (b) 2 and 3 are correct
(c) 1 and 3 are correct   (d) 1 and 2 are correct

31. A single-phase two pulse converter feeds RL load with sufficient smoothing so that the conduction is continuous. If the resistance of the load circuit is increased
    (a) the ripple content of the load current will remain the same
    (b) the ripple content of the load current will increase
    (c) the ripple content of the load current will decrease
    (d) there is a possibility of discoutinuous conduction.

32. Neglecting drops across SCRs and the circuit resistance except that in the load, the regulation of a converter, ΔV_d, is given by $I_d.R$ where R is an equivalent output resistance. It is proportional to
    (a) input line voltage   (b) trigger angle α
    (c) back emf in the dc circuit   (d) number of commutations per second

33. Reactive loading of supply lines by a converter is directly dependent on
    (a) displacement angle only   (b) displacement angle and distortion factor
    (c) back emf in the load circuit   (d) circuit configuration

34. A single-phase full converter operates as an inverter, when
    (a) $0° \le \alpha \le 90°$
    (b) $90° \le \alpha \le 180°$
    (c) it supplies to a back-emf load
    (d) $90° \le \alpha \le 180°$ and there is a suitable dc source in the load circuit.

35. An inductance is inserted in the load circuit of SCR. With this
    1. the turn-on time of SCR is increased
    2. dc output voltage is reduced for the same firing angle
    3. conduction continues even after reversal of phase of input voltage
    4. α by-pass diode is connected in such circuits.
    From above, the correct statements are
    (a) 1, 2, 3, 4   (b) 1, 3, 4   (c) 2, 3, 4   (d) 1, 2, 3

36. Overlap in a phase-controlled converter, under continuous conduction, does not depend on
    (a) frequency   (b) applied voltage   (c) load current   (d) load inductance.

37. Commutation overlap in the phase-controlled ac to dc converters is due to
    (a) load inductance
    (b) harmonic content of load current
    (c) switching operation in the converter
    (d) source inductance

38. A single-phase full converter would lose its controllability if it is feeding a load having
    (a) resistance and a current source in it
    (b) parallel combination of a resistance and capacitance
    (c) resistance and inductance in series
    (d) none of the above

39. Modern ac to dc converters employ GTOs instead of SCRs in order to have
    (a) low reactive volt-ampere flow   (b) reliable commutation
    (c) low switching loss   (d) smaller heat sink.

40. Each diode of a 3-phase half-wave diode rectifier conducts for
   (a) 60°   (b) 120°   (c) 180°   (d) 90°.

41. Each diode of a 3-phase, 6-pulse bridge diode rectifier conducts for
   (a) 60°   (b) 120°   (c) 180°   (d) 90°.

42. In a 3-phase half-wave diode rectifier, if per phase input voltage is 200 V, then the average output voltage is
   (a) 233.91 V   (b) 116.95 V   (c) 202.56 V   (d) 101.28 V.

43. In a 3-phase half-wave diode rectifier, the ratio of average output voltage to per-phase maximum ac voltage is
   (a) 0.955   (b) 0.827   (c) 1.654   (d) 1.169.

44. For a 3-phase, six-pulse diode rectifier, the average output voltage in terms of maximum value of line voltage V_m is
   (a) $\dfrac{3\sqrt{2}}{\pi} V_{ml}$ (b) $\dfrac{3 V_{ml}}{\pi}$ (c) $\dfrac{3\sqrt{3}}{2\pi} V_{ml}$ (d) $\dfrac{3\sqrt{3}}{\pi} V_{ml}$.

45. In a 3-phase half-wave rectifier, dc output voltage is 230 V. The peak inverse voltage across each diode is
   (a) 481.7 V   (b) 460 V   (c) 345 V   (d) 230 V.

46. In a 3-phase full-wave diode rectifier, the peak inverse voltage in terms of average output voltage is
   (a) 1.571   (b) 0.955   (c) 1.047   (d) 2.094.

47. In a 3-phase half-wave diode rectifier, if V_{mp} is the maximum value of per phase voltage, then each diode is subjected to a peak inverse voltage of
   (a) V_{mp} (b) $\sqrt{3}\, V_{mp}$ (c) $2V_{mp}$ (d) $3V_{mp}$.

48. In a 3-phase full-wave diode rectifier, if V_{ml} is the maximum value of line voltage, then each diode is subjected to a peak inverse voltage of
   (a) V_{ml} (b) $\sqrt{3}\, V_{ml}$ (c) $2V_{ml}$ (d) $3V_{ml}$

49. In a 3-phase full-wave diode rectifier, if V is the per phase input voltage, then average output voltage is given by
   (a) 0.955 V   (b) 1.35 V   (c) 2.34 V   (d) 3 V.

50. A converter which can operate in both 3-pulse and 6-pulse modes is a
   (a) 1-phase full converter   (b) 3-phase half-wave converter
   (c) 3-phase semiconverter   (d) 3-phase full converter.

51. In a 3-phase semi-converter, for firing angle less than or equal to 60°, each thyristor and diode conduct, respectively, for
   (a) 60°, 60°   (b) 90°, 30°
   (c) 120°, 120°   (d) 180°, 180°.

52. In a 3-phase semiconverter, for firing angle less than or equal to 60°, freewheeling diode conducts for
   (a) 30°   (b) 60°
   (c) 90°   (d) zero degree

53. In a 3-phase semiconverter, for a firing angle equal to 90° and for continuous conduction, each SCR and diode conduct, respectively, for
   (a) 30°, 60°   (b) 60°, 30°
   (c) 90°, 90°   (d) 30°, 30°.

54. In a 3-phase semiconverter, for a firing angle equal to 90° and for continuous conduction, freewheeling diode conducts for
    (a) 30°   (b) 60°   (c) 90°   (d) zero degree.

55. In a 3-phase semiconverter, for firing angle equal to 120° and extinction angle equal to 110°, each SCR and diode conduct, respectively, for
    (a) 30°, 60°   (b) 60°, 60°   (c) 90°, 30°   (d) 110°, 30°.

56. In a 3-phase semiconverter, for firing angle equal to 120° and extinction angle equal to 110°, freewheeling diode conducts for
    (a) 10°   (b) 30°   (c) 50°   (d) 110°.

57. In a 3-phase semiconverter, for firing angle equal to 120° and extinction angle equal to 100°, none of the bridge elements conduct for
    (a) 10°   (b) 20°   (c) 30°   (d) 60°.

58. A 3-phase semiconverter can work as
    (a) converter for $\alpha = 0°$ to 180°   (b) converter for $\alpha = 0°$ to 90°
    (c) inverter for $\alpha = 90°$ to 180°   (d) inverter for $\alpha = 0°$ to 90°.

59. In a 3-phase semiconverter, the three SCRs are triggered at an interval of
    (a) 60°   (b) 90°   (c) 120°   (d) 180°.

60. In a 3-phase full converter, the six SCRs are fired at an interval of
    (a) 30°   (b) 60°   (c) 90°   (d) 120°.

61. In a 3-phase full converter, three SCRs pertaining to one group are fired at an interval of
    (a) 30°   (b) 60°   (c) 90°   (d) 120°.

62. For a single-phase two-pulse phase-controlled rectifier, with a freewheeling diode across RL load,
    (a) the instantaneous output voltage v_0 is always positive
    (b) v_0 may be positive or zero
    (c) v_0 may be positive, zero or negative
    (d) v_0 is always zero or negative.

63. The frequency of the ripple in the output voltage of a 3-phase semiconverter depends upon
    (a) firing angle and load resistance   (b) firing angle and load inductance
    (c) the supply frequency   (d) firing angle and the supply frequency.

64. In a single-phase full converter, if load current is I and ripple free, then average and rms values of thyristor current are
    (a) $\frac{1}{2}I, \frac{I}{\sqrt{2}}$ (b) $\frac{1}{3}I, \frac{I}{\sqrt{3}}$ (c) $\frac{1}{4}I, \frac{I}{2}$ (d) $I, \frac{I}{\sqrt{2}}$.

65. In a 3-phase full converter, if load current is I and ripple free, then average and rms values of thyristor current are
    (a) $\frac{1}{2}I, \frac{I}{\sqrt{2}}$ (b) $\frac{1}{3}I, \frac{I}{\sqrt{3}}$ (c) $\frac{1}{4}I, \frac{I}{2}$ (d) $I, \frac{I}{\sqrt{3}}$.

66. The effect of source inductance on the performance of single-phase and three-phase full converters is to
    (a) reduce the ripples in the load current
    (b) make discontinuous current as continuous

(c) reduce the output voltage

(d) increase the load voltage.

67. In a single-phase full converter, the number of SCRs conducting during overlap is

(a) 1   (b) 2   (c) 3   (d) 4.

68. In a single-phase full converter, the output voltage during overlap is equal to

(a) zero

(b) source voltage

(c) source voltage minus the inductance drop

(d) inductance drop.

69. In a 3-phase full converter, the output voltage during overlap is equal to

(a) zero

(b) source voltage

(c) source voltage minus the inductance drop

(d) average value of the conducting-phase voltages.

70. The total number of SCRs conducting simultaneously in 3-phase full converter with overlap considered has the sequence of

(a) 3, 3, 2, 2   (b) 3, 3, 3, 2   (c) 3, 2, 3, 2   (d) 2, 2, 2, 3.

71. A 3-phase full converter has an average output voltage of 200 V for zero degree firing angle and for resistive load. For a firing angle of 90°, the output voltage would be

(a) zero   (b) 50 V   (c) 100 V   (d) 26.8 V.

72. A four quadrant operation requires

(a) two full converters in series

(b) two full converters connected back to back

(c) two full converters connected in parallel

(d) two semiconverters connected back to back.

73. In a 3-phase full converter, the output voltage pulsates at a frequency equal to

(a) supply frequency, f   (b) 2 f

(c) 3f   (d) 6f.

74. In circulating-current type of dual converter, the nature of voltage across reactor is

(a) alternating   (b) pulsating   (c) direct   (d) triangular.

75. For the same ac input voltage, the peak inverse voltage in ac to dc converter systems is highest in

(a) single-phase full wave M-2 converter

(b) single-phase full converter

(c) 3-phase bridge converter

(d) 3-phase M-3 converter.

76. The three-phase ac to dc converter which requires neutral point connection is

(a) 3-phase semiconverter   (b) 3-phase full converter

(c) 3-phase half-wave converter   (d) 3-phase full converter with diodes.

77. A 3-pulse converter feeds RLE load. The source has a definite inductance causing overlap. The thyristors are ideal. It has an overlap angle μ of 20° at the minimum firing angle 'α'. The current remains constant in the complete range of firing angle. The range of firing angle for the converter would be

(a) 0 < α < 180°   (b) 20 < α < 180°

(c) 20 < α < 160°   (d) 0° < α < 160°

78. The range of firing angle for a 3-phase, 3-pulse converter feeding a resistive load is
(a) 0° to 180°   (b) 0° to 150°   (c) 30° to 150°   (d) 30° to 180°

79. A 3-pulse converter has a freewheeling diode across its load. The operating range of the converter is
(a) $0° \leq \alpha \leq 150°$ (b) $60° \leq \alpha \leq 120°$ (c) $30° < \alpha \leq 150°$ (d) $180° \leq \alpha \leq 360°$

80. If the commutation angle of a diode rectifier (due to source inductance) is μ, then inductive voltage regulation is
(a) $\dfrac{1 + \cos \mu}{2}$ (b) $1 + \dfrac{\cos \mu}{2}$ (c) $1 - \dfrac{\cos \mu}{2}$ (d) $\dfrac{1 - \cos \mu}{2}$

81. In a 3-phase bridge rectifier fed from star-connected secondary winding of a transformer, let the voltage to the neutral of A-phase (phase sequence A, B, C) be $V_m \sin \omega t$. At the instant, when voltage of A-phase is maximum, the output voltage at the rectifier terminals would be
(a) $V_m / \sqrt{2}$ (b) V_m (c) $1.5\, V_m$ (d) $\sqrt{3} \cdot V_m$

82. Three-phase voltages v_a, v_b, v_c are applied to 3-phase M-3 diode rectifier. When v_a is passing through zero and becoming positive, the load voltage v_o would be (V_{mp} = maximum value of phase voltage)
(a) $v_b = 0.5\, V_{mp}$ (b) $v_c = 0.866\, V_{mp}$ (c) $v_b = 0.866\, V_{mp}$ (d) $v_c = 0.5\, V_{mp}$

83. Line voltages $v_{ab}, v_{ac}, v_{bc}, v_{ba}$ etc. are applied to 3-phase, six-pulse diode bridge rectifier. When v_{ab} is zero and becoming positive, the output voltage v_o would be (V_{ml} = maximum value of line voltage),
(a) $v_{cb} = V_{ml}$ (b) $v_{bc} = \dfrac{\sqrt{3}}{2} V_{ml}$ (c) $v_{cb} = \dfrac{\sqrt{3}}{2} V_{ml}$ (d) $v_{ac} = \dfrac{\sqrt{3}}{2} V_{ml}$

84. Line voltages $v_{ab}, v_{ac}, v_{bc}, v_{ba}$ etc are applied to 3-phase semiconverter. For firing angle equal to 15°, when v_{ab} is zero and becoming negative, the load voltage would be (V_{ml} = maximum value of line voltage)
(a) $v_{ac} = \dfrac{\sqrt{3}}{2} V_{ml}$ (b) $v_{bc} = 0.866\, V_{ml}$ (c) $v_{ca} = 0.866\, V_{ml}$ (d) $v_{ac} = -0.866\, V_{ml}$

85. When firing angle α of a single-phase full converter feeding constant dc current into a load is 30°, the displacement factor of the rectifier is
(a) 1   (b) 0.5   (c) $\dfrac{1}{\sqrt{3}}$ (d) $\dfrac{\sqrt{3}}{2}$

86. A 3-phase fully-controlled converter is feeding power into a dc load at a constant current of 150 A. The rms current through each thyristor of the converter is
(a) 50 A   (b) 100 A   (c) $100 \cdot \sqrt{2/3}$ A   (d) $50 \cdot \sqrt{3}$ A

87. In a 3-phase full converter, the ratio of average voltage to maximum line voltage is
(a) $0.9549 \cos \alpha$ (b) $0.9549 \sin \alpha$ (c) $0.4775 \cos \alpha$ (d) $0.9549\,(1 + \cos \alpha)$

In phase-controlled converters feeding RL load, the ripple content of load current is decided by
(a) load resistance alone   (b) load inductance alone
(c) both R and L   (d) neither R nor L

89. The function of centre-tapping on the secondary in a full-wave rectifier is to
(a) step-up the voltage   (b) step-down the voltage
(c) isolate the load from ground   (d) cause the diodes to conduct alternately.

90. The peak inverse voltage rating of a diode in M-2 rectifier is 'X' time larger than that of a bridge rectifier yielding the same dc output voltage, where the value of 'X' is

(a) 0.5   (b) 1.0   (c) $\sqrt{2}$ (d) 2.0

91. PIV rating of a diode in M-2 rectifier is 'X' times that of a B-2 rectifier yielding the same dc output voltage, where the value of 'X' is

(a) 0.5   (b) 1.0   (c) $\sqrt{2}$ (d) 2.0

92. PIV rating of a diode in M-3 rectifier is 'X' times that of a 3-phase full converter yielding the same dc output voltage, where the value of 'X' is

(a) 0.5   (b) 1.0   (c) $\sqrt{2}$ (d) 2.0

93. For the same ac input voltage, PIV rating of a diode in M-3 rectifier is 'X' times that of a 3-phase B-6 rectifier, where the value of 'X' is

(a) 0.5   (b) 1.0   (c) $\sqrt{3}$ (d) 2.0

94. A six-pulse thyristor rectifier bridge is connected to a balanced 50 Hz three-phase ac source. Assuming that the dc output current of the rectifier is constant, the lowest frequency harmonic component of the ac source line current is

(a) 100 Hz   (b) 150 Hz   (c) 150 Hz   (d) 300 Hz

95. The output of a single-phase full-wave rectifier contains

(a) dc plus even harmonics

(b) dc plus odd harmonics

(c) dc plus both odd and even harmonics

(d) dc and no harmonics

96. A single-phase half-wave converter feeds RLE load from 220 V mains. The value of E may lie between 0 and $\sqrt{2}\,V_s$. The PIV stress on the SCR is

(a) $220\sqrt{2}$ V   (b) $220/\sqrt{2}$ V   (c) $2 \times 220 \times \sqrt{2}$ V   (d) 2×220 V

97. A 3-pulse converter supplies from an ideal transformer, a load current I_o, which is ripple free, to RLFD load. The trigger angle is such that the freewheeling diode **FD** conducts each time for the same duration as each SCR. The rms values of **load** current, SCR current and FD current, respectively, are

(a) $1 : 1/\sqrt{6} : 1/\sqrt{6}$ (b) $1 : 1/\sqrt{3} : 1/\sqrt{3}$

(c) $1 : 1/\sqrt{6} : 1/\sqrt{2}$ (d) $1 : 1/6 : 1/2$

98. A 2-pulse converter supplies RLE load with $R = 5\Omega$, $L = 20$ H, $E = 160$V from a 220 V, 50 Hz supply. Load draws an average current I_o of 7.614 A. If the value of E is changed to 155 V, the new value of I_o will be

(a) 7.914 A   (b) 8.614 A   (c) 7.214 A   (d) 8.414 A

99. Inter-group reactor in a dual converter is needed

(a) to absorb the instantaneous inequalities between output voltages of the converters

(b) to absorb the regenerated energy from motor and load inertia

(c) to eliminate interconverter circulating current

(d) to avoid abrupt transfer of current from converter to another.

100. A thyristorised, three phase, fully controlled converter feeds a dc load that draws a constant current. Then the input ac line current of the converter has

(a) an rms value equal to the dc load current

(b) an average value equal to the dc load current

(c) a peak value equal to the dc load current

(d) a fundamental frequency component, whose rms value is equal to the dc load current.
[GATE, 2000]

101. A fully controlled natural commutated 3-phase bridge rectifier is operating with a firing angle $\alpha = 30°$. The peak to peak voltage ripple expressed as a ratio of the peak output dc voltage at the output of the converter bridge is

(a) 0.5 (b) $\dfrac{\sqrt{3}}{2}$ (c) $\left(1 - \dfrac{\sqrt{3}}{2}\right)$ (d) $\sqrt{3} - 1$
[GATE, 2003]

102. A rectifier type ac voltmeter consists of a series resistance R_s, an ideal full-wave rectifier bridge and a PMMC instrument as shown in Fig. C.35. The internal resistance of the instrument is 100 Ω and a full-scale deflection is produced by a dc current of 1 mA. The value of R_s required to obtain full scale deflection with an ac voltage of 100 V (rms) applied to the input terminals is

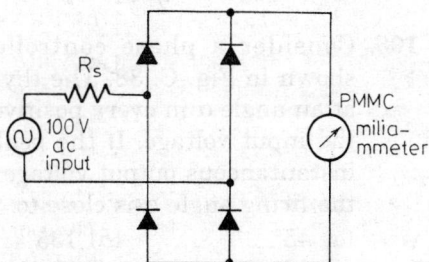

(a) 63.56 Ω (b) 89.93 Ω
(c) 89.93 kΩ (d) 141.3 kΩ
[GATE, 2003]

Fig. C.35

103. A phase-controlled half-controlled single-phase converter is shown in Fig. C.36. The control angle $\alpha = 30°$. The output dc voltage waveshape will be as shown in

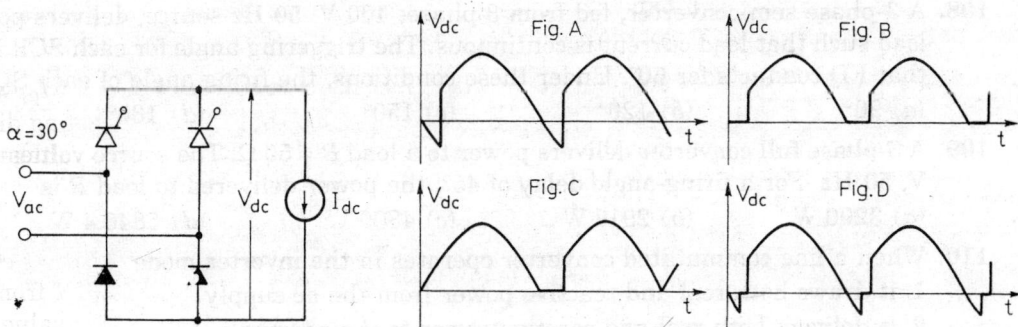

Fig. C.36

(a) Fig. A (b) Fig. B
(c) Fig. C (d) Fig. D
[GATE, 2003]

104. Analysis of voltage waveform of a single-phase bridge converter shows that it contains $x\%$ of 6th harmonic. The 6th harmonic content of the voltage waveform of a 3-phase bridge converter would be

(a) less than $x\%$ due to an increase in the number of pulses
(b) equal to $x\%$, the same as that of the single-phase converter
(c) greater than $x\%$ due to changes in the input and output voltages of the converter
(d) difficult to predict as the analysis of converters is not governed by any generalised theory

105. Fig. C. 37 shows the voltage across a power semiconductor device and the current through the device during a switching transition. Is the transition a turn ON transition or a turn OFF transition ? What is the energy loss during the transition ?

(a) Turn ON, $\dfrac{VI}{2}(t_1 + t_2)$

(b) Turn OFF, $VI(t_1 + t_2)$

(c) Turn ON, $VI(t_1 + t_2)$

(d) Turn OFF, $\dfrac{VI}{2}(t_1 + t_2)$ [GATE, 2005]

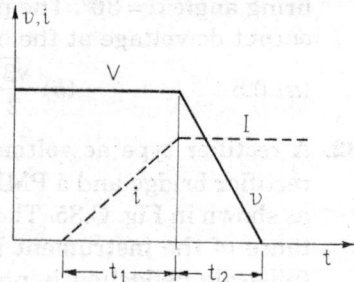

Fig. C.37

106. Consider a phase controlled converter shown in Fig. C. 38. The thyristor is fired at an angle α in every positive half cycle of the input voltage. If the peak value of the instantaneous output voltage equals 230 V, the firing angle α is close to

(a) 45°   (b) 135°

(c) 90°   (d) 83.6°   [GATE, 2005]

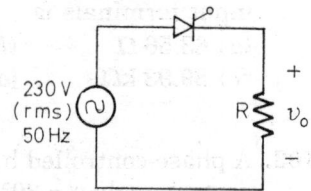

Fig. C.38

107. A 3-phase full converter, fed from 3-phase, 400 V, 50 Hz source, delivers power to load R. Each SCR is triggered sequentially. If the peak value of the instantaneous output voltage is 400 V, the firing angle of 3-phase full converter would be

(a) 30°   (b) 45°   (c) 60°   (d) 75°

108. A 3-phase semiconverter, fed from 3-phase, 400 V, 50 Hz source, delivers power to load such that load current is continuous. The triggering angle for each SCR is such that FD conducts for 60°. Under these conditions, the firing angle of each SCR is

(a) 90°   (b) 120°   (c) 150°   (d) 160°

109. A 3-phase full converter delivers power to a load $R = 50\ \Omega$. The source voltage is 400 V, 50 Hz. For a firing-angle delay of 45°, the power delivered to load R is

(a) 3200 W   (b) 2918 W   (c) 4800 W   (d) 5846.4 W

110. When a line commutated converter operates in the inverter mode
1. it draws both real and reactive power from the ac supply
2. it delivers both real and reactive power to the ac supply
3. it delivers real power to the ac supply
4. it draws reactive power from the ac supply.
From these, the correct statements are

(a) 1 only   (b) 2 only   (c) 3 only   (d) 3, 4

111. In a 3-phase semiconverter, frequency of the ripple in the output voltage wave may be
1. 3 times the supply frequency f for firing angle α < 60°
2. $3f$ for α > 60°
3. $6f$ for α < 60°
4. $6f$ for α > 60°
From above, the correct statements are

(a) 1, 3   (b) 1, 4   (c) 2, 3   (d) 2, 4

112. For the same ac voltage and load impedance, read the following statements about rectifiers :

1. The average load current in a full-wave rectifier is twice that in a half-wave rectifier.
2. The average load current in a full-wave rectifier is π times that in a half-wave rectifier.
3. Half-wave rectifier will have bigger sized transformer compared to full-wave rectifier.
4. Half-wave rectifier will have a smaller tansformer compared to a full-wave rectifier.

From these, the correct statements are

(*a*) 1, 4   (*b*) 2, 4
(*c*) 1, 3   (*d*) 2, 3

113. For the triangular waveform shown in Fig. C. 39, the rms value of the voltage is equal to

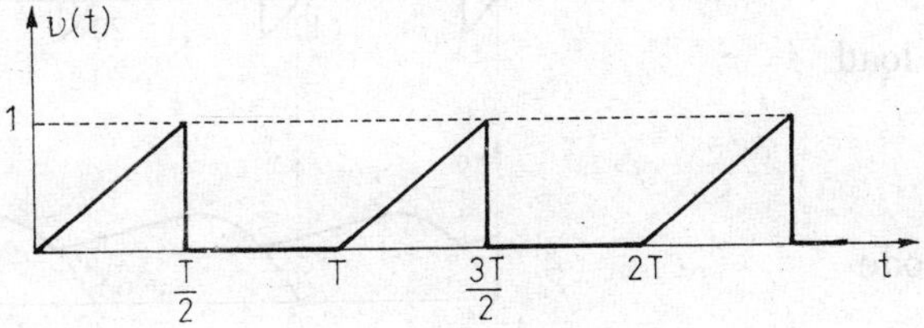

Fig. C.39

(*a*) $\sqrt{1/6}$ (*b*) $\sqrt{1/3}$
(*c*) $\dfrac{1}{3}$ (*d*) $\sqrt{2/3}$

(GATE, 2005)

114. A three-phase diode bridge rectifier is fed from a 400 V, 50 Hz, three-phase ac source. If the load is purely resistive, the peak instantaneous output voltage is equal to

(*a*) 400 V   (*b*) 400 $\sqrt{2}$ V
(*c*) $400 . \sqrt{2/3}$ (*d*) $\dfrac{400}{\sqrt{3}}$ V

(GATE, 2005)

115. For the current waveform shown in Fig. C. 40, the rms value of the current is equal to

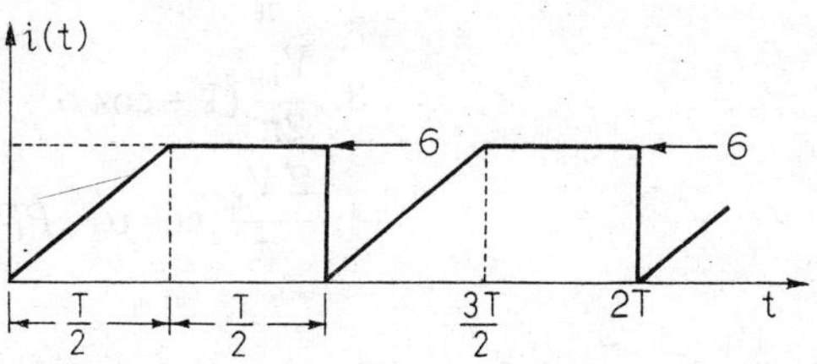

Fig. C.40

(*a*) $2\sqrt{3}$ (*b*) $3\sqrt{3}$ (*c*) $2\sqrt{5}$ (*d*) $2\sqrt{6}$

116. Match List-I with List II and select the correct answer by using the codes given below :

List I	*List II*
Single phase full converter with 50 Hz supply	*Output voltage waveform*

A. RC (parallel) load

B. Continuous conduction mode

C. Resistance load

D. Inverter mode

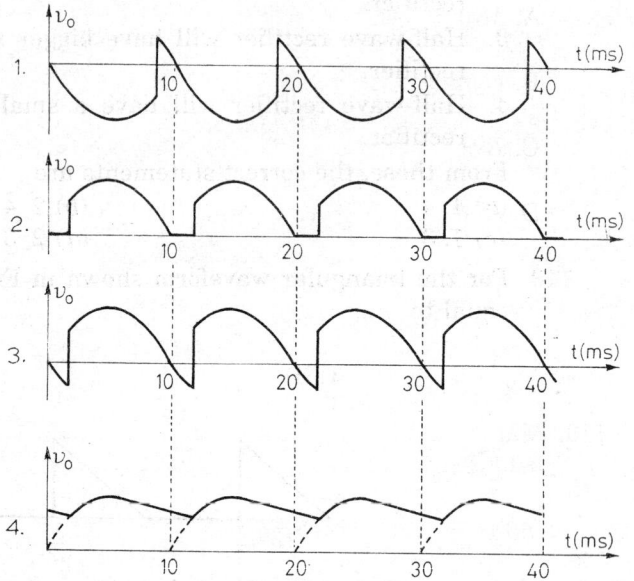

1.

2.

3.

4.

Codes :

	A	B	C	D		A	B	C	D
(a)	4	3	1	2	(b)	3	4	2	1
(c)	4	3	2	1	(d)	3	4	1	2

117. Match List I with List II and select the correct answer by using the codes given below (V_m = maximum value of source voltage) :

List I	*List II*
Single-phase converters	*Output voltage and PIV*

A. B-2

B. Half-wave

C. Semiconverter

D. M-2

1. $\dfrac{V_m}{\pi}(1+\cos\alpha)$, $PIV = V_m$

2. $\dfrac{2V_m}{\pi}\cos\alpha$, $PIV = 2V_m$

3. $\dfrac{V_m}{2\pi}(1+\cos\alpha)$, $PIV = V_m$

4. $\dfrac{2V_m}{\pi}\cos\alpha$, $PIV = V_m$

Codes :

	A	B	C	D		A	B	C	D
(a)	4	3	2	1	(b)	4	3	1	2
(c)	2	3	1	4	(d)	3	4	1	2

118. Match List I with List II and select the correct answer by using the codes given below (V_{mp} = maximum value of per-phase supply voltage) :

List I	List II
3-phase converters	*Output voltage*

A. M-3 $\qquad$ 1. $\dfrac{3V_{mp}}{\pi} \cos \alpha$

B. semiconverter $\qquad$ 2. $\dfrac{3\cdot\sqrt{3}}{2\pi} \cdot V_{mp} \ (1 + \cos \alpha)$

C. B-6 $\qquad$ 3. $\dfrac{3\cdot\sqrt{3}}{2\pi} V_{mp} \cos \alpha$

D. M-6 $\qquad$ 4. $\dfrac{3\sqrt{3}}{\pi} V_{mp} \cos \alpha$

Codes

	A	B	C	D			A	B	C	D
(a)	3	2	4	1		(b)	3	2	1	4
(c)	1	2	4	3		(d)	3	4	2	1

119. Match List-I with List-II and select the correct answer by using the codes given below the lists :

List I	List II
50-Hz system measurements :	*Waveforms*

A. Voltage across an *R-C*. (parallel) load connected through a full-wave bridge $\qquad$ 1.

B. Instantaneous power consumed by a resistor. $\qquad$ 2.

C. Output voltage of a positive clamped circuit. $\qquad$ 3.

D. Instantaneous power consumed by an *R-L* circuit. $\qquad$ 4.

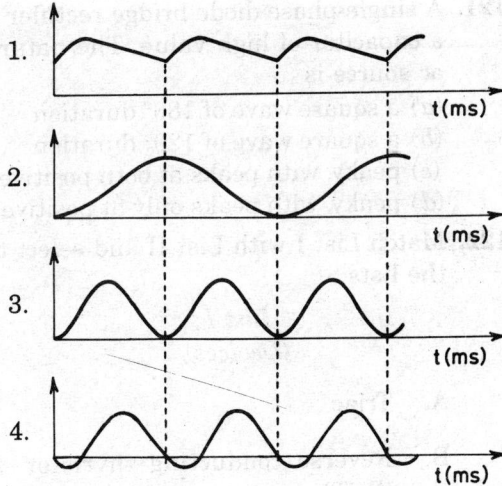

Codes :

	A	B	C	D			A	B	C	D
(a)	1	2	3	4		(b)	1	3	4	2
(c)	3	1	2	4		(d)	1	3	2	4

120. Match List I with List II and select the correct answer from the codes given below the lists :

	List I		List II	
	Controlled rectifiers with 50 Hz supply		*Output voltage waveforms*	

A. 1-phase full converter with   1.
   source inductance

B. 3-phase full converter   2.

C. 3-phase semiconverter   3.

D. 3-phase half-wave converter.   4.

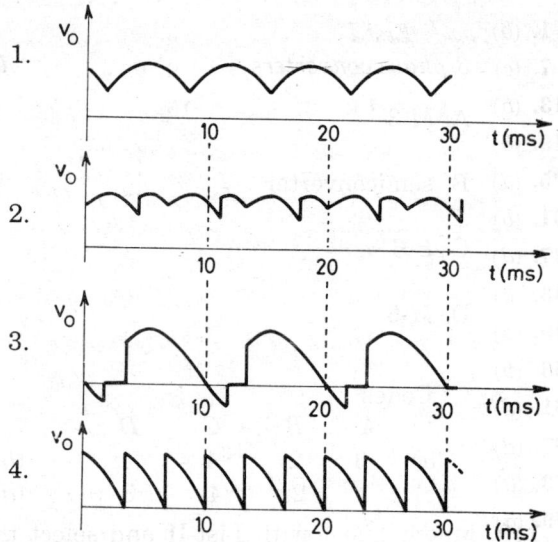

Codes

	A	B	C	D			A	B	C	D
(a)	3	4	2	1		(b)	3	4	1	2
(c)	4	3	2	1		(d)	4	3	1.	2

121. A single-phase diode bridge rectifier is feeding a parallel combination of R load and a capacitor of high value. The nature of input current drawn by the rectifier from ac source is

(a) a square wave of 180° duration

(b) a square wave of 120° duration

(c) peaky, with peaks at both positive and negative half cycles of the input voltage

(d) peaky, with peaks only at positive half cycle of the input voltage.

122. Match List I with List II and select the correct answer using the codes given below the lists :

	List I		List II	
	(Devices)		*(Properties)*	

A. Triac   1. Good $\dfrac{di}{dt}$ behaviour even at low gate currents

B. Reverse conducting thyristor (RCT)   2. Normally provided with a small continuous negative gate pulse during off state

C. Gate turn-off thyristor (G.T.O.)   3. Negative gate current for reverse conduction

D. Amplifying gate thyristor   4. No gate pulse for reverse conduction

Codes :

	A	B	C	D			A	B	C	D
(a)	4	3	1	2		(b)	3	4	2	1
(c)	3	4	1	2		(d)	4	3	2	1

ANSWERS

1. (b)	**2.** (a)	**3.** (c)	**4.** (a)	**5.** (d)	**6.** (b)
7. (c)	**8.** (c)	**9.** (b)	**10.** (b)	**11.** (d)	**12.** (a)
13. (b)	**14.** (a)	**15.** (b)	**16.** (d)	**17.** (a)	**18.** (d)
19. (b)	**20.** (c)	**21.** (c)	**22.** (b)	**23.** (d)	**24.** (a)
25. (d)	**26.** (d)	**27.** (c)	**28.** (a)	**29.** (c)	**30.** (b)
31. (b)	**32.** (d)	**33.** (d)	**34.** (d)	**35.** (b)	**36.** (d)
37. (d)	**38.** (a)	**39.** (b)	**40.** (b)	**41.** (b)	**42.** (a)
43. (b)	**44.** (b)	**45.** (a)	**46.** (c)	**47.** (b)	**48.** (a)
49. (a)	**50.** (c)	**51.** (c)	**52.** (d)	**53.** (c)	**54.** (a)
55. (b)	**56.** (c)	**57.** (b)	**58.** (a)	**59.** (c)	**60.** (b)
61. (d)	**62.** (b)	**63.** (d)	**64.** (a)	**65.** (b)	**66.** (c)
67. (d)	**68.** (a)	**69.** (d)	**70.** (c)	**71.** (d)	**72.** (b)
73. (d)	**74.** (a)	**75.** (a)	**76.** (c)	**77.** (d)	**78.** (b)
79. (a)	**80.** (d)	**81.** (c)	**82.** (b)	**83.** (c)	**84.** (a)
85. (d)	**86.** (d)	**87.** (a)	**88.** (c)	**89.** (d)	**90.** (b)
91. (d)	**92.** (d)	**93.** (b)	**94.** (b)	**95.** (a)	**96.** (c)
97. (c)	**98.** (b)	**99.** (a)	**100.** (c)	**101.** (a)	**102.** (c)
103. (b)	**104.** (b)	**105.** (a)	**106.** (b)	**107.** (d)	**108.** (b)
109. (a)	**110.** (d)	**111.** (c)	**112.** (a)	**113.** (a)	**114.** (b)
115. (d)	**116.** (c)	**117.** (b)	**118.** (b)	**119.** (d)	**120.** (a)
121. (d)	**122.** (b)				

CHOPPERS

1. In dc choppers, if T_{on} is the on-period and f is the chopping frequency, then output voltage in terms of input voltage V_s is given by
   (a) $V_s \cdot T_{on}/f$ (b) $V_s \cdot f/T_{on}$ (c) $V_s/f \cdot T_{on}$ (d) $V_s \cdot f \cdot T_{on}$

2. In dc choppers, the waveforms for input and output voltages are respectively
   (a) discontinuous, continuous   (b) both continuous
   (c) both discontinuous   (d) continuous, discontinuous.

3. A chopper can be used on
   (a) pulse-width modulation only   (b) frequency modulation only
   (c) amplitude modulation only   (d) both PWM and FM

4. In PWM method of controlling the average output voltage in a chopper,
   1. on-time T_{on} is varied and chopping frequency f is kept constant
   2. T_{on} is kept constant and f is varied
   3. both T_{on} and off-time T_{off} are varied and f is kept constant
   4. T_{off} is varied and T is kept constant.
   From above, the correct statements are
   (a) 1, 3   (b) 1, 3, 4   (c) 2, 3, 4   (d) 3, 4

5. In FM method of controlling the average output voltage in a chopper,
   1. on-time T_{cn} is kept constant and chopping period T is varied
   2. turn-off time T_{off} is kept constant and T is varied
   3. T_{on} is kept constant and T_{off} is varied
   4. T_{off} is kept constant and T_{on} is varied
   From these the correct statements are
   (a) 1, 3, 4 (b) 2, 3, 4 (c) 1, 2, 3, 4 (d) 1, 2, 3

6. A step-down chopper is operated in the continuous conduction mode in steady state
   with a constant duty ratio D. If V_o is the magnitude of the dc output voltage and if
   V_s is the magnitude of the dc input voltage, the ratio V_o/V_s is given by

   (a) D (b) $1 - D$ (c) $\dfrac{1}{1-D}$ (d) $\dfrac{D}{1-D}$

   (GATE, 2002)

7. A step-down chopper is operated in the discontinuous conduction mode in steady
   state with a constant duty ratio α. If t_x = extinction time, V_s = dc source voltage, T
   = chopping period and E = constant dc load voltage, then the magnitude of the
   average output voltage is given by

   (a) $\alpha V_s - \left(1 - \dfrac{t_x}{T}\right) E$ (b) $\alpha V_s + \left(1 - \dfrac{t_x}{T}\right) E$

   (c) $\alpha V_s + \left(1 + \dfrac{t_x}{T}\right) E$ (d) $\alpha V_s + \left(\dfrac{T}{t_x} - 1\right) E$

8. For type-A chopper, V_s is the source voltage, R is the load resistance and α is the
   duty cycle. The average output voltage and current for this chopper are respectively
   (a) $\alpha V_s, \alpha \cdot (V_s/R)$ (b) $(1 - \alpha) V_s, (1 - \alpha) V_s/R$
   (c) $V_s/\alpha, V_s/\alpha R$ (d) $V_s/(1 - \alpha), V_s/(1 - \alpha)R$.

9. A chopper has V_s as the source voltage, R as the load resistance and α as the duty
   cycle. For this chopper, rms value of output voltage is
   (a) αV_s (b) $\sqrt{\alpha} \cdot V_s$ (c) $V_s/\sqrt{\alpha}$ (d) $\sqrt{1 - \alpha} \cdot V_s$.

10. For a chopper, V_s is the source voltage, R is the load resistance and α is the duty
    cycle. Rms and average values of thyristor currents for this chopper are

    (a) $\alpha \cdot \dfrac{V_s}{R}, \sqrt{\alpha} \cdot \dfrac{V_s}{R}$ (b) $\sqrt{\alpha} \cdot \dfrac{V_s}{R}, \sqrt{\alpha} \cdot \dfrac{V_s}{R}$

    (c) $\sqrt{\alpha} \dfrac{V_s}{R}, \alpha \dfrac{V_s}{R}$ (d) $\sqrt{1 - \alpha} \cdot \dfrac{V_s}{R}, (1 - \alpha) V_s/R$.

11. In dc choppers, per unit ripple is maximum when duty cycle α is
    (a) 0.2 (b) 0.5
    (c) 0.7 (d) 0.9.

12. In the circuit shown in Fig. C. 41, $L = 5\,\mu H$ and
    $C = 20\,\mu F$. C is initially charged to 200 V. After the
    switch S is closed at $t = 0$, the maximum value of
    current and the time at which it reaches this value
    are, respectively,
    (a) 400 A, 15.707 μs (b) 50 A, 30 μs
    (c) 100 A, 62.828 μs (d) 400 A, 31.414 μs.

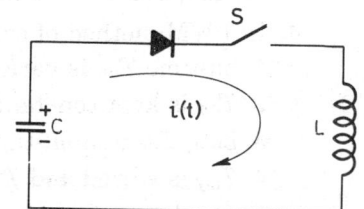

Fig. C.41

13. A voltage commutated chopper has the following parameters :

V_s = 200 V, Load circuit parameter : 1 Ω, 2 mH, 50 V

Commutation circuit parameters, L = 25 μH, C = 50 μF

T_{on} = 500 μs, T = 2000 μs

For a constant load current of 100 A, the effective on period and peak current through the main thyristor are respectively

(a) 1000 μs, 200 A   (b) 700 μs, 382.8 A

(c) 700 μs, 282.8 A   (d) 1000 μs, 382.8 A.

14. For the voltage-commutated chopper of Prob. 13, the turn-off times for main and auxiliary thyristors are, respectively,

(a) 120 μs, 60 μs   (b) 100 μs, 0.5 μs

(c) 120 μs, 55 μs   (d) 100 μs, 55.54 μs.

15. In the current-commutated chopper shown in Fig. C. 42, thyristor T1 is conducting a load current I_0. When thyristor TA is turned on, with capacitor polarity as shown, the capacitor current i_c would flow through.

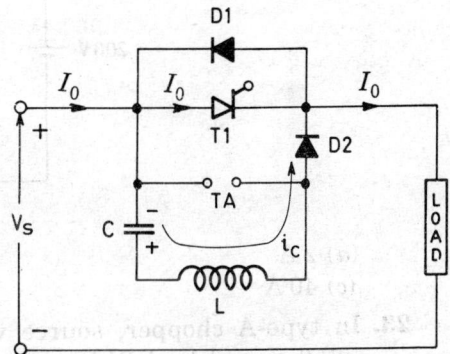

(a) diode D1 because it provides an easy path.

(b) thyristor T1 because it is already conducting

(c) diode D1 because thyristor T1 is unidirectional device and therefore current i_c cannot flow from cathode to anode

(d) SCR T1 because diode D1 is reverse biased by voltage drop across T1.

Fig. C.42

16. A load commutated chopper, fed from 200 V dc source, has a constant load current of 50 A. For a duty cycle of 0.4 and a chopping frequency of 2 kHz, the value of commutating capacitor and the turn-off time for one thyristor pair are respectively

(a) 25 μF, 100 μs   (b) 50 μF, 50 μs

(c) 25 μF, 25 μs   (d) 50 μF, 25 μs.

17. A dc battery is charged from a constant dc source of 200 V through a chopper. The dc battery is to be charged from its internal emf of 90 to 120 V. The battery has internal resistance of 1 Ω. For a constant charging current of 10 A, the range of duty cycle is

(a) 0.45 to 0.6   (b) 0.5 to 0.65   (c) 0.4 to 0.55   (d) 0.5 to 0.6

18. For type-A chopper ; V_s, R, I_0 and α are respectively the dc source voltage, load resistance, constant load current and duty cycle. For this chopper, average and rms values of freewheeling diode currents are

(a) $\alpha I_0, \sqrt{\alpha} \cdot I_0$ (b) $(1 - \alpha) I_0, \sqrt{1 - \alpha} \cdot I_0$

(c) $\alpha \cdot V_s/R, \sqrt{\alpha} \cdot V_s/R$ (d) $(1 - \alpha) I_0, \sqrt{\alpha} \cdot I_0$.

19. A step-up chopper has V_s as the source voltage and α as the duty cycle. The output voltage for this chopper is given by

(a) $V_s (1 + \alpha)$ (b) $V_s/(1 - \alpha)$ (c) $V_s (1 - \alpha)$ (d) $V_s/(1 + \alpha)$.

20. A dc chopper is fed from 100 V dc. Its load voltage consists of rectangular pulses of duration 1 msec in an overall cycle time of 3 msec. The average output voltage and ripple factor for this chopper are respectively

(a) 25 V, 1   (b) 50 V, 1

(c) 33.33 V, $\sqrt{2}$ (d) 33.33 V, 1

21. When a series LC circuit is connected to a dc supply of V volts through a thyristor, then the peak current through thyristor is

(a) $V \cdot \sqrt{LC}$ (b) $V/\sqrt{CL}$

(c) $V \cdot \sqrt{C/L}$ (d) $V \cdot \sqrt{L/C}$

22. For the arrangement shown in Fig. C. 43, the circuit is initially in steady state with thyristor T off. After thyristor T is turned on, the peak thyristor current would be

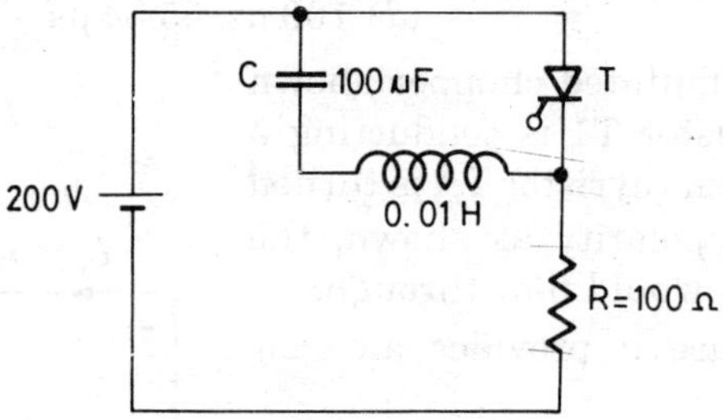

Fig. C.43

(a) 2 A   (b) 22 A

(c) 40 A   (d) 42 A.

23. In type-A chopper, source voltage is 100 V dc, on-period = 100 μs, off-period = 150 μs and load RLE consists of $R = 2\,\Omega$, $L = 5$ mH, $E = 10$ V. For continuous conduction, average output voltage and average output current for this chopper are respectively:

(a) 40 V, 15 A   (b) 66.66 V, 28.33 A

(c) 60 V, 25 A   (d) 40 V, 20 A.

24. Refer to the circuit in Fig. C. 44. The maximum current in the main SCR M can be

(a) 200 A   (b) 170.7 A

(c) 141.4 A   (d) 70.7 A.

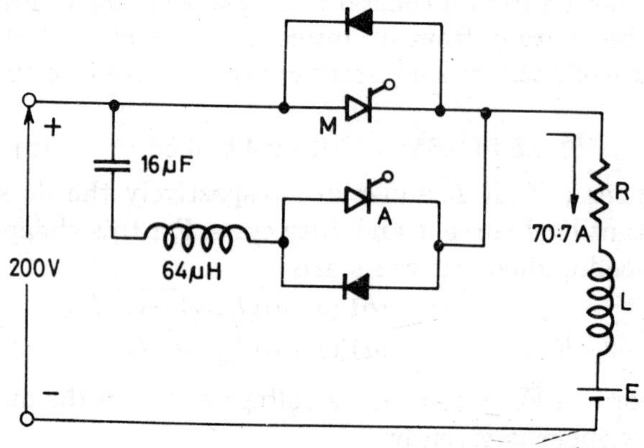

Fig. C.44

25. Refer to the circuit in Fig. C. 44. The maximum turn-off time of the main SCR M to ensure its proper commutation, in µs, is

(a) 2π (b) 4π

(c) 6π (d) 8π

26. Match List I with List II and give the correct answer by using the codes given below :

List I		List II
Chopper configurations		*Output Voltage Waveforms*

A. Voltage-commutated chopper   1.

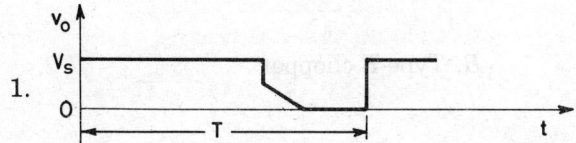

B. Load-commutated chopper   2.

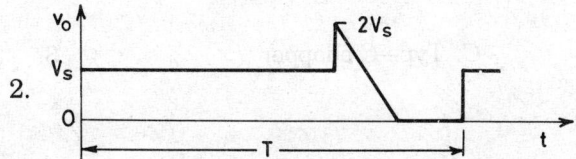

C. Current-commutated chopper   3.

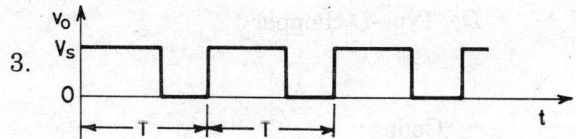

D. Ideal dc chopper   4.

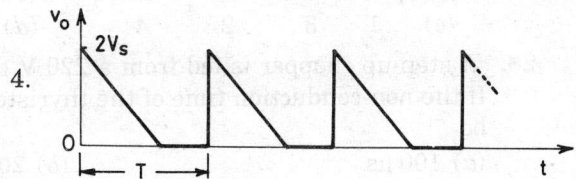

Codes :

	A	B	C	D		A	B	C	D
(a)	2	4	1	3	(b)	2	4	3	1
(c)	4	2	1	3	(d)	2	1	4	3

27. Match List I with List II and give the correct answer by using the codes given below the lists.

| | List I
(Types of choppers) | | List II
(Circuit Configurations) |

A. Type-A chopper   1.

B. Type-B chopper   2.

C. Type-C chopper   3.

D. Type-D chopper   4.

Codes :

	A	B	C	D		A	B	C	D
(a)	3	1	2	4	(b)	1	3	4	2
(c)	1	3	2	4	(d)	3	1	2	4

28. A step-up chopper is fed from a 220 V dc source to deliver a load voltage of 660 V. If the non-conduction time of the thyristor is 100 μs, the required pulse width would be
   (a) 100 μs   (b) 200 μs
   (c) 220 μs   (d) 660 μs

29. A chopper, where voltage as well as current remain negative, is known as
   (a) type-A   (b) type-B
   (c) type-C   (d) type-D

30. A chopper, in which current remains positive but voltage may be positive or negative, is known as
   (a) type-A   (b) type-B
   (c) type-C   (d) type-D

31. The freewheeling diode is subjected to double the source voltage in the following chopper configurations :
   1. Voltage commutated chopper
   2. Current commutated chopper

3. Load commutated chopper

4. Jone's chopper

From these, the correct statements are

(a) 1, 3, 4   (b) 1, 2, 3   (c) 1, 2, 4   (d) 2, 3, 4

32. The effective on period in a voltage commutated chopper

    (a) increases with load current I_o as well as with the commutating capacitance C

    (b) decreases with I_o as well as C

    (c) decreases with I_o but increases with C

    (d) increases with I_o but decreases with C

33. A step-down chopper operates from a dc voltage source V_s and feeds a dc motor armature with a back emf E_b. From oscilloscope traces, it is found that the current increases for time t_r, falls to zero over time t_f and remains zero for time t_o, in every chopping cycle. Than the average dc voltage across the freewheeling diode is

    (a) $V_s . t_r/(t_r + t_f + t_o)$

    (b) $(V_s . t_r + E_b . t_f)/(t_r + t_f + t_o)$

    (c) $(V_s . t_r + E_b . t_o)/(t_r + t_f + t_o)$

    (d) $[V_s . t_r + E_b (t_f + t_o)]/(t_r + t_f + t_o)$ [GATE, 2000]

34. Fig. C. 45 shows a step-down chopper switched at 1 kHz with a duty ratio $D = 0.5$. The peak-peak ripple in the load current is close to

    (a) 10 A   (b) 0.5 A

    (c) 0.125 A   (d) 0.25 A

                       [GATE, 2005]

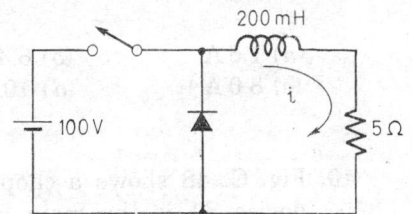

Fig. C.45

35. A dc chopper is fed from constant voltage mains. The duty ratio α of the chopper is progressively increased while the chopper feeds RL load. The per unit current ripple would

    (a) increase progressively

    (b) decrease progressively

    (c) decrease to a minimum value at $\alpha = 0.5$ and then increase

    (d) increase to a maximum value at $\alpha = 0.5$ and then decrease

36. In a two-quadrant dc to dc chopper, the load voltage is varied from positive maximum to negative maximum by varying the time-ratio of the chopper from

    (a) zero to unity   (b) unity to zero

    (c) zero to 0.5   (d) 0.5 to zero

37. A chopper circuit is operating on TRC principle at a frequency of 2 kHz on a 220 V dc supply. If the load voltage is 170 V, then the conduction period of thyristor in each cycle is

    (a) 3.86 ms   (b) 7.72 ms

    (c) 0.772 ms   (d) 0.386 ms

38. A chopper is employed to charge a battery as shown in Fig. C. 46. The charging current is 5 A. The duty ratio is 0.2. The chopper outut voltage is also shown in Fig. C. 46. The peak to peak ripple current in the charging current is

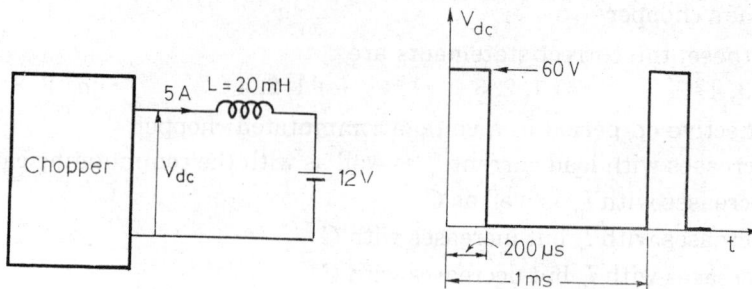

Fig. C.46

(a) 0.48 A   (b) 1.2 A
(c) 2.4 A   (d) 1 A

[GATE, 2003]

39. Fig. C. 47 shows a chopper operating from a 100 V dc input. The duty ratio of the main switch S is 0.8. The load is sufficiently inductive so that load current is ripple free. The average current through the diode D under steady state is

(a) 1.6 A   (b) 6.4 A
(c) 8.0 A   (d) 10.0 A

[GATE, 2005]

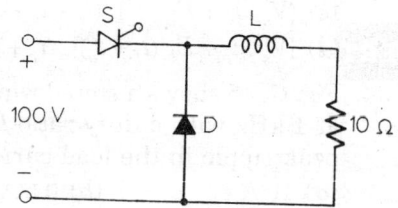

Fig. C.47

40. Fig. C. 48 shows a chopper. The device S1 is the main switching device. S2 is the auxiliary commutation device. S1 is rated for 400 V, 60 A. S2 is rated for 400 V, 30 A. The load current is 20 A. The main device operates with a duty ratio of 0.5. The peak current through S1 is

(a) 10 A   (b) 20 A
(c) 30 A   (d) 40 A

[GATE, 2004]

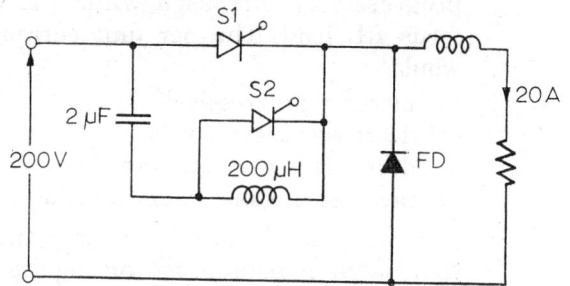

Fig. C.48

41. For eliminating fifth harmonic from the output voltage wave of a dc chopper, the ripple factor should be

(a) 1   (b) 2   (c) 3   (d) 4

42. In a chopper, for eliminating third harmonic from the output voltage wave, the duty cycle should be equal to

(a) 1/5   (b) 1/4   (c) 1/3   (d) 1/2

43. A chopper circuit, feed from an input voltage of 20 V dc, delivers a load power of 16 watts. For a chopper efficiency of 0.8, the input current is

(a) 0.64 A   (b) 0.8 A   (c) 1 A   (d) 1.25 A

44. Type-A chopper, fed from 200 V dc, is connected to load $R = 5\Omega$. This chopper operates with on and off periods of 2 ms and 3 ms respectively. The peak value of load current and ripple factor are

   (a) 16 A, 1.225   (b) 25.3 A, 1.225   (c) 24 A, 0.8165   (d) 40 A, 1.225

ANSWERS

1. (d)	2. (d)	3. (d)	4. (b)	5. (c)	6. (a)
7. (b)	8. (a)	9. (b)	10. (c)	11. (b)	12. (a)
13. (b)	14. (d)	15. (d)	16. (a)	17. (b)	18. (b)
19. (b)	20. (c)	21. (c)	22. (b)	23. (a)	24. (b)
25. (d)	26. (a)	27. (d)	28. (b)	29. (b)	30. (d)
31. (a)	32. (c)	33. (c)	34. (c)	35. (d)	36. (b)
37. (d)	38. (a)	39. (a)	40. (d)	41. (b)	42. (c)
43. (c)	44. (d)				

INVERTERS

1. If, for a single-phase half-bridge inverter, the amplitude of output voltage is V_s and the output power is P, then their corresponding values for a single-phase full-bridge inverter are

   (a) V_s, P (b) $2V_s, P$ (c) $2V_s, 2P$ (d) $2V_s, 4P$.

2. In voltage source inverters
   (a) load voltage waveform v_0 depends on load impedance Z, whereas load current waveform i_0 does not depend on Z
   (b) Both v_0 and i_0 depend on Z
   (c) v_0 does not depend on Z whereas i_0 depends on Z
   (d) both v_0 and i_0 do not depend upon Z.

3. A single-phase full bridge inverter can operate in load-commutation mode in case load consists of
   (a) RL (b) RLC underdamped
   (c) RLC overdamped   (d) RLC critically damped.

4. A single-phase bridge inverter delivers power to a series connected RLC load with $R = 2\ \Omega$, $\omega L = 8\ \Omega$. For this inverter-load combination, load commutation is possible in case the magnitude of $1/\omega C$ in ohms is
   (a) 10   (b) 8   (c) 6   (d) zero.

5. In the half-bridge inverter of Fig. C. 49, main thyristor T1 is conducting a load current. With polarity of the capacitor voltage as shown, when auxiliary thyristor TA1 is turned on, capacitor current i_c
   1. would flow through D1, because capacitor voltage v_c forward biases diode D1
   2. cannot flow through D1, because voltage drop across T1 reverse biases D1
   3. cannot flow through T1, because thyristor is unidirectional device
   4. would flow through T1, such that load current minus i_c flows from anode to cathode.

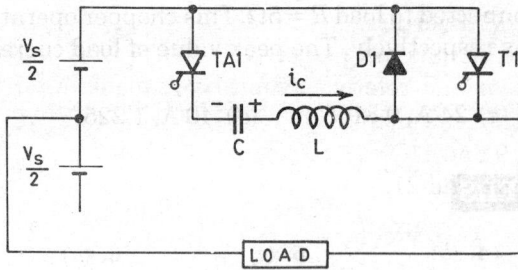

Fig. C.49

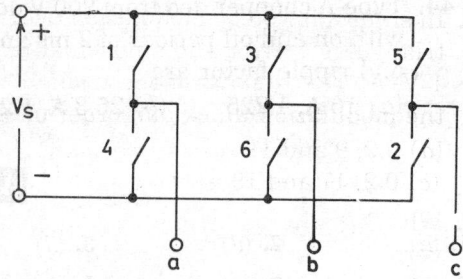

Fig. C.50

From these, the correct statements are

 (*a*) 1, 3   (*b*) 2, 4   (*c*) 3 only   (*d*) 4 only

6. For a 3-phase bridge inverter in 180° conduction mode, Fig. C. 50, the sequence of SCR conduction in the first two steps, beginning with the initiation of thyristor 1, is

 (*a*) 6, 1, 2 and 2, 3, 1   (*b*) 2, 3, 1 and 3, 4, 5
 (*c*) 3, 4, 5 and 5, 6, 1   (*d*) 5, 6, 1 and 6, 1, 2.

7. For a 3-phase bridge inverter in 120° conduction mode, Fig. C. 50, the sequence of SCR conduction in the first two steps, beginning with the initiation of thyristor 1, is

 (*a*) 6, 1 and 1, 2   (*b*) 1, 2 and 2, 3   (*c*) 1, 6 and 5, 6   (*d*) 1, 3 and 3, 4.

8. In single-pulse modulation of PWM inverters, third harmonic can be eliminated if pulse width is equal to

 (*a*) 30°   (*b*) 60°   (*c*) 120°   (*d*) 150°

9. In single-pulse modulation of PWM inverters, fifth harmonic can be eliminated if pulse width is equal to

 (*a*) 30°   (*b*) 72°   (*c*) 36°   (*d*) 108°.

10. In single-pulse modulation of PWM inverters, the pulse width is 120°. For an input voltage of 220 V dc, the r.m.s. value of output voltage is

 (*a*) 179.63 V   (*b*) 254.04 V
 (*c*) 127.02 V   (*d*) 185.04 V.

11. In single-pulse modulation used in PWM inverters, V_s is the input dc voltage. For eliminating third harmonic, the magnitude of rms value of fundamental component of output voltage and pulse width are respectively

 (*a*) $\dfrac{2\sqrt{2}}{\pi} V_s$, 120°   (*b*) $\dfrac{\sqrt{6}}{\pi} V_s$, 60°

 (*c*) $\dfrac{2\sqrt{2}}{\pi} V_s$, 60°   (*d*) $\dfrac{\sqrt{6}}{\pi} V_s$, 120°.

12. In multiple-pulse modulation used in PWM inverters, the amplitudes of reference square wave and triangular carrier wave are respectively 1 V and 2 V. For generating 5 pulses per half cycle, the pulse width should be

 (*a*) 36°   (*b*) 24°   (*c*) 18°   (*d*) 12°.

13. In multiple-pulse modulation used in PWM inverters, the amplitude and frequency for triangular carrier and square reference signals are respectively 4 V, 6 kHz and 1 V, 1 kHz. The number of pulses per half cycle and pulse width are respectively

 (*a*) 6, 90°   (*b*) 3, 45°   (*c*) 4, 60°   (*d*) 3, 40°.

14. In sinusoidal-pulse modulation used in PWM inverters, amplitude and frequency for triangular carrier and sinusoidal reference signals are respectively 5 V, 1 kHz and 1 V, 50 Hz. If zeros of the triangular carrier and reference sinusoid coincide, then the modulation index and order of significant harmonics are respectively
    (a) 0.2, 9 and 11   (b) 0.4, 9 and 11
    (c) 0.2, 17 and 19   (d) 0.2, 19 and 21.

15. Which of the following statement/statements is/are correct in connection with inverters:
    (a) VSI and CSI both require feedback diodes
    (b) Only CSI requires feedback diodes
    (c) GTOs can be used in CSI
    (d) Only VSI requires feedback diodes.

16. In a CSI, if frequency of output voltage is f Hz, then frequency of voltage input to CSI is
    (a) f (b) $2f$ (c) $f/2$ (d) $3f$.

17. In sinusoidal-pulse modulation used in PWM inverters, amplitude and frequency of triangular carrier and sinusoidal reference signals are respectively 5 V, 1 kHz and 1 V, 50 Hz. If peak of the triangular carrier coincides with the zero of the reference sinusoid, then the modulation index and order of significant harmonics are
    (a) 0.2, 9 and 11   (b) 0.4, 9 and 11
    (c) 0.2, 17 and 19   (d) 0.2, 19 and 21.

18. In sinusoidal PWM, there are 'm' cycles of the triangular carrier wave in the half cycle of reference sinusoidal signal. If zero of the reference sinusoid coincides with zero/peak of the triangular carrier wave, then number of pulses generated in each half cycle are respectively
    (a) $(m-1)/m$ (b) $(m-1)/(m-1)$
    (c) m/m (d) $m/(m-1)$.

19. In an inverter with fundamental output frequency of 50 Hz, if third harmonic is eliminated, then frequencies of other components in the output voltage wave, in Hz, would be
    (a) 250, 350, 450, high frequencies (b) 50, 250, 350, 450
    (c) 50, 250, 350, 550   (d) 50, 100, 200, 250.

20. A single-phase CSI has capacitor C as the load. For a constant source current, the voltage across the capacitor is
    (a) square wave   (b) triangular wave
    (c) step function   (d) pulsed wave.

21. A single-phase full bridge VSI has inductor L as the load. For a constant source voltage, the current through the inductor is
    (a) square wave   (b) triangular wave
    (c) sine wave   (d) pulsed wave.

22. A VSI will have better performance if its
    (a) load inductance is small and source inductance is large
    (b) both load inductance and source inductance are small
    (c) both load inductance and source inductance are large
    (d) load inductance is large and source inductance is small.

23. A series capacitor commutated inverter can operate satisfactorily if

(a) $\dfrac{1}{LC} > \dfrac{R^2}{4L^2}$ (b) $\dfrac{1}{LC} = \left(\dfrac{R}{2L}\right)^2$ (c) $\dfrac{1}{LC} < \left(\dfrac{R}{2L}\right)^2$

(d) irrespective of the values of R, L and C

24. Simplest method of eliminating third harmonic from the output voltage waveform of a single-phase bridge inverter is to use

(a) inverters in series (b) single-pulse modulation
(c) stepped-wave inverters (d) multiple-pulse modulation.

25. A PWM switching scheme is used in single-phase inverters to

(a) reduce the total harmonic distortion with modest filtering
(b) minimise the load on the dc side
(c) increase the life of the batteries
(d) reduce low-order harmonics and increase high-order harmonics.

26. In three-phase 180°-mode bridge inverter, the lowest order harmonic in the line to neutral output voltage (fundamental frequency output = 50 Hz) is

(a) 100 Hz (b) 150 Hz (c) 200 Hz (d) 250 Hz

27. A single-phase inverter has square wave output voltage. What is the percentage of the fifth harmonic component in relation to the fundamental component ?

(a) 40% (b) 30% (c) 20% (d) 10%

28. A time-margin for series inverter ensures

(a) low power loss (b) safety of the device
(c) improved power factor (d) absence of harmonics

29. A single-phase full-bridge VSI operating in square-wave mode supplies a purely inductive load. If the inverter time period is T, then the time duration for which each of the feedback diodes conduct in a cycle is

(a) T (b) $T/2$ (c) $T/4$ (d) $T/8$

30. Consider the following statements :

1. Inherent short-circuit operation 2. Regeneration capability
3. Need for inverter grade thyristors 4. Voltage spikes across the load

Which of these features are associated with CSI ?

(a) 1, 2 and 3 (b) 2, 3 and 4 (c) 1, 3 and 4 (d) 1, 2 and 4

31. The output voltages e_1 and e_2 of two full-bridge inverters are added using output transformers. In order to eliminate the fifth harmonic from the output voltage, the phase angle between e_1 and e_2 should be

(a) $\dfrac{\pi}{3}$ rad (b) $\dfrac{\pi}{4}$ rad

(c) $\dfrac{\pi}{5}$ rad (d) $\dfrac{\pi}{6}$ rad

32. Full bridge inverter is shown in Fig. C. 51. The maximum rms output voltage V_{01} at fundamental frequency is

(a) 24 V (b) 21.61 V
(c) 43.22 V (d) 48 V

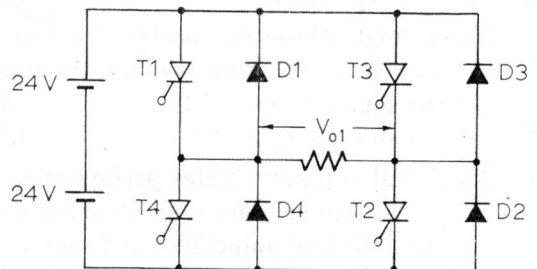

Fig. C.51

33. In a 1-phase bridge inverter, the maximum value of fundamental component of load current is I. For a load which is highly inductive in nature, the maximum value of nth harmonic component of load current would be

    (a) $\dfrac{I}{n}$ (b) $\dfrac{I}{n\sqrt{n}}$ (c) $\dfrac{I}{n^2}$ (d) I

34. In a 1-phase bridge inverter, the maximum value of fundamental component of load current is I. For a load which is highly capacitive in nature, the maximum value of nth harmonic component of load current would be

    (a) $\dfrac{I}{n}$ (b) $\dfrac{I}{n\sqrt{n}}$ (c) $\dfrac{I}{n^2}$ (d) I

35. In a 1-phase bridge inverter, the maximum value of fundamental component of load current is I. For a load which is highly resistive in nature, the maximum value of nth harmonic component of load current would be

    (a) $\dfrac{I}{n}$ (b) $\dfrac{I}{n\sqrt{n}}$ (c) $\dfrac{I}{n^2}$ (d) I

36. Output voltage of a single-phase bridge inverter, fed from a fixed dc source, is varied by
    (a) varying the switching frequency   (b) pulse-width modulation
    (c) pulse amplitude modulation   (d) all of the above.

37. A 3-phase VSI supplies a purely inductive three-phase load. Upon Fourier analysis, the output voltage waveform is found to have an h-th order harmonic of magnitude α_h times that of the fundamental component ($\alpha_h < 1$). The load current would then have an h-th order harmonic of magnitude
    (a) zero
    (b) α_h times the fundamental frequency component
    (c) $h \cdot \alpha_h$ times the fundamental frequency component
    (d) α_h/h times the fundamental frequency component.   [GATE, 2000]

38. A 3-phase VSI supplies a purely capacitive 3-phase load. Upon Fourier analysis, the output voltage waveform is found to have an h-th order harmonic of magnitude α_h times that of the fundamental component ($\alpha_h < 1$). The load current would then have an h-th order harmonic of magnitude
    (a) zero
    (b) α_h times the fundamental frequency component
    (c) $h \cdot \alpha_h$ times the fundamental frequency component
    (d) α_h/h times the fundamental frequency component.

39. A 3-phase VSI supplies a purely resistive three-phase load. Upon Fourier analysis, the output voltage waveform is found to have an h-th order harmonic of magnitude α_h times that of the fundamental component ($\alpha_h < 1$). The load current would then have an h-th order harmonic of magnitude
    (a) zero
    (b) α_h times the fundamental frequency component
    (c) $h \cdot \alpha_h$ times the fundamental frequency component
    (d) α_h/h times the fundamental frequency component.

40. What is the rms value of the voltage waveform shown in Fig. C. 52 ?

(a) $\dfrac{200}{\pi}$ V   (b) $\dfrac{100}{\pi}$ V

(c) 200 V   (d) 100 V

[GATE, 2002]

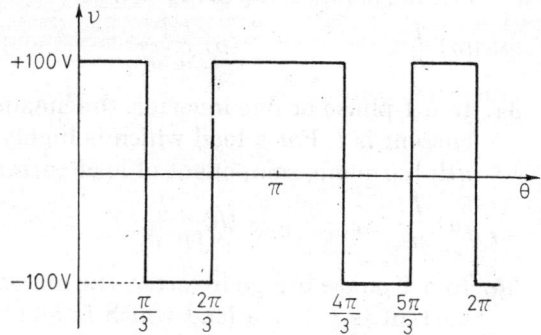

Fig. C.52

41. The output voltage waveform of a 3-phase square-wave inverter contains

(a) only even harmonics
(b) both odd and even harmonics
(c) only odd harmonics
(d) only triplan harmonics

42. Fig. C. 53 (a) shows an inverter circuit with a dc source voltage V_s. The semiconductor switches of the inverter are operated in such a manner that the pole voltages v_{10} and v_{20} are as shown in Fig. C. 53 (b). What is the rms value of the pole to pole voltage v_{12} ?

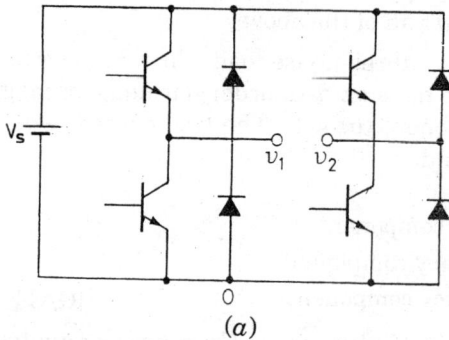

(a)

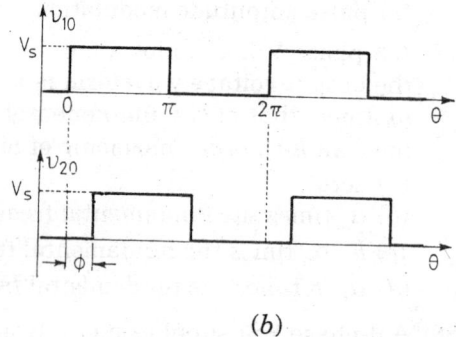

(b)

Fig. C.53

(a) $\dfrac{V_s \cdot \phi}{\pi \sqrt{2}}$ (b) $V_s \sqrt{\dfrac{\phi}{\pi}}$ (c) $V_s \cdot \sqrt{\dfrac{\phi}{2\pi}}$ (d) $\dfrac{V_s}{\pi}$ [GATE, 2002]

[**Hint.** Here $v_{12} = v_{10} - v_{20}$]

43. An inverter has a periodic output voltage with the output waveform as shown in Fig. C. 54. When conduction angle $\alpha = 120°$, the rms fundamental component of the output voltage is

(a) 0.78 V
(b) 1.10 V
(c) 0.9 V
(d) 1.27 V

[GATE, 2003]

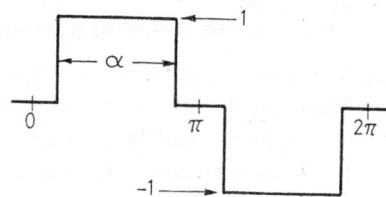

Fig. C.54

44. With reference to the output voltage waveform given in Fig. C. 54, the output of the converter will be free from 5th harmonic when

(a) $\alpha = 72°$ (b) $\alpha = 36°$ (c) $\alpha = 120°$ (d) $\alpha = 150°$

[GATE, 2003]

45. The output voltage waveform of a 3-phase square-wave inverter contains no third harmonics in
    1. line voltages in 180° mode
    2. phase voltages in 120° mode
    3. lirre voltages in 120° mode
    4. phase voltages in 180° mode

    From these, the correct statements are
    (a) 1, 3, 4   (b) 1, 2, 3   (c) 2, 3, 4   (d) 1, 2, 3, 4

46. McMurray commutation is superior to parallel capacitor commutation in respect of
    (a) number of components
    (b) overvoltage spike at the output
    (c) instantaneous reduction in SCR current
    (d) trigger circuit

47. A single-phase bridge inverter can be designed by having thyristors without forced commutation circuitry if the load it is handling is
    (a) series combination of resistance and a large inductance
    (b) series combination of resistance and a large capacitance
    (c) series combination of resistance, capacitance and inductance with resonant frequency of the circuit being lower than the inverter switching frequency
    (d) series combination of resistance, inductance and capacitance with resonant frequency of the circuit being higher than the inverter switching frequency.

48. The power delivered to a star-connected load of $R \, \Omega$ per phase, from a 3-phase bridge inverter fed from fixed dc source, is 10 kW for 180° mode. For 120° mode, the power delivered to load would be
    (a) 10 kW   (b) 5 kW   (c) 6.667 kW   (d) 7.5 kW

49. Match List I with List II and select the correct answer using the codes given below the lists :

List I	List II
A. Freewheeling diode	1. Voltage spikes in the output voltage
B. Feedback diode	2. Peaks in the inverter current
C. Current source inverter	3. Inductive loads of phase-controlled converters
D. Voltage source inverter	4. Inductive loads of dc to ac inverters

    Codes

	A	B	C	D		A	B	C	D
(a)	4	3	1	2	(b)	3	4	1	2
(c)	3	4	2	1	(d)	4	3	2	1

50. Control of frequency and control of voltage in 3-phase inverters operating in 120° mode or 180° mode of conduction is
    (a) possible only through inverter control circuit
    (b) possible through the control circuit of inverter and converter simultaneously
    (c) possible through inverter control for frequency and through converter control for voltage
    (d) possible through converter control only

51. In a series resonant inverter
    (a) the load current has square waveform
    (b) trigger frequency is higher than damped resonant frequency
    (c) change of frequency does not alter transferred power
    (d) output voltage depends upon damping factor of the load

52. In McMurray commutation circuit, the circuit turn-off time is

(a) dependent on load current and independent of operating frequency

(b) dependent on load current and also on load power factor

(c) independent of load current and dependent on operating frequency

(d) independent of load current and dependent on recovery period

53. In a 3-phase bridge inverter, the line to line voltage waveform is

1. square wave for 180° mode   2. square wave for 120° mode

3. stepped wave for 180° mode   4. stepped wave for 120° mode

From these, the correct statements are

(a) 1, 3   (b) 2, 3   (c) 1, 4   (d) 2, 4

54. In a 3-phase 180° mode bridge inverter, feeding a star-connected load with open neutral, the third harmonic component will be present in

(a) voltage of each inverter phase with respect to the mid-point of the dc source

(b) line to line output voltage

(c) line currents

(d) none of the above

55. A single-phase bridge inverter shown in Fig. C. 55. has an ideal transformer with primary turns equal to 10. For obtaining a fundamental frequency output voltage of 240 V, the number of secondary turns in transformer should be equal to (take $\pi = 3$)

(a) 120   (b) 150   (c) 150 $\sqrt{2}$ (d) 150/$\sqrt{2}$

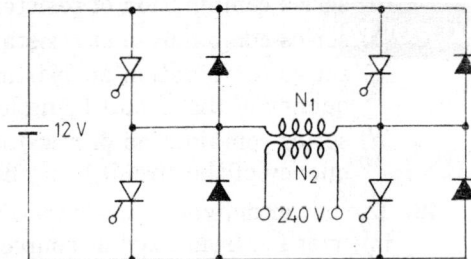

Fig. C.55

56. The operating frequency of a self-oscillating inverter using a saturable core is dependent upon

(a) battery voltage only

(b) battery voltage and saturation flux density

(c) battery voltage, saturation flux density and number of turns on primary winding

(d) load circuit power factor

57. Match List I with List II and select the correct answer using the codes given below the lists.

List I	*List II*
A. Phase-controlled rectifier feeding RL load with perfect smoothing	1. Depends on the values of R and L of the load
B. Single-pulse converter feeding RL load	2. Depends on firing angle
C. A constant dc voltage fed dc to ac inverter feeding RL load	3. Constant and independent of R and L of the load
D. A constant dc current-fed dc to ac inverter feeding RL load	4. Depends on firing angle and also impedance angle of the load.

Codes

	A	B	C	D			A	B	C	D
(a)	2	3	4	1		(b)	1	4	3	2
(c)	1	3	4	2		(d)	2	4	3	1

1. (d)	**2.** (c)	**3.** (b)	**4.** (a)	**5.** (b)	**6.** (d)
7. (a)	**8.** (c)	**9.** (b)	**10.** (a)	**11.** (d)	**12.** (c)
13. (b)	**14.** (c)	**15.** (d)	**16.** (b)	**17.** (d)	**18.** (a)
19. (c)	**20.** (b)	**21.** (b)	**22.** (b)	**23.** (a)	**24.** (b)
25. (d)	**26.** (d)	**27.** (c)	**28.** (b)	**29.** (c)	**30.** (d)
31. (c)	**32.** (c)	**33.** (c)	**34.** (d)	**35.** (a)	**36.** (b)
37. (d)	**38.** (c)	**39.** (b)	**40.** (d)	**41.** (c)	**42.** (b)
43. (a)	**44.** (a)	**45.** (d)	**46.** (b)	**47.** (d)	**48.** (d)
49. (b)	**50.** (c)	**51.** (d)	**52.** (a)	**53.** (c)	**54.** (a)
55. (c)	**56.** (d)	**57.** (d)			

AC VOLTAGE CONTROLLERS

1. A single-phase voltage controller feeds an induction motor (A) and a heater (B)
   (a) In both the loads, fundamental and harmonics are useful
   (b) In A only fundamental and in B only harmonics are useful
   (c) In A only fundamental and in B harmonics as well as fundamental are useful
   (d) In A only harmonics and in B only fundamental are useful.

2. A load resistance of 10 Ω is fed through a 1-phase voltage controller from a voltage source of 200 sin 314 t. For a firing angle delay of 90°, the power delivered to load in kW, is
   (a) 0.5   (b) 0.75   (c) 1   (d) 2.

3. A single-phase voltage controller is employed for controlling the power flow from 260 V, 50 Hz source into a load consisting of $R = 5$ Ω and $\omega L = 12$ Ω. The value of maximum rms load current and the firing angle are respectively
   (a) 20 A, 0°   (b) $\dfrac{260}{10.91}$ A, 0°   (c) 20 A, 90°   (d) $\dfrac{260}{10.91}$ A, 90°.

4. A load, consisting of $R = 10$ Ω and $\omega L = 10$ Ω, is being fed from 230 V, 50 Hz source through a 1-phase voltage controller. For a firing angle delay of 30°, the rms value of load current would be
   (a) 23 A   (b) $\dfrac{23}{\sqrt{2}}$ A   (c) $> \dfrac{23}{\sqrt{2}}$ A   (d) $< \dfrac{23}{\sqrt{2}}$ A.

5. In a single-phase voltage controller with RL load, ac output power can be controlled if
   (a) firing angle $\alpha > \phi$ (load phase angle) and conduction angle $\gamma = \pi$
   (b) $\alpha > \phi$ and $\gamma < \pi$
   (c) $\alpha < \phi$ and $\gamma = \pi$
   (d) $\alpha < \phi$ and $\gamma > \pi$.

6. A single-phase voltage controller feeds power to a resistance of 10 Ω. The source voltage is 200 V rms. For a firing angle of 90°, the rms value of thyristor current in amperes is
   (a) 20   (b) 15   (c) 10   (d) 5.

7. A single-phase voltage controller is connected to a load of resistance 10 Ω and a supply of 200 sin 314*t* volts. For a firing angle of 90°, the average thyristor current in amperes is

(*a*) 10   (*b*) 10/π
(*c*) 5√2/π   (*d*) 5√2.

8. A single-phase voltage controller, using two SCRs in antiparallel, is found to be operating as a controlled rectifier. This is because

(*a*) load is *R* and pulse gating is used
(*b*) load is *R* and high-frequency carrier gating is used
(*c*) load is *RL* and pulse gating is used
(*d*) load is *RL* and continuous gating is used.

9. A single-phase ac voltage controller (or regulator) fed from 50 Hz system supplies a load having resistance and inductance of 2.0 Ω and 6.36 mH respectively. The control range of firing angle for this regulator is

(*a*) 0° < α < 180°   (*b*) 45° < α < 180°
(*c*) 90° < α < 180°   (*d*) 0° < α < 45°.

10. Two identical SCRs are connected back to back in series with a load. If each SCR is fired at 90°, a PMMC voltmeter across the load would read

(*a*) peak voltage   (*b*) $\dfrac{1}{\pi} \times$ peak voltage

(*c*) zero   (*d*) $\dfrac{1}{2} \times$ peak voltage

11. Two identical SCRs, connected back to back, feed a load *R*. If each SCR is fired at 90°, a MI voltmeter across the load would read

(*a*) $\dfrac{1}{2} \times$ peak voltage   (*b*) peak voltage

(*c*) $\dfrac{1}{\pi} \times$ peak voltage   (*d*) zero

12. A purely inductive load is controlled by a single-phase ac voltage controller using back to back connected SCRs. If firing angle of each SCR is 75°, the current through two SCRs will flow for

(*a*) 285° and 0°   (*b*) 210° and 0°
(*c*) 105° and 105°   (*d*) 180° and 180°

13. A purely inductive load is controlled by a single-phase ac voltage controller using back to back connected SCRs. If firing angle of each SCR is 100°, the current through two SCRs will flow for

(*a*) 180° and 180°   (*b*) 160° and 160°
(*c*) 100° and 100°   (*d*) 160° and 0°

14. A single-phase voltage controller, using one SCR in antiparallel with a diode, feeds a load *R* and the supply is 230 V, 50 Hz. For a firing angle of 90° for the SCR, a PMMC voltmeter connected across the load would read

(*a*) zero   (*b*) – 51.8 V   (*c*) 51.8 V   (*d*) – 36.63 V

15. A single-phase voltage controller, using back to back connected an SCR and a diode, feeds a load *R* from 200 V, 50 Hz source. For a firing angle of 90° for the SCR, a MI voltmeter across the load would read

(*a*) 230 V   (*b*) 173.2 V   (*c*) – 173.2 V   (*d*) 51.8 V

16. A single-phase half wave ac voltage controller feeds a load R. For a firing angle of 180°, a PMMC voltmeter across the load would read

(a) $\frac{1}{2} \times$ peak voltage

(b) $-\frac{1}{\pi} \times$ peak voltage

(c) $-\frac{1}{2\pi} \times$ peak voltage

(d) zero

17. A single-phase half-wave ac voltage controller feeds a load R. For a firing angle of 180°, a MI voltmeter across the load would read

(a) $\frac{1}{2} \times$ peak voltage

(b) $-\frac{1}{\pi} \times$ peak voltage

(c) $-\frac{1}{2\pi} \times$ peak voltage

(d) peak voltage

18. A single-phase voltage controller has input voltage of 240V, 50Hz and a load $R = 5\,\Omega$. For three cycles on and two cycles off, a PMMC voltmeter across the load would read

(a) 160 V

(b) 80 V

(c) zero

(d) 195.96 V

19. A single-phase voltage controller has input voltage of 240 V, 50 Hz and a load $R = 5\,\Omega$. For three cycles on and two cycles off, a MI voltmeter across the load would read

(a) 144 V

(b) 151.79 V

(c) 185.9 V

(d) 96 V

20. A single-phase voltage controller has input voltage of 240V, 50Hz and a load of $R = 6\,\Omega$. For 3 cycles on and two cycles off, the load would consume a power of

(a) 2880 W

(b) 5760 W

(c) 3456 W

(d) 11520 W

21. In a single-phase voltage controller feeding RL load, when

1. firing angle $\alpha < \phi$ (load phase angle), load voltage v_o is sinusoidal

2. $\alpha > \phi$, v_o is non-sinusoidal

3. $\alpha < \phi$, v_o is non-sinusoidal

4. $\alpha = \phi$, v_o is sinusoidal

From these, the correct statements are

(a) 2, 3, 4

(b) 1, 3, 4

(c) 1, 2, 4

(d) 1, 4

22. In a single-phase voltage controller with RL load, α is the firing angle, ϕ is the load phase angle and β is the extinction angle. For this voltage controller, output power can be controlled if $\alpha > \phi$ and

1. $(\beta - \alpha) = \pi$

2. $(\beta - \alpha) < \pi$

3. $\beta > \pi$

4. $\beta < \pi$

5. $\beta = \pi$

From these, the correct statements are

(a) 2, 3

(b) 2, 5

(c) 2, 4

(d) 1, 5

23. A single-phase voltage controller, using a triac, controls the ac output power to the resistive load $R = 10\ \Omega$. The input voltage is $230\ \sqrt{2}\ \sin \omega t$ and firing angle of the triac is 45°. The peak power dissipation in the load is

(a) 3968 W   (b) 5290 W

(c) 7935 W·   (d) 10580 W   [GATE, 2004]

24. A single-phase voltage controller, using a triac, controls the ac output power to the resistive load $R = 10\ \Omega$. The input voltage is $230\ \sqrt{2}\ \sin \omega t$ and firing angle of the triac is 120°. The peak power dissipation in the load is

(a) 3968 W   (b) 5290 W

(c) 7935 W   (d) 10580 W

25. Output voltage wave from a converter is distorted and is found to have fundamental and third harmonic components. The distorted voltage wave differs from fundamental voltage as under :

1. Peak value of distorted wave (DW) > Peak value of fundamental wave (FW)

2. Rms value of DW > Rms value of FW

3. Average value of DW > Average value of FW

From these, the correct statements are

(a) 1, 2   (b) 2, 3

(c) 2 only   (d) 1, 2, 3

26. Output voltage wave from a converter is distorted and is found to have fundamental and second harmonic components. The distorted voltage wave differs from fundamental voltage as under :

1. Peak value of distorted wave (DW) > Peak value of fundamental wave (FW)

2. Rms value of DW > Rms value of FW

3. Average value of DW > Average value of FW

4. Average value of DW over half cycle > Average value of FW over half cycle

5. Average value of DW over half cycle = Average value of FW over half cycle

From above, the correct statements are

(a) 1, 2, 5   (b) 1, 2

(c) 2, 3, 5   (d) 1, 2, 4

27. When a single-phase ac voltage controller supplies power to an inductive load, control is lost if :

(a) $\alpha < \beta - \pi$ (b) $\alpha = \beta$

(c) $\alpha > \beta - \phi$ (d) $\alpha = \beta - \phi$

where α = firing angle, β = extinction angle with triggering of one of the SCRs and ϕ = load power factor angle.

28. Integral cycle control

(a) is very fast in action

(b) does not introduce sub-harmonics in the supply lines which are difficult to filter

(c) cannot be used on inductive loads

(d) can be advised only for loads with high time constants and limited range control.

29. Match List-I with List-II and select the correct answer using the codes given below the lists :

List I
(Power circuit diagrams)

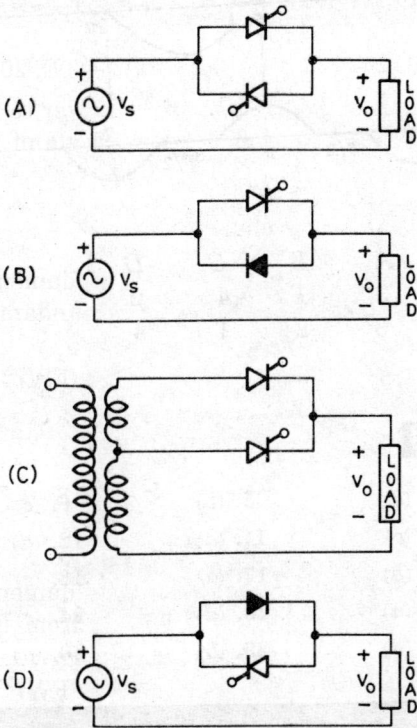

List II
(Output voltage waveforms)

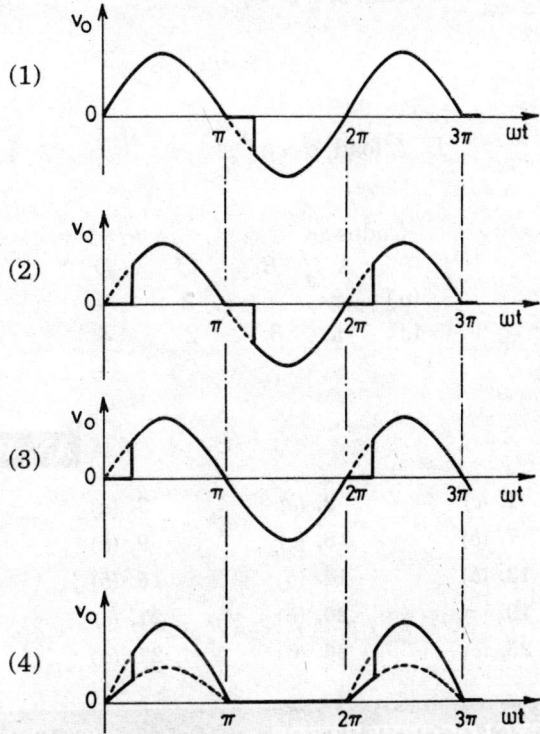

(A)

(1)

(B)

(2)

(C)

(3)

(D)

(4)

Codes :

	A	B	C	D		A	B	C	D
(a)	2	3	1	4	(b)	2	3	4	1
(c)	3	2	4	1	(d)	2	4	3	1

30. Match List I with List II and select the correct answer using the codes given below the lists :

List I
(Firing angle = α, load phase
angle = φ. Nature of load on single-
phase voltage controllers)

List II
(Load Current Waveforms)

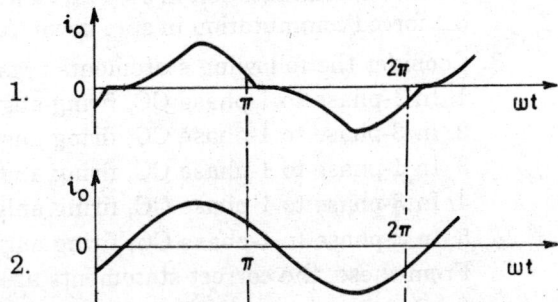

A. R load

1.

B. RL load, α > φ

2.

C. RL load, $\alpha < \phi$ 3.

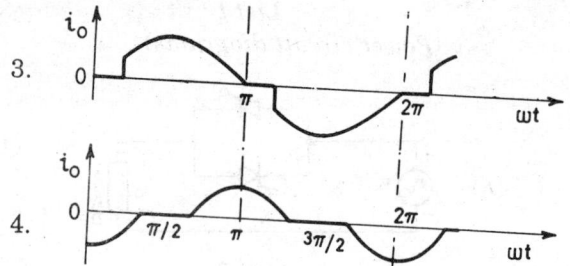

D. L load, $\alpha > \phi$ 4.

Codes :

	A	B	C	D			A	B	C	D
(a)	3	1	2	4		(b)	3	1	4	2
(c)	1	3	2	4		(d)	3	2	1	4

ANSWERS

1. (c)	**2.** (c)	**3.** (a)	**4.** (b)	**5.** (b)	**6.** (c)
7. (b)	**8.** (c)	**9.** (b)	**10.** (c)	**11.** (a)	**12.** (d)
13. (b)	**14.** (b)	**15.** (b)	**16.** (b)	**17.** (a)	**18.** (c)
19. (c)	**20.** (b)	**21.** (c)	**22.** (a)	**23.** (d)	**24.** (c)
25. (c)	**26.** (a)	**27.** (a)	**28.** (d)	**29.** (b)	**30.** (a)

CYCLOCONVERTERS

1. A cycloconverter is a frequency converter from
   1. higher to lower frequency with one-stage conversion
   2. higher to lower frequency with two-stage conversion
   3. lower to higher frequency with one-stage conversion
   4. ac at one frequency to dc and then dc to ac at a different frequency
   From these, the correct statements are
   (a) 2, 4   (b) 1 only   (c) 2, 3   (d) 1, 3

2. The cycloconverters (CCs) require natural or forced commutation as under :
   (a) natural commutation in both step-up and step-down CCs
   (b) forced commutation in both step-up and step-down CCs
   (c) forced commutation in step-up CCs
   (d) forced commutation in step-down CCs.

3. Consider the following statements regarding cycloconverters :
   1. In 1-phase to 1-phase CC, firing angle may be varied
   2. In 3-phase to 1-phase CC, firing angle may be kept constant
   3. In 1-phase to 1-phase CC, firing angle may be kept constant
   4. In 3-phase to 1-phase CC, firing anlge may be varied
   5. In 3-phase to 1-phase CC, firing angle must be varied.
   From these, the correct statements are
   (a) 2, 4, 5   (b) 1, 3, 5   (c) 2, 3, 5   (d) 2, 3, 4

4. Three-phase to three-phase cycloconverters employing 18 SCRs and 36 SCRs have the same voltage and current ratings for their component thyristors. The ratio of VA rating of 36-SCR device to that of 18-SCR device is
   (a) 1/2   (b) 1
   (c) 2   (d) 4.

5. Three-phase to 3-phase cycloconverters employing 18 SCRs and 36 SCRs have the same voltage and current ratings for their component thyristors. The ratio of power output from 36-SCR converter to that outpatted by 18-SCR converter is
   (a) 4   (b) 2
   (c) 1   (d) 1/2.

6. The number of thyristors required for single-phase to single-phase cycloconverter of the mid-point type and for three phase to three-phase 3-pulse type cycloconverter are respectively
   (a) 4, 6   (b) 8, 18
   (c) 4, 18   (d) 4, 36.

7. A 3-phase to single-phase conversion device employs a 6-pulse bridge cycloconverter. For an input voltage of 200 V per phase, the fundamental rms value of output voltage is
   (a) $600/\pi$ V   (b) $300\sqrt{3}/\pi$ V
   (c) $300/\pi$ V   (d) $600\sqrt{3}/\pi$ V.

8. A three-phase to single-phase cycloconverter consists of positive and negative group of converters. In this device one of the two component converters would operate as a
   1. rectifier if the output voltage V_0 and output current I_0 have the same polarity
   2. inverter if V_0 and I_0 have the same polarity
   3. rectifier if V_0 and I_0 are of opposite polarity
   4. inverter if V_0 and I_0 are of opposite polarity.
   From above, the correct statements are
   (a) 1, 4   (b) 2, 3
   (c) 3, 4   (d) 1, 2

9. A 3-phase to 3-phase cycloconverter requires
   1. 18 SCRs for 3-pulse device
   2. 18 SCRs for 6-pulse device
   3. 36 SCRs for 3-pulse device
   4. 36 SCRs for 6-pulse device.
   From these, the correct statements are
   (a) 1, 3   (b) 2, 3
   (c) 2, 4   (d) 1, 4

10. Which of the following statements are correct for cycloconverters ?
    1. Step-down cycloconverter (cc) works on natural commutation
    2. Step-up cc requires forced commutation
    3. Load commutated cc works on line commutation
    4. Load commutated cc requires a generated emf in the load circuit.

    From above, the correct statements are
    (a) 1, 2   (b) 1, 2, 4
    (c) 2, 3, 4   (d) 1, 2, 3

11. Match List I with List II and select the correct answer using the codes given below the lists :

List I	List II
(Power electronic controller)	(Applications)
A. Controlled rectifier	1. Aircraft supplies
B. Chopper	2. Electric car
C. Cycloconverter	3. Induction heating
D. Inverter	4. Rolling mill drive.

Codes :

	A	B	C	D			A	B	C	D
(a)	4	2	3	1		(b)	2	4	1	3
(c)	4	2	1	3		(d)	4	1	2	3

12. Match List I with List II and select the correct answer using the codes given below the lists :

List I	List II
(Power electronic controller)	(Applications)
A. Controlled rectifier	1. High-power ac drive
B. Voltage controller	2. Solar cells
C. Cycloconverter	3. Ceiling fan drive
D. Inverter	4. Magnet power supply.

Codes :

	A	B	C	D			A	B	C	D
(a)	4	3	1	2		(b)	4	3	2	1
(c)	3	4	1	2		(d)	4	1	3	2

13. Match List I with List II and select the correct answer using the codes given below the lists :

List I	List II
(Power electronic controller)	(Applications)
A. Inverter	1. Fork-lift truck
B. Controlled rectifier	2. Illumination control
C. Voltage controller	3. Uninterruptible power supply
D. Chopper	4. Hydrogen production.

Codes :

	A	B	C	D			A	B	C	D
(a)	4	3	2	1		(b)	3	4	2	1
(c)	3	4	1	2		(d)	3	2	4	1

'14. Match List I with List II and select the correct answer using the codes given below the lists :

List I
(Types of cycloconverters)

A. 1-phase to 1-phase with continuous conduction

B. 1-phase to 1-phase with discontinuous conduction

C. Step-up device

D. 3-phase to 1-phase device

List II
(Output Voltage Waveforms)

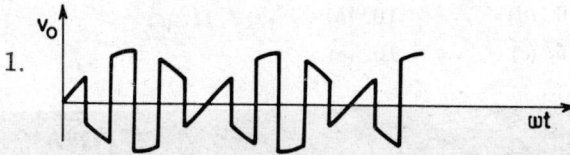

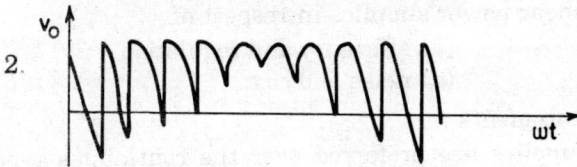

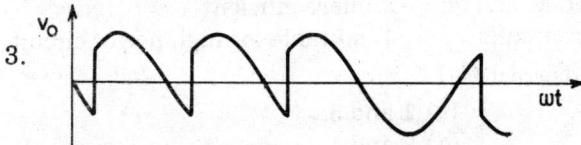

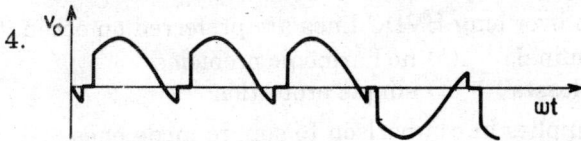

Codes :

	A	B	C	D		A	B	C	D
(a)	3	4	2	1	(b)	4	3	1	2
(c)	3	1	4	2	(d)	3	4	1	2

15. In a 3-phase to 1-phase cycloconverter employing 3-pulse positive and negative group converters, if the input voltage is 200 V per phase, the fundamental rms value of output voltage would be

(a) $\dfrac{600}{\pi}$ V   (b) $300\sqrt{3}$ V   (c) $\dfrac{300\sqrt{3}}{\pi}$ V   (d) $\dfrac{300}{\pi}$ V

16. In a single-phase cycloconverter arrangement shown in Fig. C.56, the positive direction of currents $i_{p1}, i_{p2}, i_{n1}, i_{n2}$ is as indicated. The turns ratio from primary to each secondary is unity. The source current i_s is given by

(a) $i_s = i_{p1} + i_{p2} + i_{n1} + i_{n2}$

(b) $i_s = i_{p2} - i_{p1} - i_{n1} + i_{n2}$

(c) $i_s = i_{p1} - i_{p2} - i_{n1} + i_{n2}$

(d) $i_s = i_{p1} + i_{p2} - i_{n1} - i_{n2}$

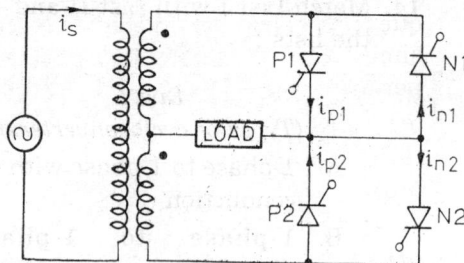

Fig. C.56

ANSWERS

1. (d)	2. (c)	3. (b)	4. (c)	5. (a)	6. (c)
7. (d)	8. (a)	9. (d)	10. (b)	11. (c)	12. (a)
13. (b)	14. (d)	15. (c)	16. (c)		

SOME APPLICATIONS

1. SMPSs are superior to linear power supplies in respect of
   (a) size and efficiency
   (b) efficiency and regulation
   (c) regulation and noise
   (d) noise and cost.

2. Consider the following statements :
   Switched mode power supplies are preferred over the continuous types, because these are
   1. suitable for use in both ac and dc
   2. more efficient
   3. suitable for low power circuits
   4. suitable for high power circuits

   Of these statements, the correct is
   (a) 1 and 2
   (b) 1 and 3
   (c) 2 and 3
   (d) 2 and 4

3. Bulk power transmission over long HVDC lines are preferred on account of
   (a) low cost of HVDC terminals
   (b) no harmonic problems
   (c) minimum line power losses
   (d) simple protection

4. Resonant mode power supplies in comparison to square mode ones
   (a) have smaller component count
   (b) have negligible power loss
   (c) do not cause overvoltages
   (d) slower in control action

5. Resonant converters are basically used to
   (a) generate large peak voltages
   (b) reduce the switching losses
   (c) eliminate harmonics
   (d) convert a square wave into a sine wave.

6. HVDC transmission is preferred to EHV-AC because
   (a) HVDC terminal equipment are expensive
   (b) VAr compensation is not required for HVDC systems
   (c) system stability can be improved
   (d) harmonic problem is avoided.

7. In a 3-phase converter used in HVDC transmission, the three anodes conduct sequentially. Due to overlap caused by the circuit inductances, two anodes conduct simultaneously during the overlap period. The output voltage waveform during this period is the
   (a) voltage of the 1st anode, because the 2nd anode has not completely taken over
   (b) mean of the two anode voltages, as they conduct together
   (c) voltage of the 2nd anode, because the voltage of this anode is greater than that of the 1st
   (d) sum of the 1st and the 2nd anode voltages, because both the anode are conducting.

8. Which one of the following statements in respect of HVDC transmission line is not correct ?
   (a) The power transmission capability of bipolar line is almost the same as that of single circuit ac line.
   (b) HVDC link line can operate between two ac systems whose frequencies need not be equal.
   (c) There is no distance limitation for HVDC transmission by UG cable
   (d) Corona loss is much higher in HVDC transmission line.

ANSWERS

1. (a)   2. (c)   3. (c)   4. (b)   5. (b)   6. (b)
7. (b)   8. (d)

ELECTRIC DRIVES

1. A separately-excited dc motor is required to be controlled from a 3-phase source for operation in the first quadrant only. The most preferred converter would be
   (a) fully-controlled converter
   (b) fully-controlled converter with freewheeling diode
   (c) half-controlled converter
   (d) sequential control of two series connected fully-controlled converters.

2. A separately-excited dc motor, when fed from 1-phase full converter with firing angle α, runs at a speed of N rpm. When this motor is fed from 1-phase semiconverter but with the same firing angle as for full-converter, the motor speed is found to be 2N rpm. The value of firing angle is
   (a) 70.528°   (b) 75.572°   (c) 70°   (d) 69.88°

3. A separately-excited dc motor, when fed from 1-phase full converter with firing angle 60° runs at 1000 rpm. If this motor is connected to 1-phase semiconverter with the same firing angle of 60°, the motor would now run at
   (a) 2000 rpm   (b) 1500 rpm   (c) 1450 rpm   (d) 1000 rpm

4. A single-phase half-wave converter with freewheeling diode, drives a separately-excited dc motor at 900 rpm with firing angle 60°. When this motor is fed from 1-phase full converter with α = 60°, the motor speed would be
   (a) 600 rpm   (b) 900 rpm   (c) 1200 rpm   (d) 1800 rpm

5. A single-phase half-wave converter with freewheeling diode, drives a separately-excited dc motor at 900 rpm with firing angle 60°. When this motor is fed from 1-phase semiconverter with $\alpha = 60°$, the motor speed would be
   (a) 1800 rpm   (b) 1200 rpm   (c) 900 rpm   (d) 1500 rpm

6. A separately-excited dc motor, when fed from 1-phase full converter, runs at a speed of 1200 rpm. Load current remains continuous. If one of the four SCRs gets open-circuited, the motor speed will reduce to
   (a) 900 rpm   (b) 800 rpm   (c) 600 rpm   (d) 400 rpm

7. A 1-phase semiconverter delivers power to a separately-excited dc motor. Armature current is ripple free at 20 A. For a firing angle delay of 45°, the average and rms values of freewheeling diode current would respectively be
   (a) 10 A, 5 A   (b) 5 A, 10 A   (c) 15 A, 17.32 A   (d) 17.32 A, 15 A

8. A single-phase half-controlled rectifier is driving a separately-excited dc motor. The dc motor has back emf constant of 0.5 V/rpm. The armature current is 5 A without any ripple. The armature resistance is 2Ω. The converter is working from a 230 V, 1-phase ac source with a firing angle of 30°. Under this operating condition, the speed of the motor will be
   (a) 339 rpm   (b) 359 rpm   (c) 366 rpm   (d) 386 rpm

   [GATE, 2004]

9. A 3-phase semiconverter feeds the armature of a separately-excited dc motor, supplying a non-zero torque. For steady-state operation, the motor armature current is found to drop to zero at certain instances of time. At such instances, the voltage assumes a value that is
   (a) equal to the instantaneous value of the ac phase voltage
   (b) equal to the instantaneous value of the motor back emf
   (c) arbitrary
   (d) zero   [GATE, 2000]

10. A 3-phase bridge inverter is used for controlling the speed of a squirrel cage induction motor. If frequency of supply voltage is decreased with
    1. constant supply voltage V_1, starting torque T_{est} decreases
    2. constant V_1, T_{est} increases
    3. constant V_1/f_1, T_{est} increases
    4. constant V_1/f_1, maximum torque $T_{e.m}$ must remain constant
    5. constant V_1/f_1, T_{est} decreases
    6. constant V_1/f_1 and at very low frequencies, $T_{e.m}$ may decrease.
    From these, the correct statements are
    (a) 1, 4, 5   (b) 2, 3, 4, 6   (c) 1, 3, 6   (d) 2, 3, 6

11. For an ac voltage controller fed induction motor drive negotiating a load whose torque requirement does not vary with speed
    (a) the fundamental component of current drawn from the supply decreases as speed is reduced
    (b) the fundamental component of current drawn from the supply increases as speed is reduced
    (c) the fundamental component of current drawn from the supply is independent of speed
    (d) none of the above

12. A delta-connected induction motor being fed by a 3-phase ac to dc inverter and operated in constant V/f control mode requires during starting a
    (a) star-delta starter
    (b) DOL starter
    (c) auto-transformer starter
    (d) none of the above

13. An ac induction motor is used for speed control application. It is driven from an inverter with a constant V/f control. The motor name-plate details are as follows :
    $V : 415V$ Ph : 3 $f : 50$ Hz $N : 2850$ rpm
    The motor is run with the inverter output frequency set at 40 Hz and with half the rated slip. The running speed of the motor is
    (a) 2400 rpm
    (b) 2280 rpm
    (c) 2340 rpm
    (d) 2790 rpm
    [GATE, 2003]

14. A variable speed drive rated for 1500 rpm, 40 Nm is reversing under no load. Figure C. 57 shows the reversing torque and the speed during the transient. The moment of inertia of the drive is
    (a) 0.048 kgm^2
    (b) 0.064 kgm^2
    (c) 0.096 kgm^2
    (d) 0.128 kgm^2
    [GATE, 2004]

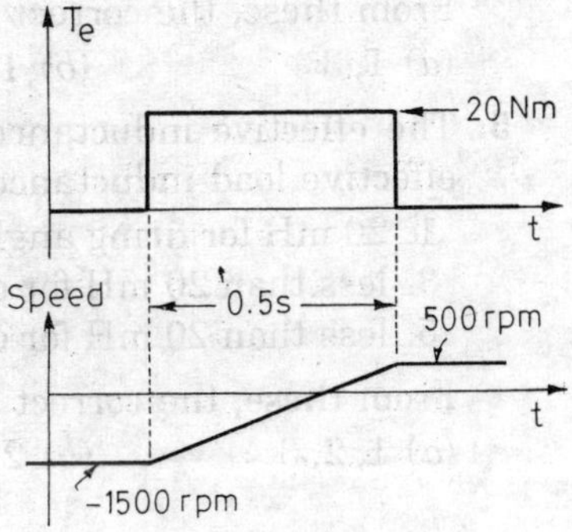

Fig. C.57

15. Slip-power control schemes provide a range of speed control of a 3-phase induction motor. The range is
    (a) 0 to N_s
    (b) $- N_s$ to N_s
    (c) 0 to $2N_s$
    (d) $- 2N_s$ to $2N_s$

    where N_s is the synchronous speed.

ANSWERS

1. (c)	2. (a)	3. (b)	4. (c)	5. (a)	6. (c)
7. (b)	8. (c)	9. (b)	10. (d)	11. (c)	12. (b)
13. (c)	14. (a)	15. (c)			

POWER FACTOR IMPROVEMENT

1. The most accurate and versatile method of achieving reactive power compensation is by using
   (a) switched capacitors
   (b) fixed capacitor with controlled reactor
   (c) saturable reactor with capacitor bank
   (d) saturable reactor with controlled reactor

2. Thyristor controlled compensator is usually designed to operate at
   (a) a slightly lagging pf
   (b) a slightly leading pf
   (c) unity pf
   (d) zero pf leading

3. Read the following statements regarding thyristor controlled reactor :
   1. It takes maximum reactive power at firing angle $\alpha = 90°$
   2. It delivers maximum reactive power at $\alpha = 180°$
   3. It delivers no reactive power at $\alpha = 0°$
   4. It takes maximum reactive power at $\alpha = 45°$
   From these, the correct statements are
   (a) 1 only   (b) 2, 4   (c) 1, 4   (d) 1, 3

4. The use of static VAr compensator in an ac system
   1. improves the supply power factor
   2. reduces the source current
   3. improves the load voltage profile
   4. reduces the load reactive power
   From these, the correct statements are
   (a) 1, 3   (b) 1, 2, 3   (c) 1, 2, 3, 4   (d) 1, 2, 4

5. The effective inductance offered by a thyristor controlled inductor is 20 mH. The effective load inductance, seen by the source, would be
   1. 20 mH for firing angle $\alpha = 60°$ 2. 20 mH for $\alpha = 90°$
   3. less than 20 mH for $\alpha = 120°$ 4. more than 20 mH for $\alpha = 120°$
   5. less than 20 mH for $\alpha = 60°$

   From these, the correct statements are
   (a) 1, 2, 3   (b) 2, 4, 5   (c) 1, 2, 4, 5   (d) 1, 2, 4

ANSWERS

1. (b)   **2.** (a)   **3.** (c)   **4.** (b)   **5.** (d)

Appendix : D

References

1. SCR Manual, 5th Edition, N.Y., General Electric Company, 1972.
2. F. Csaki et. al., 'Power Electronics', Budapest : Akademiai Kiado, 1975.
3. F.E. Gentry et. al., 'Semiconductor Controlled Rectifiers', Prentice-Hall of India New Delhi, 1964.
4. B.K. Bose, 'Power Electronics and AC Drives', Prentice-Hall, Englewood Cliffs, New Jersey 07632, 1986.
5. P.C. Sen, 'Thyristorised DC Drive', New York : Wiley Interscience, 1981.
6. J.M.D. Murphy and F.G. Turnbull 'Power Electronic Control of AC Motors', Pergamon Press, Oxford, 1988.
7. B.K. Bose, 'Evaluation of Modern Power Semiconductor Devices and Future Trends of Converters', IEEE Trans. Industry Applications, vol. 28, No. 2, pp. 403 – 413, March/April, 1992.
8. NED MOHAN et. al. , 'Power Electronics', John Wiley and Sons, 1989.
9. M.H. Rashid, 'Power Electronics', Prentice-Hall of India, New Delhi, 1993.
10. M. Chilikan, 'Electric Drive', Mir Publishers, Moscow, 1970.
11. G.K. Dubey and C.R. Kasarabada, "Power Electronics and Drives", IETE Book Series, Vol. 1, TM HILL P.C. Ltd., New Delhi - 1993.
12. B.K. Bose, "Energy, environment and advances in Power Electronics", IEEE Trans. on P.E. Vol 15, No. 4, July 2000.

INDEX

Index